Lecture Notes in Electrical Engineering

Volume 349

About this Series

"Lecture Notes in Electrical Engineering (LNEE)" is a book series which reports the latest research and developments in Electrical Engineering, namely:

- Communication, Networks, and Information Theory
- Computer Engineering
- Signal, Image, Speech and Information Processing
- Circuits and Systems
- Bioengineering

LNEE publishes authored monographs and contributed volumes which present cutting edge research information as well as new perspectives on classical fields, while maintaining Springer's high standards of academic excellence. Also considered for publication are lecture materials, proceedings, and other related materials of exceptionally high quality and interest. The subject matter should be original and timely, reporting the latest research and developments in all areas of electrical engineering.

The audience for the books in LNEE consists of advanced level students, researchers, and industry professionals working at the forefront of their fields. Much like Springer's other Lecture Notes series, LNEE will be distributed through Springer's print and electronic publishing channels.

More information about this series at http://www.springer.com/series/7818

Mitsuo Gen · Kuinam J. Kim
Xiaoxia Huang · Yabe Hiroshi
Editors

Industrial Engineering, Management Science and Applications 2015

Springer

Editors
Mitsuo Gen
Tokyo University of Science
Tokyo
Japan

Kuinam J. Kim
Department of Convergence Security
Inst. of Creative Advanced Tech., Sci. and Engg.
Kyonggi University
Suwon-si
Korea

Xiaoxia Huang
University of Science and Technology Beijing (USTB)
Beijing
China

Yabe Hiroshi
Tokyo University of Science
Tokyo
Japan

ISSN 1876-1100 ISSN 1876-1119 (electronic)
Lecture Notes in Electrical Engineering
ISBN 978-3-662-47199-9 ISBN 978-3-662-47200-2 (eBook)
DOI 10.1007/978-3-662-47200-2

Library of Congress Control Number: 2015938793

Springer Heidelberg New York Dordrecht London

Printed on acid-free paper

Springer-Verlag GmbH Berlin Heidelberg is part of Springer Science+Business Media
(www.springer.com)

Preface

This LNEE volume contains the papers presented at the International Conference on Industrial Engineering, Management Science and Applications (ICIMSA2015) which was held in Tokyo, Japan on May 26-28, 2015.

ICIMSA2015 received over 350 paper submissions from various countries. After a rigorous peer-review process, 114 full-length papers were accepted for presentation at the conference. This is intended for maintaining the high standards of the conference proceedings.

The conference is intended to bring together the researchers and technologists working in different aspects of Industrial Engineering, Management Science and Applications. In addition to the contributed papers, internationally known experts from several countries were invited to deliver Keynote speeches at ICIMSA2015.

Much of the credit of the success of the conference is due to the topic coordinators who have devoted their expertise and experience in promoting and in general coordination of the activities for the organization and operation of the conference. The coordinators of various session topics have devoted a considerable time and energy in soliciting papers from relevant researchers for presentation at the conference. The Session Chairs of the different session played important role in conducting the proceedings of the session in a timely and efficient manner.

On behalf of the Organizing Committee, we would like to thank Springer LNEE for publishing the proceedings of ICIMSA2015. We also would like to express our sincere and grateful thanks to our Program Committee and Reviewers for providing extra help in the review process. The quality of a refereed volume depends mainly on the expertise and dedication of the reviewers.

Our sincere thanks to the Institute of Creative Advanced Technology, Engineering and Science (iCatse) for designing the conference web page and also spending countless days in preparing the final conference program in time for printing. We would also like to thank the ICIMSA2015 Secretariat and Staff for arranging a large number of the invitation letters and assisting in the various stages of the editorial work. Finally we would like to thank our organization committee for their several months of hard work in sorting out manuscripts from our authors.

We look forward to seeing all of you next year at ICIMSA2016 in Korea.

Mitsuo Gen
Fuzzy Logic Systems Institute and Tokyo University of Science, Japan
Xiaoxia Huang
University of Science and Technology Beijing, China
Kuinam J. Kim
Kyonggi University, Republic of Korea
Yabe Hiroshi
Tokyo University of Science, Japan

Organizing Committee

General Chairs

Mitsuo Gen	Fuzzy Logic Systems Institute and Tokyo University of Science, Japan
Xiaoxia Huang	University of Science and Technology Beijing, China
Kuinam J. Kim	Institute of Creative Advanced Technologies, Science and Engineering, Republic of Korea
Yabe Hiroshi	Tokyo University of Science, Japan

Steering Committee

Nikolai Joukov	Chair of IEEE CS STCOS, USA
Borko Furht	Florida Atlantic University, USA
Bezalel Gavish	Southern Methodist University, USA
Kin Fun Li	University of Victoria, Canada

Publicity Chairs

Dan (Dong-Seong) Kim	University of Canterbury, New Zealand

Workshop Chairs

Kuinam J. Kim	Kyonggi University, Republic of Korea
Donghwi Lee	University of Colorado, USA

Financial Chairs

Kyoungho Choi — Institute of Creative Advanced Technologies, Science and Engineering, Republic of Korea

Program Chairs

Adel Hejaaji — Engineering Services Management Limited [ESM LTD] ESSEX, UK

Organizers and Supporters

Institute of Creative Advanced Technologies, Science and Engineering (iCatse)
Chinese Management Science Society (CMSS)
Korean Industry Security Forum (KISF)
Korea Information Assurance Society (KIAS)
Kyonggi University
University of Science and Technology Beijing
Tokyo University of Science
River Publishers

Program Committee

Chil-Chyuan Kuo	Ming Chi University of Technology, Taiwan
Suksan Prombanpong	King Mongkut's University of Technology Thonburi, Thailand
Ramayah Thurasamy	Universiti Sains Malaysia, Malaysia
Siana Halim	Petra Christian University, Indonesia
Yves De Smet	Université Libre de Bruxelles, Belgium
Marco Aiello	University of Groningen, The Netherlands
Catalina Lucía Alberto	Universidad Nacional de Córdoba, Argentina
Llewellyn C.M. TANG	University of Nottingham Ningbo China, China
M. Birasnav	New York Institute of Technology, USA
Enslin Johannes van Rooyen	Tshwane University of Technology, South Africa
Supachart Iamratanakul	Kasetsart University, Thailand
Luciana Hazin Alencar	Universidade Federal de Pernambuco, Brazil
Jui-Sheng Chou	National Taiwan University of Science and Technology, Taiwan
Shimpei Matsumoto	Hiroshima Institute of Technology, Japan
Ahm Shamsuzzoha	University of Vaasa, Finland
Minghai Jiao	Northeastern University, China
Masaharu Tsujimoto	Tokyo Institute of Technology, Japan

Ahm shamsuzzoha	Sultan Qaboos University, Oman
Seren OZMEHMET TASAN	Dokuz Eylul University, Turkey
Yoshinobu Tamura	Yamaguchi University, Japan
António Grilo	Faculdade de Ciências e Tecnologia da Universidade Nova de Lisboa, Portugal
Fabrizio Maria Maggi	University of Tartu, Estonia
V. Cruz Machado	Universidade Nova de Lisboa, Portugal
Md Mamun Habib	Universiti Utara Malaysia, Malaysia
Kittisak Jermsittiparsert	Rangsit University, Thailand
Adel Hejaaji	Engineering Services Management Limited [ESM LTD] ESSEX, UK
Ulrich Reimer	University of applied sciences St. Gallen, Switzerland
Hardeep Singh	Ferozepur College of Engg & Technology, India
Fatemeh Almasi	Amirkabir University of Technology (Tehran polytechnic), Iran
Yiliu Liu	Norwegian University of Science and Technology, Norway
Ryo HARUNA	Kanazawa Gakuin University, Japan
Ilias Santouridis	Technological Educational Institute (TEI) of Thessaly, Greece
Purit Thanakijkasem	King Mongkut's University of Technology Thonburi, Thailand
Chun-Cheng Lin	National Chiao Tung University, Taiwan
Michel ALDANONDO	Toulouse University - Mines Albi -, France
Jinho Lee	Korea Naval Academy, Republic of Korea
Mojahid F. Saeed Osman	King Fahd University of Petroleum and Minerals, Saudi Arabia
Wasawat Nakkiew	Advanced Manufacturing Technology Research Center (AMTech), Chiang Mai University, Thailand
Somlak Wannarumon KIELAROVA	Naresuan University, Thailand
Jaekyung Yang	Chonbuk National University, Republic of Korea
Suprakash Gupta	Indian Institute of Technology (B H U), India
Alejandro Escudero-Santana	Universidad de Sevilla, Spain
RIKA AMPUH HADIGUNA	Andalas University, Indonesia
Andreas Dewald	University of Erlangen, Germany
Kit Fai Pun	The University of the West Indies, West Indies
Abdol S. SOOFI	University of Wisconsin-Platteville, USA
El-Houssaine AGHEZZAF	Ghent University, Belgium
Chompoonoot Kasemset	Chiang Mai University, Thailand
Ana Paula Ferreira Barroso	Universidade Nova de Lisboa, Portugal
Virgínia Helena Arimateia de Campos Machado	Universidade Nova de Lisboa, Portugal
Antonio Ruiz Molina	University of Malaga, Spain

Contents

Regional Location Decision for Thai Garment Industry: An AEC Perspective

Walailak Atthirawong[1,*] and Wariya Panprung[2]

[1] School of Statistics, King Mongkut's Institute of Technology Ladkrabang, Bangkok 10520, Thailand
walailaknoi@gmail.com
[2] Management Science Department, Phranakhon Rajabhat University, Bangkok 10220, Thailand
nwariya@yahoo.com

Abstract. Thailand may encounter both opportunities and crises when the ASEAN Economic Community (AEC) is implemented by the end of 2015. AEC will open more opportunities to enterprises in both trade and investment, which will create a competitive market of over 600 million people, as well as enhance the competitive advantage. As such, it may be necessary for enterprises to redesign their supply chains via finding a new production base which has low labor cost, high labor skill and numerous of labor available. This strategy will help companies especially for garment industry in order to maintain a competitive position in the global market. Therefore, location decision play a key role in achieving corporate competitiveness, and as a result of this, selecting the right country is a critical issue of these new strategies. Hence, this research aims to 1) investigate factors affecting location decision for garment industry, 2) select countries in ASEAN region which has potential for establishing or expanding a production base of Thai garment industry, and 3) propose a generic model for location selection decision using Fuzzy Analytical Network Process (FANP). The results from the empirical study reveal that according to the entrepreneurs' opinion, key criteria affecting location decision for garment industry are 1) raw materials factors, 2) labor factors, 3) logistics system, 4) infrastructure, 5) government policies and 6) economy factors, respectively. The top four sub-criteria of those main factors are also taken in consideration in this research as their important details. Three outstanding countries i.e. Myanmar, Cambodia and Vietnam are mentioned for alternative location decision from interviewing expert persons. Due to both tangible and intangible criteria are included in the problem, Multi Criteria Decision Making Approach (MCDM) is then employed to this research. The generic model for regional location selection decision is proposed using Fuzzy Analytical Network Process (FANP) technique. This model contributes to planning practice for top management by suggesting a more comprehensive decision making tool in location selection.

Keywords: Garment Industry, ASEAN Economic Community, Location Decision, Multiple Criteria Decision Making (MCDM), Fuzzy Analytical Network Process (FANP).

* Corresponding author.

M. Gen et al. (eds.), *Industrial Engineering, Management Science and Applications 2015*,
Lecture Notes in Electrical Engineering 349, DOI: 10.1007/978-3-662-47200-2_1

1 Introduction

The garment industry is one of the important industries for a handful of countries. The industry is fully integrated into a whole supply chain, including an upstream sector (i.e. synthetic fiber and yarn manufacturing), an intermediate sector (i.e. manufacturing fabric, spinning, weaving, knitting, bleaching and dyeing) and a downstream sector (i.e. garment manufacturing). In addition, garment sector has played the largest sector in terms of production, GDP and employment as well as export values. Nevertheless, the industry faces some serious challenges in the middle of rising competition in the ASEAN region subject to fierce competition, as well as the increased cost of labor and raw materials and shortage of labor with in the country. Moreover, export values have been declined as key customers i.e. US and Europe have faced with economic recession. These will be a severe situation for most of garment companies which have their no brand name and innovation.

By the end of 2015, the Association of Southeast Asian Nations (ASEAN) countries proclaim to transform into the next stage of an economic era by having a single market and production base called ASEAN Economic Community (AEC). The objective of AEC is to achieve "a stable, prosperous and highly competitive ASEAN economic" [1]. The principle of AEC is based on four pillars, i.e. 1) a single market and production base, 2) highly competitive economic region, 3) equitable economic development and 4) full integration in global economy. In addition, there will be an elimination of tariffs in the sense of having free flow of goods/services/investment capitals among ASEAN countries. Some sophisticated procedures related to customs will also be streamlined. The implementation of a single market and production base will direct ASEAN countries to the five core elements which are 1) free flow of goods, 2) free flow of services, 3) free flow of investment, 4) free flow of capital, and 5) free flow of skilled labor. These will lead to a transformation in the way of doing business in ASEAN region and therefore, there is a need to reconsider the logistics and supply chain system for ASEAN countries in general and in specific, for each particular country in AEC [2]. Despite more challenges in high-wage rate and shortage labors issues and the increase of minimum wage to 300 baht per day, Thai garment industry should not only prepare to overcome these challenges but also should to take advantage of AEC liberalizations by shifting or expanding investment to a neighbor country which has lower manufacturing cost, as well as an abundance and fruitful resources.

Accordingly, this research aims to 1) investigate factors affecting location decision for garment industry, 2) select countries in ASEAN region which has potential for establishing or expanding a production base of Thai garment industry, and 3) propose a generic model for location selection decision using Fuzzy Analytical Network Process (FANP). Hence, this paper is organized as follows. The first section provides short information about garment industry in Thailand and some benefit of AEC. In Section 2, research methodologies and FANP are explained. In the following section, the results of this research and a proposed model will be mentioned. Finally, the conclusion and further work will be highlighted in the last section.

2 Research Methodologies

2.1 Interview and Questionnaire

The initial part of this primary research was conducted via site visits and in-depth interviews. Several key parties, for instance, government offices, Thailand Textile Institute (THTI), Thai Garment Manufacturers Association (TGMA), top management in garment industry were interviewed over a three month period at the beginning of 2014. Principally, the main objective of this stage was to understand the current situation and problems within the industry as, effect of AEC, as well as factors affecting location decision. Then a questionnaire was developed to investigate key success factors and sub-factors which garment companies have considered locating facilities elsewhere. Queries were collected from theories and interview results, with all 8 main factors and 57 sub-factors. Each variable gauges according to Likert Scale. Ratings are divided into 5 levels [3]. Level 1 means that factor has minimal influence on the decision making while level 5 means that factor has maximum influence on the decision making.

2.2 Alternative Location

In order to expand or enlarge a production base of garment industry in AEC country, it is necessary to further investigate by looking at statistical information, investment situation, import, and export products of each country. Furthermore, in-depth interviews with senior industry executives of Thai and CLMV garment companies in Lao, Myanmar, Vietnam and Cambodia, as well as site visits with some potential location countries were also conducted.

2.3 Generic Model for Location Decision

Multi Criteria Decision Making (MCDM), an advanced field of operations research, was employed to this study as it can provides decision makers to investigate a number of alternatives in light of conflicting priorities [4-6]. Multiple criteria analysis (MCA) provides a framework for breaking a problem into its constituent parts which considered both tangible and intangible factors and this attribute fits to the subjectivity feature of economical decision problems [7].

a) The Analytic Network Process (ANP)

The ANP, introduced by Saaty [8], is a generalization of the AHP. However, ANP is employed whenever the problem cannot be structured hierarchically. ANP allows both interaction and feedback within clusters of element (inner dependence) and between clusters (outer dependence). It is a multi-criteria theory of measurement used to derive relative priority scales of scales of absolute numbers from individual judgments that also belong to a fundamental scale of absolute numbers. These judgments represent the relative influence, of one of two elements over the other in the system, with respect to an underlying control criterion. Through its supermatrix, whose entries are themselves matrixes of column weights, and the ANP draws the result of dependence and feedback within and between elements clusters [9]. ANP models have two parts. One is a control hierarchy or network of objectives and criteria that control the interactions in the system under study. Another is there are many sub-networks of influences among the elements and clusters of the problem, one for each control criterion [8-9].

b) Fuzzy Analytic Network Process

Zadeh [10] introduced the fuzzy set theory to deal with the uncertainty due to imprecision and vagueness. A major contribution of fuzzy set theory is its capability of representing vague data. The theory also allows mathematical operators and programming to apply to the fuzzy domain [11]. Generally, a fuzzy set is defined by a membership function, which represents the grade of any element x of X that have the partial membership to M. The degree to which an element belongs to a set is defined by the value between zero and one. If an element x really belongs to M , $\mu_M(x) = 1$ and clearly not $\mu_M(x) = 0$

A triangular fuzzy number is defined as (l,m,u) , where $l \leq m \leq u$. The parameters l, m and u respectively, denote the smallest possible value, the most promising value, and the largest possible value that describe a fuzzy event. (l,m,u) has the following triangular type membership function.

$$\mu M(x) = \begin{cases} (x-l)/(m-l) & l \leq x \leq m \\ (u-x)/(u-m) & m \leq x \leq u \\ 0 & otherwise \end{cases} \tag{1}$$

The decision makers could be uncertain about their own level of preference, due to incomplete information or knowledge, complexity and uncertainty within the decision environment, or a lack of an appropriate measurement units and scale. Therefore, the fuzzy ANP has been used to solve the problem of location selection. The decision-makers use the linguistic variables shown in Fig. 1 to evaluate the importance of the criteria and the ratings of alternatives with respect to qualitative criteria. The algebraic operations with fuzzy numbers used in this paper can be found in Fig. 1.

The membership functions of triangular fuzzy number M_1, M_3, M_5, M_7 and M_9 are used to represent the pair-wise comparison of decision variables from "Very bad" to "Excellent", and TFNs M_2, M_4, M_6 and M_8 represent the middle preference values between them. Then according to the concept of extent analysis, each object is taken and extent analysis for each goal g_i is performed, respectively [11].

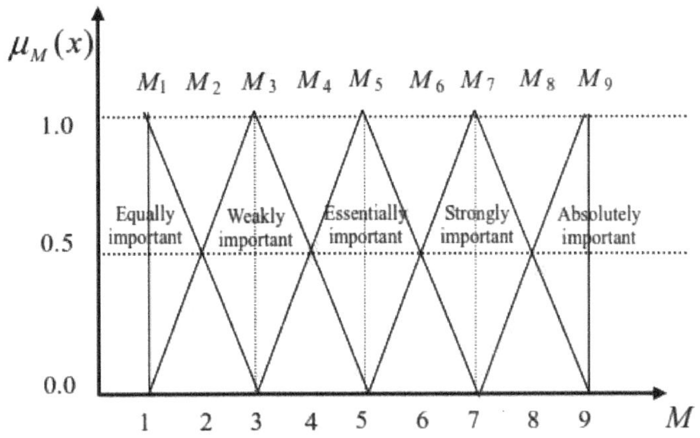

Fig. 1. The membership functions of triangular fuzzy number [11]

3 A Proposed Model for Location Decision in an AEC Perspective

3.1 Factors Influencing Decision-Making

For identification of location decision criteria and sub-criteria, questionnaires were distributed to 119 garment companies, both Thailand and the CLMV countries. Weight from average score, mean score of eight criteria and fifty-seven sub-criteria were computed. Then, Table 1 provided a process of ranking main criteria from companies' point of views as follows.

Table 1. Rank of main criteria

Criteria (1)	Weight from AVG (%) (2)	Weight from AVG Score (3)	(4)= (2) * (3)	Percent (5)	Rank (6)
Labor	0.19	3.83	0.73	18.86	1
Raw Materials	0.16	4.12	0.66	17.05	2
Economy	0.14	3.84	0.54	13.95	3
Logistics System	0.12	3.92	0.47	12.14	4
Infrastructure	0.12	3.83	0.46	12.04	5
Government Policies	0.09	3.82	0.35	8.93	6
Location Facilities	0.09	3.78	0.34	8.78	7
Risk	0.09	3.58	0.32	8.27	8
Total	**1.00**	**30.72**	**3.87**	**100**	

According to Table 1, six main criteria which can mean score more than 3.80 (see Column 3) were employed to further study. The top four ranks sub-criteria of each main criterion were also selected in order to structure a location decision modeling in the next step. The criteria and sub-criteria, which are claimed as key success factors for locating a garment company are displayed in Table 2.

Table 2. Criteria and their sub-criteria affecting location decision

Criteria	Sub-criteria
Labor	• Low labor cost • Labor availability • Flexibility of wage determination • High labor skill
Raw Materials	• Material quality • Supplier quality • Material availability • Supplier reliability
Logistics System	• Low cost transport • Transport lead time • Ease of transport document • Low investment in import/export
Infrastructure	• Available of road transport • Reliability of road transport • Cost of vehicle transport • Connectivity
Government Policies	• Government security • High government potential • Law enforcement • Funding for infrastructures project
Economy	• Exchange rate • High rate inflation • Interest rate spread • Domestic market size

3.2 Decision Alternatives

According to statistical information [12] along with in-depth interviews with senior industry executives of Thai and CLMV garment companies and site visits with some potential countries, it is revealed that the best trade and investment opportunities for Thai garment companies are the CLMV countries, especially Myanmar, Vietnam and Cambodia. The statistical information reveals that there are a number of investors from elsewhere, for instance; Europe, USA, Canada and Korea, seeking more investment in a garment production in Myanmar, Vietnam and Cambodia. This information can be assumed that labors have skillful for making clothes. Furthermore, labor cost is lower than Thailand 3-4 times. Investment climate related to governance of each country has a remarkable policy to support garment industry for local and foreigner investors as well. Since 2012, Vietnam has also had to increase minimum wage, making 1.5 times higher than Cambodia and Myanmar. However, Vietnam will have larger market than the others [12]. Cambodia and Myanmar belong to the group of least developed countries (LDC); therefore, Cambodia can be benefit from the Generalized System of Preference (GSP) of US, EU and Japan while Myanmar can benefit of the GSP of Japan [13]. Lao PDR, which is a member of the CLMV country, can also benefit of the GSP privileges; however, according to experts' opinion, Lao PDR may not be a suitable country for locating or expanding a new manufacturing base in a long run. The main concern is that a number of populations of Lao PDR are lowest among four countries. The foregoing discussion also reveals that Lao men do not like to work in garment industry. Therefore, in the research, Myanmar, Vietnam and Cambodia are selected as alternative countries in this proposed model.

3.3 A Proposed Model of Decision-Making Process

One of the primary investigations of this study is to explore the complexity of the decision-making process at the industry level. As such, it is important to understand how the decisions are made and the various levels of decision-making involved. The study reveals that the decision making for this problem is hierarchy involving with various criteria and sub-criteria. Also, these factors are interrelated with each other. In order to allow all factors to be evaluated and taken place in the same network; therefore, the application of the Analytic Network Process (ANP), a multi attribute approach for decision making, can be adopted as a tool for a top management to make a more proper and acceptable sense. However, due to the complexity of the problem in reality, the decision maker may feel more confident about making a fuzzy judgment rather than in making a crisp evaluation [14]. Consequently, in this research Fuzzy Analytical Network Process (FANP) is employed to handle interdependency among evaluation criteria and integrate the divergent judgments of experts in a location selection senior exclusive.

The ANP model is formed by criteria and sub-criteria obtained from an empirical study as discussed above. The proposed model is comprised of four stages. The ultimate goal of this study, which is to choose the best location in AEC for expanding a new production base, is placed in the first level. Six criteria are structured in the second stage which has two way direction arrows. The arrow in this stage represents the inner-dependence among the factors. Sub-criteria related to their criteria are in the third stage of the proposed model. These sub-criteria have also interrelationships among each other. The last stage is location alternatives. According to this study, three countries i.e. Myanmar, Vietnam and Cambodia are selected into this model. There is interdependence among the items and the alternatives as well. Fig. 2 presented the shamanic structure of the proposed model.

Next step, data will be obtained via 5 senior garment industry executives by pair-wise comparisons among factors, sub-factors and alternatives based on the relationship motioned above by using triangular fuzzy scales. FANP computations using Mathlab [15] will continue, with the final results reported by middle of 2015.

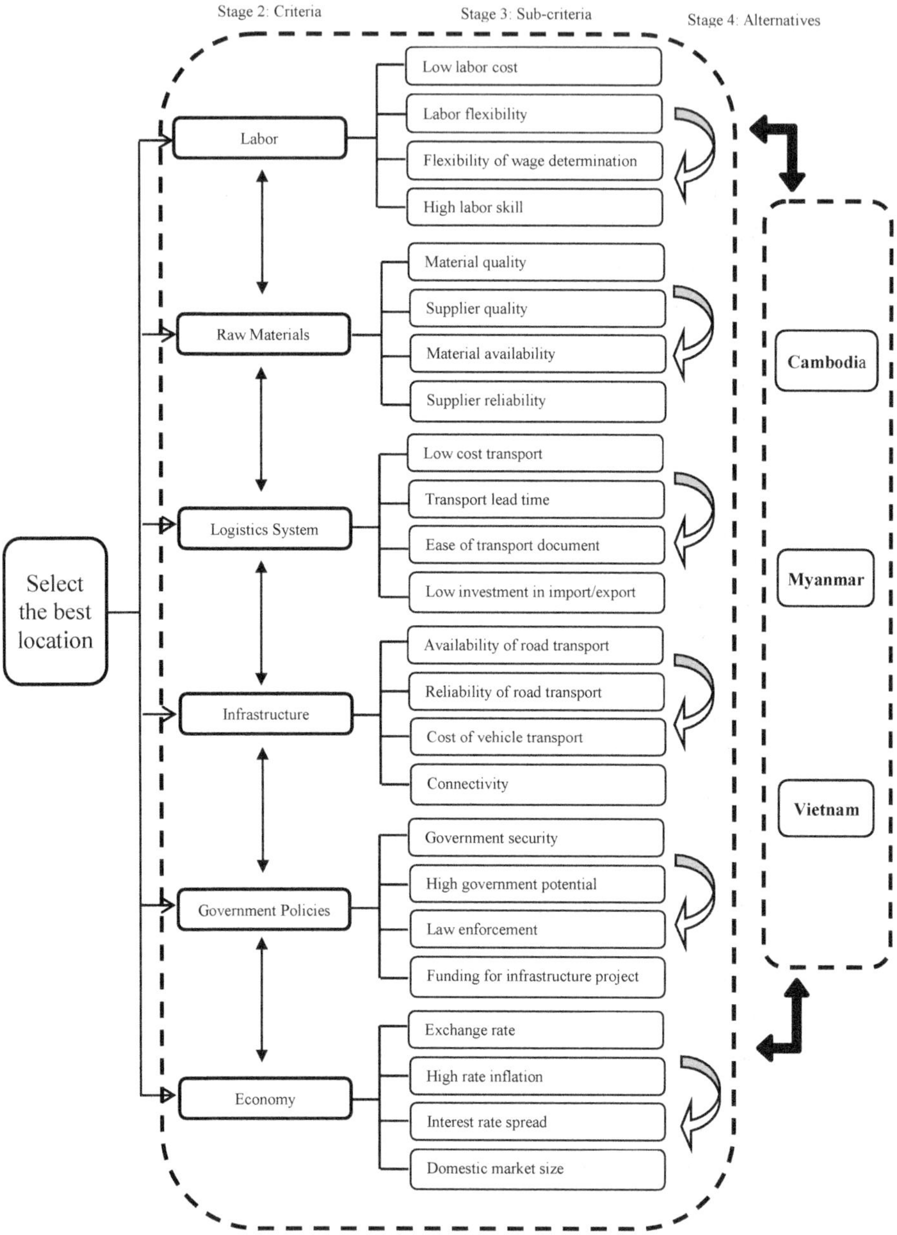

Fig. 2. The proposed model for location decision

4 Conclusion and Further Work

From the theoretical perspective, this paper has developed a FANP-based location decision modeling for garment companies to employ as a tool in order to select the best country when they are considering a new production base. The prototype of the proposed model is presented step by step. The process methodology has started from investigating main criteria and their criteria for international location decision by empirical study using in-depth interview and questionnaire. The process of seeking suitable countries for locating a new manufacturing plant for garment industry in ASEAN region was also mentioned in this paper. In the next stage of this research, the validity and effectiveness of the proposed model will be tested the proposed model through senior exclusives' view point from Thai garment companies.

Acknowledgements. The authors of this study would like to acknowledge the National Research Council of Thailand (NRCT) for financial sponsorship of this research. The authors are also grateful to all the members who were kindly involved and provided valuable and fruitful information for this study.

References

1. Association of Southeast Asian Nations (ASEAN): ASEAN Economic Scorecard. The ASEAN Secretariat, Jakarta (2012)
2. Banomyong, R.: ASEAN Economic Community (AEC) Logistics Connectivity Development Framework (2011)
3. Wolfer, L.: Real Research: Conducting and Evaluating Research in the Social Sciences. Pearson/Allyn and Bacon, Boston (2007)
4. Kaplinski, O., Tupenaite, L.: Review of the Multiple Criteria Decision Making Meth-ods, Intelligent and Biometric Systems Applied in Modern Construction Economics. Transformations in Business & Economics 10(1), 166–181 (2011)
5. Kapliński, O., Tamosaitiene, J.: Game Theory Applications in Construction Engineering and Management. Technological and Economic Development of Economy 16(2), 348–363 (2010)
6. Tamosaitiene, J., Bartkiene, L., Vilutiene, T.: The New Development Trend of Operational Research in Civil Engineering and Sustainable Development as A Result of Collaboration between German–Lithuanian–Polish Scientific Triangle. Journal of Business Economics and Management 11(2), 316–340 (2010)
7. Erdoğmus, Ş., Aras, H., Koç, E.: Evaluation of alternative fuels for residential heating in Turkey using analytic network process (ANP) with group decision-making. Renewable and Sustainable Energy Reviews 10, 269–279 (2006)
8. Saaty, T.L.: Decision Making with Dependence and Feedback: The Analytic Network Process. RWS Publications, Pittsburgh (1996)
9. Saaty, T.L.: Making with Dependence and Feedback. 2nd edn. RWS Publication (2001)
10. Zadeh, L.A.: Fuzzy Sets Information Control 8, 338–353 (1965)
11. Rezaeiniya, N., Ghadikolaei, A.H.: Fuzzy ANP Approach for New Application: Greenhouse Location Selection; a Case in Iran. Journal of Mathematics and Computer Science 8, 1–20 (2014)

12. Masami, I.: Comparing Investment Climates among Major Cities in CLMV, retrieved from `http://www.ide.go.jp/English/Publish/Download/Brc/pdf/04_chapter1.pdf`
13. Vivain, C.J.: Generalized System of Preferences: Background and Renewal Debate. Congressional Research Service Report (2014)
14. Tah, J.H., Carr, V.: A proposal for construction project risk assessment using fuzzy logic. Construction Management and Economics 18, 491–500 (2000)
15. Gilat, A.: MATLAB: An Introduction with Applications. 2nd edn. John Wiley & Sons (2004)

Key Performance Indicators for Sustainable Campus Assessment: A Case of Andalas University

Elita Amrina* and Febriza Imansuri

Department of Industrial Engineering, Andalas University, Padang, Indonesia
elita@ft.unand.ac.id, febrizaimansuri14@gmail.com

Abstract. Sustainable campus has became an important issue amongst universities around the world. Universities can generate a significant impacts to environment due to the high usage of energy, extensive transportation, massive waste, high consumption of materials, and extensive development of buildings and facilities. Thus, there is a need to assess the sustainable campus performance. This paper proposes a set of key performance indicators (KPIs) for sustainable campus assessment consisting of six categories divided into a total of 35 indicators. Analytical Hierarchy Process (AHP) method is applied to determine the importance weight of the KPIs. The results indicated the most important category for the sustainable campus assessment is education with an importance weight of 0.2665, while energy and climate change is regarded as the least important category. It is hoped the proposed KPIs can assist the universities to achieve the higher performance in sustainable campus.

Keywords: Analytic hierarchy process, key performance indicators, performance, sustainable campus, university.

1 Introduction

Nowadays, campus sustainability has become an increasingly issue of global concern for university policy makers and planners as a result of the realization of the impacts the activities and operations of universities have on the environment [1]. Like manufacturing, an university can also generate a significant impact to environment. It might becaused of the high usage of energy, extensive transportation, massive waste, high consumption of materials, and extensive development of buildings and facilities [2]. Increasing concerns to sustainability have forced universities to consider sustainability into their strategies and activities.

A sustainable university defined as a higher educational institution, as a whole or as a part, that addresses, involves and promotes, on a regional or global level, the minimization of negative environmental, economic, societal, and health effects generated in the use of their resources in order to fulfill its functions of teaching, research, outreach and partnership, and stewarship in ways to help society make the

* Corresponding author.

M. Gen et al. (eds.), *Industrial Engineering, Management Science and Applications 2015*,
Lecture Notes in Electrical Engineering 349, DOI: 10.1007/978-3-662-47200-2_2

transition to sustainable lifestyle [3]. According to the definition, sustainable campus must address the integration all the three aspects of sustainability of environmental, economic, and social in a better balance. University has several activities and complex operations which potentially generate significant environmental impacts. Sustainability must affects every sphere of a university, from the classrooms and laboratories, to housing, transportation and other services, and to the entire campus [1]. Therefore, assessing the sustainable campus has become a necessity.

In this study, a literature review was carried out in an attempt to identify key performance indicators (KPIs) used to assess the sustainable campus. One of the most commonly used indicators for the sustainable campus assessment is referred to the UI Greenmetric World University Ranking consisting of six categories and a total of 33 indicators [4]. It is a world university ranking for universities to assess and compare campus sustainability efforts [4]. The UI Greenmetric World University Ranking is the first attempt to make a global ranking of universities' sustainable behavior [5]. This paper proposes a set of Key Performance Indicators (KPIs) to assess the sustainable campus. The Analytical Hierarchy Process (AHP) methodology is applied to weighting the KPIs. It is believed that the proposed KPIs can aid universities to improve their sustainable campus performance.

2 Methodology

The methodology has two main stages. First, the key performance indicators (KPIs) for sustainable campus assessment were identified and derived from the literature. The KPIs were then validated to a case of university. Second, the importance weight of the KPIs is determined using Analytic Hierarchy Process (AHP) methodology.

Analytic Hierarchy Process (AHP) first introduced by Thomas L. Saaty in 1971 has become one of the most widely used methods for multiple criteria decision making (MCDM) problems. It is a decision approach designed to aid in making the solution of complex multiple criteria problems to a number of application domains [6]. It has been known as an essential tool for both practitioner and academics to conduct researches in decisions making and examining management theories [7]. AHP as a problem solving method is flexible and systematic that can represent the elements of a complex problem [8].

AHP method has several benefits [7]. First, it helps to decompose an unstructured problem into a rational decision hierarchy. Second, it can elicit more information from the experts or decision makers by employing the pair-wise comparison of individual groups of elements. Third, it sets the computations to assign weights to the elements. Fourth, it uses the consistency measure to validate the consistency of the rating from the experts and decision makers.

3 Identification of KPIs

This study starts with the development of key performance indicators (KPIs) for sustainable campus assessment through the literature review. The KPIs have been mostly adopted from the UI Greenmetric World University Ranking [4]. Besides, the

KPIs were also taken from the Alshuwaikhat and Abubakar's campus sustainability framework [1], sustainable UKM programme's framework [9], University of Nottingham's campus sustainability indicators [10], and University of Connecticut's campus sustainability indicators [11]. All the six categories and 33 of a total 35 indicators of the proposed KPIs are identified and derived from the UI Greenmetric World University Ranking. Another two indicators of category of the energy and climate change were taken from the other literatures. As a result, the KPIs of sustainable campus assessment consist of six categories divided into a total of 35 indicators were identified as shown in Table 1.

Table 1. The KPIs of sustainable campus assessment

Categories	Indicators
1. Setting and Infrastructure	1. Open space area/total area
	2. Open space area/total people
	3. Area on campus covered in forested vegetation
	4. Area on campus covered in planted vegetation
	5. Non-retentive surfaces/total area
	6. Sustainability budget/total university budget
2. Energy and Climate Change	7. Energy efficient appliances usage
	8. Renewable energy usage policy
	9. Total electricity use/total people
	10. Energy conservation program
	11. Green Building
	12. Climate change adaptation and mitigation program
	13. Greenhouse gas emission reduction policy
	14. Smooking area policy on campus
	15. Sustainable food program on campus
3. Waste	16. Recycling program for university waste
	17. Toxic waste recycling
	18. Organic waste teatment (garbage)
	19. Inorganic waste teatment (rubbish)
	20. Sewerage disposal
	21. Policy to reduce the use of paper and plastic on campus
4. Water	22. Water conservation program
	23. Piped water
5. Transportation	24. Total cars entering/total people
	25. Total bicycles/total people
	26. Transportation policy on limiting vehicles on campus
	27. Transportation policy on limiting parking space
	28. Campus buses
	29. Bicycle and pedestrian policy
6. Education	30. Sustainability courses / total courses
	31. Sustainability research funding/total research funding
	32. Sustainability publications
	33. Sustainability events
	34. Sustainability organizations (student)
	35. Sustainability website

The KPIs of sustainable campus assessment are then validated to a case of university located in Padang, West Sumatra, Indonesia. Established in 1956, Andalas University is the oldest university outside of Java Island, and the fourth oldest university in Indonesia. Currently, Andalas University has 15 faculies and about 25,000 students. In 2014, Andalas University has been accredited by National Accreditation Board for Higher Education with rank A (excellent). In term of sustainable campus, Andalas University has placed rank 146th and become ranked 8th of Indonesian universities in UI Greenmetric World University Ranking 2014.

A total of 5 members of green campus team from the university were consulted to validated the KPIs. The experts suggest that all categories and indicators of the KPIs are highly important. Thus, proposed as the KPIs to assess the sustainable campus.

4 Determining the Importance Weight of KPIs

Analytic Hierarchy Process (AHP) methodology was applied to determine the importance weight of the KPIs of sustainable campus assessment. The methodology consists of constructing the hierarchy, conducting the pairwise comparisons, constructing the pairwise comparisons matrix, computing the consistency ratio, and calculating the importance weight. Details are given as follows.

4.1 Constructing the Hierarchy

The proposed key performance indicators (KPIs) for sustainable campus assessment are then used in constructing a hierarchy. The three groups were defined and constructed in the hierarchy including goal, categories, and indicators. In the

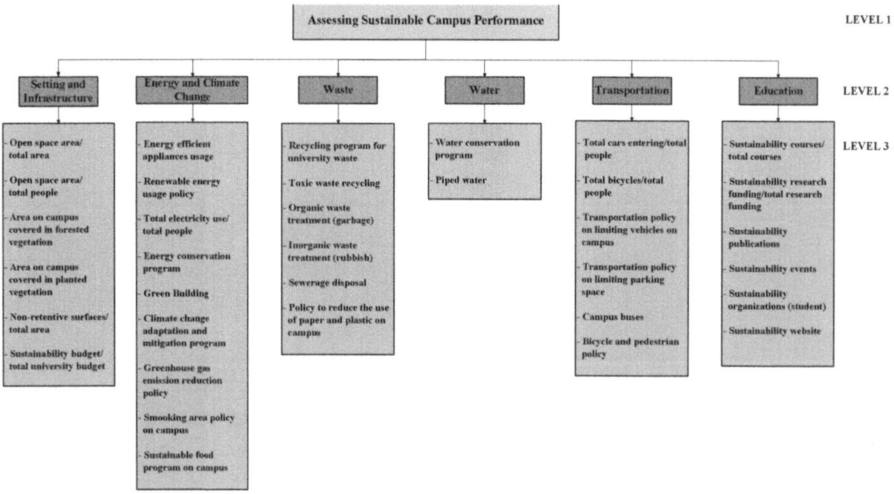

Fig. 1. The hierarchy structure of KPIs

hierarchy, assessing sustainable campus performance is set to be the goal. The next level consists of six categories of setting and infrastructure, energy and climate change, waste, water, transportation, and education. The third level consists of the indicators that described each of categories with a total of 35 indicators. The hierarchy is depicted in Fig. 1.

4.2 Conducting the Pairwise Comparisons

Once the hierarchy has been constructed, the importance weight of the KPIs should be calculated. A pairwise comparison questionnaire was then designed. A total of 30 experts from Andalas University were consulted to give their preferences on the KPIs. Those experts consist of dean and vice dean of each faculty in Andalas University. The pairwise comparisons were determined between categories, and indicators within each category of the KPIs. A Saaty' scale of 1 to 9 (1= equally, 3= moderate, 5= strong, 7= very strong, 9= extreme) was used to reflect these preferences. The consistency ratio (CR) was used to check the consistency of the pairwise comparisons for each expert. The CR values are less than 0.1 which means it matches the consistency test. If it is not yet consistent, the comparison has to be repeated again.

4.3 Constructing the Pairwise Comparisons Matrix

The preferences from the 30 experts were geometrically averaged and the pairwise comparisons matrices were then constructed. For example, the pairwise comparison matrix of the categories of sustainable campus assessment as below:

	Setting & infrastructure	*Energy & climate change*	*Waste*	*Water*	*Transportation*	*Education*
Setting & infrastructure	1	1.196	0.708	0.716	0.746	0.472
Energy & climate change	0.836	1	0.687	0.519	1.041	0.473
Waste	1.412	1.456	1	0.825	1.198	0.581
Water	1.397	1.928	1.212	1	1.775	0.689
Transportation	1.340	0.961	0.835	0.563	1	0.497
Education	2.121	2.113	1.720	1.452	2.010	1

All the diagonal elements of the matrix are equal to 1 as the elements are compared with themselves. The values of elements in the upper triangular matrix are obtained from the averaged preferences of pairwise comparisons and the reciprocals of these values are presented in the lower triangular matrix.

4.4 Computing the Consistency Ratio

The consistency ratio (CR) is used to check the consistency of pairwise comparisons and a value of less than 0.1 is acceptable [8]. The consistency test was performed to all the combined pairwise comparison matrixes. The results show that the consistency ratio (CR) values ranged from 0.0000 to 0.0081, which means that all the pairwise comparisons are consistent since the values are within the acceptable level recommended by Saaty [8]. It indicates that the experts have assigned their preferences consistently in determining the importance weights of the KPIs to assess sustainable campus performance.

4.5 Calculating the Importance Weight

The importance weight of KPIs are then calculated using the Expert Choice software. Table 2 presents a summary of the results of the importance weights of the KPIs of sustainable campus assessment. The importance weights show the importance value of one indicator over other indicators. In term of categories, education is the highest importance weight with a value of 0.2665. It is not suprisingly since the main function of an university in education field. Universities have responsibility in sustainable development to promote the sustainability culture to its students, staff, and community [2]. It followed by water with an importance weight of 0.2005. Clean water has become one of the main problems faced by any people in any place of the world. Universities with a high number students, staffs, and communities should be consider the need of clean water for their activities.

Table 2. The importance weights of KPIs

Categories	Weight	Indicators	Weight
1. Setting and Infrastructure	0.1234	1. Open space area/total area	0.0150
		2. Open space area/total people	0.0134
		3. Area on campus covered in forested vegetation	0.0237
		4. Area on campus covered in planted vegetation	0.0170
		5. Non-retentive surfaces/total area	0.0145
		6. Sustainability budget/total university budget	0.0392
2. Energy and Climate Change	0.1156	7. Energy efficient appliances usage	0.0145
		8. Renewable energy usage policy	0.0174
		9. Total electricity use/total people	0.0084
		10. Energy conservation program	0.0172
		11. Green Building	0.0157
		12. Climate change adaptation and mitigation program	0.0121
		13. Greenhouse gas emission reduction policy	0.0131
		14. Smooking area policy on campus	0.0081
		15. Sustainable food program on campus	0.0096
3. Waste	0.1630	16. Recycling program for university waste	0.0398
		17. Toxic waste recycling	0.0191
		18. Organic waste teatment (garbage)	0.0306
		19. Inorganic waste teatment (rubbish)	0.0284
		20. Sewerage disposal	0.0202
		21. Policy to reduce the use of paper and plastic on campus	0.0248
4. Water	0.2005	22. Water conservation program	0.1490
		23. Piped water	0.0510
5. Transportation	0.1309	24. Total cars entering/total people	0.0151
		25. Total bicycles/total people	0.0106
		26. Transportation policy on limiting vehicles on campus	0.0248
		27. Transportation policy on limiting parking space	0.0206
		28. Campus buses	0.0376
		29. Bicycle and pedestrian policy	0.0224
6. Education	0.2665	30. Sustainability courses / total courses	0.0299
		31. Sustainability research funding/total research funding	0.0272
		32. Sustainability publications	0.0510
		33. Sustainability events	0.0513
		34. Sustainability organizations (student)	0.0654
		35. Sustainability website	0.0422

The third category is waste with an importance value of 0.1630. This category is most related to environmental. As mention earlier, universities generate massive waste as a result of their activities and operations. Waste management is needed to solve this problem and to promote sustainability in campus environment. The next categories are transportation with a value of 0.1309, setting and infrastructure (0.1234), and energy and climate change (0.1156).

In term of indicators, water conservation program (0.1490) is regarded to the most important indicator. This indicator is of water category which suggested as the second highest important category. It followed by sustainability organizations (students) with an importance value of 0.0654, sustainability events (0.0513), piped water (0.0510), and sustainability publications (0.0510). Those indicators are categorized in education, and water category of the KPIs. Of all the indicators of KPIs of sustainable campus assessment, smoking area policy on campus with an importance weight of 0.0081 is suggested as the least important indicators.

5 Conclusions

An university can generate a significant environmental impacts due to the high usage of energy, extensive transportation, massive waste, high consumption of materials, and extensive development of buildings and facilities. Thus, it is essential to assess the sustainable campus performance. This paper has developed a set of Key Performance Indicators (KPIs) for sustainable campus assessment. The KPIs are identified and derived from the literature and then validated to a case of university. Based on the results, six categories divided into a total of 35 indicators are proposed as the KPIs of sustainable campus assessment. The importance weight of the KPIs then determined using Analytic Hierarchy Process (AHP) methodology. First, the hierarchy structure is established based on the proposed KPIs of sustainable campus assessment. Next, the pairwise comparisons conducted to the policy makers from the case of university using Saaty's scale of 1-9. The pairwise comparisons matrix are then contructed and the consistency ratio (CR) is computed. Finally, the importance weights of the KPIs is calculated.

The results show the importance value of one indicator over other indicators. Category of education is regarded as the most important category of the KPIs, followed by water, and waste. In term of indicators, water conservation program is suggested to the highest important indicator, followed by sustainability organizations (students), sustainability events, piped water, and sustainability publications. It is hoped the KPIs can aid the policy makers and planners of university to achieve a higher performance in the context of sustainable campus. Future research will focus on developing a tool to assess sustainable campus performance.

Acknowledgments. The authors would like to thanks to Andalas University, Padang, Indonesia and Directorate General of Higher Education, Indonesia for research funding with scheme of Hibah Bersaing No: 52/H.16/HB/LPPM/2015.

References

1. Alshuwaikhat, H.M., Abubakar, I.: An Integrated Approach to Achieving Campus Sustainability: Assessment of the Current Campus Environmental Management Practices. Journal of Cleaner Production 16, 1777–1785 (2008)
2. Gunawan, T.E., Prayogo, D.N., Mardiono, L.: Eco-sustainable Campus Initiatives: A Web Content Analysis. In: Proceedings of the 3rd International Conference on Technology and Operations Management (2012)
3. Velazquez, L., Munguia, N., Platt, A., Taddei, J.: Sustainable University: What Can be the Matter? Journal of Cleaner Production 14, 810–819 (2006)
4. Universitas Indonesia.: Guidelines of UI GreenMetric World University Ranking (2014), `http://www.greenmetric.ui.ac.id`
5. Grinsted, T.S.: Sustainable Universities – from Declarations on Sustainability in Higher Education to National Law. Journal of Environmental Economics 2, 29–36 (2011)
6. Saaty, T.L.: The Analytic Hierarchy and Analytic Network Measurement Processes: Application to Decisions Under Risk. European Journal of Pure and Applied Mathematics 1, 122–196 (2008)
7. Cheng, E.W.L., Li, H., Ho, D.C.K.: Analytic Hierarchy Process: A Defective Tool When Used Improperly. Measuring Business Excellence 6, 33–37 (2002)
8. Grinsted, T.S.: Sustainable Universities – from Declarations on Sustainability in Higher Education to National Law. Journal of Environmental Economics 2, 29–36 (2011)
9. Chan, F.T.S., Chan, H.K., Lau, H.C.W., Ip, R.W.L.: An AHP Approach in Benchmarking Logistics Performance of the Postal Industry. Benchmarking: An International Journal 13, 636–661 (2006)
10. Fadzil, Z.F., Hashim, H.S., Che-Ani, A.I., Aziz, S.: Developing a Campus Sustainability Assessment Framework for the National University of Malaysia. Int. J. Environmental, Ecological, Geological and Mining Engineering 6(6), 44–48 (2012)
11. University of Nottingham.: Sustainability Report 2012-13 (2013), `http://www.nottingham.ac.uk/sustainability`
12. University of Connecticut.: Campus Sustainability Design Guidelines (2004), `http://www.ecohusky.uconn.edu/pcc/sustainabledevelopment.html`

Minimum Partial Encryption for JPEG/JPEG2000 Medical Image Protection

Seong Min Yoo*, Jun Hoo Park, and Jae Cheol Ryou

Department of Computer Engineering, Chungnam National University, Daejeon, Korea
{ysm8521,ascndl}@naver.com,
jcryou@home.cnu.ac.kr

Abstract. A digitalized medical image data has personal information such as name and phone number or medical image that contains patient's physical characteristics, so it requires protective measure. The simplest way for meeting this demand is encrypting that, but, in recently, performance problem occurs owing to a considerable increase of data size with the development of medical imaging equipment. Thus, we need the way that provides protection function by transforming only some elements of medical image information. In this paper, we propose the method of obfuscation and integrity verification by encrypting at the least part of medical image that is stored in JPEG or JPEG2000 format.

1 Introduction

As Picture Archiving and Communication System (PACS) is introduced, digitized medical image information such as MRI, CT, X-ray is widely used. Such medical image information should comply with Digital Imaging and Communications in Medicine (DICOM) standard for compatibility between systems [1]. This medical image information according to DICOM standard includes many medical images with personal information of patient. For this reason, if medical imaging information is managed without appropriate protection, there may problem such as personal information leakage or reusing medical image that patients do not agree [2].

To solve this problem, encrypting whole file is very effective in terms of safety. But recently, a single size of medical image reached 6MB by performance enhancement of medical recording equipment. Also, because DICOM file contains many image files, file size becomes more and more massive. It is difficult to apply encrypting the whole DICOM file because of matter of performance.

Therefore it is necessary that a method is capable of providing protection to transform only the partial element of DICOM file. In this paper, to solve this problem, distinguishing a minimum part of medical image in JPEG/JPEG2000 format and encrypt it, we propose a method (MPEM-JPEG/JPEG2000) that has obfuscation and integrity check.

2 Related Study

M. Ulutas proposed a technique which adds further authentication procedure in managing the data on the EPR medical image based on 'Secret sharing' presented by

* Corresponding author.

M. Gen et al. (eds.), *Industrial Engineering, Management Science and Applications 2015*,
Lecture Notes in Electrical Engineering 349, DOI: 10.1007/978-3-662-47200-2_3

Shamir [3]. Secret sharing divides the secret key used for encryption into n pieces of key based on the proposed polynomial, they are used for decryption after assembled [4]. If secret sharing applies to encryption of the medical data, increase the reliability by storing to divide the key, we can expect an effect of enhancing the confidentiality, since it needs a number of key in decryption. But this method is very effective in terms of safety because of encrypting personal information and medical image, in the case of large files, it is difficult universally to use it because the dependence of the computing power is very large.

Giakoumak proposed a multiple watermarking method based on wavelet transform for copyright protection and integrity verification of the medical image [5]. The method performs wavelet transform on an image to protect and selects low frequency region that is key information and inserts the digital signed watermark. The watermark image is inserted in the frequency domain. This method very effects on the integrity verification of medical image and copyright protection due to robust against techniques that bypass the watermarking such as Cut-and-paste attack or counterfeiting attack. However, such watermarking techniques don't exist or are insufficient obfuscation effect, if a watermarked medical image has been collected without consent of the patient, there may be a risk that can exposes like physical defects and pathological records of patients.

3 JPEG Medical Image Protection Using DQT Encryption

3.1 DQT (Define Quantization Table)

An image compression process of JPEG format is composed of color transform, DCT process, quantization, Huffman coding process like the Fig 1. DQT and the quantization is a key process in the lossy compression of JPEG standard, a quantization table which plays an important role in image restoration is generated in two processes. DCT is a process that reduces the size as minimizing the loss of quality of the image by storing the image frequency data obtained by the discrete cosine transform.

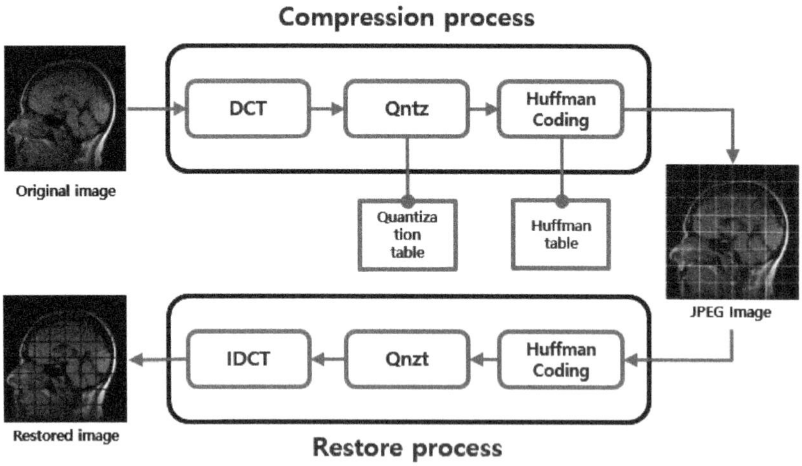

Fig. 1. Compression and Decompression Process of JPEG Image

(1) is a representation of discrete cosine transform process by formula. The amplitude C of cosine is called DCT coefficients, DCT will obtain new array F using the existing array f. At this time, DCT coefficients data is divided into 8x8 quantization table and stored. If this quantization table is abnormally corrupted, since DCT coefficient restoration not be achieved successfully in image restoration, it is not restored to the original image. (2) is expressed as a formula that array T is to be output. Array T is different from array f because the DCT coefficients E is abnormally restored due to corrupted quantization table [6].

$$F(i,j) = C(i,j) \sum_{n}^{N-1} \sum_{m}^{N-1} f(n,m) \cos\frac{(2n+1)j\pi}{2N} \cos\frac{(2m+1)i\pi}{2N}$$
$$C(0,0) = \frac{1}{N}, C(i,j) = \frac{2}{N} (i \neq 0, j \neq 0) \quad (1)$$

$$T(n,m) = \sum_{i}^{N-1} \sum_{j}^{N-1} E(i,j) F(i,j) \cos\frac{(2n+1)j\pi}{2N} \cos\frac{(2m+1)i\pi}{2N} \quad (2)$$

3.2 MPEM-JPEG (Minimum Partial Encryption Method for JPEG)

As described in section 3.1, a quantization table corruption can be used for image obfuscation. In particular, the quantization table has the size only up to 256bytes in least 64bytes regardless of the image size. MPEM-JPEG proposed in this paper is how to perform image obfuscation to encrypt only the quantization table.

MPEM-JPEG has the following four kinds of advantages. Firstly, encrypting the quantization table of a fixed size less than 256 bytes regardless of the size, it can significantly reduce the computing power dependence such as memory usage. Secondly, although such obfuscation method can't ensure 100% confidentiality of medical image, it achieves an obfuscation level image can't be read out. Thirdly, it can prove the integrity without any padding bits by using CBC-MAC because size of DQT is a multiple of 128bit [7]. Finally, encrypting by extracting only the quantization table in order to apply the MPEM-JPEG, it is not necessary to change the conventional JPEG image compression process.

3.3 Testing Result

Fig. 2 is a result that a medical image is obfuscated using PEM and restored. (B) and (E) is to encrypt quantization table having a size of 128 bytes that is included in the original image (A) and (D) by using an AES-CBC algorithm. As shown below, although the overall outline can be recognized, it is not possible to recognize morbid records or part related to the physical characteristics of the patient because of obfuscation. (C) and (F) is to decrypt the quantization table, we can see that being restored to the same state as the original image.

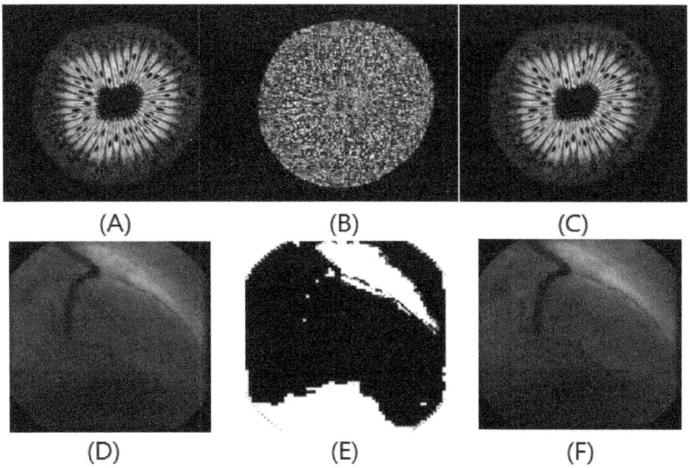

Fig. 2. Original Image-(A),(D), Encrypted image(B),(E), Restored Image-(C),(F)

4 JPEG2000 medical image protection using DWT

4.1 DWT (Discrete Wavelet Transform)

Image compression process of JPEG2000 format is consists of DWT, quantization, entropy coding process such as Fig. 3, there is a differences as compared with the JPEG. That is a quantization table is not generated in quantization process and DCT process has turned into a DWT process [8]. Medical image protection scheme using DQT encryption proposed in chapter 3 can't be applied to the image of the JPEG2000 format because there is no quantization table.

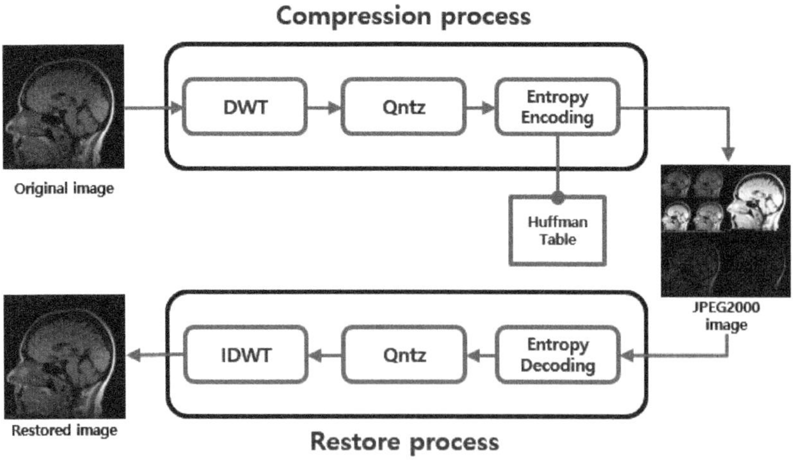

Fig. 3. Compression and Decompression Process of JPEG2000 Image

Critical part of image compression in JPEG2000 is DWT process. Since DCT process of JPEG format allocates less bits or removes high-frequency components that a human eye does not recognize, it has a disadvantage that is to occur blocking artifacts in the compressed image with high magnification. On the other hand, DWT process of JPEG2000 does not cause the blocking artifact by splitting the image each frequency domain. An original image is decomposed into LL, LH, HL, HH image data from filtering by using high pass filter (H) and low pass filter (L) in DWT process [9] [10].

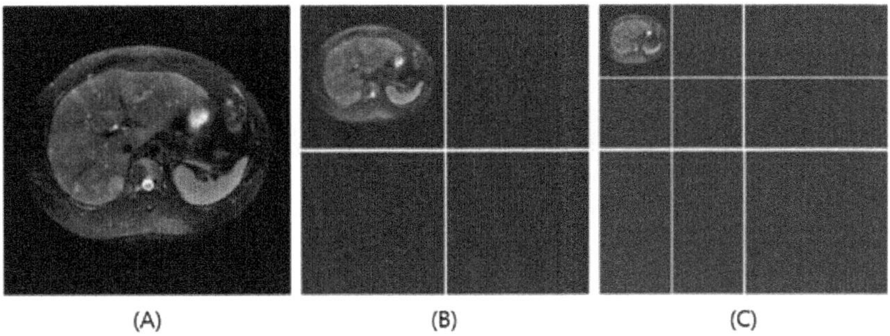

Fig. 4. (A) Original Image, (B) 1 Level DWT, (C) 2 Level DWT

Fig. 4 shows the DWT process. It is called 1 level DWT transformation that filtering is completed once, after filtering the LL image once again, this is 2 levels DWT transformation. The more the number of DWT, the higher the compression ratio. Small brain imaging data (C) located on the left top in Fig 3 is a resolution level R_0 as the image data created after the last DWT process. Since the R_0 region includes a sensitive data of the low frequency band to a human eye, it can be seen to appear in a similar form to the original image. If this R_0 region is abnormally corrupted, it is not restored to the original image because the low-frequency data of the original image can't be obtained exactly.

4.2 MPEM-JPEG2000(Minimum Partial Encryption Method for JPEG2000)

As described in section 4.1, it can be used for image obfuscation to corrupt LL image data in R_0 region. The size of R_0 region is variable depending on the original image and resolution level, it can be obtained from the header of the packet including R_0 region. Packet headers have a size of 3~4 bytes, it can be found through storing the size of data included in packet.

The proposed MPEM-JPEG2000 in this paper is how to perform the image obfuscation by encrypting only image data in this R_0 region. As MPEM-JPEG2000 is encrypting very small data compared with the original image, it can significantly reduce the computing power dependence. Also since MPEM-2000 encrypts to extract only R_0 region in JPEG2000 image, it does not change an existing JPEG image compression process.

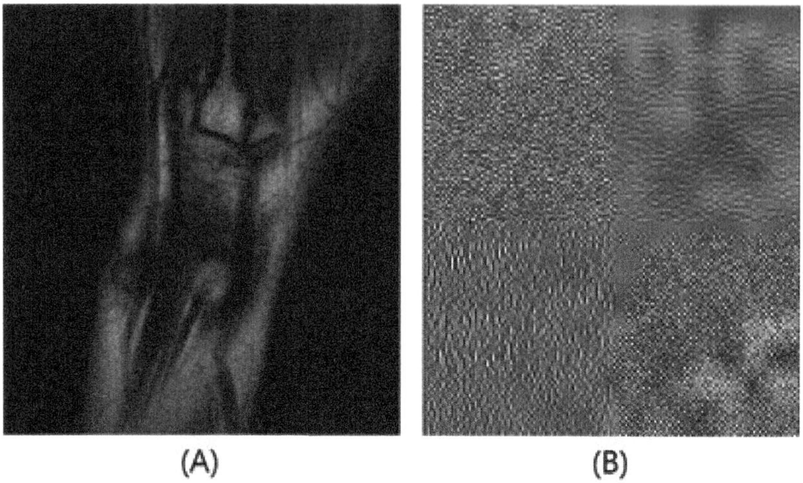

Fig. 5. (A) Original JPEG2000 Image, (B) Encrypted JPEG2000 Image

4.3 Testing Result

Fig. 5 shows the result of medical image obfuscation by using MPEM-JPEG2000. (A) is JPEG2000 original image (1 level DWT) that has size of 87.9KB extracted from the actual DICOM files, (B) is a result of encrypting the entire packet 222bytes including the R_0 region. As shown in (B), despite encrypting 222bytes which is 0.25% of the original image 87.9KB, we can see that the original image is completely impossible to distinguish.

5 Conclusion

In this paper, we proposed a minimum partial encryption for medical image in JPEG and JPEG2000 to manage safety and efficiently. First, we introduced encrypting a quantization table to protect medical image in JPEG format. This method can significantly reduce the computing power dependence such as memory usage because it just encrypt a quantization table with a minimum of 64bytes and a maximum of 256bytes regardless of size. Next we proposed to encrypt LL image data that is generated after the last DWT process for medical images protection in JPEG2000 format. Although it encrypts only very small partial compared with an original image, there is an advantage that to recognize entire image is impossible since the proposed method is to encrypt only the data of the low frequency band sensitive to a human eye. Two proposed method have an advantage that is not require change about compression process because only some of the data needed for obfuscation is encrypt.

Acknowledgments. This work was supported by the IT R&D program of MSIP/KEIT. [10041579, Development the Personal Information Security service solution using tokenization technology].

References

1. Parisot, C.: The DICOM Standard. The International Journal of Cardiac Imaging
2. Lee, S.M.: The medical treatment informatization and medical treatment information protection. Journal of Korean Law (2008)
3. Ulutas, M., Ulutas, G.: VV. Nabiyev.: Medical image security and EPR hiding
4. Shamir, A.: How to share a secret. Communications of ACM 22, 612–613 (1979)
5. Giakoumaki, A.: A medical image watermarking scheme based on wavelet transform. Engineering in Medicine and Biology Society (2003)
6. Wallace, G.K.: The JPEG Still Picture Compression Standard. IEEE Transactions on Consumer Electronics (1991)
7. Daemen, J., Rijmen, V.: AES Proposal: Rijndael. Version2. NIST (1999)
8. Lee, W.-Y., Ou, Y., Rhee, K.H.: Selective Encryption Scheme Based on Region of Interest for Medical Images. Journal of Korea Multimedia Society 11, 588–596 (2008)
9. Jung, G.-H., Han, W.-J., Yoo, D.-S., Choi, S.-C., Kim, E.-K.: Evaluation of compression ratios using JPEG 2000 on diagnostic images in dentistry. Korean Journal of Oral and Maxillofacial Radiology 35, 157–165 (2005)
10. Lim, J.M., Yoo, J.: Super-resolution Algorithm using Discrete Wavelet Transform for Single-image. Journal of Broadcast Engineering 17, 344–353 (2012)

The Optimal Hedging Strategy for Commodity Processors in Supply Chain

Ehsan Bolandifar and Zhong Chen

The Chinese University of Hong Kong Business School, Hong Kong
{ehsan,zhongchen}@baf.cuhk.edu.hk

Abstract. This paper studies the optimal hedging strategies for risk neutral players in a competitive supply chain setting. We consider two processors procuring two commodities from spot markets to process and sell through index-based contracts to a retailer. Commodity prices are stochastic and correlated. The retailer is facing price sensitive demand curves so it controls its market demands through market pricing. First, we characterize the optimal index-based contracts for processors which implies that processors prefer to be exposed to commodity price risks. We show that the processor's optimal contract consists of a processing margin which is independent of its financial hedging decisions and a hedge ratio which is a function of commodity price volatility. We show that processors can benefit from market pricing, when these prices are linked to input commodity prices and index-based contracts are a means to link these prices. We also explore how commodity and product market parameters affect optimal hedging decisions.

Keywords: Supply chain hedging, commodity procurement, index-based contracts.

1 Introduction

Nowadays, commodities are exposed to even more unexpected price fluctuations due to the drastic uncertainty in supply-demand and potential global political and financial crises. A recent example of unexpected drastic changes in commodity prices is the oil price which dropped from more than 100 dollars per barrel to about 40 in the second half of 2014. This fluctuation creates unprecedented challenges for commodity processors in procurement and process of their outputs. Usually, since commodity processors have limited market power to pass any shock in their input commodity prices to their downstream firms and customers, they are more susceptible to downside effect of high input commodity prices. They are usually exposed not only to input price risks but also to demand risks since their demand is also correlated with commodity prices. This highlights the importance of hedging in mitigating price risk and maximizing a firm's expected value.

Hedging, as Van Mieghem (2003) defines, "involves taking counterbalancing actions so that, loosely speaking, the future value varies less over the possible states of nature". One way to do hedging is to trade financial instruments, like futures,

M. Gen et al. (eds.), *Industrial Engineering, Management Science and Applications 2015*,
Lecture Notes in Electrical Engineering 349, DOI: 10.1007/978-3-662-47200-2_4

options and other financial derivatives which is referred to as *financial hedging*. In this paper, we study index-based contracts as a tool that processors use to reduce volatility of their profits. In an index-based contract, transaction price is determined as a function of spot and forward prices which is a weighted function of forward prices observed at the time of contracting for the ordering time and the spot price to be observed at the time of ordering. The weight on the forward contract versus the one on the spot price determines the volatility of the processor's transaction prices and therefore its profit.

Index-based price contract is commonly used in many industries dealing with commodities like agribusinesses and petroleum industries. For example, Kuwornu et al., (2009) studies a food supply chain agency problem where providing price is correlated with the retail value, which is random. In the petroleum industry, the largest natural gas supplier in the world, Russia, was negotiating with China as one of the largest consumers of natural gas in the world for about 10 years to price the future supply for the next 30 years. Since gas price is quite volatile and might change drastically during the next 30 years, it is believed that the transaction price is in the form of index-based contract as a function of the oil price (Wall Street Journal, 2014).

In our model, the index-based contract between supply chain firms is correlated with the spot price of the processor's input commodity. It's similar to trading in an over-the-counter market[1], where firms are free to negotiate any mutual trading in the future, and the contracts do not have to be specified as in the exchange-traded markets. After supply chain firms agree on the index-based contract, unit price would be realized at the time of contract execution. This form of contract helps the processor link its future prices to its input prices, thus makes its profit less volatile, following Van Mieghem (2003), we refer to this contract as financial hedging.

A commodity processor that produces a commodity based product to sell through a downstream firm is not only exposed to its own input price risks but it might be also exposed to substitutable commodity prices through its downstream orders for substitutable products. For example, a beef processor is not only exposed to the beef spot prices but it is also exposed to the pork spot prices since consumers may consider substituting beef products with pork in the case of sudden increase in the beef products prices. While we expect that beef and pork spot prices be correlated but still this correlation is partial and consumers may substitute one for the other.

We study a supply chain where two commodity processors that procure raw material from commodity spot markets process and sell their substitutable outputs to a common retailer. Spot prices for these input commodities are stochastic and correlated. We assume that retailer's demand is price sensitive so the retailer controls its market demand through market pricing. While all firms are risk neutral, we show that processors partially hedge their exposure to the input commodity price risks. Unlike processors, under certain conditions, retailer has a strong preference to be fully exposed to the commodity price risks. For example, when commodities are not close substitutes, the retailer's profit is maximized under cost plus contracts where processors completely pass their exposure to the input commodity prices to the retailer. This is due to the fact that the retailer has a natural hedge against any changes in its input prices, i.e., any increase in retailer's input costs can be reflected in its

[1] Options, Futures, and Other Derivatives, John C. Hull, 8th edition.

market prices. This not only increases the retailer's margin but it also results in a smaller customer orders when input costs are high. Similarly, any decrease in the input commodity prices would result in lower market prices but it brings higher orders which again benefits the retailer. We also show that how index-based contracts perform as a mechanism which allows processors to benefit from retailer's natural hedge.

While we show that partial hedging is optimal for processors, we also show that it decreases supply chain profit. We also investigate how commodity and product market parameters affect optimal hedging strategies. We show that more volatility is favorable for all firms since it increases the value of the natural hedge for the whole supply chain. Similarly, any strong correlation between commodity prices or high degree of substitution decrease the value of the natural hedge and therefore processor's optimal hedge ratios.

The rest of the paper is organized as follows. In section 2, we set up our model. The optimal partial hedging strategy is analyzed in Section 3. And, finally, we conclude the paper in section 4.

2 The Model

We consider a retailer procuring two substituable products (e.g., packed meat like beef and pork) from two processors. These two processors, denoted by P1 and P2 procure two input commodities from two spot markets, and after processing these commodities they sell their outputs to the retailer. An important feature of our model is that the processors' products are not commodity anymore, i.e., the retailer faces two demand curves which allows it to control its demand by pricing its products in the market. We consider a linear duopoly demand model as follows.

$$q_1 = 1 - p_1 + \gamma p_2, \quad (1)$$

$$q_2 = \alpha - p_2 + \gamma p_1. \quad (2)$$

This model has been widely used to model price competition in the market (see, e.g., Feng and Lu, 2013). In this model, α denotes P2's market size while P1's market size has been normalized to 1. Without loss of generality we assume that $\alpha \geq 1$. The parameter γ captures the demand interdependence for the products and satisfies $0 \leq \gamma < 1$. When γ is equal to zero, the retailer's demands for its products are independent and there is no market competition between these products. As γ increases, products become more substitutable, and one product's price has a larger impact on the other product's demand. Thus γ also reflects the intensity of competition between the processors' products. $\alpha > 1$ implies that P2's product has a stronger position than P1's product in the market, i.e., P2 enjoys a higher demand if the retailer charge the same market price for both of these products. All firms' production costs are normalized to zero to simplify our exposition without affecting the qualitative findings. Each firm's objective is to maximize its own profit.

We model this problem as a one period model. At the beginning of the period, $t = 0$, both processors offer their index-based contracts to the retailer as Stackelberg leaders in the pricing game and at the end of the period, $t = 1$, the retailer places its orders at both processors, after realization of commodity and therefore transaction prices. We assume that there is a time lag between firms' decisions on priceing structure and ordering, which is in-line with what we observe in practice. It is straightforward to show that without this assumption there is no value for hedging through index-based contracts. Processors are exposed to input commodity price risks. Since they are offering index based contracts to the retailer, retailer is also exposed to commodity price risks. Let q_i and w_i denote the order quantity and contracted index-based price between processor i and the retailer, for $i = 1,2$. We denote index-based price for processor Pi as:

$$w_i = b_i F_i + (1 - b_i) S_i + d_i, \tag{3}$$

where S_i and F_i denote commodity i's spot and forward prices for $t = 1$ and d_i is the fixed processing margin charged by processor i. In this contract, b_i denotes the hedging ratio for the retailer, since it fixes part of the contract price at $b_i F_i$. This indicates that $100 \times b_i$ percent of q_i is bought at F_i while the rest needs to be bought at the spot price, S_i. By reducing the hedge ratio, b_i, processors reduce their exposure to the input commodity price risk while this increases the retailer's exposure to commodity price risks. For example, when b_1=0, processor P1 guarantees its profit margin to be d_1 since it is offering a cost plus contract (i.e., $w_i = S_i + d_i$) to the retailer, and when $b_1 = 1$, the retailer is not exposed to any price risks since the contract price is fixed (i.e., $w_1 = F_1 + d_1$). The second scenario is indeed the *perfect hedging* in Van Mieghem (2003)[2]. Figure 1 provides a schematic view of the supply chain.

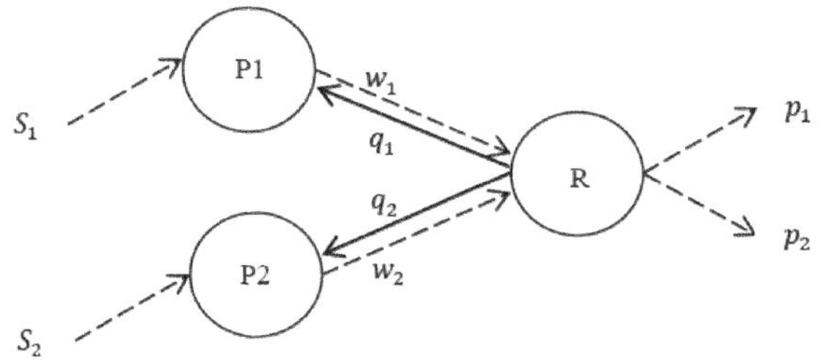

Fig. 1. Schematic view of the supply chain

To facilitate our analysis, we are making the following standard assumptions.

Assumption 1. There is no arbitrage opportunity in the spot market, and risk premium for forward markets is normalized to zero, i.e., $E[S_i] = F_i$, for $i = 1,2$.

We assume that the commodity market is competitive with the same information available to all players in the market, so if the expected spot price is not equal to the forward price, there will exist profit making opportunities without risk, so risk-neutral

[2] Perfecting hedging is a portfolio that provides a constant future value in any state of nature.

speculators will enter the market which induce the equality of forward price and expected spot prices (Popescu and Seshadri 2013). We also assume that both input commodity prices are positively correlated.

Assumption 2. Input commodity prices are positively correlated, i.e.,$0 < \rho < 1$.

Since we are assuming that final products are substitutable, it is natural to assume that this assumption is also satisfied which indicates that processors' input prices are positively correlated. In practice, when processors offer index-based contracts they choose a ratio between 0 and 1. We formalize this assumption as follows.

Assumption 3. The hedge ratios for both processors satisfy $0 \leq b_i \leq 1$, for $i = 1,2$.

This assumption helps to ensure the price of the index contract is always positive, and the processors can fully or partially hedge their exposures to commodity price risks.

In what follows, we characterize the subgame perfect equilibrium of the procurement game, so we start our analysis by investigating the retailer's optimal decisions which is its market price after realization of spot prices for input commodities.

We formulate the retailer's pricing decisions in the market as:

$$max_{p_1,p_2} \pi_R = q_1(p_1 - w_1) + q_2(p_2 - w_2), \tag{4}$$

where order quantities are given by (1) and (2) and index-based prices are given by (3). To simplify our exposition, we assume that poduction cost of both processors are normalized to zero and there is no inventory or managerial cost for the retailer.

Anticipating the retailer's optimal ordering, processors solve for their optimal index-based contract (b_i, d_i) at the beginning of the period to maximize their expected profits, $E[\pi_{Pi}]$, where expectations are taken with respect to both commodity prices joint distribution at the end of the period. We can write processor i's problem as:

$$max_{b_i,d_i} E[\pi_{Pi}] = E[q_i(w_i - S_i)]. \tag{5}$$

In the next section we try to shed some light on the processors' preference for partial hedging in a supply chain setting.

3 Analysis

In the previous section, we formulate our problem. After solving (5), our analysis indicates that processors need to partially hedge their exposure to their input commodity price risks by offering $0 \leq b_i < \frac{1}{2}$, for $i = 1,2$. Notice that processors are risk neutral in our settings but still they ask for index-based contracts. To better understand this preference, we start by investigating the retailer's profit function. It is straightforward to show that for any given index-based contract, the retailer's profit is convex in its input prices. This indicates that as a direct result of Jensen's inequality, the retailer does not have any incentive to hedge its exposure to input price risks when it is offered index-based contracts by its processors.

As offered hedge ratios by processors increase, the input price of the retailer becomes less volatile, i.e., the input commodity prices affect retailer's input prices less. As a result of less volatile input prices, the retailer has less flexibility to manipulate its market prices and therefore demands. This reduces the retailer's expected profit. As a result, the retailer prefers to be fully unhedged (i.e., $b_1 = b_2 = 0$) (when substitute rate is not very large)but since processors are the Stackelberg leaders in the pricing stage, the retailer's preference for cost-plus contract needs to be inline with processors' interest, which we show is not.

Proposition 1. The processor Pi's expected profits are joint concave in it's own hedge ratio and processing margin. The optimal hedge ratios are given by:

$$b_1^* = \frac{2 - \rho^2\gamma^2 - \gamma\rho\sigma_{2/}\sigma_1}{4 - \rho^2\gamma^2},$$
$$b_2^* = \frac{2 - \rho^2\gamma^2 - \gamma\rho\sigma_{1/}\sigma_2}{4 - \rho^2\gamma^2},$$

and optimal processing margins are given by:

$$d_1^* = \frac{2 + \alpha\gamma + \gamma^2 F_1 + \gamma F_2 - 2F_1}{4 - \gamma^2},$$
$$d_2^* = \frac{2\alpha + \gamma + \gamma^2 F_2 + \gamma F_1 - 2F_2}{4 - \gamma^2},$$

where σ_1 and σ_2 denote standard deviations for the first and the second commodity prices, respectively.

In our model, processors are Stackelberg leaders in the pricing stage but it is the retailer that determines market prices and therefore order quantities. This is in-line with what we usually observe in practice where firms negotiate over their transaction price structure which would be realized at the time of contract execution and processors are required to produce and deliver retailer's order quantities (i.e., contracts are enforceable). Even though processors do not have the power to determine market prices and order quantities, still they can pass part of their exposure to the input price risks to the retailer to affect their market prices and order quantities. Index-based contracts, is a means that processors use to benefit from retailer's natural hedge. High input costs results in high transaction prices which would be accompanied by lower orders from the retailer. Similarly, low input costs reduce transaction prices which brings larger order to processors. But, processor's profit does not monotonically increase in hedge ratios. As hedge ratios increase, initially processors benefit from sharing extra value generated through retailer's market pricing but at the same time, it reduces the value that retailer's market price brings to the supply chain (Since it reduces retailer's input price volatility). This results in concave profit functions for processors in their hedge ratios. In summary, processors can benefit from market pricing, when these prices are linked to input commodity prices. Index-price contract is a means to link them.

After characterizing the optimal hedge ratios for processors, we can investigate how different final product and commodity market parameters affect optimal contract terms. The following proposition characterizes the effect of input commodity price variances (σ_i^2), correlation (ρ) and market substitution rate (γ), on the optimal hedge ratios (b_i^*).

Proposition 2. a) Commodity Price Variances: The hedging ratio for each commodity increases in its price variance, while it decreases in the competing commodity price variance.
b) Commodity Price Correlation: The hedging ratio for each commodity decreases in ρ.
c) Demand Substitution factor: The hedging ratio for each commodity decreases in γ.

Part (a) indicates that for a processor, if its input commodity price volatility increases, it would ask for larger hedge ratio. As volatility of the input commodity price increases, the value of the retailer's option to price its product in the market increases, therefore the processor would be able to ask for a larger share of this option's value. On the other hand, if competing commodity price becomes more volatile, the processor needs to ask for smaller margins since competing processor would be able to increase its requested margin.

If the correlation of the two commodity prices increases, the retailer's option to price its products would become less valuable which indicates that both processors need to reduce their hedge ratios and margind. Similarly, when the retailer's products are more substitutable, the retailer's option to price its products would be less valuable and again both processors need to reduce their resuested margins and hedge ratios.

And finally, it is interesting to know how the whole supply chain profit would change as a function of the processors' partial hedging policies.

Proposition 3. The whole supply chain's expected profit is concave in b_i and maximizes at $b_1 = b_2 = 0$.

This observation is inline with our previous findings. Any partial hedging policy by processors reduces the volatility in input prices for the retailer, which reduces the value of the whole supply chain's option to customize market prices based on input prices. This results in reduction in the whole supply chain profit. Therefore, to maximize supply chain profit, cost plus contract are the favorite form of contracts.

4 Conclusion

In this paper, we explore a new form of financial hedging in supply chain settings through index-based contracts. Index-base contract is a popular mechanism that commodity processors usually use to pass part of their exposure to commodity price risks to their customers in commodity-based supply chains. We study a general framework in which all firms are risk neutral and still we show that commodity processors are interested to partially hedge their exposure to commodity price risks. In particular, we show that even though processors can ask for full hedge by offering cost-plus contracts to their retailer, they still prefer to be exposed to the commodity price risk. While processors are exposed to commodity price risks, they are also exposed to the demand risk when they pass part of their exposure to commodity price risks to the retailer through index-based contracts. We show that for the retailer, their ability to charge different prices based on the realized input prices to control their market demand works as a natural hedge against the volatile input prices which brings extra value to the retailer compared to a retailer which is fully hedged. Since processors do not have the capability to affect their production quantity after spot market realization, they choose to partially hedge their exposure, which indicates that the retailer is also partially exposed to the commodity price risk. In summary, processors can benefit from market pricing, when these prices are linked to input commodity prices. Index-price contract is a means to link them.We show that processor's optimal contract consists of a margin charged for production and also a margin that it charges to benefit from retailer's market pricing power. We show that how different commodity and product market parameters affect processors optimal hedging decisions. For example, all firms prefer more volatile commodity prices which decreases processors' hedge ratios.

Our paper provides another explanation for the existence of partial hedging and index-based contracts in commodity processing supply chains.

References

1. Feng, Q., Lu, L.X.: The role of contract negotiation and industry structure in production outsourcing. Production and Operations Management 22(5), 1299–1319 (2013)
2. Hull, J.C.: Options, Futures, and Other Derivatives. Prentice Hall (2012)
3. Kuwornu, J.K.M., Kuiper, W., Pennings, J.M.E.: Agency problem and hedging in agri-food chains: Model and application. Journal of Marketing Channels 16(3), 265–289 (2009)
4. Mock, V., Spegele, B., Ma, W.: China and Russia Sign Natural Gas Deal. The Wall Street Journal (May 21, 2014)
5. Popescu, D.G., Seshadri, S.: Demand uncertainty and excess supply in commodity contracting demand uncertainty and excess supply in commodity contracting. Management Science 59(9), 2135–2152 (2013)
6. Van Mieghem, J.A.: Capacity management, investment and hedging: review and recent developments. Manufacturing and Service Operations Managment 5(4), 269–302 (2003)

Application of MFCA and Dynamic Programming in Operations Improvement: A Case Study

Atchara Songkham and Chompoonoot Kasemset*

Department of Industrial Engineering, Faculty of Engineering,
Chiang Mai University, Chiang Mai, Thailand
{atchara_s,chompoonoot.kasemset}@cmu.ac.th

Abstract. This study aimed to present an application of Material Flow Cost Accounting (MFCA) to discover loss in the process and dynamic programming in decision making of improvement solutions. The research focused on internal supply chain operations of the case study. Based on MFCA analysis, the overall cost can be classified as the cost of positive product as 46.97% and the cost of negative product as 53.03%. The highest cost of negative product was material cost (MC) 45.52% following by system cost (SC) 6.41%, and energy cost (EC) 1.10% of the total product cost. The improvement solutions were proposed for MC reduction. Industrial engineering problem solving tools; cause and effect diagram, pareto diagram and why-why analysis, were used to investigate root causes for negative MC. Seven improvement solutions were proposed and evaluated based on MFCA calculation. The decision making was carried out using dynamic programming when the incursion was formulated using the cost of positive product increased by each solution. Finally, only the 1st, 2nd, 6th and 7th solution were selected and the cost of positive product can be increased from 46.97% to 56.08% and the cost of negative product reduced from 53.03% to 43.92%

1 Introduction

Material Flow Cost Accounting (MFCA) is one of environment management tools that aim to reduce the environmental impact and cost at the same time because wastes from operations mean loss of input resources. The key concept of MFCA is to present the cost of production in term of the cost of positive and negative products. The cost of negative product is the main problem that should be reduced as small as possible. To reduce the cost of negative product, there is always more than one solution that can be considered. In many cases, it is very difficult to make decision on which solutions are the optimal plan for the improvement.

This study aims to apply the concept of MFCA to improve the production line of the case study company. The improvement solutions are proposed based on MFCA analysis and industrial engineering tools, i.e. ECRS concept, cause and effect diagram and pareto chart, to reduce input material and the cost of negative product and increase the cost of positive product. During decision making for improvement, the concept of MFCA is adopted in dynamic programming to select the best improvement solution among all solutions.

* Corresponding author.

M. Gen et al. (eds.), *Industrial Engineering, Management Science and Applications 2015*,
Lecture Notes in Electrical Engineering 349, DOI: 10.1007/978-3-662-47200-2_5

2 Preliminaries

MFCA is a concept that interprets waste and environmental impacts in term of cost to allow management person to easily understand. Based on MFCA concept, cost is classified as "cost of positive product", cost of quantity of materials made into company products, and "cost of negative product" which is overall materials wasted. Both costs can be identified into four categories of cost as material cost (MC), system cost (SC), energy cost (EC) and waste management cost (WC) [1, 2]. The total cost of negative product from all four parts is considered to be reduced during the step of improvement solution implementation.

To implement MFCA, there are 7 steps as presented in [2] that are; (i) Select the target product and process, (ii) Collect concerned data and information, (iii) Perform MFCA calculation, (iv) Identify points for improvement, (v) Introduce improvement methods, (vi) Implement improvement methods and (vii) Evaluate improvement methods by performing MFCA calculation again and comparing the results.

There were many research works applied MFCA in various case studies: plastic packaging company [3], brewery manufacturer [4], textile producers [5, 6], adhesive tape factory [7] and household drain boards company [7].

MFCA can be used as an identification tool for pointing the location of waste along a production line. To reduce the cost of negative product, other industrial engineering techniques should be employed. ECRS's principle was applied after MFCA analysis in order to reduce material wastes from plastic packaging production line in [3]. Design and experiment (DOE) was applied with MFCA to reduced wastes from wooden product in [8]. Plant layout and design was implemented after MFCA analysis in [9] in food industry to reduce the transportation distance in the case study production line to reduce defect product during the internal transportation.

After MFCA analysis, there was certainly more than one solution for improvement. When there were many improvement solutions, decision making techniques should be considered. Dynamic programming is one widely used tool for find the best solutions among many solutions with different effects. The concept of dynamic programming is to solve large problem by breaking into small problems and solve small problem until all problems are solved [10]. The application of Dynamic programming and MFCA was firstly proposed in [11] for multistage serial production processes using the cost of positive product and the investment cost for each improvement solution to form the recursion equation.

3 Research Methodology

The research methodology was presented as follows:

3.1 Perform MFCA Procedure

During this step, 7 steps of MFCA implementation were carried out. Firstly, Target product and process were selected. Then, data of the target product and process including process steps, machine list, raw material used in process, raw material cost, labor cost, electricity charge, energy consumption and loss in process, were collected. Secondly, these data were used to create the material flow chart to represent the flow and loss of material along the whole processes. The cost of positive and negative product can be calculated using material balance concept and MC, SC, EC and WC were also classified during the third step. Then, during the fourth and fifth steps, the results from MFCA analysis were presented in from of Pareto chart to identify the main negative product cost and problem analysis tools, cause and effect diagram and why – why analysis, were used to find out the root cause of the main negative product cost. Subsequently, improvement methods were proposed based on ECRS concept. The sixth and seventh steps were to implement the improvement solution and evaluated the results using MFCA calculation again. Finally, there were many improvement solutions with different effects that were carried on to the next step of dynamic programming to make the decision on the practical improvement solutions.

3.2 Decision Making Using Dynamic Programming Technique

The characteristic of dynamic programming is the problem can break into the small problems or stage and there are the decision choices in each stage or state. The solution was separated in to T stages and must repeat the equation for returning the calculation from stage *t, t+1,...,T* to stage *t+1, t+2,...,T*. This research identified the process as stage and the improvement method as Fig 1.

Each stage (t) was set as each process or quality center (QC). In each stage, there can be more than one improvement method defined as node (i).

The incursion can be formulated as equation (1) and (2)

$$f_t^*(s) = max f_t(s, x_t) \tag{1}$$

$$f_t(s, x_t) = \left(\frac{PC_{st} + PC_{st-1}^*}{TC_{st} + TC_{st-1}^*} \right) * 100\% \tag{2}$$

Given:

f_t = Positive Product Cost of stage *t* in %
x_t = Selected method after stage *t*
PC_{st} = Total Positive Product Cost of stage *t* in THB
TC_{st} = Total Cost of stage *t* in THB

3.3 Improvement Solution Evaluation Using MFCA Analysis

The set of improvement solutions selected based on dynamic programming was evaluated again based on MFCA analysis to present the overall improvement gain from these set selected solutions.

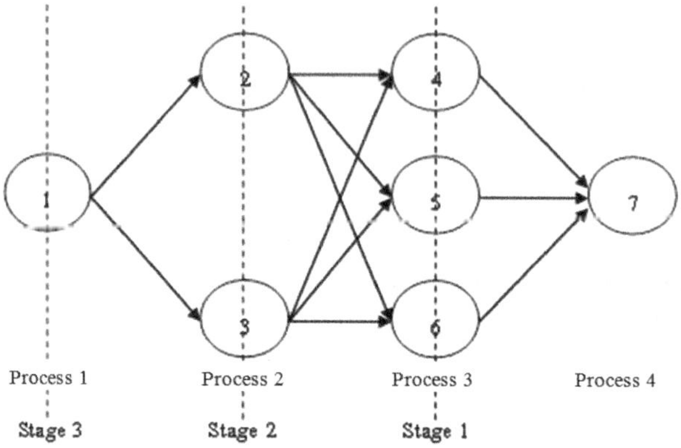

Fig. 1. Dynamic Programming Formulation for MFCA

4 Results

4.1 MFCA Analysis

The comfort chair was selected to be the target product because it was the main product of the case study company. The production process of this product consists of 6 processes or QC presented as material flow chart as Fig 2. This flow chart was created to show the input material and waste of this production line.

From Fig. 2, it was found that the maximum loss occurred at QC3 as 20.36%. All of the waste from wood can be used as fuel in factory's boilers and no rework process found in operations so there was no WC occurred along this production line and WC was ignored for this case study.

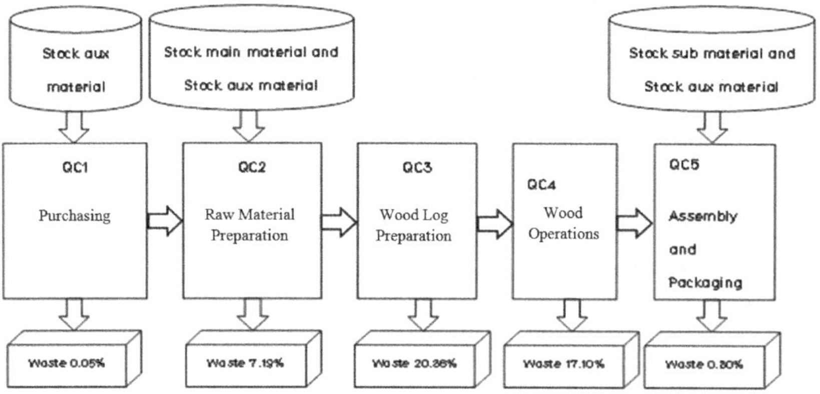

Fig. 2. Material flow chart of comfort chair

The MFCA analysis displayed the detail of newly input total cost, total cost handed over from previous process, process total of input cost, percent quantity of negative product, percent quantity of positive product, positive product cost and negative product cost as shown in Fig 3.

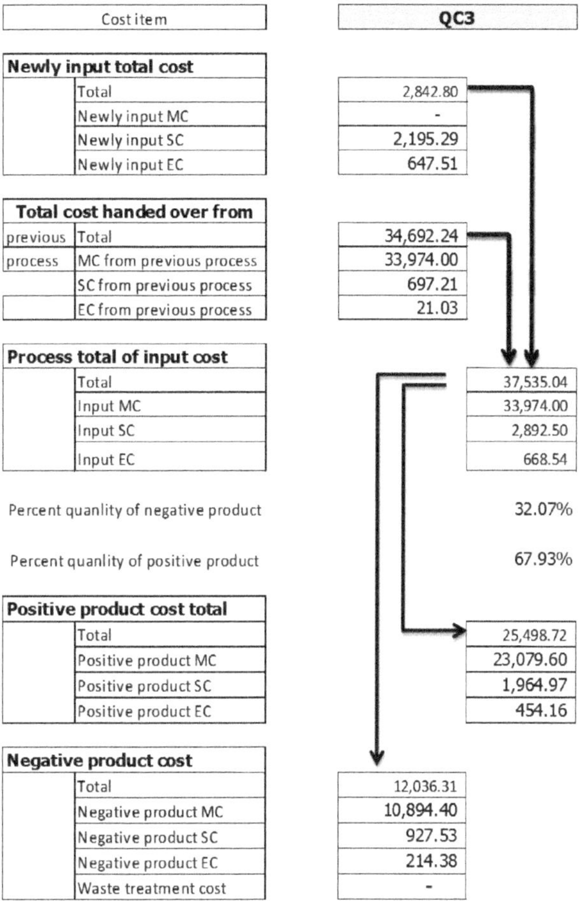

Fig. 3. Example of MFCA analysis for QC3

The results from MFCA analysis showed total cost of one production lot of comfort chair (50 sets) was 53,495.82 THB[1] divided as the cost of positive product 25,128.77 THB (46.97%) classified as MC 20,598.06 THB (38.50%), SC 3,986.59 THB (7.45%) and EC 544.11 THB (1.02%) relatively, and for the cost of negative product was 28,367.05 THB (53.03%) classified as MC 24,349.66 THB (45.52%), SC 3,427.61 THB (6.41%) and EC 589.79 THB (1.10%) as shown in Table 1. From these results, the maximum negative product cost was MC as 84% comparing with the total cost of negative product.

[1] Approximately 32.40 THB = 1 USD as of January 2015.

Based on the data from MFCA analysis, the Pareto chart was developed as Fig 4 to present the process with negative product cost of MC. From the chart, the maximum negative product cost occurred at QC3 as 44.74% following by QC4 as 37.57%.

Table 1. MFCA calculation (Before improvement) (cost unit: THB)

	MC	SC	EC	Total Cost
Total	44,947.72	7,414.20	1,133.89	53,495.82
	84.02%	13.86%	2.12%	100.00%
Positive	20,598.06	3,986.59	544.11	5,128.77
	38.50%	7.45%	1.02%	46.97%
Negative	24,349.66	3,427.61	589.79	28,367.05
	45.52%	6.41%	1.10%	53.03%

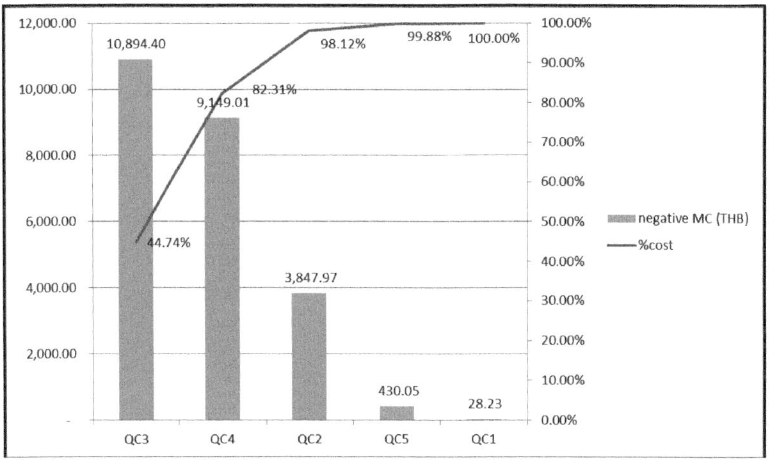

Fig. 4. Pareto chart for costs of negative product for MC

Cause and effect diagram and why – why analysis were applied to find out root causes of the negative MC and the result showed that the maximum loss occurred in process of wood log preparation for operation and sawing because the log size was not fit to the size of the final part. ECRS concept was used to design the improvement methods. The objective of MFCA was to maximize the benefit from material usage that reflected to reducing input material and increasing positive product cost. There were 7 improvement methods proposed as follows.

1) Reducing unnecessary paper usage (QC1): To reduce input material.

2) Separating starting wood log length before performing operations (QC 3): To increase positive product cost.

3) Cutting wood log length before operations (QC 3): To increase positive product cost.

4) Using one wood log for making 2 back legs (QC 3+4): To reduce input material and increase positive product cost. (see Fig.5a)

5) Using one wood log for making back leg and back stand (QC 3+4): To reduce input material and increase positive product cost. (see Fig.5b)

6) Using one wood sheet for making 2 back leg and back stand (QC 3+4): To reduce input material and increase positive product cost. (see Fig.5c)

7) Reducing used quantity of painting color at (QC 5). Reduce input.

The results of all improvement solutions from MFCA calculation were presented as Table 2.

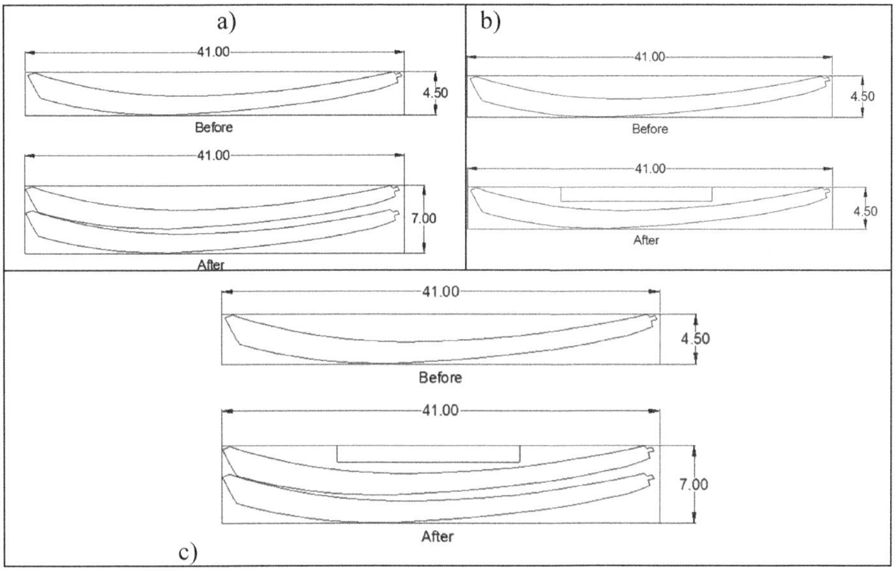

Fig. 5. Drawing for a) 4th solution, b) 5th solution and c) 6th solution

Table 2. MFCA calculations for improvement methods (cost unit: THB)

Method	Current	1	2	3	4	5	6	7
Total	53,495.82	53,490.64	48,687.07	48,814.75	51,233.82	52,914.14	50,942.89	53,304.14
	100.00%	100.00%	100.00%	100.00%	100.00%	100.00%	100.00%	100.00%
Positive	25,128.77	25,128.76	25,351.96	25,412.25	25,306.06	25,190.62	25,456.42	25,170.44
	46.97%	46.98%	52.07%	52.06%	49.39%	47.61%	49.97%	47.22%
Negative	28,367.05	28,361.87	23,335.11	23,402.50	25,927.76	27,723.52	25,486.46	28,133.69
	53.03%	53.02%	47.93%	47.94%	50.61%	52.39%	50.03%	52.78%

4.2 Dynamic Programming

To set up dynamic programming for this case problem, stage and node were defined as in Fig 6. Each node represent each improvement solution and each stage represented process or QC that the solution was implemented.

The incursion addressed as equation (1) and (2) were formalized the working-backward procedure to find the optimal decision for each stage that implied to maximize the cost of positive product for each method at each stage. The results of dynamic programming were presented as Table 3 to 5 and the final decision was presented as Fig 7. The optimal solution was to implement solution 1, 2, 6 and 7.

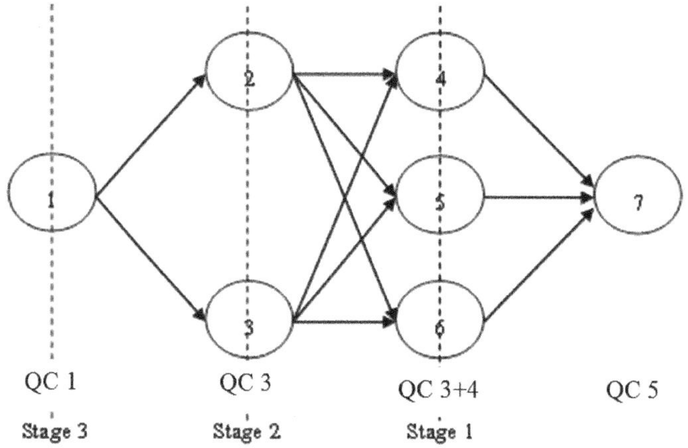

Fig. 6. Dynamic programming for the case study

Table 3. Dynamic programming operation for stage 1

State, s	x_1	PC_1^*	TC_1^*	$f(x)$	x_1^*
4	7	25,349.37	51,042.14	49.66%	7
5	7	25,232.87	52,722.46	47.86%	7
***6**	**7**	**25,501.12**	**50,751.21**	**50.25%**	**7**

Table 4. Dynamic programming operation for stage 2

State, s	x_2			PC_2^*	TC_2^*	$f_2^*(x_2)$	x_2^*
	4	5	6				
***2**	**55.38%**	**53.15%**	**56.07%**	**25,760.86**	**45,942.46**	**56.07%**	**6**
3	55.37%	53.14%	56.06%	25,826.67	46,070.14	56.06%	6

Table 5. Dynamic programming operation for stage 3

State, s	x_3		PC_3^*	TC_3^*	$f_3^*(x_3)$	x_3^*
	2	3				
***1**	**56.08%**	**56.07%**	**25,760.86**	**45,937.28**	**56.08%**	**2**

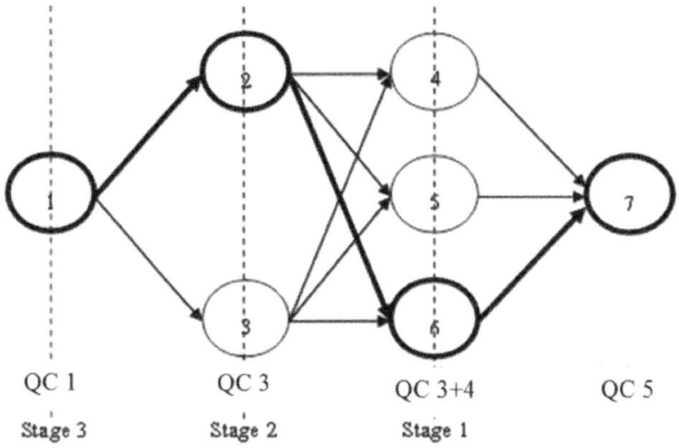

Fig. 7. Final solution from dynamic programming

4.3 Solution Evaluation Using MFCA Calculation

The results of MFCA calculation for the decision making from dynamic programming was presented in Table 6. The total input cost was reduced from 53,495.82 THB to 45,937.28 THB and the cost of positive product was increased from 46.97% to 56.08% and the cost of negative product reduced from 53.03% to 43.92% as presented.

Table 6. Improvement solutions evaluation based on MFCA (cost unit: THB)

	MC	SC	EC	Total Cost
Total	37,429.79	7,384.96	1,122.52	45,937.28
	81.48%	16.08%	2.44%	100.00%
Positive	20,598.06	4,514.39	648.40	25,760.86
	44.84%	9.83%	1.41%	56.08%
Negative	16,831.73	2,870.57	474.12	20,176.42
	36.64%	6.25%	1.03%	43.92%

5 Conclusion

This study aimed to present the application of MFCA and dynamic programming in supply chain operations improvement. The case study results showed that when MFCA was implemented the cost of positive and negative products can be clearly identified. To improve operations of the case study, one main product was selected to study and the source of negative product cost can be identified as material waste from the production line. Thus, seven improvement solutions were proposed and evaluated based on MFCA concept. The positive cost from each solution was used to formulate incursion equation for dynamic programming. The results from dynamic programming showed that only the 1st, 2nd, 6th and 7th solution should be implemented to obtain the maximum value of total positive product cost comparing with current situation. The benefit of dynamic programming when applied with MFCA was in making decision on the optimal set of solution when there was more than one alternative for operations improvement.

References

1. Nakajima, M.: Environmental Management Accounting for Better Eco-Efficiency - Close Look at Material Flow Cost Accounting. In: The International Symposium on Environmental Accounting, Japan (2003)
2. Environmental Industries Office. Environmental Policy Division. Industrial Science and Technology Policy and Environment Bureau. Ministry of Economy. Trade and Industry. Japan.: Guide for Material Flow Cost Accounting (2011), http://www.meti.go.jp
3. Kasemset, C., Sasiopars, S., Suwiphat, S.: The Application of MFCA Analysis in Process Improvement: A Case Study of Plastics Packaging Factory in Thailand. In: Proceedings of the Institute of Industrial Engineers Asian Conference 2013, pp. 353–361 (2013)
4. Fakoya, M.B., Poll, B.: The feasibility of applying Material Flow Cost Accounting as an integrative approach to Brewery Waste-reduction Decisions. African Journal of Business Management 35(6), 9783–9789 (2012)
5. Viere, T., Stock, M., Genest, A.: How to achieve energy and resource efficiency: Material and energy flow modeling and costing for a small and medium-sized company. Paper presented at the Wissenschaftliche Scripten, Auerbach. Germany (2013), Retrieved from http://www.qucosa.de/fileadmin/data/qucosa/documents/10523/218_Material_and_energy_flow_modelling_and_costing.pdf
6. Kasemset, C., Chernsupornchai, J., Pala-ud, W.: The Application of MFCA in Textile Factory: A Case Study. Journal of Cleaner Production (in press, 2014)
7. Ministry of Economy. Trade and Industry.: Material Flow Cost Accounting MFCA Case Examples, Tokyo, Japan (2010)
8. Chompu-Inwai. R., Jaimjit, B., Premsuriyanunt, P.: A combination of Material Flow Cost Accounting and design of experiments techniques in an SME: the case of a wood products manufacturing company in northern Thailand. Journal of Cleaner Production (in press (2014)
9. Nakkiew, W.: Application of Material Flow Cost Accounting in Dried Longan Manufacturer: A Case Study of Small-to-Medium Enterprise Company in Thailand. In: Proceeding of EMAN-EU 2013 conference on Material Flow Cost Accounting Conference Proceedings, Germany, pp. 34–37 (2013)
10. Shukr, S.M., Alwan, N.A.S., Ibraheem, I.K.: The Multi- Constrained Dynamic Programming Problem in View of Routing Strategies in Wireless Mesh Networks. International Journal of Information and Communication Technology Research 6(2), 471–476 (2012)
11. Chattinnawat, W.: Identification of Improvement for Multistage Serial Processes with respect to Material Flow Cost Accounting via Dynamic Programming. In: Paper presented EMAN-EU 2013 Conference on Material Flow Cost Accounting Conference Proceedings, Germany, pp. 30–33 (2013)

Value Analysis of Coco Board for Production Sustainability

Marianne B. Calayag

Department of Industrial Engineering, Bulacan State University,
Philippines City of Malolos, Bulacan, Philippines
mariannecalayag18@gmail.com, marianne_calayag@yahoo.com

Abstract. The principle of Value Engineering is applied to make an innovative product that can help the public and the environment. Value Engineering is a technique in which the value of a system's output is optimized by improving the function or reducing the cost. Innovation is the application or creation of new product to satisfy new requirements and market needs. Wood has been an important construction material since humans began building shelters. It has always been used extensively for furniture, such as chairs, tables and beds. But there will come a time that the consumption of woods is more than the amount of trees that are growing back. For these purpose a composite board that can be an alternative to wood using waste to reduce the need to consume natural resources is created. The coconut coir fibers, plastic resin, and hardener are used as the materials for the composite board. In order to come up with the product, the techniques including coir pounding and washing, mold waxing, molding, pressing, drying, and polishing were executed. Upon completion of the project, a naturally wood looking product is manufactured because of the coir strands which even made the product stronger than ordinary plywood.

Keywords: Value engineering, coconut coir fiber, wood, composite board.

1 Problem and Its Background

Demand for plywood in 2013 reached 436,251 cubic meters which was only satisfied through importation.[1] There is also an Executive Order No. 23 Sec. 2011, declaring a national moratorium on logging natural and residual forests.[2] Cost of plywood is inevitably increasing because of these known facts.

An idea for an alternative product of wood came up which is a Coco Board. Coconut tree is abundant in the Philippines.[3] Considering that it is a tree of life[4] because of the many purposes of every part of this tree.[5] The leaves are used for

[1] GNP, 2013.

[2] Bendijo, 2014.

[3] Coconut - Tree of Life, 2005.

[4] Ramos, 2013.

[5] 165 + Uses for the Coconut Tree - There's More To It Than Coconut Oil, July 23, 2012.

M. Gen et al. (eds.), *Industrial Engineering, Management Science and Applications 2015*,
Lecture Notes in Electrical Engineering 349, DOI: 10.1007/978-3-662-47200-2_6

broomstick, its fruits are also exported for coconut juice which is considered a fountain of youth, and also for virgin coconut oil and its fruits' shell is used for floor polishing among many others.

The coco husk[6] is one of its parts where we can still look into. It is the rough external covering of a coconut used for fuel and are a source of energy. While coco husk fiber which is also called coir[7] is the fiber removed from the coconut and it is used for making floor mats, doormats, brushes and mattresses.

Using Value Analysis, the possibility of developing a coco board can be studied. Value Analysis[8] is a systematic analysis that identifies and selects the best value alternatives for designs, materials, processes, and systems. It proceeds by repeatedly asking "can the cost of this item or step be reduced or eliminated, without diminishing the effectiveness, required quality or customer satisfaction?" Its objectives are to distinguish between the incurred costs which are the actual use of resources and the costs inherent which is the locked-in cost in a particular design and to minimize the locked-in costs.

2 Statement of the problem

The general problem of the study is: How can the values of coco coir are analyzed for the production of coco board?

Specifically, the study seeks answers to the following questions:

1. Where can we see the abundant supply of coco coir in the Philippines?
2. What is the sustainability of coco board if developed in terms of
   2.1. materials;
   2.2. process; and
   2.3. cost
3. What is the demand for plywood that can be supplied by coco board?
4. How to manufacture a coco board?
5. What are the different tests that will be implemented in the coco board?

3 Significance of the Study

The study will look into the analysis of developing coco board using Value Analysis. It aims to find out the possibility of manufacturing coco board in the Philippines.

The study may also be beneficial for the Department of Environment and Natural Resources for the preservation of our forestry. It will lessen cutting of trees as the main source of plywood. And likewise reduce waste of coco husk will be the primary source of coco board.

This will also be beneficial for the present manufacturer and seller of plywood like hardware stores for this will be an alternative for plywood which has an increase in

---

[6] Coconut Husk is a Cutting Edge Technology, 2012.

[7] What is Coco Coir, 2012.

[8] Value Analysis, 2014.

demand but low in supply. Instead of resulting for importation, developing coco board is a great alternative for meeting the demand for plywood.

4 Scopes and Delimitation

The scope of the study is solely for the manufacture of coco board using coco coir as its raw material. It aims to find out the sustainability of the product in terms of materials, process and cost. It also aims to determine the demand of existing plywood that can be catered by the manufacture of coco board.

The marketing of the product is not included in the study, as well as the other parts of the coconut tree. It only focuses on the manufacture of coco board and coco coir as its primary raw material.

5 Population of the Study

A survey is conducted to the primary and secondary users of plywood which is similar to the product being developed which is coco board. The primary users considered are the home owners, furniture shops and interior designers who use the product most often, while the secondary users are schools and government agencies.

The number of respondents to the survey is 100 persons from the province of Bulacan only. Interview is also given to the respondents after they answer the questionnaire to better understand what they need and want for the product being developed.

6 Research Method and Instrumentation

Value Analysis is the focus of this research. It starts with the survey of what the customers want to improve in the product. The product is plywood that is used in building houses and creating furniture and fixtures. The improvements that the customers want are then converted into Needs Statements and then arranging it into hierarchy. Based from the survey, the customers want durable, cheap or economical and decorative material for plywood that can be used for creating furniture. The secondary needs are convenience, and resistance to water and termites.

A Need Metric Matrix is then created to see how the needs of the customers can be formulated in the improvements for the new product which is coco board.

Benchmarking from local brands of plywood is also done to find out the ideal and marginal value of the product specification.

Using Combination Table, the final concepts for the product is set. Several composition of mixture of percentage by weight of coco coir and adhesive can be tested.

The durability of the product being developed must be tested before it can be manufactured and then distributed in the market.

A mixture of coco coir and adhesive can produce a board that is called coco board. Product can be produced by adding the amount in weight needed of coco coir and adhesive.

7 Board Process Design

The process of producing a coco board starts by separating the coconut coir from the outer husk of the coconut. Coconut coir is properly washed in tap water to remove the coco peat. Pound the coconut coir until some of it are separated and then remove the coco peat. The coconut coir is dried in direct sunlight. Then, prepare the binders which are polyester resin and hardener. Set the coir into the mold to attain its desired shape. Remove the coir into the mold. Apply wax in the surface of the mold to have a shiny look after it has been made and to easily remove from the molder. Pour the binder into the mold with the desired amount. Lay the contoured coir into the mold with the binder. Press the composite material to remove the excess binder and to flatten the composite material to attain its designed durability. Let it dry for1440 minutes or until it reaches its desired hardness. The composite material will be removed from the mold when the composite material achieves its desired hardness. Lastly, polish it for the final process.

Table 1. Time of Production

PROCESS	**TIME (mins.)**
Separate coir from husk	3
Wash the coir	3
Pound the coir	3
Set aside	1
Prepare the binder	2
Set the coir in the mold	2
Remove the coir in the mold	3
Apply wax in the surface of the mold	3
Pour the binder into the mold	3
Lay the contoured coir to the mold	2
Press the composite material	2
Remove the composite material from the mold	3
Drying	1440
Polish	3
TOTAL	**1473**

7.1 Experimental System

Testing of strength, durability, and loading capacity of the product can be done by experts. Suitable machine and equipment are only available in the laboratory and can also be utilized only by the experts in the field. The objective of testing the product is to find out the strength, durability and loading capacity of the product being developed and compared it to existing product in the market today.

8 Presentation, Analysis and Interpretation of Data

The abundant supply of Coconut production can be found in the provinces of Quezon and Davao Oriental. There is an increasing supply of coconut all over the country, including the province of Bulacan wherein the study is focused.

The cost of producing Coco board is also sustainable since coir only costs 10 pesos per kilo, resin costs 400 pesos per gallon and hardener costs 200 pesos per gallon only.

Table 2. Costing of Coco Board

Year	Hardener	Resin	Coir	TOTAL
2014	200	400	10	610
2015	210	420	10	640
2016	222	440	10	672
2017	236	460	10	706
2018	248	480	10	738
2019	260	500	10	770

A forecast on the costs of raw materials is also presented in Table 4 below. Using linear regression, the total cost of raw materials for the next five years is computed.

Studies also show that accumulation of coir pith near coir factories causes solid waste pollution problem plus the fact that it undergoes a very slow decomposition rate because of its low carbon to lignin ration of less than 0.5%. In this regard, using coco coir as main material of the board can also solve the waste management disposal of the coir. And can only prove the sustainability of the cost of coco board.

The increasing demand of plywood can be supplied with coco board. Forecasting showed an increase of 11.1% in the demand for plywood. While there will be a steady supply of 297,482 cu. m. because of the implementation of EO 23. This means that there is a gap between the demand and supply of plywood, which gives an opportunity for alternative materials such as coco board to get into the market.

8.1 Manufacturing Process of Coco Board

The Product Life Cycle of the Coco board begins with the requirement definition followed by the design stage then manufacturing and finally testing stage.

Fig. 1. Product Life Cycle

Requirement definition stage fits into other stages in manufacture of Coco board. In this stage the output from one process become inputs to other processes. This includes the raw materials which are coco coir, resin and hardener. The design stage is where the concept of manufacturing is created taking in consideration the costing, appearance, functionality, strength and durability of the product being developed. The manufacture stage is where the process took place. The mixture of coco coir, resin and hardener produced the end product which is the coco board. The testing stage is where the final product undergone test that proved the strength and durability of coco board.

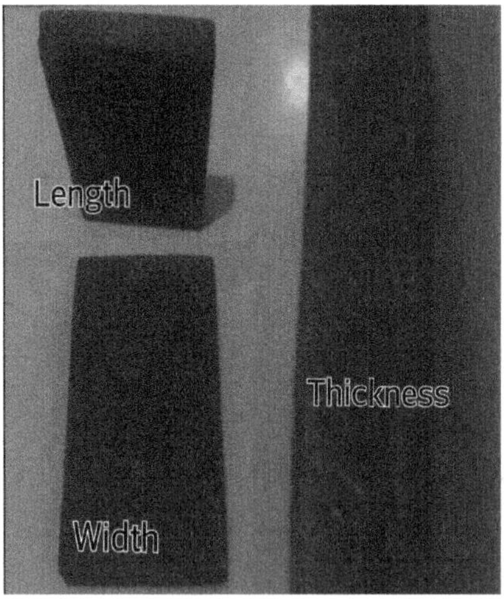

Fig. 2. Image of Coco Board

9 The Different Tests that will be Implemented in the Coco Board

Conducting a comprehensive strength test for the block showed that the compressive strength of the block having 70/30 combination (meaning 30% of the composition is coco coir) has 8188psi and 56mpa, a 60/40 combination has 8731psi and 60mpa while a 50/50 combination has 9520psi and 66mpa as shown in Table 6. Coco board is more durable compared to plywood which only has 3011psi.

Studies on granite environmental coconut products as presented in the related studies can also back up this research. Coconut fiber logs, blocks, blankets and wattles are ideal for erosion control, soil erosion prevention, bank restoration and stream reinforcement.

Using the Factorial Statistical treatment with the weight of Coco Coir as variable, it is found out that as the amount of coco coir increases the better the strength and durability of the product.

Std	Run	Block	Factor 1 A:Coir grams	Response 1 Strength MPa
1	5	Block 1	30	56
2	8	Block 1	30	58
3	3	Block 1	30	54
4	9	Block 1	40	60
5	7	Block 1	40	61
6	4	Block 1	40	59
7	1	Block 1	50	66
8	2	Block 1	50	69
9	6	Block 1	50	63

Fig. 3. ANOVA of Test Result

Unlike in the traditional manufacturing of plywood wherein the first step is to cut tree, coco board only uses waste material of coconut.

With the results presented in the ANOVA, coir is significant to be a good composition of board with a probability of 0.0038 (f value = 16.29) that implies a good impact with regards to the tested material. Also, a low in probability means that a material is more reliable to be used. Thus, it has a good characteristic tobe used as a board.

Conducting a comprehensive strength test for the block showed that the compressive strength of the block having 70/30 combination (meaning 30% of the composition is coco coir) has a 8188psi and 56mpa, a 60/40 combination has 8731psi and 60mpa while a 50/50 combination has 9520psi and 66mpa. Coco board is more durable compared to plywood which only has 3011psi. Studies on granite environmental coconut products as presented in the related studies can also back up this research. Coconut fiber logs, blocks, blankets and wattles are ideal for erosion control, soil erosion prevention, bank restoration and stream reinforcement.

10 Conclusions

In view of the foregoing findings, the following conclusions were drawn:

1. There is abundant supply of Coco coir in the Philippines.
2. The development of Coco board is sustainable in terms of materials, process and cost.
3. There is an increasing demand of plywood that can be catered by Coco board if developed.
4. Manufacturing of Coco board is practical and economical.
5. Strength and Durability & Loading Capacity tests proved the acceptability of Coco board.

11 Recommendations

In light of the findings and conclusion of the study, the following recommendations will be drawn:

1. Further study on the sources of the coco coir all over the Philippines to maximize the production of coco board.
2. Machineries for compression and drying of coco board can also be studied to it make easier and faster.
3. Different variations of coco board as to the thickness, length and width can also be set into study depending on the usage of the board.
4. Additional components for the production of coco coir can also be studied so as to lessen the price of production more.
5. Curing time of the coco board can also take into consideration to find out if it affects the strength and durability of the product.

References

http://www.pca.da.gov.ph (2005), retrieved from http://www.pca.da.gov.ph/tol.html

http://www.be-healthy-with-coconuts.com (2010), retrieved from http://www.be-healthy-with-coconuts.com/coconut-husk.html

http://www.tradingeconomics.com (2013), retrieved from http://www.tradingeconomics.com/philippines/gross-national-product

http://www.businessdictionary.com (2014), retrieved from http://www.businessdictionary.com/definition/value-analysis.html

Bendijo, L. (2014), retrieved from http://www.denr.gov.ph, http://www.denr.gov.ph/news-and-features/features/93-eo-23-renewing-hopes-for-sustainable-forestry-in-the-philippines-.html

Ramos, R. (June 21, 2013), retrieved from http://www.dailyroar.com, http://thedailyroar.com/culture/the-coconut-tree-the-philippines-tree-of-life/#.U9fhkPldXEA

Effect of Temperature on the Colour and TSS Removal of Batik Dye Wastes in an Integrated Biological and Filtration Treatment System

N.A. Ramlee*, M.N. Mohd Rodhi, S.F. Abdul Manaf, F. Hamzah, A. Anuar, and A. Datu Brandah

Faculty of Chemical Engineering, Universiti Teknologi MARA, 40450 Shah Alam, Malaysia
azrini@salam.uitm.edu.my

Abstract. In this study, conventional filter system (with a polypropylene membrane) was used with *L. delbrueckii* to treat dye waste from textile waste water. The purpose of this study was to determine the effect of temperature on the colour and TSS removal of dye wastes in an integrated biological and filtration treatment system. Temperature was varied in a range of 33°C, 35 °C, 37 °C and 39 °C and pH was controlled at 6 for 50 wt% dilution of dye. Fourier Transform Infrared Sprectroscopy (FTIR), Inductively Coupled Plasma-Optical Emission Spectroscopy (ICP-OES), color removal, pH profile and total suspended solid (TSS) analyses were done on the dye waste before and after the treatment. The optimum temperature and pH identified for dye degradation were at 37°C and 6 with initial color removal of 22.6% and final removal of 75%. Initial total suspended solid removed was 5% whereas the final removal was 75%.

1 Introduction

Nowadays, there are many industries in Malaysia use dyes as one of their tools to produce variety of products such as textile industries, paper printing industries, rubber industries, and also plastic industries. However, among these industries, textile industries use dyes the most in their production [1]. Even though Malaysia is developing rapidly in these few decades, there are still some traditional methods preserved to produce a good textile art such as batik canting. Batik is widely produced in Kelantan and Terengganu that are located at the north-east of Peninsular Malaysia and has been divided into small and medium enterprises (SMEs) [2]. Although the batik canting industry is beneficial to Malaysia's economic growth, there are also some negative impacts caused by this industry mainly on dye and chemical waste production.

Since there are many types of dyes and pigments used in batik production, the waste from this process contains dye compounds. This type of waste causes negative impact to the environment especially to the water bodies such as rivers and streams.

* Corresponding author.

M. Gen et al. (eds.), *Industrial Engineering, Management Science and Applications 2015*, Lecture Notes in Electrical Engineering 349, DOI: 10.1007/978-3-662-47200-2_7

In Kelantan, some of the industrial units are built alongside of river and also at the manufacturer's home backyard [3]. Unfortunately, the effluents from the process, especially in Kelantan, are discharged directly into the stream without any or with minimum treatment. It was found that the discharges of batik industry in Kelantan have only 5 percent level of compliance with Environmental Quality Act 1974 between January and September 2010 compared to other industrial effluents [3].

It has been proven that different synthetic compounds in dye waste shows different biological activity in the water. Unfortunately, traditional wastewater treatment that is currently being used is not capable of treating dye waste due to its chemical stability and many kind of synthetic compound present in the dye [4]. Hence, new technology must be developed to overcome this problem.

Recent studies have found that anaerobic bacteria such as *L. delbrueckii* are able to treat dye waste up to 60 percent of colour removal [2]. It is also stated that the use of anaerobic membrane bioreactors is advantageous since it can generate energy which is methane gas, during the treatment process [5]. However, the study of the use of membrane with *L. delbrueckii* is less found. Thus, a further study should be done on the effectiveness of the dye treatment by using *L. delbrueckii* utilizing membrane filter.

In this study, *L. delbrueckii*, anaerobic bacteria, was used to treat the dye waste from batik canting process in an integrated biological and filtration treatment system in which a polypropylene (PP) membrane was applied. Previous studies have shown that *L. delbrueckii* can be used to treat the dye waste by removing up to 45-60% of colour in the waste water. Commercial membrane filter was used in this research to study the effectiveness of the dye waste treatment by using this integrated biological and filtration treatment system. The objectives of this study were to determine the effect of temperature on the colour and TSS removal of batik dye wastes after treated with *L. delbrueckii* and further filtration system using PP membrane.

2 Methodology

2.1 Inoculation and Cultivation of *L. delbrueckii*

Throughout this study, *Lactobacillus delbrueckii subsp. Lactis* ATCC: 12315 were used. The bacteria strains were obtained from Bioprocess Laboratory, UiTM Shah Alam, Malaysia. The inoculation of *L. delbrueckii* was done on agar plate. The method used was streaking techniques under aseptic conditions. The microorganism was incubated at 37^0C for 24 hours. Then, the microorganism was transferred to MRS(Man de Rose) broth. In order to maintain the concentration of *L. delbrueckii* during degradation process, serial dilution was performed and incubated.

2.2 Characterization of Dyes Wastewater

The dye waste was diluted to 50% concentration. Firstly, using a UV-vis Spectrophotometer (Uviline 9400, SECOMAM), the maximum absorbance of the sample (300 nm) was measured. The functional group existed in the sample was determined by using Fourier Transform Infrared Spectroscopy (FTIR). The application of Inductively Coupled Plasma Atomic Emission Spectroscopy (ICP-OES) was used to determine metal element in the sample.

2.3 Experimental Set-up and Degradation Process

A conventional filter (with a polypropylene membrane type) was installed in the water filter casing. Dyes wastewater was added to the water filter casing for 300 ml. The bacteria, *L. delbrueckii* was fed in and the degradation process was started. The wastewater penetrated through the membrane filter and was drawn out into the vacuum Erlenmeyer flask under vacuum pressure [18]. The sample was collected from the flask.

This study was done under different temperature and at controlled pH in 72 hours. Table 1 shows the conditions used for dye degradation. The pH adjustment was made by using 1 mol sodium hydroxide (NaOH) or 1 mol hydrochloric acid (HCl) according to the required pH. The ratio for the inoculums and composition of dye waste water are 3:50 or 0.06 [2]. Degradation process was conducted under anaerobic condition in an incubator. In every 12 hours of interval time, the treated sample was taken and analyzed. The completed time for degradation process was 72 hours. The sample was being analyzed at hour 0, 12, 24, 36, 48, 60 and 72 hours for each parameter.

Table 1. Control and variables condition

Parameter	pH	Temperature
Operating Condition	6	33°C
		35°C
		37°C
		39°C

2.4 Analytical Methods

pH was measured to determine any changes of pH after the process. The colour changes of the treated water were also identified by using UV-vis spectrophotometer at maximum wavelength. TSS was also measured. Colony forming unit (CFU) measurement of the treated sample was analyzed. FTIR spectrophotometer was used again to detect the functional group that presence in the treated water sample. The determination of the trace metals after the experiment was made by applying the ICP test.

3 Results and Discussion

3.1 Characterization of Dyes Waste

ICP-OES Analysis

Referring to [6] dye that contains metal ions does affect the efficiency of dye degradation by anaerobic bacteria. It is known that metal-complex dyes might cause toxicity to the degradation of bacteria [7]. Table 2 shows the metal elements present in dye waste in this study.

As presented in Table 2, the metal elements present in the dye waste before degradation process were aluminium (Al), arsenic (As), calcium (Ca), cobalt (Co), copper (Cu), potassium (K), magnesium (Mg), manganese (Mn), lead (Pb) and vanadium (V). After the degradation took place, Ca, K, Mg and Mn were degraded. Ca, Mg and Mn were completely degraded whereas K was degraded into very low concentration. A study done by Umesh et. al. [8] on effect of metals on decolorization of Reactive Blue HERD by Comamonas sp. UVS showed that the supplementation of 1.0 g L^{-1} Mg^{2+} to the growth medium reduced the time required for decolorization of dye from 6 to 3 h. However, the lowest concentration of Cu^{2+} inhibited decolorization [9]. Bakshi et. al. [3] also reported that decolorization of dye increase to 6% and 9% with the presence of Zn^{2+} and Mg^{2+}. Therefore, the presence of Mg in the dye waste in this study helps in increasing the degradation of dye. However, Cu was also present and might reduce the degradation effectiveness since it can inhibit the anaerobic bacteria.

Table 2. Metals Element in Dye Wastes

Symbol	Elements	Concentration	
		Before	After
Al	Aluminium	+	+
As	Arsenic	+	+
Ba	Barium	-	-
Ca	Calcium	+*	-
Cd	Cadmium	-	-
Co	Cobalt	+*	+*
Cu	Copper	+*	+*
Fe	Iron	-	-
K	Potassium	+	+*
Li	Lithium	-	-
Mg	Magnesium	+*	-

+ = Present - = Absent +* = Present with very low concentration

FTIR

Figure 1 shows the spectrum of FTIR analyses on the dye waste for the temperature of 33°C, 35°C, 37°C and 39°C at initial and final samples. Based on the graph, the peaks observed were generally similar for all conditions. However, there were slight different at temperature of 37°C and 39°C. The peaks were similar at the initial and final sample and indicated that there was no degradation of the functional group occurred during the treatment. At time zero, the peaks observed at the temperature of 37°C were at 3296.98 cm^{-1} with broad peak which indicated hydroxyl group (-OH), 2107.10 cm^{-1} with stretch peak indicated alkynes (C≡C), 1635.74 cm^{-1} with bend peak showed primary amines (N-H), 1365.52 cm^{-1} with bend peak mean nitro group (N=O) and 1217.03 cm^{-1} with stretch peak indicated ethers (C-O). After the treatment at hour 72, the functional groups identified were similar with functional group in the initial sample. However, the wavelength numbers were slightly different for hydroxyl group and alkynes.

FTIR spectrum is a plot of percentage transmittance versus wave number or wavelength. As the readings moves along the x-axis on the graph, the bond vibration energy varies. Therefore, the x-axis of absorption band on the graph represents the different energy for different molecules tested [11]. A slight reduction of alkynes from 2107.10 cm^{-1} to 2103.34 cm^{-1} showed a slight changes on the alkynes molecules structure after the treatment by *L. delbruekii*.

Growth Curve of *L. delbruekii*

Fig. 2 shows the growth curve of *L. delbruekii* in temperature of 37°C and pH 6. From the graph, it took 36 hours for the bacteria to enter the exponential phase which occurred for 12 hours during the treatment. After 48 hours, the bacteria entered stationary phase until hour 72. However, Zuraidah et. al. [12] had studied the growth of *L. delbruekii* and found that the exponential phase of the bacteria started at hour 25 and entered the stationary phase at hour 70. However, as stated by Umesh et. al. [8], the type of dye is very important since some of the dye might cause toxicity to the bacteria used in the degradation. The presence of copper in dye waste in this study might influence the growth of the anaerobic.

3.2 Effect of Temperature on the Colour Removal

Temperature is a critical factor that can influence the cell growth and also the rate of color removal during the degradation process [13]. Fig. 3 showed the result of color removal percentage for different temperature of 33°C, 35 °C, 37 °C and 39 °C. Based on the result of this study, the maximum color removal observed was 75.5% for the temperature of 37°C whereas the color removals for other temperature conditions were between 48% and 52%.

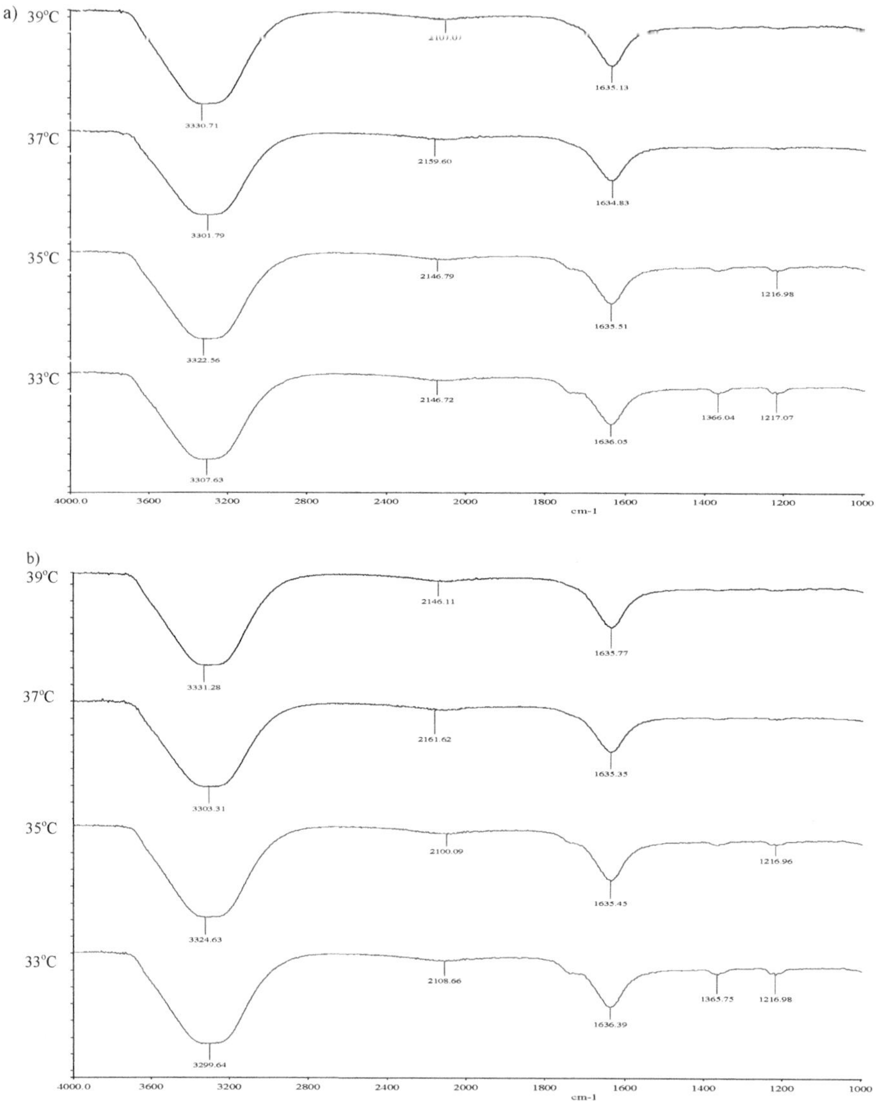

Fig. 1. FTIR: a) initial and b) final (different temperature) at 33°C, 35 °C, 37 °C and 39 °C

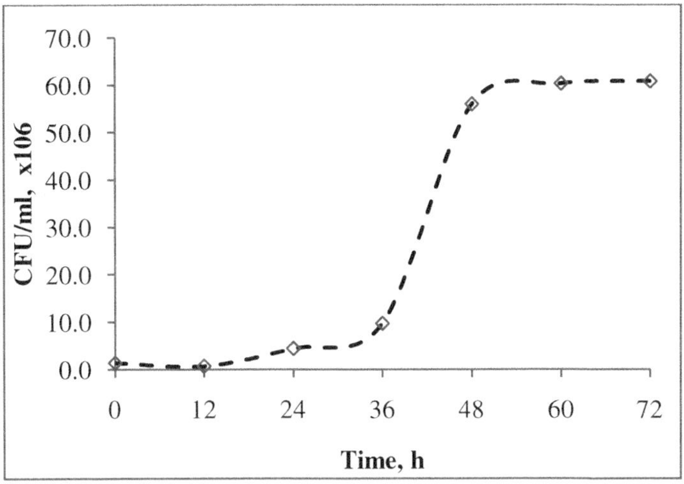

Fig. 2. Profile of *L. delbrueckii* growing rate during degradation of dye waste

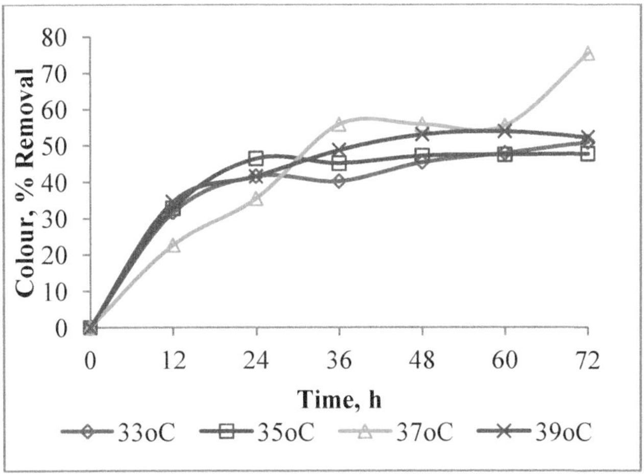

Fig. 3. Colour removal percentage for temperature of 33°C, 35 °C, 37 °C and 39 °C.

Zuraida et. al. [12] has studied on color degradation of textile dye using *L. delbruekii* in a shake flask and reported that the optimum condition for the color removal was 37°C with pH 6 and the maximum color removal observed was 46% [2]. Another study performed by You et. al. [11], on anaerobic membrane bioreactor using *Bacillus cereus* showed that the maximum color removal of Reactive Black 5 was 62.5% using polytetraflouroethylene (PTFE) membranes. According to Sandhya [14], the enzyme that anaerobically degrades dye is azoreductase enzyme.

Based on the result of FTIR analyses, there was no sulfonate group present in the dye sample. Based on work done by Hitz et. al. [7], acid dyes has lower colour removal due to the presence of sulfonate group whereas direct dye exhibit higher colour removal since there were no sulfonate groups present in the dye sample [15]. Nigam et. al. (1996) stated that azo dye compound consist of hydroxyl group or amino group are more likely to be degraded compared to dye containing methyl, methoxy, sulfo, or nitro group [16].

3.3 Effect of Temperature on the TSS

The maximum total suspended solid removed in this study was 70% at the temperature 37°C and pH 6. Fig. 4 shows the turbidity removal percentage for all of the conditions used in this study. Due to the presence of polypropylene membrane in the bioreactor, the suspended solids were retained causing the TSS to decrease after the treatment [10]. Since there was no additional carbon sources such as glucose or lactose supplied to the *L. delbrueki* during the degradation, the bacteria use an alternative carbon source to acquire energy [17]. Therefore, the degradation of suspended solid in this study might also occur because the bacteria consume the carbon source that might present dye waste. Moreover, since the degradation process occurred in 3 days (72 hours), the suspended particles inside the dye waste settled at the bottom of the filter casing which increased the removal of total suspended solid in the dye waste.

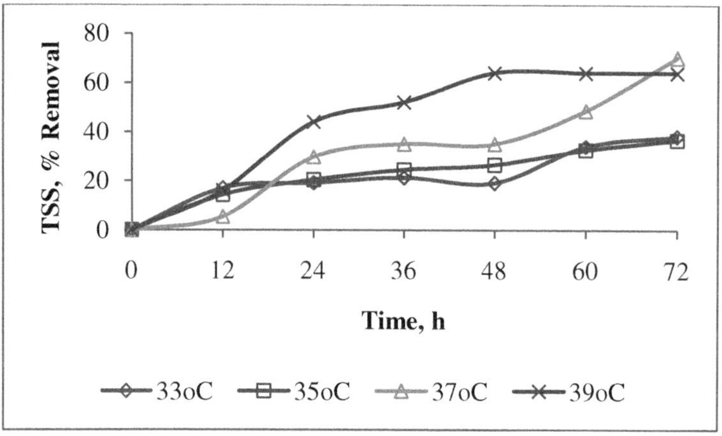

Fig. 4. TSS removal percentage for temperature of 33°C, 35 °C, 37 °C and 39 °C

3.4 Temperature and pH Profile

The changes in pH during the study are shown in Fig. 5. Based on the result, it was indicated that the pH in all of the conditions were increased towards alkaline condition.

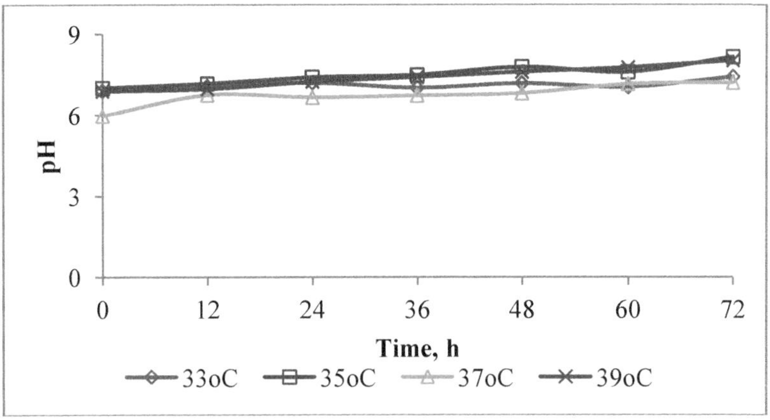

Fig. 5. Effect of degradation on pH in the dye waste for temperature of 33°C, 35°C, 37°C, and 39°C

Based on FTIR result, the functional groups present in the dye after the treatment were all in base condition. Since there was no carboxylic acid present in the dye waste, most of the components were in basic conditions causing the pH to change towards basicity during the treatment process. Ali et. al. [9] had studied on starvation of *Lactobacillus casei* with absence or limited lactose and found out that the bacteria produce less than 3mM of lactic acid during the fermentation without lactose as the carbon source[18]. This production of lactic acid might not significantly affect the pH of the dye waste during the treatment process.

4 Conclusion

From the study, it is proven that at temperature of 37°C and pH 7 the *L. delbruekii* in an integrated biological and filtration treatment system is able to degrade dye waste from textile industries. At the optimum condition, it has shown up to 75.5% of colour removal and 70% of TSS removal.

Acknowledgment. The authors acknowledge the Universiti Teknologi MARA (UiTM) for granting the Research Acculturation Fund, 600-RMI/RAGS 5/3 (72/2013) to carry out this project. The support from all staff of Faculty of Chemical Engineering, UiTM, involves in this research is highly appreciated. Thank you also to Faculty of Art and Design, UiTM in providing the samples of dyes waste to be used in this research.

References

1. Asgher, M., Bhatti, H.N.: Evaluation of Thermodynamics and Effect of Chemical Treatments on Sorption Potential of Citrus Waste Biomass for Removal of Anionic Dyes from Aqueous Solutions. Ecological Engineering, 79–85 (2012)

2. Siti Zuraida, M., Nurhaslina, C.R., Ku Halim, K.H.: Influence of Agitation, pH and Temperature on Growth and Decolorization of Batik Wastewater by Bacteria Lactobacillus Delbrueckii. IJRRAS 14(2) (2013)
3. Bakshi, D., Gupta, K., Sharma, P.: Enhanced biodecolorization of Synthetic Textile Dye Effluent by Phanerochaete Chrysosporium under Improved Culture Conditions. World Journal of Microbiology & Biotechnology 15, 507–509 (1999)
4. George, S., Daphne, H., Patricio, L., Carlos, N., Ángeles, B.: Anaerobic Membrane Bioreactors for Wastewater Treatment: A Review. Chemical Engineering Journal (2012)
5. Maredia, S.: A Comparison of Biofilters, Biotrickling Filters and Membrane Bioreactors for Degrading Volatile Organic Compounds, MMG 445 Basic Biotechnol, eJournal (2015), `http://www.msu.edu/course/mmg/445`
6. Çinar, Ö., Demiröz, K.: Biodegradation of Azo Dyes in Anaerobic–Aerobic Sequencing Batch Reactors. Biodegradation of Azo Dyes 9, 59–72 (2010)
7. Hitz, H.R., Huber, W., Reed, R.H.: The Absorption of Dyes on Activated Sludge. J. Soc. Dyers. Colorists. 94, 71–76 (1978)
8. Umesh, U.J., Vishal, V.D., Anuradha, N.K., Sanjay, P.G.: Effect of Metals on Decolorization of Reactive Blue HERD by Comamonas sp. UVS. Water Air Soil Pollut. 216, 621–631 (2011)
9. Ali, A., John, P.B., Richard, W., Rolf, E.N., Margaret, L.B.: Impact of Lactose Starvation on the Physiology of Lactobacillus casei GCRL163 in the Presence or Absence of Tween 80. Journal of Proteome Research 12, 5313–5322 (2013)
10. Lin, H., Peng, W., Zhang, M., Chen, J., Hong, H., Zhang, Y.: A Review on Anaerobic Membrane Bioreactors: Applications, Membrane Fouling and Future Perspectives. Desalination 314, 169–188 (2013)
11. You, S.J., Rahul, A.D., Hou, S.C.: Degradation of Reactive Black 5 Dye Using Anaerobic/Aerobic Membrane Bioreactor (MBR) and Photochemical Membrane Reactor. Journal of Hazardous Materials 177(1-3), 1112–1118 (2010)
12. Spagni, A., Casu, S., Selene, G.: Decolourisation of Textile Wastewater in A Submerged Anaerobic Membrane Bioreactor. Bioresource Technology 117, 180–185 (2012)
13. Siti Zuraida, M., Nurhaslina, C.R., Ku Halim, K.H.: Removal of Synthetic Dyes from Wastewater by Using Bacteria, Lactobacillus delbruckii. International Refereed Journal of Engineering and Science (IRJES) 2(5) (2012)
14. Pearce, C.I., Lloyd, J.R., Guthrie, J.T.: The Removal of Colour From Textile Wastewater using Whole Bacterial Cells: A Review. Dyes Pigments 58(3), 179–196 (2003)
15. Sandhya, S.: Biodegradation of Azo Dyes Under Anaerobic Condition: Role of Azoreductase. Biodegradation of Azo Dyes 9, 38–57 (2010)
16. Seesuriyachan, P., Takenaka, S., Kuntiya, A.: Metabolism of Azo Dyes by Lactobacillus casei TISTR 1500 and Effects of Various Factors on Decolorization. Wat Res. 41, 985–992 (2007)
17. Azhar, S.S., Liew, A.G., Suhardy, D., Hafiz, K.H., Irfan, M.D.H.: Dye Removal from Aqueous Solution by using Adsorption on Treated Sugarcane Bagasse. American Journal of Applied Sciences 2(11), 1499–1503 (2005)
18. Kaya, Y., Ersan, G., Vergili, I., Gonder, Z.B., Yilmaz, G., Dizge, N., Aydiner, C.: The treatment of Pharmaceutical Wastewater Using in A Submerged Membrane Bioreactor Under Different Sludge Retention Times. Journal of Menbrane Science 442, 72–82 (2013)
19. Ramlee, N.A., Mohd Rodhi, M.N., Anak Brandah, A.D., Anuar, A., Alias, N.H., Tengku Mohd, T.A.: Potential of Integrated Membrane Bioreactor in Batik Dye Degradation - A Review. Journal of Applied Mechanic and Materials 575, 50–54 (2014)

A Multi-agent Approach for Production Management

Ana Paula M. Tanajura [1,2,*], Valdir Leanderson C. Oliveira[2], and Herman Lepikson[1,2]

[1] SENAI CIMATEC, Salvador, Bahia, Brazil
[2] Program of Industrial Engineering, Federal University of Bahia, Salvador, Bahia, Brazil
{ana.tanajura,herman.lepikson}@fieb.org.br,
valdirleanderson@yahoo.com.br

Abstract. Multi-agent systems have been successfully used to represent distributed manufacturing systems. Each part or characteristic of the system can be represented by an agent which acts independently and in a cooperative way. A better integrated operation can emerge as the result of each agent following its own rules and communicating with the others. Better decisions from a holistic viewpoint can be achieved when manufacturing asset management is supported by a multi-agent approach. A model to manage distributed manufacturing is proposed and applied to on-shore oil fields. The distributed characteristics of oil field units, such as wells, collecting stations, compressing stations and supplies mean that they have to cooperate to reach production targets. The integrated management model carries out a cost analysis and helps to identify unprofitable assets and to support decision making processes. Agents help to reduce the load of information for the operator, giving more time to focus on situations that require greater attention. Example applications are presented in which the proposed multi-agent model, its ontology and rules, could help oil field surveillance and support decision-making process.

1 Introduction

An agent is a computational system that is situated in an environment and that is capable of acting independently in this environment in order to achieve its design goals. In multi-agent systems, the ability for interaction means that each agent may be affected by other agents, or by humans, and perform tasks in pursuit of their goals [1]. This is typical in supply chain situation, where agents represent different links in this chain such as suppliers and customers.

Some authors present successful applications of multi-agent systems to problems in supply chain integration [2, 3, 4, 5]. According to [2], multi-agent systems seem appropriate for supply chains because the different business units involved can be modeled as autonomous agents. For [3, 5], the potential benefits of this approach are: the establishment of an effective mechanism for trading; the adoption of a technology for multi-attribute negotiation; the offer of an effective mechanism to improve performance across the supply chain, thus supporting decision-making. To achieve the desired level of collaboration, [4] presents a decentralized approach to the supply chain

* Corresponding author.

M. Gen et al. (eds.), *Industrial Engineering, Management Science and Applications 2015*,
Lecture Notes in Electrical Engineering 349, DOI: 10.1007/978-3-662-47200-2_8

through three subsystems based on agents. The first is the communication subsystems within the plants (which will manage unforeseen events that may cause the need for a new production schedule of the plant or part of it). The second is the interplant communication subsystem, which will manage events occurring in a plant that will affect other plants. Finally, the third is the communication subsystem of the supply chain, which will manage events in the plant that affect suppliers and /or external customers.

An oil field can be considered as a set of distributed systems consisting of surface facilities, wells, reservoirs, logistic systems, suppliers and. The supply chain approach is useful to study oil fields because it allows an extensive analysis of the production process, similar to a network of integrated operations. The proposed model seeks to represent each of the production functions of the field as an agent, as well as their relationships with service providers, equipment and the maintenance staff.

The independence of agents is also a desired characteristic for oil fields, since production systems may evolve over time, adding new equipment and capacities. For example, a well may change its artificial lift technology or even change its function from producer to water injection well, or a new collection station can be incorporated into the model. Modelling the oil field as a collection of independent agents allows changes to be handled without any modification of the model.

Internally, the operatives seek the best result through cooperation rules that lead to greater profitability for the asset. The operatives are connected through a flow value where there is an increase in value at each stage as the raw material moves along the transformation process. Management involves the administration of the manufacturing value stream (monitoring resources used at each stage of the process) as well as investments for improved system performance.

Externally, maintenance and provision service suppliers compete for work on the oil field. Therefore a natural progression of service improvement can be achieved through a careful selection of service providers on the oil field. The independence of the agents and their capacity to act and react also enables using them to monitor the facilities, minimizing some of activities previously undertaken by operators. This condition can lead the oil field to function as an unassisted plant.

The paper is organized as follows. Section 1 is this introduction. Section 2 synthesizes some challenges and opportunities in oil field management to support the proposed model. Section 3 presents the multi-agent model in the following aspects: framework, ontology (the agent knowledge environment), structure of agents, and applications and system interfaces, to show how the system acts during production surveillance activities and profitability analyses. Finally, Section 4 provides conclusions and future work.

2 Challenges in Oil Field Management

According to [6], there is huge complexity involved in managing an asset operating in the oil and gas industry. For the authors, the management of these assets is a complex activity due to several factors, among them: the need for physical control while physical and chemical rules of procedure are respected; frequent changes in operation according to economic, social and strategic needs; equipment reliability and criteria of integrity, and safety of staff working on assets. The authors consider the management of operating assets as the joint work of several disciplines and hierarchies. It requires the coordination of people and their knowledge of procedures incorporated in the work and the coordination of business processes. This can only be successful if it is guided by a clear strategy with stakeholders who are mobilized and aligned to the strategic objectives.

Traditionally, according to [7], efforts to increase the ultimate capacity of oil recovery have resulted in competition. This has often resulted in incompatible plans and proposals from various disciplines of the asset management team. Each one focuses on issues in their respective areas, but they have no ability to understand the ramifications for other disciplines or the asset as a whole.

The complexity of the problem is further increased by restrictions arising from the current workforce. The retirement of senior professionals reduces the number of experienced workers in the area of oil exploration and production. A small team means that fewer people understand the complexity and implications of competing proposals, and few people are qualified to evaluate these proposals and make recommendations. The authors also point out that the ultimate goal of all team members of a group of assets is to make decisions that benefit the entire business, not only their respective domains [7].

[8] corroborate [7] with regard to the difficulty of forming a technical body. From the authors' experience at Shell Exploration and Production, they mention the difficulty of transmitting knowledge and lack of methods for sharing information in the technical area. This makes it imperative to implement systems that standardize the work, capture knowledge and automate surveillance activities.

In the opinion of [9], the areas that compose assets, such as wells, reservoirs and facilities, are well developed but still treated separately. There is a need for engineers to understand the systems in an integrated manner and they should be able to perform cost estimates for the facilities. This competence will enable better assessments at the various stages involved in the development of active installations, reservoirs and wells.

[10] have had positive results with optimization and integrated operation of oil fields including the reservoir, the surface facilities and an economic model. They believe that the current industry still has a very segmented supply chain. Each part of the chain is treated separately from the other. The authors attribute this to the fact that in different parts of the supply chain people are recruited with specific knowledge and use different tools to support decision-making. This limits integration, even in situations where integration has clear potential.

Many technologies for digital oil fields have been developed [11] and applied [12,13]: Remote Real-Time Facility Monitoring and Control, Real-Time Production Surveillance, Intelligent Wells, 4D Visualization and Modelling, Remote Communications, Workflow and Knowledge Management Systems. These models represent an evolution from existing production workflows and applications in a collaborative decision making environment that brings together reservoir, production and process engineering domains [13]. All of these help to reach useful Integrated Asset Models. Nevertheless, these models are almost static, based on pre-determined configurations and, in general, have a hierarchical configuration, which results in useful but limited deterministic solutions.

The authors [6, 7, 8, 9, 10] highlight the possibilities of gaining advantage with integrated asset management in the oil and gas industry, and [11] and [12] suggest some solutions in this direction. However, in spite of the technological development, there is much work still to be done in old assets, to solve the lack of communication between different areas, integration between systems, standardized methods and procedures, or even trained professionals [13].

There is a gap in the oil industry: the need for decision-making taking into account not only the production systems, but also other elements that constitute the supply chain of this industry. There is the possibility to offer the engineers and managers an integrated view of an oil field including service providers and cost analysis of each part of the field. This situation is even more critical in mature oil fields, where the yield is not as high as more recent ones and operation investments may not be justified [14]. Thus, this research aims to improve decision-making in onshore mature oil fields through an analysis of economic variables and techniques. It considers the oil field as a set of wells, reservoirs, surface facilities, suppliers and teams which act to release the oil and pipe it to the refinery in processing condition.

3 The Multi-agent Model

Because the elements that compose the oil field are distributed in nature, a multi-agent approach presents a suitable model, as noted by previous researchers [2, 3, 4]. The independence of agents is also a desired characteristic for oil fields. The production systems can be modeled separately with specific rules and production targets. In oil fields, the production systems may change over time, e.g. the mode of production of a well may change, its role may change from oil producer to water injection well, or a new collecting station may be incorporated into the model. In the multi-agent model, these changes can be done without modifying the whole model. The model can provide the platform that integrates simulations and economic models, to provide a total asset solution with full transparency and capabilities to seek optimization along the entire value chain. Each agent in the model can access different existing systems, with their own local culture, to reach a better global solution.

The independence and capacity of action and reaction of the agents also reflects the possibility of using them to monitor facilities, minimizing some of the activities previously undertaken by the operator. When the agents are intelligent, in addition to monitoring, they can be used to analyze the situation, to suggest action and to act, resulting in better conditions for the entire field.

3.1 Framework

The proposed model considered three types of agents: operational, intervention and supply agents (Figure 1). These agents are classified according to their role in the supply chain. The *operational agents* are those directly related to production. They do the processing of the product from extraction of oil and gas from the reservoir to the refinery. This category of agent includes: wells and other smaller stations; collecting station; carrier systems; compression unit; oil transfer unit; water handling and injection units; storage unit. *Intervention agents* are those who perform preventive or corrective maintenance in industrial facilities, as well as small repairs and even larger investments. There are two types of intervention agents: internal and external intervention agents. The latter are partner companies that perform services and are managed by internal intervention agents. The external *supply agents* procure inputs (spare parts, chemicals and other necessary items to conduct operational processes) for production and are driven and coordinated by the internal supply agent.

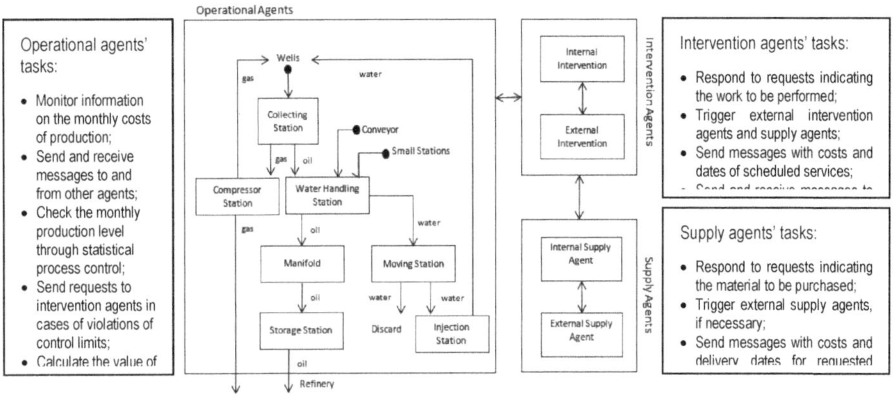

Fig. 1. Multi-agent framework for an oil field

The costs incurred by each operating agent are the expenses needed to perform their functions, such as labor, energy, supplies and maintenance service requests. The latter are made by the internal intervention agent who may require the supply agent to procure services, parts and equipments. The agent that performs the service will have a price paid by the agent who requested the service. The operational agent, that provides its product with a cost to other agents, will also have an income, paid by the one who purchased it. Thus, incomes are amounts paid for services rendered or products purchased by other agents. The profit function can be obtained from the difference between incomes and costs for the entire system as well as the agents individually.

3.2 Ontology

Another aspect to be considered in the construction of the model concerns the ontology. The ontology is the representation of abstract concepts such as events, time, physical objects and beliefs. In multi-agent systems, this is the knowledge representation of the agent. All the actors involved in the same environment and that relate to each other must share the same ontology. The possibilities for status display of industrial control systems based on ontology are mentioned by [15]. According to the authors, the use of ontologies for an explicit description of the world has many advantages, including ease of integration and extensibility. Some authors suggest the use of ontology for geological interpretation [16], oil exploitation [17], exploration [18], and process surveillance [19].

Table 1. Set of terms and concepts of the ontology.

Term	*Concept*
Well	Equipment used to link the surface and the reservoir.
Extraction Well	Equipment used to extract oil from the reservoir.
Water Injection Well	Equipment used to pressurize the reservoir and improve its production.
Production Mechanism	Techniques used to achieve a better production quality of the extraction well.
Collecting Station	System used to collect oil from wells and separate gaseous and liquid phases.
Treatment Station	Equipment used in the treatment of the produced oil, including separating all the water present in the production.
Compression Station	System used to compress the produced gas in the exploration process, sending it to extraction wells and refineries.
Manifold	System of pumps and pipes used to transport the treated oil from treatment station to the storage station.
Moving Station	System of pumps and pipes used to transport produced water from the treatment station to the injection stations and controlled disposal.
Storage Station	Large tanks that hold the produced oil in the oilfield before be sent to refineries.
Water Injection Station	Equipment which performs reinjection of produced water into the reservoir, aiming to pressurize it.
Refinery	Consumer of oil and gas produced in the field.
Internal Intervention Team	Human resources group of the company responsible for performing services within the oilfield.
External Intervention Team	Outsourced human resources group responsible for conducting services in the field.
Services	Interventions with the purpose of corrective or preventive maintenance, or investments such as installing new equipment to expand the oilfield capacity.
Supply	Acquired materials for the performance of services in the oilfield.
Oil	Main product of the oilfield.
Crude Oil	Emulsion of oil, gas and water. Product extracted from the reservoir through the wells
Pre Treated Oil	Emulsion of oil and water. Product of the action of the separators in the collection stations.
Treated Oil	Treated product containing oil only.
Water	Liquid separated from the oil during the treatment process.
Gas	Gaseous phase of the production.
Gas LP	Compressed gaseous phase of the production at a low pressure.
Gas HP	Compressed gaseous phase of the production at a high pressure.
Chemical	Inputs used in the oilfield.
Manpower	Specialized staff to carry out general and specific services within the oilfield.
Eletric Power	Power used to make the equipment of the field work.
Problem	Unexpected situation which happens in some component of the oilfield.
Sensor	Equipment that collects useful information on production and costs.

During the ontology construction process, the Protégé 4.2 software was chosen for formatting and editing diagrams. Protégé is an open-source software, which is easy to learn and to use, and well suited to academic work. The integration of the ontology with the agents was realized with the class Ontology.java, one of the components of the JADE framework [20]. Following [1], we enumerate all the relevant terms of the ontology in Table 1.

The classes in the ontology are defined based on the kind of agents used in the model: operational agent, intervention agent, supply agent. After the definition of the classes it is necessary to set up properties representing the relationships among classes (Object Properties), or to represent the classes' attributes (Data Properties). Some relationship properties are common to all the classes, for example hasCost, hasSensor and isPartOf. Other relationships are specific, like isInjectorOf, which is specific to water injector wells. The following examples describe a few of the many properties of classes in the ontology:

- The hasCost relationship links some component of the oilfield to all the costs related to the consumption of chemicals, electric power, maintenance, investments, suppliers and services.
- hasSensor links each component to the sensor. This sensor has all the necessary information on production and costs.
- isPartOf indicates that all components are parts of the oilfield.
- isInjectorOf is a property of the water injection well, because it is a type of well used to inject produced water in the reservoir to improve the production through pressurization.

4 Structure of Agents

There are four basic kinds of agent programs that embody the principles underlying almost all intelligent systems [21]: Simple reflex agents; Model-based reflex agents; Goal-based agents; and Utility-based agents. Each kind of agent combines particular components in particular ways to generate actions.

The agents in this model were proposed according to model-based reflex agents. This agent type maintains an internal state that depends on the agents' perception history and thereby reflects at least some of the observed aspects of the current state [21]. Figure 2 describes the structure of the reflex agents.

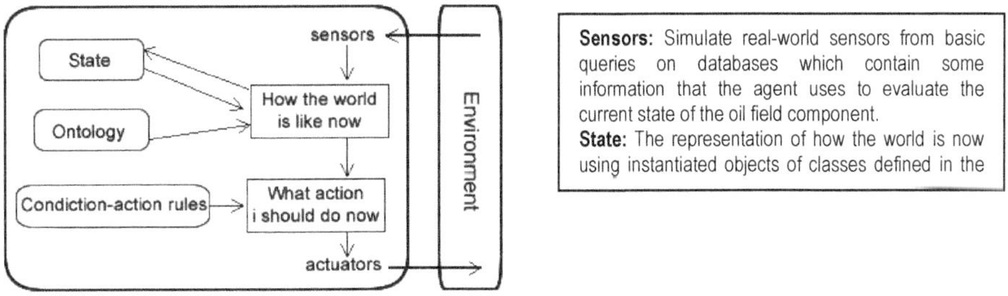

Fig. 2. Model-based *reflex* agents. Based on Russel and Norvig (2009)

The platform *JADE—Java Agent DEvelopment Environment Framework* [20] —was used for the implementation of the multi-agent system. JADE is based on the JAVA programming language and provides very useful resources for developing multi-agent applications. The system is connected to a MySQL 5.0 database in which the environment information is stored during the simulation run. The agent behaviours were implemented based on the JADE standard behaviour classes *CyclicBehaviour.java* and *OneShotBehavior.java*. Transmission of information from one agent to another is accomplished through FIPA-ACL, a message-asynchronous mechanism, and these messages use FIPA-ACL communicative act, such as follow, inform, call for proposal (CFP), propose, accept-proposal, reject-proposal, refuse, agree, failure and request.

Intervention and supply agents have their activities registered in files called yellow pages [20]. These files facilitate the search for services when there is some unexpected event. When more than one intervention or supply agent provides the same service type, the choice of a supplier is made based on criteria of cost and service quality through the cooperation protocols Contract-Net and Request. The agent can trigger alerts whenever the cost or performance is not compatible with its function. For example, if the cost of a well exceeds the revenue from the volume of oil transferred to the collection station, the agent can send a warning message about this.

5 Applications and System Interfaces

In an abnormal situation, when unexpected events occur, operational agents start to interact with intervention agents, and intervention agents start to interact with supply agents. In this scenario, unexpected events are characterised by alerts that are generated every time the production level exceeds the upper or lower control bounds, according to statistical process control rules. The monitoring process is continuous. When the production flow is not satisfactory, the system initiates a process to identify possible problems. The operational agents exchange messages with intervention agents informing them about the event. The intervention agents receive messages and direct them to supply agents to request materials. These agents confirm the sending of the required equipment and materials, and the intervention agents send messages concerning service programming. As soon as the problem is solved, the supply agents send messages with costs to intervention agents, who in turn pass on their costs to the operational agents.

Figure 3a, the main interface, designed to represent all the oil field agents on a single screen, shows the interface at a moment when an alert is generated, informing the user about the occurrence of some unexpected event. The intervention and supply agents post messages to the user interface so that the operator has information about events that occur (Figure 3b) and the necessary mechanisms to solve possible problems, rather than raw data.

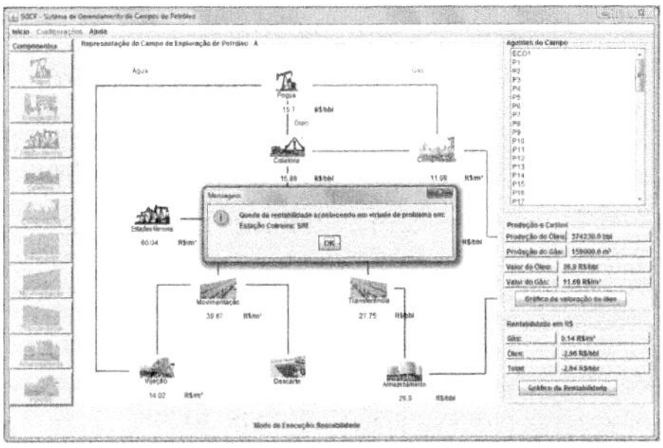

Fig. 3a. User interface emitting an alert about an unexpected event.

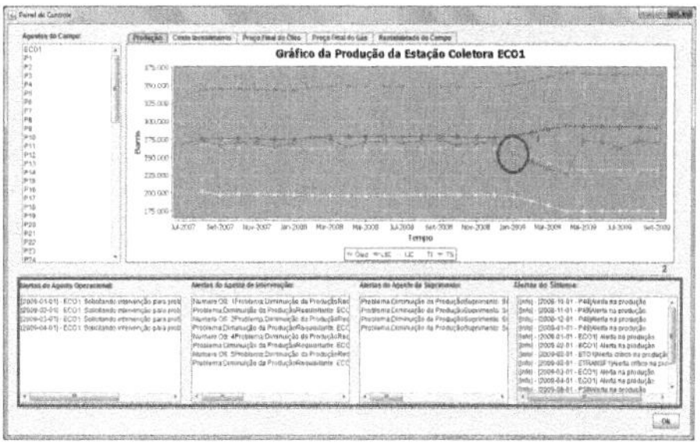

Fig. 3b. User interface showing a production graph and messages exchanged between agents.

Fig. 3. Multi-agent System interface

6 Conclusion

The model suggested herein contributes to decision-making processes, by supplying operators and managers with integrated information about the entire system and alternatives for solving specific problems that have been identified by a global analysis. In addition, the proposed cost model enables profitability analysis per agent or process step and for the entire oil field. Modelling with multi-agent systems offers advantages such as easy inclusion or exclusion of elements in the system, which often occurs in the field when wells are activated or deactivated or when the lift methods are modified. Through artificial intelligence techniques, rather than simply detecting points of inefficiency, the model can be extended to integrate solution detection and implementation. An intelligent agent can indicate a better production strategy, a more reliable supplier or other interventions directing the oil field strategy towards profitability.

The proposed model was developed based on an oil field located on-shore in Brazil, but the proposed approach can be applied also to other applications, including off-shore oil fields. Each operation on the platform, wells and other ground support systems can be represented by agents. These intelligent agents modeled with cost information and production data for each production system can help managers to make more rational decisions.

The proposed model integrates the different systems in an oil field and each production step is modeled as an intelligent agent. Each stage of production has individual goals and performance metrics and they interact with other agents in a collaborative way to achieve the overall goal of the field. The model offers an effective mechanism to improve oil field performance and decision-making processes, as well as offering a systemic view for operators, supervisors and managers, providing new perspectives for oil field management. We believe that the proposed model can be successfully applied to oil field management on a daily basis and further evaluation will be carried out to validate its effectiveness in gathering intelligence from real operators in real applications.

References

1. Wooldridge, M.: Intelligent Agents. In Multiagent Systems – A Modern Approach to Distributed Artificial Intelligence. MIT Press (1999) edited by Gerhard Weiss
2. Ferreira, L., Borenstein, D.: Supply chain based agent simulation: towards a normative approach. In: 13th IFAC Symposium on Information Control Problems in Manufacturing, Moscow, Russia (2009)
3. Jin, C.H., Li, Q.M.: Study on a Multiagent Construction Supply Chain Management System. In: 4th International Conference on Wireless Communications, Networking and Mobile Computing, pp. 1–5. IEEE (2008), doi:10.1109/WiCom.2008.1606
4. Álvarez, E., Diaz, F.: A web-based approach for exceptions management in the supply chain. Robot Comput Integr Manuf doi:10.1016/j.rcim.2010.12.004 (2011)
5. Wang, G., Wong, T.N., Wang, X.: An ontology based approach to organize multi-agent assisted supply chain negociations. Computers & Industrial Engineering 65, 2–15 (2013)
6. Al Meshabi, O., Khazandar, M.: Attitude of Collaboration, Real-time Decision Making in Operated Asset Management. In: SPE Intelligent Energy Conference and Exhibition, Utrecht, The Netherlands, 23-25 March (2010)
7. Cabrera, B., Musri, D., Selva, C., Lardone, C.: Developing a Holistic Global Approach to Asset Management. In: SPE Annual Technical Conference and Exhibition, Anaheim, California, U.S.A, November 11-14 (2007)
8. Yero, J., Moroney, T.A.: Exception Based Surveillance. In: SPE Intelligent Energy Conference and Exhibition, Utrecht, The Netherlands, March 23-25 (2010)
9. Boschee, P.A.M.: Challenges of accurate cost estimation. Oil and Gas Facilities (2012)
10. Rahmawati, S.D., Whitson, C.H., Foss, B., Kuntadi, A.: Integrated field operation and optimization. Journal of Petroleum Science and Engineering 81, 161–170 (2012)
11. Steinhubl, A., Klimchuk, G., Click, C., Morawski, P.: Unleashing Productivity: The Digital Oil Field Advantage. Booz&co Viewpoint, Summer (2008)
12. Al-Khamis, M.N., Zorbalas, K.I., Al-Matouq, H.M., Almahamed, S.M.: Revitalization of Old Asset Oil Fields into I-Fields. In: SPE Saudi Arabia Section Technical Symposium and Exhibition, Al-Khobar, Saudi Arabia, May 9-11 (2009)
13. Lima, C.B.C., Sobreira, G., Rossi, D., Kumar, A., Sauvé, B.: State-of-Art Digital Oilfield Implementation in Petrobras Campos Basin. In: SPE Intelligent Energy Conference and Exhibition, Utrecht, The Netherlands, March 23-25 (2010)
14. Babadagli, T.: Development of mature oil fields – A review. Journal of Petroleum Science and Engineering (57), 221–246 (2007)
15. Obitko, M., Vrba, P., Kadera, P., Jirkovský, V.: Visualization of ontologies in multi-agent industrial systems. IEEE Symposium on Emerging Technologies and Factory Automation, ETFA. In: Proceedings of 2011 IEEE 16th Conference on Emerging Technologies and Factory Automation (2011)
16. Nimmagadda, S.L., Dreher, H.: Petroleum Ontology: An effective data integration and mining methodology aiding exploration of commercial petroleum plays. In: 6th IEEE International Conference on Industrial Informatics, INDIN 2008, July 13-16, pp. 1289–1295 (2008) ISSN:1935-4576
17. Du, R., Li, Y., Shang, F., Wu, Y.: Study on Ontology-based Knowledge Construction of Petroleum Exploitation Domain. In: International Conference on Artificial Intelligence and Computational Intelligence (2010)
18. Ge, J., Li, Z., Li, T., Qiang, B.: Petroleum Exploration Domain Ontology-based Knowledge Integration and Sharing System Construction. In: International Conference on Network Computing and Information Security (2011)

19. Natarajana, S., Ghosha, K., Srinivasana, R.: An ontology for distributed process supervision of large-scale chemical plants. Computers and Chemical Engineering 46, 124–140 (2012)
20. Bellifemine, F., Caire, G., Greenwood, D.: Developing Multi-Agent Systems with JADE. John Wiley & Sons, England (2007) ISBN: 978-0-470-05747-6, HB
21. Norvig, P., Russel, S.: Artificial Intelligence: A Modern Approach (3rd edn) Hardcover (2009)

Design of Experiment for Predicting Residual Stresses in Gas Tungsten Arc Welding Process

Adirek Baisukhan, Wasawat Nakkiew*, and Siwasit Pitjamit

Advanced Manufacturing Technology Research Center (AMTech),
Department of Industrial Engineering, Faculty of Engineering,
Chiang Mai University, Chiang Mai, Thailand, 50200
wasawat@eng.cmu.ac.th,
{adirek_b,siwasit_pitjamit}@cmu.ac.th

Abstract. Gas Tungsten Arc Welding (GTAW) requires high heat input for welding. After welding, the workpiece cools down non-uniformly which creates residual stresses. Generally tensile residual stress weaken the strength of the welded joints and shorten the fatigue life of the material. It is very important to identify appropriate welding conditions in order to prevent negative effects from the residual stresses. In this research, the full factorial design of experiment technique was constructed for predicting residual stress occurred after the GTAW welding process. The values of the residual stresses were obtained from the results from the Finite Element Analysis (FEA) model, which had been verified by the residual stress measurement from the X-ray diffraction (XRD) machine. Effects of four welding process parameters: efficiency, arc voltage, arc current, and welding speed on residual stresses at the center of the welded joint and at the heated affected zone (HAZ) were investigated. Each factor had 2 levels (2^4 Full Factorial Design). The regression equation for predicting the residual stress from significant factors was constructed. The results showed that the arc voltage, arc current, welding speed and interaction effects between efficiency and welding speed were significant factors, with the most compressive residual stress value of about -50.97 MPa.

1 Introduction

Residual stress in the welded bead is caused by non-uniform cooling rate of the workpiece after welding. The generated residual stress creates changes in both macro and micro-structures. Residual stress in the workpiece creates permanent deformation which affects the strength as well as the fatigue life of the materials. Determination of the residual stress can be done in several ways: (I) Non-destructive method such as X-ray diffraction (XRD) [6], which is a commonly used method, Neutrons diffraction, Ultrasonic and so on. (II) Destructive method such as Deep-hole drilling, Contour method, Slitting and so on. (III) Using a mathematical model based on the finite element analysis (FEA) program such as ANSYS, ABAQUS, MSC/MARC, etc.

---

* Corresponding author.

M. Gen et al. (eds.), *Industrial Engineering, Management Science and Applications 2015*,
Lecture Notes in Electrical Engineering 349, DOI: 10.1007/978-3-662-47200-2_9

The results from the XRD method compared with the FEA model was consistent [3][5]; therefore, the FEA model is commonly used for predicting the residual stress values. The welding process parameters that affect to the residual stress can be identified by using a statistical method. Design of experiment [4][8] such as full factorial design, fractional factorial design, Taguchi method and so on, can be used for determining significant factors, before using response surface method (RSM) such as Central composite design (CCD) for obtaining the optimal response. In this research, 2^4 full factorial design was used for determining significant factors of four GTAW welding parameters: arc current, arc voltage, welding speed, and efficiency, to the residual stresses occurred after the welding process. Linear regression technique was used for determining an equation for predicting residual stress values from given significant factors.

2 Experiment

Specimen used in this research (shown in Figure 1) was stainless steel grade 304, size width x length x thickness = 150 x 200 x 3 mm., with butt joint welded, single pass by GTAW robotic arm welding. Chemical compositions, mechanical and physical properties for the stainless steel grade 304 are shown in Tables 1-3 [1].

Table 1. Composition ranges for 304 grade stainless steel

Grade 304	C	Mn	Si	P	S	Cr	Mo	Ni	N
Minimum	-	-	-	-	-	18.0	-	8.0	-
Maximum	0.08	2.0	0.75	0.045	0.030	20.0	-	10.5	0.10

Table 2. Mechanical properties of 304 grade stainless steel

Grade	Tensile strength [MPa] minimum	Yield strength 0.2% proof [MPa] minimum	Elongation (% in 50 mm) minimum	Hardness	
				Rockwell B [HRB] maximum	Brinell [HB] maximum
304	515	205	40	92	201

Table 3. Physical properties of 304 grade stainless steel in the annealed condition

Grade	Density [kg/m^3]	Elastic modulus [GPa]	Mean coefficient of thermal Expansion [μm/m/^{0}C]		Thermal conductivity [W/mK]		Specific heat 0-100^{0}C [J/kgK]
			0-100^{0}C	0-538^{0}C	100^{0}C	500^{0}C	
304/L/H	8000	193	17.2	18.4	16.2	21.5	500

Fig. 1. Butt joint welded workpiece from a single pass

3 Finite Element Analysis Model

The FEA model was constructed on simulation software, ANSYS version 15 for predicting residual stress from butt joint single pass by GTAW welding of stainless steel grade 304. The heat flux input during welding used in this FEA model was defined as Eq. 1 [7].

$$Q = \eta \frac{60UI}{V} \tag{1}$$

Where η is the welding efficiency, V is the welding speed (mm/min.), U is the arc voltage (volts) and I is the arc current (ampere). This FEA model using SOLID90 element for analysis [1] and constructed as a moving heat flux on the surface of the element with constant volume from one element to the next element until the welding process was finished with the application of birth-death technique as shown in Figure 2. At the end of the welding process, the material was instructed to be cooled down to room temperature. Then, the FEA model was switched to the structure analysis mode in order to determine the residual stress.

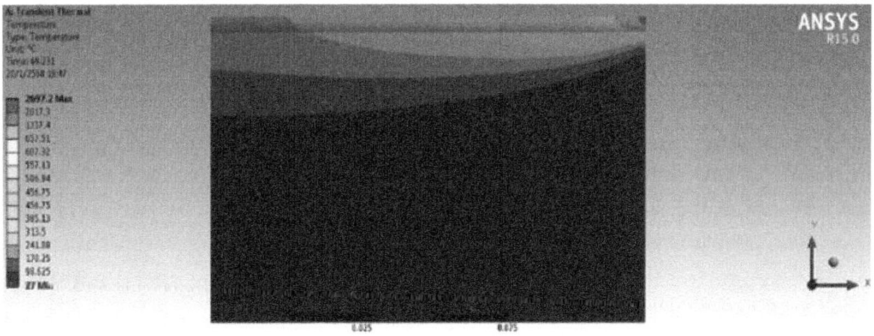

Fig. 2. Welding simulation with birth-death technique

4 Residual Stress Measurement

After the two specimen was welded together, non-uniform cooling rate inside the materials caused residual stresses. The result of residual stress obtained from a non-destructive measurement by X-ray diffraction (XRD) method was compared with the results from the FEA model on a given condition. Positions for measurement by XRD and FEA model were at the following distances from the middle of the welding line: 0, 5, 10, 20, 30, 40 mm. (Transverse direction). The FEA model results showed that the residual stress profile is consistent with the XRD measurement as shown in Figure 3.

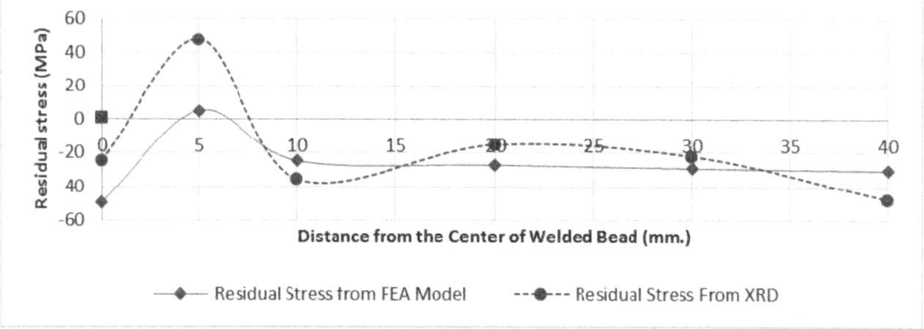

Fig. 3. Residual stress from FEA model and XRD

After the FEA was verified, the design of experiment was firstly used for determining experimental conditions. It was more cost effective for using data from a verified FEA model than obtained directly from measurements. Therefore, in this research the values of residual stresses were obtained from varying four main effects conditions in the FEA model, the output results were then analyzed statistically by MINITAB software.

5 Design of Experiment (DOE)

In this research, design of experiment (DOE) of full factorial designs (2^4 full factorial) was used for determining significant factors that affect the values of the residual stress. This full factorial designs had four factors: welding efficiency, arc voltage, arc current and welding speed. Each factor consists of a low and high levels as shown in Table 4 [8]. There were 16 trials per 1 replicate from the 2^4 full factorial designs and the response was the values of compressive residual stress as shown in Table 5. The analysis was performed by the statistical software, MINITAB. The results showed in Figure 4-7 suggested that the arc voltage, arc current, welding speed and interaction between efficiency and welding speed were statistically significant factors (p-value $<$ 0.05). The coefficient of determination (R^2) of 94.39% indicated that the model was sufficient to fit the data. The result was then used to create a predictive equation for determining residual stress from given significant factors.

Table 4. GTAW welding process parameter

Factors	Unit	Level	
		Low	High
Efficiency	%	50	84
Arc voltage	V	15	28
Arc current	A	90	140
Welding speed	mm/min	100	200

Table 5. Residual stress result

Standard Order	Run Order	Efficiency	Arc voltage	Arc current	Welding speed	Residual stress (MPa)
16	1	84	28	140	200	-40.5310
4	2	84	28	90	100	-48.5460
13	3	50	15	140	200	-44.8625
11	4	50	28	90	200	-46.1390
6	5	84	15	140	100	-48.8850
12	6	84	28	90	200	-44.5870
1	7	50	15	90	100	-48.7280
10	8	84	15	90	200	-47.7470
14	9	84	15	140	200	-46.0500
5	10	50	15	140	100	-45.4120
7	11	50	28	140	100	-43.1215
9	12	50	15	90	200	-47.6750
8	13	84	28	140	100	-42.9810
15	14	50	28	140	200	-44.3450
2	15	84	15	90	100	-50.9730
3	16	50	28	90	100	-44.8390

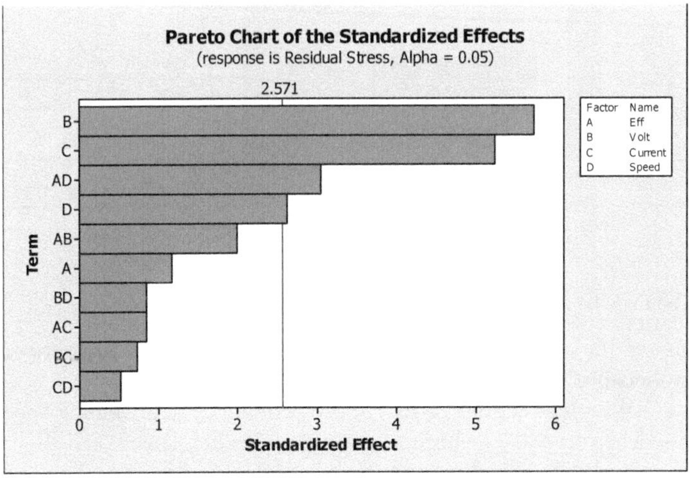

Fig. 4. Pareto chart of the standardized effects

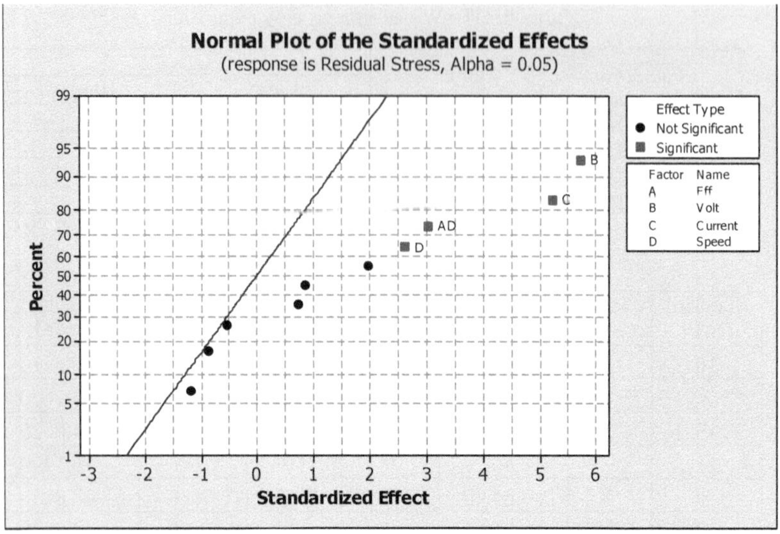

Fig. 5. Normal plot of the standardized effects

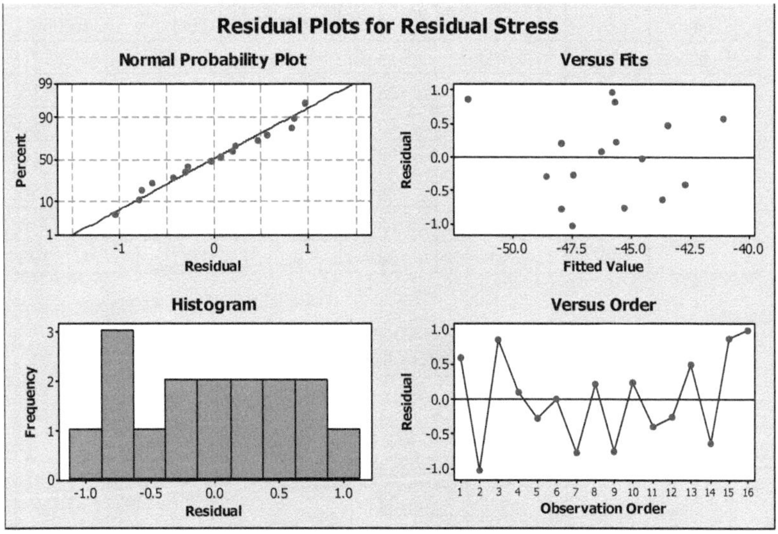

Fig. 6. Residual plots for residual stress

5.1 ANOVA (Analysis of Variance)

The results of the ANOVA showed that the p-value (with the level of significance α=0.05) of main effects: Volt (arc voltage), Current (arc current), Speed (welding speed) and 2-Way Interactions of Eff*Speed (efficiency and welding speed) were less than 0.05. Therefore, arc voltage, arc current, welding speed, and the interaction of efficiency and welding speeds were significant factors which then be used for creating a regression-equation for predicting residual stress values.

Analysis of Variance for Residual Stress

Source	DF	Seq SS	Adj SS	Adj MS	F	P
Main Effects	4	83.032	83.0324	20.7581	17.15	0.004
Eff	1	1.676	1.6757	1.6757	1.38	0.292
Volt	1	39.826	39.8256	39.8256	32.91	0.002
Current	1	33.195	33.1949	33.1949	27.43	0.003
Speed	1	8.336	8.3362	8.3362	6.89	0.047
2-Way Interactions	6	18.787	18.7867	3.1311	2.59	0.158
Eff*Volt	1	4.815	4.8147	4.8147	3.98	0.103
Eff*Current	1	0.886	0.8864	0.8864	0.73	0.431
Eff*Speed	1	11.207	11.2074	11.2074	9.26	0.029
Volt*Current	1	0.648	0.6476	0.6476	0.54	0.497
Volt*Speed	1	0.892	0.8921	0.8921	0.74	0.430
Current*Speed	1	0.338	0.3384	0.3384	0.28	0.620
Residual Error	5	6.051	6.0508	1.2102		
Total	15	107.870				

Fig. 7. Results of full factorial analysis

5.2 Residual Stress Predictive Equation

The predictive equation of residual stress was obtained from the regression analysis of significant factors mentioned in previous section (shown in Eq. 2)

$$Residual\ Stress = -46.0 - 0.324\ Eff + 1.58\ Volt + 1.44\ Current + 0.722\ Speed \quad (2)$$

6 Conclusion

The 2^4 full factorial experimental design was used for determining significant GTAW welding process parameters to the occurred residual stresses after the process. The residual stress values in this research were obtained from the constructed FEA predictive model, which was verified by XRD measurement on a given condition. This FEA model applied the birth-death technique which can simulate the GTAW welding process more realistically. The full factorial results showed that the arc voltage, arc current, welding speed, and the interaction of efficiency and welding speeds were significant factors. These significant factors were then used for creating a regression equation for predicting residual stress values.

References

1. Akbari, M.S.A.A., Miresmaeili, R.: Experimental and numerical analyses of residual stress distributions in TIG welding process for 304L stainless steel. J. Mat. Pro. Tech. 208, 383–394 (2008)
2. Bappa, A., Arunanshu, S.K., Souren, M., Dipten, M.: Modeling of laser transmission contour welding process using FEA and DoE. Opt. Laser Tech. 44, 1281–1289 (2012)

3. Brickstad, B., Josefson, B.L.: A parametric study of residual stresses in multi-pass butt-welded stainless steel pipes. Int. J. Pres. Vess. Pip. 75, 11–25 (1998)
4. Khan, M.M.A., Romoli, L., Fiaschi, M., Dini, G., Sarri, F.: Experimental design approach to the process parameter optimization for laser welding of martensitic stainless steels in a constrained overlap configuration. Opt. Laser Tech. 43, 158–172 (2011)
5. Martins, J.A., et al.: Analyses of residual stresses on stamped valves by X-ray diffraction and finite elements method. J. Mat. Pro. Tech. 179, 30–35 (2006)
6. Rossini, N.S., Dassisti, M., Benyounis, K.Y., Olabi, A.G.: Methods of measuring residual stresses in components. Mat. Des. 35, 572–588 (2012)
7. Stamenković, D., Vasović, I.: Finite element analysis of residual stress in butt welding two similar plates. Sci. Tech. Rev. 59 (2009)
8. Worrayottrakul, P.: Appropriate parameters in gas tungsten arc welding using robotic arm welder. M.Eng. Thesis. Chiang Mai University (2013)

Hands-on Industrial Process Modelling Using the MATLAB System Identification Toolbox

Abubakar Sadiq Bappah

Faculty of Technology Education, Abubakar Tafawa Balewa University, Bauchi, Nigeria
asbappah@gmail.com

Abstract. This paper employs a MATLAB based system identification toolbox (SID) to obtain a mathematical model of a loss-in-weight (LIW) feeder. In this study a pseudorandom binary sequence (PRBS) was used to excite a loss-in-weight feeder used in a cement manufacturing process. The input and output data were then processed using a MATLAB system identification toolbox that yielded an ARMAX model of the process. The identified model was sufficiently validated in the MATLAB control toolbox. Details of data capture, system identification, open-loop simulation, and validation tests were presented in this paper. The results are well assuring to control practitioners and students that MATLAB SID toolbox is a perfect candidate for offline process identification.

Keywords: Loss-In-Weight Systems, System Identification, Modeling, Raw Cement Feeder, MATLAB Control Toolbox.

1 Introduction

A model is the result systematic and purposeful mapping of relationships between a system's physical variables onto mathematical structures such as algebraic equations, differential equations or systems of differential equations. Mathematical models can be developed using different approaches. Theoretical approaches which employ physical laws of nature and empirical approaches that are based on experimentation with the existing systems. Some other approaches may combine both theoretical and empirical methods. It is often beneficial, therefore, to combine mathematical model building with experiments [1]. However, if a plant cannot be experimented on or if the plant does not exist at all then theoretical modeling has to be employed to arrive at a mathematical model suitable for control algorithm synthesis and system simulation.

Experimental modelling of a plant or system identification requires accurate measurement of input and output signals. The input signals can either be the operating signals of the plant or artificial test signals. This involves careful planning of the inputs to be applied so that sufficient information about the system dynamics is obtained. This is imperative in obtaining good quality data that contains sufficient information about the system behavior [2]. There are two ways to carry out this process; by applying an excitation rich in desired frequencies and directly identifying a parametric model from input-output data or by finding a non-parametric frequency

M. Gen et al. (eds.), *Industrial Engineering, Management Science and Applications 2015*,
Lecture Notes in Electrical Engineering 349, DOI: 10.1007/978-3-662-47200-2_10

response of plant by performing one or more experiments with periodic inputs. Then finding a parametric model based on frequency response samples. The measurements are processed in an identification procedure that yields the plant's mathematical model. The result of the identification is an experimental model. In this paper an experimental model is obtained for a loss-in-weight feeder used in a cement manufacturing industry. Input and output data from the plant are processed using the MATLAB system identification toolbox (SID) to yield an ARMAX model of the system. The procedure elucidated the suitability of using MATLAB SID toolbox in process modeling.

2 System Description

In the cement manufacture process, the rate at which the raw meal is fed into the kiln is one of the critical factors required to ensure product consistency and quality. The purpose of any feeder is to deliver material at a setpoint or desired rate. This rate or 'feedrate' is measured as units of weight delivered per unit of time, such as pounds per minute, tons per hour, etc. The feeder controller is the device that controls this operation by constantly measuring the actual feedrate of the material being delivered and comparing it to the desired feedrate and making corrections as necessary. The desired feedrate entered into the controller is called the 'setpoint'. It can be entered directly into the controller by the operator or it may be imported from an external source, such as a remote analogue signal coming from the process. The output of the feeder controller is called the control signal which manipulates the rate at which the feeder delivers material. A Loss-In-Weight (LIW) feeder system consists of a hopper and feeder mounted on load cells commonly used to deliver raw cement at a desired feedrate. When operated in a continuous discharge mode, accurate gravimetric operation is achieved by controlling the speed of the feeder in order to provide a constant decrease in the weight of the feed hopper [3]. Feedrate regulation is in the main based on closed-loop control to deliver powders and other bulk solids at a desired feedrate.

Many industrial processes are designed to handle bulk solids, often in granular or powdered form. Various devices such as conveyors, screw augers, pneumatic tubes, vibrating platforms, bucket elevators, and so on can be used for transporting and metering these materials. The screw conveyor varies the flow by varying the speed of the screw. A variable speed motor drives the feed screw that discharges the material. As the material is discharged from the hopper, a feeder controller keeps track of the weight; constantly comparing it to how much the weight should have decreased based on the set-point. If the weight is not decreasing fast enough, the controller increases the control signal to speed it up. If the weight is decreasing too fast on the other hand, the control signal will be decreased accordingly. To measure the feedrate, the hopper is mounted on load cells. The hopper is supported in such a way that the weight of the hopper and its contents is sensed by three load cells. An analog electrical signal from the load cells which corresponds to the mean weight on the load cells is used for feedback purposes. Two screw feeders actuated by thyristor driven variable speed motors are used to transfer material from the hopper at a specified rate [3]. The advantage of this method is that there is no lag between the time the material is weighed and when it actually leaves the weighing device.

Automatic control requires that the transport device be controllable over a reasonable range by means of some type of control variable. As material leaves the hopper, the weight will go down, showing a loss-in-weight to the controller. A microprocessor-based controller MC^2 is employed in the control loop's error channel. It compares the feedrate set-point value with the actual feedrate and employs a digital PID control algorithm to generate the control signals that actuate the thyristor regulating the speed of the motors. The paper addresses the issue of designing and simulating an industrial process. A model of a LIW feeder is developed from actual open loop LIW process data using the MATLAB SID toolbox. The problem generally consists of data acquisition, estimation, characterization, and verification [4]. The procedure was followed to acquire input-output data necessary for system identification from a 45ton loss-in-weight raw cement feeder. Simulation results based on the identified process are then presented.

3 Data Acquisition

The first and most important step is to acquire the input - output data of the system to be identified. Open loop input-output data relating PV to CO was extracted for the purpose of this study. The average operating SP for the feeder under investigation, Feeder 1 (main) was 90 tons per hour (TPH). At this constant SP, the CO was shifted between two levels by ±20% for the period of this experiment. The procedure was facilitated by the microprocessor based MC^2 process controller in the "manual %" mode of operation. This is the first setpoint mode available in the MC^2 feeder controller. This mode allows the user to control the controller output directly. There is no closed loop control of feedrate in this mode. The output to the motor speed controller can simply be regulated by setting this parameter which is adjustable from 0 to 100%. The method was both safe and realizable. The following algorithm shows the order in which identification data was obtained.

```
Step 1: Log on SP @ 90TPH %feedrate set at 90 tph
Step 2: Set to manual %manual mode @ ±20% of CO
Step 3: While time < specified duration %read and store
    Read PV
    Read CO
Step 4: Read SP %confirm SP
        Read clock %read and store in tie
go to step 1. %loop
```

Fig. 1 shows the LIW Feeder identification data set when imported and preprocessed. The upper sub - plot being the measured feedrate in tons per hour (output signal) and the lower sub - plot represents the input signal as % change in CO. The output data vector is a scaled measured feedrate in tons per hour (PV) while the input data vector is in form of percentage change in CO. That measured data set was exported into the MATLAB workspace as `Fedat.sid` and examined to determine the time delay. The data set is then preprocessed by means of filtering and detrending to remove offsets noise, and other spurious phenomena that may affect the process description. This de-trended working data, `Fedatdd`, is shared into two equal places through the action of select range prop-ups. The first range 0 to 500 samples, `Fedatdde` is the final working data for system identification. While the second range 501 to 1000 samples, `Fedatddv` is the validation data.

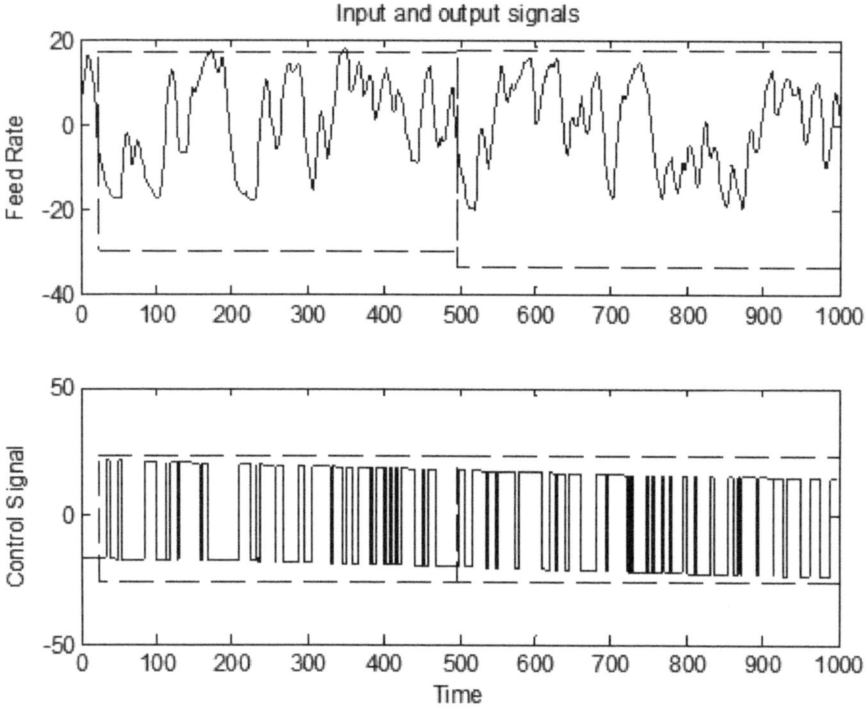

Fig. 1. Data Views with SID Toolbox

4 Model Estimates

The second step is estimation, which involves determining the numerical values of the structural parameters, which minimize the error between the system to be identified, and its model. Parameter estimation can be formulated as an optimizing problem, where the best model is the one that best fits the data according to the given criterion. There is a large number of different estimation methods. Interestingly, however, the choice of method is not crucial as there is no method that is universally the best [5]. Common approaches to parameter estimation are the prediction error method (PEM), maximum likelihood (MLE) and instrument-variable (IV). Parameterizations in the MATLAB toolbox are preferably done with the PEM estimation method, which consistently estimates models from the data set. Although called conventional by many researchers, the PEM method will consistently estimate a system if the data set is informative and the model set contains the true system, irrespective of whether or not data have been collected under feedback [4]. All these technical provisions are therefore on the side of experiments in open-loop.

5 Model Structure

The third step defines the structure of the system, for example, type and order of the differential equation relating the input to the output. The set of systems from which a model is chosen is usually defined by a model structure. Once the structure is chosen, an appropriate excitation is applied to the true plant and input-output data is measured [6]. Appropriateness here refers to the ability to outweigh the effect of disturbances and to excite the desirable system dynamics.

The measured input - output data is then mapped into a model which best explains this data. Let $\{e(t)\}$ be a pseudorandom sequence which is similar to white noise in the sense that

$$\frac{1}{N}\sum_{t=1}^{N} e(t)e(t-\tau) \to 0 \text{ as } N \to \infty \; (\tau \neq 0) \; . \tag{1}$$

This relation is to hold for τ at least as large as the dominating time constant of the unknown system. Let $y(t)$ and $u(t)$ be scalar signals and consider the model structure

$$A(q^{-1})y(t) = B(q^{-1})u(t) + C(q^{-1})e(t). \tag{2}$$

where

$$\begin{aligned} A(q^{-1}) &= 1 + a_1 q^{-1} + \ldots + a_{na} q^{-na}. \\ B(q^{-1}) &= 1 + b_1 q^{-1} + \ldots + b_{nb} q^{-nb}. \\ C(q^{-1}) &= 1 + c_1 q^{-1} + \ldots + c_{nc} q^{-nc}. \end{aligned} \tag{3}$$

The parameter vector is taken as

$$\theta \;=\; (a_1 \ldots a_{na} \quad b_1 \ldots b_{nb} \quad c_1 \ldots c_{nc})^T. \tag{4}$$

The model in Eqn. (2) can be written explicitly as the difference equation

$$\begin{aligned} y(k) + a_1 y(t-1) + \cdots + a_{na} y(t-na) &= b_1 u(t-1) + \cdots \\ + b_{nb} u(t-nb) \; + e(t) + c_1 e(t-1) &+ \cdots + c_{nc} e(t-nc). \end{aligned} \tag{5}$$

But the previous form using the polynomial formalism will be more convenient. This model is called an ARMAX model, which is short for an autoregressive moving average with an exogenous signal (a control variable $u(t)$). When all $b_i = 0$, it is called an MA (moving average) process, while for an autoregressive (AR) process all $c_i = 0$ The ARMAX model structure is chosen and MATLAB is required to select the best model order in the range of 1 10. The result indicates that a second order ARMA model best describes the system under investigation. Using a sampling interval of 1secs, the state space model for the identified process is:

$$\begin{aligned} x_{k+1} &= Ax_k + Bu_k . \\ y_k &= Cx_k + Du_k . \end{aligned} \tag{6}$$

where

$$\mathbf{A} = \begin{bmatrix} -0.8754 & -1.4060 \\ 1 & 0 \end{bmatrix}.$$

$$\mathbf{B} = \begin{bmatrix} 1 \\ 0 \end{bmatrix}.$$

$$\mathbf{C} = [0.0830 \quad 1.2654].$$

and

$$\mathbf{D} = [0].$$

u is the control signal, y is the federate, and x is the state vector.

The corresponding pulse transfer function is:

$$\frac{PV}{CO} = \frac{Y(z)}{U(z)} = \frac{0.4750z + 0.2750}{z^2 - 0.5833z + 0.4167}. \tag{7}$$

The process model can be re-written in the difference equation form as

$$\begin{aligned} y(k) - 0.5833\,y(k-1) + 0.4167\,y(k-2) = \\ 0.4750\,u(k-1) + 0.2750\,u(k-2). \end{aligned} \tag{8}$$

Fig. 2 reveals that the LIW is a self-regulating process with a considerably high overshoot of around 20% and settling period of about 0.80secs at known sampling rate. This configuration results into a steady state offset of about 0.1. These are indicators to an open loop stable second order process with trivial delay. The overshoot should be reduced to its barest minimum and the offset removed without jeopardy to the settling time. The process is therefore validated with the remaining one half of the data that were not used for model estimation (i.e. *Fedatddv*). Standard validation tools are residual analysis and cross – validation [7].

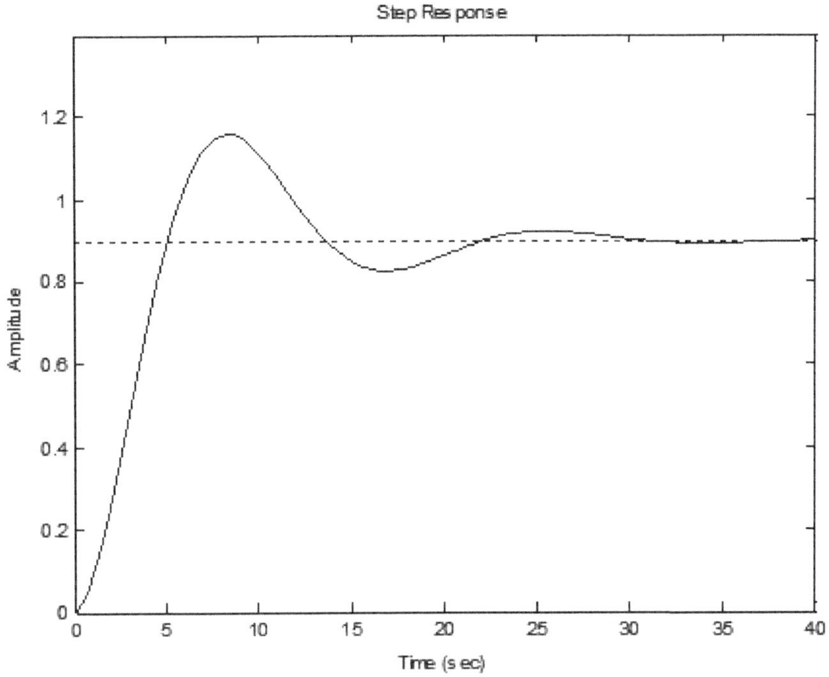

Fig. 2. Time Response of the Identified Process

6 Model Validation

The final step, verification, consists of relating the system to the identified model responses in time or frequency domain to instill confidence in the obtained model. Residual (correlation) analysis, Bode plots and cross-validation tests are generally employed for model validation. In the residual analysis models are statistically evaluated by way of correlation tests. The auto–correlation and cross-correlation functions of the errors with the outputs do not go significantly outside the 99% confidence bounds as shown in Fig. 3. While, on the other hand, cross–validation test was administered to establish the percentage fit between the predicted and the measured output from the SID toolbox. Fig. 4 shows that the simulated outputs follow the measured output closely, since relatively very high percentage of fit was obtained. Though the process is self-regulating, its step response with a peak overshoot of around 20% and an offset of 10%, indicates the need for tight closed-loop control.

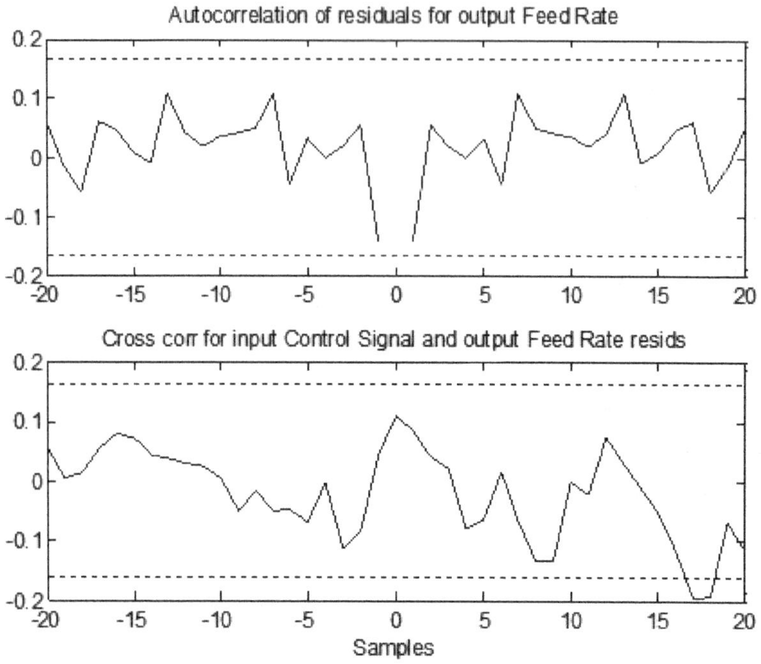

Fig. 3. The Auto-correlation and Cross-correlation Functions for Open-loop Feeder Model

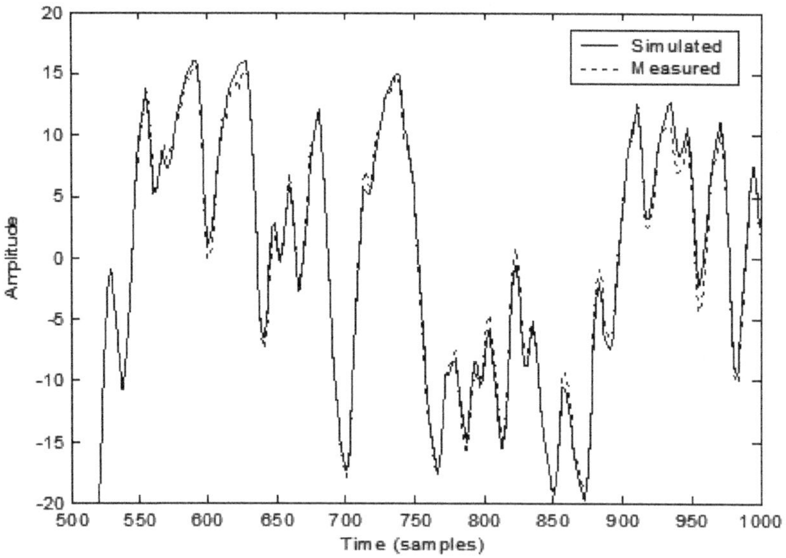

Fig. 4. Cross Validation Test for the Feeder Model

7 Conclusion

This paper presented details of the determination of an experimental model a 45 ton loss-in-weight raw cement feeder. The complete hardware setup and the algorithm followed for data acquisition, analysis and interpretation have been explained. Details of the procedure used to identify the model of a loss-in-weight raw meal feedrate process have been presented. The approach employed the microcontroller in the feedrate control loop to capture the process input and output experimental data with the control loop in manual mode. From the data the MATLAB system identification toolbox produced pulse transfer function for the process. Cross-validation test based on the comparative analysis between the actual and simulated open loop transient response of the model identified yielded a very high correlation. Simulation results indicate a great deal of flexibility and straightforward industrial process modelling using the MATLAB SID toolbox. Though the process is self-regulating, its step response with a peak overshoot of around 20% and an offset of 10%, indicates the need for tight closed-loop control.

References

1. Majhi, S., Atherton, D.P.: Obtaining Controller Parameters for a New Smith Predictor Using Autotuning. Automatica 36(11), 1651–1658 (2000)
2. Ljung, L.: System Identification: Theory for the User, 2 edn. Preniice Hall, New Jersey (1999)
3. Merricks Inc. Operation and Maintenance Instructions: MC2 Merricks Micro-computer Controller Software 30.00. HP. H & H Services Ltd, Nottinghamshire (1996)
4. Åström, K.J., Eykhoff, P.: System Identification—A Survey. Automatica 7(2), 123–162 (1971)
5. Kozin, F., Natke, H.: System Identification Techniques. Structural Safety 3(3), 269–316 (1986)
6. Balchen, J.G., Mummé, K.I.: Process Control: Structures and Applications. Van Nostrand Reinhold, New York (1988)
7. Dunn, P.F.: Measurement and Data Analysis for Engineering and Science, 3rd edn. CRC Press, Florida (2014)

Model Based Design of Finger Exoskeleton for Post Stroke Rehabilitation Using a Slotted Link Cam with Lead Screw Mechanism

Mohd Nor Azmi Bin Ab Patar*, Takashi Komeda, Jamaluddin Mahmud, and Cheng Yee Low

Research Organization for Advance Engineering, Shibaura Institute of Technology, Japan,
Centre of Excellence for Humanoid Robotics and Bio-Sensing,
Universiti Teknologi MARA, Malaysia
{azmipatar,jm,chengyee.low}@salam.uitm.edu.my,
komeda@se.shibaura-it.ac.jp

Abstract. Post stroke rehabilitation consumes a huge amount of health care resources in terms of costs related to hospital and home assistance. Recently, robot-assisted rehabilitation has been adapted to support physiotherapists in providing high-intensity and repetitive rehabilitation sessions. It has been observed that robotics offers an objective and reliable instrumented tool to monitor patient's progress and accurately assess their motor function. Each finger is attached to an instrumented mechanism which allowing force control and a mostly linear displacement. This paper presents a novel finger rehabilitation approach for acute paralyzed stroke survivors using a wearable robotic interface for hand motor function recovery. The device designed based on biomechanics measurements, able to assist the subject in opening and closing movements. It capable to adapt with various hand shapes and finger sizes. Main features of the interface include a differential system, and a lead screw mechanism which allows independent movement of the five fingers with actuators. The device is safe, easily transportable, and offers multiple training possibilities. The prototype deployment was carried out to determine the requirements for a finger rehabilitation device, the design and characterization of the complete system. Offering ease of use and affordability, the device has great potential to be deployed for individualized rehabilitation session for patients who have to undergo therapy in their home.

1 Introduction

Post stroke rehabilitation at the acute stage usually starts with one-to-one therapies conducted by physiotherapists in acute-care clinics [1]. To reduce the total cost of the treatment, patients are usually sent back to their home when their ability to walk is improved, even though they have not fully recovered the function of upper extremity, especially the distal segments, such as hands and fingers. In many cases, it will take a

* Corresponding author.

M. Gen et al. (eds.), *Industrial Engineering, Management Science and Applications 2015*,
Lecture Notes in Electrical Engineering 349, DOI: 10.1007/978-3-662-47200-2_11

long period of time to recover the function of flexion, extension, abduction, and adduction of the fingers [2]. Thus, leaving the fingers in a flexed or extended position leads to difficulties in activities of daily living (ADL), such as feeding, dressing, grooming and personal hygiene.

One of the approaches in solving the finger disabilities and injuries is undergoing a finger rehabilitation [3]. The finger rehabilitation is a physiotherapy approach which aims to partially or entirely recover the finger motor function of the patient. The physiotherapy approach is based on how to manipulate the paretic limb which is supported by a physiotherapist. The approach may be accomplished with daily repetitive rehabilitation frequently for up to several months [4]. It is depending on the severity of the finger and the condition of the patient. In order to recover back as a normal life, the patient needs time and undertaking the consistent rehabilitation assisted by a physiotherapist [5]. However, since the number of the physiotherapists is limited, it will not be easy for the patient to do the rehabilitation which needs to be supported by a physiotherapist all the time. Due to the limited numbers of physiotherapists, there are needs to develop a rehabilitation system where the patient can have their own rehabilitation exercises without aids from a therapist.

In this study, we investigated to develop and fabricate a new type of a robust hand and finger rehabilitation device which can control a human hand to do flexion and an extension motion. Our hypothesis by enforcing the correct flexion and extension motion, it can help patients with hand and finger muscle problems to close their hand and open hand correctly and improve healing. Most hand and finger devices for rehabilitation available on the market uses the passive control system. Unfortunately, the active control systems are costly and need a bigger space to install, not portable and not suitable to use at home. Therefore, the current study for the first time attempts to produce a robust, low cost device employing an active control system with a novel slotted link cam integrated with sliding lead screw mechanism.

Furthermore, most of the literature reviews on hand rehabilitation robotic devices focus on the recovery of motor functions, specifically the extension and flexion movements of the hand. However, there are no established approaches or publications available on the hand sensory functions restoration. In other words, the recovery of the sensory functions of the hand has yet to be explored by researchers. Therefore, improvements in the sensory functions of the hand are just as crucial to the recovery of the motor functions of the hand.

In this paper will discuss on the development process of the design concept, simulation and the fabrication of the device. The initial prototype of the device is also included. Since the exoskeleton only performs a flexion and extension thru the mechanism, the modification of the exoskeleton will be conducted in order to qualify as an index finger rehabilitation device. The design concept determined by these specifications is described in Section 2. The choice of materials, the actuation system, the implemented control schemes and evaluation experiments are described in Section 3.

2 Hand and Finger Exoskeletons: Functionality System

This section describes the specific functions [6] of the robotic system in rehabilitating the motor function of muscle in each fingers of human hand. The key function of a hand exoskeleton device is the ability to decrease the stiffness of the contracture finger. The stiffness of the muscle in human finger need to reduce according to normal human finger orientation [7], thus the robotic exoskeleton must be able to reproduce the flexion and extension of the finger movement repetitively [8]. Besides, the device must be able to detect angle of the flexion and extension in order to measure the trajectories for index, middle, ring and small fingers while performing the movement [9]. As shown in Figure 1, the sub-functions of the robotics exoskeleton consist of the ability to control angular velocity and producing a normal range of motion for the finger depend on each input angle on finger joint. The point of view from the occupational physiotherapist and the feedback from the healthy subjects are important to avoid incongruity during training session with real patients.

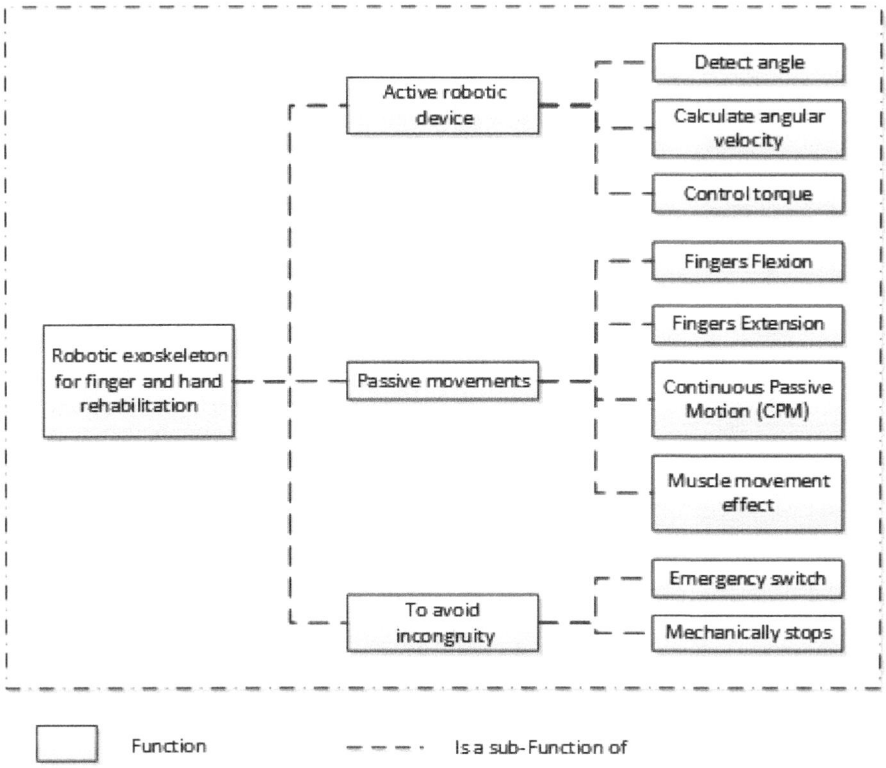

Fig. 1. Functions of robotic exoskeleton for finger and hand rehabilitation

3 Robotic Exoskeleton Prototype

This project is a pilot study to improve the finger rehabilitation. The project starts with a mechanical design of exoskeleton for an index finger. The main idea of the design is to perform an extension and flexion of the finger based on mechanisms that can transmit the force from the actuators.

In this study, we investigated a new type of a robust hand and finger rehabilitation device which can control a human hand to do flexion and an extension motion. Our hypothesis by enforcing the correct flexion and extension motion, it can help patients with hand and finger muscle problems to close their hand and open hand correctly and improve healing. Most hand and finger devices for rehabilitation available on the market uses the passive control system [10]. Unfortunately, the active control systems are costly and need a bigger space to install, not portable and not suitable to use at home. Therefore, the current study for the first time attempts to produce a robust, low cost device employing an active control system with a DC motor integrated with lead screw mechanism.

A hybrid actuation system consist of a DC servo motor integrated with lead screw mechanism has been developed to realize the functions in aforementioned criteria. The architecture of the actuation system is illustrated in Figure 2. The prototype is shown in Figure 4.

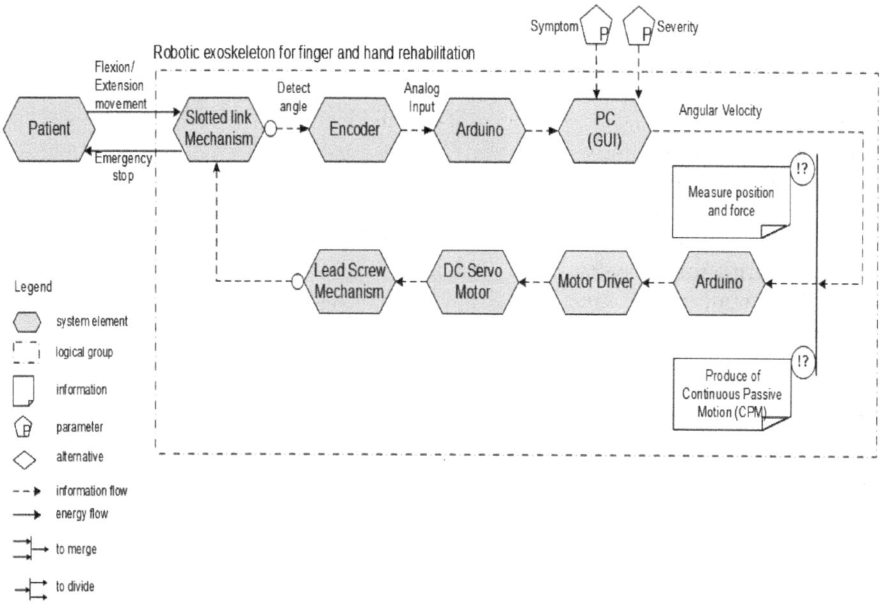

Fig. 2. System architecture of a DC servo motor integrated with lead screw mechanism in robotic exoskeleton for finger and hand rehabilitation

3.1 DC Servo Motor

A DC servo motor acts as an actuator to drive the lead nut in lead screw mechanism in order to repetitively flexion and extension a human finger. When the actuator actuate the mechanism, lead screw will convert rotary input motion to linear output motion. The nut is constrained from rotating with the screw, thus as the screw is rotated the nut travels back and forth along the length of the shaft. Depending on the level of severity, the DC servo motor provides a constant stiffness against the force given by the subjects' finger.

To further realize the real training session, variations in stiffness and angular velocity is added by applying the torque control via DC servo motor to provide continuous passive motion (CPM) helping subjects reduce joint stiffness of the fingers together and individually.

3.2 Lead Screw Mechanism

A lead screw typically is a linear actuator based mechanism that converts an oscillating input torque in the form of an angular displacement into a desired linear displacement. The major benefits of using a lead screw mechanism in linear actuators are inherent mechanical advantage, high stiffness, high strength, and a cost-effective package. Lead screws fall under the category of power screws and can be classified into ball screw, acme/trapezoidal screw and roller screw. A ball screw mechanism, consists of a ball screw and a ball nut with recirculating balls providing rolling contact between the nut and the screw. An acme or trapezoidal screw, which hereafter will be addressed as lead screw, consists of a screw and a nut that are in sliding contact with each other. The screw is generally made up of alloy steel with a trapezoidal thread form, and the nut is typically made of an engineering polymer or bronze. The contact between the nut and the screw is a sliding contact. Therefore, friction plays a very important role in the performance and efficiency of the mechanism. These screws offer low efficiencies due to the relatively greater coefficient of friction in sliding. Figure 3 illustrates trapezoidal thread profile of lead screw mechanism. Consider that a single thread of the screw is unrolled for exactly one turn. When determining the amount of input torque required to produce an amount of output linear force, there are many factors to consider. The following equations provide a practical approach in making force and torque calculation in lead screw mechanism. Equation (1) was used to approximate the total force involving in the system.

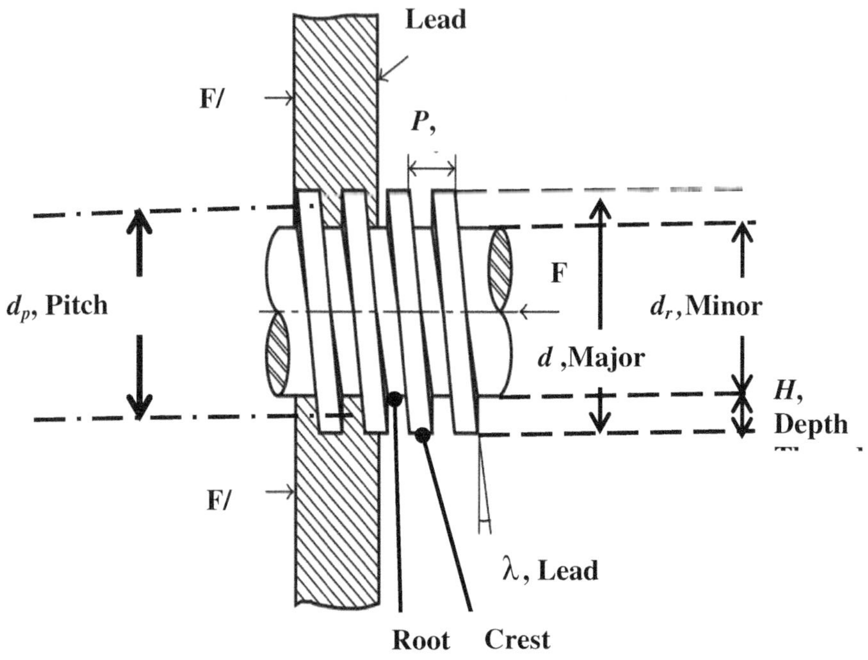

Fig. 3. Trapezoidal thread profile, Legends: *d*: Nominal or major diameter of screw, d_r: minor diameter of screw, d_p: pitch diameter, *P*: pitch, *L*: Lead, λ: Lead angle, *H*: depth thread of screw

$$F_T = F_A + F_E + F_F \tag{1}$$

Here F_T is total force, F_A represents acceleration force, F_E is external force, and F_F is friction force. External force due to clockwise (CW) and counter clockwise (CCW) motion of DC servo motor shaft direct connection with coupling in horizontal applications which is the requirements in extension and flexion of the finger. Friction force required to overcome all of the friction in the load bearing system with a low friction bearing system, this can be negligible. The total force must be below the compressive trust rating of the lead screw chosen. A modest factor of safety should be added to the total force. Thus, unexpected dynamic loads are safely handled by the lead screw mechanism system. Pitch and lead are closely related as shown in Figure 4.

Pitch, *P* is the distance between the crests or troughs of two consecutive threads. Lead, *L* is the distance navigated along the axis of the screw, by one complete rotation of the screw.

The crucial function of a lead screw is to transmit a load force axially, through a specified linear distance, *x*, called travel. It is a one degree of freedom (DOF) mechanism with the travel constrained between a full extent and full retract position.

The nut, which is engaged with the screw, is generally the linearly traveling member, whereas the screw is the rotating member driven by DC servo motor. Assuming that the screw is driven by a DC servo motor at an angular velocity of n, linear velocity v of the nut is given by Equation (2).

$$v = nx \tag{2}$$

Here v is in the unit of mm/s, n is in the unit of revolutions/second, and x is in the unit of mm.

The torque, T required to move the mechanism system can be approximated by Equation (3), where F_T is total force, L represents lead and, e is efficiency of lead screw assembly.

$$T = F_T \frac{L}{2\pi e} \tag{3}$$

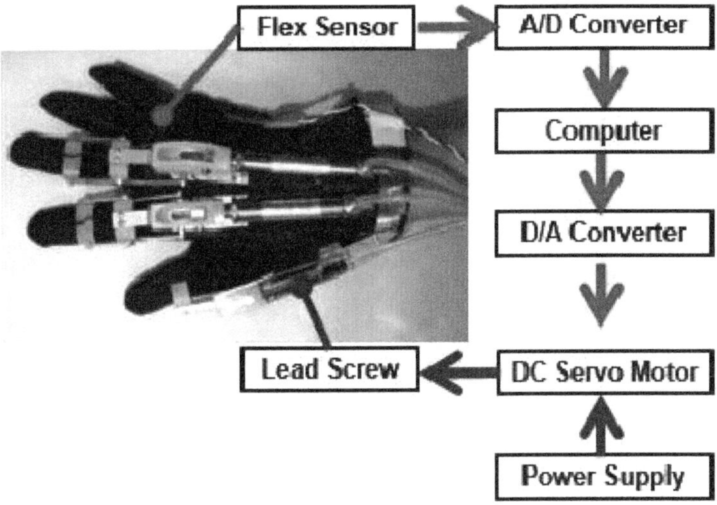

Fig. 4. Prototype of robotic exoskeleton for finger and hand rehabilitation

3.3 Control System

The prototype was developed with three operation modes for each individual finger. Each mode has different control system. In order to explain the hybrid actuation strategy implemented in the system, the control system design for different level of spasticity which is explained as illustrate in Figure 2.

Figure 5 shows the lead screw mechanism allows three operation modes for every single finger. Rest mode occurred where the finger cannot move. Passive mode is when the driven mechanism allow rotate freely which make the finger move freely. Active mode started where the lead screw is driven by the DC servo motor's shaft, which moves the finger according the range of motion of normal human.

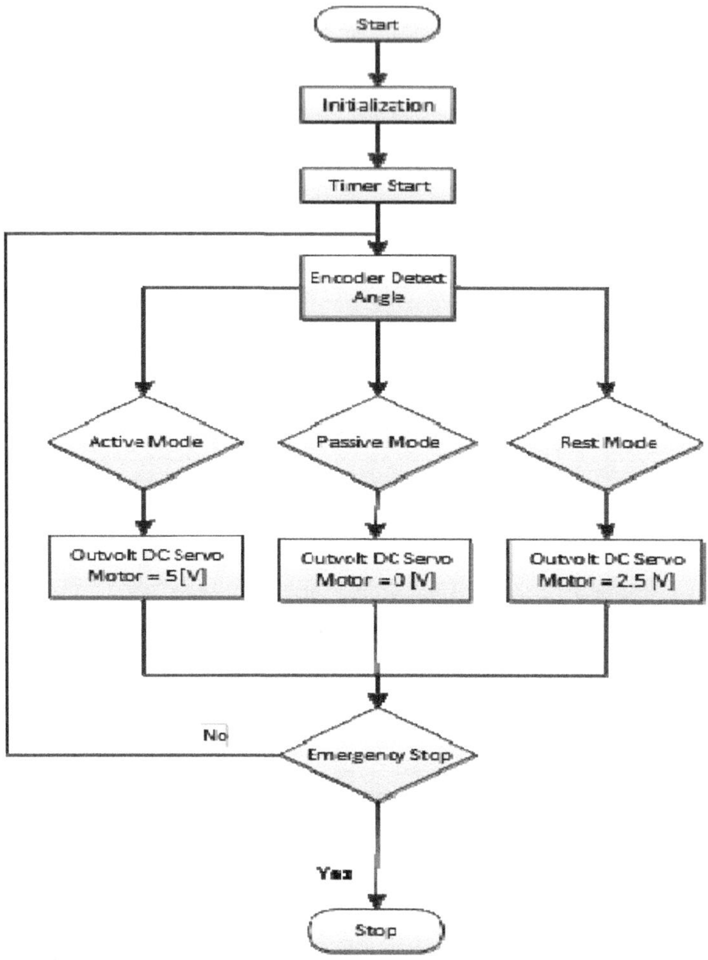

Fig. 5. Flowchart described control programming pseudo code of robotic exoskeleton for finger and hand rehabilitation

4 System Performance

This section presents measurement results in terms of joint angle produced by the hybrid actuation system of the lead screw mechanism in the robotic exoskeleton for hand and finger. These results shows plot of the experimental measurement of joint angle obtained from device versus time. The device joint angle increased according to the voltage supplied to the encoder. We observed that there had been a linear relationship between the voltages supplied to the encoder with joint angle produced by the device. Each operation mode can be distinguished through observation of the level of slope and speed that contribute to their discrete characteristics.

5 Conclusion

A hybrid actuation control strategy integrated with DC servo motor and lead screw mechanism has been developed to actuate the flexion and extension of the finger in rehabilitation. By independently control the DC servo motor and utilizing the behavior of lead screw mechanism, the system is provide with a range of bandwidth to operate in different modes.

Acknowledgments. The authors would like to thank the Research Management Institute (RMI) of Universiti Teknologi MARA (UiTM) for their support in managing the research funds.This research was initiated by the Research Organization for Advanced Engineering in Shibaura Institute of Technology, Japan and later, a part of this study was funded by Ministry of Science Technology and Innovation (MOSTI), Malaysia via Science Fund Scheme (Grant No. 100-RMI/SF 16/6/2 (7/2012).

References

1. Mohd Nor Azmi, A.P., Komeda, T., Mori, T., Seki, T., Saito, Y., Mahmud, J., Low, C.Y.: Hand rehabilitation device system (HRDS) for therapeutic applications. In: 5th IEEE RAS/EMBS Int. Conf. Biomed. Robot. Biomechatronics, pp. 381–386 (August 2014)
2. Lambercy, O., Metzger, J.-C., Santello, M., Gassert, R.: A method to study precision grip control in viscoelastic force fields using a robotic gripper. IEEE Trans. Biomed. Eng. 9294(1), 1–10 (2015)
3. Takahashi, C.D., Der-Yeghiaian, L., Le, V.H., Cramer, S.C.: A Robotic Device for Hand Motor Therapy after Stroke. In: 9th Int. Conf. Rehabil. Robot, ICORR 2005, pp. 17–20 (2005)
4. Lambercy, O., Dovat, L., Yun, H., Wee, S.K., Kuah, C.W.K., Chua, K.S.G., Gassert, R., Milner, T.E., Teo, C.L., Burdet, E.: Effects of a robot-assisted training of grasp and pronation/supination in chronic stroke: a pilot study. J. Neuroeng. Rehabil. 8(1), 63 (2011)
5. Mohd Nor Azmi, A.P., Komeda, T., Low, C.Y.: System Integration and Control of Finger Orthosis for Post Stroke Rehabilitation. Procedia Eng. 15, 787–796 (2014)
6. Yee, C., Kasim, M.A.A., Koch, T., Dumitrescu, R., Yussof, H., Jaafar, R., Jaffar, A., Aqilah, A., Mun, K.: Hybrid-Actuated Finger Prosthesis with Tactile Sensing. Int. J. Adv. Robot. Syst. 1 (2013)
7. Dovat, L., Lambercy, O., Gassert, R., Maeder, T., Milner, T., Leong, T.C., Burdet, E.: HandCARE: a cable-actuated rehabilitation system to train hand function after stroke. IEEE Trans. Neural Syst. Rehabil. Eng. 16(6), 582–591 (2008)
8. Patar, M.N.A.B.A., Mali, M., Ramli, M.H.M., Makhtar, A.K., Mahmud, J.: Simulation and performance evaluation of a new type of powered Dynamic Ankle Foot Orthosis. In: 2011 IEEE Colloquium on Humanities, Science and Engineering (CHUSER), pp. 167–171 (2011)
9. Seki, T., Nakamura, T., Kato, R., Morishita, S., Yokoi, H.: Development of five-finger multi-DoF myoelectric hands with a power allocation mechanism. In: 2013 IEEE Int. Conf. Robot. Autom., pp. 2054–2059 (May 2013)
10. Arata, J., Ohmoto, K., Gassert, R., Lambercy, O., Fujimoto, H., Wada, I.: A new hand exoskeleton device for rehabilitation using a three-layered sliding spring mechanism. In: IEEE Int. Conf. Robot. Autom., pp. 3902–3907 (May 2013)

Quantile Estimation Using a Combination of Stratified Sampling and Control Variates

Marvin K. Nakayama

Computer Science Department
New Jersey Institute of Technology
Newark, NJ 07102, USA
marvin@njit.edu
http://web.njit.edu/~marvin

Abstract. Quantiles are used to measure risk in many application areas. We consider simulation methods for estimating a quantile using a variance-reduction technique that combines stratified sampling and control variates. We provide an asymptotically valid confidence interval for the quantile.

Keywords: Risk, quantile, value-at-risk, Monte Carlo simulation, variance reduction, confidence interval.

1 Introduction

For fixed $0 < p < 1$, the p-quantile of a continuous random variable Y is the (smallest) constant ξ such that $P(Y \leq \xi) = p$. For example, the median is the 0.5-quantile. Also known as the $100p$ percentile, the p-quantile can alternatively be expressed as $F^{-1}(p)$, where F is the *cumulative distribution function (CDF)* of Y.

Quantiles are used in many application settings to measure risk [11]. For example, in finance, a 0.99-quantile, which is known as a value-at-risk, is often employed to assess portfolio risk and capital adequacy; e.g., see Chapter 9 of [7]. Acceptable risk levels for nuclear power plants (NPPs) are expressed in terms of a 0.95-quantile; see [17].

A common approach for estimating a quantile via (Monte Carlo) simulation is to first estimate the CDF from multiple runs of the simulation model, and then invert the estimated CDF. In addition to constructing a point estimator for the p-quantile ξ, we also want a confidence interval (CI) to provide a measure of the statistical error in the point estimator. Indeed, the U.S. Nuclear Regulatory Commission (NRC) requires NPP licensees to satisfy a so-called *95/95 criterion*, which entails establishing with 95% confidence that the 0.95-quantile lies below a mandated safety limit.

In the case when the simulation applies *simple random sampling (SRS)*, there are several methods for constructing a CI for a quantile. One approach exploits a binomial property to build an exact CI; e.g., see Setion 2.6.1 of [16]. Asymptotic

M. Gen et al. (eds.), *Industrial Engineering, Management Science and Applications 2015*
Lecture Notes in Electrical Engineering 349, DOI: 10.1007/978-3-662-47200-2_12

techniques include using a finite difference [4], kernel methods [6], batching (e.g., p. 491 of [7]), and sectioning (Section III.5a of [1]).

When applying SRS, the resulting CI for a quantile may be unusably wide, especially when examining an extreme quantile, i.e., $p \approx 0$ or $p \approx 1$. In this case, one can employ a *variance-reduction technique (VRT)* to try to obtain a more efficient quantile estimator. VRTs developed for quantile estimation include control variates (CV) [10,12], importance sampling (IS) [9], a combination of IS and stratified sampling (SS) [8], and Latin hypercube sampling [2,13]. The general approach is to apply the VRT to estimate the CDF, and then invert to CDF estimator to obtain a quantile estimator. While the above papers develop VRT point estimators for a quantile ξ and prove they obey central limit theorems, most do not provide a CI for ξ. To address this issue, [5] develops a general approach for constructing a CI for ξ based on a finite difference, and specializes it to apply for combined IS and SS (IS+SS), CV, and antithetic variates. The asymptotic validity of the technique is established by first showing that the VRT quantile estimator satisfies a so-called *Bahadur representation* [3], which approximates the quantile estimator as the sum of the true quantile and a linear transformation of the VRT CDF estimator, with a remainder that vanishes faster than $n^{-1/2}$, where n is the sample size. Also by exploiting a Bahadur representation, [15] develops asymptotically valid CIs for ξ using sectioning for the VRTs IS and CV.

The current paper describes how to estimate a quantile with a combination of SS and CV, which we denote by SS+CV. We also provide a sectioning CI for the SS+CV approach. Similar to batching, sectioning divides the n outputs into $b \geq 2$ equal-sized nonoverlapping batches. We compute a p-quantile estimator from each batch, and batching constructs a CI using the sample average and sample variance of the b quantile estimators. Sectioning replaces the sample average of the b quantile estimators, which we call the *batching point estimator*, with the overall quantile estimator based on all n outputs. An advantage of sectioning over batching arises since quantile estimators are generally biased. While the bias vanishes as the sample size grows large, it can be significant for small samples. The bias of the batching point estimator is determined by the batch size $m = n/b$, which is strictly smaller than the overall sample size n. Hence, the sectioning CI is centered at a point that may be less biased than the batching point estimator, and this can lead to the sectioning CI having better coverage than the batching CI, especially for small sample sizes.

The rest of the paper unfolds as follows. We first describe the mathematical framework in Section 2. We review stratified sampling and control variates in Sections 3 and 4, respectively. Section 5 shows how to combine the two VRTs, and we provide the sectioning CI for a quantile in Section 6. We provide some concluding remarks in Section 7.

2 Mathematical Framework

Consider a random variable Y representing the output of a simulation model, and let F denote the CDF of Y, i.e., $F(y) = P(Y \leq y)$ for each constant y. For a fixed

$0 < p < 1$, our goal is to estimate $\xi \equiv F^{-1}(p)$, where $G^{-1}(q) = \inf\{y : G(y) \geq q\}$ for any function G and $0 < q < 1$. We call ξ the p-quantile of F (or equivalently of Y).

We can estimate ξ when applying simple random sampling as follows. Let $Y_1, Y_2, \ldots, Y_n$ be a sample of n independent and identically distributed (i.i.d.) replicates of Y. We can express the CDF of Y as $F(y) = E[I(Y \leq y)]$, where $I(\cdot)$ denotes the indicator function, which assumes value 1 (resp., 0) when its argument is true (resp., false). By estimating the expectation through a sample average, we obtain the *empirical distribution function* F_n defined by

$$F_n(y) = \frac{1}{n}\sum_{i=1}^{n} I(Y_i \leq y). \tag{1}$$

Since the true p-quantile $\xi = F^{-1}(p)$, a natural point estimator of ξ is then

$$\xi_n = F_n^{-1}(p).$$

We can equivalently express $\xi_n = Y_{(\lceil np \rceil)}$, where $Y_{(1)} \leq Y_{(2)} \leq \cdots \leq Y_{(n)}$ are the ordered values of $Y_1, Y_2, \ldots, Y_n$, and $\lceil \cdot \rceil$ is the ceiling function. Sections 2.3–2.6 of [16] provide an extensive overview of SRS quantile estimation.

3 Stratified Sampling

We now review stratified sampling for estimating ξ; Section 4.3 of [7] gives a thorough introduction of SS for estimating a mean. Suppose that Z is an auxiliary random variable that is generated in the same simulation as the output Y. One possibility is that $Z = Y$, but we allow for other choices for Z. To use Z as a stratification variable, partition the support S of Z as $S = \cup_{s=1}^{t} R_s$, where $R_s \cap R_{s'} = \emptyset$ for $s \neq s'$, and we call R_s the sth *stratum*. Let $\lambda_s = P(Z \in R_s)$, so $\sum_{s=1}^{t} \lambda_s = 1$. We assume that the λ_s are known.

Example 1. Suppose that $Y = \sum_{k=1}^{Z} M_k$, where Z is a positive discrete random variable, and $M_1, M_2, \ldots$ are i.i.d. random variables, independent of Z. Further, suppose that Z has finite support $S = \{1, 2, \ldots, t\}$ and known probability mass function with $P(Z = m) = p_m$, $m \in S$. Then we can define each individual value $s \in S$ as a stratum, i.e., $R_s = \{s\}$, and $\lambda_s = P(Z = s)$.

In general, SS exploits the following representation for $F(y)$:

$$F(y) = \sum_{s=1}^{t} \lambda_s P(Y \leq y \,|\, Z \in R_s). \tag{2}$$

Thus, we may estimate F by estimating each $F'_s(y) \equiv P(Y \leq y \,|\, Z \in R_s)$ and combining the estimators as in (2). We finally obtain the SS estimator of ξ by inverting the resulting estimator of F. We next provide the details.

Define γ_s, $s = 1, 2, \ldots, t$, as user-specified positive constants summing to 1, and let $n_s = \lfloor \gamma_s n \rfloor$ be the sample size for the sth stratum, where n is the overall sample size and $\lfloor \cdot \rfloor$ is the floor function. To simplify the development, we will assume that $\gamma_s n$ is integer-valued, so that $n_s = \gamma_s n$ and $n = \sum_{s=1}^{t} n_s$. Let (Y'_s, Z'_s) denote a pair of random variables having the conditional distribution of (Y, Z) given $Z \in R_s$. We assume that we have a method for generating (Y'_s, Z'_s); Section 4.3 of [7] describes approaches for doing this. Let $(Y'_{s,i}, Z'_{s,i})$, $i = 1, 2, \ldots, n_s$, be a sample of n_s i.i.d. observations of (Y'_s, Z'_s). Also, assume that samples across strata are independent. Then an estimator of $F'_s(y)$ is

$$\hat{F}'_{s,n_s}(y) = \frac{1}{n_s} \sum_{i=1}^{n_s} I(Y'_{s,i} \leq y). \tag{3}$$

From (2), we obtain the SS estimator $\hat{F}_{\mathrm{SS},n}$ of F defined by

$$\hat{F}_{\mathrm{SS},n}(y) = \sum_{s=1}^{t} \lambda_s \hat{F}'_{s,n_s}(y), \tag{4}$$

and an estimator of the p-quantile ξ is then $\hat{\xi}_n = \hat{F}_{\mathrm{SS},n}^{-1}(p)$. The SS-only p-quantile estimator $\hat{\xi}_n$ is a special case of the IS+SS quantile estimator given in [8,5]. Thus, the finite-difference method in [5] can be used to construct a CI for ξ when applying SS-alone.

Example 1 (continued). We have that $Y'_s = \sum_{k=1}^{s} M_k$ and $F'_s(y) = P(\sum_{k=1}^{Z} M_k \leq y | Z = s) = P(\sum_{k=1}^{s} M_k \leq y)$ by the assumed independence of Z and $M_1, M_2, \ldots$. Thus, (3) becomes

$$\hat{F}'_{s,n_s}(y) = \frac{1}{n_s} \sum_{i=1}^{n_s} I\left(\sum_{k=1}^{s} M_{s,k} \leq y\right),$$

where $M_{s,k}$, $s = 1, 2, \ldots, t$, $k = 1, 2, \ldots, s$, are i.i.d.

By adapting an algorithm in [5] for inverting the IS+SS CDF estimator, we can compute $\hat{F}_{\mathrm{SS},n}^{-1}(p)$ as follows. We are given λ_s, $s = 1, 2, \ldots, t$, and after completing all simulations, we have $n = \sum_{s=1}^{t} n_s$ total outputs $Y'_{s,i}$, $s = 1, 2, \ldots, t$, $i = 1, 2, \ldots, n_s$. For each $s = 1, 2, \ldots, t$, and $i = 1, 2, \ldots, n_s$, define $A_k = Y'_{s,i}$ and $B_k = \lambda_s / n_s$, where $k = \sum_{l=1}^{s-1} n_l + i$. Then sort A_k, $k = 1, 2, \ldots, n$, in ascending order as $A_{(1)} \leq A_{(2)} \leq \cdots \leq A_{(n)}$, and let $B^{(k)}$ be the concomitant variable corresponding to $A_{(k)}$. Finally set $\hat{F}_{\mathrm{SS},n}^{-1}(p) = A_{(i_p)}$, where i_p is the smallest integer for which $\sum_{k=1}^{i_p} B^{(k)} \geq p$.

4 Control Variates

Let X be another auxiliary variable generated in the same simulation as the original output Y, and we now review how to use X as a control variate to estimate ξ. We assume that we know the mean ν of X. For example, in a simulation of a $G/G/1$ queue with Y denoting the sum of the waiting times of the first r customers, we often know the mean τ of an individual service time. Thus, let V be the sum of the service times of the first $r-1$ customers and we can define the CV $X = V$, where $\nu = (r-1)\tau$. Another possibility is to let $X = I(V \leq \eta)$, where η is the p-quantile of V, which we assume can be computed, so $\nu = p$.

Let (Y_i, X_i), $i = 1, 2, \ldots, n$, be n i.i.d. copies of the pair (Y, X). Let β be any constant, and we can express the CDF F of Y as

$$F(y) = E[I(Y \leq y)] - \beta E[(X - \nu)]$$

since $E[X] = \nu$. Replacing the expectations with sample averages leads to an estimator of the CDF F as

$$\hat{F}_{n,\beta}(y) = F_n(y) - \beta(\bar{X}_n - \nu),$$

where F_n is defined in (1) and $\bar{X}_n = (1/n)\sum_{i=1}^{n} X_i$. The variance of $\hat{F}_{n,\beta}(y)$ depends on β and y, and it is well known (e.g., see p. 184 of [7]) that the value of β minimizing the variance for a given y is $\beta^*(y) = \mathrm{Cov}[I(Y \leq y), X]/\mathrm{Var}[X]$. A consistent estimator of $\beta^*(y)$ is

$$\hat{\beta}_n(y) = \frac{[(1/n)\sum_{i=1}^{n} I(Y_i \leq y)X_i] - F_n(y)\bar{X}_n}{(1/n)\sum_{i=1}^{n}(X_i - \bar{X}_n)^2},$$

which leads to the CV estimator of $F(y)$ as

$$\hat{F}_{\mathrm{CV},n}(y) = F_n(y) - \hat{\beta}_n(y)(\bar{X}_n - \nu).$$

We then obtain $\hat{\xi}_{\mathrm{CV},n} = \hat{F}_{\mathrm{CV},n}^{-1}(p)$ as the CV estimator of the p-quantile ξ. To build a CI for ξ when applying CV, [5] uses a finite difference and [15] employs sectioning.

Inverting $\hat{F}_{\mathrm{CV},n}$ initially appears to be complicated by the fact that $\hat{\beta}_n(y)$ depends on y. But [10] provides an alternative representation of $\hat{F}_{\mathrm{CV},n}$ that circumvents this issue, which we now describe. It can be shown that

$$\hat{F}_{\mathrm{CV},n}(y) = \sum_{i=1}^{n} W_i I(Y_i \leq y), \tag{5}$$

where

$$W_i = \frac{1}{n} + \frac{(\bar{X}_n - X_i)(\bar{X}_n - \nu)}{\sum_{k=1}^{n}(X_k - \bar{X}_n)^2}.$$

Note that W_i does not depend on y, and $\sum_{i=1}^{n} W_i = 1$. Thus, we can invert $\hat{F}_{\mathrm{CV},n}$ as follows. Sort Y_i, $i = 1, 2, \ldots, n$, in ascending order as $Y_{(1)} \leq Y_{(2)} \leq \cdots \leq Y_{(n)}$, and let $W^{(i)}$ be the concomitant variable corresponding to $Y_{(i)}$. Then $\hat{F}_{\mathrm{CV},n}^{-1}(p) = Y_{(j_p)}$, where j_p is the smallest integer such that $\sum_{i=1}^{j_p} W^{(i)} \geq p$.

Both SS and CV use an auxiliary variable from a simulation to reduce variance. If we can use only one auxiliary variable, is it better to employ it as a stratification variable, as in Section 3, or as a control? An asymptotic analysis on p. 220 of [7] provides a partial answer when the auxiliary variable has a continuous distribution. Suppose we apply SS in which the number t of strata grows large in an equiprobable stratification (i.e., each $\lambda_s = 1/t$) that becomes increasingly refined. Then in the limit as $t \to \infty$, the SS variance is smaller than the CV variance, showing the superiority of SS in this asymptotic context. But an infinitely fine stratification requires knowing the auxiliary variable's full distribution, instead of just its mean, as in CV.

5 Combining SS with CV

Now we describe how to combine SS with CV to estimate the quantile ξ, and we denote this approach SS+CV. Again, let Z be a stratification variable as in Section 3, and given $Z \in R_s$, we use a random variable X'_s as a control. We assume that we know $\nu_s = E[X'_s]$. One possibility for X'_s is to let it be the control X from Section 4 conditional on $Z \in R_s$, assuming that $\nu_s = E[X|Z \in R_s]$ is known, but we do not require this choice for X'_s. Let $(Y'_{s,i}, X'_{s,i}, Z'_{s,i})$, $i = 1, 2, \ldots, n_s$, be a sample of n_s i.i.d. observations of (Y'_s, X'_s, Z'_s) for stratum R_s. For each stratum R_s, we modify the estimator $\hat{F}'_{s,n_s}(y)$ in (3) to incorporate CV to obtain

$$\hat{F}''_{s,n_s,\beta_s}(y) = \hat{F}'_{s,n_s}(y) - \beta_s(\bar{X}'_{s,n_s} - \nu_s),$$

where $\hat{F}'_{s,n_s}(y)$ is defined in (3), β_s is any constant, and $\bar{X}'_{s,n_s} = (1/n_s)\sum_{i=1}^{n_s} X'_{s,i}$. For each y, the variance of $\hat{F}''_{s,n_s,\beta_s}(y)$ depends on β_s, and the optimal value of β_s for a given y is

$$\beta^*_s(y) = \mathrm{Cov}[I(Y'_s \le y), X'_s]/\mathrm{Var}[X'_s], \tag{6}$$

which can be consistently estimated by

$$\hat{\beta}_{s,n_s}(y) = \frac{[(1/n_s)\sum_{i=1}^{n_s} I(Y'_{s,i} \le y)X'_{s,i}] - \hat{F}'_{s,n_s}(y)\bar{X}'_{s,n_s}}{(1/n_s)\sum_{i=1}^{n_s}(X'_{s,i} - \bar{X}'_{s,n_s})^2}.$$

Then our SS+CV estimator of $F(y)$ is

$$\hat{F}_{\mathrm{SS+CV},n}(y) = \sum_{s=1}^{t} \lambda_s[\hat{F}'_{s,n_s}(y) - \hat{\beta}_{s,n_s}(y)(\bar{X}'_{s,n_s} - \nu_s)],$$

and the corresponding SS+CV p-quantile estimator is

$$\hat{\xi}_{\mathrm{SS+CV},n} = \hat{F}^{-1}_{\mathrm{SS+CV},n}(p). \tag{7}$$

As in (5), we can compute $\hat{F}^{-1}_{\mathrm{SS+CV},n}(p)$ by first expressing

$$\hat{F}'_{s,n_s}(y) - \hat{\beta}_{s,n_s}(y)(\bar{X}'_{s,n_s} - \nu_s) = \sum_{i=1}^{n_s} W_{s,i} I(Y'_{s,i} \le y)$$

for each $s = 1, 2, \ldots, t$, where

$$W_{s,i} = \frac{1}{n_s} + \frac{(\bar{X}'_{s,n_s} - X'_{s,i})(\bar{X}'_{s,n_s} - \nu_s)}{\sum_{k=1}^{n}(X'_{s,k} - \bar{X}'_{s,n_s})^2}$$

for $i = 1, 2, \ldots, n_s$. Then for each $s = 1, 2, \ldots, t$, and $i = 1, 2, \ldots, n_s$, define $C_k = Y'_{s,i}$ and $D_k = \lambda_s W_{s,i}$, where $k = \sum_{l=1}^{s-1} n_l + i$. Then sort the $n = \sum_{s=1}^{t} n_s$ total outputs C_k, $k = 1, 2, \ldots, n$, in ascending order as $C_{(1)} \leq C_{(2)} \leq \cdots \leq C_{(n)}$, and let $D^{(k)}$ be the concomitant variable corresponding to $C_{(k)}$. Then $\hat{F}^{-1}_{\mathrm{SS+CV},n}(p) = C_{(l_p)}$, where l_p is the smallest integer such that $\sum_{i=1}^{l_p} D^{(i)} \geq p$.

6 Confidence Interval with SS+CV

We now describe one method to construct a confidence interval for the p-quantile when applying SS+CV. The approach we discuss is known as sectioning, which was originally developed for simple random sampling by [1] (Section III.5a) and extended to some variance-reduction techniques in [15]. Similar to batching, sectioning partitions the data into $b \geq 2$ nonoverlapping batches, and computes a p-quantile estimator from each batch. (Numerical results in [15] suggest selecting $b = 10$ batches for sectioning can work well for SRS, IS-alone and CV-alone, and this choice is probably reasonable for SS+CV.) Batching constructs a CI for ξ using the sample average and the sample variance of the b quantile estimators. Sectioning instead centers the CI around the overall quantile estimator from all of the data directly instead of the sample average of the b quantile estimators. We now provide more details.

Fix an integer $b \geq 2$ as the number of batches. For each stratum R_s, $s = 1, 2, \ldots, t$, divide the n_s triples $(Y'_{s,i}, X'_{s,i}, Z'_{s,i})$, $i = 1, 2, \ldots, n_s$, from the stratum into b nonoverlapping batches, each of size $m_s = n_s/b$, assumed to be an integer. Specifically, for each $j = 1, 2, \ldots, b$, the jth batch for stratum R_s consists of data $(Y'_{s,i}, X'_{s,i}, Z'_{s,i})$ for $(j-1)m_s + 1 \leq i \leq jm_s$. For each batch j and stratum R_s, let $\hat{F}'_{s,m_s,j}(y) = (1/m_s)\sum_{i=(j-1)m_s+1}^{jm_s} I(Y'_{s,i} \leq y)$ be an estimator of $F'_s(y)$, and let $\bar{X}'_{s,m_s,j} = (1/m_s)\sum_{i=(j-1)m_s+1}^{jm_s} X'_{s,i}$ be the sample average of the m_s i.i.d. controls for the jth batch in stratum R_s. Then define

$$\hat{\beta}_{s,m_s,j}(y) = \frac{[(1/m_s)\sum_{i=(j-1)m_s+1}^{jm_s} I(Y'_{s,i} \leq y)X'_{s,i}] - \hat{F}'_{s,m_s,j}(y)\bar{X}'_{s,m_s,j}}{(1/m_s)\sum_{i=(j-1)m_s+1}^{jm_s}(X'_{s,i} - \bar{X}'_{s,m_s,j})^2}$$

as the batch-j estimator of the optimal CV multiplier $\beta^*_s(y)$ in (6) for stratum R_s. Define the total batch size across all t strata to be $m = \sum_{s=1}^{t} m_s = \sum_{s=1}^{t} n_s/b = n/b$. Then our SS+CV estimator of F for batch j across all strata is

$$\hat{F}_{\mathrm{SS+CV},m,j}(y) = \sum_{s=1}^{t} \lambda_s[\hat{F}'_{s,m_s,j}(y) - \hat{\beta}_{s,m_s,j}(y)(\bar{X}'_{s,m_s,j} - \nu_s)],$$

leading to $\hat{\xi}_{SS+CV,m,j} = \hat{F}^{-1}_{SS+CV,m,j}(p)$ as the corresponding SS+CV p-quantile estimator from batch j. The b batch SS+CV quantile estimators $\hat{\xi}_{SS+CV,m,j}$, $j = 1, 2, \ldots, b$, are i.i.d., and we define their sample variance

$$S^2_{m,b} = \frac{1}{b-1}\sum_{j=1}^{b}(\hat{\xi}_{SS+CV,m,j} - \bar{\xi}_{m,b})^2, \tag{8}$$

where $\bar{\xi}_{m,b} = (1/b)\sum_{j=1}^{b}\hat{\xi}_{SS+CV,m,j}$. The $(1-\alpha)$-level batching CI for ξ using SS+CV is

$$I_{n,b} \equiv \left(\bar{\xi}_{m,b} \pm \tau_{b-1,\alpha/2}S_{m,b}/\sqrt{b}\right), \tag{9}$$

where $\tau_{b-1,\alpha/2}$ is the $\alpha/2$-level critical point of a Student t random variable T with $b-1$ degrees of freedom; i.e., $P(T \geq \tau_{b-1,\alpha/2}) = \alpha/2$.

While the batching CI in (9) is asymptotically valid in the sense that $P(\xi \in I_{n,b}) \to 1-\alpha$ as $n = bm \to \infty$ for fixed $b \geq 2$, the CI may have poor coverage for small n. This arises because quantile estimators are generally biased. Although the bias vanishes as the sample size grows large, it can be significant for small sample sizes. Moreover, the bias of the batching point estimator $\bar{\xi}_{m,b}$ is determined by the total batch size $m = n/b$, which is smaller than the overall sample size n. Hence, the batching CI is centered at a point, $\bar{\xi}_{m,b}$, that can have considerable bias, which may cause poor coverage; i.e., $P(\xi \in I_{n,b})$ may differ significantly from $1-\alpha$ when n is small.

Sectioning addresses this issue by replacing the batching point estimator $\bar{\xi}_{m,b}$ with the overall quantile estimator $\hat{\xi}_{SS+CV,n}$ from (7), which is typically less biased than $\bar{\xi}_{m,b}$. We then obtain the sectioning CI as

$$\hat{I}_{n,b} \equiv \left(\hat{\xi}_{SS+CV,n} \pm \tau_{b-1,\alpha/2}\hat{S}_{m,b}/\sqrt{b}\right), \tag{10}$$

where

$$\hat{S}^2_{m,b} = \frac{1}{b-1}\sum_{j=1}^{b}(\hat{\xi}_{SS+CV,m,j} - \hat{\xi}_{SS+CV,n})^2.$$

This CI will often have better coverage than the batching CI in (9). The following result can be established by combining ideas from [5] and [15]; see [14] for details.

Theorem 1. *The SS+CV sectioning $(1-\alpha)$-level CI $\hat{I}_{n,b}$ in (10) is asymptotically valid in the sense that*

$$P(\xi \in \hat{I}_{n,b}) \to 1-\alpha$$

as $n = bm \to \infty$ for any fixed number $b \geq 2$ of batches.

Another CI described in [15] centers the interval at the overall quantile estimator $\hat{\xi}_{SS+CV,n}$ from (7) but uses $S_{m,b}$ from (8) to determine its width. This CI is also asymptotically valid. Numerical results in [15] with SRS, IS-alone and CV-alone show this approach performs similarly to (10).

7 Concluding Remarks

We considered quantile estimation through a combination of stratified sampling and control variates. This can provide a more efficient point estimator than using either method by itself or just simple random sampling. We also described how to apply sectioning to build an asymptotically valid confidence interval for the quantile.

Key factors that determine how much variance reduction is obtained by SS+CV are the choices of the stratification variable Z, the strata R_s, $s = 1, 2, \ldots, t$, and the CV X'_s for each stratum. In each stratum R_s, CV will significantly reduce variance when there is strong correlation (either negative or positive) between $I(Y'_s \leq \xi)$ and X'_s. The paper [10] examines ways to obtain effective controls when applying CV-alone, and it would be interesting to investigate similar ideas for SS+CV.

Acknowledgments. This work has been supported in part by the National Science Foundation under Grants No. CMMI-0926949, CMMI-1200065, and DMS-1331010. Any opinions, findings, and conclusions or recommendations expressed in this material are those of the author and do not necessarily reflect the views of the National Science Foundation.

References

1. Asmussen, S., Glynn, P.: Stochastic Simulation: Algorithms and Analysis. Springer, New York (2007)
2. Avramidis, A.N., Wilson, J.R.: Correlation-induction techniques for estimating quantiles in simulation. Operations Research 46, 574–591 (1998)
3. Bahadur, R.R.: A note on quantiles in large samples. Annals of Mathematical Statistics 37, 577–580 (1966)
4. Bloch, D.A., Gastwirth, J.L.: On a simple estimate of the reciprocal of the density function. Annals of Mathematical Statistics 39, 1083–1085 (1968)
5. Chu, F., Nakayama, M.K.: Confidence intervals for quantiles when applying variance-reduction techniques. ACM Transactions on Modeling and Computer Simulation 36, Article 7 (25 pages plus 12–page online–only appendix) (2012)
6. Falk, M.: On the estimation of the quantile density function. Statistics & Probability Letters 4, 69–73 (1986)
7. Glasserman, P.: Monte Carlo Methods in Financial Engineering. Springer, New York (2004)
8. Glasserman, P., Heidelberger, P., Shahabuddin, P.: Variance reduction techniques for estimating value-at-risk. Management Science 46, 1349–1364 (2000)
9. Glynn, P.W.: Importance sampling for Monte Carlo estimation of quantiles. In: Mathematical Methods in Stochastic Simulation and Experimental Design: Proceedings of the 2nd St.Petersburg Workshop on Simulation, pp. 180–185. Publishing House of St. Petersburg Univ., St.Petersburg, Russia (1996)
10. Hesterberg, T.C., Nelson, B.L.: Control variates for probability and quantile estimation. Management Science 44, 1295–1312 (1998)
11. Hong, L.J., Hu, Z., Liu, G.: Monte Carlo methods for value-at-risk and conditional value-at-risk: A review. ACM Trans. Mod. Comp. Sim. 24, Article 22 (37 pages) (2014)

12. Hsu, J.C., Nelson, B.L.: Control variates for quantile estimation. Management Science 36, 835–851 (1990)
13. Jin, X., Fu, M.C., Xiong, X.: Probabilistic error bounds for simulation quantile estimation. Management Science 49, 230–246 (2003)
14. Nakayama, M.K.: Quantile estimation with stratified sampling and control variates (in preparation)
15. Nakayama, M.K.: Confidence intervals using sectioning for quantiles when applying variance-reduction techniques. ACM Transactions on Modeling and Computer Simulation 24, Article 19 (2014)
16. Serfling, R.J.: Approximation Theorems of Mathematical Statistics. John Wiley & Sons, New York (1980)
17. U.S. Nuclear Regulatory Commission: Acceptance criteria for emergency core cooling systems for light-water nuclear power reactors. Title 10, Code of Federal Regulations Section 50.46 (10CFR50.46), U.S. Nuclear Regulatory Commission, Washington, DC (2010)

A Simulation-Based Analysis for Inter Release Problem in Airport Baggage Handling Systems

James T. Lin, Irene Liou, and Chun-Chih Chiu

Department of Industrial Engineering and Engineering Management,
National Tsing Hua University, Hsin-Chu, Taiwan, R. O. C.
jtlin@ie.nthu.edu.tw, liouirene@yahoo.com, vhasb3210@gmail.com

Abstract. Airport baggage handling systems (BHS) are a crucial infrastructure in major airports to ensure smooth baggage transfer while avoiding congestion. Therefore, it is important to operate the BHS efficiently.

Recently, the number of tourists increases year after year in Taiwan, leading to an insufficient capacity of airport baggage storage. Hence, the parameter setting in BHS for baggage transfer becomes more important and challenging regarding future system reconstructions. For experiment efficiency and cost reduction, a widely used method to evaluate system performances and compare alternatives is to apply discrete event simulation models. This paper presents an empirical study of interval release time by using simulation with full factorial design of the BHS. The goals are to determine the best combination of the interval release time, understand their interactions, and minimize congestion time. The results demonstrate a 23% time reduction in peak-time congestion while the optimized parameter setting achieves smooth system operation.

Keywords: Baggage handling system, Airport, Simulation, Full factorial design.

1 Introduction

The growing number of passengers traveling to/from Taiwan has brought a capacity issue to Taiwan's major airport, Taiwan Taoyuan International Airport (TPE), which commenced operation in 1979. In 2013, with 200,000 aircraft movements (ACMs), growing at a rate of around 6% [1] per year since 2009 (shown in Figure 1), TPE handled a total of over 30,700,000 passengers, indicating a matching number of baggage amount. (Statistic figures for 2014 have not been published as of this date.) In terms of international passenger number, TPE has become the 11th busiest airport worldwide in 2014 [2, 3]. Therefore, the baggage handling system (BHS) plays an important role in delivering the baggage accurately and punctually.

Detailed process of the baggage handling system is introduced as follows: (1) Checked-in and transferred baggage is carried by the conveyors to go through x-ray screening for security check. (2) The baggage is then sent to the barcode scanner where each piece of baggage is identified and assigned a corresponding destination by the control system. (3) The baggage finally enters the main sorter for destination

M. Gen et al. (eds.), *Industrial Engineering, Management Science and Applications 2015*,
Lecture Notes in Electrical Engineering 349, DOI: 10.1007/978-3-662-47200-2_13

sorting and dispatch based on its scanned information. The destination is either a buffer zone for a later release for re-sorting or an unloading zone. The unloading zone for a specific flight is usually assigned two hours before the departure time. If the baggage arrives before the unloading zone is assigned, it will be transported to the buffer zone and wait until the controller releases it back to the main sorter for re-sorting.

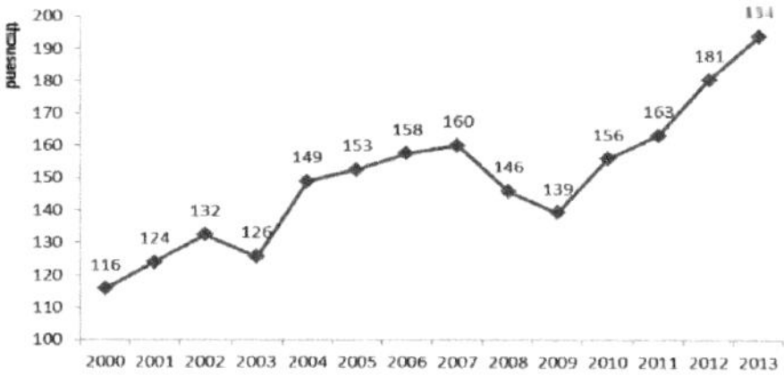

Fig. 1. Aircraft movements from 2000 to 2013

Presently TPE is facing a capacity issue of buffer zone shortage in BHS due to the annual large increase in passenger number. Therefore, an overall BHS upgrade and a future reconstruction are under consideration with the main terminal in the early phase of such projects. As a result, its large amount of baggage will have to be handled by Terminal 2 North Concourse, a BHS that is far away from flight gates and rarely used, for temporary baggage storage and unloading. Terminal 2 North Concourse is connected to the main terminal by two conveyors, on which ends locate a barcode scanner respectively before entering the sorter in the Concourse, for transporting checked-in baggage. Also before entering the sorter, there locates a third barcode scanner on the conveyor for transporting buffer-released baggage. The fourth barcode scanner locates on the conveyor, connected to the sorter as well, for the baggage that misses scanning. After being scanned by these four scanners, baggage enters the sorter for sorting. However, when the sorter is busy in operation, the scanned baggage may need to wait at the scanners for a turn to enter it. Therefore, the respective setting of interval release time for baggage to enter the sorter at these four scanners will significantly influence the operation in the Concourse. Without proper parameter settings, baggage congestion may occur on the conveyors. In the worst situation, blockage may reverse to the check-in counters, leading to poor level of service.

The purpose of this paper is to find the best combination of interval release time parameter settings for the baggage waiting at the four correlated scanners to enter the sorter. The best combination may contribute to a smooth operation and the reduction of congestion in the Concourse system. Since the four conveyors work with interaction, blockage may occur at the merging points when they are busy in operation. Due to the difficulty in achieving an accurate prediction of BHS in real world, we propose a simulation model to test the designed alternatives of different interval release time settings in order to obtain the best combination of the four interval release time.

This paper is organized as follows: Section 2 presents research literature relating to BHS. Section 3 briefly introduces BHS and its components and describes the problem using a BHS simulation model. Section 4 conducts experiments with different control rules; the experiment and test results for estimating performance are included. Section 5 concludes with a summary of work as well as guidelines for future research.

2 Related Literature

In recent years, issues on BHS have been focused on the effect of passenger arrival rate [4], check-in counter arrangements [5], unloading zone arrangements for flights [6], conveyors-merging in the analytic system [7], etc. A great many important effects influencing BHS performances are also discussed [8]. BHS is the processing system that transports baggage from check-in counters to aircrafts; hence, it is highly important that the system operates steadily and smoothly.

Ayman [9] analyzed the causes of congestion through data-mining with acquired baggage information. To detect the probable congestion conveyors, the author used three visual patterns to analyze and summarize system operation, more frequently used conveyors, and the usage rate for each conveyor.

Gary [7] considered merging factors in the system. The study considered possibilities of conveyor merges as the merging rule had an effect on system performance. Le and Johnstone [8] proposed problems encountered in current baggage transportation system. While solving problems using discrete simulation, they proposed the factors considered such as baggage sizes, system recovery time, length of delay time, etc. They described situations of speed reductions or stoppages in baggage transportation when merging occurred in the system.

BHS is one of the most complex systems in real world operation at the airport that includes many processes, machines, and operators. It is not a straightforward task to achieve an accurate prediction on site; hence, applying simulation models is necessary. Simulation models are widely used for analyzing BHS performances [8][10]. They can be used for targeted systems not even physically existing. For BHS, Chun and Mak [11] developed intelligent resource simulation system (IRSS) to minimize number of counters while maintaining a given level of service at the same time.

BHS is affected by many important performance factors; there are also difficulties for system operation planners to predict performances of a complex system. A simulation model is therefore developed in order to understand BHS performances by testing different parameter settings in the operation system.

3 Problem Description and Simulation Modeling

The problem under consideration may be summarized as the interaction among the four operating conditions with different settings of interval release time.

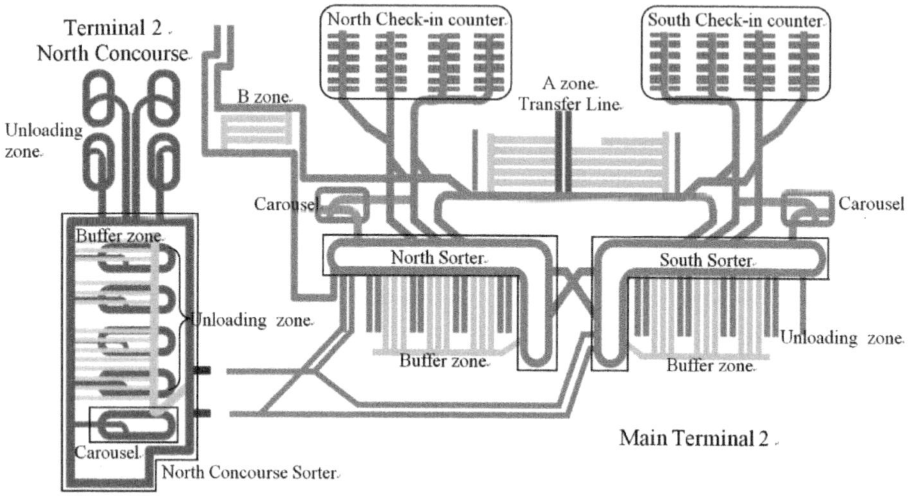

Fig. 2. The TPE BHS sketch

Figure 2 provides a layout of the BHS of TPE, comprising mainly of North/South check-in counters, barcode scanners, sorters, buffer zones, unloading zones, carousels, and Terminal 2 North Concourse. The simulation model is constructed using Flexsim™, a commercial off-the-shelf (COTS) tool and a fully 3D simulation software environment that provides excellent visually settings for users. This simulation model is also validated and used for assessing the underlying system performance and conducting analyses. Detailed construction of the simulation model can be referred to Shih et al. [12]. The simulation model is shown in Figure 3.

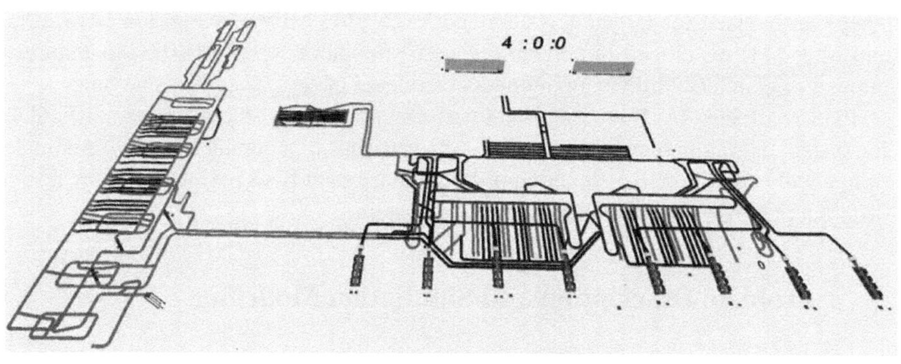

Fig. 3. Flexsim simulation model

The Operation Condition of the Four Barcode Scanners

Terminal 2 North Concourse is presently facing a problem of balancing the incoming baggage input. Therefore, this paper focuses on its operation condition.

Figure 4 shows an enlargement of Terminal 2 North Concourse in which there are four barcode scanners, namely: "A", "B", "C", and "D". "A" and "B" locate at the ends of the two conveyors connected to the main terminal. After being scanned by "A" or "B", the baggage is sent to the sorter for destination sorting based on its scanned information. The destination is either a buffer zone or an unloading zone.

Since baggage flows in the direction of the arrow, shown in Figure 4, barcode scanner "B" is in domination. When dominant conveyor is busy in operation, baggage from "A" may accumulate at the merging point. Consequently, the blockage may then reverse to the main terminal, causing passengers queuing up at the check-in counters, which produces problems for the airport and leads to poor system performances.

Baggage released from the buffer zones goes through "C" and re-enters the sorter. Such re-entering process affects the merging of the input baggage from "A" and is influenced by the baggage coming from "B". Either of the two situations may cause queuing delay and blockage in the system.

The baggage missing scanning is sent to "D" for a rescan. This process also adversely affects the system performance and causes baggage queuing in the system.

The four above-mentioned barcode scanners work with interaction. Therefore, it is very important to set the most suitable interval release time for baggage to merge. In addition, the conveyor to the route where scanner "A" locates has a heavy traffic of excess baggage amount. To avoid baggage congestion on such route and in the Concourse system, this paper aims at the smooth operation of the conveyors and the reduction of congestion in Terminal 2 North Concourse by setting up appropriate interval release time for baggage to enter the sorter at the four barcode scanners.

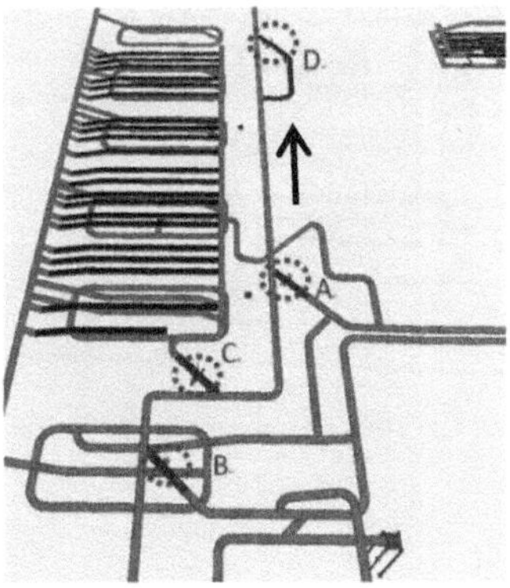

Fig. 4. Enlargement of Terminal 2 North Concourse in Flexsim model

4 Simulation and Analysis

In this section, different control rules are compared, and their impacts on system performance are analyzed. The congestion time of baggage queue on the four conveyors is used as a performance indicator. The congestion time of baggage queue is defined as the length of time while the number of queuing baggage exceeds the specific higher bound on each conveyor, detailed in Section 4.2. The analysis result is stated in Section 4.3. The model constraints and assumptions are listed in Section 4.1.

4.1 Model Constraints and Assumptions

Constraints:

- System behavior in North and South Areas is not considered.
- Accommodation for each buffer zone is at most 24 pieces of baggage.
- There are 9 buffer zones, 10 unloading zones, and one carousel.
- There is limited baggage amount for flights; conveyor routes are constant.

Assumptions:

- Scanning time for barcode scanners is constant.
- Distance between each piece of baggage is at least 2 kilometers.
- Speed rate for all conveyors and sorters is constant.
- The size of all baggage is equal.
- The hypothetical rate for baggage going to "D" is 20%.

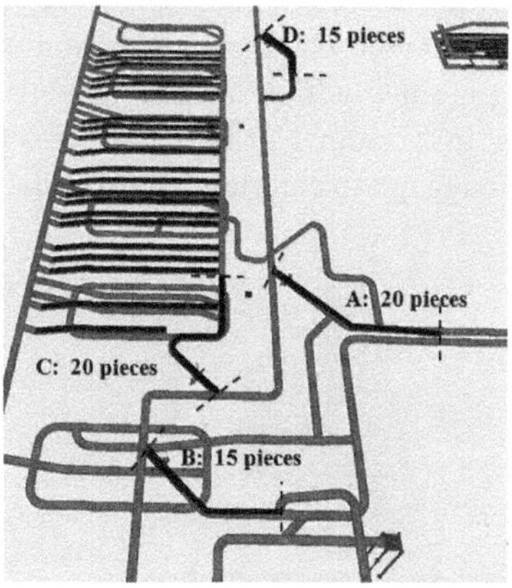

Fig. 5. The paths of the four conveyors in Terminal 2 North Concourse

4.2 System Performance

The performance effect focuses on the recorded congestion time of baggage queue on the four conveyors where the four scanners, A, B, C, and D, respectively locate.

It is considered an occurring of congestion in BHS when individual conveyor's baggage queue reaches a specific amount on its path, defined as the distance between the barcode scanner and a specific point on the conveyor. The distance is, shown in Fig. 5, the length between the two dotted lines on each conveyor. The details of each specific amount of baggage queue are described as follows:

"A" intakes an excess amount of baggage; also, during non-congestion time, it is observed that the baggage amount is fewer than 20 pieces. Hence, a baggage amount of over 20 pieces on the path where "A" locates is considered congestion.

"B" has a lighter baggage load than "A". With a baggage amount of over 15 pieces, it is considered congestion on the path where "B" locates.

Baggage storage for each buffer is limit to 24 pieces at most. It is considered congestion when the released baggage on the path going through "C" exceeds 20 pieces.

"D" intakes a baggage amount of at most 30 pieces. It is considered congestion on this path with a baggage amount of over 15 pieces.

Follow-up analysis is done through the recording of the length of congestion time.

4.3 Experimental Analysis

In this section, investigation outcomes of the combination of interval release time for the baggage waiting at the four barcode scanners are presented. In this system, when baggage is unable to enter the sorter, it is allowed to wait on the conveyors based on the control rules of interval release time at the scanners. The control rule of interval release time at a barcode scanner is to determine the time for baggage to enter the sorter as well as to control congestion time on each conveyor. In the situation that the setting of interval release time for all four scanners is too small, there may be insufficient space for baggage coming from " A" or "C" , since "B" is in domination, to enter the sorter due to the constraint of at least 2-kilometer distance between each piece of baggage. Moreover, when large amount of baggage comes at the same time from different conveyors, blockage may occur at "D", leading to longer baggage queuing time for all four conveyors. In contrast, when the interval release time is too large, there may be excess space for baggage to enter the sorter, causing extending baggage queue on each conveyor in the long run and leading to poor system performance. Different interval release time at the four scanners produces different scenarios and performances. Hence, the setting of the system control rules of interval release time at the four barcode scanners is of critical importance as it has a strong impact on baggage flow and operation. It is also essential to evaluate the interaction among the four barcode scanners.

The Analysis of Full Factorial Design

The simulation model processes the four factorial designs of interval release time parameter settings to determine which among them significantly affects the performance as well as to detect their interaction on the performance.

Full factorial design is a well-known technique based on a statistical method that analyzes and collects all possible combinations of factors of a specific problem [13].

The four parameters, A, B, C, and D, affect simulation performance. Therefore, different levels are considered for these parameters (presented in Tables 1) in order to prove that the parameter setting contributes significantly to system performance. The full factorial design requires 81 experiments (3^4−81); five sets of simulation runs are given, resulting in a total simulation number of 405. The parameter experiment is analyzed using Minitab 16.

Table 1. Four factors and 3 different levels (in seconds)

Factor	Level 1	Level 2	Level 3
A	1.5	2	2.5
B	1.5	2	2.5
C	1.5	2	2.5
D	1.5	2	2.5

The simulation model is used to simulate one single-day operation in BHS with a real dataset from TPE. The results of the full factorial design using ANOVA is presented in Table 2. Such results indicate that all four main effects of the interval release time parameters significantly affect the performance at α=0.05, and so do the interactions between A and B, A and D, and B and D.

During the process of simulation, it is observed that the large amount of baggage coming from A or B coincidently misses the baggage released from the buffer zone for re-sorting but increases the amount of baggage for re-scanning at scanner D, dominated by A and B due to its location . Therefore, the interaction between A and D is significant, and so is it between B and D. The interaction between A and B is also significant since B dominates A as shown in Figure 4.

Table 1. ANOVA of full factorial design

Source	DF	Seq SS	F	P
A	2	7245215727	9460.25	0.000*
B	2	8357667482	10912.81	0.000*
C	2	13602087	17.76	0.000*
D	2	45034659	58.80	0.000*
A*B	4	3356484638	2191.32	0.000*
A*C	4	3396595	2.22	0.067
A*D	4	17734746	11.58	0.000*
B*C	4	2070631	1.35	0.251
B*D	4	50043610	32.67	0.000*
C*D	4	1552958	1.01	0.400
Error	324	124069075		
Total	404	19287042503		

Note: DF: Degrees of freedom, Seq SS: Sequential Sum of squares, F: F Value, P: P Value.
* indicates that the effect is significant at α=0.05.

As analyzed in the full factorial design, when A=2, B=2.5, C=2 and D=1.5, the best simulation performance is obtained, with an average congestion time of approximately 1.94 hours/day for the four conveyors. In contrast, the worst parameter combination appears when A=2.5, B=2, C=1.5, and D=2.5, resulting in an average congestion time of 3.57 hours/day; the difference of 1.64 hours/day contributes to an improvement gap of about 84.38%. Using these two sets of parameter combination, the congestion time in each time window from the simulations is collected. Figure 6 presents the comparison of congestion time on the four paths, in both the best and the worst cases, in each time window. As observed in the figure, in the worst case, a one-hour congestion time in each time window starts from 7:00 to 19:00 without intermissions for "A". In contrast, in the best case, such congestion time only occurs in two time windows: 7:00~8:00 and 8:00~9:00.

The congestion patterns in the two above-mentioned cases for "B" are similar but with a congestion time of under 0.43 hour in each time window in the worst case and a 0.88 hour congestion time in 15:00~16:00 in the best case. By the same token, differences in congestion are insignificant for "C" and "D" respectively, except that there is no congestion for "D" in the best case.

As observed, the peak-time congestion occurs in the 15:00~16:00 time window in both cases. The average congestion time is 0.2940 hour in the best case while 0.3626 in the worst case; the difference of 0.069 hour results in an improvement gap of 23%.

The above experiment results indicate that the parameter setting would significantly affect the system performance.

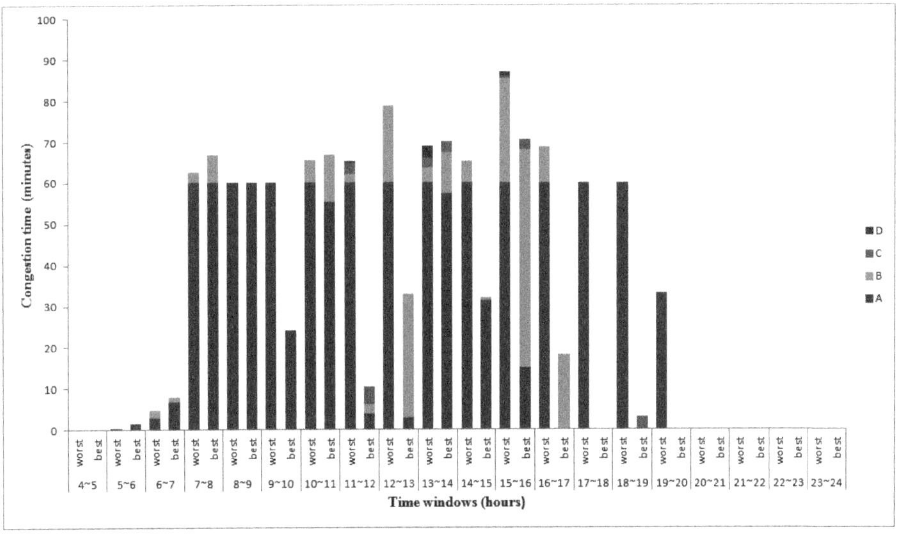

Fig. 6. Comparison of congestion time on four paths in the best and the worst cases

5 Conclusion

This paper addresses the empirical study of the interval release time in the airport baggage handling system. In this paper, object-oriented graphical user interface (OOGUI) simulation model with full factorial design is developed for the reduction of congestion time. The experimental results demonstrate that the best combination reduces 1.64 hours of congestion time per day, with an improvement gap of 23% at peak hour, in comparison with the worst situation. Consequently, it enhances smooth system operation while promoting the service level at Taiwan Taoyuan Airport.

In this study, the simulation time for each alternative is about 5.32 minutes, depicting a required total time of about 35.9 hours (near 1 and a half days) to perform all the designed alternatives, which is time-consuming. Future research directions intend to consider other analysis approaches based on this simulation framework. For example, the Taguchi method or fraction factorial design may require fewer experiment times than the full factorial design and achieve similar best parameter selection.

Acknowledgment. The research in this paper is partially supported by the Advanced Control & Systems Inc. of Taiwan. We also thank Taiwan Taoyuan International Airport for providing research data.

References

1. International Air Transport Association (IATA): Airlines Expect 31% Rise in Passenger Demand by 2017, http://www.iata.org/pressroom/pr/pages/2013-12-10-01.aspx
2. Taiwan Taoyuan International Airport Co. Ltd. Taiwan Taoyuan International Airport 2013 Annual Report (2013), http://www.taoyuanairport.com/uploads/20140630/20140630014441_TTIA%202013%20Annual%20Report.pdf
3. International Passenger Rankings, International Passenger Traffic for past 12 months ending April 2014 (2014), http://www.aci.aero/Data-Centre/Monthly-Traffic-Data/International-Passenger-Rankings/12-months (retrieved August 19, 2014)
4. Le, V.T., Creighton, D., Nahavandi, S.: Simulation-based Input Loading Condition Optimisation of Airport Baggage Handling Systems. In: Proceedings of the IEEE Intelligent Transportation Systems Conference, USA, pp. 574–579 (2007)
5. Joustra, P.E., Van Dijk, N.M.: Simulation of check-in at airports. In: IEEE International Conference on Simulation, vol. 2, pp. 1023–1028 (2001)
6. Abdelghany, A., Abdelghany, K., Narasimhan, R.: Scheduling Baggage-Handling Facilities in Congested Airports. Journal of Air Transport Management, 76–81 (2006)
7. Jing, G.G., Kelton, W.D., Arantes, J.C., Houshmand, A.A.: Modeling a controlled conveyor network with merging configuration. In: Proceedings of the 30th Conference on Winter Simulation, pp. 1041–1048 (1998)
8. Le, V.T., Zhang, J., Johnstone, M., Nahavandi, S., Creighton, D.: A Generalised Data Analysis Approach for Baggage Handling Systems Simulation. In: Proceedings of the IEEE International Conference on Systems, Korea, pp. 1681–1687 (2012)
9. Al-Serafi, A., Elragal, A.: Visual Trajectory Pattern Mining: An Exploratory Study in Baggage Handling Systems. Springer International Publishing Switzerland, 159–173 (2014)
10. Johnstone, M., Creighton, D., Nahavandi, S.: Enabling Industrial Scale Simulation / Emulation Models. In: Proceedings of the Winter Simulation Conference, US, pp. 1028–1034 (2007)
11. Chun, H.W., Mak, R.W.T.: Intelligent Resource Simulation for an Airport Check-In Counter Allocation System. IEEE Transactions on Systems, Man, and Cybernetics 29(3), 325–335 (1999)
12. Lin, J.T., Huang, E., Shih, P.H., Chiu, C.C.: Airport Baggage Handling System Simulation Modeling Using SysML. In: Proceedings of the International Conference on Industrial Engineering and Operation Management (2015)
13. Montgomery, D.: Design and Analysis of Experiments. 5th ed. John Wiley & Sons New York (2000)

A New Two-Phase Approach for Petri Net Based Modeling of Scheduling Problems

Reggie Davidrajuh

University of Stavanger
Stavanger, Norway
reggie.davidrajuh@uis.no

Abstract. This paper presents a new two-phase approach for Petri Net based modeling of scheduling problems. Though Petri Nets have been used as heuristic approach for modeling scheduling problems, literature study reveals two major difficulties: 1) the large size of Petri Net models, and 2) the inability to differentiate workstations. In this paper, these two difficulties are avoided by a two-phase approach known as Activity-Oriented Petri Nets (AOPN). General Purpose Petri Net Simulator (GPenSIM) is a new Petri Net simulator that implements AOPN on MATLAB platform. This paper introduces AOPN and GPenSIM in a tutorial style, working through an example on job scheduling in grid computing. This example shows the usability of the AOPN approach for the modeling of scheduling problems and the easiness of GPenSIM for coding and simulation.

Keywords: Scheduling problems, Activity-Oriented Petri Net (AOPN), GPenSIM, Petri Nets, grid computing.

1 Introduction

Scheduling is a process to manage time and costs effectively when a sequence of tasks are to be performed using a set of workstations; the sequence of tasks have a precedence order, meaning some tasks must have been completed before a particular task can start [1].

Scheduling is an important issue in all branches of engineering. In production engineering, scheduling can minimize the production time and costs, by optimal use of the resources such as man and machines; assembly line balancing problems (ALBP) is a special class of scheduling problems in production engineering [2].

Literature study reveals a large number of works that use Petri Nets for scheduling problems; literature also reveals the huge size of resulting Petri Net models and the difficulties in differentiating workstations as two main problems associated with the use of Petri Nets [3; 4, 5]. This problem is especially true when modeling resource-scheduling problems, where a large number of resources is usually involved [5]. However, the two problems are solved using advanced techniques: for example, Jeng and DiCesare [6] uses control nets and Wu and Zhou [7] uses Resource-oriented Petri Nets to overcome these difficulties.

M. Gen et al. (eds.), *Industrial Engineering, Management Science and Applications 2015*,
Lecture Notes in Electrical Engineering 349, DOI: 10.1007/978-3-662-47200-2_14

The aim of this paper is to introduce Activity-Oriented Petri Nets (AOPN) as a simple yet effective technique to model scheduling problems; this paper is written in a tutorial style to achieve this aim, taking job scheduling in grid computing as an example. As the goal of this paper is to show how simple and easy it is to use AOPN for scheduling problems, discussion is kept to a minimum about the various advanced techniques and complex approaches for scheduling problems, inclusive advanced Petri Nets based approaches.

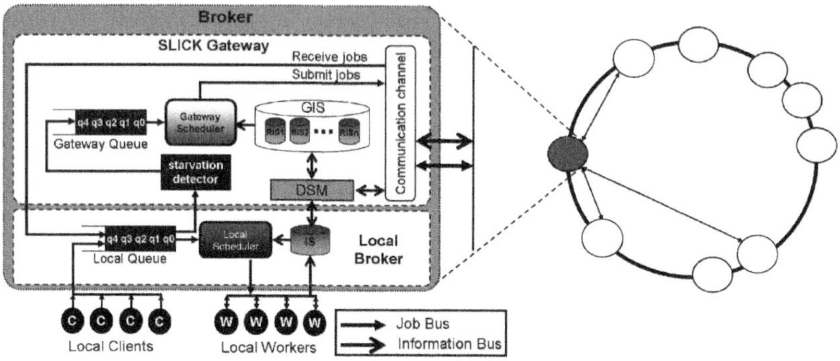

Fig. 1. Grid computing environment [8]

2 A Scheduling Problem in Grid Computing

This section presents a scheduling problem in grid computing environment.

Figure 1 shows a grid where the broker plays a central role in accepting jobs (multi-task jobs comprised of multiple tasks) from the local clients and delegating these tasks to the local workers [8]. Figure 1 also shows that the broker is a node in the ring of many brokers; thus, the broker may also accept tasks from other brokers and also route local tasks to external workers via the brokers in the ring.

Let us assume that the broker accepts a job from a local client; the job consists of 19 tasks as shown in Figure 2. The tasks are shown as nodes with labels n_1 to n_{19}. The diagram shown in Figure 2 is known as the precedence graph, as it shows the number of tasks involved in the job, and the sequential relationship between the tasks.

Table 1 shows the types of the tasks, based on the required software and hardware to perform the task (requirements). Table 2 lists the tasks n_1 to n_{19}, along with their type and the expected processing time of the respective task, if it was to be run on a standard desktop system. The processing times are given as stochastic timing, with Gaussian (normal) distribution with mean and standard deviation pair. Table 3 shows the types of the eight local workers based on their software and hardware configuration. The broker can only assign a task with specific requirements to a worker with the matching configuration; e.g. task n_1 can be assigned to workers w_1 to w_5, whereas task n_6 cannot be assigned to any local worker as the local workers are not equipped with enough main memory (minimum 40GB); thus, n_6 must be routed to an external worker. Worker w_8 cannot be assigned to perform any task of this job, as it possesses too little memory.

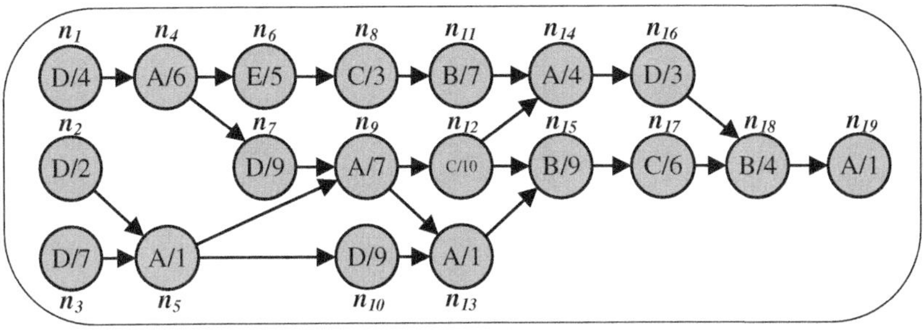

Fig. 2. The precedence graph of the tasks

Table 1. Type of the tasks.

Type	Requirements
A	environment=java; OS=LINUX; memory=15GB
B	environment=java; OS=Win-7; memory=10GB
C	environment=.NET; OS=Win-7; memory=10GB
D	environment=java; OS=(don't care); memory=10GB
E	environment=java; OS=LINUX; memory=40GB

Table 2. Type and processing time (PT) of the tasks

Task	*Attributes*		*Task*	*Attributes*		*Task*	*Attributes*	
	Type	*PT*		*Type*	*PT*		*Type*	*PT*
n_1	*D*	(4,0.7)	n_8	*C*	(3,0.5)	n_{15}	*B*	(9,2)
n_2	*D*	(2,0.1)	n_9	*A*	(7,1)	n_{16}	*D*	(3,0.5)
n_3	*D*	(7,1.1)	n_{10}	*D*	(9,1)	n_{17}	*C*	(6,1)
n_4	*A*	(6,1.2)	n_{11}	*B*	(7,1)	n_{18}	*B*	(4,0.3)
n_5	*A*	(1,0.1)	n_{12}	*C*	(10,2)	n_{19}	*A*	(1,0.2)
n_6	*E*	(5,1)	n_{13}	*A*	(1,0.2)	-	-	-
n_7	*D*	(9,1)	n_{14}	*A*	(4,0.4)	-	-	-

In the next section, using the AOPN approach, a Petri Net model is created for the job to be performed by the grid. AOPN is a two-phase modeling approach, in which the phase-I is for creating a static Petri Net graph and the phase-II is for adding run-time dynamics into the model.

Table 3. Configuration of the workers.

Type	Requirements
w_1, w_2, w_3	environment=java; OS=LINUX; memory=20GB
w_4, w_5	environment=java; OS=Win-7; memory=20GB
w_6, w_7	environment=.NET; OS=Win-7; memory=20GB
w_8	environment=java; OS=Win-7; memory=8GB

3 Phase-I: Creating the Static Petri Net Graph

The first phase of AOPN approach is to create the static Petri Net graph. In this phase, taking an "activity-oriented view", only the tasks and their precedence are considered; workers are not considered as they are resources used by the tasks. Because of the absence of workers as places in the Petri net graph along with all their connections (arcs) with the tasks, the resulting Petri Net graph becomes much smaller than the usual Petri Net models for scheduling presented in literature.

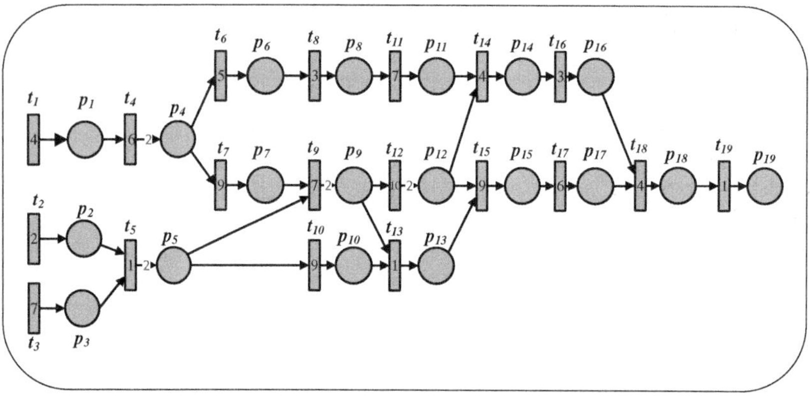

Fig. 3. The Petri Net model of the job

The static Petri Net graph for the job shown in Figure 3 can be easily obtained by taking an "activity-oriented view" (only considering the tasks of the job and their precedence); the procedure for creating the Petri Net graph from the precedence graph consists of the following two steps [9]:

1. In the static Petri Net graph, transition-place pair (t_i, p_i) represents the node n_i of the dependency graph.
2. Unless the task n_i is a sink (no outputs for it), the weight of the arc a_{ii} between the transition t_i and the place p_i is equal to the number of outputs of n_i; if n_i is a sink then the weight of the arc a_{ii} becomes (default) one.

Figure 3 shows the static Petri Net graph obtained by transforming the precedence graph shown in Figure 2. The Petri Net graph shown in Figure 3 has 19 transition-place pairs representing the 19 tasks of the job; there is no other elements (either place or transition) to represent the workers.

4 Phase-II: Adding the Run-Time Dynamics

The second phase of the AOPN approach is mainly about adding the run-time details that cannot be coded (or cannot be coded easily) in the static Petri Net graph.

For example, tasks requesting, using, and releasing of a variety of workers cannot be easily shown in the Petri Net graph; these details are added in the second phase [9].

The phase-II is divided into two steps, step-1 is to add initial dynamics, such as firing times of the transitions, initial tokens in places, etc. and step-2 is about the resource usage: like requesting a worker to start firing of a task, locking a worker during its execution, and releasing a worker after completion of work.

4.1 Step-1: Adding Initial Dynamics

Firing times of the transitions: The task processing time of a task n_i becomes the firing time of the corresponding transition t_i; firing times of the transitions (processing times of the corresponding tasks) are shown inside each transition in Figure 3. In addition, the life of a transition is rather limited: each transition in the Petri Net has a lifetime limited to one firing only.

Initial tokens: as clearly visible from the Figure 3, there is no need for initial tokens as the Petri Net is started by 'cold transitions' (self-starting, with no enabling input places with enough tokens) such as t_1, t_2, and t_3.

4.2 Step-2: Adding Resource Usage

The tasks require workers that are configured with adequate software and hardware. From the Table 2 and Table 3, we can match tasks with workers as shown in Table 4.

Table 4 indicates that the tasks that falls within type-D can be easily assigned to a worker as many workers can perform this type of task. However, no local worker can perform the task n_6 of type-E, thus this task must be assigned to an external worker.

5 Simulation with GPenSIM

In this section on coding and simulations, we assume the following simplification: since no local worker can perform the task n_6, we simply assume that it is performed by an external worker. Due to the additional overheads involved in dispatching the task to an external worker and retrieving the results from the other grid, the resulting processing time has increased by a larger amount to 20 Time Units.

Table 4. Matching the tasks with the workers.

Task type	Tasks	Matching Workers
A	n_4, n_5, n_9, n_{13}, n_{14}, n_{19}	w_1, w_2, w_3
B	n_{11}, n_{15}, n_{18}	w_4, w_5
C	n_8, n_{12}, n_{17}	w_6, w_7
D	n_1, n_2, n_3, n_7, n_{10}, n_{16}	w_1, w_2, w_3, w_4, w_5
E	n_6	*(no match)*

5.1 Some Issues in Scheduling

There is a number of issues in resource management:

Instances of a resource: This means a resource can have many indistinguishable copies ('instances'). For example, there are three cashiers in a bank (resource cashier has three instances), all of them can process all kind of transactions. For a casual customer, it does not make sense to prefer on cashier over the others. In this case, we could generalize all the cashiers into a group, name this resource as 'cashier', and say that cashier is a single resource with three instances.

Generic resources: this means, there is a number of 'named' resources available, but all of them are same for some specific applications. For example, in a car workshop, we have three mechanics called 'Alfred', 'Bob' and 'Chuck'; though they are specialist in some works, when it comes to engine repair, they all are the same. Thus for engine repair, we can pick up a generic mechanic, without naming anyone.

Specific resources: When the resources are named, we may request specific (named) resources. For example, in the car workshop, 'Chuck' is a painter. Hence, when we need a painting job, we would prefer using the specific resource 'Chuck'. Only if Chuck is not available, we may go for another resource.

Write Access: When a resource has many instances, a transition may try to acquire one or many of these instances. Write access means, the resource will be locked and all the instances will be made available to a requesting transition.

Table 5 shows a number of GPenSIM functions available for resource management.

Table 5. GPenSIM functions for resource management

Function	Description
availableInst	checking whether any instances is available in a resource
availableRes	checking whether any resources is available (not on use)
requestsSR	request a number of instances from specific ('named') resources
requestGR	request a number of resource instances, without naming any resource
requestAR	request a number of resource instances among many alternatives
requestWR	request all the instances of a specific resource (either all or none)
release	release all the resources and resource instances held by a transition
prnschedule	prints information on resources usage

5.2 Code Implementation with GPenSIM

Using GPenSIM, a Petri net model of a discrete event system usually consists of four files [10]: the main simulation file (MSF), the Petri Net Definition file (PND), the COMMON_PRE file, and the COMMON_POST file.

The MSF is shown in Figure 4. In the MSF, firstly, the initial dynamics (e.g. initial tokens in places, firing times of transitions, available systems resources) are declared. Then, the simulation iterations are started. When the simulation iterations are complete, the results can be plotted (graphics) or displayed.

PND is the code implementation of the static Petri Net graph. The PND file shown in Figure 5 is the GPenSIM code for the Petri Net graph shown in Figure 3. The PND file declares the sets of places, transitions, and arcs.

```
% 1: declare the firing times
dyn.ft = {'t01','normrnd(4,0.7)', 't02','normrnd(2,0.1)', ...
't03','normrnd(7,1.1)','t04','normrnd(6,1.2)','t05','normrnd(1,0.1)',
't06','normrnd(5,1)','t07','normrnd(9,1)','t08','normrnd(3,0.5)',
't09','normrnd(7,1)','t10','normrnd(9,1)','t11','normrnd(7,1)',
't12','normrnd(10,2)','t13','normrnd(1,0.2)','t14','normrnd(4,0.4)',
't15','normrnd(9,2)','t16','normrnd(3,0.5)','t17','normrnd(6,1)',
't18', 'normrnd(4,0.3)', 't19','normrnd(1,0.2)'};

% 2: declare the workers ("resources")
dyn.re = {'w1',1,inf, 'w2',1,inf, 'w3',1,inf, 'w4',1,inf, ...
        'w5',1,inf, 'w6',1,inf, 'w7',1,inf, 'w8',1,inf};
% 3: create the Petri-Net run-time
pni = initialdynamics(pns, dyn);

sim = gpensim(pni); % 4: run the simulation iterations
prnschedule(sim);   % 5: print the results
```

Fig. 4. The main simulation file (MSF)

```
function [png] = pnd()
png.PN_name = 'Model of a Job in grid computing';
png.set_of_Places = {'p01','p02','p03',...
    'p04','p05','p06','p07','p08','p09','p10',...
    'p11','p12','p13','p14','p15',...
    'p16','p17','p18','p19'};
png.set_of_Transitions = {'t01','t02','t03',...
    't04','t05','t06','t07','t08','t09',...
    't10', 't11','t12','t13','t14','t15',...
    't16','t17','t18','t19'};
png.set_of_Arcs={'t01','p01',1,'t02','p02',1, ...
   't03','p03',1,'p01','t04',1, 't04','p04',2, ...
   'p02','t05',1,'p03','t05',1, 't05','p05',2, ...
   'p04','t06',1, 't06','p06',1, ...
   'p04','t07',1, 't07','p07',1, ...
   'p06','t08',1, 't08','p08',1, ...
   'p07','t09',1,'p05','t09',1, 't09','p09',2, ...
   'p05','t10',1, 't10','p10',1, ...
   'p08','t11',1, 't11','p11',1, ...
   'p09','t12',1, 't12','p12',2, ...
   'p09','t13',1,'p10','t13',1, 't13','p13',1,...
   'p11','t14',1,'p12','t14',1, 't14','p14',1,...
   'p12','t15',1,'p13','t15',1, 't15','p15',1,...
   'p14','t16',1, 't16','p16',1, ...
   'p15','t17',1, 't17','p17',1, ...
   'p16','t18',1, 'p17','t18',1, 't18','p18',1,...
   'p18','t19',1, 't19','p19',1};
```

Fig. 5. The Petri Net Definition file (PND)

```
function [] = COMMON_POST(transition)

% (if used) release all resources used by transition
release();
```

Fig. 6. The COMMON_POST file

The COMMON_PRE file (Figure 7) is for coding the conditions for the enabled transitions to start firing ('guard conditions'), e.g. assigning a worker to a task. In this file, two operations are done: Firstly, the enabled transition is checked to see if it had already fired once before. If it did, then it is prevented from firing again, as the transitions are allowed to fire once only. Secondly, using the Table 4, a suitable worker is selected for the transition (task).

In the COMMON_POST file (Figure 6), the post-firing actions of the transitions are coded. This file is small, as it only performs one operation: if any transition (task) used any worker, then the worker is released at the completion of the transition.

6 Simulation Results

The following cases are studied with the simulations:

- Case-I: no grid computing: the job (all the tasks) is completely performed by a single worker (computer node).
- Case-II: Using grid computing: there is no differentiation between the workers; all the workers are capable of performing all the tasks.
- Case-III: Using grid computing: as shown in the Tables 1-4, the workers are configured differently, thus can perform only the matching tasks.

```
function [fire, transition] = COMMON_PRE (transition)

tn = transition.name;
tx = get_trans(tn);
if tx.times_fired,
   fire = 0; return;
end;

% case 3
if ismember (tn, {'t04','t05','t09','t13','t14','t19'}),
   granted = requestAR({'w1','w2','w3'}, 1);
elseif ismember (tn, {'t11','t15','t18'}),
   granted = requestAR({'w4','w5'}, 1);
elseif ismember (tn, {'t08','t12','t17'}),
   granted = requestAR({'w6','w7'}, 1);
elseif ismember (tn, {'t01','t02','t03','t07','t10','t16'}),
   granted = requestAR({'w1','w2','w3','w4','w5'}, 1);
else
   granted = 1;
end;

fire = granted;
```

Fig. 7. The COMMON_PRE file

Simulation codes presented in Figures 4-7 are for Case-III. The code for the other two cases, Case-I and II are very similar to that of Case-III, thus not shown here. Interested readers can download the code from the website cited in the reference as [11]. The results of the simulations are presented in Table 6.

Table 6. Completion time (CT) under different cases.

Case	CT
Case-I: No grid computing	96.8
Case-II: homogenous workers.	55.2
Case-III: heterogeneous workers	54.9

The results show that by using grid computing, the job can be done faster, completing in 56 TU, if the stochastic parts of the processing times are dropped (processing times are deterministic and equal to the mean). This means, the use of grid computing boosts performance by 42.8%, provided that the workers are available at that time (they are not busy executing tasks from other jobs).

The results also show that whether or not the task n_6 was performed locally (taking 5 TU) or done externally (taking 20 TU), it will not influence the completion time of the job (56 TU in both cases). This is because, task n_6 falls outside the critical path that is decisive for the overall completion time of the job.

7 Conclusion

This paper introduces a new approach known as Activity-Oriented Petri Nets (AOPN) and its MATLAB implementation known as General Purpose Petri Net Simulator (GPenSIM). This paper is written in a tutorial style to achieve this aim, taking job scheduling in grid computing as an example; the scheduling example is also purposely chosen as a simple one and the advanced topics in scheduling (e.g. loops) are not touched in this example.

The modeling approach presented in this paper shows how simple the AOPN approach is and the complete code (Figures 4-7) shows how easy it is to code in GPenSIM. In addition, GPenSIM allows imposing costs to transitions and resources, thus the total operational costs of the job can be estimated too; this potential makes GPenSIM very useful for industrial applications.

References

1. Scholl, A.: Balancing and sequencing of assembly lines. Physica-Verlag, Heidelberg (1995)
2. Grzechca, W.: Idle Times Analysis in Two-Sided Assembly Line Balancing Problem. Manufacturing Modelling, Management, and Control 7(Part 1), 1720–1725 (2013)
3. Kim, C., Lee, T.E.: Modeling and simulation of Petri nets for complex scheduling rules of automated manufacturing systems. Manufacturing Systems 3(3-4), 275–307 (2011)

4. Piera, M.A., Mušič, G.: Coloured Petri net scheduling models: Timed state space exploration shortages. Mathematics and Computers in Simulation 82(3), 428–441 (2011)
5. Davidrajuh, R.: Activity-Oriented Petri Net for Scheduling of Resources. In: Proc. IEEE International Conference on Systems, Man, and Cybernetics (IEEE SMC 2012), Seoul, Korea, October 14-17 (2012)
6. Jeng, M.D., DiCesare, F.: Synthesis using resource control nets for modeling shared resource systems. IEEE Transactions on Robotics and Automation 11, 317–327 (1995)
7. Wu, N., Zhou, M.: System Modeling and Control with Resource-Oriented Petri Nets. CRC Press, NY (2010)
8. Azab, A., Meling, H., Davidrajuh, R.: A fuzzy-logic based coordinated scheduling technique for inter-grid architectures. In: Magoutis, K., Pietzuch, P. (eds.) DAIS 2014. LNCS, vol. 8460, pp. 171–185. Springer, Heidelberg (2014)
9. Davidrajuh, R.: Modeling Resource Management Problems with Activity-Oriented Petri Nets. In: 2012 Sixth UKSim/AMSS European Symposium on. Computer Modeling and Simulation (EMS). IEEE (2012)
10. Davidrajuh, R.: Developing a New Petri Net Tool for Simulation of Discrete Event Systems. In: Proc. Second Asia International Conference on Modeling & Simulation (AMS), pp. 861–866. IEEE (2008)
11. GPenSIM code for Job Scheduling in Grid Computing, `http://www.davidrajuh.net/gpensim/2015-ICIMSA-JobScheduling`

Simulation and Analysis of Impulse Faults in Power Transformer

Kaveri Bhuyan[1], Saibal Chatterjee[2], Anwesa Yadav[3], Sarita Bansal[4], and Sanat Kumar Paul[5]

[1] Department of Electric Power Engineering, Norwegian University of Science and Technology, Trondheim, Norway
kaveri.bhuyan@ntnu.no
[2] Department of Electrical Engineering, North Eastern Regional Institute of Science and Technology, Arunachal Pradesh, India
saibalda@ieee.org
[3] Zindal Steel and Power Ltd., Raigarh, India
anwesa.ee@gmail.com
[4] Department of Management Studies, Malaviya National Institute of Technology Jaipur, India
saritabansal86@gmail.com
[5] Tripura Satte Electricity Corp. Ltd., Agartala, India
sanatkumarpaul@gmail.com

Abstract. The effects of impulse faults of different nature and different locations of occurrence in the high voltage winding of a transformer are investigated in this work. During the investigation, a high frequency transformer model developed in MATLAB Simulink was subjected to different lightning and switching impulse waveforms. Turn-to-earth and turn-to-turn faults were simulated at different locations of the windings due to applied impulse. The transformer response is studied owing to the variation of the input impulse waveforms, fault types and their location along the windings. This study is important in the purview of fault characterization which provides an important basis for developing more reliable and sensitive methods to detect internal faults in the transformer windings.

Keywords: Impulse, Faults, MATLAB, Modelling, Power transformers, Simulation.

1 Introduction

Transformers are an integral part of electrical transmission and distribution network. Transformers are often excited by lightning or switching impulses comprising of high frequency oscillatory voltages, which degrades the insulation system [1]. Impulse testing of power equipments is generally done according to IEC/IEEE standards in high voltage laboratories for determining the insulation strength against the application of impulse waveforms [2]. Insulation failure in power transformer can occur between coils, turns or layers, and winding to earth. The accurate diagnoses of these

M. Gen et al. (eds.), *Industrial Engineering, Management Science and Applications 2015*, Lecture Notes in Electrical Engineering 349, DOI: 10.1007/978-3-662-47200-2_15

faults are significant to assess insulation condition of windings, which will eventually help to improve the reliability.

The main objective in this work is identification of impulse fault characteristics in power transformers under lightning and switching impulse voltages. The fault characteristics are useful to identify the type, and location of internal impulse faults. When an impulse voltage is applied in the transformer, a part of the current flow through the neutral wire. The nature of neutral current changes for different types of fault and this sensitivity demonstrates the possibility to extract some special features for fault characteristics [3]. Therefore, in this work it is of interest to study the variation of magnitude and frequency of transformer neutral current due to simulated series and shunt faults at different locations i.e fault in line-end, the mid-winding, and the earth-end sections under different impulse voltages.

2 Investigation Approach

Impulse faults are simulated on a high frequency power transformer model of 3 MVA, 33/11 kV, 3-phase, 50 Hz, Dyn 11 transformer [4]. The transformer main winding constitutes of 80 coils and 8 extra coils are used as tap coils. The main winding is divided into three sections namely line-end section (coils numbers 1 to 27), mid-winding (28 to 54) and earth-end section (coil no. 55 to 80). The design data of the transformer model are:

Power rating of transformer	3 MVA
Number of coils in main winding	80
Number of tap coils	8
Average number of turns per disc	19
Voltage rating of transformer	33/11 kV
Axial height of 33 kV disc	6.6 mm
Outer diameter of HV winding	534 mm
Inner diameter of HV winding	424 mm
Mean radius of the disc	237 mm
Inductance of each coil	0.324 mH
Resistance of each coil	0.151 Ω
Ground capacitance per coil (C_g)	24.63 pF
Series capacitance per coil (C_s)	934 pF

The mathematical model is shown in Fig.1 [4]. Details of modeling and validation of transformer model are presented in [4, 5]. Two different types of fault i.e. shunt and series fault are considered for fault analysis. For simulating shunt fault, the terminals are grounded and for series fault, the windings are shorted. Fig. 1 shows the equivalent circuit of transformer model with induced shunt and series faults [4].

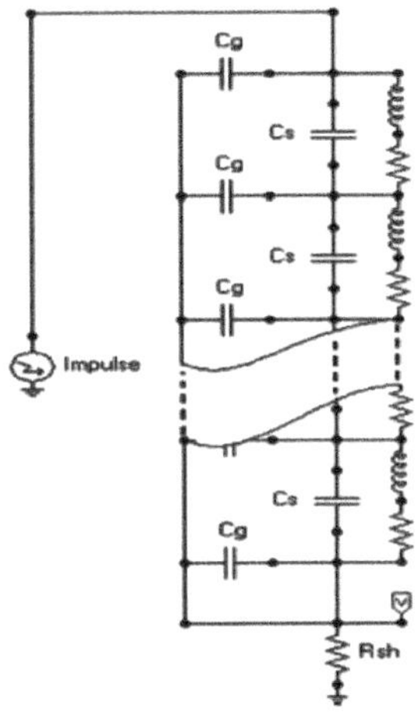

a) Transformer model under no-fault condition

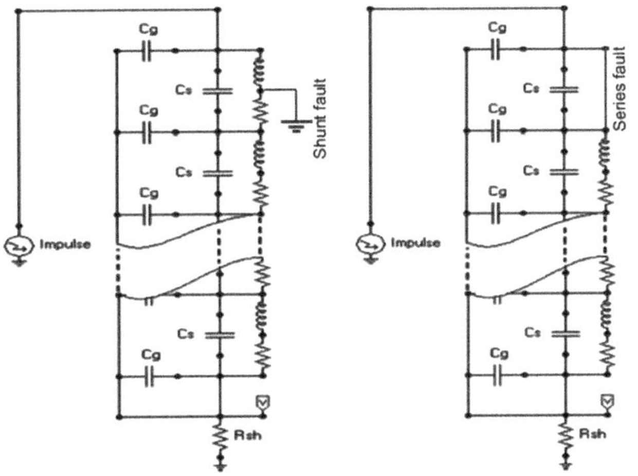

b) Shunt fault c) Series fault

Fig. 1. Digital model of 3 MVA, 33/11 kV, 3 phase power transformer [4]

Full and chopped lightning and full switching impulse voltages are applied to the transformer model for simulating impulse faults. As per IEC 60060-1[2], the full standard lightning impulse voltage rises to its peak value in 1.2 μs and the tail of the wave decays to a level of 50 percent of the peak in 50 μs. In addition to standard full lightning impulse wave, tail-chopped lightning impulse voltage wave has been used to indicate failure due to voltage collapse. The chopped wave is similar to the full wave lightning impulse except that its crest value is 10% greater than that of full impulse wave and the wave is chopped at 3-6 μs. In this work, tail chopped impulses at 7 μs and 20 μs has been used to know the effect of non-standard chopped impulses. The full standard switching impulse voltage rises to its peak value at 250 μs and the tail of the wave decays to a level of 50 percent of the peak in 2500 μs. The magnitude and frequency of maximum value of neutral current under series and shunt faults, different input impulse voltages and locations of fault are calculated by Fast Fourier Transform (FFT).

3 Observations

The maximum frequency components of neutral current obtained against application of a particular type of impulse during series and shunt fault of the transformer in different locations is shown in Fig. 2 and Fig. 3. It is observed that when there is a series fault in the transformer the neutral current magnitude increases gradually and when there is a shunt fault in the transformer, the current magnitude gradually decreases. It is seen that the value of neutral current depends on the input voltage, fault type and it's location. The frequency component of neutral current for different types of fault, applied impulse and location of fault are given in Table I.

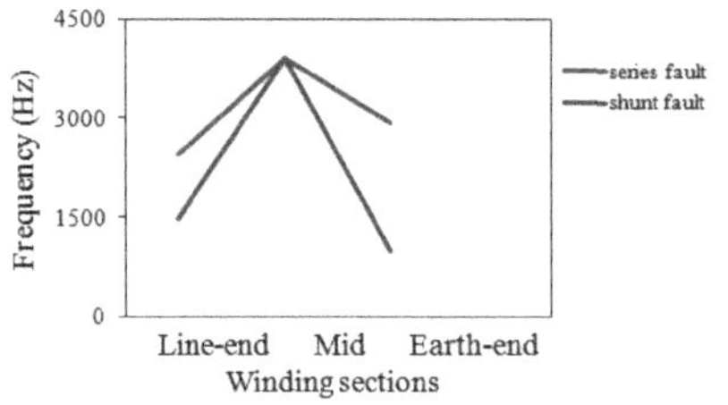

(a) Maximum frequency of fault current against standard lightning impulse (1.2/50 μs)

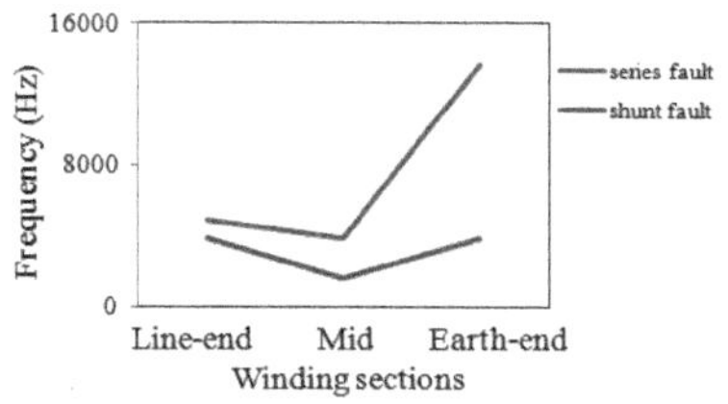

(b) Maximum frequency of fault current against chopped impulse (7μs)

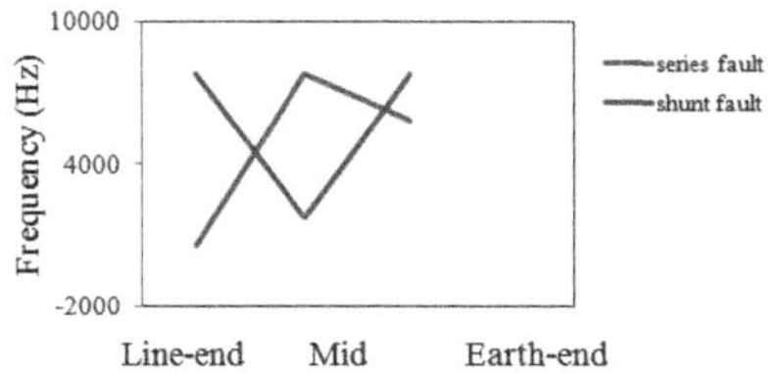

(c) Maximum frequency of fault current against chopped impulse (20 μs)

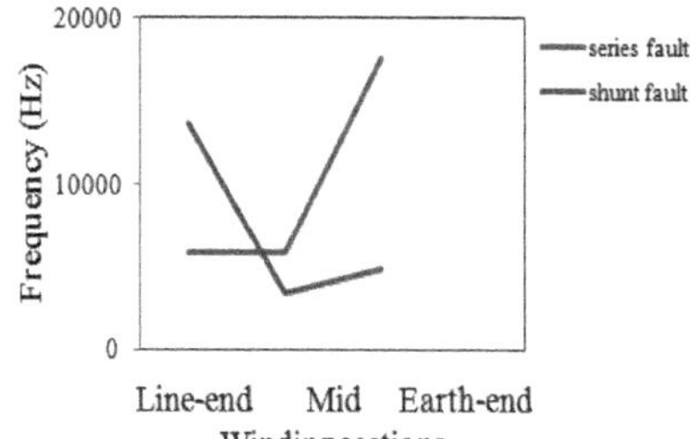

(d) Maximum frequency of fault current against standard switching impulse (250/2500 μs)

Fig. 2. Maximum frequency of series and shunt fault current against standard lightning impulse (1.2/50 μs), chopped impulse (7μs), chopped impulse (20 μs), switching impulse (250/2500 μs)

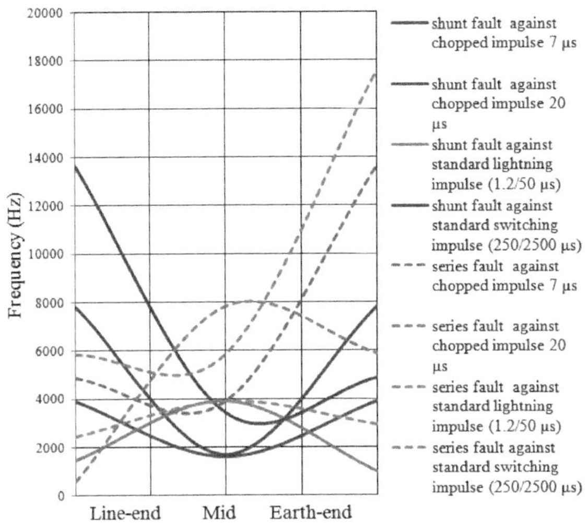

Fig. 3. Maximum frequency component in neutral current versus fault location along the transformer winding during series and shunt faults

Table 1. Frequency component of neutral current for different types of fault, applied impulse and location of fault

Impulse voltage waveforms	Type of fault	Location of fault	Max. frequency of neutral current (Hz)	Magnitude of neutral current (A) at max. frequency
Std. full lightning impulse	Series	line-end	2441	3.75 $X10^{-15}$
		mid	3906	2.39 $X10^{-15}$
		earth-end	2930	1.57 $X10^{-15}$
	Shunt	line-end	1465	3.78 $X10^{-15}$
		mid	3906	4.80 $X10^{-16}$
		earth-end	976.6	6.35 $X10^{-15}$
Std. full switching impulse	Series	line-end	5859	5.45 $X10^{-17}$
		mid	5859	2.19 $X10^{-17}$
		earth-end	1.76 $X10^{4}$	5.37 $X10^{-17}$
	Shunt	line-end	1.37$X10^{4}$	6.32 $X10^{-18}$
		mid	3418	9.68 $X10^{-17}$
		earth-end	4883	5.04 $X10^{-17}$
Impulse chopped at 7μs	Shunt	line-end	3906	3.4 $X10^{-19}$
		mid	1587	7.61 $X10^{-20}$
		earth-end	3906	1.16 $X10^{-19}$
	Series	line-end	4883	2.06 $X10^{-19}$
		mid	3906	5.43 $X10^{-20}$
		earth-end	13670	6.7 $X10^{-20}$
Impulse chopped at 20 μs	Shunt	line-end	7813	4.61 $X10^{-20}$
		mid	1709	7.18 $X10^{-20}$
		earth-end	7813	5.3 $X10^{-20}$
	Series	line-end	585.9	2.15 $X10^{-19}$
		mid	7813	3 $X10^{-20}$
		carth end	5859	1.29 $X10^{-19}$

4 Inferences

- Irrespective of fault type, maximum frequency of neutral current is obtained against full lightning impulse in the mid-winding section of the transformer. The frequency components are dominant during the series fault as compared to that of the shunt fault.
- Irrespective of fault type, minimum frequency of neutral current is obtained against chopped lightning impulse (7μs) in the mid-winding section of the transformer. The frequency components are dominant during the series fault as compared to that of the shunt fault.
- Irrespective of fault type, minimum frequency of neutral current is obtained against switching impulse in the mid-winding section of the transformer. The frequency components are dominant during the series fault as compared to that of the shunt fault.
- When shunt fault occurs in the mid-winding of the transformer against chopped lightning impulse (20 μs), minimum frequency of neutral current is. During series fault, the maximum frequency of neutral current is obtained in the mid- winding section of the transformer.
- The frequency of the shunt fault current, when the applied voltage is a switching impulse is more than that of the fault due to the full lightning impulse.
- The frequency of the series fault current against switching impulse is more dominant compared to the full lightning impulse.

5 Conclusion

Series and shunt faults were simulated in a 3 MVA transformer model against the application of standard lightning impulse (1.2/50 μs), chopped impulse at 7μs and 20 μs and standard switching impulse (250/2500μs) at three different locations i.e. line-end, the mid-winding, and the earth-end sections of the winding. The transient response due to faults of different natures and different physical locations of occurrence in the winding were investigated. Using FFT some typical features of the impulse fault currents was identified. These fault current characteristics are useful to develop intelligent transformer fault detection techniques. In future work, the identified fault characteristics derived from the current simulation results will be used with an intelligent method such as wavelet networks to develop reliable and sensitive method for detection and localization of the multiple and simultaneous faults on the windings of the power transformers.

Acknowledgment. The authors acknowledge the support of DST-FIST Program project number ETI-211/2007, dated 28-04-2008 for this work.

References

1. Popov, M., van der Sluis, L., Smeets, R.P.P.: Evaluation of Surge Transferred Overvoltages in Distribution Transformers. Electric Power Systems Research, 441–449 (May 2007)
2. High-Voltage Test Techniques Part 1: General Definitions and Test Requirements, IEC 60060-1 (September 2010)
3. Baral, A., Lahiri, A.: Applicability of Stockwell Transform in Diagnosing Impulse Faults of a Transformer. In: IEEE 1st International Conference on Power and Energy (ICPEN), NERIST, India (December 2012)
4. Bhuyan, K., Chatterjee, S.: Surge Modelling of Transformer Using Matlab-Simulink. In: Annual IEEE India Conference (INDICON), Gujarat, India (December 2009)
5. Purkait, P., Chakravorti, S.: Can Fractal Techniques be used for Impulse Fault Pattern Discrimination in Distribution Transformers? Electric Power Systems Research 68, 258–267 (2004)

The Online Study Design for Different Study Location Environment, Using ICT and Social Methodology Tool

Platt Pangthong and Phudinan Singkamfu

College of Arts, Media, and Technologies. Chiang Mai University.
239 Huaewkeaw Road. Suthep Sub District Muang Chiang Mai District
plattpangthong@gmail.com, mr_pu@hotmail.com

Abstract. The influence of study environment is one of the significant impacts to study efficiency. This research examine the problem of two different group of student which study in the same course but different study environment, The samples are divided into 2 groups. The one is study in middle of the town campus, the other one is study in the far away in rural area. The primary concern problem is to reduce the gap of study result between both of campus, due to the study result of faraway campus was less than the other. One of our assumptions is from the different amount of campus's facility such as library and daily life. Although, this research introduce solution by using tools to help both of study group connect to each other and also share their study environment together. The tool simulates social network status to deliver short lesson to student's mobile phone. Notification will send to every one in the class to shows if any of students has read short lesson, or post any comment. There are approximately 40 students of faraway classroom and 120 students in in town class participated. Social network application for study has been experimented in other research and produces interested result. This research will show the approach that will reduce location gap between to different classroom by sharing assignment and connects them by use social network methodology to update their mate activity and duce bounder of the distance. After deployed tools students are response positive way for both, during the study period, and examination result.

Keywords: Social Network Application, Education Data Exchange, Mobile Application.

1 Introduction

With the increasing number of students in several universities in the city, there is a need for a separate campus for each university. However, these newly built campuses are still ill-equipped in some areas. For example, some libraries lack atmosphere that encourages students to study, resulting in less attention span and lower grades for students. The following table shows the details of two campuses of Chiang Mai University in several aspects.

M. Gen et al. (eds.), *Industrial Engineering, Management Science and Applications 2015*,
Lecture Notes in Electrical Engineering 349, DOI: 10.1007/978-3-662-47200-2_16

Table 1. Factors affecting the study environment of the students from 2 different campuses

	Chiang Mai University (Chiang Mai Campus)	**Chiang Mai University (Lamphun Campus)**
Courses	Same courses	Same courses
Number of Students	120 students	40 students
Teachers	Same teachers	Same teachers
Teaching Tools	Ms Office	Ms Office
Teaching Hours	4 hours per week	4 hours per week
Studying Resources	Main library, Faculty libraries, ITSC corner, etc.	Faculty libraries at the campus
Campus Atmosphere	Friends are both in the same faculty and other faculties. There are some competition and some exchange of educational data.	Friends are in the same faculty. There are low competition and low exchange of educational data.
Average score of the Same course	68.79 Point	59.22 Point

It can be seen from the table that both campuses share the same courses, the same teachers, the same teaching tools and the same teaching hours; whereas, the number of students, the studying resources and the campus atmospheres are different. The studying resources at Chiang Mai campus exceed in number and diversity. In addition, the campus atmosphere in Chiang Mai constitutes academic competition and educational data transfer while students enjoy having friends in the same faculty and other faculties as well. Whereas in Lamphun, there is only one studying resource available for students and there are less academic competition with low exchange of educational data.

In the past years, information and communication technologies (ICTs) have delivered culture of range new definition of communication potentials. ICTs provide technological tools and resources used to communicate, create, broadcast, store, and manage information. These technologies tools and resources include computers, tablets, smartphones, the Internet, broadcasting technologies (radio and television), and telephony. ICT in Education means "Teaching and Learning with ICTs tools and resources. [1] Nowadays, Information Technology plays an important role in daily routines in various forms such as search engines, plane ticket online reservations and accommodation online reservations, fitting with the lifestyles of university students as can be seen with smart phone usage and social networking. Therefore, there is an initiative to exercise an IT tool to give students more and easier access the educational data by classroom activities control via Information Technology. Technology is responsible for distorting the concept of distance between learner and instructor, and enabling learners to access education at any time and from any place [2]. This research was start from ideal of Kim Munich, who success to contribute experiment nursing students' ability and motivation to learn online to enhance their online learning experience and promote completion of their programs [3].

2 Methodology

This experiment will concentrate on educational data exchange among students. One of which is ICT classroom activity sharing which can help gather up course details and educational data for exchange via Information Technology.

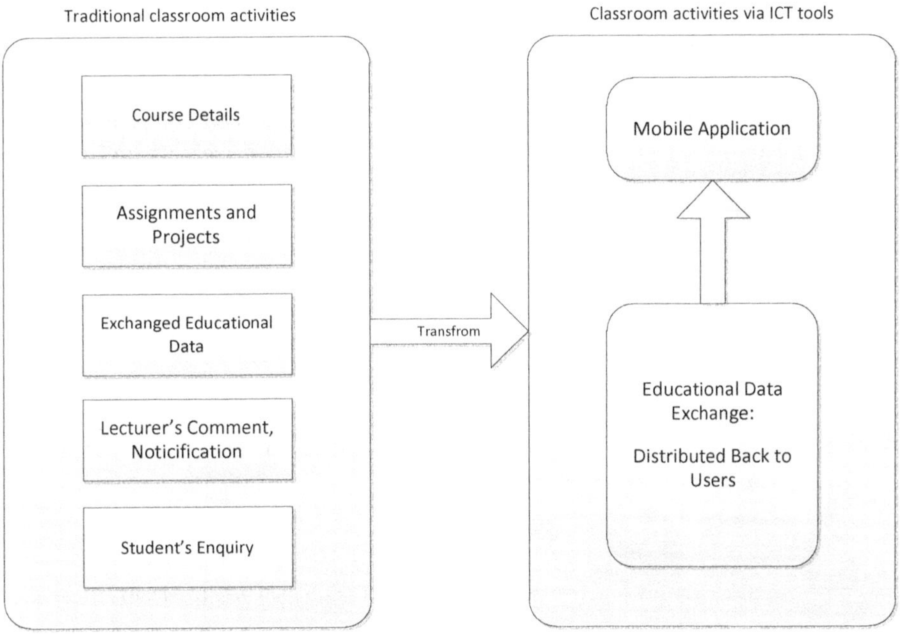

Fig. 1. Transform normal study flow to information technology study center

The figure above shows how course details, assignments and projects and exchanged educational data can be stored in the Information Technology tool. Students and teachers can log in to add, edit or delete data to an extent designated according to each user's access rights. These data will be shown on the website or the application for later use.

Knowledge Management can help store data in the Information Technology tool. The next part will help explain how the Information Technology tool can be used. Teachers will put in course details under Course Details and create assignments or projects under Assignments for students to download. Also, both teachers and students can exchange data and answer questions regarding course details in the provided blog.

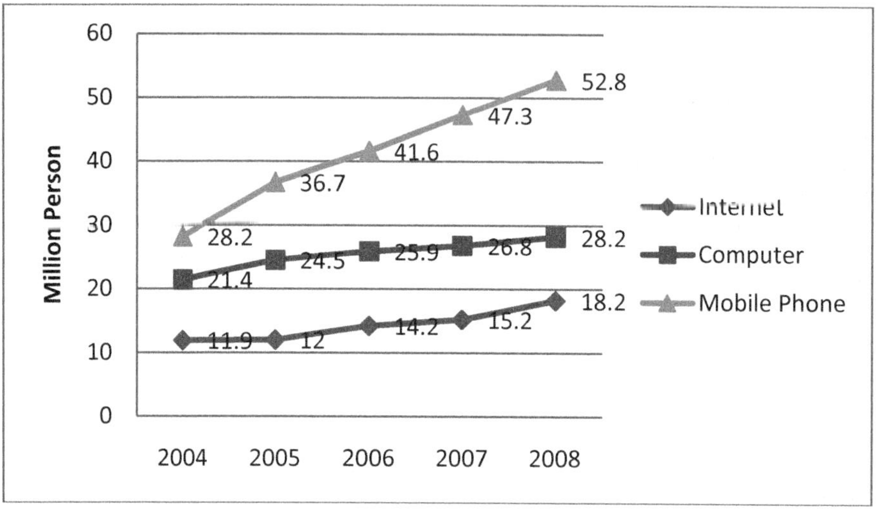

Fig. 2. Internet, computer, and mobile phone frequently use in population between year 2004-2008 [4]

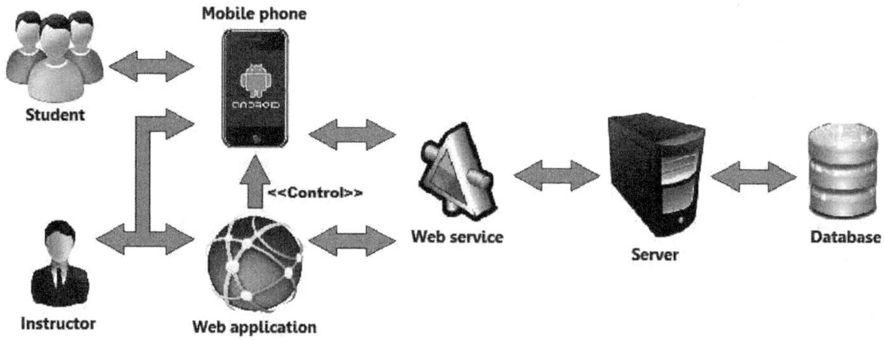

Fig. 3. System Architecture of application [5]

Figure 2. Illustrated comparison of Internet, computer, and mobile phone usage in population, we can see clearly that there are significant number of mobile phone is greater than the other. The smart phone seems to be necessary gadget for individuals particularly university student. There are verities of software which intend to assist study process buy use mobile environment. The mobile social study application is one application among them and will select as experiment tools in the paper. The software perception gaining motivation of student in class to read study content what posted from lecturer. An observation shows that the student frequently access Facebook page when they have free time. The software will send short class room lesson in to mobile phone in order to make it similar Facebook status message. To reduce strained of lesson reading the shot lesson message will concurrently sending to student patrician to share.

Figure 3. Teachers can put the course details on the website, which will also be shown on the application on the students' smart phones. Consequently, students will immediately know of the update and can interact via the application on the smart phones. In addition, these data can also be exchanged with and shared to other devices. And the system will store all data in the database for future use.

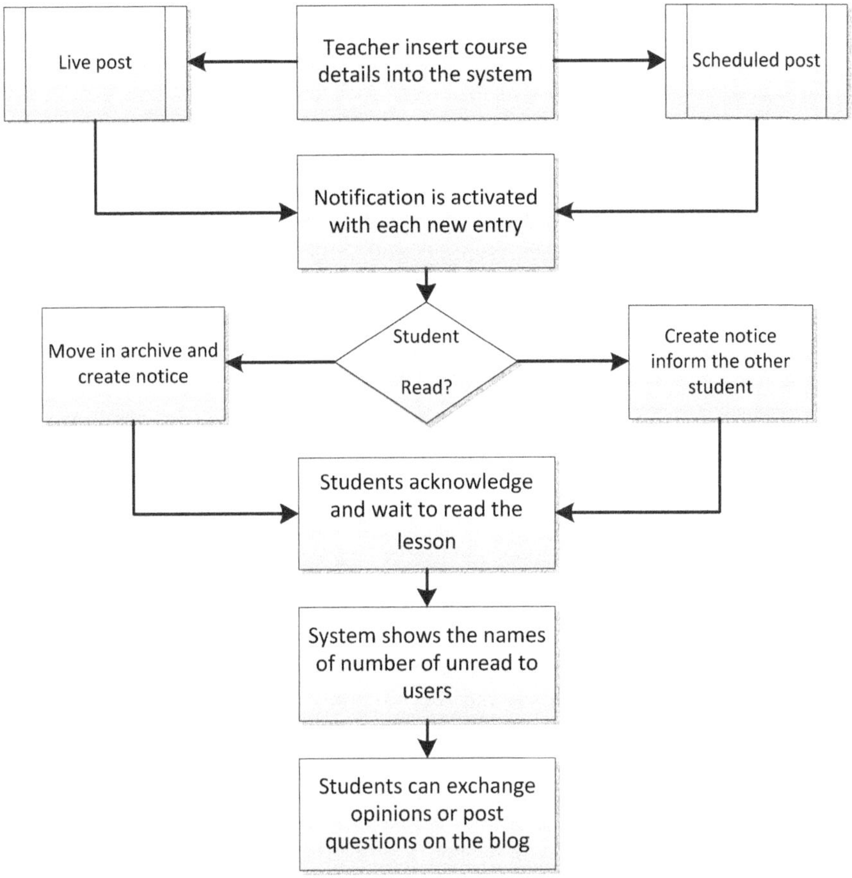

Fig. 4. The working process of the system

From the figure, after a teacher creates a new course detail or a new assignment, the system will send out a notification alert, and students have to press "Accept" to acknowledge it. After that, students can then download the course details or the assignment posted. The system will also let the teacher know which students have downloaded the data. Furthermore, if any students have questions regarding the course details or the assignment, the questions can be posted in the blog available for educational data exchange.

3 Conclusion

The expecting results from student who participate the experiment that they will dramatically increase more attention the study and gain more awareness in any of class mate study activity. With this strategy the student who studies in both campuses will be able to share same study resource and reduce study gap of both classes. Noticeably, those education students about their online learning tools and new ICT tools study methods can help them increase more effective lesson understanding. Also get narrow score gap between two different study environments, and use improved study abilities.

Students can cooperate in employing Information Technology in class. Teachers can also know which students still have not read and downloaded the course details. In addition, the notification function of the system will keep the students up-to-date at all times on their smart phones, suiting well with their lifestyles. Also students expecting the good responses with the system, pay more attention to the course, and ultimately, can comprehend the course details more effectively.

Since the utilization of the system is still in an early stage, some functions are still incomplete and there can be some problems with the communication.

1. Students interesting in logging into read on the lesson and download the course details.
2. There will an exchange of educational data between the students at the Chiang Mai campus and the students at the Lamphun campus.
3. Students have better scores and grades.

References

1. Mbodila, M., Kikunga, M., Serote, M.: Factors Contributing to the Successful Implementation of Information and Communications Technology in the Classroom. In: The Asian Conference on Technology in the Classroom (2014)
2. Beldarrain, Y.: Distance Education Trends: Integrating new technologies to foster student interaction and collaboration. Distance Education 27(2), 139–153 (August 2006)
3. Munich, K.: Social Support for Online Learning: Perspectives of Nursing Students. International Journal of E-Learning & Distance Education 29(2), 1–12 (2014)
4. Compare Mobile & Fixed Subscribers, Bangkok: The National Broadcasting and Telecommunications Commission (2012)
5. Singkhamfu, P., Inkong, P., Suraphruk, C.: Study Assistance by Social methodology application for mobile device. In: Conference on Global Trends in Academic Research (2014)

Telematics Technology Development Forecasting: The Patent Analysis and Technology Life Cycle Perspective

Shu-Hao Chang and Chin-Yuan Fan

Science & Technology Policy Research and Information Center,
National Applied Research Laboratories
14F., No. 106, Sec. 2, Heping E. Rd., Da'an Dist., Taipei 10636, Taiwan
{shchang,cyfan}@narlabs.org.tw

Abstract. Telematics technology is the most popular in recent years, the field of intelligent vehicles issues, was becoming increasingly relevant research. However, the research is still lack in the technology development forecasting. The purpose of this research is to study the technology life cycle of the telematics, and investigates the major technologies in different technology life cycle to point out technical change trends. We used logistic growth curve model and patent analysis to explore future development time phase and technology development saturation. The results indicate that telematics technology is currently in a growth phase, and until the saturation phase, the technology will have eighteen years to grow and develop. Telematics is an emerging technology and will continue to integrate the services or facilities adapted for wireless communication networks and traffic control systems in the future. We assessed telematics development phase and proposed technology development recommendations to telematics industry and government.

1 Introduction

As the technology and computing power keep on development, the transport is no longer just simple to solve the problem between the point-to-point distance, how to effectively increase the security and efficiency of the transport process, and even increased transport effects in the entertainment, and reduce anxiety and impatience during transport, which is the most important issue for manufacturers. Telematics system driven by the automotive industry digital change user traffic environment, improve road safety, and combination of vehicles, traffic, real-time vehicle location information and network communication services, leading the automotive industry from manufacturing-oriented to service-based to knowledge-based economy into a new direction. The purpose of this study, a preliminary understanding is that the status of the patent development for Telematics and further reflection on the major motor manufacturers in the overall activity trends of this technology, as an important evidence for thinking about the future patent portfolio strategies and R & D direction.

In the recent years, as intelligent vehicles development, intelligent automotive driver assistance device has become a promising area of research, one of the key development in each vehicle manufactories, which encompasses a wide range of quite

M. Gen et al. (eds.), *Industrial Engineering, Management Science and Applications 2015*,
Lecture Notes in Electrical Engineering 349, DOI: 10.1007/978-3-662-47200-2_17

diverse, including driving safety or ease of operation, etc. Related technologies have been widely encouraged by the national government in the world, coupled with the recent advances in computer technology, supported by appropriate algorithms for lane detection and monitoring vehicle barriers, are able to provide better analysis. According to the United Nations survey found that about 1.24 million road traffic deaths occur annually on the world's roads [16]. Telematics can provide immediate lane information to help drivers react quickly. The main purpose is to make use of Telematics combines information, communications and automotive electronics technology, providing the driver a wide range of services to meet the person driving the car when the safety, recreation, driving the demand for aid.

In this paper, we adopt patent analysis to understand the life cycle of technology development, mainly to explore the fitting pattern of patent growth through a database search and statistical methods, and through a variety of statistical information related to the patent document, analyze and compare to understand countries' technology development. Previous studies pointed out through patent analysis to understand industry trends, competitors dynamic, strategic planning, [3], [15], etc. For a particular technology, conduct effective analysis for the life cycle, can master the correct technology trends, forecast the level of technology maturity, particularly in high growth technology, the information of technology life cycle is more important.

This study uses patent analysis to explore telematics development. First, we perform major countries telematics patent technical analysis to understand the current status of the technology development; secondly, this study is executed technology life cycle to analysis the technology development stage via logistic growth model and provides practice suggestions.

2 Literature Review

2.1 Definition of Telematics

In this era of digital convergence, telematics has been hailed as a field with unlimited market potential. Telematics involves the integrated use of telecommunications and informatics and thus can provide drivers and passengers with high-value services with a number of functions, including location traffic information, internet access, and vehicle testing [18]. "Telematics" derived from the telecommunication and information of typographical ligature, such as the integration of information and communication technology and automotive industry to develop a platform suitable for telematics driving environment (telematics platform), in order to provide a wide range of vehicle drivers and user demand for access to information by individuals. Telematics industry, including hardware and software systems industry, communications industry, the back-end content providers, to form a chain of industrial chain relationships, and system architecture shown in Figure 1, which provides real-time traffic information, point of information, the return traffic, weather information, emergency and other services.

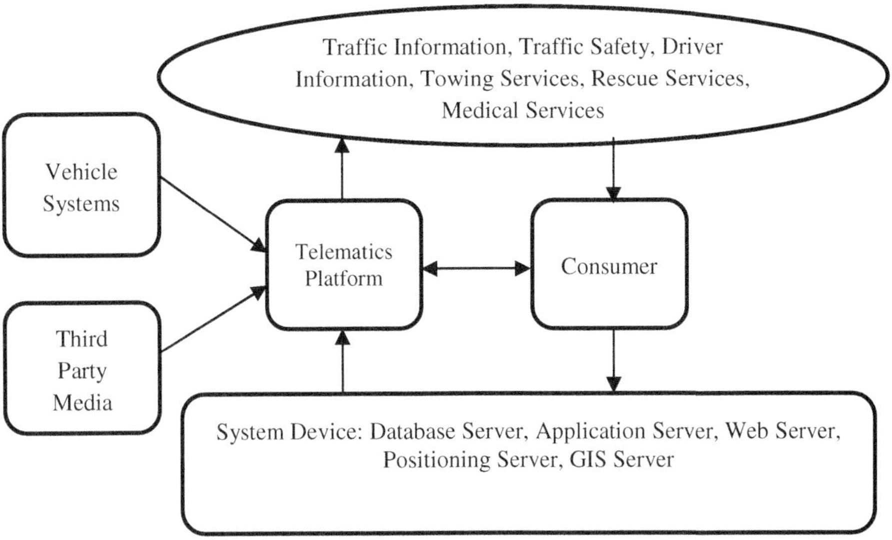

Fig. 1. Telematics system services framework

In recent years, the popularity of Internet, motor manufacturers focus on the development of energy saving and new energy exploiting, but has even more insight on e-car, to provide users with the required information and communication and other services. Telematics systems integrate information and communication systems in vehicles, so that the driver can always contact the support center (call center), to obtain services and information content. In other words, telematics system can be considered as the driver-side contact with the external service pipe.

2.2 Telematics Application and Development

The global commercial vehicle telematics industry has witnessed high growth during last years (2006-2011) and is expected to continue its growth momentum and reach an estimated US $13.5 billion by 2017 with a CAGR of 9.3% over the next five years [11]. According to the definition of telematics, as long as information services and communications functions with the system, can be called the telematics system. There are a variety of telematics combined communication and information systems services in current market. On Telematics hardware can be roughly divided into two, one for the car's computer system in conjunction with the vehicle (also known as the monitor type), when the original factory embedded systems, and the other was detachable mobile devices, such as PDA and so on, this device can be divided into non-monitoring device type and mobile integration, the former to provide a navigation system, while the latter without the navigation system, only integrated media communications interface and provides hands-free function. Telematics services can be divided into personal communications systems, lifestyle information queries, driver assistance, the remote vehicle control, safety and preservation, mobile commerce, mobile entertainment and other. Telematics Services as shown in Table 1.

Table 1. Service categories of telematics

Category	Types	Examples
Drivers	Traffic	Route information, traffic jam, parking
	Security	Anti-theft systems, remote open/close control, vehicle location tracing, stolen vehicle tracking
	Safety	Driving logs, driving surveillance/warning, emergency support systems, remote car accident forwarding, emergency calls
	Personal records	Miscellaneous information, personal information, voice mail/analyses, call records
	Lifestyle	Weather information, events/occasions, local information, points of interest
Unit Installation	Customer relationship management	Diagnostic services, driving records, vehicle maintenance, the auto manufacturer's A/S, auto insurance
Passengers	Entertainment	Music, movies, games, the internet, remote education

Source: [17]

Telematics system is a comprehensive many technical achievements in industry, therefore, a need for the development of telematics systems in conjunction with the application of the technology in different industries. Telematics is mainly applied in the moving car driving environment, taking into account the safety and feasibility of providing the information consumers need it, so that is the key technology in the IT and communications technology, and information technology such as display technology, computing ability, human-machine interface and software platforms.

From the above, the development of telematics received considerable attention, which is closely related to human life, especially as Internet of Things and cloud technology on development, and therefore forecast telematics technologies were necessary.

3 Methodology

3.1 Technology Life Cycle

The life cycle models have been used in different management fields (strategy, marketing, production) in order to represent the evolution of industries, products, brands, [10], etc. Using the conceptual framework of the life cycle model, some authors have defined different models that are representative of the evolutions of technologies [12, 14, 20]. [2] believes that the technology life cycle is a technology from the appear to be replaced or eliminated all the time on the market. That is when the technology reaches a natural limit, it will be easily replaced and recession. Technology development go through emerging, growth, maturity and saturation stage, and technology life cycle curve is like S-shaped curve, so technology life cycle curve can also be called the S-curve [4].

Logistic growth curve model is an S-shaped curve. The starting growth is exponential, but when the inner and outer resistance is noticed, a slowdown appears, and then we can notice the logistic curve growth. It increases a little slow in the beginning. At a certain point, it speeds up exponentially. After passing the 'midpoint' (where the growth rate inflects), the growth slows down and eventually reaches a growth limit. The growth curve analysis fits the given time series data with a logistic function so that one is able to predict the growth limit, the midpoint, and the life cycle of an S-shaped curve [8]. Mathematically, a logistic growth curve model can be represented with a logistic function [9]. The classical logistic curve is given by:

$$\mathrm{L(t)} = \frac{k}{1+exp[-\alpha(t-\beta)]} \tag{1}$$

where k is the growth limit or the asymptotic maximum, often called the carrying capacity; α, the rate of change of growth; and β, the inflection point or mid-point of the curve. Eq. (1) produces three parameters are needed to fully specify the curve, α, β, k. The growth rate parameter, α, specifies the steepness of the logistic growth curve. It is often helpful replacing α with a variable that specifies the time required for the trajectory to grow from 10% to 90% of the limit k. a period which is called the characteristic duration, or Δt. [9]. The characteristic duration is related to α by $\Delta t = \frac{\ln(81)}{\alpha}$. The parameter d_t is usually more useful than α, because the units are easier to appreciate. The parameter β specifies the time when the curve reaches 1/2 k, or the midpoint of the growth trajectory, often re-labeled t_m. The adjusted logistic growth model define as Eq. (2):

$$\mathrm{L(t)} = \frac{k}{1+exp[-(\ln(81)/\Delta t)(t-t_m)]} \tag{2}$$

where Δt is the characteristic duration of the curve, i.e., the time needed for L to grow from 10% to 90% of k. t_m is the midpoint of the curve at which 50% of k is reached.

3.2 Patent Analysis and Data Sources

In recent years, patent analysis has been adopted as the analytical tool for evaluation of technology forecasting [5, 7, 13]. Patent analysis can reveal the current field of technological development, and stimulate new technology solutions, also shows the relationship between technology or stimulate investment policy. [6] and [19] consider that patent analysis can be provided through government public database for technology trends. Based on the above mentioned, we use the patent database of the US government disclosed, that is the United States Patent and Trademark Office (USPTO). Due to the United States is the largest commercial market, system development and data can be traced to 1975, thereby enabling the exhibition of historical trends. In the analysis of global technology, systems from the United States exhibit universal representation [1]. Therefore, this study used the United States patent database for conducting patent analysis and collected patents available to the public from the US Patent and Trademark Office database. Patent data were limited to the US patents publicly available from 1976 to December 31, 2014. The search conditions were as follows: (TTL/telematics) or (ABST/telematics) or (ACLM/telematics). Initially, 813 patents were obtained.

4 Research Results

4.1 Verification of the Growth Curve Model

The most common mode of growth curve model is linear growth, exponential growth and logistic growth. Linear growth model means that it grows by the same amount in each year; exponential growth refers to increases at a consistent rate, and it occurs when a quantity increases by the same proportion in each year; logistic growth model is an S-curve that can be used to model functions that increase gradually at first, more rapidly in the middle growth period, and slowly at the end, leveling off at a maximum value after some period of time. In order to accurately forecast the development of telematics technology curve, this study uses curve estimation method to verify the curve fitting of the data. Curve estimation results shown in Table 2.

Table 2. Curve estimation results

Growth curve model	R square	F
Linear growth	0.895	110.492***
Exponential growth	0.905	123.868***
Logistic growth	0.905	123.868***

Note: p***<0.001

We conduct a logistic growth curve model on the accumulated number of patents from 1976 to 2014, and use the Loglet Lab software and patent publications data for telematics technologies. We can see from the R square, the explanatory power of exponential growth and logistic growth are higher than linear growth model, which represents the patent growth patterns is more fit with exponential growth and logistic growth. However, the patent growth of specific technology has its limitations (technology development saturated), which does not meet the exponential growth assumptions. Finally, this study adopts logistic growth curve model to verify the patent growth of telematics technology, and utilize to make projections about the technology trend.

4.2 Logistic Growth Curve Model

This study used logistic growth curve model to conduct technology forecasting. Figures 2 presents logistic growth curves based on actual data and fitted models for the period 1998-2014 (first patent appeared in the 1998). The figure supper-left corner lists the estimates of the growth limit, the midpoint, and the life cycle. The midpoint of the logistic growth curve was at the year 2021, the duration of growth time was 18.926 years; the drop in publication activity in 2021 may also indicate that the technology growth curve passes an inflection point. Because growth time or the time during is the curve grows from 10% to 90% of its saturation level, the emerging phase of the telematics patent growth began at 1998 and ended around 2012. In addition, growth time includes growth phase and maturity phase and midpoint is the cutoff point. For the periods 2013-2021 and 2022-2030, the growth curve was divided into growth and mature stages. The saturation number of publications of the cumulative telematics patents might be attained to 1133.951. Table 3 summarizes the results for technology life cycle.

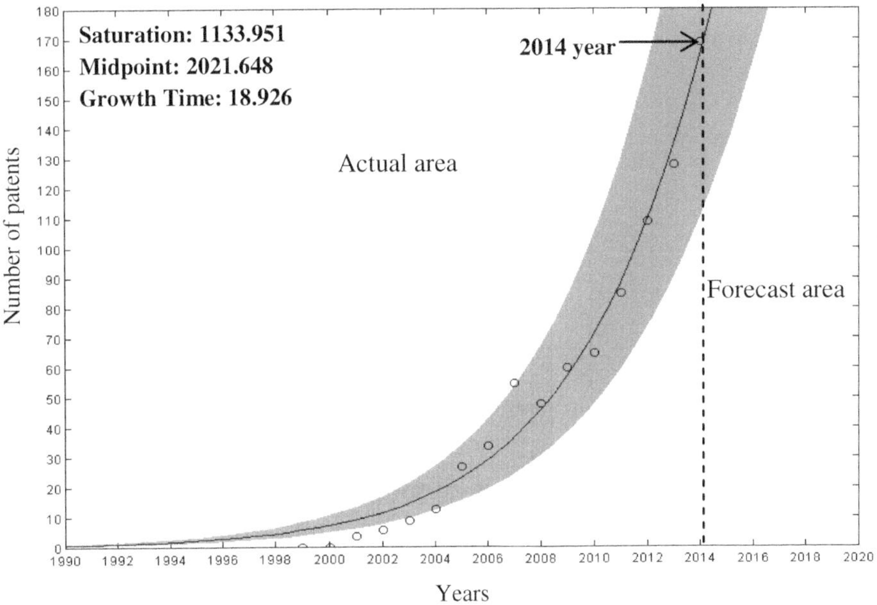

Fig. 2. Telematics patents growth trend

Table 3. Telematics technology life cycle

Telematics technology forecasting	Year				Saturation level
	Emerging phase	Growth phase	Maturity phase	Saturation phase	
	1998~2012	2013~2021	2022~2030	2031~	1133.951

Note: Saturation level is the saturation amount of the number of patents

4.3 Technology Portfolios of Technology Life Cycle

This study conducted technology portfolios of technology life cycle on 813 patent data. Due to telematics technology is currently in the growth phase, it is divided into emerging phase and after the growth phase. Table 4 shows the distribution of the 5 most numerous IPC (International Patent Classification) categories. The results shown in Table 4 indicate that telematics technology is mostly concentrated in G06F007, H04M001, and G01C021, whether in the emerging phase or after the growth phase. According to the definitions of the IPC, the G06F007 category represents methods or arrangements for processing data by operating upon the order or content of the data handled, H04M001 denotes substation equipment, and G01C021 denotes navigational instruments. After emerging phase, some technologies have developed gradually slow, for example, H04M003 and H04B001. H04M001 represents automatic or semi-automatic exchanges, and H04B001 denotes details of transmission systems. Instead, some technical fields have been began to receive attention, such as H04W004 (It is defined as the services or facilities specially adapted for wireless communication networks) and G08G001 (It denotes traffic control systems for road vehicles).

Table 4. Distribution of the Top 5 IPC Categories

Rank	Emerging phase (n=516)	After the growth phase (n=297)
1	G06F007 (n=111)	H04W004 (n=43)
2	H04M001 (n=78)	G06F007 (n=42)
3	H04M003 (n=68)	H04M001 (n=40)
4	G01C021 (n=62)	G01C021 (n=39)
5	H04B001 (n=54)	G08G001 (n=32)

Note: The numbers in the parenthesis are the number of patents

5 Conclusion and Recommendations

The objective of this study was to use relevant patent analysis and technology life cycle analysis to forecast technological development trends. The results can provide a reference for relevant industries or government when applying for telematics patents, or when devising patent-portfolio strategies. Telematics now play important roles in intelligent vehicle industry, and many countries vigorously promote the projects, which can be said to be the next star industry. However, few studies have forecasted telematics development. This study applied the logistic growth curve model to investigate the technology development of telematics, and draw out the possibility of the development time and phase.

The results show that telematics technology has been an emerging technology field, and the strong growth of relevant patents in recent years has increased. Telematics technology is still currently in the growth phase, and the timelines for telematics technology will be long before reaching the mature stage. In addition, the technology portfolios of technology life cycle analysis show that the emerging development directions of telematics are H04W004 and G08G001. It implied that the direction of the series developed by the original equipment development, progress to a wireless network, the traffic platform application and value-added services. And this trend is also consistent with current hot topics such as Machine to Machine (M to M) and Internet of Things (IOT) a considerable degree of correlation and related applications development. As the wireless information technologies keep on development, we can connect to the Internet is no longer just computers, but also cell phones or other electronic products such as iPad. The 'M' of M2M represents gradually no longer refers only to Machine it may be a Mobile even a Man. Currently common wireless network transmission protocol, such as Wireless LAN (Wi-Fi), WiMAX, GPRS, Global System for Mobile Communications (GSM), Code Division Multiple Access (CDMA), Radio Frequency Identification System (RFID), Bluetooth, ZigBee and other agreements have already been added to the conceptual framework, so the application level of telematics and has been with the progress of wireless network have become a growing.

Finally, the technology life cycle analysis was proposed as the simple and efficient tool to link the science and technology activities and to obtain quantitative and historical data for helping researchers in technology forecasting, especially in rare historical data available fields, such as the emerging technology fields. However, there are still several limitations to this study. First, the dataset is taken from the USPTO database. Although it is one of the most reliable patent databases, there are, however, some telematics patents published not included in the USPTO. Presentation and interpretation of the results should be accompanied by a warning on the limitation of the data source. Second, albeit much effort has been made to select correct telematics patents from the database, two situations may still exist: missing telematics patents and an incorrect inclusion of non-telematics patents. Since the search criteria of this study are very strict and narrow, we believe that these patents are a very small percentage of the total patents and do not change the major analysis results. Third, decision makers often need specific information as the basis for future decision. This study adopts simulation approach to provide one of the quantitative forecast techniques. Thus, one needs to be more careful in interpreting the results close to the forecasting development. Another way to overcome this issue is taking more multiple methodologies to let more of the true develop paths appear.

References

1. Bass, S.D., Kurgan, L.A.: Discovery of Factors Influencing Patent Value Based on Machine Learning in Patents in the Field of Nanotechnology. Scientometrics 82, 217–241 (2010)
2. Betz, F.: Strategic Technology Management: Technology and Economic Goals. McGraw-Hill, Singapore (1994)
3. Ernst, H.: Patent Information for Strategic Technology Management. World Patent Information 25, 233–242 (2003)
4. Foster, R.N.: Assessing Technological Threats. Research Management 29, 17–20 (1986)
5. Jun, S., Park, S.S., Jang, D.S.: Technology Forecasting Using Matrix Map and Patent Clustering. Industrial Management & Data Systems 112, 786–807 (2012)
6. Jun, S., Uhm, D.: Technology Forecasting Using Frequency Time Series Model: Bio-technology Patent Analysis. Journal of Modern Mathematics & Statistics 4, 101–104 (2010)

7. Kim, Y.G., Suh, J.H., Park, S.C.: Visualization of Patent Analysis for Emerging Technology. Expert Systems with Applications 34, 1804–1812 (2008)
8. Liu, J.S., Lu, L.Y.Y., Lu, W.M., Lin, B.J.Y.: Data Envelopment Analysis 1978-2010: A Citation-based Literature Survey. Omega 41, 3–15 (2013)
9. Meyer, P.S., Yung, J.W., Ausubel, J.H.: A Primer on Logistic Growth and Substitution: The Mathematics of the Loglet Lab Software. Technological Forecasting and Social Chang 61, 247–271 (1999)
10. Nieto, M., Lopéz, F., Cruz, F.: Performance Analysis of Technology Using the S Curve Model: The Case of Digital Signal Processing (DSP) Technologies. Technovation 18, 439–457 (1998)
11. Research and Markets: Global Commercial Vehicle Telematics Industry 2012-2017: Trends, Profits and Forecast Analysis (2012), http://www.researchandmarkets.com/reports/2773153/global_commercial_vehicle_telematics_industry.pdf (retrieved January 15, 2014)
12. Roussel, P.A., Saad, K.N., Erickson, T.J.: Third Generation R&D. McGraw-Hill, New York (1991)
13. Shin, J., Coh, B.Y., Lee, C.: Robust Future-oriented Technology Portfolios: Black-Litterman Approach. R&D Management 43, 409–419 (2013)
14. Stolwijk, C.C.M., Vanhaverbeke, W.P.M., Ortt, J.R., Pieters, M.W., den Hartigh, E., van Beers, C.: The Effect of Internal and External Technology Sourcing on Firm Performance throughout the Technology Life Cycle. Technology Analysis and Strategic Management 24, 1013–1028 (2012)
15. Tseng, C.Y., Ting, P.H.: Patent Analysis for Technology Development of Artificial Intelligence: A Country-level Comparative Study. Innovation: Management, Policy & Practice 15, 463–475 (2013)
16. World Health Organization: Global Status Report on Road Safety 2013. World Health Organization, Geneva Switzerland (2013)
17. SK Telecom: Internal Document on the Telematics Industry (2011) (as cited in Song, Y.W., Kim, J.D., Yu, L., Lee, H.K., Lee, H.S.: A Compapative Study of the Telematics Industry in Korea and China. Journal of Internet Banking & Commerce 17, 1–13 (2012))
18. Song, Y.W., Kim, J.D., Yu, L., Lee, H.K., Lee, H.S.: A Compapative Study of the Telematics Industry in Korea and China. Journal of Internet Banking & Commerce 17, 1–13 (2012)
19. Trappey, C.V., Wang, T.M., Hoang, S., Trappey, A.J.C.: Constructing a Dental Implant Ontology for Domain Specific Clustering and Life Span Analysis. Advanced Engineering Informatics 27, 346–357 (2013)
20. Taylor, M., Taylor, A.: The Technology Life Cycle: Conceptualization and Managerial Implications. International Journal of Production Economics 140, 541–553 (2012)

Empirical Study of Collaborative Learning Knowledge Management System for Thai Students

Krittawaya Thongkoo[1,*] and Chiraporn Thongkhu[2]

[1] Chiang Mai University, Chiang Mai, Thailand
krittawaya.t@cmu.ac.th
[2] Lukhamhanwarinchamrab School, Ubon Ratchathani, Thailand
chiraporn.ubn@gmail.com

Abstract. This paper deliberates about the design of the prototype model of the online collaborative learning system for introductory Structural Analysis and Design course. The model of collaborative learning system has been developed in pursuance of supporting the collaborative learning activities in virtual environment. A design of a structure model for an online collaborative learning system has been constructed for the sake of sustaining the online collaborative learning activities in a virtual environment. The methodology of this study consists of three processes which are data collection, system analysis & design and system implementation. Initially, a set of questionnaire was conducted to a sample group including one hundred of second- year students from Modern Management and Information Technology program at Chiang Mai University (CMU), Chiang Mai, Thailand. The students were selected to be respondents during the data collection process. The analysis results from the questionnaires will reveal the students' interests, learning styles and preferences in learning Structural Analysis and Design course for the online system development. The results have illustrated significant correlation proving that the online application has been essential and helpful for the students to achieve better results in Structural Analysis and Design course.

Keywords: collaborative learning, knowledge management system, distance learning education.

1 Introduction

Structural Analysis and Design (SA) is claimed to be a challenging intellectual task. Collaborative learning is generally described as coordinated and synchronous activity resulting from a continued attempt to construct and maintain a shared conception of a problem.

The goal of this study is to compare two sections of a structural analysis and design course – The first one is dominated by lecture and the second one is used the constructivist model. The need for the development of a knowledge management

* Corresponding author.

M. Gen et al. (eds.), *Industrial Engineering, Management Science and Applications 2015*,
Lecture Notes in Electrical Engineering 349, DOI: 10.1007/978-3-662-47200-2_18

system using concept maps for structural analysis and design is apparent. The first step is to characterize the problems which are typically asked to solve by the students.

Although a number of web-based systems have been developed to support learning process of SA courses, the impact of applying web-based systems on students' performances has remained elusive. Moreover, it is vital for further investigations on how the web-based systems used as the medium to teach SA collaboratively and how the effect of online collaboration to the students' performances in SA courses is significant. Therefore, the aim of this study is to introduce the concept of online collaborative learning system to support the teaching and learning of SA. Also, it aims to study the impacts which are analyzed to determine whether it could have benefits for the students or not in term of their achievements in courses.

2 Literature Review

2.1 Knowledge Management

Knowledge management [1] was initially defined as the process of applying a systematic approach to the capture, structuring, management, and dissemination of knowledge throughout an organization to work faster, reuse best practices, and reduce costly rework from project to project.

The goal of a knowledge management system [2][3] is to provide managers with the ability to organize and locate relevant content and the expertise required to address specific business tasks and projects. The relationships between content, people, topics and activity can analyzed by some knowledge management systems and produce a knowledge map report.

2.2 Collaborative Learning

Collaboration can be referred [4] as coordinated, synchronous activity emerging from constant attempt to construct and retain a consistent conception of a problem. A great number of collaborative learning methods such as the "Round Robin", "Numbered-Heads-Together", "Jigsaw", "Think-Pair-Share" and many others are generally employed in a classroom-based environment. Additionally, they have also been proven to be practical for enhancing the students' performances. With collaborative learning, it will emphasize on assisting students to plan and design their problem solving strategy and guide them to evaluate the problem during the learning process.

2.3 Computer-Supported Collaborative Learning (CSCL)

An online collaborative system is meant to be supported by specific tools [5] in contemplation of teaching and learning collaboratively. For this reason, it is meticulously associated with Computer-Supported Collaborative Learning or CSCL. CSCL has been seen as the most promising innovations to progress teaching and learning in a virtual environment. Supported by modern information and communication technologies, online collaborative learning is now possible and able to be implemented more efficiently. CSCL is also described as how collaborative activities and their coordination can be sustained via computer systems. In previous studies, CSCL has also been known as a "groupware" or "collaborative software". The online computer-supported collaborative learning expresses two significant notions which are the idea of learning collaboratively online with other community members and also in a group. In addition, CSCL can be applied to promote peer interaction and working together. Also the collaboration and technology will facilitate the sharing and distribution of knowledge and expertise among community members. Therefore, more interactive online learning environment will be provided by any online collaborative learning resulting from the concept of CSCL. Furthermore, it also encourages and facilitates interaction, negotiation and collaboration amongst and between students and their tutors.

3 Methodology

3.1 Data Collection

Firstly, a set of questionnaire was conducted to a sample group which accounted for one hundred of second-year students from Modern Management and Information Technology program in Chiang Mai University (CMU), Chiang Mai, Thailand to inspect the students' interest, learning styles and preferences in learning structural analysis and design subject. The study analysis resulted from the questionnaires will reveal whether there are needs for the development of the online system for learning and sharing knowledge of the system analysis course.

3.2 System Analysis & Design

The Collaborative Learning Knowledge Management System or CLKMS was promoted by adopting the concept of collaborative learning using concept map technique. Three main stakeholders designed for the system are combined; namely, the administrator, lecturer and student.

For the system design of the CLKMS, the system requirements are comprised of two parts which are the functional requirements and non-functional requirements. Both conditions are represented using case diagram. Finally, the physical design of CLKMS is being constructed by using the Web-based technology which includes MySQL, PHP and Apache web server.

3.3 System Implementation

CLKMS was implemented with a sample group of the second- year students from Modern Management and Information Technology program at Chiang Mai University (CMU), Chiang Mai, Thailand in one section. In order to analyze the effectiveness of using the CLKMS in structural analysis and design classes, another group of the second-year students was compared - one that was dominated by lecture named "Group A"and one that used the application with the constructivist model named "Group B". Both groups consisted of 30 students for each class and were taught by the same lecturer.

4 Experimental Results

4.1 Data Collection

The results form data collection phase consists of questionnaires to inspect the students' interest and learning styles in learning structural analysis and design subject.

Table 1. The interest of students

Question	Percent
Are you interested in structural analysis and design course?	68%
Do you always update your knowledge about structural analysis and design?	60%
Are you expert in structural analysis and design?	15%
Are you interested in working with a group while learning structural analysis and design?	76%
Are you interested in using e-learning for knowledge sharing?	87%
Are you interested in working alone on structural analysis and design homework?	18%
Do you prefer to learn and explore using e-learning?	89%

Table 1 above shows the results of the students' interests in learning structural analysis and design subject where the analysis has showed that about 76% of the students are more interested working in groups while learning structural analysis and design. Meanwhile, it has also revealed that about 87% of the students are interested to use the e-learning for knowledge sharing. And also 89% of the students preferred to learn and explore using e-learning.

4.2 System Analysis & Design

The Collaborative Learning Knowledge Management System or CLKMS was developed by adopting the concept of collaborative learning using concept map technique. There are three main stakeholders designed for the system which are the administrator, lecturer and student.

First step, the lecturer will prepare the general topic of the concept map and considers the related ideas. Then, the lecturer will select appropriate words for the topic that support the main ideas of system analysis and draw these words and connect to the main topic with a line referring to the relationship. Furthermore, the lecturer will repeat all processes in the subtopics until students understand the concept of system analysis.

For student phase, students will prepare the general topic of the structural analysis and design curriculum and brainstorm to design the words that are related to the topic. Besides, the selected words from each brainstorming sessions are ranked which will support the main ideas of structural analysis and design. In additional, the students will compare group similarities and differences; a separate map is created for each of the groups. The final concept map demonstrates the importance of the items according to the rankings. Finally, the concept map is presents the information which identifies into themes. The answers are graded by the lecturers and can be downloaded by the entire collaborative class.

For the system design of the CLKMS, the system requirements comprise of two parts that are the functional requirements and non-functional requirements. Both conditions are represented using the use case diagram. Finally, the physical design of CLKMS is being constructed using the Web-based technology which includes MySQL, PHP and Apache web server.

4.3 System Implementation

This part is an analysis to measure the effectiveness of CLKMS towards the students' performances is being conducted. Table 2 shows the result of the analysis to see the correlation between CLKMS and the students' performances. As mentioned earlier, two groups of students, where each group consists of 30 students from the second-year of Modern Management and Information Technology program was being selected. "Group A" is being exposed to CLKMS in structural analysis and design class where the lecture uses the application. Meanwhile, "Group B" is being taught using the traditional method, where the lecturer will be teaching only by using PowerPoint and examples were given to them in the course without having them to download it via the CLKMS application. Before we elaborate more about the impact of the online collaborative system to the students' performance, we first take a look at Table 2 below to see the frequency of students' achievements in their test results for structural analysis and design course.

From the Table 2 below, as depicted in the "Group A", the results were higher achievements in students' performances and there were no students who failed in the test compared to "Group B" where there were few students who have scored D for structural analysis and design course.

Table 2. The students'test results for structural analysis and design course

Grade	*Sample groups*		*Total*
	Group A	*Group B*	
A	3	1	4
B+	7	3	10
B	10	4	14
C+	5	9	14
C	3	4	7
D+	2	6	8
D	0	3	3
F	0	0	0
Total	**30**	**30**	**60**

5 Conclusion and Discussion

The classroom-based collaborative learning has been demonstrated the efficiency in enhancing students' performances and individual self-esteem. Such benefits can be easily achieved as the interactions among and between the community members. These are not only visible and can be directly monitored but also be incapable of being evaluated by the instructors. Designing the virtual collaborative learning system, on the other hands, involves complex issues and challenging tasks in furtherance of creating the virtual environment suiting every community members ideally. The developer needs to consider the selections of the right collaborative learning technique, tools and technologies which can be ideally used in a virtual environment. In this study, series of fact-finding processes have been implemented for further understanding in the current situations and most importantly to gain users' needs and requirements. The results have illustrated that the students prefer to work in small groups so that their understandings are being enhanced in system analysis and they also prefer to search for learning materials from various internet resources such as from the e-learning portals.

Further analysis has been done to measure the effectiveness of CLKMS implementation towards students' performances. Consequently, the results have declared significant correlation ensuring that the online application is crucial and applicable for students to accomplish better results in Structural Analysis and Design course.

References

1. Nonaka, I., Takeuchi, H.: The Knowledge-Creating Company, vol. 1. Oxford University Press, New York (1995)
2. Morey, D., Maybury, M., Thuraisingham, B. (eds.): Knowledge Management. MIT Press, Cambridge (2001)
3. Liebowitz, J.: Knowledge Management Handbook. CRC Press LLC (1999)
4. Roschelle, J., Teasley, S.D.: The Construction of Shared Knowledge in Collaborative Problem Solving in Computer Communication and Cognition Vygotskian Perspectives, pp. 67-197. Cambridge University Press, Cambridge (1995)
5. Stahl, G., Koschmann, T., Suthers, D.: Computer-supported collaborative learning: An historical perspective. In: Sawyer, R.K. (ed.) Cambridge Handbook of the Learning Sciences, pp. 409–426. Cambridge University Press, Cambridge (2006)

Factors for Enterprise Resource Planning System Selection to Support Information Management of Manufacturers

Porntida Kaewkamol

College of Arts, Media and Technology, Chiang Mai University
Huay Kaew Rd., Muang, Chiang Mai, Thailand
Porntida.k@cmu.ac.th

Abstract. In the age of globalisation, many firms are being operated by Information Technology (IT) to improve and streamline their business processes. IT systems usually play an important role to support internal information management. An Enterprise Resource Planning system (ERP system), which can be regarded as one of IT solutions for business, is able to integrate business process and information flow across the enterprise to support decision-making process. However, the selection of ERP system might be complicated as it is affected by various factors. This paper, therefore, aims to gather and identify key factors to select the ERP system for manufacturers. Methodology of this paper is mainly based on a combination of literature reviews from related papers. As a result, this paper provides a conceptual framework illustrating a set of major factors that should be considered to select ERP system for manufacturers.

1 Introduction

Enterprise Resource Planning system (ERP system) has become a widespread solution for companies in many industries. It synchronises business processes and also integrates company-wide information using a single database (Monk and Wagner, 2013). However, with various types of this software solution from different vendors, combines with the failure or unsatisfied implementation stories (Momoh, Roy, and Shehab, 2010). The firms should consequently consider related factors for an ability to evaluate, select and implement the most appropriate system for them. This paper, hence, aims at gathering and indentifying key factors for ERP system selection to particularly support information management of manufacturers.

2 Literature Reviews

Literature reviews of this paper consist of two main parts: Enterprise Resource Planning system selection and information management in organisation. The former topic focuses on general concepts of ERP system and its impacts on business process of an enterprise. Key factors that should be considered for ERP system selection are

M. Gen et al. (eds.), *Industrial Engineering, Management Science and Applications 2015*,
Lecture Notes in Electrical Engineering 349, DOI: 10.1007/978 3 662-47200-2_19

also reviewed in this section. Additionally, the latter topic, information management in organisation, is also mentioned concerning business activities and information system categories, as well as, types of information that should be collected to support system selection process.

2.1 Enterprise Resource Planning System Selection

Enterprise Resource Planning system is a systematic approach to streamline internal value chain (Norris et al., 2000). It integrates cross-functional business activities from Sales, Marketing, Manufacturing, through Accounting and Human Capital Management (Monk and Wagner, 2013). This leads to all-level relationship improvement and a competitive advantage development of an enterprise (Al-Mashari and Al-Mudimigh, 2003).

After deciding to adopt an ERP system, an enterprise should carefully gather and analyse all involved information to select the most suitable system. ERP system selection relates to more than an interview with vendors, since the selected vendor is likely to become a long-term business partner (Travis, 1999 cited in Baki and Cakar, 2005: 76). Furthermore, the system is also likely to be implemented in long period regarding to its initial high investment (Verville and Hallingten, 2002).

In order to select an appropriate ERP solution, there are some key factors that should be considered for the subsequent successful implementation, Reuther and Chattopadhyay (2004: 852) proposed a critical selection factors from the survey results regarding ERP system for manufacturing SMEs in Brisbane, Australia. There are six major categories ordered by the most rating from project managers who directly responsible for ERP implementation projects. These key factors can be shown in Table 1.

Table 1. Critical selection factors (Reuther and Chattopadhyay, 2004: 852)

Critical selection factors	Description
System Functionality Requirements	Requirements of the system to suit business.
Business Drivers	Financial benefit to the company of the selected system.
Cost Drivers	Direct cost of the implementation in terms of outlay and resources.
Scalability	Size of the system to suit the business and ability to grow with the business.
Flexibility	Ability to optimise the system to meet unique requirement of the company.
Others	Specific factors critical to the target business.

Besides, Baki and Cakar (2005) conducted a questionnaire-based survey and the interviews with senior executives and managers of ERP projects in Turkish manufacturers to investigate ERP selection criteria. Table 2 shows their criteria in descending order of importance.

Table 2. ERP selection criteria (summarised from Baki and Cakar, 2005)

ERP selection criteria	Description
Fit with parent and/or allied organization systems	Provide compatibility with parent and/or allied systems.
Cross-module integration	Facilitate full integration between business activities.
Compatibility with other systems	Can be integrated with in-house system or other specialised software.
References of the vendor	Have some success stories about system implementation for customers.
Vision	Consider the vendor's vision in terms of modification plan for three to five years.
Functionality	Provide sufficient modules to support core business process.
System reliability	Provide satisfied and reliable system.
Consultancy	Have experience to support both selection and implementation process
Technical aspect	Address current or future trends of hardware and software architecture.
Implementation time	Provide appropriate implementation time
Methodology of the software	Have clear and effective activities on how, when to implement the system and with which resources
Market position of the vendor	Provide global best practice to the company.
Ease of customization	Can be customised for specific needs of an enterprise.
Better fit with organizational structure	Provide compatibility with the current organisational and human resources structure.
Service and support	Include pre and post-implementation support.
Cost	Consider overall cost of the system: software, hardware, consulting, training, maintenance and upgrades.
Domain knowledge of the vendor	Specialise and have experiences in the industry

Similarly, Perera and Costa (2008) investigated the mostly considerable criteria for ERP selection of Sri Lankan manufacturers. They distributed questionnaires to companies that either already implemented or would be going to implement ERP systems. Their results can be shown in Table 3.

Table 3. ERP selection criteria (summarised from Perera and Costa, 2008)

ERP Selection Criteria	Description
Business strategy	Profit margins, supply chain management, and customer support.
Implementability	Implementation period, multi site implementability.
Risk	Requirement achievement, financial risk, technology obsolescence, and project management problems.
Functional fit	Need for customisation, user friendliness, and compatible with prosper business needs.
After sales agreement	Supportive staffs, training, and on-site maintenance.
Technology	Hardware requirement, further development limitations.
Vendor's position	Potential of local agent.
Cost and benefits	Licenses, training, version upgrades and infrastructure cost.
Change management	Need for process modifications.

Moreover, Mexus, Quelhas, and Costa (2012: 345), developed the ERP systems selection criteria tree which includes the aspects of finance, business, software, technology, and vendor. In terms of business criteria, it refers to the alignment of ERP systems with business processes of the company. Besides, each category has a number of sub-criteria as can be shown in Figure 1.

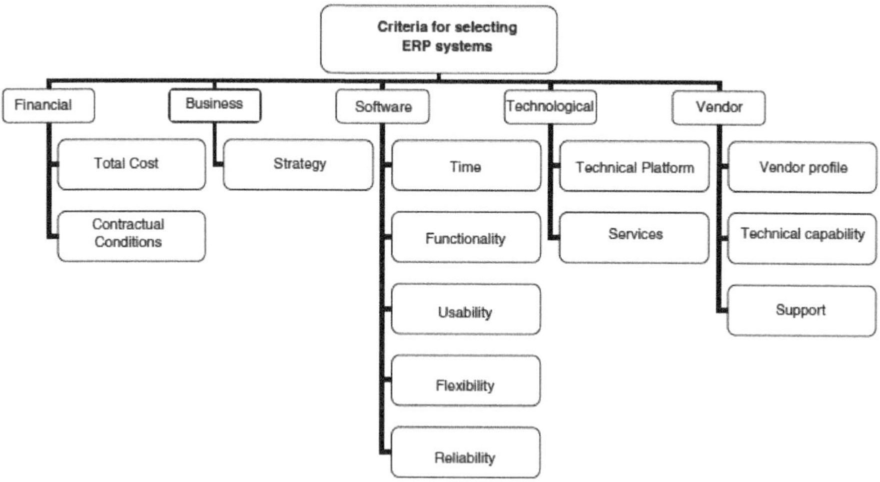

Fig. 1. Criteria tree for ERP systems selection (Mexus, Quelhas and Costa, 2012: 345)

2.2 Information Management in Organisation

In this paper, information management in organization is specifically focused on topics that are related to system selection. To begin with, information management in terms of business activities and information system can be categorised into two main groups, which are primary business processes and information system, and secondary business processes and information system. (Porter and Millar,1985 cited in Reuther and Chattopadhyay, 2004: 851)

Primary Business Processes and Information System
They are business processes and information system that have a large impact on critical operations of an organization, for instance, financial activities, and logistics management process. In case of information system failure, an organization is likely to have serious financial effects.

Secondary Business Processes and Information System
Secondary business processes and information system can support and facilitate day-to-day activities of a particular operation within an organisation.

From the aforementioned business process and information system categories, they have a relationship between each other and both are also related to information management in organisation. More importantly, Lech (2012) mentioned that there are two types of information that need to be managed before enterprise system selection process. They are organisational needs and system functionality.

Organizational Needs
Organisational needs, also known as organizational requirements, are important information that should be primarily collect and define. Organisational requirements can be regarded as a main framework to search and select an appropriate system for an enterprise.

System Functionality
After gathering organizational needs, an enterprise should use them as a guideline to subsequently analyse functionality of the new system. Additionally, system functions of the selected system should cover and support all of organizational needs.

3 Methodology

The methodology of this paper is initially a combination of literature reviews from related papers. The reviews are in terms of factors that manufacturers should be considered for ERP system selection. Acquired data is afterward gathered and analysed. A conceptual framework, illustrating a set of major factors of ERP system selection for manufacturers, is subsequently developed. Lastly, the proposed conceptual framework is summarized and discussed.

4 Result

With regard to literature reviews, the proposed conceptual framework which summarises major factors of ERP system selection for manufacturers can be shown in Figure 2.

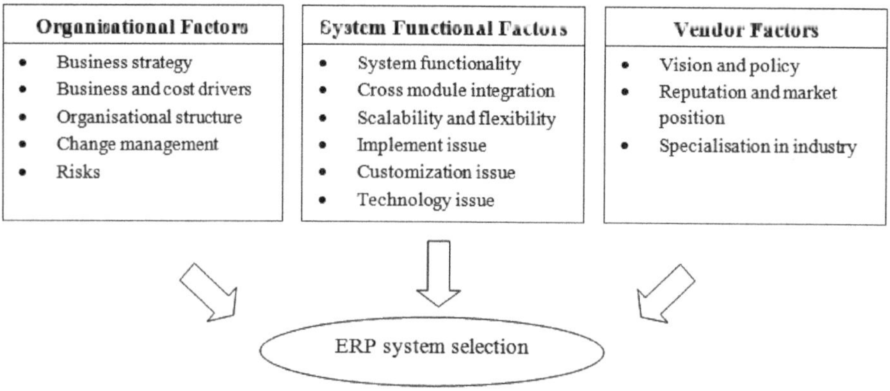

Fig. 2. The proposed conceptual framework

5 Discussion and Conclusion

The advent and rapid development of new technology have driven enterprises to adopt IT solutions to establish their competitive advantages. ERP system, as one of those kind of solutions, have also been implemented in many enterprises since it provides an integration of internal business process, transactional information, analytical information, and information flowing through a company (Momoh, Roy, and Shehab, 2010; Leger, Riedl and Brocke, 2014).

However, the different ERP systems from various vendors and the reported failure or unsatisfied implementation (Momoh, Roy, and Shehab, 2010) have also led to the difficulty and complexity in system selection. There are, hence, a set of factors that should be considered before selecting the system. According to the proposed conceptual framework, those key factors are categorised into three groups: organisational factors, system functional factors, and vendor factors.

Organisational Factors
Business strategy, business and cost drivers, and organizational structure are the factors that should initially be considered for ERP system selection to support information management. These factors provide a guideline or a framework to eliminate ERP systems that cannot meet business requirements. Additionally, cost drivers are factors that need a careful consideration, as it involves cost of installation, user training program, system maintenance, system upgrades and other ongoing costs. Besides, Change management and risks are the factors that should not be overlooked. Implementing an ERP system results in some internal changes. Obviously, employees

are supposed to work with the new system that they may not acquaint with. Manufacturers should, therefore, consider and develop an effective plan to manage these changes and risks for a successful system implementation.

System Functional Factors
ERP systems should provide functions that cover all core business processes and cross-module integration. This is essential since all transactional operations are likely to be executed via the system. If it is unable to support all key processes, there might be a mistake in operations, especially cross-functional processes, which may cause critical problems to an organisation afterwards.

ERP system should also compatibly function with organizational structure, parent and/or allied systems, and partner systems. The reason is that ERP system should be able to manage physical, process, and information flow across all business activities which involve information from other related systems.

Scalability and flexibility are other factors that should be included in selection process. Manufacturers, and almost all business, have to develop and increase their operational performance over a period of time. This contributes to the need of scalable and flexible system that an organisation can improve and upgrade along with their business growth.

Implement issue refers to both methodology and implementation time of the ERP system. Manufacturer should consider this issue, as implementation process of different system leads to a variety of time and resource needed. The following implementation plan may be more effective when considering these factors since selection process.

Technology issue includes aspects of software and hardware trends. The latest technology may not be essential to consider. However, technology used to develop and run an ERP system should be, at least, able to support current technology architecture of an organisation and their partners.

Apart from above, ease of customisation is another factor that should not be overlooked. Although manufacturers are in the same industry, business processes of them are more likely to be different in details. Manufacturers should, therefore, adopt system customisation function as one of selection factors so as to tailor the system to fit with their business needs.

Vendor Factors
A profile of ERP system vendor is also a major factor that manufacturer should consider. Vision and policy of vendor concerning consultancy, after sales agreement, maintenance plan, provided support and service, have a large impact on system functionality and reliability in long-term period. Furthermore, reputation and market positioning of vendor may be considered for effective and success of system implementation. More importantly, specialisation in the industry is another significant factor, It reflects that vendor have experiences and is likely to assist manufacturers to select and implement a suitable ERP system.

In conclusion, manufacturers should gather, analyse, and consider factors in the proposed conceptual framework in order to select an appropriate ERP system. Moreover, other related factors for specific needs of a particular manufacture should also be identified and considered to ensure that the selected ERP system is able to support all core business processes in the long run.

References

1. Al-Mashari, M., Al-Mudimigh, A.: ERP implementation: lessons from a case study. Information Technology & People 16(1), 21–33 (2003)
2. Baki, B., Cakar, K.: Determining the ERP package-selecting criteria: The case of Turkish manufacturing companies. Business Process Management Journal 11(1), 75–86 (2005)
3. Lech, P.: Information gathering during enterprise system selection: insight from practice. Industrial Management & Data Systems 112(6), 964–981 (2012)
4. Leger, P., Riedl, R., Brocke, J.: Emotions and ERP information sourcing: the moderating role of expertise. Industrial Management & Data Systems 114(3), 456–471 (2014)
5. Mexas, M., Quelhas, O., Costa, H.: Prioritization of enterprise resource planning systems criteria: Focusing on construction industry. International Journal of Production Economics 139(1), 340–350 (2012)
6. Momoh, A., Roy, R., Shehab, E.: Challenges in enterprise resource planning implementation: state-of-the-art. Business Process Management Journal 16(4), 537–565 (2010)
7. Monk, E., Wagner, B.: Concepts in Enterprise Resource Planning Fourth Edition, Course Technology, Cengage Learning (2013)
8. Norris, G., Hurley, J., Hartley, K., Dunleavy, J., Balls, J.: E-Business and ERP: Transforming the Enterprise. John Wiley & Sons Inc., Canada (2000)
9. Perera, H.: Analytic Hierarchy Process for Selection of ERP Software for Manufacturing Companies. The Journal of Business Perspective 12(4), 1–11 (2008)
10. Reuther, D., Chattopadhyay, G.: Critical Factors for Enterprise Resources Planning System Selection and Implementation Projects within Small to Medium Enterprises. In: IEEE International Engineering Management Conference, pp. 851–955 (2004)
11. Verville, J., Hallingten, A.: An investigation of the decision process for selecting an ERP software: the case of ESC. Management Decision 114(3), 206–216 (2002)

Low Level of Licensing Activities by Universities in Japan

Shigenori Hata and Kumiko Miyazaki

Tokyo Institute of Technology, 2-12-1, Ookayama, Meguro-ku, Tokyo 152-8550, Japan
{hata,miyazaki}@mot.titech.ac.jp

Abstract. With a series of institutional arrangements University-Industry Collaboration (UIC) activities in Japan have emerged since the late 1990s. This study examines the low licensing activities by universities in Japan because it is one of the serious problems universities are now facing. Interviews with the relevant technology transfer officers in major universities in Japan were conducted. Results show that co-ownership of patents, attitude of industry, and human resource and financial constraints exist behind the low licensing activities.

1 Introduction

A series of policy measures has been carried out in Japan since THE late 1990s with a hope to spur prolonged economic recession by University Industry Collaboration (UIC) such as Technology Licensing Office (TLO) Act in 1998, Japanese "Bayh-Dole" Act in 1999, and incorporation of national universities in 2004(Table 1). With more than a decade OF experience and relevant data available, it is necessary to review the results of UIC policies in Japan.

Reference [6] analyzes the background of UIC promotion in Japan. According to the analysis, as industry in Japan achieved parity with western firms in the late 1990s in most technologies and forged ahead in several very important areas such as consumer electronics, automobiles, and ceramics, and U.S. firms responded to competition by reducing their manufacturing costs, shortening product cycles, and improving quality and responsiveness to customers, it became clear that further Japanese progress depended on creating an independent base for rapid innovation. In addition, in face of prolonged economic recession in the late 1990s, Ministry of international Trade and Industry (METI) actively called for technology transfer from university to industry as one of effective tools for revitalizing the domestic economy. With this background, a series of institutional arrangements were conducted such as TLO Act and Japanese Bayh-Dole Act. The Japanese Bayh-Dole Act was enacted in 1999 as the part of Industry Revitalization Special Law ([3], [4], and [5]).

Like Bayh-Dole Act in US, the Japanese Bayh-Dole Act aims to give ownership of patents to universities and research institutions and promote transfer of their technology to industry for commercialization. This Japanese Bayh-Dole act successfully

M. Gen et al. (eds.), *Industrial Engineering, Management Science and Applications 2015*,
Lecture Notes in Electrical Engineering 349, DOI: 10.1007/978-3-662-47200-2_20

Table 1. Recent developments in UIC in Japan

	Important measures	Subsidiesto universities, TLOs
1998	Technology licensing office(TLO) act	
1999	Japanese "Bayh-Dole" act	
2000	Relaxing business on the side for faculty members at national universities	Subsidies for TLO
2001	Reorganization of governmental agencies	
2002		
2003		
2004	Incorporation of national universities	Subsidies for IP management at universities
2005		
2006	Redefinition of the role of universities	
2007		
2008		
2009		Subsidies for university-Industry collaboration at universities
2010		
2011		
2012		

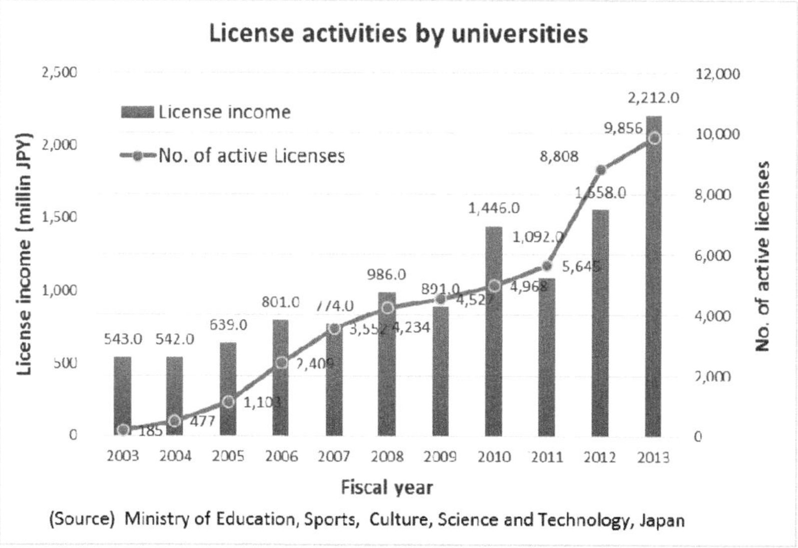

Fig. 1. License activities by universities in Japan

increased the number of applications from the universities. TLO Act aims to promote (1) the progress of industry and creation of new technology, and (2) research activities through technology transfer of research results developed by universities and public research institutions to industry. Special privileges were provided to the TLOs which were approved under Acts such as operational subsides, discount of patent annual fees and patent examination official fees ([12]).

From the perspective of technology management, TLOs receive a great deal of attention because TLOs are thought to be one of key mediators to bridge universities and industry in technology transfer. License income has increased rapidly in the last decade. There were only 477 licenses executed with license income of 532 million yen in 2004. The number of license executed and the total amount of license income grew to 9,856 cases and 2,212 million yen, respectively in 2013(Fig.1).

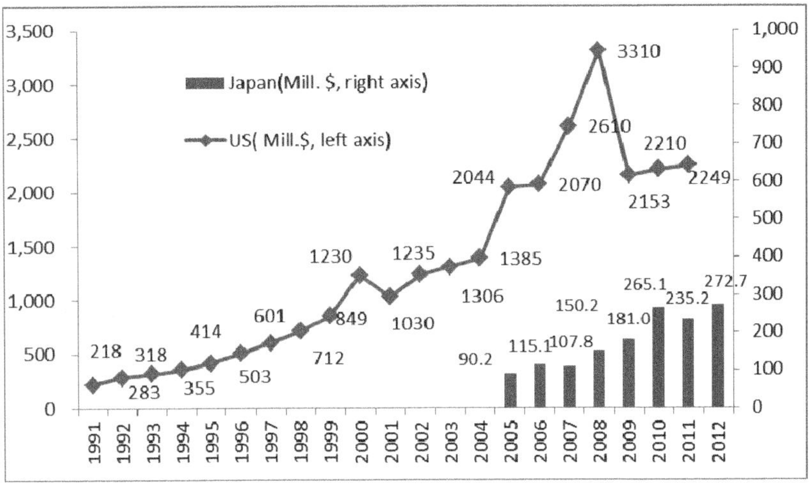

Fig. 2. Adjusted license income received by universities. *Source* AUTM Licensing Survey, UNITT Survey of University Technology Transfer in Japan.

Table 2. Comparison of the average license income

	Number of current and active licenses	Adjusted license income		Average income per license(Thousand US$)
U.S.	38,600	2.25	billion US$	58.3
Japan	6,883	2.28	billion JPY	4.3

With the expanding UIC in Japan, however, licensing statistics does not necessarily indicate active technology transfer via licensing compared to that of U.S. License income by universities in U.S. was 2,249 Million US$ in 2011 while that of Japan was only 235 Million U.S.$. License income in U.S. is ten times as much as that of Japan (Fig.2). In order to eliminate the effect of size of economy of the country, the comparison of the average license income between U.S. and Japan was calculated in Table 2. The result indicates that the average income per license by Universities in U.S. is about fifteen times as much as that of universities in Japan.

In addition, the low level of license income poses serious problems on TLOs in Japan. Major national universities such as the University of Tokyo established their own TLOs in response to TLO act in 1998. 38 approved TLOs are under operation as of the end of September, 2014. However, it is widely recognized that except for a few TLOs, most TLOs are under unprofitable business operation because of insufficient license income to cover the operational costs of TLOs.

We examine the reasons licensing activities are not so active in Japan by interviews with relevant technology transfer officers at universities. Several factors are thought to be the reasons such as legislative aspect relating to Patent Law and private companies' attitude to universities. This paper is organized as follows. In the next section we review the literature regarding TLOs in Japan as well as theoretical framework. Methodologies and research questions are presented in section 3, followed by the findings in section 4. Section 5 summarizes the conclusions.

2 Literature Survey

It is widely recognized that the theoretical framework on which UIC is based is "the cooperative technology policy paradigm" ([1]). This paradigm is a fundamental framework which underpins US policy to promote technology transfer in the 1980s such as "Bayh-Dole" Act. Under this paradigm, government laboratories and universities can play a role in developing technology, especially pre-competitive technology, for use in the private sector.

There are numerous literatures regarding UIC in Japan. We focus on two types of study :(1) examination of the effects of UIC promoting policies and (2) survey analysis to observe the UIC activities in detail. Reference [8] examines the role of new technology-based firms in UIC activities in Japan, concluding that smaller firms achieve higher productivity through UIC than large firms. In light of the findings of this paper, it appears that UIC are gaining momentum and are likely to play a strong role in reducing the dependence of Japan's system of innovation on in-house R&D conducted within large corporations. Reference [9] examines the changes in the nature and the outcome of UIC following the enactment of UIC policies in the late 1990s. Their analysis indicates that these policies increased the number of UIC patents in the late 1990s, while the quality of UIC patents does not fall over this period. It also indicates that Japanese UIC policies in the late 1990s provided favorable results in general. However, strong IP policies pursued by universities may reduce the incentive for firms to commercialize inventions resulting from UIC collaborations, so that patent policies at university should be re-examined. Reference [11] analyzes "Industrial Cluster Projects" carried out in the early 2000s to promote UIC in local area in Japan. The study reveals that cluster participants apply for more patents than others without reducing patent quality when they collaborate with national universities in the same cluster region.

Reference [10] analyzes the UIC activities based on the survey to inventors both in academia and industry in Japan. They point out that university researchers tend to participate in collaboration with industry in order to conduct practical applications of their research results. In addition, they recognize the importance of firms' research funds as valuable source of fund in light of tightening universities' own research funding. On the other hand, researchers in industry tend to seek a network formation with university as well as practical solutions in their research agenda.

3 Research Question and Methodologies

Why are licensing activities by universities in Japan so low compared to U.S.? It is misleading to compare situations in two countries directly without considering the backgrounds of each country. Private universities with ample funds are main actors in UIC in the U.S. while in Japan national universities which used to be branches of Ministry of Education play dominant roles. Some of the differences of license activities between U.S. and Japan may be attributable to this different situation. For instance, it is possible that technology transfer officers at private universities tend to be more active in licensing than the counterparts in national universities in Japan. Due to limitation of research method, we focused on managerial aspects of license activities in Japan to make clear the problems they face.

In this study semi-structured interviews were conducted with 9 technology transfer offices at universities, 2 technology transfer offices at national research institutes as well as one TLO (Table. 3). Some of the interviewed universities established TLOs as independent entities outside universities (e.g. The University of Tokyo), other universities have TLOs as internal department (e.g. Tokyo Institute of Technology). The interviewees included top national universities as well as leading research institutes in Japan. The interview covered issues including: the overview of the UIC activity, licensing policy, the barriers to expand licensing, and problems in the process.

Table 3. A list of interviewees

The University of Tokyo
Kyoto University
Tohoku University
Osaka University
Hokkaido University
Nagoya University
Tokyo Institute of Technology
Nagoya Institute of Technology
Nihon University
National Institute of Materials Science
National Institute of Advanced Industrial Science and Technology
Kansai TLO

4 Results

Interviewees' comments which hint the factors behind low licensing activities by universities in Japan can be summarized as follows: (1) co-ownership of patents with companies, (2) attitude of industry toward university, and (3) human and financial resource constraints.

4.1 Co-Ownership of Patents with Companies

It is common that a university and a company apply jointly and co-own patents arising from the results of collaborative research between them. Most interviewees emphasized co-ownership of patents with companies is a constraint for them to license patents. Statistical data shows that about 70% of university-owned patents are co-owned by private companies ([13]). The article 73 of Patent Law articulates that the assignee of a patent shall obtain the consent of the other party when the assignee licenses the co-owned patent to a third party. Private companies are in general reluctant to license their co-owned patents to other companies from the view point of operational strategies. Thus, potential licensees of the co-owned patents are limited to the co-owners themselves or affiliated companies of the co-owners. Because co-owners are legally entitled to exercise the patent rights without licensing fees, it is quite difficult for universities to receive license fees from co-owners.

There is a new movement regarding license activities. Some TLOs are now making effort to increase sole applications of patents despite the results from the collaborative research with private companies. The objective of this effort is to own basic patents covering multi-purpose technology and to license them non-exclusively to third parties.

4.2 Attitude of Industry Toward Universities

Most of interviewees emphasize that attitude of industry to UIC is one of negative factors. As pointed out by reference [2] companies in Japan have been used close ties with universities as channels for recruiting than as sources of new technology. Therefore, it is presumable that companies do not necessarily pay strict attention to the results and patents arising from their collaborative research with universities. The fact that average size of collaborative research fund between a university and a private company is about 2 Million JPY ([7]) shows limited resource is provided to university from industry, resulting in low quality of patents.

In addition, it is widely acknowledged that companies tend to do collaborative research with universities in the near field where they have strong patents, aiming at defensive patent applications. Although no specific data are available about defensive patents co-owned by universities and industry, some interviewees have the impression that defensive patents account for considerable portion of their patents co-owned with industry. The more defensive patents a university co-own with companies, the lower the average license fees it receive from license because these defensive patents basically do not generate license incomes.

Another reason also relates to the attitude of companies but more fundamental R&D strategies. As reference [8] points out, Japanese companies tend to rely on "in-house R&D" and that they are reluctant to utilize external resources of R&D.

In practice, license income received by universities in Japan heavily depends on large companies and this situation may lead to unfavorable licensing contracts for universities. According to UNITT survey, in Japan 62.5% of licensees of intellectual properties of universities are large companies, 37.3% are SMEs, and 0.3% are newly-established companies. In U.S. the proportion of large companies, SMEs, and newly-established companies are 35.4%, 47.4% and 17.2%, respectively ([17]).

In addition, the proportion of running royalties to the total license fees is around 40% in Japan while in U.S. the proportion is around 70%. In Japan around 60 % of licenses fees come from onetime payment when a contract is signed ([17]). This means that majority of patents licensed from universities in Japan are not yet utilized in the commercialization phase.

4.3 Human and Financial Constraints

Almost all of interviewees stressed that they are suffering from human and financial constraints. As for human constraint, shortage of staff members is a serious issue. This situation has close relation with financial constraints. As general funds provided from government have been decreasing continuously from the incorporation of national universities in 2004, national universities are facing serious budgetary constraints. Therefore, the profitability of Intellectual Property Management Division is strictly examined.

Severe financial constraints lead to avoidance of maintenance cost of patents. It is common for universities interviewed to transfer patents arising from collaborative research to partner companies in order to avoid maintenance cost associated with patents.

5 Conclusions and Implications

This study examined the low licensing activities by universities in Japan, by interviews with universities and national institutes.

Results from the interviews point out there are major factors behind the low licensing activities by universities in Japan: (1) co-ownership of patents with companies, (2) attitude of industry toward university, and (3) human and financial resource constraints.

Most interviewees emphasized co-ownership of patents with companies is one of the constraints for them to license patents to third parties. Patent Law requires the assignee to obtain the consent of the other party when licensing co-owned patents to a third party. In practice, it is quite difficult to license their co-owned patents with companies because companies are reluctant to license their co-owned patents to third parties.

Attitude of industry to universities is another issue to hinder licensing from universities to industry. It is presumable from the historical perspective that companies have not necessarily paid strict attention so far to the results and patents arising from their collaborative research with universities. In addition, industry tends to heavily rely on their "In-house" R&D.

TLOs are in general suffering from human resource as well as financial constraints. With decreasing general funds from government, universities are at the crossroads regarding in UIC, in particular, licensing activities.

Funding to universities by government is increasingly tightening in face to rapidly aging society in Japan. Universities should take necessary measures to expand UIC in Japan as possible funding sources.

References

1. Bozeman, B.: Technology transfer and public policy: a review of research and theory. Research Policy 29, 627–655 (2000)
2. Collins, S., Wakoh, H.: Universities and Technology Transfer in Japan: Recent Reforms in Historical Perspective. Journal of Technology Transfer 25, 213–222 (2000)
3. Fujisue, K.: Promotion of academia-industry cooperation in Japan—establishing the "law of promoting technology transfer from university to industry" in Japan. Technovation 18(6/7), 371–381 (1998)
4. Kneller, R.: The beginning of university entrepreneurship in Japan: TLOs and bioventures lead the way. Journal of Technology Transfer 32, 435–456 (2007)
5. Harayama, Y.: Industry-University Linkage in Japan, A paper presented at International Conference on "New Trends and Challenges of Science and Technological Innovation in a Critical Era" Taipei, R.O.C. (October 31, 2003)
6. Kodama, F., Branscomb, L.: University Research as an Engine for Growth: How realistic Is the Vision. In: Branscomb, L., Kodama, F., Florida, R. (eds.) Industrializing Knowledge. The MIT Press (1999)
7. MEXT, Survey of University Industry Collaboration (2014) (in Japanese)
8. Motohashi, K., Muramatsu, S.: Examining the university industry collaboration policy in Japan: Patent analysis. Technology in Society 34(2012), 149–162 (2012)
9. Motohashi, K.: University–industry collaborations in Japan: The role of new technology-based firms in transforming the National Innovation System. Research Policy 34(2005), 583–594 (2005)
10. Nagaoka, S., Hosono, M., Akaike, S., Nishimura, J.: Knowledge Creation and Innovation by University-Industry Collaboration: Basic findings from the large-scale survey of academic and corporate researchers (in Japanese), National Institute of Science and Technology Policy (NISTEP), Ministry of Education, Culture, Sports, Science and Technology (MEXT) and Institute of Innovation Research (IIR), Hitotsubashi University (2013)
11. Nishimura, J., Okamuro, H.: R&D productivity and the organization of cluster policy: an empirical evaluation of the Industrial Cluster Project in Japan. J. Technology Transfer 36, 117–144 (2011)
12. Takenaka, T.: Technology licensing and university research in Japan. International Journal of Intellectual Property 1, 27–36 (2005)
13. UNITT, Survey of technology transfer from universities (2014) (in Japanese)

Analysing Industry Clustering to Develop Competitive Advantage for Wualai Silver Handicraft

Khanita Tumphasuwan*

Department of College of Arts Media and Technology,
Univeristy Chiang Mai University, Huaykeaw Rd,
239, Muang, Chiang Mai, 50200
khanita.t@gmail.com

Abstract. The Wualai community of Northern Thailand represents important intellectual and social capital and their silver handicraft products are desirable tourist souvenirs within Chiang Mai Province. This community has been in danger of losing this social and intellectual capital due to apply an improper tool, the Scottish Enterprise model of clustering. This research aims to analyse and increase its competitive advantages for preventing the loss of social and intellectual capital. To improve the Wualai's competitive advantage, analysis is undertaken using a Porterian cluster approach, including the diamond model, five forces model and cluster mapping. Research results suggest that utilizing the community's Buddhist beliefs can foster collaboration between community members and is the only way to improve cluster effectiveness, increase competitive advantage, and in turn conserve the Wualai community.

1 Introduction

Wualai Community is the small community located in Chiang Mai Province, the Nothern Part of Thailand. It has been the community that has produced silver handcraft for more than 700 years. The ancester of Wualai Community migrated from Hongsawadee, Mynmar and settled down the South Outer Gate of Chiang Mai. The production of silver handcraft has been passed on nowadays.

In Wualai Communty, there have been the valued-add process by producing many types of products which are ornaments, i.e, bracelet, ring, bangel, hairpin, belt, bag and utensils, i.e, cup, tray, frame, cardholder, decorated aluminium picture, vase, silver flower, lamp.

In the past, Wualai Silverware has been well-known by tourists in terms of product itself and the process of production. Every home in Wualai Community had its own factory to produce silverware by using silver stove to mold the product and carving to beatify the product. The clanking was the evidence of this production process and it intrested tourists to visit its community and was considered to be the major income of the community.

* Corresponding author.

M. Gen et al. (eds.), *Industrial Engineering, Management Science and Applications 2015*,
Lecture Notes in Electrical Engineering 349, DOI: 10.1007/978-3-662-47200-2_21

In 1996, Thailand faced with economic crises resulting in creating the new tourist attraction in Bosang-Sunkumpang. The craftmen from Wualai Community was drawn to this Bosang Community. Wualai Community had been stagnant due to the few number of craftsmen. Some had died and the new generations show no interest in learing the production of silverware, nowadays, there are only 13 silverware shops and from this 13 shops there are only 4 shops that has their own factories. It is predictable that in the next 3 years, Wualai Community will become the history unless the new generation continues the production of silverware. Chiang Mai will lose one of the major source of income to attract tourist. This has led govermental office and private company including educational institution to step in to support Wualai Community in various acitivies. For example, budget provision, knowledge creation to preserve the community, the development of product to differentiate the products, provision of fund source, traning of new craftsmen to preserve the arts of production.

2 Methodology

The foundation of Wualai Cluster has followed the Scottish Enterprise Cluster in Scotland which comprises of mobilization, diagnosis, collaborations and implementation. The approach to develop cluster by Michael E. Porter was utilized, i.e., the Five Forces Model to find the 5 effects to industry which led to pricing strategy, profit, returns on invesment. The Diamond Model is used to find the level of competitive advantage including Cluster Mapping to find connections in silver cluster and other stakeholders. The details mentioned will be discussed as follows:

2.1 Cluster Analysis

Considerting Wualai competitive advantage by using Scottish Enterprise Cluster Model, the study can be concluded that

Mobilization
Mobilization of stakeholders in silver industry from all supply chain activities. The party involved are person or shops directly related to silverwre business, and govermental office. There are 19 members which shows that the supply chain of cluster is relatively small. The president was chosen and so was the vice president. The working committee consists of the business development committee, the production committee, the financial committee and the secretary. In the mobilization process, members was unlikely to cooperate in joining group activities which impeded the unity of cluster. This may be caused by lack of trust among members in cluster, no evidence of success of cluster, accompanied by the conflict among members themselves before this cluster settling up. Some member did not joined the meeting as planned, denied of and the difference of the meeting concensus.

Diagnosis
Analysis of the data and status of silver industry found that the market is highly competitive as there are other silver producers widespread in Chiang Mai Province

especially in the famous tourist attraction like Night Bazaar, Bosang-Sunkampang District, Tapae Walking Street, and Doi Suthep. In addition, there are competitors from other parts of Thailand and in the foreign countries; for instance, silver product from Lumphun, Surin, Konkhaen, France, China, India. By using the pro and con, it is found that Wualai Silverware is the household industry. The production of each piece of the product requires time and craftsmanship to produce the unique and beautiful product. The material used is costly and fluctuates with the price of gold, thus the craftsmen produce the product only to get by, unable to expand the production. The design and quality are required to improved; simultaneously, the uniqueness and the standard of production must be preserved. Though there is no trademark of Wualai product, the history of each product can be created to valued-add itself. The strategic plan has been devleoped by brainstorming ideas from cluster members and other stakeholders. Projects were proposed in order of importance and urgency.

Collaboration
Collaboration is vital in the creation of cluster. Wualai cluster set up "The Production of Original Products" which invited all members to join, exchange of idea, work together, hoping to build trust to one another. Meeting was set up to follow up on the progression, division of responsibility, for example, designing, production, selling, and income allocation. In this process, it is found that collaboration level is low, as members did not willingly share information, shows no intentions to work together as a team and the evidence of conflict stood out during this process. Personal benefits was focused rather than that of common benefits. Copy of design and price war were observed.

Implementation
There are three steps to optimize the development of cluster. Members' products was shown in the exhibition on behalf of cluster. Cluster branding has been created to build trust in Wualai product. The expansion plan was developed to ensure the equality of distribution of income among members. The working committee are to motivate members continuously. In this setp, it is found that only few cooperation has been found and they were in the same group before joining cluster. The excution of each activity requried strong request and foudn that members were unwillingly cooperate, resulting in the inefficiency of clustering.

2.2 5 Forces Model

When considering the cluster of Wualai Silver Handcraft using the analysis tool like Diamond and 5-force Models, it can be concluded that

1. Obstacle against industry entering: Wualai Silver industry is the household industry. The industry entering is difficult if there is no accumulated knowledge and wisdom which is the Intellectual capital from the ancestors. Although there is a support from the government, there is a few numbers of new entrepreneurs to enter the industry. This is because the high costs of raw materials and labor, especially when the new entrepreneurs have not enough investment money. There is also a tendency in the future that the existing entrepreneurs will terminate Their businesses if there is no successor. In

addition, Wualai has no signature goods and thus there is not sufficient competitive power.

2. Driving force from the manufacturers or competitors in the industry: Wualai is the important source of silver handcraft manufacturers in Chiang Mai. It focuses on manufacturing and retails in their community. Each product in Wualai has similar form and same amount of silver compounds. It thus makes the sellers employ the price strategy in attracting customers and gaining market share. In other communities like Mae Yoi, or Sang, sun and Patong, there is silver manufacturing as well. However, there are not so many manufacturers and the manufacturing are not complete as in Wualai. Although there are competitors both in the country and abroad, it does not however affect the silver industry of Wualai. This is because the Wualai markets are directly at their shops, regional, national, and international goods exhibitions. The customers that buy Wualai silver handcraft is satisfied with the forms and quality of the silver craft.

They are proud of the products that have identify and directly selected from the manufacturing source. The other competitors in the industry therefore less affect the Wualai silver craft than the competition in the community, which employs the price strategy.

3. Bargaining power of sellers: The problem of raw material results in the high prices of silver craft because the manufacturers will set the price based on the silver granules used, which is calculated in gram. The ordering of raw material will be done to the shops of gold sellers, in a small amount of 30-50 grams per day, and to the manufacturing orders only. If there is no order, there will not be the raw material ordering because there are already available products in the shops. The cost of silver raw material is high and fluctuating daily with reference to gold price. There is a way of cost reduction by using silver and aluminum-coated brass in manufacturing accessories and wares.
4. Bargaining power from buyers: Since the products are of craft type, beautiful appearance, worth, and high-valued collection, the buyers purchase them in a small number and therefore make the bargaining power from buyers low.
5. Driving force from substitutes: There are substitutes of souvenirs and wares from other materials, ranging from low-price goods to gold and alloy of gold and copper which are higher prices for future sale than silver. This is a factor that influences the decision of customers.

2.3 Diamond Model

When the diamond model is used as an analysis tool of competitiveness of Wualai silver craft cluster, the following results are obtained from analyzing 4 conditions that affect the cluster productivity.

1. Factor Conditions find that the Wualai silver craft demands high investment because the price of silver granules as the raw materials is high and varied according to the price of gold. The labor needs to have high intellectual capital and skill. All of these make the products high prices.

2. Demand Conditions reveal that the demanded form change from the past that the silver craft was used in daily life as the accessory showing social status of Thai. However, there is less demand in using silver wares is smaller, the maintenance is difficult, and the prices of products become higher but no value when further selling in the present. This results in the shrinkage of silver ware market. The forms of the products are thus directed towards the accessories, souvenirs, and ordered works. These products are worth for both givers and receivers, thus maintaining the market shares both in the country and abroad.
3. Related and Supporting Industries show that the cluster of Wualai silver craft small but highly competitive among members. The recognition and trust among the members has not yet existed. The demand of benefit obtained from collaborating among the members makes the driving of the cluster slow. The cluster still lacks of other environments such as financial institution, suppliers of raw materials, etc. in seriously supporting it. The effective operation is thus affected by several sensitive factors like lacking of member collaboration, bargaining power in buying raw materials, price setting due to substitutes.
4. Firm Strategy, Structure, and Revelry reveal that the Wualai silver craft is beautiful, fine, and of high quality. There is still a lack of brand as a signature showing the identity of Wualai silver craft that can be found here only in the world.

2.4 Cluster Mapping

Cluster mapping is the analysis of the factors of the Wualai silver craft cluster. It is used for evaluating relation level, role, responsibility, and relation that links with the cluster. The number of organizations is enough for supporting the cluster towards its success. It is found from the cluster mapping that the cluster is comprised of local wisdom, villagers, high skill workers, shops, and government organizations, academic institutes, and churches in the community. These factors are important in driving the formation of cluster, but not enough in supporting the cluster towards its success. It is necessary to be driven by members that profoundly understand the principle of cluster development and fully and sincerely collaborate among the others. This is because the cluster development will result in the the systematic development of products, tourism, and quality of life in community.

It requires time in incubating which cannot be seen in a very short time. The Wualai cluster is a small cluster, has a small number of members when comparing to other clusters thus having less power of bargaining, and lack of links with raw material suppliers. It is thus not completely capable of controlling the product price and lacks of links with financial sources for business development. The nature falls into household business and small shops. However, this is a charm of Wualai community as a niche market.

3 Discussion

From the study above using different analysis tools for the Wualai silver craft cluster, it can be concluded that

1. The Wualai silver craft industry is a small and household industry. The manufacturing of silver craft is made to order at a small scale because it is craft and thus consumes time in manufacturing. The raw material is of high cost. The small manufacturing leads to less bargaining power. The new generation does not succeed the wisdom of their ancestors. The development of silver craft and community should therefore be in the manner of creating values for products and exploiting the history in adding the product values than the development as general industry.
2. The establishment of cluster for creating competitiveness, reviving culture, and developing business for the Wualai silver craft industry is the guideline of the government in assisting industries of the country in the same way. Such a directive towards boosting sale rate and incomes of the shops for the purpose of maintaining the Wualai as a sustainable silver craft community according to the strategy of industry development may not be an appropriate direction for the silver craft industry of household business that is of small-scale production an lack ability of achieving the real objective of cluster. The objective that requires all parts in the supply chain to support among another for the benefit of business operation and increase of competitiveness.
3. The Wualai cluster still lacks of strong leaders, dedication, unity, as well as strong motivation if driving the cluster that can be positively developed. It is the cluster which is under the pressure from different opinions of the members. The introduction of cluster to Wualai cannot solve the problem and real desires of the community and results in bad responses and supports from the members. These are the important factor for the cluster success.
4. The entrepreneurs in Wualai expect the ready-to-go support from the government and need the support in terms of finance more than seeing the real benefits of forming cluster. This is due to the household business nature that lacks of good management system and therefore cannot see the clustering as a way of increasing competitiveness.
5. The crucial and urgent problem is that the new generation lacks of the recognition of the silver craft manufacturing as a cultural heritage. They cannot recognize the importance of the conservation and extension. Yet, they want to enter other careers. The culture of silver craft manufacturing at Wualai has the tendency of decrease and may disappear in 30 years.
6. The observation from the study is that the people in the community have a separating and conflicting opinions. They however strongly believe in Buddhism. The assessment to the people sometimes need to use the church which is the morale center of them in creating trust and motivation for the clustering.

4 Suggestions and Remedies

1. Use the church, the morale center of the people, as an archetype for the study of the church management form that can transfer the culture from a generation to another. This can be carried out in form of Time Capsule since the past to the present.
2. Use the church as an archetype because the church has a social capital which is comparable with the intellectual capital of the silver craft manufacturing of Wualai community.
3. Use "Wat Srisupan" study case which is the church supporting activities of Wualai community. It is the center of ten types of skill labors, and the source of selling, distributing, conserving, and transferring the culture of silver craft manufacturing of Wualai community.

References

1. Bianchi, S.: The Wall Street Journal (November 9, 2009), retrieved January 9, 2010 from http://wsj.com, http://online.wsj.com/article/SB125723890596024879.html
2. Bresnahan, A., Gambardella, A.: Building high-tech clusters: Silicon Valley and beyond. University of Cambridge, Cambridge (2004)
3. Cook, G.A., Pandit, N.R., Beaverstock, J.V., Taylor, P.L., Pain, K.: The cluster of financial services in London. Liverpool Managment School, Liverpool (2004)
4. Feser, E.J.: Old and new theories of industry cluster. In: Steiner, M.: Clusters and regional specialistion: On geography, technology, and networks, pp. 18–40. Pion, London
5. Martin, R., Subley, P.: Deconstructing clusters: Chaotic or policy panacea? Journal of Economic Geography, 1-52 (2002)
6. Raines, P.: Regional and Industrial Policy Research Paper. Local or National Competitive Advantage? The Tensions in Cluster Development Policy, 1-31 (2001)
7. Porter, M.E.: Location, competition, and economic development: Local cluster in global economy. Economic Development Quarterly 14, 14–34 (2000)
8. Porter, M.E.: On competition. Harvard Business School Press, Boston (1998)
9. Porter, M. E.: The competitive advantage of nations. Free Press, New York (1990)
10. Porter, M. E., & Ketels, C.: Hong Kong competiveness and its financial services cluster. Harvard Business School, Boston (2008)
11. Rosenfeld, S.A.: Bringing business clusters into the mainstream of economic development. European Planning Studies 5(1), 3–23 (1997)
12. Simmie, J., Wood, P., Sennett, J.: Iinnovative clusters and competitive cities in the UK and Europe, pp. 76–89. Brookes School of Planning, Oxford

University-Industry Linkages (UILs) and Research Collaborations: Case of Thailand's National Research Universities (NRUs)

Naparat Siripitakchai and Kumiko Miyazaki

Graduate School of Innovation Management, Tokyo Institute of Technology, Tokyo, Japan
{siripitakchai.n.aa,miyazaki.k.ae}@m.titech.ac.jp

Abstract. The Thai government have launched the National Research University (NRU) project since 2009 to boost the country's research activities and the University-Industry Linkages (UILs). The objective of this study is to investigate the joint research activities between research intensive universities and domestic firms, in the case of Thailand-NRUs. We used co-patents and co-publications as indicators representing the successful joint research outputs. The two sets of data were collected from DIP (Thailand's patent database) and ISI database. The result shows that a large portion of firms' co-patenting and co-publishing with the NRUs were from low-tech and medium-low tech sectors including food and agriculture. The NRU project seems to gain success in creating and sustaining the collaboration in low-tech and medium-low tech sectors. However, it has strengthened only the existing collaborations in the case of high-tech sector and did not create new collaboration as much as the policy makers pursued. Another concern is that a number of co-publications from high-tech sectors were from few large incumbents not from SMEs.

Keywords: Research University, University-Industry-Linkage (UIL), Co-publication, Joint research, Thailand.

1 Introduction

University-Industry Linkage (UIL) is becoming an increasingly important component in National Innovation System (NIS) for driving a country's long-term competitiveness [1]–[5]. Several types of UILs ranging from informal to formal interactions having different objectives, scopes, and institutional arrangements exist. Among different modes of interactions, joint research channel requires the highest intensity of interaction, in which both agents provide knowledge resources, dealing with long-term targets of knowledge creation. Joint research is considered to be a formal type of R&D cooperation with sophisticated bi-directional knowledge transfer channels between the actors.

Following the global trend, the Thai government have encouraged firms to engage in setting up joint research programs with universities by policies including the provision of R&D research grants, matching grants, and tax-incentives. At the same time,

M. Gen et al. (eds.), *Industrial Engineering, Management Science and Applications 2015*,
Lecture Notes in Electrical Engineering 349, DOI: 10.1007/978-3-662-47200-2_22

the National Research University (NRU) project has been launched in 2009 to leverage the country's R&D activities and to strengthen the UIL. The researchers in the research universities should actively undertake the third mission "community engagement of economic development" [6]. However, there is a lack of longitudinal study that identifies how the Thailand-NRUs have engaged in joint research activities with domestic firms. Therefore, the objective of this study is to explore the pattern of joint research collaborations to analyze the existing UILs.

In this study, we employed co-patents and co-publications as indicators representing the successful joint research output between the Thailand-NRUs and domestic firms. The empirical evidence of the existing pattern of the collaborations should provide an insight on the cooperative research areas of focus and the existing patterns of collaborations for policy makers, university administrators and management of private firms to adjust and redesign their cooperative strategies together with enhancing the research networks.

In the next section, the background related to Thai UILs and NRUs is illustrated. Section 3 explains the different aspects of two dissemination tools: patents and scientific publications. Section 4 illustrates the methodology and data according to the analysis of co-publications. The findings section is shown in section 5. The last section is conclusion and implications.

2 Thai Universities Industry Linkages (UILs) and National Research Universities (NRUs)

Studies at the national level indicated that Thailand failed to upgrade its UIL within its NIS. The evidence of weakness in Thai UIL was due to low levels of innovation resulting in few private sectors collaborating with universities. The state of Thai UILs is a reflection of weak downstream demand by indigenous Thai firms, limited commercial focus and capacity within the university system and bureaucratic fragmentation. Thailand is less successful in technological catching-up because there was a mismatch of NIS with its economic structural development level [7], [8].

Realizing these issues, the Thai government has initiated several policy measures, including the establishment of the NRU project, organized by the Office of the Higher Education Commission (OHEC) announced in 2009. This plan aimed at developing academic excellences to enhance the country's research activities and to promote better UILs for national competitiveness. The nine selected NRUs are acknowledged by their remarkably good reputation and outstanding research achievements. The nine universities include Chiang Mai University (CMU), Chulalongkorn University (CU), Kasetsart University (KU), Khon Kaen University (KKU), King Mongkut's University of Technology Thonburi (KMUTT), Mahidol University (MU), Suranaree University of Technology (SUT), Prince of Songkla University (PSU) and Thammasat University (TU). The OHEC's selection criteria were mainly based on the ranking system conducted by Times Higher Education-Quacquarelli Symonds (THE-QS) and the impact factor of publications published on Scopus database. OHEC has granted extra financial support within three consecutive years (2010-2012) to encourage the NRUs to dedicate themselves towards the intended objectives [9].

Strengthening domestic interactions and knowledge transfer processes between the key actors in the NIS are important factor for the country's endogenous development. However, the limitations on resources of universities, companies and public research institutes including funding and staffs, coupled with simultaneous increases in social needs, have forced universities, private sectors and government to pay more attention to the efficiency and effectiveness of their resource allocations for project eligibility. Therefore, the study on existing pattern of NRUs' cooperative research areas should provide better understanding for policy makers, university administrators and management of private firms to adjust and redesign their cooperative strategies based on their limited resources. It can explore how specialized the NRUs are in the joint research programs. In this study, we employed co-patents and co-publications as indicators of R&D collaborations. The different types of knowledge dissemination tools: patenting and publishing activities, are explained in the next section.

3 Patenting and Publishing

The more involvement of the universities in socio-economic development, the greater emphasis on exploiting research results in line with: an increase in patent and licensing activities; the spin-off activities; and managerial and attitudinal changes among academics related to collaborative projects with industry. Moreover, there are major concerns related to dissemination of research results, derived from the fundamentally different reward and incentive systems of academic and private sector research. These include (1) the relationship between disclosure versus secrecy; and (2) the complementarities and substitution effects between public and private R&D expenditures [10].

The main tools for dissemination of technical and scientific knowledge are patent applications and scientific publications. Scientific publication is one of the most common and rapid instruments for knowledge dissemination. The protection granted by the IP system to an article or publication is copyright, which arises automatically when the researcher writes it. However, copyright only protects expression of the words and its originality, but not the ideas underlying the research findings. It does not ensure the protection of the technical information of the invention [11].

Patents grant inventors the exclusive right to make, use or sell their inventions for a limited time period. The patent application which contains the disclosure is normally published after eighteen month from the filing date. Filing patent can be based on two strategies: commercial exploitation or knowledge dissemination. In the first case, researcher may consider applying for a patent in order to commercially exploit the invention and have financial gain in return. On the other hand, when organizations are not interested in acquiring a monopoly, they can use patenting process to publish their technology to obstruct the patenting ambitions of competitors later. It means that researchers can file a patent application and later abandon it to save costs. The application will be published and appear in patent database forever. Since the findings are in the public domain and can be accessed by everyone, it will be useful in attracting partners who use the patent database as part of a commercial intelligence search [11].

Actually, both patents and scientific publications are not very different in terms of their intellectual challenge and nature, for example, creativity, originality and novelty. These dissemination tools serve different purposes and can also be used simultaneously. For example, firms often patent and then immediately publish on the same materials in order to preclude other firms from patenting in the same area. In addition, the practice of combining patenting and publishing is more and more used in the research organizations. The researchers can take advantage of prospective patent profits and be recognized at academic level for the quality of their papers [10], [12].

In this study, co-patents (Co-pat) and co-publications (Co-pub) by multiple affiliations are used as indicators of research collaboration. Co-pat and Co-pub data represents successful joint research projects and indicates diffusion of knowledge and skills. The university-industry co-patent or co-publication (UIC-Pat or UIC-Pub) is a patent application or a scientific publication with the applicants/authors' addresses that include affiliations referring to at least one university and one private sector organization. It provides a unique source of information for developing aggregate-level proxy measures of the magnitude and intensity of university-industry research collaboration. Using UIC-Pat and UIC-Pub data as an indicator has an advantage of wide availability and international comparability [13], [14].

UIC-Pat and UIC-Pub should reveal the mutual agreements between universities and private firms as well as their strategies and purposes regarding the disclosure of their research results. However, UIC-Pat and UIC-Pub data should be handled with care because there are many cases when collaboration will not result in co-patenting or co-publishing [13], [14]. In the next section, the methodology and data for exploring the patterns of UIC-Pat and UIC-Pub to analyze the Thai UILs is explained.

4 Methodology and Data

Both patent applications and scientific publications of the nine Thailand-NRUs were extracted using the same search strategy. The university name was used as a search criteria via "Advanced Search" of two databases. Patent applications filed in the DIP[1] database during 2004-20013 were retrieved in December 2014. Scientific publications from major industrially relevant scientific fields published during 2004-2013 was collected from ISI Web of Science (WoS) database in the first week of August 2014. The extracted data from each database was divided into two subsets based on the publication year in order to represent the cooperative R&D achievements in 2004-2008 and 2009-2013, before and after the announcement of the NRU project in 2009.

In the next step, we extracted the UIC-Pat and UIC-Pub that are defined as the patent applications and scientific publications jointly co-authored by university researchers and employees of private firms. First, the patents and publications with the addresses of organizations that contained "company excluding terms", for example,

[1] The Department of Intellectual Property (DIP) is a search system provided by the Thailand Intellectual Property Information Center. All text in the website is in Thai.

addresses including the terms “Univ”, “Inst”, “Hosp”, etc., were excluded. Then, the left addresses were manually assessed. The private firms, co-patenting/co-authoring within these UIC-Pat and UIC-Pub were divided into two broad geographical classes: domestic or foreign. In this study, we focused on only the case of domestic firms. The domestic firms included Thai-owned firms, subsidiary companies of Thai-owned firms, and branches and/or subsidiary companies of foreign firms located in Thailand. A domestic co-pat/co-pub is defined as a patent-application/publication that are co-patented/co-published by the NRUs and firms located within the same country (in this case, Thailand).

The aggregated data according to the NRUs is presented in Table 1. The left column illustrates the detail of patent applications from DIP database while the right column shows publication information. The characteristics of domestic firms co-patenting or co-publishing with the NRUs is shown in Table 2. The firms were classified according to their industries and the industries' technology levels. A number of firms and a number of UIC-Pat/UIC-Pub were aggregated and shown in two periods. Firms that have continuously engaged in the R&D collaborations in both periods were extracted and shown in the separated sub column. In the next section, the findings are explained.

5 Findings

During 2004-2013, the nine NRUs held about 67% of Thailand total scientific publications from the ISI database, while they held only 4% of Thailand total patent applications from the DIP database. The average UIC-Pub of the nine NRUs is about 1.5% of the NRUs' total publications during 2004-2008 and 2.2% during 2009-2013, respectively. After the announcement of NRU project together with several concurrent S&T policies, the average percentage of domestic UIC-Pub of the nine NRUs increased from 28.3% of the NRUs' total UIC-Pub during 2004-2008 to 44.4% during 2009-2013. There are a variety of firms co-authoring in scientific publications including Thai SMEs, Thai incumbents and subsidiaries of foreign enterprises whose parent companies are in developed countries while most of the patents with corporate co-applicants were from Thai SMEs, Thai incumbents and subsidiaries of Japanese companies.

According to Table 1, the number of patent applications and publications increased drastically during the two observed periods. The NRUs have been following the global trend in commercialization activities based on an increase in number of patent applications. However, co-patenting activity was a rare case and it is not a common selected dissemination tool for joint research output between the NRUs and domestic firms. Only CU have actively involved in the co-patenting activities with the firms. Although the co-publishing activities were a more common channel for knowledge dissemination, the average number of UIC-Pub were lower than world average [2]

[2] According to the world's tops 500 universities during 2008-2011 from UIRC2013, average UIC-Pub per university in natural science and engineering technology was 6.3% with the 46% of domestic UIC-Pub.

The collaborating firms may concern about the high degree of secrecy over disclosure and don't want to disseminate their research results.

In the case of domestic UIC-Pub, CU, KKU, KMUTT, KU and MU have actively involved in these activities rather than the rest of NRUs. Their domestic UIC-Pub were increasing significantly in the second period. CU, KMUTT and MU were a few Thai universities that have been ranked in the international university ranking systems like Times Higher Education, QS and CWTS Leiden ranking systems. It can be implied that the leading Thai universities with high research performance seems to be more attractive to domestic firms in collaborating in joint research projects.

Considering the characteristic of firms that co-patent or co-publish with the NRUs, the types of domestic firms were classified by industry and R&D level as shown in Table 2. We found that a large portion of firms that co-patent and co-publish with the NRUs were from low-tech and medium-low tech sectors during the two observed periods. We also observed that UIC-Pat in high-tech sectors and UIC-Pub in medium-high tech and high-tech sectors were significantly increasing in number during 2009-2013. Unlike the developed countries and the majority of world's top universities, whose contents of UIC-Pub were relevant to pharmaceutical and biomedical; electrical engineering and telecommunication sub-fields [13], [15], the Thai NRUs tend to have UIC-Pub with contents related to food and agriculture, and advanced materials.

The highest number of co-patenting and co-publishing firms was from Food and agricultural industry including rice, shrimp, rubber, swine, etc. The average UIC-Pub per firm in low-tech and medium-low tech sectors was lower than that in medium-high tech and high tech sectors. It indicates that only a few firms in medium-high and high tech sectors have actively co-published. In the case of energy and electronics industries, the firms co-patenting and co-publishing are from a few large enterprises which have continuously collaborated with the NRUs from 2004 to 2013. In case of electronics industry, the only firm that has continuously engaged in co-publishing activities is Seagate Technology (Thailand), a subsidiary of U.S. multinational data-storage company. While the firm that has continuously engaged in co-patenting is Mektec Manufacturing Corporation (Thailand), a subsidiary of Japanese consumer electronics company. The energy firm consistently engaging in both activities was PTT group, a Thai state-owned enterprise. Its publications are related to green energy, petrochemicals and advanced materials.

NRU project announced in 2009 seems to have succeeded in creating and sustaining the collaboration between the NRUs and firms from low-tech and medium-low tech sectors, especially, food and agricultural industry. This was in accordance with the competitive advantage of Thailand based on Food and agricultural industry which are its major economic sectors. The industry and the NRUs have a long-term, reciprocal relationships under the same context of the country's economic development. In the case of high tech sector, the NRU project only enhanced the collaboration between the firms which already collaborated with NRUs, however, it did not create new collaboration as much as the policy makers expected. Moreover, a number of UIC-Pub from high-tech sectors were from few large incumbents not from SMEs. Interestingly, automotive and transportation industry have been a strategic sub-cluster of Thailand since 2006, however, we observed that they are not collaborating to co-patent and co-publish with the NRUs.

Table 1. Patents and academic publications from industrially relevant science fields of the Thailand's National Rearch Universities during 2004-2013

NRU	DIP Patent applications						WoS Publications					
	Years: 2004-2008			Years: 2009-2013			Years: 2004-2008			Years: 2009-2013		
	Total Patent Applications	Number of Domestic UIC- Pat	Number of domestic co -patenting firms	Total Patent Applications	Number of domestic UIC- Pat	Number of Domestic co-patenting firms	Total Publications	Number of Domestic UIC- Pub	Number of domestic co -publishing firms	Total Publications	Number of Domestic UIC- Pub	Number of domestic co -publishing firms
CMU	49	1	1	101	2	2	1,753	4	5	2,914	10	6
CU	137	10	6	202	23	7	3,722	23	12	5,237	68	20
KKU	21	-	-	55	4	3	622	8	1	1,522	25	11
KMUTT	51	3	3	89	3	2	889	6	4	2,002	38	11
KU	68	1	1	62	2	2	1,286	3	3	2,244	23	10
MU	70	2	2	131	3	2	2,248	11	11	3,169	19	16
PSU	42	-	-	74	-	-	1,139	1	1	2,042	5	4
SUT	48	1	1	60	3	6	592	1	1	1,244	5	4
TU	11	-	-	41	-	-	688	3	2	1,015	6	4
NRU Total	480	18	15	805	40	20	12,209	55	33	19,142	189	61
Thailand Total	13,768	NA	NA	18,025	NA	NA	18,844	NA	NA	29,277	NA	NA

Source:
Thailand patents wereaggregated from DIP database (http://www.ipthailand.go.th/)
Academic publications from industrially relevant science fields were aggregated from ISI database (https://apps.webofknowledge.

Table 2. Characteristics of domestic firms co-patenting/co-publishing during 2004-2013

Classification of manufacturing industries based on R&D intensities	Industry	DIP Patent applications						WoS Publications					
		2004-2008 (Before NRU announcement)		2009-2013 (After NRU announcement)		2004-2013 (Both periods)		2004-2008 (Before NRU announcement)		2009-2013 (After NRU announcement)		2004-2013 (Both periods)	
		Number of domestic co-patenting firms	Number of JIC-Pat	Number of domestic co-patenting firms	Number of JIC-Pat	Number of domestic co-patenting firms	Number of JIC-Pat	Number of domestic co-publishing Firms	Number of UIC-Pub	Number of domestic co-publishing Firms	Number of UIC-Pub	Number of domestic co-publishing Firms	Number of UIC-Pub
Low-Tech	Food and Agriculture	2	2	7	12	-	-	10	11	23	32	3	10
Medium-Low-Tech	Architecture and Construction	1	1	3	5	-	-	3	4	9	32	-	-
	Energy	1	1	3	15	1	14	5	10	6	48	1	45
	Metal and Mineral							3	6	2	4		
	Polymer	4	6	-	-	-	-	4	6	2	2	1	2
Medium-High Tech	Chemical	1	1	2	2	-	-	-	-	6	32	-	-
	Environment	1	1	-	-	-	-	1	1	2	4	-	-
	Automotive and Transportation	1	1	-	-	-	-	1	1	1	1	-	-
	Machinery Instrument	-	-	-	1	-	-	-	-	2	6	-	-
High-Tech	Electronics	3	4	3	7	1	7	3	11	5	24	1	14
	Medicine and Public Health	1	1	-	1			3	5	3	4	1	2
Total		15	18	20	43	2	21	33	55	61	189	7	73

Source:
Thailand patents were aggregated from DIP database (http://www.ipthailand.go.th/)
Academic publications from industrially relevant science fields were aggregated from ISI database (https://apps.webofknowledge.

In addition, the results of domestic UIC-Pub from medium-high-tech and high-tech sectors may be interpreted in two ways. First, firms from medium-high-tech and high-tech sectors which mostly are subsidiaries of MNCs still do not considered Thailand-NRUs as attractive partner for researches. Or second, these firms' production activities level may be still low related to only the improving the existing production technologies not R&D activities, so they may prefer the other modes of interaction with universities, for example, consultancy, training and testing.

6 Conclusion and Implications

In this study, we analyzed the pattern of joint research collaborations between the Thailand-NRUs and domestic firms. We employed the UIC-Pat and UIC-Pub as indicators representing the successful joint research outputs. We found that the co-patenting activity between the NRUs and domestic firms was a rare case and only be found in the leading Thai universities with high research performance. A large portion of firms co-patenting and co-publishing with the NRUs were from low-tech and medium-low tech sectors. In the case of energy and electronics industries, the firms co-patenting and co-publishing were from a few large enterprises which have continuously collaborated with the NRUs from 2004 to 2013.

The NRU project seems to have succeeded in creating and sustaining the collaboration between the NRUs and firms from low-tech and medium-low tech sectors. Nevertheless, in the case of high tech sector, the NRU project only strengthened the existing collaborations and a number of UIC-Pub from high-tech sectors were from a few large incumbents not from SMEs. Policy makers and the NRUs should pay attentions on this issue and should use proactive strategy to promote collaboration with firms in high tech industries and also to sustain the collaboration within the other sectors.

The finding related to the cooperative research areas reflects the needs and opportunities for promoting further research projects. The Thailand-NRUs have to actively monitor their research strengths together with the industry's needs to promote the cooperative researches. Policy makers have to monitor the technological capabilities of firms and their needs in upgrading their capacities. They have to act as intermediary in matching the cooperative partners for industries and universities in order to create and expand new collaborations for the country's long-term development.

References

1. Inzelt, A.: The evolution of university–industry–government relationships during transition. Res. Policy 33(6-7), 975–995 (2004)
2. Kodama, T.: The role of intermediation and absorptive capacity in facilitating university–industry linkages—An empirical study of TAMA in Japan. Res. Policy 37(8), 1224–1240 (2008)
3. Eom, B.-Y., Lee, K.: Determinants of industry–academy linkages and, their impact on firm performance: The case of Korea as a latecomer in knowledge industrialization. Res. Policy 39(5), 625–639 (2010)

4. Fernandes, A.C., de Souza, B.C., da Silva, A.S., Suzigan, W., Chaves, C.V., Albuquerque, E.: Academy—industry links in Brazil: evidence about channels and benefits for firms and researchers. Sci. Public Policy 37(7), 485–498 (2010)
5. Bruneel, J., D'Este, P., Salter, A.: Investigating the factors that diminish the barriers to university–industry collaboration. Res. Policy 39(7), 858–868 (2010)
6. Altbach, P.G.: Advancing the national and global knowledge economy: the role of research universities in developing countries. Stud. High. Educ. 38(3), 316–330 (2013)
7. Brimble, P., Doner, R.F.: University–Industry Linkages and Economic Development: The Case of Thailand. World Dev. 35(6), 1021–1036 (2007)
8. Intarakumnerd, P., Schiller, D.: University-Industry Linkages in Thailand: Successes, Failures, and Lessons Learned for Other Developing Countries. 22(4) (2009)
9. OHEC, Thailand's National Research Universities: Enhancing Academic Excellence and Competitiveness, Bangkok, Thailand (2010)
10. Van Looy, B., Callaert, J., Debackere, K.: "Publication and patent behavior of academic researchers: Conflicting, reinforcing or merely co-existing? Res. Policy 35(4), 596–608 (2006)
11. The Eurpean IPR Helpdesk, Publishing v. patenting (2013), https://www.iprhelpdesk.eu/sites/default/files/newsdocuments/Patenting_v._publishing_0.pdf (accessed: January 19, 2015)
12. Nelson, A.J.: Measuring knowledge spillovers: What patents, licenses and publications reveal about innovation diffusion. Res. Policy 38(6), 994–1005 (2009)
13. Tijssen, R.J.W., Van Leeuwen, T.N., Van Wijk, E.: Benchmarking university-industry research cooperation worldwide: performance measurements and indicators based on co-authorship data for the world' s largest universities. Res. Eval. 18, 13–24 (2009)
14. Hoekman, J., Frenken, K., Tijssen, R.J.W.: Research collaboration at a distance: Changing spatial patterns of scientific collaboration within Europe. Res. Policy 39(5), 662–673 (2010)
15. Tijssen, R.J.W.: Co-authored research publications and strategic analysis of public-private collaboration. Res. Eval. 21(3), 204–215 (2012)

Factors Affecting the Use of Information Technology for Collaboration among Government, Educational and Tourism Small Business Sectors

Kannika Daungcharone

College of Arts Media and Technology, Chiang Mai University,
239 Huaykeaw Rd, Suthep, Muang, Chiang Mai, Thailand, 50200
kannika.d@cmu.ac.th

Abstract. Tourism business is recognized as one of the main income source for Thailand for several decades. This is evident by the numbers of tourist visitors and the revenue from them has increased for several years. Besides, in the year 2015, Thailand will enter to the Asian Economic Community (AEC); Thailand then will be recognized as the main country providing tourism and aviation business within this community. As a result, such will enlarge the tourism industry in Thailand. This paper investigates the factors influencing the use of information technology for combining the operations among government, educational and tourism small business sectors in Chiang Mai. In particular, it measures the attitude of small tourism operators on how the information technology can be employed to collaborate among such government, educational and tourism small business sectors so as to sustain the business. With the use of questionnaire, the data is analyzed by using the Structural Equation Modeling to clarify the relationship between all related variables. The result is then used to provide appropriate recommendations for all parties on how information technology can be augmented the sustainability of tourism industry in Thailand.

Keywords: Small business, Tourism business, Collaboration, Sustainable.

1 Introduction

Nowadays, technologies have a big role for supporting the business processes. It can be seen that most of businesses use computer and other equipment in their business processes. Furthermore, information technology give an advantageous way to integrate all activities within networks such as generates efficiencies, reduction in complexity, improved organizational intelligence, develop stronger relationships and decrease in costs [1]. The results from the use of technology to support the business functions are flexible, reduce errors and increase profit.

Thailand is developing country and be a country that has various types of business. Tourism business is recognized as one of the main income source for Thailand for several decades. This is evident by the numbers of tourist visitors and the revenue

M. Gen et al. (eds.), *Industrial Engineering, Management Science and Applications 2015*,
Lecture Notes in Electrical Engineering 349, DOI: 10.1007/978-3-662-47200-2_23

from them has increased for several years. Besides, in the year 2015, Thailand will enter to the Asian Economic Community (AEC); Thailand then will be recognized as the main country providing tourism and aviation business within this community [2]. As a result, such will enlarge the tourism industry in Thailand. Therefore, many involved parties have an interesting in joining the development practices in tourism industry in Thailand in particular on the small tourism business in the urban district for distributing income to the community.

Although many foreign tourists come for visiting Bangkok in Thailand, they still visit other places such as Pattaya, Phuket, and Chiang Mai. By observing, any small tourism businesses in Chiang Mai, it shows that some tourism business especially small size businesses are not success. Some of them closed and others have lost their investment. Those problems are occurred by various causes. In particular there are a lot of errors happening in the information system; it needs a long-time to access to the useful information; and there is no supported system for encouraging the businesses. Besides, the small businesses usually are the stand alone business, it means they will not be able to contact with other related business easily.

Therefore, in this paper, it proposes the idea of how to investigate the factors influencing the use of information technology for combining the operations among government, educational and tourism small business sectors in Chiang Mai. In particular, it measures the attitude of small tourism operators on how the information technology can be employed to collaborate among such government, educational and tourism small business sectors so as to sustain the business. The result is then used to provide appropriate recommendations for all parties on how information technology can be augmented the sustainability of tourism industry in Thailand.

2 Literature Review

To support this paper there are various theories and researches to explain and support this concept they are explained as follows.

2.1 Collaboration among Various Actors

In these days, collaboration has been widely applied in various areas including sociology, psychology, marketing, management, and supply chain management [3]. Base on the Triple Helix model, collaboration may involve not only trading partners, but also educational institutes and governments [4]. The model highlights that the outcome from collaboration is expected to be greater when partners from industry, academia and government work together on study and development to create a common value as shown in Figure 1.

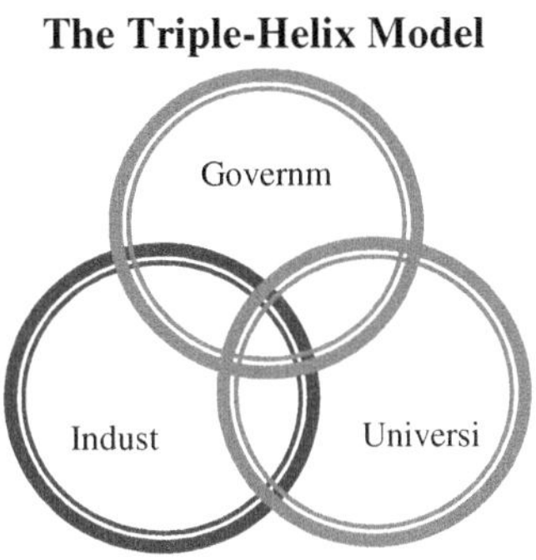

Fig. 1. The Triple-Helix model

2.2 Collaborative Process and Sustainable Development

The participation of stakeholders' effect to the collaborative methods can lead to sustainable development [5]. Sustainable development is also considered as an important goal in various businesses; its main objective is to encourage the enterprises achieving benefit in three areas: the economic benefits, the social benefits and the environmental benefits [6], which can be summarized as follows.

The Economic Benefits
Businesses can be successful in economic benefit by reducing the cost which then results in increasing the profit. To observe this matter, it can be done by considering both of direct cost/revenue and indirect cost/revenue.

The Social Benefits
To achieve this benefit, organizations have to work with the community and the people who live nearby without any problem. This can be observed with the wellbeing of the people in the organizations.

The Environment Benefits
In order to achieve sustainable development in terms of the environment, the business must be aware of the use of materials and/or devices that are safe for the environment and the reusing waste of their productions.

2.3 The Role of Information Technology in Business

Information technologies play an important role over all functions of strategic and operational management because it is the main function of tourism. It provides both opportunities and challenges for the industry. Besides, developments in information technology revolutionize both economies and enterprises which it is used for the acquisition, processing analysis, storage, retrieval, dissemination and application of information [1].

Information technology (IT) has dramatically transformed travel and tourism. From the development of Computer Reservation Systems (CRSs) in the 1960s to the Global Distribution Systems (GDSs) in the 1980s and the advent of the Internet in early 1990s, the tourism industry has always been confronted with the rise of new technological developments that posed both opportunities and challenges. There are several important trends in travelers' use of the Internet that firms should understand in order to sustain success in the online that include [7]:

- Trend 1: Internet use for travel planning is well adopted and saturated
- Trend 2: Contrasts and commonalities between generations
- Trend 3: Social media, mobile devices, and emergent channels support new behaviors

End-user computing can be defined as an end-user directly interacts with computer application software and computing systems in order to perform his or her assigned tasks. End-user computing may be defined as an end-user's applying computing knowledge, solutions, and computing systems to his or her tasks in order to efficiently perform these tasks in a business environment, which relies heavily on computing and computer operators. There composed of the computing mind, knowledge of computing technology, capability of computing application, and the potential of computing capability from a competency perspective [8] as shown in Figure 2.

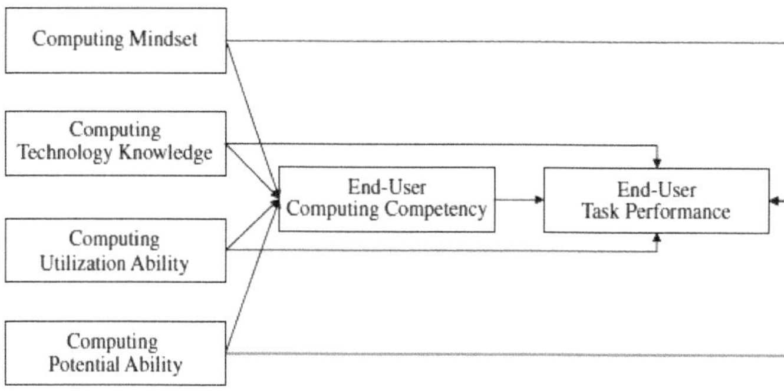

Fig. 2. The effect factors of end-user task performance on computing competency

2.4 E-Service and E-Government

Information technology (IT) capability can assist firms in collecting and categorizing information considered necessary for developing innovative services, thereby enabling them to better understand their customers' needs and preferences. Moreover, IT capability can potentially enable firms to reduce the "fuzziness" in online innovation. E-service capability emphasizes a firm's ability to access and integrate

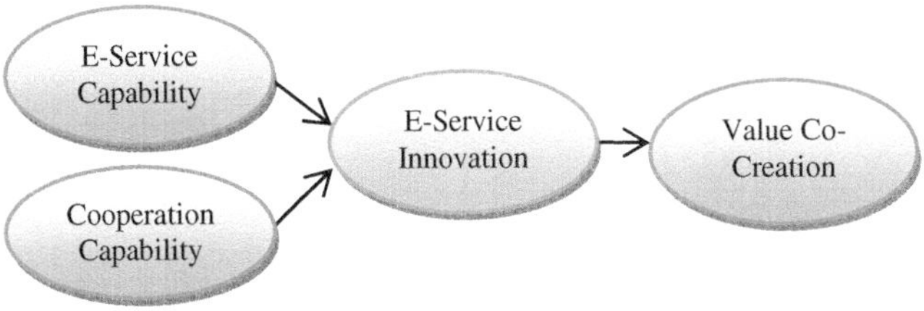

Fig. 3. The factors affect E-service innovation
(Adapted from Chuanga and Lin framework, 2015)

customer information through customer-facing technology and processes; therefore, firms possessing such a capability are more effective in leveraging this information to improve service delivery and match customer needs [9] as it shows in Figure 3.

The information of government tourism websites is very important for promoting a tourist destination's exciting such as topography, culture and tourist attraction. Furthermore, the capacities of government websites affect introduce and advertise countries [10]. Moreover, E-Government is playing its part in this digital economy by providing e-services through various ministry/department to its e-Citizen [11].

2.5 The Effect of Education on the Business

Education is an essential part of the company's development. Therefore, important factor of the business development is a high efficient information technology. Possibilities of its application in the country depend on the funds and educational preparedness of people in the area. In general, each employee has a unique potential, which needs to be stimulated in order to be productive as much as possible. Education is an important factor for this stimulation because only well-educated employees represent a real competitive advantage for the company [12].

In order to compete globally, businesses must be active and innovative in their ability to utilize evolving technologies for both product and process improvement. While over the last several decades the workers' skill are not enough with the knowledge and abilities necessary to operate and maintain modern technologies. Therefore, curriculum development and defined occupational standard are important for solving this problem because it will prepare student to entrance to labor market in various fields. The definitions of graduate employability in the academic literature focused on employment outcomes, for example; getting a job after the completion of higher education [13]. Sustainable tourism and sustainable tourism education are important characteristics of the tourism research and education curriculum [14].

3 Conceptual Framework

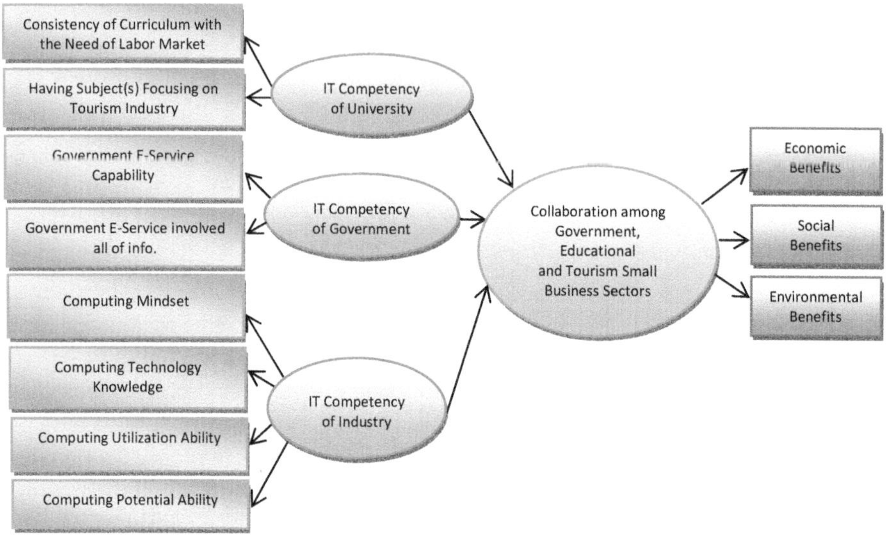

Fig. 4. The conceptual framework

4 Methodology

To achieve the objective, this study was designed to be quantity research which study and analysis about the information technology factors that affect the collaboration among government, educational and tourism small business sectors in Chiang Mai. The details of processes are as follows:

1) Study of business conduct and analyze the present problems. By observation and questioning with small tourism businesses in Chiang Mai.
2) Researching and studying theories and related researches for applying to design the questionnaire.
3) Inquiring and collecting information for 400 samples from the small tourism businesses in Chiang Mai with the questionnaire was designed.
4) Analysis of data collected from the questionnaire by using the Structural Equation Modeling, and then conclude the information technology factors of educational, government and tourism small business that effect to the efficiency and benefit of business.

5 Conclusion

From the framework of this study, it shows that the effect of collaboration among government, educational and tourism small business sectors which base on the capability of information technology will support the benefit in three areas as follows:

1) The educational can use the result of research for applying with the curriculum development then to enable students to meet the needs of the business sector.
2) The Government can use the result of research to define the policy in the related area.
3) The businesses can apply research results to improve their business processes to be the sustainable business.

References

1. Buhalis, D.: Strategic use of information technologies in the tourism industry. Tourism Management 19(5), 409–421 (1998)
2. Thai-AEC, Strengths - Weaknesses of countries in the AEC, `http://www.thai-aec.com/140` (searching on December 22, 2014)
3. Min, S., Roath, A.S., Daugherty, P.J., Genchev, S.E., Chen, H., Arndl, A.D., Richey, R.G.: Supply Chain Collaboration: What's happening. The International Journal of Logistics Management 16, 237–256 (2005)
4. Leydesdorff, L., Etzkowitz, H.: Emergence of a Triple Helix of University- Industry-Government Relations. Science and Public Policy 23, 279–286 (1996)
5. Council of Great Lakes Industries, The Boulder Principles for Effective Collaborative Processes on Sustainable Development and Environmental Policy. Effective Collaborative Processes on Sustainable Development and Environmental Policy. Boulder. Colorado: Council of Great Lakes Industries (1998)
6. Colbert, B.A., Kurucz, E.C.: Three Conceptions of Triple Bottom Line Business Sustainability and the Role for HRM. Human Resource Planning 30(1), 21–29 (2007)
7. Xiang, Z., Magnini, V., Fesenmaier, D.: Information technology and consumer behavior in travel and tourism: Insights from travel planning using the internet. Journal of Retailing and Consumer Services 22, 244–249 (2015)
8. Yoon, C.Y.: The effect factors of end-user task performance in a business environment: Focusing on computing competency. Computers in Human Behavior 25, 1207–1212 (2009)
9. Chuanga, S.H., Lin, H.N.: Co-creating e-service innovations: Theory, practice, and impaction firm performance. International Journal of Information Management 35, 277–291 (2015)
10. Horng, J.S., Tsai, C.T.: Government websites for promoting East Asian culinary tourism: A cross-national analysis. Tourism Management 31, 74–85 (2010)
11. Hojeghan, S.B., Esfangareh, A.N.: Digital economy and tourism impacts, influences and challenges. Procedia Social and Behavioral Sciences 19, 308–316 (2011)
12. Krsak, B., Tobisova, A., Sehnalkova, M.: Education in Information Technology as a tool for tourism Development. Procedia - Social and Behavioral Sciences 116, 1096–1100 (2014)
13. Azevedo, A., Apfelthaler, G., Hurst, D.: Competency development in business graduates: An industry-driven approach for examining the alignment of undergraduate business education with industry requirements. The International Journal of Management Education 10, 12–28 (2012)
14. Schweinsberg, S., Stephen, L., McManus, P.: Exploring sustainable tourism education in business schools: The honours program. Journal of Hospitality and Tourism Management 20, 53–60 (2013)

How Managerial Capabilities of Entrepreneur Leverage Innovative Capability of SMEs: A Perspective of TIM

Qingrui Xu, Siyu Liu, and Zhiyan Wu

School of Management, Zhejiang University, Hangzhou, China
{sbaxuqr,liusiyu,wuzhiyan}@zju.edu.cn

Abstract. As economic development increasingly relies on the innovation of the firm, firms are required to improve innovative capability to survive in the market. Small and medium-sized enterprises (SME) play a critical role in the transition of a developing country. It is essential for SME to leverage innovative capability in long-term growth and survival, and the role of the firm's entrepreneurs play the key role for leveraging innovative capability. Yet at present little is known about the relationship between the managerial capabilities of entrepreneurs and the innovative capability of the firm. This paper contributes to the understanding of the innovative capability leveraging of an organization by developing theory about the interplay between managerial capabilities of the entrepreneur and the firm's TIM capability.

Keywords: total innovation management (TIM), managerial capabilities of entrepreneurs, innovative capability.

1 Introduction

1.1 The Issues of Technology Innovation in SMEs in China

As economic development relies on the innovation of the firm, firms are required to conduct innovation to survive in the market. The manufacturing industry perform the main driving force of economic growth and social development, and small and medium-sized enterprise (SME) play a key role in the transition of a developing country[1]. In other words, it is essential for SME to leverage innovative capability in long-term growth and survival. However, it is difficult for SME to leverage innovative capability, because the capability is a comprehensive system that consisting of many elements, such as strategy, organization, technology, culture, institution, and entrepreneurship. Therefore, SME tend to undertake TIM (Total innovation management) to adjust the new complex and changeable environment [2].

1.2 The Structure of Innovative Capability

According to the resource-based theory, firms perform heterogeneously, because the resources they control are different [3]. Organizational resources consist of the tangible

M. Gen et al. (eds.), *Industrial Engineering, Management Science and Applications 2015*,
Lecture Notes in Electrical Engineering 349, DOI: 10.1007/978-3-662-47200-2_24

resources like funds, knowledge, talents and intangible resources liked the institution, culture, strategy [4]. Since resource is the foundation of competitive advantage, the more scarce resource the firm controls, the more innovative capability the firm performs [5]. It is also important for the firm that how the resource allocated and utilized in the organization. However, even though with the same quality and quantity resource, enterprises perform differently. One the most important resource, entrepreneurs, should be considered specially. Moreover, the entrepreneurs are directly and indirectly responsible for the resource allocation, so they play a critical role in leveraging innovative capability of the organization.

1.3 The Important Roles of Entrepreneurs in Leveraging Innovative Capabilities

The entrepreneurs reside at the strategic apexes of its organization, and their talent, ability, and action are ultimately responsible for key resource utilization [6]. Therefore, there is a relationship between the entrepreneur and the firm's capability. The nature of capability is configuration and reconfiguration of resource the firm has. The resource in the organization is allocated and configured by the person's decision. And the entrepreneur is the person who is responsible for the plan making. Some scholars explored the impact of the characteristic of entrepreneur on organizational innovation [7]. Yet at present little is known about the relationship between entrepreneur's capability and innovative capability of the firm. This study, therefore, will explore the interplay between managerial capabilities of the entrepreneur and the firm's TIM capability.

2 The Path of Innovation in Chinese Enterprise

2.1 Introduction the Path of Innovation in China: Secondary Innovation, Integrated Innovation, and TIM Innovation

A typical catch-up path of technology innovation has emerged in China today which plays an important role in national economic growth. The innovation path in China can be divided as three stags "secondary innovation-- integrated innovation --TIM innovation." Firstly, secondary innovation is the process of foreign technology acquisition and endogenous technological innovation, which is based on imported technology but fundamentally different from imitation [8]. Actually, the essence of the secondary innovation is the replacement of capability of imported technology, and it is the first step of technology innovation in national economy in China. Secondly, the integrative innovation is an integrated system consisting of two and more innovative elements. The integrated innovation emphasize that it is synergistically composed but not randomly[9]. The nature of integrative innovation is the change of capability. Thirdly, total innovation management (TIM) is a theory contributing to promote competitive advantage at this stage [10]. TIM derives from synergy theory, and it consists three all and one synergy, including all–involvement, all-element, all-time place, and total synergy [11]. All-involvement innovation refers to innovation by all individuals. All-elements innovation includes the integration among technological and non-technological factors—strategy, culture, organization, institution, and market. All time-space innovation means innovating all the time and in all space. TIM emphasize the totally synergy process of enterprise innovation combining different innovative factor [10]. As an emerging theory based on Chinese innovation characteristics, TIM is defined as a system for enterprise to promote sustainable innovative capability and competitive advantage.

2.2 How TIM Leverage Innovation Capabilities of Firms

Based on the TIM theory, innovation elements of innovation are distinguished in the process of capability leveraging. According to the research of 44 SMEs located in Zhejiang, Fujian and Shandong province, Strategy, technology, organization, market, culture, alliance, basic, learning, and creativity have been recognized as very important elements of innovative capability among the innovative factors. In terms of the enabler of each element, the NPD (new product development), entrepreneur and innovator play more important role on factors than other enabler. Specially, the entrepreneur has the most important influence on most of the factors--strategy, organization, market, culture, alliance, basic and learning--and ultimately affects the innovative capability leveraging (see Table 1)[12]. In one word, the capability of entrepreneur in the organization plays critical role in enhancing the innovation performance and promoting the innovative capability. However, how the capability of the organization leverages over time by the entrepreneur and how to describe such development appear to be open questions. Adner and Helfat (2003) identify the three main elements supporting managerial capabilities as managerial human capital, managerial social capital, and managerial cognition [13]. In this study, we use these elements to describe the managerial capabilities of the entrepreneur and explore how they improve the capability of the organization in different stages.

Table 1. Influence of different innovative enabler on the innovative factors

Factor / Enabler	Strategy	Technology	Organization	Market	Culture	Alliance	Basic	Learning	Creativity
Environment	*	*	*	**	*	*		*	*
NPD	*	***	*	***	*	*	*	*	*
Entrepreneur	**	*	**	***	***	**	**	**	*
Innovator	*	***	*	**	*	*	*	*	*
Employee				***	*	*		*	

*** very important, **important, * commonly

Note: NPD-new product development

3 Case Study: The Process of Innovative Capability Leveraging of Firm and the Capability Development of Entrepreneur in SMEs

3.1 Research Setting

Exploring the underlying processes of capability leveraging of entrepreneur requires a research setting that allowed an analysis of the role of the entrepreneur, innovative capability of the firm and interactions, at various points in time. Two cases-- Loope Co. Limited and Enjoy Co. Limited-- are selected because they had several features suitable for this purpose. First, they were both well-run SEMs, providing 10 years of data, and important leaders of the firm could be interviewed, including the founder. Second, they were both environmental protection industry, as a high-technology industry environmental protection industry heavily rely on organizational innovative capability and entrepreneur's talent. Third, comparing with matured industry like manufacturing industry, environmental protection industry in China is an emerging industry, and it is clear to observe the path of innovative capability.

3.2 The Main Role of the Entrepreneur in Different Stage: THE CASE OF LOOPE

The process of capability development in Loope includes three periods, including the formative start-up stage (2005-2008), the developing stage (2009-2011) and the murmured stage (2012-today). The detailed information is displayed in Table 2.

In the start-up stage Loope was a small enterprise ran by the founder Mr. Xia. The company introduced the advanced technology of ground-source heat pump from foreign company and the main problem was that market was urgently needed to be expanded. The founder of the company developed the market by visiting the potential customers himself.

In the developing stage, the business demand increase and the organizational structure developed fast. With the increase of business, the organization expanded its size and a team was gradually built. A horizontal organization is gradually established during this period, and function departments were formed. The main issue in this stage is the cash flow problem because of the beginning of the depression of real estate industry. With the plenty of the social capital, the entrepreneur success in financing activities and overcome the problem.

In the matured stage, the company faced the cash problem again, beside of financing activities, the executors decided to select potential customer to guarantee the program quality. During this stage, another main problem was the managerial capability didn't match its business requirement. The CEO suggested the board chairman to adjust the governance structure to branch office system.

3.3 The Main Role of the Entrepreneur in Different Stage: THE CASE OF ENJOY

The innovative capability development in Enjoy can be divided into three periods, including the formative start-up stage (2005-2006), the developing stage (2007-2012) and the murmured stage (2013-today). The detailed information is displayed in Table 3.

In the formative start-up stage Enjoy was such a small enterprise that Mr. Ye, who is the company's founder was the only staff in the organization and was responsible for all the issues in company at that time. Not only a manager, but also a salesman and a finance staff did Mr. Ye played. However, the main problem in this stag was the low rate of return on investment, and the development of the company was hindered. In order to deal with this problem, Mr. Ye made a significant decision that he adjusted its main business as offering sewage services in real estate industry at the end of 2006, because the real estate industry was flourished. With the keen insight of market, the entrepreneur noticed the booming of real estate industry and the new market of treatment equipment production.

The main subject the company faced in the developing stage was cash flow problem because of the long Mat endowment time in real estate industry. To address this problem, the team decided to reduce the project number of real estate, and cultivate new booming industry liker rural sewage disposal and river channel ecosystem governance.

In the matured stage, the company was merged and acquired by a public limited company named Xingyuan Filter Technology Co., Ltd. Then seven big center were found, including engineering management center, financial management center, enterprises development management center, technological management center and human resource management center. The leader in each center was directly response

Table 2. The managerial capabilities of entrepreneurs of loope in different stage

	Development stage		
Content	2005-2008	2009-2011	2012-today
Main business	Ground source heat pump central air-conditioning system;	ground source heat pump program; energy storage systems	district energy system planning;
Main problem the company faces	market expansion;	Cash flow problem;	managerial capability didn't match its business requirement
The way to solve the problem	visiting the potential customers and expand the market;	financing activities by the founder;	The CEO suggested the board chairman to adjust the governance structure to branch office system;
the role played by entrepreneur	CEO is the executor and the decision maker;	CEO is the coach of the TMT	CEO is the leader of the TMT; also influenced by the TMT
managerial capabilities of entrepreneurs	managerial human capital, cognition	managerial human capital, social capital, cognition	social capital, cognition

Table 3. The managerial capabilities of entrepreneurs of Enjoy in different stage

Content	Development stage		
	2005-2006	2007-2012	2013-today
Main business	Producing the water - treatment equipment of central air conditioner;	Sewage services project in real estate industry;	Rural sewage disposal; river channel ecosystem governance;
Main problem the company faces	Low rate of return on investment; Slow development speed;	Long Mat endowment time in real estate industry ; Cash flow problem;	Development of technology and integration of knowledge
The way to solve the problem	Adjust strategy ; Explore and expand the market;	Adjust strategy to focus on the ecological governance project;	Cultivate talents in organization; complete the institution;
the role played by entrepreneur	The decision maker and the executor; The CEO's capability transforms from his own resource and capability to the firm's capability;	The decision maker and strategy planner; The CEO lead the team to leverage the firm's capability;	The strategy planner; The TMT led by CEO governance the organization;
managerial capabilities of entrepreneurs	managerial human capital, cognition	managerial human capital, social capital, and cognition	social capital, and cognition

for the issue that enhanced the work efficiency of CEO. In order to overcome the development problem of technology integration with knowledge, the entrepreneur started to tend to cultivate talents in organization.

3.4 The Relationship between the Managerial Capability of the Entrepreneur and the Firm's Innovation Capability

During the three development period in these two cases, we can see the entrepreneur play a critical but different role in every stage. In the first stage, the company is an individual household and CEO is the only entrepreneur. The main managerial capability of the entrepreneur is cognition. CEO's capability transforms from his own resource and capability to the firm's capability. In the second stage, there is a team composed by key staffs, with CEO as the core of the team. The managerial capabilities of the entrepreneur are enriched and managerial human capital, social capital and cognition are considered as the main capability. In the third stage, the firm develops as a highly structured organization, and the entrepreneur is top management team. The main problem an entrepreneur focus on is how to leverage the innovative capability. The TMT led by CEO enhance the organization capability. The main managerial capabilities are social capital and cognition. The internal mechanism of the capability leveraging is that, to fit the dynamic environment the managerial capabilities the entrepreneurs keep learning. And the external mechanism is that the dynamic environment pushes the evolution of managerial capability of the entrepreneur gradually.

The relationship between entrepreneur and the firm's innovation capability are shown in Figure 1.

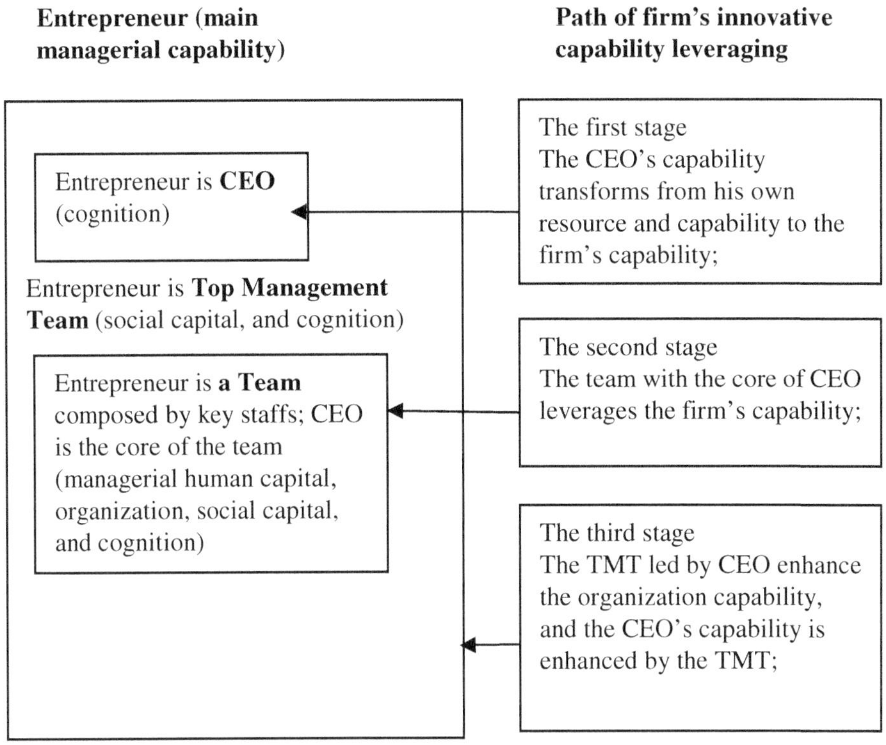

Fig. 1. The relationship between entrepreneur and the firm's innovation capability

4 Conclusion

Based on the case of Loope and Enjoy, this paper focuses on the relationship between entrepreneur's capability and the firm's innovation capability.

First, the roles entrepreneurs play and the main managerial capability in different development stages are described. At initial stage, the cognition capability of the founder plays an important role for the firm's survival. At the developing stage, all the managerial capability (managerial human capital, social capital, and cognition) are developed to fit the requirement of organization and the external environment. At the matured stage, the CEO focus on the strategic decision making and platform building.

Second, the mechanisms of the capability leveraging are researched. The internal mechanism of the capability leveraging is that entrepreneurs keep learning to fit the dynamic environment. And the external mechanism is the dynamic environment pushes the evolution of managerial capability gradually and in turn leverages the organizational capability.

Third, the relationship between managerial capability of the entrepreneur and the firm's innovation capability are explored. In the first stage, CEO's capability transforms from his own resource and capability to the firm's capability. In the second stage, CEO leads the team to enrich the managerial capability and leverage the firm's capability. In the third stage, the CEO's ability cultivates the TMT's ability to combine specialized knowledge streams and strategy, and the TMT reinforces the CEO's own dynamic managerial capabilities. Both of them enhance the organizational capability.

Acknowledgment. This research is supported by National Natural Science Foundation-"Evolution and Leverage Mechanism of Enterprise Innovation Capability in China" (Grant No.: 71172115) and consulting projects under the grant of Chinese Academy of Engineering（Grant No.： 2014-XY-40, 2013-XZ-38-2-2）.

The paper has also benefitted immensely from the comments of the reviewers and editors, and we are particularly thankful for their valuable suggestions.

References

1. Love, J.H., Roper, S.: SME innovation, exporting and growth: A review of existing evidence. International Small Business Journal 33(1), 28–48 (2015)
2. Xu, Q., Chen, J., Xie, Z., et al.: Total Innovation Management: a novel paradigm of innovation management in the 21st century. The Journal of Technology Transfer 32(1-2), 9–25 (2007)
3. Barney, J., Wright, M., Ketchen, D.J.: The resource-based view of the firm: Ten years after 1991. Journal of Management 27(6), 625–641 (2001)
4. Greco, M., Cricelli, L., Grimaldi, M.: A strategic management framework of tangible and intangible assets. European Management Journal 31(1), 55–66 (2013)
5. Wernerfelt, B.: On the role of the RBV in marketing. Journal of the Academy of Marketing Science 42(1), 22–23 (2014)
6. Balkin, D.B., Markman, G.D., Gomez-Mejia, L.R.: Is CEO pay in high-technology firms related to innovation? Academy of Management Journal 43(6), 1118–1129 (2000)
7. Sirivanh, T., Chaikeaw, A., Sateeraroj, M.: Characteristic of Entrepreneur, Business Alliance and Growth of Small and Medium Enterprises (SMEs): Case in Lao People's Democratic Republic (2014)
8. Xiaobo, W.: The evolutionary process of secondary innovation. In: Technology Management-The Key to Global Leadership. PICMET 1997: Portland International Conference on Management and Technology, p. 183. IEEE (1997)
9. Zhang, H., Xue, L.: New Technology Innovation Management Paradigm–Integrative Innovation. China Soft Science 12, 002 (2002)
10. Chen, J., Shou, Y.: Leverage Innovation Capability: Application of Total Innovation Management in China's SMEs' Study. World Scientific (2012)
11. Wu, Z., Mei, L.: Toward innovative firm under Total Innovation Management. In: 2012 IEEE International Conference on Management of Innovation and Technology (ICMIT), pp. 471–474. IEEE (2012)
12. Chen, J., Shou, Y., Liu, J.: Leverage Innovation Capability: Application of Total Innovation Management in China's SMEs' Study. World Scientific (2012)
13. Adner, R., Helfat, C.E.: Corporate effects and dynamic managerial capabilities. Strategic Management Journal 24(10), 1011–1025 (2003)

Definition of Complex Hurst and Fractional Analysis for Stock Market Fluctuation

Qing Zou[1], Yufan Hu[2], and Jun Steed Huang[3,*]

[1] Suqian College, 399 South Huanghe, Jiangsu 223800, P.R. China
[2] Loughborough University, Leicestershire, LE11 3TU, UK
[3] Suqian College, Jiangsu University, 399 South Huanghe, 223800, P.R. China
{roronoaz,yufanhu0918,15862930930}@163.com

Abstract. Hurst model has received significant interest in recent years and are being increasingly used to explain the stochastic phenomenon with long term dependency such as stock market fluctuations. Different from existing methods in traditional integer dimension construction, this paper proposes a novel fractional dimension derivation along with the exact algorithm involving the fractional norm definition and the fractional center moment extension, which ends up a complex Hurst parameter. The proposed algorithm provides more granularities in terms of the norm and the center moment that eventually leading to the detail revealing of the underline stock marketing drivers from both real and imaginary angles. The calculation results demonstrate that the complex model is able to distinguish the subtle difference between the stock market performances for the same field that the real model may overlook. We take the e-commence cluster of online related companies as an example.

Keywords: Fractional norm, stock exchange, complex Hurst, electronic commence.

1 Introduction

Hurst algorithms are used in analyzing real-world implementations, i.e. it comes as the trusted analytic solution, but typically tends to have challenges in software implementation cost and time, that leads to a demand for simpler software solution by using complex theory. Hurst algorithms have received significant interest in recent years and are being increasingly used to solve real-world problems. Among them are combination of two or more algorithms involving numerical algorithms, analytic calculation [1], and other computational techniques, such as, artificial intelligence [2], [3] and [4], fuzzy systems [5] or simulation [6], etc.

Fractal Market Hypothesis (FMH) was first introduced by Edgar E. Peters (1994) [7]. Being different from the traditional Efficient Market Hypothesis (EMH), FMH is based on a nonlinear dynamic system, which makes it more reasonable according to the reality. Especially, it emphasizes the influences on the behaviors of investors

* Corresponding author.

M. Gen et al. (eds.), *Industrial Engineering, Management Science and Applications 2015*,
Lecture Notes in Electrical Engineering 349, DOI: 10.1007/978-3-662-47200-2_25

caused by the difference of information reception and lengths of investing time horizons, or states the existence of fractal structure in stable markets.

FMH states that the capital market is formed by large amounts of investors with various beginning investment points, so the same market information may have different impacts on the individuals: Some might response at the moment when they receive the information, while others, perhaps the majority, may choose doing nothing until the information is confirmed and obvious market trends appear. Moreover, the existence of irrational investors cannot be ignored. Thus the number of daily, weekly or monthly transactions may not stay uniform. Starting from the normality test on the price change of capital market, Peters has proved that the assets' prices and returns in the capital market follow the fractional Brownian motion by using standard R/S analysis.

After the phase space reconstruction of time series, he also succeeded in calculating the fractal dimension and the Lyapunov exponent of the capital market [8], the dynamical system of which, in this way, has been successfully analyzed. However, the spread and practice of the FMH are hold back because of the difficulty in mathematical modeling. This paper will simplify the process of modeling by amending thoroughly the basis of fractal theory and applying "Complex Hurst"(which is equivalent to R/S analysis in Real domain, and beyond R/S for additional imaginary value), so that the analysis can be carried out rapidly by running the MATLAB tool.

Traditional R/S method has its own weaknesses. On one hand, researchers have to put particular emphasis on either modeling [9],[10],[11],[12] or calculation [13], [14], since these two steps cannot synchronize perfectly. Although numerical analysis is precise in theory, however, on the other hand, computers are unable to realize the theoretical accuracy, where uncontrollable error accumulation occurs. Not only the artificial intelligence, fuzzy systems, but also simulation, modeling, all have above drawbacks caused by the error accumulation or inaccuracy of modeling. Therefore, a new method called complex Hurst based on fractional norm is presented in this article to give an easy and fractional precise way to solve the real-world problems. Based on the thoughts of functional analysis [15],[16],[17],[18], it gives definitions of fractional norm and fractional center moments, and consequently ensures the precision while simplifying the calculation. In general Matlab is preferred, others like Excel is not impossible.

2 Review of Hurst Model and Fractional Analysis

The Hurst parameter is used as a measure of long-term memory of time series. It relates to the autocorrelations of the time series, and the rate at which these decrease as the lag (grouping) level between pairs of values increases. Studies involving the Hurst parameter were originally developed by Hurst himself, in hydrology for the practical matter of determining optimum dam sizing, for the Nile river's volatile rain and drought conditions, that had been observed over a hundred years. The name "Hurst coefficient", derives from Harold Edwin Hurst (1880–1978), who was the lead researcher in these studies.

In fractal geometry [19],[20], the generalized Hurst exponent has been denoted by H in honor of both Harold Edwin Hurst and Ludwig Otto Hölder (1859–1937), who set up the theoretical frame work for Hurst, by Benoît Mandelbrot (1924–2010). H is directly related to fractal dimension, *D*. However, so far, the H remains as a pure real number. We are going to extend it to a complex number in this paper, for the first time, and along with some application samples.

The Hurst parameter is referred to as the "index of dependence" or "index of long-range dependence". It quantifies the relative tendency of a time series either to regress strongly to the mean or to cluster in a direction. A value H in the range 0.5–1 indicates a time series with long-term positive autocorrelation, meaning both that a high value in the series will probably be followed by another high value, and that the values for a long time into the future will also tend to be high. A value in the range 0 – 0.5 indicates a time series with long-term switching between high and low values in adjacent pairs, meaning that a single high value will probably be followed by a low value, and that the value after that will tend to be high, with this tendency to switch between high and low values lasting a long time into the future. A value of H=0.5 can indicate a completely uncorrelated memory less series, such as Poisson point process [21],[22].

The slope of the IDC (Index of Dispersion for Count – ratio of variance) curve (Fig. 1) gives the Hurst parameter from equation 1. As shown from the curve, the Fractal effect [23] calms down, but does not disappear, for this reason we call it persistent Hurst phenomenon [24].

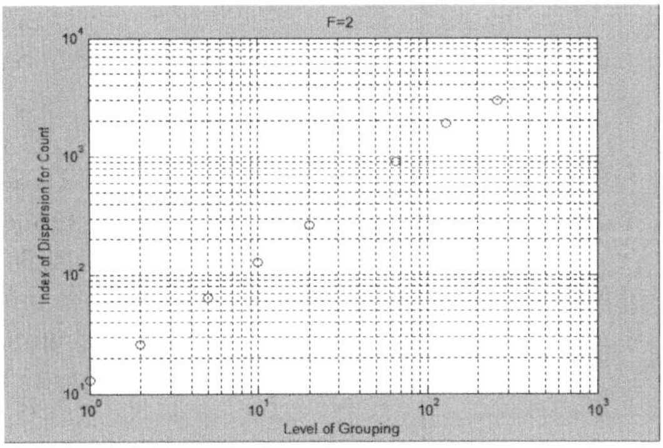

Fig. 1. IDC over Level of Grouping

$$H = \frac{1}{2}(1 + slope) \tag{1}$$

3 Definition of Fractional Norm

In order to define a complex Hurst value, we need to establish the Fractional norm and Fractional moment in sequence.

3.1 The Lemmas

Lemma 1. (Hölder Inequality) p,q are real number. Suppose $p>1,\frac{1}{p}+\frac{1}{q}=1$, for all $x=(\xi_i)\in l^p$ and $y=(\eta_i)\in l^p$, one has

$$\sum_{i=1}^{\infty}|\xi_i\eta_i|\le\left(\sum_{i=1}^{\infty}|\xi_i|^p\right)^{1/p}\left(\sum_{i=1}^{\infty}|\eta_i|^q\right)^{1/q}.$$

Lemma 2. (Minkowski Inequality) Suppose $p\ge1$, for all $x=(\xi_i)\in l^p$ and $y=(\eta_i)\in l^p$, one has

$$\left(\sum_{i=1}^{\infty}|\xi_i+\eta_i|^p\right)^{1/p}\le\left(\sum_{i=1}^{\infty}|\xi_i|^p\right)^{1/p}+\left(\sum_{i=1}^{\infty}|\eta_i|^p\right)^{1/p}.$$

Review of the definition of the norm: for a mapping: $\|\cdot\|:E\to E$, if it satisfies:

N.1) Nonnegative: $\|x\|\ge0$.

N.2) Scaling: $\|\alpha x\|=|\alpha|\cdot\|x\|,\alpha\in\mathbb{R}$.

N.3) Triangular inequality: $\|x+y\|\le\|x\|+\|y\|,x,y\in E$.

Then $\|\cdot\|$ is called a norm.

According to Lemma 1 and Lemma 2, we can prove the Euclidean norm

$$\|x\|=\left(\sum_{i=1}^{n}|\xi_i|^2\right)^{1/2} \tag{2}$$

It satisfied the norm axioms.

Proof. N.1) It is obvious that Euclidean norm satisfied the nonnegative property, and $\|x\|=0$ if and only if $x=(0,0,\cdots,0)$.

N.2) for all $\alpha\in\mathbb{R}$,

$$\begin{aligned}\|\alpha x\|&=\left((\alpha x_1)^2+(\alpha x_2)^2+\cdots+(\alpha x_n)^2\right)^{1/2}\\&=\left[\alpha^2(x_1^2+x_2^2+\cdots+x_n^2)\right]^{1/2}=|\alpha|\cdot\|x\|.\end{aligned}$$

N.3) According to Lemma 2, we can obtain

$$\|x+y\|=\left(\sum_{i=1}^{n}|\xi_i+\eta_i|^2\right)^{1/2}\le\left(\sum_{i=1}^{n}|\xi_i+|^2\right)^{1/2}+\left(\sum_{i=1}^{\infty}|\eta_i|^2\right)^{1/2}=\|x\|+\|y\|.$$

So, the definition of Euclidean norm is reasonable. □

3.2 Precursor for the Definition of Fractional Norm

Definition 1. For all $a>0,a\ne1,x\in\mathbb{R}$, define

$$a^x=\begin{cases}\sup\limits_{r\le x}\{a^r:r\in\mathbb{Q}\},a>1,\\\inf\limits_{r\le x}\{a^r:r\in\mathbb{Q}\},0<a<1.\end{cases} \tag{3}$$

Now, we explain the rationality of this definition.

If $x \in \mathbb{Q}^C$ (irrational number), there exists a rational number r_0 such that $x < r_0$, so when rational number $r < x$, one has $r < r_0$. According to the properties of rational power: when $a > 1$, one has $a^r < a^{r_0}$, it shows that nonempty set

$$\{a^r : r < x, r \in \mathbb{Q}\}$$

It has upper bound, from the supremum and infimum principle, the set has supremum and the supremum is unique. So, the right side of the first expression in equation 3 is a certain number. In the same way, when $0 < a < 1$, the right side of the second expression in equation 3 is a certain number too. Consequently, for $x \in \mathbb{Q}^C$, the definition is rational and meaningful.

Below we show that the right side number of (3) is a^x.

When $a > 1$ for any $\alpha < a^x$, if $\alpha \le 0$, then $a^r \in \{a^r : r \in \mathbb{Q}, r < x\}$, so $a^r > \alpha$; if $\alpha > 0$, according to rational density property there exist $r_0 \in \mathbb{Q}$, that

$$\log_a \alpha < r_0 < x (\log_a \alpha < \log_a a^x = x),$$

Which means $a^{r_0} > \alpha$ and $a^{r_0} \in \{a^r : r \in \mathbb{Q}, r < x\}$, thus

$$\sup_{r<x}\{a^r : r \in \mathbb{Q}\} = a^x.$$

For $0 < a < 1$ the same arguments hold.

If $x \in \mathbb{Q}$, the definition is "$r \le x$", so it is right the rational power. For this, we can see Definition 1 contains rational power. Consequently, for $x \in \mathbb{Q}$, the definition is rational and meaningful too.

In summary, the norm from Definition 1 is rational and meaningful for the dimension with both rational number and irrational number.

3.3 The Generalization of the Norm Definition

Let's review the integer norm definition first:

$$\|x\|^m = \left(\sum_{i=1}^{n} |x_i|^2\right)^{\frac{m}{2}}, m \in \mathbb{Z}.$$

On Euclidean norm we define $\|x\|^{1 \pm \frac{1}{r}}$.

Definition 2. $\forall x \in l^p$, define

$$\|x\| \cdot \sqrt[r]{\|x\|} = \|x\|^{1+\frac{1}{r}} = \left(\sum_{i=1}^{n} |x_i|^2\right)^{\frac{r+1}{2r}}.$$

Where $r \in [1, +\infty)$.

Since $\left\|\cdot\right\|$ is Euclidean norm, where Section 3.1 has shown that it is meaningful, as such Definition 2 is reasonable as well.

Definition 3. $\forall x \in l^p$, define

$$\|x\|^{1+\frac{1}{r}} = \left(\sum_{i=1}^{n}|x_i|^2\right)^{\frac{r+1}{2r}} = \begin{cases} \sup\limits_{y\le\frac{r+1}{2r}}\left\{\left(\sum_{i=1}^{n}|x_i|^2\right)^y : y\in\mathbb{Q}\right\}, (\sum_{i=1}^{n}|x_i|^2 > 1) \\ \inf\limits_{y\le\frac{r+1}{2r}}\left\{\left(\sum_{i=1}^{n}|x_i|^2\right)^y : y\in\mathbb{Q}\right\}, (0 < \sum_{i=1}^{n}|x_i|^2 < 1) \end{cases}$$

Since the current norm definition is derived from previous norm definition, as such it stands well too.

Definition 4. $\forall x \in l^p$, define

$$\|x\| \Big/ \sqrt[r]{\|x\|} = \|x\|^{1-\frac{1}{r}} = \left(\sum_{i=1}^{n}|x_i|^2\right)^{\frac{r-1}{2r}},$$

Where $r \in [1,+\infty)$.

Definition 5. $\forall x \in l^p$, define

$$\|x\|^{1-\frac{1}{r}} = \left(\sum_{i=1}^{n}|x_i|^2\right)^{\frac{r-1}{2r}} = \begin{cases} \sup\limits_{y\le\frac{r-1}{2r}}\left\{\left(\sum_{i=1}^{n}|x_i|^2\right)^y : y\in\mathbb{Q}\right\}, (\sum_{i=1}^{n}|x_i|^2 > 1) \\ \inf\limits_{y\le\frac{r-1}{2r}}\left\{\left(\sum_{i=1}^{n}|x_i|^2\right)^y : y\in\mathbb{Q}\right\}, (0 < \sum_{i=1}^{n}|x_i|^2 < 1) \end{cases}$$

Now, we have established the concept of the Fractional norm, we can move on to the Fractional moment, of particular interests to us here is the Fractional center moment.

4 Definition of Fractional Moments

In order to calculate the Fractional center moment, we need first to generalize the Binomial formula from integer domain to real domain.

4.1 Generalized Binomial Formula

Definition 6. $\forall r \in \mathbb{R}$, Let's define

$$(x+y)^r = \sum_{k=0}^{\infty}\binom{r}{k}x^{r-k}y^k,$$

Where $\begin{pmatrix} r \\ k \end{pmatrix} = \frac{r(r-1)\cdots(r-k+1)}{k!} = \frac{(r)_k}{k!}$ and Pochhammer symbol $(\circ)_k$ represents the falling factorial.

If $r \in \mathbb{Z}$,

$$(r)_k = r(r-1)\cdots(r-k+1).$$

If $r \in \mathbb{R}$,

$$(r)_k = r(r-1)\cdots(r-k+1) = \frac{\Gamma(r+1)}{\Gamma(r-k+1)}.$$

$\Gamma(\circ)$ is Gamma Function.

4.2 Backward Compatibility of Generalized Binomial Formula

For integer binomial formula, when $k > n$, we have $\begin{pmatrix} n \\ k \end{pmatrix} = 0$. For the generalized real binomial we then have $\begin{pmatrix} x \\ k \end{pmatrix}_{(x \in \mathbb{R}, k \in \mathbb{N})}$, for $x < k$ situation, if x is natural, $x, (x-1), \cdots, (x-k+1)$ may contains zero, say

$$\begin{pmatrix} 3 \\ 5 \end{pmatrix} = \frac{3 \cdot 2 \cdot 1 \cdot 0 \cdot (-1)}{5!} = 0,$$

Thus $\begin{pmatrix} x \\ k \end{pmatrix} = 0$. If x is not natural, $x, (x-1), \cdots, (x-k+1)$ will contain no zero, thus $\begin{pmatrix} x \\ k \end{pmatrix} \neq 0$.

For $x > k$ situation, if x is natural, then generalized real binomial and traditional integer has no difference, in another word, the traditional binomial formula is a special case of our generalized real formula, the new formula is backward compatible, but it allows us do much more, as shown below.

4.3 The Main Formula for Real Binomial Coefficient

Theorem 1. $\begin{pmatrix} -x \\ k \end{pmatrix} = (-1)^k \begin{pmatrix} x+k-1 \\ k \end{pmatrix}$, $x \in \mathbb{R}, k \in \mathbb{N}$.

Proof.

$$\begin{pmatrix} -x \\ k \end{pmatrix} = \frac{(-x)(-x-1)\cdots(-x-k+1)}{k!}$$
$$= (-1)^k \frac{x(x+1)\cdots(x+k-1)}{k!}$$
$$= (-1)^k \begin{pmatrix} x+k-1 \\ k \end{pmatrix}.$$

□

4.4 The Definition of Fractional Center Moment

Definition 7. Given ξ as a discrete random variable (D.R.V), $E\xi$ exist, define k order center moment as

$$\mu_k = E(\xi - E\xi)^k = E\left(\sum_{i=0}^{\infty}\binom{k}{i}(-1)^i \xi^{k-i}(E\xi)^i\right)$$

$$= \sum_{i=0}^{\infty}\binom{k}{i}(-1)^i E(\xi^{i-k}) E^i \xi \cdot \tag{4}$$

Where $1 < k < 3$.

According to section 4.2 if we let k be the integer in (4), then (4) is traditional center moment, as such we conclude the Fractional center moment is also backward compatible.

5 Definition of Complex Hurst

5.1 Fractional Dimension of Fractal Market

Judean mathematician F.Hausdorff, the founder of topology, argued that instead of discrete changes, the dimension of a continuous space changes continuously. Hausdorff believed in the existence of fractional dimension, which has been proved to be true. For example, the fractional dimensions of Cantor set, Sierpinski triangle [25] and Menger sponge [26],[27] are 0.63, 1.58, and 2.72 respectively. But their lengths, areas and volumes are all equal to 0 under the sense of integer dimension. Koch curve has fractional dimension 1.26 and infinite length while its area has a finite limit, sounds quite contradictive. Consequently, the system of the integer dimension has got many limitations, it is necessary to introduce some new concepts here.

This non-integer dimension, denoted by D, is called Hausdorff dimension, or fractal dimension. It does not equal to an integer in most cases.

Hurst was a hydrologist who had controlled reservoirs for a long time. During years of work, he found that most natural phenomena follow some kind of "biased random walk", i.e. there is a trend with noise terms. In 1940s, Hurst made an intensive study of this biased random walk and proposed a new statistical coefficient: Hurst exponent, a real number. It was first used to measure the time-dependent strength of tendency and noise, but nowadays, it has been widely applied in analyzing chaos and fractal of the capital market.

In 1972, Mandelbort [28],[29] applied R/S analysis, which was made use of by Hurst when studying Hurst exponent, to study the changes of stock returns for investment in American Securities Market. Peters [7] considered R/S analysis as the most important tool in the FMH, so he has conducted detailed discussions and has carried out plentiful empirical verification studies.

However, the difficulty in R/S calculation for Hurst results in the fact that this method can only be applied to a large amount of representative data, which are usually obtained from daily, weekly, monthly and yearly data, in order to gain some insight view.

Another problem is that most people only focus on the analysis of Hurst exponent while they ignore the numeration of the fractional dimension of the fractal market itself. Thus, it is important to find the relationship between Hurst exponent and the fractal dimension, so that distinguishing quickly between the trend of developed unified market and that of emerging fragmented market becomes possible.

FM dimension D and its Complex Hurst parameter CH (the definition of CH will be given in the next subsection) have following relationship:

$$D = \frac{\log N}{\log 1/r},$$

Where N is Fractal number, r is segmentation ratio.

5.2 Complex Hurst Parameter

Based on the functional theory, applying above mentioned relationships, we put forward a Complex Hurst parameter, as below.

Definition 8. Define Complex Hurst parameter as

$$CH = \frac{1 + \log \frac{\mathrm{IDC}(m)}{\mathrm{IDC}(1)} \cdot \log \frac{\mathrm{Level}(1)}{\mathrm{Level}(m)}}{2},$$

Where IDC means the Index of Dispersion for Count [30] [31],

$$\mathrm{IDC} = \frac{E(\xi - E\xi)^k}{\mu},$$

μ represents the mean of ξ. Grouping Level is the X-axis in Fig 1. m represents the last point.

In a similar way Hurst used to confirm Hurst Parameter [32] [33], we can obtain that $-0.5 < \mathrm{Re}(CH), \mathrm{Im}(CH) < 1.5$, which means that $-0.5 < \mathrm{Re}(BH), \mathrm{Im}(BH), \mathrm{Re}(IH), \mathrm{Im}(IH) < 1.5$. BH represents Backward Hurst and FH represents Forward Hurst.

In Forward Hurst, we used the way of forward accumulation, with $F>2$; it contains all the forward information. In Backward Hurst, we used the way of backward accumulation, with $F<2$; it contains all the backward information.

The physical meaning of the real part of H is explained as the traditional Hurst model, the imaginary part is explained as the uncertainty of the real part, it can also be understood as a measure of the amount of multi-Fractal of the point process. The high the imaginary part of stock means more underline diversity of a company, in terms of number of locations, employee culture mix or delivered product lines span etc. Which can be observed below to some extends.

5.3 Complex Irreversibility Hurst Parameter

Based on the reverse time theory, applying above mentioned complex Hurst formula, we put forward a Complex Irreversibility parameter, as below.

$$IH = \frac{FH - BH}{2}.$$

Where $IH \in (-0.5, 1.5)$ represents Irreversibility Hurst.

The irreversibility is used to judge if the stock market will repeat itself in future, the lower the value, the better chance it will repeat, the high the value, unlikely it will.

When $-0.5 < \mathrm{Re}(CH) < 1.5$, the fractional dimension of fractal market $D = 2 - \mathrm{Re}(CH)$ [34].

6 MATLAB Calculations

Figure 2-4 are the calculation results for online business related companies' recent stock fluctuation, using 1.5th, 2.0th and 2.5th order of the "variance", where 2.0th is the normal variance:

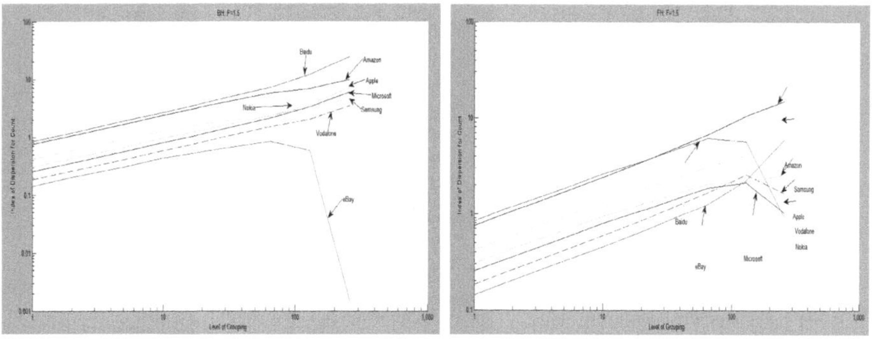

Fig. 2. Forward and Backward IDC Graphs (F=1.5, half way to Mean)

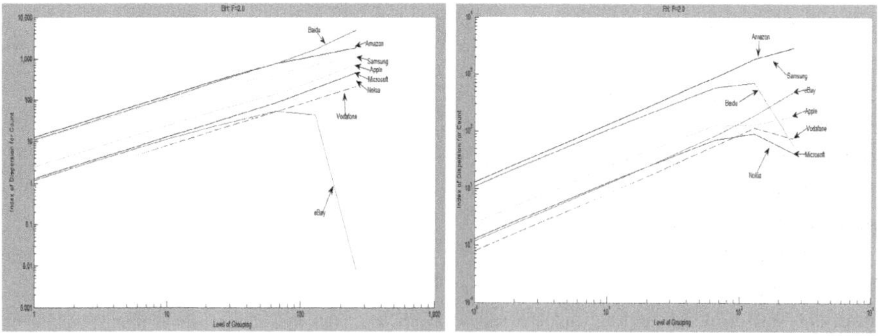

Fig. 3. Forward and Backward IDC Graphs (F=2.0, normal Variance)

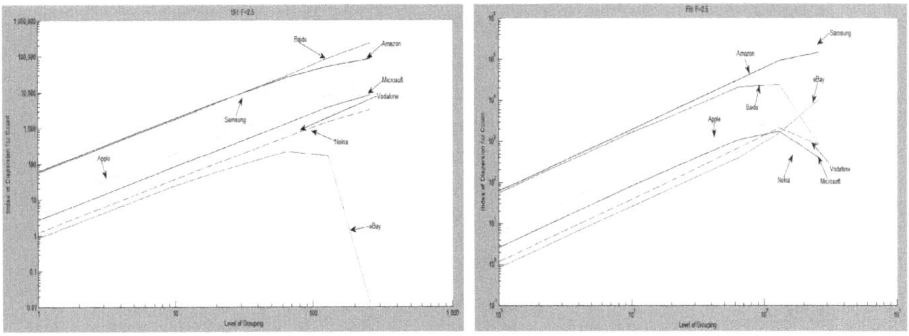

Fig. 4. Forward and Backward IDC Graphs (F=2.5, near 3rd moment)

From which we can see, each company may stay together or apart from each other, depends on which Fractional "angle" F you examine them.

Table 1. Diversity and Reversibility Comparison

Company	Re(FH)	Im(FH)	Re(BH)	Im(BH)	Re(IH)	Im(IH)
	F=2.5	F=2.5	F=1.5	F=1.5	ΔF=0.5	ΔF=0.5
eBay	1.2529	-0.0477	0.0645	0.0196	0.5942	-0.0337
Microsoft	0.9756	0.0308	0.7929	-0.0103	0.0914	0.0206
Baidu	0.7681	0.0285	0.8119	-0.0088	-0.0219	0.0186
Vodafone	1.1049	0.0213	0.7720	-0.0070	0.1665	0.0142
Nokia	0.9958	0.0151	0.7891	-0.0059	0.1034	0.0105
Apple	1.0746	0.0137	0.7870	-0.0042	0.1438	0.0090
Samsung	1.2304	-0.0108	0.7128	0.0022	0.2588	-0.0065
Amazon	1.2035	0.0052	0.7332	-0.0015	0.2352	0.0034

From above table, we can see that eBay is one time stock, if you missed the chance to be rich, it will never come back to you! The imaginary part of eBay is high, since the number of M&A went through is on top of all other companies, it has great diversity of products and capitalism culture as well. Baidu is the one in China, influenced by local socialism policy, its reversibility is negative. Samsung is the one from Korea, influenced by tight-controlled family-oriented operation strategy, its diversity is also negative.

7 Conclusion

This study has been able to construct a complex Hurst for fractional analysis of fractal stock market fluctuation using fractional norm basis with simple algorithm coded in Matlab, other similar tools can be used too. This novel complex Hurst model reveals extra performance index over the traditional real Hurst model. The study also provided graphical comparisons of the traditional calculation of the integer dimension with that of the fractional dimension. The results of the complex Hurst calculation show the difference between well established unified stock and new emerging fragmented stock, that traditional Hurst definition and calculation may overlook.

It is envisioned that the complex model is a good alternative in the stock market or risk analysis and will go a long way to help developing Complex Fractal theory and easier practical applications. The calculation is fairly simple, accurate enough and backward compatible with traditional model and calculations. The example code is available from Matlab file exchange server.

References

1. Zunino, L., Tabak, B.M., Figliola, A., Pérez, D.G., Garavaglia, M., Rosso, O.A.: A multifractal approach for stock market inefficiency. Physica A 387, 6558–6566 (2008)
2. Brooks, R.: Elephants Don't Play Chess. Robotics and Autonomous Systems 6, 3–15 (1990)
3. Buchanan, B.G.: A (Very) Brief History of Artificial Intelligence. AI Magazine, 53–60 (2005)
4. Hernandez-Orallo, J., Dowe, D.L.: Measuring Universal Intelligence: Towards an Anytime Intelligence Test. Artificial Intelligence Journal 174(18), 1508–1539 (2010)
5. Gerla, G.: Fuzzy Logic Programming and fuzzy control. Studia Logica 79, 231–254 (2005)
6. Davidovitch, L., Parush, A., Shtub, A.: Simulation-based Learning: The Learning-Forgetting-Relearning Process and Impact of Learning History. Computers & Education 50(3), 866–880 (2008)
7. Peters, E.E.: Fractal Market Analysis: Applying Chaos Theory to Investment and Economics. Wiley (1994)
8. Chisholm, A.M.: An Introduction to International Capital Markets: Products, Strategies, Participants. Wiley (2009)
9. Heath, M.T.: Scientific Computing: An Introductory Survey. McGraw-Hill (1997)
10. Li, M., Zhao, W., Jia, W., Long, D.-Y., Chi, C.-H.: Modeling autocorrelation functions of self-similar teletraffic in communication networks based on optimal approximation in Hilbert space. Applied Mathematical Modeling 27(3), 155–168 (2003)
11. Geniting, T., Schlather, M.: Stochastic Models That Separate Fractal Dimension and Hurst Effect. SIAM Review (Print) 46 22, 269–282 (2004)
12. Gneiting, T., Sasvári, Z., Schlather, M.: Analogies and Correspondences Between Variograms and Covariance Functions. NRCSE. Technical Report Series. NRCSE-TRS No. 056, pp. 617–630 (October 12, 2000)
13. Kincaid, D., Cheney, W.: Numerical Analysis. American Mathematical Society (2009)
14. Atkinson, K., Han, W.: Theoretical Numerical Analysis: A Functional Analysis Framework. Springer (2001)
15. Yosida, K.: Functional Analysis, 6th edn. Springer (1980)
16. Zeidler, E.: Applied Functional Analysis: Applications to Mathematical Physics. Applied Mathematical Sciences 108 (1995)
17. Lax, P.: Functional Analysis. Wiley-Interscience, New York (2002)
18. Reed, M., Simon, B.: Methods of Modern Mathematical Physics. Functional Analysis, vol. 1. Academic Press (1980)
19. Falconer, K.J.: Fractal Geometry-Mathematical Foundations And Applications. John Wiley (1990)
20. Feder, J.: Fractals. Plenum Press, London (1988)
21. Kallenberg, O.: Random Measures, 4th edn. Academic Press, New York (1986)

22. Daley, D.J., Vere-Jones, D.: An Introduction to the Theory of Point Processes. Springer, New York (1988)
23. Mandelbrot, B.B.: The Fractal Geometry of Nature, New York (1982)
24. Mehrvar, H.R., Le-Ngove, T., Huang, J.: Performance Evaluation of Bursty Traffic Using Neural Networks. In: Canadian Conference Electrical and Computer Engineers, vol. 2, pp. 955–958 (1996)
25. Tedeschini Lalli, E.: Sierpinski Triangles in Stone on Medieval Floors in Rome. APLIMAT Journal of Applied Mathematics 4, 114–122 (2011)
26. Menger, K., Dimensionstheorie, B.G Teubner Publishers (1928)
27. Menger, K.: Allgemeine Räume, Cartesische Räume. I., Communications to the Amsterdam Academy of Sciences (1926)
28. Mandelbrot, B.B.: Self-affine fractals and fractal dimension. Physical Script 32, 257–260 (1985)
29. Mandelbrot, B.B.: The Random Character of Stock Prices, pp. 165–179. Massachusetts Institute of Technology Press, Cambridge (1964)
30. Hoel, P.G.: On Indices of Dispersion. Annals of Mathematical Statistics 14(2), 155–162 (1943)
31. Gusella, R.: Characterizing the Variability of Arrival Process with Index of Dispersion. IEEE JSAC 9(2), 203–211 (1991)
32. Hurst, H.E.: Trans. Am. Soc. Civ. Eng. 116, 770 (1951)
33. Hurst, H.E., Black, R.P., Simaika, Y.M.: Long-term storage: an experimental study. Constable, London (1965)
34. Huang, J.S., Huang, J.M., Lu, G.Q.: An Interpretation of Quantum Entanglement and the Nature of the Universe. In: ICSS 2014 (2014)

Water Cycle and Artificial Bee Colony Based Algorithms for Optimal Order Allocation Problem with Mixed Quantity Discount Scheme

Chanikarn Praepanichawat*, Charoenchai Khompatraporn, Chorkaew Jaturanonda, and Chiranya Chotyakul

Department of Production Engineering, Faculty of Engineering, King Mongkut's University of Technology Thonburi (KMUTT), Bangkok, Thailand
{Chanikarn,enjoy.chot}@gmail.com,
{charoenchai.kho,chorkaew.jat}@kmutt.ac.th

Abstract. Supplier selection is one of the most important activities in purchasing management. Once the suppliers are determined, the proper allocation of the order among the suppliers can greatly help the company to reduce the raw material and production costs. In this paper, the order allocation with quantity discount of a single product is considered. The product can be offered with either an all unit discount model or an incremental discount one. Since the problem is NP-hard, three metaheuristics are applied to solve the problem. The metaheuristics are water cycle algorithm, artificial bee colony algorithm and hybrid water cycle-artificial bee colony algorithm. The results obtained from these algorithms are then compared.

Keywords: Water cycle algorithm, Hybrid water cycle and artificial bee colony algorithm, Order allocation problem, Quantity discount.

1 Introduction

A large portion of the product cost in many manufacturing industries is from the cost of raw materials. In some cases, this cost can be accounted for 70% of the product cost, and may go up to 80% in hi-tech firms. It is thus important for the management to recognize the importance of reducing the cost of material procurement to improve the competitiveness of the firms [1-4].

Basically, there are two types of the supplier selection problem: single sourcing where only one supplier can satisfy all buyer's requirements and multiple sourcing where one supplier cannot satisfy all buyer's requirements due to supplier capacity limitation or sometimes to reduce the risk of supply interuption [4]. The situation becomes more complicated when the suppliers motivate their customers to buy more by offering quantity discount. Two popular quantity discount schemes are called all unit discount and incremental discount. Moreover, some studies in this area may be

* Corresponding author.

M. Gen et al. (eds.), *Industrial Engineering, Management Science and Applications 2015*,
Lecture Notes in Electrical Engineering 349, DOI: 10.1007/978-3-662-47200-2_26

interested in cases with pure all unit discount or pure incremental discount. It is possible in an actual setting that different suppliers may offer different discount models (or a mixed discount scheme). In addition, the suppliers may impose a fine if the buyer purchases less than a prespecified amount in exchange for preferably allotting their capacities for the buyer as explored in Chotyakul et al. (2012).

The supplier selection and quantity allocation decision problem under quantity discount environment is an NP-Hard problem [5]. In previous studies, there are several methods used in solving this problem, and may be classified into three groups: (i) exact methods, (ii) heuristics, and (iii) metaheuristics algorithm such as genetic algorithm (GA) [1] and artificial bee colony (ABC) [7]. The first two methods sometimes require a long computational time to find the optimal solution, whereas metaheuristics may obtain near optimal solutions within a reasonable amount of time. Thus, this paper adopts metaheuristics to solve the problem [5,6].

Recently, a new algorithm called water cycle algorithm (WCA) was developed and tested on truss structure design problems. Their results showed that the WCA performed better than standard GA and particle swarm optimization [8,9]. Inspired by their success, we apply the WCA on an order allocation problem with mixed quantity discount. We also develop a hybrid water cycle-artificial bee colony algorithm (HWAA) and test it on the same problem. These results are then compared with those from ABC.

This paper is organized as follows. Section 2 presents a mathematical model of the order allocation problem with a mixed quantity discount scheme. Section 3 briefly describes the WCA and HWAA. The numerical experiment is explained in Section 4. Section 5 discusses the results. Finally, Section 6 concludes the paper.

2 Mathematical Model

2.1 Assumptions and Notation

Assumptions:

- The demand for the product is known and may be fulfilled by the suppliers.
- Each supplier offers one of the following quantity discount schemes: all unit discount or incremental discount.
- Each supplier imposes a minimal monetary value (MMV) constraint.
- The supply capacity of each supplier is finite.
- The buyer purchases to fulfill the demand of merely one period.
- The product must be purchased in whole units.

Notation:

j Index of suppliers ; $j = 1, 2, 3,\ldots,S$
k Index of discount intervals ; $k = 1, 2, 3,\ldots,K_j$
S Total number of suppliers
s_1 Number of suppliers who offer all unit discount scheme

s_2 Number of suppliers who offer incremental discount scheme
K_j Total number of discount intervals of supplier j
x_j Purchased quantity from supplier j who offers all unit discount
y_j Purchased quantity from supplier j who offers incremental discount
$P_k(x_j)$ Cost function of the purchased quantity from supplier j who offers all unit discount in discount interval k
$P_k(y_j)$ Cost function of the purchased quantity from supplier j who offers incremental discount in discount interval k
p_j Unit price of product offered by supplier j before the discount is applied
u_{jk} Upper bound of the quantity discount interval k offered by supplier j
l_{jk} Lower bound of the quantity discount interval k offered by supplier j
d_{jk} Unit price discount rate on discount interval k offered by supplier j
C_j Maximum supply capacity of supplier j
D Total demand of the product
V_j^m Minimum purchase agreed with supplier j
f_j Fine rate applied to the unpurchased amount from supplier j in comparison to the minimum purchase agreed

2.2 Mathematical Formulation

Using the above notation, the problem is formulated as the following.

Objective Function. The objective is to minimize the total purchase cost as shown in Eq. (1). The first term in the objective function consists of the costs incurred by the all unit discount and by the incremental discount, respectively. The second term in the objective function is the cost due to the penalty imposed on the buyer (the fine) if the MMV is not met.

$$\text{Minimize Total Purchase Cost} = \sum_{j=1}^{s_1}\left(\sum_{k=1}^{K_j} P_k(x_j) + f_j\, max\left\{V_j^m - \sum_{k=1}^{K_j} P_k(x_j), 0\right\}\right) + \sum_{j=s_1+1}^{S}\left(\sum_{k=1}^{K_j} P_k(y_j) + f_j\, max\left\{V_j^m - \sum_{k=1}^{K_j} P_k(y_j), 0\right\}\right) \quad (1)$$

Constraints. *Capacity constraint.* The purchased quantity from each supplier is less than or equal to the supplier's production capacity.

$$x_j \le C_j \text{ where } j = 1,2,\dots,s_1 \text{ and } y_j \le C_j \text{ where } j = s_1+1, s_1+2,\dots,S \quad (2)$$

Demand constraint. The sum of the quantities purchased from all suppliers must satisfy the demand of the product.

$$\sum_{j=1}^{S} x_j \ge D \quad (3)$$

Discount constraints. The purchased quantity from each supplier who offers all unit discount x_i and incremental discount y_i must be equal to or between there lower and upper bound of the discount interval k. Therefore, the only one of the discount interval k that the purchased quantity from the selected supplier falls within must be selected. The purchase cost of the purchase quantity to be charged by the total price under all-unit, $P_k(x_i)$ and incremental, $P_k(y_i)$ discount can be formulated as:

- All unit discount

$$P_k(x_j) = \begin{cases} x_j p_j(1-d_{jk}) & ; l_{jk} \le x_j \le u_{jk} \; ; 1 \le j \le s_1, \forall k \\ 0 & ; Otherwise \end{cases} \tag{4}$$

- Incremental discount

$$P_k(y_j) = \begin{cases} \left(\left(\sum_{m=1}^{k-1}(u_{jm} - u_{j,m-1})(1-d_{jm}) + (y_j - u_{j,k-1})(1-d_{jk})\right) p_j \right. \\ \qquad ; l_{jk} \le y_j \le u_{jk} \; if \; s_1 + 1 \le j \le S, \forall k \\ 0 \qquad ; Otherwise \end{cases} \tag{5}$$

Nonnegativity constraint. Eq.(6) specifies that the decision variables are nonnegative integers.

$$x_j \ge 0 \; where \; j = 1, 2, \ldots, s_1 \; and \; y_j \ge 0 \; where \; j = s_1 + 1, s_1 + 2, \ldots, S \tag{6}$$

3 Algorithms

3.1 Water Cycle Algorithm

The water cycle algorithm (WCA) is a relatively new algorithm proposed by Eskandar et al. in 2012. The algorithm is inspired by observation of water cycle and how rivers and streams flow downhill towards the sea in the real world. The initial population in the algorithm is called raindrops. The best (in term of the objective function value) of the raindrops is chosen as the sea. Then some top raindrops are chosen as rivers and the rest are streams. Like in the nature, streams are created from raindrops and flow downhill from one place to another and join each other to form new rivers and end up in the sea (the best point). The WCA is adapted to solve the optimal order allocation problem under a mixed quantity discount scheme with the following details:

Step 1: Randomize the initial raindrops by Eq. (7)

$$x_{ij} = rand(0,1) * x_j^{max} \; ; i = 1, 2, 3, \ldots, N_{pop} \; and \; j = 1, 2, 3, \ldots, N_{var} \tag{7}$$

Where x_{ij} is the product quantity purchased from supplier *j* at raindrop number *i*, x_j^{max} is the maximum capacity of supplier *j* and *rand(0,1)* is a random number between 0 and 1. We denote N_{pop} as the initial population of raindrops and $N_{var} = S$ or the number of suppliers.

Step 2: Calculate the objective function value (total purchase cost) of each raindrop

$$Cost_i = f(x_{ij}) \tag{8}$$

The best of the raindrops is chosen as the sea, some few top raindrops are chosen as the rivers and the rest are streams.

Step 3: Calculate the intensity of flow (NS_n) using Eq. (9) to assign the streams to rivers and rivers to sea

$$NS_n = round\left\{\left|\frac{Cost_n}{\sum_{i=1}^{Nsr} Cost_i}\right| \times N_{Raindrops}\right\} ; \ n = 1,2,\dots,N_{sr} \tag{9}$$

Where N_{sr} is the number of rivers and sea

Step 4: Streams are created from raindrops and flow to join each other to form new rivers and sea. The new positions of streams and rivers is determined by Eqs. (10) and (11).

$$v_{ij}^{Stream} = x_{ij}^{Stream} + rand(0,1) \times C \times \left(x_{ij}^{River} - x_{ij}^{Stream}\right) \tag{10}$$

$$v_{ij}^{River} = x_{ij}^{River} + rand(0,1) \times C \times \left(x_{ij}^{Sea} - x_{ij}^{River}\right) \tag{11}$$

Where C is the value between 1 and 2 (near to 2). The best value for C may be chosen as 2. If the solution given by a stream is better than its connecting river, the positions of river and stream are exchanged (i.e. the stream becomes the river and vice versa). Such exchange can similarly happen for rivers and sea.

Step 5: The evaporation process is proposed to avoid getting trapped in local optima. The following pseudo code shows how to determine whether or not river flows to the sea.

$if \quad \left|x_i^{Sea} - x_i^{River}\right| < d_i^{max} \qquad i = 1,2,3,\dots,N_{sr} - 1$

Evaporation and raining process
End if

Where d_i^{max} is a small number close to zero and use for control the search intensity near the sea. Reduce the value of d_i^{max} by Eq. (12)

$$d_{i+1}^{max} = d_i^{max} - \frac{d_i^{max}}{max\,iteration} \tag{12}$$

Step 6: After the evaporation process, the raining process is executed by:

$$x_{new}^{Stream} = x_j^{min} + rand(0,1) \times \left(x_j^{max} - x_j^{min}\right) \tag{13}$$

3.2 Hybrid Water Cycle-Artificial Bee Colony Algorithm

The hybrid water cycle-artificial bee colony algorithm (HWAA) integrates two nature-inspired metaheuristics: WCA and artificial bee colony (ABC) algorithm. More details of the ABC algorithm can be found in the literature such as [11,12]. The main structure of HWAA is based on WCA except for the step of finding new positions of streams in Eq. (10). We replace this step by a step when a new candidate food source position in ABC algorithm is determined. The new step is:

$$v_{ij}^{Stream} = x_{ij}^{River} + rand(0,1) \times \left(x_{ij}^{River} - x_{rj}^{Stream}\right) \quad (14)$$

Where r is new random candidate of existing food source

4 Numerical Experiment

From the model in Section 2, several parameters are set with the following values.

1. The total demand is 35,000 units
2. Possible discount rates that determine discount intervals (d_k): 0% 3% 10%, 0% 5% 10%, 0% 5% 7%
3. The penalty rate (f_j) of the unmet MMV: 20%
4. There are two cases considered here. The first one is composed of 10 suppliers and the other 20 suppliers
5. Six cases of different ratios of the number of suppliers who offer the all unit discount vs. the incremental discount over the total number of suppliers are: 100:0, 80:20, 60:40, 40:60, 20:80 and 0:100

Details of the discount intervals of the suppliers are shown in Table 1 and 2.

Table 1. Lower bound and upper bound of quantity discount for 10 suppliers

k		Supplier (j)									
		1	**2**	**3**	**4**	**5**	**6**	**7**	**8**	**9**	**10**
1	l_{ik}	0	0	0	0	0	0	0	0	0	0
	u_{ik}	799	499	1,679	1,719	579	859	1,099	1,639	419	699
2	l_{ik}	800	500	1,680	1,720	580	860	1,100	1,640	420	700
	u_{ik}	2,799	1,749	5,879	6,019	2,029	3,009	3,829	5,739	1,469	2,449
3	l_{ik}	2,800	1,750	5,880	1,020	2,030	3,010	3,850	5,740	1,470	2,450
	u_{ik}	4,000	2,500	8,400	8,600	2,900	4,300	5,500	8,200	2,100	3,500

Table 2. Lower bound and upper bound of quantity discount for 20 suppliers

k		Supplier (*j*)									
		1	**2**	**3**	**4**	**5**	**6**	**7**	**8**	**9**	**10**
1	l_{ik}	0	0	0	0	0	0	0	0	0	0
	u_{ik}	380	480	860	740	280	360	300	760	440	340
2	l_{ik}	381	481	861	741	281	361	301	761	441	341
	u_{ik}	1,330	1,680	3,010	2,590	980	1,260	1,050	2,660	1,540	1,190
3	l_{ik}	1,331	1,681	3,011	2,591	981	1,261	1,051	2,661	1,541	1,191
	u_{ik}	1,900	2,400	4,300	3,700	1,400	1,800	1,500	3,800	2,200	1,700
k		**11**	**12**	**13**	**14**	**15**	**16**	**17**	**18**	**19**	**20**
1	l_{ik}	0	0	0	0	0	0	0	0	0	0
	u_{ik}	420	400	800	300	260	240	960	320	460	900
2	l_{ik}	421	401	801	301	261	241	961	321	461	901
	u_{ik}	1,470	1,400	2,800	1,050	910	840	3,360	1,120	1,610	3,150
3	l_{ik}	1,471	1,401	2,801	1,051	911	841	3,361	1,121	1,611	3,151
	u_{ik}	2,100	2,000	4,000	1,500	1,300	1,200	4,800	1,600	2,300	4,500

Table 3. Supply information of the 10 suppliers case

Supplier (*j*)	Capacity C_j (units)	Unit Price P_j ($)	MMV V_j^m ($)	Supplier (*j*)	Capacity C_j (units)	Unit Price P_j ($)	MMV V_j^m ($)
1	4,000	106	84,800	6	4,300	94	80,840
2	2,500	91	45,500	7	5,500	107	117,700
3	8,400	92	154,560	8	8,200	99	162,360
4	8,600	106	182,320	9	2,100	91	38,220
5	2,900	109	63,220	10	3,500	95	66,500

Table 4. Supply information of the 20 suppliers case

Supplier (*j*)	Capacity C_j (units)	Unit Price P_j ($)	MMV V_j^m ($)	Supplier (*j*)	Capacity C_j (units)	Unit Price P_j ($)	MMV V_j^m ($)
1	1,900	106	40,280	11	2,100	98	41,160
2	2,400	91	43,680	12	2,000	93	37,200
3	4,300	92	79,120	13	4,000	104	83,200
4	3,700	106	78,440	14	1,500	107	32,100
5	1,400	109	30,520	15	1,300	103	26,780
6	1,800	94	33,840	16	1,200	105	25,200
7	1,500	107	32100	17	4,800	100	96000
8	3,800	99	75,240	18	1,600	105	33,600
9	2,200	91	40,040	19	2,300	95	43700
10	1,700	95	32,300	20	4,500	93	83,700

The WCA and HWAA described in the previous section were then applied to solve the above problem. Both algorithms were written in C++ programming language and run on a computer with Intel® Core™ i5 2.50GHz and 4GB RAM. We set N_{pop}, N_{sr} and d^{max} to be equaled to 100, 5 and 1×10-5 for the WCA, and to 100, 25 and 1×10-5 for HWAA. Each experiment setting was run 30 times (30 trials). The results are compared with those obtained by Chotyakul et al. (2012).

5 Results and Discussion

The results obtained from all three algorithms can be shown in Table 5 to Table 8. For the case of 10 suppliers, the best and the mean value of the objective function are shown in Table 5 and 6. Table 7 and 8 show the results for the case of 20 suppliers. In each table, the first column gives the percentage ratio of the number of supplier offering all unit discount and incremental discount. The next column shows the discount rate of the initial unit prices given in Table 3 and 4. The next part of the table shows the best or the mean value of the objective function obtained in 30 trials from each of the algorithms. Taking the difference between the results of any pair of the algorithms we obtain the results as shown in the last part the table. The computation time required for all 30 trials is less than 2 minutes for the case of 10 suppliers, and for 20 suppliers the time needed is less than 11 minutes.

For the case of 10 suppliers when consider the best objective function value, the HWAA tends to perform better than the other algorithms in the scenarios of 100:0 and 80:20 ratios of the number of suppliers offering the all unit discount and incremental discount, and yields similar results to the ABC algorithm for the 20:80 ratios. The overall performance in this case, the WCA gives better results than the ABC algorithm from the best value perspective, but the ABC is better from the mean value perspective.

When consider the case of the best values for 20 suppliers, the HWAA gives the best results when the ratios of the number of suppliers offering all unit discount and incremental discount are 100:0, 80:20, 60:40, 40:60, and 20:80. For the scenarios under the 0:100 ratio, HWAA gives the same results to those of the ABC algorithm. For this particular case the ABC algorithm finds better results than the WCA in most scenarios. Similar results can be seen when there are 20 suppliers involve and consider the mean values from the 30 trials. However, the ABC algorithm yields slightly little better results than the HWAA in four scenarios. The WCA performs the worst among the three algorithms in all scenarios.

Table 5. Summary of the best value of the total purchase cost in 10 suppliers case

% All Unit Discount : % Incremental Discount	Discount	Best (USD)			Difference (USD)		
		ABC	WCA	HWAA	WCA - ABC	WCA-HWAA	ABC - HWAA
100:0	0%, 3%, 10%	3,069,090	3,067,620	3,067,604	-1469	16	1485
	0%, 5%, 10%	3,062,521	3,060,344	3,060,333	-2177	11	2188
	0%, 5%, 7%	3,152,498	3,150,833	3,150,831	-1665	1	1667
80:20	0%, 3%, 10%	3,097,888	3,096,419	3,096,402	-1469	17	1485
	0%, 5%, 10%	3,086,083	3,083,899	3,083,895	-2184	4	2188
	0%, 5%, 7%	3,165,064	3,163,402	3,163,398	-1662	4	1667
60:40	0%, 3%, 10%	3,142,916	3,142,919	3,142,916	3	3	-1
	0%, 5%, 10%	3,127,771	3,126,299	3,126,298	-1472	1	1473
	0%, 5%, 7%	3,189,877	3,188,759	3,188,756	-1118	2	1121
40:60	0%, 3%, 10%	3,167,551	3,167,552	3,167,551	1	1	0
	0%, 5%, 10%	3,148,677	3,147,639	3,147,634	-1038	5	1043
	0%, 5%, 7%	3,202,383	3,201,608	3,201,607	-775	1	777
20:80	0%, 3%, 10%	3,216,638	3,216,639	3,216,638	0	0	0
	0%, 5%, 10%	3,191,514	3,191,515	3,191,514	1	1	0
	0%, 5%, 7%	3,229,261	3,229,262	3,229,261	0	0	0
0:100	0%, 3%, 10%	3,249,721	3,249,721	3,249,721	0	0	0
	0%, 5%, 10%	3,218,506	3,218,506	3,218,506	0	0	0
	0%, 5%, 7%	3,243,165	3,243,165	3,243,165	0	0	0

Table 6. Summary of the mean value of the total purchase cost in 10 suppliers case

% All Unit Discount : % Incremental Discount	Discount	Mean (USD)			Difference (USD)		
		ABC	WCA	HWAA	WCA - ABC	WCA-HWAA	ABC - HWAA
100:0	0%, 3%, 10%	3,069,090	3,075,163	3,070,108	6073	5055	-1018
	0%, 5%, 10%	3,062,521	3,063,755	3,062,079	1234	1676	442
	0%, 5%, 7%	3,152,531	3,153,600	3,151,667	1069	1933	864
80:20	0%, 3%, 10%	3,097,888	3,103,786	3,099,657	5898	4129	-1769
	0%, 5%, 10%	3,086,083	3,089,034	3,085,583	2951	3452	500
	0%, 5%, 7%	3,165,160	3,165,958	3,164,094	797	1863	1066
60:40	0%, 3%, 10%	3,143,052	3,143,336	3,143,229	284	108	-176
	0%, 5%, 10%	3,127,771	3,131,452	3,128,207	3681	3245	-435
	0%, 5%, 7%	3,189,885	3,190,975	3,188,927	1090	2048	957
40:60	0%, 3%, 10%	3,167,551	3,167,579	3,167,552	28	27	-1
	0%, 5%, 10%	3,148,677	3,152,427	3,148,892	3751	3535	-215
	0%, 5%, 7%	3,202,420	3,202,541	3,201,607	121	935	814
20:80	0%, 3%, 10%	3,216,638	3,216,770	3,216,638	132	132	0
	0%, 5%, 10%	3,191,514	3,191,661	3,191,514	146	146	0
	0%, 5%, 7%	3,229,261	3,229,578	3,229,261	317	317	0
0:100	0%, 3%, 10%	3,249,721	3,249,724	3,249,721	3	3	0
	0%, 5%, 10%	3,218,506	3,218,507	3,218,506	1	1	-1
	0%, 5%, 7%	3,243,165	3,243,168	3,243,165	3	2	-1

From the overall performance in all 36 scenarios each consider from the best value and the mean value perspectives, or 72 subcases, the HWAA gives better or similar results compared to the other algorithms in 57 subcases. When compare the WCA with the ABC algorithm, the WCA tend to yield better solutions from the best value perspective, whereas the ABC algorithm is better if the mean value criterion is used.

Table 7. Summary of the best value of the total purchase cost in 20 suppliers case

% All Unit Discount : % Incremental Discount	Discount	Best (USD)			Difference (USD)		
		ABC	WCA	HWAA	WCA - ABC	WCA-HWAA	ABC - HWAA
100:0	0%, 3%, 10%	3,052,077	3,052,532	3,050,855	456	1677	1222
	0%, 5%, 10%	3,045,200	3,044,425	3,043,314	-776	1111	1887
	0%, 5%, 7%	3,133,961	3,133,143	3,132,626	-819	517	1335
80:20	0%, 3%, 10%	3,099,573	3,099,291	3,098,949	-282	342	624
	0%, 5%, 10%	3,089,313	3,088,466	3,088,301	-848	165	1012
	0%, 5%, 7%	3,162,335	3,161,783	3,161,105	-552	678	1230
60:40	0%, 3%, 10%	3,111,186	3,111,286	3,111,056	100	230	130
	0%, 5%, 10%	3,103,547	3,103,461	3,103,357	-86	104	190
	0%, 5%, 7%	3,170,159	3,169,679	3,169,443	-480	236	716
40:60	0%, 3%, 10%	3,152,629	3,152,670	3,152,499	41	171	130
	0%, 5%, 10%	3,137,455	3,137,472	3,137,257	17	214	198
	0%, 5%, 7%	3,188,243	3,187,734	3,187,527	-509	208	716
20:80	0%, 3%, 10%	3,184,461	3,184,509	3,184,461	48	47	-1
	0%, 5%, 10%	3,164,894	3,164,998	3,164,907	104	91	-13
	0%, 5%, 7%	3,204,167	3,203,882	3,203,728	-285	154	439
0:100	0%, 3%, 10%	3,228,975	3,229,067	3,228,975	92	92	0
	0%, 5%, 10%	3,198,676	3,198,758	3,198,676	83	83	0
	0%, 5%, 7%	3,224,385	3,224,498	3,224,385	114	114	0

Table 8. Summary of the mean value of the total purchase cost in 20 suppliers case

% All Unit Discount : % Incremental Discount	Discount	Mean (USD)			Difference (USD)		
		ABC	WCA	HWAA	WCA - ABC	WCA-HWAA	ABC - HWAA
100:0	0%, 3%, 10%	3,052,079	3,058,430	3,052,089	6351	6341	-10
	0%, 5%, 10%	3,045,200	3,049,655	3,043,664	4455	5991	1536
	0%, 5%, 7%	3,133,961	3,134,726	3,132,638	764	2088	1323
80:20	0%, 3%, 10%	3,099,595	3,105,104	3,099,450	5509	5655	146
	0%, 5%, 10%	3,089,439	3,092,244	3,088,734	2806	3510	704
	0%, 5%, 7%	3,162,335	3,163,696	3,161,149	1361	2546	1185
60:40	0%, 3%, 10%	3,111,293	3,111,970	3,111,163	677	807	130
	0%, 5%, 10%	3,103,737	3,105,203	3,103,506	1466	1697	231
	0%, 5%, 7%	3,170,159	3,170,264	3,169,443	105	821	716
40:60	0%, 3%, 10%	3,152,706	3,154,795	3,152,563	2089	2232	143
	0%, 5%, 10%	3,137,660	3,139,015	3,137,517	1355	1498	143
	0%, 5%, 7%	3,188,243	3,188,478	3,187,527	235	951	716
20:80	0%, 3%, 10%	3,184,461	3,186,784	3,184,466	2323	2318	-6
	0%, 5%, 10%	3,165,086	3,166,259	3,164,977	1173	1282	109
	0%, 5%, 7%	3,204,167	3,204,502	3,203,728	335	774	439
0:100	0%, 3%, 10%	3,228,975	3,229,546	3,228,975	571	570	-1
	0%, 5%, 10%	3,198,676	3,199,259	3,198,676	584	583	-1
	0%, 5%, 7%	3,224,385	3,224,943	3,224,385	559	559	0

6 Conclusion

In this paper, the order allocation with mixed quantity discount schemes is considered. The problem is formulated and a numerical example is set for cases of 10 and 20 suppliers with different mixes of all unit discount and incremental discout, as well as discount rates. Since the problem is NP-hard, we choose to apply the water cycle algorithm (WCA) and hybrid water cycle-artificial bee colony algorithm (HWAA) to find the optimal solutions. To HWAA developed in this research utilizes the core methodology from the WCA but replacing a step in the WCA with an exploration step (finding a new candidate food source) of the ABC algorithm. The results obtained from the ABC algorithm of this silimar problem appeared in the literature are also considered. Among the three algorithms, the results show that the HWAA tends to find the best solutions in most scenarios. When compare only the WCA and ABC algorithm, the WCA yields better results if the best value criterion is used, but the ABC is better from the mean value perspective.

References

1. Setak, M., Sharifi, S., Alimohammadian, A.: Supplier Selection and Order Allocation Models in Supply Chain Management: A Review. World Applied Sciences Journal 18, 55–72 (2012)
2. Ghodsypour, S.H., Brien, C.O.: The Total Cost of Logistics in Supplier Selection, under Conditions of Multiple Sourcing. Multiple Criteria and Capacity Constraint, International Journal of Production Economics 73, 15–27 (2001)
3. Charles, A.W., Current, J.R., Benton, W.C.: Vendor Selection Criteria and Methods. European Journal of Operational Research 50, 2–18 (1991)
4. Xia, W., Wu, Z.: Supplier Selection with Multiple Criteria in Volume. Omega The International Journal of Management Science 35, 494–504 (2007)

5. Burke, G.J., Carrillo, J., Vakharia, A.J.: Heuristics for Sourcing from Multiple Suppliers with Alternative Quantity Discounts. European Journal of Operational Research 186, 317–329 (2008)
6. Goossens, D.R., Maas, A.J.T., Spieksma, F.C.R., van de Klundert, J.J.: Exact Algorithms for Procurement Problems under a Total Quantity Discount Structure. European Journal of Operational Research 178, 603–626 (2007)
7. Chotyakul, C., Khompatraporn, C., Somboonwiwat, T., Jaturanonda, C.: Optimal Order Allocation Decision with Multiple Quantity Discount Schemes and Minimum Monetary Value Penalty by Artificial Bee Colony Algorithm. In: Proceedings of the 7th International Congress on Logistics and SCM Systems (ICLS 2012), Seoul, Korea (2012)
8. Eskandar, H., Sadollah, A., Bahreininejad, A.: Weight Optimization of Truss Structures Using Water Cycle Algorithm. International Journal of Optimization in Civil Engineering 3, 115–129 (2013)
9. Eskandar, H., Sadollah, A., Bahreininejad, A., Hamdi, M.: Water Cycle Algorithm - a Novel Metaheuristic Optimization Method for Solving Constrained Engineering Optimization Problems. Computer and Structure 110-111, 151–166 (2012)
10. Blum, C., Roli, A.: Hybrid Metaheuristics: An Introduction. In: Blum, C., Aguilera, M.J.B., Roli, A., Sampels, M. (eds.) Hybrid Metaheuristics an Emerging Approach to Optimization. SCI, vol. 114, pp. 1–30. Springer, Heidelberg (2008)
11. Akay, B., Karaboga, D.: Solving integer programming problems by using artificial bee colony algorithm. In: Serra, R., Cucchiara, R. (eds.) AI*IA 2009. LNCS, vol. 5883, pp. 355–364. Springer, Heidelberg (2009)
12. Akay, B., Karaboga, D.: A modified artificial bee colony algorithm for real-parameter optimization. Information Sciences 192, 120–142 (2012)

The Management of Assessment and Allocation of Marshalling Yards and Designation Their Catchment Areas

Juraj Camaj[*], Jana Lalinská, and Jaroslav Masek

Faculty of Operation and Economics of Transport and Communication / University of Zilina
Univerzitna 8215/1, 010 26 Zilina, Slovakia
{juraj.camaj,jana.lalinska,jaroslav.masek}@fpedas.uniza.sk

Abstract. The paper is focused on the operational model for transport the single wagon consignments on Slovak railway network using time-discrete train formation. The paper gives an overview of evaluation of railway stations by multi-criteria approach. The resulting indicators characterize the station based on evaluation scale. This model is one of the effective ways to meet the development strategy of the EU transport policy. By applying this model in practice, the transport company can guarantee a higher quality of service and expect increasing of transport performance. The model is also applicable to other railway networks. This model supplements a theoretical problem of train formation by new ways of looking at evaluation of factors affecting the organization of wagon flows.

1 Introduction

European Commission and government of European countries declared in the many official documents aim to shift part of the volume of freight transport from road to railway. This change should occur in particular to decrease pollution, reduce road fatal accidents and eliminate congestion on the road network in metropolitan areas.

One of the way, how to fulfil these goals is improving the flexibility and quality of services provided by railway cargo operators in the transport services segment of single wagon consignments. It is necessary that for this purpose create the right conditions and get back lost customers by offering innovative services.

However, the providing of services does not mean an automatic change of the volume of freight transport from road to rail. It need to be change the whole technological process, so that the service should be acceptable for the operators from the perspective of the customer for acceptable cost.

Requirement for the quickly, safe and affordable transport time may offer for the customers the model of Just in time technology. As a starting point to improve the current technology is use of time-discrete train formation ("night jump").

---

[*] Corresponding author.

M. Gen et al. (eds.), *Industrial Engineering, Management Science and Applications 2015*,
Lecture Notes in Electrical Engineering 349, DOI: 10.1007/978-3-662-47200-2_27

2 Formulation of Train Formation Problem

The train formation plan is prognostic model for determining the best organization of wagons flows to the trains. The problem is the variability of transport options of organization of wagon flows of freight trains (direction and movement from the forwarding station to the destination station) and the best combination gives the final solution.

The main idea of this model is based on a three-level hierarchical structure stations. Namely:

- stations authorized for sending cargo wagons – base station of the core network;
- satellite marshalling station (station of first train formation);
- main marshalling yards.

A schematic diagram of the model is in figure 1.

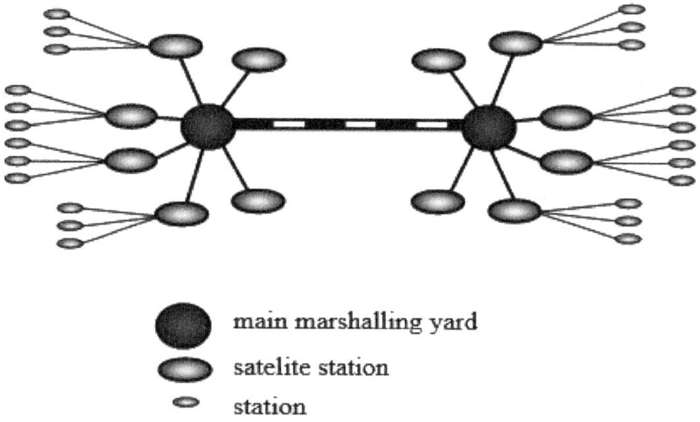

Fig. 1. The basic scheme of a three-level hierarchy stations

The station of core network belongs to the attraction area of only one satellite railway station. Each of satellite station belongs to the attraction area of only one main marshalling yard. Single wagons consignments are taken from the stations of the core network by pick-up trains to the nearest satellite station or directly to marshalling yard. In the satellite station are wagons connected to the train sets from pick-up trains to the sectional trains (for transport to the main marshalling yard). The main marshalling yard forms the direct trains to other major marshalling yards to their attraction area. The advantage is reducing redeployment of wagons (max. 2) during the transport. This technology reduces the average transport time of the wagon on the network. Described train formation technology can be called as a time-continuous or discrete time technology.

2.1 Time-Continuous Train Formation Technology

The model assumes full day train formation by collecting of wagons to train according norms of weight or length of train set. Criterion of optimization in model which was proposed is the average transport time of the wagon or time losses of wagon. Time loss due to simplification consist of transport time of the wagon in the sectional train between the satellite and the main marshalling yard and conversely – t1 and the residence time of the wagon while direct train is collecting in the first major marshalling yard – t2.

These times can be calculated by equation

$$t_1(x)=\frac{2d}{v} \quad [hours\ per\ wagon] \tag{1}$$

$$d=q\sqrt{\frac{p}{x}} \quad [\mathrm{km}] \tag{2}$$

$$t_2(x)=\frac{24\,mx\,(x-1)}{2n} \quad [hours\ per\ wagon] \tag{3}$$

To calculate the average time loss of the wagon is used the function:

$$t(x)=t_1(x)+t_2(x) \quad [hours\ per\ wagon] \tag{4}$$

where:

- x number of main marshalling yards;
- d average distance of train run from the satellite to the main yard [*km*]
- v average technical speed of train;[$km.h^{-1}$]
- p area of state [km^2];
- q average conversion factor square root of area attraction area of the main marshalling yard in length d;
- m the average number of wagons per train running between the main marshalling yard;
- n the average daily number of loaded and empty wagons moved between the main marshalling yards.

The optimal number of main marshalling yards is calculated of the function:

$$t'(x)=0 \tag{5}$$

This equation cannot be solved analytically. The value of "x" can be calculated using one of the methods of numerical mathematics. When omission of the definite article can be analytically calculated initial approximate solution:

$$x_0 = \sqrt[5]{\frac{q^2 n^2 p}{576 m^2 v^2}} \tag{6}$$

The biggest problem of the model is to determine the coefficient "q". Author of the normal forms of the attraction area is recommended $q = 4$.

Designed relations are the simplification of the problem and do not include other parts of the wagon stay on the network and the capacity of the main marshalling yards.

If the network as "s" marshalling yards, each marshalling yard needs on average $\frac{(s-x)}{x}$ the sorting tracks for train formation to own satellite station and "$x - 1$" the sorting tracks for train formation to other main marshalling yards.

This model requires a large number of sorting tracks and causes an increase of delays in the collecting process.

2.2 Time Discrete Train Formation Technology – "Night Jump"

This model is a qualitative upgrading time-continuous train formation. It is based on the same principles, i.e. three-level hierarchy stations, one-way transport service and creation of one group direct trains between all main marshalling yards.

Trains are created and runs only for a limited period during the day with compliance with timetables. This condition is based on customer requirements for the implementation of goods manipulations in daylight hours on working days.

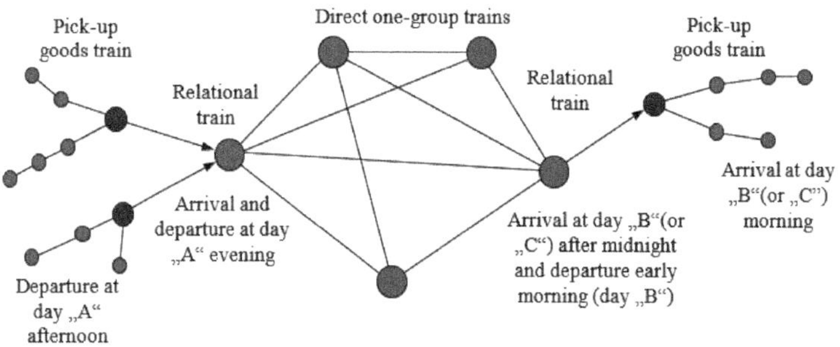

Fig. 2. Implementation of time-discrete train formation technology

After loading the goods to wagons in the afternoon of the first day (day "A") are wagons transported to the nearest main marshalling yard via satellite stations. Subsequently, it starts processing the target sets of trains and creation of direct one-group trains to all other main marshalling yards. These trains leave the marshalling yard in the evening hours. At the earliest way trains to more distant relations, taking care to ensure the transfer of all wagons from arriving trains to the departing trains. Wagons are collected separately according the target station in the own attraction area. They await the arrival of trains from other major marshalling yards. Processes in main marshalling yard during night hours is just focused on sorting wagons whose destination station in the attraction area of the marshalling yard.

The next day (day "B") for longer distant relations, exceptionally, on the third day (day "C"), from midnight come for sorting the one group train to the main marshalling yard from other major marshalling yards. Their wagons pass through to sectional trains to the satellite station or pick-up trains for stations of the core network. Activities in main marshalling yards finish after leaving the trains until late afternoon. In the satellite stations the trains are divided to distribution trains to operate the stations of the core network. These wagons are transported to their destination station in the morning.

The benefits of time discrete train formation:

- Shortening the period of transport according necessary technological time;
- Reducing the transport time of wagon - loading in the day "A", unloading in the day "B" (exceptionally day "C");
- Delivery of goods within logistic technology "just in time";
- Optimization of the rotations locomotives;
- Temporal separation of passenger and freight transport on railway network;
- High productivity of workers operating in the marshalling yard concentrated in the night hours, while the day-time activities marshalling yards are minimized and use of operational staff is minimal;
- During daylight hours can be realized the repairs and maintenance of equipment in the marshalling yards

3 Location of Main Marshalling Yards and Determination of Each Attraction Area

In this technology is one of fundamental problem the determining the location of the main marshalling yards. The main marshalling yards are in a model referred to as the station with the highest priority.

One of the main objectives is already determine the optimal number of main marshalling yards of the optimization model in time discrete train formation technology. The results of this optimization are determined the optimal variant of concentration and organization of wagon flows in the network.

3.1 The Background Information of the Mathematical Model for Allocation

The task of the mathematical model is selection from more station (theoretically suitable stations) the best one suited for placing the marshalling yard. In view of the foregoing factors appears to be the most preferred using a multi-criteria evaluation.

Multi-criteria evaluation methods can be used generally for comparison and subsequent selection of any objects on the basis of several indicators. Because of its ability to synthesize several different variables are particularly suitable into a quantitative statement of comprehensive indicator for the analysis of the position of the object (the marshalling yard) on the network.

The method allows to compare a set of several objects based on multiple characteristics of their activities and also to determine the order of placing the object being analysed.

3.2 The Criteria for Main Marshalling Yards

Selection can be performed for any point on the transport network, where is the possibility of building the new marshalling yard, ("building from scratch") or to determine directly choice the marshalling yard and therefore "only" determine the attraction area of for the specified marshalling yard.

The objective of the methods of multi-criteria evaluation is the transformation and synthesis values of different variables to a single summary indicator (the resulting characteristics). This indicator shows the comprehensive level of individual objects in the surveyed stations. It is a summary indicator representation of the overall level (importance) of the station or suitability for placement the marshalling yard of the station.

$$Kj = \frac{\sum_{i=1}^{n} a_{ji}}{\sum_{i=1}^{n} a_{mi}}.100 \quad ; j \in 1,\dots, m-1 \tag{7}$$

where:

Kj – coefficient of the j-th station;

$\sum_{i=1}^{n} a_{ji}$ – sum of the evaluation j-th station;

$\sum_{i=1}^{n} a_{mi}$ – sum of the evaluation the model the station

"Model station" is marked the fictional station that all evaluation factors meets in full. Thus, the value xm1, ... , xmn are the maximum. Summary of indicators reflects the importance of the point. The indicators will have a substantial impact in deciding where allocate the marshalling yard. The basis of multi-criteria evaluation is processing the initial matrix of objects and their characteristics. The objects represent all points (the station) on the network, which in each case meet the criteria for inclusion in a specific choice.

In the construction of the initial matrix of objects is necessary to observe the following steps:

1. Selection of objects included in the analysis file.
2. The assignment the "model station" in the list of objects.
3. Selection of parameters characterizing every object.
4. Election of weights of indicators.
5. The preparation of the initial matrix.

Table 1. Initial matrix

Station \ Indicator	a_1	a_2	...	a_i	...	a_n	$\sum$
1	x_{11}	x_{12}				x_{1n}	$\sum_{i=1}^{n} a_{1i}$
2	x_{21}	x_{22}				x_{2n}	$\sum_{i=1}^{n} a_{2i}$
...	...						...
j				x_{ji}			$\sum_{i=1}^{n} a_{ji}$
...	...						...
m-1	$x_{m-1,1}$	$x_{m-1,2}$				$x_{m-1,n}$	$\sum_{i=1}^{n} a_{m-1,i}$
m (the "model station")	x_{m1}	x_{m2}				$x_{m\,n}$	$\sum_{i=1}^{n} a_{mi}$

Legend of the matrix:
$a_{1,\ldots,n}$ – evaluation indicators;
x_{ij} – value of i-th indicator in the j-th object;
n – number of indicators;
m -1 – number of objects included to the initial matrix;
m – the model station.

4 Application the Evaluation of Stations on the Network of Slovak Railways

For the solution to this problem is Slovak railways network as essential input data. The original model of this network where are given all stations, loading yards and siding. Applying the model of the network movement results to the reduction of the items listed in the original model. Thus, the total number of registered points were 824, after reduction were enrolled to the base map total of 404 points.

4.1 The Methodology of Evaluation of Points

Individual assessment criteria were processed under traffic and transport limitations and the extent of the loading of wagons. To balance the point were included the following evaluation factors:

- The number of wagons manipulated per week.
- Type of station interlocking equipment.
- Number of running tracks that are in the station designated for freight trains.
- Other evaluation factors as technical inspection of station, depot locomotives and other equipment evaluated stations of technical and operational perspective.

4.2 Methodology for Evaluating of Railway Tracks

In process of solution the location problem should be evaluated not only the station but also the importance of the individual railway lines. For simplification this assessment to be made only on the basis of mutual distances. However is not problem the distances substitute for example by the price of used transport route, technical speed of track or time demands of passing of individual lines.

4.3 The Software Solution of Location Problem

For this task has been processed a simple software application shown in figure 4. The software contains menu, graphical part, textual part and a part of the window for specifying the criteria for solving task allocation on the Slovak railways network.

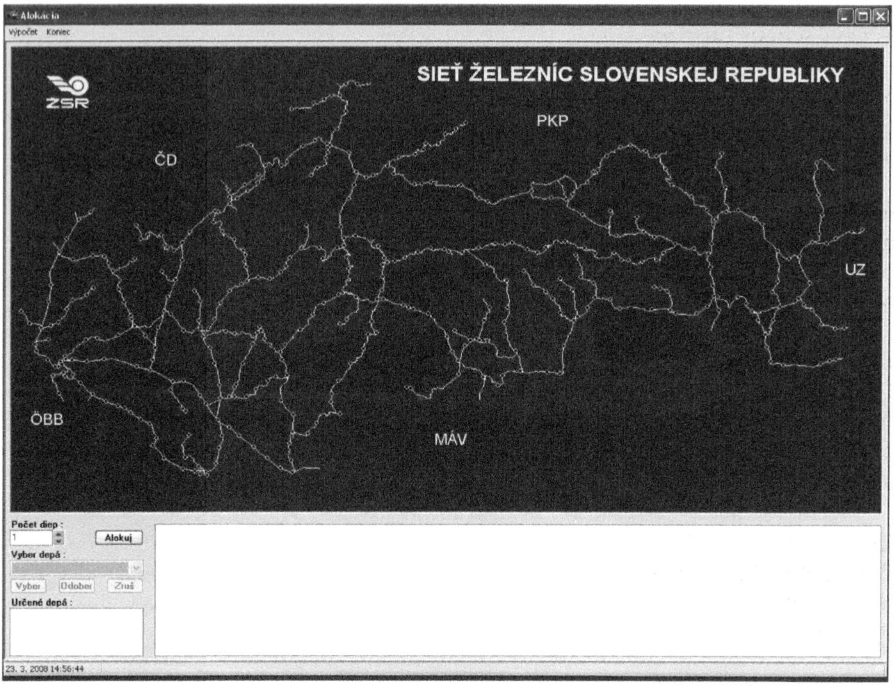

Fig. 3. The initial program window

In the menu of the "calculation" is the possibility to choose "general calculation", where the user only selects the number of depots on the network to be created with the respective attraction area. Wingspan search depots on the network's program set 1-50 depots. This variant calculation is the predefined as a source.

A second option program provides is the possibility of calculating attraction area for the desired station directly. This option can be involved in the program menu offer "enter depots". In this case, the user enters directly stations where stations are located.

The subsequent calculation is carried out to determine the individual attraction area of a entered station.

Range of depots at program (1-50) allows the use of this program not only for the allocation of marshalling yards, but then also the allocation of satellite stations and the attraction of oven sections in each attraction area of marshalling yards.

After performing the allocation results are relevant allocation exported to files for further processing options.

4.4 Practical Application of Locations

In practical use of the program is the given a sample of a direct award depots, which are selected as the main marshalling yards in accordance with the concept of strategy Slovak railways. The incorporated station are Bratislava east, Zilina - Teplicka, Zvolen and Kosice Freight Station (for a total of 4 stations). Solve this problem is shown in figure 5

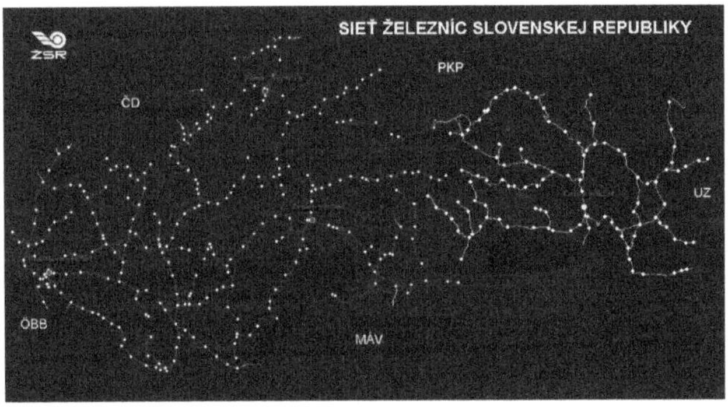

Fig. 4. Graphical solution of layout in attraction area with application of selected stations

5 Conclusion

Notwithstanding the guaranteed support rail funding by European Union is necessary that railway undertakings to provide transport services to be competitive with other transport trade unions offer.

The method of single wagon loads using a time-discrete model of train formation offers solutions for network Slovak railways. This model represents one of the effective ways to meet the development strategy of the European Community's railways. By applying this model in practice Transport Company, which has offered to bring the carriage of a single wagon load can guarantee a higher quality of service and then expect an increase in performance. The model is also applicable to other rail networks of infrastructure managers.

Acknowledgements. Táto publikácia vznikla v nadväznosti na riešený projekt spolufinancovaný zo zdrojov EÚ s názvom „Kvalita vzdelávania a rozvoj ľudských zdrojov ako piliere vedomostnej spoločnosti na Fakulte PEDAS", ITMS kód projektu 26110230083"riešeného na Žilinskej univerzite v Žiline.

This paper is prepared with the support of the project "The quality of education and development of the human resources as pillars of the knowledge society at the Faculty PEDAS", ITMS project code 26110230083, University of Zilina.

Operačný program VZDELÁVANIE

Moderné vzdelávanie pre vedomostnú spoločnosť/Projekt je spolufinancovaný zo zdrojov EÚ

Modern education for the kowledge society/ Project is co-financed by funds from the EC

Európska únia
Európsky sociálny fond

References

1. Camaj, J., Masek, J.: The train formation as important tool for increasing the freight transport services. In: Horizons of Railway Transport. Scientific papers vol. 2(1), pp. 27–32 (2011) ISSN 1338-287X
2. Camaj, J., Masek, J.: Location of marshalling yards as basic problem for time discreet train formation. In: Horizons of Railway Transport 2011. International Scientific Conference, Terchová, Slovak Republic, September 29-30, pp. 36–42 (2011) ISBN 978-80-554-0426-4
3. Camaj, J., Gasparik, J.: Railway station evaluation for systematic train forming on the railway network. Scientific Papers of the University of Pardubice, Series B – The Perner Transport Faculty 14, 257–263 (2008)
4. Camaj, J.: Optimization of train formation on transport network. Dissertation theses. University of Zilina, Faculty of Operation and Economics of Transport and Communications, p. 107 s.
5. Cerny, J.: The capacity problems of marshalling yard for time-discrete time formation. In: Vědecko Technicky Zborník ČD No. 10, Czech Republic, Praha. General department of Czech railways (2000)
6. Kendra, M., Lalinska, J., Camaj, J.: Optimization of transport and logistic processes by simulation. In: ISTEC: 3rd International Science, Technology and Engineering Conference, Dubai, United Arab Emirates (UAE), December 13-14-15, pp. 886–892 (2012) ISSN 2116-7383
7. Kendra, M., Masek, I., Babin, M.: Impact of railway infrastructure parameters on safety of goods transportation by railways. Machines, Technologies, Materials: International Virtual Journal 8(3), 44–47 (2014) ISSN 1313-0226
8. Kendra, M., Babin, M., Sulko, P.: Interaction between railway infrastructure parameters and quality of transportation services. In: BulTrans-2013: Anniversary Scientific Conference on Aeronautics, Automotive and Railway Engineering and Technologies: Proceeding Book, Sofia, Bulgaria, October 16-18, pp. 95–97. Izdatelstvo na Techničeskija Universitet, Sofija (2013) ISSN 1313-955X
9. Nedeliaková, E., Nedeliak, I., Blinova, E.: Dynamic models of service quality as a new trend in quality management. In: Economics and Business Management: 8th International Scientific Conference, September 19-20. Proceeding Book, pp. 273–279. Technical University, Zvolen (2013) ISBN 978-80-228-2565-8

Improving Vehicle Routing Decision for Travel Agency in Chonburi, Thailand

Tantikorn Pichpibul

Faculty of Business Administration, Panyapiwat Institute of Management,
85/1 Moo 2 Chaengwattana Road, Bangtalad, Pakkred, Nonthaburi, Thailand
pichpibul@gmail.com

Abstract. This paper focuses on vehicle routing decision, a real-world problem proposed by a travel agency company operating in Chonburi, Thailand. The objective is to minimize the total number of vehicles used and the total travelling distance. Currently, the company's planners have to spend hours on manually planning with spreadsheet software. Therefore, the decision-making software was developed by using an approach based on golden ball algorithm to increase the efficiency of the route assignment and especially suitable for the practical use. Computational results on a set of benchmark problems show that the proposed approach is competitive when compared it with the best existing approaches in the literature. Moreover, when the proposed approach is applied to solve the company's problems, the obtained solutions were able to significantly reduce the total number of vehicles used and the total travelling distance for supporting the planners' decision.

Keywords: Vehicle routing, Golden ball algorithm, Transportation, Logistics.

1 Introduction

This paper is mainly focused on a vehicle routing decision about daily traveler delivery of a travel agency company in Chonburi, Thailand. Fig. 1 graphically illustrates the locations of port (in black) and hotels/resorts (in grey) on the map. The company operates both managing vehicles and planning to deliver the travelers from 68 hotels/resorts to the Pattaya port located in the same city for doing company's activities such as snorkeling trips, diving trips and sightseeing tour. The company does not own the vehicles but outsources the service. In every morning, the vehicles are assigned to pick up travelers from their hotels/resorts and deliver them to the Pattaya port. Each vehicle can carry maximally 15 persons. When the vehicle arrives to the hotel/resort, the travelers have to register their tickets with the vehicle driver. After finish the trips, each vehicle has to deliver the same travelers from the Pattaya port back to their hotels/resorts in the reverse order.

Currently, the company's planners have to spend hours on manually planning with spreadsheet software. Therefore, the decision-making software was developed by using an approach based on golden ball algorithm. The objective of this research is to minimize the total number of vehicles used and the total travelling distance.

M. Gen et al. (eds.), *Industrial Engineering, Management Science and Applications 2015*,
Lecture Notes in Electrical Engineering 349, DOI: 10.1007/978-3-662-47200-2_28

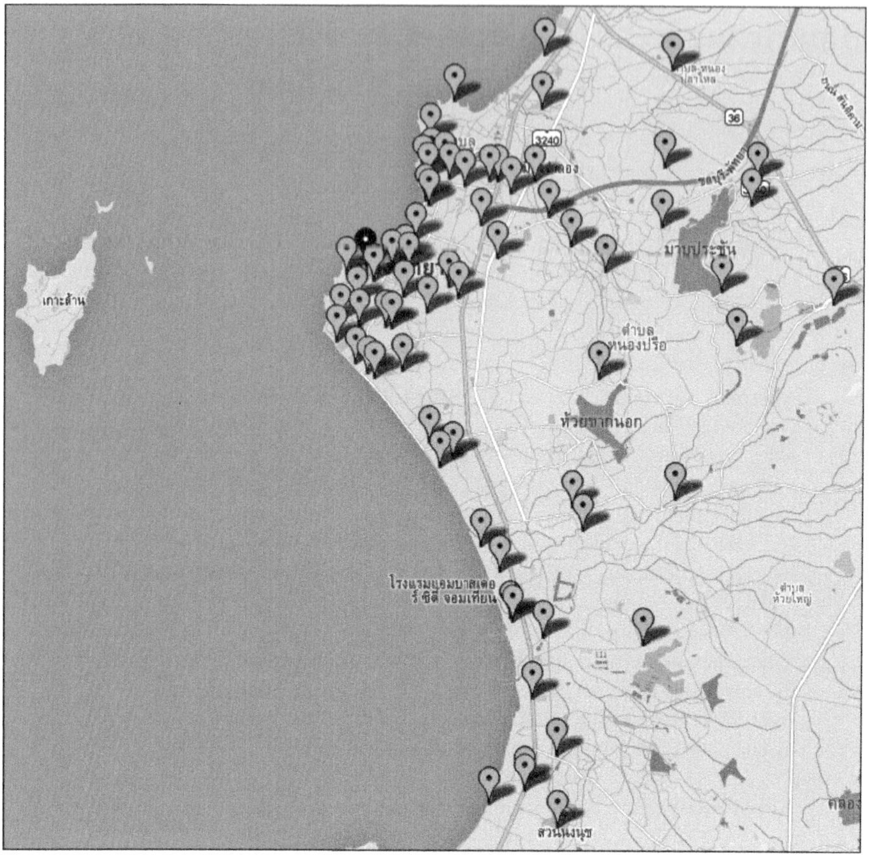

Fig. 1. Locations of port (in black) and hotels/resorts (in grey)

Golden ball algorithm is a relatively new meta-heuristic algorithm which can be used to solve the vehicle routing and other problems [1] [2] [3]. However, there is no previously published research that presents golden ball algorithm to solve this kind of problem. Therefore, this is major contribution to develop the proposed approach which can efficiently solve the problem in terms of solution quality.

2 Problem Description

Given those characteristics from previous section, the problem can be modeled as a variant of the well-known vehicle routing problem (VRP) called open VRP (OVRP). The OVRP was firstly solved by Sariklis and Powell [4] in their article on distribution management problems.

The mathematical model of the OVRP is based on the classical VRP formulation presented in Bodin et al. [5]. It includes notations, an objective function, and constraints. The following notation will be used in the mathematical calculations:

N	total number of hotels/resorts
V	total number of vehicles
d_{ij}	traveling distance between hotel/resort i and hotel/resort j
r_i	hotel/resort i 's request
Q	vehicle capacity
x_{ij}^{v}	binary variable, where $x_{ij}^{v} = 1$ if vehicle v travels from hotel/resort i to hotel/resort j and $x_{ij}^{v} = 0$ otherwise.
y_i^{v}	binary variable, where $y_i^{v} = 1$ if vehicle v visits hotel/resort i and $y_i^{v} = 0$ otherwise.

The VRP can be stated as a problem where V vehicles based at a port are required to serve N hotels/resorts in order to satisfy the known request. All of the vehicles have the same capacity, denoted by Q. All hotels/resorts have a non-negative request, denoted by r_i, which is associated with each hotel/resort i. The traveling distance between hotels/resorts i and j, denoted by d_{ij}, is symmetric. The objective function of the VRP can be written as:

$$\text{Minimize} \sum_{i\in N}\sum_{j\in N}\sum_{v\in V} d_{ij}x_{ij}^{v} \tag{1}$$

The VRP objective function is subject to the following constraints:

$$\sum_{v\in V} y_i^{v} = 1 \quad \text{for } i\in N \tag{2}$$

$$\sum_{i\in N} x_{ij}^{v} = y_j^{v} \quad \text{for } j\in N \text{ and } v\in V \tag{3}$$

$$\sum_{j\in N} x_{ij}^{v} = y_i^{v} \quad \text{for } i\in N \text{ and } v\in V \tag{4}$$

$$\sum_{i\in N} r_i y_i^{v} \le Q \quad \text{for } v\in V \tag{5}$$

$$\sum_{j\in N} x_{1j}^{v} \le 1 \quad \text{for } v\subset V \tag{6}$$

$$\sum_{i\in N} x_{i1}^{v} \le 1 \quad \text{for } v\in V \tag{7}$$

In this formulation, the objective function is expressed in Equation (1), which states that the total travelling distance of all of the vehicles is to be minimized. Constraint (2) states that each hotel/resort must be visited once by vehicle v. Constraint (2) is guaranteed by Constraints (3) and (4), where each hotel/resort is visited and left with vehicle v. Constraint (5) illustrates that the total requests of vehicle v do not exceed the vehicle capacity. Constraint (6) guarantees that the vehicle v leaves the port and arrives at hotel/resort j. Constraint (7) guarantees that vehicle v leaves hotel/resort i and returns to the port. Both Constraints (6) and (7) ensure that vehicle v is used no more than once.

The characteristics of the OVRP are quite similar to that of the classical VRP. The major difference between both problems is that, for the VRP, all vehicles have to return to the port after completing their service; for the OVRP, all vehicles do not have to return to the port. Therefore, the d_{i0}, the traveling distance from hotel/resort i to port 0, is set equal to zero.

3 Solution Approach

The development of the solution procedures which are based on golden ball algorithm is proposed, and the program code of the proposed approach is shown as follows.

```
For end_iter = 1 To 1000
   team(1) = CreateSpecialTeam()
   For t = 2 To 4
      team(t) = CreateRandomTeam()
   Next t
   For seasons = 1 To 10
    teampoint = 0
    For match = 1 To 3
       For t = 1 To 4
          For p = 1 To k
             For basictrain_iter = 1 To 10
                team(t) = BasicTraining(player(p))
             Next basictrain_iter
          Next p
       Next t
       For t = 1 To 4
          For tactictrain_iter = 1 To 10
               team(t) = TacticTraining()
          Next tactictrain_iter
       Next t
       teampoint = MatchPlay(match)
    Next match
    team = PlayerTransfer(teampoint)
   Next seasons
Next end_iter
```

3.1 Special and Random Teams

The population of golden ball algorithm is represented by soccer teams that each team represents one problem solution. Each team (*t*) has the number of players that each player (*p*) represents one vehicle route. The example of soccer team encoding is shown in Fig. 2.

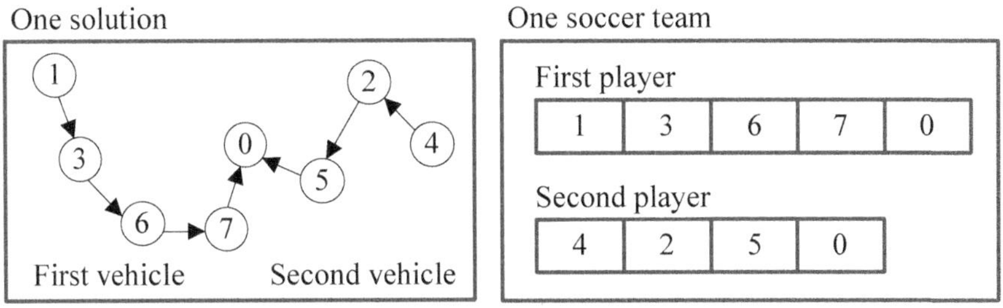

Fig. 2. Example of soccer team encoding

For the first player, the vehicle departs from its parking space, picks up travelers from hotels/resorts (1, 3, 6, 7), and deliver them to the Pattaya port (0). For the second player, the vehicle departs from its parking space, picks up travelers from hotels/resorts (4, 2, 5), and deliver them to the same port.

In this research, all teams are randomly created except one special team which is created by using a modified Clarke-Wright algorithm proposed by Pichpibul and Kawtummachai [6].

3.2 Basic Training

The basic training is an individual player development plan to improve the quality of each player in each team by using intra-route moves [7] composed of shift, swap, 2-opt and 3-opt moves. The obtained route is accepted only if its quality is better than the existing route.

3.3 Tactic Training

The tactic training is for all players in each team to improve their quality by using inter-route moves [7] composed of shift, swap, 2-opt and 3-opt moves. The obtained solution is accepted only if its quality is better than the existing solution.

3.4 Match Play

When the season starts, each match is played between two teams. All teams have to play matches against each other. In each match, the team with better solution quality is the winner of the match, and obtains three points. The loser obtains zero point. In case of a tie match, both teams get one point.

3.5 Player Transfer

The player transfer is a procedure that the players are moved between the teams. In every season, all teams are ranked depending on their points. The top two teams are chosen to move their players by using partial-mapped crossover [8]. The obtained solutions are accepted only if their quality is better than the existing solutions. The proposed approach is stopped when the total number of iterations is reached.

4 Computational Results

The results of the computational testing conducted on the proposed approach are discussed. First, a set of benchmark problems proposed in the literature was considered. Then the real-world problems from the travel agency company that motivated this research were examined.

The proposed approach was implemented in Visual Basic 6.0, and was run on an Intel Core i7 PC with a 2.80 GHz CPU and a 1.99 GB RAM. The values of the parameters used are: t=4, p=k, *specialt*=1, *specialt_iter*=1000, *basictrain_iter*=10, *tactictrain_iter*=10, *match*=3, *seasons*=10, and *end_iter*=1000.

4.1 Benchmark Problems

The performance of the proposed approach is first evaluated on the benchmark problems proposed by Christofides and Eilon [9] (dataset E) and Fisher [10] (dataset F). All problems used the same nomenclature consisting of a dataset identifier. This is followed by n, which represents the number of customers (including depot), and k, which represents the number of available vehicles. The problems also include the vehicle capacity constraints. In problems 1-3, the customer locations are randomly distributed in the plane and the depot is either in the center or near to it. Problems 4 and 5 are the real-world problems that the customer locations are clustered. The depot is not centered in both problems.

Table 1. Computational results for benchmark problems

No.	Problem	Vehicle Capacity	Solution			
			[11]	[12]	[13]	Proposed Approach
1	E-n51-k5	160	**416.06**	**416.06**	**416.06**	**416.06**
2	E-n76-k10	140	**567.14**	**567.14**	**567.14**	**567.14**
3	E-n101-k8	200	**639.74**	**639.74**	**639.74**	**639.74**
4	F-n72-k4	30000	**177.00**	**177.00**	178.09	**177.00**
5	F-n135-k7	2210	**769.66**	**769.66**	**769.66**	**769.66**

Bold number indicates the best solution was found.

Table 1 demonstrates the effectiveness of the proposed approach for the solution of the benchmark problems. It obtains competitive solutions in comparison to the best existing solutions obtained by a branch-and-cut algorithm [11], a record-to-record travel algorithm [12] and a variable neighborhood search algorithm [13]. The proposed approach is able to find all the best solutions, reported by these approaches. Therefore, it consistently outperforms.

4.2 Real-World Problems

This section shows the obtained results by using real data provided by the company. In this experiment, 30 problems correspond to 30 days of January 2015. For all problems, the company provided the characteristics of real data, as shown in Table 2.

Table 2. Characteristics of company's data

Item	Detail
Number of hotels/resorts	68
Number of ports	1
Number of hotel/resort requests per problem	30 – 60
Number of travelers per hotel/resort	2 – 6
Vehicle capacity	15 persons/trip

Table 3. Computational results for company's problems

Item	Company's approach [a]	Proposed approach [b]	Value difference a – b	Percent difference a – b
Total number of problems	30	30	-	-
Total number of vehicles used (unit)	588	401	-187	-31.80%
Total number of travelers (person)	5,926	5,926	-	-
Total travelling distance (km.)	6,120	4,472	-1,648	-26.93%

From the results in Table 3, it can be shown that the proposed approach is much better than the company's approach in all directions. The total number of vehicles used reduces by 31.80% and the total travelling distance reduces by 26.93%. This would result in the reduction in the transportation cost of the whole chain and would increase the competitiveness of the company. Moreover, the developed software can be easily used by the company's planners that will increase the efficiency of their process.

5 Conclusions

The research about vehicle routing decision for a travel agency has been done. The model studied in this paper concerns with the assignment of the vehicles to deliver the travelers from 67 hotels/resorts to the Pattaya port. The proposed approach based on golden ball algorithm was tested by using the developed software. From the computational results, the proposed approach gave competitive results compared with the best existing approaches, and illustrated much better performance in comparison with the company's approach. It is found that the proposed model, approach and the software can be used for the involved companies' planners to solve this kind of problem.

References

1. Osaba, E., Diaz, F., Onieva, E.: Golden ball: a novel meta-heuristic to solve combinatorial optimization problems based on soccer concepts. Applied Intelligence 41, 145–166 (2014)
2. Ruttanateerawichien, K., Kurutach, W., Pichpibul, T.: An Improved Golden Ball Algorithm for the Capacitated Vehicle Routing Problem. In: Pan, L., Păun, G., Pérez-Jiménez, M.J., Song, T., et al. (eds.) BIC-TA 2014. CCIS, vol. 472, pp. 341–356. Springer, Heidelberg (2014)
3. Osaba, E., Diaz, F., Carballedo, R., Onieva, E., Perallos, A.: Focusing on the Golden Ball Metaheuristic: An Extended Study on a Wider Set of Problems. The Scientific World Journal 2014, 1–17 (2014)
4. Sariklis, D., Powell, S.: A Heuristic Method for the Open Vehicle Routing Problem. The Journal of the Operational Research Society 51, 564–573 (2000)
5. Bodin, L., Golden, B., Assad, A., Ball, M.: Routing and scheduling of vehicles and crews: The state of the art. Computers and Operations Research 10, 63–211 (1983)
6. Pichpibul, T., Kawtummachai, R.: A Heuristic Approach Based on Clarke-Wright Algorithm for Open Vehicle Routing Problem. The Scientific World Journal 2013, 1–11 (2013)
7. Groër, C., Golden, B., Wasil, E.: A library of local search heuristics for the vehicle routing problem. Mathematical Programming Computation 2, 79–101 (2010)
8. Gen, M., Cheng, R., Lin, L.: Network models and optimization: Multiobjective genetic algorithm approach. Springer (2008)
9. Christofides, N., Eilon, S.: An algorithm for the vehicle dispatching problem. Operational Research Quarterly 20, 309–318 (1969)
10. Fisher, M.L.: Optimal solution of vehicle routing problems using minimum K-trees. Operations Research 42, 626–642 (1994)
11. Letchford, A., Lysgaard, J., Eglese, R.W.: A Branch-and-Cut Algorithm for the Capacitated Open Vehicle Routing Problem. The Journal of Operational Research Society 58, 1624–1651 (2007)
12. Li, F., Golden, B., Wasil, E.: The open vehicle routing problem: Algorithms, large-scale test problems, and computational results. Computers & Operations Research 34, 2918–2930 (2007)
13. Fleszar, K., Osman, I.H., Hindi, K.S.: A variable neighbourhood search algorithm for the open vehicle routing problem. European Journal of Operational Research 195, 803–809 (2009)

Differential Evolution Algorithm for Storage Location Assignment Problem

Warisa Wisittipanich* and Pongsakorn Meesuk

Industrial Engineering Department, Faculty of Engineering,
Chiang Mai University 239 Huay Kaew Road, Muang District, Chiang Mai, Thailand, 50200
warisa.o@gmail.com, enjoyneering@hotmail.com

Abstract. This study addresses warehouse storage location assignment problems (SLAP) where the traveling distance in an order-picking process is considered with three-axis traveling distance; two-horizontal and one-vertical distance. A mathematical model of the problem is first presented, then LINGO is used to find optimal solutions for a set of generated problems. However, as the problem size increases, computing time increases rapidly, and eventually the solution could not be found when the problem size is very large. Thus, this study presents an application of Differential Evolution (DE) algorithm to solve SLAP. The performance of proposed DE is evaluated on a set of generated problems, and the experimental results shows that the algorithm is able to provide good solutions especially for the large-size problems with relatively shorter computing time.

1 Introduction

Warehouse management is an art of storage and movement of inventory throughout the warehouse. Typical operations in warehouse management are comprised of receiving, storing, picking and delivering. Among these operations, product storage and retrieval are considered as one of the most critical resource-consuming activities [1]. Storage location assignment problem (SLAP) in warehouse management is a problem of assigning goods to storage locations that aims to satisfy one or more objectives i.e. space utilization, total transfer time, total transportation distance.

The storage management system can be classified into three main policies; dedicates storage policy, random storage policy, and class based storage policy. Brynzer and Johansson [2] stated the different between dedicated storage policy in which each stock-keeping unit (SKU) has a set of certain designed location, random storage policy in which any SKU can occupy any storage location, and class-based storage policy in which a group of storage location is allocated to a class of SKUs and random storage is allowed within the group of storage locations.

Many researchers have proposed several solution techniques to solve SLAP. Heragu et al. [3] considered a warehouse with five functional areas and proposed a

* Corresponding author.

M. Gen et al. (eds.), *Industrial Engineering, Management Science and Applications 2015*,
Lecture Notes in Electrical Engineering 349, DOI: 10.1007/978-3-662-47200-2_29

heuristic algorithm to determine SKUs' s allocation to different storage areas and the size of each functional area in order to minimize the total cost of material management. Chen and He [4] presented warehouse assignment strategies for storage systems with automation and developed a mathematical model for warehouse assignment optimization. Then, they applied particle swarm optimization with Pareto concept to overcome big-size problems. Muppani and Adil [5] studied the integer programing of a storage system with classed-based storage policy, and developed the simulated annealing (SA) algorithm to solve storage assignment and forming the classes. Next, they [6] proposed a non-linear integer programing for class-based storage system considering area decrease, handling costs and storage area cost, and used the brand and bound (B&B) algorithm for solving the developed nonlinear model. Hsu, Chen and Chen [7] presented a batching approach based on genetic algorithm (GA), which directly minimize the total travel distance. Roodbergen and De Koster [8] proposed heuristic methods for solving order picking routing problem in warehouses where two or more aisles exist and random storage is used.

This paper presents an implementation of a novel evolutionary algorithm called differential evolution (DE) to solve large-scale storage location assignment problems (SLAP) considering three-axis traveling distance. The storage management used in this study is based on a dedicated storage policy. The remainder of this paper is organized as follows. The problem description and model specification of SLAP are provided in section 2. Section 3 describes the proposed DE algorithm and its application to the problem. Experimental results are reported in section 4. Finally, conclusion and further research are provided in section 5.

2 Problem Description

The classical SLAP is to assign each product to a storage location in order to obtain certain objectives subjected to the constraints. In this paper, the objective of the problem is to minimize the total traveling distance along three-axis traveling distance; two horizontal axis and one vertical axis.

In this study, the warehouse layout used is based on the work in [9]. Warehouse layout is assumed to be symmetric. Therefore, the width of all storage locations are the same. The input/output (I/O) point is located at one corner of the warehouse. The number of storage blocks is limited and one storage block can be assigned for one product only. Fig.1 illustrates an example of a warehouse layout with four columns, two racks, three levels, and two rows, viewed from the top and the side of storage rack.

The notation and variable used in this model are listed as follows:

D_x	:	Distance from I/O point to origin point along X-axis
D_y	:	Distance from I/O point to origin point along Y-axis
W_a	:	Width of aisle
W_r	:	Width of storage row (equal to two times of storage block width)
L_s	:	Length of storage block

H	:	Height of storage block
T_p	:	Number of picking product p ($p = 1,2,3,\ldots, n$)
S_p	:	Number of storage block required for product p
I	:	Order of storage block position i ($i = 1,2,3,\ldots, m$)
J	:	Order of storage rack j ($j = 1,2,3,\ldots, q$)
K	:	Order of level k ($k = 1,2,3,\ldots, k$)
R	:	Order of row r ($r = 1,2,3,\ldots, r$)

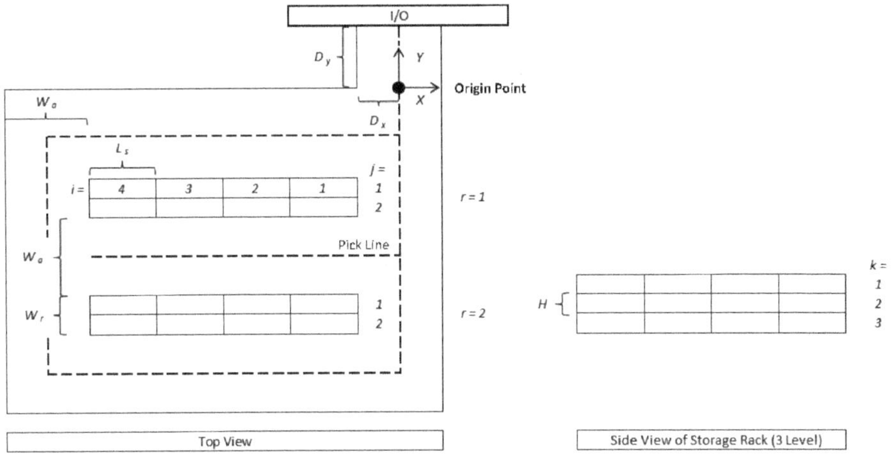

Fig. 1. A warehouse layout with four columns, two racks, three levels, and two rows

Decision variable:

$$X_{pijkr} = \begin{cases} 1, \text{if product } p \text{ is assigned to storage position } i \text{ rack } j \text{ level } k \text{ row } r \\ 0\text{, otherwise} \end{cases}$$

The mathematical model of the problem is formulated as follows:
Minimization of Total traveling distance:

$$f_2 = \sum_{p=1}^{n}\sum_{i=1}^{m}\sum_{j=1}^{q}\sum_{k=1}^{k}\sum_{r=1}^{r} \left(\frac{Tp}{Sp}\right) X_{pijkr} D_{ijkr} \tag{1}$$

Subjected to constraints:

$$D_{ijkr} = TD_x + TD_y + TD_z \tag{2}$$

$$TD_x = D_x + (I - 0.5)L_s \tag{3}$$

$$TD_y = D_y + (0.5J\, W_a) + [(J-1)(W_r + 0.5W_a)] + [(W_r + W_a)(R-1)] \tag{4}$$

$$TD_z = H(K-1) \tag{5}$$

$$\sum_{p=1}^{n} X_{pijkr} \leq 1 \qquad \forall\, i,j,k,r \tag{6}$$

$$\sum_{i=1}^{m}\sum_{j=1}^{q}\sum_{k=1}^{k}\sum_{r=1}^{r} X_{pijkr} = S_p \qquad \forall\, p \tag{7}$$

The objective function of the model is expressed in equation (1); minimization of the total traveling distance. It is noted that traveling distances are measured along the aisle centerline and centerline of storage block, and horizontal traveling distance considered by the picker moved along the aisle floor and vertical distance considered by height of rack shelf. Equation (2) to (5) illustrates the calculation of three-axis traveling distance. Equation (6) ensure that no more than one product is assign to one storage unit. Equation (7) is to guarantee that each product stored in the storage units as equal to the number of storage needed.

3 Adaptation of Differential Evolution Algorithm

Differential Evolution (DE), proposed by Storn and Price in 1995 [10], is one of the latest Evolutionary Algorithms (EAs) for global optimization over continuous search space. Due to its advantage of relatively few control variables but performing well in search ability and convergence, DE has been recently applied to solve many combinatorial *NP*-hard problems.

3.1 Differential Evolution Algorithm

As a population-based search method, DE begin with randomly generate initial population of size *N*. Each population is represented as a *D*-dimensional vector, and each variable's value in the *D*-dimensional space is represented as the real number. The main idea which makes DE different from other EAs is its mechanism for generating a new solution by mutation and crossover operation. At initialization stage ($g = 0$), the jth value of the ith vector is generated as equation (8).

$$x_{j,i,0} = u_j.(b_{j,U} - b_{j,L}) + b_{j,L} \tag{8}$$

The lower bound, b_L, and upper bound, b_U, for the value in each dimension j^{th} (j=1,2,…,D) must be specified. A uniform random number, u_j, is in the range [0, 1]. DE performs mutation operation by combining randomly selected vectors to produce a mutant vector. For each target vector, $X_{i,g}$, at generation g, the mutant vector, $V_{i,g}$, is generated equation (9).

$$V_{i,g} = X_{r1,g} + F(X_{r2,g} - X_{r3,g}) \tag{9}$$

It is noted that X_{r1}, X_{r2}, and X_{r3} are vectors randomly chosen from the current population. They are mutually exclusive and different from the target vector, $X_{i,g}$. F is a scale factor which controls the scale of the difference vector between X_{r2}, and X_{r3} , added to the base vector, X_{r1} . DE applies crossover operator on $X_{i,g}$ and $V_{i,g}$ to generate the trial vector $Z_{i,g}$. In the classic DE, the binomial crossover is employed and the trial vector is generated by equation (10).

$$z_{j,i,g} = \begin{cases} v_{j,i,g}, & \text{if } u_j \leq C_r \text{ or } j = j_u \\ x_{j,i,g}, & \text{otherwise} \end{cases} \quad (10)$$

Where C_r is crossover probability in the range [0, 1], and j_u is a random chosen index ($j_u \in \{1, 2, ... D\}$). C_r value controls the probability of selecting the value in each dimension from a mutant vector over its corresponding target vector. Then, the selection or replacement of an individual occurs only if the trial vector outperforms its corresponding vector. As a result, all individuals in the next generation are as good as or better than their counterparts in the current generation. The evolution of DE population continues through repeated cycles of three main operations; mutation, crossover, and selection until stopping criterion are met.

3.2 Solution Mapping to SLAP

In this study, a solution of the problem were represented using a DE vector with dimensions equal to the total number of storage blocks. Consider an example of a warehouse with three columns (i = 1, 2, 3), two racks (j = 1, 2), two levels (k = 1, 2), and one rows (r = 1). The warehouse store two product types; A and B. The value of Tp and Sp value each product type is shown in Table 1.

Table 1. Data for product type A and product type B

	Tp	*Sp*	*Tp/Sp*
Product Type A	15	4	3.75
Product Type B	20	6	3.33

The number of vector dimension is equal the number of storage units which is 12. Fig.2. illustrates a random key representation encoding scheme where each value in a vector dimension is initially generated with a uniform random number in range [0, 1].

Dimension d	1	2	3	4	5	6	7	8	9	10	11	12
	0.03	0.55	0.62	0.48	0.86	0.25	0.19	0.97	0.46	0.81	0.35	0.23

Fig. 2. Random key representation encoding scheme

Next, this study adopts the permutation of n-repetition of n jobs [11] with a sorting list rule to determine the assignment of each product unit to a storage location. According to the data in Table 1, product type A has a higher value of movement ration (*Tp*/*Sp*) than product type B. Therefore product type A are first allocated to the dimension with sorted values until the last unit of product type A has been assigned, and the product B unit are allocated next. Since the number of storage blocks is 12 and the number of total product units is 10, there are two storage blocks that are not assigned to store any product. The advantage of this approach is that it always provides a feasible storage allocation. This procedure results in completed storage location assignment as shown in Fig. 3.

Dimension d	*1*	*7*	*12*	*6*	*11*	*9*	*4*	*2*	*3*	*10*	*5*	*8*
	0.03	0.19	0.23	0.25	0.35	0.46	0.48	0.55	0.62	0.81	0.86	0.97
Product Type	*A*	*A*	*A*	*A*	*B*	*B*	*B*	*B*	*B*	*B*	-	-

Dimension d	*1*	*2*	*3*	*4*	*5*	*6*	*7*	*8*	*9*	*10*	*11*	*12*
	0.03	0.55	0.62	0.48	0.86	0.25	0.19	0.97	0.46	0.81	0.35	0.23
Product Type	*A*	*B*	*B*	*B*	-	*A*	*A*	-	*B*	*B*	*B*	*A*
Storage location	*1*	*2*	*3*	*4*	*5*	*6*	*7*	*8*	*9*	*10*	*11*	*12*

Fig. 3. Decoding scheme for SLAP

4 Computational Experiments

4.1 Parameter Setting

In this study, the DE population size (*N*) and number of iterations are set as 10 and 20 respectively to provide an adequate number of function evaluations used in the search procedures. After some preliminary experiments, the value of *F* is set to be uniformly randomized between 1 and 1.5 to retain population diversity throughout the search process. Based on some preliminary experiments, the use of binomial crossover yields better results than exponential crossover in SLAP. Thus, binomial crossover operation is used in this experiment with constant crossover rate (*Cr*) at 0.5.

4.2 Experimental Results

The performance of proposed DE algorithm is evaluated using seven generated data sets. Each instance is characterized by problem size: (number of product types) x (total number of products) x (total number of slots). Tables 2 shows the comparison of the traveling distance obtained by the proposed DE and those obtained by LINGO optimization software. It is noted the computational time of the proposed DE is also reported and the best result of the proposed DE for each instance is obtained from 5 independent runs.

Table 2. Comparison of total traveling distances on a set of generate instances

Instance	Problem Size	LINGO	Time (sec.)	Proposed DE	Time (sec.)
WH1	3x38x48	588.46	4	588.46	2.7
WH2	60x259x300	47273.04	8	47273.04	4.4
WH3	135x840x1000	112705.19	240	112705.19	6.9
WH4	270x1675x2000	204283.57	1200	204283.57	15.5
WH5	500x2825x3360	290062.88	3600	290062.88	45.2
WH6	500x2825x4992	-	-	271333.12	64.7
WH7	500x2825x6300	-	-	259022.57	83.5

According to the results from Table 2, it can be easily seen that the proposed DE is an effective solution technique to SLAP in this study. DE is able to find optimal solutions equal to those obtained by LINGO in small-size problem. Although solutions from LINGO is guaranteed to be optimal, when the problem size increases, computing time increases rapidly, and eventually the solution could not be found when the problem size is very large. On the other hand, for large-size problems, DE is able to obtain solution with relatively faster computing time.

5 Conclusion

This paper presents an implementation of differential evolution (DE) algorithm for solving storage location assignment problem (SLAP). The proposed DE employs the classic DE mutation scheme with binomial crossover operation. The random key representation and permutation of n-job repetition is applied to assign products to storage locations. The performance of proposed method is evaluated on a set of generated instances and compared with results from LINGO optimization program.

The experimental results indicates that DE can be used as efficient alternative approach for solving SLAP as it yields competitive solutions in term of quality and computing time especially for the large size problems. The ongoing researches are under investigation to improve DE performance and apply DE to deal with other aspects of combinatorial optimization problems.

References

1. Van Den Berg, J., Zijm, W.: Models for warehouse management: Classification and examples. Int. J. Prod. Econ. 59(1), 519–528 (1999)
2. Brynzer, H., Johansson, M.I.: Storage location assignment: using the product structure to reduce order picking times. Int. J. Oper. Res. 47, 595–603 (1996)
3. Heragu, S.S., Du, L., Mantel, R.J., Schuur, P.C.: Mathematical model for warehouse design and product allocation. Int. J. Pro. Res. 43(2), 327–338 (2005)
4. Chen, Y., He, F.: Research on particle swarm optimization in location assignment optimization. In: Proceedings of the 7th World Congress on Intelligent Control and Automation (2008)

5. Muppani, V.R., Adil, G.: Efficient formation of storage classes for warehouse storage location assignment: A simulated annealing approach. Int. J. Manag. Sci. 36, 609–618 (2008)
6. Muppani, V.R., Adil, G.K.: A branch and bound algorithm for class-based storage-location assignment. Eur. J. Oper. Res. 189, 492–507 (2008)
7. Hsu, C.M., Chen, K.Y., Chen, M.Y.: Batching orders in warehouse by minimizing travel distance with genetic algorithms. Comput. Ind. 56, 169–178 (2005)
8. Roodbergen, K.J., De Koster, R.: Routing methods for warehouses with multiple cross aisle. Int. J. Prod. Res. 39(9), 1865–1883 (2001)
9. Kasemset, C., Meesuk, P.: Storage location assignment considering three-axis traveling distance: A mathematical model. In: Golinska, P. (ed.) Logistics Operations, Supply Chain Management and Sustainability, EcoProduction, pp. 499–506. Springer, Heidelberg (2014)
10. Storn, R., Price, K.: Differential evolution – a simple and efficient adaptive scheme for global optimization over continuous spaces. Technical Report TR-95-012, International Computer Science, Berkeley, CA (1995)
11. Bierwirth, C.: A generalized permutation approach to job shop scheduling with genetic algorithms. In: Pesch, E., Vo, S. (eds.) OR-Spektrum. Special issue: Applied Local Search, vol. 17(213), pp. 87–92 (1995)

Master Production Scheduling for the Production Planning in the Pharmaceutical Industry

Sivinee Wattitham, Tuanjai Somboonwiwat, and Suksan Prombanpong

Department of Production Engineering, Faculty of Engineering,
King Mongkut's University of Technology Thonburi (KMUTT), Bangkok, Thailand
sivinee.vi@gpo.or.th, {tuanjai.som,suksan.pro}@gmail.com

Abstract. Pharmaceutical products are indispensable for human life. Typically, a manufacturer should well plan the production in order to meet the customer demand. However, it is due to the fact that pharmaceutical industry typically produces mixed products in different dosage forms; therefore, a good plan is hardly obtained. Therefore, this research presents the business process redesign of the Master Production Schedule MPS for the tablet manufacturing process. The objective of this study is to plan synchronized master production schedule of tableting, coating and packing line utilizing the developed heuristic algorithm. As a result, practical MPS can be created from starting to finish date scheduling in each process of tablet production. This methodology is then applied to sample data sets, attempting to develop MPS with an aim for on-time delivery.

Keywords: Production planning, Pharmaceutical, MPS.

1 Introduction

The Government Pharmaceutical Organization (GPO), a state enterprise under the Ministry of Public Health is the main generic drug manufacturer in Thailand and it is responsible for producing various products used in both government and private hospitals and health care centers. Its main product is tablet dosage form which accounts the main portion of its production. The process of tablet manufacturing consists of compression, coating and packing. Therefore, the production planning is of important to efficiently respond to customer demand [1]. Due to ever increasing of the customer demand, The GPO has faced with backorder problems and must strive to meet the quantity and delivery date required [2]. A production planner in the pharmaceutical manufacture usually is a pharmacist without engineering background. Since the production planning task is sophisticated as a consequent, there is no simple algorithm available to use. [3,4]. By using the starting inventory and demand forecast for a particular item, a planner can calculate the amount of production needed per period to meet anticipated customer demand [5]. MPS can be briefly described as what to produce and also time in which the production is to be completed [6].

M. Gen et al. (eds.), *Industrial Engineering, Management Science and Applications 2015*,
Lecture Notes in Electrical Engineering 349, DOI: 10.1007/978-3-662-47200-2_30

There are several literatures in the area of production planning related to MPS. Hill et al. proposed the modified design of the MPS and the use of a sequence-dependent scheduling heuristic for a chemical manufacturing [7]. The proposed method can provide improvements in changeover time and total shortages with sequence-dependent changeovers. Sonklin et al. developed the MPS of the production planning and scheduling for the plastic forming products in order to plan the production quantity and time schedule for reducing split order and its transportation surcharge from late delivery [8]. Venkataraman and Nathan established a weighted integer goal-programming model for the development of a rolling horizon master production schedule for a paint industry environment with multiple production lines, under demand certainty conditions, and minimum batch-size production restrictions [9].

Heuristics have been applied by a number of researchers to generate a valid and realistic MPS. Guilhereme et al. proposed simulated annealing where overcoming the local optimum [10]. Guilhereme and Favaretto proposed a practical heuristic for the MPS creation which strongly impacts final product costs [11]. Chern, and Hsieh proposed multi-objective master planning algorithm (MOMPA), for a supply chain network with multiple finished products [12]. Most of papers have been attempted so far in the industry out of pharmaceutical. Unlike, this paper aims at presenting heuristic algorithm to generate MPS for tablet manufacturing process at GPO for case study.

2 Methodology

2.1 Problem Description

The process flow of tablet production can be divided into 4 stages i.e. mixing, compression, coating and packing respectively as shown in Fig. 1. Tablet dosage form is classified into two different types i.e. a plain tablet and coated one (plain tablet with coating). The task starts from the compression process which the right ingredient of materials is done by dispensing center. Here the materials will be re-mixed to ensure they are in the right condition for compression. Once the materials are well amalgamated, the tablet will be formed by the punching machine and they will be coated and subsequently transferred for packing and then packed into a box or container. Note that each processing step must follow GMP/PICs (Good Manufacturing Product / Pharmaceutical Inspection Co-Operation Scheme).

Previously, the production planning of tablet manufacturing process is usually conducted by expert pharmacists. The planner generates a monthly aggregate plan which is consisted of quantity and product name. This plan is distributed to all production units as shown in Fig. 2. The non-linkage problem of each production planning is frequently occurred and is typically difficult to prevent due to lack of efficient MPS.

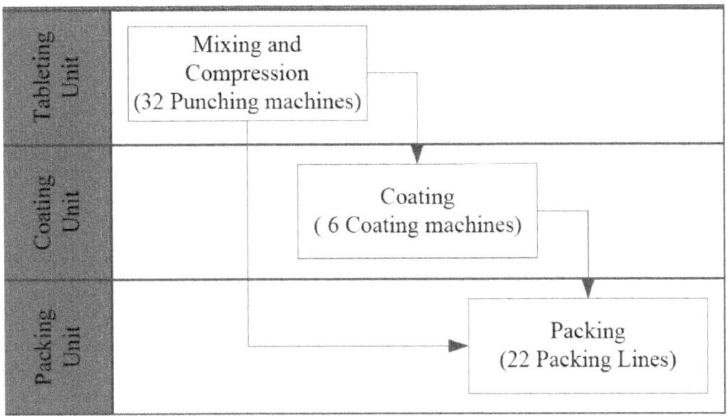

Fig. 1. Flow Chart of tablet manufacturing process

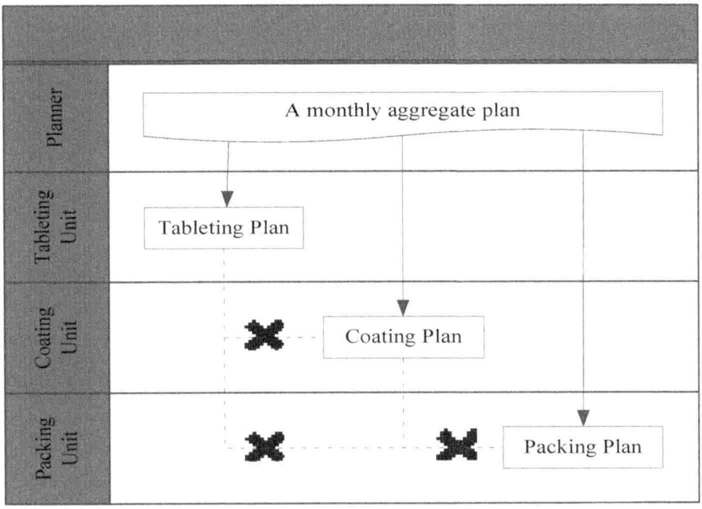

Fig. 2. The non-linkage problem of as-is plan

In this paper, tablet manufacturing at GPO which produces a total of 142 items, its ranging between 36,000 to 897,500,000 tablets per year and on the average order of 30-40 items per month. The MPS procedure is shown in Fig. 3. It begins with collecting production data of each product and also required processing steps. Then a heuristic algorithm will be developed based on the priority setting. Next, the latest start and latest finish will be calculated using backward scheduling. The main concept of the algorithm is to classify each product whether it is Make to Order (MTO) or Make to Stock (MTS) and then order the product based on the priority setting scenario. The flow chart of the algorithm is depicted in Fig. 4. The procedure steps of the algorithm are as follows:

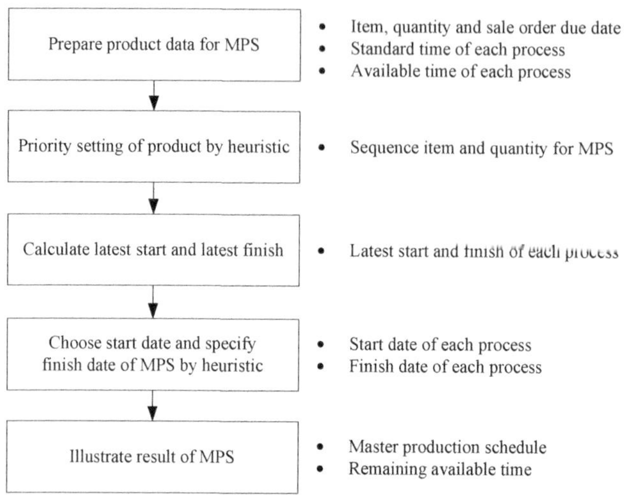

Fig. 3. The procedure of Master Production Scheduling

- Step 1. Check all concerned product information
- Step 2. Determine each product item whether (MTO), named Group 1 or (MTS), named Group 2.
- Step 3. Sort the items in Group 1 by the respective sale order due date.
- Step 4. Check the date having more than one product item. If Yes, go to step 5. If No, go to step 6.
- Step 5. If tie occurs, the importance level obtained from the priority setting scenario will be used to break the tie. The first priority is with the product in the previous list, The second priority is the lesser remaining number of production batch. The third one is the higher net profit margin.
- Step 6. Select the candidate and assign to the sequence items for the MPS.
- Step 7.Check whether all products in Group 1 are completely ordered. If Yes, go to step 8; otherwise go to step 3.
- Step 8. Sort the items in Group 2 in the ascending Inventory Usage Ratio (IUR).
- Step 9. Check IUR whether the product has the same ratio. If Yes, go to step 10. If No, go to step 11.
- Step 10. If tie occurs, the importance level obtained from the priority setting scenario will be used to break the tie again.
- Step 11. Select the candidate and assign to the sequence items for the MPS.
- Step 12. Check whether all product items in Group 2 are completely ordered. If Yes, finish; otherwise go back to step 8.

2.2 Calculation of the Latest Start and Latest Finish

The latest start date will be calculated for each process of tablet production to ensure that the latest finish of each step still comply with the order due date. In this study the delivery lead time of all products is 3 days. The latest start and latest finish can be calculated from (1) and (2) respectively.

$$LF_{ij} = LS_{(i+1)j} - 1 \tag{1}$$

$$LS_{ij} = LF_{ij} - LT_{ij} + 1 \tag{2}$$

Where

I = process; i = 1, 2, 3, 4 whereas 1 = compression, 2 = coating, 3 = packing, 4 = delivery.

J = product sequence ; j = 1, 2, 3,…J

LS_{ij} is the latest start date of process i at sequence j, LF_{ij} is the latest finished date of process i at sequence j whereas; LF_{4j} = Sale order due date. LT_{ij} is lead time (Day/batch)

2.3 Heuristic Algorithm for Selecting the Start Date and Finished Date

The sorted items based on the priority setting rule must be determined the start date and finished date. The start date of each process must start before the latest start date. Also the required production time must be checked against the available time. If the available time is enough to accommodate this item, the finished date of this one will be in the same day as start date. If it is not enough, the next available time from the next day will be considered. All the sorted product must be assigned finished date using the abovementioned procedure as shown in Fig. 5.

3 Results

In this section a generated MPS is presented. The algorithm is applied to a case study for the production planning of tablets manufacturing in June 2014. The as-is production plan indicated required products and number of batches is shown in Table 1. We prioritize the 33 products (434 lots) to 175 sequences by priority setting rule as shown in Table 2. After that, the synchronized master production schedule of tableting, coating and packing are created the start date to finished date scheduling of each product in each process as shown in Table 3.

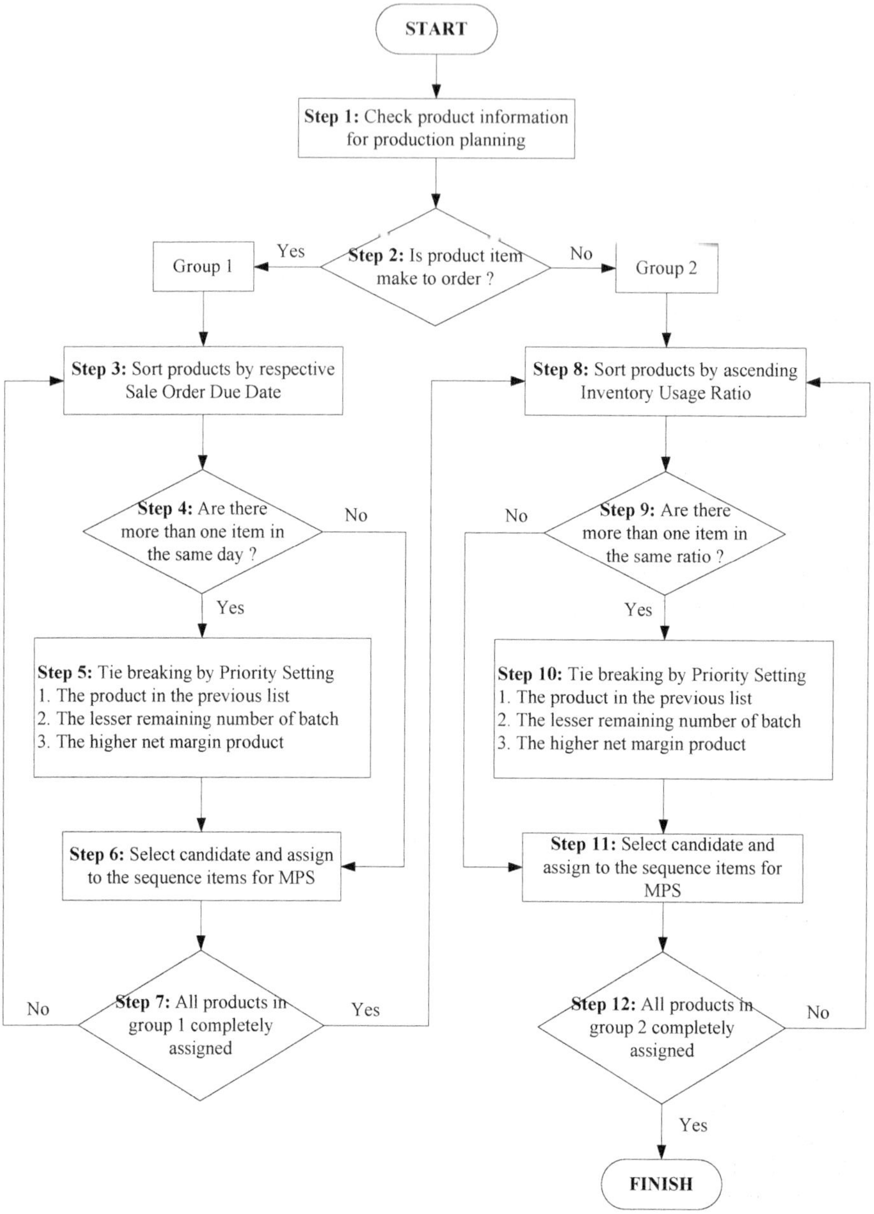

Fig. 4. Flow chart of the priority setting algorithm

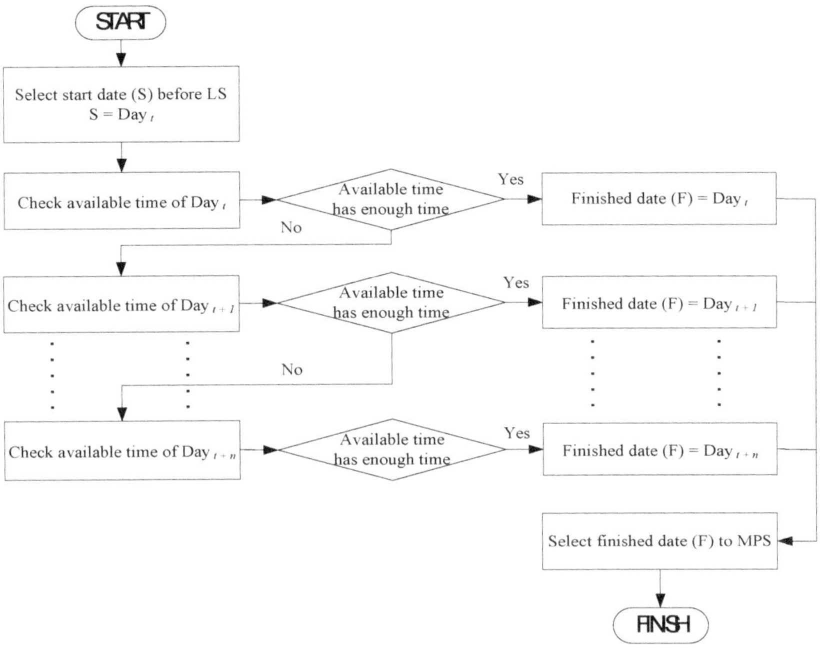

Fig. 5. The algorithm to choose start and finished date

Table 1. The as-is production plan in June 2014

No.	Product code	No. of lots	No.	Product code	No. of lots
1	P001	3	**18**	P072	4
2	P002	10	**19**	P074	13
3	P003	32	**20**	P081	5
4	P004	70	**21**	P097	10
5	P005	8	**22**	P099	8
6	P006	10	**23**	P102	6
7	P024	8	**24**	P106	1
8	P037	8	**25**	P107	2
9	P042	13	**26**	P113	4
10	P043	2	**27**	P114	2
11	P049	15	**28**	P115	2
12	P050	20	**29**	P116	15
13	P054	9	**30**	P117	30
14	P055	5	**31**	P118	10
15	P059	18	**32**	P123	30
16	P060	8	**33**	P131	20
17	P071	33	**Total**	-	***434***

Table 2. The product sequence on June 2014

Sequence	Product code	No. of lots	Sale order due date
1	P055	1	09-Jun-14
2	P118	1	11- Jun-14
3	P002	1	13- Jun-14
4	P074	1	13- Jun-14
5	P074	1	15- Jun-14
6	P117	2	15- Jun-14
7	P042	1	16- Jun-14
8	P102	1	16- Jun-14
9	P097	1	16- Jun-14
.	.	.	.
.	.	.	.
.	.	.	.
175	P054	2	31- Jul-14

Table 3. The example of the master production schedule on June 2014 (to-be)

Sequence	Tableting		Coating		Packing		Due date
	S	F	S	F	S	F	
1	2-Jun-14	2-Jun-14	-	-	5-Jun-14	5-Jun-14	09-Jun-14
2	2-Jun-14	2-Jun-14	4-Jun-14	4-Jun-14	5-Jun-14	5-Jun-14	11-Jun-14
3	2-Jun-14	2-Jun-14	-	-	5-Jun-14	5-Jun-14	13-Jun-14
4	2-Jun-14	2-Jun-14	5-Jun-14	5-Jun-14	8-Jun-14	8-Jun-14	13-Jun-14
5	3-Jun-14	3-Jun-14	8-Jun-14	8-Jun-14	9-Jun-14	9-Jun-14	15-Jun-14
6	2-Jun-14	2-Jun-14	4-Jun-14	4-Jun-14	5-Jun-14	5-Jun-14	15-Jun-14
7	4-Jun-14	4-Jun-14	-	-	5-Jun-14	5-Jun-14	16-Jun-14
8	4-Jun-14	4-Jun-14	-	-	5-Jun-14	5-Jun-14	16-Jun-14
9	2-Jun-14	2-Jun-14	5-Jun-14	5-Jun-14	8-Jun-14	8-Jun-14	16-Jun-14
.	.	.	.	.	.	.	.
.	.	.	.	.	.	.	.
.	.	.	.	.	.	.	.
.	.	.	.	.	.	.	.
175	25-Jun-14	25-Jun-14	-	-	26-Jun-14	26-Jun-14	31-Jul-14

4 Discussion

The master production schedule (to-be) in June 2014 can integrate planning of all processes with all products and as a result all products can be delivered on time. Whereas the previous planning can achieve on time only 80.65% of 434 batches. Moreover, the study reveals that the total available time from the master production scheduling is decreased by 3,755.40 hour or 26.30% and the manufacturer can foresee the production problem because of this synchronized plan.

5 Conclusion

This research paper shows an attempt to apply an industrial engineering technique in pharmaceutical industry. It is shown that the simplified algorithm is one of the potential and useful methods for problem solving. Although master production scheduling (MPS) has been studied and used by both academia and industries for quite a long time, the real complexity involved in making a master plan under constraints is not so easy. These constraints include production flexibility, quality issue, efficacy and safety and so on. Thus, the proposed heuristic algorithm will simplify this task and make it more efficient.

Acknowledgments. We would like to thank the Government Pharmaceutical Organization for the financial support of this research and Production planning division for the support data.

References

1. Chaoleam, S., Somboonwiwat, T., Prombanpong, S.: The Production Planning of Pharmaceutical Production under Multi variables. In: Proceedings of the 2013 IEEE Industrial Engineering and Engineering Management, Bangkok (2013)
2. Asgerirsson, E.I., Axelsdottir, G.S., Stefansson, H.: Automating a Manual Production Scheduling Process at a Pharmaceutical Company. In: Proc. IEEE, Computation Intelligence in Production and Logistics System, Paris, pp. 1–8 (2011)
3. Engudomnukul, S., Prombanpong, S., Somboonwiwat, T.: A Heuristic Approach to Capacity Assignment in Pharmaceutical Manufacturing Production Line. Applied Mechanics and Materials 548-549, 1928–1933 (2014)
4. Segerstedt, A.: Master Production Scheduling and a Comparison of Material Requirements Planning and Cover-time Planning. International Journal of Production Research 44(18), 3585–3606 (2006)
5. Feng, K., Rao, U.S., Raturi, A.: Setting Planned Orders in Master Production Scheduling under Demand Uncertainty. International Journal of Production Research 49(13) (2011)
6. Radhika, S., Rao, C.S.: A New Multi-Objective Optimization of Master Production Scheduling Problems Using Differential Evolution. International Journal of Applied Science and Engineering 12(1), 75–86 (2014)
7. Hill, J.A., Berry, W.L., Leong, G.K., Schilling, D.A.: Master Production Scheduling in Capacitated Sequence-dependent Process Industry. International Journal of Production Research 38(18) (2000)
8. Sonklin, A., Somboonwiwat, T.: Schedule Creation of Molded Plastic Production for Parallel Machinery Identical under Different Production Ratio. Presented at National Operations Research Conference, Thailand, pp. 89–96 (2009)
9. Venkataraman, R., Nathan, J.: Master Production Scheduling for a Process Industry Environment. International Journal of Operations and Production Management 14(10), 44–53 (1994)

10. Vieiraa, G.E., Ribas, P.C.: A New Multi-objective Optimization Method for Master Production Scheduling Problems using Simulated Annealing. International Journal of Production Research 42, 4609–4622 (2004)
11. Vieira, G.E., Favaretto, F.: A New and Practical Heuristic for Master Production Scheduling Creation. International Journal of Production Research 44(18-19), 3607–3625 (2006)
12. Chern, C.C., Hsieh, J.S.: A Heuristic Algorithm for Master Planning that Satisfies Multiple Objectives. Computers and Operations Research 34, 3491–3513 (2007)

A Hybrid Genetic Algorithm for Simultaneous Scheduling of Machines and AGVs in FMS

James T. Lin, Chun-Chih Chiu, Yu-Hsiang Chang, and Hung-Ming Chen

Department of Industrial Engineering and Engineering Management
National Tsing-Hua University, Hsinchu, Taiwan, ROC
jtlin@ie.nthu.edu.tw,
{vhasb3210,andyworld123}@gmail.com,
k2478417@hotmail.com

Abstract. In this paper, simulation optimization approach for simultaneous scheduling of machines and identical automated guided vehicles with minimizing the makespan in flexible manufacturing system is presented. In the past time, this problem has been solved by using heuristic algorithm such as genetic algorithms (GA), particle swarm optimization (PSO) etc. Actually, many factors should be considered in real situation such as deadlock or blockage of AGV and processing time with uncertainty in FMS. This will impact the system performance significantly. Hence, discrete event simulation model is used to evaluate the system performances which consider those random factors and to compare alternatives. In addition, optimal computing budget allocation (OCBA) embedded with GA is used to reduce simulation replications and provide reliable evaluations and identified for ranking chromosomes of the GA procedure. As a result, we prove those random factors affect system performance significantly. The numerical experiment results demonstrate the superiority of the hybrid approach in terms of computing cost for this problem.

Keywords: Flexible manufacturing system, simulation optimization, genetic algorithm, optimal computing budget allocation.

1 Introduction

Flexible manufacturing systems (FMSs) are a high degree of flexibility manufacturing system which can be used to increase machines utilization and productivity. The systems components have intricate relationships and interactions. Operating such complex system is challenging. Several operating rules such as alternative machine selection, vehicle dispatching and routing can have a strong impact to the system performance.

Due to the complexity of the system, scheduling is a tough task which can affect the system performance directly. Most research only considers the machine scheduling [8][13][14], but in FMSs, the AGVs also plays a crucial role which can effective impact the system performance. Some research emphasized that both the scheduling of operations on machine centers as well as the scheduling of AGVs are essential

M. Gen et al. (eds.), *Industrial Engineering, Management Science and Applications 2015*,
Lecture Notes in Electrical Engineering 349, DOI: 10.1007/978-3-662-47200-2_31

factors contributing to the efficiency of the overall system [1][12]. But only a few researchers aim to simultaneous scheduling of jobs and AGVs. Raman et al. [10] developed a deterministic off-line scheduling model and formulated the simultaneous scheduling problem with resource constrained as an integer programming problem (IP) then solve it. The principle assumption of this problem is that AGVs have to return to the U/L after handling a product, this assumption will decrease the system flexibility.

Bilge and Ulusoy [2] constructed a non-linear mixed integer programming problem (MIP) with a time-window approach for the simultaneous scheduling of machines and AGVs, and the target is to minimize makespan.

Ulusoy et al. [12] advanced an appropriate encoded method to present the job sequence and AGVs assignment with chromosomes and proposed a genetic algorithm (GA) for the simultaneous scheduling problem. Jerald J. et al. [6] addressed an adaptive genetic algorithm (Adaptive GA) to solve this simultaneous scheduling problem, the crossover and mutation rate changed during the adaptive GA procedure, and the system performance is to minimize a combination of total machine idle time and the penalty cost for not meeting the delivery date.

Some research only evaluate simultaneous scheduling problem of FMSs with single objective (e.g., minimize makespan, maximize utilization, maximize system throughput), but most of real-world scheduling problems, multiple objectives should be considered to optimize at the same time, B. S. P. Reddy and C. S. P. Rao [1] proposed a hybrid multi-objective genetic algorithm (Hybrid-MOGA) to optimize the job sequence for simultaneous scheduling of machines and AGVs problem in FMS, which objectives are to minimize makespan, minimize mean flow time and minimize tardiness simultaneously.

Even there are some research addresses the simultaneous scheduling of jobs and AGVs, their proposed model is still deterministic. Without considering uncertainty in FMS, it can be inapplicable to practice. For example, the blocking and congestion between vehicles may be rare in a deterministic model. The random factors like car dispatching, zone-control, alternative routing and etc., can make. To evaluate the system performance under uncertainty, Monte Carle simulations are widely used [3][11]. Computer simulation technology provides an effective tool to help plan for solving, analyzing and evaluating different alternatives.

Evaluating a decision using simulation can be time-consuming. The performance of a decision cannot be evaluated in a single run of the simulation. The replications are needed to robust the optimal solution, but certainly it requires more computational resources, we need to allocate these resources more effectively, in this study, we employed Optimal Computing Budget Allocation (OCBA) technique which was proposed by Chen et al. [4] as resource allocation method to enhance simulation efficiency with a given computing budget. In FMS, the solution space, i.e., all possible decisions we can choose, can be large. Then, evaluating all possible decisions under uncertainty can become infeasible. Machine scheduling and AGVs dispatching are well-known NP-hard problems [2][9], in this research, we proposed a hybrid framework approach for simultaneous scheduling of machines and AGVs in FMS and demonstrate the efficiency of this framework.

This paper is structured as follows. Section 2 presents the FMS problem description and formulates the problem with the non-linear programming model. The proposed a simulation optimization based methodology for the simultaneous scheduling problem in this research will be introduced in section 3. In section 4, two numerical experiments and the results were presented, which include two experiments, one is the comparison of FMS problem used zone-control technique or not, and, the another one is the comparison of GA_{EQUAL} with the proposed simulation optimization framework-GA_{OCBA}, the conclusions are given in section 5.

2 Problem Description

The FMS consists a group of computer numerically controlled (CNC) machines, automated guided vehicles and computer control system are connected together to process different products. An example of FMS can be presented as figure 1.

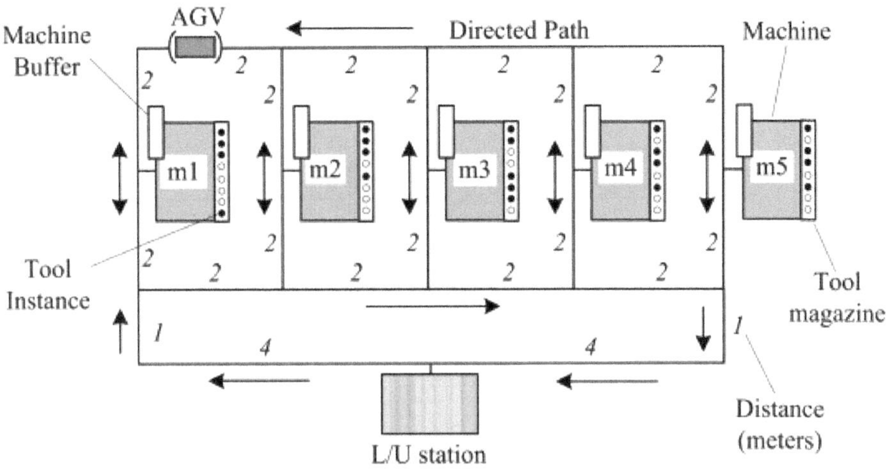

Fig. 1. Example of FMS [7]

As shown in Fig. 1. There is a load/unload (L/UL) station, an automated guided vehicle and eight machines, local buffers (input/output buffer) are in front of each machine, and a tool store behind each machine.

If the problem is a deterministic problem, the objective function of alternative x is $Z = f(x)$. But if we considered some factors are randomness in this problem, for example, processing time, machine setup time, MTTR, MTTF etc., thus, the objective function of x will have some randomness, we will use expectation value to decrease the variability, and the objective function will become to $Z = E[f(x,\omega)] = \frac{1}{n}\sum_{i=1}^{n} f(x,\omega)$, where n denotes the replication time and ω denotes the randomness. If the replication time getting larger, the sample mean will converge to it actual value, the expectation value will be similar to the real performance, therefore, an intelligent resource allocation method is needed to increase the simulation efficiency and robust the optimal solution.

Zone Control

Most of literatures, the AGV handling time is equals to the handling distance divided by the handling speed, in this way, a situation of an AGV may pass through to the other AGVs is ignored. But in real case, handling behavior may suffer blocking and congestion, further to make the AGV deadlock, and then to cause the system performance even worse. In this research, to avoid blocking and congestion cause AGVs deadlock, we use zone-control [5] to ensure AGVs will not collide by separating the track into several disjoint regions, when there's an AGV enter a zone, the other AGVs cannot enter the same zone at the same time and have to wait until the zone becomes empty. The zone-control technique can prevent AGVs deadlock but it will cause to increase the makespan. A comparison of a system with/without zone-control will be present later.

3 Introduction of Proposed Framework

The conceptual framework of the proposed hybrid approach is shown in Figure 2. The proposed framework is composed of a meta-heuristic, which generates design alternatives to exploit the solution space, and a discrete event simulation to estimate performance by OCBA procedure to allocate multiple sampling.

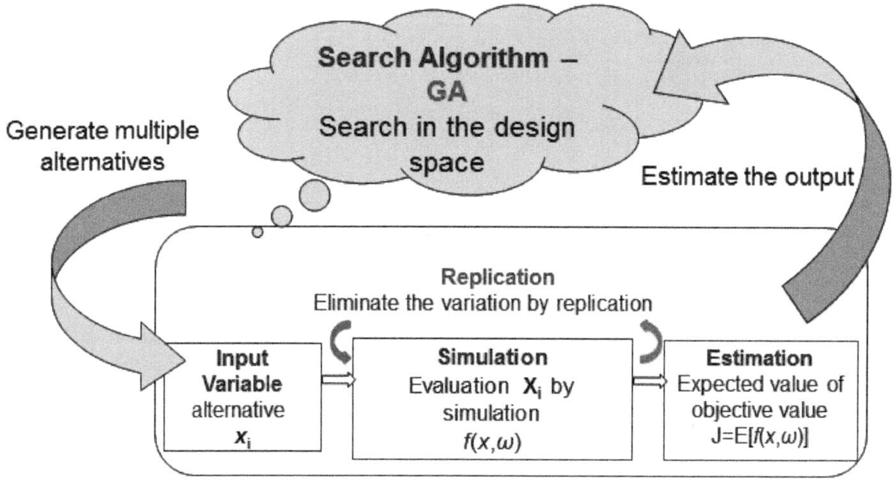

Fig. 2. Simulation optimization framework

3.1 Genetic algorithm

Genetic algorithm in this framework is applied to well exploration and exploitation the scheduling of machine in design space. In addition, we also embedded vehicle assignment algorithm to control job handling in simulation model. The GA procedure can be discussed by the following 6 steps: Chromosome encoding, population initialization, legalization, chromosome decoding and evaluation, crossover and mutation.

1. *Chromosome representation* Chromosomes represented the operation sequence, and each gene means a specific job and its processing step. The chromosome length is equal to the total number of operations of all the jobs. For example, there is a scheduling situation with 4 machines and 3 jobs, each job has 3 operations, which shown in table 1, the chromosome length is 12.

Table 1. A jobset and the correspond genes

Jobs	1			2			3		
Operation	1	2	3	1	2	3	1	2	3
Machine	M1	M2	M4	M1	M3	M2	M3	M4	M1
Gene	11	12	13	21	22	23	31	32	33

2. *Population initialization.* We randomly generate a set of chromosomes as population initially. A chromosome is shown as an example in table 2.

Table 2. A randomly generated chromosome

Operation sequence	1	2	3	4	5	6	7	8	9
Gene	22	31	21	11	33	23	12	13	32

3. *Legalization* Because of operations are nonpreemptive, a legalization is needed when a chromosome is infeasible. Table 2 presents a randomly generated chromosome, because the job operation sequence cannot be preempted, so the chromosome is infeasible, so a legalization is employed to make chromosomes feasible.

Operation sequence	1	2	3	4	5	6	7	8	9
Gene	22	31	21	11	33	23	12	13	32

Operation sequence	1	2	3	4	5	6	7	8	9
Gene	21	31	22	11	32	23	12	13	33

Fig. 3. Chromosome legalization

4. *Chromosome decoding and evaluation* Fig.4 contains a set of job operation sequence which has been legalized, the first gene is 21, which means the first step of operations is to move the job 2 to its first operation machine- M1, and the second gene is 31 means to move the job 3 to its first operation machine- M3, and so on. After decoding, the operation sequence will be put into the simulation model to evaluate the performance.
5. *Crossover.* Because of each gene in chromosome is unique and cannot be repeated, a "two-point order crossover method" has been quoted for this feature in this study. The steps are as follow:

(a) Select two chromosomes randomly from population as parents. The two chromosomes' fitness value should be in the top 10% of the population.
(b) Randomly select 2 positions of chromosome.
(c) Let the genes within the 2 positions be replaced by another chromosome's genes which are same as the original genes, thus, the order of genes within the 2 positions will be change.

An example is shown in figure 5.

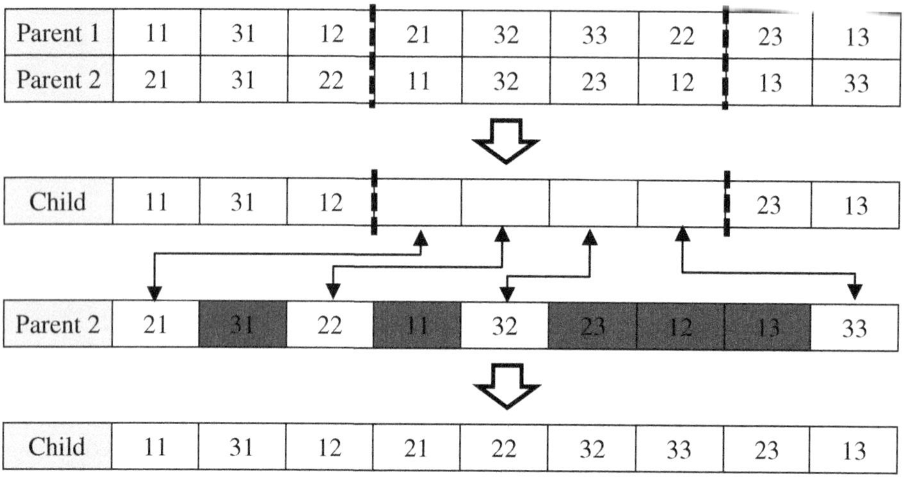

Fig. 4. Two-point crossover method

6. *Mutation.* In this research, we adopted "shift mutation" as the mutation method, the steps are as follow:
   (a) Select a gene randomly.
   (b) Select a position randomly, and insert the selected gene into the selected position, then the genes behind the position have to shift for one grid.

3.2 Vehicle Assignment Algorithm

To dispatch the AGVs handling, we employed a vehicle assignment algorithm [1]. Assume there are two AGVs to provide material handling, at the beginning, the first two operations will be assigned to these two AGVs randomly, and then the other operation will be assigned to AGVs by vehicle assignment algorithm.

3.3 Optimal Computing Budget Allocation (OCBA)

In this study, we use an intelligence procedure, OCBA, which assigns a number of replications to each particle. The OCBA approach was introduced by Chen et al. [4] to determine the optimal design alternative x within a reasonable amount of time.

There are some notations should be defined, T is the total replication and P(CS)* is the expectation to reach P(CS) by user setting. OCBA allocates simulation budget n_0 to each alternatives, and then allocate Δ each iteration according to (1), where n_0 is the initial simulation replications and Δ is the total simulation budget for each iteration. When OCBA is integrated with GA, the formulation of the maximized simulation efficient is chosen when the P(CS)* is satisfied. The function P(CS) is indicated in (2).

Minimize $n_1+n_2+,\ldots,+n_k=T$ such that P(CS)>=P(CS)*

Where T is the total replication and P(CS)* is the expectation to reach P(CS) by user setting.

$$\begin{cases} \dfrac{N_i^{l+1}}{N_j^{l+1}} = \left(\dfrac{s_i\left(\bar{J}_b - \bar{J}_j\right)}{s_j\left(\bar{J}_b - \bar{J}_i\right)} \right)^2 \\ N_b^{l+1} = s_b \sqrt{\displaystyle\sum_{i=1, i \neq b}^{k} \left(\dfrac{N_i^{l+1}}{s_i} \right)^2} \end{cases} \tag{1}$$

$$PCS \equiv 1 - \sum_{i=1, i \neq b}^{k} P\left\{\tilde{J}_b > \tilde{J}_i\right\} \tag{2}$$

where

k is number of design alternatives.

N_i^{l+1} and N_j^{l+1} denotes i-th or j-th alternatives' simulation replications in iteration l+1.

N_b^{l+1} is the simulation times for the best object in iteration $l+1$.

s_i and s_j denote the standard deviation of the i-th or j-th design alternative.

s_b denotes the standard deviation for the best design alternative.

$\bar{J}_i$ and $\bar{J}_j$ denotes the expected value for the i-th and j-th design alternative.

$\bar{J}_b$ represents the expected value for the best alternative.

3.4 Hybrid Approach Procedure

The implementation of this hybrid approach, as shown in Table 3.

Table 3. Procedure of GA_{OCBA}

Initialization		Initialization of GA and OCBA (parameter setting) The chromosomes are generated by uniform randomly and simulate n0 times for each chromosome.
Loop while stopping criterions is met (GA).		

Updating (GA)		For each chromosome i do
		Update chromosome i using crossover and mutation;
		End for

Loop while P(CS)<P(CS)* or Tl<T do (OCBA)		Calculate sample means sample variances and P(CS)* of particles based on the simulation outputs;
	Allocation	Let Tl+1 =Tl +Δ and allocate the new simulation replications to chromosomes m $N^l = (N_1^l, N_2^l, \ldots, N_m^l)$ based on equation (1).
	Simulation	Perform max $= (0, N_m^{l+1} - N_m^l)$ numbers of replication for each chromosome m; let l=l+1.
End of loop (OCBA)		
End of loop (Stopping GA procedure.)		If the stopping criteria is satisfied, stop and output the best solution and its performance. Otherwise, go to Updating(GA).

4 Numerical Results

4.1 Benchmark Case in Simultaneous Scheduling

In this experiment, Machine processing time is considered as a random factor in the simulation model, we assumed machine processing time is following *Normal*(μ= *Pt*, σ^2= *3*). A comparison of two simulation budget allocation methods (OCBA and EQUAL) in stochastic problem is represented in this experiment.

Parameter setting. GA and OCBA parameter setting is shown in table 4.

Table 4. GA and OCBA parameter setting

GA			
Population size	Crossover rate	Mutation rate	Generation
50	0.6	0.4	50
OCBA			
Initial replication (n0)	Simulation budget per OCBA iteration (Δ)	Total simulation budget (T)	Probability of correct selection (P{CS})
10	100	1500	80% 58.71%

Experiment result. The convergence plot is shown as figure 5. In Fig. 5. GA_{OCBA} shows faster converge than GA_{EQUAL}. In table 6, GA_{OCBA} can save 65.95% of simulation replication than GA_{EQUAL}, which implies GA_{OCBA} is more efficient than GA_{EQUAL}.

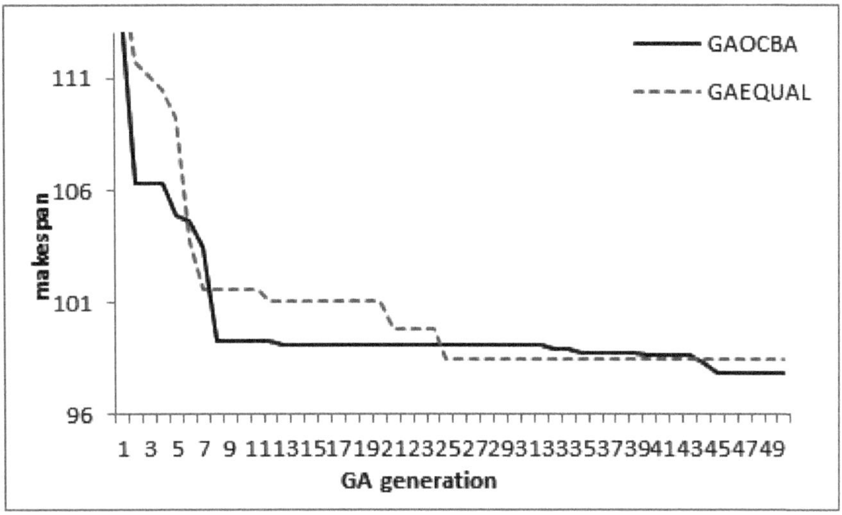

Fig. 5. Convergence plot of EQUAL and OCBA

Table 5. Comparison of GA_{EQUAL} and GA_{OCBA}

Search method	Makespan	Simulation replication	Saving
GA_{EQUAL}	98.479	195000	
GA_{OCBA}	97.875	66400	65.95%

5 Conclusion

In this study, we proposed a framework based on simulation optimization which integrate genetic algorithm with OCBA in simultaneous scheduling problem of machines and AGVs of FMS. This approach can effective and efficacy for this problem. Traditionally, simultaneous scheduling problems of machines and AGVs of FMS are considered a deterministic problem and solved by using meta-heuristic algorithm. Actually, there are many noise factors in this system, such as uncertain processing time, congestion in order handling, etc. Those factors are contained in real system and strongly impact system performance. Hence, stochastic simulation is used to evaluate the system performance in this problem. For avoiding congestion of AGVs, we prove deadlock situation of two AGVs in the system and use zone control technique to improve it. Finally, the numerical results demonstrate GA_{OCBA} is superiority of solution quality and reducing computing cost. Future research can also consider parameter setting in this hybrid approach such as fraction factorial design, Taguchi method and other proper methods can embed with genetic algorithm for enhancing solution quality.

References

1. Reddy, B.S.P., Rao, C.S.P.: A hybrid multi-objective GA for simultaneous scheduling of machines and AGVs in FMS. The International Journal of Advanced Manufacturing Technology 31(5-6), 602–613 (2006)
2. Bilge, U., Ulusoy, G.: A time window approach to simultaneous scheduling of machines and material handling system in an FMS. Institute for Operations Research and the Management Sciences 43(6), 1058–1070 (1995)
3. Billinton, R.: Peng Wang: Teaching distribution system reliability evaluation using Monte Carlo simulation. IEEE Transactions on Power Systems 14(2), 397–403 (1999)
4. Chen, C.H., Lin, J.Y., Ucesan, E., Chick, S.E.: Simulation budget allocation for further enhancing the efficiency of ordinal optimization. Discrete Event Dynamic Systems 10(3), 251–270 (2000)
5. Ho, Y.-C., Wang, C.-R.: A Shifting-Zone Control Strategy for Vehicle-Collision Prevention in a Multiple-Vehicle AGV System. In: Proceedings of the 14th International Conference on Production Research, OSAKA, Japan
6. Jerald, J., Asokan, P., Saravanan, P., Rani, D.C.: Simultaneous scheduling of parts and automated guided vehicles in an FMS environment using adaptive genetic algorithm. International Journal Advanced Manufacturing Technology 29, 584–589 (2006)
7. Novas, J.M., Henning, G.P.: Integrated scheduling of resource-constrained flexible manufacturing systems using constraint programming. Expert Systems with Applications 41(5), 2286–2299 (2014)
8. Li, K., Leung, J.Y.-T., Cheng, B.-Y.: An agent-based intelligent algorithm for uniform machine scheduling to minimize total completion time. Applied Soft Computing 25, 277–284 (2014)
9. Lacommea, P., Moukrimb, A., Tchernev, N.: Simultaneous job input sequencing and vehicle dispatching in a single-vehicle automated guided vehicle system: a heuristic branch-and-bound approach coupled with a discrete events simulation model. International Journal of Production Research 43(9), 1911–1942 (2005)
10. Raman, N., Talbot, F.B., Rachamadugu, R.V.: Simultaneous scheduling machines and material handling devices in automated manufacturing. In: Proceedings of Second ORSA/TIMS Conference on Flexible Manufacturing System, pp. 455–466 (1986)
11. Shanmugam, K.S., Balaban, P.: A Modified Monte-Carlo Simulation Technique for the Evaluation of Error Rate in Digital Communication Systems. IEEE Trans. Commun. 28(11), 1916–1924 (1980)
12. Ulusoy, G., Srvnkaya, F., Umit, B.: A genetic algorithm approach to the simultaneous scheduling of machines and automated guided vehicle systems. Computers & Operations Research 24(4), 335–351 (1997)
13. Wang, W.-L., Wang, H.-Y., Zhao, Y.-W., Zhang, L.-P., Xu, X.-L.: Parallel machine scheduling with splitting jobs by a hybrid differential evolution algorithm. Computers & Operations Research 40(5), 1196–1206 (2013)
14. Wang, X., Cheng, T.C.E.: A heuristic for scheduling jobs on two identical parallel machines with a machine availability constraint. International Journal of Production Economics 161, 74–82 (2015)

Collaborative Agents Supporting Tactical Planning Activities – An Industrial Application

Ana Paula M. Tanajura[1,2,*], Pinar Öztürk[3], and Herman Lepikson[1,2]

[1] SENAI CIMATEC, Salvador, Bahia, Brazil
[2] Program of Industrial Engineering, Federal University of Bahia, Salvador, Bahia, Brazil
[3] NTNU, Norwegian University of Science and Technology, Trondheim 7491, Norway
{ana.tanajura,herman.lepikson}@fieb.org.br,
pinar@idi.ntnu.no

Abstract. This paper presents a collaborative agents model for Sales and Operation Planning (S&OP) in industry. We show how an S&OP system can be used in a multi-agent approach considering existing legacy software and how people and software agents interact in a planning environment in a process industry, such as the petrochemical industry. The model helps to describe the process, determine the pre-existing software integrations, define responsibilities, and promote communication and learning. The solution adopted is crucial for a high performance S&OP. It is also important to consider the interaction among human agents and software agents, which are required for its success. The multi-agent system (MAS) paradigm is useful for helping industries handle distributed information sources and interactions between a number of actors and teams. We examine activities of an S&OP in a petrochemical company and how these activities can be more efficiently performed through MAS.

1 Introduction

Many research studies have focused on the adoption of multi-agent systems (MASs) in industrial environments [1, 2]. Inside manufacturing systems, MASs provide solutions for manufacturing enterprise integration [3], enterprise collaboration [4], manufacturing process planning and scheduling [5, 6], shop floor control and holonic manufacturing [7]. In the production planning and control field, MASs are applied to order quotations, demand forecasts, order management, purchasing, master production scheduling, monitoring and distribution [8].

In all of these examples, MASs are useful for data collection, data analysis, communication and information exchange. At the tactical level of production planning in Sales and Operation Planning (S&OP) [9], these agents help in the reduction of planning time, activities coordination, control information and retrieve learned-lessons that are relevant for computer reasoning.

S&OP is a business process that links the corporate strategic plan to daily operation plans [10]. It includes representatives from operations, supply chain,

* Corresponding author.

M. Gen et al. (eds.), *Industrial Engineering, Management Science and Applications 2015*,
Lecture Notes in Electrical Engineering 349, DOI: 10.1007/978-3-662-47200-2_32

marketing, human resources, and production planning from middle management to the executive level. To accomplish this complex activity, S&OP requires the involvement of a number of experts working on different parts of a problem to provide the right information, define the work processes and support the technology. The process is a series of pre-planning sessions that promote team work, as well as the involvement and communication among executives and the mid-level operational management [11].

According to [12], S&OP gained recognition in 1990s in operations planning and control evolution with the switch in focus from Manufacturing Resource Planning (MRP) to Enterprise Resource Planning (ERP). Some workflows [13, 14] and maturity models [15, 16, 17, 18, 19, 20] were proposed, but S&OP is still a challenge, and few companies reach the advanced S&OP level. For S&OP to succeed, functional teams must communicate, share data, strive for consensus, and make plans to achieve common sets of objectives with adequate technology [21]. Without technology, a process such as S&OP cannot support the scale needed to realize all of the benefits. There are technologies for demand and supply planning, but S&OP workbenches that integrate with ERP, MRP, MES and Legacy Systems [22] are less available.

An S&OP workbench is an automated tool for sharing information among team members [10] related to the what-if analysis of the potential changes to the supply chain and/or the demands plan and Key Performance Indicators (KPI) for performance measurement [22]. Some suggested that new technologies for Integrated Business Planning (IBP) should be considered an advanced S&OP essential for incorporating the integration of targets and cooperation among teams and sub-processes into the strategy-to-operation cycle in a company [23].

This paper proposes an S&OP workbench that relies on an MAS approach. In our approach, a *task decomposition method* [24] is used to create a plan for accomplishing a complex task, and a multi-agent framework [25] is used to determine the system service and agents in the S&OP process. As a result, our S&OP process may incorporate the potential benefits from the integration of the best utilities from ERP and MRP to provide an effective, useful, powerful and intelligent system. This is an effective way to embed intelligence into ERP and MRP applications without interfering with their structure.

The paper is organized as follows. Section 2 synthesizes the S&OP general concepts and activities. Section 3 describes the framework and the problem solving method. In Section 4, we describe the problem solving method in detail and apply it to S&OP in a model industry. Finally, Section 5 provides conclusions.

2 Sales and Operation Planning

S&OP is about bringing together various plans prepared within a company, e.g., sales plans, marketing plans, research and development plans, production plans, purchasing plans, and financial plans, to generate an integrated set of plans. It is a process for creating tactical plans to ensure the company's competitive advantage based on the continuous integration of client-oriented marketing plans and supply chain management.

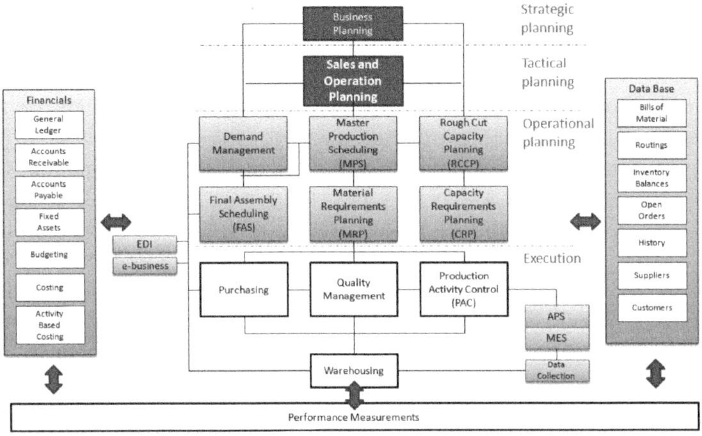

Fig. 1. The S&OP and the 3 different levels of general planning, based on APICS [27]

The plan is prepared once a month at the level of product groups. S&OP is both a combination and a confrontation of objectives that fulfill the requirements for medium-term planning of resources and one-year business planning. This plan also combines strategic plans with the implementation, performance measurements and continuous improvement of operations [25].

As a tactical approach, S&OP connects the strategic plan with the operational plan in a company (Figure 1). In terms of ERP, it connects the Final Assembly Scheduling (FAS), MRP and Capacity Requirements Planning (CRP) to Business Planning where the company strategy is defined [27].

According to [28], the main purposes of S&OP are the following: to build bridges between the business or strategic plan and the operational plans of the enterprise and to balance supply and demand, [29] also included some other main objectives such as evaluating the performance of the planning process by gathering measurements. The performance evaluation helps to identify gaps in the communication and interactions between different teams and among the team members, thereby improving the teamwork capabilities.

The S&OP cycle has 5 main stages [29, 10, 14, 15]. It starts with gathering data on past sales for sales forecasting. Next, demand planning validation is conducted using forecasts of the sources of demand and inventory and customer service policies. In the third stage, supply planning (the ability to meet demand) is assessed by reviewing the available capacity and the required scheduling for production operations. The production and logistic plans that best fit the proposed demand plan are then selected. A preliminary meeting (Pre-S&OP) is then conducted to match the supply and demand plans with financial considerations (impact and constraints). Finally, after adjustments are made, the S&OP objectives are released for implementation.

More recently, S&OP has been recognized under the term Integrated Business Planning (IBP), which "prioritizes the strategic planning and more robust product and portfolio review over the balancing supply and demand process" [23, p.28]. In essence, IBP can be considered advanced S&OP with targets and KPIs adapted to the type of company strategy, e.g., the segmentation strategy or cost leadership and differentiation strategy, as proposed by Porter in 1980, in his book *Competitive Strategy*.

We consider computer support to be essential for a successful IBP. In particular, the MAS paradigm seems to be appropriate for this purpose. The MAS provides a suitable environment for collaborative software that may be applied to S&OP using the MAS framework proposed by [25]. In the next section, we describe the MAS framework and provide details about the correspondence between S&OP and MAS-based computer support.

3 MAS Framework and the Task Decomposition

One of the main questions in collaborative agent systems relates to what the agents should collaborate on. In the MAS approach we adopted, anything the agents do in collaboration is defined as a "task". A task can denote an instance of a problem, a problem class or both a problem class and an abstract description of a method for solving the problem [30]. In our case, the S&OP process is a complex task, i.e., an instance of a problem to be solved. A task is described in terms of a goal and refers to the "problem solving method" through which it can be accomplished. Hence, a task describes what to do, while the problem-solving methods describe how to do it [31].

Problem solving methods decompose a higher level task (or problem) into subtasks and follow certain steps and rules to construct an unambiguous tree that represents the Task Structure [24]. The subtasks are decomposed into smaller tasks iteratively until reaching the "leaf-node tasks" that can be directly accomplished by agents without further decomposition. The Task Structure model can be integrated into the MAS framework [25]. The MAS includes a number of different software agents, each of which has a different role and responsibility in terms of the tasks in the system [32]. The types of agents are defined according to the requirements of the domain and the characteristics of the problem at hand (i.e., the S&OP).

In the MAS, each agent has a set of capabilities based on their knowledge (e.g., ontology, or a rule base) and reasoning skills (i.e., algorithms). A Task Structure tree maps the capabilities of the agents in the system. An agent should have the knowledge and information (i.e., input) required for the task to which it is mapped in order to be able to accomplish/execute the task (activity). That is, an agent can be allocated to a task (i.e., mapped) only if it is has the computational skills for producing the type of output from the input, as well as the ontology that the task specifies.

In the industrial context, the capabilities of agents are represented by the capabilities of a person or a computer system that is currently in use. The plan generated through such a problem decomposition process is useful for finding the best sequence of activities ("tasks" in MAS terminology) to be performed as well as the best team (a set of "agents" in MAS terminology) for executing these activities.

In the Coordinated Problem Solving Framework (CoPSF) proposed by [25], task decomposition occurs in the second stage of the five stage problem solving process: (i) problem identification, (ii) problem analysis, (iii) team formation, (iv) team work, and (v) presentation of the solution. There are five different types of agents in the CoPS framework that have different roles and responsibilities: Personal Assistant (PA), Matchmaker (MAT), Decomposer (DEC), Problem Solvers (PS), and Task Administer (TA). A PA is a bridge between the user and the system, as shown in Figure 4. The MAT keeps a record of the actors and their abilities. The DEC decomposes the tasks. The PSs either are workers that possess certain domain specific skills or are software platforms.

S&OP involves a complex set of activities due to the amount of distributed information, the number and heterogeneity of the actors, uncertainties and the limited time to deliver the S&OP objectives. Some activities require special attention with respect to time and availability of people, who are usually busy. In the next section, we describe how task decomposition and the CoPSF framework are suitable and useful for structuring the S&OP process.

4 The Problem Solving Method Applied to S&OP in an Industrial Context

This section synthesizes some concepts, targets and activities for S&OP in a petrochemical company and describes the problem solving method in this context. The S&OP context is important for problem identification and task decomposition. After we describe the task decomposition, we will detail the interactions among different teams for each task and propose an intelligent S&OP System.

4.1 S&OP Context

The petrochemical company used for this case study is one of the largest companies in its sector. It has 36 industrial units: 29 in Brazil, 5 in the United States and 2 in Germany. In our example, we consider S&OP in one unit of the company, the Basic Petrochemical Unit (BPU) located in Bahia, Brazil. The S&OP considers 26 products: ethylene, propylene, benzene, toluene, gasoline, solvents, butadiene and isoprene, among others. Ethylene and propylene are the two main products, accounting for approx. 40% of all production in BPU. These products go to an internal customer, the Polymer Unit. The main raw material is Nafta, which is produced by the oil and gas industry. Figure 5 shows where the BPU is placed in the supply chain.

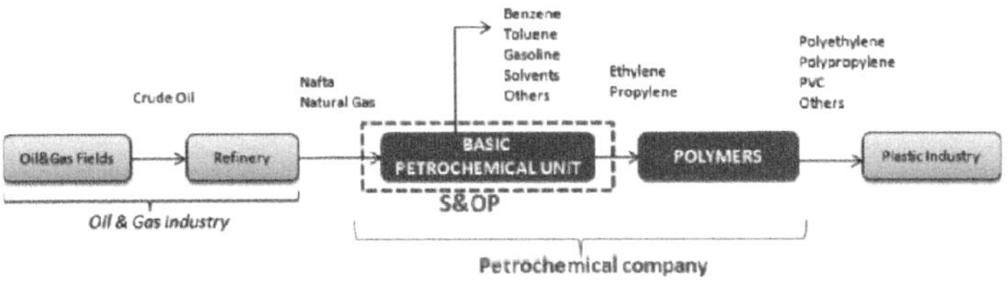

Fig. 2. Petrochemical supply chain stages and the focus of S&OP process

One of the aspects to be considered in the petrochemical company scenario is the existence of many international competitors with no local competitors. This requires a team with expertise in international marketing to provide information for the S&OP process. Another important aspect is that it is not possible to make different arrangements for equipment, e.g., assigning some manufacturing companies to produce different final products. And there are no modifications to the final product specifications because the products are commodities, but it is possible to make arrangements with internal products to produce others like gasoline.

All of these aspects should be defined to develop the MAS and the S&OP context (Table 1). The S&OP context considers multiple *business aspects* (type of industry, location, logistics, environmental restrictions, availability of new business and the market) and multiple *planning aspects* (strategic plan, the maintenance plan, planning horizon, available products, products to launch, production capacity and logistics).

Table 1. The elements of the S&OP context

S&OP context
Type of industry
Location
Products available
Products to launch
Production capacity
Logistics (trucks, pipelines, vessels, people)
Planning horizon
Maintenance plan
Environment restriction
New business available
Market share
Strategic plan

There are four main areas involved in the S&OP process: Marketing, Commercial, Industrial and Supply Chain.

- *Marketing* gives information about the competitor's actions and the premises of oil price that influence the final product prices.
- *Commercial* defines volumes and prices for each costumer based on the premises of the market.
- *Industrial* is responsible for production and covers health and environment restrictions and maintenance coordination.
- *Supply Chain* coordinates three teams in the company: *Planning* – responsible for S&OP execution; *Raw Material* – responsible for Nafta acquisition; and *Logistic* - responsible for distribution of the final products (gases and liquids).

4.2 Legacy Systems

The company has an ERP and an optimization tool called SCMart®. The ERP delivers planned production orders to be executed at the control rooms. These orders come from SCMart®, which has planned production using S&OP processes in the company since 2007. The SCMart® is a mixed integer nonlinear programming (MINLP) solver that helps with scenario development [33]. SCMart constructs several scenarios from demand, supply and production data [34]. During its execution, the system highlights many constraints and restrictions, such as lack of raw material, lack of ethylene demand, lack of benzene demand, availability and seasonal fluctuations due to lack of natural gas or high raw material prices. The type of information that can be found in SCMart includes Finance (profit, shadow prices, and costs), Supply, Demand, Inventory (raw materials, intermediates, and products) and Production data.

The main goal of SCMart is to improve profitability. The software defines the best-case scenario, the volumes to be produced and the sales that provide the most profit for the company. SCMart is a good optimization tool that fits the company purpose. However, for intelligent S&OP processes, it is necessary to use a holistic view that considers other aspects, including the gathering of information and the possibility of learning from analyses realized during the S&OP process.

4.3 S&OP Task Decomposition

For the task decomposition process, we start with the problem definition:

How to perform S&OP in an effective way?

The main goal of S&OP is to optimize the assertiveness of prices and volumes. The assertiveness measures the deviation between the planned and the real data to guide the improvement of the S&OP process. The assertiveness is targeted because S&OP includes the elaboration of a plan to produce the highest profit. The highest profit plan defines the volumes and prices that should be reached to provide the highest profit for the company.

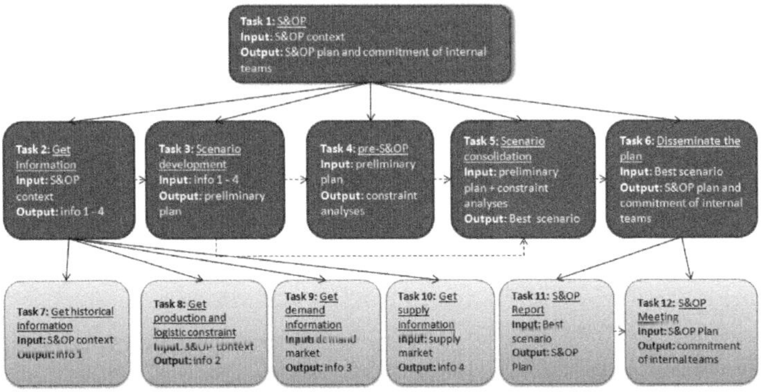

Fig. 3. S&OP plan after the Task decomposition process

The activity starts with supertask S&OP (Task 1). The input for that task is the S&OP context, as defined in Section 4.1. The output of the S&OP is the commitment of the company to execute the best plan defined through the whole planning process. If the best plan is elaborated, but there is not commitment to follow this plan, then the S&OP task is not successful. The commitment is measured through the assertiveness KPI.

The S&OP *Task 1* is divided into five subtasks: Get information (Task 2), Develop the scenario (Task 3), pre-S&OP (Task 4), Consolidate the scenario (Task 5), and Disseminate the plan (Task 6). The planning team is involved in all of these tasks. They perform a coordination function in horizontal alignment and should have good interactions with the other teams.

The input of *Task 2* is the S&OP context, which defines which information is relevant for the S&OP activity. This Task combines information from its four subtasks: Get historic information (Task 7) – important for initial inventory forecasting, Get production and logistic constraints (Task 8), Get demand information (Task 9), and Get supply information (Task 10). Each of these subtasks addresses different types of information, e.g., historical information about the product and raw material storage, sales, deliveries, production and interventions in the last month. Tasks 8, 9 and 10 refer to the forecast data and the operational constraints.

The output of Task 2 is necessary for the execution of Task 3. *Task 3* needs a lot of expertise to understand the information it receives from the different teams in the company. All of the information is gathered and alternative scenarios are created to find the highest profit scenario. The constraints are analyzed and some of the opportunities are highlighted to compare with alternative plans.

The output of Task 3 is a preliminary plan to drive the work meeting, i.e., the pre-S&OP in *Task 4*. The people responsible for obtaining information through tasks 7, 8, 9 and 10 have to participate in Task 4 to validate the plan premises. The opportunities identified in Task 3 are analyzed in a holistic group during the execution of Task 4. The coordinators or managers responsible for any process with constraints should participate to eliminate the process or propose alternatives. As proposed by the Theory of Constraints (TOC), it is necessary to identify and eliminate constraints to improve the results in a company [35]. The TOC is considered in the company and in the software used for scenario development to provide a list of constraints and shadow prices to help find opportunities. The output of Task 4 is a confirmation about the constraints that helps in the scenario consolidation in Task 5.

In *Task 5*, the final scenario is defined by planners with the complete Sales and Operation Plan. The inputs are the preliminary plan from Task 3 and the constraint analyses from Task 4. The output is the best scenario for the company after considering all of the possibilities in Task 3 and the plan validation in Task 4.

Task 6 addresses the dissemination of the plan for the main teams involved in the S&OP process: industrial, commercial, marketing, logistic, raw material and planning teams. It has two subtasks: S&OP Report (Task 11) and S&OP Meeting (Task 12). The Report is drafted by a planning team and delivered to a group of people who need this information for their daily work. It can be communicated via email, on paper, through an order in the ERP system or all three of these. The S&OP meeting is an opportunity to share challenges and opportunities for the business, analyze the performance of the planning activity thought KPI and mobilize people to engage.

4.4 Intelligent S&OP System

The MAS based on the CoPSF framework is used to support the S&OP task. One of the important properties of the MAS is to promote the interaction between agents through special agent-based languages such as KLM [36] or FIPA-ACL, both of which rely on the speech act theory developed by Searle [37]. A set of standardized interaction protocols have been developed that capture some typical situations that require interactions between agents working on the same task. One such example is the "contract net protocol" [38], which is the standard protocol for the task allocation process. It enables interactions between agents where a call has been broadcasted to a set of agents to allocate a certain task to the most suitable agents. CoPSF uses the contract net protocol in addition to its own more custom-made protocols. CoPSF uses FIPA-ACL as the agent language.

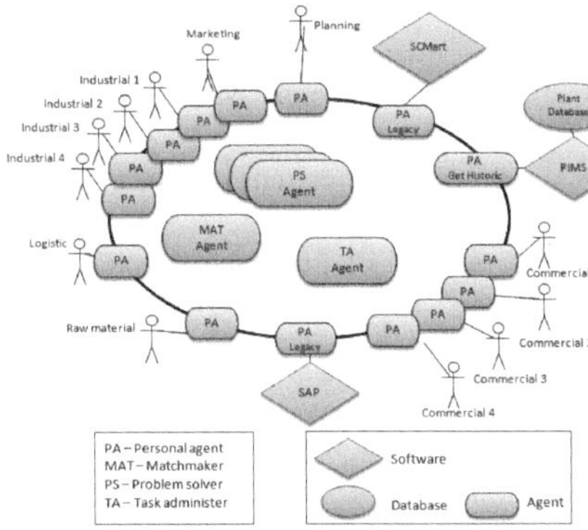

Fig. 4. The Sales and Operation Planning System

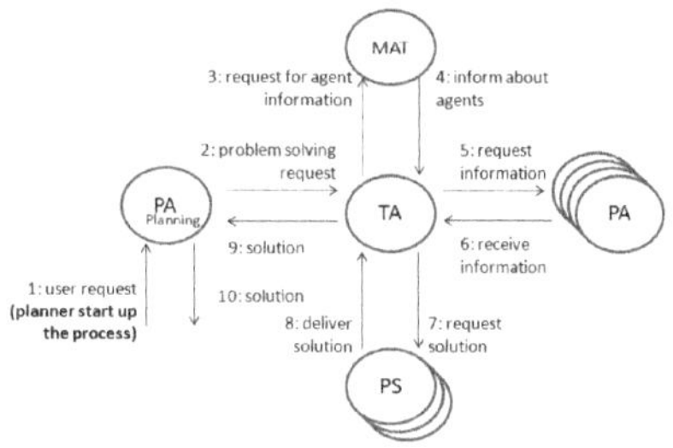

Fig. 5. Overview of the interactions involved in S&OP process

The plan obtained through task decomposition in Section 4.3 (Figure 6) will be accomplished by the CoPSF agents. In the system, each entity (people, database, or agent) responsible for providing the information is represented by one PA agent. Thus, in Figure 7, it is possible to see many PA agents interacting with others and the legacy software.

The SAP APO Advanced Planner and Optimizer is a planning tool that is used to plan and optimize the supply chain processes by making use of various modules. SAP APO is also used in the system to connect the S&OP process to other processes such as MRP, which request material based on the production orders. Demand Planning is one of the SAP APO modules and is used for statistical forecasting. The Process Information Management System (PIMS) provides data from the Data Acquisition Level to a PA Get Historic. Useful information is obtained when checking if the production, inventory and sales are proceeding according to the planning.

As described above, CoPSF has five types of agents: Task Administer agent (TA), Personal Assistant agent (PA), Matchmaker agent (MAT), Problem Solver agent (PS) and Decomposer agent (DEC). During the S&OP process, the agents execute their roles as described below:

- The *Task Administer* (TA) agent coordinates the S&OP Process. It resolves dependencies and distributes the tasks to the other agents, controls the deadline, and ensures the quality of the process.
- The *Personal Assistant* (PA) agents interact with human agents when there is a need for communication of a question/information between the system and the persons in the company. In the S&OP process some of the information gathered by the agents in the MAS needs to be confirmed by a human before being used in the later stages of the process. The PA agents collect the information and display it for the users to approve or refuse. In this way, each piece of information that is used in S&OP is ensured to have a human responsible for it.
- The *Matchmaker* (MAT) agent keeps a record of agents in the system and their capabilities regarding knowledge and reasoning skills. This agent knows which agents are suitable for certain tasks. Every MAS agent has a matchmaker [39].
- *Problem Solver* (PS) agents are the planners that work in the planning team to analyze scenarios (Tasks 3 and 5 in Figure 6) to reach the best sales and operations plan based on the embedded intelligence. Some of the PS agents can also serve as dummy agents that receive an output from the optimization software outside of the MAS, e.g., a legacy software system that the company already uses.

The Decomposer (DEC) agent is not considered in our model. The S&OP process is a pre-defined activity and the task plan in Figure 6 will not change with the S&OP process execution. Figure 9 shows an overview of the interactions involved in the S&OP process.

The user, via the PA agent, starts the S&OP process. The PA communicates the user's request about creating an S&OP to the TA agent. The TA gets the requisition and asks the MAT which agents in the system match the tasks in the decomposed task tree. The MAT informs the agents and people who are responsible for each piece of information inside the company. The TA collects the required information (i.e., the commercial, supply, production and historical data) that the human actors need to provide using the PAs. Each PA sends the requested information to the TA in a timely manner. The TA requests an integrated solution from the set of PSs it allocated to the tasks in the task structure tree. The PSs deliver a solution to the TA. The TA then integrates these and delivers the results to the relevant PAs, who in turn deliver it to the human actors they are connected to. Interaction 8 and 7 between the TA and the PS agent will happen at least twice during scenario development and consolidation.

The flowchart in Figure 6 shows the communication for data acquisition (Task 2). The flowchart in Figure 7 shows the negotiation during scenario development and consolidation (Tasks 3 to 6). Each PA (Planning, Marketing, Industrial, Logistic, Raw Material and Commercial) is responsible for specific information. For example, only the

Raw Material PA can define the naphtha strategy (origin, quality, price, and minimum and maximum lot size) and only the Commercial PA who is responsible for Customer 1 can define the price and minimum and maximum volume for that customer.

There are several Commercial PAs and Industrial PAs. Each group of products produced in the same industrial plant has one Industrial PA. Each group of products negotiated by a commercial manager has one commercial PA. There are approximately 8 Industrial PAs and 5 Commercial PAs.

The TA requests information from all of the PAs. If the requested information is not available, the TA designates the Planning PA for filling in the information gaps based on the PA's experience. After gathering all of the information necessary for S&OP, the TA asks the PS agents to find the best S&OP plan (II) and to send it to the SCMart PA legacy (I) (Figure 7).

If necessary, the PS agent may access other software to provide the best S&OP. In our case, only the SCMart is used for optimization, as shown in Figure 11. During the what-if analyses, the pre-S&OP meeting is held to formalize the modifications suggested by the TA. In this meeting, human agents join a holistic group to collectively analyze the alternatives. For example, if the Raw Material PA indicates that there is not enough naphtha available, the Industrial PA can reduce the GN consumer for energy production and increase the capacity of the gas furnaces that produce ethylene. Similarly, the Commercial PA can reduce the sales to less profitable customers. The Logistic PA can review the inventory policy and transfer as much products as possible to avoid customer shut down.

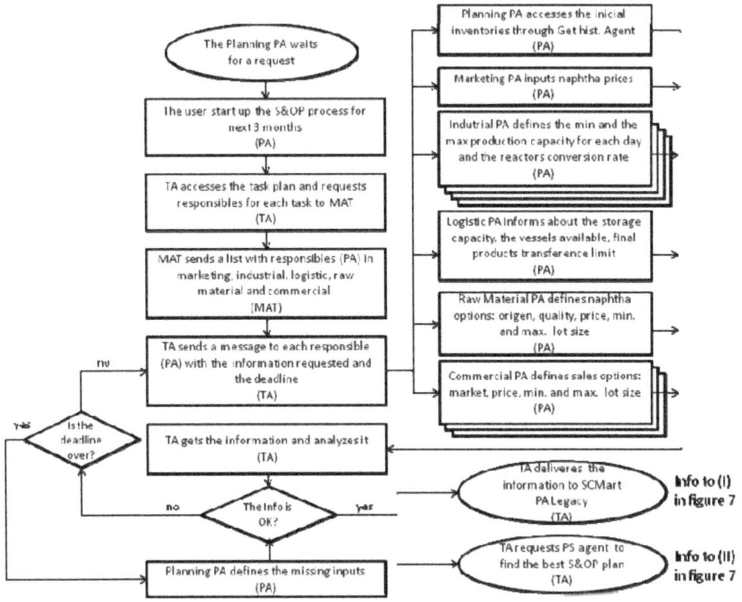

Fig. 6. Communication flow to get information (Task 2 in Figure 3)

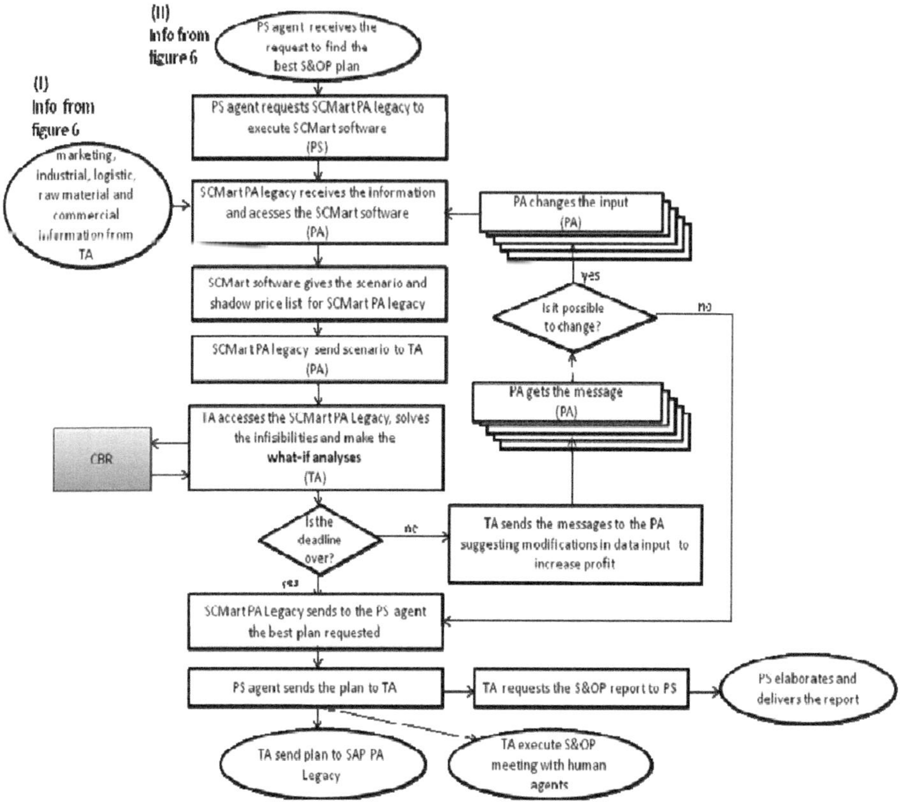

Fig. 7. Communication flow for elaborating the S&OP objectives (Tasks 3 to 6 in Figure 3)

The capabilities of the workbench can be extended through the addition of a Case Based Reasoning (CBR) skill. CBR is an artificial intelligence method in which the principal idea is solving new cases using solutions to similar cases that were solved in the past. It has been applied in diverse industrial applications for reuse of experiences [40]. A CBR system learns new "cases" after solving a new problem to be used in future. The TA in our system can be connected with an agent with this capability (i.e., CBR software) that can learn from analyses executed during the scenario development and scenario consolidation (Task 3 and 4) and suggest solutions to maximize the results based on learned cases. The TA registers the what-if analyses and gains suggested by SCMart. The analysis of the constraints and opportunities needed to maximize the results in a holistic scenario is complex and requires a lot of expertise from the planning team. A CBR module can help new professionals during the execution of the S&OP process, thereby reducing the learning period.

5 Conclusion and Future Work

The paper presented a framework for a powerful intelligent S&OP system developed using a multi-agent approach. The system used the collaborative environment proposed in CoPSF to show how agents can interact and communicate to develop the best S&OP. S&OP is a tactical planning strategy that is important for connecting business and operation plans. S&OP gathers the relevant information at the right time and joins several software platforms and people with different knowledge and expertise. The MAS improved the S&OP execution by reducing planning time, improving activity coordination, and providing control over information and learning. The people responsible for each piece of information should be involved in the S&OP process to guarantee that the plan is coherent and feasible. The MAS seems to be the right bridge to link people and systems in distributed environments in a more efficient way. The use of CBR connected to the system will be detailed in future work. CBR should be formalized for computer reasoning in a four-step process: retrieving, reusing, revising and retaining. The reuse step maps the solution from the previous case to the target problem. This may involve adapting the solution as needed to fit the new situation. Having mapped the previous solution, it can be tested in the real world or by simulation and, if necessary, it can be revised. After the solution has been successfully adapted to the target problem, the experience is retained as a new case in the computer memory.

References

1. Vrba, P.: Review of Industrial Applications of Multi-agent Technologies. In: Borangiu, T., Thomas, A., Trentesaux, D. (eds.) Service Orientation in Holonic and Multi agent, SCI, vol. 472, pp. 327–338. Springer, Heidelberg (2013)
2. Shen, W., Hao, Q., Yoon, H.J., Norrie, D.H.: Applications of agent-based systems in intelligent manufacturing: An updated review. Advanced Engineering Informatics 20, 415–431 (2006), doi:10.1016/j.aei.2006.05.004.
3. He, N., Zhang, D.Z., Li, Q.: Agent-based hierarchical production planning and scheduling in make-to-order manufacturing system. International Journal of Production Economics (2013), http://dx.doi.org/10.1016/j.ijpe.2013.08.022
4. Hernández, J.E., Lyons, A.C., Mula, J., Poler, R., Ismai, H.: Production Planning & Control: Supporting the collaborative decision-making process in an automotive supply chain with a multi-agent system. Production Planning & Control: The Management of Operations (2013), doi:10.1080/09537287.2013.798086
5. Tichý, P., Slechta, P., Maturana, F., Balasubramanian, S.: Industrial MAS for Planning and Control. In: Marik, V., et al. (eds.) Proceedings of the 9th ECCAI-ACAI/EASSS 2001, AEMAS 2001, HoloMAS 2001 on Multi-Agent-Systems and Applications II-Selected Revised Papers, pp. 280–295 (2002)
6. Jindal, K., Srinivasan, S., Sharma, M.: Review of Decision Support System Based on Multi Agent in Production Scheduling. International Journal of Engineering and Social Science 3(10), 33–36 (2013)
7. Leitão, P.: Agent-based distributed manufacturing control: A state-of-the-art survey. Engineering Applications of Artificial Intelligence 22(7), 979–991 (2009)

8. Caridi, M., Cavalieri, S.: Multi-agent systems in production planning and control: an overview. Production Planning & Control: The Management of Operations 15(2), 106–118 (2004), doi:10.1080/09537280410001662556
9. Vollmann, T.E., Berry, W.L., Whybark, D.C., Jacobs, F.R.: Manufacturing Planning and Control Systems for Supply Chain. Irwin, NY (2005)
10. Grimson, J.A., Pyke, D.F.: Sales and operations planning: an exploratory study and framework. International Journal of Logistics Management 18(3), 322–346 (2007)
11. Schlegel, G.L., Murray, P.: Next Generation of S&OP: Scenario Planning with Predictive Analytics & Digital Modeling. Journal of Business Forecasting (Fall 2010)
12. Olhager, J.: Evolution of operations planning and control: from production to supply chains. International Journal of Production Research 51, 6836–6843 (2013), doi:10.1080/00207543.2012.761363.
13. Wallace, T.F.: Sales & Operations Planning: The How-To Handbook, USA
14. Lapide, L.: Sales and operations planning Part I: the process. The Journal of Business Forecasting 23(3), 17–19 (2004a)
15. Wing, L.G.P.: Toward twenty-first-century pharmaceutical sales and operations planning. Pharmaceutical Technology North America, 20–26 (2001)
16. Lapide, L.: Sales & operations planning part III: a diagnostic model. Journal of Business Forecasting 24(1), 13–16 (2005)
17. Ventana Research. Sales and operations Planning: Measuring Maturity and Opportunity for Operational Performance Management. Ventana Research, San Mateo, CA, USA (2006)
18. Feng, Y., Sophie D'Amours, S., Beauregard, R.: The value of sales and operations planning in oriented strand board industry with make-to-order manufacturing system: cross functional integration under deterministic demand and spot market recourse. International Journal of Production Economics 115(1), 189–209 (2008), doi:10.1016/j.ijpe.2008.06.002.
19. Viswanathan, N.: Sales and operations Planning: Integrate with Finance and Improve Revenue. Aberdeen Group, Boston (2009)
20. Cacere, L., Barret, J., Mooraj, H.: Sales and Operations Planning: Transformation from Tradition. Industry Value Chain Strategies. AMR Research, Boston (2009)
21. Wells, A.M., Schorr, J.: Sales and Operations Planning: The key to continuous demand satisfaction. SAP INSIGHT Business Process Innovations (2007), `http://arnoldmarkwells.com/images/Sales_and_Operations_Planning_The_Key_to_Continuous_Demand_Satisfaction.pdf`
22. Lapide, L.: Sales and operations planning Part II: enabling technology. The Journal of Business Forecasting 23(3), 18–20 (2004)
23. Jurečka, P.: Strategy and Portfolio Management Aspects of Integrated Business Planning. Central European Business Review 2(1), 28–36 (2013)
24. Gundersen, O.E., Kofod-Petersen, A.: Multiagent Based Problem-solving in a Mobile Enviroment. In: Norsk Informatik Konferanse, Bergen (2005)
25. Öztürk, P., Rossland, K., Gundersen, O.E.: A multiagent framework for coordinated parallel problem solving. Appl. Intell. 33(2), 132–143 (2010), doi:10.1007/s10489-008-0154-7
26. Adamczak, M., Domański, R., Cyplik, P.: Use of sales and operations planning in small and medium-sized enterprises. LogForum 9(1), 11–19 (2013)
27. APICS. Sales & Operations Planning. Presented at the Region IV Meeting for the Association for Operation Management, New Orleans (April 14, 2007)

28. Thomé, A.M.T., Scavarda, L.P., Fernandez, N.S., Scavarda, A.J.: Sales and Operations Planning: A Research Synthesis. International Journal of Production Economics 138(1), 1–13 (2012)
29. Corrêa, H.L., Gianesi, I.G.N., Caon, M.: Planejamento, Programação e Controle da Produção. Editora Atlas, São Paulo (2001)
30. Chandrasekaran, B., Johnson, T.R.: Generic tasks and Task structures: History, critique and new directions. In: David, J.M., Krivine, J.P., Simmons, R. (eds.) Second Generation Expert Systems, pp. 232–272. Springer, Berlin (1993)
31. Fensel, D., Motta, E., Benjamins, V.R., Crubezy, M., Decker, S., Gaspari, M., Groenboon, R., Grosso, W., Van Harmelen, F., Musen, M., Plaza, E., Schreiber, G., Studer, R., Wielinga, B.: The unified problem-solving method development language UPML. Knowledge and Information Systems (1999)
32. Wooldridge, M.: Intelligent Agents. In: Weiss, G. (ed.) Multiagent Systems – A Modern Approach to Distributed Artificial Intelligence. MIT Press (1999)
33. Tanajura, A.P.M., Cabral, S.: Sales and Operations Planning (S&OP) in a Petrochemical Company. TAC, Curitiba 1(2), 55–67 (2011)
34. Rossi, M.C., Bandoni, J.A.: Planning of an integrated Petrochemical Complex Using SCMart®. Paper presented at the 2nd Mercosur Congress on Chemical Engineering 4th Mercosur Congress on Process Systems Engineering, Costa Verde, August 14-18 (2005)
35. Cox, J., Goldratt, E.M.: The goal: a process of ongoing improvement. North River Press, Great Barrington (1986)
36. Finin, T., Fritzson, R., McKay, D., McEntire, R.: KQML as an agent communication language. In: Proceedings of the Third International Conference on Information and Knowledge Management, CIKM 1994, New York, USA, pp. 456–463 (1994)
37. Searle, J.: Indirect speech acts. In: Syntax and Semantics, vol. 3: Speech Acts, pp. 59–82. Academic Press, New York (1975)
38. Smith, R.G.: The Contract Net Protocol: High-Level Communication and Control in a Distributed Problem Solver. IEEE Transactions on Computers C-29(12), 1104–1113 (1980)
39. Foner, L.N., Crabtree, I.B.: Multi-Agent Matchmaking. In: Nwana, H.S., Azarmi, N. (eds.) Software Agents and Soft Computing: Towards Enhancing Machine Intelligence. LNCS, vol. 1198, pp. 100–115. Springer, Heidelberg (1997)
40. Öztürk, P., Aamodt, A.: A context model for knowledge-intensive case-based reasoning. International Journal of Human Computer Studies 48, 331–355 (1998)

Minimizing Makespan Using Node Based Coincidence Algorithm in the Permutation Flowshop Scheduling Problem

Ornrumpha Srimongkolkul and Prabhas Chongstitvatana

Department of Computer Engineering, Faculty of Engineering, Chulalongkorn University Bangkok, Thailand
o.srimongkolkul@gmail.com, Prabhas.C@chula.ac.th

Abstract. This paper proposes a Node-Based Coincidence Algorithm (NB-COIN) for the permutation flowshop scheduling problems (PFSP) aimed at Makespan minimization. For almost half a century, a variety of complex algorithms have been introduced to solve the problems. Nevertheless, these algorithms will be useless if they fail to implement in practice where computational time and complexity of algorithm become an important issue of concern. NB-COIN is proved to be an effective algorithm and it is extremely fast. Based on the bench-mark data sets of Taillard, the presented algorithm provides acceptable solutions within a very short period of time. More importantly, the results generated by NB-COIN are also better than other well-known algorithms in consideration.

Keywords: Permutation flow shop, Makespan, Coincidence algorithm, Scheduling.

1 Introduction

In the highly competitive industrial market, production speed and operation cost become important factors for a manufacturing process. Producing wide variety of goods using the same production line is a key solution for many manufacturing companies. This technique is called flowshop.

In the permutation flow shop scheduling problem (PFSP), there are n jobs and m machines. All jobs have to be processed on every machine in the same order. Over the production period, all machines are ready and only one job can enter to the machines at a specific time. The pre-emption and interruption is not allowed. In order to reduce the search space, passing any jobs is prohibited in PFSP. In general, the performance is measured by two main objectives, makespan minimization or flowtime minimization. The makespan criterion is well-known to lead to rapid turn-around of jobs, uniform utilization of resources and minimization of work-in-process inventory.

The permutation flow shop scheduling problem (PFSP) has become an interesting research topic for many researchers since Johnson [1] introduced it in

M. Gen et al. (eds.), *Industrial Engineering, Management Science and Applications 2015*
Lecture Notes in Electrical Engineering 349, DOI: 10.1007/978-3-662-47200-2_33

the 1950s. Later, the complexity of PFSP is proved to be NP-hard by Garey et al [2] and Rinnooy Kan [3]. Many heuristic optimization methods have been developed to achieve high quality solutions in a reasonable computational time such as Nawaz et al. [4], Palmer [5], Campbell et al. [6], Dannenbring[7], Taillard [8], Framinan et al. [9] and Framinan and Leisten [10]. The results given by the most powerful heuristics, NEH, proposed by Nawaz et al. [4] are still far, at almost 7%, from the optimal value. Using only heuristics may not be capable enough to reach the optimum solution for the PFSP, many researchers developed more complex methods, metaheuristics, such as tabu search [11-14], genetic algorithms (GAs) [15-16], ant colony optimization [17-19], particle swarm optimization [20], iterated local search (ILS) [21] or the Estimation of Distribution Algorithm (EDA) [22]. Although these methods provide better results, they need to trade-off with long computational time or a lot of resources. Later, the algorithms are even enhanced by integrating two or more metaheuristics, called the hybrid metaheuristics. This technique was used by G.I. Zobolas [23] and H. Liu [24] to achieve optimal solution. However a simple algorithm that can provide a reasonable solution in a short period of time is more practical in the real world. Node-Based Coincidence Algorithm (NB-COIN) is a new metaheuristic tool improved from Coincidence Algorithm (COIN) [25]. This method is proved to be an effective algorithm for flowshop scheduling problems in terms of total flow time minimization[26]. Moreover, NB-COIN is easy to implement, using very few user-defined parameters.

This paper is organized as follows: Section 2 illustrates the Permutation Flowshop Scheduling Problem. Section 3 determines the related work. The Node-Based Coincidence Algorithm (NB-COIN) is described in Section 4. Section 5 presents the computational result and the conclusion is shown in Section 6.

2 Permutation Flowshop

The makespan is the finished time of the last job in the schedule. The makespan minimization is described as $n/m/P/C_{max}$. It consists of a set J of n jobs, $J = \{j_1, \ldots, j_n\}$, and set K of m machines, $K = \{k_1, \ldots, k_m\}$. Let $t_{k,j}$ denotes as the processing times of job J on machine K and $C(k, j)$ be the completion time of job J on machine K. Thus, $C(k, j)$ can be calculated as follows:

$$C(1, 1) = t_{(1,1)} \tag{1}$$
$$C(1, j) = C(1, j - 1) + t_{1,j} \text{ where } j = 2, \ldots, n \tag{2}$$
$$C(k, 1) = C(k - 1, 1) + t_{k,1}, \text{ where } k = 2, \ldots, m \tag{3}$$
$$C(k, j) = \max\{C(k, j - 1), C(k - 1, j)\} + t_{k,j} \tag{4}$$

3 The Related Work

In this section, a brief overview of four well-known methods for the PFSP is provided. In addition, the strength and weakness of these algorithms are pointed out. The methods include three heuristics; the NEH [4], the constructive greedy (CG)and the stochastic greedy (SG) [29], two metaheuristics; ant colony system [18] and the hybrid metaheuristic by G.I. Zobolas [23].

The performance of the NEH Heuristic (NEH) [4] has been confirmed by Park et al [27] since 1984. In addition, Turner and Booth [28], and Taillard [8], also came to a conclusion that NEH is an efficient tool for minimizing makespan in flow shop scheduling problem. The main idea of this heuristic is that the high priority should be given to the job with more total processing time on all machines. There are 3 main steps of NEH. Step 1 order the jobs by decreasing sums of processing time, $T_j = \sum_{k=1}^{m} P_{kj}$ where P_{kj} defines as the processing time of job j in machine i. In step 2, schedule the first two jobs to minimize partial makespan. Finally, The Kth jobs are inserted individually at the position with the shortest makespan.

The greedy concept has been adapted to the PFSP by M. Ancau [29]. The author proposed two heuristics; the constructive greedy heuristic (CG) and the stochastic greedy heuristic (SG) and compares them against the NEH. As the result, these algorithms provide a better result in the makespan criterion, however they consume longer computational time than the well-known NEH heuristic.

The constructive heuristic algorithm (CG) generates a job's sequence using two lists called job list and optimal schedule. A job list consist of n elements $(j_1, j_2, ..., j_n)$. Firstly, a pair of jobs from the job list will be selected and arranged to find the minimum completion time passing to the optimal schedule. Then, repeat the first step, however either increase the selected elements to $k(n-k-1)$, k is the number of rounds, or pass to the optimal schedule in the relative position that minimize completion time.

In the stochastic heuristic (SG), the job list consists of n random job's elements. The first pair from the job list will be selected and finds an optimal completed time. Other jobs in the job list will be selected individually and find the best position in the optimal schedule.

K.C. Ying presented an Ant Colony System (ACS) [18] for the PFSP. This method was first introduced by Dorigo [30]. It is inspired from real ant behavior, finding shortest path using the relevant pheromones. The algorithm consists of four steps. In the first step, the method generates a set of artificial ants. Each ant employs a stochastic greedy to create a path (in the PFSF is job's sequence). The amount of pheromone is updated when the ants build a tour. Then, after all ants have completed their paths, the pheromones are modified again. The ACS is an effective algorithm that generates a solution within a small amount of time. Nevertheless, this method is sensitive to the user-defined parameters.

The hybrid metaheuristic was proposed by G.I. Zobolas[23] in 2009. This method combines different techniques and concepts from four construction heuristics and two metaheuristics to expand the solution space search and overcome the limitation of a single metaheuristic. The combination includes the heuristic proposed by Nawaz et al[4], Campbell et al. [6], Palmer's [5], Gupta's [32], and the metaheuristic algorithms such as the well-established GA [15-16] and variable neighbourhood search (VNS) [31]. In the initialization stage, the algorithm employs four well-known heuristics (NEH, Gupta, CDS, and Palmer) to generate the population. Then, the GA that adopts a special variation of the operator from Murata et al. [33] is applied for improving the population. In the third step, the VNS is used to avoid the trap of local optima. Finally, the populations are updated by replacing the old population with the new one. The results obtained from this algorithm are achieved all optimum solution when the number of job are lower than 50.

4 Node Based Coincidence Algorithm

NB-COIN is a permutation based Estimation of Distribution Algorithm (EDA). It generates solution strings in sequences, ensuring that only valid permutations are sampled. It uses a data structure called coincidence matrix H to model substructures from absolute positions. The matrix H_{xy} represents the probability of y found in the absolute position x. The update equation of NB-COIN is

$$H_{xy}(t+1) = H_{xy}(t) + \frac{k}{n}(r_{xy}(t+1) - p_{xy}(t+1)) + \frac{k}{(n-1)^2}\left(\sum_{j=1}^{n} p_{xj}(t+1) - r_{xj}(t+1)\right) \tag{5}$$

where k denotes the learning step, n is the problem size, r_{xy} is the number of xy found in the good solutions, and p_{xj} is the number of xy found in the not-good solutions. The term $\frac{k}{(n-1)^2}\left(\sum_{j=1}^{n} p_{xj}(t+1) - r_{xj}(t+1)\right)$ represents the adjustment of all other H_{xj} where $j \neq x$ and $j \neq y$.

After each population was evaluated and ranked, two groups of candidates are selected according to their fitness values: better-group and worse-group. The better-group is selected from the top $c\%$ of the rank and is used as a reward, and H_{xy} is increased for every pair of xy found in this group. The punishment is a decrease in H_{xy} for every pair of xy found in the worse group of the bottom $c\%$ of the population rank.

The pseudo code of NB-COIN is simplified as follows:

```
Step 1: Initialize the model
Step 2: Sample the population
Step 3: Evaluate the population
Step 4: Select candidates
Step 5: Update the model
Step 6: Repeat steps 2 to 5 until terminated.
```

5 Computational Result

The proposed algorithm, Node Based Coincidence Algorithm(NB-COIN), was coded in C++ and run on MS Windows 7 using Intel Core i5 450M, 2.40GHz and 4GB of RAM. 40 instances of Taillard benchmark[8] where $n \in \{20, 50\}$ and

$m \in \{5, 10, 20\}$ were selected and represented in four sets; 20×5, 20×10, 20×20 and 50×5 to determine the efficiency and performance of NB-COIN in the PFSP. Each set consist of 10 instances. Moreover, the proposed algorithm was tested according to two different criteria; computational time and performance.

5.1 Computational Time

The CPU time obtained from NB-COIN were compared against the powerful metaheuristics such as ant colony system [18] and the hybrid metaheuristic proposed by G.I. Zobolas[23] in 2009 with the allocation of CPU time for 5, 15, 25 and 100 seconds to four problem sets; 20×5, 20×10, 20×20 and 50×5. In Table 1, the results obtained from all groups of instances are summarized. The computational time of NB-COIN is superior when the number of jobs is twenty, especially in the 20×5 Taillard instance. It is twice as fast as the hybrid metaheuristic and the ACS in 20×5 problem. Furthermore, the speed of hybrid metaheuristic is slower than NB-COIN by 5 and 15 seconds in 20×10 and 20×20 problems while NB-COIN is slightly slower than the ACS by 3 seconds and 9 seconds. However, the computational speed of NB-COIN decreases when the number of jobs exceeds 50.

Table 1. The computational speed

Instances	CPU time (Seconds)		
	Hybrid Metaheuristic	ACS	NB-COIN
20×5	10	11	5
20×10	20	12	15
20×20	40	16	25
50×5	25	44	100

5.2 The Performance Analysis

In this section, the solutions acquired from NB-COIN were tested on the Taillard benchmark against the upper bound. Although NB-COIN achieved the upper bound only in a few solutions, it is essential to mention that NB-COIN was run on the PC. It finds the high quality solution very fast while usually the upper bounds are generally generated by branch and bound techniques and run on more powerful workstations for extended time periods.

To compare the quality of solutions, the percentage gap between the makespan from our algorithm and the upper bound (UB) of Taillard. Each instance was run 5 times. To calculate the percentage gap, the equation is presented as follows:

$$Gap(\%) = \frac{C_{max} - UB}{UB} \times 100 \qquad (6)$$

The results of three methods; NEH, CG and SG are adopted from the original paper proposed by Nawaz et al. [4] and M. Ancau[29] and compared against NB-COIN. Overall, we found that NB-COIN performs far better than the NEH and the constructive greedy (CG) in all problem sizes while it is slightly superior compared to the stochastic greedy (SG) in the small size problem (20×5). Moreover, NB-COIN provides a wide variety of solutions that share the same quality.

Table 2 shows the result of Taillard's 20×5 instance. NB-COIN not only found many optimum solutions the average gap is also a lot lower than both NEH and CG. Comparing with SG algorithm, the average gap is slightly higher. However NB-COIN is better in terms of the number of good solutions.

Table 2. Performance comparison of Taillard's 20×5 instances

Instance	UB	NEH	CG	SG	NB-COIN	Gap%			
						NEH	CG	SG	NB-COIN
Ta001	1278	1286	1286	1278	1294	0.626	0.626	0	1.252
Ta002	1359	1365	1367	1366	1363	0.442	0.589	0.515	0.294
Ta003	1081	1159	1141	1097	1090	7.216	5.550	1.480	0.833
Ta004	1293	1325	1358	1306	1304	2.475	5.027	1.005	0.851
Ta005	1235	1305	1301	1244	1244	5.669	5.344	0.729	0.729
Ta006	1195	1228	1224	1210	1210	2.762	2.427	1.255	1.255
Ta007	1239	1278	1264	1251	1251	3.148	2.018	0.968	0.968
Ta008	1206	1223	1268	1206	1206	1.410	5.141	0	0
Ta009	1230	1291	1277	1253	1253	4.959	3.821	1.870	1.870
Ta010	1108	1151	1144	1117	1120	3.880	3.250	0.812	1.083
Average						3.258	3.379	0.863	0.913

The quality of solutions in the 20×10 and 20×20 problem are shown in Table 3 and Table 4. Since the performance of CG and SG algorithm for the instance where $m \in \{10, 20\}$ are not reported by M. Ancau [29]; NB-COIN is solely tested with the NEH. The results show that the average gap of NB-COIN is over three times better than the NEH in both sizes of problem.

As seen in Table 5, NB-COIN performs very well in Taillard's 50×5 instance. It is clear that the SG provides slightly better results in this size of problem. However, NB-COIN found more optimum solutions than SG and has a lower average gap than both the NEH and CG algorithm. In addition, in each instance, although the average gap of SG algorithm is slightly lower than NB-COIN, the SG consumes more CPU time, at almost double.

Overall, the quality of solutions as measured by gap averaging over all test instances (Table 2-5) is 0.96% from the upper bound.

Table 3. Performance comparison of Taillard's 20×10 instances

Instance	UB	NEH	NB-COIN	Gap%	
				NEH	NB-COIN
Ta011	1582	1680	1599	6.195	1.074
Ta012	1659	1729	1679	4.219	1.205
Ta013	1496	1557	1518	4.077	1.471
Ta014	1377	1439	1392	4.502	1.089
Ta015	1419	1502	1433	5.850	0.987
Ta016	1397	1453	1417	4.008	1.432
Ta017	1484	1562	1513	5.256	1.954
Ta018	1538	1609	1575	4.616	2.406
Ta019	1593	1647	1608	3.390	0.942
Ta020	1591	1653	1617	3.897	1.634
Average				4.601	1.419

Table 4. Performance comparison of Taillard's 20×20 instances

Instance	UB	NEH	NB-COIN	Gap%	
				NEH	NB-COIN
Ta021	2297	2410	2323	4.919	1.132
Ta022	2099	2150	2119	2.430	0.953
Ta023	2326	2411	2349	3.654	0.989
Ta024	2223	2262	2242	1.754	0.855
Ta025	2291	2397	2314	4.627	1.004
Ta026	2226	2349	2243	5.526	0.764
Ta027	2273	2362	2300	3.915	1.188
Ta028	2200	2249	2235	2.227	1.591
Ta029	2237	2320	2276	3.710	1.743
Ta030	2178	2277	2200	4.545	1.010
Average				3.731	1.123

Table 5. Performance comparison of Taillard's 50×5 instances

Instance	UB	NEH	CG	SG	NB-COIN	Gap%			
						NEH	CG	SG	NB-COIN
Ta031	2724	2733	2761	2724	2724	0.330	1.358	0	0
Ta032	2834	2843	2889	2848	2848	0.317	1.941	0.494	0.494
Ta033	2621	2640	2674	2622	2640	0.725	2.022	0.038	0.725
Ta034	2751	2782	2782	2782	2771	1.127	1.127	1.127	0.727
Ta035	2863	2868	2908	2863	2863	0.175	1.572	0	0
Ta036	2829	2850	2863	2840	2835	0.742	1.202	0.389	0.212
Ta037	2725	2758	2781	2732	2739	1.211	2.055	0.257	0.514
Ta038	2683	2721	2780	2701	2704	1.416	3.615	0.671	0.783
Ta039	2552	2576	2595	2562	2565	0.940	1.685	0.392	0.510
Ta040	2782	2790	2787	2784	2782	0.287	0.180	0.072	0
Average						0.727	1.676	0.343	0.396

6 Conclusion

In this paper, we present a new method to solve the permutation flowshop scheduling problem called the Node Based Coincidence Algorithm (NB-COIN). This algorithm makes use of positive and negative knowledge to rapidly improve the solution. The proposed method was tested on a set of 40 Taillard instances. The experiment shows that the solution of NB-COIN is very close to the optimal value, at only 0.96% from the upper bound on average. Moreover, the proposed algorithm not only provides acceptable results very fast, it has few user-defined parameters. Hence, NB-COIN is a highly appropriate method that is easy to apply to real world situations where lower computational time and higher quality solutions are preferred.

References

1. Johnson, S.M.: Optimal two-and three-stage production schedules with setup times included. Naval Research Logistics Quarterly 1(1), 61–68 (1954)
2. Garey, M.R., Johnson, D.S., Sethi, R.: The complexity of flowshop and jobshop scheduling. Mathematics of Operations Research 1(2), 117–129 (1976)
3. Lenstra, J.K., Kan, A.R., Brucker, P.: Complexity of machine scheduling problems. Annals of Discrete Mathematics 1, 343–362 (1977)
4. Nawaz, M., Enscore, E.E., Ham, I.: A heuristic algorithm for the m-machine, n-job flow-shop sequencing problem. Omega 11(1), 91–95 (1983)
5. Palmer, D.: Sequencing jobs through a multi-stage process in the minimum total time–a quick method of obtaining a near optimum. OR, 101–107 (1965)
6. Campbell, H.G., Dudek, R.A., Smith, M.L.: A heuristic algorithm for the n job, m machine sequencing problem. Management Science 16(10), B-630–B-637 (1970)
7. Dannenbring, D.G.: An evaluation of flow shop sequencing heuristics. Management Science 23(11), 1174–1182 (1977)
8. Taillard, E.: Some efficient heuristic methods for the flow shop sequencing problem. European Journal of Operational Research 47(1), 65–74 (1990)
9. Framinan, J.M., Leisten, R., Ruiz-Usano, R.: Efficient heuristics for flowshop sequencing with the objectives of makespan and flowtime minimisation. European Journal of Operational Research 141(3), 559–569 (2002)
10. Framinan, J., Leisten, R.: An efficient constructive heuristic for flowtime minimisation in permutation flow shops. Omega 31(4), 311–317 (2003)
11. Grabowski, J., Wodecki, M.: A very fast tabu search algorithm for the permutation flow shop problem with makespan criterion. Computers & Operations Research 31(11), 1891–1909 (2004)
12. Nowicki, E., Smutnicki, C.: A fast tabu search algorithm for the permutation flowshop problem. European Journal of Operational Research 91(1), 160–175 (1996)
13. Reeves, C.R.: Improving the efficiency of tabu search for machine sequencing problems. Journal of the Operational Research Society, 375–382 (1993)
14. Watson, J.-P., Barbulescu, L., Whitley, L.D., Howe, A.E.: Contrasting structured and random permutation flow-shop scheduling problems: search-space topology and algorithm performance. INFORMS Journal on Computing 14(2), 98–123 (2002)
15. Reeves, C.R.: A genetic algorithm for flowshop sequencing. Computers & Operations Research 22(1), 5–13 (1995)

16. Reeves, C.R., Yamada, T.: Genetic algorithms, path relinking, and the flowshop sequencing problem. Evolutionary Computation 6(1), 45–60 (1998)
17. Stützle, T.: An ant approach to the flow shop problem, pp. 1560–1564 (1998)
18. Rajendran, C., Ziegler, H.: Ant-colony algorithms for permutation flowshop scheduling to minimize makespan/total flowtime of jobs. European Journal of Operational Research 155(2), 426–438 (2004)
19. Ahmadizar, F., Barzinpour, F., Arkat, J.: Solving permutation flow shop sequencing using ant colony optimization, pp. 753–757. IEEE (2007)
20. Tasgetiren, M.F., Liang, Y.-C., Sevkli, M., Gencyilmaz, G.: A particle swarm optimization algorithm for makespan and total flowtime minimization in the permutation flowshop sequencing problem. European Journal of Operational Research 177(3), 1930–1947 (2007)
21. Stützle, T.: Applying iterated local search to the permutation flow shop problem. FG Intellektik, TU Darmstadt, Darmstadt, Germany (1998)
22. Jarboui, B., Eddaly, M., Siarry, P.: An estimation of distribution algorithm for minimizing the total flowtime in permutation flowshop scheduling problems. Computers & Operations Research 36(9), 2638–2646 (2009)
23. Zobolas, G., Tarantilis, C.D., Ioannou, G.: Minimizing makespan in permutation flow shop scheduling problems using a hybrid metaheuristic algorithm. Computers & Operations Research 36(4), 1249–1267 (2009)
24. Liu, H., Gao, L., Pan, Q.: A hybrid particle swarm optimization with estimation of distribution algorithm for solving permutation flowshop scheduling problem. Expert Systems with Applications 38(4), 4348–4360 (2011)
25. Wattanapornprom, W., Olanviwitchai, P., Chutima, P., Chongstitvatana, P.: Multi-objective Combinatorial Optimisation with Coincidence algorithm, pp. 1675–1682 (2009)
26. Srimongkolkul, O., Chongstitvatana, P.: Application of Node Based Coincidence algorithm for flow shop scheduling problems, pp. 49–52 (2013)
27. Park, Y.B., Pegden, C.D., Enscore, E.E.: A survey and evaluation of static flowshop scheduling heuristics. The International Journal of Production Research 22(1), 127–141 (1984)
28. Turner, S., Booth, D.: Comparison of heuristics for flow shop sequencing. Omega 15(1), 75–78 (1987)
29. Ancău, M.: On Solving Flowshop Scheduling Problems. Proceedings of the Romanian Academy. Series A 13(1), 71–79 (2012)
30. Dorigo, M.: Optimization, learning and natural algorithm. Politecnico di Milano (1992)
31. Hansen, P., Mladenović, N.: Variable neighborhood search: Principles and applications. European Journal of Operational Research 130(3), 449–467 (2001)
32. Gupta, J.N.: A functional heuristic algorithm for the flowshop scheduling problem. Operational Research Quarterly, 39–47 (1971)
33. Murata, T., Ishibuchi, H., Tanaka, H.: Genetic algorithms for flowshop scheduling problems. Computers & Industrial Engineering 30(4), 1061–1071 (1996)

An Inventory System of Packaging Materials: Case Study at PT. Djambi Waras Jujuhan

Nilda Tri Putri, Jonrinaldi, and Y.R. Risa Noviani

Department of Industrial Engineering, Faculty of Engineering,
Andalas University, Padang 25163, Indonesia
nildatriputri@gmail.com, jonrinaldi@ft.unand.ac.id

Abstract. PT Djambi Waras Jujuhan is one of Crumb Rubber Factory which is inseparable of inventory problems. It needs packaging materials for production process, especially in packaging process. Currently, PT Djambi Waras Jujuhan has the problems in controlling inventory system of packaging materials because of one supplier only supplying materials, variation of lead time, high minimum stocks of packaging materials and pile up in the storage. The aim of the research is to propose an inventory system of packaging materials to improve inventory performance considering variations of lead time. The inventory system we proposed consists of four steps. The first step is calculating Inventory Turnover (ITO) to identify how the performance level of the existing inventory system. The second step is determining aggregate planning of packaging materials, classifying packaging materials using ABC analysis and testing distribution of lead time. The third step is determining economic order quantity and order interval, safety stock and reorder point. The fourth step is calculating total annual inventory cost considering safety stock and variation of lead time. Sensitivity analysis is performed to see the effects of changes of input parameters to the decision variables and total inventory cost. Based on analysis we have done, current inventory system of packaging materials at PT Djambi Waras Jujuhan has not been efficient yet that we can see from the value of inventory turnover. From 38 types of packaging materials, 18 types of packaging materials have the value of inventory turnover more than 1 and 20 types have the value of inventory turnover less than 1. The inventory system we proposes in this paper can save 78,46% of total current inventory cost that is Rp.320.728.188,-. Therefore, the inventory system considering variation of lead time we proposed at PT Djambi Waras Jujuhan has increased the inventory performance.

1 Introduction

Inventory is one of the most important factors in logistics management. Inventory planning will determine the smoothness of the production process in the company. Deficiency or excess inventories may cause problems in the company. So, inventory management seeks to achieve balancing between deficiency and excess inventory in a period of planning that involve risks and uncertainties [2]. In addition, each company must be able to maintain the optimum amount of inventory so as to ensure the

M. Gen et al. (eds.), *Industrial Engineering, Management Science and Applications 2015*,
Lecture Notes in Electrical Engineering 349, DOI: 10.1007/978-3-662-47200-2_34

smoothness whole operation of the company in the right quantity, right quality and lowest possible cost.

Inventory problems can be influenced by several parameters such as demand, lead time, holding costs, ordering cost, back order costs and price, which often vary in the real situation. Deterministic models are not sensitive for these changes. To cope their variations, especially variations of lead time and demand, the probabilistic model is characterized by the presence of safety stock that is an important part of the company inventory policies. Safety stock be able to meet the demand during lead time [1,2].

PT Djambi Waras Jujuhan is one of Crumb Rubber Factory that produces natural rubber balles into products such as: SIR-10, SIR-20 and SIR-20 CV. The company needs packaging materials for packaging process. And the company had difficulty to control the packaging materials inventory because the company has only one supplier, varying demand and lead time which varies from 1 day, 3 days, 5 days until more than 1 month, the packaging materials ordered the last period came periodically, minimum stock of packaging materials in warehouses is quite high. Because the existing inventory system for packaging materials in the company has not yet precise and optimal, so the packaging material is damaged and pill up in warehouses. This situation happened because of the size specification of the packaging materials that has been holding in previous period by the company no longer fit the size specifications set by the consumer (buyer), so that situation has explained before happened and the company is loss. Figure 1, 2, and 3 showed the variation of the finished product demand, variation of lead time, and one of the packaging materials inventory.

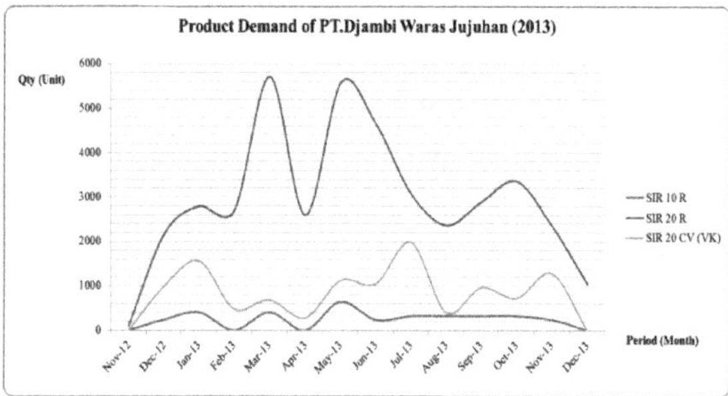

Fig. 1. Graph of Product Demand (SIR) in PT Djambi Waras Jujuhan

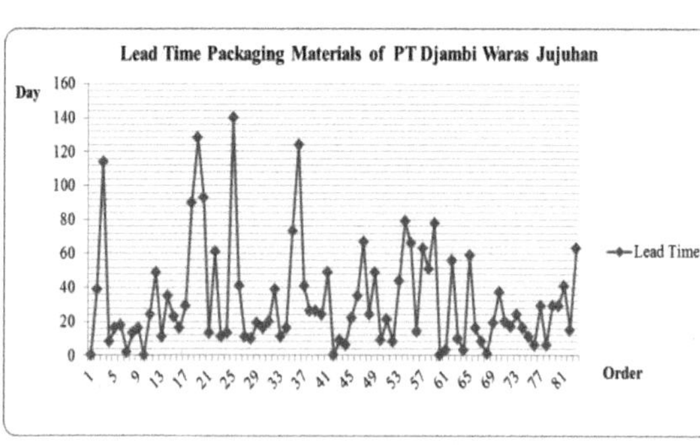

Fig. 2. Graph of Lead Time Packaging Materials in PT Djambi Waras Jujuhan

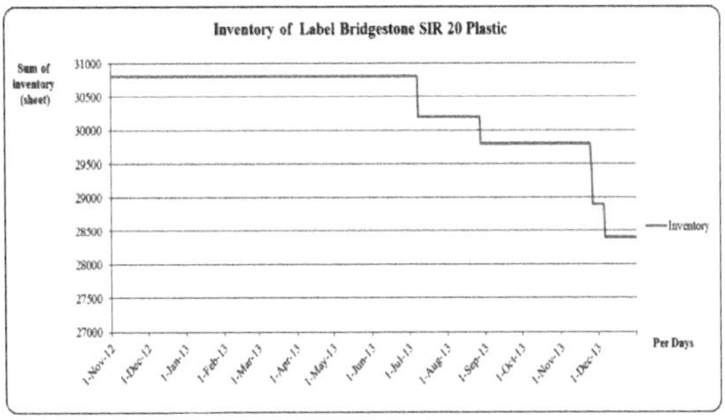

Fig. 3. Graph of Label Bridgestone SIR 20 Plastic Inventory

Examples for safety stock of Label Bridgestone SIR 20 Plastic is 30,000 sheets and changed in October 2013 to be 7000 sheets, though this type of packaging material start in November 2012 until in July 2013 are not used. That type of packaging material used on August 28th, 2013 was 400 sheets. It was concluded that this type of packaging material has a lot of stock and long stored in the warehouse. As a result of too much storage, there are some packaging material that discarded due to damage such as Plastic bags SIR R 20/35 SEU damaged amount of 654 sheets or approximately 27.25 kg and removed from the warehouse on April 10, 2013. In addition, there are some packaging material have not been established safety stock to be provided by the company to cope with demand fluctuations or variations. For example: Label SMPT C 20 UG230 SIR Plastic, Red Plastic, etc.

Consequences that must be accepted by the company because inventory system still doesn't optimal or appropriate is the high inventory cost and inventory policy for safety stock is not accurate that will increase the risk of losses in the future. Now, company management has not been too considering the consequences. Based on the issues that have been disclosed and supporting data of this research to be done so this research to improve the existing inventory system is needs by evaluate inventory system with considering variations of lead time, so the company can make appropriate planning of when order time, optimal quantity order and how much safety stock should be provided for a certain period to anticipate variations of lead time so that inventory cost can be minimized.

2 Literature

2.1 Inventory

According to reference [1] inventory is a number of materials, spare-parts, work in process, finished goods/products or components supplied to meet consumer demand at all times and use an asset that is waiting in a production process. So, basically this inventory is intended to facilitate or expedite the process of the operation in the company.

2.1.1 Inventory Costs

According to reference [1] the elements contained in the inventory can be divided into four groups, that is:

1. Ordering Costs
2. Inventory carrying costs/stock holding cost (%)
   According to Tersine (1994), holding costs are typically at intervals of 20-40% of the investment companies [8].
3. Shortage costs/out of stock cost).
4. Capacity associated costs

2.1.2 ABC Analysis

ABC Analysis is the method to classification material into group based on materials usage during the period of time (price per unit multiplied by the volume of material usage in period of time). ABC classification is using principle 80-20 or Pareto law which about 80% of the total inventory of material represented or represented by 20% of material inventory. Use of ABC analysis is to establish the engineering priority, purchase priority, security, recharging system (replenishment systems), and investment decisions and also can be applied in the company that have various types of inventory materials with different using value.

2.1.3 Probabilistic Models

This model considers all variables have values that are uncertain and one or more variable is a random variable. Parameters such as demand, lead time, inventory costs are varies in the real condition. And deterministic models are not sensitive to such things. The equation used to determine the total inventory cost is using equations Economic Order Interval (EOI)-Multiple Items. The equation is as follows:

1. Total Inventory Cost

$$TC(T) = \sum_{i=1}^{n} P_i R_i + \frac{C+nc}{T} + \frac{TF}{2} \sum_{i=1}^{n} P_i R_i \tag{1}$$

2. Optimum Order Interval (T*)

$$T^* = \sqrt{\frac{2(C+nc)}{F \sum_{i=1}^{n} P_i R_i}} \tag{2}$$

3. Maximum Inventory each items (E_i)

$$E_i = \frac{R_i T}{N} + \frac{R_i L}{N} = \frac{R_i (T+L)}{N} \tag{3}$$

where:

R_i = Demand or requirement per item
P_i = Purchasing cost per item
n = number of items are ordered at the same time
c = additional ordering cost for each item

When the lead time varies, reorder point policies that may set the lead time is the time that the minimum lead time, the average lead time or maximum lead time. With a minimum or maximum limit, then the reorder point should be different. If the reorder point based on a minimum lead time then tend not sufficient, and if the maximum lead time based on the results obtained is excessive inventory levels, so it can not be evaluated statistically, so in practice the reorder point based on the average lead time. The solution in this case is similar to variation demand and constant lead time. The basic difference is the demand during lead time probability distribution is obtained by adding a constant demand during lead time. This case can be seen in Figure 4

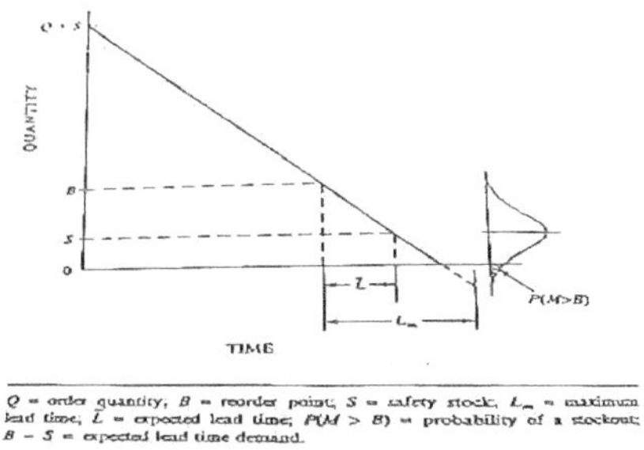

Fig. 4. Constant Demand and Variable Lead Time (Source: Tersine, 1994)

If the lead time is normal distribution, the formulations are used to optimize the reorder point is as follows:

$$B = \bar{M} + Z_{\sigma} = D\bar{L} + ZD\sigma_L \quad \textbf{(4)}$$

where:

B = reorder point
D = rate constant demand per period
σ = standard deviation of demand during lead time
σ_L = standard deviation of lead time
$\bar{L}$ = average lead time in the period

2.2 Sensitivity Analysis

Sensitivity analysis is one step of the modeling system in validation the inventory model built or developed. Two important issues in the sensitivity analysis are as follows [3]:

1. Find response of optimal solutions generated to changes in the input values.
2. Find out how the big error occurs (loss of profits or savings).

3 Research Methodology

Research methodology is structured for problems can be resolved with better and more focused. The steps are shown in the flowchart in Figure 5.

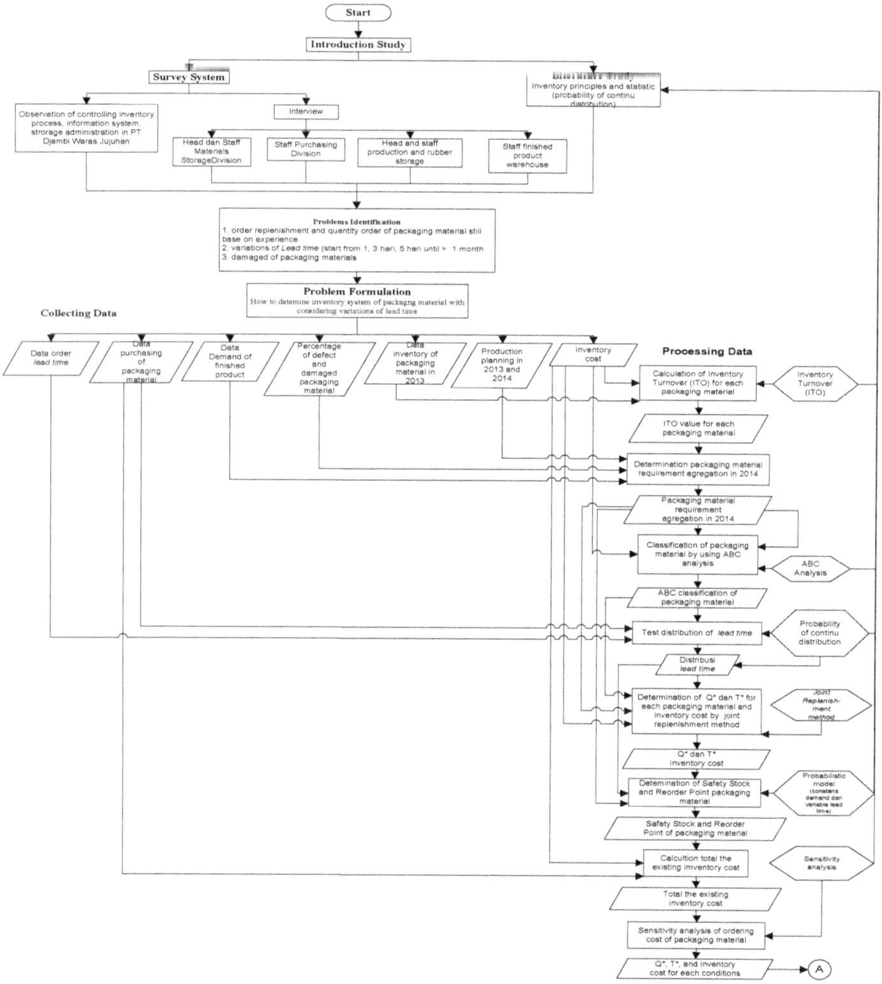

Fig. 5. Flowchart of Research Methodology

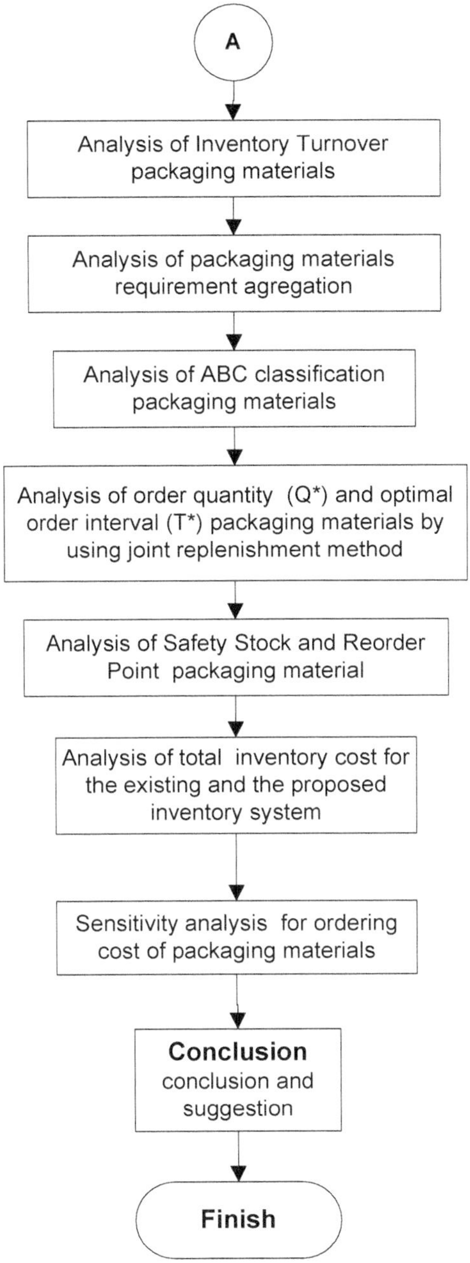

Fig. 5. Flowchart of Research Methodology (continue)

4 Results and Discussion

4.1 Data Collection

The data collected are as follows:

1. Data demand of finished products in 2013
2. Plan production in 2013 and 2014
3. Data supplies packaging materials in 2013
4. Component inventory cost of packaging materials
5. Percentage of defect or damaged packaging materials
6. Data oerder lead time of packaging materials
7. Data purchase of packaging materials.

Ordering costs of packaging material can be seen in Table 1, and transportation cost and PPN can be seen in Table 2, the percentage of defect packaging materials can be seen in Table 3, and data lead time can be seen in Table 4.

Table 1. Ordering Costs of Packaging Materials

Number	Component	Cost
1	Make PR from storage and buy sheet (@Rp. 1000)	Rp 6.000
2	Make an coverring letter (3 pieces) @Rp. 1000)	Rp 3.000
3	Goods Receiving Report 6 pieces (@Rp. 1000)	Rp 6.000
4	Communication Fee via Telephone and email	Rp 30.925
5	Administration Fee	Rp 72.565
6	Goods Receiving Fee	Rp 206.360
Total		Rp 324.850

(Source: Materials Storage and Purchasing Division of PT Djambi Waras Jujuhan)

Table 2. Transportation Costs of Packaging Materials

Number	Component	Cost
1	Transportasi Cost Per Kg	Rp 600
2	Delivery Cost Per Kg	Rp 500
3	Value Added Tax (PPN) % Per Unit	10%

(Source: Purchasing Division of PT Djambi Waras Jujuhan)

Table 3. Percentage Defect of Packaging Material

Number	Code	Type of Packaging Material	Unit	Percentage (%)
1	PLS-0001	Plastic Bag SIR 10 SEUR	Kg	3%
2	PLS-0002	Plastic Bag SIR 20 SEUR	Kg	3%
3	PLS-0003	Plastic Bag SIR 20 SEUVK	Kg	3%
4	PLS-0071	Bridgeston Plastic SIR 20	Sheet	3%
5	PLS-0074	Good Year Plastic SIR 20 NOLO	Sheet	3%
6	PLS-0075	Good Year Plastic SIR 20 SEUNOVOLO	Sheet	3%
7	PLS-0076	Good Year Plastic SIR 10 NIMBO	Sheet	3%
8	PLS-0077	Good Year Plastic SIR 20 SEUCVNIVCO	Sheet	3%
9	PLS-0093	Plastic SIR 10	Sheet	3%
10	PLS-0099	SMPT Plastic C UG230SIR20	Sheet	3%
11	PLS-0163	Shrink Wrapped Plastic 0,25mm x 165 mm x252 mm	Kg	5%
12	PLS-0189	Blank Plastic SIR 20 SEU	Sheet	5%
13	PLS-0195	Blueprint Plastic 0,14mm x 145 x 870 cm	Kg	5%
14	PLS-0199	Blueprint Plastic 0,10 mm x 140 x 160 cm	Kg	3%
15	PLS-0202	Gyo Plastic SR 20 SEUCOOPER TIRE	Sheet	3%
16	PLS-0223	Blank Plastic 0,1 x 140 x 160 cm	Kg	3%
17	PLS-0225	White Blank Plastic 0,1 mm x 125 x 160 cm	Kg	3%
18	PLS-0230	Trapping Band Plastic SMPT	Rol	3%
19	PLS-0253	Wire Plastic MB 0,2mm x 90 x 271 cm	Kg	3%
20	PLS-0296	Red Plastic 0,14 mm x 146 x 340 cm	Kg	3%
21	PLS-0297	Red Plastic 0,14 mm x 146 x 370 cm	Kg	3%
22	PLS-0302	Blue Plastic 0,1 mm x 140 x 170 cm	Kg	3%
23	PLS-0312	Blue Plastic 0,1 mm x 110 x 150 cm	Kg	3%
24	PLS-0326	Sumtomom Plastic	Sheet	3%

(Source: Finished Product Warehouse of PT Djambi Waras Jujuhan)

Table 4. Order Lead Time of Packaging Materials

Order	Lead Time (Days)	Order	Lead Time (Days)	Order	Lead Time (Days)	Order	Lead Time (Days)
1	8	15	13	29	22	43	37
2	16	16	11	30	35	44	20
3	18	17	10	31	24	45	17
4	2	18	19	32	9	46	24
5	13	19	16	33	21	47	16
6	16	20	20	34	8	48	11
7	24	21	11	35	14	49	6
8	11	22	16	36	3	50	29
9	35	23	26	37	10	51	6
10	23	24	26	38	3	52	29
11	16	25	24	39	16	53	29
12	29	26	1	40	8	54	15
13	13	27	9	41	1		
14	11	28	6	42	19		

(Source: Purchasing Division of PT Djambi Waras Jujuhan)

4.2 Results

The following are the results of the calculations have been carrying out.

Table 5. Summary of Packaging Material Requirements 2014

Number	Code	Type of Packaging Material	Unit	Total Requirement (2014)
1	PLS-0001	Plastic Bag SIR 10 SEUR	Kg	4201,73
2	PLS-0002	Plastic Bag SIR 20 SEUR	Kg	59969,25
3	PLS-0003	Plastic Bag SIR 20 SEUVK	Kg	14833,31
4	PLS-0071	Bridgeston Plastic SIR 20	Sheet	2294
5	PLS-0074	Good Year Plastic SIR 20 NOLO	Sheet	17955
6	PLS-0075	Good Year Plastic SIR 20 SEUNOVOLO	Sheet	32398
7	PLS-0076	Good Year Plastic SIR 10 NIMBO	Sheet	9515
8	PLS-0077	Good Year Plastic SIR 20 SEUCVNIVCO	Sheet	43951
9	PLS-0093	Plastic SIR 10	Sheet	1158
10	PLS-0099	SMPT Plastic C UG230SIR20	Sheet	25891
11	PLS-0163	Shrink Wrapped Plastic 0,25mm x 165 mm x252 mm	Kg	25772
12	PLS-0189	Blank Plastic SIR 20 SEU	Sheet	21665
13	PLS-0195	Blueprint Plastic 0,14mm x 145 x 870 cm	Kg	20045,2
14	PLS-0199	Blueprint Plastic 0,10 mm x 140 x 160 cm	Kg	17522,44
15	PLS-0202	Gyo Plastic SR 20 SEUCOOPER TIRE	Sheet	7632
16	PLS-0223	Blank Plastic 0,1 x 140 x 160 cm	Kg	4008,9
17	PLS-0225	White Blank Plastic 0,1 mm x 125 x 160 cm	Kg	1068,48
18	PLS-0230	Trapping Band Plastic SMPT	Rol	1295
19	PLS-0253	Wire Plastic MB 0,2mm x 90 x 271 cm	Kg	41346,84
20	PLS-0296	Red Plastic 0,14 mm x 146 x 340 cm	Kg	3107,52
21	PLS-0297	Red Plastic 0,14 mm x 146 x 370 cm	Kg	3301,74
22	PLS-0302	Blue Plastic 0,1 mm x 140 x 170 cm	Kg	3365,96
23	PLS-0312	Blue Plastic 0,1 mm x 110 x 150 cm	Kg	30789,64
24	PLS-0326	Sumtomom Plastic	Sheet	34491

Table 6. Summary of ABC Classification Based on Usage Value / Demand Packaging Materials

Class	Usage Value	Amount of items	Percentage of items	Total usage value/demand	Percentage usage value/demand
A	>360 million	5	20,83%	Rp 4.226.892.496	79.41 %
B	70 million < usage value < 360 million	4	16,67%	Rp 782.532.795	14,56%
C	< 70 million	15	62,50%	Rp 323.674.462	6,02%
Total			100%	Rp 5.373.099.753	100%

Data collected amount of 80 data and after testing data distribution, data can only 54 data. This happened because there are some data that is too extreme like ordering lead time is more than 1 month and out of the control limits I-MR control chart. Distribution of lead time obtained is a normal distribution with a mean = 16.2 days (0,044 years) and standard deviation = 8.97 days (0,025 years).

Determination of Q * and T * packaging material the proposed system (2014) by using joint replenishment method

Code material = PLS-0002
Type of material = Plastic Bag SIR 20 SEU R
Unit = kg
Class = A
Total requirement (Ri) = 59.969,25 kg
Price per unit (Pi) = Rp. 27.940, -
S = ordering cost for 1x order = Rp. 324.850, -
k = fraction of holding cost = 0,24

Step to determination Q * and T * are as follows:

Step 1: Calculate annual demand (PiRi)
PiRi = Rp. 27 940, - x 59969.25 kg = Rp. 1.675.540.845, -

Step 2: Total annual demand = Rp. 4.894.941.013, -
Step 3: Economic Order Interval (EOI)

$$T^* = \sqrt{\frac{2(C+nc)}{F\sum_{i=1}^{n} P_i R_i}} = \sqrt{\frac{2(Rp.324.850 + (7x0))}{0{,}24\ x\ Rp.4.894.941.013}} = 0{,}02\ years = 8{,}58\ days$$

That included in Group AB.

Step 4: Optimal Order Quantity (Q *)
Q * = Ri x Q * = 59.969,25 kg x 0,02
= 1.410,28 kg per order

Step 5: Ordering cost (TC_p) Group AB

$$TC_p = \frac{C+nc}{T} = \frac{Rp.324.850 + (7x0)}{0{,}01} = Rp.13.968.562,-$$

Step 6: Holding cost (TC_s) Group AB

$$TC_s = \frac{TF}{2}\sum_{i=1}^{n} P_i R_i = \frac{0{,}01x0{,}24}{2} xRp.4.894.941.013 = Rp.13.813.571,-$$

Step 7: Total Inventory Cost

TC = TCP + TCS
= Rp.13.968.562,- + Rp.13.813.571,-
= Rp. 27.782.133, -

Determination of SS and B for the proposed system (2014)

Code material = PLS-0001
Type of material = Plastik Kantong SIR 10 SEU R
Class = B

Total demand in 2014 (D) = 4201,73 kg
$\bar{L}$ = 16.2 days = 0,044 years
σ_L = 8.97 day = 0,025 years

SS $= ZD\sigma_L$
= 2,33 x 4.201,73 kg x 0,025
= 240,59 kg

B $= \bar{M} + Z_\sigma = D\bar{L} + ZD\sigma_L$
= (4201,73 kg x 0,044 years) + 240,59 kg
= 427,08 kg

Table 7. Summary of Inventory Cost Comparison between The Existing and The Proposed Inventory System in 2013

component of inventory system	existing inventory system	proposed inventory system
ordering cost	Rp 5,522,455	Rp 16,892,214
holding cost	Rp 403,234,152	Rp 71,136,205
total inventory cost	Rp 408,756,607	Rp 88,028,419
save	Rp 320,728,188	
% save	78.46%	

Table 8. Summary Inventory Costs of The Proposed System with Considering Safety Stock (SS) in 2014

component of inventory cost	proposed system 2014
ordering cost	Rp 19,815,867
holding cost	Rp 94,877,098
total inventory cost	Rp 114,692,965

Sensitivity analysis performed two conditions of ordering cost are:
Initial condition = Rp. 324 850, -
Condition 1 booking fee up to 30% = Rp. 422 305, -
Condition 2 booking fee rise 50% = Rp. 487 275, -

Table 9. Summary of Inventory Cost Comparison between Initial Condition, Condition 1 and 2

component of inventory cost	initial condition (proposed 2014)		sensitivity analysis			
			condition 1		condition 2	
ordering cost	Rp	19,815,867	Rp	22,804,489	Rp	24,363,771
holding cost	Rp	19,368,350	Rp	22,083,316	Rp	23,721,287
total inventory cost	Rp	39,184,216	Rp	44,887,805	Rp	48,085,057
% increasing TC				14.56%		22.72%

4.3 Analysis

4.3.1 Analysis of Inventory Turnover (ITO) of Packaging Materials

The purpose of determination ITO value was to know how quickly turnover of packaging materials in the company and how much the company invested to inventory. Inventory management of the company will be efficient if ITO value more higher. Based on the results that have been obtained from 38 types of packaging material, just only 18 types of packaging material that has ITO value > 1, its means that 18 types of packaging material is changed at least once time for a year. For example, Plastic Bag SIR 20 SEU Plastic have ITO value = 10.28 x turn of the year,

Table 10. Inventory Turnover Value (ITO) of Proposed Inventory System

Number	Code	Type of Packaging Material	Unit	Class	ITO
1	PLS-0002	Plastic Bag SIR 20 SEUR	Kg	A	58.46%
2	PLS-0253	Wire Plastic MB 0,2mm x 90 x 271 cm	Kg	A	58.46%
3	PLS-0163	Shrink Wrapped Plastic 0,25mm x 165 mm x252 mm	Kg	A	58.46%
4	PLS-0312	Blue Plastic 0,1 mm x 110 x 150 cm	Kg	A	58.46%
5	PLS-0195	Blueprint Plastic 0,14mm x 145 x 870 cm	Kg	A	58.46%
6	PLS-0199	Blueprint Plastic 0,10 mm x 140 x 160 cm	Kg	B	58.46%
7	PLS-0003	Plastic Bag SIR 20 SEUVK	Kg	B	58.46%
8	PLS-0001	Plastic Bag SIR 10 SEUR	Kg	B	38.55%
9	PLS-0223	Blank Plastic 0,1 x 140 x 160 cm	Kg	B	38.55%
10	PLS-0302	Blue Plastic 0,1 mm x 140 x 170 cm	Kg	C	38.55%
11	PLS-0297	Red Plastic 0,14 mm x 146 x 370 cm	Kg	C	38.55%
12	PLS-0296	Red Plastic 0,14 mm x 146 x 340 cm	Kg	C	38.55%
13	PLS-0077	Good Year Plastic SIR 20 SEUCVNIVCO	Sheet	C	27.69%
14	PLS-0230	Trapping Band Plastic SMPT	Rol	C	27.69%
15	PLS-0326	Sumtomom Plastic	Sheet	C	27.69%
16	PLS-0075	Good Year Plastic SIR 20 SEUNOVOLO	Sheet	C	27.69%
17	PLS-0225	White Blank Plastic 0,1 mm x 125 x 160 cm	Kg	C	27.69%
18	PLS-0099	SMPT Plastic C UG230SIR20	Sheet	C	24.54%
19	PLS-0189	Blank Plastic SIR 20 SEU	Sheet	C	24.54%
20	PLS-0074	Good Year Plastic SIR 20 NOLO	Sheet	C	24.54%
21	PLS-0076	Good Year Plastic SIR 10 NIMBO	Sheet	C	24.54%
22	PLS-0202	Gyo Plastic SR 20 SEUCOOPER TIRE	Sheet	C	24.54%
23	PLS-0071	Bridgeston Plastic SIR 20	Sheet	C	8.53%
24	PLS-0093	Plastic SIR 10	Sheet	C	8.53%

its mean that packing material turn over 10 times a year and inventory control for this packaging materials quite good compared with other types of packaging materials. Meanwhile, 20 types of packaging material has ITO value <1 and ITO value = 0 as Label Good Year 10 VK / Nevo Plastic, Pink Plastic 0.1 x 150 x 880 cm, etc. That is happened because packaging materials not changed in one year.

If ITO value of the existing inventory system compared with the proposed inventory system in Table 5.1, so ITO value of the existing inventory system is still not optimal or it can be concluded that inventory turnover is still relatively slow if it compared with the proposed inventory system has been considering with variations of lead time. Based on these conditions, the company should improve the existing inventory system and implement the proposed inventory system that can reduce average inventory of packaging materials in warehouses. So turnover of packaging materials can be faster than the previous inventory system.

4.3.2 Analysis of Packaging Material Requirements Aggregation

Existing packaging material requirements planning is still not well arranged and has not been integrated with the demand of finished product, it still based on previous usage, so the risk of stock out or over stock becomes larger. To solve this problem, the company must plan for packaging material in 2014 so that the risks can be minimized.

Packaging material requirements planning is done based on production plan in 2014. The result of this packaging material requirements planning is aggregation of packaging materials. Packaging material requirements planning has been undertaken to consider the percentage of defects or damage of the packaging material during the production process or after production during for a year. Production planning of packaging material in 2014 use for another step for evaluation and give recommendations inventory system for this company.

4.3.3 Analysis of ABC Classification Packaging Materials

Condition of packaging material inventory systems in PT Djambi Waras Jujuhan still not grouping and ordering process in the company is done if inventory of packaging material has reached the minimum stock. So, if there are several types of packaging material has reached the minimum stock of packaging materials will be immediately ordered to suppliers. In this research, packaging materials will be grouped into three classes using the ABC analysis.

Data packaging materials in that company amounted to 38 types. But just only 24 types are used and classify by ABC Analysis because only 24 types of packaging material that has required in the year 2014. Packaging materials are not grouped into packaging material requirements planning in 2014, the inventory system is specific policy based on historical data of the company.

ABC analysis performed in this research is based on value usage criteria of the packaging material and the result showed that 5 types of packaging materials are in A class, 4 types of packaging materials are in B class, and 15 types of packaging materials are in C class. Packaging materials including into A class with usage percentage 79.41% and its annual usage value more than Rp. 360 million, B class with usage percentage of 14.56% and its annual usage value is above Rp.70 million - under Rp.360 million, and C class with usage percentage 6.02% and its annual usage valua is under Rp.70 million.

The high demand value of packaging material is influenced by price per unit and the number of demand for a year. If price and amount of usage per year increase, so packaging material can classify into A class. And if one of two factors that mentioned before is low, so possibility of the packaging material can classify into B or C class. The situation also happened of packaging material in PT Djambi Waras Jujuhan. For example, Plastic Bag SIR 10 SEU has a high price but low annual usage for a year, so this type classify into C class.

The results of ABC classification will be using for packaging material inventory system in PT Djambi Waras Jujuhan in 2014. A class of packaging materials should be given the main focus in the Materials Storage and Purchasing Division. Although B and C class are focus of attention quite normal and enough, but for certain types of packaging material still needs to be given more attention than the other types that are same class because there is a packaging material that has a rapid rate of usage in the class. The results of ABC classification is also used as the basis for ordering process the packaging material to suppliers.

4.3.4 Analysis of Optimal Order Quantity (Q*) Order Interval (T*) of Packaging Materials by Using Joint Replenishment Models

Order interval for every group is different. The differences happened are because of the influence from purchasing cost, ordering cost, fraction of holding cost and requirement of packaging material. The higher of demand and purchasing cost, so order interval will be smaller. This means that company is increasingly often an order to the supplier. For example, the actual order interval C order 1 that is 59 days so actual order frequency is 11 times not 10.95 times as the result of the calculation. This is done so that demand can be met during the year.

Thus, in the decision-making process should consider order interval and order frequency so that a more optimal decisions compared just only one factor to consider. For example, when a decision is taken only consider order interval there was a risk that order made by the optimal quantity will exceed demand for a year, and it will increase order and holding cost. This problem happens if packaging materials has great demand. Optimal order quantity (Q *) is different for every item in the group, depending on the needs of the packaging materials and purchasing cost. Q * will be greater if demand is high and purchasing cost of packaging materials is low. Conversely, if purchasing cost of packaging materials is high so Q * will become smaller. Recommended for the proposed system, the company would choose the actual order frequency because the company only adds one order.

4.3.5 Analysis of Safety Stock (SS) and Reorder Point (B) Packaging Materials

Based on the results of the calculations have been done SS obtained is smaller than the SS set by companies today. It is shown that the the proposed system has a better inventory system inventory systems inventory now because a given proposal already has the optimal ordering quantity (Q*) with optimal ordering interval (T*) for each packaging material, although given the the proposed inventory system has messaging costs are greater because the frequency of ordering frequently. However, packaging materials stored in the warehouse a little more and have faster inventory turnover compared to the current inventory system. So the risk of damage or loss due to specification changes the size of the packaging material can be reduced in the the proposed inventory system.

Safety stock provided by the company is still greater when compared with the results of calculations performed, but this time the SS policy is much better than the company's policy before the month of October 2013 as the SS set is much larger than the current inventory system. Reorder Point (B) packaging materials are obtained from calculations carried out amounted to two times of the safety stock is obtained. This occurs because the service level that is 99% larger.

4.3.6 Analysis of Current and Proposed Inventory System

Calculation of total inventory cost for the the existing and the proposed inventory system 2013 was conducted to determine the proposed inventory system is feasible to be implemented or not. The proposed inventory system that given was feasible to implement because: a) it can save total inventory costs up to 78.46% or Rp.320.728.188,- from total inventory cost of the existing inventory system, b) ordering costs of the proposed inventory system is greater than the existing inventory system because ordering frecuency more often done by considering the optimal quantity of packaging material, purchasing of packaging materials, ordering costs holding costs, c) holding costs of the proposed inventory system is smaller than the existing inventory system because average inventory and safety stock packaging materials is smaller than the existing inventory system. In the existing inventory system, average inventory and safety stock is higher when it compared with the the proposed system, its meaning that company save many packaging materials that resulted turnover of stock to be slow.

4.3.7 Sensitivity Analysis for Ordering Cost of Packaging Materials

Based on result from calculation, it showed that condition 1 when ordering costs increased up to 30%, so total inventory costs increased up to 14.56%. Meanwhile, the condition 2 ordering costs increased up to 50% so the total inventory cost increased up to 22.72%. It can be concluded that the response given the optimal solution or the total inventory costs if ordering costs chaged so total inventory costs chages less than increase of ordering costs, this can be interpreted that propose or recommendations inventory system feasible to implement in the company. In addition, if ordering costs increase so order interval and order quantity will be increasing too. That mean, time period between the previous order to the next order increases because ordering quantity increase or it can be said ordering frequency reduced for a year.

5 Conclusions and Suggestions

5.1 Conclusions

Conclusions derived from this research include the following:

1. Packaging materials inventory systems in PT Djambi Waras Jujuhan still not optimal if their compared with the the proposed inventory system because of from 38 types of packaging materials holding in warehouse just only 18 types of packaging material that has a value of ITO > 1 and 20 types of packaging material has a value of ITO < 1 and ITO = 0. While the value of

the the proposed inventory system of order class AB has a value of ITO = 58.46 x turn of the year, order class BC has a value of ITO = 38.55 x turn of the year, order class C 1 has a value of ITO = 27.69 x turn of the year, order class C 2 has a value of ITO = 24.54 x turn of the year, and order class C 3 has a value of ITO = 8.53 x turn of the year. This happens because the company has an average inventory and safety stock is high. Solution for this problem is the company can reduce the average inventory in the warehouse so risk of damage or change the size specifications of packaging materials can be minimized.

2. Proposal or recommendation packaging materials inventory systems will give for the the existing inventory system, are:

   a. Planning or aggregation of packaging material needs for a year using a production plan that has been made by considering the percentage of defect packaging materials.
   b. Grouping of packaging materials using ABC analysis based on rate of using criteria ordering process and the controlling.
   c. Determination of the optimal order quantity (Q*), optimal order interval (T*), and frequency ordering (f) by using Joint Replenishment Method.
   d. Determination of safety stock and reorder point of packaging materials by considering variations of lead time.
   e. Calculate the total inventory costs during for a year.

5.2 Suggestions

Suggestions given to the company and further research are:

1. The company may implement the packing material inventory systems is the proposed that the the existing inventory system becomes more accurate and optimal.
2. Future research could design an application or information system for inventory planning of packaging materials, so that company can plan packaging materials inventory system for the next period to more easily and quickly and also can be used by stakeholders. In addition, data inventory or inventory reports more accessible.
3. Future research may develop or make a new model inventory system of packaging materials which more appropriate and more sensitive if one of the components inventory system in the company changes, so optimal solutions that result by the model more optimal.

Acknowledgement. This research was financial supported by Andalas University, West Sumatera Indonesia.

References

1. Assauri, S.: Production and Operations Management. Revised Edition 2008. Economic Faculty of Indonesia University, Jakarta (2008)
2. Bowersox, D.J.: Logistics Management. Systems Integration - Management Physical Distribution Systems and Materials Management, vol. 1. Bumi Aksara, Jakarta (2002)
3. Daellenbach, H.G.: Systems and Decision Making: A Management Science Approach. John Wiley & Sons Ltd., England (1995)
4. Fogarty, et al.: Production and Inventory Management, 2nd edn. South - Western Publishing Co., Ohio (1991)
5. Gaspers, V.: Production Planning and Inventory Control: Based Integrated Systems Approach Towards MRP II and JIT Manufacturing 21. PT Gramedia Pustaka Utama, Jakarta (1998)
6. Isaac, A.: Operation Manajemen, 1st edn. Graha Ilmu, Yogyakarta (2010)
7. Walpole, R.E., Myers, R.H.: Probability & Statistics for Engineers and Scientists, 4th edn. Bandung Institute of Technology, Bandung (1995)
8. Tersine, R.J.: Principles of Inventory and Materials Management, 4th edn. Prentice Hall International, New Jersey (1994)

A Pattern in Formgiving Design: Giving Priority to a Principle Solution in Industrial Design Situation

Rusmadiah Anwar*, Shahriman Zainal Abidin, and Oskar Hasdinor Hassan

Formgiving Design Research Group, Faculty of Art & Design Universiti Teknologi Mara
40450 Shah Alam, Malaysia
{rusma935,shahriman.z.a,oskar}@salam.uitm.edu.my

Abstract. This research discusses on proposing a new design methodology for industrial ceramic based on "formgiving" perspective. A significant finding of formgiving between two different domains which, industrial design and engineering design has defined the potential new design segment introduced in ceramic industries. However, with the predisposed of aesthetics and human sensation, it insists to investigate the principle of design requirements among the designer, artifacts and users, in order to aesthetically solve the appearance and consequences of form development. The aim of this study is to propose a principle of solution embarks with ablution tub as sub-sanitary ware design rooted in user's expectations. Finally, the study concludes the most important pattern of formgiving used in between designer and design situations.

1 Introduction

In ceramic sanitary ware design (CSWD), the most critical part in design development is finalizing the form requirements between industrial designers and engineering designers [1]. The understandings of each role build a constraint to this field towards introducing a new sub-sanitary ware product segment. Akin of crucial design process faced during the form creation [2]. For over a decade ceramic industries been introduced in Malaysia, the methodology of design practice was not clearly carried out beside, lack of research findings ever been reported. It was an agreeable that we happenstance to identify the most design investigation practicing in this country towards technical and material design [3,4,5]. Based on these findings, we attempt to come out with a design guideline for any industrial ceramic practice. Resolving the divisions of design phase that can be performed align with the manufacturing processes. However, this sort of practice can be considered as part of engineer's task which mainly pertaining their scientific and engineering knowledge to the solution of technical problems.

The constraint of materials, technological, economic, legal, environmental and human-related is considered by optimizing those solutions [6]. Appreciating the role of formgiving in design seems to be a solution addressing entirely misunderstanding

* Corresponding author.

M. Gen et al. (eds.), *Industrial Engineering, Management Science and Applications 2015*,
Lecture Notes in Electrical Engineering 349, DOI: 10.1007/978-3-662-47200-2_35

design flew among industrial ceramic sanitary ware designers. According to Abidin [7], the words formgiving has been adapted among design authors when discussing any design practiced related. In engineering design, it was related to a specific phase in design process, well thought-out as a part of principle-solution. The emphasis was on the determination of form and material, as well as the process of bringing both in line with each other's. In this paper, we intend to endow with some standpoint about the direction of introducing a sub-sanitary ware design, which ablution becomes a subject being considered. The viewpoints about this study will base on the following structure; (1) introduction; (2) idea generation from the principle of ablution; (3) the influence of formgiving in ceramic design practices; (4) identifying pattern language in design situation; (5) conclusion. The objective of this study is to present an approach to analyze a principle solution in the next stage for industrial ceramic design methodology. This study then presented to demonstrate the terms and understanding of formgiving design to contribute with introducing a principle of design for CSWD development.

2 Idea Generation from the Principle of Ablution to CSWD

The idea to classifying the important role of formgiving in product development for CSWD following the problem raised by the Johari's analysis, who manage to explain on the subject related to important discipline, requirements and procedure for Muslims to perform a *wudhu'* (ablution) [8]. A kind of individual's self-cleaning as ritual, it has became as a massive issues among Muslims in order to improve and discover the most suitable practice to complete the ablution. The main concern is about the determination of form and materials, ergonomic and human behavior [9]. On our perspective, lack of solution discussed to achieve the ablution performance due to the deficiency on design thinking. This problem has generated new possibilities in introducing an intricate design in CSWD as new segment [3]. In order to complete the ideation, three fundamental factors of ablution principles should put in practice as important design development [1];

2.1 The Act of Ablution

The ablution required each person performed four obligatory areas on human body to be wash. Any ablution's procedure must perform in sequence beginning with; face, wrist to elbows, forehead to the crown of head finally both feet up to ankle [9].

2.2 Water Consumption

The success of ablution procedures were not measured by washed the sequence important obligatory areas. It is also depending on the discipline of water used while performing the ablution. Before any ablution begins, each Muslim should able to identify the water requirements with only pure water is permitted such; water came from rain, spring, sea or river; water from melting snow or any water of a big tank or pond. In connection with the requirements, Faruqui [10, 11] explained about the water consumption is about one *mudd*. As conversion to a scientific measurement, it equal to 2/3 liter.

2.3 Ergonomic Precautions

Human posture became another issue been discussed among scholar in practicing the ablution procedures. Ironically, ergonomic design has not been discussed to overcome the issues. Signify the arguments on the ablution area or space in the mosque is more important than the product itself. The coordination of ablution space neither ignoring the ergonomic perspective even though it clearly mentioned about the ablution always performed in an option either sit, stand nor hunchbacked [9].

3 The Role of Formgiving in Industrial Design Practices

Typical investigation in design studies have shown that design activities is a momentous factor in designer's success, the field of design theory and methodology has produced much useful research to aid engineers in designing both classical machines and more modern artefacts [12]. With a clear description on the role of formgiving in design [7] able to indirectly affect the evolvement of an industrial ceramic product form. As principles, Quinn in the most part explained about the ceramic design process mainly stakes on the technical approach. Ceramic designers often translate the idea effectively using imaginations, experimental development and association of maquette or scale model explorations [13]. These models clearly closed to the principle solution such as the problem-solving process [14] and in the state of the arts, became as a booster to the creative process with widespread applications [2]. Comparative study of formgiving between Industrial Design (ID) and Engineering Design (ED) based on two experts (see Table 1) chosen by Abidin [7] became a significant tool to integrate between the needs of ID and ED in industrial CSWD. Obviously, the form evaluation through several stages was expended by the manipulation of visual element. Akner-Koler able to describe the transitional of form development begin with geometric until the transformation completed as organic forms.

This process answered evidently when ceramic designer has to use their imaginations to embark with new design ideas. A construction development of form based on Akner-Koler pictures still need to bring in practice to control the un-limitation of ceramic designer's creativity and aesthetics. Most important, the right time to conclude an aesthetical form development. The experimental development and maquette exploration in ceramic design phase can be associated with Muller judgment. He believes the beginning of form generation phase, designers have point out the important of design is founding the transition of function into form and then the transition marks the form creation phase through the evolution process. The quantified phase of form evaluation defined by criteria; from a basic structure to a structural concept, formal concept and material concept. The different between each form composition perform as the beginning point in form integration. This study shows another method to tame a wild experimental development and maquette exploration initiated in industrial ceramic practices. In the other view, the adoption of formgiving features in modern ceramic design especially for the product form, people are more likely to perceive the ceramic design as elegant, efficient with a good function of performance.

Table 1. Comparative study of formgiving based on difference design approaches [7]

Cheryl Akner-Koler (Industrial Design)	Wim Muller (Engineering Design)
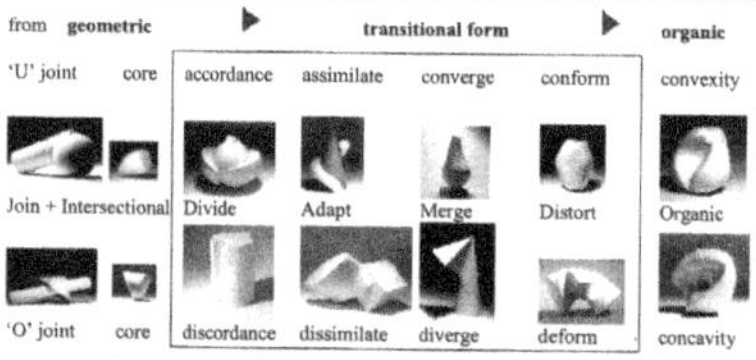	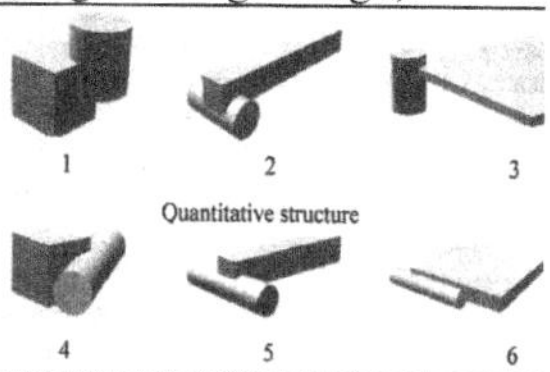

Research on formgiving for both academic and industrial firm should address issues on genetic construction, contextual, and affective elements [15]. It is because, most CSWD industry not able to introduce a new segment for their products. In recent industrial practice, they only continue on a new or improvement product development that always comes along with the establishment of company's brand image and identity.

4 Identifying Principle-Solution in Design Situation

4.1 Alexander's Design Theory and Approaches

Alexander, well-known architect and researcher, has found the interaction between human activity and designed artifacts and space [16]. The Alexander's Pattern Language has been formulated and described precisely by Denef [17]. Instead of designing a building, he found that Alexander aims to design overall configuration that are compatible with and beneficial for human life. Denef also describe about Alexander's believes where the design process should be iterative, gradually molding the new shape. In view of that, he characterizes this design process as careful and personal, a humane process between designers and users that is; not guided by the grasp for a goal, but guided by the minute-to-minute necessity of caring, dynamically, for the feelings and well-being of another.

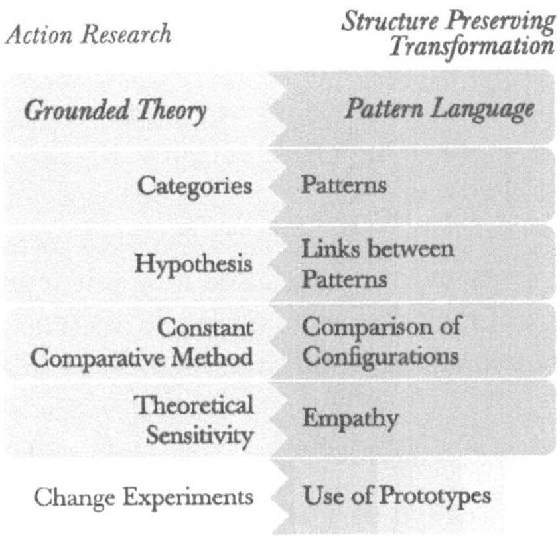

Fig. 1. Alexander's concepts of structure preserving transformations and pattern languages transposed to action research and grounded theory [16]

Figure 1 shows the Adaptation of Alexander's approaches (structure preserving transformation) to be compared with grounded theory (action research) where both methods obviously derive pattern by comparing phenomena. On the other views, the same transposition can easily convert to the formgiving design; as a part of principle solution. Denef also mentioned about most of the organizational action researchers focus on research and data driven decision-making. Alexander's greatest conclusion has some agreements to the one of Jonathan [12]; he reported, they need a concept that will enable them to address the interactions between designers, artifacts, and users in any design situations. Figure 2 shows the nature, which explained the relationship between artifacts and users is that, artifacts was used by users (which is obvious), but it is the affordances of the artifacts that determine how the artifacts can be used (which is not obvious).

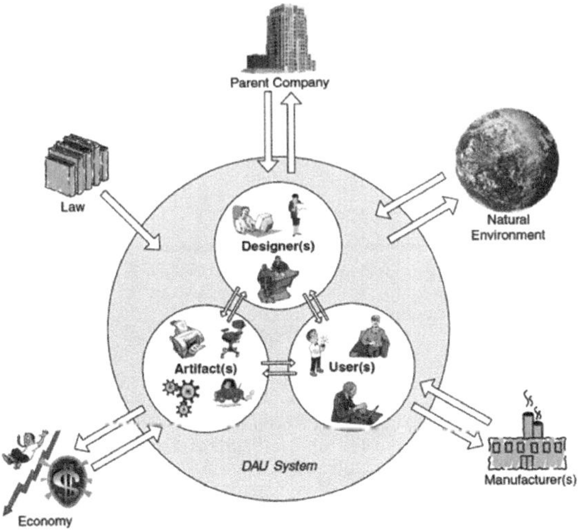

Fig. 2. Generic situated designer-artefact-user (DAU) system [12]

As mentioned on the second structure of this paper, three fundamental factors of ablution principles stated as important design development. In relation to this theory, the act of ablution (human activity), water consumption (designed artifacts) and ergonomic precaution (space). In order to build a relational theory for industrial design situation, it's also required a rational concept for CSWD. Based on both theories' point of views, pattern languages between designer-artefact-user mainly need to be systematized to make possible the new sub-sanitary ware introduced. The configuration of formgiving design in humane process (designers and users) can grasp not only a goal, but include thoughtful and enthusiastically, for the transcendent of ablution design.

4.2 Constructing a Pattern of Formgiving Design Principle (FDP)

As a part formgiving required design-inspired approaches, understanding engineering principle solution can frame the design process easier [7]. It's come to a potentially results (FDP) in constructing ablution design grounded by the pattern language theory. The integration is base on Andreason [18] and Pahl [6] product development process model. The transposition also will bring action research into design decision-making. Comparing grounded theory with Alexander's pattern language contribute to the introductory of CSWD (ablution design) methodology (see Table 2). Three methods applied a same driving force imagination data, pre-defined and develop the theory from the research topic bottoms-up from the empirical data. As shown, Alexander's pattern appears from the observation of human behaviour, and the same approach used on grounded theory as well as ablution principles. Those link between pattern and lead to hypothesis bring a design proposition, are important stage to conclude overall FDP. Finally, Alexander's model of makes use human intuition align with theoretical sensitivity that brings design intelligence as final phase in ablution design theory. Based on pattern described, we defined link between Andreason and Pahl model (Figure 3) with seven patterns (bold with seven different colours) which describe most important design activities and should developed accordingly with FDP (ID and ED) to establish ablution design research model.

Table 2. The Construction Of Formgiving Design Principle For Ablution Design Research Model (ADRM)

Structure Preserving Transformation	Action Research	Formgiving in ADRM
Pattern Language	Grounded Theory	Ablution Theory
Pattern	Categories	Principle
Link Between Pattern	Hypothesis	Design Proposition
Comparison Of Configuration	Constant Comparative Method	Quantitative Structure Organization
Empathy	Theoretical Sensitivity	Design Intelligence
Use Of Prototypes	Change Experiments	Behaviour Alterations

4.3 Discussion on Case Examples on Industrial Ceramic Design Practice

In order to establish the ADRM to any related industrial design field, we investigated the existing of formgiving in industrial ceramic design based on five different design research and activities [19-23]. All design and manufacturing decision among designers also revealed to activity of product development model defined by Andreason and Pahl. As results, the designers manage to give priority to a FDP. However, it does not clearly mentioned about the methodology in a principle-solution manner. ADRM, translating action research into industrial design activity, identifying patterns of design process must put in practice, as principle-solution to design needs. Figure 3 indicates the principle-solution in development process used among designers, in order to fulfill the need between designer, artifacts and users. Here, we summarized seven (7) patterns in the case examples as a important practiced (by designers) in which; the need (red); problem solving (yellow); principle of design (blue); product design structure (green); design specification (orange); manufacturing design (dark blue); and continues improvement (pink). The organizational of selected principle-solution then structured as relation to ADRM (see Table 3).

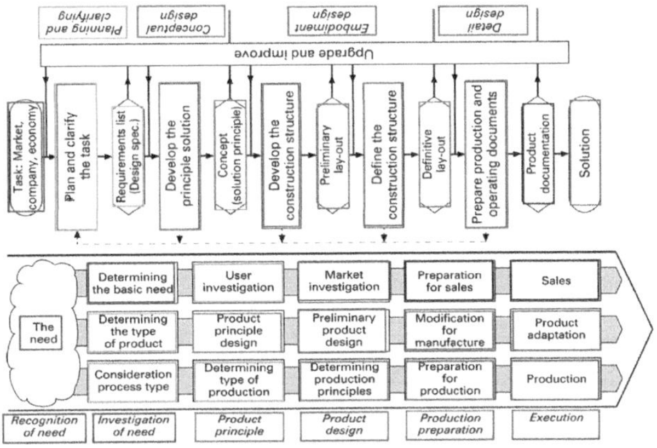

Fig. 3. Identifying paterns in Adreason and Pahl product development process

Table 3. The Details Summary of Principle-Solution based on Seven Patterns

Formgiving in ADRM	Principle-Solution
Ablution Theory	The need; Problem solving
Principle	Principle of design
Design Proposition	Product design structure
Quantitative Structure	Design specification;
Organization	Manufacturing design
Design Intelligence	Continues improvement
Behaviour Alterations	

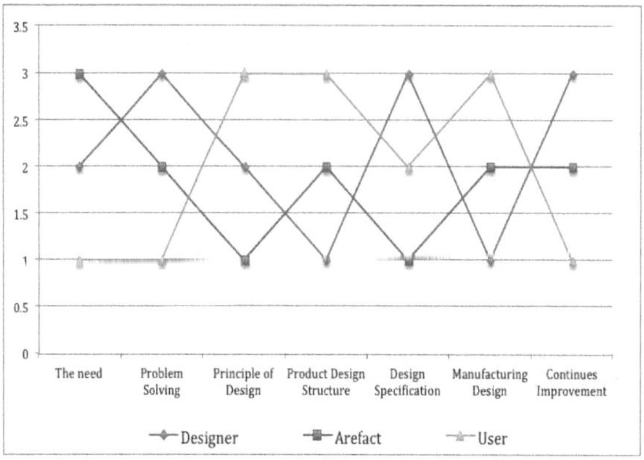

Fig. 4. The relation between Designer, Artifact and User with the 7 patterns in principle-solution of product development

Based on design situation faced by the designer [19-23], the formgiving design activities clearly identified became the main issues highlighted by those designers. The principle-solution also seems to be reflected with subject and set to be important to discuss by each product development phase (see Fig. 4). Here, designer, artifact and user described the roles in each development phase. However, the designer chose to concentrate to solve the users need on the semi-concrete level in design phase compare to the abstract level. By the end of the product development stage, the designer still play the important role of decision making followed by the artifact then the users need. This final stage shows the production designer change the early discussion of product development by positioning the artifact as important subject to discuss followed by designer and again the artifact still became the last issue to discuss.

5 Conclusion and Future Work

In this paper, we have presented our approach of identifying design thinking for industrial design, in its close to identify the important three major element as design research framework: (1) The early stage, meeting requirement of ablution as subject matter; (2) The middle stage, make clear on industrial ceramic design by appreciating role of formgiving in design, and (3) The final stage, identifying a ADRM to described design behavior based on two famous model of product development process. These stages became a main key in order to generate new transition of conventional to intricate design evolutions for ceramic industries. The finding of pattern language introduced by Alexander comes to conclusion on vital connection between "human activity-designed artifacts-space" towards action research. On the perspective of measurement of making practices available for new design, the achievement of balancing the needs among those vital connections will benefit the design results.

Making a case for the principle-solution, we present an example in which we were on shift to tackle the task of designing ablution product. A project idea sets which until this day, many scholars despite to uncover the existing problems. For future work, we will discover a work that specifically focused on ablution practices; consider the ablution act, water consumption and ergonomic influence. Preserving the spiritual needs of ablution is crucial to ensure that Muslim can cleanse particular body mentioned. The experiment undergoes Protocol Analysis to seeks answers to questions, which were formulated on literature reviews and on what is often practiced by designers in product industry and academia by focusing on three elements such as product design, form development and design thinking. These three different levels involve: 1) Different level of form development level such as explorative, explanatory and persuasive; 2) Different level of career development such as expert, senior, intermediate, novice and student; and 3) Different level of learning/work such as product design art-based and science-based.

Acknowledgement. We would like to acknowledge the generous participation of the interaction designers in the research. This study was conducted in Formgiving Design Research Lab established by Research Management Institute, Universiti Teknologi MARA (UiTM). Fully appreciation to Malaysia Ministry of Higher Education for the financial support under RAGS grant and UiTM for the under Research Excellent Fund Scheme (RIF).

References

1. Anwar, R., Hassan, O.H., Abidin, S.Z.: A Framework of Empirical Study Through Design Practice for Industrial Ceramic Sanitary Ware Design. In: Hassan, O.H., Abidin, S.Z., Legino, R., Anwar, R., Kamaruzaman, M.F. (eds.) Springer, Singapore (2014)
2. Pahl, G., Beitz, W.: Engineering design: A systematic approach. Springer, London (1996)
3. Anwar, R., Kamarun, H.R., Vermol, V.V., Hassan, O.H.: Marble dust incorporate in standard local ceramic body as enhancement in sanitary ware products. In: IEEE Colloquium on Humanities, Science and Engineering, CHUSER 2011, Penang, pp. 355–357 (December 2011)
4. Vermol, V.V., Kamsah, K., Hassan, O.H., Anwar, R.: A study on porcelain anti slip tile design. In: IEEE Colloquium on Humanities, Science and Engineering, CHUSER 2011, Penang, pp. 121–124 (December 2011)
5. Zainuddin, N.M., Rahim, Z.A., Anwar, R., Mujir, M.S., Hassan, O.H.: Conceptual framework of hydroxyapatite for damaged skull through design approach. In: IEEE Business Engineering and Industrial Applications Colloquium, BEIAC 2012, Kuala Lumpur, pp. 455–459 (April 2012)
6. Pahl, G., Beitz, W., Feldhusen, J., Heinrich Grote, K.: Engineering Design: A Systematic Approach, 3rd edn. Springer, London (2007)
7. Abidin, S.Z., Sigurjónsson, J., Liem, A., Keitsch, M.: On the role of form giving in design. In: 10th International Conference on Engineering and Product Design Education - New Perspective in Design Education, E&PDE 2008, Barcelona, pp. 365–370 (2008)

8. Johari, N.H., Anwar, R., Hassan, O.H.: Design Framework of Ceramic Ablution Tub. In: IEEE Symposium on Business, Engineering and Industrial Applications, ISBEIA 2012, Bandung, pp. 608–610 (September 2012)
9. Johari, N.H., Anwar, R., Hassan, O.H., Kamaruzaman, M.F.: Human Behaviours Influence Frameworrk of the Ablution Tub Design. In: IEEE Business Engineering and Industrial Applications Colloquium, BEIAC 2013, Langkawi, pp. 750–752 (April 2013)
10. Faruqui, N.I., Biswas, A.K., Bino, M.J.: Water Management in Islam. The United Nations University Press, USA (2001)
11. Faruqui, N.I.: Water, Human Rights, and Economic Instruments the Islamic Perspective. Journal of Water Resources Development 9/10 (2003) (Format Printing Press Kalopul, Kathmandu)
12. Maier, J.R.A., Fadel, G.M.: Affordance based design: a relational theory for design. Research in Engineering Design 20(1), 13–27 (2009)
13. Quinn, A.: The Ceramics Design Course. Thames & Hudsons Ltd., London (2007)
14. Simon, H.A.: The new science of management. Harper, New York (1961)
15. Abidin, S.Z., Othman, A., Shamsuddin, Z., Samsudin, Z., Hassan, H.: The Challenges of Developing Styling DNA Design Methodologies for Car Design DS 78. In: Proceedings of the E&PDE 2014 16th International conference on Engineering and Product Design, University of Twente, The Netherlands, pp. 738–743 (2014)
16. Alexander, C.: Notes on the synthesis of form. Harvard University Press, Cambridge (1964)
17. Denef, S., Oppermann, R., Keyson, D.V.: Designing for social configurations: Pattern languages to inform the design of ubiquitous computing. International Journal of Design 5(3), 49–65 (2011)
18. Andreasen, M.M., Hein, L.: Integrated Product Development. Springer, New York (1987)
19. Noordin, S.N.A., Sanusi, S.A., Anwar, R., Hassan, O.H., Kamaruzaman, M.F.: A Fusion Design Study Evolving a Malay Modern Teapot. In: 2013 IEEE Business Engineering and Industrial Applications Colloquium, Langkawi, pp. 199–201 (2013)
20. Zainuddin, N.M., Yusof, N.A., Anwar, R., Hassan, O.H., Jalil, A.R.: Humanistic study in ceramic cereal breakfast set as children learning tool. In: Business Engineering and Industrial Applications Colloquium (BEIAC), Langkawi, pp. 195–198 (April 2013)
21. Ali, A., Jusoh, S.S., Anwar, R., Hassan, O.H., Jalil, M.F.A.: Study on human posture and gesture elements for industrial ceramic robotic artware. In: Business Engineering and Industrial Applications Colloquium (BEIAC), Langkawi, pp. 772–775 (April 2013)
22. Raif, D.M., Anwar, R., Ahmad, N.A., Zakaria, Z., Jalil, M.F.A.: Revision on cartoon character integrate with chess concept for industrial ceramic artware. In: Business Engineering and Industrial Applications Colloquium (BEIAC), pp. 776–779 (April 2013)
23. Johari, N.H., Anwar, R., Hassan, O.H., Kamaruzaman, M.F.: Human behaviours influence framework of the ablution tub design. In: 2013 IEEE Business Engineering and Industrial Applications Colloquium (BEIAC), pp. 750–752 (2013)

Simulation of Logistic Operations

Jana Lalinská, Kendra Martin, and Čamaj Juraj

Department of Railway Transport, University of Žilina/Slovakia
Univerzitná 8215/1, 010 26 Žilina, Slovakia
{jana.lalinska,martin.kendra,juraj.camaj}@fpedas.uniza.sk

Abstract. Inquiries and transfer of innovative knowledge and technologies would be a one of the main goal of an each company. Through these innovations get company a new view of solving problems and an opportunity to be better than its competitors. In this paper we would like to describe a project, which was oriented to transfer of innovation knowledge and technologies in the logistics and transport processes. As a result of the project is a simulation module of technology progress of transfer and logistics processes. The simulation module is consisted of three separated parts which are railway classification yard, storehouse and intermodal terminal.

1 Introduction

Modernization of departments and their integration to the national and international research centres as a general part university activities are included to development of scientific and research activities at the University of Žilina. Lot of this modernization was possible with the aid of structural funds from European Union.

One of the modernizations was solved in the Department of railway transport. This project was oriented to the Transport laboratory. Thanks this modernizations we are able to simulate processes in logistic centres complex or separately in marshalling yard, storehouse and intermodal terminal.[1]

2 Time Harmonogram Project Realization

At the end of 2009, Department of railway transport, Faculty of operation and economics, transport and Communication University of Žilina gained project supported by European Union. The title of project was „"Transfer of innovative knowledge and technology in logistics and transport processes", ITMS project code 26220220006.

Main project target was "Developing simulation model of technological process in transport and logistics processes". This target was supported by two specific targets:

- Research support of new progressive technology in managing transport operation and logistic processes,

M. Gen et al. (eds.), *Industrial Engineering, Management Science and Applications 2015*,
Lecture Notes in Electrical Engineering 349, DOI: 10.1007/978-3-662-47200-2_36

- The application of newly acquired research knowledge into practice in order to ensure sustainable transport.

  Project was divided to five interconnected activities:

- ***1.1 The Development of Advanced Technologies in the Management of Traffic Operations and Logistics Processes***
  - aim of the activity - Development of advanced technologies in the management of traffic operations and logistics processes,
- ***2.1 Implementation of technology in the management of traffic operations and logistics processes in the simulation model***
  - aim of the activity - Implementation of the developed technologies in the management of traffic operations and logistic processes to a specific software tool,
- ***2.2 Simulation Model Function Testing in the Management of Traffic Operations and Logistics Processes***
  - aim of the activity - Implementation of simulation technology procedures,
- ***2.3 Developing a Manual for Handling Simulation Model***
  - aim of the activity - Developing a manual for simulation model,
- ***2.4 Supplying of technical equipment with a focus on IKT technologies***
  - aim of the activity - Ensuring the technical conditions and equipment for the operation, application and subsequent publicity simulation model.

In addition to these five activities, in the project were defined two supporting activities:

- Project control,
- Publicity and unforcedness.

Expected time frame of the project realization was specified from October 2009 till December 2011. Consequential duration of particular activities is illustrated in table 1 and chart 1, in which is illustrated actual time period project solution:

Table 1. Time period project realization

Activity name	Beginning of activity realization	Ending of activity realization
1.1 The Development of Advanced Technologies in the Management of Traffic Operations and Logistics Processes	10/2009	01/2011
2.1 Implementation of technology in the management of traffic operations and logistics processes in the simulation model	04/2010	09/2012
2.2 Simulation Model Function Testing in the Management of Traffic Operations and Logistics Processes	06/2010	09/2012
2.3 Developing a Manual for Handling Simulation Model	01/2011	12/2012
2.4 Supplying of technical equipment with a focus on IKT technologies	10/2009	03/2011
Supporting activities		
Project control	10/2009	12/2012
Publicity and informedness	10/2009	12/2012

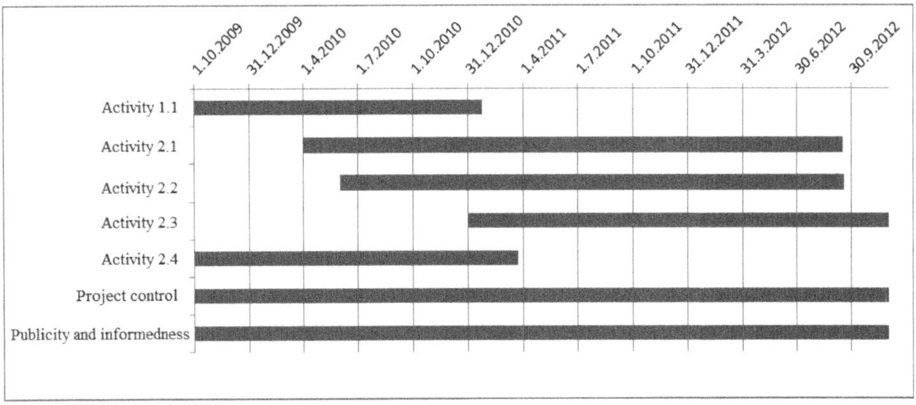

Chart 1: Gantt chart project realization

3 Project Outputs

As a result of this project is developed a Simulation model of logistic centre. Simulation model is able to simulate different technological practices in one application using its three independently modules (railway classification yard, storehouse and intermodal terminal).

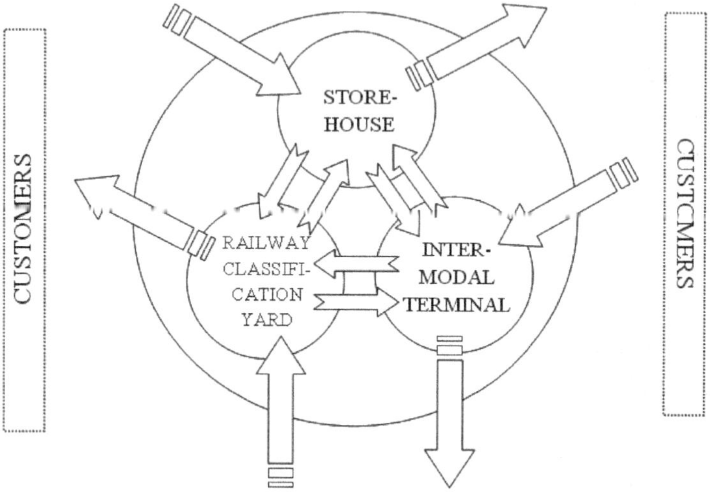

Fig. 1. Simplified scheme of the logistics center

Every part (storehouse, railway classification yard, intermodal terminal) can operate independently and it is possible to monitor all the processes in these partial components. It is necessary to follow the accouplements and the flows between storehouses, railway classification yard and intermodal terminal to optimize logistic processes of the entire logistic center. Optimizing of the processes in partial parts of the logistics center and between them can achieve the great synergies.

Application includes Base model data that are valid for all modules or for every module independently. Base model data containing baseline standard attributes of parameters as well as different files, for example list of destination station. Any specified values of parameters in base model data is possible to change other are given by simulation. List of data in simulation are definite and editable, some of them is possible redefine by simulation needs and requirements. [3]

The basic output characteristics obtained from the simulation could be:

- average queue length of the requirements in the systems,
- maximum queue length of the requirements in the systems,
- average waiting time of the requirements in the systems,
- maximum waiting time of the requirements in the systems,
- average time which spent requirements in the systems,
- maximum time which spent requirements in the systems.

4 Characterization of Individual Modules Logistics Centre

4.1 Railway Classification Yard

Configuration of railway classification yard allows choosing from standard arrangement of stock groups (serial or parallel). Consequently can user verify –

structure of the rail-yard, number of the tracks in the rail-yard groups, number of employees and locomotives, technology of the primary and secondary splitting, operating efficiency.

4.2 Intermodal Terminal

Configuration of intermodal terminal allows to choose from two basic type of internal infrastructure, which are defined by handling equipments (bridge cranes, reach-stacker – their abilities can be affected by parameters settings). Consequently can user verify – type and number of handling equipment, size of the handling areas, technology of the trains-shipment cargo units, staffing demand, process times, load carrying capacity.

4.3 Storehouse

Configuration of internal storehouse infrastructure allows selection from two primary infrastructure types whit handling equipments (electric stacker, forklift stacker - their abilities can be affected by parameters settings. Handling equipment affects arrangement of storage area whereby has to be observing minimal building module. After that it is define storage building module and storage layer. Consequently can user verify – number and capacity, type and number of the handling equipment, size of the handling areas, storage technology, staffing demand, process times, load carrying capacity.

5 Conclusion

Simulation modelling provides a comprehensive and dynamic view on the whole technological process and can provide the necessary information about its behaviour. In second place, utilizing simulation modelling can optimize the number of handling equipment, the handling area size, the number of transport, the range of transport infrastructure. Using designated conditions eliminate any negative impact on the environment.

By using this application can user simply simulate processes and identify the weaknesses of every part logistic centre in complex or independently in modules. Application offers complex view of logistics operations.

Acknowledgement. This paper is prepared with the support of the project "The quality of education and development of the human resources as pillars of the knowledge society at the Faculty PEDAS University of Žilina.", ITMS project code 26110230083, University of Žilina.

Operačný program VZDELÁVANIE

Modern education for the knowledge society/
Project is co-financed by funds from the EC

Agentúra
Ministerstva školstva, vedy, výskumu a športu SR
pre štrukturálne fondy EÚ

Európska únia
Európsky sociálny fond

References

1. Buková, B., a kol.: Zasielateľstvo a logistické činnosti. IURA Edition (2008) ISBN: 9788080782320
2. Lábaj, J., Patsch, M., Barta, D.: Combustion of alternative fuels. In: TRANSCOM 2009: 8th European Conference of Young Research and Scientific Workers: Žilina, Slovak Republic, June 22-24. Section 6: Machines and equipments. Applied mechanics, pp. 67–76. University of Žilina, Žilina (2009) ISBN 978-80-554-0031-0
3. External Costs of Transport. Update Study. Final Report. Zurich/Karlsruhe (October 2004) ISBN 2-7461-0891-7
4. Dolinayová, A., Čamaj, J., Průša, P.: Economic effects of realization logistics centres. In: Logistic Centres, Institut Jana Pernera, o.p.s., Pardubice (2008) ISBN 978-80-86530-52-9

Contracting Decisions in Project Management – An Outline of the Dedicated Decision Support System

Tomasz Błaszczyk[1] and Paweł Błaszczyk[2]

[1] Department of Operations Research, University of Economics in Katowice,
1 Maja 50, 40-287 Katowice, Poland
[2] Institute of Mathematics, University of Silesia, Bankowa 14, 40-007 Katowice, Poland
Tomasz.blaszczyk@ue.katowice.pl, pawel.blaszczyk@us.edu.pl

Abstract. In this paper we describe partial results of the research focused on the improvement of the decision-making quality and control in the project contracts management in concurrent environment. Bearing in mind the diversity of interdependences between the stakeholders involved in the project and their decisions, our considerations are embedded in the complex systemic process model introduced by one of the most popular project management methodologies. Here we propose the improved process of project contractors selection associated with the contract type selection and work distribution. The model of this process is developed with the ability to use in the environment with multiple contractors and their sub- contractors on the one of more levels.

1 Introduction

The Project Management Body of Knowledge (PMBOK Guide) developed by the Project Management Institute [19] is a project management methodology that is known and used worldwide. This methodology describes the decision-making processes in the 10 knowledge areas with a group of project processes connected with procurement management. In a situation in which the agreement seems to be beneficial to the contractor, but he or she is not able to provide sufficient resources for its implementation or the knowledge, experience, and technology is not sufficient, the contractor will be considered for the purchase of selected scopes of work with external contractors. The structure of the PMBOK Guide processes is based on a systemic approach that determines the range of the input data in the sphere of implementation tools of individual processes and their results. Issues tools are formulated in general terms and provide guidance for project managers and project teams to show the general methodology of the proceedings. However, they do not contain precise indications of the methods and tools that can be used to implement these processes. Procurement management issues have been also discussed in many papers. For example Bajari, McMillan, & Radelis [1] compared the advantages and disadvantages resulting from the use of auctions and negotiations. They concluded that when ex ante information is valuable and when ex post change is anticipated, the use of auctions, which often require fixed-price contracts, may be inefficient. Eriksson and Laan [7] investigated how construction clients deal with procurement and analyze how the choices that are made during the buying stages affect the combination

M. Gen et al. (eds.), *Industrial Engineering, Management Science and Applications 2015*,
Lecture Notes in Electrical Engineering 349, DOI: 10.1007/978-3-662-47200-2_37

of governance mechanisms and control types in client-contractor relationships. Currently there also is research work being conducted on the architecture of procurement decision support systems, including web-based applications. One of the recent publications was provided by Liu, Sun, Wang, & Zhao [15]. As a result of their work, they proposed a framework for a multi-agent based system for e-procurement. Let's begin our consideration when the prospective contractor of the project receives an inquiry from the customer, i.e., the owner of the project. Certainly, the reasonably progressive contracting before making the procurement decision will recognize the market in order to identify potential contractors and their offers. Therefore, every potential contractor must decide whether to make an offer or not. If a potential contractor decided to make an offer, he or she must, acting in a situation of uncertainty, determine the parameters of the offer. This uncertainty stems from both a lack of knowledge of competitive offers and the actual conditions in which the project will be implemented. Therefore, the contractor cannot measure the cost of service in a way that will cover any consequences arising from the identified risks, because such an offer would certainly be worse than other competing offers in which the authors estimated the expected level of costs under conditions of incomplete information. We propose the implementation of a decision support system (DSS) that facilitates the analysis of the decision-making process, while allowing full control over the collection of data, filtering the data, and making decisions. Such a system also should offer the possibility of taking into account the opinions of experts (also postulated by the PMBOK Guide), because their opinions are supported by knowledge, experience from similar projects, and intuition. However, DSSs are used extensively in many professional fields to make decisions about what to do or what to buy, among other things (see [10],[11]). In this article, we show the design and functionality of a system to support the management decision-making process. The system will address the current gap between the functionality of programs for project management and the functionality of advanced decision support systems.

2 Theory

The planning phase of the project usually must be initiated before notification of the contractor selection has been received, because of the need to reserve resources, provide funds, and maintain compliance with other projects in the portfolio. During this period of time, the bidder, before becoming the contractor, must work conceptually in the solutions for the bid that was requested by the client while simultaneously working organizationally at the beginning of the planning phase when the procurement issues for the future realization generally are prepared. A client who is seeking a potential contractor can maintain a list of qualified contractors that already have positive evaluations and send an RFx inquiry to them (where RFx represents various types of requests, e.g., Request For Bid, Request For Quotation, and Request for Proposal). If such a list of qualified contractors is not available, clients must make a full assessment of the contractor and her or his bid/proposal/... Maintaining the list of qualified contractors is an efficient solution for clients who buy goods or outsource services regularly. It improves the selection process by making it faster, and it minimizes the risks associated with an unknown contractor. However, the list of qualified contractors requires periodic verification and updating, but it also can be automated, as demonstrated in the paper of [2]. From the bidder's perspective, efforts must be made to become included on such a list, which allows both RFx sourcing and reduces the risk of sending an uncompetitive offer for the job. The problem with choosing the winner offer is that the decision usually is a multi-criteria decision-making (MCDM) problem. Therefore, except in special cases, when the potential qualified contractors are indifferent in terms of the client preferences (only if the offer price is the single criterion for selection) for its solution tools of the multi-criteria analysis should be used. In terms of contract management and the bidding procedure, the bidder must submit a good offer, indicating a thorough understanding of the project's feasibility. The bidder also must be cognizant of contractual risks and the profit that the contract likely will produce. With the intention of preserving the universality of the proposed DSS, it cannot be assumed that the potential customer uses the same system of evaluation and selection of contractors. It is possible that the client uses an interactive approach, i.e., negotiation or auction, rather than a formalized procedure. Due to the detailed descriptions in the literature (see [18], [17] and [14]), interactive approaches and public procurement issues (see [21]) are not analyzed in this paper. For the purposes of the proposed system, it was assumed that the process of selecting a contractor for external client follows the procedure described in Figure 1.

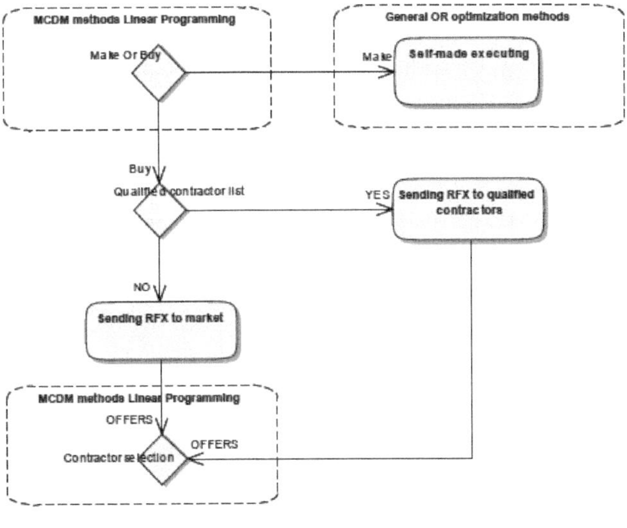

Fig. 1. Contractor selection process

From the methodical perspective, this process will fit therein indescribable process "contract acquisition" and each area of the PMBoK Guide "procurement plan management" and "conduct procurement". Thus, exploratory action will be undertaken, aimed at determining the feasibility and profitability of the contract, and any arrangements with potential contractors will be conditional in nature. For the purposes of conducting the process of decision-making by the bidder, the following data and actions are required: external data (from the client) such as scope statement, project Schedule, quality requirements, time limitations, enterprises' environmental factors, templates for the contract documents, External data (from bidders/potential sub-contractors) such as bids, bidders' data, internal data such as organizational Process Assets, previous clients' cooperation reports (if any), previous sub-contractors' qualifications, previous sub-contractors' performance reports. Bidders must make a set of decisions based on the data listed above, i.e.,: design of the work breakdown structure, method of scheduling (CPM/PERT/CCPM/...), make-or-buy analysis for the entire project or parts of the scope/WBS, Acquisition of staff, selection of the contract type (with sub-contractors), weighted selection criteria for subcontractors, weighted criteria for bid selection, definitions of risk probability and impact, strategies for risks (proactive and reactive), methods of quality assurance, tender quotation or decision not to tender a quotation. In the contract acquisition process, several well-known and extensively used Operations Research methods and tools can be applied to improve the decision-making process and to make optimal or near-optimal decisions. In the described DSS, some of the decisions are planned to be modeled in a software supporting individual and group decision-making. The data and problem structure in the project contract/procurement process make MCDM methods suitable for aiding most of the required decisions. One of the methods that can be used is introduced by Saaty [20]the Analytical Hierarchy Process (AHP). This is a very good method for supporting multi-criterial decision making or problems with quantitative and qualitative criteria with multiple decision makers. It has particular application in decision making and is used around the world in a wide variety of decision situations in various fields. This approach also is used extensively for procurement issues ([6], [13]). To support achieving optimal decisions by negotiation or auction procedures one can use the TOPSIS method developed by Hwang and Yoon in 1981 ([12]). The idea of TOPSIS is that the chosen alternative should have the shortest geometric distance from the "positive ideal solution" and the longest geometric distance from the "negative ideal solution". TOPSIS allows trade-offs between criteria. A poor result in one criterion can be negated by a good result in another criterion. This method already has been used as an auctions supporting tool ([22]) and as an interactive procedure for negotiations ([17]). Turn to portfolio optimization one can use mathematical programming ([16]). This method allow to achieve the best outcome (such as maximum profit or lowest cost) in a mathematical model the requirements of which are make-or-buy analysis. This analysis compares the costs and benefits of manufacturing a product with the costs and benefits of purchasing the product. If the purchase price is higher than the manufacturing cost or if the manufacturer's suppliers are unreliable, the manufacturer may choose to make the product. To support the list of qualified contractors, the statistical approach of Partial Least Squares in the modified version ([3]) will be used. To finalize the process of contract acquisition, the decision maker must have the following outputs: a list of the client's qualified contractors (if provided), contract acquisition, make-or-buy decisions, preliminary (conditional) agreement with potential sub-contractors. After successful bidding and a contract has been obtained, the other project procurement management processes in the PMBoK Guide may be activated, i.e.,: conduct procurements (update and finalize), control procurements, close procurements. In the section 2.1 of this article we will present methods which can be used to reduce the risk of delays and contractual penalties. Our method is an implementation of Goldratt's (1997) Critical Chain-based approaches (described by [2]). Moreover, in section 2.2 we show how to perform the procedure for selecting the type of contract.

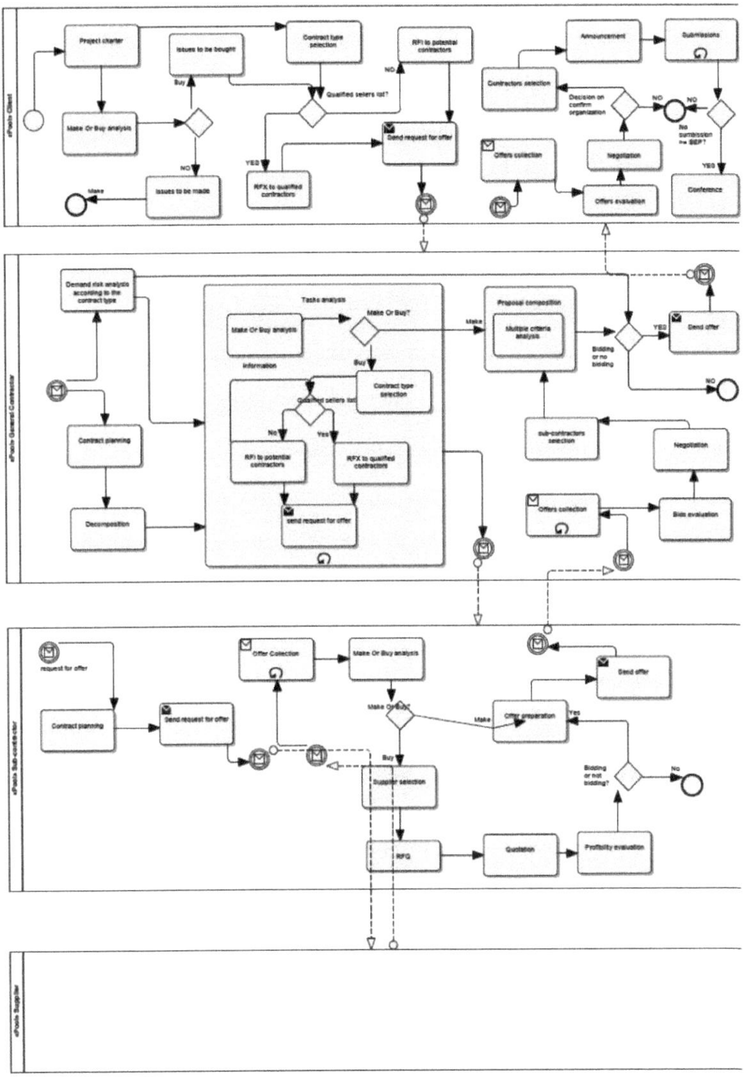

Fig. 2. Business Process Model for contractor-aiding DSS [5]

2.1 Cost and Time Buffers

We consider project which consist $x_1, \dots, x_n$ activities characterized by cost and time criteria. We assume that only q factors has any influence on the cost and the time of the project. Let us consider the following matrix X:

$$X = \begin{bmatrix} x_{11} & \cdots & x_{1q} \\ \vdots & \ddots & \vdots \\ x_{n1} & \cdots & x_{nq} \end{bmatrix} \tag{1}$$

Elements of the matrix *X* equals 0 or 1. If x_{ij} equals 1 it means that factor j has influence on the completion of task x_i. In the other case there is no influence of factor j on task x_i. The matrix *X* we will call the factor's matrix. Let $K = \left[k_{ij}\right]$ to be the matrix of cost's ratios of all q factors for all tasks and $W^m = \left[w_1^m, \dots, w_n^m\right]$ to be the vector of minimal amounts of work for the tasks $x_1, \dots, x_n$. On the basis of matrix *X* and vector W^m for task x_i we can calculate the total amount of work w_i by $w_i = f_{w_i}\left(x_{i1}, \dots, x_{iq}, w_i^m\right)$ where f_{w_i} is a work assigning function. Moreover we assume that there is vector $R = \left[r_1, \dots, r_q\right]$ describing the restrictions of accessibility of factors for whole project. Let $T = \left[t_{ij}\right]$ be the matrix of amounts of work for each factor in each task. On the basis of the matrix X, T and K we calculate the cost and the duration of each task by: $k_i = f_{i_k}\left(x_{i1}, \dots, x_{iq}, t_{i1}, \dots, t_{iq}, k_{i1}, \dots, k_{iq}\right)$ and $t_i = f_{i_t}\left(x_{i1}, \dots, x_{iq}, t_{i1}, \dots, t_{iq}\right)$ where f_{i_k}, f_{i_t} are some functions. We called this functions the *cost* and the *time* functions, respectively. Thus the total cost and the total duration of the project are given by $K_c = \sum_{i=1}^{n} k_i$ and $T_c = \max_{i=1,\dots,q}\left(ES_i + t_i\right)$ where ES_i is the earliest start of task x_i. Under the following assumptions we minimize total cost of the project. It leads to find the optimal work assignments for every factor in each activities. From the set of alternate optimal solutions we choose this one, for which the total duration of project is minimal. In this way we obtain the optimal solution in safe case. According to the contractors' safe estimations the amount of work could be overestimated. It leads up to overestimations of the activities' cost and duration expected values and afterwards the total cost and the total duration of the whole project. That means that cost (duration) of the active is the sum of the reasonable cost k_i^e (duration t_i^e) for activity x_i and some buffers of budget k_i^B and time t_i^B for activity x_i i.e $k_i = k_i^e + k_i^B$ and $t_i = t_i^e + t_i^B$. Then the total cost (total duration) of project is the sum of the reasonable cost K^e (reasonable duration T^e) and the buffers of budget K^B (time T^B) i.e. $K_c = K^e + K^B$ and $T_c = T^e + T^B$. To set the buffers K^B, T^B up we must estimate the most probable amounts of work. We do that by changing appropriate elements x_{ij} in matrix X. It means that some factors which had influence on activity x_i in safe estimation case does not have it in real estimation case and vice versa. Than we using the function w_i for each activity x_i. In this way we get the new factor's matrix and the new vector of amounts of work. Then we execute the same procedure for the most probable amount of work but under additional condition that the assigned amount of work in the safe case cannot be smaller than the new one for each factor in each activity. Because that is unlikelihood that all factors will occur, we can reduce the buffers for project by: $K_r^B = \alpha K^B$ and $T_r^B = \beta T^B$ where $\alpha, \beta \in [0,1]$ are the ratios revising amount of buffers. The total cost and duration for project are given by the following formulas $K^P = K^e + K_r^B$ and $T^P = T^e + T_r^B$. Part of saved money can generate bonus pool B and be divided between the factors. Let us introduce the weight of importance of activities $S = [s_i]$ where $s_i \in [0,1]$. To share the bonus pool we define function which depends on saved amount of work, importance of activity x_i and if the activity is critical or not and on the reduced buffers of cost and time. In the general case that factor i can receive the amount of money $b_i = f_{i_b}(s_i, D_i^W, c, D_B^K, D_B^T)$ where s_i is the importance of task xi, D_i^W is the saved amount of work for task x_i, $c = 1$ if the task is on critical path or $c = 1$ if is not on the critical path, D_B^K is the amount of saved cost, D_B^T is the amount of saved time and f_{i_b} is some function.

2.2 Main Types of Contracts

We consider project which consist $x_1, \dots, x_n$ tasks characterized by cost and time criteria. As the consequence of the contract between company and principal there are budget $K_{\max}$ and duration $T_{\max}$ of project. Moreover we can in the contract with the fixed price there are defined price I_p, success fee S_p and penalty fee P_p. The success fee and penalty fee can be defined as follow: $S_p = r_s \cdot I_p / day$, $P_p = r_p \cdot I_p / day$ where r_s and r_p are success rate and penalty rate respectively. We assume that we have q potential subcontractors for n tasks in the project.

The Fixed Price Contract. In our first model we assume that between project owner and the contractor there is contract with the fixed price. Let $A = [a_{ij}]$ where $i = 1,\dots,n;\ j = 1,\dots,q$ to be the subcontractors assign matrix. Elements of the matrix A equal 0 or 1. If a_{ij} equals 1 it means that subcontractor j will be perform task x_i. Moreover, let $M = [m_{ij}]$ denote the preference matrix. Elements of the matrix M equal 0 or 1. If m_{ij} equals 1 it means that the task x_i should be realized together with task x_j by the same subcontractors. Of course there are ones on the main diagonal. On the other hand let $D=\left[d_{ij}\right]$ denote the restriction for tasks in project. Elements of the matrix D equal 0 or 1. If d_{ij} equals 1 it means that the task x_i not be realized together with task x_j by the same subcontractors. In our case we the following optimization problem. In order to simplified the calculation let us also introduce the following vectors $m = MI + I$ and $d = DI$ where $I=\left[i_{ij}\right]$ is identity matrix. Under this assumptions we maximize the total benefits of the project. We have the following optimization problem

$$I_p + S_p - P_p - \sum_{i=1}^{n} a_{ij} x_{ij} \left(k_{ij} + s_{ij} - p_{ij}\right) \rightarrow \max$$

$$\forall_{i=1,\ldots,n} \sum_{j=1}^{n} a_{ij} = 1$$

$$\forall_{i=1,\ldots,n} \sum_{j=1}^{n} a_{ij} x_{ij} = 1$$

$$T_c = \max_{i=1,\ldots,n} \left\{ES_i + t_i\right\} < T_{\max}$$

$$K_c = \sum_{i=1}^{n} \sum_{j=1}^{q} a_{ij} k_{ij} < K_{\max}$$

$$\forall_{j=1,\ldots,q} \sum_{i=1}^{n} \sum_{k=1}^{n} a_{ij} x_{ij} m_{ik} = m_i$$

$$\forall_{j=1,\ldots,q} \sum_{i=1}^{n} \sum_{k=1}^{n} a_{ij} x_{ij} d_{ik} = d_i$$

$T_{\max}$, $K_{\max}$ denote maximum duration and cost for the project respectively. The $T_{\max}$, $K_{\max}$ are results of the project requirements. It leads to find the optimal work assignments for every task. From the set of alternate optimal solutions we choose this one, for which the total duration of project is minimal.

The Cost-Plus Contract. In second model we assume that the project will be settled by the cost-plus formula on the basis of the quantity survey. Let cost and duration matrices be given in the section 2.1. We treat cost from matrix as the cost of actual implementation of each task for each subcontractors. Let $G=\left[g_i\right]$ be the vector of profit margins for all subcontractors. The values g_i belongs to the interval $[0,1]$. To protect against the uncontrolled growth of the cost of the task x_i in such type of contracts the so-called ceiling price is used. So let $C=\left[c_i\right]$ be the vector of ceiling price for each tasks. Like in previous case denote by k_i the cost and by t_i duration of the task x_i. Thus the total cost of the project is given by $K_c = \sum_{i=1}^{n} k_i g_i$ where g_i is the profit margin of subcontractor who will perform the task x_i. Under this assumptions we maximize the total benefits of the project. In our case we have the following optimization problem

$$I_p + S_p - P_p - \sum_{i=1}^{n} a_{ij} x_{ij} \left(k_{ij} + (1 + g_i) s_{ij} - p_{ij}\right) \to \max$$

$$\forall_{i=1,\dots,n} \sum_{j=1}^{n} a_{ij} = 1$$

$$\forall_{i=1,\dots,n} \sum_{j=1}^{n} a_{ij} x_{ij} = 1$$

$$T_c = \max_{i=1,\dots,n} \{ES_i + t_i\} < T_{\max}$$

$$K_c = \sum_{i=1}^{n} \sum_{j=1}^{q} a_{ij} k_{ij} < K_{\max}$$

$$\forall_{j=1,\dots,q} \sum_{i=1}^{n} \sum_{k=1}^{n} a_{ij} x_{ij} m_{ik} = \{0, m_i\}$$

$$\forall_{j=1,\dots,q} \sum_{i=1}^{n} \sum_{k=1}^{n} a_{ij} x_{ij} d_{ik} = \{0, d_i\}$$

$$\forall_{j=1,\dots,q} \sum_{i=1}^{n} \sum_{k=1}^{n} a_{ij} x_{ij} k_{ij} (1 + g_i) < c_i$$

From the set of alternate optimal solutions we choose this one, for which the total duration of project is minimal.

3 Conclusion

The general project management methodologies describes processes and decision-making problems in a general manner. In the approach described in this paper we tried to maintain the versatility and compatibility with commonly used methodologies while introducing elements of the model adaptable to the actual decision-making situations in specific projects. Just as in real-life cases, we have introduced the possibility of interaction between the project owner and its potential and actual performers, who may also contract the services of sub-contractors. The proposed decision models are easy to implement and to solve in a plain spreadsheet.

Acknowledgments. The project was funded by the Polish National Science Centre on the basis of the decision number DEC-2011/03/D/HS4/01649.

References

1. Bajari, P., McMillan, R., Radelis, S.: Auctions Versus Negotiations in Procurement: An Empirical Analysis. Journal of Law, Economics, and Organization 25, 372399 (2009)
2. Blaszczyk, P., Blaszczyk, T.: Project subcontractors selection in fixed price and cost-plus contracts. Lectures Notes on Engineering and Computer Science, pp. 1131–1136 (2013)

3. Blaszczyk, P., Blaszczyk, T.: Assessment of the risk of the project's contractor bankruptcy using the Partial Least Squares approach. Procedia Technology 9, 777–786 (2013)
4. Blaszczyk, P., Blaszczyk, T.: The Draft of the System Supporting Purchasing Decision Making in Project Management Processes. In: GIBA 2014 Conference Proceedings, pp. 827–833 (2014)
5. Blaszczyk, P., Blaszczyk, T.: On the support of the general contractors decisions in the processes of event project management. In New perspectives on the challenges and future developments of global enterprise management, ch. 11. Palgrave Macmillan (in press)
6. Cheung, S.-O., Lam, T.-I., Leung, M.-Y., Wan, Y.-W.: An analytical hierarchy process base, New perspectives on the challenges and future developments of global enterprise management procurement selection method. Construction Management and Economics 19 (2001)
7. Eriksson, P.E., Laan, A.: Procurement effects on trust and control in client contractor relationships. Engineering, Construction and Architectural Management 14, 387–399 (2006)
8. Figueira, J., Greco, S., Ehrgott, M.: Multiple criteria decision analysis: state of the art surveys. Kluwer Academic Publishers (2005)
9. Goldratt, E.: Critical Chain. North River Press (1997)
10. Humphreys, P., McIvor, R., Huang, G.: An expert system for evaluating the make or buy decision. Computers & Industrial Engineering 42, 567–585 (2002)
11. Humphreys, P., McIvor, R.: Knowledge-based system technology in the make or buy decision in manufacturing strategy. In: Intelligent Knowledge-Based Systems, pp. 83–108. Springer (2005)
12. Hwang, C.L., Yoon, K.: Multiple Attribute Decision Making: Methods and Applications. Springer, New York (1981)
13. Kahraman, C., Cebeci, U., Ulukan, Z.: Multi-criteria supplier selection using fuzzy AHP. Logistics Information Management 16, 382–394 (2003)
14. Kersten, G.E., Pontrandolfo, P., Vahidov, R., Gimon, D.: Negotiation and Auction Mechanisms: Two Systems and Two Experiments. In: Shaw, M.J., Zhang, D., Yue, W.T. (eds.) WEB 2011. LNBIP, vol. 108, pp. 399–412. Springer, Heidelberg (2012)
15. Liu, Q., Sun, S., Wang, H., Zhao, J.: A multi-agent based system for e-procurement exception management. Knowledge Based System 24, 49–57 (2011)
16. Ogryczak, W.: Multiple criteria linear programming model for portfolio selection. Annals of Operation Research 97, 143–162 (2000)
17. Nowak, M., Blaszczyk, T.: The Time-Cost Trade-Off Analysis in Construction Project Using Computer Simulation and Interactive Procedure. Technological and Economic Development of Economy 4/2009, 523–539 (2009)
18. Nowak, M.: Interactive Multicriteria Decision Aiding Under Risk Methods and Applications. Journal of Business Economics and Management 12, 69–91 (2011)
19. PMI. A Guide to the Project Management Body of Knowledge, 5th edn. Project Management Institute, USA (2012)
20. Saaty, T.: The analytic hierarchy process: planning, priority setting, resources allocation. McGraw-Hill, USA (1980)
21. Thai, K.V.: Public procurement re-examined. Journal of Public Procurement 1, 9–50 (2001)
22. Wachowicz, T., Blaszczyk, P.: TOPSIS Based Negotiation Offers Scoring Approach For Negotiation Support Systems. Group Decision and Negotiation, 1–30 (2012)

Factory Logistics Improvement Projects: Case Northern Thailand

Sakgasem Ramingwong*, Apichat Sopadang, and Korrakot Yaibuathet Tippayawong

Center of Excellence in Logistics and Supply Chain Management
Chiang Mai University, Thailand
sakgasem@gmail.com

Abstract. The paper analyses and discusses the interest and results from 3-years factory logistics improvement projects in 75 factories in Northern Thailand region. Funded by government agency, the consultant had conducted man-days of consultancy for the factory, aiming at improving logistics efficiency or reduce logistics cost. However, with limited resource, only top priority problem can be chosen. Where the size and type of industry varies, the problem, approach and result are different as expected. Therefore, it is interesting to understand the nature of the factory and their selected project type.

1 Introduction

Thailand is regarding a major manufacturing hub of Asia. As the world's 31st biggest economy, 31st in World Economic Forum's the Global Competitiveness Report and 18th in 2014 World Bank's Ease of Doing Business, Thailand is global supply for hard disk drive, rubber, sugar, rice, food, fishery and automobile. [1, 2]

However, focusing on logistics performance of Thailand, there are room for improvement. Study by Office of the National Economic and Social Development Board (Thailand) suggested that Thailand logistics cost per GDP lies at 18.8% which is considerably high, compared to those in advances such as EU, Japan, Singapore, US. Thailand is weak on managing logistics. Efficiency and effectiveness of the logistics activities in terms of cost, time, quality, agility, traceability and reliability are low. [3]

The issue is well aware by Thai government. Several logistics improvement projects as well as budgets have been issued by several agencies. Among the government agencies, Logistics Bureau, Ministry of Industry is the dedicated key policy driver. Aiming at reducing logistics cost, improve logistics efficiency, building capacity, more than 30 projects and 3 million USD has been placed under the bureau actions per year. In 2014, more than 500 factories and 3,500 logistics staffs have been involved with these projects.

* Corresponding author.

M. Gen et al. (eds.), *Industrial Engineering, Management Science and Applications 2015*,
Lecture Notes in Electrical Engineering 349, DOI: 10.1007/978-3-662-47200-2_38

2 Factory Logistics Improvement Projects

Pattern of consultancy for improving logistics activities is familiar with Thai industry eco-system. Working together with the factory, the budget is funded by the government agencies and the consultancy is conducted by expertise, e.g., university professors or professional consultants. The duration, man-day can be varied from the resource and readiness.

In specific, factory logistics Improvement projects is the continuously funded by Logistics Bureau, aiming at improving logistics efficiency or reducing logistics cost in participated factories. The projects have been scoped in both area-based and industry-based. For example, the northern Thailand project aims at improving factories in northern Thailand, regardless of industry type. The Food project aims at improving food factories, regardless of location and area. Here, the paper focuses on the northern Thailand project which has been continuously addressed to 75 factories in northern Thailand for the past 3 years.

Learning from best practice and know-how of the nature of the problem, the project is planned. [4, 5] The objectives of the project are to (1) improve logistics efficiency, (2) build logistics capacity and (3) outreach the outcome of the projects. The duration of the project is basically 6 months. The process of the project is consistent to other factory logistics improvement projects, ie., in-depth training for prospective participants, selecting participating factories (based on sizing, readiness, team support, etc.). Then, 6 man-day of consultancy was assigned to each participating factory. Each factory must conduct at least 2 projects, bywhich 1 project with inventory/ warehouse management is compulsory. The aim is either the factory can reduce logistics cost by 15% or the factory can improve logistics efficiency by 15%.

3 Result Discussion

Fig. 1 firstly illustrates distribution on type of industry involving in projects in 2012-2014, 75 factories in total. It can be seen that the food industry is very dominant

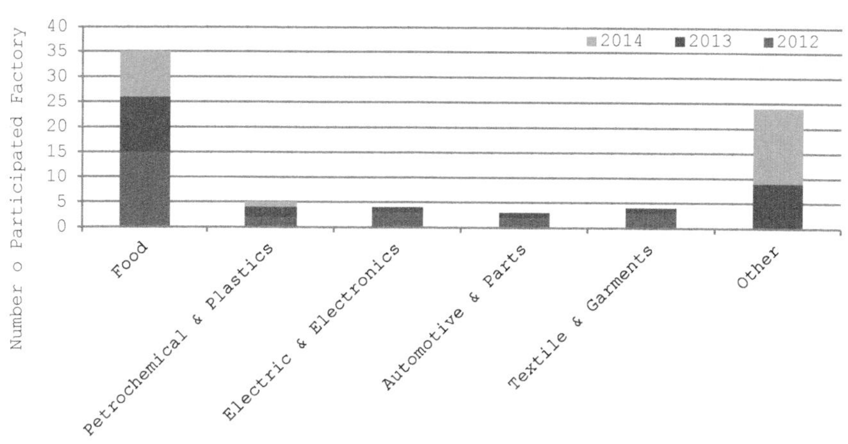

Fig. 1. Distribution on Type of Industry Involving in Projects

in terms of number of participants as the industry supply chain is very strong in northern Thailand. Other industry here includes ceramics, wooden products, glass, fertilizer, concrete, metal parts, packaging and mining. Size of the factory varies from small to large enterprise, number of staffs from 10 to 1,600, sales from 10 to 14,000 million THB.

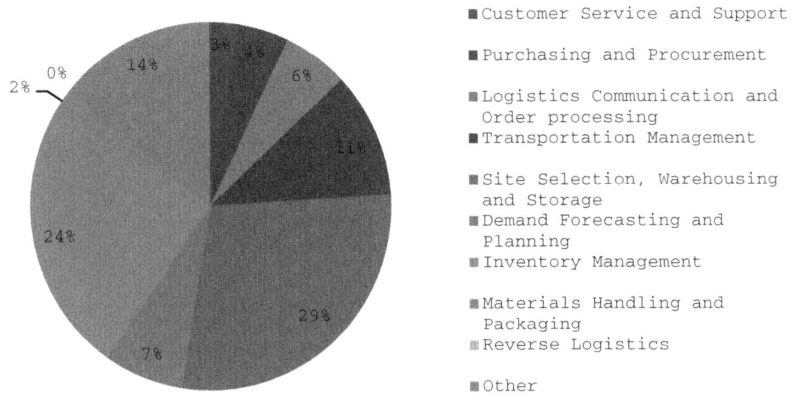

Fig. 2. Distribution of Project Type Based on Logistics Activities

Dividing the logistics activities into 9 followed the logistics management masterpiece of Grant et.al., [6] with other improvement project is optional, further inspection is conducted in order to understand the nature of the factory and their needs.

Apart from the inventory- and warehouse-related management projects with are of mandatory, the factories and consultants must choose at least one another logistics improvement project. It can be seen from Fig. 2 that transportation-related and other improvements project are among those were chosen at total of 25% of all projects.

Focusing on food industry, which is the majority of the factories, layer of enterprise size is of interest here. Where small, medium and large size is here based on number of labor, following the definition by Office of Small and Medium Enterprises Promotion, of those 35 food factories, there are 10 small, 12 medium and 13 large enterprises.

To inspect the nature of the enterprise based on their selected project topics, the paper choose to see projects that is non-compulsory.

Here, large enterprise seems focus on transport management and demand forecasting and planning. The projects are for example, distribution and network design, vehicle routing, warehouse investment decision, using Neural Network for multi-factor demand forecasting, supply prediction and planning. Most of the project requires advance engineering technique to approach the large-scale, complicated problem.

Where medium-size firms focus on logistics communication and order processing perspective, other production efficiency improvement are of interests. The projects are for example, using Lean concept in logistics communication and documentation, using Value Stream Mapping to reduce production lose, using line balancing for labor productivity. The technique and complication are rather simple than those of the large-firms.

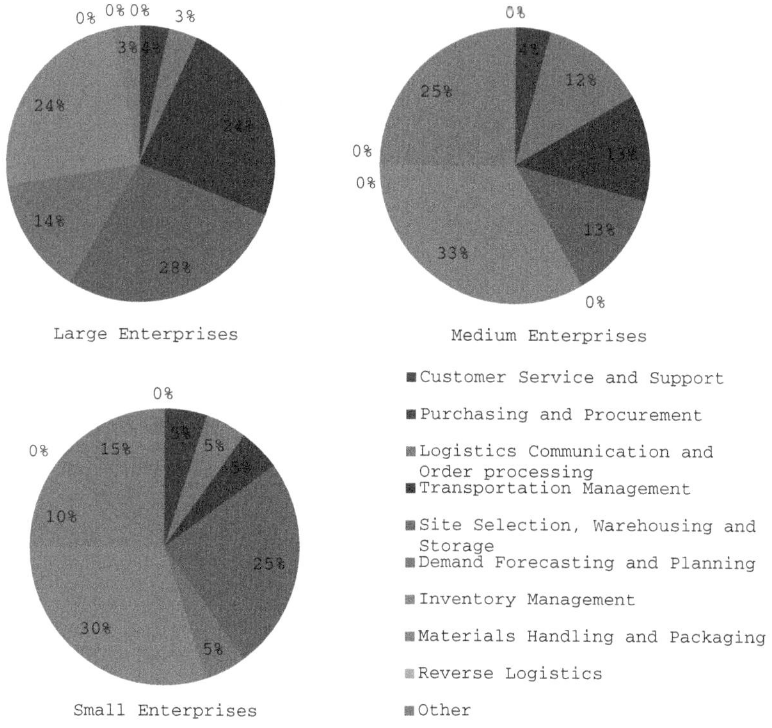

Fig. 3. Distribution of Project Type Based on Logistics Activities for Food Industry: Large, Medium and Small Enterprises

Small enterprises on the other hands give priorities to material handling and packaging as well as other production efficiency improvement. Using a simple logistics and engineering tools, the projects are for example, production improvement using QC tools, production process analysis, packaging design for logistics, material handling equipment design using Ergonomics and safety concept.

Focusing on the success of the project, the improvement KPIs in terms of (1) cost reduction, (2) time reduction, (3) area utilization, (4) accuracy improvement, (5) delivery improvement and (6) efficiency improvement can be discussed as follow (see Fig. 4):

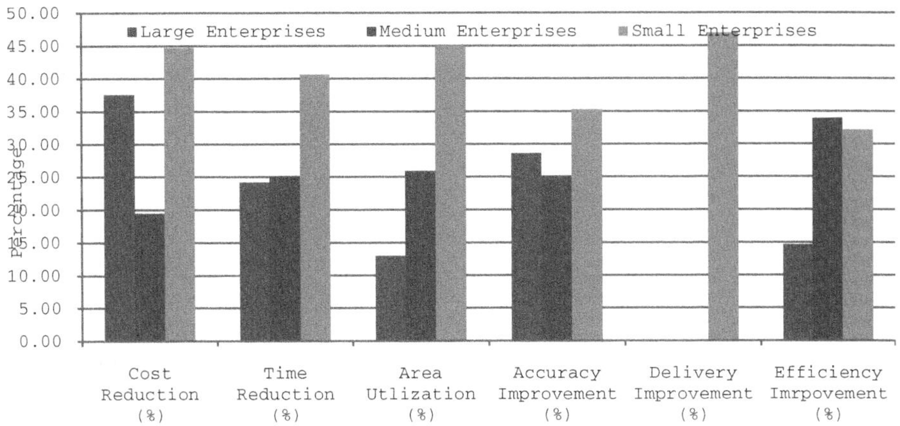

Fig. 4. Percentage Improvements in Food Industry

In overview, it can be seen that small enterprises enjoy their benefit of the project since the improvement is comparably high. Third party consultants have successfully approached opportunity for improvement. Whilst, the staffs, budget and knowledge are often limited, thus decision making is quick due to short organizational hierarchy. The challenge is to sustain the development and transfer the knowledge and practice for these factories.

4 Conclusion

The paper does not intend to conclude any statistical findings due to limited samples. However, it is the objectives to share a little but useful information for such project. Where factory are required to conduct the inventory- and warehouse-related project, which are successfully obtained, their interest on other logistics improvement is reflected by their project selection. Where of food industry, the majority of the industry in northern Thailand, the large-, medium- and small-firms approach the problem differently. The scope and level of improvement are inspected and suggestive for further similar projects. It is indeed to state that such projects are beneficial to the factories.

Acknowledgement. The Factory Logistics Improvement projects (Northern Thailand) had been conducted by Center of Excellence in Logistics and Supply Chain Management, Chiang Mai University. The projects had been funded by Logistics Bureau, Ministry of Industry - THAILAND.

References

1. Board of Investment Thailand: Thailand's Economic Fact Sheet (2013)
2. KPMG: Investment in Thailand (2008)
3. Office of the National Economic and Social Development Board: Thailand Logistics Report 2011 (2011)
4. Lenny Koh, S.C., Demirbag, M., Bayraktar, E., Tatoglu, E., Zaim, S.: The impact of supply chain management practices on performance of SMEs. Industrial Management & Data Systems 107(1), 103–124 (2007)
5. Banomyong, R., Supatn, N.: Developing a supply chain performance tool for SMEs in Thailand. Supply Chain Management: An International Journal 16(1), 20–31 (2011)
6. Grant, D.B., Lambert, D., Stock, J.R., Ellram, L.M.: Fundamentals of Logistics Management. McGraw-Hill Higher Education (2006)

Enhancing Project Funding Decision Quality

Ofer Zwikael and Ying-Yi Chih

The Australian National University, Australia

Abstract. Successfully implementing projects accelerate organizational growth. Given the limited organizational resources available, project funding is a critical strategic decision to ensure the most appropriate projects are selected. Recent research suggests that a proposed project's target benefits—those anticipated to be realized at project completion—are important to consider when making decisions for project funding. Yet, the literature provides few measures to appraise the quality of target benefits. Consequently, sub-optimal projects are funded due to estimations errors, optimism bias about the future ("planning fallacy") and strategic misrepresentation of target benefits. This paper presents three studies to develop and validate a new "Quality of Target Benefits" (QTB) scale consisting of three dimensions: specificity, attainability, and comprehensiveness. This scale can be used practically—that is, to objectively appraise a proposed project's target benefits and to support an informed funding decision. It can also be used theoretically—as an instrument to facilitate theory development in areas of decision making, organizational strategy implementation, and project performance.

1 Introduction

Organizations implement strategies through projects (Pellegrinelli and Bowman 1994). Organizational performance is thus strongly associated with the successful realization of project benefits—"the flows of value that arise from a project" (Zwikael and Smyrk 2012: p.11). For example, benefits from a new product development project may include "increased company market share" and "increased product portfolio." Because organizations have limited resources, they need to ensure those available are allocated to projects that best delivers benefits and supports organizational strategies. In other words, choosing which projects to fund is a critical strategic organizational decision.

In making project funding decisions, senior managers take into account the proposed projects' estimated cost, duration, level of risk, and target benefits—those benefits expected to be realized at a project's completion (Zwikael and Smyrk 2012). To support such decision-making processes, prior research has discussed numerous methodologies ranging from financial-focused to multi-criteria approaches (Shapira and Shaver 2013). Examples of financial-focused approaches include Net Present Value (NPV) and Internal Rate of Return (IRR). These approaches are limited in their capability to integrate non-financial factors. Alternatively, multi-criteria approaches, such as the Multi-Attribute Utility/Value Theories (MAUT/MAVT) and Analytical Hierarchical Process (AHP) can enable a more comprehensive consideration of multiple aspects (Grushka-Cockayne, De Reyck and Degraeve 2008). Although information inputs significantly affect the quality of decision making (Raghunathan

M. Gen et al. (eds.), *Industrial Engineering, Management Science and Applications 2015*,
Lecture Notes in Electrical Engineering 349, DOI: 10.1007/978-3-662-47200-2_39

1999, Mihm 2010), the quality of the information is seldom discussed in this context. Instead, it is taken for granted that the information required for decision making is readily available and of good quality. Unfortunately, this is not the common case in the project funding decision-making process.

Funding decisions are made at an early stage of a project's life when uncertainty is substantial. Recognizing that poor quality or lack of information may lead to inadequate project decisions (Flyvbjerg, Holm and Buhl 2005), researchers have investigated ways to estimate project cost, duration, and risk level to ensure information inputs are of a high quality (e.g., Jørgensen, Halkjelsvik and Kitchenham 2012, Zeng, An and Smith 2007). However, the literature offers little information on ensuring the quality of another important piece of information: the proposed project's target benefits. Consequently, target benefits of proposed projects are often incomplete (Lin and Pervan 2003) and vaguely defined (Norris 1996). In some cases, under uncertainty, decision makers' forecasts are adversely biased (Ansar, Flyvbjerg, Budzier and Lunn 2014), as explained by the work that lead to Daniel Kahneman's 2002 Nobel Prize in Economics (Kahneman and Tversky, 1979). The phenomenon of being systematically and predictably too optimistic about the future and hence overestimating target benefits and underestimating cost is called "planning fallacy" (Weick and Guinote 2010, Kahneman 2011). In addition to these candid psychological biases, some target benefits are deliberately inflated to increase the chance of project funding (Jenner 2009, Flyvbjerg 2007). As a result, suboptimal and unsuitable projects are endorsed, whereas more suitable projects are excluded (Flyvbjerg 2013).

Following the call for clear project goals (Cats-Baril and Thompson 1995), research has provided initial wisdom regarding the appraisal of proposed projects' target benefits. For example, Jenner (2009) suggests that target benefits should be "robust and realizable"; yet, how to measure these two dimensions and whether they are conclusive remains vague. This stream of research is incomplete and lacks empirical support. Accordingly, this paper aims to address this gap in the literature by developing a new "Quality of Target Benefits" (QTB) scale. The proposed scale is expected to contribute to both practice and research. In practice, this scale can be used to appraise and compare alternative proposed projects' target benefits to support informed funding decisions and investment prioritizations. In research, it can be used to further develop theory in the areas of strategy implementation, decision making, goal setting, and project performance. To ensure the rigor of the scale development and validation, we followed a well-establish approach suggested by Churchill (1979) and Hinkin (1998). We started with the definition of the domain of QTB construct. We then presented three studies to develop and validate the new QTB scale. Finally, we discuss the implications and limitations of this research and suggest future research directions.

2 Defining the Domain of the Quality of Target Benefits (QTB) Construct

The first step in developing a new scale is to define it (Churchill 1979). Because QTB is a quality-related construct, we looked at the quality management literature and found that the term "quality" is defined inconsistently. Examples of definitions include excellence, value, conformance to specifications, and meeting expectations (Reeves and Bednar 1994). In this paper, we build upon Crosby (1979) and operationalize "quality" as "conformance to specifications"—this definition provides an objective basis for the initial appraisal of target benefits and allows their monitoring and evaluation. For example, when target benefits are proposed for a new project, their quality can be appraised against a set of specifications. The remainder of this section aims to establish such a definition.

Next, to derive potential dimensions for appraising QTB, we draw upon goal-setting theory, because target benefits are strategic goals of a project. Goal-setting theory suggests that goals need to be SMART, namely: Specific, Measureable, Attainable, Relevant, and Time targeted (Doran 1981). Such SMART goals can enhance performance by directing attention towards goal-relevant activities, energizing effort, encouraging persistence, and supporting strategy development (Locke and Latham 2002, Verbeeten 2008). We conceptualized QTB as a five dimensional construct and identified potential items from the literature for their assessment. In the next section, we present three studies to further develop and validate the QTB scale.

3 Studies

This paper develops and validates a scale to measure QTB of proposed projects. Drawing on the SMART principle in goal-setting and other relevant theories, we conceptualized QTB as a five-dimensional construct and identified potential items for each dimension. According to the structured scale development process (Churchill 1979, Hinkin 1998), we further conducted three studies to refine the initial pool of items and test the validity and reliability of the scale. To ensure the generalizability of our findings, in these studies we recruited participants with relevant experience in forming target benefits and reviews across different countries, including the United States, Australia, and China.

In Study 1, we conducted semi-structured interviews with senior managers in Australia, drawing on their experience to validate the initial QTB dimensions and identify missing ones. This study ensures our scale is comprehensive and captures both theoretical and practical viewpoints. In Study 2, we examined the content validity of the dimensions and items using a group of international experts from academia and industry. In Study 3, using a survey approach, we conducted an Exploratory Factor Analysis (EFA) to validate the proposed items and scale structure. The three studies are summarized in Table 1.

Table 1. Research Overview

Study	Scale Development Step	Study Description
Study 1	Item generation and refinement	Interviews with 15 senior managers in Australia to supplement the literature review analysis
Study 2	Content validity	Survey of 21 experts from the United States, Australia, Israel, and New Zealand to validate the proposed construct structure
Study 3	Initial item reduction	A survey of 132 senior managers in China to confirm and refine the construct structure

3.1 Study 1 – Item Generation and Refinement

3.1.1 Participants and Procedure

To refine further the preliminary list of QTB dimensions derived from the literature, we conducted a series of interviews with senior managers from eight Australian government agencies. We focused on the Australian government because of its strong emphasis on project benefit management practices. For example, benefit realization is a core practice in assessing projects in the Australian Department of Finance and Deregulation's Gateway Review (2012) and is used in the majority of Australian government agencies (Gershon 2008).

Participants were selected through a purposive sampling strategy (Kerlinger 1986) based on their involvement in formulating and appraising project target benefits. Potential participants were identified through the Australian government website and invited for interviews via emails and follow-up phone calls. Fifteen participants were chosen based on the relevance of their roles and work experience. Participants comprised of two directors, one assistant director, one executive secretary, seven national managers, three general managers, and one program manager. The Australian Public Service (APS) classifications of these participants ranged from Executive Level 1 (EL1) to Senior Executive Service (SES) level 3, which are senior ranks. The majority of participants were at SES level 1 with an average of 16 years work experience. All participants played an active role in formulating and appraising target benefits and managing benefit realization processes for high-level government projects. The heterogeneous knowledge of these participants allowed us to develop a comprehensive understanding about practices of appraising QTB.

A semi-structured interview approach was chosen to keep participants focused on the interview questions while also allowing them flexibility to discuss additional relevant practices. The interviews were divided into three parts. In the first part, we began with "ice-breaking" questions about the participants' roles and responsibilities in formulating and apprising target benefits. In the second part, we asked participants to describe how QTB was determined in their agencies.

3.1.2 Data Analysis

The interview transcripts were coded and analyzed with reference to our initial list of dimensions. This was achieved through an iterative process of data analysis from different perspectives at different times, as recommended by Miles and Huberman (1994). First, each transcript was read several times to grasp each participant's view on the QTB. Second, all transcripts were re-read and systematically reviewed for consistency and variations in participants' responses. Then, we amalgamated the interview results and cross-checked them with the findings derived from the literature.

3.1.3 Results

Our interviews confirmed the five dimensions derived from the goal-setting theory, which suggests that QTB can be appraised based on whether the benefits are specific, have clearly defined target values, have measureable, attainable and given organizational constraints, fit into organizational strategic goals, and have target dates. In addition to confirming the five proposed dimensions, our participants mentioned two additional ones: "accountability" and "comprehensiveness." The additional two dimensions are discussed below.

Accountability. Our participants emphasized the need to establish clear lines of accountability in realizing target benefits for two major reasons. First, accountable managers can ensure that target benefits are formulated through consultation with a diverse group of stakeholders, and consideration of the project's strategic fit. Second, accountability can also help organizations investigate how they may effectively realize the benefits. This input from the interviews aligned with the literature. For example, Ashurst et al. (2008) suggested assigning a project owner—the person held accountable for securing the project's target benefits (Zwikael and Smyrk 2012)—as the accountable person for their realization. This clear and visible line of accountability can enhance the achievability of target benefits (Breese 2012, Lin and Pervan 2003).

Comprehensiveness. Our participants considered it important to include a complete web of benefits to meet the objectives of various stakeholder groups, which aligns with the literature (Breese 2012, Jenner 2009). Although there is no consensus in the literature on what can be considered "comprehensive", Henderson and Ruikar (2010) suggested that benefit formulation should rely on diverse categories, including financial/non-financial, direct/indirect, short/long term, internal/inter-organizational, and economical/cultural benefits. Similarly, Irani and Love (2001) suggested that target benefits should comprise of operational, tactical, and strategic natures.

The above two additional dimensions were added to those five derived from the literature, leading to a total of seven dimensions. Proposed items representing these dimensions were captured from both the literature and interviews. For example, items suggested to assess the "Specific" dimension include: "target benefits were assigned a target value (e.g. 10% decrease in road fatalities)" adopted and modified from Breese (2012), Ward and Daniel (2006), and Zwikael and Smyrk (2012); and "target benefits were assigned measures that are consistent with those measuring similar benefits across the organization", adopted and modified from Nicholson-Crotty et al. (2006). However, due to the large number of dimensions and the fact that two dimensions consisted of only a single item, factors were consolidated (Zikmund, Babin, Carr and Griffin 2010). The "Measurable" dimension was integrated into the similar dimension of "Specific", which created a new dimension called "Specificity", defined as "the extent to which target benefits are clearly defined and measurable." "Accountability" and "Time-targeted" were integrated into the "Attainability" dimension, defined as "the extent to which target benefits are realistic, given the context in where the organization is managing the benefit realization process and the constraints it has." Finally, "Relevant" was integrated into the "Comprehensiveness" dimension, defined as "the extent to which target benefits reflect organizational strategies and the objectives of various stakeholder groups." This exercise resulted in a three-dimensional 17-item scale.

3.2 Study 2 – Content Validity

3.2.1 Participants and Procedure

Study 2 aimed to examine the content validity of the initial pool of items and their corresponding dimensions developed in Study 1 (Anderson and Gerbing 1991).

This was achieved by asking participants to sort the 17 items (presented in random order) into one of the three dimensions where they fitted best and provide feedback on the clarity and readability of each item (Hinkin 1998). Participants were provided with a definition for each of the three proposed dimensions and the list of items, but were not given any indication as to which dimension we expected the items to fall into. Prior to the main study, a pilot study was conducted involving one senior academic in the area of management and one research assistant to ensure the clarity and readability of the questionnaire. Participants for the full study were experts from both academia and industry. A total of 24 requests for participation were sent. Twenty-one experts (88 percent)—16 senior academics with a PhD in management and related areas from leading universities, and five senior practitioners in the area of project management—returned complete questionnaires. Among these responses, nine were received from Australia, seven from the United States, three from Israel, and two from New Zealand. This number of respondents falls within the recommended 12 to 30 experts suggested by Anderson and Gerbing (1991) and Sendjaya, Sarros and Santora (2008).

3.2.2 Data Analysis

The average rate of answers that fit our hypothesized dimension from an expert was 77 percent with a standard deviation of 14 percent. On average, an expert marked the same dimension from Study 1 for 13 items, ranging from nine to the maximum of 17 "correct" items. All experts had a "correct" answer rate within two standard deviations of the mean; hence, none was removed from the analysis. The 21 experts also made 48 suggestions to improve the clarity and readability of the items.

We used two measures suggested by Anderson and Gerbing (1991) and used by Linderbaum and Levy (2010) to test the level of agreement of the experts with the proposed two dimensions: (1) substantive agreement, meaning the proportion of respondents who assign an item to its intended dimension and with a minimum cut off value of 0.7; (2) substantive validity coefficient, meaning the extent to which respondents assign an item to its posited construct than other construct, with a cut off value of 0.5.

3.2.3 Results

13 of the 17 proposed items received values higher than the thresholds required for both measures. These items were confirmed in their proposed dimensions. Changes were made for items with low scores based on suggestions from experts. We also revised slightly the wording of some other items, and improved the clarity of the definitions for the three dimensions. These results led to a revised three-dimensional 15-item scale.

3.3 Study 3 – Initial Item Reduction

3.3.1 Participants and Procedure

Study 3 aimed to validate the three QTB dimensions and 15 items proposed at the end of Study 2. It involved a cross-sectional survey of senior managers who were engaged in formulating and reviewing target benefits as part of their regular organizational duties. Participants of this study consisted of Master of Management part-time students at Tsinghua University in China. All students held full-time management positions in various industries, such as finance, manufacturing, service, and engineering. Survey questionnaires were distributed and collected by the researchers, who also conducted briefing on the questionnaire at the start of the session and addressed individual questions throughout data collection. To assure equivalence between the questionnaire in Chinese and the original English version, a standard translation and back-translation procedure was performed (Brislin 1980). First, an English version of the questionnaire was translated into Chinese by a bi-lingual Chinese-English researcher. Then, an additional person proficient in both languages compared the original questionnaire with the translated questionnaire. This process was repeated until satisfactory results were reached.

In total, 180 questionnaires were distributed in three classes of the same program. All three classes were analyzed as one sample, following a confirmation that participants from the various groups shared similar gender, age, and work experience characteristics, as well as working in organizations of similar types and sizes. Participants were asked to recall a recent project proposal they had proposed or reviewed, and mark their level of agreement according to the 15 items derived from Study 2 for that proposal (the unit of analysis). Proposed items (presented in random order) were phrased in the form of statements on seven-point Likert scales, from "Strongly disagree" (1) to "Strongly agree" (7). Such a technique is widely regarded as appropriate for measuring attitudes (Nunnally 1978). Only participants with relevant job experience in formulating and reviewing target benefits were allowed to complete the questionnaire, and therefore 48 questionnaires were dropped. This resulted in a total of 132 completed questionnaires for analysis, with a response rate of 73 percent. Seventy-one percent of participants were male, 29 percent were under 30 years old, 39 percent were between 30 to 40 years old, 30 percent were between 40 to 50 years old, and two percent were between 50 to 60 years old. The majority of participants hold managerial organizational positions (eight percent were non-supervisory staff, six percent were junior managers, 63 percent were middle managers, and 23 percent were top managers), with an average work experience of 13.2 years. Participants hold positions in both private (36 percent) and government (64 percent) organizations, with an average organization size of 2,667 employees.

3.3.2 Data Analysis

As suggested by Churchill (1979), we purified the proposed measure by conducting an Exploratory Factor Analysis (EFA) to determine the dimension structure of the QTB construct. To examine whether the items load onto the specified dimensions, we performed a rotated Varimax factor analysis with Kaiser Normalization.

3.3.3 Results

EFA results (loading of .4 and above) for all 15 items included in the questionnaire confirm the belonging of all items in the three proposed dimensions. These three dimensions explained 63 percent of the variance. Twelve of the 15 items fell into the same dimensions confirmed at the end of Study 2. Three other items fell into different dimensions suggested in previous studies. Due to these contradictory results and to ensure we captured correctly the valued views of our interviewees and experts, we have again rephrased the wording of these items. The final QTB scale is presented in Table 2.

Table 2. The Final Version of the "Quality of Target Benefits (QTB)" Scale

Dimension	Item Title
Specificity (7 items)	Target benefits are assigned a specific target value (e.g. 10 percent increase in market share)
	Target benefits are explicitly defined to leave no other interpretation (e.g. clear title and description)
	Target benefits are assigned specific measures that will enable the evaluation of their realization
	Target benefits are assigned measures that are defined consistently
	Target benefits have clear units of measurement
	The source of data to measure the target benefits is clear
	Target benefits have a dedicated person accountable for their realization
Attainability (3 items)	Target benefits are achievable given the context of the organization
	The organization has the capacity to realize the target benefits
	Timeframes set for target benefit realization are realistic
Comprehensiveness (5 items)	Target benefits are aligned with the organization's current strategy
	Target benefits are relevant to the organization's long term vision
	Target benefits comprehensively comprise multiple categories (e.g. both financial and non-financial benefits)
	Target benefits are the result of intensive consultation with various stakeholders
	Target benefits reflect the views of key stakeholders

4 Discussion

Projects play an important role in implementing organizational strategy (Pellegrinelli and Bowman 1994), particularly in the context of limited organization resources. Therefore, choosing which projects to fund is a critical strategic decision. Prior studies suggested that decision quality improves with the quality of the information available for decision-makers (Raghunathan 1999). Specifically, a proposed project's target benefits are essential pieces of information for making high quality decisions (Zwikael and Smyrk 2012). Accordingly, in this research, we developed and validated a new Quality of Target Benefits (QTB) scale to appraise the quality of this information. We first reviewed the literature on related theories to theoretically conceptualize the QTB construct. We then conducted three studies to further develop, refine, and validate the scale. Across these studies, substantial support was found for a three-dimensional 15-item scale (see Table 2 for the final scale). The three dimensions are: specificity, attainability, and comprehensiveness. Results from our studies confirmed the reliability and validity of this scale.

Construct development is often considered the first step toward future research in a new area. The new QTB construct encourages us to study its relationship with constructs, specifically those related to decision making, strategy management, and project performance. For example, because the strategy management literature focusses on the management of strategic initiatives (e.g., new product development) (Papadakis 1998, Lechner and Floyd 2012, Carr et al. 2010), research can be conducted on improving the process of transforming strategic goals into proposed projects' target benefits. Further, in decision making, the QTB scale can be used to investigate how the target benefits of strategic initiatives affect the decision-making process and the final decisions. Finally, it is well-recognized in the literature that target benefit formulation is critical for successful benefit realization—they are used as a baseline for monitoring and evaluating ongoing project performance. The QTB scale can thus be used to examine the relationships between the formulated target benefits and the final performance of the strategic initiatives.

The QTB scale also has significant practical implications. At the most fundamental level, the dimensions and items of QTB can serve as a checklist to guide managers in formulating target benefits for new project proposals. Instead of relying on the "inside view", which focuses on the constituents of the specific planned action, the QTB scale encourages the use of the "outside view", which is based on previous actual objective and comparable data from similar actions already completed (Kahneman and Tversky 1979, Flyvbjerg 2013). In particular, the QTB scale suggests that a proposed project's target benefits should be comprehensive (for example, reflect the views of key external stakeholders), specific (for example, use measures defined consistently across the organization), and attainable given the organization's context and external constraints. As an illustrative example, we can refer to the comprehensive procurement process improvement case study discussed by Zwikael and Smyrk (2011). In this project, a well-formulated target benefit may be "Aligned with the strategy to enhance operational efficiency, National Procurement Manager is accountable for reduced procurement costs by 25 percent per order by one year from project approval, measured by the company's cost accounting information system." For comprehensiveness, this target benefit should also consider organizational capabilities and constraints following consultation with diverse groups of stakeholders, and be part of a web of target benefits, which also includes non-financial target benefits. The challenges in formulating measurable non-financial target benefits are recognized. Despite this, organizations should still identify measures and proxies to allow these benefits to be managed. This quality assurance, validated using the QTB scale, can resolve common problems stemming from unclear project objectives, such as misunderstood or ignored users' needs and conflicts over strategic project objectives (Lin and Pervan 2003, Norris 1996). Finally, evaluating the quality of project proposals using QTB can improve funding decisions. Senior managers can develop an understanding of what target benefits are expected at a project's completion, how these benefits should be measured, and who is accountable for their realization. This practice can not only reduce "planning fallacy" (Kahneman 2011) but also increase executives' confidence in their funding decisions (Gimbert, Bisbe and Mendoza 2010).

5 Conclusions

Across three studies, substantial support was found for the reliability and validity of a three-dimensional 15-item QTB scale. This new scale encourages further research into the influence of projects' target benefits on strategic decision-making, as well as on project performance monitoring and evaluation. The new construct also provides opportunities for improving organizational target benefit formulation practices, which can result in better decision making, reduced cases of funding unsuitable projects, and enhanced resource allocation.

Our studies have several limitations that need to be acknowledged. First, participants across the three studies were recruited from five countries only. Expanding our studies to organizations in a range of contexts and countries can enrich the validity and generalizability of the construct. Second, whereas our research assumes all QTB dimensions are of equal weight, future research can look into alternative weighting schemes. Finally, it is widely recognized that projects vary (Shenhar 2001). Analyzing whether different scales are required for various project types and industries was outside the scope of this paper. Future studies can investigate this issue and further develop a dataset of norms for the QTB scale for different types of projects and across industries.

References

Ansar, A., Flyvbjerg, B., Budzier, A., Lunn, D.: Should we build more large dams? The actual costs of hydropower megaproject development. Energy Policy (2014) (forthcoming), http://dx.doi.org/10.1016/j.enpol.2013.10.069i

Anderson, J.C., Gerbing, D.W.: Predicting the performance of measures in a confirmatory factor analysis with a pretest assessment of their substantive validities. Journal of Applied Psychology 76(5), 732 (1991)

Ashurst, C., Doherty, N.F., Peppard, J.: Improving the impact of IT development projects: The benefits realization capability model. European Journal of Information Systems 17, 352–370 (2008)

Australian Department of Finance and Deregulation, "Assurance Review Process - Lessons Learned Benefits Realisation Management". Australian Government (2012)

Breese, R.: Benefit realisation management: Panacea or false dawn? International Journal of Project Management 30, 341–351 (2012)

Brislin, R.W.: Translation and content analysis of oral and written material. In: Triandis, H.C., Berry, J.W. (eds.) Methodology, vol. 2. Handbook of Cross-Cultural Psychology, pp. 349–444. Allyn and Bacon, Boston (1980)

Carr, C., Kolehmainen, K., Mitchell, F.: Strategic investment decision making practices: A contextual approach. Management Accounting Research 21(3), 167–184 (2010)

Cats-Baril, W., Thompson, R.: Managing Information Technology Projects in the Public Sector. Public Administration Review 55(6), 559–566 (1995)

Churchill, G.A.: A Paradigm for Developing Better Measures of Marketing Constructs. Journal of Marketing Research 16, 64–73 (1979)

Crosby, P.B.: Quality Is Free: The Art of Making Quality Certain. New American Library, New York (1979)

Doran, G.: There's a SMART way to write management's goals and objectives. Management Review 70(11), 35–36 (1981)

Flyvbjerg, B., Holm, M.K.S., Buhl, S.L.: How (In)accurate are demand forecasts in public works projects? Journal of the American Planning Association 71(2), 131–146 (2005)

Flyvbjerg, B.: Policy and planning for large-infrastructure projects: problems, causes, cures. Environment and Planning B: Planning and Design 34, 578–597 (2007)

Flyvbjerg, B.: Quality control and due diligence in project management: Getting decisions right by taking the outside view. International Journal of Project Management 31, 760–774 (2013)

Gershon, P.: Review of Australian Government's Use of Information and Communication Technology. Department of Finance and Deregulation, Australian Government, Parkes (2008)
Gimbert, X., Bisbe, J., Mendoza, X.: The role of performance measurement systems in strategy formulation processes. Long Range Planning 43(4), 477–497 (2010)
Grushka-Cockayne, Y., De Reyck, B., Degraeve, Z.: An integrated decision-making approach for improving European air traffic management. Management Science 54(8), 1395–1409 (2008)
Henderson, J., Ruikar, K.: Technology implementation strategies for construction organisations. Engineering, Construction and Architectural Management 17(3), 309–327 (2010)
Hinkin, T.R.: A Brief Tutorial on the Development of Measures for Use in Survey Questionnaires. Organizational Research Methods 1(1), 104–121 (1998)
Irani, Z., Love, P.: The propagation of technology management taxonomies for evaluating investments in information systems. Journal of Management Information System 17, 161–177 (2001)
Jenner, S.: Realising Benefits from Government ICT Investment—AFool's Errand? Academic Publishing, Reading (2009)
Jørgensen, M., Halkjelsvik, T., Kitchenham, B.: How does project size affect cost estimation error? Statistical artifacts and methodological challenges. International Journal of Project Management 30(7), 839–849 (2012)
Kahneman, D.: Thinking, Fast and Slow, 1st edn. Farrar, Straus and Giroux, New York (2011)
Kahneman, D., Tversky, A.: Prospect theory: an analysis of decision under risk. Econometrica, 263–291 (1979)
Kerlinger, F.N.: Foundations of behavioral research. CBS Publishing Limited, Japan (1986)
Lechner, C., Floyd, S.W.: Group influence activities and the performance of strategic initiatives. Strategic Management Journal 33(5), 478–495 (2012)
Li, S., Rao, S.S., Ragu-Nathan, T.S., Ragu-Nathan, B.: Development and validation of a measurement instrument for studying supply chain management practices. Journal of Operations Management 23(6), 618–641 (2005)
Lin, C., Pervan, G.: The practice of IS/IT benefits management in large Australian organizations. Information Management 41, 13–24 (2003)
Linderbaum, B.A., Levy, P.E.: The development and validation of the Feedback Orientation Scale (FOS). Journal of Management 36(6), 1372–1405 (2010)
Locke, E.A., Latham, G.P.: Building a practically useful theory of goal setting and task motivation: a 35-year odyssey. American Psychologist 57(9), 705–717 (2002)
Mihm, J.: Incentives in new product development projects and the role of target costing. Management Science 56(8), 1324–1344 (2010)
Miles, M.B., Huberman, M.A.: Qualitative Data Analysis: An Expanded Sourcebook, 2nd edn. Sage Publications, Newbury Park (1994)
Norris, G.D.: Post-investment appraisal. In: Willcocks, L. (ed.) Investing in Information Systems: Evaluation and Management, pp. 193–223. Chapman & Hall, Boca Raton (1996)
Nicholson-Crotty, S., Theobald, N., Nicholson-Crotty, J.: Disparate Measures: Public Managers and Performance-Measurement Strategies. Public Administration Review 66(1), 101–113 (2006)
Nunnally, J.C.: Psychometric theory, 2nd edn. McGraw-Hill, New York (1978)
Pellegrinelli, S., Bowman, C.: Implementing strategy through projects. Long Range Planning 27, 125–132 (1994)
Papadakis, V.M.: Strategic investment decision processes and organizational performance: an empirical examination. British Journal of Management 9(2), 115–132 (1998)

Raghunathan, S.: Impact of information quality and decision-maker quality on decision quality: a theoretical model and simulation analysis. Decision Support Systems 26(4), 275–286 (1999)

Reeves, C.A., Bednar, D.A.: Defining quality: alternatives and implications. Academy of Management Review 19(3), 419–445 (1994)

Sendjaya, S., Sarros, J.C., Santora, J.C.: Defining and measuring servant leadership behaviour in organizations. Journal of Management Studies 45(2), 402–424 (2008)

Shapira, Z., Shaver, J.M.: Confounding changes in averages with marginal effects: How anchoring can destroy economic value in strategic investment assessments. Strategic Management Journal (2013), doi:10.1002/smj.2165

Shenhar, A.J.: One size does not fit all projects: exploring classical contingency domains. Management Science 47, 394–414 (2001)

Verbeeten, F.H.M.: Performance management practices in public sector organizations. Accounting, Auditing & Accountability Journal 21(3), 427–454 (2008)

Ward, J., Daniel, E.: Benefits Management: Delivering Value from IS & IT Investments. John Wiley & Sons Inc., New York (2006)

Weick, M., Guinote, A.: How long will it take? Power biases time predictions. Journal of Experimental Social Psychology 46, 595–604 (2010)

ZengJ, A.M., Smith, N.J.: Application of a fuzzy based decision making methodology to construction project risk assessment. International Journal of Project Management 25(6), 589–600 (2007)

Zikmund, W.G., Babin, B.J., Carr, J.C., Griffin, M.: Business Research Methods, 8th edn. South-Western Cengage Learning (2010)

Zwikael, O., Smyrk, J.R.: A general framework for gauging the performance of initiatives to enhance organizational value. British Journal of Management 23, S6–S22 (2012)

Zwikael, O., Smyrk, J.R.: Project Management for the Creation of Organisational Value. Springer, London (2011)

Software Project Team Selection Based on Enterprise Social Networks

Panos Fitsilis, Vassilios Gerogiannis, and Leonidas Anthopoulos

Technological Educational Institute of Thessaly, 41110 Larissa, Greece
{fitsilis,gerogian,lanthopo}@teilar.com

Abstract. Project management is a well understood management method, widely adopted today, in order to give predictable results to complex problems. It is based on the assumption that unique undertakings require flexible organizational structures and different skill sets, in order to be implemented successfully. It is evident that matching the required, by each project, skills with the skills offered from the available human capital, is not a trivial process. This need of building high performing, effective teams with the appropriate skill set, is quite important since it is a prerequisite for successful project completion. Furthermore the recent shift to agile project management, together with the fact that projects today do not only require technical skills but also behavioral and contextual skills increases the importance of project team selection. In this paper, we present the ONSOCIAL system, a system that attempts to capitalize knowledge accumulated over social networks and to support the project team selection process. For achieving this objective, the ONSOCIAL system is using information stored in social network systems, analyses the stored profiles and offers recommendations for candidate team members.

Keywords: Project team selection, expert location, social networks, project management

1 Introduction

Projects are by definition temporary and unique endeavors performed in a dynamic environment [1]. According to Antoniadis [2] the definition of project is "*The management of transient, dynamic and complex adaptive systems/agents, so as to deliver the expected change within certain parameters that are established by seemingly ordered and stable environments*". According to this definition a shift is required towards to the behavioral paradigm that emphasizes on issues such as collaborative organization structures, team selection, behavioral skills and competence management.

Cohen and Bailey [3] defined a team as "*a collection of individuals who are independent in their tasks, who share responsibility for outcomes, who see themselves and are seen by others as an intact social entity embedded in one or more larger social systems, and who manage their relationships across organizational*

M. Gen et al. (eds.), *Industrial Engineering, Management Science and Applications 2015*,
Lecture Notes in Electrical Engineering 349, DOI: 10.1007/978-3-662-47200-2_40

boundaries". As such project teams require of its members extensive collaboration not only within the project team but as well as with other agents/actors from whom the team is dependent. As a result organizations need to select team members that are able to work effectively in collaborative environments. This implies that organizations need to manage competences in a systematic way. Generally speaking, *competence management* is the process in which organizations manage the competencies of the *corporation*, the *groups* and the *individuals* [4].

Today, social networks and specifically enterprise social networks can be used to seek and retrieve individual technical competences [5] but more importantly to mine behavioral competences that are implicitly found within the social networks and can be used in the context of project management in order to answer questions, such as [6]:

- How to locate experts with specific technical and behavioral skills?
- How to select the best performing project team that can collectively perform a given task?

The above questions correspond to the expert location and expert team formation problem respectively, research topics that attract significant research lately [7]. For addressing the above research questions, we present an approach which is based on four different disciplines and attempts to combine them in a single model/system. Namely these disciplines are: software project management, enterprise social networks, ontologies-knowledge management and fuzzy recommendations. The work presented is currently implemented in ONSOCIAL research project (http://onsocial.teilar.gr/). The remainder of the paper is structured as follows. Section 2 provides very briefly the background of the project: project management methodologies, competence management, social networks, and ontological engineering. Section 3 presents the proposed system use cases, architecture and how the system can be used in order to derive results and finally, section 4 presents conclusions and our future research plans.

2 Related Work

According to Project Management Body Of Knowledge (PMBOK) [8] developed by Project Management Institute (www.pmi.org), human resource management is one out of ten subject areas that a project manager should manage for successfully executing a project. Within this subject area four are the main processes:

1) Planning Human Resource management focusing on defining the needs of a project in terms of required roles, skills, responsibilities and job descriptions.
2) Acquiring the project team which implies locating and selecting project team members the
3) Developing the project team and
4) Managing the project team

Both processes of HR planning and team acquisition are related with the problem of expert location. As such, expert location is the task of identifying a set of candidates that have a specific set of skills in order to perform a set of project tasks.

The expert finding problem was introduced by Craswell, et al. [9]. and since then it has generated a lot of interest and a number of different approaches have been developed. The common questions are: a) what constitutes expertise evidence and b) how to identify expertise when documents are analysed. Traditional expert finding systems quite often use information retrieval techniques by estimating the probability that a candidate satisfies a set of skills required by a vacant project position. The probability calculation is done either by modelling the expertise of each candidate, a strategy called user profile-centric approach, or by identifying document relevant to a query and then establishing the link between the document and the expert [10, 11].

Further, the success of a project depends on the behavioural skills of the team members, on how efficiently team members are communicating, collaborate and work closely. This trend, to give emphasis on skills and behavioural issues is followed by many international organizations and groups. As an example, we can refer to International Project Management Association (IPMA) that developed IPMA Competence Baseline (ICB) [12] that defines a set of behavioural skills such as leadership, liability and motivation; self-control; self-confidence; relaxation; openness; creativity; outcome orientation; efficiency; consultation; negotiation; conflicts and crisis; reliability, etc. Similarly, agile methods like Scrum [13] and XP [14] give emphasis on people and interactions rather than process and tools, on daily communication and customer satisfaction.

Software Project Management is a knowledge intensive process that can benefit substantially from ontology development and ontology engineering. Since the emergence of the "semantic web" ontologies attract much attention and more specifically on issues such as on ontology representation, machine-processing, and ontology sharing. At the same time, a large number of ontologies have been developed attempting to address various software engineering aspects, such as requirements engineering, components reuse, domain modeling, human resource management, etc. In the problem area of "expert location", Mochol and Oldakowski [15] have developed a Human Resource Ontology to support the Recruitment Process. This HR-ontology is mainly based on the German version of the HR-XML standard developed by the HR-XML Consortium[1]. HR-XML is a library of more than 75 interdependent XML schemes which define data components for various HR transactions, as well as options and constraints governing the use of these components. The ontology was adapted so the Industry sub-ontology to use both the German Classification of the Industry Sector (WZ2003) and the North American Industry Classification System (NAICS). For the occupational classification the German Occupation Codes (BKZ) are used along the Standard Occupational Classification (SOC) System.

Dorn et al. [16] developed an ontology containing concepts of HRM for two different projects: a meta-search engine for job searching in job portals and for a university competence management system. In their paper, they presented the requirements derived from these two projects and the design of the ontology is described. This ontology is characterized by its integration of job descriptions, concepts for evaluating competencies on different levels and evidences for

[1] `http://www.hr-xml.org/`

competencies. The definition is also aligned to HR-XML for the definition of competence profiles. Schmidt and Kunzmann [17] describe an ontology that integrates concepts from skill management and learning. Their approach is similar to Dorn et al. [16], however, they do not consider job descriptions composed of required competencies. In the same paper a holistic view of the organization and of the HR problem area is presented showing how the ontology is covering different organizational aspects. Gómez-Pérez et al. developed an ontology for HR based on standards [18]. More specifically, they developed SEEMP ontology as a common "language" as a controlled vocabulary in order to describe the details of a job posting and the CV of a job seeker. SEEMP Ontology is composed of thirteen sub-ontologies: *Competence, Compensation, Driving License, Economic Activity, Education, Geography, Job Offer, Job Seeker, Labor Regulatory, Language, Occupation, Skill and Time*. An elaborate study on the software project management ontologies available today in the literature is presented in [19].

The relations between team members and with possible team candidates are commonly represented by a social network, with links between nodes to indicate easiness of communication. According to Lappas et al [20] the problem of identifying a team of experts that can communicate effortlessly is defined as the team formation problem. The team formation problem has been used either as an integer linear programming problem, or by using fuzzy sets [21].

A number of methodologies have been used in order to measure the personality traits of an individual or of team performance. The Myers-Briggs Type Indicator (MBTI): MBTI is probable the most frequently employed personality assessment tool. It matches the personality of an individual with job descriptions [22]. Similarly Kolbe Conative Index (KCI) was used to predict team's performance [23].

Social Network Analysis (SNA) offers a powerful tool that provides the means for analyzing informal networks [24]. Social network analysis aims to understand the relationships between people, groups, organizations, and other types of social entities [25], and has been used extensively in fields such as sociology and management. A social network is modeled as a graph with nodes representing the individual actors in the network and ties representing the relationships between the actors.

The concept of using data collected in social networks has become increasingly popular is a number of different disciplines such as marketing, political science, economy, organizational science, human resource management etc. Coleman's [26] definition of social capital considers it as one of the resources that somebody can use besides other resources such as human or cultural capital (their own skills and expertise), physical capital (equipment), or economic capital.

As such, in the area of human resource management, finding the available social capital and using it within a group e.g. the group of employees in a large organization or within informal networks is quite interesting lately. However, in order to use social capital we should find ways to define and measure it more formally. Bourdieu's [27] definition for measuring social capital is based on number of network connections somebody can effectively mobilize and on the volume of the capital (economic or other) possessed by each of those to whom he is connected.

Cross et al. [28] identified consistent behaviours within "communities of practice" for encouraging information flow, knowledge reuse, and learning amongst employees.

Interestingly enough they have identified patterns on behaviour such as: Central Connectors, brokers, peripheral players, and external connectors. For example, Central connects have the maximum number of direct connections in a network, playing a critical role in the community effectiveness. Oppositely, peripheral players are loosely connected or isolated employees that represent under-utilized resources of a community. Brokers are employees that are disproportionately important for community connectivity. This type of people can be found on the shortest path between two different community members.

Abbasi et al. [29] in their attempt to measure social capital proposed a set of metrices as assets in networks. More specifically they proposed:

- *Ego network size index* for measuring the diversity of contacts. Ego network size index is measures as the degree centrality within the social network and it is the number of different contact each node of the social network has.
- *Ego average tie strength index* for measuring the tie strengths. Tie strength is measuring the frequency of communication or collaboration between two actors. The sum of tie strengths of an author is the total number of collaborations an actor has.
- *Ego betweenness centrality index* for measuring the structural position. Betweenness centrality is a measure of the potential of an employee to control the communication flow within the group of people. When an actor is consider central within a group this implies that this individual exhibits bridging behavior. Ego-betweenness centrality is computed as the sum of the times an actor is within the shortest path between other actors of the network.
- *Individual effectiveness index* for measuring brokerage and diversity. Effectiveness of an actor is defined as the number of non-redundant (not connected) contacts. It is measured as the number of contacts that an actor has, minus the average number of links that each contact has to other contacts of the specific actor.
- *Contact status (power)* for measuring the embeddedness of resources. Power is measured either a) by degree centrality: actors who have more ties to other actors may be advantaged positions, or b) by degree centrality: measures the immediate ties that an actor has, or the ties of the actor's neighbors, rather than indirect ties to all others, or c) by betweenness centrality where an actor is being in a favored position to the extent that the actor falls on the geodesic paths between other pairs of actors in the network [30].

A recent trend in the area of social networks technology is an attempt made by large number of researchers and especially companies to use social networks as collaboration management systems within enterprises. This results in a new category of systems so called Enterprise Social Networking (ESN) systems. According to Riemer and Richter [31], ESN is "the application of Internet platforms for relationship building and short message exchanges in the context of workplace communication". Examples of ESN are Yammer system (http://www.yammer.com), Jive software (http://www.jivesoftware.com), IBM Connections (http://www-03.ibm.com/software/products/en/conn), Socialcast (http:// socialcast.com), etc.

3 Our Approach

It is evident that project management success depends heavily from human capital management. This factor is even more important in the case of software project management since software development is based mainly on intellectual effort.

Our approach, as it was implemented with ONSOCIAL system is facilitating the process of project team selection by combining in an intuitive manner ontologies, social networks, and social network analysis.

In our proposed approach [32], four are the basic steps involved:

a) **Building the enterprise corpus** of data based on data extracted from commercial social networks. The basic scenario used is that all enterprise employees will authorize ONSOCIAL system to crawl on their own data collecting profiles and established social links in order to build an *enterprise social corpus*. According to this view, the enterprise social corpus is considered to be the sum of the profiles of all employees.

b) **Modelling the competences** and the knowledge required for successfully implementing a software project using custom developed software project management ontology. The ontology has been developed using protégé system.

c) **Analyzing the social network** for extracting behavioral competences, by calculating specific network metrics, such as

d) **Locating and recommending experts/project team members.** This is a two steps process that includes definition of project team requirements and querying the enterprise data corpus for finding matches for the empty project positions.

In Figure 1, we present the high level use case diagram of the ONSOCIAL system. The system is interacting with five different actors, namely:

- the *Employees* that *Donate Own Social Network Data* to the enterprise data corpus,
- the *Administrator* that initiates the use case of *Constructing the Enterprise Data Corpus* (operation of the crawler) and the use case *Analyze Social Network* than computes network metrics
- the *Human Resource (HR) manager* that is *constructing/maintaining the ontology* and finally
- the *Project Manager* that defines the *project team requirements* as a set of competences/skills in accordance with the ontology and *searches within enterprise data corpus* for possible matches.

The ONSOCIAL Crawler is the component of the system that constantly runs in the background, with the task of gathering information from the supported social networks, translating it to an internal common representation and storing it in a form that is appropriate for querying. The *crawler* is using an incremental approach for building the group's corpus of data, containing all profiles of the persons belonging to the specific group that have accepted to invitation issued by ONSOCIAL system. The internal representation is based on HR-XML in order to maximize the data interoperability with similar systems.

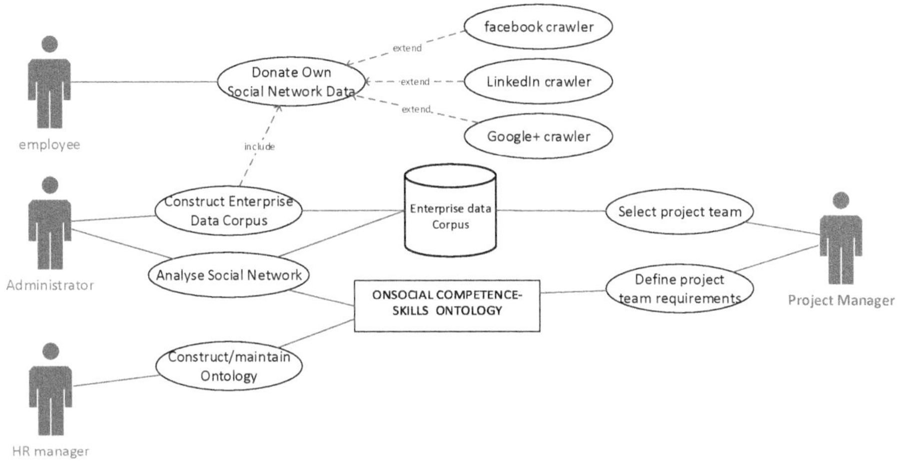

Fig. 1. ONSOCIAL system use cases

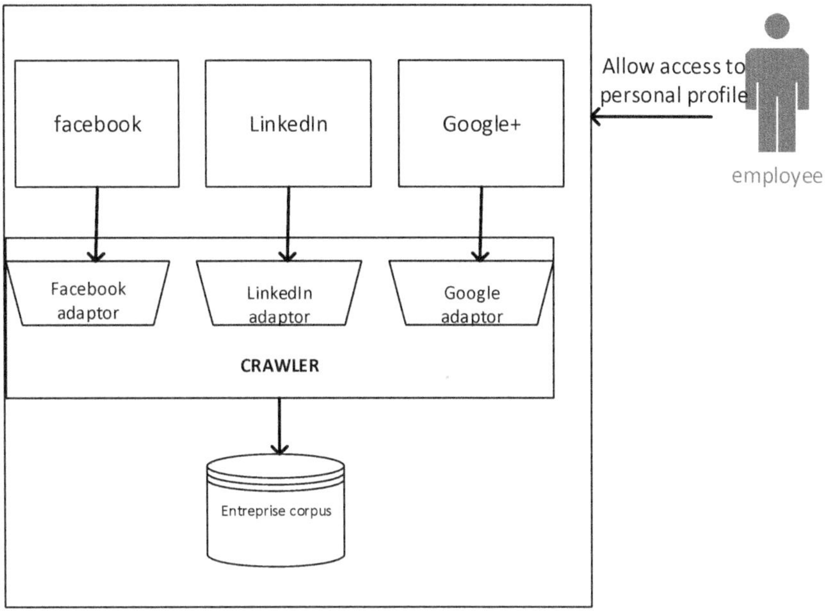

Fig. 2. ONSOCIAL building enterprise corpus

The ontology was developed using Protégé system and it models competencies which a human resource may possess for the performance of a project activity. Furthermore, the ontology includes concepts that enable to associate employees with sector of activity where they acquired the competences, seniority level, location where they worked etc.

A brief description of ontology classes is presented in the Table 1.

Table 1. ONSOCIAL ontology main classes

Class name	Description
Education Type	The type of education of a person (High School, Graduate School, College)
Gender	The gender of a person (Male or Female)
Job	A job may be related to a company and one or more projects.
Person	A person may be seen as a current, past or potential software project team member.
Project	The project type
School	A school where a person was educated.
School Type	The type of a school (e.g. College, University etc.)
Seniority	The seniority on a technical skill. A seniority refers to a technical skill, may be required by a task and has a seniority level.
Seniority Level	A level of seniority (Junior, Moderate or Senior)
Set of tasks	A set of tasks related to a project.
Task	A task is part of a project. It may belong to a set of tasks, it may require some competence, it may require some seniority and it may be performed at one or more locations.
Team	A project team has one or more persons as members and is related to a project.
Competence	A competence is a technical, social or other type of skill.
Education	The education of a person. The class is defined in HRM Education Ontology and extended in OnSocial Ontology. Education can have a begin date, an education type, be related to a school, be on an education field and have an education level.
Education Field	An education field. HRM Education Ontology provides a detailed categorization of education fields.
Education Level	HRM Education Ontology provides seven education levels from level 0 to level 6.
Location	A geographical location (Continent, Country or Region)
Sector	HRM Economic Activity Ontology provides a detailed categorization of industry sectors.

When searching for team members, a project manager or a representative of human resource department can ask for all profiles that match the profile needed. In this matching, we are interested in determining whether or not a profile satisfies a set of requirements. Since the requested requirements could be mandatory or optional the matching in many cases is partial. This implies that a fuzzy approach for the recommendation is the most appropriate in our case [21].

4 Conclusions

ONSOCIAL system explores the process of project management human resource management by assisting large distributed project organizations in their efforts to recruit project team members by using social networks profiles and associations between members. For achieving this objective ONSOCIAL system is using social networks to create an enterprise corpus of data, data are classified according to an enhanced competence ontology and finally the system is using a recommendation engine for proposing candidate team members. Currently ONSOCIAL system is implementing the “Project Team Selection” use case, while development is under way for implementing more project management use cases, such as “supplier selection”.

Acknowledgments. This research has been co-financed by the European Union (European Social Fund - ESF) and Greek national funds through the Operational Program "Education and Lifelong Learning" of the National Strategic Reference Framework (NSRF) - Research Funding Program: ARCHIMEDES III. Investing in knowledge society through the European Social Fund.

References

1. Rose, K.H.: A Guide to the Project Management Body of Knowledge (PMBOK® Guide)—Fifth Edition. Project Management Journal 44(3) (2013)
2. Antoniadis, D.N.: Managing complexity in project teams. PhD Thesis, Loughborough University, UK (2009)
3. Cohen, S.G., Bailey, D.E.: What makes teams work: Group effectiveness research from the shop floor to the executive suite. Journal of Management 23(3), 239–290 (1997)
4. Braun, S., Kunzmann, C., Schmidt, A.: People tagging and ontology maturing: Toward collaborative competence management. In: From CSCW to Web 2.0: European Developments in Collaborative Design, pp. 133–154 (2010)
5. Fitsilis, P., Gerogiannis, V., Anthopoulos, L., Kameas, A.: Using Social Network Analysis for Software Project Management. In: Proceedings of the 2009 International Conference on the Current Trends in Information Technology. IEEE Press, Dubai (2009)
6. Fitsilis, P., Gerogiannis, V., Anthopoulos, L.: Project team selection based on social networks. In: IEEE International Technology Management Conference (ITMC), pp. 1–4 (2014)
7. Lappas, T., Liu, K., Terzi, E.: A survey of algorithms and systems for expert location in social networks. In: Social Network Data Analytics, pp. 215–241. Springer US (2011)
8. Rose, K.H.: A Guide to the Project Management Body of Knowledge (PMBOK® Guide)—Fifth Edition. Project Management Journal 44(3) (2013)
9. Craswell, N., de Vries, A., Soboroff, I.: Overview of the trec 2005 enterprise track. In: The 14th Text REtrieval Conference Proceedings, TREC 2005 (2006)
10. Balog, K., Fang, Y., de Rijke, M., Serdyukov, P., Si, L.: Expertise Retrieval. Foundations and Trends in Information Retrieval 6(2-3), 127–256 (2012)
11. Balog, K., Azzopardi, L., de Rijke, M.: A language modeling framework for expert finding. Information Processing & Management 45(1), 1–19 (2009)
12. International Project Management Association: ICB-IPMA competence baseline version 3.0. International Project Management Association, Nijkerk, The Netherlands (2006)
13. Schwaber, K.: Agile Project Management with Scrum. Microsoft Press (2004)
14. Beck, K.: Planning extreme programming. Addison-Wesley Professional (2001)
15. Mochol, M., Oldakowski, R., Heese, R.: Ontology based Recruitment Process. In: GI Jahrestagung (2), pp. 198–202 (2004)

16. Dorn, J., Naz, T., Pichlmair, M.: Ontology development for human resource management. In: 4th International Conference on Knowledge Management, pp. 109–120 (2007)
17. Schmidt, A., Kunzmann, C.: Towards a Human Resource Development Ontology for Combining Competence Management and Technology-Enhanced Workplace Learning. In: Meersman, R., Tari, Z., Herrero, P. (eds.) OTM 2006 Workshops. LNCS, vol. 4278, pp. 1078–1087. Springer, Heidelberg (2006)
18. Gómez-Pérez, A., Ramírez, J., Villazón-Terrazas, B.: An ontology for modelling human resources management based on standards. In: Apolloni, B., Howlett, R.J., Jain, L. (eds.) KES 2007, Part I. LNCS (LNAI), vol. 4692, pp. 534–541. Springer, Heidelberg (2007)
19. Fitsilis, P., Gerogiannis, V., Anthopoulos, L.: Ontologies for Software Project Management: A Review. Journal of Software Engineering and Applications 7(13), 1096 (2014)
20. Lappas, T., Liu, K., Terzi, E.: Finding a team of experts in social networks. In: KDD, pp. 467–476 (2009)
21. Gerogiannis, V., Rapti, E., Karageorgos, A., Fitsilis, P.: A Fuzzy Linguistic Approach for Human Resource Evaluation and Selection in Software Projects. In: Proceedings of the 2015 International Conference on Industrial Engineering and Operations Management, Dubai, United Arab Emirates, UAE (2015)
22. Hammer, A.L., Huszczo, G.E.: Teams. Consulting Psychologists Press (1996)
23. Kolbe, K.: The Conative Connection: Acting on Instinct. Addison-Wesley (1997)
24. Cook, K.S., Whitmeyer, J.M.: Two approaches to social structure: Exchange theory and network analysis. Annual Review of Sociology, 109–129 (1992)
25. Bird, C., Gourley, A., Devanbu, P., Gertz, M., Swaminathan, A.: Mining email social networks. In: Proceedings of the 3rd International Workshop on Mining Software Repositories (2006)
26. Coleman, J.S.: Social capital in the creation of human capital. American Journal of Sociology 94(1), 95–120 (1988)
27. Bourdieu, P.: The forms of capital. In: Woolsey Biggart, N. (ed.) Readings in Economic Sociology, pp. 280–291. Blackwell, Oxford (1986)
28. Cross, R., Laseter, T., Parker, A., Velasquez, G.: Using social network analysis to improve communities of practice. California Management Review 49(1) (2006)
29. Abbasi, A., Wigand, R.T., Hossain, L.: Measuring social capital through network analysis and its influence on individual performance. Library & Information Science Research 36(1), 66–73 (2014)
30. Hanneman, R.A., Riddle, M.: Introduction to social network methods. University of California, Riverside (2005)
31. Riemer, K., Richter, A.: S.O.C.I.A.L. - Emergent Enterprise Social Networking Use Cases: A Multi Case Study Comparison (2012), http://hdl.handle.net/2123/8845
32. Fitsilis, P., Gerogiannis, V., Anthopoulos, L.: Project team selection based on social networks. In: 2014 IEEE International Technology Management Conference (ITMC), pp. 1–4 (2014)

Study on the Agriculture Information Cloud Architecture and Application

Peng Qing[1], Ming Ye[1,2], and Guangyuan Liu[1,2]

[1] School of Computer and Information Science, Southwest University, Chongqing400715, China
[2] College of Electronic and Information Engineering, Southwest University, Chongqing 400715, China
zmxym@swu.edu.cn

Abstract. Cloud computing refers to subscription-based, fee-for-service utilization of computer hardware and software over the Internet. The model is gaining acceptance for business information technology (IT) applications because it allows capacity and functionality to increase on the fly without major investment in infrastructure, personnel or licensing fees. This paper propose a agriculture information cloud solution to provide for agriculture information computing of a mass of heterogeneous, distributed giant collection of information systems. At the same time, this paper also proposes agriculture information cloud architecture that is optimized. This cloud architecture mainly is constituted of hardware layer, data service, virtual layer, middleware layer, application layer. To effectively manage agriculture information platform, this paper also provides insights on market-based resource management strategies that encompass both customer-drive service and application management. Furthermore this paper finally evaluate the performance of CPU and internet-based service workload in the environment of proposed agriculture information cloud architecture and management service strategies. Experiments shows that the proposed agriculture information cloud infrastructure and management service strategies is effective and essential for large-scale agriculture information resource computing of a mass of heterogeneous, distributed giant collection of information systems which carry environmental resources, agricultural production, agricultural products, rural synthesize management and farmers' education and training.

Keywords: Cloud computing, Virtualization computing, Agriculture information, Cloud platform architecture.

1 Introduction

Cloud computing is a model for enabling convenient, on-demand network access to a shared pool of configurable computing resources (e.g., networks, servers, storage,

M. Gen et al. (eds.), *Industrial Engineering, Management Science and Applications 2015*,
Lecture Notes in Electrical Engineering 349, DOI: 10.1007/978-3-662-47200-2_41

applications, and services) that can be rapidly provisioned and released with minimal management effort or service provider interaction. This cloud model promotes availability and is composed of five essential characteristics, three service models, and four deployment models. Although the concept of cloud computing is emerging a few years ago, but the accumulation of technology is concerned, already up to two to three decade[1-5]. In essence, cloud computing is to a large number of distributed and high-cost computer, networking equipment and storage resource management to support a variety of applications over the Internet to provide customers with high-quality low-cost service[6-9].

Do not confuse cloud computing with the term data center, as it typically sits on top of the latter. Viewing the cloud as logical rather than a physical, you can see it object describes it better.

As learned from past events, computing in its purest form, has changed hands multiple times. First from near the beginning when mainframes were predicted to be the future of computing. Indeed mainframes and large scale machines were built and used, and in some circumstances are used similarly today[14-15]. The trend, however, turned from bigger and more expensive, to smaller and more affordable commodity PCs and servers.

Most of our data is stored on local networks with servers that may be clustered and sharing storage. This approach has had time to be developed into stable architecture[10-13], and provide decent redundancy when deployed right. A newer emerging technology, cloud computing, has shown up demanding attention and quickly is changing the direction of the technology landscape. Whether it is Google's unique and scalable Google File System, or Amazon's robust Amazon S3 cloud storage model, it is clear that cloud computing has arrived with much to be gleaned from.

In dealing with the abstract term, "the cloud", it is easy to misunderstand what makes up the structure and function. The basic function is what comes from "the cloud". This is primarily output, however, not only. Input is what makes the cloud tick. Do not confuse cloud computing with the term data center, as it typically sits on top of the latter. Viewing the cloud as logical rather than a physical, you can see it object describes it better. The proposed benefits of using cloud architecture to build applications on top of, are immense. SmugMug, a successful photography community, saved an estimated 21 Million dollars in 2007 by using Amazon S3 cloud storage. Organizations and individuals can benefit from mass computing and storage, provided by large companies with stable and strong cloud architectures[16-18]. On the other hand, companies that desire to build massive, scalable environments, utilizing virtualization and cloud computing will increase their future margin of success greatly.

As new technologies emerge, they often tend to build on the success of previous developments. Cloud computing and storage, benefit from years of development and testing of large scale infrastructure. The most important take away is cloud storage is for everyone and every organization. From big to small, groups to individual, the use of grid infrastructure can be deployed for maximum return and efficiency[23].

2 Relate Work

2.1 Agriculture Information Cloud Solution

"95","15" and "115" period, Chinese government the Ministry of Science and Technology, Ministry of Education, the Chinese Academy of Sciences, Ministry of Agriculture and other departments, local governments have implemented a large number of rural agricultural information projects and works in rural agricultural information infrastructure[19-21], and agriculture-related information resources[24]. Information service system has begun to take shape. By the end of 2009, most of china area have established physical network architecture and information service organizations of the covering over provincial, municipal, county, township, village, 97 percent of prefectures and cities, 77 percent of the county set up a special agricultural information management and service institutions, 50 percent of the township was set up information kiosks. In the first half of 2010,most of the provinces and cities in china have formed a mass of heterogeneous, distributed giant collection of information systems which carry environmental resources, agricultural production, agricultural products, rural synthetize management and farmers' education and training, and so on large-scale database/knowledge repository. In the three-year(2007-2010)period, remote sensing and geographical information of the National rural , plants and animals, agricultural supply chain information, multimedia information resources, business data can be growed from 1012 (in bytes) to 1014 (in bytes) magnitude, the total reach hundreds of terabytes. To effectively improve the rural construction of agricultural information and application level, the Ministry of Science and Technology, the National Remote Office and the Ministry of Industry jointly have launched the demonstration province construction work of National Rural agricultural information and has gradually expanded to all Chinese era.

In accordance with the principle of "platform to move upward, the service extension", we make unified planning the top-level design of the Chongqing Municipal agricultural and rural information service platform to overcome the common key technologies and the underlying kernel to build the agricultural and rural information cloud service platform system. To support cloud computing technology, we mainly rely on government departments and enterprises of information system resources. To build the ecological chain of agricultural and rural information services, we integrated construction of agricultural and rural cloud services platform and support the construction of the sub-platform by science and technology correspondent information. At same time, this agriculture information cloud platform also support science and technology correspondent for "a machine with two wings".

We make unified planning and design of agricultural rural information of the top-level framework of the integrated service platform to develop the agricultural and rural information cloud services platform [22] for top-level design specifications and technical standards to build agricultural rural information service platform architecture. "Agricultural Information notify", "12396" three agriculture service hotline, rural party members and distance education platform integrated together to form a "cloud". This agriculture information cloud reference solution see Fig.1.

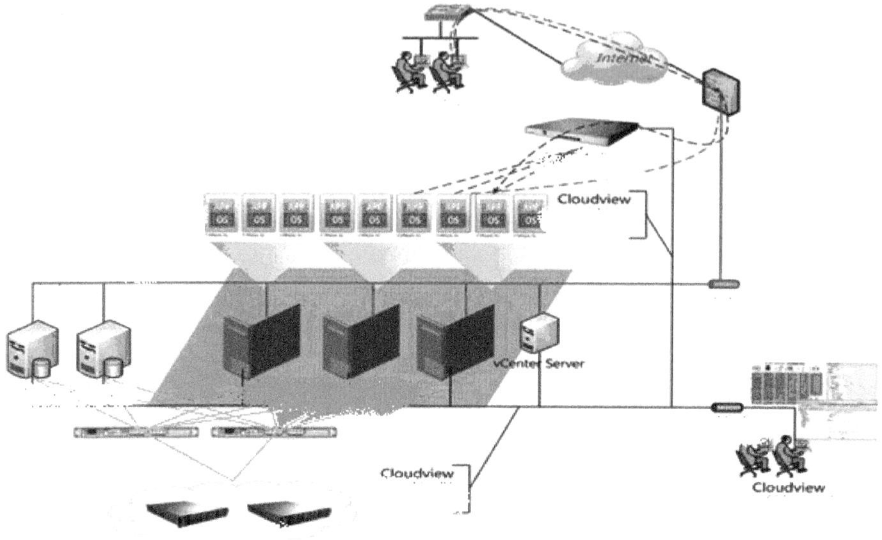

Fig. 1. Agriculture Information Cloud Solution

2.2 Agriculture Information Cloud Architecture

The underlying cloud architecture and environment must be designed and implemented to be flexible and scalable. Unfortunately, the history of designing, delivering, and managing very large scale federally-developed systems does not offer many success stories to build upon. If not implemented properly, the government risks significant challenges and costs in migrating information to different technologies as the third-party vendor upgrades its processing and storage environment. If this type of upgrade is managed in-house, resident IT professionals can more readily manage migration and harmonization of data, users, and processes. But the procedures that a cloud vendor executes in scaling its environment is managed without the input of its customers, and may change services the customer requires. Customers require the ability to increase bandwidth, speed, and response time. In some cases, the cost to move data to a cloud infrastructure has proven quite costly in terms of time (bandwidth) and money. Some cloud users have resorted to using physical media to send data in order to expedite changes in their business needs. All IT systems are subject to regular considerations of their life spans and durability. The question arises as to how long certain technologies will exist. The need for interoperability, the ability to switch providers, compatibility between vendors, and avoidance of migration issues will all be demanded by customers. As the government approaches the cloud, this will likely be very problematic as there are no universal, ratified standards within the industry or through NIST that would govern these issues. So it is important for users how to design the integrated solving strategy for cloud computing platform.

In this paper, we present infrastructure and management strategy for cloud computing to meet users' demand. Cloud computing platform integrated in together by certain relatively independent subsystems, include data server, computing server, business intelligence application. Users can make use of cloud computing infrastructures for their serve, for example, data storage, information sharing and remote computing and so on.seeFigure.2.

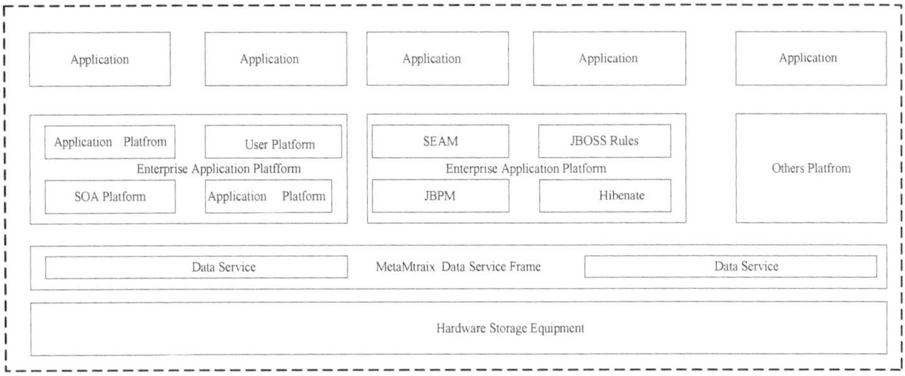

Fig. 2. Cloud Platform Architecture

3 Performance Evaluation

We conduct extensive experiments on the performance of our proposed cloud Infrastructure. Fig.3shows the Agriculture information Cloud environment setup used for performance evaluation. The Agriculture information Cloud contains 30 personal computers (PCs) with 1 master node and 32 execution nodes located across 3 student computer laboratories in the Department of Computer Science and Software Engineering. This setup demonstrates that the Agriculture information Cloud is able to present a unified resource to the users/brokers by harnessing computing resources located physically apart in the 3 laboratories. Synthetic workloads are created by utilizing trace data of HPC applications. The experiments utilize 238 reservation requests.

Experiment result shows that CPU workload and internet workload of this work model of the agriculture information cloud are acceptant. But this work model in the standard distribution is unable to perform well in heterogeneous loud computing infrastructure. Though agriculture information cloud is successful in homogeneous computing environments, experimental observations reveal that the homogeneity assumptions of reduce can cause wrong and often unnecessary speculative execution in heterogeneous environments, sometimes resulting even worse performance than with speculation disabled. This valuation and performance results demonstrate that Cloud execution management systems need to be designed to handle heterogeneity that is present n workloads, applications, and computing infrastructure.

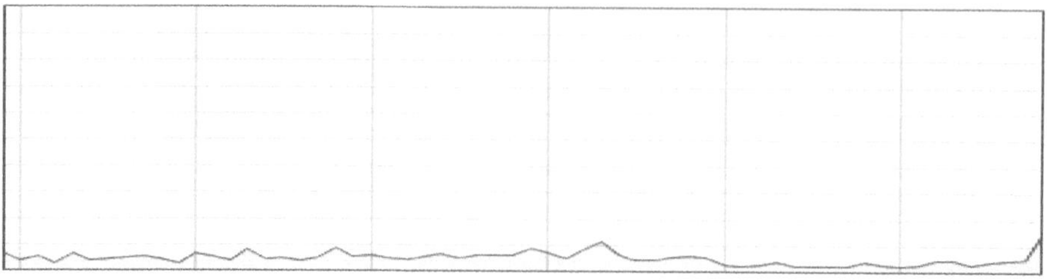

(a)

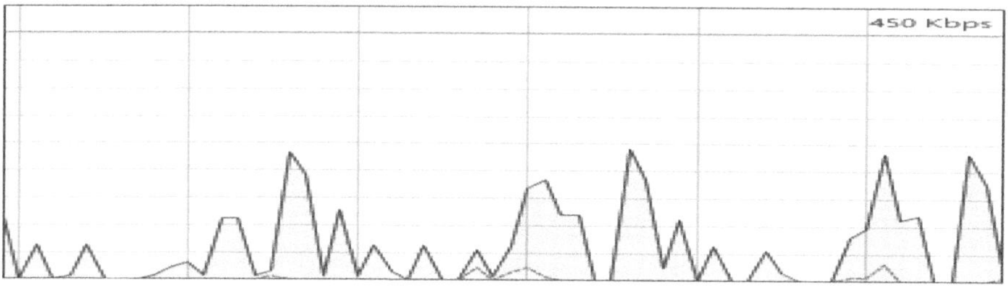

(b)

Fig. 3. Cloud platform workload (a)CPU workload(b)Network workload

4 Conclusions and Outlook

Cloud computing refers to subscription-based, fee-for-service utilization of computer hardware and software over the Internet. As Clouds are designed to provide services to external users, providers need to be compensated for sharing their data resources, computing and storage capabilities. In this paper, we have proposed architecture for market-oriented allocation of resources within Clouds. We have also presented a management service strategy to manage agriculture information cloud platform. Moreover, we have evaluated this agriculture information cloud system based on this infrastructure and system architecture. The result proves CPU workload and internet workload are acceptant.

In particular, we have presented various Cloud efforts in practice from the market-oriented perspective to reveal its emerging potential for the creation of third-party services to enable the successful adoption of Cloud computing. In addition, we need programming environments and tools that allow rapid creation of Cloud applications. Data Centers are known to be expensive to operate and they consume huge amounts of electric power. As Clouds are emerging as next-generation data centers and aim to support ubiquitous service-oriented applications, it is important that they are designed to be energy efficient to reduce both their power bill and carbon footprint on the environment. To achieve this software systems level, we need to investigate new techniques for allocation of resources to applications depending on quality of service expectations of users and service contracts established between consumers and providers. Finally, we need to address regulatory and legal issues, which go beyond technical issues. Some of these issues are explored in related paradigms such as Grids and service-oriented computing systems. Hence, rather than competing, these past developments need to be leveraged for advancing Cloud computing. Also, Cloud computing and other related paradigms need to converge so as to produce unified and interoperable platforms for delivering IT services as the 5th utility to individuals, organizations, and corporations.

Acknowledgements. The authors thank the editors and the anonymous reviewers for their helpful comments and suggestions. This work is supported by"the Fundamental Research Funds for the Central Universities (XDJK2012C023).

References

1. Schweitzer, E.J.: Reconciliation of the cloud computing model with US federal electronic health record regulations. Journal of the American Medical Informatics Association: JAMIA 19(2), 161–165 (2012)
2. Lin, Y.-K., Chang, P.-C.: Evaluation of system reliability for a cloud computing system with imperfect nodes. Systems Engineering 15(1), 83–94 (2012)
3. Pieter, G., Paul, G.: Supporting the internet-based evaluation of research software with cloud infrastructure. Software and Systems Modeling 11(1), 11–28 (2012)
4. Rajkumar, B., Shin, Y.C., Srikumar, V.: Cloud computing and emerging IT platforms: Vision, hype, and reality for delivering computing as the 5th utility. Future Generation Computer Systems-The International Journal of Grid Computing-Theory Methods and Applications 25(6), 599–616 (2009)
5. Burroughs, J.M.: What users want: assessing government information preferences to drive information services. Government Information Quarterly 26, 203–218 (2009)
6. Jaeger, P.T., Lin, J., Grimes, J.M.: Cloud computing and information policy:computing in the policy cloud. Journal of Information Technology & Politics, 269–283 (2008)
7. Jaeger, P.T., Lin, J., Grimes, J.M., Simmons, S.N.: Where is the cloud Geography, economics, environment, and jurisdiction in cloud computing. First Monday 14(5) (2009)
8. Jaeger, P.T., McClure, C.R., Bertot, J.: The e-rate program and libraries and library consortia. Information Technology and Libraries 24(2), 57–67 (2005)
9. Wen, G.Y., Marshak, A., Cahalan, R.F., et al.: 3-D aerosol-cloud radiative interaction observed in collocated MODIS and ASTER images of cumulus cloud fields. Journal of Geophysical Research-Atmospheres 112(D13) (2007)
10. Grossman, R.L., Gu, Y.H., Sabala, M., Zhang, W.Z.: Compute and storage clouds using wide area high performance networks. Future Generation Computer Systems-The International Journal of Grid Computing Theory Methods and Application 25(2), 179–183 (2009)
11. Buyya, R., Abramson, D., Venugopal, S.: The grid economy. Proceedings of the IEEE 93(3), 698–714 (2005)
12. Stuer, G., Vanmechelena, K., Broeckhovea, J.: A commodity market algorithm for pricing substitutable grid resources. Future Generation Computer Systems 23(5), 688–701 (2007)
13. Venugopal, S., Chu, X., Buyya, R.: A negotiation mechanism for advance resource reservation using the alternate offers protocol. In: Proc. 16th Int. Workshop on Quality of Service, Twente, The Netherlands (June 2008)
14. Van Looy, B., Gemmel, P., Van Dierdonck, R. (eds.): Services Management: An Integrated Approach. Financial Times, Prentice Hall, Harlow, England (2003)

15. Schneider, B., White, S.S.: Service Quality: Research Perspectives. Sage Publications, Thousand Oaks (2004)
16. Yeo, C.S., Buyya, R.: Integrated risk analysis for a commercial computing service. In: Proc. 21st IEEE Int. Parallel and Distributed Processing Symposium, IPDPS 2007, Long Beach, USA (March 2007)
17. Crouhy, M., Galai, D., Mark, R.: The Essentials of Risk Management. McGraw-Hill, New York (2006)
18. Moeller, R.R.: COSO Enterprise Risk Management: Understanding the New Integrated ERM Framework. John Wiley and Sons, Hoboken (2007)
19. Barham, P., Dragovic, B., Fraser, K., Hand, S., Harris, T., Ho, A., Neugebauer, R., Pratt, I., Warfield, A.: Xen and the art of virtualization. In: Proc. 19th ACM Symposium on Operating Systems Principles, SOSP 2003, Bolton Landing, USA. Integrated Solving Strategy for Cloud Computing. Applied Mechanics and Materials, vols. 44-47 (October 2011)

Enhanced Value Stream Mapping: Potentials and Feasibility of IT Support through Manufacturing Execution Systems

Markus Philipp Roessler[1,*], Ina Kleeberg[1], Moritz Kreder[1], Joachim Metternich[1], and Klaus Schuetzer[2]

[1] Institute of Production Management, Technology and Machine Tools, Technische Universitaet Darmstadt, Otto-Berndt-Str. 2, 64287 Darmstadt, Germany
[2] Lab. for Computer Integrated Design and Manufacturing, Methodist University of Piracicaba, Rod. Luis Ometto (SP 306), 13451-900 Santa Barbara d'Oeste, Brazil
roessler@ptw.tu-darmstadt.de

Abstract. Value stream mapping (VSM) is widely spread both in research and industrial application. In some industries it is established as a standard approach and starting point for improving production systems. This is because of the relatively distinct ease of use and the high effectiveness of the method as well as the standardized visualization. Nowadays production systems are more and more characterized by a high volatility concerning volume and variants of produced parts. Here the method of VSM reaches its limits because of its snapshot character and the rather manual data collection. In order to cope with this volatility quick data collections and the possibility to view historical data regarding e.g. inventory, process characteristics or energy consumption as a factor for sustainability are getting more important. Thus in this paper the use of manufacturing execution systems (MES) is examined to support the main elements of VSM: the value stream analysis (VSA) leading to current state maps and the value stream design (VSD) of future states. Further potentials and inhibitions are discussed using a case study.

1 Introduction

Due to a constant increase of international competition through globalization of markets and a dynamic environment, producing companies are forced to increase their productivity and this at lower costs while rising quality. In order to strengthen their competitiveness and to cope with global mega trends, modern production systems have to be lean and flexible. To put this into practice, the methods of lean manufacturing are most suitable. [1]

The main principles of lean manufacturing, which bases on the work of Ohno at Toyota [2], are the increase of production flow, the elimination of waste and variability as well as the involvement of employees. [3] By eliminating waste and reducing

* Corresponding author.

M. Gen et al. (eds.), *Industrial Engineering, Management Science and Applications 2015*, Lecture Notes in Electrical Engineering 349, DOI: 10.1007/978-3-662-47200-2_42

lead time, the productivity of companies, work in progress levels and the flexibility regarding volume and variants can be systematically improved [4]. In order to meet the claim of a holistic improvement on production system level, the method of VSM was developed as a tool for the implementation of lean thinking. The focus of this method is not on optimizing single (production) processes, but rather entire process chains from receipt to delivery of goods. [5] An optimization of material flow systems is of crucial importance for the performance of production systems [6]. VSM is widely accepted in industry and already established as a standard approach especially in the automotive sector [7]. Projects to improve production systems require a comprehensive analysis of the existing value stream. Information on shop floor level needs to be collected, analyzed and interpreted. The mapping as well as the design process is typically performed with paper and pen [8], the representation of complex production systems however is difficult to achieve by this method [9].

Nowadays MES are increasingly in use for the control and optimization of production, supporting order control on the one hand and data acquisition and analysis on the other hand. A large amount of production relevant data is available in such systems, so it seems reasonable to use this data to support value stream improvement projects. Originally MES don't offer the possibility to help mapping value streams because the systems are not able to picture material flows and are centrally controlled. So the two principles can't be combined that easy at the first glance. [10] However, lean production systems can benefit from MES in order to react quickly to problems and deviations in production [11]. In this paper inhibitions for the combination of both MES and VSM are uncovered, potentials are pointed out and discussed.

2 Methodology and Study Design

To show the potentials of a connection between VSM and MES it's recommendable to discuss a procedure model for VSM in the first step, see therefore figure 1. Such a model can be used to describe the sequence of actions in industrial improvement projects using VSM.

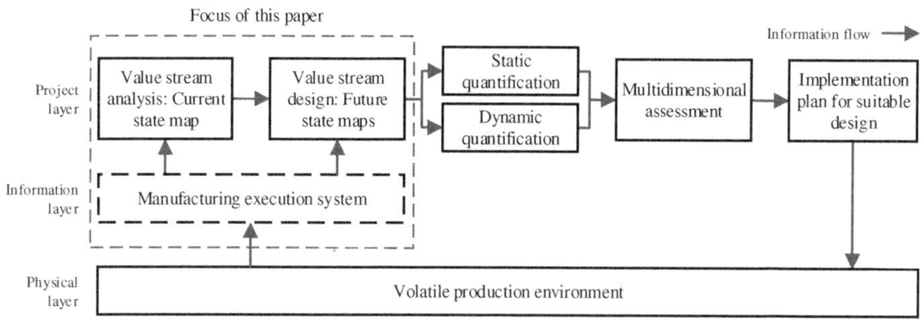

Fig. 1. Process model for VSM enriched by an information layer with MES

This procedure model shows an enhanced VSM approach based on the one developed by Rother [12]. It was set up with the goal of developing multiple future states and includes quantification and an assessment step of developed future states before actual implementation. As pointed out here, information about the real production environment contained in MES can be potentially used in two ways. At first information out of MES could be used to derive a current state map, or at least facilitate the data collection process. A second potential field of application is the support of the design phase.

3 Feasibility of VSM Supported by MES

3.1 VSA Supported by MES

This paragraph describes synthesis and evaluation of a feasibility study, whether MES are capable of providing support during a VSA or if they can be used to automatically generate a current state map out of the included data. Before proceeding to examine this issue, it is necessary to prepare a brief overview of VSA followed by the feasibility assessment.

VSA is an instrument often used as a starting point for the implementation of a lean production concept [9]. Objectives of this step are to transparently represent the production process, to visualize areas of wastes and to ensure a system-wide perspective, whereby a good understanding between material and information flow is given. Another key aspect is the symbolism, whereby a common language is created. [8]

VSA is the representation of an existing value chain, including the material and information flow. A systematic process representation is the result. Each production process can be mapped using VSA; afterwards it can be optimized by an appropriate VSD [3]. The requirements of the customers and the suppliers are integrated into the analysis, whereby the circuit is mapped completely from incoming orders to the dispatch of the products [8]. The method can reveal deficits in production planning and control, as well as manufacturing processes.

In terms of creating current state value stream maps out of MES data, the information content required for a VSA has to be identified and structured in a first step. Afterwards the core functions of MES will be analyzed and examined, to what extent they are able to provide the necessary data for VSA. Currently a vast amount of providers for MES exist on the market. Therefore a market analysis was performed to identify the main functions which are available by the most systems and then to assign them to the main tasks of MES [13]. 75 systems were identified, which fulfil the defined functions. Table 1 visualizes these functions and the VSA data demand.

Table 1. Functions of MES and data demand for VSA

MES function	Data demand for value stream analyses
Planning of production program	product variants, sales per product, work days per year, no. of identical parts, planned production time, demand per period, raw material
Steps and sequence	product variants
Detailed production planning	production processes, sales per product, parts produced, delivery period, work days per year, working time per day, no. of workers per shift, no. of alternative equipment, process time, lot size, no. of identical parts, planned production time per production process, no. of produced parts, no. of scrap parts, no. of rework parts, no. of variants per production process, transport relations, storage location, storage capacity, raw material, replenishment lead time
Order management	no. of produced parts, customer, demand per period, delivery period, real delivery time, business processes, no. of orders, information flow relations
Production control	product variants, production processes, real delivery time, work days per year, no. of workers per shift, process time, no. of identical parts, no. of alternative equipment, changeover time, lot size, no. of produced parts, no. of scrap parts, no. of rework parts, transport relations, storage location, storage capacity, stock count, demand per period, raw material, replenishment lead time
Tool management	equipment, no. of alternative equipment
Maintenance	planned production time
Production logistics	transport relations, storage location, storage capacity, demand per period
Material management	product variants, no. of produced parts, identical parts, lot size, storage location, storage capacity, demand per period, raw material, replenishment lead time
HRM	work days per year, working time per day, no. of workers per shift
Plant and machine data collection	process time, machining time, lot size, changeover time, machine operating time, availability, planned operating time, no. of produced parts, no. of scrap parts, no. of rework parts, stock count
Analysis / reporting	no. of produced parts, real delivery time, process time, machining time, changeover time, planned production time, planned operating time, real operating time, availability, transport relations
Quality management	no. of produced parts, customer, delivery period, real delivery time, no. of produced parts, no. of scrap parts, no. of rework parts, error rate, delivery reliability
Optimization	production processes
Documentation and history	sales per product, no. of produced parts, demand per period, real delivery time, transport relations, replenishment lead time, error rate, delivery reliability, no. of employees
Work flow	business processes, information flow relations
Tracking / tracing	equipment

The survey shows that it is theoretically possible to collect most of the necessary indicators by using commercially available MES. In practice the actual data collection process is not that easy, e. g. when it comes to mapping stocks. A permanent inventory management enables access to accurate inventory data of each stock at any time. To this end, an existing warehouse management is required which accurately picks up the data. [14] However, the book inventory represents only the stock levels derived out of inventory and order transactions. Schoensleben describes that despite accurate inventory management deviations of the actual and set inventory may occur [15].

The theoretical influences to the stock amount are quite high, in order to cope with these influences highly sophisticated systems and sensor implementations are necessary. In practice this is mostly linked to high financial investments.

Another main inhibition regarding the data demand of a current state map is the detection and mapping of in- and outbound orders like orders from customers or orders for raw material to suppliers. This level of information is usually only available at ERP (enterprise resource planning) systems because they include the corporate processes of purchasing and distribution.

The most striking observation to emerge from the feasibility study is that it is easy to receive information about internal (production) processes and orders, but not about the actual material flow between the process steps. Here MES don't offer the function of parameterizing logistics functions, like if processes are connected via push principle, FIFO-lanes (first-in-first-out), supermarkets, directly via flow, etc.

The results of this investigation showed that currently MES are capable of supporting existing current state maps with up-to-date information, the automatic and complete generation of current state maps is not possible with the main functions provided by available MES solutions.

3.2 VSD Supported by MES

In this section the results of another feasibility study are shown, if MES are capable of providing support during a VSD process or if they can be used to automatically generate a future state map on the basis of included data. Before proceeding to examine this issue, a brief report about VSD will be given, followed by the assessment.

VSD is the second step of VSM. Based on the current state of the production system, one or multiple better future states have to be derived [8]. Today's production systems attempt to align with the Toyota production system (TPS). This production system is considered as being an ideal system for the total elimination of waste. The basic idea of the TPS is the same as for VSD: the elimination of all non-value adding activities. The advantage of VSM is the capability of detecting waste and its causes [8]. The latter can be located precisely by using a VSA and the offered graphical representation. The main focus of VSD is on the observation and improvement of the entire production system. Waste identified at individual process steps can be resolved by single lean methods such as "single minute exchange of die" (SMED) or "poka yoke" (error avoidance strategies).

For improving production systems with the VSD approach eight design guidelines are available [8]. By following these guidelines, one or multiple future state maps can be derived out of the current state map. In order to find out if MES are able to offer recommendations to improve a production system the defined MES functions (see table 1) were analyzed and compared to the needed information for fulfilment of the VSD guidelines. The results of the feasibility study are shown in table 2.

The results show that currently MES are capable to support decision makers in the process of designing improved value streams with additional data. For example data to draw tact diagrams and visualize capacities as well as live data allocated to acquisitions at processes (like changeover times, availability, process times, etc.) are available out of MES databases. However, the same inhibitions regarding the missing function of mapping material flows exist, that makes it impossible to automatically create suggestions by the system. Recommendations to rearrange or link processes cannot be realized by the systems.

Table 2. Guidelines for VSD, resulting data demand and evaluation of feasibility based on MES main functions

Design guideline	Data demand for design guideline	Feasibility
Orientation at customer's tact	cycle times, process times, lot sizes, no. of identical parts, no. of identical processes, working time, demand per period, customer's tact, capacities, machining times, changeover times, produced parts, OEE	Data available for creating tact diagram, visualization using cycle times and customer's tact not included in most MES. Suggestions could be made from the system to raise the working time or improve effectiveness of processes. Visualization of over-capacity theoretically possible
Process integration	work sequences for products, layout and position of processes, changeover times, availability, process times, size of products	A similar work sequence for all products over the processes is required. MES mostly don't use graphical representations of factory layout, or product sizes, so a suggestion for process integration only could be theoretically made considering process characteristics
FIFO connection	no. of variants, volumes for each variant, process times, container sizes, transport times, working time, demand, lot sizes, changeover times, availability	Calculation of buffer sizes and theoretical feasibility is possible, technical suggestions, however are difficult to derive because of layout issues. Use of simulation software plugin possible
Kanban control	no. of variants, volumes for each variant, process times, container sizes, transport times, working time, demand per period, lot sizes, changeover times, availability, WIP, no. of kanbans, safety stock	Calculation of buffer sizes and theoretical feasibility is possible, technical suggestions, however are difficult to derive because of layout issues. When already using kanban, the dynamic kanban and volume calculation is possible to integrate into MES
Pacemaker process	material flow types between processes	Last supermarket has to be set as pacemaker. Identifying material flow types is not possible in existing systems. Pacemakers have to be allocated manually
Smoothing of production	lot sizes, process and machining times, changeover times, no. of identical processes, EPEI indicator, working time per day	Possibility of manual sequencing is integrated by most MES, however a module for automatic order sequencing only is partially available. MES can suggest to lower the EPEI indicator to gain flexibility
Production order entity	lot sizes, working time, demand per period, customer's tact, container sizes	Volume released for each production order can be calculated using customer's tact and container sizes. Such information is distributed from an ERP system which is not scope of the MES
Bottleneck control	cycle times, process times, lot sizes, no. of identical parts / processes, working time, demand, customer's tact, capacities, machining times, changeover times, produced parts, OEE	Calculation of bottlenecks is possible using existing data. A suggestion could be offered by the MES to release orders directly at this process. However technical properties and material flow types are to be considered as well, which currently is not possible

3.3 General Requirements

In addition to both feasibility studies some further, general requirements could be identified out of interviews with two automotive value stream experts. These requirements serve as prerequisites and have to be fulfilled by the used MES in order to be able to contribute purposefully to VSM:

- Timeliness: The collected data must be up to date and assigned to one time period.
- Correctness: Data used for creating a value stream map or derive improvement potential has to be accurate, otherwise decisions based on that can be faulty.
- Completeness: It must be possible to map the chosen value stream completely. If process steps are missing or have no connection to the MES, VSM cannot be applied correctly. This also requires interfaces to processes and storages.

4 Case study

In order to validate the theoretical assessment of the combination of MES and VSM a case study was conducted at the Center for industrial Productivity (CiP) at the Technische Universitaet Darmstadt. The CiP represents a real industrial production environment with machining and assembly areas for educational and staff training purposes. The production system produces pneumatic cylinders in four variants for various applications and is controlled using the MES Hydra 8 from MPDV Mikrolab. This system offers the standard functions (see table 1). The fulfilment of the general requirements was given during this study.

4.1 Practical Feasibility of VSA Supported by MES

In a first step the necessary criteria (see table 1) could be identified out of the implemented system and a current state map could be set up manually. The MES does not offer a possibility to create and visualize current state maps automatically.

The case study has confirmed the findings and underlines the identified barriers of the feasibility study (see paragraph 3.1). These are mainly inhibitions regarding the acquisition of present inventory and top-level order information. The other barrier of mapping material flows is also present. To map all processes and material flows it is required to set up a complete order network in the MES. The possibility to add order networks does already exist, only the links must be added e. g. with the help of a BOM (bill of material). The information about the control principle of the respective production segment must be communicated to every connection to map the whole value stream; this problem could be solved with little effort. Information regarding the supplier management would have to be provided by the ERP system. Figure 2 visualizes the derived, manually created current state map and shows easy and difficult realization areas. A bright dotted pattern is used where no problems occurred during data acquisition, grey diagonal stripes are used when additional acquisition effort (above functions of standard MES) is necessary and dark dotted is used for areas, where it was not feasible to support the concept. This only is the case for in- and outbound orders on top-level, because the used ERP offers no suitable interface for acquiring necessary data.

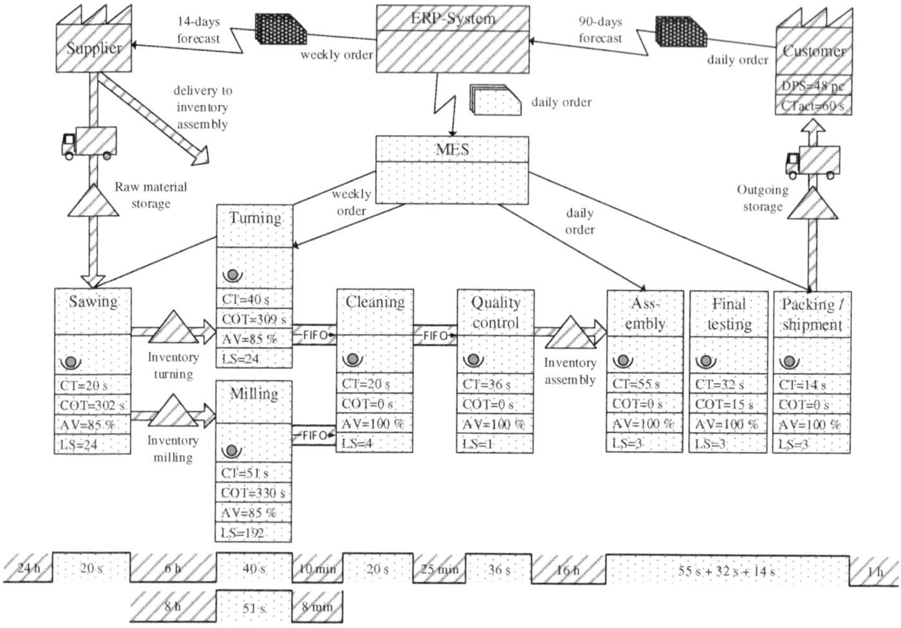

Fig. 2. Current state map of the CiP value stream

The results of the theoretical as well as the practical feasibility study can be summarized with following aspects:

- Necessary data can be automatically derived, if system requirements (necessary functions of the used MES) are covered.
- Mapping of the whole information flow is only possible with additional effort.
- To fulfil the general requirements mapping of inventory requires also additional effort (e. g. through automatic stock-taking using sensor systems).
- All data have to be up-to-date, correct and complete.
- Necessary interfaces have to be available (e. g. machine connections, order reporting systems).
- Visual representation of current state maps has to be implemented via interface or directly with plugins into MES.

4.2 Practical Feasibility of VSD Supported by MES

Based on a previously set up current state map during the case study the design principles (see table 2) were worked through and the necessary information for making explicit statements were tried to identify out of the MES. If no automatic functions are included in the MES this step has to be performed manually.

The case study confirms the inhibitions mentioned in paragraph 3.2. A majority of the criteria for implementing the design guidelines could be covered by the existing data base of the MES. Important elements like the visualization of a tact diagram, the possibility for selecting an appropriate pull-system and the selection of a pacemaker process are not implemented in the MES. There is no direct possibility in the system to define the material flow between the processes. This could be solved as stated in the previous paragraph. Additional inhibitions to the theoretical ones identified could not be discovered.

The findings of this study show that for performing a VSD using MES still some (technical) extensions and enhancements are to be implemented. It is recommended to integrate simulation software into MES in order to solve some of the stated issues.

5 Conclusions

In this paper the use of manufacturing execution systems (MES) in value stream mapping (VSM) projects is examined to give practitioners direct access to information about the current value stream or to automatically derive suggestions to improve the production system.

The work is split in two parts, on the one hand the examination of value stream analysis (VSA) and on the other hand the process of value stream design (VSD). The first feasibility assessment has shown that in principle MES would be capable of carrying out VSA, with some modifications of the system. However as a first step it is recommendable to use data available in MES to enrich existing current state maps with live information. An automatic deduction of current state maps would be linked to investments in software adaptions (e. g. new plugins) as well as intelligent sensor systems and an enhancement of interfaces to the ERP level.

The results regarding the feasibility of MES to automatically suggest value stream improvements or to deduce future state maps are not that encouraging. Automatic recommendations regarding most of the design principles suggested during VSD currently cannot be made by MES because of missing functions, interfaces and visualization options. The creation of complete future state maps with MES is also not possible at the present time. Furthermore, it seems fundamentally questionable whether some of those listed barriers can be solved and thus a complete value stream can be designed by current MES.

The feasibility studies have shown that despite the shortcomings at a material flow level, MES can contribute to improvements through creation of transparency at production processes. The current state mapping of value streams can be partially supported by MES and their valuable live data.

Acknowledgements. The authors want to express their kindest thanks to the German BMBF and the Brazilian CNPq, which enabled and supported this work by funding.

References

1. Abele, E., Reinhart, G.: Zukunft der Produktion: Herausforderungen, For-schungsfelder, Chancen. [Future of production: Challenges, fields of research, opportunities]. Carl Hanser Verlag, Muenchen (2011)

2. Ohno, T.: How the Toyota production system was created. Japanese Economic Studies 4, 83–101 (1982)
3. Womack, J.P., Jones, D.T., Roos, D.: The machine that changed the world. The story of lean production. Free Press, New York (1990)
4. Oeltjenbruns, H.: Organisation der Produktion nach dem Vorbild Toyotas. [Organization of production systems according to the Toyota model]. Dissertation, Shaker Verlag (2000)
5. Bokranz, R., Landau, K.: Produktivitaetsmanagement von Arbeitssyste-men: MTM Handbuch. [Productivity management of work systems: MTM field book]. Schaefer-Poeschel Verlag, Stuttgart (2006)
6. Negahban, A., Smith, J.S.: Simulation for manufacturing system design and operation. Literatur review and analysis. Journal of Manufacturing Systems 33(2), 241–261 (2014), doi:10.1016/j.jmsy.2013.12.007
7. Brueggemann, H., Mueller, P.: Digitales Wertstromdesign. [Digital value stream mapping]. In: Rabe, M. (ed.) Advances in Simulation for Production and Logistics Applications, ASIM Simulation in Production and Logistics, Berlin, October 01-02, vol. 13, pp. 575–584. Fraunhofer IRB Verlag, Stuttgart (2008)
8. Rother, M., Shook, J.: Learning to see: Value-stream mapping to create value and eliminate muda. Lean Enterprise Institute (1999)
9. Braglia, M., Carmignani, G., Zammori, F.: A new value stream mapping approach for complex production systems. International Journal of Production Research 44(18-19), 3929–3952 (2006)
10. Steger-Jensen, K., Hvolby, H.H.: Review of an ERP system supporting lean manufacturing. In: Koch, T. (ed.) Lean Business Systems and Beyond. IFIP, vol. 257, pp. 67–74. Springer, Boston (2008)
11. Yang, T., Lu, J.C.: The use of a multiple attribute decision-making method and value stream mapping in solving the pacemaker location problem. International Journal of Production Research 49(10), 2793–2817 (2011)
12. Roessler, M.P., Metternich, J., Abele, E.: Learning to see clear. Quantification and multidimensional assessment of value stream mapping alternatives considering variability. Business and Management Research 3(2), 93–109 (2014), doi:10.5430/bmr.v3n2p93
13. Verein Deutscher Ingenieure: Manufacturing execution systems (MES). VDI 5600 Part 1 (2007)
14. ten Hompel, M., Heidenblut, V.: Taschenlexikon Logistik. [Encyclopedia logistics]. Springer, Berlin (2011)
15. Schoensleben, P.: Integrales Logistikmanagement. [Integrative logistics management]. Springer, Berlin (2007)

Automatic Oil Palm Detection and Identification from Multi-scale Clustering and Normalized Cross Correlation

Teerawut Wong-in, Tonphong Kaewkongka, Nagul Cooharojananone, and Rajalida Lipikorn

Machine Intelligence and Multimedia Information Technology Laboratory
Department of Mathematics and Computer Science, Faculty of Science
Chulalongkorn University, Bangkok, Thailand
Teerawut.Wo@student.chula.ac.th,
{Tonphong.K,Nagul.C,Rajalida.L}@chula.ac.th

Abstract. Oil palm cultivation is one of the most important occupation in South East Asia. Since oil palm plantations cover the wide range of area, it is difficult to count the population of oil palm manually. This paper presents a new method to detect and identify oil palms over the plantation from aerial images regardless of their sizes using features such as shape, size, and texture. The proposed method can handle the problem of identifying oil palms from an aerial image when oil palms are too close to each other which cause them to be detected as a single stand. The process consists of removing non-tree components from an image, distinguishing oil palms from other components, identifying individual oil palm, and counting the number of oil palms. In this paper, oil palm can be detected and distinguished from other components using ideal low-pass filter and normalized cross correlation. Then a proposed multi-scale clustering method and erosion are used to identify individual oil palm from a bush. The method was evaluated on a set of 21 aerial images taken over oil palm plantations from different regions of Thailand by attaching a digital camera to a remote airplane. The experimental results reveal that our proposed method can detect most of the oil palms with the accuracy as high as 90%.

Keywords: Normalized cross correlation, multi-scale clustering, oil palm identification, template matching.

1 Introduction

Oil palms are economic plants that grow in warm and humid weather with fertile soils close to rivers. The oil palm is a plant that has only trunk and leaves without any branches. It has semispherical shape. Palm oil becomes raw material in many product and is needed in great demand which causes the development of oil palm plantation to increase rapidly during the past decades. This situation can lead to supply over demand situation. Thus, the exploration of oil palm population can be very helpful in forecasting of palm oil supply and demand. However, the areas of oil palm plantation are very wide and some plantations are located in the rainforest areas which make

M. Gen et al. (eds.), *Industrial Engineering, Management Science and Applications 2015*, Lecture Notes in Electrical Engineering 349, DOI: 10.1007/978-3-662-47200-2_43

them difficult to be explored. Even though, it is suggested that 143 oil palms per hectare is the best density but many plantations may plant at more or less density.

This paper thus proposes a new method to detect and identify oil palms from aerial images using special features of oil palms and the planting pattern. Once the oil palms can be identified, the proposed method can estimate the population of the oil palms by converting an image to the actual measurement.

This paper is organized as follows: section 1 presents a brief introduction about oil palms, section 2 gives some literature reviews, section 3 proposes a new method to detect and identify oil palms from an aerial image, section 4 presents the experimental results, and section 5 gives discussion and conclusions.

2 Literature Reviews

From the past decades, several methods on tree detection have been proposed in the literatures. Chen, et al. [1] proposed the scalable approach to tree detection in large urban landscapes using aerial LiDAR data by projecting their 3D point cloud onto a 2.5D depth image and then applying the Robert cross gradient operator and seed- based region growing segmentation to segment objects from an image. They then used a random forest classifier [2] to classify objects as tree or non-tree. They report- ed that precision and recall rates are over 95%. This proposed method still has prob- lem about discriminating building edges and trees. Another method was proposed to classify 3D aerial LiDAR height data using a supervised parametric classification algorithm with the mixture of Gaussian models [3]. This method used five features for classification: normalized height, height variation, multiple returns, luminance, and intensity. They observed that using more features produced better results in some cases, but not always.

Arasato, et al. [4] proposed a method to detect individual palm tree (Arecaceae family) in the Amazon rainforest using high resolution image classification. They used real color composition with region growing algorithm for segmentation and used four different unsupervised classifications. The drawback of this method is that the input images has to be color images. Yang, et al. [5] presented a method for tree detection from aerial imagery. The proposed method can automatically segment the region of trees, locate each individual tree and provide an estimate of its crowd size. It segmented tree/non-tree regions by applying a pixel-level tree classification based on a set of visual features and a partitioning algorithm. The classifier needs to be trained by only 1% of data to yield more than 90% accuracy. The problem with this method is that it used color as one of the features for classification.

Kattenborn, et al. [6] proposed the UAV-based photogrammetric point clouds for single palm tree detection using pouring algorithms. Their proposed method yielded high accuracy but it needed color images as input. Malek et al. [7], on the other hand, proposed to use SIFT to extract a set of keypoints and then use extreme learning classifier to detect palm trees. Then the shape of each tree is captured by merging the keypoints with active contour method based on level sets and using the local binary patterns to distinguish palm trees from other vegetations. Pouliot et al. [8] proposed to use local maximum refinement and delineation algorithm to automatically detect conifer trees which yielded 91% accuracy rate.

3 Proposed Methodology

Before proceeding to oil palm detection and identification, all color aerial images are converted to gray scale and histogram equalization is applied to adjust the contrast of the images. Then the proposed method, which consists of three main steps, is applied to each aerial image. These steps are noise filtering, non-oil palm elimination, and individual oil palm identification as shown in Fig. 1.

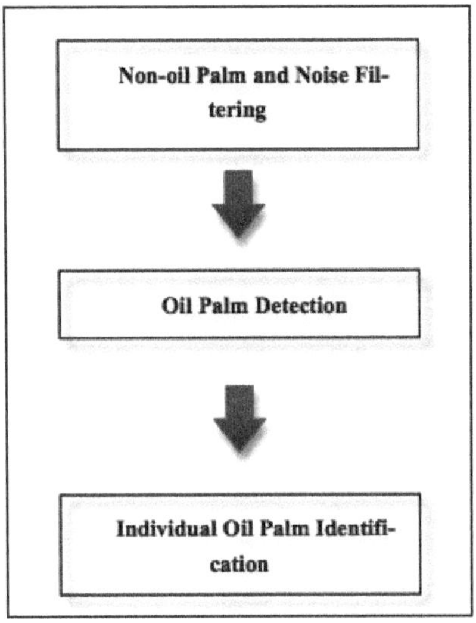

Fig. 1. The process of oil palm detection and identification

3.1 Noise Filtering

In this step, an image, $f(x,y)$, of size $M \times N$ is transformed to Fourier spectrum, $F(u,v)$, and Butterworth low-pass filter, $H(u,v)$, with cut-off, D, is applied to an image in order to remove speckle noise with high frequency spectrum. The size of the cut-off depends on the elevation of the airplane. The higher the elevation is, the larger the cut-off will be. The filtered image, $G(u,v)$, can be obtained from

$$G(u,v) = H(u,v)F(u,v) \tag{1}$$

where

$$H(u,v) = \frac{1}{1+(D(u,v)/D)^{2n}} \tag{2}$$

$$D(u,v) = \left[(u-\tfrac{M}{2})^2 + (v-\tfrac{N}{2})^2\right]^{\frac{1}{2}} \tag{3}$$

and $D(u,v)$ is the distance from any pixel to the origin.

A filtered image is converted back to spatial domain and Gaussian smoothing is applied to an image to remove young oil palms and other small bushes. The Gaussian smoothing function used in this step is calculated from.

$$G(x,y) = \frac{1}{2\pi\sigma^2} e^{-\frac{x^2+y^2}{2\sigma^2}} \tag{4}$$

where $\sigma = 1$. The size of Gaussian kernel also depends on the elevation of an airplane. The higher the elevation is, the smaller the Gaussian kernel will be.

3.2 Non-oil Palm Elimination

A gaussian blurred image is then converted to a binary image using adaptive local thresholding with the initial size of 3x3 mask and the threshold value is set equal to the average of intensities under the mask. This adaptive local thresholding is used to remove grass, weed, background, road, and fence from an image. After that, 8-connected component labeling is used to identify the remaining components in an image. An example of a binary image is shown in Fig. (2).

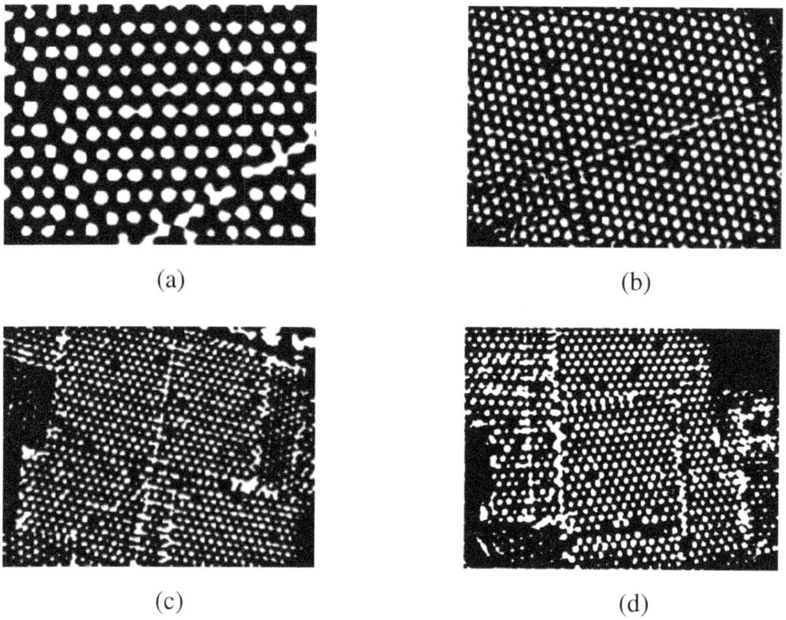

Fig. 2. The result images from using radius of 15, 20, 40, and 60 for ButterWorth low pass filter are shown in (a) – (d), respectively

However, there is still a problem with a binary image obtained from the previous step. Pond and swamp cannot be removed from an image because its intensity value is very close to the intensity of the trees. This problem can be solved by using normalized cross correlation template matching to match the texture of a template with each component as expressed in Equation (5).

$$R(x,y) = \frac{\sum_{i=0}^{m-1}\sum_{j=0}^{n-1}(T(i,j)-C(x+i,y+j))^2}{\sqrt{\sum_{i=0}^{m-1}\sum_{j=0}^{n-1}T(i,j)^2}.\sqrt{\sum_{i=0}^{m-1}\sum_{j=0}^{n-1}C(x+i,y+j)^2}} \tag{5}$$

where $T\ (i,j)$ represents a template of size $m \times n$ and $C(x,y)$ represents each component. If the size of a component is not equal to the size of a template, its size will be scaled to $m \times n$. As a result, only oil palm components are remained in an image as shown in Fig. 3.

Fig. 3. The result of non-oil palm elimination

3.3 Individual Oil Palm Identification

Now that an image contains only oil palm components, however, some components still contain oil palm stands; that is they contain two or more oil palms. The last step is to separate oil palm stands into individual oil palms. At first, k-means clustering is used to classify oil palms in an image into three clusters due to the fact that the sizes of oil palms in each image are different. Some are large and some are small. Thus, these three clusters represent small trees, large trees, and tree stands. The seed points for these three clusters are set to 30, 60, and 95 percentiles of the component's size, respectively. From the clustering results shown in Fig. 4 and Fig. 5, the clusters of small and large oil palms are marked on the original image as individual trees, while the cluster of tree stands is forward to the proposed multi-scale clustering which is used to separate individual trees from each tree stand.

The multi-scale clustering algorithm for separating individual trees from each tree stand is as follows:

For each component of tree stands
- *Erode a component until it splits into two or more components or until the size of a component is smaller than an oil palm*
- *Compare the size of each new component to the average size of an oil palm*
  - *If it is smaller than the average oil palm*
    - *Mark each component as a single oil palm on the original image*
  - *Else*
    - *Go back to the beginning*
- *Repeat the process until each component is smaller than the average oil palm*

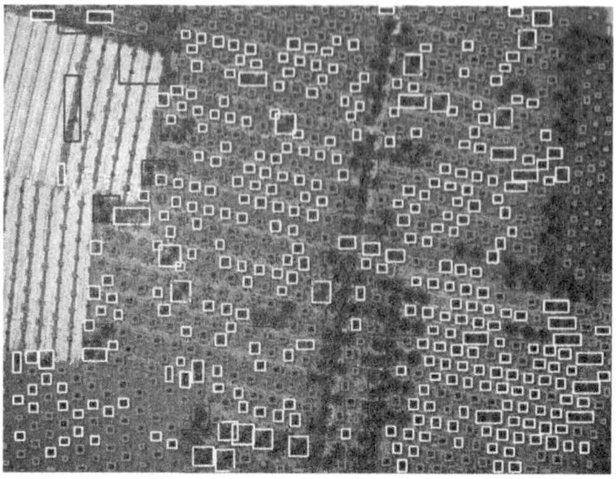

Fig. 4. The identification result from k-means clustering

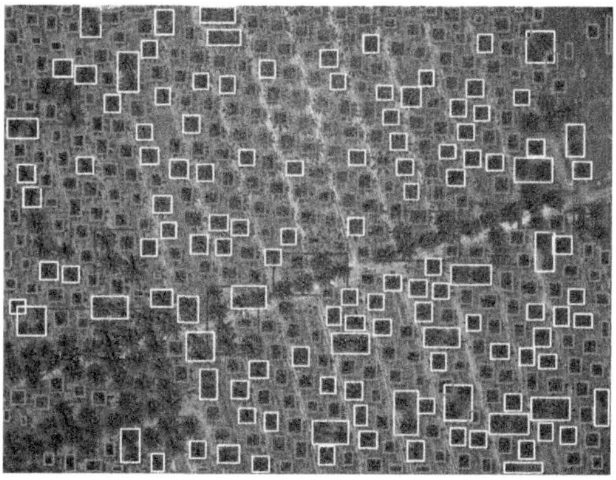

(a) An image prior to multi-scale clustering

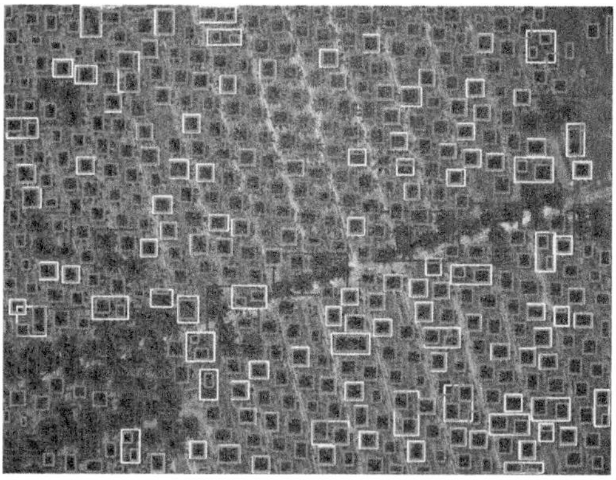

(b) An image after performing multi-scale clustering

Fig. 5. The result of multi-scale clustering

4 Experimental Results

The proposed method was evaluated on a set of 21 digital aerial images captured by a digital video camera mounted to a remote-control model airplane. The airplane was controlled to fly over the oil palm plantations in the Southern part of Thailand. The actual size of each image is 3648×2736 pixels. The elevation of the airplane was not constant, thus some images were taken from far distance and some images were taken from near distance. They can be seen that most of the images do not contain only oil palms but also other components, such as road, swamp, pond, and weed which are needed to be removed.

After obtaining the original images, they were converted to gray scale and were reduced to 912×684 pixels. The contrast of all the images was enhanced by histogram equalization to make all trees more distinguishable and was then proceeded to oil palm detection and identification by the proposed method. The cut-off radius of Butterworth low-pass filter, which was used to remove noise from an image, depended on the distance between the airplane and the ground. In our experiments, the cut-off radius was set to 15 if the distance was short and was set to 40 or 50 or 60 if the distance was long. For normalized cross correlation template matching, the size of the mask depended on the size of the template. The size of each component was adjusted to be equal to $m \times n$ before the matching.

From the experimental results, most of the oil palms can be identified correctly. However, there are some problems that cause misidentification. The first problem is when the ground is covered with thick grass which has intensity in the same range as the oil palms. The second problem is when other shrubs are similar to oil palms as shown in Fig. 6.

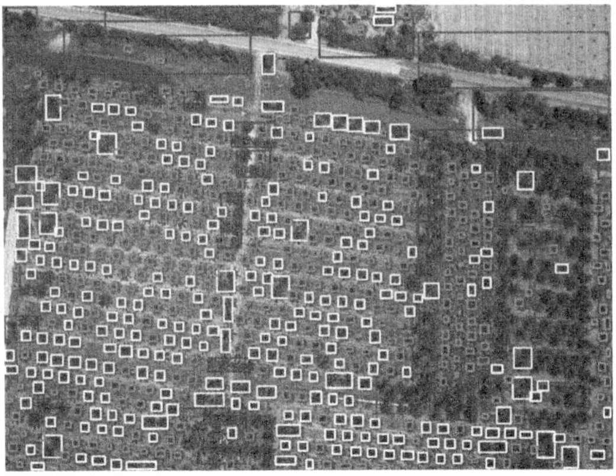

Fig. 6. An image with shrubs that caused misidentification

5 Conclusion

This paper represents the method for individual oil palm detection and identification from aerial images using the proposed multi-scale clustering and normalized cross correlation. From a dataset of 21 aerial images, there are both far-distance and near-distance images. The proposed method can detect and identify individual oil palms with the accuracy rate over 90%. In the future, we are trying to distinguish oil palms from thick grass and other shrubs in order to improve the accuracy rate.

References

1. Chen, G., Zakhor, A.: 2D tree Detection in Large Urban Landscapes using Aerial LiDar Data, Video and Image Processing Lab, EECS Department, University of California, Berkeley
2. Breiman, L.: Random forests. In: Machine Learning, pp. 5–32 (2001)
3. Charaniya, A.P., Manduchi, R., Lodha, S.K.: Supervised Parametric Classification of Aerial LiDar Data. In: 2004 IEEE Computer Society Conference on Computer Vision and Pattern Recognition Workshops, CVPRW 2004 (2004)
4. Arasato, L.S., Amaral, S., Renno, C.D.: Detecting Individual Palm Tree (Arecacear Family) in the Amazon Rainforest using High Resolution Image Classification
5. Yang, L., Wu, X., Praun, E., Ma, X.: Tree Detection from Aerial Imagery. In: ACM GIS 2009, Seattle, WA, USA © 2009 ACM, November 4-6 (2009)
6. Kattenborn, T., Sperlich, M., Bataua, K., Koch, B.: Automatic Single Palm Tree Detection in Plantation using UAV-based Photogrammetric Point Clouds. In: Remote Sensing and Spatial Information Sciences, vol. XL-3 (2014)
7. Malek, S., Bazi, Y., Alajlan, N., Alhichri, H., Melgani, F.: Efficient Frame-work for Palm Tree Detection in UAV Images. 1939-1404 © 2014 IEEE Journal of Selected Topics in Applied Earth Observations And Remote Sensing (2014)
8. Pouliot, D.A., King, D.J., Bell, F.W., Pitt, D.G.: Automated Tree Crown Detection and Delineation in High-Resolution Digital Camera Imagery of Coniferous Forest Regeneration. Remote Sensing of Environment 82, 322–334 (2002)

A Comparison Approach for Accuracy Feature of Requirements Prioritization Models

Jenjira Jaimunk and Pradorn Sureephong

Department of Software Engineering, College of Arts, Media and Technology, Chiang Mai University, Chiang Mai, Thailand 50200
janjira.se@gmail.com, dorn@camt.info

Abstract. As the quality of software defined by the users satisfaction, in order to maximize the users satisfaction under constrains of budget, time and resource, the software engineer need to prioritize the requirements. Requirement prioritization involves identifying the core requirements used for project planning. This article proposes a method to compare the accuracy of requirement prioritization model. The subjects include customers, project leaders, users and the representatives of developer such as programmers and testers. They have to select a related list of requirements among two sets of requirements. Those pair lists of requirements have to be the same order from difference prioritizing models. Whereas, the subjects don't realize those which requirements are belong to which practical model. The main objective of this study is to assess the accuracy of a proposed model with a reference model. The results show that the proposed approach provides precise measurement of accuracy for prioritizing model.

1 Introduction

Requirements prioritization would not been mentioned if time box was not tight with large number of user's requirements and every requirements have identical priority.

Requirement Prioritization involves the process to evaluate the sequence of requirements. The main objective of requirements prioritization is to identify the core requirements including important and urgent requirements (M. Aasem et al., 2010). It facilitates the scheduling of requirements implementation. (Firesmith, 2004) Moreover, requirements prioritization deals with requirement management under resources constrains. It can minimize the capital and resolves the conflicts among stakeholders. (Sommerville, 2004; Wiegers, 2003)

The success in the execution of requirement prioritization affects the correctness of requirement choosing for product release. (Karlsson, 2006) Regarding to Stanislav (Stanislav, 2012), inaccuracy of requirement prioritization is one of the reasons causes software development projects fail. Because requirement is the fundamental for the rest of construction phases, if the identifying of requirement candidates for a single product release is wrong, there is a high chance to design the incorrect software

M. Gen et al. (eds.), *Industrial Engineering, Management Science and Applications 2015*,
Lecture Notes in Electrical Engineering 349, DOI: 10.1007/978-3-662-47200-2_44

structure. This leads the erroneous in final product and high cost of development. The user satisfaction for a particular product is going to be abated.

There is a number of requirement prioritization models have been employed to get the right set of requirements. In order to select the proper one, software engineer need an effective approach to compare the requirements prioritization models. Such an approach also provides comparison results precisely.

According to Karlsson's article, the process to evaluate the rightness of the requirements succession has not been well defined. Nevertheless, she performed an experimental studied on testing 2 prioritization techniques: pair-wise and planning game. The approach used to compare the accuracy of between models is to ask the subjective which prioritization technique provides the most accurate results. The experiments are executed based on blind test which means the subjective don't know on those lists of requirement are related to which techniques.

While, Perini (Perini, 2006) had conducted an empirical study on comparison between 2 prioritization models which are AHP and CBRanking. The aim of her comparison approach is to resolves the difficult of comparing. Perini also adopted the procedure to measure the accuracy which proposed by Karlsson. The experimental results showed that the AHP outperforms CBRanking in accuracy aspect. Hence, this study will follow and develop the comparison approach based on Karlsson's and Perini's procedures.

In this paper, we discuss on an empirical study to assess the requirements prioritization model for accuracy aspect using a comparison approach. A deployed approach must provide precise results on accuracy comparison. Moreover, such an approach must minimize the difficulty of comparison as well. The paper consists of 4 discussing topics as follows. Section 2 describes the experiment design, covering the experimental tools, subjects, objects, and experimental execution. The experimental results are explained in Section 3. Lastly, the conclusion is discussed in Section 4.

2 The Experimental Designs

The main aim of this investigation is to evaluate the following hypothesis: 2 Practical requirements prioritization models: the simple ranking method and the combination method provide equal accuracy.

There are 4 discussing topic in the design of experiment containing the experimental tools, subjects, objects and experimental execution

2.1 Experimental Tools

This section explains the deployed experimental tools. Topics include the questionnaires and the level of accuracy for grading of prioritization models. Both of them are elaborated as follows:

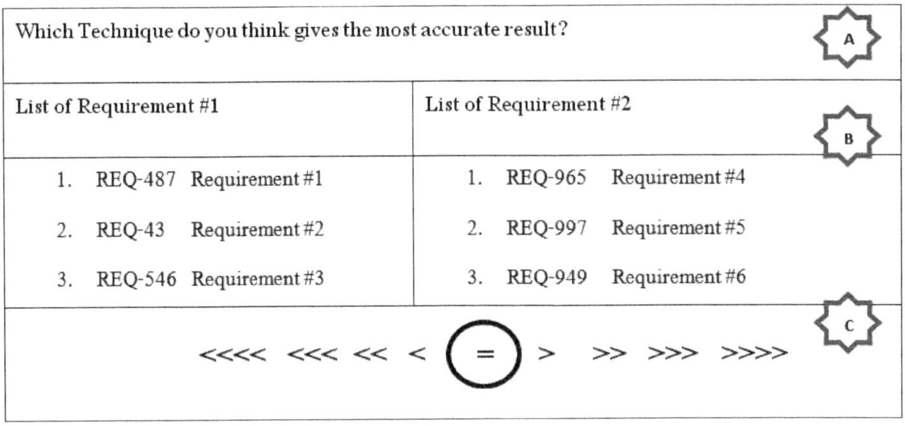

Fig. 1. Components of the employed questionnaire

The questionnaire includes three sections: Section A, Section B and Section C: question, requirements lists, and accurate levels, respectively. The Fig.1 illustrates the sample of questionnaires.

Firstly, the Section A is the question of the questionnaire. The subjects are requests to select which list of requirements are the most accurate.

Secondly, the Section B composes by two lists of requirements. Each consists of 10 requirements. These requirements lists are arranged by difference prioritization methods, even though their orders are similar. In addition, the positions of requirements lists are random. This means the lists of requirements from a particular method can be either on the left-hand side or the right-hand side for different questionnaires.

Table 1. The table of accurate levels

The symbols represent levels of accuracy	**Descriptions**
<<<<	Lists of requirements on the left-hand side are more accurate than the right ones extremely.
<<<	Lists of requirements on the left-hand side are more accurate than the right ones majorly.
<<	Lists of requirements on the left-hand side are more accurate than the right ones essentially.
<	Lists of requirements on the left-hand side are more accurate than the right ones moderately.
=	Two lists of requirements are accurate equally.
>	Lists of requirements on the right-hand side are more accurate than the left ones moderately.
>>	Lists of requirements on the right-hand side are more accurate than the left ones essentially.
>>>	Lists of requirements on the right-hand side are more accurate than the left ones majorly.
>>>>	Lists of requirements on the right-hand side are more accurate than the left ones extremely.

And lastly, the Section C is the level of accuracy section. Subjects have to compare the requirement lists regarding to defined value of accuracy and then select the proper symbol of accuracy. Table 1 presents all 9 symbols of accurate level and their descriptions.

2.2 Subjects

The subjects comprise of following people participated in the experiments:

1) Customers
2) Team Leaders
3) The representatives of Developer including designer programmers and testers. The numbers of them are counted on equal ratio.
4) Users

The distribution of subjects is difference for each project. It depends on the size of project. However, the ratio of customers, team leaders, developer, and users is 1:1:2:2, respectively. The number of developers and users are double comparing to customers and team leaders because the study considers most for maximization of user's satisfaction while the possibility of implementation is high as well. So, we need more opinion of these kinds of people.

2.3 Objects

This section describes two topics: two developing projects and their requirements and two prioritization methods which are going to be evaluated for their correctness.

Table 2. Detail of projects

Details	Project 1	Project 2
Contract Type	Long-term contract	Long-term contract
Consideration	Mobile-application support user's scheduling	Web-application support employee evaluation
Among of Requirements	118 requirements	102 requirements

All requirements are from two different projects including mobile application development project and web application development project. There are 118 requirements and 102 requirements for the first and the second project, respectively. The subjects are also stakeholders of these projects.

The first project concerns with developing mobile application base on the iOS platform. The application is designed to support user for scheduling. This project is a long-term contract which consists of 118 requirements. It has been developed for two years. Its users are all around the world, mainly in European countries.

The second project is web application development project. The application functions for evaluating employees. It's a long-term contract project as well. There are 102 requirements composed in this project. This project has been constructed for 4 years. The main users of this project are team leader and directors.

Both projects have deployed the similar requirement prioritization method which is a simple-ranking method.

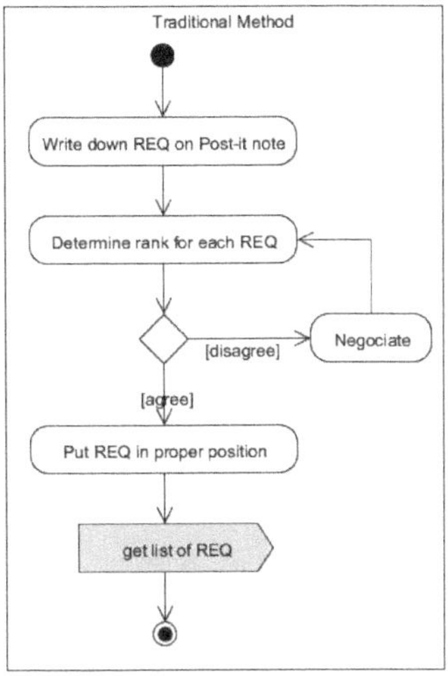

Fig. 2. The process of a simple-ranking method

Fig.2 shows the process to prioritize requirements using a simple-ranking prioritization method. The first step in the requirement prioritization is to white down requirement (REQ) on post-it notes. Then the project stakeholders have to determine rank for each requirement. Afterwards, all projects stakeholder judge on the sequence of requirements whether they are position on the right order. If some people disagreed on the requirements succession, negotiation will be carried out. At the end of the process, they will get the list of requirement production as shown in Fig.3.

Fig. 3. The sequence of requirements produced by a simple ranking prioritizing method

The second prioritizing method which is going to be accessed for accuracy is a combination method.

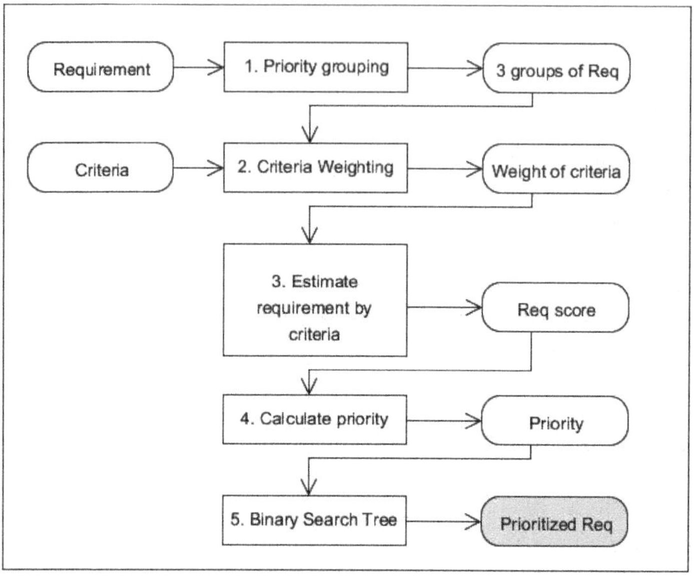

Fig. 4. The processes of the combination prioritizing method

The diagram in Fig.4 shows the procedures of a combination prioritizing method. The process begins with categorization of requirements into three groups: essential requirements, useful capabilities and desirable capabilities and weighing criteria which present in ratio. These 2 elements (requirements in separated groups and weight of criteria) are introduced into a requirement prioritization. After this, requirements in the first and second group (essential requirements and useful capabilities) are estimated their value by the defined criteria. The estimated value of each requirement and weight of criteria are used to calculate for priority value using following formula:

$$priority = \frac{\text{value\%}}{(cost\% * cost\ weight) + (risk\% * risk\ weight)}$$

For example,

Requirement	benefit	penalty	Total Value	Value %	cost	st % [C]	risk	Risk % [E]
	1	1.5			1		0.5	
REQ-43	3	3	7.5	17.44 [A]	6	21.43 [B]	2	28.57 [D]

Referring to the formula mentioned above,

$$priority = \frac{A}{(B * C) + (D * E)}$$

So, the priority value of REQ-43 is

$$priority = \frac{17.44}{(21.43 * 1) + (28.57 * 0.5)} = 0.49$$

These priority values will be prioritized in the last step by using Binary-search tree method where the prioritized requirements come out.

2.4 Experimental Execution

The designed questionnaires are distributes to all subjects. The experiment starts with a little talk aims to introduce the objectives of the experiment and step to respond the questionnaire. Regarding to the questionnaire, the subjects have to select the listed of requirement which supposed to be the most accurate. These lists of requirement are generated by unknown prioritization method. The subjects cycle on the symbol of accurate level in the questionnaire shown in Figure 1. The symbols of accurate level have been defined in Table 1. At the end of the experiment, all subjects return their questionnaires to researcher.

3 Experimental Results

The results from the experiment are show in Fig.5.

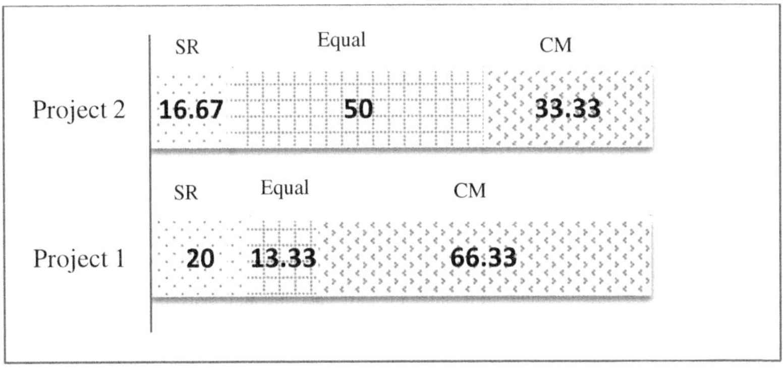

Fig. 5. The experimental results

The graph shows percentage of accuracy for the simple ranking method (SR) and combination method (CM) of Project 1 and Project2. In the first project, the greatest accurate level was on the combination prioritizing method having as much as 66.33 %. In contrast, the second project shown SR and CM are accurate equally, account for 50%. The SR was minimal for both project, with only 20 % and 16.67 % for Project 1 and Project 2, respectively.

Overall, it is clear that CM has the highest overall percentage of Project 1 while SR has the lowest overall percentage of both projects in accuracy aspect.

4 Conclusions

In conclusion, we conduct an experiment between 2 practical requirement prioritization methods including simple ranking method and combination method. This experiment focuses on assessing the accuracy of priority sequence of requirements. All interviewee have to choose the most accurate sequence of requirements without realizing on which sequence is from which method (Blind Test).

The experimental results indicate that two practical prioritizing methods provide insignificantly differ in accuracy aspect. Level of accuracy for simple ranking method in Project 1 is minimal. Similarly, the simple ranking method accounts for the least accuracy in Project 2. The difference between both projects is that the combination method has greatest accuracy on the Project 1 while the combination and simple ranking method are similar accurate in Project 2.

Acknowledgments. The authors would like to express deepest appreciation to all staffs from Chiang Mai Lanna Business Services Co.,Ltd. for participation on experimental execution and data collection. This study cannot be accomplished without their help. Besides, we gratefully acknowledge to the Graduate School, Chiang Mai University for financial support.

References

1. Hatton, S.: Choosing the Right Prioritization Method. In: Proceedings of the Australian Conference on Software Engineering, pp. 517–526 (2008)
2. Carod, N., Cechich, A.: Requirements Prioritization Techniques. In: Encyclopedia of Information Science and Technology, 2nd edn., pp. 3283–3291. IGI Global (2009)
3. Hasan, M.S., et al.: An Evaluation of Software Requirement Prioritization Techniques. International Journal of Computer Science and Information Security, 83–94 (2010)
4. Firesmith, D.: Prioritizing Requirements. Journal of Object Technology, 35–47 (2004)
5. Karlsson, L., Thelin, T., Regnell, B., Berander, P., Wohlin, C.: Pair-wise comparisons versus planning game partitioning–experiments on requirements prioritisation techniques. Journal of Empirical Software Engineering 12(1), 3–33 (2007)
6. Perini, A., Ricca, F., Susi, A., Bazzanella, C.: An Empirical Study to Compare the Accuracy of AHP and CBRanking Techniques for Requirements Prioritization. In: Fifth International Workshops on Comparative Evaluation in Requirements Engineering (2010)

A Novel Approach on Operation and Maintenance Guideline Using Semantic Processing and Clustering

Ki Hoon Jang[1], Gyeong-June Hahm[1], Heejung Lee[2], and Hyo-Won Suh[1]

[1] Dept. of Industrial&Systems Engineering, KAIST, Dae-jeon 305-701, Republic of Korea
[2] Dept. of Applied Systems, Hanyang University, Seoul, 133-791, Republic of Korea
{jangki,hamstein,hw_suh}@kaist.ac.kr,
stdream@hanyang.ac.kr

Abstract. The rapid progress in the industry increases the number of plant, and a myriad of operation and maintenance manuals is required for the plant safety activities. Moreover, major challenge in safety prevention and maintenance is that manual search and retrieval requires extensive labor-intensive and time-consuming process incurred by high degree of semantic ambiguity and linguistic complexities among the various operation and maintenance manuals. In this paper, we propose a novel approach on plant operation and maintenance guideline using semantic processing and clustering techniques for the plant safety activities. Specifically, the proposed method can provide the engineers with the more enhanced guidelines by information propagation.

1 Introduction

With industries making rapid progress, the number of plants and facilities has been considerably increased. Thereby plants and facilities failure emphasize the importance of effective and efficient operation & maintenance activities. Major challenge in the safety related activities is that O&M manual search and retrieval requires extensive labor-intensive and time-consuming process incurred by high degree of semantic ambiguity and linguistic complexities among the various O&M manuals[1]. When an accident occurs, it is solved by using experience or judges from skilled engineers instead of finding associative informative guidelines because most of documents are non-structured and not easy to exploit them. Frequently, a myriad of O&M manuals has further exacerbated engineers efforts for efficient mechanisms for information selection and retrieval[2].

To address the above issues, we propose a novel approach on plant operation and maintenance guideline using semantic processing and clustering techniques for the plant safety activities. The main contents of this paper are as follows: (1) From O&M manuals, important terms are extracted using criteria based on syllable unit and then experts construct the domain-specific semantic model, (2) Applying Term-Frequency & Inverse Document Frequency (TF-IDF) and K-medoid clustering method to find the semantic relationships in the O&M manual, (3) Finally, the failure reports are analyzed and the engineers will be provided with the more enhanced O&M guidelines through the information propagation.

M. Gen et al. (eds.), *Industrial Engineering, Management Science and Applications 2015*,
Lecture Notes in Electrical Engineering 349, DOI: 10.1007/978-3-662-47200-2_45

2 Backgrouds

2.1 Information Extraction by Text-Mining Techniques

Information Extraction (IE) methods bring natural language processing (NLP) tools with domain knowledge to extract meaningful sentence constituents from unstructured texts for retrieval or for knowledge mining purposes[3]. It has been widely explored by using a variety of documents such as CAD drawings[4], [5], [6], sensor data[3]-[9], web data[10], [11] and so on. The domain knowledge can either be formalized as expression patterns by experts[12] or learned from a large training corpus[13]. Several systems have been reported in the literature[9], [14], [15], which uses the text-mining techniques to analyze the diagnosis data. Buddhakulsomsiri et el.[14] developed the association rule mining algorithms based on *A priori* and elementary set concept techniques to identify the relationships between the product attributes (such as production date, repair date, mileage) and the failure causes (labor code) from the automotive warranty data. In [9], the text mining has been used to map the problem description to their appropriate diagnostic categories, such as engine, electrical, brake and transmission. The techniques, such as text document categorization and term weighting schemes, similarity functions, and latent semantic indexing are used to cluster the data and to identify the similarities between a problem description and a diagnostic categorization. In [15], the knowledge discovery in database along with the case-based reasoning, neural network and rule-based reasoning techniques are used to extract the fault conditions and the checkpoint information.

2.2 Ontology-Based Semantic Representation

The structured and semantics-based representation of designs has been studied in such fields as product modeling and ontology modeling[5]. The use of ontology to overcome the limitations of keyword-based search has been put forward as one of the motivations of Semantic Web since its emergence in the late 1990s[6]. For understanding information which is extracted, the semantic representation is usually used to find out meaning and relationships. Therefore it is carried out based on domain knowledge. However, constructing an ontology semantic representation is a challenging task. So the work engineers must have detailed and specific knowledge.

Velardi et el.[1] constructed an ontology model that has an interoperability infrastructure for small and medium European enterprise. Choi[16] proposed semantic model based on FMEA structure and engineering domain. It also proposed model expansion by FMEA concepts and instance. So it realized the integrated representation model. A few models on integrating ontology with Information Retrieval have been proposed as well[14], [15]. Rajpathak[7] proposed a novel ontology-based text mining system to achieve the aforementioned objectives by automatically analyzing the repair verbatim data. Through finding hidden meaning of abbreviations it offers important, accurate and specific diagnosis information.

In this paper, we propose the hybrid approaches of an ontology-based semantic model and text mining method. These incorporated methods enable both skilled and novice engineers to find necessary information in more effective and efficient way.

2.3 Sample Facility

We studied a Korean engineering company which consists of a variety of facilities to apply our approach, and selected a certain target facility, PUMP #201 as shown in Fig. 1 and related O&M manuals. We will use this sample facility throughout this paper, and define the semantic models by consulting O&M manuals and finally provide the enhanced O&M guidelines for PUMP #201. In this paper, some pieces of O&M manuals are selected for the simplicity.

Fig. 1. Sample facility

3 Semantic Model

3.1 O&M Manual

This section addresses the specific words used in O&M manual and their special meaning which is used throughout this paper. Our O&M manual has two types: <failure-cause documents> and <maintenance guidelines> as shown in Fig. 2.

<failure-cause-documents>

Doc. ID	Failure	Cause
D_1	...	...
D_2 ...	...	...
D_N	...	...

<maintenance guidelines>

Part ID	To-Check	Action
P_1	...	...
P_2 ...	...	...
P_N	...	...

Fig. 2. O&M manual template

Failure means the problem, concern, error, or challenge, which is the state of inability of the system, design, process, service, or subsystem to perform based on the design intent. Cause means what is the root cause of failure. When failure occurs you have to check the related information and do appropriate action in a right way.

3.2 Concept Extraction

A concept can be a single word such as "plant" or "oil", or a phrase such as "hole missing". The word "concept" can be used with different meanings in different situations or communities. In this paper, concept is the basic unit of text used in O&M manuals, and can be extracted by a series of operations. In general O&M manuals are represented by a variety of types. However, we mainly focus on the only text-based manual so that word-based information, i.e., concept, can be extracted by parser and analyzed. We used the *HanNanum* Korean Morphological Analysis System which is a morphological analyzer and a POS(part-of-speech) tagger. In *HanNanum*, tokenizer reads and divides a sentence into tokens when there are '\t', '\n' and special characters. The tokenized terms will be POS-tagged based on the following simple statistic technique: $P(W_{1,n}) = arg\ max\ P(C_{1,N} \mid W_{1,N})$. $W_{1,n}$ means a sentence which consists of n words and $C_{1,N}$ means the number of possible POS. The overall morphological analysis workflow is depicted in Fig. 3. For instance, if a word in a sentence was frequently used as a certain type of POS then the probability values stored and the word is going to be tagged later. For example, a word *fly* can be used as noun or verb.

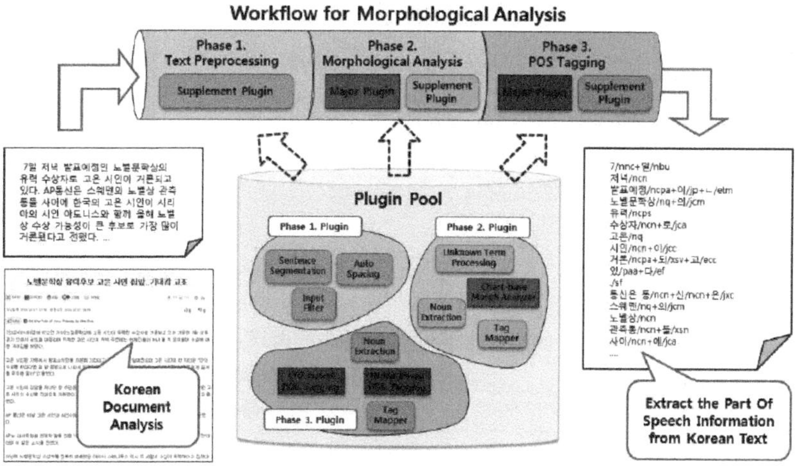

Fig. 3. Workflow example for morphological analysis

In this paper, eight failure-cause-documents are selected and used for the concept extraction test, and we obtained a total number of 104 concepts such as: MOTOR, SPEED, PUMP, etc.

3.3 Semantic Model Construction

In this study, ontology and engineering experts was selected and developed the semantic model using the following steps: defining reference model → extracting concepts from the O&M manuals → identifying the relationships among these concepts and integrating all concepts and relationships. Firstly we built the high-level reference model by consulting general function of O&M operations. The terminology in this reference model has been found by securing the literature [17], [18], and we define the O&M domain concepts and relationships as shown in Fig. 4(a). Secondly, all the sentences in O&M manuals are morphologically analyzed and classified into several concepts as shown in Fig. 4(b). Finally, we can construct the semantic model by matching the obtained concepts and reference model step by step. Fig. 5 shows how the semantic model can be constructed by matching and linking using the predefined reference model.

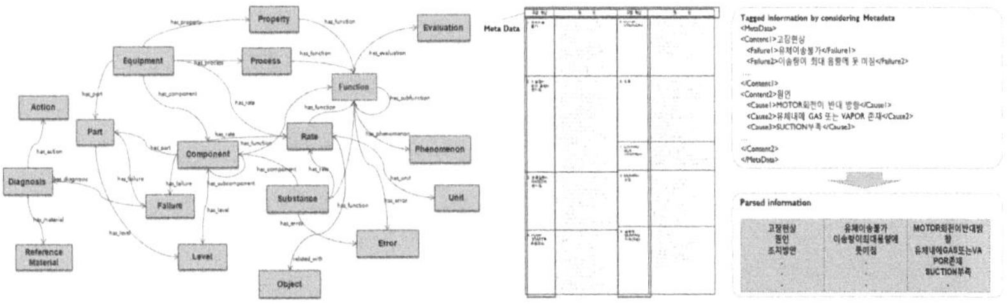

(a) O&M reference model (b) concept extraction from O&M manuals

Fig. 4. Reference model and concept extraction

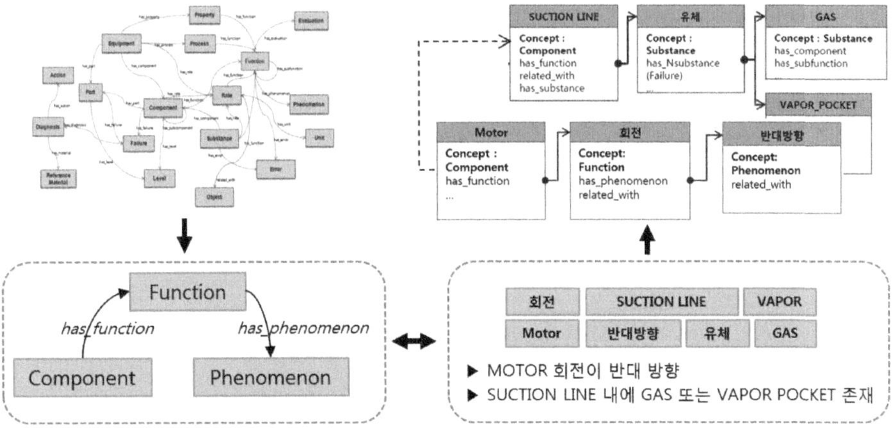

Fig. 5. Procedure of O&M semantic construction

4 Clustering

4.1 Concept Clustering by K-Medoid

When it comes to concepts extracted from O&M manuals, we can expect that that some concepts are more similar each other than other concepts. There are many ways to assess the similarity between two concepts, and the most common way to measure the similarity is as follows: similarity is a real function from a pair of concepts to a real number expressing the similarity between two concepts. In this paper, we used the number of intermediate edges between two classes in the O&M semantic model as the similarity measure. Fig. 6 shows the distance calculation (i.e. dis-similarity) example. For instance, the distance between 회전 and MOTOR is 1 since there are only one intermediate edge, but the distance between 회전 and PUMP is 2 since there are two intermediate edges. Based on this distance or dis-similarity, we applied K-medoid algorithms for concept clustering and four clusters was obtained as shown in Fig. 7.

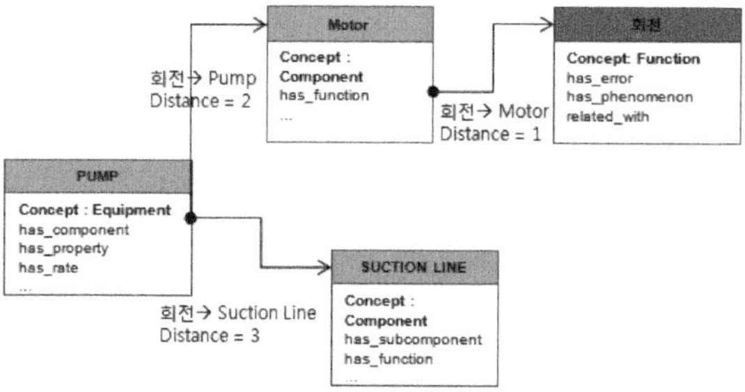

Fig. 6. Distance calculation example

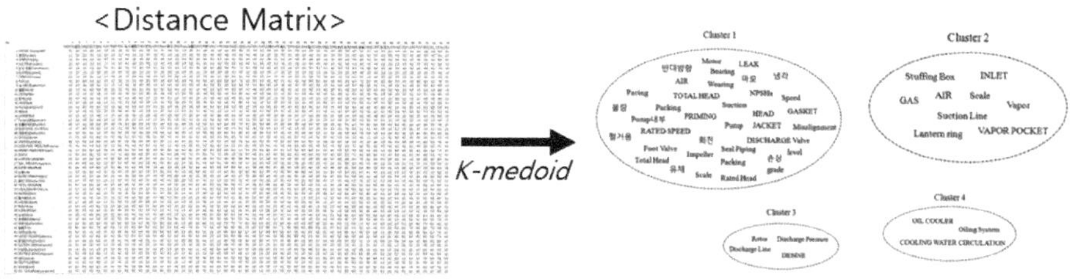

Fig. 7. Concept Clustering by K-medoid

4.2 Documents Clustering by TF-IDF

TF-IDF(Term Frequency and Inverse Document Frequency) is a numerical statistic that is intended to reflect how important a word is to a document in a collection[19]. The TF-IDF value increases proportionally to the number of times a word appears in the document. The TF is the value which indicates how a particular word appears in a document, i.e., the number of times that term t occurs in documents d, and the IDF is the inverse value of number of document that a particular word appears, that is, whether the term is common or rare across all documents, i.e., $log((N-n)/n)$. Hence, the criticality of extracted terms is calculated by using the equation such as:

$$TF\text{-}IDF(t) = TF \times log((N-n)/n).$$

Where TF is the term frequency (concept frequency) extracted from O&M document. N is number of total documents and n is number of document including a particular word. TF-IDF result can be also represented by matrix as follows.

$$W(w_{i,j})_{M\times N} = \begin{array}{c|ccccc} & d_1 & d_2 & d_3 & \cdots & d_N \\ c_1 & w_{1,1} & w_{1,2} & w_{1,3} & \cdots & w_{1,N} \\ c_2 & w_{2,1} & w_{2,2} & w_{2,3} & \cdots & w_{2,N} \\ c_3 & w_{3,1} & w_{3,2} & w_{3,3} & \cdots & w_{3,N} \\ \vdots & \vdots & \vdots & \vdots & \ddots & \vdots \\ c_M & w_{M,1} & w_{M,2} & w_{M,3} & \cdots & w_{M,N} \end{array}$$

The d_j means *Doc ID*, in <failure cause documents>, c_i means extracted concepts, and w_{ij} means TF-IDF values. We can more easily understand the relationship of document and concepts using this matrix, and also produce the documents clusters. With the 104 concepts acquired from concept extraction method and 8 failure-cause-documents we performed the TF-IDF clustering procedure, and the results is as shown in Fig. C. The documents 1, 2 are included in cluster A, and documents 4, 6 in cluster B, and finally documents 5, 7, 8, and 9 in cluster C.

5 Enhanced O&M Framework

The proposed enhanced O&M system is designed for providing additional O&M guidelines to the both skilled and novice engineers based on the concept extraction and clustering techniques. Fig. 8 shows the overall procedure of the proposed approach which is compared with the as-is one. In enhanced O&M system, occurred failure will be analyzed and parsed into the concepts, and the matching and extension procedure is invoked to provide the engineers with additional helpful guidelines.

For example, assume that O&M system detect failure and activates warning. The text formatted failure-cause "LANTERN RING 으로의 SEAL PIPING 막힘 will be morphologically analyzed and classified into concepts: LANTERN_RING/part, SEAL_PIPING/component as shown in Fig. 9.

Fig. 10 shows how we can propagate related information using TF-IDF matrix. We can find that LANTERN RING has the highest TF-IDF value (1.91) in document 4 (Assume that the TF-IDF threshold value is 0.8, so SEAL_PIPING with TF-IDF value (0.65) is not considered here), and we will expand our document space into documents 4 and 6, since these are in the same document cluster B. In document 6, we will search for the concept in the same concept cluster 1 which has the highest TF-IDF value, and find that BEARING has the highest TF-IDF value (0.95).

Applying this enhanced framework, we can find not only guidelines related to LANTERN RING but also BEARIING as shown in Fig. 11. It is meaningful result that engineers can get necessary information about primary cause as well as extra one.

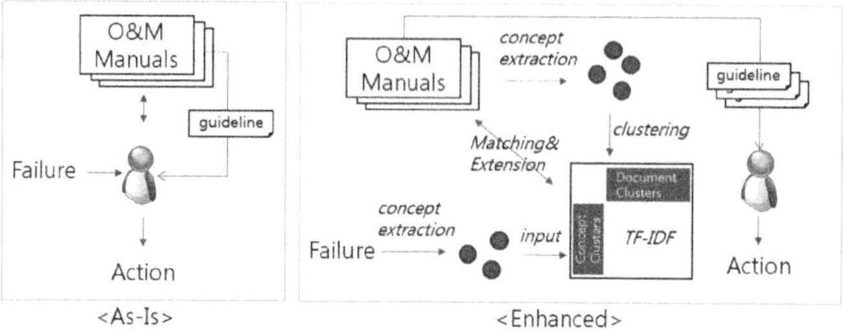

Fig. 8. Outline of enhanced O&M framework

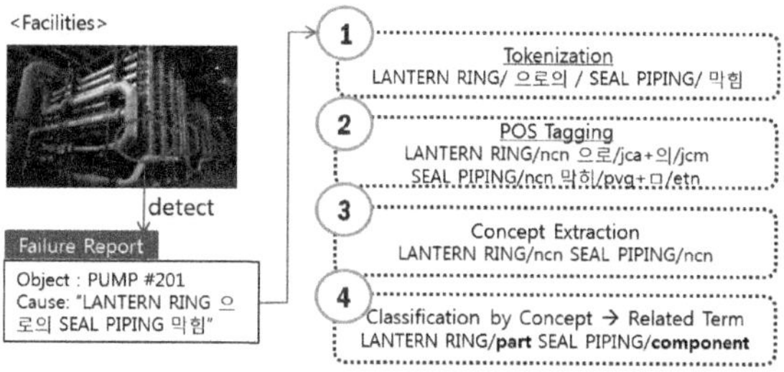

Fig. 9. Concept extraction from the failure report example

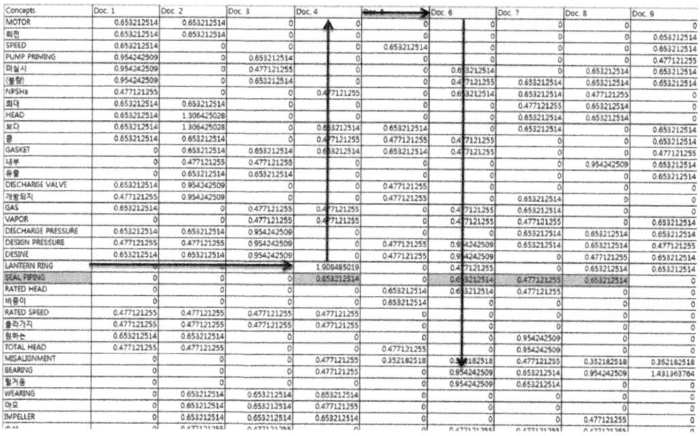

Fig. 10. TF-IDF matrix and information propagation example

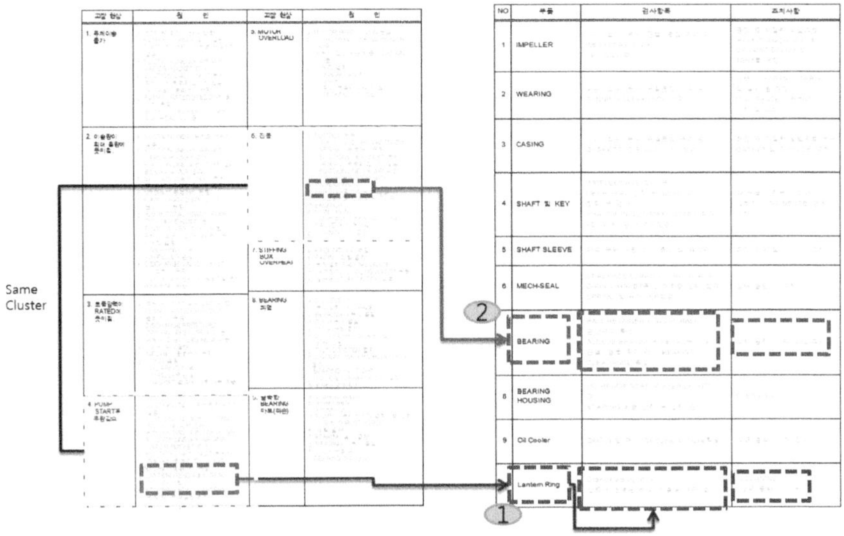

Fig. 11. Guideline expansion example

6 Conclusion

This research proposed a semantic retrieval approach for knowledge searching and retrieval in operations & maintenance areas. The reference model was defined and we built clusters and calculate the TF-IDF value for checking important concepts or term. When the input queries or texts as failure reports are entered into the system, the system can find a variety of information by using the our approach. So engineers can get necessary action information about primary cause as well as extra action information. When this study is compared to existing researches, it has several advantages. First, it is an integrated approach of text-mining and semantic processing. Second, it offers meaning-based and necessary information to engineers. So when the abnormal state is detected from sensor data, engineers can fix and check the state with more accurate and efficient approaches.

But the approach has some limitations. First, the reference model is manually constructed by domain experts. Second, expected result is too dependent on reference model. In case that the experts constructed the reference model incorrectly, the whole process and result might be negatively affected.

In this paper, we did not discuss the concrete implementation for proposed approach. Also we manually constructed the reference model, which possibly causes some troubles such as incorrect information, subjective information and so on. All these implementation issues are the big theme, so we will secure how the series of manual operations can be implemented in the future works.

Acknowledgments. This research was supported by the Industrial Core Technology Development Program (Project ID: 10048341) funded by the Ministry of Trade, Industry and Energy of the Korean government. This research was also supported by a grant(Project ID: 14IFIP-B091004-01) from Plant Research Project funded by Ministry of Land, Infrastructure and Transport of the Korean government.

References

1. Missikoff, M., Velardi, P., Fabriani, P.: Text Mining Techniques to Automatically Enrich a Domain Ontology. Appl. Intell. 18(3), 323–340 (2003)
2. Nayak, R., Tao, X., Li, Y.: A knowledge retrieval model using ontology mining and user profiling. Integrated Computer-Aided Engineering 15(4), 1 (2008)
3. Ferrucci, D., Lally, A.: Building an example application with the Unstructured Information Management Architecture. IBM Syst. J. 43(3), 455–475 (2004)
4. Zhang, X., Hou, X., Chen, X., Zhuang, T.: Ontology-based semantic retrieval for engineering domain knowledge. Neurocomputing 116, 382–391 (2013)
5. Li, Z., Raskin, V., Ramani, K.: Developing Engineering Ontology for Information Retrieval. J. Comput. Inf. Sci. Eng. 8(1), 7 (2008)
6. Li, Z., Ramani, K.: Ontology-based design information extraction and retrieval. AI EDAM 21(2), 137–154 (2007)
7. Rajpathak, D.G.: An ontology based text mining system for knowledge discovery from the diagnosis data in the automotive domain. Comput. Ind. 64(5), 565–580 (2013)
8. Castells, P., Fernandez, M., Vallet, D.: An Adaptation of the Vector-Space Model for Ontology-Based Information Retrieval. IEEE Trans. Knowl. Data Eng. 19(2), 261–272 (2007)
9. Huang, L., Murphey, Y.L.: Text Mining with Application to Engineering Diagnostics, pp. 1309–1317 (2006)
10. Pang, B., Lee, L.: A sentimental education: Sentiment analysis using subjectivity summarization based on minimum cuts. In: Proceedings of the 42nd Annual Meeting on Association for Computational Linguistics, ACL 2004, p. 10 (2004)
11. Wilson, T., Wiebe, J., Hoffmann, P.: Recognizing contextual polarity in phrase-level sentiment analysis. In: Proceedings of the Conference on Human Language Technology and Empirical Methods in Natural Language Processing, HLT 2005, pp. 347–354 (2005)
12. Hobbs, J.R., Appelt, D., Bear, J., Israel, D., Kameyama, M., Stickel, M., Tyson, M.: FASTUS: A Cascaded Finite-State Transducer for Extracting Information from Natural-Language Text, pp. 1–25 (1997)
13. Riloff, E.: Automatically Generating Extraction Patterns from Untagged Text (1996)
14. Buddhakulsomsiri, J., Zakarian, A.: Sequential pattern mining algorithm for automotive warranty data. Comput. Ind. Eng. 57(1), 137–147 (2009)
15. Hui, S.C., Jha, G.: Data mining for customer service support. Inf. Manag. 38(1), 1–13 (2000)
16. Choi, S., Suh, H.: A method of Creating Semantic Models of FMEA Documents for Reuse of Failure Knowledge in Product Development. Korea Advanced Institute of Science & Technology (2013)
17. Hirtz, J., Stone, R.B., McAdams, D.A., Szykman, S., Wood, K.L.: A functional basis for engineering design: Reconciling and evolving previous efforts. Res. Eng. Des. 13(2), 65–82 (2002)
18. Be, D.G., Tumer, I.Y., Stone, R.B.: Requirements for a Failure Mode Taxonomy for Use in Conceptual Design, pp. 563–564 (2003)
19. Hiemstra, D.: A probabilistic justification for using tf-idf term weighting in information retrieval. Int. J. Digit. Libr. 3(2), 131–139 (2000)

Rutting Load Equivalency Factors of Heavy Vehicles Operating in the Sothern Part of Malaysian Peninsula

Osama Mahmoud Yassenn[1], Intan Rohani Endut[2], Mohamed Ahmed Hafez[3], and Siti Zaharah Ishak[1]

[1] Universiti Teknologi MARA, Malaysia Institute of Transport, 40450 Shah Alam, Selangor, Malaysia
[2] Universiti Teknologi MARA, Faculty of Civil Engineering, 40450 Shah Alam, Selangor, Malaysia
[3] Universiti Tenaga Nasional, Department of Civil Engineering, 43000 Kajang, Selangor, Malaysia
eng.osama.alhusyni@gmail.com

Abstract. In this study, load equivalency factors based on rutting criteria of heavy vehicles operating at the southern portion of Malaysian peninsula were developed. The study was based on the traffic volume and weight collected by three weight-in-motion stations for more than 6 million heavy vehicles. It was also based on the analysis of the effect of their weight on more than three hundred and fifty kilometers of flexible pavements. Based on the analysis made in this study, the average rutting load equivalency factors for each heavy vehicles type were developed. The developed rutting load equivalency factors could be used in flexible pavement design and rehabilitation purposes in the southern portion of Malaysian peninsula.

Keywords: Flexible pavements, Heavy Vehicles, load equivalency factors, Malaysia, and rutting.

1 Introduction

The economic development of any country requires an effective transport infrastructure to help in movement of people and goods with great ease and speed. Thus it is highly important to understand and solve all types of problems faced by transportation sector in any country.

In Malaysia the economy is developing rapidly and thus it is highly important to keep the transportation system at its highest possible quality. One of the problems facing transportation infrastructure in Malaysia is the overloading of heavy vehicles [1-4]. This problem causes premature damages and fast deterioration of pavements and bridges. Noteworthy that, Malaysia and many other countries have implemented load limits [5-7] to control transport infrastructure deterioration due to overloading problem. However, a lot of studies have shown that overloading is a common problem around the world [8-11].

M. Gen et al. (eds.), *Industrial Engineering, Management Science and Applications 2015*, Lecture Notes in Electrical Engineering 349, DOI: 10.1007/978-3-662-47200-2_46

Although material properties, design considerations, quality of construction, environment may accelerate pavement deterioration, the most important factor causes pavement deterioration is the traffic loading, especially the heavy vehicles traffic loading [12, 13]. [14] reported that loading is one of the important factors that could increase the pavement damage by 20 times. Studies in US and South Africa reported that overloading could increase pavement damage out of proportion, for instance, an axle loaded double the legal limits may cause the pavement damage to increase up to 60 times [5].

Load equivalency factor based on rutting , also known as rutting damage ratio, is a number comparing the rutting damage caused by an arbitrary axle loading and configuration with the rutting damage caused by the single axle, four wheels standard axle loading (8.16 tons). For example, if the rutting damage ratio of specific axle loading and configuration is 2, it means that it causes 2 times as much as rutting damage the standard axle. In case of underestimating of these ratios, for instance due to overloading, the pavement will deteriorate faster and premature rutting will occur along the road network. On the other hand, if these ratios were overestimated the construction cost will be higher and this will cause unnecessary expenditures due to over-designing.

The main aim of this paper is to develop rutting damage ratios for heavy vehicles operating along the southern portion of the Malaysian peninsula based on their actual loading collected by weight-in-motion stations. These developed factors could be used for new pavements and overlay designs purposes.

2 Methodology

The heavy vehicles traffic volume and weights were collected by PLUS expressway Berhad from several weighting stations spread along the longest Malaysian expressway (North-South-Expressway). For the purpose of this study, the data collected by three weighting stations spread along the southern part of Malaysian peninsula.

The counting of heavy vehicles traffic volume and weighting was carried out using weight-in-motion systems. . The weight-in-motion systems are very reliable and accurate systems installed on the pavement of the selected sites. This weighting systems are capable of determining the heavy vehicles gross vehicle weight and axles loading at normal traffic speed without the need of these vehicles to stop.

There are many advantages of the Weight-In-Motion system. One of these advantages is the high processing rate it can weigh the vehicles at normal speed which allows it to process a large number of vehicles comparing to the static weighting system. Another advantage is the ability of the system to have a continuous data processing rather than sampling technique which is used in the static weighting system, which eliminates any bias in the collected data. One more advantage is that since this system reduces the accumulation of the heavy vehicles leading to the weigh station, the weighting process is safer. A further advantage is that the Weight-In-Motion system can weigh the vehicle without alerting the driver, which results in a more truthful data since the drivers will not try to avoid the weighting process.

Although there are many advantages for using the Weight-In-Motion system, there are also some disadvantages. The fact that trucks do not stop at the weighting station reduces the number of different types of data collected in the normal static weighting station. These data include fuel type, origin, destination, and loaded or unloaded status. However, as these data are not required in the study objective, this makes the weight-in-motion system a very good choice for this study (Figure 1).

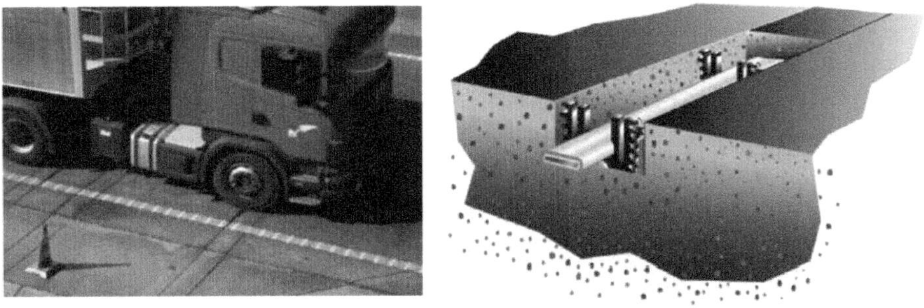

Fig. 1. Weight-In-Motion System [15]

In this study, the data collected were used to determine the traffic volume for each heavy vehicles classes, average gross vehicle weight, and average axles loading. Although, the data were collected along the Malaysian Expressway, it is possible to believe that the average heavy vehicles load along the road network is similar in the sense that the same trucks which use the expressways will use the rest of the road network.

In this study, data collected from the following weight-in-motion stations (Figure 2) were used:

1. Tangkak
2. Yong Peng
3. Sedenak

The heavy vehicles investigated by this study, which forms the extreme majority of heavy vehicles in Malaysia, are as follow:

a) Single unit trucks with two axles
b) Single unit trucks with three axles
c) Single unit trucks with four axles
d) Single trailer trucks with four axles
e) Single trailer trucks with five axles
f) Single trailer trucks with six axles
g) Multi trailer trucks with five axles
h) Multi trailer trucks with six axles
i) Multi trailer trucks with seven axles
j) Busses

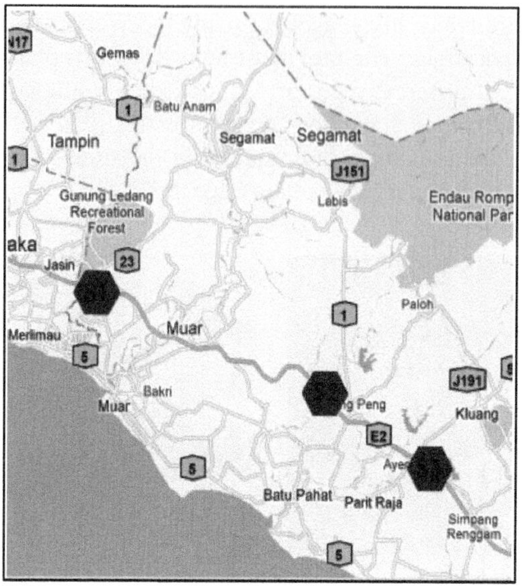

Fig. 2. Weight-In-Motion Approximate Locations. S1At Tangkak, S2 At Yong Peng, S3 At Sedenak.

After collecting the traffic volume and weight data, flexible pavement of the North-South Expressway was analyzed using finite element analysis to estimate the rutting life of the pavement under the current loading, and to compare it with its rutting life under standard axle loading. To perform this analysis, it is necessary to identify pavement layer thicknesses and elasticity moduli. These two types of data were collected by PLUS Expressway Berhad every 25 meters. The data were collected using Falling Weight Deflectometer (FWD), Ground Penetration Radar (GPR), and Coring and Dynamic Cone Penetration (DCP).

The pavement cross-sections were analyzed to determine the maximum compressive strain at the top of each subgrade layer. The compressive strains were used to determine the pavement rutting life as shown by the following formula:

$$N_R = f_4(\varepsilon_c)^{-f_5}$$

Where:

N_f : Allowable number of load repetition before rutting failure.
ε_t : Maximum compressive strain at the top of subgrade
f_4, f_5 : Constants equal to 1.365 × 10^{-9} and 4.477 based on American Asphalt Institute [16].

The same analysis was done on pavement cross-sections under standard axle load (8.16 tons). Next, rutting damage ratios were determined by dividing the pavement expected rutting life under a specific heavy vehicle loading by the pavement rutting life under standard axle loading. This was done for more than three hundred and fifty kilometers of the north-south expressway, in which the stations are located. Then, the average rutting damage ratio of each heavy vehicle type under this study was calculated.

3 Results

3.1 Tangkak

Figure 3 shows a comparison between the percentages of weighted heavy vehicles at Tangkak Weight-In-Motion Station. A total of 2,074,116 heavy vehicles were weight at Tangkak Weight-In-Motion Station. It could be clearly seen that the majority of heavy vehicles traffic volumes (89.3%) is made of merely three types: single unit trucks with two axles (27.0%), single trailer truck with 4 axles (31.8%) and buses (30.5%).

All types of multi-trailer vehicles forms only 1.1%. The multi-trailers with five axles forms 1.1% and only 48 multi-trailer truck with six axles passed the weighting station (0.002%), also there is no records for any multi trailer vehicles with more than 6 axles passed the weighting station. All other types of heavy vehicles form 9.5%.

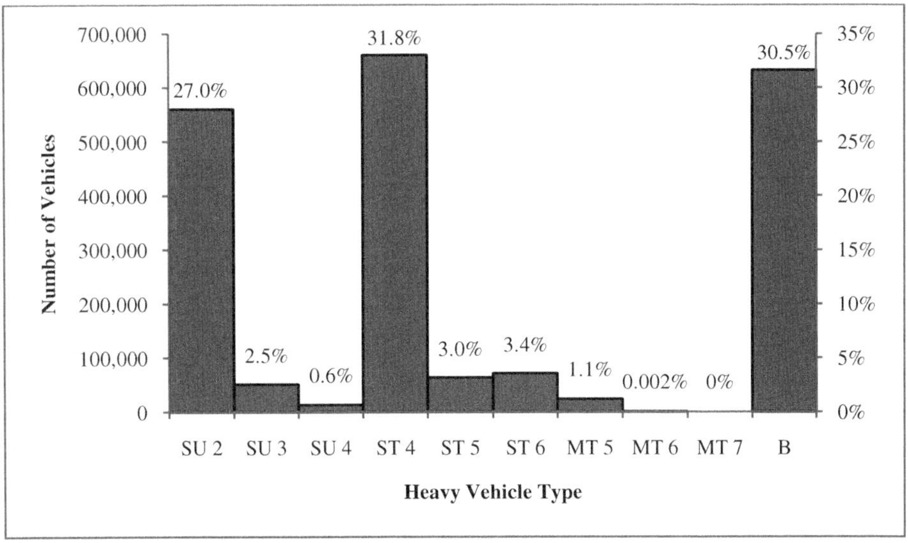

Fig. 3. Heavy Vehicles Traffic Volume at Tangkak WIM Station

Figure 4 compares the rutting damage ratios calculated for each heavy vehicle classes. It can be clearly seen that three types of heavy vehicles causes the highest rutting damage ratio. The highest rutting damage ratio was caused by single trailer trucks with six axles (19.12). This indicates that, one pass of this vehicle causes a rutting damage approximately 19 times higher than the rutting damage caused by standard axle load, this is mainly caused by overloading. The second heights rutting damage ratio was caused by single trailer trucks with five axles (12.75), followed by single trailer trucks with four axles which causes 9.41 rutting damage ratio.

The rest of the heavy vehicles have a damage ratio ranged between a minimum of 1.83 for single unit trucks with two axles to a maximum of 8.09 for the multi-trailer trucks with five axles.

Overall, single unit trucks causes a rutting damage ratio ranged between 1.83 and 3.84, single trailer tucks causes a rutting damage ratio ranged between 9.41 and 19.12, and multi trailer trucks causes a rutting damage ranging between 7.15 and 8.09.

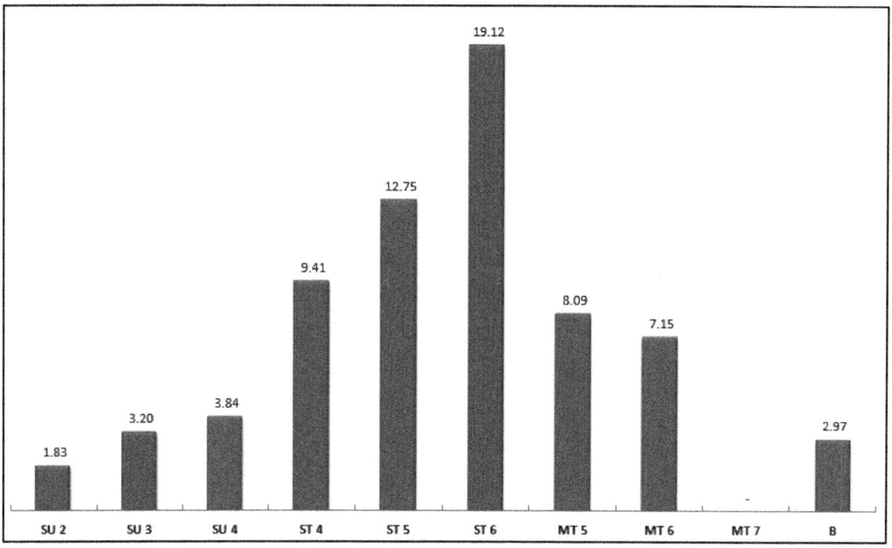

Fig. 4. Calculated Rutting Damage Ratios for Heavy Vehicles at Sedenak WIM Station

3.2 Yong Peng

Figure 5 shows a comparison between the percentages of weighted heavy vehicles at Yong Peng Weight-In-Motion Station. A total of 1,850,915 heavy vehicles were weight at Yong Peng Weight-In-Motion Station. It could be clearly seen that the majority of heavy vehicles traffic volumes (87.5%) is made of merely three types: single unit trucks with two axles (17.0%), single trailer truck with 4 axles (35.2%) and buses (35.3%).

All types of multi-trailer vehicles forms only 1.6%. The multi-trailers with five axles forms 1.1% and the multi-trailer with seven axles forms only 0.5%. Also, only 33 multi-trailer truck with six axles passed the weighting station. All other types of heavy vehicles form 14.6%.

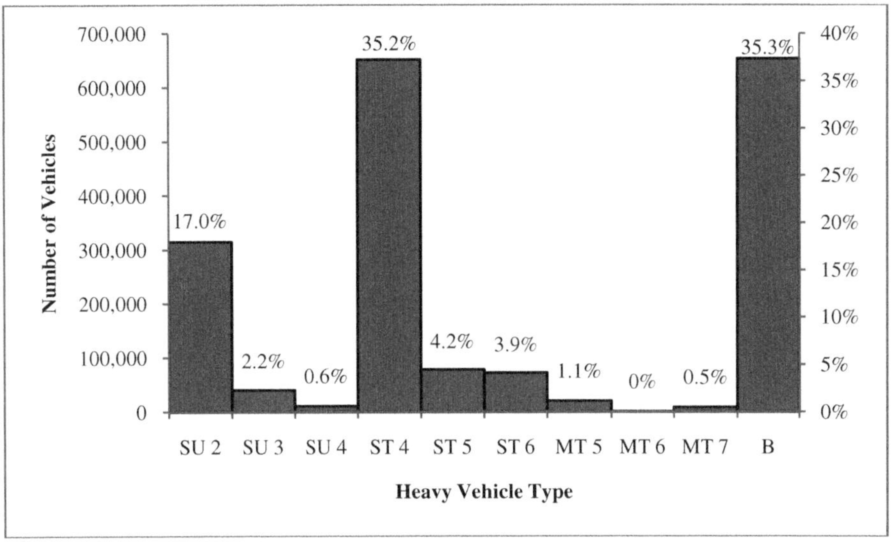

Fig. 5. Heavy Vehicles Traffic Volume at Yong Peng WIM station

Figure 6 compares the rutting damage ratios calculated for each heavy vehicle classes. It can be clearly seen that, the highest rutting damage ratio was caused by single trailer trucks with six axles (17.22). This indicates that, one pass of this vehicle causes rutting damage approximately 17 times higher than the rutting damage caused by standard axle load. On the other hand, the lowest rutting damage ratio is caused by single unit trucks with two axles (1.89).

Overall, single unit trucks causes a rutting damage ratio ranged between 1.89 and 4.35, single trailer tucks causes a rutting damage ratio ranged between 6.74 and 17.22, and multi trailer trucks causes a rutting damage ranging between 5.22 and 7.74.

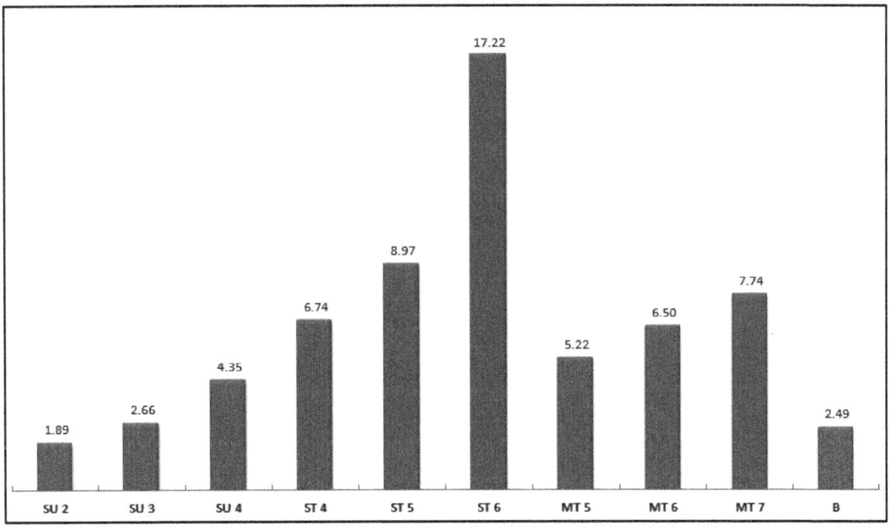

Fig. 6. Calculated Rutting Damage Ratios for Heavy Vehicles at Yong Peng WIM Station

3.3 Sedenak

Figure 7 shows a comparison between the percentages of weighted heavy vehicles at Sedenak Weight-In-Motion Station. A total of 2,557,805 heavy vehicles were weight at Sedenak Weight-In-Motion Station. It could be clearly seen that the majority of heavy vehicles traffic volumes (86.6%) is made of merely three types: single unit trucks with two axles (20.0%), single trailer truck with 4 axles (35.5%) and buses (31.0%).

All types of multi-trailer vehicles forms only 1.5%. The multi-trailers with five axles forms 1.0% and the multi-trailer with seven axles forms only 0.5%. Also, only 36 multi-trailer truck with six axles passed the weighting station (0.001%). All other types of heavy vehicles form 11.9%.

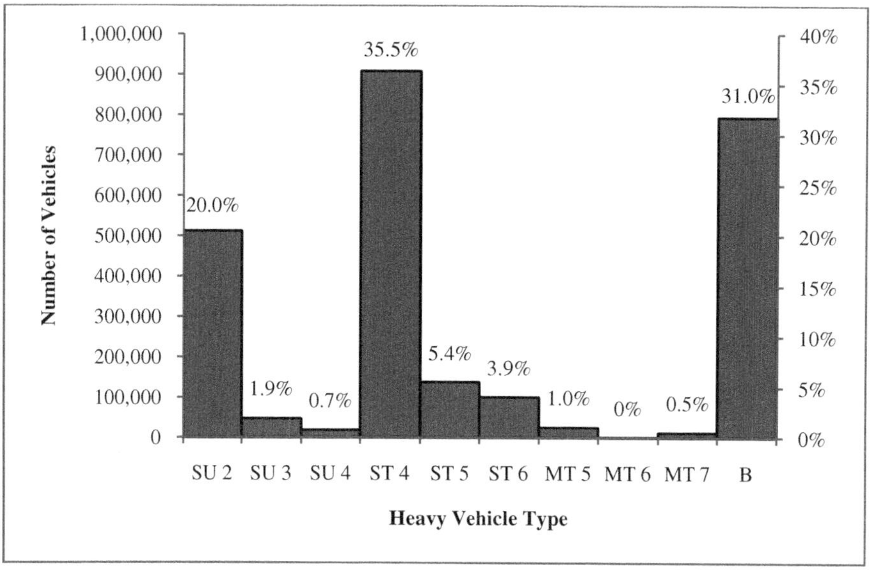

Fig. 7. Heavy Vehicles Traffic Volume at Sedenak WIM station

Figure 8 compares the rutting damage ratios calculated for each heavy vehicle classes. It can be clearly seen that, the highest rutting damage ratio was caused by single trailer trucks with six axles (12.92). This indicates that, one pass of this vehicle causes rutting damage approximately 13 times higher than the rutting damage caused by standard axle load. On the other hand, the lowest rutting damage ratio is caused by busses (2.29).

Overall, single unit trucks causes a rutting damage ratio ranged between 2.59 and 4.80, single trailer tucks causes a rutting damage ratio ranged between 5.84 and 12.92, and multi trailer trucks causes a rutting damage ranging between 4.58 and 7.01.

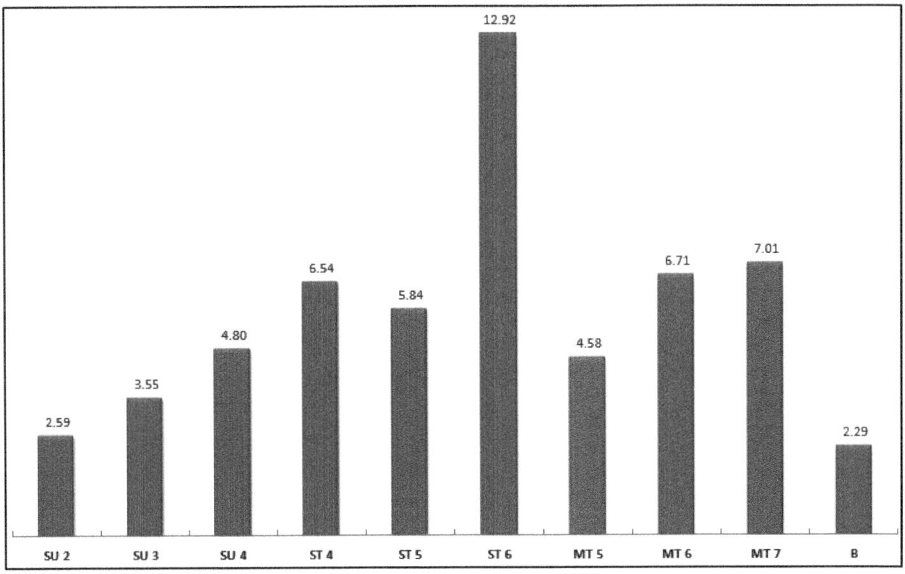

Fig. 8. Calculated Rutting Damage Ratios for Heavy Vehicles at Sedenak WIM Station

4 Conclusion

For the pavements' new design and overlay design at the southern region of the Malaysian Peninsula, the average rutting damage ratios based on actual weights for each vehicle type were calculated as follows:

1. Single unit trucks with two axles = 2.51
2. Single unit trucks with three axles = 3.98
3. Single unit trucks with four axles = 4.73
4. Single unit trailers with four axles = 6.01
5. Single unit trailers with five axles = 7.07
6. Single unit trailers with six axles = 8.18
7. Multi-trailer trucks with five axles = 4.81
8. Multi-trailer trucks with six axles = 8.56
9. Multi-trailer trucks with seven axles = 8.67
10. Buses = 2.90

In case of inaccurate traffic data, a weighted average of 4.34 can be used to represent rutting damage ratio of heavy vehicles.

Acknowledgements. The authors would like to express their gratitude to the Fundamental Research Grant Scheme for initiating the grant awarded by the Ministry of Education (MOE), and also greatest support from Malaysia Institute of Transport (MITRANS), Research Management Institute (RMI), and Universiti Teknologi MARA (UiTM).

The writers would also like to express their gratitude to PLUS Expressway Bhd. and Soil Central Lab Sdn. Bhd. for providing WIM and FWD data.

References

[1] Yassenn, O.M., Hafez, M.A., Endut, I.R., Baharom, B., Wahab, M.Y.: Overloading at the Northern Part of the Malaysian Expressway. Presented at the 2011 IEEE Colloquium on Humanities, Science and Engineering Research, Penang, Malaysia (2011)

[2] Yassenn, O.M., Hafez, M.A., Endut, I.R., Baharom, B., Wahab, M.Y.: Axle Load Distribution and Overloading at the Central Part of the North-South Expressway in Malaysia. European Journal of Scientific Research 79 (2012)

[3] Yassenn, O.M., Hafez, M.A., Endut, I.R., Baharom, B., Wahab, M.Y.: Axle Overloading at the Southern Part of the North-South Expressway in Malaysia. Presented at the 2012 IEEE Business Engineering and Industrial Applications Colloquium, Malaysia, (2012)

[4] Abdullah, A.S.: Development of Integrated Weigh-in-motion System and Analysis of Traffic Flow Characteristics considering Vehicle Weight. The University of Tokushima (2011)

[5] CSIR, Roads and Transport Technology. The Damaging Effects of Overloaded Heavy Vehicles on Roads (Pad27 - revised). Republic of South Africa (1997)

[6] P.U., (A) 275 Weight Restrictions (Amendment) Order (2003)

[7] Thailand Infrastructure Annual Report (2008),

[8] http://siteresources.worldbank.org/INTTHAILAND/Resources/333200-1177475763598/3714275-1234408023295/5826366-1234408105311/chapter2-transport-sector.pdf (accessed August 13, 2011)

[9] Wen, H., Jie, H., Peng, J., Xuhong, L.: Site survey and analysis of highway trucks overloading status quo in Anhui (2005), http://eprints.qut.edu.au/27727/

[10] Chan, Y.C.M.: Truck overloading study in developing countries and strategies to minimize its impact (2008)

[11] Chaudry, R., Memon, A.B.: Effects of Variation in Truck Factor on Pavement Performance in Pakistan (2000)

[12] Chou, C.-P.J.: Effect of overloaded heavy vehicles on pavement and bridge design. Transportation Research Record: Journal of the Transportation Research Board 1539, 58–65 (1996)

[13] Salama, H.K., Chatti, K., Lyles, R.W.: Effect of Heavy Multiple Axle Trucks on Flexible Pavement Damage Using In-Service Pavement Performance Data. Journal of Transportation Engineering 132, 763–770 (2006)

[14] Embong, H.M.B.H.: Kajian Kerosakan Permukaan Jalan Raya Dan Kerja-Kerja Membaiki Di Bawah Jabatan Kerja Raya Kuala Terengganu. Bachelor of Civil Engineering, Faculty of Civil Engineering, Universiti Teknologi Malaysia, Kuala Terengganu (2004)

[15] Gillespie, T.D.: Effects of heavy-vehicle characteristics on pavement response and performance: Transportation Research Board National Research (1993)

[16] I. R. D. I. (IRD). Weigh-in-Motion (WIM) (August 5, 2012), http://www.irdinc.com/systems/wim/index.php

[17] Huang, Y.H.: Pavement analysis and design. Pearson/Prentice Hall, United States of America (2004)

Enhancing Virtual Manipulatives for After-School Tutoring in the Subtraction Unit

Wen-Chung Shih

Department of Computer Science and Information Engineering,
Asia University
500, Lioufeng Rd., Wufeng, Taichung 41354, Taiwan
wjshih@asia.edu.tw

Abstract. The learning of subtraction is essential for elementary students. However, common subtraction misconceptions hinder mathematics underachievers from learning subtraction. Existing remedial instruction focuses on teacher-oriented or self-learning approaches. However, we think that the tutor-based approach with virtual manipulatives can provide students with an opportunity to correct their misconceptions. In this study, a simple tutoring method based on misconception diagnosis is proposed, and supportive virtual manipulatives are prepared for voluntary tutors to use. First, the tutor identifies the tutee's misconception by conducting a diagnostic test. Then, the tutor applies corresponding virtual manipulatives to help the tutee to understand the concept. For the first case manipulatives from the National Laboratory of Virtual Manipulatives (NLVM's) are used, and the tutor-tutee interaction patterns are identified by sequential analysis. An enhanced virtual manipulative was then developed according to the aforementioned analysis, and applied to the second case to investigate its performance. Evaluation results indicate that despite the need for further improvement, the tutoring method with the enhanced virtual manipulative is effective for correcting subtraction misconceptions. Finally, recommendations for future research are proposed.

1 Introduction

The learning of subtraction is essential for elementary students, yet at the level of elementary students, misconceptions are common. Furthermore, helping elementary students to develop subtraction literacy is not an easy task. These misconceptions conflict with scientifically accepted subtraction concepts and seriously hinder the comprehension and application of subtraction. Vanlehn's [29] study of the systematic errors that students make while attempting to perform subtraction came to similar conclusions. Chang et al. [6] reported on misconceptions in mathematics and showed that students always had misunderstandings in processing problem solving, even for simple skills such as addition or subtraction. The forming of misunderstandings may be due to experience in real life or come from their learning experiences in the classroom [26]. Brown and Burton [4] found that some misconceptions occurred in the solving procedures of the subtraction exercises. Among these misconceptions of

M. Gen et al. (eds.), *Industrial Engineering, Management Science and Applications 2015*,
Lecture Notes in Electrical Engineering 349, DOI: 10.1007/978-3-662-47200-2_47

the subtraction unit, there are six types to be frequently found in students. For example, students might think that the answer of (32 – 19) is 27. The students might, by mistake, subtract a smaller digit from a larger digit, but the smaller one may be a minuend and the larger one a subtrahend.

Most previous studies of remedial instruction have focused on self-learning methods, and researchers have assumed that low-achievers would be able to perform well. Chen [7] proposed a personalized diagnosis and remedial learning system, and showed that learners who received personalized remedial learning guidance achieved improved learning performance. Huang et al. [13] have developed a computer-assisted mathematics learning system to serve as a supplementary tool that helps teachers conduct remedial instruction. Heh et al. [11] proposed a full-loop learning system to identify the students' misconceptions by using a knowledge map, and then selecting suitable learning materials, according to the misconceptions of individual students, to aid their remedial learning. Volunteer tutoring emerges as one kind of possibility to solve this problem. The instructional models of tutoring in small groups and in one-to-one settings have been investigated in the literature [3, 30]. The tutoring model has been characterized by the main advantage of taking care of the tutee's learning needs [9].

Recent studies have proposed virtual manipulatives to help students learn mathematics. The learner can move the computer mouse to operate dynamic visual objects [2, 22, 23]. The virtual manipulatives mentioned in the literature are not simply physical copies of the teaching aids. Virtual manipulatives and physical teaching aids are different in function and effect, such as interactivity and availability [12, 23, 25, 27]. Despite the recognized potential of virtual manipulatives for helping students to understand mathematics, the design and development of virtual manipulatives for learning mathematics still requires further study.

In this study, a simple tutoring method is proposed, and supporting virtual manipulatives are prepared to assist voluntary tutors. First, the tutor identifies the tutee's misconception by conducting a diagnostic test. Then, the tutor applies several kinds of virtual manipulatives to allow the tutee to understand the concept. A formative evaluation is conducted by using a case study to explore the effects and limitations of the proposed approach. Evaluation results indicate that despite the need for further improvement, two-phased tutoring is effective for correcting subtraction misconceptions.

2 Related Work

The proposed tutoring method integrates a tutor-based approach and a computer-assisted self-learning approach in which the tutor follows a pre-defined work-flow to help the tutee learn mathematics. The theory of multiple representations is employed to develop virtual manipulatives for remedial instruction. This section introduces related work about an instruction model for tutoring, a learning model for simulation-based computer assisted learning and the theory of mathematical representation.

Karsenty [16] proposed an instructional model for preparing tutors. This instructional model consisted of a sequence of stages. In the first two stages, mathematical concepts and representations were discussed. In the next stage of instruction, the counselor discussed with tutors possible difficulties that students might encounter with these concepts. In the fourth stage of the instruction sequence, tutors were introduced to teaching approaches that might assist students in overcoming the difficulties discussed.

Liu [19] has developed simulation-based computer assisted learning to correct students' statistical misconceptions based on cognitive conflict theory. This model and its corresponding system is a good approach for students to learn by themselves. However, most underachievers need teachers or tutors to accompany them and provide adaptive instruction. Moreover, if the system is not intelligent enough, human supervisors are needed to standby and provide suitable support to the tutees.

Mathematical representation means that learners represent an external concept in different cognitive symbols when solving a problem. Mathematical representation plays an important role in the learning of mathematics and learning to construct knowledge [14, 28]. Bruner [5] proposed that the progress of learning consists of three stages: operations, images and symbols. Heddens [10] divided the students' learning process into four stages: Concrete, Semi-concrete, Semi-abstract and Abstract, advocating that learners must establish a good link between the real world and the abstract world, thereby ameliorating the difficulties in the understanding of mathematical concepts. Kaput [15] presented four representations to model the relationship between mathematical representation and mathematical learning, which included three internal representations and one external representation. Lesh et al. [17] pointed out five external representations used in mathematics education including real world object representation, concrete representation, arithmetic symbol representation, spoken-language representation, and picture or graphic representation. Willis and Fuson [31], who explored the use of picture-characterization with elementary school sophomores, found that the effectiveness of their approach was enhanced. Based on the theory of mathematical representation, Hwang et al. [14] developed a Virtual Manipulatives and Whiteboard (VMW) system, which allowed users to manipulate virtual objects in 3D space and find clues to solve geometry problems. In this work, we apply mathematical representation to remedial instruction by providing operational virtual manipulatives.

3 Methodology of Phase 1: Analysis of Tutor-Tutee Interaction Patterns

To explore the tutor-tutee interactions in applying virtual manipulatives to tutoring this study conducted an exploratory case study with observation and analysis of a videotaped tutoring case. The participants in the first phase of the study were one third-grade boy and a graduate student as a tutor. "Base blocks", the virtual manipulative available from the website of National Laboratory of Virtual Manipulatives (http://nlvm.usu.edu/), was used as a tutorial resource.

Base blocks consist of individual "units," "longs," "flats," and "blocks" (ten of each set for base 10). They can be used to show place value for numbers and to increase understanding of addition and subtraction. Blocks representing positive numbers are colored blue. Negative numbers are displayed using red blocks. If a user places a blue block on top of a red block of the same size (click-hold-drag with the mouse) they will disappear or "cancel" each other out because they add to zero. The same thing happens when a red block is placed on top of a blue block of the same size, as shown in Fig. 1.

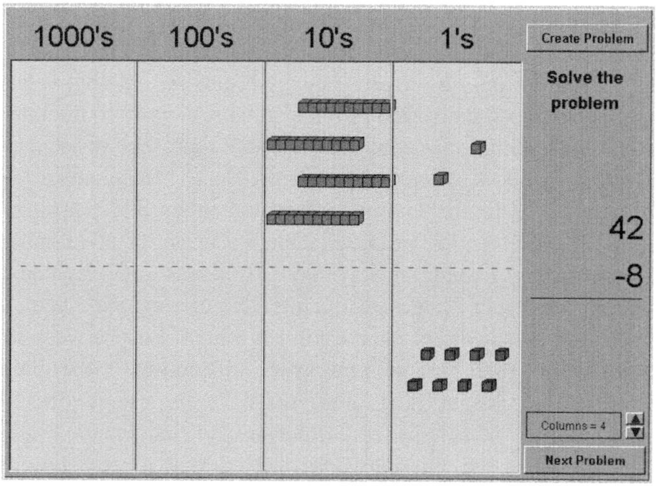

Fig. 1. The virtual manipulative: "Base blocks"

The tutor-tutee interaction coding scheme was based on a design from related studies [8, 20, 24]. Six codes were included in this coding scheme: T1 (providing guidance), T2 (correcting mistakes), T3 (asking questions), S1 (practicing the VM), S2 (solving problems with the VM), and S3 (sharing and expressing ideas), as shown in Table 1.

Table 1. Coding scheme for tutor-tutee interaction in the tutoring activities of Phase 1

Code	Actor	Dimension	Description
T1	Tutor	providing guidance	The tutor explains how to use the VM.
T2	Tutor	correcting mistakes	The tutor identifies the student's mistake and provides direction.
T3	Tutor	asking questions	The teacher asks questions to guide the tutee.
S1	Tutee	practicing the VM	The tutee practices with the VM independently.
S2	Tutee	solving problems with the VM	The tutee operates the VM to solve the problems.
S3	Tutee	sharing and expressing ideas	The tutee shares her/his experience and ideas with the tutor.

For content analysis, each interaction behavior was coded by three educational researchers. Coding the one-hour tutoring activities produced 134 codes. The value of the Kappa coefficient was .91 (p<.001), indicating that the data had very good inter-rater reliability. Lag sequential analysis was used to explore whether significant tutor-tutee interaction patterns existed [1]. The sequential analysis software Multiple Episode Protocol Analysis 4.10 (MEPA 4.10, http://edugate.fss.uu.nl/mepa/), designed by Gijsbert Erkens, was used in this study.

The distribution of coded interaction behaviors during the one-hour tutoring activities is shown in Fig. 2. The results indicate that S2 (solving problems) was the most frequent behavior (n=35, 26.12%), followed by T1 (guidance: n=31, 23.13%) and S1 (practicing: n=24, 17.91%). The frequencies of the other three codes (T2, T3, and S3) were lower than the top three frequencies. S3 (sharing and expressing ideas) constituted only 8.21% of all interactional behaviors during the tutoring activities. It is worth noting that in the tutoring activities, the tutee rarely shared and expressed ideas, and the tutee tended to practice the VM (17.91%) or solve problems with the VM (26.12%). Also, T1 (guidance) was an important interaction behavior for the tutee, which was helpful for the tutee to learn during the tutoring activities.

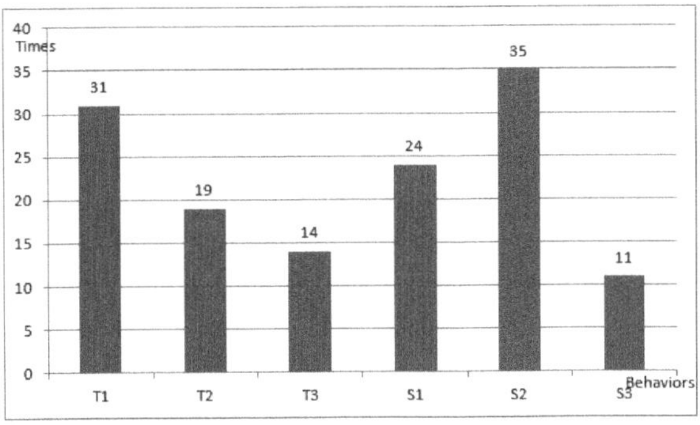

Fig. 2. Distribution of quantitative content analysis of interaction behaviors in the tutoring process

The adjusted residual table of sequential analysis was then inferred (as shown in Table 2).

Table 2. Adjusted residuals table of sequential analysis (lag=1)

	T1	T2	T3	S1	S2	S3
T1	-1.71	0.29	1.01	1.83	1.22	0.73
T2	1.52	-2.08	0.56	1.96	2.47*	1.38
T3	-0.67	-2.15	-2.20	0.36	0.24	2.28*
S1	1.76	-3.62	-2.84	-0.98	3.55*	0.08
S2	3.20*	1.87	1.42	1.27	-1.14	1.88
S3	-1.26	-2.30	-3.05	-2.63	-1.92	-3.22

*p<0.05

The first case study found that the student's behavior in practicing the VM was significantly followed by his behavior in solving problems with the VM (S1→S2). This result indicated that the tutee tended to practice before he began to solve problems. Also, it was found that the tutor's correcting behaviors were significantly followed by the tutee's behaviors of solving problems with the VM (T2→S2). This result indicated that the tutee tended to go to solve problems right after mistakes were corrected. The following excerpts of dialogue from the content analysis of the video demonstrate that the questions provided the opportunity for the tutee to explain the motivation behind his behavior.

Tutor: There doesn't seem to be enough blocks. Right?
Tutee: Yes!
Tutor What can we do?
Tutee: We can borrow more blocks from the left!
Tutor: Go ahead.
(20111230-S3-1-T003)

It is noteworthy that the sequence S2→T1 achieved significance ($z=3.20$, $p<0.05$), indicating that the tutor's guidance followed the tutee's problem-solving. To clarify the reason behind this behavior we interviewed the tutor and the tutee and found that the main reason is that the tutee needed the tutor's guidance to solve the problem (22 from the total of 35 cases). Furthermore, as indicated by the tutee, it was not easy for the tutee to learn the VM and then use it to solve the math problem without the tutor's help. This sequence suggests that providing suitable support (such as giving a hint to the next step) helps the tutee learn to solve problems independently. However, the existing VM lacks design about guidance. This result motivates us to enhance the VM by incorporating new guidance features.

4 Methodology of Phase 2: Enhancing the Virtual Manipulative

According to the analysis of Phase 1, an enhanced VM is developed to improve the tutoring process. Phase 2 of this study uses the case-study method [18, 21], which is considered to be "an empirical inquiry that investigates a contemporary phenomenon within its real-life context, especially when the boundaries between phenomenon and context are not clearly evident" [32]. This means that a case study is an empirical and holistic inquiry that explores a social unit, a single instance, or a phenomenon within a natural setting [21]. Therefore, this approach is particularly appropriate for exploring the possible effects that the proposed diagnostic and remedial methods have on the tutoring sessions [18, 28].

One third-grade girl and a graduate student, different from those in Phase 1, participated in this case study during January 2012. The subject of this research is a third-grade student from an elementary school in Taichung, Taiwan. She was selected from a class which had eight students. The performance ranking of this child fell within the bottom 25% to 40%, according to the midterm assessment of the school. The selection criteria for the student included parental agreement to participate in this study. It should be noted that students with distinct learning disabilities and/or evidence of violent behavior were excluded from this study.

During the tutoring sessions, the tutor sat with the tutee in a separate area. Tutoring was conducted according to a concept of a "fresh start", where the key was not students' previous knowledge but rather their willingness to think and keep an open mind about the material discussed. Sessions were usually guided by the virtual manipulatives. The procedure for this study is composed of five stages:

- Preparation. In this stage, common misconceptions of third graders' "subtraction" are collected according to related literature. Next, diagnostic items and remedial materials are designed and implemented in terms of these conceptions. In addition, test items for a pre-test are designed.
- Case Selection. A misconception test is administered to the class as a pre-test. The students who are underachievers and willing to participate in this tutoring program are selected as the cases. The behavior of the selected cases is analyzed with respect to their performance in school and at home.
- Misconception Diagnosis. Diagnostic tests are administered to the selected cases. The tutor interviews the tutee to identify their misconception.
- Remedial Instruction. The tutor uses virtual manipulatives to conduct remedial instruction.
- Evaluation. A misconception test is administered to the selected cases as a post-test. Also, the tutor and the tutee are interviewed by the researcher to understand the effect of the proposed approach.

The current study utilizes a case study to evaluate whether the proposed approach can correct subtraction misconceptions and what the cases' perceptions are. A comparison of pre-test and post-test scores on the subtraction unit showed that this tutoring process could reduce misconceptions about subtraction concepts in both students. Three of the four misconceptions held by Student B were corrected. To further explore how the proposed approach benefited the tutee, the following sections present the results and discussions about the effects and limitations of this tutoring method.

The tutoring process includes two steps: Diagnosis and Remedial Instruction. This section describes the tutoring process observed in the misconception diagnostic phase. In addition, the subsequent interview results are presented to show how the tutee benefited from the diagnostic method. For brevity, the following results and discussion focus on the unit "subtraction of three-digit numbers". The items of pre-test are designed to reflect misconceptions in the options. For example, an item has four options. One is the correct answer, and the other three represent three misconceptions.

According to the pre-test results, the tutee fails in the unit of "subtraction of three-digit numbers". First, the tutor needs to verify that the tutee really has misconceptions in this unit. Hence, the tutor downloads a diagnostic item from the repository and helps the tutee to do it. In the first item, case A drags the option "156" to the answer area. In the second item, case A drags the option "233" to the answer area. These two actions show that the tutee has the misconception of "using a large number to subtract a small number".

This section describes the tutoring processes observed in the remedial instruction phase. In the misconception diagnostic phase, the tutor found that the tutee has the misconception of "using a large number to subtract a small number". Therefore, the tutor downloads an operational virtual manipulative about the concept of "borrowing" from the repository. Then the tutee practices it to conduct remedial instruction. This operational virtual manipulative guides the tutee to use the mouse to show the correct sequence for subtracting three-digit numbers.

After the remedial instructions, the tutor asks the tutee to do another item: "502 – 379 = ?" The tutee wrote the right process and answer. Seeing the tutee modify his answer in the process, the tutor asks, "Why did you replace 127 by 123?" The tutee answered, "I used to subtract the small number from the large number. But I immediately recall from the virtual manipulative that I should borrow from the higher digit. So I modify the answer. " This interview illustrates that the tutee has corrected the misconception.

The researcher interviewed the tutee to understand their perceptions about the proposed approach. For the diagnostic process, the tutee agreed that the process was helpful to identify her misconception. She reported, "The diagnostic tool can identify my misconception, and it is fun."

Regarding the remedial instruction phase, the tutee also agreed the virtual manipulatives are interesting. She said, "I think the interactive tool is like video games, which attracts me to learn mathematics." In addition, the tutee thought the remedial approach can help her understand mathematic concepts. In the tutoring process, she felt focused. More important, mathematics seemed not to be as difficult as it usually was. That is, the tutee was more confident about learning mathematics.

The remedial instruction proposed in this work was used to promote the learning performance of underachievers in mathematics. While the previous sections show how the proposed method works and its effectiveness, this section discusses why it works. First, the interview data revealed that the feature of "operational guidance" enhanced the learning experience. For example, Student A stated in the interview, "The virtual manipulative showed and pointed out to me the next step in the subtraction process, which makes it easy to understand and memorize". Additionally, the misconception-based diagnostic test can help students identify their misconceptions. For example, Student B said, "The result of the diagnostic test is very helpful. I can learn what I really don't understand". Finally, the virtual manipulatives can improve students' motivation. For example, Student A said in the interview, "I like the animation and the virtual manipulatives. When doing this, I don't feel that learning is hard. "

Although the evaluation results showed that the two students corrected their misconceptions and that the designs of the proposed tutoring method could achieve their objectives, some limitations of this method were revealed from interview data and learning behaviors. For example, the tutors indicated that, for some misconceptions, the virtual manipulatives that are retrieved from the repository are not appropriate for the tutoring process. For example, the three existing virtual manipulatives still could not help case A to understand some difficult concepts. This limitation can be improved by creating more educational virtual manipulatives for the repository. After the tutor has helped students to resolve most misconceptions, it would be possible for the underachievers to return to classes for formal instruction. Although the operational guide in the virtual manipulatives provided clear hints for helping students learn the correct procedures for subtraction by following the animation's directions, the researcher found that, for some virtual manipulatives, Student B could not understand how representations from her manipulation results are related to the correct concepts of subtraction. This may have affected the student's subsequent learning, and probably thus failed to correct the misconceptions. These problems should be considered in future work.

5 Conclusion

The pilot study reported in this paper focused on tutoring mathematics to low-achieving students. It is well known from the literature that professional and paraprofessional tutoring both have very positive effects on students' achievements [16]. The novelty of the study is that nonprofessional tutoring can still provide a certain positive effect on students' achievements in elementary school mathematics.

Although understanding and applying concepts of subtraction are essential abilities, students often have a number of misconceptions. This study proposed a two-phase tutoring method, consisting of diagnosis followed by remediation, to eliminate misconceptions about the basic concept of subtraction. Two important elements were considered in the design and development of the proposed approach. One is the organization of virtual manipulatives based on a taxonomy of common misconceptions, which facilitates efficient retrieval of manipulatives. Another is the introduction of operational "virtual manipulatives" into the tutoring sessions, which makes the learning process interesting. The results of the evaluation indicate that students substantially reduced their misconceptions after the tutoring sessions.

Acknowledgments. This research was supported by Ministry of Science and Technology of Republic of China under the number of NSC100-2628-S-468-001-MY2, NSC101-2511-S-468-003, NSC102-2511-S-468-001 and MOST 103-2511-S-468 -003.

References

1. Bakeman, R., Gottman, J.M.: Observing interaction: An introduction to sequential analysis, 2nd edn. Cambridge University Press, Cambridge (1997)
2. Baker, M.: Merging technology & mathematics instruction: The case of virtual manipulatives & geometric concepts. In: Richards, G. (ed.) Proceedings of World Conference on E-Learning in Corporate, Government, Healthcare, and Higher Education, pp. 566–573. AACE, Chesapeake (2008)
3. Bloom, B.S.: The search for methods of group instruction as effective as one-to-one tutoring. Educational Leadership 41, 4–17 (1984)

4. Brown, J.S., Burton, R.R.: Diagnostic models for procedural bugs in basic mathematic skills. Cognitive Science 2, 155–192 (1978)
5. Bruner, J.S.: Toward a theory of instruction. Harvard University, Cambridge (1966)
6. Chang, K.-E., Lin, M.-L., Chen, S.-W.: Application of the Socratic dialogue on corrective learning of subtraction. Computers & Education 31(1), 55–68 (1998)
7. Chen, L.H.: Enhancement of student learning performance using personalized diagnosis and remedial learning system. Computers & Education 56(1), 289–299 (2011)
8. Flanders, N.A.: Analyzing teaching behavior. Addison-Wesley, Reading (1970)
9. Gaustad, J.: Tutoring for at-risk students. Oregon School Study Council, Eugene (1992)
10. Heddens, J.W.: Today's mathematics, 5th edn. Science Research Associates, Chicago (1984)
11. Heh, J.-S., Li, S.-C., Chang, A., Chang, M., Liu, T.-C.: Diagnosis Mechanism and Feedback System to Accomplish the Full-Loop Learning Architecture. Educational Technology & Society 11(1), 29–44 (2008)
12. Highfield, K., Mulligan, J.T.: The role of dynamic interactive technological tools in preschooler's mathematical patterning. In: Watson, J., Beswick, K. (eds.) Proceedings of the 30th Annual Conference of the Mathematics Education Research Group of Australasia, Hobart, vol. 1, pp. 372–381. MERGA, Adelaide (2007)
13. Huang, T.-H., Liu, Y.-C., Chang, H.-C.: Learning Achievement in Solving Word-Based Mathematical Questions through a Computer-Assisted Learning System. Educational Technology & Society 15(1), 248–259 (2012)
14. Hwang, W.-Y., Su, J.-H., Huang, Y.-M., Dong, J.-J.: A Study of Multi-Representation of Geometry Problem Solving with Virtual Manipulatives and Whiteboard System. Educational Technology & Society 12(3), 229–247 (2009)
15. Kaput, J.J.: Representation systems and mathematics. In: Janvier, C. (ed.) Problems of Representation in Teaching and Learning of Mathematics, pp. 159–195. Lawrence Erlbaum, Hillsdale (1987)
16. Karsenty, R.: Nonprofessional mathematics tutoring for low-achieving students in secondary schools: A case study. Educ. Stud. Math. 74, 1–21 (2010)
17. Lesh, R., Post, T., Behr, M.: Representations and translations among representations in mathematics learning and problem solving. In: Janvier, C. (ed.) Problems of Representation in the Teaching and Learning of Mathematics, pp. 33–40. Lawrence Erlbaum, Mahwah (1987)
18. Liu, T.C.: Teaching in a wireless learning environment: A case study. Educational Technology & Society 10(1), 107–123 (2007)
19. Liu, T.-C.: Developing Simulation-based Computer Assisted Learning to Correct Students' Statistical Misconceptions based on Cognitive Conflict Theory, using "Correlation" as an Example. Educational Technology & Society 13(2), 180–192 (2010)
20. Liu, E.Z.F., Chang, M.: Effects of improvement on selective attention: Developing appropriate somatosensory video game interventions for institutional-dwelling elderly with disabilities. Turkish Online Journal of Educational Technology 12(1), 9–18 (2013)
21. Merriam, S.B.: Qualitative Research and Case Study Applications in Education. Jossey-Bass, San Francisco (1998)
22. Moyer, P.S., Bolyard, J.J., Spikell, M.A.: What are virtual manipulatives? Teaching Children Mathematics 8(6), 372–377 (2002)
23. Moyer, P.S., Salkind, G., Bolyard, J.J.: Virtual manipulatives used by K-8 teachers for mathematics instruction: Considering mathematical, cognitive, and pedagogical fidelity. Contemporary Issues in Technology and Teacher Education 8(3), 202–218 (2008)

24. Perraton, H.: The role of theory and generation in the practice of distance education (Report No. CE049375). Zentrales Institut fur Fernstudienforschung, Hagen (1987) (ERIC Document Reproduction Service No. ED290015)
25. Reimer, K., Moyer, P.S.: Third graders learn about fractions using virtual manipulatives: A classroom study. Journal of Computers in Mathematics and Science Teaching 24(1), 5–25 (2005)
26. Schoenfeld, A.H.: Cognitive science and mathematics education. Erlbaum, Hillsdale (1987)
27. Steen, K., Brooks, D., Lyon, T.: The impact of virtual manipulatives on first grade geometry instruction and learning. Journal of Computers in Mathematics and Science Teaching 25(4), 373–391 (2006)
28. Tsui, C.Y., Treagust, D.F.: Genetics reasoning with multiple external representations. Research in Science Education 33(1), 111–135 (2003)
29. VanLehn, K.: Mind bug: The origins of procedural misconceptions. MIT Press, Cambridge (1990)
30. Wasik, B.A., Slavin, R.E.: Preventing early reading failure with one-to-one tutoring: A review of five programs. Reading Research Quarterly 28(2), 178–200 (1993)
31. Willis, G.B., Fuson, K.C.: Teaching children to use schematic drawings to solve addition and subtraction word problems. Journal of Educational Psychology 80, 192–201 (1988)
32. Yin, R.: Case study research: Design and methods, 2nd edn. Sage Publishing, Beverly Hills (1994)

Implementing an Information System Development Simulation in an Industrial Engineering Class: A Case Study

Sakgasit Ramingwong[1,*] and Lachana Ramingwong[2]

[1] Excellence Center in Logistics and Supply Chain Management, and
Department of Computer Engineering, Faculty of Engineering, Chiang Mai University,
Chiang Mai, Thailand
sakgasit@eng.cmuac.th

[2] Department of Computer Engineering, Faculty of Engineering, Chiang Mai University,
Chiang Mai, Thailand
lachana@gmail.com

Abstract. Development of an information system is a challenging task due to many of its unique and sensitive nature. As stakeholders, it is important for both managerial and technical counterparts to understand the challenges in information system development. This paper explores the efficiency of implementing a simulation workshop on information system development in an industrial engineering class as well as report lesson leant from the industrial points of view.

Keywords: Simulation, Education, Information system development, Case study.

1 Introduction

Information systems have become a critical foundation for modern organizations. Indeed, technology can increase efficiency and effectiveness of virtually all business processes of a supply chain. Appropriate implementation and utilization of information systems can yield major competitive edges. Yet, due to many unique characteristics of information systems, the development could be challenging. Human factor is one of the most vulnerable issues in information system development [1][2][3][4]. Any forms of applications that can help improve mutual understanding of stakeholders could well increase the chance of successful implementation [5].

The supply chain is a system of necessary elements that are involved in moving products or services from producers to consumers. This includes people, documents and processes [6]. Different type of business could have different supply chain elements and sequences. Basic implementation of information system in supply chain involves digitizing of documentation, information flows and all forms of machinery assistances. Higher precision, higher efficiency, scalability and lower lead-time are the most common benefits of information systems [7].

* Corresponding author.

M. Gen et al. (eds.), *Industrial Engineering, Management Science and Applications 2015*,
Lecture Notes in Electrical Engineering 349, DOI: 10.1007/978-3-662-47200-2_48

Information systems can be classified into a few categories based on their complexity and capability. The higher the competency they have, the more difficult they can be developed [8]. The most complete proprietary information system might not be as efficient as a small tailor-made application in certain business environment.

It is reported that the success rate of information system implementation is interestingly low. Despite of all modern technology, techniques and professional approaches, global information system projects struggle at approximately 50% chance of success [9]. Intangible parts of information systems, especially software, make the project progress extremely difficult to be tracked. Lack of understanding from management level is reported to be a critical factor for failure [10]. Various other issues, such as requirement creeps, communication gap, and user resistance, are some of the key challenges in information system development [10]. The more the stakeholders realize these constraints, the higher chance of success the projects have [11][12].

This paper reports a result of implementing Engineering Construction for Software Engineers (ECSE), an educational simulation on software/information system development, in a class of graduate multi-cultural students. The second section of this paper describes the challenges in information system development and the necessary of simulation. The third and fourth sections respectively discuss the research methodology and discussion of the result. Finally, the fifth section concludes the paper.

2 Information System Development and ECSE Simulation

Information systems play critical roles in modern businesses. It supports both basic functions such as accounting and inventory and advance competitive activities such as forecasting and decision making. Major benefits of information systems involve increased efficiency, higher accuracy, faster processing time, scalability and portability. Implementing an information system has become increasingly cheaper due to the advancement of computer technology.

However, the development of an information system is challenging. During the past decade, it is reported that more than half of global information system projects neither achieved planned requirement, budget nor schedule [9]. Some of the projects were even cancelled before its completion. Several reasons are claimed to be the factors behind this unfortunate phenomenon. Examples of these issues include lack of executive support, lack of technical expertise and lack of user involvement.

Information systems consist of two main components, hardware and software [13]. Installation of hardware and networking could be costly but not generally complicated. This is however different on the software side. Due to its intangible form, planning and tracking of their progress can be challenging [14]. Unstable requirements, changes, technical difficulties and many more issues are also expectable in software projects. For example, everyone realizes that it is illogical to ask the engineers to add two more floors or make a large change on the structure of a half-finished building. In contrast, it is rather common that the stakeholder ask for more features during the implementation of a software project. In some cases, additional requirements are made during the acceptance or integration testing [15]. Some of these changes might be completely different from the initial requirements. Proceeding without a strong understanding towards such characteristics could lead to numerous problems afterwards.

In educational level, simulations are frequently implemented in order to provide comprehensive insight to students [16]. There are several simulations that focus on system/software development processes [17][18][19][20]. However, most of them highlight technical side of the project and are suitable for skilled personnel. Engineering Construction for Software Engineers (ECSE) [21], however, is a system development simulation that is designed to educate general participants on information system development from the technical perspective. Since no actual technical skills are needed, this workshop is suitable to any relevant stakeholders.

In ECSE, the participants are to form a group of engineers who are responsible for constructing a toy house. Since the house is made from corrugated plastic board and common office materials, it is more similar to software than an actual house in the sense that every component can be swapped, changed or modified with some efforts. With regards to a number of changing specifications, only several features are deemed as "true requirements". Within 90 minutes and given resources, the participants strive to make the product as close these "true requirements" as possible. Together with the introduction and debriefing, the total time required for the workshop is 150 minutes. Two main stakeholders, which are engineers and customers, are portrayed by the participants and instructors, respectively. Some of the main key learning points achieved from ECSE include:

- *Focusing on the objectives of the system* – In information system development, the stakeholders need to mainly focus to the main objectives of the system rather than trivial features. For example, the key question the development team should ask the customer is what the main objective of this house is. Important issues such as function and usability must also be prioritized over cosmetics such as color and alignment. In real-life scenario, overlooking over this key point can ultimately lead to disaster. In ECSE, the "true requirements" represents these objectives.
- *Importance of communication* – In information system development, where the product is largely intangible during the development, communication is critical for the success of the project. Frequent communication can efficiently help bridging the gaps of understanding between stakeholders. Design is a very important document which can further clarify the requirements. Indeed, the longer time the team spends on discussing and working on the design of the main functions, the shorter time it needs to make changes to the product afterwards.
- *Unstable requirements* – Changes are very common in information system development. As long as the changes lead to better services of the system, the stakeholders should not resist them. Certain development methodologies are more appropriate for continuous requirement change. One practical strategy for this workshop is to develop the system as modules and, similar to an aforementioned point, prioritize the development of the core features. In this way, when requested, the team needs to work on only parts of the product that is relevant to changes.

- *Time management* – Time is always an essential essence. Controlling the schedule in information system development could be more difficult than other engineering projects since it is usually difficult to measure the progress of parts of the system, notably the software. In order to manage time efficiently, the developer should break the system into small measurable modules and, obviously, prioritize the more important features.
- *Investment in appropriate tools* – Two types of tools are available to be purchased in ECSE. Either of them is suitable for certain tasks but not for other. Since the tools are rather expensive, the participants are forced to plan ahead and decide earlier what building strategy and techniques they use. Inadequate tools can result in severe delay and low quality products.

3 Research Methodology

This research implemented ECSE in "255766 Information Systems for Logistics" course for master of engineering in Logistics and Supply Chain Management, Faculty of Engineering, Chiang Mai University, Thailand. Eighteen students participated in this 150 minutes simulation. Among them, there were 13 Thai and 5 German students. Most of them did not have experience in information system development. Only a few of these students roughly participated as consultants in implementation of certain enterprise resource planning (ERP) packages. The students were divided into 4 groups of 4-5 persons. They were equally given development resources at the beginning of the workshop. The students were allowed to form their own team and this resulted in 3 all-Thai and 1 all-German teams. The lecturer role-played as a customer who wanted a house but did not have much engineering knowledge and, without a clear guidance from the developers, could be confused by his own requirements.

Four features were selected as the "true requirements" for this workshop. Firstly, the customer expected to open a grocery store at the ground floor. However, due to certain reasons, the customer did not clearly specify this objective in the beginning of the project. As a result, this requirement would be revealed only when the development team asked for the main objectives of the house. Participants who did not ask such question would likely to be trapped in their common sense that they are building a general accommodation. The second "true requirement" was the requirement of a "hip roof". The hip roof was the modern type of roof which each side slope downwards to the walls. This requirement imitated a scenario when the customer actually knows his needs but was unable to express accurately. Design documents could become a major help for this learning point. Thirdly, a backdoor was selected as another "true requirement" in order to highlight the importance of simple mistake. Indeed, a back door is an essential part of a house due to safety reasons. However, it is highly possible that after being crowded with specifications, some of these critical foundations could be overlooked. The final "true requirement" involved the number of floors. During the simulation, certain scenario happened to the customer. This caused two major changes of this specification during the workshop. For instance, the customer stated the initial requirement that he wanted a three-floor house and changed to two-floor during the last thirty minute of implementation. This forced a scenario of reusing and reassembling parts of the product that is another important strategy in information system development.

After the activities, the students were asked to anonymously answer a set of questionnaire on their opinions towards the simulation. The results of this questionnaire are discussed in the next section.

4 Results and Discussion

Within the given time and resources, all groups of student completed the products. The German students won the simulation by achieving three "true requirements" while the other three Thai groups attain only half of them. Table 1 displays the final achievement of the participants.

Table 1. Final achievement of the participants

Team	Requirements			
	Grocery	Hip Roof	Backdoor	Two floors
#1 (Thai)			X	X
#2 (Thai)		X		X
#3 (Thai)			X	X
#4 (German)	X		X	X

Due to their lack of experiences in information system development, all teams naturally followed the classic Waterfall system development, i.e. they distinctively divided the development into planning, designing, implementation and testing, and processed them in a sequential fashion. All teams attempted to communication with the customer using designs. However, all of these designs were 2-dimensional top view of each floor of the house. No perspective views of the finished house were illustrated to the customer.

Based on observation, the German students were much more communicative than the Thais. They set up a meeting with the customer almost every 10-15 minutes, resulting in 8 meetings throughout the simulation. Their questions to the customer ranged from decorative to structural and objective issues. This helped the German team to identify final floor specification as well as the most difficult hidden requirement, i.e. the needs of grocery space on the ground floor. The backdoor was installed based on a suggestion of a team member without any specification from the customer. On the other hand, the Thai teams spent significantly less time to the customer. One of them made only 3 contacts during the entire workshop. Yet, they all successfully finalized the floor requirement within the time limit. One of the team queried the customer about the backdoor and accordingly implemented this feature while another team, similar to the Germans, installed the backdoor based on their member's suggestion. The fulfillment of the hip roof of the second team was actually done by chance, however.

After the simulation finished, the participants anonymously filled in the questionnaire. On a Likert scale of five where 3 means neutral, the enjoyability of this workshop are rated 3.78. The average score of 4.12 on appropriateness indicated that the students agreed that it is important for managerial stakeholders to understand the nature of information system development. The participants gave an average score of 3.78 when asked if this workshop gave them new knowledge on information system development. ECSE was scored 4.35 which is fairly high when being asked if it should be implemented for the next batch of students or not. Figure 1 summarizes feedbacks from the students.

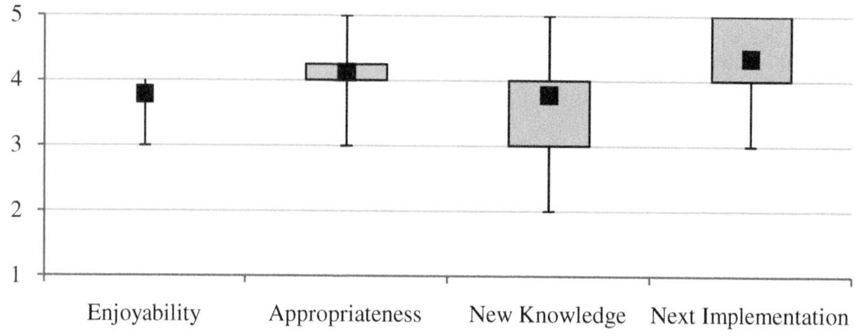

Fig. 1. Feedback of ECSE on a 5-scale Likert

The students were asked to give their opinions on satisfactory and unlikeable parts of the workshop. According to the answers, the most satisfactory facet of ECSE was reported to be the designing of the product. Additionally, the fun of implementation was also signified as another positive element of the simulation. Other noteworthy addresses included the enjoyment of teamworking and communication.

On the other hand, the participants commented that the workshop could be improved if more time is given to the developers. They also complaint that the continuously changing requirements are disturbing. The final important suggestion included the difficulty of finding the "true requirements".

When being asked for what new knowledge were gained from ECSE, the students revealed that this simulation made them realize the importance of the control of fluctuate requirements. Furthermore, they indicated that they recognized the importance of communication, both internal and external. Some students added that the criticality of teamworking, planning process and decision making were highlighted. Others indicated that they considerably experienced general challenges in information system development.

5 Conclusion

Information system is a key element in modern businesses. However, development of an information system can be challenging. Realizing their unique characteristics is one of the promising success strategies. Indeed, it is a responsibility of all stakeholders to make the project success. In order to increase chance of success, both managerial and technical counterparts need to realize unique challenges in information system development. This research attempts to implement and investigate the efficiency of a workshop, ECSE, to the management side of the stakeholders. In ECSE, the participants imitate the activities in information system projects by building a house from corrugate plastic board and basic office equipment. Due to many characteristics of the materials, the participants are expected to experience several unique nature of information system development.

Eighteen master students in Logistics and Supply Chain Management, Faculty of Engineering, Chiang Mai University participated the simulation. After spending 150 minutes in ECSE, they successfully submitted their product. Although none of the teams were unable to fulfill all essential requirements, some of them managed to get the most difficult feature right. Afterward, the students anonymously evaluated the workshop.

The evaluation of this case study revealed satisfying results. The high average score on appropriateness indicating that ECSE is suitable for any stakeholders. It gives the participants informative insights on key challenges of information system development while providing them teamworking and fun factors. As the materials can be easily purchased and the activities are not sophisticate, this workshop could be considered an option to transfer knowledge in information system development to any stakeholders.

References

1. Beckworth, G., Garner, B.: A Framework for the Success Factors in the Information System Development Process. Deakin University (1994)
2. Snodgrass, C.R.: Human Factors in Information Systems. IGI Global (2001)
3. Yeo, K.T.: Critical Failure Factors in Information System Projects. International Journal of Project Management 20(3), 241–246 (2002)
4. Martinsons, M.G., Chong, P.K.C.: The Influence of Human Factors and Specialist Involvement on Information Systems Success. Human Relations 52(1), 123–152 (1999)
5. Offen, R.: Domain Understanding is the Key to Successful System Development. Requirements Engineering 7(3), 172–175 (2002)
6. Shirley, R.: Practices for Managing Supply Chain Risks to Protect Federal Information Systems, ITL Bulletin (2012)
7. Auramo, J., et al.: The Roles of Information Technology in Supply Chain Management. In: 17th Annual NOFOMA Conference (2005)
8. Xia, W., Lee, G.: Complexity of Information Systems Development Projects: Conceptualization and Measurement Development. Journal of Management Information Systems 22(1), 45–83 (2003)
9. The Standish Group.: CHAOS Manifesto. The Standish Group (2013)

10. Nelson, R.R.: IT Project Management: Infamous Failures, Classic mistakes, and Best Practices. MIS Quarterly Executive 6(2), 67–78 (2007)
11. Hofmann, H.F., Lehner, F.: Requirements Engineering as a Success Factor in Software Projects. IEEE Software 18(4), 58–66 (2001)
12. Sjöström, J., Goldkuhl, G.: A Stakeholder-centric Approach to Information Systems Design. eChallenge (2010)
13. O'brien, J.A., Marakas, G.M.: Introduction to Information Systems. McGraw-Hill (2005)
14. Phillips, J.: IT Project Management: On Track from Start to Finish. McGraw-Hill (2002)
15. Williams, B.J., Carver, J., Vaughn, R.B.: Change Risk Assessment: Understanding Risks Involved in Changing Software Requirements. Software Engineering Research and Practice, 966–971 (2006)
16. Oh, E., van den Hoek, A.: Towards Game-based Simulation as a Method of Teaching Software Engineering. Frontiers in Education 3, S2G-13 (2002)
17. Zhu, Q., Wang, T., Tan, S.: Adapting Game Technology to Support Software Engineering Process Teaching: From SimSE to MO-SEProcess. In: The Third International Conference on Natural Computation, pp. 777–780. IEEE (2007)
18. Navarro, E.O., van den Hoek, A.: Comprehensive Evaluation of an Educational Software Engineering Simulation Environment. In: 20th Conference in Software Engineering Education & Training, pp. 195–202. IEEE (2007)
19. Blake, M.B.: A Student-enacted Simulation Approach to Software Engineering Education. IEEE Transactions on Education 46(1), 124–132 (2003)
20. Connolly, T.M., Stansfield, M., Hainey, T.: An Application of Games-based Learning within Software Engineering. British Journal of Educational Technology 38(3), 416–428 (2007)
21. Ramingwong, S., Ramingwong, L.: ECSE: A Pseudo-SDLC Game for Software Engineering Class. In: Overcoming Challenges in Software Engineering Education: Delivering Non-Technical Knowledge and Skills, pp. 296–309. IGI Global (2014)

Exploring the ISO 14001 Environmental Management System (EMS) towards SMEs Organizational Performance: Case Study of Southern Malaysia Furniture Manufacturers

A.H. Nor Aziati*, Ng Seow Chian, Abdul Talib Bon, Y. Ngadiman, and M.F. Ahmad

Department of Production & Operation Management
Faculty of Technology Management & Business
Universiti Tun Hussein Onn Malaysia, Johor, Malaysia
{aziati,talib,yunos,mohdfauzi}@uthm.edu.my,
ngseowchian@gmail.com

Abstract. Environmental issues associated with industrial activities such as gas emissions and toxic chemicals waste caused the nature imbalance effects. Therefore, ISO14001 standard was introduced and employed globally. Previous studies related to ISO 14001 in Malaysia are focused at high technologies and construction sectors. Thus, this research aims to determine the correlation between ISO 14001 orientation factor and environmental management capability (EMC) of suppliers towards SMEs organizational performance. The surveys were distributed to various ISO 14001 certified furniture manufacturing companies registered under Federation of Malaysian Manufacturers in southern region of Malaysia. The data were collected among 70 respondents at the managerial and production level. Efficient production and EMC of suppliers' factors are significantly correlated to organizational performance. Thus, companies are recommended to employed green supply chain practices and eco-efficiency production processes. This quantitative case study is important to SMEs in the specific area in order to provide knowledge on the impact of ISO14001 into production and operation.

1 Introduction

The importance of preserving the environment while increasing production requires a difficult balance to achieve. In Malaysia context, the government's response to the worldwide suggestion to implement sustainable production, environmental management systems play an important role. The implementation of ISO14001 Environmental Management System (EMS) has gained much interest by manufacturers due to the ability of minimizing environmental impact and simultaneously obtaining various benefits from the adoption (88% of the studies showed positive impact on organizational performance)[1]. Thus, it is vital for manufacturers to join in the

* Corresponding author.

M. Gen et al. (eds.), *Industrial Engineering, Management Science and Applications 2015*,
Lecture Notes in Electrical Engineering 349, DOI: 10.1007/978-3-662-47200-2_49

bandwagon because manufacturing sector has contributed 29.4% to Malaysia GDP index in year 2013[2] the second highest sector after the service sector. SMEs in general could contribute 70% of all industrial pollution [3]. Thus, improving the environmental performance of SMEs is vital, irrespective of their size and sector, because they are a crucial part of the enterprise society that collectively can contribute to sustainable development.

As the demand for sustainable development in growth, there is only a few companies especially small manufacturing companies obtain ISO14001 certification. Among reasons often discussed in previous studies are the companies still doubtful with the impact of ISO on organizational overall performance and financial performance[1]. Based on ISO survey in 2011; in total, there were only 1,934 companies in Malaysia are accredited with ISO 14001 with total increment of 16% compared to 2010. This report implies increased awareness among companies in Malaysia on the importance of well-organized environmental management systems. Nonetheless, there is a lack of study from southern area in Peninsular Malaysia specifically furniture industries. This study attempts to bridge the gap by identifying the effect of ISO14001 EMS implementation towards furniture industries organizational performance. In contrast with the study carried out by [4], previous study was conducted in Klang Valley. Meanwhile, the study focuses on office based furniture product. Hence, the need for this study is imperatively important. Additionally, the number of companies in the furniture sector with ISO 14001 accreditation is very low when compared with the major furniture exporter such as Europe. Therefore, the next section discusses some issues and awareness of the importance of EMS among furniture manufacturers in Malaysia.

1.1 Malaysia Furniture Industry

The Malaysian furniture industry today has transformed into a technologically advanced multi-billion ringgit industry from a traditional, domestic cottage-based production in the beginning years in 1980s. With the furniture industry's adoption of high technology production manufacturing capabilities coupled with greater emphasis on design, market expansion and aggressive promotional efforts, the Malaysian furniture industry has the potential to increase exports for a bigger global market share. The furniture industry is highly export-oriented with more than half of its production are destined for overseas market which accounted 3% value of global exports. In 2013 for instance, Malaysia exported RM7.4 billion worth of furniture. Malaysia ranked as the eighth largest exporter of furniture in the world and third in Asia with the exports destined to more than 180 countries. According to Malaysian Timber Corporation, in 2012, Malaysia exports more than RM2.0 million worth of furniture to United State with wooden furniture were recorded as the highest exported furniture. Besides these traditional markets, Malaysian furniture has also gained access to the markets in Russia, New Zealand, South America, Middle East and Africa.

Furniture manufacturing has been receptive to management tools that could increase productivity. According to [4] factors that hinder implementation of ISO 14001 among SMEs in Malaysia is due to a very high cost, lack of customer demand, lack of government incentives and lack of experts to implement and maintain the system. This is also supported by the varied demands by different market segments. However, with the country's MyHijau agenda and the need for companies to provide eco-friendly product resulted in companies adopting environmental management systems. Most of the companies implement ISO 14001 on regulation bases as required by the parent company. Companies usually adhere to the EMS because the requirements and policies set forth by the parent company and not done on a voluntary basis. The implementation of ISO 14001 in improving the environmental performance of SMEs is vital for the survival irrespective of their size and sector because SMEs are the enterprise society that collectively can contribute to sustainable development. ISO14001 can be applied to businesses of all sizes and shapes. As for SMEs, the benefits of using ISO standards include opening up of export markets, increased efficiency, and increased credibility as well as confidence. Research done by [5] supported that ISO14001 EMS implementation has make improvement on the Malaysian SMEs performance. In the context of this study, the concentration is the Southern region SMEs. Based on SME Corporation Malaysia, it is recorded that there are 4804 SMEs in Johor area with 3201 companies of the total number are manufacturing companies while 1603 are from service sector. To further minimize the scope of research, the focus is mainly concentrated on furniture manufacturing industry. According to Federation of Malaysian Manufacturers (FMM) 2012, there only 82 companies certified with ISO14001 in southern area in which only 5 companies from Kluang and 6 companies from Batu Pahat have obtained ISO14001 certification.

Although a number of studies had proven that ISO14001 certification shows a positive impact on company's performance, organizations are still contemplating and argue about the impact of ISO14001 on environmental situation [6] as well as firm's performance, one of the reason is because the deployment of ISO14001 standard is diffused in a different way across the globe. Furthermore, there is no international standard guideline to cope with the diverse environmental policies in various countries. ISO14001 concentrates more on the processes neglecting to establish performance standard to measure actual impact [7].

Nowadays in Malaysia, consumers are increasingly concern about the environmental friendly products or services in the marketplace. Consequently, it becomes a trend to take part in the bandwagon of implementing environmental management system such as ISO14001 in the business in order to stay competitive in current market. Despite the fact that, [8] argued the ISO14000 principle only provides framework to improve environment performance within the firm's operation boundaries instead of through the supply chain. Additionally, [9] reinforced that organization often concentrates merely on the issues within the organization but neglected the negative spill-over effects from the poor environmental performance of its supply chain partners. Managers realized a large and growing amount of environmental risk can be found in company's supply chain, particularly from small- and medium-sized enterprise (SME) suppliers [10]. Furthermore, community stakeholders rarely discern between an organization and supplier's poor environmental practices [11]. From the literatures, it is clearly seen that there is a lack of concern about the environmental management capability (EMC) of suppliers.

Hence, an international guideline is suggested to be executed universally in order to solve orientation problem. While, firms are encouraged to put more attention on EMC of suppliers such as green purchasing in order to enhance environmental performance throughout the supply chain by working with high level EMC of suppliers. Because purchasing is at the beginning of a green supply chain, a firm's environmental efforts cannot be successful without integrating environmental goals into purchasing activities [12]. This study attempts to answer two questions as follows; what is the impact of ISO14001 orientation and EMC of suppliers' factors towards organizational performance? The selections of factors are based on the significant results of previous studies in environmental management. Therefore, this study is vital to SMEs of furniture manufacturing industry in Malaysia in order to better implement environmental system to their production and operation. Moreover, academicians have additional knowledge on the relationship of orientation factors and EMC of suppliers towards organizational performance. Thus, it can be a foundation for further study on how orientation factors and EMC of suppliers could be enhanced to improve organizational performance.

2 Malaysian Standards (MS) ISO 14001:2004

ISO 14000 is an environmental management system standard, environmental auditing, labeling, environmental performance evaluation and life cycle analysis. ISO 14001 comprises four general areas: EMS, auditing, performance evaluation, labeling, life cycle assessment, and product standards [13]. The standards are divided into two general categories; organizational evaluation and product. It is a voluntary standard that intended to be used worldwide as a catalyst for create a favorable global environment and to improve the business and international trade. When implemented, it is expected to be able to standardize environmental control efforts in every country in form a global framework. These standards should be practical, useful and can be used by all organizations regardless the firm's size. In context of this study, the concentrate is MS ISO 14001:2004 Environmental Management System (EMS) because it is one of the most popular standard in Malaysia. This standard enables businesses of any sizes and sectors around the world to continuously minimize negative impact towards environment. A framework is provided for a company's environmental policy, plans and actions in its production activities, products or services. However, the company is required to commit to continual improvement and compliance with the applicable environmental legislation and regulations.

2.1 Organizational Performance

Organizational performance is the dependent variable (DV) in this study which is also the main concentrations of our findings. It is an analysis of a company's performance as compared to goals and objectives. Verweire & Berghe [14] defined organizational performance in terms of the value that an organization creates using its productive assets in comparison with the value that the owners of these assets expect to obtain. Organizational performance is always been the dependent variable of interest for researchers in any area of management to evaluate organizations, actions and environments [15]. In recent years, organizational performance is being measured by large and small companies in various dimensions specifically in financial and non-financial performance [16]. Generally, financial performance refers to financial results of a company in monetary terms such as return on investment (ROI) and return on assets (ROA), while non-financial performance measures how well a company performs that is not in expressed in monetary units such as customer satisfaction and employee satisfaction. In the context of this study, only nonfinancial performance is discussed due to some constraints such as limited time to collect data and confidentiality of financial information.

2.2 Orientation Factor

In this research, orientation factor which is also one of the independent variables (IV) bring the meaning that how is the company oriented towards the implementation of ISO14001. Ann et al. [6] supported that company must committed to the plan of action on environmental management which indicates what the company is going to do, how it is going to go about doing it, and when it is to be done. The orientation factor is measured by previous scholars in multi-dimensional perspective. They perceived orientation factor as a combination of various elements such as waste management, product quality, and efficient production. According to [17], adoption of waste management was more prevalent to be described as cost-reducing practices. Besides, [18] supported that development of systematic approach for improving environmental performances resulted in an improvement of product quality. Dües et al. [19] stated in their research that green practices would also improve efficiency in the production process and could achieve better leaner results than those companies which do not. Therefore, these factors are investigated in this research in order to find out whether they influence organizational performance in terms of non-financial performance after the implementation of ISO14001.

2.3 Environmental Management Capabilities of Suppliers Factor

The concept of environmental management capabilities (EMC) is concerned with conserving natural capital, whereby firms reduce their environmental impact by such means as reducing waste in operations, using renewable inputs, and continuously improving their operations to sustain yield with minimum adverse impact to the environment [20]. EMC of suppliers is about suppliers' ability to perform business activities and respond to their buying firms in an environmentally friendly manner while attaining financial gains [21]. Based on [22], it is insufficient to focus intensely on improving the environment while sppliers provide harmful materials. EMC of suppliers is important to the implementation of green practices because 87% of customers would accuse firms of environmental negligence when their suppliers are environmentally irresponsible, e.g., use harmful chemicals, and refuse product recycling [21]. Corporate image and reputation will be affected because of the careless made by their suppliers. Wong et al. [21] further support that negligent behaviors of suppliers can devastate the green practice of their downstream partners. Therefore, this study intends to find out the relationship between EMC of suppliers towards organizational performance.

3 Methodology

The aim of this study is to identify the relationship between orientation factors and EMC of suppliers factors toward organizational performance among furniture manufacturers in Southern region of Malaysia. Data collection for this study was collected using cross-sectional survey design. The survey based approach allowed the research team to collect data pertaining to the attitudes of the respondents towards environmental business activities and their plant's environmental management system. The survey was also used to identify the most influential factors that impact organizational performance. The sampling technique used in the research is purposive sampling in which researchers select subjects who are judged to be the representatives of the population [23] based on some criterions. The questionnaire consists of four sections which are demographic of respondents, orientation factors with 14 items and EMC of suppliers with 5 items towards organizational performance.

In this research, selected respondents are among top management, quality assurance staff and knowledgeable in the area of ISO14001 standard. The reason is because employees from basic level might not have the information about company's performance and they might not have deep understanding about ISO14001 standard that is applied in their companies [13]. A total of 110 questionnaires were self administered to the respondents. With a close follow up by telephone call, email, first reminder letter and personally meet up with the respondent to hand in the questionnaires, 70 completed questionnaires was received. The usable response rate is approximately 42%. The response rate is deemed to be exceptionally good as responses expected from mail survey are usually low [24]. The response rate of the study is considered to be good when comparing to other similar studies from previous research (47.4-57.7% response rate).

The developed survey has undergone pre-testing phase prior the distribution. Pre-testing of questionnaire is carried out to improve the readability and understandability of the target respondents [25]. According to [26] there are three of the methods that are used to pretest questionnaires: cognitive interviewing, respondent debriefing, and behavior coding of respondent/interviewer interaction. For the purpose of this study, the authors employed respondent's debriefing method. Respondent debriefing is incorporated into the actual data collection method of the survey. It can be included as part of a survey pretest, to provide input for revision for the production survey, or it can be included in the actual survey to provide input for the next administration of a continuing survey. In this research, 5 experts from the related area of the topic had been invited to participate in this questionnaire pretesting. Once the process is complete, the improved questionnaires were distributed to the appropriate respondent.

3.1 Conceptual Framework and Hypotheses

A conceptual framework in Fig. 1 is formed based on the literature review discussed in problem statement and literature. In addition, these hypotheses are generated to answer the two identified research questions. Total 12 hypotheses were developed in the study.

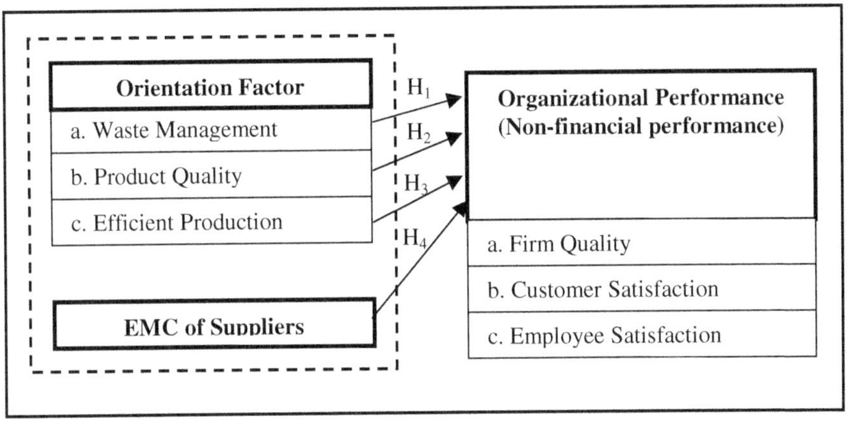

Fig. 1. Conceptual Framework

To understand the relationship of each ISO 14001 on furniture companies' performance, the following hypotheses were set up to be tested. Based on the literature review, these hypotheses will be stated based on a numbering system from H_1 until H_4. This style of hypotheses statement was chosen due to the nature of answering hypotheses using multivariate data analysis. The respondents were then asked to access their firm's systems using a five-point scale (0 = strongly disagree; 5 = strongly agree). A five-point scale has been widely used in research studies, as this Likert scale is designed to examine how strongly respondents agree or disagree with statements in questionnaire. The developed hypotheses are;

H_{1a}: Waste management has positive relationship with firm quality.
H_{1b}: Waste management has positive relationship with customer satisfaction.
H_{1c}: Waste management has positive relationship with employee satisfaction.
H_{2a}: Quality of product has positive relationship with firm quality.
H_{2b}: Quality of product has positive relationship with customer satisfaction.
H_{2c}: Quality of product has positive relationship with employee satisfaction.
H_{3a}: Efficient production has positive relationship with firm quality.
H_{3b}: Efficient production has positive relationship with customer satisfaction.
H_{3c}: Efficient production has positive relationship with employee satisfaction.
H_{4a}: EMC of suppliers has positive relationship with firm quality.
H_{4b}: EMC of suppliers has positive relationship with customer satisfaction.
H_{4c}. EMC of suppliers has positive relationship with employee satisfaction.

3.2 Data Collection Instrument and Measurement

In the study each item measured were adopted from previous studies in related area. Table 1. until Table 3. exhibits the complete measurement items of dependent and independent variables.

Table 1. Organizational Performance Item Measurement and data type

No	Variables	No. Items	Sources	Data Categories
1.	Firm Quality	(1) Produce effective products. (2) Improve of products manufacturing formulas. (3) Produce consistent products with low defect rate. (4) Produce durable products with long expiration dates. (5) Gaining new technology. (6) Gaining new expertise. (7) Provide better service quality compared to competitors. (8) Gaining good reputations.	Adopted from [27]	Data type: Scale Data scaling: Five-point Likert Scale
2.	Customer Satisfaction	(1) Customers are always satisfied with products. (2) Purchase frequency of customers has increased. (3) The number of customer complaints has decreased.	Adopted from [28]	Data type: Scale Data scaling: Five-point Likert Scale
3.	Employee Satisfaction	(1) Employees are satisfied with the implementation of ISO14001 standard. (2) Low employee absenteeism. (3) Low employee turnover.	Adopted from [27]	Data type: Scale Data scaling: Five-point Likert Scale

Table 2. Measurement Items of Orientation Factor (IV_1)

No	Variables	No. Items	Sources	Data Categories
1.	Waste Management	(1) Dispose of hazardous waste appropriately (2) Have a recycling program (3) Use re-usable packaging (4) Minimize product packaging (5) Set measurable targets for waste reduction (6) Take back packaging (7) Take back end-of-life products	Adapted from [29]	Data type: Scale Data scaling: Five-point Likert Scale
2.	Product Quality	(1) Use non-hazardous materials (2) Design products to be easy to repair and/or last longer (3) Design products to be easy to disassemble and/or recycle (4) Replace virgin materials with recycled materials	Adapted from [17]	Data type: Scale Data scaling: Five-point Likert Scale
3.	Efficient production	(1) Application of high-technology equipment to reduce waste and energy consumption (2) Production planning and control focused on optimization and maximizing use of resource (3) Processing methods focused on waste, pollutants and energy reduction	Adapted from [4]	Data type: Scale Data scaling: Five-point Likert Scale

Table 3. Measurement items of EMC of suppliers (IV_2)

No	Variables	No. Items	Sources	Data Categories
1.	EMC of supplier	(1) Company's suppliers are ISO14001 certified. (2) Second-tier supplier environmental evaluations are conducted by suppliers. (3) Suppliers are able to provide ecological proof of their products. (4) Specific environmental management guidelines are provided to upstream suppliers. (5) Suppliers cooperate with company to reduce environmental impact in the manufacturing processes.	Adopted from [21]	Data type: Scale Data scaling: Five-point Likert Scale

4 Data Analysis and Results

4.1 Demographic Profiles

Demographics information provides data about the respondents and it is essential to determine whether the individuals in a particular study are the representative sample of the target population in order to generalize the research. The first question for demographic is job position whereby majority of the respondents are managers which constituted to 45.71%. Besides that, 41.43% of the respondents have been working in the organization for 1 to 5 years while 48.57% of the respondents have been taking their current position for 1 to 5 years. The result shows that on average the companies have been established for more than 10 years. Most of the companies have been operated from 20 to 29 years. In terms of number of staff, 37.14% of the companies involved have a number of employees between 151-200 employees. This result indicates more medium enterprises adopt ISO14001 standard compare to small enterprises.

There are a total of 15 companies involved in this research. Majority of the respondents, which contributed to 71.43% of the overall percentage, claimed that the export number of product per year is approximately more than 50,000. As for the market penetration, 81.43% of the respondents answered that their companies enter 1 to 5 new markets in a year. Furthermore, 71.43% of them perceived that the growth in export has become much higher after deploying ISO14001 standard. Lastly, 72.86% of the respondents stated that the improvement of productivity has been higher after the adoption of ISO14001 standard.

4.2 Reliability Test

Reliability of a measuring instrument is defined as its ability to provide similar results in repetitive measurements under identical conditions. Cronbach's coefficient alpha is used to measure internal consistency of the scale whereby the coefficient values indicate the average correlation among all of the items that make up the scale. The higher the values of Cronbach's alpha, the greater is the reability of scale [30]. According to [31], 0.6 is the common threshold for sufficient values of Cronbach's alpha. The results appear to be all of the variables are reliable with Cronbach's alpha greater than 0.6. Table 4. below displays the reliability statistics for each variable.

Table 4. Reliability Statistics

Constructs	Cronbach's Alpha (α)	N of Items	No. of Item Deleted
Waste Management	0.726	7	None
Product Quality	0.863	4	None
Efficient Production	0.843	3	None
EMC of Suppliers	0.779	5	None
Firm Quality	0.854	8	None
Customer Satisfaction	0.724	3	None
Employee Satisfaction	0.830	3	None

4.3 Descriptive Statistics and Hypotheses Testing

The "Mean" obtained from SPSS gives the average score of their responses for each of the items that they answered [32]. Whereas, standard deviation is used to measure the dispersion of the data in which how close the entire set of data is to the average value. The lower the value of standard deviation, the closer is the data to the average value. The majority of the items obtained high mean score. The results here provide an indication that respondents do perceive there are benefits to certification as also reported by [6, 33]. Besides that, the values of standard deviation ranged from 0.721 to 1.227 indicate a general consensus amongst the respondents. Similarly, majority of EMC of achieved high mean score. In addition, the standard deviation is only slightly different with the range of 0.679 to 0.728. Additionally, respondents agreed with all the elements in non-financial performance, thus, a high mean score is achieved. The value of standard deviation that is ranging from 0.647 to 0.845 shows the data points are closely distributed around the mean. To conclude, the data points are clustered closely to the mean.

Kolmogorov-Smirnov is used in this research because the dataset for this research is 77 [34] which are more than 50 elements. For sample size less than 50, Shapiro-Wilk test is more appropriate [35] because Shapiro-Wilk test will rejects the hypothesis that data is normal when the sample size is equal to 40. If p value of Kolmogorov-Smirnov is greater than 0.05, null hypothesis is accepted and the data is assumed to be approximately normally distributed. On the other hand, if p value is less than 0.05, the null hypothesis is rejected and the data is said to be deviated from

Table 5. Results of Hypotheses Tests

	Hypothesis	Correlation Coefficient	Inference
H_{1a}	Waste management has positive relationship with firm quality.	0.075	rejected
H_{1b}	Waste management has positive relationship with customer satisfaction.	-0.099	rejected
H_{1c}	Waste management has positive relationship with employee satisfaction.	-0.011	rejected
H_{2a}	Quality of product has positive relationship with firm quality.	0.128	rejected
H_{2b}	Quality of product has positive relationship with customer satisfaction.	0.063	rejected
H_{2c}	Quality of product has positive relationship with employee satisfaction.	-0.119	rejected
H_{3a}	Efficient production has positive relationship with firm quality.	0.651	accepted
H_{3b}	Efficient production has positive relationship with customer satisfaction.	0.563	accepted
H_{3c}	Efficient production has positive relationship with employee satisfaction.	0.535	accepted
H_{4a}	EMC of suppliers has positive relationship with firm quality.	0.661	accepted
H_{4b}	EMC of suppliers has positive relationship with customer satisfaction.	0.561	accepted
H_{4c}	EMC of suppliers has positive relationship with employee satisfaction.	0.571	accepted

the normal distribution [36]. In the study, all six constructs are not normal. Thus the data is not distributed normally. Hence, the study employed non-parametric analysis for hypotheses testing. Table 6. shows hypotheses testing results.

5 Discussions and Conclusion

This research study has been conducted to determine the relationship between the independent variables (orientation factors of ISO14001 and environmental management capability (EMC) of suppliers) and dependent variable (organizational performance).

The results obtained from previous chapter showed that not all of the factors are significantly affected the organizational performance specifically on firm's quality, customer satisfaction as well as employee satisfaction. The effect of waste management on firm quality, customer satisfaction, and employee satisfaction were found to be insignificant (correlation coefficient = 0.075, -0.099, -0.011, $p > 0.05$). Based on the collected questionnaires, majority of the companies in the research areas do not emphasis on waste management in their production and operation. As a result, there is no relationship between the variables.

The effect of product quality on firm quality, customer satisfaction, and employee satisfaction was found to be insignificant (correlation coefficient = 0.128, 0.063, -0.119, $p > 0.05$). This proved that product quality is less important by the companies involved. Majority of the companies do not design their products to be easily repaired and last longer. Besides that, they rarely focus on replacing original materials with recycled materials. This finding is supported by the study of [6] which has stated that quality of products ware not significantly related to firm's performance.

The effect of efficient production on firm quality, customer satisfaction and employee satisfaction was found to be significant (correlation coefficient =0.651, 0.563, 0.535, $p < 0.01$). According to [4], adoption of environmental management practices has contributed to a more cost-efficient production process that ensure built in quality while at the same time maximize the use of resources, reduce waste and energy consumption. Based on the results, efficient production does have a positive relationship with nonfinancial performance of a company.

The effect of EMC of suppliers on organizational performance was found to be significant (correlation coefficient = 0.661, 0.561, 0.571, $p < 0.01$). Based on the data in Chapter 4, companies' suppliers are said to have strong EMC. Consequently, when EMC of suppliers is strong, it produces a better firm quality. Study carried out by [21] provided evidence about suppliers with high EMC can bring positive impact on financial performance of a company. While, the results of this research show that EMC of suppliers is the most influential factor that affect nonfinancial performance of a company.

From the data above, it is clear that EMC of suppliers has the highest coefficicient for the effect on organizational performance. In another words, EMC of suppliers is the most significant factor that impact organizational performance. Based on the research, EMC of suppliers is obviously an essential factor to the application of ISO14001 in order to provide nonhazardous materials to the company and indirectly improve organizational performance. The second influential factor is efficient production in which green practices on the production process would improve organizational performance of a company.

In conclusion, this research attempts to identify the impacts of ISO14001 towards organizational performance with respect to orientation factors and EMC of suppliers. In this quantitative research study, data is collected by distributing survey questionnaires to targeted population. Data analysis was then being carried out. Reliability test was conducted to test on the data before analysing the postulated hypotheses. Next, the hypothese were tested with bivariate approach and Shapiro-Wilk correlation coefficient by using SPSS software. The results showed that only 3 out of 9 of the hypothses in orientation factors which is efficient production ($H_{3a,}$ H_{3b}, H_{3c}) as well as EMC of suppliers ($H_{4a,}$ H_{4b}, H_{4c}) were found to be significantly supported. Last but not least, this research could be improved by expand the population to have more sample in order to obtain a more dependable results. This research study is expected to assist SMEs in Johor area in order to better implement ISO14001 standard into their production and operation.

Acknowledgments. Researchers want to thank the students involved in the study, namely Ng Seow Chian and support provided by ORRIC, Universiti Tun Hussein Onn Malaysia when conducting research.

References

1. de Vries, H.J., Bayramoglu, D.K., van der Wiele, T.: Business and environmental impact of ISO 14001. Int. J. Qual. Reliab. Manag. 29, 425–435 (2012)
2. Malaysia Economic Planning Unit: The Malaysian Economy in Figures 2013 - Economic Planning Unit, Prime Minister's Department (2013)
3. Hillary, R.: Environmental management systems and the smaller enterprise. J. Clean. Prod. 12, 561–569 (2004)
4. Ratnasingam, J.: The Impact of ISO 14001 on Production Management Practices: A Survey of Malaysian Wooden Furniture Manufacturers. J. Appl. .... (2009)
5. Nee, G., Wahid, N.: The Effect of ISO 14001 Environmental Management System Implementation on SMEs Performance: An Empirical Study in Malaysia. J. Sustain. Dev. 3, 215–220 (2010)
6. Ann, G.E., Zailani, S., Wahid, N.A.: A study on the impact of environmental management system (EMS) certification towards firms' performance in Malaysia. Manag. Environ. Qual. An Int. J. 17, 73–93 (2006)
7. Perrini, F., Tencati, A.: Why Do Standardized ISO 14001 Environmental Management Systems Lead to Heterogeneous Environmental Outcomes? Bus. Strateg. Environ. 308, 296–308 (2006)
8. Wooi, G., Zailani, S.: Green supply chain initiatives: investigation on the barriers in the context of SMEs in Malaysia. Int. Bus. Manag. (2010)
9. Hsu, C.-C., Tan, K.C., Zailani, S.H.M., Jayaraman, V.: Supply chain drivers that foster the development of green initiatives in an emerging economy. Int. J. Oper. Prod. Manag. 33, 656–688 (2013)
10. Kamaruddin, N.K., Kovalan, V., Adi, M.N.M., Ahmad, A.R., Noor, H.M.: Adoption of green supply chain initiatives among small and medium sized suppliers. Entrep. Vis. 2020 Innov. Dev. Sustain. Econ. Growth., 527–536 (2013)
11. Eltayeb, T.K., Zailani, S., Ramayah, T.: Green supply chain initiatives among certified companies in Malaysia and environmental sustainability: Investigating the outcomes. Resour. Conserv. Recycl. 55, 495–506 (2011)
12. Mariana, N.: Green Purchasing and Supply: A Literature Review and Future Research, pp. 1–23 (2011)

13. Melnyk, S.A., Sroufe, R.P., Calantone, R.: Assessing the impact of environmental management systems on corporate and environmental performance. J. Oper. Manag. 21, 329–351 (2003)
14. Verweire, K., Van den Berghe, L.: Integrated Performance Managment: A Guide to Strategy Implementation (2004)
15. Richard, P.J., Johnson, G.: Measuring Organizational Performance as a Dependent Variable: Towards Methodological Best Practice. 1901 (2008)
16. Richard, P.J., Devinney, T.M., Yip, G.S., Johnson, G.: Measuring Organizational Performance: Towards Methodological Best Practice. J. Manage. 35, 718–804 (2009)
17. Yacob, P., Nur Syaheeda, A., Mohamad Fared, M.M., Adi Wira, M.Z.: The policies and green practices of Malaysian SMEs. Glob. Bus. Econ. Res. J. 2, 52–74 (2013)
18. Arena, M., Azzone, G., Platti, M.: ISO14001: Motivations and benefits in the italian metal industry. Int. J. Eng. Bus. Manag. 4, 1–9 (2012)
19. Dües, C.M., Tan, K.H., Lim, M.: Green as the new Lean: How to use Lean practices as a catalyst to greening your supply chain. J. Clean. Prod. 40, 93–100 (2013)
20. Wong, C.: Leveraging Environmental Information Integration to Enable Environmental Management Capability and Performance. J. Supply Chain Manag. 49 (2013)
21. Wong, C.W.Y., Lai, K., Shang, K.-C., Lu, C.-S., Leung, T.K.P.: Green operations and the moderating role of environmental management capability of suppliers on manufacturing firm performance. Int. J. Prod. Econ. 140, 283–294 (2012)
22. Hajmohammad, S., Vachon, S., Klassen, R.D., Gavronski, I.: Reprint of Lean management and supply management: Their role in green practices and performance. J. Clean. Prod. 56, 86–93 (2013)
23. Tongco, M.D.C.: Purposive sampling as a tool for informant selection. Ethnobot. Res. Appl. 5, 147–158 (2007)
24. Sekaran, U.: Research Methods For Business. John Wiley & Sons, Inc. (2003).
25. Collins, D.: Pretesting survey instruments: An overview of cognitive methods, pp. 229–238 (2003)
26. Couper, M.P.: Methods for Testing and Evaluating Survey Questionnaires (2005)
27. Chow, C., Van, W.: The Use and Usefulness (2006)
28. Nilsson, L., Johnson, M.D., Gustafsson, A.: The impact of quality practices on customer satisfaction and business results: product versus service organizations. J. Qual. Manag. 6, 5–27 (2001)
29. Yacob, P.: The policies and green practices of Malaysian SMEs. Glob. Bus. .... 2, 52–74 (2013).
30. Pallant, J.: SPSS Survival Manual: A step by step guide to data analysis using SPSS. Allen & Unwin (2011)
31. Vinzi, V.E., Trinchera, L., Amato, S.: PLS PathModeling: From Foundations to Recent Developments and Open Issues forModel Assessment and Improvement. In: Esposito Vinzi, V., Chin, W.W., Henseler, J., Wang, H. (eds.) Handbook of Partial LEast Squares, pp. 47–82. Springer, Heidelberg (2010)
32. Tabachnick, B.G., Fidell, L.S.: Using Multivariate Statistics. Allyn and Bacon (2007)
33. Boiral, O., Henri, J.-F.: Modelling the impact of ISO 14001 on environmental performance: A comparative approach. J. Environ. Manage. 99, 84–97 (2012)
34. Ahad, N.A., Yin, T.S., Othman, A.R., Yaacob, C.R.: Sensitivity of Normality Tests to Non-normal Data. Sains Malaysiana 40, 637–641 (2011)
35. Razali, N.M., Wah, Y.B., Sciences, M.: Power comparisons of Shapiro-Wilk, Kolmogorov-Smirnov, Lilliefors and Anderson-Darling tests. J. Stat. Model. Anal. 2, 21–33 (2011)
36. Slaughter, S.A., Kirsch, L.J.: The Effectiveness of Knowledge Transfer Portfolios in Software Process Improvement: A Field Study. Inf. Syst. Res. 17, 301–320 (2006)

Happy Workers Work Happy? The Perspective of Frontline Service Workers

Wan-Jung Hsiao

Graduate Institute of Adult & Continuing Education, National Chi Nan University, 1 University Rd., Puli, Nantou Hsien, Taiwan
wjhsiao@ncnu.edu.tw

Abstract. Frontline service workers provide service to customers on behalf of the organization. Many recent studies have confirmed that customer service performances are positively linked with customer's feelings. Frontline service workers in contact with customers have to conceal their true emotions while at work, instead displaying emotions appropriate to those required by the organization. However, the everyday emotions of members in the organization affect individual behavior. The display of day-to-day behavior by frontline service workers may be influenced by personal emotions. This study is thus designed as a longitudinal study to observe the effect of daily emotional variations on the emotional labor of service personnel.

1 Introduction

Frontline service workers provide service to customers on behalf of the organization. Many recent studies have confirmed that customer service performances are positively linked with customer's feelings (Spencer & Rupp, 2009). Thus, organizations are increasingly attentive to the behaviors and conversations between service providers and customers to create positive and memorable affects in every consumer experience. Frontline service workers in contact with customers have to conceal their true emotions while at work, instead displaying emotions appropriate to those required by the organization (Hochschild, 1983). The need to display appropriate emotions in public in accordance with company requirements is termed "emotional labor."

However, the everyday emotions of members in the organization affect individual behavior (Ashforth & Humphrey, 1995). The display of day-to-day behavior by frontline service workers may be influenced by personal emotions. During the empirical study on affective events theory (AET), Grandey, Tam, and Brauburger (2004) discovered that the positive and negative affects of employees will be reflected in individual work behavior and attitude.

The AET indicates that the employee emotions can be influenced by certain events, which in turn influence the display of personal behavior. Thus, the various events or contextual factors which the frontline service workers experience every day can influence personal positive or negative affect. This further causes different strategies to be adopted when facing the appropriate display of emotions as required by

M. Gen et al. (eds.), *Industrial Engineering, Management Science and Applications 2015*, Lecture Notes in Electrical Engineering 349, DOI: 10.1007/978-3-662-47200-2_50

organizational regulations. However, relatively few studies have explored the day-to-day variability in the emotional levels of frontline service workers and the variations in service behavior caused by the emotional labor involved. This study is thus designed as a longitudinal study to observe the effect of daily emotional variations on the emotional labor of service workers.

2 Design/Methodology/Approach

This study observed the influence of day-to-day emotional states on behavior, which required the subjects to record their emotions and behavior on a daily basis. Information was gathered via online questionnaire during the recording period. The research topic and methodology were explained before the tests. The subjects were asked to record their emotional states and behaviors over 25 working days in detail using a method referred to the research design of Yang and Diefendorff (2009), and the questionnaire was distributed online. An email was sent to each subject at 4 pm every day to remind them to fill out the questionnaires. To prevent the subjects from filling in the records in the morning and thus creating bias, the online questionnaire system remained closed before 4 pm each day. Even if a subject had missed a day or two of records, they were encouraged to continue participating in the study. If the completed day-to-day records of an individual totaled fewer than 10 working days, the records were omitted from further analysis.

We used convenience sampling and invited frontline workers of the service industry to participate in this study. We also ask the supervisors of the study subjects to fill in information on customer-oriented behavior to avoid common method variance. The results included 124 effective subjects after removal of those individuals who had recorded fewer than 10 working days of available data. A total of 3,261 day-to-day records were collected. Among the subjects included in the final analysis, 76 were male (61.3 %), and 48 were female (38.7 %). Eighty-four of them had experiences in sales, with 4.1 years of experience on average (ranging between one month and 25.9 years). Regarding education levels, 32.5 % had a high school level education or below, 39.8 % had graduated from vocational schools, and 37.7 % held university or advanced degrees.

3 Findings

This longitudinal study explores the relationship between the positive and negative affects of the individual and the emotional labor strategy adopted, and incorporates the moderating effect from self-monitoring. After a period of tracking, the relationship between emotional labor strategies and work customer orientation behavior are explored. The major findings and the implications are discussed as follows: first, day-to-day emotional changes affect the emotional labor strategy. When negative affect is higher on a particular day, service workers will "put-on" a mask to disguise their emotions to match the expectations and requirements placed upon them by the environment (adopting the surface acting strategy). Less cognitive changes occur to make internal feelings congruent with the external environmental demands on emotion (adopting the deep acting strategy). On days when the positive affect of frontline service workers is higher, it is easier for them to see things in an enthusiastic and positive manner. Thus, it is easier for them bring their internal feelings in line with external expressions by conscious control, and sincerely display the emotional requirements enforced by the organization (adopting the deep acting strategy).

Second, self-monitoring characteristics moderate the relationships between negative affect and emotional labor. When an individual experiences negative affect, higher levels of fear, anxiety, and stress are sensed. Thus, it is more difficult for individuals to convert their true feelings in the face of organizational requirements for emotional displays through the emotional regulation process. One is able to satisfy the organizational requirements only through surface acting. However, if an individual possess a higher level of self-monitoring, he or she would be more concerned about the appropriateness of their behavior in the social context. Thus, compared to those with lower self-monitoring, individuals who are highly self-monitoring are able to sense environmental demands and expectations. If personal affect is negative at this moment, the individual will still strive to moderate his or her emotional cognition so that internal feelings become congruent with external environmental demands, which indicates a higher level of deep acting.

Third, in the long run, deep acting is beneficial to work customer orientation behavior. When an employee adopts the deep acting strategy, he or she is able to more sincerely accord with the required display of emotions, which helps generate customer-oriented behavior and increase individual job performance.

4 Research Implications and Limitations

With the arrival of service economy, frontline service workers affect the overall impression of customers toward the organization. Such emotional experiences determine customer retention and impacts organizational performance. Thus, frontline service workers occupy a critical role in the age of the service economy (Rafaeli, 1989). Day-to-day emotional regulation is dynamically affected by affective states. This places more importance on the emotional management of frontline service workers. The behavioral school belief that "happy employees bring higher productivity" has been constantly questioned in the field of Industrial Organizational (IO) psychology (Kaplan et al., 2009). However, this perspective seems to be even more strongly grounded in the age of the service economy. The results of this study show that the positive or negative affect of frontline service workers influences emotional labor strategy, which affects personal well-being in the long run. Consequently, making employees happy seems to be another focal point of management in the age of the service economy.

From the viewpoint of organizational management, however, "making employees happy" and "emotional management training for employees" appear to consume more resources from the organization. The results from this study show that service businesses can first screen for workers with higher self-monitoring during the hiring process, which helps in moderating the relationship between the negative affect of employees and deep acting. When under higher negative affect, the levels of deep acting of those who are highly self-monitoring are still higher than those with low self-monitoring. Over extended periods of time, this bolsters personal well-being and organizational performance. Thus, "self-monitoring" should be included as one of the key factors during the hiring process of service industry workers.

This study has limitations in methods and context as follows. First, AET illustrate that the affective state is influenced by affective events. We did not include considerations for affective cues in this study, which is a limitation. We recommend that future studies include emotional cues to completely validate the AET and the emotional regulation process. Second, samples for this empirical study mostly come from frontline service workers from the finance, telecom, and funeral services industries. It remains to be seen whether these results can be generalized over all types of service industry, such as department store sales, food and beverages, and medical services. We recommend that future studies enlarge their sample size and conduct longitudinal studies on units with different organization types. Third, self-monitoring was the only moderating factor used to explore the relationship between positive and negative affect and emotional labor. However, other moderating factors may exist, such as emotional intelligence and stress. Even environmental factors such as supervisor support, coworker support, and organizational/unit service atmosphere may generate moderating effects (Grizzle et al., 2009). We suggest that future studies consider other moderating factors, and further elucidate upon the relationship between positive and negative affect and emotional labor.

References

1. Ashforth, B.E., Humphrey, R.H.: Emotion in the workplace: A reappraisal. Human Relations 48(2), 97–125 (1995)
2. Grandey, A.A., Dickter, D.N., Sin, H.P.: The Customer is not always right: Customer aggression and emotion regulation of service employess. Journal of Organizational Behavior 25, 397–418 (2004)
3. Grizzle, J.W., Zablah, A.R., Brown, T.J., Mowen, J., Lee, J.M.: Employee customer orientation in context: How the environment moderates the influence of customer orientation on performance outcomes. Journal of Applied Psychology 94(5), 1227–1242 (2009)
4. Hochschild, A.R.: The managed heart: Commercialization of human feeling. University of California Press, Berkeley (1983)
5. Kaplan, S., Bradley, J.C., Luchman, J.N., Haynes, D.: On the role of positive and negative affectivity in job performance: a meta- analytic investigation. Journal of Applied Psychology 94(1), 162–176 (2009)
6. Rafaeli, A.: When clerks meet customers: A test of variables related to emotional expressions on the job. Journal of Applied Psychology 74(3), 385–393 (1989)
7. Spencer, S., Rupp, D.E.: Angry, guilty, and conflicted: Injustice toward coworkers heightens emotional labor through cognitive and emotional mechanisms. Journal of Applied Psychology 94(2), 429–444 (2009)
8. Yang, J., Diefendorff, J.M.: The relations of daily counterproductive workplace behavior with emotions, situational antecedents, and personality moderators: a diary study in Hong Kong. Personnel Psychology 62(2), 259–295 (2009)

Analyzing Cargo Loss Severity of Electronics Products with Decision Tree

Mu-Chen Chen[1], Pei-Ju Wu[2], and Chih-Kai Tsau[3]

[1] Department of Transportation and Logistics Management, National Chiao Tung University, 4F, No. 118, Sec. 1, Chung-Hsiao W. Rd., Taipei, Taiwan
[2] Department of Transportation Technology and Management, Feng Chia University, No.100, Wenhwa Road, Seatwen, Taichung, Taiwan
[3] 2F., No.3, Ln. 31, Yifang St., Beitou Dist., Taipei City 112, Taiwan
ittchen@mail.nctu.edu.tw, wupj@fcu.edu.tw, tsaukai@gmail.com

Abstract. Supply chain risk management has been an essential issue in recent years. Cargo loss in supply chain and logistics activities has been the major cause of delays and supply chain disruption; however, rarely do studies provide comprehensive studies focusing on cargo loss analysis and prevention in various modes of transportation. Hence, this study aims to investigate the cargo loss severity of an electronics company. Decision tree analysis is adapted to develop classification models for cargo loss severity of electronics products. The empirical results with the classification rules can be utilized as a cargo loss prediction tool to help managers to make an effective plan on cargo loss prevention.

Keywords: Supply chain risk management, Cargo loss severity, Cargo loss prevention, Decision tree.

1 Introduction

Cargo loss has been considered by logistics operators, insurance companies, and surveyors to involve cargo damage and loss, a risk that persists in global supply chains [1]. Factors causing cargo loss are complex and difficult to identify; these factors can be classified as controllable factors (i.e., management-related factors such as improper cargo packaging or poorly designed workflow) or uncontrollable factors (i.e., external environmental factors such as adverse weather or accidents during cargo transports, low-quality workers of transit stations, negligence during operational processes, or loose management). Among the types of cargo loss, planned cargo theft conspired by theft groups all over the world is one of the most common cause of severe financial loss; specifically in recent years, electronic products have become the target most preferred among theft groups worldwide [2].

Cargo loss has been extensively investigated by insurance companies, with the aim of reducing cargo loss incidents and insurance compensations. In particular, international large insurance companies often invite cargo loss prevention teams and experts to conduct relevant risks analyses. Nonetheless, few studies have investigated cargo

M. Gen et al. (eds.), *Industrial Engineering, Management Science and Applications 2015*, Lecture Notes in Electrical Engineering 349, DOI: 10.1007/978-3-662-47200-2_51

losses due to varying types of transportation modes, the effects of cargo loss on global supply chains, and recommendations for cargo loss prevention.

This study aims to utilize data mining techniques to tackle the causality of cargo loss severity. Specifically, this study establishes a classification model to extract essential patterns of the cargo loss incidents of the case company that arose during sea, air, and ground international logistics activities. Based on the analytical results of the cargo loss classification model, managers can determine which logistics combination will lead to cargo loss incidents of varying severity, and subsequently concentrate managerial resources into preventing moderately and highly severe incidents.

The rest of this paper is organized as follows. Section 2 conducts a literature review related to cargo loss issues. Section 3 introduces the decision tree analysis for cargo loss severity. Section 4 then analyzes a case of electronic products. Next, Section 5 provides managerial implications for cargo loss prevention. Finally, Section 6 draws conclusions.

2 Cargo Loss Issues

Regarding the logistics packaging–cost trade-offs concept, although increasing packaging protection decreases damage and theft in transit, it increases package weight and cost, thereby elevating transportation costs [3]. In response to the cost-oriented competitive strategy persistently advocated by the electronic industry and the increasing demand for green environmental protection, numerous cargo owners typically endeavor to reduce cost and actualize environmental protection by reducing cargo packaging to the minimum requirements of cargo protection standards. However, such practice increases the risk of damaging electronic products during transit.

The conditions of international cargo transportations is the predominant cause of cargo loss should not be overlooked. Taking airfreight forwarding, the modes of transport on which electronic industry supply chain is heavily dependent, as an example, causes of cargo loss during ground handlings and air transport are complex. According to NipponExpress [4], six types of operating procedures during air cargo transport lead to cargo loss: cargo pallet loading and unloading at airports, cargo cabin handling, flight processes, airport warehouse tally operations, temporarily storing cargos in outdoor areas, and improper cargo shipping and packaging.

According to the UK P&I Club [5], among the cargo losses incurred during sea container transport, 25% of cargo losses involved physical damages, 14% were related to temperature changes, 11% were damages due to water leakage, 9% were cases of theft, and 8% were associated with shortage of goods. In addition, numerous factors cause cargo loss in sea container transport.

Cargo loss prevention is a persisting problem that requires the collective focus of every employee in a company. This problem is particularly crucial when the cargo involves high-priced, fragile products that can be easily stolen and, therefore, all logistics operations must be performed according to a standard operating procedure. For example, cargo delivery time must be predefined so that when the cargo reaches its destination, it can be collected immediately, eliminating the need to temporarily store expensive products at a location exposing them to the risk of being stolen [6]. Moreover, Inbound Logistics [7] highlights the two directions to avoid cargo losses and damages: 1. On the supplier or shipper side: First, shipments must be packaged in a container that is suitable for its contents. Cargos should never exceed the container's maximum gross weight, as indicated by the packaging vendor. Containers or boxes should be rigid with no dents, tears, or rips. 2. On the carrier side: Partnering with a reliable carrier is critical. Cargo owners must frequently inspect the carrier's transport vehicles and inspect their cargo operating procedures onsite to ensure cargo safety.

The literature review found that cargo losses have occurred in various transportation processes. Therefore, after gaining an insight into these issues, cargo owners should pay attention to the effects of cargo losses and how such losses can be eliminated.

3 The Decision Tree Analysis for Cargo Loss Severity

This research utilized decision tree analysis in data mining to generate a cargo loss model. The primary process to establish a decision tree model is as follows.

A. Select target and input attributes: According to the research purpose, the target attribute of cargo loss severity were selected from all attributes. Then, the attributes related to the target attributes were then chosen as input attributes to construct the classification model.
B. Separate the dataset: Before developing a data mining model, the chosen dataset should be divided into training data, validation data, and test data. The percentages of training data, validation data, and test data will be 60%, 30%, 10%, respectively.
C. Choose a classification model for data mining: The decision tree model, a data mining technique, was selected to develop the classification model.
D. Set up the split rule of decision tree development: Decision tree algorithms and split parameters, such as split rule, maximum branch, and maximum tree depth, were determined. This research adopted default values for split rules in SAS Enterprise Miner to generate the model. Based on default value results, this research examined and adjusted the parameters to the accuracy of classification and meaningful managerial implications.

4 A Case of Electronic Products

4.1 The Case Background

The case company is a brand electronic manufacturer for global logistics covering consumer electronic market around the world. The electronic products sold by the case company involve laptops, desktop computers, tablets, smart phones, mainboards, display cards, LCDs, etc. These products are featured for their high precision, high price, and high market liquidity. Accordingly, these products are too vulnerable to cause damage by shaking or compress, and even to be robbed by robbers. Cargo damage or robbery are more likely to engender firms' severe financial loss and the risk of not delivering on time to their customers.

The case company has large scale business with competitive global logistics network involving air freight forwarders, sea freight forwarders, truck companies, global express. Hundreds of flights are shipped around the world every day. Due to such frequent and complicated intermodal shipments, the case company should deal with cargo loss incidents every day worldwide. The only approach that case company handles cargo loss issues is to purchase global cargo insurance as its main logistics risk management. Nonetheless, cargo insurance is not the fundamental way to prevent cargo loss. Furthermore, as the rate of cargo loss increases by year, the case company should pay more cargo insurance fee for the next year correspondingly.

This research collected three years cargo loss data from the insurance reimbursement claim database of the case company's insurance company in the time frame of 2011-2013 covering 4,372 cargo loss cases. It implies that nearly four cargo loss cases happened per day worldwide. In this study, input attributes were set to be "Product Category", "Forwarder Type", "On Board Date", "Ship From", "Ship To Area", and "Transit Type." Moreover, target attribute was set to be "Financial Impact" representing cargo loss severity. In practice, the cargo loss severity is usually evaluated by the amount allowable for loss. Hence, three levels of cargo loss severity are classified as follows:

- Low severity: A loss amount of less than US$200 is directly compensated by the insurance company. The case company perceives this loss to exert minimal financial influence and thus defined the level of severity for this loss as "low."
- Medium severity: For a loss amount of US$201–US$5,000, the insurance company must conduct extensive investigations according to stringent claim criteria; therefore, the resulting communication and personnel costs are high. The level of severity for this degree of loss is defined as "medium."
- High severity: A loss amount of US$5,001 and higher is perceived by the insurance company to be a large sum. The insurance company will delegate notary agencies to conduct onsite investigation, a process that the case company considers to be time-consuming and complex. In addition, a high compensation amount directly influences the insurance premium for the subsequent year, generating a strong economic impact, costs, and risks. Therefore, the case company viewed the level of severity for this type of loss as "high."

Consequently, this research aims to clarify the case company's cargo loss situation and to create corresponding prevention strategies of cargo loss. Specifically, what scenarios or logistics conditions are more likely to trigger what levels of cargo loss severity are investigated in this research? Moreover, the prevention strategies of cargo loss are devised by systematically drawing the data mining results of different logistics conditions.

4.2 The Empirical Results

The severity model exhibited an average accuracy rate of 71% (the accuracy rate of training data is 70.59%; the accuracy rate of validation data is 71.54%, the accuracy rate of test data is 72.19%), demonstrating favorable classification accuracy. Furthermore, a total of 7 rules were summarized from the tree branches of the loss severity classification model, as elaborated in Table 1.

Table 1. A summary of the loss severity classification results

Rules	Tree nodes	Rule descriptions	Accurate rate of leaf nodes	Coverage rate of observations
S1	2	if Transit_Type IS ONE OF: AIR then Number of Observations = 474 Predicted: Financial_impact=Low = 0.70 Predicted: Financial_impact=Medium = 0.27 Predicted: Financial_impact=High = 0.03	70%	60.6%
S2	8	if Transit_Type IS ONE OF: SEA AND Ship_To_Area IS ONE OF: ASIAN, NORTH_AMERICA, AFRICA then Number of Observations = 77 Predicted: Financial_impact=Low = 0.82 Predicted: Financial_impact=Medium = 0.18 Predicted: Financial_impact=High = 0.00	82%	9.8%
S3	9	if Transit_Type IS ONE OF: SEA AND Ship_To_Area IS ONE OF: EUROPE then Number of Observations = 68 Predicted: Financial_impact=Low = 0.59 Predicted: Financial_impact=Medium = 0.29 Predicted: Financial_impact=High = 0.12	59%	8.7%
S4	10	if Transit_Type IS ONE OF: SEA AND Ship_To_Area IS ONE OF: OCEANIA, MIDDLE_EAST then Number of Observations = 21 Predicted: Financial_impact=Low = 0.29 Predicted: Financial_impact=Medium = 0.57 Predicted: Financial_impact=High = 0.14	57%	2.7%

Table 1. *(continued)*

Rules	Tree nodes	Rule descriptions	Accurate rate of leaf nodes	Coverage rate of observations
S5	11	if Transit_Type IS ONE OF: TRUCK AND Cargo IS ONE OF: NB or MISSING then Number of Observations = 98 Predicted: Financial_impact=Low = 0.17 Predicted: Financial_impact=Medium = 0.83 Predicted: Financial_impact=High = 0.00	83%	12.5%
S6	12	if Transit_Type IS ONE OF: TRUCK AND Cargo IS ONE OF: MB_VGA then Number of Observations = 29 Predicted: Financial_impact=Low = 0.34 Predicted: Financial_impact=Medium = 0.48 Predicted: Financial_impact=High = 0.17	48%	3.7%
S7	13	if Transit_Type IS ONE OF: TRUCK AND Cargo IS ONE OF: OTHERS, DESKTOP, LCD MONITOR then Number of Observations = 15 Predicted: Financial_impact=Low = 0.67 Predicted: Financial_impact=Medium = 0.33 Predicted: Financial_impact=High = 0.00	67%	1.9%

Note: Coverage rate was calculated by dividing the number of node training data by 780 pieces of root node original training data. For example, Rule S1 involves 474 pieces of leaf node training data, with a coverage rate of 474/780 = 60.6%.

Based on the results of cargo loss severity, managers can predict cargo loss incidents by inputting logistics components relevant to new shipping operations into a decision tree model, and prioritize management tasks to avert cargo loss risks encompassed in critical tasks, thereby effectively utilizing management resources and reducing relevant costs.

5 Managerial Implications for Cargo Loss Prevention

This study identified the following key classification rules for cargo loss severity, from which corresponding cargo loss prevention strategies were derived.

A. According to Rule S4, when products are shipped using sea transport from Australia or Middle East regions, cargo loss of medium severity is likely to occur, with an accuracy rate of 57%. Therefore, managers making management decisions should pay much attention to cargo loss risks when shipments are made via sea transport to Australian or Middle East regions. They should strengthen cargo packaging to enforce freight forwarders to assume their rightful duty as managers in reinforcing tight control at every point of sea transit. Particularly, emphasis should be paid to cargo collection procedures to avoid fictitious pickups at collection ports. Furthermore, examining the decision tree rules from a probability perspective reveals that Rule S4 resulted in 14.3% highly severe cargo loss incidents, which is substantially higher than root nodes (4.1%). This result indicates that extremely severe financial loss is highly likely to occur under such logistics condition and, therefore, managers must strengthen its control for the condition outlined in Rule S4.
B. According to Rule S5, when laptops are shipped via land transportation modes, cargo loss of medium severity is most likely to occur, exhibiting an accuracy rate of 83%. Logistics mangers should strictly regulate their control of land forwarders and extensively understand the potential risks of cargo theft at each component of land transit. Since electronic products such as laptops are small in size, expensive, and highly mobile, making them the main target of theft during land transits. Hence, regularly inspecting the service contract signed between clients and outsourced logistics companies and formulating rigorous standard operating procedures and penalty policies for specific transportation modes are a vial loss prevention measure.
C. According to Rule B6, when electronic card products are shipped on land, cargo loss of medium severity is most likely to occur, exhibiting an accuracy rate of 48%. However, noticeably, the probability of highly severe cargo loss occurring under Rule B6 was 17%, the highest percentage among all lead nodes generated in this loss severity model. Rule B6 (land transit involving electronic card products) exhibited a possibility of engendering high degree of loss greater than did Rule S5 (land transport shipping laptop products). For managers, moderately or highly severe losses are risks warranting attention. Prevention plans for land transported product theft should be devised synchronously.
D. According to Rule S3, when cargos are sea transported from European regions, cargo loss of low severity is most likely to occur, with an accuracy of 58.8%. However, the probability of moderately and highly severe cargo loss occurring under this condition is relatively high, at 29.4% and 11.8%, respectively. This information serves as a reminder for managers to take extra precaution in averting the latent financial risks engendered from cargo loss incidents during sea transit in European regions.

6 Conclusions

Cargo loss incidents not only cause financial losses to an enterprise but also disrupt supply chain operations. This study gives academics and managers a holistic view of cargo loss in global operations through data mining techniques. The empirical results can be used by managers as a cargo loss prediction tool for devising systematic cargo loss prevention plans.

This study differs from previous investigations by addressing the supply chain risk management of cargo loss in numerous ways. First, this study systematically absorbs the valuable insights of both academics and practitioners for the issues of cargo loss analysis and prevention. Second, data mining techniques was adopted to analyze cargo loss severity, collecting cargo loss data from the insurance reimbursement claim database of the case company's insurance company. Third, the proposed decision tree model yields a satisfactory classification accuracy rate for cargo loss issues, thus providing a foundation for supply chain risk management. Finally, the essential classification rules for cargo loss severity allow managers to predict potential cargo loss results and thereby execute effective loss prevention strategies during global operations. Analyzing cargo loss severity of electronics products can help not only prevent financial losses incurred by cargo loss incidents, but also avoid major losses to an enterprise or jeopardizing its competitiveness in the supply chain.

Acknowledgments. This work is partially supported by Ministry of Science and Technology, Taiwan, R.O.C. under grant MOST 103-2221-E-009 -135 -MY3.

References

1. Hung, W.-C.: A study on the compensation responsibility of the International Multimodal Transport operator, shipper, and insurer underwriter. National Taiwan Ocean University, Taiwan (2005)
2. FreightWatch: 2013 FreightWatch International Global Threat Assessment (2013)
3. Lambert, D.M., Stock, J.R., Ellram, L.M.: Fundamentals of logistics management. Irwin/McGraw-Hill, Mass (1998)
4. NipponExpress: Damage prevention measures and the change of freight forwarder. A Course of Lectures in Business Administration and Ethics, Taiwan, National Chiao Tung University (2013)
5. Carefully to Carry Committee: Cargo Damage–the Causes (2013), `http://www.ukpandi.com/fileadmin/uploads/uk-pi/LP%20Documents/Carefully_to_Carry/Cargo%20Damage.pdf` (accessed October 19, 2013)
6. Fennelly, L.J.: Handbook of loss prevention and crime prevention. Butterworth-Heinemann, MA (2012)
7. Cortese, L.: Reducing the Risk of Damage, Loss and Claims (2003), `http://www.inboundlogistics.com/cms/article/reducing-the-risk-of-damage-loss-and-claims/` (accessed January 15, 2014)

A Comparison of Inventory Management between Decentralized and Centralized Distribution Networks with Backorder

Kanokwan Singha[1], Parthana Parthanadee[2], and Jirachai Buddhakulsomsiri[3]

[1,3] School of Manufacturing Systems and Mechanical Engineering, Sirindhorn International Institute of Technology, Thammasat University, P.O. Box 22, Pathum Thani 12121, Thailand
[2] Department of Agro-Industrial Technology, Faculty of Agro-Industry, Kasetsart University, 50 Ngam Wong Wan Rd, Lat Yao Chatuchak Bangkok 10900, Thailand
kanokwan.singha@gmail.com, parthana.p@ku.ac.th, jirachai@siit.tu.ac.th

Abstract. This paper investigates the effect of information sharing on inventory management of two-echelon distribution network, consisting of a distribution center and retailers. Shortages that occur at DC and at retailers are assumed backlogged and lost sales. Performance of a decentralized system, where inventory decisions are made locally, is compared with a centralized system, where echelon inventory position and lead time are used through information sharing. An iterative procedure to determine inventory policy parameters is used in both systems. The purpose is to minimize the total inventory management cost, including costs of ordering, inventory holding, and shortage. A numerical example is provided.

1 Introduction

A distribution network in a supply chain system generally consists of at least two stages: a distribution center (DC) and retailers. One of the most important aspects of managing distribution networks is inventory management. Many research studies have developed techniques to manage inventory distribution networks (e.g., [1], [6]). A simple technique is an order quantity-reorder point (Q, R) policy. Typically, inventory management problem in a distribution network that uses the (Q, R) policy as shown in Fig. 1.

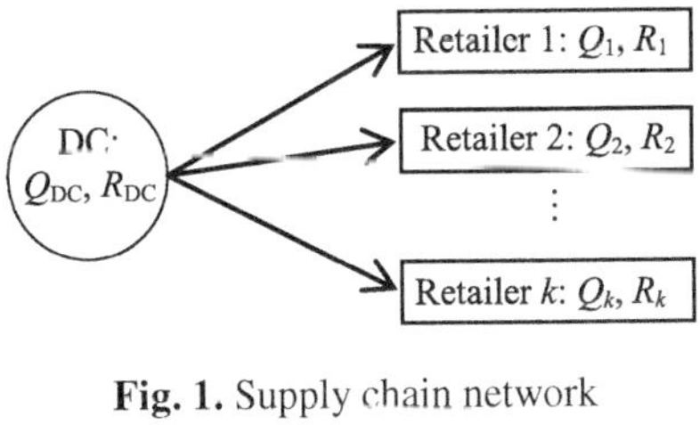

Fig. 1. Supply chain network

M. Gen et al. (eds.), *Industrial Engineering, Management Science and Applications 2015*, Lecture Notes in Electrical Engineering 349, DOI: 10.1007/978-3-662-47200-2_52

From the figure, each node carries inventories of finished goods to satisfy its customers' demand. When the inventory level falls on or below *R*, an order of *Q* for replenishment will be placed. There are two main systems: decentralized and centralized systems, for distribution network [7]. In the decentralized system where retailers and DC do not share information, the inventory parameter decisions are calculated independently based on their local information. The order quantity is calculated from their demand, i.e. retailers use end customers' demand, and DC uses orders placed by retailers as its demand. *R* is calculated from demand, its variation, and lead time from its supplier, i.e. retailers use lead time from DC and DC uses lead time from manufacturers.

In the centralized system, all nodes in the retailer stage calculate both parameters similar to that of the previous system. However, the DC would use the aggregated demand of end customers that occur at the retailers, instead of using retailers' orders, as demand. The main advantage is that the demand variability at the DC will be reduced, because the end customers usually place order of smaller quantity and more frequently than the retailers. In addition, the DC would use the echelon inventory position, which includes inventory positions at the DC and all retailers, as well as in-transit inventories between them. Moreover, this system considers echelon lead time, which consists of lead time between DC and retailers and lead time between DC and its supplier. Both echelon inventory positions and lead time can be implemented through information sharing. This can be achieved with enabling information technology, and when different stages in the supply chain are under a single ownership, or trust between stages with different ownerships are well established.

Axsäter (2005) studied a two-echelon decentralized distribution system with a central warehouse and multi-retailer. The system is controlled by (*Q*, *R*) continuous review under stochastic demand. A simple approximate technique is proposed for determining the backorder cost at the warehouse and the warehouse calculate its reorder point with the objective function to minimize the sum of expected holding and backorder cost. The performance between decentralized and centralized inventory system with a single DC and multi-retailer are compared in [5]. Normal and Poisson demand distributions are investigated with no lead time. If the DC has enough inventory, all retailers' order are satisfied. Otherwise, two strategies are examined: proportional allocation based on the size of retailers' orders, or batch-size allocation priority rule. Important finding is that sharing information does not always benefit to system inventory cost when using proportional allocation rule. On the other hand, if batch-size allocation rule is used, sharing information provide much higher benefit than not sharing one. Duan and Liao (2013) studied a single DC multi-retailer supply chain system. The inventory level is reviewed periodically while all nodes adopt the (*s*, *S*) inventory policy. They conclude that centralized control or decision generates fewer inventories cost than decentralized system. Ye and Xu (2010) considered a single product supply chain with a vendor and a buyer. Inventory models with controllable lead time under centralized and decentralized system are considered. They conclude that if all demand cases during the shortage period, the cost of supply chain under centralized system is less than that of decentralized system.

This paper considers a single product and a two-echelon distribution network. The objective is to investigate the benefit of information sharing by comparing the total inventory management cost between decentralized and centralized systems. Shortages that occur at DC and at retailers are assumed backlogged and lost sales, respectively. The (Q, R) policy is used in both systems where the parameters are determined by an iterative approach. The objective function is to minimize the total inventory management cost, including the costs of ordering, inventory holding, and shortage. Demand is assumed normally distributed and lead time is constant. A small numerical example is provided to demonstrate the benefit of information sharing.

2 Iterative Approach to Determine Inventory Parameters

The method for determining Q and R parameters is adopted from the method first developed by Felter and Dalleck (1961) for continuous review policy. The method searches to obtain Q and R that result in a balance among ordering cost, shortage cost, and inventory holding cost. The shortage cost is subject to the expected number of shortage, which is the function of R, which in turn is the function of Q. Thus, the algorithm proceeds iteratively until the values of Q and R converge. After the solutions for Q and R are obtained, the total inventory cost is calculated using Eq. (1).

$$TC = \frac{D}{Q}[C_P + C_S E\{s\}_R] + C_H\left[\frac{Q}{2} + (R - \bar{d})\right] \quad (1)$$

Where Q = replenishment order quantity, R = reorder level, D = demand rate per period, C_P = fixed reordering cost, C_S = shortage cost per unit, $E\{s\}_R$ = expected number of shortage units when the reorder level is R, C_H = unit inventory holding cost per period, and $\bar{d}$ = average demand during the replenishment lead time. The approach has five steps as follows, with additional notation: L = lead time, d = random demand during the replenishment lead time, and $P(d)$ = probability that demand during lead time will be equal to d.

Step 1: Initialize $Q = \sqrt{\dfrac{2DC_P}{C_H}}$.

Step 2: Find R that satisfies $P(d > R) = \dfrac{QC_H}{DC_S}$ from the cumulative probability of demand during lead time distribution.

Step 3: Compute $E\{s\}_R = \sum_{d=R+1}^{d_{\max}} (d - R)P(d)$.

Step 4: Compute $Q = \sqrt{\dfrac{2D[C_P + C_S E\{s\}_R]}{C_H}}$.

Step 5: Repeat Steps 2-4 until Q and R converge.

The method starts with setting an initial value of Q to the economic order quantity (EOQ) in Step 1. In Step 2, the probability of shortage, $P(d > R)$, for a given value of R is computed by setting the additional inventory carrying cost from a unit increase in R to be equal to the savings in the shortage cost. Then, find R that is associated with the probability of shortage from the cumulative probability distribution of demand during lead time. Step 3 computes the expected number of shortage for the value of R obtained from Step 2. Then, Step 4 computes Q by adding the total shortage cost to the computation of EOQ. Finally, the method iterates from Steps 2-4 until the solutions of Q and R converge.

3 Numerical Example

The numerical example considers a distribution network under one ownership consisting of one DC and five retailers for a single product. Demand of each retailer is assumed to be normally distributed with the mean (*avg*), standard deviation (*s*), and coefficient of variation (*cv*) as shown in Figure 1. These setting are chosen to systematically vary the level of end-customer demand as well as their variation, so as to investigate their effects on the performance of the system. One year of retailers' daily demand are randomly generated using these demand parameters. The iterative procedure described in the previous section is used to determine the inventory parameters. The cost parameters are actual data provided by an industrial user (a product's distributor) as follows: $C_p = 9.6$ THB/order, $C_H = 0.01189$ THB/unit/day, $C_S = 8$ THB/unit. The values of Q and R for each retailer are shown in Fig. 1.

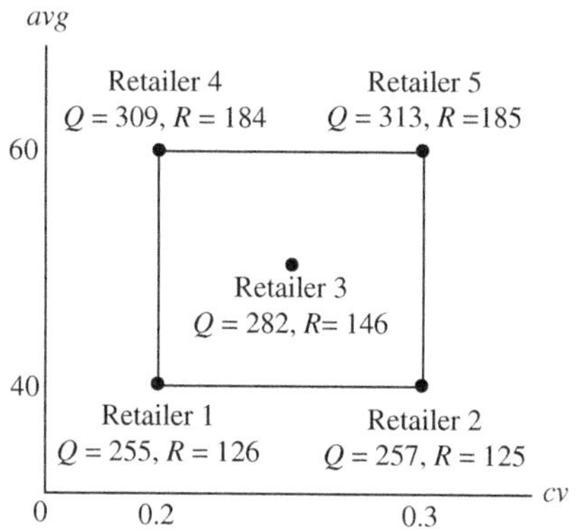

Fig. 2. Demand and inventory parameters at retailer stage

Lead time from supplier to DC, and from DC to each retailer is four and two days, respectively. Initial inventory on-hands is 3,600 units at the DC and 720 units at each retailer. For the centralized system, the demand data from five retailers are aggregated and treated as demand for the DC. The echelon lead time is six days and initial echelon inventory position of the DC is 7,200 units. For the decentralized system, DC's demand are the orders placed by the retailers. The orders are simulated from the retailers' day-to-day ordering process, based on the randomly generated daily demands at the retailers, their initial inventory positions, and the inventory parameters from Table 1. The resulting retailers' orders are then used to compute the *avg* and *s* of demands that arrive at the DC. The summary statistics of DC's demand data for both systems are shown in Table 1. The apparent difference in *s* between both systems contributes to the difference in *R*.

Table 1. Summary statistics of demand and inventory parameters at DC stage

DC	*avg*	*s*	*Q*	*R*
Centralized system	250	28.7	657	1,685
Decentralized system	234	246.6	614	2,358

Using the demand and inventory parameter data from Fig.1 and Table 1, ten sets of one-year demand data are generated. The day-to-day inventory management process is simulated based on these data in order to evaluate and compare the performance, i.e. total inventory management cost, between the decentralized and centralized systems. The simulation process records the amount of inventory on-hand, level of inventory position, number of replenishment orders, and shortage amounts. Unmet demands at retailers are lost sales with unit shortage cost of 8 THB.

Unmet demands at the DC are backordered without incurring additional cost. This is assumed because in the real situation, there are multiple products in the distribution network such that deliveries of some products are made on a daily basis. As a result, the backlogged amount of a product can be delivered along with other items. When backlog occurs, the DC will fulfill partial demand to the retailer up to its inventory on-hands and deliver the backlogged amount with express delivery lead time of one day. When the DC cannot satisfy two or more orders on the same day, each retailer will receive the amount proportional to their order quantity before the backlogged amounts are delivered.

Simulation results from 10 replications are summarized in Table 2. First, the paired t-test is used to analyze the results. Based on the simulation results, there is a strong evidence that the average total cost of the centralized system is significantly lower than that of the decentralized system, a reduction of 16% in the total cost, p-val<0.001. A closer look reveals that the centralized system gains significant savings in its inventory holding cost from using echelon inventory position, while sacrificing the lost sale cost at the retailers due to backlogged orders that occur.

Table 2. Numerical example results

	Decentralized system				Centralized system			
	Ordering	Holding	Shortage	Total	Ordering	Holding	Shortage	Total
Retail 1	530	800	328	1,658	529	786	578	1,893
Retail 2	523	806	980	2,309	521	788	1,270	2,579
Retail 3	602	853	1,458	2,913	600	835	1,740	3,175
Retail 4	663	978	870	2,511	661	948	1,325	2,934
Retail 5	654	994	1,710	3,358	652	969	2,150	3,771
All retails	2,972	4,431	5,346	12,749	2,963	4,326	7,063	14,352
DC	1,363	7,496	-	8,859	1,256	2,487	-	3,743
System	4,335	11,927	5,346	21,608	4,219	6,813	7,063	18,095

Another statistical analysis is performed to investigate the effect of demand (i.e. *avg*) and its variation (i.e. *cv*) on all three cost components of the retailers. Analysis results, see Fig. 2, show that *avg* demand has a strong positive effect on all three cost components, which explains the increase in all three costs as demand increases. In addition, *cv* has a strong positive effect on the shortage cost only, with slight effects on ordering and inventory holding costs. The effect of *cv* can be seen from the increase of the shortage costs between Retailer 1 and Retailer 2, and between Retailer 4 and Retailer 5 in both systems.

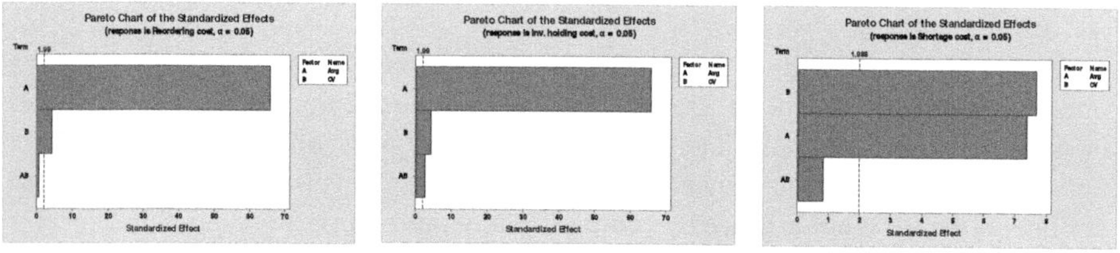

Fig. 3. Effects of average and variation of demand on retailers' costs

4 Conclusion

This study focuses on inventory management of a single product in two-echelon distribution network. The total inventory management cost of the centralized system is compared with that of the decentralized system. A numerical example shows a promising results in cost savings through information sharing in the centralized system. Further study will be conducted that considers limited capacity of shelf space, especially at the retailers, multiple products sharing the same space, other patterns of demand, and distribution networks with more than two echelons.

References

1. Abdul-Jalbar, B., Gutiérrez, J.M., Sicilia, J.: Single cycle policies for the one-warehouse N-retailer inventory/distribution system. Omega 34, 196–208 (2006)
2. Axsäter, S.: A simple decision rule for decentralized two-echelon inventory control. Int. J. Production Economics 93-94, 53–59 (2005)
3. Duan, Q., Liao, W.: Optimization of replenishment policies for decentralized and centralized capacitated supply chains under various demands. Int. J. Production Economics 142, 194–204 (2013)
4. Felter, R.B., Dalleck, W.C.: Decision models for inventory management. Irwin, Inc., Homewood (1961)
5. Geng, W., Qiu, M., Zhao, X.: An inventory system with single distributor and multiple retailers: Operating scenarios and performance comparison. Int. J. Production Economics 128, 434–444 (2010)
6. Monthatipkul, C., Yenradee, P.: Inventory/distribution control system in a one-warehouse/multi-retailer supply chain. Int. J. Production Economics 114, 119–133 (2008)
7. Simchi-Levi, D., Kaminsky, P., Simchi-Levi, E.: Designing and managing the supply chain: concepts, strategies and case studies, 3rd edn. McGraw-Hill (2007)
8. Ye, F., Xu, X.: Cost allocation model for optimizing supply chain inventory with controllable lead time. Computers & Industrial Engineering 59, 93–99 (2010)

A Study on Hong Kong Rice Supply Chain Risk Management with Value Chain Analysis

Anthony Lam[1], Tao Zhang[2,1], and Kin Keung Lai[3,1,*]

[1] Department of Management Science, City University of Hong Kong, Tat Chee Avenue, Kowloon, Hong Kong
[2] School of Management, Xi'an Jiaotong University, Xi'an, 710049, China
[3] International Business School, Shaanxi Normal University, Xi'an, 710049, China
anthony.lam@grdil.com, zhangtao05@stu.xjtu.edu.cn, mskklai@cityu.edu.hk

Abstract. Hong Kong rice supply chain is characterized as totally importing from overseas countries and huge local demand. There exist numerous risk issues in different stages of the supply chain. Using the technique of value stream analysis, this paper investigates the supply chain structure and identifies the relevant risk factors on the supply side and demand side of the chain. Findings show that the rice supply chain risk may arise from the process of rice production and milling, storing and distribution, government policy, demand fluctuation and so on. All risk factors are assessed based on value chain analysis and the related mitigation measures are proposed.

1 Introduction

Since rice is a staple food in Hong Kong and rice consumption is around 44 Kg per person per year, ensuring supply of rice is very important for the people and the government of Hong Kong. From past evidence, many events in the rice supply chain can result in negative supply performance and even supply chain disruption. For example, the Thailand political situation in 1973, SARS in 2002, rice commodity price increasing in 2003 and Thailand rice policy changing in 2011 all led to the supply crisis for Hong Kong rice market, accompanying with rice firms' bad operations and financial performance.

Systemic identification of Hong Kong rice supply chain risk issues is full of value, as it is the basis of risk mitigation to guarantee better supply chain performance. Previous studies on rice supply chain management have mainly focused upon the production and distribution of rice in a single market, which is not suitable for the specific business mode of Hong Kong. To analyze the complex rice supply chain structure, including export and import markets, and to identify the risk issues in each stage, we conduct the method of value stream mapping and analysis. Based on the value chain analysis, risk issues are assessed and related mitigation measures are proposed.

---

* Corresponding author.

M. Gen et al. (eds.), *Industrial Engineering, Management Science and Applications 2015*,
Lecture Notes in Electrical Engineering 349, DOI: 10.1007/978-3-662-47200-2_53

2 Literature Review

2.1 Supply Chain Risk Management

Supply chain risk management has been a hot research topic for many years. In most literatures, supply chain risk is defined as issues that are related to negative consequences or impact[1~3]. The risk factors in supply chain are grouped with material flow, financial flow and information flow. For the risk in material flow, sourcing flexibility[4], supplier selection[5], transportation lead time[6] and supply monitoring[7] are found to be relevant with supply chain risk. In the financial flow, the exchange rate in global supply chain[8], firm's after tax profit[9], financial crises[10], control and visibility of procure-to-pay process[11] are the risk related factors. The risk in information flow is about information inaccuracy caused by problems in information accessibility[12], information efficiency[13] and data accuracy[14]. Supply chain risk mitigation measures mainly consist of ex-ante and ex-post measures.

2.2 Rice Supply Chain Risk Management

Rice is an agricultural product, and its supply chain is characterized with biological production and special distribution. In the production process, the quality of seed and fertilizer, temperature change, rainfall and flood, diseases and pests are playing important roles in determining rice yield and output[15,16]. According to different risk factors, the rice supply chain risk is divided into weather and natural disasters related risk, biological and environmental risk, market related risk, logistical and infrastructural risk, managerial and operational risk, public policy and institutional risk and the order of risk magnitude.

2.3 Value Chain of Rice Industry

Value chain refers to the value added process from production to final consumption[17]. Focusing on the process that materials and information are delivered to customer, the method of value stream mapping (VSM) and value stream analysis (VSA) turns out to be one of the most popular techniques for supply chain analysis. Using time as an indicator, all related activities are categorized into value-added activities (VA), non-value-added activities (NVA) and necessary-non-value-added activities (NNVA). With value stream mapping, potential risk factors in each stage of the supply chain can be identified and assessed. Even though there are many studies on agricultural supply chains such as Jasmine rice supply chain and parboiled rice supply chain, which take advantages of VSM and VSA[18~20], there's hardly any research on Hong Kong rice supply chain using the above analytical method, to the best of our knowledge. This paper investigates Hong Kong rice supply chain structure and depicts it in a value chain map for the first time. Based on the analysis results, we provide comprehensive identification and assessment of risk factors that arise from export and import markets. Some risk mitigation measures are also provided in this paper.

3 Value Chain Analysis

3.1 Hong Kong Rice Industry

The Hong Kong rice market is characterized by total dependence on imports and a huge demand in the local marketplace. Data from the Trade and Industry Department of Hong Kong[1] show that annual rice consumption per person in Hong Kong is 44 Kg and the total annual consumption in Hong Kong is 320,000 MT. According to the law of Hong Kong, importers and re-exporters of rice have to obtain license from the government, stockholders must fulfill the import amount standard and put the imported rice into government approved warehouses. The newest statistics by September 2014 illustrate that Hong Kong has 187 registered stockholders and about 10 of them are active stockholders. In the time interval from July 1, 2014 to September 30, 2014, the total imported quantity is up to 75,800 MT, and the consumed quantity is 78,200 MT. In the same period, the stockholders have 15,100 MT closing stock and 13,900 MT reserve stock, note that stockholders must keep 17% of the total stock as reserve stock for government.

The main imports are from Thailand, Vietnam, China, Australia and so on. As illustrated in Figure 1, it is obvious that Thailand is the largest exporting country for Hong Kong, though its share in total imports has declined from 67.45% to less than 50% in recent years. Meanwhile the share of Vietnam has increased to about 40% of the total imports. Mainland China is the third largest rice exporter to Hong Kong.

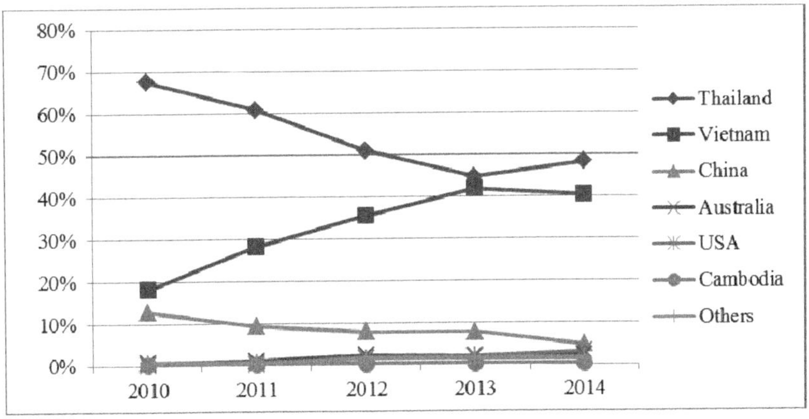

Fig. 1. The share of Hong Kong rice imports from different exporters

From the supply side of Hong Kong rice market, we mainly consider exporters from Thailand and Vietnam as the suppliers. Thailand exports about 10 million tons of rice to different countries and districts, including jasmine white rice, glutinous rice and brown rice, etc., among which the parboiled rice accounts for nearly 40% of the total volume; Thailand accounts for more than half of world's parboiled rice

---

[1] The website is `http://www.tid.gov.hk/eindex.html`

consumption. For Vietnam, rice is the most important crop that supports more than half of the people's livelihood and contributes about $1.5 billion in export revenue annually. However, many factors, such as weather and climate change, pest disaster and political instability strongly influence production, processing and distribution of rice in these countries. We then give an inspection on the entire supply chain of Hong Kong rice industry.

3.2 Rice Supply Chain Structure

Considering that Thailand is the largest exporter of rice to Hong Kong, we take the rice supply chain of Thailand as the upstream of the entire supply chain of Hong Kong rice market. The upstream and downstream of supply chain are connected by export and import trade. In Hong Kong, the main process of rice market is re-milling and distributing to customers.

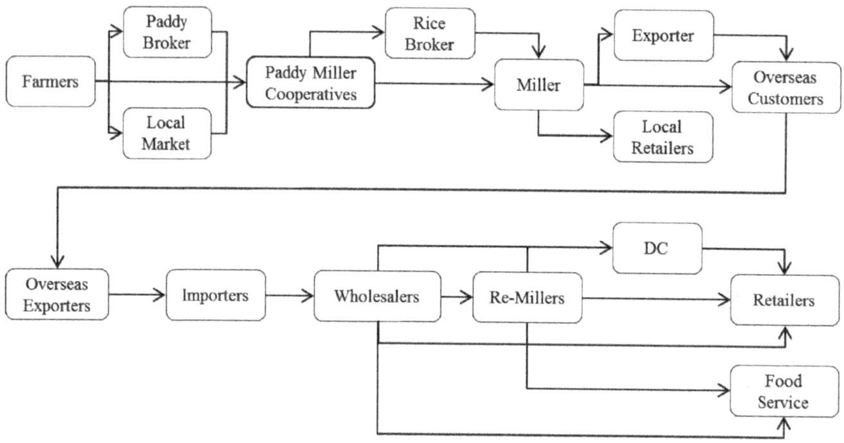

Fig. 2. The structure of Hong Kong rice supply chain

As showed in Figure 2, we pay the same attention to rice supply chain structures both in export country and in Hong Kong, because the problems in every stage of upstream supply chain may disrupt the entire chain and fluctuate Hong Kong rice market.

The entire supply chain starts from farmers in Thailand. After harvest, farmers sell their paddy to brokers or the local market, which play role as middle man. Through the first level of distribution, middle man sells paddy to rice millers. Millers receive semi-manufactured paddy and then process it into rice by milling. The primary rice can be sold to the local market immediately through retailers, and can also be exported to overseas markets through exporters. In the downstream, importers in Hong Kong sell rice to wholesalers. Many wholesalers choose to re-mill the primary rice to enhance the quality. With distribution centers (DC) in different areas, wholesalers, as well as re-millers, provide rice directly to retailers and various food servers, which at last reach customers.

3.3 Value Stream Mapping and Analysis

By interviewing and questionnaire survey, we depict the value stream map of Hong Kong rice industry as shown in Figure 3. The value stream map consists of 8 parts, which are the main stages of the whole supply chain. Using value stream analysis, we divide the 8 parts into 22 activities, and estimate the average time that every activity takes.

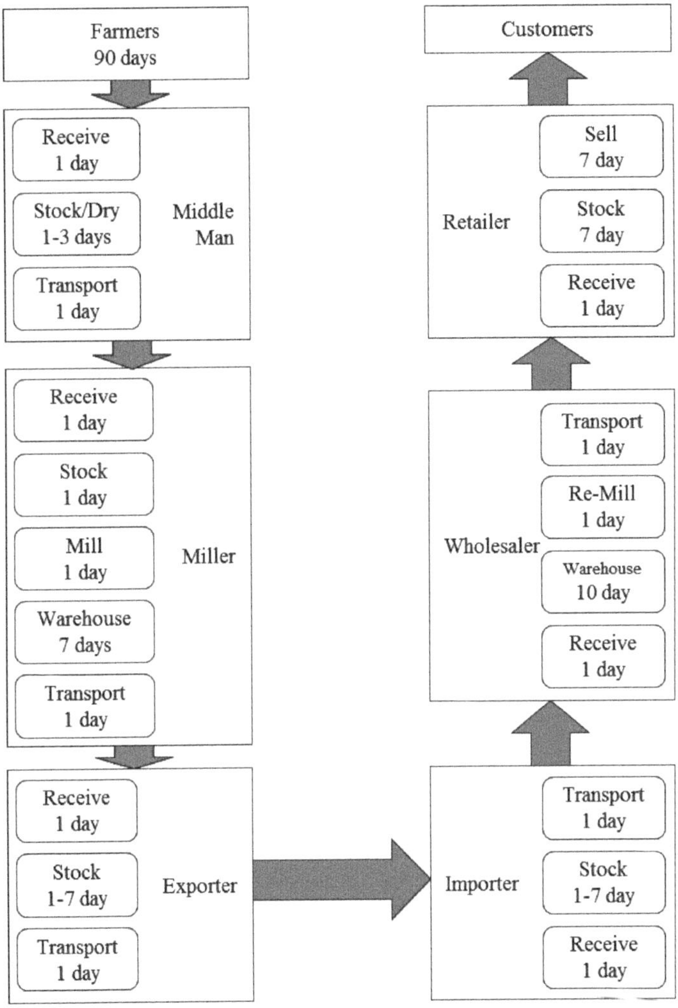

Fig. 3. Value stream map of Hong Kong rice supply chain

Some straightforward results can be drawn from the value stream map in Figure 3. It takes 151 days from farmers to overseas customers, which includes 114 days in the upstream supply chain and 37 days in the downstream supply chain. From sewing to harvesting of rice, it takes up to 90 days in the whole rice lifecycle. A further step of value stream analysis is given in Table 1.

Table 1. Value stream analysis of Hong Kong rice supply chain

Position	Value of activity	Activities		Time	
		No.	%	Day	%
Thailand (upstream)	Value-Added Activities (VA)	2	9.09	91	60.26
	Non-Value-Added Activities (NVA)	4	18.18	17	11.26
	Necessary-Non-Value-Added Activities (NNVA)	6	27.27	6	3.97
Hong Kong (downstream)	Value-Added Activities (VA)	1	4.55	1	0.66
	Non-Value-Added Activities (NVA)	3	13.64	24	15.89
	Necessary-Non-Value-Added Activities (NNVA)	6	27.27	12	7.95
Total		22	100	151	100

Table 1 shows that about 61% of the activity time goes toward value-added activities. Farmers spend 90 days for producing rice. However, they are not dominant in the supply chain, especially in pricing, bargaining and negotiation power. In the period of production, farmers are not capable of handling the numerous risk factors. Therefore, the upstream rice production process risk management needs to be addressed intensively.

Another type of activity that needs to be considered in risk management is necessary-non-value-added activities such as receiving, transportation and selling. These count for 12% of the total time with 3.97% in export country and 7.95% in Hong Kong. These activities are necessary and easily suffered from risk factors like natural disasters, political instability and demand volatility.

For the non-value-added activities such as stock and warehouse, we should optimize our supply chain operations to minimize the time length and decrease the probability of disruption from risk events.

4 Risk Management Based on Value Chain Analysis

4.1 Supply Side Risk Management

The upstream of Hong Kong rice supply chain is related to rice production, processing and distribution in exporting countries. Because rice is an agricultural product, there exist numerous uncontrollable natural forces that may decrease rice yield, leading to significant supply fluctuations. The two largest suppliers, Thailand and Vietnam, frequently suffer from floods, temperature events, pest disasters and rice disease, which directly affect the agricultural system and output. As shown in the value chain analysis, production is conducted by farmers and it lasts for a long period. In this time interval, farmers lack the ability to mitigate risk factors and deal with risk-related consequences. Risk factors in production lead to supply chain disruptions. Risk mitigation in rice production should involve supply chain partners and governments. More of the value generated by the entire supply chain should be transferred to farmers by pricing and contract design. Governments can expand their investment in agricultural infrastructure to facilitate safe production of rice.

Processing and distribution system is another aspect that needs to be considered in the rice supply chain of Hong Kong. Different members in the supply chain face a variety of processes. The millers must handle the milling process, storage, quality control and packing, while the exporters must deal with their own storage and packing processes. All of these processes can vary with quantity, quality and external conditions. Meanwhile, any failure in the various processes may induce a systemic fluctuation. To mitigate the risk factors, usage of information technology can effectively control the task planning and reduce uncertainty. For this reason, IT tools should be used widely in exporting countries and Hong Kong rice firms.

An important environmental risk factor in exporting countries is government policy. Because agriculture is vital for Thailand and Vietnam and failure of rice production may lead to starvation and political instability, governments of these countries intervene in the rice industry through policies on production, trade and export. As exemplified, Thailand government has intervened in rice industry by setting export tax. Moreover, policies in developing countries are turbulent and unpredictable, which brings up uncertainty for firms' operations in the supply chain. In this situation, coordination between upstream and downstream firms can alleviate the negative effects of policy changes.

4.2 Demand Side Risk Management

Hong Kong rice market is the demand side of the entire supply chain. The customers have different requirements in terms of quantity, rice types and purchase time. Different seasons and social events impact the demand of rice in Hong Kong, making it difficult to predict precisely, especially for a certain type of rice. Nowadays, the information revolution has provided customers a wide range of choices. Preferences of customers change rapidly, which makes the prediction of rice demand even harder. As a consequence, firms in rice industry often run out of stock when customers want to buy their products. The increasing risk from demand side requires the members of rice supply chain take more integration practices to share demand information among partners.

Apart from the real fluctuation of rice demand in Hong Kong market, rice supply chain members' perceptions of demand fluctuations are exaggerated from the downstream to the upstream, known as the 'bullwhip effect'. Effects of risk from demand uncertainty expand with the bullwhip effect. To mitigate the possible negative effects, upstream firms are advised to coordinate with downstream firms in terms of demand information sharing. That is to say, risk information disclosure, collection and processing are important tasks for rice firms' risk management teams.

5 Conclusion

The rice industry in Hong Kong is particular about the modes of supply and demand. There exist numerous unpredictable and uncontrollable factors that may lead to supply chain disruption. In this paper, we take advantage of value stream analysis to identify and assess the rice supply chain risk issues.

In the value stream mapping and analysis, we find that the upstream of Hong Kong rice supply chain is dominant to the success of the entire business. The risk factors may arise in the process of production, milling, storing and distribution. Moreover, government policies in exporting countries also play a vital role. On the demand side, the preference of customers is increasingly changing and demand fluctuation can be exaggerated through the supply chain, which makes risk management of rice supply chain even harder. For many of the risk factors we have identified with value chain analysis, some mitigation measures are proposed. However, the systemic risk mitigation measures still need further studies.

Acknowledgement. The work described in the paper is financially supported by the NSFC grant (project number 71390335.

References

1. Christopher, M., Lee, H.: Mitigating supply chain risk through improved confidence. International Journal of Physical Distribution & Logistics Management 34(5), 388–396 (2004)
2. Spekman, R.E., Davis, E.W.: Risky business: expanding the discussion on risk and the extended enterprise. International Journal of Physical Distribution & Logistics Management 34(5), 414–433 (2004)
3. Wagner, S.M., Bode, C.: An empirical investigation into supply chain vulnerability. Journal of Purchasing and Supply Management 12(6), 301–312 (2006)
4. Kamrad, B., Siddique, A.: Supply contracts, profit sharing, switching, and reaction options. Management Science 50(1), 64–82 (2004)
5. Levary, R.R.: Ranking foreign suppliers based on supply risk. Supply Chain Management: An International Journal 12(6), 392–394 (2007)
6. Stalk, G.: The costly secret of China sourcing. Harvard Business Review 84(2), 64–66 (2006)
7. Khan, O., Christopher, M., Burnes, B.: The impact of product design on supply chain risk: a case study. International Journal of Physical Distribution & Logistics Management 38(5), 412–432 (2008)
8. Kouvelis, P.: Global sourcing strategies under exchange rate uncertainty. In: Quantitative Models for Supply Chain Management, pp. 625–667. Springer US (1999)
9. Goh, M., Lim, J., Meng, F.: A stochastic model for risk management in global supply chain networks. European Journal of Operational Research 182(1), 164–173 (2007)
10. Hartley-Urquhart, R.: Managing the financial supply chain. Supply Chain Management Review 10(6) (2006)
11. Kerr, J.: Streamlining the cash flow. Supply Chain Management Review 10(7), 25–31 (2006)

12. Lee, H.L.: The triple-A supply chain. Harvard Business Review 82(10), 102–113 (2004)
13. Faisal, M.N., Banwet, D.K., Shankar, R.: Information risks management in supply chains: an assessment and mitigation framework. Journal of Enterprise Information Management 20(6), 677–699 (2007)
14. Barry, J.: Supply chain risk in an uncertain global supply chain environment. International Journal of Physical Distribution & Logistics Management 34(9), 695–697 (2004)
15. Matthews, R., Wassmann, R.: Modelling the impacts of climate change and methane emission reductions on rice production: a review. European Journal of Agronomy 19(4), 573–598 (2003)
16. Zhang, T., Zhu, J., Wassmann, R.: Responses of rice yields to recent climate change in China: an empirical assessment based on long-term observations at different spatial scales (1981–2005). Agricultural and Forest Meteorology 150(7), 1128–1137 (2010)
17. Kaplinsky, R., Morris, M.: A handbook for value chain research, vol. 113. IDRC, Ottawa (2001)
18. Ramingwong, S., Wattanutchariya, W., Laosiritaworn, W.: The Study of Parboiled Rice's Supply Chain in Thailand (2010)
19. Francis, M.: Application of the food value chain analysis method in the UK red meat industry. In: Proceedings of the 9th International Symposium on Logistics, vol. 104 (2004)
20. Simons, D., Francis, M., Jones, D.T.: Food value chain analysis. In: Consumer Driven Electronic Transformation, pp. 179–192. Springer, Heidelberg (2005)

The Role of Product Development to Drive Product Success: An Updated Review and Meta-Analysis

Yosephine Suharyanti[1,2,*], Subagyo[1], Nur Aini Masruroh[1], and Indra Bastian[3]

[1] Department of Mechanical and Industrial Engineering, Universitas Gadjah Mada, Indonesia
[2] Department of Industrial Engineering, Universitas Atma Jaya Yogyakarta, Indonesia
[3] Department of Accounting, Universitas Gadjah Mada, Indonesia
yosephine@mail.uajy.ac.id, {subagyo,aini}@ugm.ac.id,
indra_bastian@yahoo.com

Abstract. Researches on product development since 1970s have discussed many aspects, yet studies on the role of product development in driving product success, as believed in SCM concept, are rarely available. The available ones do not clearly describe the way product development affects product success. This paper presents a model development of the relationship between several aspects including product development, and product success. Six variables are constructed based on a review, and the relationships are evaluated using meta-analysis on 166 studies spanning 35 years, carried out in four continents, covering large scale industries and SMEs, manufacturers and services, profit and non-profit organization. The result shows that five of the six variables are clearly correlated to product success, while one of them could not be proven. The future work then is model modification based on this result.

1 Introduction

Society economic growth in a country is driven by the growth of its industry. Manufacturing industries, as the engine for the industries chain, take role as the strongest stimulator for the growth of the other sectors. A comparison on some countries shows that the higher the manufacturing output is, the higher the gross domestic product (GDP) becomes. For example, the manufacturing output in 2012 of the United States (US), China, Japan, Germany, and Indonesia is 1,912; 1,623; 960; 632; and 107 billions of US$ respectively [1], in line with their GDP in 2013 i.e. 16.768; 9,240; 4.920; 3.730; and 868 billions of US$ respectively [2].

Nowadays, two of some basic issues about manufacturing industries are customization and supply chain. These are triggered by the increase of sophistication of customers' needs altogether with the rapid change in every aspect of life.

Customization drives the shortening of product life cycle. The consumer products like electronic products, computers, and apparels are the example of the products which are changed rapidly. The first two are related to the rapid change of technology, while the last changes rapidly because of fashion trend. In this condition, companies must give their effort for product development process, by which the

* Corresponding author.

M. Gen et al. (eds.), *Industrial Engineering, Management Science and Applications 2015*,
Lecture Notes in Electrical Engineering 349, DOI: 10.1007/978-3-662-47200-2_54

product success will be supported. Some researches show that the companies with good product development process tend to be more successful [3, 4].

Supply chain point of view places the product development as the beginning of all. Fig. 1 shows the supply chain scheme in the context of the way a company realizing the product success, starting from product development (PD) stage, product realization (PR) stage, product delivery (PDv) stage, to product success (PS).

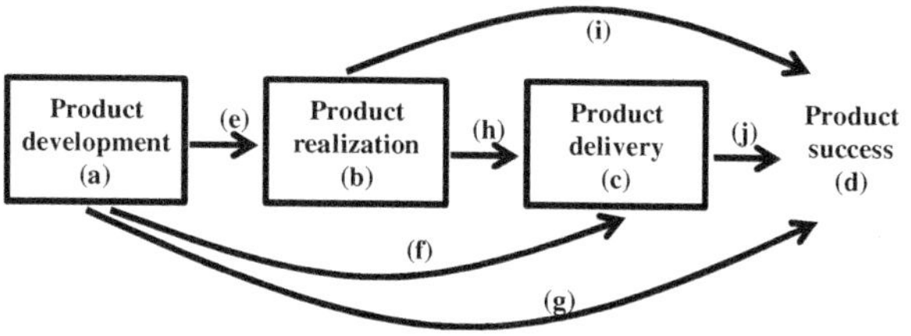

Fig. 1. Product development and product success under supply chain subsystem viewpoint

Major researches related to PD (a) cover frameworks, strategies, processes, and product data management. Studies on PR (b) spread from material preparation to packaging, including soft aspects like production managements and manufacturing information systems. The PDv problems (c) lead to studies on distribution and transportation management, while researches on PS (d) cover product success factors and brand managements.

The relationship between PD and PR (e) is dominated by researches on design-to-manufacture and concurrent engineering, while the relationship between PD and PDv (f) encourage the researches on design-to-supply-chain. Specific researches related to the relationship between PD and PS (g) are rarely found. Therefore, the way the product development affecting the product success becomes an opportunity to study further. PR and PDv relationship (h) encourages researches on logistic management and supply chain management, while the relationship between PR to PS (i) gives the birth of the studies on quality and reliability. Last, PDv and PS relationship (j) drives the customer relationship management discussions.

Fig. 2 shows the map of some researches related to product development (1979 to 2014) categorized according to Fig. 1 and the focus of the researches. The size of the block represents the intensity of researches on the related field.

	PD	PS	PD-PR	PD-PDv	PD-PS
Concept	[5, 6, 7, 8, 9]	[6, 17, 18]	[22, 23, 24, 25]	[3, 12, 24, 28, 33]	[8, 10, 11, 21, 34, 35]
Technique	[9, 10, 11, 12]	[19]	[26, 27, 28, 29, 30]	[25, 28, 33]	[18, 24, 29]
Cost	[13, 14, 15, 16, 17]	[6, 18, 20, 21]	[9, 17, 22, 23, 26, 27, 31, 32]	[28, 30, 33]	[12]

Fig. 2. Researches on product development

Based on the review illustrated in Fig. 2, the research opportunity in product development is related to product success and costing. Some researches discuss the relationship between product development and product success [17, 18, 24, 29], yet none of them clearly discusses the contribution of product development to product success. Up to now, their discussions are generally around product success factors and product life-cycle management. A discussion about cost in the term of a bench mark study reviews the relationship between product development effort and economic recovery [12]. However, some experimental studies [36, 37] state that there is no significant correlation between product development effort and product successes.

On the other hand, some evidence shows that superiority in product development lead companies to gain success. For instance is the evidence in Indonesia, a developing country located in Southeast Asia, in which many emerging manufacturing industries are developing and growing. Most of the large scale manufacturing companies in Indonesia are operated under foreign enterprise's licenses. The medium scale manufacturing companies become the subsidiaries of the large ones. One of the weaknesses of this condition is that those companies are inferior in product development. They become fragile; the uncertainty of the environment deteriorates them. However, there are some companies already major in product development. They are proven to be strong in facing the uncertain surrounding and relatively successful. A company, say, Company A, for example, produces electronic consumer products which are able to get large market share because of their uniqueness. This company is one of the ten best electronic manufacturers, together with some international brands [38].

2 The Question

Some researches discuss the effect of product development to product success. Yet, the way the product development affects the product success is still not clear. Some researches, somehow, show the opposite result. On the other hand, companies' experiences indicate that the company major in product development gains success in business. The question then, what actually the relationship between product development and product success is. This paper presents a preliminary analysis to answer that question.

3 Model Development

Product success is affected by many aspects. Evaluation of the product development effect on product success must be conducted altogether with other aspects. Here, all the aspects are classified into six groups, i.e. product characteristic, technological advance, management, market environment, product development, and product development cost. These six groups are built by collecting, from published researches, all the aspects believed to have influence on product success, and categorizing them based on production system elements and, specifically, the aim of this study.

In order to evaluate further the relationship between product development and product success as mentioned above, a simple model describing the relationship between those six aspects and the product success is developed. The product success (S) is assigned as the dependent variable, and the six aspects are assigned as the independent variables affecting the product success. Figure 3 shows the diagram of the model.

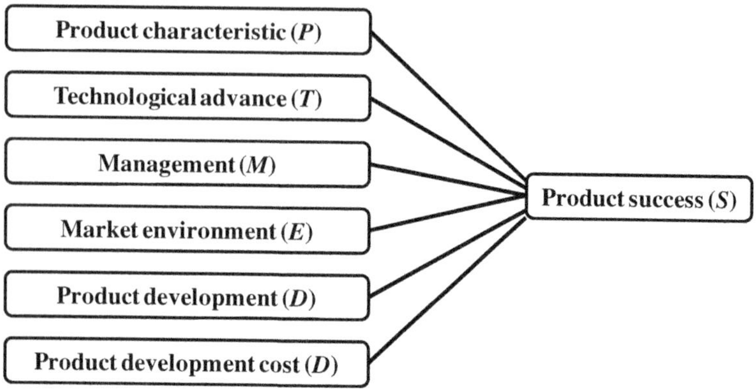

Fig. 3. A model of relationship between product success and the six variables

Product success (*S*) here is defined as the ability of product to create profit for the company. It can be measured by sales volume, market share, market life, cash flow, revenue, profit, or payback [12, 17, 19, 20, 32, 33, 39].

The first independent variable, product characteristic (*P*), is defined as the properties of product from the customers' viewpoint. It covers, for instance, product superiority, product innovativeness, product uniqueness, relative price, relative quality, product variation, product flexibility, and conformance to standard [18, 20, 40, 41, 42, 43, 44, 45, 46, 47, 48, 49, 50, 51, 52, 53, 54, 55, 56, 57].

The second variable, technological advance (*T*), defines the character of technology owned or implemented by the company. It describes the technology ownership, the synergy or compatibility of technology with the organization, the proficiency of technology, the technology sophistication related to the product innovativeness, and technological flexibility to response the rapid change, as elaborated through some experimental researches about the relationship between T to S [18, 20, 40, 41, 45, 54, 58, 59].

The management of the overall manufacturing system, from material procurement to product delivery is involved in the third independent variable, management (*M*). Production planning and control, logistic management, suppliers involvement, resources compatibility, manufacturing and marketing synergy, marketing and sales management, management support, organization support, integration of the organization, and networking, are some of the aspects covered in this variable [18, 20, 40, 41, 42, 44, 45, 47, 48, 49, 50, 52, 54, 56, 57, 60].

The uncontrollable aspects outside the company are presented by the forth variable, market environment (*E*). It explains, for instance, market dynamic, market need, market growth, market size, market maturity, brand, early success with customers, and customers' knowledge on product [20, 41, 42, 44, 46, 47, 55, 61].

Product development (*D*) variable covers all the aspects of product development, involves its power, its resources, its management, and its strategy. Some of the examples could be mentioned are R&D excellence, team competency, resources compatibility, resources flexibility, activities involved in product development, and process flexibility [6, 37, 40, 42, 47, 49, 51, 52, 59].

Cost of product development (*C*) represents the effort, the money spent, or the intensity of resources used in product development activities. R&D spending, R&D to asset ratio, and investment magnitude are some terms mentioned in some related researches [20, 36, 37, 54, 57, 62, 63].

4 Meta-Analysis

The effect of the six independent variables on product success will be evaluated by meta-analysis on the result of other studies previously conducted. This meta-analysis is applied on the correlation (r) as the most available and common data presented in research papers. Some research papers present no correlation data but mention the value of F statistic, t statistic, or R^2 for simple linear regression model. In this case, those values are converted to correlation. Several other papers discussing structural equation model present the value of the effect size. It will be converted also to correlation value [64].

The studies involved in this meta-analysis are published from 1979 to 2014. The location of observed cases covers almost all the five continents, i.e. the United States including Alabama specifically, Canada, Europe including London, Hungary, and Turkey specifically, Brazil, China, Taiwan, Korea, Thailand, Indonesia, Pakistan, India, Kenya, and Nigeria. The industry sectors spread from manufacturing industries to services, large scale industries to SMEs, profit oriented companies to non-profit organizations. The manufacturing industries involved are manufacturing industries in general, electronics, computers, control equipments, telecommunication equipments, medical equipments, foods, beverages, and textiles. The services industry involved are service industries in general and insurances.

The total number of studies (K) observed for all the six variables is 166 and the total number of sample (N) is 30,759, thus the average sample size per study (N_K) is around 185. The reliability (α) of the variables on those studies is ranged from 0.52 to 0.964, exceeding the minimum reasonable α value of 0.5 [64]. For every variable, the number of studies observed is ranged from 10 to 56 and the total number of sample is ranged from 3056 to 9820. It has exceeded 3000, the rule of thumb of minimum number of sample required for meta-analysis [64].

Table 1 and Table 2 show the summary of the data used and the result of this analysis. The r presents the observed correlation value mentioned by the studies, while ρ is the representative correlation value came from the meta-analysis. The σ_ρ is the standard deviation of ρ, and σ_r is the standard deviation of r. The ratio σ_ρ/ρ represents the feasibility of the result of the meta-analysis. Under 5% confidence interval, the minimum value of σ_ρ/ρ is $1.96 \cong 2$ [64].

Table 1. Meta-analysis data summary of the variables

Relationship between	Data					
	K	N	N_K	R	σ_r	α
P and S	34	5945	24-800	-0.760-0.870	0.222	0.520-0.964
T and S	14	3120	24-800	0.113-0.720	0.208	0.590-0.867
M and S	33	5604	24-800	-0.680-0.822	0.216	0.590-0.964
E and S	19	3214	24-367	0.128-0.810	0.205	0.860-0.964
D and S	56	9820	30-332	0.026-0.703	0.077	0.590-0.964
C and S	10	3056	51-1260	-0.154-0.843	0.274	0.600-0.911

Table 2. Meta-analysis result of the correlation of the variables

Relationship	Meta-analysis result		
	ρ	σ_ρ/ρ	Conclusion
P and S	0.583	2.123	Feasible
T and S	0.623	2.099	Feasible
M and S	0.636	2.443	Feasible
E and S	0.494	2.151	Feasible
D and S	0.320	3.392	Feasible
C and S	0.752	1.778	Not feasible

5 Discussion

As presented in Table 2, the correlation between cost of product development and product success generated from meta-analysis is not feasible. Technically, there are two reasons could be identified to explain it. First, the range of the observed data is too wide, and the second, the data is not normally distributed (Fig. 4). The other variables may have wide range of data too, as product characteristic and management. However, the extreme data is not significant and most of the data is normally distributed. The data related to technology advance variable may be not normally distributed perfectly, but there is no extreme data involved.

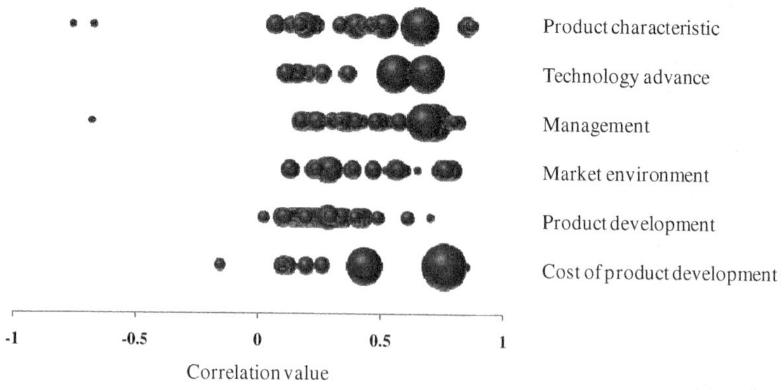

Fig. 4. Correlation value distribution of the observed studies

6 Conclusion

One conclusion generated from the result of the meta-analysis is that the cost of product development is not proven to be correlated to product success. Some studies (31%) say that cost product development has low, even negative, correlation to product success ($-0.154 \leq \rho \leq 0.260$), while the others (69%) say that the cost of product development is highly correlated to product success ($0.437 \leq \rho \leq 0.843$).

Nonetheless, correlation could only explain the linear relationship between two variables [65]. The low correlation does not mean that there is no relationship between two variables. The nonlinear relationship still becomes the possibility. In this case, cost of product development itself, in the terms of overall cost amount, may have no impact on the product success, like what the 31% of studies has observed. However, the way the cost is utilized in product development process may have relationship with the product success. Cost, nowadays, is not just a tool to measure the level of activities. It takes a role as performance accelerator, in facilitating activities to be conducted properly [66]. For example, the fallacy in utilizing the cost for resources procurement will lead to non-effective activities.

7 Future Work

Hence, combining the variable *D* and *C* into one variable becomes the reasonable follow-up to be evaluated. The meta-analysis result shows the correlation between product development and product success. The product development here means the way the product is developed, hence it is related to the resources utilized, beside the arrangement of the activities. Therefore, it is related to the cost utilization.

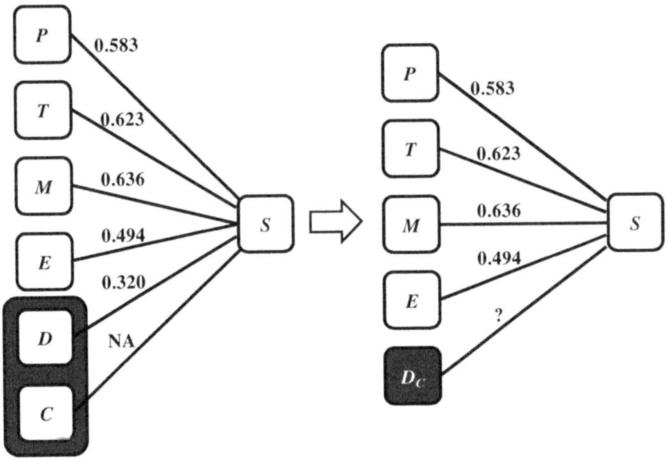

Fig. 5. The future work: model improvement

Following-up the result of the meta-analysis, the future work of this study is to reconstruct the model and reevaluate the effect among variables. Fig. 5 gives the scheme of the model modification. The new variable, the scheme of product development cost D_C, presents the strategy or scenario of how the product development is managed to contribute the product success. Finally, the correlation among the independent variables will also be evaluated, in order to find the alternative of improvement.

References

** used in meta-analysis

1. Rhodes, C.: Manufacturing: International Comparisons. UNCTAD (2014), `http://unctadstat.unctad.org` (accessed: January 26, 2015)
2. World Bank: Gross Domestic Product 2013, World Development Indicators Database (2014), `http://databank.worldbank.org` (accessed: January 26, 2015)
3. Griffin, A.: Product Development Cycle Time for Business-to-Business Products. Ind. Market. Manage. 31, 291–304 (2002)
4. Industry Innovation Council: Trends in Manufacturing to 2020, Dept. of Innovation, Industry, Science and Research, Canberra (2011)
5. Cooper, R.G.: Stage-gate Systems: A New Tool for Managing New Products. Business Horizons, 44–54 (May-June 1990)
6. Cooper, R.G., Kleinschmidt, E.J.: An Investigation into the New Product Process: Steps, Deficiencies, and Impact. J. Prod. Innovat. Manag. 3, 71–85 (1986)**
7. Ahmadi, R., Roemer, T.A., Wang, R.H.: Structuring Product Development Process. Eur. J. Oper. Res. 130, 539–558 (2001)
8. Kapoor, S., Bhattacharya, K., Buckley, S., Chowdhary, P.: A Technical Framework for Sense-and-Respond Business Management. IBM Syst. J. 44(1), 5–24 (2005)
9. Weustink, I.F., ten Brinke, E., Streppel, A.H.: A Generic Framework for Cost Estimation and Cost Control in Product Design. J. Mater. Process. Tech. 103, 141–148 (2000)
10. Burruss, J., Kuettner, D.: Forecasting for Short-lived Products: Hewlett-Packard's Journey. The Journal of Business Forecasting 21(4), 9–14 (2002)
11. Chang, P.T., Chang, C.H.: A Stage Characteristic — Preserving Product Life Cycle Modeling. Math. Comput. Model. 37, 1259–1269 (2003)
12. Jackson, C.: The CIO's Role in PLM: Facilitating "Great Product" Development to Drive Economic Recovery. Aberdeen Group, A Harte-Hanks Company (2010)
13. McGrath, M.E., Romeri, M.N.: The R&D Effectiveness Index: A Metric for Product Development Performance. World Class Design to Manufacture 1(4), 24–31 (1994)
14. Asiedu, Y., Gu, P.: Product Life Cycle Cost Analysis: State of the Art Review. Int. J. Prod. Res. 36(4), 883–908 (1998)
15. Ben-Arieh, D., Qian, L.: Activity-based Cost Management for Design and Development Stage. Int. J. Prod. Econ. 83, 169–183 (2003)
16. Manalo, A.J., Ramon, G.P.: The Cost of Product Development of Bt Corn Event MON810 in the Philippines. J. Agrobiotech. Manage. Econ. 10(1), 19–32 (2007)
17. Cooper, R.G., Edgett, S.J., Kleinschmidt, E.J.: Portfolio Management: Fundamental for New Product Success. Prod. Innov. Best Pract. Series, Product Development Inst. (2014)
18. Cooper, R.G.: Selecting Winning New Product Projects: Using the NewProd System. J. Prod. Innovat. Manag. 2, 34–44 (1985)**
19. Chang, T.H., Lee, J.Y., Chen, R.H.: The Effects of Customer Value on Loyalty and Profits in a Dynamic Competitive Market. Comput. Econ. 32, 317–339 (2008)
20. Cooper, R.G.: The Dimensions of Industrial Success and Failure. J. Marketing 43, 93–103 (1979)**

21. Cooper, R.G.: From Experience: The Invisible Success Factors in Product Innovation. J. Prod. Innovat. Manag. 16, 115–133 (1999)
22. Ulrich, K.T., Pearson, S.A.: Does Product Design Really Determine 80% of Manufacturing Cost? Working Paper. Massachusetts Inst. of Tech. (1993)
23. Shehab, E.M., Abdalla, H.S.: Manufacturing Cost Modelling for Concurrent Product Development. Robot. CIM-Int. Manuf. 17, 341–353 (2001)
24. Kušar, J., Duhovnik, J., Grum, J., Starbek, M.: How to Reduce New Product Development Time. Robot. CIM-Int. Manuf. 20, 1–15 (2004)
25. Tyagi, S.K., Yang, K., Tyagi, A., Dwivedi, S.N.: Development of a Fuzzy Goal Programming Model for Optimization of Lead Time and Cost in an Overlapped Product Development Project Using a Gaussian Adaptive Particle Swarm Optimization-based Approach. Eng. Appl. Artif. Intel. 24, 866–879 (2011)
26. Roemer, T.A., Ahmadi, R., Wang, R.H.: Time-cost Trade-offs in Overlapped Product Development. Oper. Res. 48(6), 858–865 (2000)
27. Hon, K.K.B., Xu, S.: Impact of Product Life Cycle on Manufacturing Systems Reconfiguration. Annals of the CIRP 56(1), 455–458 (2007)
28. Östlin, J., Sundin, E., Björkman, M.: Product Life-cycle Implications for Remanufacturing Strategies. J. Cleaner Prod. 17, 999–1009 (2009)
29. Thevenot, H.J., Simpson, T.W.: A Product Dissection-based Methodology to Benchmark Product Family Design Alternatives. J. Mech. Design 131, 1–9 (2009)
30. Chiu, M.C., Okudan, G.: An Integrative Methodology for Product and Supply Chain Design Decisions at the Product Design Stage. J. Mech. Design 133, 1–15 (2011)
31. Niazi, A., Dai, J.S., Balabani, S., Seneviratne, L.: Product Cost Estimation: Technique Classification and Methodology Review. J. Manuf. Sci. E 128, 563–575 (2006)
32. Shiau, C.S.N., Michalek, J.J.: Optimal Product Design under Price Competition. J. Mech. Design 131, 1–10 (2009)
33. Aitken, J., Childerhouse, P., Towill, D.: The Impact of Product Life Cycle on Supply Chain Strategy. Int. J. Prod. Econ. 85, 127–140 (2003)
34. Komninos, I.: Product Life Cycle Management. Urban and Regional Innovation Research Unit, Aristotle University of Thessaloniki (2002)
35. Gecevska, V., Chiabert, P., Anisic, Z.: Product Lifecycle Management through Innovative and Competitive Business Environment. J. Ind. E. Manag. 3(2), 323–336 (2010)
36. Cooper, R.G., Kleinschmidt, E.J.: Winning Businesses in Product Development: The Critical Success Factors. Res. Technol. Manage., 1–15 (2007)**
37. Sisodiya, S.R., Johnson, J.L.: Resources in NPD: An Investigation of Resource Capabilities, Int. J. Bus. Manage. 9(2), 103–113 (2014)**
38. Desain Bagus Indonesia, PT: 10 Perusahaan Elektronik Terbesar di Indonesia Perebutkan Pasar Rp 83 Triliun (2012), `http://Duniaindustri.com` (accessed: June 2, 2014)
39. Cooper, R.G.: Doing It Right: Winning with New Products. Prod. Innov. Best Pract. Series. Product Development Inst. (2000)
40. Zirger, B.J., Maidique, M.A.: A Model of New Product Development: An Empirical Test. Manage. Sci. 36(7), 867–883 (1990)**
41. Yap, C.M., Souder, W.E.: Factors Influencing New Product Success and Failure in Small Entrepreneurial High-technology Electronics Firms. J. Prod. Innovat. Manag. 11, 418–432 (1994)**
42. da Silva, S.L., de Toledo, J.C., Jugend, D., de Sousa Mendes, G.H.: Critical Success Factors on Product Development Management in Brazilian Technological Based Companies. In: Complex Syst. Concurrent E, pp. 739–747. Springer, London (2007)
43. Sethi, A.: The Stage-gate Process, Organizational Politics, and Performance of New Products. Dissertation. Carleton University, Ontario (2012)**

44. Altindag, E., Zehir, C.: Back to the Past: Re-Measuring the Levels of Strategic Orientations and Their Effects on Firm Performance in Turkish Family Firms: An Updated Empirical Study. Soc. Behav. Sci. 41, 288–295 (2014)**
45. Hassan, M.U., Shaukat, S., Nawaz, M.S., Naz, S.: Effects of Innovation Types on Firm Performance. Pakistan J. Comm. Soc. Sci. 7(2), 243–262 (2013)**
46. Turhan, G.: Risk Perceptions and Brand Extension Success: Just Another Antecedent or One that Shapes the Effects of Others? Fibres Text. East. Eur. 22, 3(105), 23–30 (2014)**
47. Thongpravati, O.: Market-Driving Innovation: Understanding the Critical Success Factors at the Front End of the Development Process. Dissertation. RMIT University (2014)**
48. Wangbenmad, C., Rashid, N.R.N.A.: Supplier Involvement in Product Development Process (PDP) and New Product Performance. J. Market. Manage. 2(2), 227–238 (2014)**
49. Onguko, M.A., Ragui, M.: The Role of Strategic Positioning on Products Performance in the Telecommunications Industry in Kenya. Int. J. Sci. Res. 3(10), 1309–1314 (2014)**
50. Choi, M.J., Song, J.W., Choi, R.H., Choi, J.S.: An Empirical Analysis on the Performance Factors of Software Firm. Int. J. Softw. Eng. Appl. 8(7), 121–132 (2014)**
51. Cortese, M.: Co-creating Products with Customers. Dissertation. Luiss Guido Carli, Rome (2014)**
52. Muriira, L.K.: Competitive Strategies Adopted by Insurance Companies in Kenya. Thesis. University of Nairobi (2014)**
53. Rachim, T., Setiawan, I.: The Effects of Product Attributes and Pricing Policy to Netbook Purchase Decision. Int. J. Sci. Res. 3(4), 522–537 (2014)**
54. Szerb, L., Ulbert, J.: The Examination of Competitiveness in the Hungarian Small Business Sector and Family Businesses. Recontres de St-Gall, Switzerland, pp. 1–25 (2014)**
55. Ojeyinka, T.O., Ajayi, O.O.: Analysis and Modelling of Brand Equity Data to Predict the Success of a Business. J. Emerg. Trends Eng. Appl. Sci. 5(7), 66–72 (2014)**
56. ul Hassan, E., Hameed, Z., Fatima, M., Shaheen, M.: Innovation is the Key Driver for Corporate Success. Middle-East J. Sci. Res. 19(6), 843–847 (2014)**
57. Zheng, H., Wan, N., Chen, D., Wang, T.: Antecedents of Project Implementation Success in Crowdfunding. In: 18th PACIS, Chengdu, China, pp. 1–13 (2014)**
58. Wang, L.W., Kuang, N.P., Shu, M.H.: Impacts of R&D-marketing Interaction: Market Performance of New Products. Int. J. Econ. Stat. 2, 137–144 (2014)**
59. Atanu, C., Saloni, D.: Impact of Product Development Efforts on Product Introduction and Product Customization Abilities. In: EUROMA, pp. 1–10 (2014)**
60. Brahmane, J.: An Empirical Study on Sales Capability and Marketing Implementation Capability of SMEs in India and Their Impact on Market Share. IOSR J. Bus. Manage. 16(11), 7–16 (2014)**
61. Yang, C.C., Tsai, K.C., Huang, P.B.: An Empirical Study of Critical Success Factors for the Manufacturing Industry in Taiwan. In: 19th ICIT, pp. P1–P6 (2014)**
62. Ma, L., Mello, A.S., Wu, Y.: Industry Competition, Winner's Advantage, and Cash Holdings, Working Paper, University of South Carolina, Columbia (2014)**
63. Park, S.: Analyzing the Efficiency of Small and Medium-sized Enterprises of a National Technology Innovation Research and Development Program. SpringerPlus 3(403), 1–12 (2014)**
64. Hunter, J.E., Schmidt, F.L.: Methods of Meta-Analysis, 2nd edn. Sage Publications, California (2004)
65. Velickovic, V.: What Everybody Should Know about Statistical Correlation. American Scientist, 1–3 (2015), `http://www.americanscientist.org` (accessed on January 19, 2015)
66. Suharyanti, Y., Bastian, I., Subagyo, M.N.A.: From Fixed Term to Quality Thinking. In: 1st ICBESS, Bali, p. 51 (2014)

Management Practices of Thai Silk Product

Kanogkan Leerojanaprapa* and Walailak Atthirawong

School of Statistics,
King Mongkut's Institute of Technology Ladkrabang, Bangkok 10520, Thailand
klkanogk@kmitl.ac.th, walailaknoi@gmail.com

Abstract. Thai silk is one of Thailand's most well-known products, which is considered as the best silk in the world. It is famous for its bearing uniqueness features, beautiful designs, bright colors and elegant quality. Silk weaving is a long-established folk craft passed down from generation to generation that resembles the region's cultural heritage. It pays an important role to the foundation of Thai local economy for centuries.Thai products produced by local communities using indigenous skills and craftsmanship combined with available natural resources and raw materials. However, it is widely accepted that there exists a managerial gap in many small or micro-sized enterprise, many activities may not be enough to make those products are sustainable. Understanding the key issues affected Thai silk entire chain is an important step to improve the competitiveness of those products. In this paper, case studies are therefore conducted, as part of a research project to examine the current stage. The SCOR model is employed to identify challenges and pinpoint weaknesses in management practices also in what processes have an effect on inefficiency within the supply chain before radical suggestions improvements. This paper intends to report two case studies of Thai silk products and concludes with a summary of key issues that are raised in the fieldwork. Finally, conclusions and recommendations for further development of the main study are also provided in the paper.

Keywords: Thai Silk, SCOR model, Supply Chain Performance.

1 Introduction

"Mai Thai" or "Thai Silk" has been worldwide recognized of its reputation for its attractiveness and unique, different from other countries' silk. Thailand is considered the leading country in ASEAN region, in terms of both researches and products that are unique, with short thread and rough knots. Cocoon is small, delicate, shiny, gentle, flexible and fluttering. For a genuine native variety, despite its floss is thin but sticky and springy. As handicraft, the silk cannot be put into the machines due to its length of about 250 - 450 meters. Silk production in household is found almost in all regions, especially in the northeast and the north where production has been at household level with establishment of more than 200 groups of textile weavers across the country,

* Corresponding author.

M. Gen et al. (eds.), *Industrial Engineering, Management Science and Applications 2015*,
Lecture Notes in Electrical Engineering 349, DOI: 10.1007/978-3-662-47200-2_55

at present, for local distribution and exporting [1]. In silk production, Thai labors, especially workers in rural area, will be used. This is deemed generation of income for the Thai workers and entrepreneurs, mostly small and medium-sized enterprises. Effective production of the genuine Thai silk will be significantly handmade. Thai silk product is one of the interesting products, with potential to be further developed to support free trade of the ASEAN Economic Community (AEC), which will expand the market. It is generally found that the demands for silk and silk products have currently been steadily growing in both domestic and foreign markets. Moreover, Thailand has the ability to produce Thai silk products in a manner that ranges from upstream to downstream.

This paper aims to investigate in which practices have the most influence in supply chain performance of Thai silk producers. Therefore, in the first stage of the study will focus on broadening an understanding in management practices of whole supply chain. Case studies are conducted, as part of a research project to examine the entire supply chain from both demand and supply sides at the current stage to understand the differences and similarity among businesses. In-depth interview via site visits, as well as the SCOR model is employed to gain more understanding insights into the chain prior to conducting the main study in the next stage. This paper presents the preliminary results of the first phase of the case studies of Thai silk producers and concludes with a summary of key issues that are raised in the fieldwork.

The rest of this paper is organized into the following sections. Methodological details are presented in section 2. The results and findings are then illustrated in section 3. Finally, conclusions of the study along with recommendations for further work will be drawn in the last section of the paper.

2 Methodology and Tools

The study is concentrate particularly on a formal group of people in a community or community producers who join together to produce Thai silk products, since they create value of the chain from raw materials through the production process to end customers. The important steps including methodologies employed to this study are discussed as follows.

2.1 Initial Interview

The initial part of this primary research was conducted via site visits and in-depth interviews. Several key parties or stakeholders i.e. suppliers, a formal group of members or community producers, government officers, network members in each village as well as end customers were interviewed over a two month period between December 2014 and January 2015. Principally, the main objective of this stage was to understand of problems and difficulties within the group of society in producing and selling their products to the market. The interview questions intend to measure business's performance using the SCOR model [2, 3]. To illustrate the point, in this paper, two case studies have been presented.

2.2 Supply Chain Operations Reference Model (SCOR Model)

SCOR model is developed in 1996 by Supply Chain Council [2, 3]. The SCOR contains standard descriptions of management processes and standard metrics to measure process performance, enabling effective communication among the supply chain partners [4]. The SCOR model provides standard descriptions of relevant management processes, a framework of relationships among the standard process performance metrics and standard alignment to features and functionality. The model is separated supply chain operations into five primary management processes - PLAN, SOURCE, MAKE, DELIVER, and RETURN [3]. The current version of SCOR model is also available on Supply Chain Council website [5]. Once a complex management process is captured using SCOR Model form, all unambiguously can be described and explained. As such, by applying a SCOR model, it will help us to depict the "AS-IS" state of a process and also provides a basis for improvement those processes in the future state. The model consists of over 150 key indicators that measure the performance of supply chain operations. These metrics derive from the contribution and experience of the council members, who are independent not-for-profit corporation [2].

2.3 Supply Chain Management Practice

Supply chain management practices based on main four decision areas on SCOR as PLAN, SOURCE, MAKE and DELIVERY were summarised by Lockamy III and McCormack [6] by excluding RETURN from their framework. They propose nine keys created from literature review and interviews with supply chain experts and practitioners. Their framework will be employed to analyse data from the Thai silk case studies and the qualitative analysis from case study will be presented in Section 3.2.

Planning process relates to the process of forecasting, planning and scheduling with suppliers, planning in manufacturing and also scheduling for delivery to customers.

Collaboration relates to the process of establishing customer and supplier relationships, sharing of planning information with suppliers, collaboration between sales, manufacturing distribution department etc. The supplier lead-times can base on collaborative information.

Teaming relates to the process of setting the strategy team from the major supply chain functions of the focal firm or identifying procurement, manufacturing and distribution team. Furthermore, particular teams should have regular meetings.

Process measures relates to the measuring of forecasting accuracy, a joint assessment of the needs among sales, manufacturing and distribution department, measuring differentiation between customer requests and actual delivery.

Process document refers to the process of documenting in the operation strategy process, procurement process, manufacturing process and distribution process. It is used to explain what is to be done within or between processes.

Process ownership refers to the process of identifying owner for supply chain planning process, procurement process planning team, manufacturing planning process team to ensure its effectiveness and working closely with other relevant teams.

Process credibility relates to the delivery process and it can be evaluated from customers' satisfaction with on-time delivery performance.

Process integration is the linking process along with the supply chain in particular process. For example, delivery process from inventory control and automatic replenishment throughout the supply chain can be integrated in order to increase the effectiveness.

IT support is generally using IT in particular process.

Furthermore, it is found that Process credibility, Process integration, IT support can impact significantly only in the Delivery process [6].

3 Results

In this research, case studies were firstly conducted mainly in the Northwest part of Thailand, for instance Ubon Ratchathani, Khon Khan and Maha Sarakham. The results of two case studies in Ubon Ratchathani province are reported below.

3.1 Thai Silk Handicraft Industry

Traditionally Thai silk is hand woven in household. The majority of Thai silk producers are farmers who spend their time out of the agricultural season for weaving for private consumption. Now a day Thai silk is famous as mentioned above so it becomes commercial fabric. As the purpose of production changes, Thai silk supply chain becomes more complicated and involves with different stakeholders. During this transition, variety types of Thai silk producers and their stakeholders are investigated in this exploratory study.

Since traditional Thai silk is hand woven, each silk fabric is unique and difficult to be duplicated through commercial means by mass production. Therefore we will scope our research in Thai silk handcraft industry and also define the focal agents in this research as individuals or group of members.

3.1.1 Individual Operators

Most individual operators for Thai Silk are originated from a small household-based which has been weaving its traditional products for several years. Treads for weaving are produced from the cocoons of Thai silkworms which had been raised the caterpillars using mulberry leaves. After a suitable time, the silk thread is separated from the caterpillar inside the cocoon. Then the threads are hand-reeled onto a wooden spindle to produce a uniform strand of raw silk. Once washed and dried the silk yarn to remove the natural yellow coloring, the silk is then woven using a traditional hand operated loom. It takes a weaver for several months to produce about 2 meter of handmade silk as most of them usually have their own jobs. They will design their own styles of weaving according to their cultures. The final products mostly use in their own families and giving as gifts for a wedding or engagement ceremony, for example. The rest of products will be then sold to customers in their provinces.

Based upon the in-depth interview results and observation of the production process, it seems that these operators run every process by themselves starting from the primary producers of the silk cocoons through producing the silk. For upstream process, they need to make their own plan to set how many of silkworms they can raise as they need to make sure that they will have time to feed silkworms and protect from other animals and the elements. Furthermore, it is needed to make sure that they have enough mulberry leaves to feed them (between 5-700 kg. per 20,000 eggs). The life cycle of a silkworm spans about 40-50 days. After the cocoons have been formed, the strands have to be reeled within two days; otherwise the cocoons will be opened by pupa and those cocoons cannot be used anymore. Thread is then further processed and spun to the desired count (in general 20-22 or 40-44 denier). The amount and quality of yarn are also depended on the quality of eggs as well. Hence, those activities have a noteworthy impact on supply chain performance. The process of planning, sourcing of raw materials (eggs and mulberry) and production of cocoons are required to make sure its effectiveness. Normally, the rearer, reeler and the weaver is the same person in the village. After finishing the process of washing and drying the silk yarn, the process of weaving silk products can be done whenever they have a free time from their normal jobs and festival. Silk product represents Thai local folk, the lives which come from fundamental ideas of the weavers and descendent from their ancestors. Figure 1 displays Thai silk products and traditional process of weaving.

Fig. 1. Thai silk and traditional process

3.1.2 A Group of Operators

Thai silk operators in handicraft industry can cooperate in their community as a local entrepreneur. An example of the group of operators is called "OTOP" (One Tambon One Product) to support local make and marketed products of each Tambon [1].

According to the case studies, members in the group produce silk individually from their house since it takes long time to complete hand-woven silk and they need to use waving machine which is big and difficult to be moved. They also use the same process to produce the product as individual operators. However, they can show and sell their products in the shop where is the general office in their community so it is easier for customers to find and know them. In addition, they can gain financial and knowledge supports from government in order to improve potential and to sustain their groups in the market.

The similarity of both agents can be found that most individual operators do not yet adopt any SCOR model areas included in this study (Plan, Source, Make, and Delivery).

3.2 Different Supply Chain Practices by SCOR Model

This research employs the framework by Lockamy III and McCormack [6] to characterize and compare two focal Thai silk operators who are individual operators and a group of members. Table 1 compares two types of focal agents according to their supply chain practices.

Table 1. Key supply chain practices of Thai silk individual operators or group of members from Ubon Ratchathani province case study

Practice	Individual operators				A group of members			
	PLAN	SOURCE	MAKE	DELIVERY	PLAN	SOURCE	MAKE	DELIVERY
Planning process	✗	✓	✗	✗	✗	✓	(✓)	(✓)
Collaboration	✗	✗	✗	✗	✗	✗	✗	✗
Teaming	✗	✗	✗	✗	✓	✗	✗	✗
Process measures	✗	✗	✗	✗	✗	✗	✗	✗
Process credibility*	-	-	-	✗	-	-	-	✗
Process integration	-	-	-	✗	-	-	-	✗
IT support	-	-	-	✗	-	-	-	(✓)
Process documentation	✗	✗	✗	✗	✗	✗	(✓)	✓
Process ownership	✗	✗	(✓)	✗	✗	✗	(✓)	✗

* Only impact significantly on DELIVERY

✗ not available, ✓ available, (✓) available in some occasion

3.2.1 Individual Operators

Generally, individual operators do not deliver products to their customers and their major customers are local people so they may not concern about any practices for delivery. Since they can work anytime and no commitment as their part-time job, they do not use forecasting to predict demand. Generally, they will produce silk by considering their available materials e.g. silk yarn rather than demands from customers. Before they start weaving, they usually try to get enough materials for a piece of silk. Once they start weaving, they may not set the deadline to finish their products.

As explained in Section 3.1.1, they can assign different tasks to their family members by their expertises in the producing process. However, there is no process ownership in other SCOR factors. Furthermore, they do not concern either relationship with other partners or the strategy of a team since they are mainly own raw materials and do not set sales target. Therefore, they do not neither set formal strategy nor predict demand that why they cannot set measuring in any process. Interestingly the know-how of Thai silk operation has been transferred within family without documenting but they implement "learning by doing" manner. They are using traditional processes, and also old style machines. Furthermore, most of them are old ladies in rural communities so the modern IT or any technology may not suitable or relevant to them.

3.2.2 A Group of Members or Society

The process of setting the strategic team is the originated requirement when they set their Thai silk group. Leaders can response in planning process for all members in the team and to link between their group and government agencies. Their community shop is a showroom and also an office. They can sell their products to walk-in customers and also make product by customers' orders.

According to the interview especially with the leader of Thai silk handicraft group, she is a key person to deal with customers orders (make to order). Once she gets the order, she can ask other members to find out the current level of resources that they have and then she can plan to sort out from suppliers until getting enough materials (e.g. color, silk yarn). After that she can split the order to members and also planning the deadline for delivery to their customers, if they required. Occasionally, some organizations want to make their uniform, they may contact and order silk from the group which is capable to deal with big orders. In this case they can assign the different tasks such as thread producing, coloring, silk weaving so sometime the process ownership is applied.

In addition the formal collaboration between suppliers or customers cannot be observed. Main customers' satisfactions depend simply on the quality of the silk. Customers normally come to buy silk from their shop so process credibility which is measured by customers' satisfaction with on-time delivery performance is out of their focuses. The process integration cannot be defined since the quantity of their product depends on available quantity of silk yarn that they can produce rather than customers' demand. Information to match demand and supply cannot be observed in the case studies so the process measures of the accuracy of demand forecasting are unavailable.

Members in the group are using similar process of silk production to the individual operators do and they transfer the knowledge without documenting. However, the group will support their members by preparing the striped pattern paper to produce striped Thai silk. Furthermore, they need to record stock, product owners and also incomes or profits which can be shared with all members. We also found that generally they do not use IT to support any process except their groups' name will be able to find on the Internet or when they get promote as OTOP of their province.

4 Conclusion and Further Research

In this paper, two case studies of Thai silk products were analyzed. Data was gathered mainly from all relevant partners along the chains. The SCOR model was used to measure and evaluate supply chain configurations of both case studies. Similar obstacles are occurred in management practices from both cases. For instance, the issues of meeting deadlines, production capacity, quality control, design preferences and marketing are common challenges for two groups. Moreover, there is no use of material planning, recording, as well as inventory management with in a business. Likewise, there is a lack of information sharing between partners and their members. They don't know exactly when to produce at each time. For the group of silk producer, every determination comes from one person, which is the president of the group. It becomes clear, though, that both businesses must improve their current coordination of operations along the chain.

The results of the study will be applied to improve and develop the Thai silk and silk products to meet the consumer needs. Although one should be very careful about drawing conclusion of such kind of these products from a small group of sample sizes, some recurring issues do become apparent from data gathered in interviews and observation process. However, it is clear, from the study that, in order to improve competitiveness, there is a need to totally integrate material flow and information flow in the whole supply chain. Meanwhile, the establishment of strong network within village communities and with other silk producers will be a benefit for communities to provide supports for each other. For instance, they can pool together to buy raw materials or take turn in bringing goods for sale in OTOP fairs which are arranged in the main cities far away from the communities. The government's focus should be shifted towards strengthening the competitiveness of Thai Silk products rather than put only a financial assistance or knowledge that sometimes is difficult to apply into the villagers.

This will help expand the market to be potentially competitive once the ASEAN Free Trade opens. A subsequent work will be driven into this direction to confirm the conclusion and classify those barriers and challenges. Finally, the main task of the research will be concerning in a strategic scope in order to match demand to supply, such as developing a distribution centre in the community area for collecting, displaying and distributing all Thai silk products from each group to market. Data collection will continue, with the final results reported by the middle of 2015.

Acknowledgements. The authors would like to acknowledge the National Research Council of Thailand (NRCT) pass to National Institute of Development Administration (NIDA) for financial sponsorship of this research. The authors are also grateful to all the members who were kindly involved and provided valuable and fruitful information for this study.

References

1. http://www.thai-otop-city.com.html (available on April 12, 2008)
2. Coyle, J.J., Bardi, E.J., Langley Jr., C.: The Management of Business Logistics: A Supply Chain Perspective. In: 7th Thomson South-Western, USA, (2002)
3. Bolstorff, P., Rosenbaum, R.: Supply Chain Excellence: A handbook for Dramatic Improvement Using the SCOR Model, 2nd edn. AMACOM-American Management Association, New York (2007)
4. http://www.isye.gatech.edu/~lfm/8851.html (available on April 12, 2008)
5. http://www.apics.org/sites/apics-supply-chain-council/frameworks/scor.html (available on January 31, 2015)
6. Lockamy, A., McCormack, K.: Linking SCOR Planning Practices to Supply Chain Performance - An Exploratory Study. International Journal of Operations & Production Management 24(11-12), 1192–1218

Fuzzy Multi-objective Supplier Selection Problem: Possibilistic Programming Approach

Dicky Fatrias*, Ahmad Syafruddin Indrapriyatna, and Difana Meilani

Dept. of Industrial Engineering, Andalas University
Kampus limau Manis, Padang (25163), West Sumatera – INDONESIA
dicky@ft.unand.ac.id, ahmadsyaf@ft.unand.ac.id,
difana@ft.unand.ac.id

Abstract. In this paper, a multi-objective mathematical model is developed in fuzzy environment in which the vagueness in aspiration level of objectives and data imprecision regarding the selection criteria and related constraints are considered simultaneously as a source of fuzziness. In the model, such data imprecision is presented based on the estimation of its possibility distribution to better capture the uncertainty. Finally, a fuzzy solution methodology is constructed by the aid of weighted additive aggregation function to derive optimal solution. As preliminary investigation, we report that the proposed model is more flexible and convenient than the previous models whose imprecise parameters are treated as a given single estimated value.

1 Introduction

To remain competitive in a dynamically global market, the need to improve efficiency has prompted enterprises to seek opportunity to reduce costs while continuously improve their operation. Within the purchasing function, one of the key activities to achieve this goal is by selecting the appropriate supplier(s). In essential, supplier selection problem is a multi-criteria decision making within which criteria may be defined in quantitative and qualitative dimensions. Dickson [1] was the first to identify several criteria which are the most considered criteria in a practical supplier selection where quality, on-time delivery, performance history, warranty policy, and production facility/capacity of supplier were on the top five ranked in the list. A recent survey by Olson and Wu [2] study reported that cost, quality, and time response are major criteria that consistently appear for supplier selection.

The issue of considering uncertainty in supplier selection problem has received a great deal of concern in the field of supply chain management. This complexity in supplier selection stems from imprecise preferences of the decision maker (DM) regarding the aspiration level of decision objective and/or the imprecise nature of decision criteria and constraints. While the usefulness of stochastic approach has been documented, it is not always applicable in coding the information regarding the

* Corresponding author.

M. Gen et al. (eds.), *Industrial Engineering, Management Science and Applications 2015*,
Lecture Notes in Electrical Engineering 349, DOI: 10.1007/978-3-662-47200-2_56

imprecision of data and vagueness of goals. To avoid this drawback, the fuzzy approach is employed for modeling uncertain parameters. Moreover, it is frequently emphasized in the literature that fuzzy approach has had a great impact in preference modeling and multi-objective problem and has helped bring optimization techniques closer to the users' needs [3].

A number of studies have been devoted to examining supplier selection methods. Quantitative techniques have become increasingly applied recently. A comprehensive review of numerous quantitative techniques used for supplier selection has been done by [4].

This paper focuses on fuzzy multi-objective linear programming (fuzzy MOLP) to deal with supplier selection problem. Kumar et al. [5] developed a fuzzy multi-objective integer programming approach for supplier selection problem subject to constraints including buyer's demand and suppliers' capacity, and derived an optimal solution using max-min operator (Zimmermann's approach). To evaluate the performance of the model, they perform sensitivity analysis on the order allocation and objective function by changing the degree of uncertainty in supplier capacity. Amid et al. [6] solved fuzzy MOLP supplier selection problem by applying weighted additive aggregation function to facilitate an asymmetric fuzzy decision making technique. Since they found the performance of such a method is not adequate to support decision making process, α-cut approach is then proposed to improve the resulted achievement level. Later on, Amid et al. [7] applied weighted max-min aggregation function in supplier selection problem and compared the performance of the proposed approach with max-min operator and weighted additive aggregation function. They found that the ratio of achievement level of objectives matches the ratio of the objectives weight. Yucel & Guneri [8] proposed a new method of weights calculation in fuzzy MOLP supplier selection. Recent study by Arikan [9] developed a modified augmented max-min aggregation function that originally proposed by Lai dan hwang ([10],[11]) to solve fuzzy multi-objective supplier selection problem by considering a preference of the decision maker(s) (DMS) in determining the desired minimum achievement level of fuzzy objectives. The performance of the proposed approach is then is compared with the original augmented max-min and the weighted additive aggregation function in solving the test problem in Yucel & Guneri [8]'s study. The study reported that while the modified augmented max-min outperforms the weighted additive in terms of the achievement level of fuzzy objectives, it shows insignificant improvement in performance when compared to the original augmented max-min.

Related to coding the imprecise data involved in supplier selection problem, all models in literatures assumed such imprecision is tackled by assigning a given single estimated value. In this paper, a solution methodology for multi-objective supplier selection problem is developed in fuzzy environment in which the data imprecision regarding the selection criteria and related constraints, and vagueness in aspiration level of objectives are considered simultaneously as a source of fuzziness. Unlike the previous models, data imprecision is generated based on its possibility distribution in order to better capture the uncertainty. A fuzzy mathematical model is then developed by the aid of weighted additive aggregation function to derive a set of optimal solution.

2 Fuzzy Multi-objective Programming Methodology

2.1 Fuzzy Multi-objective Preliminary Formulation

A fuzzy formulation of the multi-objective linear programming (MOLP) with imprecise coefficient and fuzzy aspiration level of objectives can be stated as

$$\min Z_k = \sum_{i=1}^{n} \tilde{c}_{ki} x_i \leq \sim Z_k^0, \quad k = 1,2,\dots,p \tag{1}$$

$$\max \tilde{Z}_l = \sum_{i=1}^{n} \tilde{c}_{li} x_i \geq \sim Z_l^0, \quad l = p+1, p+2, \dots, q \tag{2}$$

subject to:

$$g_s = \sum_{i=1}^{n} a_{si} x_i \leq \tilde{b}_s, s = h+1, h+2, \dots, m \tag{3}$$

$$x_i \geq 0, \ i = 1,2,\dots,n \tag{4}$$

The symbol $\leq\sim$ and $\geq\sim$ denotes the fuzzified version of $\leq$ and $\geq$, respectively. The notation of $\tilde{c}_{ki}$ and $\tilde{c}_{li}$ are imprecise coefficients, and Z_k^0 and Z_l^0 are the aspiration levels that the DM wants to reach. The above fuzzy mathematical formulation is characterized by linear membership function whose value changes between 0 and 1. The linear membership function for fuzzy objectives are given as [12]:

$$\mu_{Zk} = \frac{Z_k^{max} - Z_k(x)}{Z_k^{max} - Z_k^{min}} \tag{5}$$

$$\mu_{Zl} = \frac{Z_l(x) - Z_l^{min}}{Z_l^{max} - Z_l^{min}} \tag{6}$$

Here Z_k^{max}, Z_l^{max}, Z_k^{min}, and Z_l^{min} means the maximum value and the minimum value of Z_k and Z_l, respectively. They are obtained by solving a single objective optimization problem respectively under each objective function.

2.2 Modeling Imprecise Parameter using Possibilistic Programming

In possibilistic programming, each imprecise data (ill-known parameter) has its possibility distribution which represents the possibility degree of occurrence of possible value for each imprecise parameter. A several number of distributions exist in literature such as triangular, trapezoidal and so on. Among them, triangular are the most commonly used distributions in solving possibilistic programming problems [2]. Using triangular distribution, a possibilistic programing with imprecise parameter $\sum_{i=1}^{n} \tilde{c}_i x_i$ is redefined as

$$\max/\min \sum_{i=1}^{n} (c_i^p, c_i^m, c_i^o) x_i \tag{7}$$

where c^p, c^m and c^o are the most pessimistic, the most likely and the most optimistic value of imprecise parameter, respectively. These values are usually estimated by the DMs based on available data as well as their knowledge.

Jimenez *et al.*, [13] proposed a method to define a single crisp representation of c^p, c^m and c^o based on the concept of expected interval and expected value of fuzzy numbers. It has been proven that this method is computationally efficient to solve such problems as it can preserve its linearity and do not increase the number of objective functions and inequality constraints [14].

The crisp representation of imprecise parameters of fuzzy MOLP problem in Eq. (1) - (4) can be formulated as follows [13]:

$$\tilde{Z}_k = \sum_{i=1}^{n} \left(\frac{c_{ki}^{pes} + 2c_{ki}^{mos} + c_{ki}^{opt}}{4} \right) x_i \leq \sim Z_k^0 \tag{8}$$

$$\tilde{Z}_l = \sum_{i=1}^{n} \left(\frac{c_{li}^{pes} + 2c_{li}^{mos} + c_{li}^{opt}}{4} \right) x_i \geq \sim Z_l^0 \tag{9}$$

subject to:

$$g_s = \sum_{i=1}^{n} a_{si} x_i \leq \left[\alpha \left(\frac{b_s^{pes} + b_s^{mos}}{2} \right) + (1 - \alpha) \left(\frac{b_s^{mos} + b_s^{opt}}{2} \right) \right] \tag{10}$$

$$x_i \geq 0, \; i = 1,2, \ldots, n \tag{11}$$

where α is a minimum acceptable feasibility degree of decision vector which can be varied according to subjective preference of the DM.

2.3 Fuzzy Aggregation Function

A fuzzy aggregation function is typically used to solve fuzzy multi-objective programming problem by converting such problem into single objective formulation. Solving the aggregation function results in the efficient solution in terms of the satisfaction degree of each objective from which the DMs choose the final decision based on his/her preference (relative importance among objectives). Tiwari et al [15] proposed weighted additive aggregation function which had been widely used in vector-objective optimization problems. The function is stated as follow:

$$\text{Max} \sum_k \omega_k \lambda_k$$

subject to:

$$\lambda_k \leq \mu_{Zk} \tag{12}$$

$$\lambda_k, \mu_{Zk}, \omega_k \in [0,1]$$

$$\sum_k \omega_k = 1$$

where λ_k denote the satisfaction degree of k-th objective (individual satisfaction degree of each objective).

3 Fuzzy Multi-objective Supplier Selection Problem

3.1 Model Formulation

A typical linear model for multi-objective supplier selection problems is presented as follows [7]:

Index set

i index for suppliers, for all i = 1, 2, . . . , n

Decision variable

x_i The number of units purchased from the i-th supplier

Parameters

D Aggregate demand of the item over a fixed planning period.
n Number of suppliers competing for selection
p_i Unit net purchase cost from supplier i
f_i percentage of product quality of the supplier i
s_i Service performance of the supplier i
C_i Capacity of i-th supplier

Following the formulation of the fuzzy MOLP problem in Eqs. (1)-(4), the crisp representation of the above problem using Eqs. (8)-(11) can be stated as follows:

$$\text{Min } \tilde{Z}_1 = \sum_{i=1}^{n} \left(\frac{p_i^{pes}+2p_i^{mos}+p_i^{opt}}{4}\right) x_i \leq \sim Z_1^0 \tag{13}$$

$$\text{Max } \tilde{Z}_2 = \sum_{i=1}^{n} \left(\frac{f_i^{pes}+2f_i^{mos}+f_i^{opt}}{4}\right) x_i \geq \sim Z_2^0 \tag{14}$$

$$\text{Min } \tilde{Z}_3 = \sum_{i=1}^{n} \left(\frac{s_i^{pes}+2s_i^{mos}+s_i^{opt}}{4}\right) x_i \geq \sim Z_3^0 \tag{15}$$

subject to:

$$\sum_{i=1}^{n} x_i \geq \left[\alpha\left(\frac{D^{mos}+D^{opt}}{2}\right) + (1-\alpha)\left(\frac{D^{pes}+D^{mos}}{2}\right)\right] \tag{16}$$

$$x_i \leq \left[\alpha\left(\frac{C_i^{pes}+C_i^{most}}{2}\right) + (1-\alpha)\left(\frac{C_i^{most}+C_i^{opt}}{2}\right)\right] \tag{17}$$

$$x_i \geq 0 \tag{18}$$

Eq. (13) minimizes the net cost for ordering product to satisfy demand. Eq. (14) maximizes the quality requirement of each supplier. Eq. (15) maximized the service performance of each supplier. Eq. (16) ensures that order quantity assigned to suppliers must satisfy the total demand. Eq. (17) guarantees that the order quantity assigned to each supplier will not exceed supplier capacity limit. Eq. (18) is non-negativity constraint.

3.2 Step-by-step Solution Methodology

We propose solution methodology to facilitate the decision-making process in solving multi-objective supplier selection problem with imprecise parameters and fuzzy aspiration level of objectives. The steps are summarized as follows:

Step 1: Construct the fuzzy MOLP supplier selection problem with imprecise parameters and fuzzy aspiration level according to defined criteria and constraints.

Step 2: Transform the model into an equivalent crisp representation of multi-objective model by converting all the imprecise parameters (i.e., criteria data, aggregate demand and capacity of each supplier).

Step 3: Determine the minimum acceptable feasibility degree (α-level) and then construct membership function for each fuzzy objective function using lower and upper bounds of each objective for the desired α-level.

Step 4: Specify the weight of each objective and solve the model using weighted additive aggregation function.

Step 5: Present the optimal solution set according to predetermined α value. When the DM desires to change his/her preference in respond to uncertainty and/or the weight of each objective, change the corresponding values and repeat the procedure from step 3.

Table 1. Supplier quantitative information

Supplier n	Cost ($)	Quality (%)	Service (%)	Capacity (unit)
1	{11,13, 15.5}	{65,80,95}	{70,85,90}	{550,700,800}
2	{10,11.5, 13}	{60,70,80}	{60,75,85}	{400,600,700}
3	{13,15,16.5}	{70,80,99}	{70,80,95}	{300,500,650}

4 Numerical Example

The following example is based on Amid, Ghodsypour and O'Brien [7]' s study.

Consider one company which considers three candidates of supplier for ordering plan of one product. Management wants to improve the efficiency of the purchasing process by evaluating their suppliers using three criteria which are net price, quality and service. Based on this description the objectives are developed as minimizing net cost of purchasing a product to the suppliers, maximizing quality rate and maximizing service performance of suppliers.

To show the effectiveness of the proposed solution methodology, the original criteria data from Amid, Ghodsypour and O'Brien [7]' s study are presented as imprecise parameters, following the assumption that the data is imprecise. As a result, rather than estimates a single value for each of the data, the DM determines the estimation of its possibility distribution by deciding three prominent values (i.e., the most likely, the most pessimistic and the most optimistic values) based on their current available information and knowledge. The constraints regarding the total demand and the capacity of each supplier are also considered imprecise in nature. As a result, the estimated values of their cost, quality and delivery performance, and associated constraints of suppliers are presented in Table 1.

Several results with different α-level (i.e. $\alpha = 0.0, 0.2, 0.4, 0.6, 0.8, 1.0$) are provided in performance testing and for each α-level alternative solution sets are generated by the aid of the weighted additive aggregation function. The weight of cost, quality and service are given as $\omega_1 = 0.63$, $\omega_1 = 0.21$ and $\omega_1 = 0.16$ [16]. Due to space limitation, the detail formulations according to the step-by-step procedures of methodology are not presented in the paper.

Table 2. Different sets of optimal solution

Item	Amid et al. (2011)			Proposed model	
		$\alpha = 0.0$	$\alpha = 0.2$	$\alpha = 0.4$	$\alpha = 0.6$
Z_1	12000	10756	11264	11773	12281
Z_2	740	655	686	717	748
Z_3	807	686	717	749	780
x_1,x_2,x_3	400,600,0	250,650,0	315,650,0	380,590,0	445,560,0
μ_{z1}	1.000	1.000	1.000	1.000	1.000
μ_{z2}	0.000	0.000	0.000	0.000	0.000
μ_{z3}	0.300	0.004	0.005	0.007	0.010

According to the result provided in Table 2 the value of all objective functions increases when the minimum acceptable feasibility degree (**α**-level) is increased. In other words, when the DM decided to response to uncertainty with a higher confidence level, all objective functions are also augmented. This could be due to the need to order more quantity of product (in total) in higher **α**-level.

It is also revealed that the second and third objective (quality and service) are critical objectives as the corresponding achievement level is always in the worst possible value in any **α**-levels. This implies that the model tends to sacrifice the performance of these objectives because it is at less of cost decreasing the performance of these objectives rather than decreasing the performance of the first objective (net price). This phenomenon is directly influenced by the fact that the first objectives is the most important ones whose assigned weight is the highest, according to the DM's preference ($\omega_1 >> \omega_2 > \omega_3$).

As it was mentioned in Section 3.2, the value of minimum acceptable feasibility degree (**α**-level) and the weight of the objectives can also be varied according to the DM preferences (other than illustrated above), yielding some alternative solution sets from which the DM select the most preferred solution. Hence, besides providing a broader decision spectrum, the proposed model is also more flexible and convenient than the previous models whose imprecise parameters are treated as a single estimated value.

5 Conclusion

In this paper, a solution methodology for multi-objective supplier selection problem is developed by simultaneously considering vagueness in aspiration level of objectives as well as the imprecision nature of to criteria data and related constraints. To better capture the uncertainty embedded in selection process, the model facilitates a judgment of the DM to estimate of the possibility distribution of each criterion and constraints by deciding three prominent values based on their current available information and knowledge.

According to the preliminary investigation, the main feature of the proposed model is the ability to yield different solution set with adjusted ordering decision based on a different minimum acceptable feasibility degree ($\boldsymbol{\alpha}$-level) in order to facilitate the DM to set his/her confident level in response to the uncertainty in imprecise criteria data and related constraint (i.e., demand and capacity) in supplier selection problem. Another interesting feature were also mentioned regarding the flexibility of the model compared to the recent models whose imprecise parameters are treated as a single estimated value.

Acknowledgments. This work has been funded by Directorate General of Higher Education, INDONESIA, under a program of Fundamental Research Grant 2015.

References

1. Dickson, G.W.: An analysis of vendor selection: system and decisions. Journal of Purchasing 1, 5–17 (1966)
2. Olson, D.L., Wu, D.: Simulation of fuzzy grey relationships. European Journal of Operational Research 175(1), 111–120 (2006)
3. Selim, H., Araz, C., Ozkarahan, I.: Collaborative production-distribution planning in supply chain: A fuzzy goal programming approach. Transportation Research Part E 44, 396–419 (2008)
4. Weber, C.A., Current, J.R., Benton, W.C.: Vendor selection criteria and methods. European Journal of Operation Research 50, 2–18 (1991)
5. Kumar, M., Vrat, P., Shankar, R.: A fuzzy programming approach for vendor selection problem in a supply chain. Int. J. Production Economics 101, 273–285 (2006)
6. Amid, A., Ghodsypour, S.H., O'Brien, C.: Fuzzy multiobjective linear model for supplier selection in a supply chain. Int. J. Production Economics 104, 394–407 (2006)
7. Amid, A., Ghodsypour, S.H., O'Brien, C.: A weighted max–min model for fuzzy multi-objective supplier selection in a supply chain. Int. J. Production Economics 131, 139–145 (2011)
8. Yucel, A., Guneri, A.F.: A weighted additive fuzzy programming approach for multi-criteria supplier selection. Expert Systems with Applications 38, 6281–6286 (2011)
9. Arikan, F.: Fuzzy Solutio approach for fuzzy multi-objective supplier selection problem. Expert Systems with Applications 40, 947–952 (2013)
10. Lai, Y.J., Hwang, C.L.: Possibilistic linear programming for managing interest rate risk. Fuzzy Sets and Systems 49, 121–133 (1993)

11. Lai, Y.J., Hwang, C.L.: Fuzzy multiple objective decision making-methods and applications. Springer, Germany (1996)
12. Zimmermann, H.J.: Fuzzy programming and linear programming with several objective functions. Fuzzy sets and systems 1, 45–46 (1978)
13. Jimenez, M., Arenas, M., Bilbao, A., Rodriguez, M.V.: Linear programming with fuzzy parameters: An interactive method resolution. Eur. J. Oper. Res. 177, 1599–1609 (2007)
14. Pishvaee, S.M., Razmi, J.: Environmental supply chain network design using multi-objective fuzzy mathematical programming. Applied Mathematical Modelling 36, 3433–3446 (2012)
15. Tiwari, R.N., Dharmar, S., Rao, J.R.: Fuzzy goal programming an additive model. Fuzzy Sets and Systems 24, 27–34 (1987)
16. Ghodsypour, S.H., O'Brien, C.: A decision support system for supplier selection using an integrated analytical hierarchy process and linear programming. International Journal of Production Economics 56 57, 199–212 (1998)

Value Adding and Improving Factors of Thai Long Steel Supply Chain for ASEAN Economic Community

Apinun Chedchoosuwan*, Tuanjai Somboonwiwat, and Charoenchai Khompatraporn

Department of Production Engineering, Faculty of Engineering
King Mongkut's University of Technology Thonburi, Bangkok
apinun_pte50.kmutt@hotmail.com, tuanjai.som@gmail.com,
charoenchai.kho@kmutt.ac.th

Abstract. At the end of 2015, all the ASEAN member countries including Thailand will become one ASEAN Economic Community (AEC) to promote regional integration. A consequence of the AEC is that there will be no taxes for the products whose origins are in the ASEAN region. Local businesses in all sectors must be prepared to compete with their counterparts from other countries within the region. Steel is an important fundamental industry in developing countries. Every year, Thailand imports over two million tons of long steel. Part of this is due to the lack of a steel smelting plant in Thailand. This paper analyzes the value-added chain of long steel products in Thailand and projects the total circulating volume up to 2017. This paper also assesses important factors influencing the long steel supply chain in Thailand, including those from the AEC, in order to recommend improvement strategies for this industry in the future.

1 Introduction

The long steel industry can be considered as a primary industry, and it is important for the development of the national economy [1]. Steel is a raw material for the manufacturing and industrial sectors, such as construction, automotive, and other industries. The Thai long steel industry is affected by the ASEAN Economic Community (AEC). Moreover, the ASEAN countries are currently increasing their capability in long steel product manufacturing. Meaning that there will be higher competition from free trade, and they might use some trade barriers to protect their own domestic markets [2]. In addition, neighboring countries of Thailand, such as Cambodia, Laos, Myanmar and Vietnam (CLMV), will increase their demand for long steel products. Therefore, the development of Thai long steel supply chain should be accelerated in order to compete in the AEC market.

Previous studies regarding steel supply chain include [3] a study of the decision support in rational use of steel products for dynamic and heterogeneous global supply chain (GSC). Another study focuses on the creation of business strategies in the automotive supply chain with respect to the relationship between automakers and

* Corresponding author.

M. Gen et al. (eds.), *Industrial Engineering, Management Science and Applications 2015*,
Lecture Notes in Electrical Engineering 349, DOI: 10.1007/978-3-662-47200-2_57

steel mills. The result of the study provides information regarding production expansion in the supply chain [4]. Other research studies the competitiveness factors of steel supply chain. The study identifies the competitiveness factors for the European steel sector in order to create proper policies for the steel industry in the European economic system [5]. Another study is about improvement of logistics and supply chain system for Thai steel industry, in relation to economic factors influencing logistic costs [1]. Moreover, the study identifies service strategies that are consistent with each customer's requirements, using multiple data sources and strategies in sourcing, demand management and fit with buoys [6]. These researches demonstrate the potential of the steel supply chain from various dimensions. This research therefore aims to analyze the position and dynamics of the steel supply chain in response to the context of the AEC, in order that factors can be identified to improve Thai steel supply chain.

The rest of the paper is organized as follows. In section 2 the supply chain of Thai long steel and research methodology is presented as a framework. Section 3 presents value adding to Thai long steel supply chain, Section 4 presents the work to analyze factors for improving Thai long steel supply chain. In section5, the conclusion and suggestions for future research are given.

2 Thai Long Steel Supply Chain and Research Methodology

Thai long steel supply chain consists of 3 sections: the upstream section including pig iron, scrap steel, semi-finished products, so called bloom and billet; the midstream section that includes finished steel products, so called bar & section steel and wire, steel; and the downstream section, industries who use the finished steel products as raw materials, such as construction, automotive, and other industries. In Thailand, the upstream begins with blast furnace works which use raw materials such as scrap and pig iron for production of semi-finished products; bar &section steel and wire steel. All the pig iron is imported. Next, bar and wire steels are produced for finished products by transforming the steel into shapes and sizes, so called hot-rolled and cold-rolled steel in the midstream industries.

As Thai long steel supply chain joins the ASEAN Economic Community (AEC), a study regarding value adding and competitiveness factors would be highly useful. In performing such a study, the analysis of the current situation in terms of production capacities, and import and export volumes must be done. Then the impact of joining the AEC must be analyzed by forecasting production and import volumes to estimate quantities of scrap to be used and export volumes of Thai long steel. Then value adding in the Thai long steel supply chain must be analyzed. Later, improvement factors should also be scrutinized. Brainstorming by experts in the Thai long steel industry is required to form strategies and identify improvement factors. These strategies and improvement factors are used to increase value adding along the supply chain. Figure 1 presents the research methodology framework.

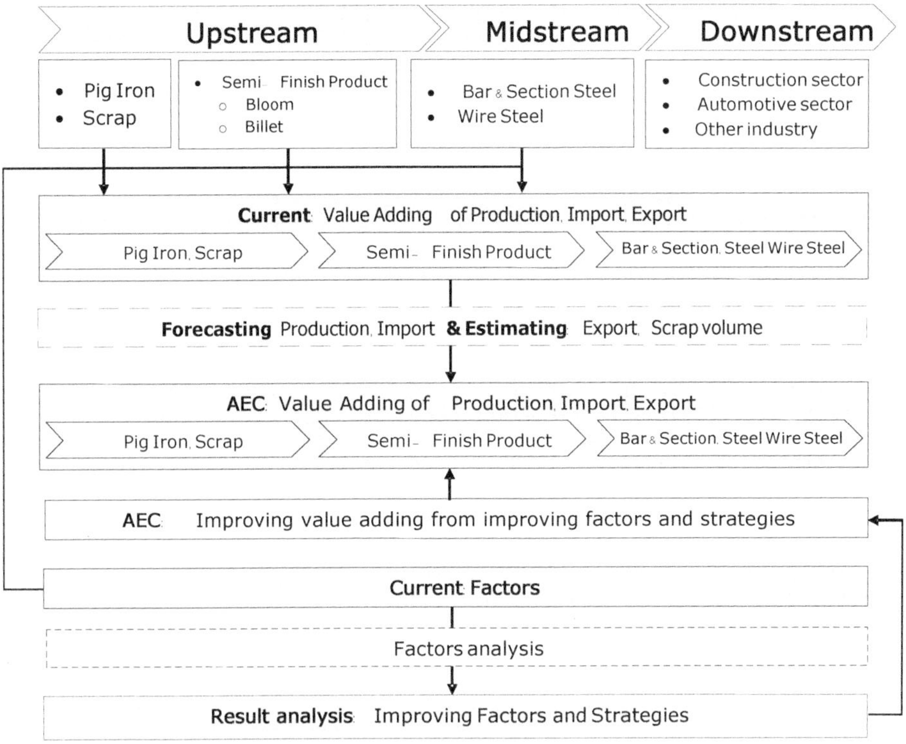

Fig. 1. Research methodology framework for value adding of Thai long steel supply chain for AEC

3 Value Adding to Thai Long Steel Supply Chain

Value adding analysis in Thai long steel supply chain can be separated into 2 parts; value added per unit of each steel product; and situation of that steel industry in the context of the AEC.

3.1 Value Added per Unit of Steel Products in Thai Long Steel Supply Chain

The analysis of value added per unit of long steel products utilizes average price of products in the global market between 2008 – 2012[7]. The data is taken from the perspective of Thai long steel supply chain structure and its statistical data [8]. The data can be classified into 4 cases by product, including 1.) Bar & section steel from pig iron, 2.) Bar & section steel from scrap steel, 3.) Wire steel from pig iron, and 4.) Wire steel from scrap steel (Fig. 2).

Fig. 2. Value added per unit of production of the long steel industry

Figure 2 shows value added per unit of product caused by differences of the cost of each product. The value added of long steel products can be classified by material flow along the supply chain as presented in Table 1.

Table 1. The value adding of long steel products by material flow along the supply chain

Case	Production process	Value adding	Total (Dollar/ton)
Cases1	Pig Iron – Semi finish Product - Bar & Section Steel	122.23+218.35	340.58
Cases2	Pig Iron – Semi finish Product - Wire Steel	122.23+110.18	232.41
Cases3	Scrap – Semi finish Product - Bar & Section Steel	225.62+218.35	443.97
Cases4	Scrap – Semi finish Product - Wire Steel	225.62+110.18	335.80

Comparison among these 4 cases indicates that the highest value added per production unit is Cases 3. Scrap – Semi finish Product - Bar & Section Steel, which needs to be studied further.

3.2 Analysis of Thai Long Steel Supply Chain

Thai long steel supply chain status is analyzed under the current situation as well as the projected situation within the AEC. The analysis includes the stages of pig iron and scrap import, production of semi-finished and finished products, and exports of the finished products.

Current Status. From SEAISI Statistical Yearbook 2013 [8], the products included in the statistical data can be categorized into pig iron, scrap steel, semi-finished products, bar & section steel, and wire steel as presented in Table 2.

The data in Table 2 shows that there is a decreasing trend of imported scrap and pig iron. The volume of import, export and consumption of semi-finished products are varied, while the imported volume of finished product; both bar &section and wire steels increases. This trend shows that the production of finished-products are inadequate and this must be improved.

Table 2. Current status of Thai long steel supply chain (Unit: thousand tons)

		2008	2009	2010	2011	2012	Total 5 Year	%
Pig Iron	Import	792	432	332	385	232	2,174	100.00%
Scrap	Import	2,480	1,735	1,973	1,877	1,701	9,766	35.42%
	Domestic	3,790	2,651	3,015	3,014	2,478	14,948	54.22%
	Export	724	507	576	567	481	2,855	10.35%
	Stock	335	234	266	86	370	1,291	4.68%
Semi-finish Product	Import	1,069	1,352	1,086	1,132	1,117	5,755	21.61%
	Production	5,211	3,646	4,145	4,238	3,328	20,568	77.22%
	Export	135	65	35	67	10	312	1.17%
	Consumption	4,250	3,604	3,663	3,653	3,763	18,933	71.08%
Bar & Section Steel	Import	605	319	551	585	725	2,785	13.73%
	Production	3,579	2,846	2,819	2,731	2,950	14,925	73.58%
	Export	433	377	544	622	598	2,573	12.69%
Wire Steel	Import	629	484	747	775	1,093	3,728	46.00%
	Production	671	758	844	921	813	4,008	49.45%
	Export	48	91	125	70	35	369	4.55%

Projected Status under AEC Context. The analysis of the projected status within the AEC context covers the production and import volumes of semi and finished products in Thai long steel supply chain, as suggested from the analysis of the current status. The imports and production are projected for a 5-year period (2013-2017). The years 2013 to 2014 represent preparation years before AEC, and the year 2015 is the year of entry. The forecasts apply Mean Absolute Percentage Error (MAPE) for the selection of forecasting equation. The results of the trend analysis can be represented as a best forecasting model according to MAPE. The equations and their MAPE for import and production of each product are shown in Table 3, 4, and 5.

Variables include
A_i = productions volume of steel type i (Unit: tons)
Im_i = imported volume steel type i (Unit: tons)
t = year of production (t = 1 is set to the year of 2008) (Unit: years).

Table 3. Forecasting equations for production and import volumes

Type		Model	Equation
Production	Semi-Finish Product	Linear	$A_1 = 5065559.1-317314.9*t$ MAPE = 9.22 %
	Bar & Section Steel	Quadratic	$A_2 = 4318610.8-927317.6*t+131664*t^2$ MAPE = 2.29 %
	Wire Steel	Linear	$A_3 = 667537.9+44686.5*t$ MAPE = 5.82%
Import	Pig Iron	Exponential	$Im_1 = 863726.644e^{-0.257*t}$ MAPE = 15.61 %
	Semi-Finish Product	Exponential	$Im_2 = 1178056.85\ e^{-0.0089*t}$ MAPE = 6.10 %
	Bar & Section Steel	Quadratic	$Im_3 = 732505.2-229902.871*t+46746.929*t^2$ MAPE = 15.42 %
	Wire Steel	Linear	$Im_4 = 379479.30+122064.49*t$ MAPE = 14.14 %

Table 4. Estimate equation for scrap volumes and export volumes

Type	%	Equation
Scrap Import	37.07	$S_1 = A_1/(77.89*100)*(37.07/100)$
Scrap Domestic	56.65	$S_2 = A_1/(77.89*100)*(56.65/100)$
Scrap Export	11.55	$S_3 = (S_1 + S_2)*(11.55/100)$
Export Semi-Finish Product	37.07	$E_1 = A_1*(1.52/100)$
Export Bar & Section Steel	56.65	$E_2 = A_2*(17.24/100)$
Export Wire Steel	11.55	$E_3 = A_3*(9.2/100)$

Table 5 shows the entire Thai long steel supply chain analysis under AEC context. The production of semi-finished products are inadequate for the demand of bar & section steels, leading to increased imports of the steels. Inadequate production results from the reduction in pig iron imports. For the wire steels, imports are increasing, and the production and exports are gradually increasing.

Table 5. Summary of steel volumes within AEC context (Unit: thousand tons)

		2013	2014	2015	2016	2017	5 Years	%
Pig Iron	Import	185	143	110	85	66	589	100.00%
Scrap	Import	1,505	1,354	1,203	1,052	901	6,013	0.04%
	Domestic	2,299	2,069	1,838	1,607	1,376	9,190	0.05%
	Export	439	395	351	307	263	1,756	0.01%
	Stock	203	183	162	142	122	812	0.00%
Semi-finish Product	Import	1,116	1,106	1,096	1,086	1,077	5,482	0.03%
	Production	3,162	2,844	2,527	2,210	1,892	12,635	0.07%
	Export	48	43	38	34	29	192	0.00%
	Consumption	4,430	5,259	6,352	7,707	9,326	33,075	0.18%
Bar & Section Steel	Import	1,036	1,414	1,885	2,450	3,108	9,893	0.02%
	Production	3,495	4,279	5,327	6,638	8,212	27,949	0.07%
	Export	602	738	918	1,144	1,416	4,818	0.01%
Wire Steel	Import	1,112	1,234	1,356	1,478	1,600	6,780	0.05%
	Production	936	980	1,025	1,070	1,114	5,125	0.04%
	Export	86	90	94	98	103	472	0.00%

Analysis of Changes in Value under AEC Context. The analysis employs the data from Thai long steel statistical data between the years 2008 to 2012 and the projection through the period 2013 – 2017. The numbers used in the analysis refer to the average price of products in the global market from 2008 – 2012 [7]. Changes in value added of imports, exports, production, and domestic trading along the long steel supply chain are analyzed. (Fig. 3).

Fig. 3 shows that there is a decreasing trend in the use of raw materials for semi-finished products production, while the consumption of bar & section steels increases. This leads to an increase of imported semi-finished products for finished product production.

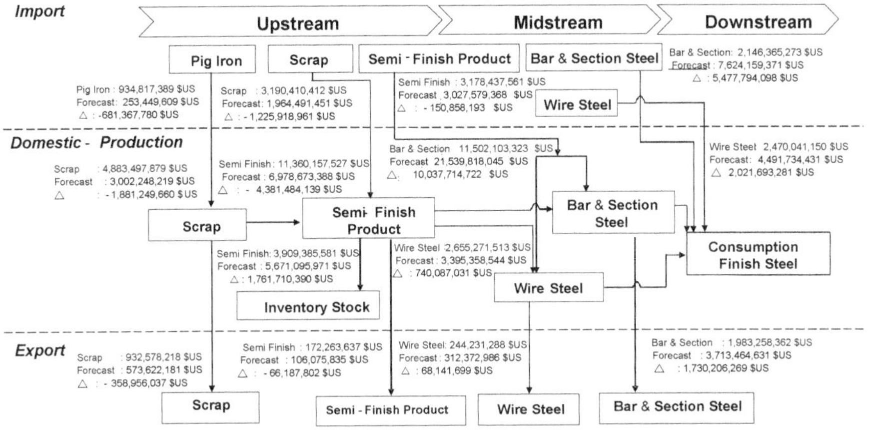

Fig. 3. Changes in the value adding of Thai long steel supply chain under AEC context

The changes along the supply chain are: changes in upstream section equal to -6.984 billion US dollars; midstream changes with a total of 20.075 billion US dollars; production changes for 6.276 billion US dollars in total; import changes equal to 5.441 billion US dollars; and export changes equal to 1.373 billion US dollars.

4 Improvement Factors to Enhance Value Adding

To enhance the value adding of Thai long steel supply chain, improvement factors must be explored. 65 questionnaires were responded by experts from Thailand, Indonesia, Malaysia and Vietnam. The factors were ranked according to their effect on market competitiveness. For each factor, Thai experts participated in the survey to propose strategies for improvement.

Ranked Improvement Factors for value adding in ASEAN Long Steel Supply Chain: The average weight of 11 improvement factors is calculated from the questionnaires. The ranked factors are as follows: Material and Production, Economy and Market, Government, Customer and Supplier Coordination, Competitiveness, Labor, Logistics, Information Technology, Risk, Utility and Work Environment.

Strategy for Improvement: Thirty nine experts particaipated in the survey to propose strategies for enhancement of value adding in each factor. The participants then brainstormed in focus groups to formulate proposed strategies.

The ranked factors and strategies for improvement are summarized in Table 6.

Table 6. Ranked improvement factors and strategies

Rank	Factor	Strategy for improvement
1	Material and Production	• Increase efficiency in production management • Adding investment in new technology
2	Economy and Market	• Seek new markets of the steel products in the AEC • Enhance the quality of product for increasing exports
3	Government	• Promote investment in advanced machinery
4	Supplier and Customer Coordination	• Increase ease of customs procedure
5	Competitiveness	• Develop technology knowledge
6	Labor	• Develop labor skills • Increase metallurgical engineering and laboratory experts
7	Logistics	• Develop logistics systems
8	Utility	• Lower energy cost
9	Risk	• Invest in cleaner technology
10	Information Technology	• Promote integration of ERP in the organization
11	Work Environment	• Create a better working environment in the factory

5 Conclusions

Typically, changes in value adding are analyzed in fragments either from downstream to midstream or from midstream to upstream. This study improves the analysis by integrating all the players from upstream to downstream in the supply chain. The study in Thai long steel supply chain shows that the most value adding to the supply chain is from the chain of Scrap-Semi finished Products-Bar & Section Steel. To improve Thai long steel supply chain requires improvements in the factors of Material and production as well as Economy and market. Accordingly, the strategies to improve Thai long steel supply chain include Strategies for increasing the supply of raw materials and Strategies to support investments in advanced technology.

Strategies for increasing the supply of raw materials include collaborations among the supply chain members, building smelting plants, increasing the number of blast furnace plants, and increasing the export tax as well as limiting the amount of exports of scrap steel. Strategy for supporting investments in advanced technology is to promote the cooperation with overseas companies for advanced technology implementation.

For future research, the optimization model should be applied to maximize value adding in long steel supply chain. Scenario analysis can be applied in changing production, import, and export at each stage of the supply chain to identify the value adding.

Acknowledgements. This research was supported by National research council of Thailand and Thailand Research Fund under grant number RDG 5650042-003 for the project entitled Supply Chain Redesign for Steel Industry in Preparation for AEC.

References

1. Suthikarnnarunai, N., Muangthong, K.: Economics Factors Influencing Logistics Cost of Thai Steel Industry. In: Proceedings of the World Congress on Engineering and Computer Science (WCECS 2009), vol. II, pp. 1203–1209 (2009)
2. ISIT.: Impacts Assessment and Guidelines Set up for Thailand Steel Industry under ASEAN Economic Community, pp. 1–3 (2011)
3. Wang, X., Wong, T.N., Fan, Z.-P.: Ontology-based supply chain decision support for steel manufacturers in China. Expert Systems with Applications 40, 7519–7533 (2013)
4. Jung, K.-H., Lee, S.-K.: New paradigm of steel mills in the supply chain of automotive sheets. Supply Chain Management: An International Journal 11(4), 328–336 (2006)
5. ECORYS SCS Group: Study on the Competitiveness of the European Steel Sector, pp. 1–149 (2008)
6. Hua, S., Chatterjee, S.-R., Jingliang, C.: Achieving competitive advantage in service supply chain: evidence from the Chinese steel industry. Chinese Management Studies 5(1), 68–81 (2011)
7. Steelonthenet.: About the Steelmaking Input Costs and Steelmaking Input Costs (November 30, 2014), http://www.steelonthenet.com/ (retrieved)
8. SEAISI.: The South East Asia Iron and Steel Institute (SEAISI) Statistical Yearbook, Statistical Yearbook 2013, pp. 1–104 (2013)

Analysis of Causal Competitive Factors of Thai Iron and Steel Supply Chain by DEMATEL Method

Supattra Rattanavijit*, Tuanjai Somboonwiwat, and Charoenchai Khompatraporn

Department of Production Engineering, Faculty of Engineering,
King Mongkut's University of Technology Thonburi (KMUTT), Bangkok, Thailand
{supattra.r.1989,tuanjai.som}@gmail.com,
charoenchai.kho@kmutt.ac.th

Abstract. The steel industry is an important basic industry to the economic growth of developing countries. Steel is a primary raw material for many products downstream in the supply chain. In this paper, we study important factors in the steel industry in Thailand. The relations among the factors are analyzed using a method called DEMATEL. There are totally 11 important factors. The analysis shows that there are three groups of factors: causal, affected, and independent ones. Since the causal factors affect other factors, they should be paid attention very closely. The analysis shows that there are four causal factors which are large domestic market size, favorable government policies, high labor skill, and well planned production schedule. The other factors, however, are also important to the industry even though their roles are different. Strategies are then formulated based on the roles of the factors to enhance the competitiveness of the Thai iron and steel supply chain.

1 Introduction

Steel industry is an important primary industry which contributes to the economic growth of Thailand and it is also an important upstream material to many industries such as construction, automotive industry, machinery, electrical and electronic industry, packaging, and many more. In 2011, the value of imported iron and steel was accounted for approximately 10,102 million USD and the exported value was about 1,340 million USD [1]. This is partly due to the fact that Thailand has no steel smelting plant, an upstream part of the steel industry. This posts a disadvantage to the steel supply chain in Thailand, particularly in terms of the cost of raw materials to the downstream industries and sometimes the quality of the materials.

ASEAN member countries (AMCs) have agreed to launch ASEAN Economic Community (AEC) at the end of 2015. Under AEC, the import/export tax among the AMCs will be reduced to zero. This is a threat to the local steel industry in Thailand since there could be inflows of similar materials from other AMCs where the cost of the production is less expensive. Therefore, in order to be competitive the steel industry in Thailand must adapt in order to enhance the industry.

* Corresponding author.

M. Gen et al. (eds.), *Industrial Engineering, Management Science and Applications 2015*,
Lecture Notes in Electrical Engineering 349, DOI: 10.1007/978-3-662-47200-2_58

There are several studies on development of the steel industry. For instance, Potter et al. (2004) described the evolution of the steel supply chain from traditional community business; Hua et al. (2011) and Sandhu et al. (2013) were interested in the management perspectives of the steel supply chain to create strategies that suit the industry. Conceicao et al. (2012) and Zadeh et al. (2013) were concerned with the facility location problem on the supply chain to improve its efficiency. Zhang (2008) estimated the impact of energy saving technologies to the Chinese steel supply chain, while Skelton and Allwood (2013) explored incentives to improve material efficiency in the industry worldwide. However, there is no study that explores the factors impacting to the steel industry.

From the literature, a methodology that has been applied in many industries and organizations is called DEMATEL. The methodology was first developed by identifies critical factors for increase economic and societal values chain. Then it has been used to identified causal factors influencing to the industry such as in automotive remanufacturing industry [9], computer recycling supply chains [10], and to an organization such as a hospital [11] so that suggestions on how to improve the industry or organization can be properly prioritized.

In this paper, we study the factors influencing the iron and steel industry in Thailand, and apply the DEMATEL method to identify the causal factors to the industry. The organization of the paper is the followings. Section 2 gives and overview of the iron and steel industry in Thailand, reviews factors mentioned in various literature, and describes the DEMATEL method in details. The method is then applied to the industry and the results are shown in Section 3. Section 4 discusses strategies to improve the industry based on the results previously obtained. Finally, Section 5 concludes the paper.

2 Literature Review

2.1 Overview of the Iron and Steel Industry in Thailand

The supply chain of steel industry in Thailand is composed of midstream and downstream industries. The midstream industries produce semi-finished products such as bloom, beam blank, billet, slab, and ingot for the downstream industries such as construction, automotive, machinery, and appliances. The industries can be classified into two types based on the shape of the semi-products which are long products and flat products. The consumption of the long steel products is 51.9% of the total national steel consumption, and the majority of the long products are used in the construction industry. The demand estimate of iron and steel in other sectors in Thailand can be broken down into automotive 22.4%, machinery 11.7%, appliances 10.8%, and packing 4.3% [12].

Since Thailand has no upstream part of the iron and steel supply chain in the country, the midstream and downstream industries must depend on the raw materials imported from oversea. The Iron and Steel Institute of Thailand has warned the iron and steel manufacturers in Thailand that there will be pressure from the products from China who has the largest excess steel production capacity in the world and can produce similar products with less expensive costs. Thus, the local manufacturers must be more efficient and effective to be able to compete in the industry.

2.2 Influencing Factors

The factors influencing the Thai iron and steel industry were identified by interviewing many stakeholders such as iron and steel companies, Iron and Steel Institute of Thailand, Thai Customs Department, the Federation of Thai Industries, Department of Foreign Trade, and Office of Industrial Economics. Many of them were agreed with those mentioned in the literature such as World Competitiveness Index by World Economic Forum [13], Logistics Performance Index by the World Bank [14], and Global Climate Risk Index by Germanwatch [15]. Then the factors were presented to a focus group composing representatives from many organizations to select only those that had a strong connection to the competitiveness of the industry. Only 11 factors remained and they are:

Table 1. Competitiveness factors of Thai iron and steel industry

Competitiveness Factor	Literature
Competitive local interest rates (F1)	[16]
Large domestic market size (F2)	[13]
Well responsive to customer's requests(F3)	[3,14,17]
Favorable government policies to the steel industry (F4)	[13,18]
Good working attitude (F5)	[13]
Product price is competitive (F6)	[19]
High labor skill (F7)	[13]
Well planned production schedule (F8)	[20]
Good company's assets utilization (F9)	[21,22]
Good customer relationship (F10)	[14,17]
High stability in local government (F11)	[13,18]

2.3 DEMATEL Method

The DEMATEL method was developed by Battelle Memorial Institute in Geneva between 1972-1976, and referred to the Decision Making Trial and Evaluation Laboratory [10]. It has been utilized in many applications to determine causal relationships between determining factors. This paper follows the DEMATEL procedure described by Lee et al. (2011) and Hsu (2012) as follows.

Step 1. Calculate the average matrix. First step is to collect assessment results from a panel of respondents and construct a score matrix for each of the respondents. Then take the average of all the data. Suppose we have Q experts and n factors. Each respondent is asked to assess the degree to which he/she believes a factor i affects factor j. The scores provided by the respondent q can be written as an $n \times n$ matrix denoted by $Z^q = \left[z_{ij}^q\right]_{n\times n}$ where q = 1,2,…,Q. The diagonal matrix of each matrix Z^q is set to zero. Then, the average scores in the average matrix $Z = \left[z_{ij}\right]_{n\times n}$ are computed from:

$$z_{ij} = \frac{\sum_{q=1}^{Q} z_{ij}^q}{Q} \tag{1}$$

Step 2. Calculate the normalized matrix Z_{nor}. The matrix is obtained by normalizing the average matrix by:

$$S = \max\left(\max \Sigma_{i=1}^{n} z_{ij}, \max \Sigma_{j=1}^{n} z_{ij}\right) \tag{2}$$

$$Z_{nor} = Z / S \tag{3}$$

Step 3. Compute the total relation matrix T. Using $\lim_{m\to\infty} Z_{nor} = [0]_{n\times n}$ and $\lim_{m\to\infty}(I + Z_{nor} + Z_{nor}^2 + \ldots + Z_{nor}^m = (I - Z_{nor})^{-1}$ where I is the identity matrix, then we have:

$$T = [t_{ij}]_{n\times n} = \sum_{i=1}^{\infty} Z_{nor}^i = Z_{nor}(I - Z_{nor})^{-1}; i,j = 1,2,\ldots,n \tag{4}$$

Step 4. Calculate the degree of influence. We compute the sum of direct and indirect effects given by factor i, r_i, and the sum of direct and indirect influence by factor j, c_j, by:

$$r_i = \sum_{j=1}^{n} t_{ij}, i = 1,2,\ldots, n \tag{5}$$

$$c_j = \sum_{i=1}^{n} t_{ij}, i = 1,2,\ldots, n \tag{6}$$

In addition, the sum (r_i+c_j) gives an index representing the total effects (both given and received) by the i factor. The difference (r_i-c_j) shows the net cause by factor i if the value is positive, and net effect if the number is negative.

Step 5. Set the threshold value & obtain the impact relations. To filter only important relations, the decision maker may set a threshold value. In doing so only the effects greater than the threshold value will be chosen and shown in a digraph. In this study, we set the threshold equaled to 1.25 x 0.3622, where 0.4527 refers to the average of the elements in matrix T.

3 Application

From the 11 competitiveness factors obtained, a focus group of 25 professionals and representatives from various organizations relating to the Thai iron and steel industry were asked to rate the factor using a scale ranging from 0 to 3, where 0 represents "no influence" and 3 indicates "strong influence". Table 2 summarizes the average scores given by the focus group, or the average matrix.

Table 2. The Average Matrix

	F1	F2	F3	F4	F5	F6	F7	F8	F9	F10	F11
F1	0.0000	1.9200	1.8000	1.8800	0.9200	3.1600	0.8800	1.0400	2.2000	1.1600	1.2400
F2	1.7200	0.0000	2.7600	2.5200	1.8400	2.7600	2.2400	2.2400	2.6800	2.1600	2.0400
F3	1.8000	2.6400	0.0000	2.2000	1.9600	2.8800	2.5200	2.7600	2.5600	3.3600	1.5600
F4	2.7600	2.6800	2.4000	0.0000	1.9600	3.0800	2.2800	1.9600	2.2800	1.6000	2.2000
F5	0.7600	1.4400	2.8800	1.4800	0.0000	1.8400	2.9200	2.3200	2.6800	2.0400	1.3600
F6	2.6400	3.0400	3.3600	2.5600	1.8800	0.0000	2.0400	2.4000	2.4800	2.5600	1.5600
F7	1.1600	1.7600	3.1600	1.5600	2.8800	2.4400	0.0000	2.9600	3.1200	2.1600	1.4400
F8	1.0400	2.0400	3.4400	1.6800	1.9600	3.0000	1.7600	0.0000	3.4000	2.5600	1.2400
F9	1.9200	2.2400	2.4800	1.8400	1.8400	2.8000	2.0400	3.0800	0.0000	1.4800	1.2400
F10	0.7200	2.1600	2.8000	1.2800	1.7200	2.4400	1.4000	2.4000	1.6800	0.0000	1.1200
F11	2.1600	2.1200	1.6800	3.0000	1.6000	1.8000	1.4400	1.4800	1.7200	1.2800	0.0000

From Table 2, we follow Steps 2 and 3 in Section 3 and obtain the total relation matrix as shown in Table 3. Then we follow Step 4 of the DEMATEL method and obtain the sum of influences given and received among the factors as shown in Table 4. The threshold in Step 5 was set to be the average from all the values in the total relation matrix, and is equal to 0.4527. The relations of the factors with values higher than the threshold show strong causal-effect affiliations and are indicated with the "*" sign in Table 3. These causal and effect relations can be depicted by a digraph as shown in Fig.1.

Table 3. The total relation matrix

	F1	F2	F3	F4	F5	F6	F7	F8	F9	F10	F11
F1	0.1890	0.3110	0.3516	0.2848	0.2376	0.3890	0.2486	0.2877	0.3427	0.2698	0.2114
F2	0.3161	0.3335	0.4917*	0.3856	0.3464	0.4812*	0.3749	0.4219	0.4597*	0.3890	0.2996
F3	0.3280	0.4385	0.4184	0.3876	0.3637	0.5031*	0.3971	0.4552*	0.4735*	0.4426	0.2939
F4	0.3517	0.4258	0.4809*	0.3016	0.3502	0.4930*	0.3764	0.4123	0.4484	0.3709	0.3060
F5	0.2489	0.3414	0.4470	0.3119	0.2484	0.4021	0.3606	0.3839	0.4148	0.3473	0.2471
F6	0.3597	0.4553*	0.5331*	0.4036	0.3630	0.4114	0.3853	0.4461	0.4748*	0.4204	0.2974
F7	0.2916	0.3905	0.5022*	0.3494	0.3779	0.4662*	0.2964	0.4437	0.4716*	0.3879	0.2761
F8	0.2857	0.3958	0.5039*	0.3494	0.3430	0.4782*	0.3527	0.3381	0.4729*	0.3958	0.2663
F9	0.3020	0.3849	0.4540*	0.3402	0.3248	0.4536*	0.3458	0.4223	0.3424	0.3444	0.2555
F10	0.2291	0.3401	0.4137	0.2838	0.2852	0.3915	0.2881	0.3579	0.3535	0.2532	0.2222
F11	0.2820	0.3410	0.3768	0.3431	0.2816	0.3748	0.2897	0.3265	0.3555	0.2958	0.1850

Table 4. The sum of influences given and received among the factors

Factors	r_i	c_i	r_i+c_i	r_i-c_i
Well responsive to customer's requests(F3)	4.5017	4.9734	9.4751	-0.4717
Product price is competitive (F6)	4.5502	4.8441	9.3944	-0.2939
Good company's assets utilization (F9)	3.9700	4.6098	8.5798	-0.6398
Well planned production schedule (F8)	4.1816	4.2956	8.4772	-0.1140
Large domestic market size (F2)	4.2997	4.1577	8.4574	0.1420
Favorable government policies to the steel industry (F4)	4.3172	3.7410	8.0581	0.5762
High labor skill (F7)	4.2534	3.7156	7.9690	0.5378
Good customer relationship (F10)	3.4182	3.9171	7.3353	-0.4989
Good working attitude (F5)	3.7534	3.5218	7.2752	0.2316
High stability in local government (F11)	3.4518	2.8604	6.3122	0.5913
Competitive local interest rates (F1)	3.1232	3.1839	6.3071	-0.0606

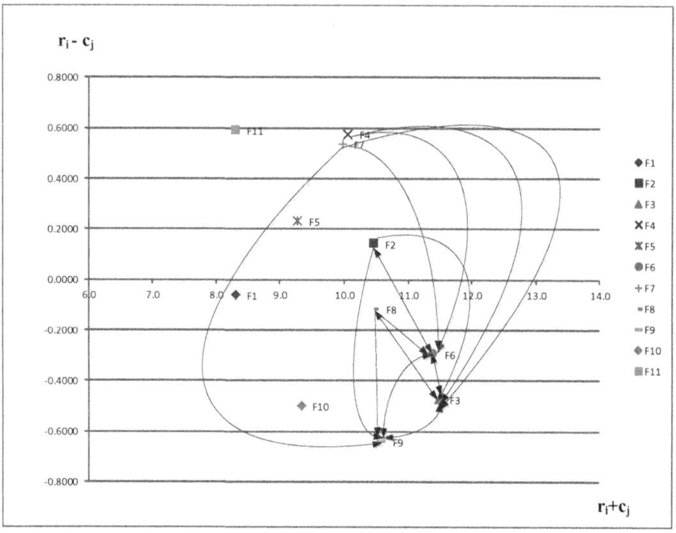

Fig. 1. The digraph of causal and effect relations among the factors

From Fig.1, the factors can be divided into three categories: causal, affected, and independent ones. The causal factors are F8 > F2 > F4 > F7 where the order implies the degree of influence from the highest to the lowest within the group. The affected ones are F3 > F6 > F9, while F10 > F5 > F11 > F1 are in the independent category. These factors are to be discussed based on their categories in the next section.

4 Discussion and Recommendations

From the results above, we can divide the factors based on their relations into the causal factors, the affected factors, and the independent ones. The causal factors are well planned production schedule (F8), large domestic market size (F2), favorable government policies to the steel industry (F4), and high labor skill (F7). To improve the competitiveness of the industry, certain strategic recommendations are made and shown in Table 5. The affected factors are well responsive to customer's requests (F3), product price is competitive (F6), good company's assets utilization (F9), and their corresponding strategic recommendations are shown in Table 6. Lastly, the independent factors which are good customer relationship (F10), good working attitude (F5), high stability in local government (F11), and competitive local interest rates (F1), and their recommendations are shown in Table 7.

Table 5. Strategic recommendations for the causal factors

Causal Factors	Strategic Recommendations
Well planned production schedule (F8)	▪ Organize training programs to improve schedulers' skills and new techniques in production scheduling. ▪ Promote cooperation with educational institutes to develop more effective and practical production planning tools.
Large domestic market size (F2)	▪ Expand oversea by taking the advantage of zero tax benefit from AEC in the ASEAN region. A challenge though is the different local standards of the iron and steel products in each of the ASEAN countries which hinder intra-regional trades.
Favorable government policies to the steel industry (F4)	▪ The government should support investments in high grade steel production ▪ The government should highly consider initiatives to improve upstream capacity locally. ▪ Review regulations relating to the customers harmonize codes of the steel products to eliminate loopholes.
High labor skill (F7)	▪ Conduct training on international standard such as ASTM, JIS or TIS. ▪ Develop competency standard of workers in the industry.

Table 6. Strategic recommendations for the affected factors

Affected Factors	Strategic Recommendations
Well responsive to customer's requests (F3)	▪ Reduce order processing procedure within the organization to shorten response time. ▪ Utilize information technology to integrate all processes to better respond to customer requests.
Product price is competitive (F6)	▪ Invest in upstream industry in neighboring countries to acquire quality raw materials with competitive price. ▪ Explore cooperation among the ASEAN member countries to negotiate on international antidumping regulations.
Good company's assets utilization (F9)	▪ Better management of inventory to reduce stock of raw materials and finished products. ▪ Better scheduling of production to fully utilize the existing machinery and assets.

Table 7. Strategic recommendations for the independent factors

Independent Factors	Strategic Recommendations
Good customer relationship (F10)	▪ Promote long-term partnership. ▪ Co-design and work closely with the customers to develop new products.
Good working attitude (F5)	▪ Improve working conditions for the employees. ▪ Use cleaner energy sources and better control of pollutions. ▪ Offer attractive welfare package to the employees.
High stability in local government (F11)	▪ This factor is a rather uncontrollable factor. The industry must at least agree on the strategic direction and supports required from the government. The consistency of the industry is a key against instability of the local government.
Competitive local interest rates (F1)	▪ The financial institutes may offer competitive interest rates to the manufacturers through the support by the government. ▪ Promote joint venture with certain manufacturers oversea to alleviate the need for acquiring a large amount of loan.

The grouping of these factors indicates that the causal factors should be paid more attention since they can affect other factors. The recommendations of these factors should be implemented with top priorities. The suggestions of the independent factors can be executed in parallel with the causal factors, whereas those of the affected factors may be carried out after the other recommendations.

5 Conclusion

The Thai iron and steel industry contributes to the growth of the country but its competitiveness factors have not been studied. This paper identifies the factors influencing to the competitiveness of the industry through a focus group. To enhance the competiveness, effective strategies are needed. The strategies can be formulated based on the important role of the factors; causal, affected and independent factors. In this study, 11 important factors are identified and their relations analyzed by using DEMATEL method. Four of the factors are classified as the causal factors, which are well planned production schedule (F8), large domestic market size (F2), favorable government policies to the steel industry (F4), and high labor skill (F7). The affected factors are well responsive to customer's requests (F3), product price is competitive (F6), good company's assets utilization (F9), and the independent factors are good customer relationship (F10), good working attitude (F5), high stability in local government (F11), and competitive local interest rates (F1). Recommendations for each of these factors are given to improve the overall competitiveness of the industry. The priorities of the recommendations should be based upon their roles. Specifically, the focus should be on these causal factors to improve the competitiveness of the industry, while the other recommendations are needed to further enhance the overall competitiveness.

Acknowledgments. This research was supported by Thailand Research Fund under grant number RDG 5650042-003.

References

1. Office of Industrial Economic.: Summary Economic Industry Report, pp. 35–37 (2011) (in Thai)
2. Potter, A., Mason, R., Naim, M., Lalwani, C.: The evolution towards an integrated steel supply chain: A case study from the UK. Int. J. Production Economics 89, 207–216 (2004)
3. Hua, S., Chatterjee, S.R., Jingliang, C.: Achieving competitive advantagein service supply chain: evidencefrom the Chinese steel industry. Chinese Management Studies 5, 68–81 (2011)
4. Sandhu, M.A., Helo, P., Kristianto, Y.: Steel supply chain managementby simulation modelling. International Journal 20, 45–61 (2013)
5. Conceicao, S.V., Pedrosa, L.H.P., Neto, A.S.C., Vinagre, M., Wolff, E.: The facility location problem in the steel industry: a case study in Latin America. Production Planning & Control 23, 26–46 (2012)
6. Zadeh, A.S., Sahraeian, R., Homayouni, S.M.: A dynamic multi-commodity inventory and facility location problem in steel supply chain network design. Int. J. Adv. Manuf. Technol. 70, 1267–1282 (2014)
7. Zhang, J., Wang, G.: Energy saving technologies and productive efficiency in the Chinese iron and steel sector. Energy 33, 525–537 (2008)
8. Skelton, A.C.H., Allwood, J.M.: The incentives for supply chain collaboration to improve material efficiency in the use of steel: An analysis using input output techniques. Ecological Economics 89, 33–42 (2013)
9. Xia, X., Govindan, K., Zhu, Q.: Analyzing internal barriers for automotive parts remanufacturers in China using grey-DEMATEL approach. Journal of Cleaner Production 87, 811–825 (2015)
10. Rahman, S., Subramanian, N.: Factors for implementing end-of-life computer recycling operations in reverse supply chains. Int. J. Production Economics 140, 239–248 (2012)
11. Shieh, J., Wu, H.H., Huang, K.K.: A DEMATEL method in identifying key success factors of hospital service quality. Knowledge-Based Systems 23, 277–282 (2010)
12. Iron and Steel Institute of Thailand: Summary Overview Steel Industry (2015), http://www.isit.or.th/modulesphp?mod=Article&file=view&ID=58 (in Thai)
13. World Economic Forum: Insight Report The Global Competitiveness Report (2014), http://reports.weforum.org/the-global-competitiveness-report-2013-2014
14. World Bank: The Logistics Performance Index and Its Indicators (2014), http://www.worldbank.org/content/dam/Worldbank/document/Trade/LPI2014.pdf
15. Germanwatch: Global Climate Risk Index (2013), http://germanwatch.org/fr/download/7170.pdf
16. Ocal, M.E., Oral, E.L., Erdis, E., Vural, G.: Industry financial ratios-application of factor analysis in Turkish construction industry. Building and Environment 42, 385–392 (2007)
17. Holmen, E., Pedersen, A.C.: How do suppliers strategise in relation to a customer's supply network initiative? Journal of Purchasing & Supply Management 16, 264–278 (2010)
18. Zhao, Z., Yan, H., Zuo, J., Tian, Y., Zillante, G.: A critical review of factors affecting the wind power generation industry in China. Renewable and Sustainable Energy Reviews 19, 499–508 (2013)

19. Zhang, J., Nault, B.R., Yang, W., Tu, Y.: Dynamic price quotation in a responsive supply chain for one-of-a-kind production. Int. J. Production Economics 139, 275–287 (2012)
20. Tang, L., Liu, J., Rong, A., Yang, Z.: A review of planing and scheduling system and methods for integrated steel production. European Journal of Operation Research 133, 1–20 (2001)
21. Bernolak, I.: Effective measurement and successful elements of company productivity: The basis of competitiveness and world prosperity. Int. J. Production Economics 52, 203–213 (1997)
22. Lee, W.S., Huang, A.Y., Chang, Y.Y., Cheng, C.M.: Analysis of decision making factors for equity investment by DEMATEL and Analytic Network Process. Expert Systems with Applications 38, 8375–8383 (2011)
23. Hsu, C.C.: Evaluation criteria for blog design and analysis of causal relationships using factor analysis and DEMATEL. Expert Systems with Applications 39, 187–193 (2012)

Combining Balanced Scorecard and Data Envelopment Analysis to Design Performance Measurement for Supply Chain Actor and Regulator: A Case Study in Innovative Product in Indonesia

Elisa kusrini[1,2,*], Subagyo[1], and Nur Aini Masruroh[1]

[1] Department of Mechanical and Industrial Engineering, Faculty of Engineering, Gadjah Mada University, Jl. Grafika No. 2, Yogyakarta, 55281, Indonesia
[2] Department of Industrial Engineering, Faculty of Industrial Technology, Indonesia Islamic University, Indonesia
1elisa_kusrini@yahoo.com,
{subagyo,aini}@ugm.ac.id

Abstract. This research is a sequel of the authors' earlier conducted researches in the fields of identifying the key performance indicator for actors and regulators. In the previous study, authors proposed a model based on Balanced Scorecard perspective, by integrating process based on SCOR into the internal business processes and incorporated the role of regulators in each perspective. This model referred as B-S-Rc model (Balanced Scorecard-SCOR-Regulator contribution). The weakness of this model is it has no standard methodology and the inability to provide a benchmark. The objective of the present study is developing a model to overcome the shortcomings of the B-S–Rc, by combining the B-S - Rc with Data Envelopment Analysis (DEA). This integrated model is validated on supply chain's SME and regulators of innovative products in Yogyakarta, Indonesia. It is found that the combination of BS-Rc-DEA provides information that enables benchmarking to overcome the weaknesses of B-S-Rc model.

Keywords: Performance Measurement, Supply Chain Actors, Supply Chain Regulator, Balanced Scorecard, Data Envelopment Analysis.

1 Introduction

Nowadays, performance measurement SC is attaining more attention from both practitioners and academics. Various supply chain measurement model have been developed by many researchers with various approaches, such as the Balanced Scorecard [1], [2], [3], [4], [5], [6], [7], [8], SCOR [9], [10], [11], [12], [13], [14], Component of measurement, such as quality, cost, delivery time etc [15], [16], [17], [18], [19], [20], Data Envelopment Analysis (DEA) [21], [22] ,[23].

* Corresponding author.

M. Gen et al. (eds.), *Industrial Engineering, Management Science and Applications 2015*,
Lecture Notes in Electrical Engineering 349, DOI: 10.1007/978-3-662-47200-2_59

Various performance measurements that mentioned above are usually just focus on the performance measurement from the internal side of supply chain's actors. Meanwhile, the supply chain performance measurements are also being necessary for the government (regulators). The government has a responsibility to facilitate an economic growth in a supply chain [24], [25]. Regulator's role is one of the important factors for the success of a supply chain especially Small Medium Enterprises (SME) [26]. However, not all government policies positively impact on the development of the economy and SME supply chain in general. It is consider that the government policy is considered will disturb the balance of the market economy because of an overprotection toward SME [27]. Therefore , it is necessary to have a model to evaluate the government policies in improving the performance of the SME .

As far as the author's concern, the model focusing on performance evaluation of government policy toward SME supply chain is still limited. Moreover, a model that integrating performance measurements between the supply chain actors and regulator (government) is not yet available. Integrated measurement model is necessary so that the government can facilitate the economic growth effectively. The current research is a sequel of the authors' earlier conducted researches in the fields of identifying the key performance indicator for actors and regulators. In the previous study, authors proposed a model based on Balanced Scorecard perspective, by integrating process based on SCOR (plan, source, make, deliver, return) into the internal business processes and incorporated the role of regulators in each perspective. This model referred as B-S-Rc model (Balanced Scorecard-SCOR-Regulator contribution) [28]. The weakness of this model is it has no standard methodology and the inability to provide a benchmark [29], [30], [31], [32]. The objective of the present study is developing a model to overcome the shortcomings of the B-S - Rc . This is done by combining the model B-S - Rc with Data Envelopment Analysis (DEA) model. DEA is a mathematical model by taking into account the level of businesses' efficiency. This model can be used to compare (benchmarking) with clear and objective methodology [33], [34], [32]. The lack of DEA is that it is only focused on the input and output variables. In addition it couldn't see the vision [35]. The combination of BSC - DEA models is combining the advantages of both models to overcome their weaknesses .

2 Literatur Review

2.1 Supply chain Performance Measurement

The model employed in this study was a model based on the BSC. Researchers proposed a model based on balanced scorecard perspective, by integrating process based on SCOR into the internal business processes and incorporating the role of regulators in each perspective. This model refers as B-S-Rc model (Balanced Scorecard-SCOR-Regulator contribution). B-S-Rc model will be combined with DEA model to address the weaknesses of the BSC model.

A model of Balanced Scorecard (BSC) was used to measure the performance of the company[36]. Later, it was developed to measure the performance of the supply chain and become a model of supply chain balanced scorecard. Model BSC supply chain (BSC-SC) is a measurement framework around the SC's actors with four perspectives i.e. financial perspective, customer perspective, business process perspective and learning and growth perspective. The whole perspectives in this framework are no longer viewed within the internal side of the company, but they are viewed from the side of the supply chain. The model is balanced with the four perspectives considering the financial and non financial factors as well as short term and long term goals. This model is also the most widely used for a variety of industries.

Data Envelopment Analysis was introduced by Charnes, Cooper and Rhodes (1978) [37]. Data Envelopment Analysis (DEA) method was created as a tool for the evaluation of the performance of an activity within a unit of entities (organizations). The principle of the DEA model is comparing the input and output data among organization called the Decision Making Unit (DMU). This comparison was done to get a value of efficiency. DEA can identify units that are inefficient and can quantify the amount of the improvement to achieve the efficiency frontier. Research on the measurement of efficiency, especially with the approach of DEA has been done in various fields, such as the measurement of SME [38], [39], public sector [40], [41], environment [42], hotel industry [43], airlines [44] and banking [45], [35].

2.2 Government (Regulator) Performance Measurement

The performance measurement of government (regulator/local authorities) has been getting enough attention from various researchers . However, empirical evidence about the benefit and performance measurement practices are still very limited. The measurement is more focusing on the measurement of inputs and the work load of certain programs run by local governments. The measurement result does not contribute in decision making[46].

A performance measurement is a method to measure the activities performance, program or service which is provided by the government. Neely et al. (2005) state that performance measurement system as the sets of metric used to quantify both the efficiency and effectiveness of action [47]. Nyhan and Martin (1999) defined the performance measurement as "The regular collection and reporting of information about the efficiency, quality and effectiveness of governmental programs" [48]. Performance measurement can be divided in 4 categories: input, output, efficiency and effectiveness [49]. In general efficiency and effectiveness is the most widely used as a measure of performance [49], [50].

There are many models to measure the government performance (regulators), such as Logic models [51], the Balanced Scorecard [52], [53], [54], PRISM [55], Malcolm Baldrige National Quality Award [56] and DEA [40], [41], [57]. But the model focusing on performance evaluation of government policy towards supply chain SME is still very limited. Moreover, the model that integrates performance measurement of SCM and regulatory policy does not exist. Therefore, this study will contain proposals for the development of performance measurement model for SCM and regulators to identify key indicators that will be used by the actors of the SCM and regulators to improve their performance.

3 Research Methodology

3.1 Model Development

As described earlier, this study is a sequel of previous research that generates valid models for supply chain's actors and regulators performance measurement. The main model is the BSC supply chain model. The weakness of this model is it has no standard methodology and the inability to provide a benchmark. In order to overcome the shortcomings of the B-S – Rc, combining the model B-S - Rc with Data Envelopment Analysis (DEA) is proposed. Performance indicators of the generated BSC model will be combined with the DEA.

The concept of DEA is to measure efficiency of DMU, so the combination model of DEA and BSC can measure the efficiency of the SCM members and regulators based on performance indicators in the BSC model. Thus the combination of this model despite able to measure the performance, it also can be used for comparison (benchmark). Each member of the SC can make improvements based on the value which has optimum efficiency [32].

3.1.1 Development of BSC Model

In the previous study, authors proposed a model based on Balanced Scorecard perspective, by integrating process based on SCOR (Plan, Source, Make, Deliver, Returns) into the internal business processes and incorporated the role of regulators in each perspective. This model referred as B-S-Rc model (Balanced Scorecard-SCOR-Regulator contribution) [28] as depicted in Figure 1. This research found that there were 31 valid key indicators consist of 19 indicators to measure actors' s performance and 12 indicators to measure regulator's performance.

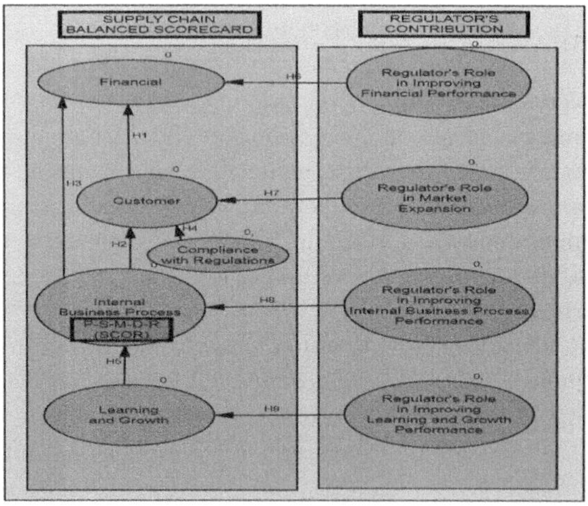

Fig. 1. BSC-SCOR- Regulator Contribution Framework [28]

3.1.2 Combining B-S-Rc Model with DEA

The development of the BSC model produced a B-S-Rc model. A valid indicator to measure the performance of the SCM and the regulators is obtained in this model. Its indicators were used as input and output variable in DEA model. The combination produces a B-S-Rc-DEA model. The concept of combining B-S-Rc-DEA model is depicted in Figure 2.

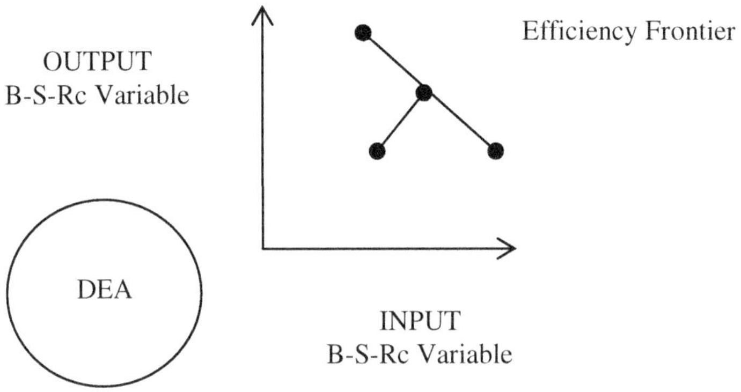

Fig. 2. Combining B-S-Rc-DEA

Input variable are divided in to 7 variables: supply chain's customer performance, Supply chain' business process performance , Supply Chain's learning and growth performance, Regulator's role in improving financial performance, Regulator's role in market expansion, Regulator's role in improving Internal business process performance and Regulator's role in improving learning and growth performance. Output variable consists of one variable, known as supply chain's financial performance.

3.2 Population and Sample

The population of the research is actors (SME) and regulator of SME's leather product in Yogjakarta, one of the province in Indonesia. Supply chain actors consists of leather craftsmen and traders. Regulators consist of The Cooperative, Industrial and Trade service of Yogyakarta and Center for Leather, Rubber and Plastics Yogyakarta, Indonesia. This leather cluster potentially produces leather craft. Some types of leather products can be either leather craft (puppets, fan, lamp cap), leather goods (leather bag, leather wallets, belts), garment leather (leather jacket, leather gloves, leather hats, vests), leather furniture (chair, sofa) and leather accessories (necklaces, bracelets, wristwatches). Random sampling technique was used in this research. The questionnaire, interview and observation were done to obtain data from supply 's actors and regulator.

4 Result and Discussion

4.1 Performance Measurement of Supply chain Actors and Regulators Using B-S-Rc Model

Based on data from The Cooperative, Industrial and Trade service of Yogyakarta of Yogyakarta city, there are about 50 actors (SME) of leather and leather products in Yogyakarta. The measurement was done by distribute questionnaires and interview. There were 40 SME (name as DMU A2 – A46) are willing to fill the data and interview. Forty SME represented the existing population (80%) of the whole population. Performance measurement was done by measuring the performance of indicators in accordance with the model B-S-Rc (as seen in Table 1). It was conducted in two year time range from 2012 and 2013. The result of each measurement indicators normalized by SNORM, which valued from a range of 0 – 100. The process of normalization (Snorm) De Boer employed scale of 0 to 100 to get the value of each metric with the following formula:

Larger is better

$$Snorm = ((Si - S\ Min)/(S\ max - S\ Min)) \times 100 \quad (1)$$

Lower is better

$$Snorm = ((S\ Max - Si)/(S\ Max\text{-}Smin)) \times 100 \quad (2)$$

Si = average value
S max = the value of achieving the best performance indicators
S min = the value of achieving the worst performance indicators

Regulator's performance measurement are based on valid indicators which has been generated in the B-S-Rc model. There are 12 valid indicators as shown in Table 1. Performance measurement was done based on the results of interview and government performance accountability report documents in year 2012. The impact of regulator's performance in year 2012 would be obtained by SME on the one year later, that is year 2013. The result of each measurement indicators had normalized by SNORM. It resulted the score of range 0-100. The total performance was obtained from the average score of the regulators and actors (SME) as presented in Table 1.

The result indicates that the contribution of the regulators impacted the actors depended on their program or support. Some actors did not get the support from regulators in some indicators. From the assessment results, there are only 4 DMU which has score above 50,i.e. A5, A17, A28 and A42. But a comparison of the results with B-S-Rc could not be exploited for improvement because the calculation B-S-Rc model had not been based on the best targets of each indicator [35]. The score was calculated based on the achievement of the highest and lowest of each SME in Yogyakarta. The achievement value used B-S-Rc model had not reflected the best management yet because was not based on mathematical optimization model [31]. A high score on the B-S-Rc model doubtly reflects the performance indicator that is better than the others, because the value of the B-S-Rc was obtained from the average of all indicators and not based on a mathematical model optimization.

Table 1. The Performance of actors and regulators using B-S-Rc model (in %)

DMU	A2	A3	A4	A5	A6	A7	A8	A9	A10	A11
Score of actors and regulators performance	35.85	34.90	35.17	53.27	49.75	35.07	44.19	36.86	37.24	42.34

DMU	A12	A13	A14	A15	A16	A17	A18	A19	A20	A21
Score of actors and regulators performance	40.55	41.17	41.11	34.00	33.21	55.36	34.53	30.67	28.06	38.06

DMU	A22	A24	A27	A28	A29	A30	A31	A32	A33	A35
Score of actors and regulators performance	37.45	32.37	28.20	50.55	36.21	32.09	33.92	33.61	44.94	35.43

DMU	A36	A37	A38	A40	A41	A42	A43	A44	A45	A46
Score of actors and regulators performance	38.93	36.47	33.74	40.61	45.19	53.63	32.43	34.11	33.40	33.14

Figure 3 shows an example of performance score comparison of three DMU (A3,A17,A28), where the score of the B-S-Rc A17 > A28 > A 3. Meanwhile the score of the internal business process performance were A28 > A 17 > A 3. This indicates that the value of the performance with the model B-S-Rc cannot be used as a basis for comparison in SME performance improvement. B-S-Rc model needs to be combined with other models which could identify the comparison (benchmarking) between actors to address the weakness.

Data Envelopment Analysis (DEA) can support this inability. This combination produces a value based on the valid indicators performance (valid measurement variables) of the B-S-Rc model and on the other hand generates the measurement score which could be compared (benchmark) mathematically. The combination of B-S-Rc-DEA measurement model will generate the performance value which would be used to improve the performance not only for actors but also the regulators in their business management.

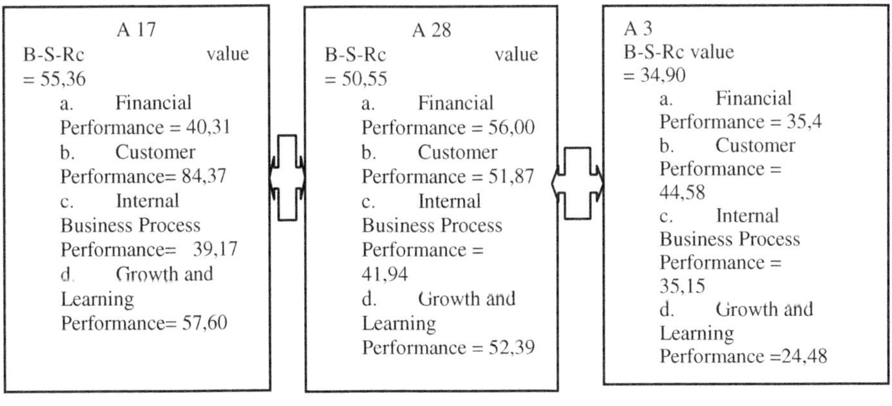

Fig. 3. An example of the performance score comparison of 3 DMU

4.2 Performance Measurement of Supply Chain Actors and Regulators Using B-S-Rc-DEA Model

Performance measurement uses B-S-Rc-DEA model beginning by identifying the inputs and outputs. The selection of input and output can be seen in Figure 1. In this early framework, the input variable consists of 7 models, they are Regulator's role in improving financial performance (X1), Regulator's role in market expansion (X2), Regulator's role in improving Internal business process performance (X3) and Regulator's role in improving learning and growth performance (X4), supply chain's customer performance (X5), Supply chain' business process performance (X6), Supply Chain's learning and growth performance (X7). The model's output consists of a single variable; it is supply chain's financial performance (Y1).

Performance measurement uses B-S-Rc-DEA solved using a CRS primal model. It was calculated with Linier Programing by the help of Software Win QS. The performance score reflecting the efficiency value is shown in Tabel 2. The efficiency's result with B-S-Rc –DEA indicates that the highest value was reached by A19 and A44 with the efficiency score of 1. It shows that both of the SME could manage their inputs to gain optimum output result [34]. The regulator's role for both of these SME was optimum with the help and the facilitation which could be effectively managed by these SME. The other SME could refer to these SME as a benchmark for their performance management.

Table 2. The performance score using B-S-Rc – DEA model (in %)

DMU	A2	A3	A4	A5	A6	A7	A8	A9	A10	A11
Score	40.81	40.81	40.81	72.30	77.72	43.29	60.02	40.81	40.81	68.02

DMU	A12	A13	A14	A15	A16	A17	A18	A19	A20	A21
Score	40.81	72.79	72.51	55.18	64.14	54.05	48.17	100.00	80.53	74.15

DMU	A22	A24	A27	A28	A29	A30	A31	A32	A33	A35
Score	40.81	45.99	49.62	96.31	61.45	63.61	42.58	85.90	83.28	45.01

DMU	A36	A37	A38	A40	A41	A42	A43	A44	A45	A46
Score	61.50	50.02	48.91	40.81	57.38	60.55	68.70	100.00	74.15	60.27

5 Conclusion

B-S-Rc model is intended as a model for measuring performance of supply chain actors and regulator which consist 19 indicators to measure the performance of the SC actors and 12 indicators to measure the regulator performance. This performance indicator can accommodate the needs of supply chain's actors and regulator simultaneously. B-S-Rc model based on Balanced Scorecard perspective and integrating process based on SCOR into the internal business processes and incorporated the role of regulators in each perspectives. The advantages of B-S_Rc model are (1) it is a comprehensive model (2) It considers the operation process thoroughly and (3) it involving the role of regulators in increasing the performance of SME. The weakness of this model is its inability to provide a benchmark. The performance score employing B-S-Rc model has not reflected the best management yet because it is not based on mathematical optimization models. The high score on the model B-S-Rc doubtly reflect the performance indicators which are better than the others, because the score of the B-S-Rc was obtained from the average of all indicators. This drawback was overcome by combining the B-S-Rc model with DEA. This combination generated a score that could be compared among DMU, so it gave opportunity for DMU to improve their performance referring the best one. It can be concluded the combination of BS-Rc-DEA provides information that enables benchmarking (comparison) to overcome the weaknesses BS-Rc models.

The deficiency of this research is that the object is limited to innovative products, so it cannot be generalized to be applied to the other product characteristics. All variables in this study assumed to have equal weight. Later, the research opportunity may develop a measurement model by considering the weight indicator, because the indicator has different importance degree of performance.

References

1. Barber, E.: How to measure the "Value" in value chains. International Journal of Physical Distribution & Logistics Management 38(9), 685–698 (2008)
2. Bhagwatt, R., Sharma, M.K.: Performance measurement of supply chain management using the analytical hierarchy process. Production Planning & Control 18(8), 666–680 (2007a)
3. Brewer, P.C., Speh, T.W.: Using balanced scorecard to measure supply chain performance. Journal of Business Logistics 21(1) (2000)
4. Chia, A., Goh, M., Hum, S.: Performance measurement in supply chain entities: balanced scorecard perspective. Benchmarking: An International Journal 16(5), 605–620 (2009)
5. Chang, H.H.: An empirical study of evaluating supply chain management integration using balanced score card in Taiwan. The Service Industries Journal 29(2), 185–201 (2009)
6. Gunasekaran, A., Patel, C., Tirtiroglu, E.: Performance measures and metrics in a supply chain environment International Journal of Operation & Production Management 21, 71–87 (2001)
7. Gunasekaran, A., Patel, C., McGaughey, R.E.: A framework for supply chain performance measurement. International Journal Production Economy 87, 333–347 (2004)
8. Park, J.H., Lee, J.S., Yoo, J.S.: A framework for designing the balanced supply chain scorecard. European Journal of Information Systems 14, 335–346 (2005)

9. Georgise, F.B., Thoben, K.D., Seifert, M.: Adapting the SCOR model to suit the different scenarios: a literature review & research agenda. International Journal of Business and Management 7(6), 1–17 (2012)
10. Chae, K.: Developing key performance indikators for supply chain: an industry perspective. Supply Chain Management: An International Journal 14(6), 422–428 (2009)
11. Huan, S.H., Sheoran, S.K., Wang, G.: A review and analysis of supply chain operations reference (SCOR) model. Supply Chain Management, ABI/INFORM 9(1), 23–29 (2004)
12. Lockamy, A., McCormack, K.: Linking SCOR planning practices to supply chain performance: an exploratory study. International Journal of Operations & Production Management 24(11/12), 1192–1217 (2004)
13. Supply Chain Council: Supply Chain Operation Reference Model (2001), http://www.supply-chain.org
14. Theeranuphattana, A., Tang, J.C.S.: A conceptual model of performance measurement for supply chains alternative considerations. Thailand Journal of Manufacturing Technology Management 19(1), 125–148 (2008)
15. Beamon, B.M.: Measuring supply chain performance. International Journal of Operation & Production Management 19, 275–292 (1999)
16. Espinoza, O., Bond, B.H., Kline, E.: Supply chain measures of performance for wood products manufacturing. Forest Production Journal 60(7/8), 700–708 (2010)
17. Farris, M.T., Hutchison, P.D.: Cash-to-cash: the new supply chain management metric. International Journal of Physical Distribution & Logistics Management 32(4), 288–298 (2002)
18. Hum, S.-H., Parlar, M.: Measurement and optimization of supply chain responsiveness. IIE Transactions, Taylor & Francis Group, Philadephia 46(1) (2014)
19. Marthin, P.R., Patterson, J.W.: On measuring company within a supply chain. International Journal of Production Research 47(9), 2449–2460 (2009)
20. Pohlen, T.L., Coleman, B.J.: Evaluating internal operations and supply chain performance using EVA and ABC. S.A.M. Advanced Management Journal 70(2), ABI/INFORM Global, 45–58 (2005)
21. Wong, W.P., Wong, K.Y.: Supply chain performance measurement system using DEA modeling. Industrial Management & Data Systems 107(3), 361–381 (2007)
22. Yang, F., Wu, D., Liang, L., Bi, G., Wu, D.D.: Supply chain DEA: production possibility set and performance evaluation model. Annals of Operations Research 185(1), 195–211 (2011)
23. Liang, L., Yang, F., Cook, W.D., Zhu, J.: DEA models for supply chain efficiency evaluation. Ann. Oper. Res. 145, 35–49 (2006)
24. Tambunan, T.: Promoting small and medium enterprises with a clustering approach: a policy experience from Indonesia. Journal of Small Business Management 43(2), 138–154 (2005)
25. Mawardi, M.K., Choi, T., Perera, N.: The factors of SME cluster developments in a developing country: the case of Indonesian clusters. In: ICSB World Conference Proceedings, pp. 1–28. International Council for Small Business (ICSB), Washington (2011)
26. Parrilli, M.D.: Collective efficiency, policy inducement and social embeddedness: drivers for the development of industrial district. Entrepreneurship & Regional Development 21(1), 1–24 (2009)
27. Nakagawa, R.: The policy approach in promoting small and medium sized enterprises in Japan. International Business & Economics Research Journal 11(10) (2012)

28. Kusrini, E.S., Masruroh, N.A.: A New Approach to Design Supply Chain Key Performance Indicator for Actors and Regulator: A Case Study in Innovative Produc. In: Indonesia. International Journal of Business Performance Management (2014) (accepted)
29. Striteska, M., Spickova, M.: Review and Comparison of Performance Measurement Systems. Journal of Organizational Management Studies 2012 (2012)
30. Banker, R.D., Potter, G., Srinivasan, D.: An empirical investigation of an incentive plan that includes nonfinancial performance measures. Accounting Review 75(1), 65–92 (2000)
31. Lee, J.Y.: Combining balanced scorecard and data envelopment analysis in kitchen employees performance measurement - An exploratory study - A dissertation Hospitality Management, Iowa (2012)
32. Rickards, R.C.: Setting benchmarks and evaluating balanced scorecards with data envelopment analysis. International Benchmarking Journal 10(5), 62–86 (2003)
33. Coelli, T.: A Guide to DEAP version 2.1. A Data Envelopment Analysis (Computer) Program. Department of Economic University of New England (1996)
34. Banker, R.D., Charness, A., Cooper, W.W.: Some models for estimating technical and scale inefficiencies in data envelopment analysis. Management Science 30(6), 1078–1092 (1984)
35. Aryanezhad, M.B., Najafib, E., Bakhshi, F.S.: A BSC-DEA approach to measure the relative efficiency of service industry: a case study of banking sector. International Journal of Industrial Engineering Computations 2(2), 273–282 (2011)
36. Kaplan, R.S., Norton, D.P.: Using the balanced scorecard as a strategic management system. Harvard Business Review 74(1), 75–85 (1996)
37. Charnes, A., Cooper, W., Rhodes, E.: Measuring the efficiency of decision making units. European Journal of Operational Research 2(6), 429–444 (1978)
38. Chen, Y.S., Chen, B.Y.: Using data envelopment analysis (DEA) to evaluate the operational performance of the wafer fabrication industry in Taiwan. Journal of Manufacturing Technology Management 20(4), 475–488 (2009)
39. Huang, S.P.: Performance analyses of traditional industries in taiwan with data envelopment analysis. The International Journal of Organizational Innovation 4(3) (Winter 2012)
40. Nooreha, H., Mokhtar, A., Suresh, K.: Evaluating publicsector efficiency with Data envelopment Analysis (DEA): A case study in road Transport deartement, Seangor Malaysia. Total Quality Management, pp. 4–6 (July 2000)
41. Meza, D.: Measuring the efficiency of a lean six sigma program in a u.s. government agency using data envelopment analysis, thesis the Faculty of University of Houston Clear Lake (2012)
42. Sarkis, J., Talluri, S.: Ecoefficiency measurement using dataenvelopment analysis: research and practitioner issues. Journal of Environmental Assessment Policy and Management 6(1), 91–123 (2004)
43. Min, H., Min, H., Joo, S.J.: A data envelopment analysis-based balanced scorecard for measuring the comparative efficiency of Korean luxury hotels. International Journal of Quality & Reliability Management 25(4), 349–365 (2008)
44. Wu, W., Liao, Y.: A balanced scorecard envelopment approach to assess airlines' performance. Industrial Management & Data Systems 114(1), 123–143 (2014)
45. Chen, T., Chen, C., Peng, S.: Firm operation performance analysis using dataenvelopment analysis and balanced scorecard: a case study of a credit cooperative bank. International Journal of Productivity and Performance Management 57(7), 523–539 (2008)
46. Poister, T.H., Streib, G.: Performance Measurement in Municipal Government: Assessing the State of the Practice. Public Administration Review 59(4), 325–335 (1999)

47. Neely, A., Gregory, M., Platts, K.: Performance measurement system design: a literature review and research agenda. International Journal of Operation and Production Management 25, 1228–1263 (2005)
48. Nyhan, R.C., Martin, L.L.: Assessing the Performance of Municipal PoliceServices Using Data Envelopment Analysis: An Exploratory Study. State and Local Government Review 31(1), 18–30 (1999)
49. Ammons, D.N.: Overcoming the Inadequacies of Performance Measurement in Local Government: the Case of Libraries and Leisure Services. Public Administration Review 55(1), 37–47 (1995)
50. Kusrini, E., Subagyo, Masruroh, N.A.: Good Criteria for Supply Chain Performance Measurement. International Journal of Engineering Business Management 6, 9 (2014), doi:10.5772/58435
51. Kellogg Foundation: Using Logic models to bring together planning, evaluation, and action Logic model development guide, One East Michigan Avenue East Battle Creek, Michigan 49017-4058 (2004), `http://www.wkkf.org` (accessed September 2, 2012)
52. Northcott, D., Taulapapa, T.M.: Using the balanced scorecard to manage performance in public sector organizations Issues and challenges. International Journal of Public Sector Management 25(3), 166–191 (2012)
53. Anthoula, K., Alexandros, H.: Designing a balanced scorecard for the evaluation of a local authority organization European Research Studies XIV(2) (2011)
54. Farneti, F., Guthrie, J.: Italian and Australian local governments: balanced scorecard practices: A research note. Journal of Human Resource Costing & Accounting 12(1), 4–13 (2008)
55. Neely, A., Adam, C., Kennerly, M.: The performance prism – the scorecard for measuring and managing business success. London Pearson Education Limited (2002)
56. Baldrige National Quality Program. Criteria for performance excellence. Administration Building, Gaithersburg (2007), `http://www.baldrige.nist.gov` (accessed November 18, 2011)
57. Deng, Z., Wu, C., Zhang, J., Wang, J.: Study on the supply chain efficiency of rural public service in china based on three stage dea model. Asian Agricultural Research 69(1), 6–13 (2014)

A Study of Consumers' Post Consumption Behaviour for Mobile Phone in Indonesia

Siti Mahsanah Budijati[1,2], Subagyo[1], Muhammad Arif Wibisono[1], and Nur Aini Masruroh[1]

[1] Faculty of Engineering, Department of Mechanical and Industrial Engineering, Gadjah Mada University, Jl. Grafika No. 2, Yogyakarta, 55281, Indonesia
[2] Faculty of Industrial Technology, Department of Industrial Engineering, Ahmad Dahlan University, Jl. Prof. Dr. Soepomo, S.H., Warungboto, Yogyakarta, 55164, Indonesia
smbudijati@ie.uad.ac.id,
{subagyo,arifwibisono,aini}@ugm.ac.id

Abstract. The Study of consumers' post consumption behaviour is important to obtain an explanation of consumers' behaviour. This information is needed to design the appropriate reverse logistics activities. This study conducted toward the mobile phone users in Indonesia, aims to solicit the data of consumers' post consumption behaviour for mobile phone. From 346 respondents, there were three post consumption behaviours identified, namely: sell to second-hand market, storage/ donated to others, and direct discarding, meanwhile there was no consumer who were willing to return to a take back program. The study showed that the behaviour with the highest frequency distribution was storage/ donated to others. The homogeneity tests revealed the differences in proportion of storage/ donated to others and direct discarding based on gender, as well as the differences in proportion of direct discarding based on the income. The independent test found that there was no relationship between demographic backgrounds with the choices of post consumption behaviour.

1 Introduction

Currently, the level of mobile phone usage is quite high and continues increasing over the years. Indonesia is one of the largest markets for cellular industry. The data from BPS Indonesia (Central Bureau of Statistics) showed that the percentage of households possessing mobile phone has increasing, i.e. 72.01% in 2010 increased to 78.96% in 2011 and also increased to 83.52% in 2012, where the number of households in 2010 was 61.1645 million with the average number of members in the household was 3.9 inhabitants [1].

The increasing use of mobile phones will lead to the increasing of numbers on the mobile phone waste known as e-waste. The accumulation of e-waste would become a problem for the environment if there is no adequate treatment, but can be profitable if it is managed correctly. As stated in [2], the increasing volume of e-waste is not only an environmental problem but also an economic opportunity.

M. Gen et al. (eds.), *Industrial Engineering, Management Science and Applications 2015*, Lecture Notes in Electrical Engineering 349, DOI: 10.1007/978-3-662-47200-2_60

There is no definitive data on the rate of mobile phone waste in Indonesia. Likewise, there is no special program aimed at addressing the mobile phone waste. Traced from government regulation of electronic waste, Indonesia does not have specific regulations to arrange the management of e-waste. [3] argued that the existing legislation is about B3 (Toxic and Hazardous Substances) waste management specifically is the Minister of Environment Regulation No.18/2009.

One of the ways for handling EOL (end of life) product, including e-waste is by reverse logistics activities. Basically the reverse logistics activities are intended to take the value of the EOL product or, if the disposal should be conducted, then do it correctly. This is confirmed in [4] which expressed that the essence of reverse logistics is to get the value of the product that is no longer used by the consumers. When a product has lost its value, reverse logistics can recover it to become a new product by recycling some parts or components of the product.

In fact, some mobile phone manufacturers have offered a take back program for the used or EOL mobile phones, as a form of reverse logistics activities. However, consumers' response to the program is still low. This is shown in a survey held by Nokia, that 70 percent of mobile phone users were not aware that their handsets and accessories can be recycled. While the 3 to 4 percent of people who already knew about it never think to recycle their phones. Approximately 88 percent of the people are Indonesian [5].

To be able to run the mobile phone reverse logistics, the authorities need to know the post consumption behaviour of the mobile phone consumers. The information of the behaviour will give an idea to design reverse logistics activities that were appropriate with mobile phone.

2 Importance of the Study

Mobile phone contains materials that are harmful to the environment and human health in case of inadequate disposal. This is confirmed in [6], [7], [8], [9], and [10]. The hazardous materials are: lead, cadmium, mercury, hexavalent chromium, and flame-retardant materials.

Meanwhile, if the handling of mobile phone waste was conducted properly, it can be profitable. As mentioned in [5] the advantage of recycling mobile phone is its ability to reduce the potential greenhouse effect that is equivalent with the gas pollution from 4 million vehicles. This can be achieved if every mobile phone users around the world are willing to give their used mobile phones and accessories for recycling.

Furthermore, according to the Environmental Protection Agency (EPA) in [11], the energy which can be saved from recycling of one mobile phone is enough to turn on a laptop for 44 hours. If American people recycle approximately 130 million used mobile phones each year, there will be enough energy to be used for more than 24 thousand houses each year. In line with that [12] stated that if the mobile phone discarded, it takes 400 million years to decompose.

Mobile phone recycling is one of the typical product recoveries in reverse logistics activities which require cooperation with consumers. Consumers in these activities take a role as suppliers. The willingness and behaviour of the consumers to participate in the collecting process of used mobile phones will affect the success of reverse logistics activities. For this reason, it is important to investigate the behaviour of consumers towards their used or EOL products.

Some authors emphasized the importance of consumers' behaviour to be accommodated in the reverse logistics activities, particularly in the collecting process. As noted by [13], the end-user characteristics affect different types of reverse logistics system. Similarly [14] expressed that the characteristics of the population differ from region to region, so the reverse logistics need a customized EOL product collection strategy to achieve a suitable reverse logistics system. Moreover, [15] suggested that consumers' behaviour in returning the used product need to be considered. Furthermore, [16] argued that the best solution to manage reverse logistics system depends on the economic and cultural context where the system operates. Meanwhile [17] stated that consumers' behaviour is one of the main components that should be considered in the disposition decision of reverse logistics. Recently [2] said that to build an effective reverse logistics, it is necessary to know the consumers' behaviour.

A way to find out the wishes and behaviour of consumers towards their used or EOL product is studying the consumers' post-consumption behaviour. Research on the consumers' post-consumption behaviour for used or EOL product is still slightly, two of whom are [18] and [2]. [18] analyzed the factors influencing the behaviour of post-consumption disposal for beverages, while [2] studied the impact of demographic variables on the choices of mobile phones waste disposal in India.

Meanwhile researches on post-consumption related to recycling behaviour were conducted by [19] and [20]. [19] discussed consumers' decision on recycling the used-fluorescent lamp in São Paulo, Brazil, while [20] investigated the consumers' willingness to pay for the recycling process of mobile phone waste in China.

Demographic factors are believed to be able to influence the behaviour of individuals. Some studies demonstrated the influences, as in [2], it is shown that the differences in some demographic background bring out the differences in disposal choices for mobile phone. [21] verified that education level and residence type have a significant impact in the intention and behaviour for recycling, while [22] demonstrated the difference in the education level correlates with the energy-efficient investment behaviour. Furthermore, [20] found that the differences in the education level and monthly income provide a difference in willingness to pay for recycling process of mobile phone.

Therefore a study of consumers' post consumption behaviour for mobile phones in Indonesia needs to be done. It is based on the differences of demographic background for decision of post-consumption behaviour. The results of the study can provide some recommendations to the authorities in handling the mobile phone waste. So, the main objectives of this paper are:

1 to identify consumers' post consumption behaviour for mobile phone in Indonesia
2 to study the choices of consumers' post consumption behaviour for mobile phone based on the demographic background
3 to study the relationship of consumers' post consumption behaviour for mobile phone with demographic background

3 Research Method

The respondents of this research were mobile phone users in Indonesia which were at least 17 years old. Meanwhile, the stages of the research included:

1 questionnaires distribution to ask the demographic data and the post consumption behaviour of the consumers for their used or EOL mobile phones
2 frequency distribution analysis of each post consumption behaviour based on demographic background
3 Homogeneity test of each post consumption behaviour based on demographic background
4 Independent test to find out the relationship between demographic background with the choices of post consumption behaviour

4 Result and Discussion

4.1 Respondent Description

The questionnaire was conducted from June to November 2013, 350 questionnaires were distributed and the numbers of questionnaire returned were 346. It meant that the percentage of questionnaires returned was 69.2%. The demographic background of the respondents can be seen at Table 1.

Table 1. Demographic background of respondent

Demographic variable		Percentage of respondents
Gender	Male	47
	Female	53
Age	17-25 years	33
	26-35 years	32
	36-45 years	20
	46-55 years	13
	56-65 years	2
Education	Elementary school	1
	Junior high school	5
	Senior high school	28
	Diploma	3
	Under graduate	51
	Master	11
	Doctoral	1
Income/ month	< 1 million Rp	33
	1-5 million Rp	53
	5-10 million Rp	8
	10-20 million Rp	2
	others (housewife or student)	5

4.2 Frequency Distribution Result

From 346 respondents, there were three post consumption behaviours identified, namely: sell to second-hand market, storage/ donated to others, and direct discarding. Meanwhile there was no consumer who were willing to return to a take back program.

The frequency distributions for each type of post consumption behaviour based on demographic background are shown in Figure 1 to Figure 4. It can be seen that the highest frequency distribution of consumers' post consumption behaviour is storage/ donated to others.

4.3 Homogeneity Test Result

The homogeneity test is intended to find out the differences in proportion based on demographic background for each type of post consumption behaviour. The homogeneity test result based on gender showed the differences in proportion of storage/ donated to others and direct discarding. Female's proportion was higher than the male's for storage/ donated to others behaviour. On the contrary, for direct discarding behaviour, male's proportion was higher than female's.

The homogeneity test result based on age and education showed no difference in the proportion of all kinds of post-consumption behaviour. As for the homogeneity test results based on income showed the differences in proportion of direct discarding, which the highest proportion was the respondents with the income of 10-20 million rupiah/ month. The detail result of homogeneity test is shown in Table 2 to Table 5.

4.4 Independent Test Result

The independent test is aimed to test whether there is a relationship between demographic backgrounds with post-consumption behaviour choices or not. The results of the independent tests showed that there was no relationship between demographic backgrounds with the choices of post consumption behaviour. The detail result of independent test is shown in Table 6.

4.5 Discussion

From the identified mobile phone post-consumption behavior, there was no consumer who answered return to a take back program. This indicates that the behavior has not become a habit for mobile phone users in Indonesia. Consequently, the company has been offering a take back program needs to campaign the program more aggressively. Possible lack of information or lack of collecting point facilities of take back program, makes consumers not to understand if there is such a program. Given that the highest frequency of behaviour is storage/ donated to others, then most likely the consumer is willing to participate in a take back program as long as supported by the availability of information and adequate facilities.

The homogeneity test result showed evidently that female has higher proportion on storage/ donated to others than male. This means that female customers were potential to participate in the take back program. So, the companies need to think a strategy to encourage the willingness of the female consumer.

Meanwhile male consumers have higher proportion in direct discarding than female. This has some implications for environmental pollution. Therefore in handling mobile phone waste, the authorities should distribute the information about the dangerous of disposal mobile phone material into the environment. The dissemination of the information about the dangerous of environmental damage is also required to address the consumers with the highest income, because they also showed the highest proportion in direct discarding than those with lower incomes.

Table 2. Homogeneity test result based on gender (H_0: p1=p2)

Behaviour	Statistical value	Acceptance value of H0	Decision
Sell to 2nd hand market	Z = 1.40102915	-1,96< Z < 1,96	H_0 accepted
Storage/donated to others	Z = -3.3591301	-1,96< Z < 1,96	H_0 rejected (female's proportion > male's)
Direct discarding	Z = 2.9600514	-1,96< Z < 1,96	H_0 rejected (male's proportion > female's)

Table 3. Homogeneity test result based on age (H_0: p1=p2=p3=p4=p5)

Behaviour	Statistical value χ^2	Value χ^2 from table (df=4; α=0.05)	Decision
Sell to 2nd hand market	1.216445	9.48773	H_0 accepted
Storage/donated to others	3.5383575	9.48773	H_0 accepted
Direct discarding	2.8890617	9.48773	H_0 accepted

Table 4. Homogeneity test result based on education (H_0: p1=p2=p3=p4=p5=p6=p7)

Behaviour	Statistical value χ^2	Value χ^2 from table (df=6; α=0.05)	Decision
Sell to 2nd hand market	5.691571	12.5916	H_0 accepted
Storage/donated to others	4.223251	12.5916	H_0 accepted
Direct discarding	9.795341	12.5916	H_0 accepted

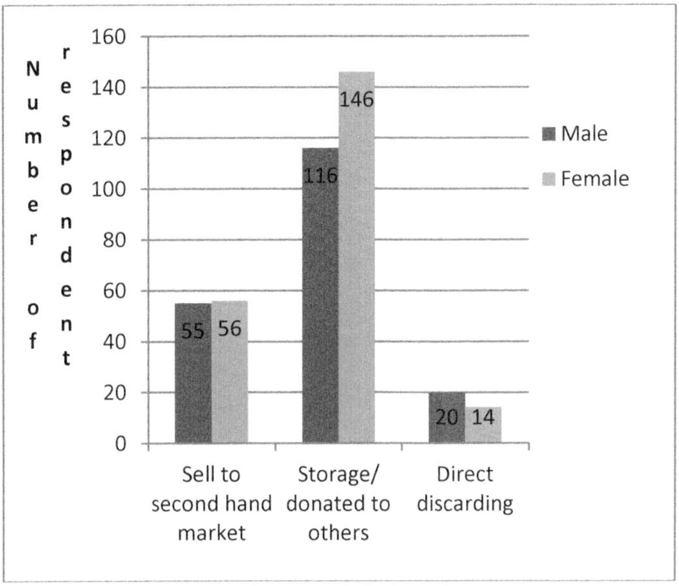

Fig. 1. Consumers' post consumption behavior based on gender

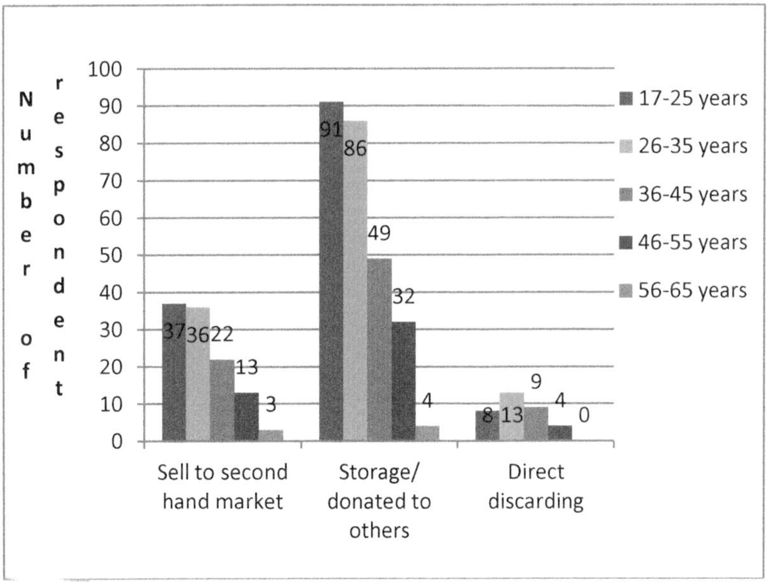

Fig. 2. Consumers' post consumption behavior based on age

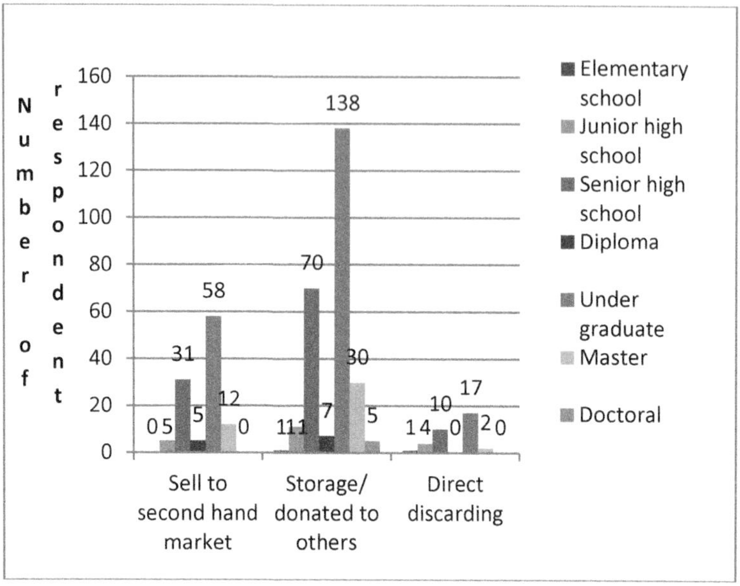

Fig. 3. Consumers' post consumption behavior based on education

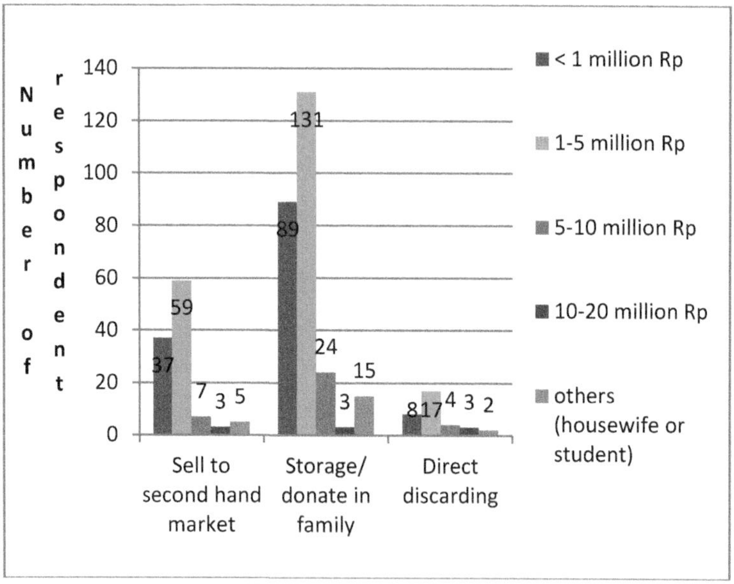

Fig. 4. Consumers' post consumption behavior based on income

Table 5. Homogeneity test result based on income (H_0: p1=p2=p3=p4=p5)

Behaviour	Statistical value χ^2	Value χ^2 from table (df=4; α=0.05)	Decision
Sell to 2nd hand market	1.426046	9.48773	H_0 accepted
Storage/donated to others	7.883309	9.48773	H_0 accepted
Direct discarding	12.82209	9.48773	H_0 rejected (the highest proportion is the income of 10-20 million Rp)

Table 6. Independent test result based on demographic background (H_0: there is no relationship between the demographic background with post consumption behaviour choices)

Demographic variable	Statistical value χ^2	Value χ^2 from table (df=(k-1)(b-1); α=0.05)	Decision
Gender	2.978559	5.99147	H_0 accepted
Age	3.631038	15.35075	H_0 accepted
Education	14.07777	21.0261	H_0 accepted
Income	10.75381	15.35075	H_0 accepted

Considering to the results of the independent tests that there is no relationship between demographic backgrounds with the choices of post-consumption behaviour, this provides an opportunity to divert the current behaviour become willing to follow the take-back program.

5 Conclusion and Recommendation

From the study, it could be identified that there are three post consumption behaviours, namely: sell to second-hand market, storage/ donated to others, and direct discarding. There was no consumer who was willing to return to a take back program. Meanwhile, the behaviour with the highest frequency was storage/ donated to others.

The homogeneity test showed the differences in proportion of storage/ donated to others and direct discarding based on gender, as well as the differences in proportion of direct discarding based on income. The independent test found that there was no relationship between demographic backgrounds with the choices of post consumption behaviour.

Looking at the whole study results, it could be inferred that there is an opportunity to offer mobile phone take back program as a collecting point for reverse logistics activities. Nevertheless the program offers must be accompanied by a clear and widespread information, and the easily accessible of collecting point facilities for consumers.

The results of the study may be different when applied to other electronic products. The forthcoming research may be developed to accommodate other variables such as the role of media, regulation, and economic incentives to support the positive post-consumption behaviour.

Acknowledgments. This paper is a part of a dissertation financially supported by a Doctoral Grant from the Directorate General of Higher Education, Ministry of Education and Culture of Indonesia which is highly appreciated.

References

1. BPS-Central Bureau of Statistics: Trends of Selected Socio-Economic Indicators of Indonesia, BPS, Jakarta, Indonesia (2013)
2. Dixit, S., Vaish, A.: Sustaining environment and organisation through e-waste management: a study of post consumption behaviour for mobile industry in India. International Journal of Logistics Systems and Management 16(1), 1–15 (2013)
3. Aslia, U.S.: Green ICT: Pengelolaan E-Waste di Indonesia, Unit Asdep Urusan Verifikasi Pengelolaan Limbah B3-Deputy Assistant Unit of Hazardous Materials Waste Management Verification, Kementerian Lingkungan Hidup-Ministry of Environment, Jakarta, Indonesia (2010)
4. Jingbo, Y.: The Coordinate Study of Reverse Logistics and Green Supply Chain (2005), `http://www.seiofbluemountain.com/upload/product/201002/12651652325cqooobk.pdf` (retrieved August 11, 2011)
5. World Wide Fund for Nature: Daur Ulang Ponsel dan Aksesoris Bekas (2009), `http://www.wwf.or.id/berita_fakta/berita_fakta/newsclimateenergy.cfm?10080/Daur-ulang-Ponsel-dan-Aksesori-Bekas` (retrieved August 11, 2011)
6. Chatterjee, S., Kumar, K.: Effective electronic waste management and recycling process involving formal and non-formal sectors. International Journal of Physical Sciences 4(13), 893–905 (2009)
7. Chi, X., Porte, M.S., Wang, M.Y.L., Reuter, M.A.: Informal electronic waste recycling: a sector review with special focus on China. Waste Management 31(4), 731–742 (2011)
8. Joseph, K.: Electronic waste management in India-Issues and strategies. In: Proceedings Sardinia 2007, Eleventh International Waste Management and Landfill Symposium. CISA, Environmental Sanitary Engineering Centre, S. Margherita di Pula, Cagliari, Italy (2007)
9. Realff, M.J., Raymond, M., Ammons, J.C.: E-waste: an opportunity. Materials Today 7(1), 40–45 (2004)
10. Schwarzer, S., De Bono, A., Giuliani, G., Kluser, S., Peduzzi, P.: E-waste, the hidden side of IT equipment's manufacturing and use. Environment Alert Bulletin (January 2005)
11. Media Indonesia: Ramai-ramai Daur Ulang Ponsel (2010), `http://digilib-ampl.net/detail/detail.php?row=&tp=kliping&ktg=sampahdalam&kode=10290` (retrieved August 23, 2011)
12. Roney: 50.000 Ponsel Didaur Ulang Bakrie (2010), `http://www.harianberita.com/50-000-ponsel-di-daur-ulang-bakrie.html` (retrieved August 23, 2011)
13. Flygansvaer, B.M., Gadde, L.E., Haugland, S.A.: Coordinated Action in Reverse Distribution Systems. International Journal of Physical Distribution & Logistics Management 38(1), 5–20 (2008)

14. Hanafi, J., Kara, S., Kaebernick, H.: Reverse Logistics Strategies for End-of-Life Products. The International Journal of Logistics Management 19(3), 367–388 (2008)
15. Poles, R., Cheong, F.: A System Dynamic Model for Reduction Uncertainty in Remanufacturing System. In: Pacific Asia Conference on Information System (PACIS) Proceedings. Association for Information System (2009)
16. Roslim, N.F., Ishak, M.B.: Electrical and electronic waste condition and handling practices in bangi Industrial Estate, Selangor. International Journal of Environmental Sciences 1(7), 1648–1658 (2011)
17. Hazen, B.T., Hall, D.J., Hanna, J.B.: Reverse logistics disposition decision-making developing a decision framework via content analysis. International Journal of Physical Distribution and Logistics Management 42(3), 244–274 (2012)
18. van Birgelen, M., Semeijn, J., Keicher, M.: Packaging and pro environmental consumption behavior: investigating purchase and disposal decisions for beverages. Environment and Behavior 41(1), 125–146 (2009)
19. Laruccia, M.M., Nascimento, J.V., Deghi, G.J., Garcia, M.G.: A study of consumer behavior on recycling of fluorescent lamps in São Paulo, Brazil. International Journal of Business Administration 2(3), 101–112 (2011)
20. Yin, J., Gao, Y., Xu, H.: Survey and analysis of consumers' behaviour of waste mobile phone recycling in China. Journal of Cleaner Production 65, 517–525 (2014)
21. Latif, S.A., Omar, M.S., Bidin, Y.H., Awang, Z.: Environmental problems and quality of life: situational factor as a predictor of recycling behaviour. Procedia – Social and Behavioral Sciences 35, 682–688 (2012)
22. Wenshun, W., Xiaohua, L., Hualong, L.: Empirical research of the environmental responsibility affected on the urban residential housing energy saving investment behaviour. Energy Procedia 5, 991–997 (2011)

Green Supply Chain Assessment to Operations Improvement for Can Packaging Industry

Tuanjai Somboonwiwat[1], Tuangyot Supeekit[2,*], and Patiyut Punta[1]

[1] Department of Production Engineering, Faculty of Engineering, King Mongkut's University of Technology, Bangkok, 10140, Thailand
[2] Department of Industrial Engineering, Faculty of Engineering, Mahidol University, Nakhonpathom, 73170, Thailand
tuanjai.som@kmutt.ac.th, tuangyot.sup@mahidol.ac.th

Abstract. To successfully operate an organization nowadays needs to improve the environmental performance due to increasing demands on environmental concerns. To identify improving area requires assessment methodology. This study applies supply chain operations assessment framework and FMEA to identify area that need to be urgently improved. The application of this assessment methodology is implemented to the can packaging industry to identify the improving area. Then the suggestion for improvement is achieved by the application of mathematical modelling. The improvement can be achieved and the performance of the can packaging manufacturer is increase both environmental and supply chain performance.

1 Introduction

Can packaging industries can be considered an environmentally unfriendly due to its production process which release and emit environmentally harmful gases, liquids and solids to the environment. Green packaging is under high competition nowadays meaning that the package to be used for the manufactured products must be environmentally friendly. Can packaging which are to be used in canned food and beverage production must, therefore, be improved to become environmental friendly. Improvement in the operations can be achieved by green supply chain management which aids managing operations in the supply chain to be effective in reducing their impacts to the environment [1]. The environmental impacts of supply chain operations should be reduces to ensure the greenness [2, 3]. If the environmental performance is improved, it is believed that the supply chain performance will be improved as well [4, 5]. The improvement in environmental performance requires green performance assessment because assessment of environmental performance can suggest the improvement area [2]. This study attempt to improve the operations in the can packaging manufacturer. The green operations assessment framework [6] incorporate with Failure Mode and Effect Analysis are chosen to assess the operations in the can packaging manufacturer because of its easiness in assessment. By the

* Corresponding author.

M. Gen et al. (eds.), *Industrial Engineering, Management Science and Applications 2015*, Lecture Notes in Electrical Engineering 349, DOI: 10.1007/978-3-662-47200-2_61

selection of operations to be improved, the improvement suggestion must be done to ensure the environmental performance of the manufacturer.

The rest of the paper is organized as follows. In section 2, assessment methodology of green operations assessment and FMEA are presented. Then the selection of operations to be improved applying green operations assessment and FMEA is presented in section 3. Section 4 provides improvement suggestion by developing the mathematical model for calculation. Finally, discussion on the performance improvement and the conclusion are presented.

2 Assessment Methodology

To identify the most urgent area for improving greenness of the supply chain, it is necessary to have a tool or methodology to assess the performance of current supply chain practice. This study employs the green supply chain operations assessment framework proposed by Somboonwiwat and Supeekit [6] to discover areas for improvement in the supply chain followed by Failure Mode and Effect Analysis (FMEA) to discover the most urgent improvement area or process. The assessment process can be presented in Fig. 1.

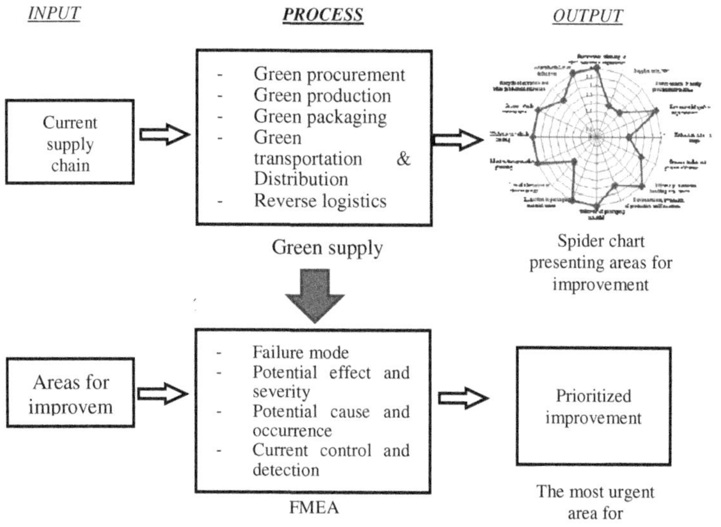

Fig. 1. Assessment methodology

2.1 Green Supply Chain Operations Assessment

The green supply chain operation assessment methodology is developed by Somboonwiwat and Supeekit [6] to assess the operations of green supply chain.

This framework is proven that these green supply chain operations can be mapped to Life Cycle Assessment (LCA) stages and can be used interchangeably [6]. Assessing the greenness of a supply chain prefers this assessment methodology to LCA due to its simplicity. The assessment can be completed by involved personnel in the assessed operations while LCA requires calculations of carbon releases, waste disposal, or usage of natural resources which has to be done by specialists. Kangang [7] believes that the easier assessment aids the organization in achieving the environment and business goals more easily. The detail of assessment involves the assessment of activities in green supply chain operations.

Green procurement includes the greenness in the processes of purchasing or procurement of raw materials and other materials and resources for production; procurement planning in accordance with the customer's requirement; inbound logistics from suppliers and storage.

Green production includes the greenness in the transformation process from raw materials to finished goods; the selection and efficiency of the production process; efficiency in material handling; and green awareness of production staff.

Green packaging considers the selection of packaging materials, size, and other characteristics; and reduction in packing material usage.

Green transportation & distribution includes the activities of storage and transportation the goods to customers as well as the choice of cleaner technology; fleet maintenance; truck loading and transportation planning.

Reverse logistics (Internal) includes the process of bringing the defects defectives and other production resources back for recycle or reproduction.

Green operation assessment framework can be summarized in table 1. After being assessed by interviewing the experts and scoring each supply chain operational practices, improvement areas can be surfaced in a spider chart (Fig 2).

Table 1. Green operation assessment framework

Green operations assessment criteria	Score
Green Procurement	
· Procurement planning to fulfill customer's requirement	
· Supplier selection	
· Environmental friendly procurement process	
· Raw material quality improvement	
Green Production	
· Reduction in raw material usage	
· Process design and process efficiency	
· Process design	
· Efficiency in material handling and internal transportation	
· Environmental awareness of production staff members	
Green Packaging	
· Selection of packaging material	
· Reduction in packaging material usage	
Green Transport & Distribution	
· Use of alternative or cleaner energy	
· Efficient in transportation planning	
Efficiency in vehicle loading	
· Proper vehicle maintenance	
Reverse Logistics (Internal)	
· Recycle of materials and other production resources	
· Reproducibility of defectives	

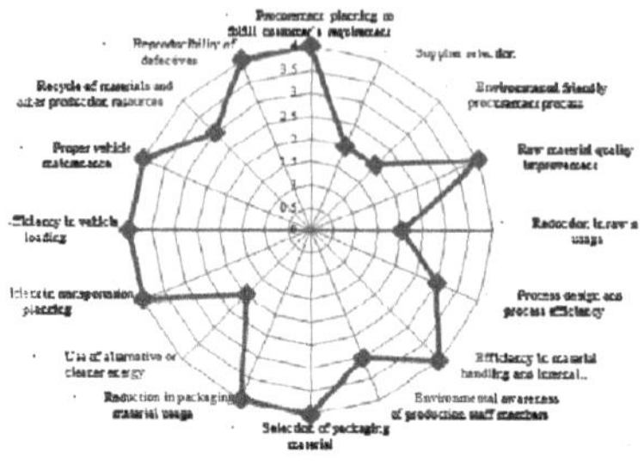

Fig. 2. Spider chart, presenting areas for improvement

2.2 Failure Mode and Effect Analysis (FMEA)

Failure Mode and Effect Analysis (FMEA) is a systematic methodology for analyzing potential product and process problems at the early stages [8]. The FMEA involves anticipating potential failure modes, their effects on operations, and actions to mitigate the failures. Despite the assertion in [8], there are 3 major types of FMEA including design, process, and service FMEA which focus on different aspects of a system. In the process FMEA, the processes or operations under investigation are analyzed for the potential failure that cause the particular process to miss the defined objectives; the effect to the system or process that the failure might cause to see how severe the effect might be; the cause of that failure mode and chance that cause might occurs; and the current control or action to detect or prevent the failure. After being analyzed, the processes are then raked by Risk Priority Number (RPN). RPN can be determined by multiplication of severity of the failure mode (S), frequency or chance of the cause of failure occurrence (O) and the ability or action to detect the cause of failure mode. The higher RPN means the higher priority for improvement. Table 2 is used for process FMEA analysis and RPN calculation

Table 2. Process FMEA

No.	Key Process	Potential Failure Mode	Potential Effect of Mode	Severity (S)	Potential Cause of Failure	Occurrence (O)	Current Process Control	Detection (D)	RPN
1									
2									
3									

3 The Case of a Can Packaging Manufacturer

The can packaging manufacturer in this study aim to maintain the competitiveness and create a good image for the company via promoting environmental concerns. Moreover, considering the highest production volume product; A-TP, the carbon footprint of this product equals to 205g which is higher than the counterparts. The executives believe that the concept of green supply chain can improve the carbon footprint as well as reduce the cost of operations. The current supply chain operations of the can packaging, therefore, must be explored and analyzed to identify areas that need to be improved and then selection of the most urgent area that need an improvement should be made. The following sections elaborate the selection process.

3.1 Supply Chain Operations in the Can Packaging Manufacturer

The operations of the can packaging manufacturer in this study are analyzed under the scope of gate-to-gate LCA which covers the operations within the manufacturer. The operations includes procurement, raw material storage, production planning, can packing production, packing, finished goods storage and goods transportation & distribution (Fig. 3).

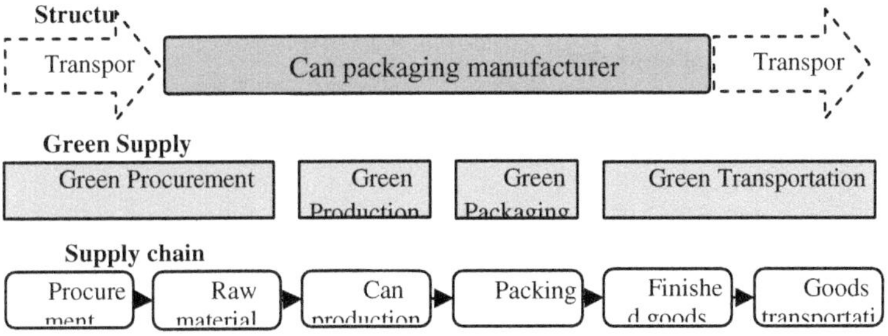

Fig. 3. Can packaging manufacturer supply chain operations

3.2 Green Operations Assessment for Can Packaging Manufacturer

The framework of green operations assessment framework (table 1) is taken to interview with the involved department managers including the following departments: Sales, Planning, Procurement department, Raw material storage, Production, and Finished goods storage and distributions. The result of the assessment is presented in table 3 and the spider chart is illustrated in fig 4.

Table 3. Green operation assessment for can packaging manufacturer

Green operations assessment criteria	Score
Green Procurement	
· Procurement planning to fulfill customer's requirement	3
· Supplier selection	2
· Environmental friendly procurement process	2
· Raw material quality improvement	4
Green Production	
· Reduction in raw material usage	4
· Process design and process efficiency	2
· Product design	2
· Efficiency in material handling	4
· Environmental awareness of production staff members	4
Green Packaging	
· Selection of packaging material	3
· Reduction in packaging material usage	3
Green Transport & Distribution	
· Use of alternative or cleaner energy	3
· Efficient in transportation planning	2
· Efficiency in vehicle loading	3
· Proper vehicle maintenance	3
Reverse Logistics (Internal)	
· Recycle of materials and other production resources	4
· Reproducibility of defectives	3

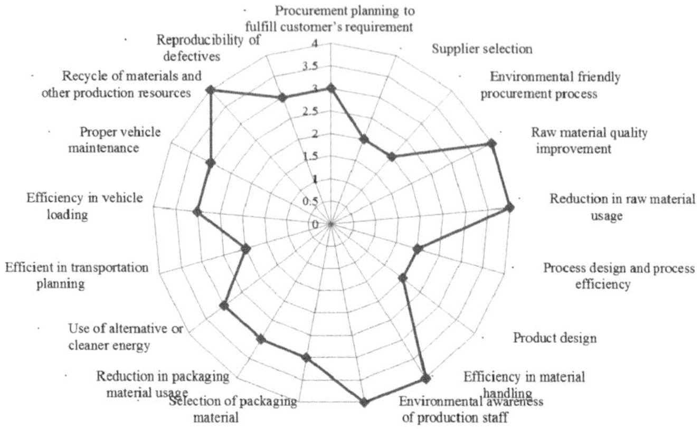

Fig. 4. Spider chart, presenting areas for improvement for can packaging manufacturer

It can be seen from the spider chart (fig 4) that the areas for improvement in this can packaging manufacturer includes supplier selection and procurement process (green procurement), process and product design (green production) and transportation planning (green transportation & distribution). Taking into account the capability for improvement, supplier selection, and procurement process are not easily improved due to the limited numbers of suppliers and uncontrollable practices of suppliers. Therefore, the area for improvement can be reduced to 1.) Process design, 2.) Product design, and 3.) Transportation planning. These three activities are further analyzed in the next section.

3.3 FMEA in Can Packaging Manufacturer

The activities of process design, product design and transportation planning are now analyzed again applying FMEA for the most urgent in need for improved activity. Each activity is investigated for the possible failure mode; effect of failure and its severity; cause of the failure and likelihood to happen; and the current control to ensure whether the cause can easily be detected. Then the severity, occurrences and detection are calculated for RPN. The following section explains the analysis of each activity.

Process Design. The failure in process design is the extensive use of energy and chemical substances. This might cause extra costs to the production processes and the use of energy and chemical substances are contributed to the much emission of Carbon dioxide. The extensive use of energy might result from the old-designed production line. To alter the process would cost a lot and require an investment analysis. The current process control for this activity is the regulation regarding the use of electricity in the factory.

Product Design. The failure is the variety of products which are customers' requirements and uncontrollable. The product varieties would result in extra uses of aluminum and chemical for coating.

Transportation Planning. The failure in this process is high transportation costs and complicated transportation routes. This would result in the waste of energy and expenses. This failure result from unplanned transportation or consideration only cost of materials. The current control is the full truck loading for transportation in every single trip.

The result of FMEA analysis is presented in table 4.

Table 4. Process FMEA of the can packaging manufacturer

No	Key Process	Potential Failure Mode	Potential Effect of Mode	Severity (S)	Potential Cause of Failure	Occurrence (O)	Current Process Control	Detection (D)	RPN
1	Process design	Extensive use of energy and chemical substances	Extra costs and CO_2 emission	8	The old design of production line	6	Regulation in the uses of electricity	6	288
2	Product design	Products varieties	Additional uses of Aluminum and chemical	5	Customer's requirement	5	No control	5	125
3	Transportation planning	High transportation costs and complex transportation routes	Waste of energy and expenses and heaps of CO_2 emission	9	No transportation route planning	9	Full truck loading for each trip	7	567

It can be seen from the RPN analysis that the most urgently in need of improvement in this can packaging manufacturer is the transportation planning. Considering the current practice of transportation in this company, it means that the transportation must be planned. The transportation planning for the can packaging manufacturer are presented in the next section.

4 Mathematical Modelling for Transportation Planning of the Can Packaging Manufacturer

Transportations for this company include raw materials transportation from suppliers to the manufacturing plant and finished goods transportation from the plant to distribution centers of the manufacturer. There are 5 suppliers from Singapore, Germany, Taiwan R.O.C and the other two from eastern Thailand. Raw materials are transported to 2 factories, one in central Thailand; the other in southern Thailand, by ship to the port located near factory then by truck to the factory for can packaging production. The can packaging products are then transported to 3 distribution centers in different locations; central, southern and northern Thailand, for distribution to customers (fig. 5).

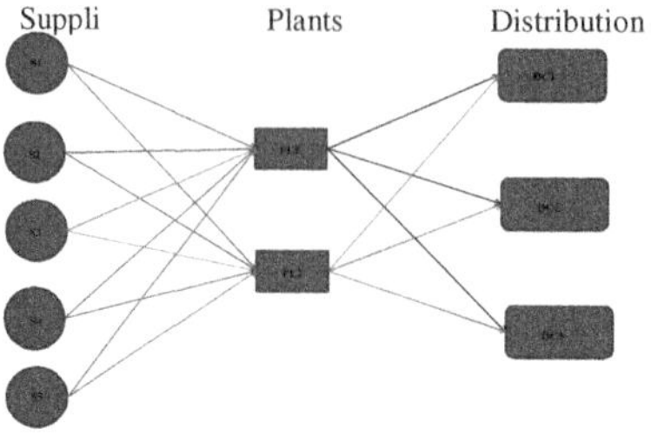

Fig. 5. Transportation system for can packaging manufacturer

Transportation planning in order to reduce the carbon dioxide emission can be achieved by creating mathematical modeling to find the amount of raw materials to be received from each supplier and the amount of finished goods to be transported to each distribution center. Assumptions of transportation planning are described as follows: factories produce adequate products for transportation; all transportations are full truck load transportations; there is no limitation in transportation capacity; amount of finished goods stored at each distribution centers are known; production rate of can packaging is consistent.

To ease the reading, the following lists of symbols summarizing the elements used for composing the model are presented.

Indices

i number of suppliers; where $i = \{1, 2, 3, 4, 5\}$
j number of plants; where $j = \{1, 2\}$
k number of distribution centers; where $k = \{1, 2, 3\}$

Determinant variables

X_{ij} Amount of raw materials shipped from supplier *i* to plant *j*
Y_{jk} Amount of products transported from plant *j* to distribution center *k*

Parameters

D_j Raw material requirement of plant *j*
D_k Product requirement of distribution center *k*
CAP_i Transportation capability of supplier *i*
CAP_j Transportation capability of supplier *j*
C_{ij} Transportation cost from supplier *i* to plant *j*
C_{jk} Transportation cost from plant *j* to distribution center *k*
CO_{ij} Carbon dioxide emission resulting from transportation from supplier *i* to plant *j*
CO_{jk} carbon dioxide emission resulting from transportation from plant *j* to distribution center *k*
E_i exchange rate of supplier's currency
SCR Social Cost Rate in carbon dioxide emission

Objective Formulation

The objective of transportation plan is that the total transportation would minimize the emission of carbon dioxide

$$Min\,Z = \left[\sum_i \sum_j X_{ij}\ C_{ij}E_i + \sum_j \sum_k Y_{jk}\ C_{jk}E_i\right] + \left[\sum_i \sum_j X_{ij}\ CO_{ij} + \sum_j \sum_k Y_{jk}\ CO_{jk}\right]SCR \quad (1)$$

Constraints

1. Raw materials transported from supplier i to all plants j must not exceed the transportation capability of supplier i

$$\sum_{j=1}^{j} X_{ij} \leq CAP_i \qquad ;\forall i,\ \ X_{ij} \geq 0 \qquad (2)$$

2. Raw materials transportation to plant j must not less than the raw material demand of plant j

$$\sum_{i=1}^{i} X_{ij} \geq D_j \qquad ;\forall j,\ \ X_{ij} \geq 0 \qquad (3)$$

3. Products transported from plant j to all distribution center k must not exceed the transportation capability of plant j

$$\sum_{k=1}^{k} Y_{jk} \leq CAP_j \qquad ;\forall j,\ \ Y_{jk} \geq 0 \qquad (4)$$

4. Products transportation to distribution center k must not less than the product demand of distribution center k

$$\sum_{j=1}^{j} Y_{jk} \geq D_k \qquad ;\forall k,\ \ Y_{jk} \geq 0 \qquad (5)$$

5. Raw material transported from suppliers to all plant must not less than products transported from plants to all distribution centers

$$\sum_{i=1}^{i} \sum_{j=1}^{j} X_{ij} \geq \sum_{j=1}^{j} \sum_{i=1}^{i} Y_{jk} \qquad (6)$$

5 Discussion and Conclusion

The study applies the concept of green supply chain to select the activity of can packaging manufacturer for improvement. Transportation planning is the activity selected from the application of green operations assessment and FMEA. The improvement in transportation planning is suggested by the mathematic model. The problem in the model is the problem of linear programming. Excel Solver is employed to find the solutions of number of raw material and can packaging products to be transported from one place to another. The result of transportation plan can yield the reduction in both carbon dioxide emission and transportation cost. Emission of carbon dioxide is reduced from 14,812 to 11,873 CO_2 kg/month for raw materials transportation and from 5,868 to 5,018 CO_2 kg/month for products transportation. Moreover, transportation planning can also decrease overall costs of transportation from over 38.2 million baht to around 37.6 million baht/month or about 1.87% per month. To conclude, green operations assessment and FMEA analysis can help the organization identify the areas for improvement with less effort than other methodologies. This methodology can be applied to other company or supply chain to select the process that need to be urgently improved. The improvement of selected process can be achieved by other application including mathematical modelling to make sure that the process is improved and the supply chain becomes greener.

References

1. Beamon, B.M.: Designing the green supply chain. Logistics Information Management 12(4), 332–342 (1999)
2. Björklund, M., Martinsen, U., Abrahamsson, M.: Performance measurements in the greening of supply chains. Supply Chain Management: An International Journal 17(1), 29–39 (2012)
3. Hervani, A.A., Helms, M.M., Sarkis, J.: Performance measurement for green supply chain management. Benchmarking: An International Journal 12(4), 330–353 (2005)
4. Green, K.W., et al.: Green supply chain management practices: impact on performance. Supply Chain Management: An International Journal 17(3), 290–305 (2012)
5. Villanueva, R., Garc'a, J.L., Adame, W.: Green Supply Chain Management; a competitive advantage. In: International Congress on Logistics & Supply Chain, Sanfandila, Queretaro, Mexico (2013)
6. Somboonwiwat, T., Supeekit, T.: Operations assessment for greening supply chain: A case of rubber glove. In: Global Conference on Applied Engineering and Environment Research, Okinawa, Japan (2014)
7. Kangang, M.T.: Green supply chain implementation: Best practices and Challenges. In: African International Business and Management. KICC, Nairobi (2011)
8. Crow, K.: Failure Modes and Effects Analysis, FMEA (2002), `http://www.npd-solutions.com/fmea.html` (cited December 29, 2014)

Selection of Digital Marketing Channels: Application of Modern Portfolio Theory

Tomás Frausto-da-Silva, António Grilo*, Virgílio Cruz-Machado

UNIDEMI, Faculdade de Ciências e Tecnologia da UNL
Campus FCT-UNL, Monte Caparica, Portugal
tda.silva@campus.fct.unl.pt,
{acbg,vcm}@fct.unl.pt

Abstract. This paper proposes the application of Markowitz's Modern Portfolio Theory to the selection of digital communication channels, in order to attain the finest trade-off between return on investment and risk. Previous application of financial portfolio theory in Marketing, as a means to optimize the portfolio, focused solely on the use of the model in decisions related to product, customer and customer segment, retail format and price promotion portfolios. The authors concluded that the Modern Portfolio Theory may be used as a decision support system in the selection of a company's Digital Marketing channels or tactics, mutatis mutandis, so to find the share of the communication budget to be allocated to each type of digital channel or tactic.

1 Introduction

Technological developments in the last two decades brought the mass worldwide adoption of smartphones and tablets, as well as the internet revolution, leading to the proliferation of e-mail, blogging and social networking sites, for instance. Due to the wide range of ways of reaching consumers, marketers currently face major challenges when planning and shaping its marketing strategies. The purpose of this paper is to develop a tool that helps improving the process of investment decisions regarding Digital Marketing and that benefits, to that end, from the marketers' intuitive component and from a comprehensible, but still reliable and efficient mathematical analysis. In this sense, the research question of this paper stands as follows:

How can we improve the efficiency of investments in Digital Marketing, combining analytical tools with the intuition and freedom of marketers?

The proposition of this paper is that Markowitz's Modern Portfolio Theory may be an helpful tool during the process of selecting Digital Marketing channels or tactics, considering (1) it relies on simple statistical measures (such as the mean and the standard deviation) to find the set of optimal solutions to an investment problem and (2) it offers the investor the opportunity to select, within a set of optimal investment choices, the solution that best suits his interests. Another advantage of Modern

* Corresponding author.

M. Gen et al. (eds.), *Industrial Engineering, Management Science and Applications 2015*,
Lecture Notes in Electrical Engineering 349, DOI: 10.1007/978-3-662-47200-2_62

Portfolio Theory is that it allows the analysis of portfolios based on a combination of the expected return and the investment risk (referred to as risk-adjusted return), and not solely on the expected return, which may lead to more robust solutions.

This paper is composed of six sections. Section 2 covers the literature review, presenting general concepts of Marketing and Digital Marketing and introducing the Modern Portfolio Theory, with regard to (1) the concepts of risk and diversification, (2) the assumptions inherent to the theory, and (3) the mathematical formulation of the theory, which consists of an optimization problem. In the final part of this section, some examples of application of the Modern Portfolio Theory to non-financial contexts are given, in order to identify key elements that may allow the extension the financial Modern Portfolio Theory to other areas, namely to Marketing. The Sharpe Ratio is also introduced on section 2. The proposed framework is described on section 3 and its proof of concept is conducted on section 5. The results and findings of the proof concept are discussed on section 5 along with the conclusions.

2 Literature Review

2.1 Digital Marketing

In the last twenty years, the world has witnessed an intensive growth in the communication and information technologies [1] [2] [3], particularly with respect to computers, mobile phones, digital televisions and, of course, the internet, which shook the foundations of Marketing and shaped what is now known as the Digital Era [4] [5]. Ryan and Jones [6] proposed the alternative term The Age of Digital Marketing, in line with [7], who state that the concept of Digital Marketing describes the practice and management of Marketing by means of electronic media, such as web, e-mail, interactive television and wireless media, along with digital data about customers' intrinsic characteristics and behaviour.

The technological growth propelled the fragmentation of communication channels, making a lot more difficult the process of selecting an effective means of delivering a Marketing message, that is, an effective communication channel [8]. When going digital, marketers have at its disposal tools like websites, display ads and search ads, search engine optimization (SEO), mobile Marketing, e-mail Marketing, affiliate Marketing and social media Marketing, [9] [10]. Considering there are millions of websites where to place display ads, several search engines where to place search ads and develop SEO strategies and also a plethora of different social media platforms (e.g., blogs and social networks), nowadays there are endless Marketing options available, making it harder for marketers to plan and shape their Marketing strategies.

The performance of a Digital Marketing activity, in general, is quite easy to quantify in terms of Marketing Metrics, thanks to *Web Analytics* tools. Web Analytics services are computer applications designed to collect and aggregate the results achieved by online Marketing campaigns, allowing the study and the subsequent understanding of its impacts. Marketers may analyze what worked and what did not, in the past, and then identify potential areas for improvement concerning future campaigns [12].

Key performance indicators to be used could be the number of clicks on an ad, the number of impressions (or views) an ad gets, the number of site visitors (often expressed in terms of the *clickthrough rate*, CTR), conversion rates (namely *clicks to leads* and *leads to sales* rates), the Cost per Click (CPC), the Cost per Mille[1] (CPM), the Cost per Action (CPA) – whether a lead (such as a file download, a newsletter subscription or a form filling) or a sale –, the number of sales and also the sales revenue [7] [12] [13].

2.2 Modern Portfolio Theory

Markowitz [14] proved, that the Expected Returns-Variance of Returns (E-V), since it reflects both the expected return on investment and its variance, which illustrates the uncertainty or risk associated with that return leads to optimal (or efficient) portfolios, in terms of the risk-return compromise, that is, it leads to the portfolios that, for a given level of risk, maximize the expected return or that, conversely, for a given level of return, ensure the lowest possible risk [15]. Since this rule is based on the principle of diversification, Markowitz's study shows that, from investing in a diversified set of financial assets, it is possible to maximize the return on a portfolio while controlling the risk it holds [16] [17] [18].

This conclusion from Markowitz further developed and explored in Markowitz [19], gave rise to the Modern Portfolio Theory [20], which remains since then as the dominant theory in the field of portfolio selection. The most relevant finding is the fact that in the presence of uncertainty (i.e., risk), assets should not be selected only on the basis of its unique features alone; rather, the investor should consider how each of the assets moves in terms of yields over time, comparatively to all other; that is, the investor should analyze how the assets' yields are correlated.

In the analysis of investment projects, the concept of risk is related to the investment's uncertainty and it can be regarded as the likelihood of obtaining from that investment a return other than the expected, whether by excess or by default. The portfolio's Overall Risk is the risk inherent to the universe of financial assets that may become part of the portfolio and it can be broken down into two major components: the Market Risk and the Specific Risk [21].

Also known as Systematic Risk, the Market Risk relates to the price sensitivity to market changes and depicts the possibility of an investor incurring losses due to factors that affect the performance of all financial markets, such as economic cycles, terrorist attacks, and changes in the inflation rate and in interest rates. This component of the overall risk cannot be eliminated through diversification.

Conversely, the Specific Risk portrays the factors affecting the performance of a specific title (or set of titles), company or industry, in particular. Being directly associated with some title, company or industry, the Specific Risk does not affect the whole lot of securities being traded in the market, being however present in all investments involving such titles, companies or industries. Price variations regarding the Specific Risk may be motivated by (1) the introduction of new products in the market with returns below expectations, (2) an unexpected increase in the number of

[1] Cost per Thousand impressions.

competitors or (3) an organizational change, for instance, thus representing a risk to which only a certain title, company or industry is subject to [21].

Simultaneously holding titles from different companies and different industries would minimize an investor's exposure to the negative consequences of events or decisions that may affect the performance of a specific company or industry. In that sense, the Specific Risk can be reduced by means of a diversification of investment, that is, through the combination of several diverse investments in a portfolio, being therefore also referred to in the literature as Diversifiable Risk. Being part of the Overall Risk, the reduction of the Specific Risk leads to the reduction of the investment's total risk. Diversification of investment is then, building up a portfolio with several titles from different natures and origins. In other words, it is the distribution of an investment by a set of distinct assets, allocating to each one a portion (not necessarily uniform) of an available budget [14] [19]. The principle of diversification is hence used due to its Specific Risk reduction capability: when a certain asset, Asset A, gets a worse return than its expected return, it is likely that there is some second asset, Asset B, which will have a better performance than expected. In his pioneering work, Markowitz proved that investing in two assets with equal variance reduces the overall variance of the investment and emphasize that the variance of a diversified portfolio may even be less than the variance of the asset (pertaining to the portfolio) with lower associated risk. Sharpe [21] states that, in general, the greater the number of assets in a portfolio, the greater the likelihood of good performances offset poor performances.

A risky asset is a financial asset whose future return is uncertain. Given this uncertainty, it is considered that the return of a risky Asset i (R_i) is a random variable that can assume a n finite number of values. When there is a history of the past n returns of that financial Asset i, for n periods of time, the expected return of the asset may be given by the mean value of the history data, that is, by the sum of all $R_i(t)$ divided by the total number of periods, n. Under the principle of diversification introduced earlier, the portfolio is a weighted combination of n financial assets with returns R_i. Accordingly, the portfolio return (R_p) will be given by the weighted sum of the returns generated by each of those assets, R_i, with the respective weights, ω_i.

Considering that the expected return of an asset is equal to the average value of its past historical data (that is, making $E(R_i) = \mu_i$), the expected return of the portfolio can be written up as:

$$E(R_p) = \sum_{i=1}^{n} \omega_i \, \mu_i \tag{1}$$

The risk of a financial asset, in turn, is determined through statistical techniques that measure the dispersion of its returns around the expected value. The dominant dispersion measure in the context of portfolio analysis and selection is the variance or, equivalently, the standard deviation [14] [19]. Thus, each financial asset can be characterized in terms of its expected return (mean) and its risk (variance or standard deviation). The risk of a portfolio is related to the risk of the assets it encompasses [21]. However, despite what happens with the expected return, the portfolio risk cannot be determined by a simple weighted sum of the individual risks, since it is also a function of the correlation coefficients and the covariance of each pair of financial instruments [14] [18].

Covariance is the statistical measure of how two random variables, such as the returns of two assets i and j, move together [22]. The covariance of the returns yielded by the assets i and j is represented by σ_{ij} and can be expressed in terms of the correlation coefficient of the assets (ρ_{ij}) and also of their standard deviations (σ_i and σ_j).

Since one of Markowitz's (1952) assumptions states that short-selling is not allowed (i.e., it is not possible to divest), it shall be taken as a constraint that $\omega_i \geq 0$, which has the mathematical formulation of the optimization problem:

Minimize

$$\sigma_P^2 = \sum_{i=1}^{n} \sum_{j=1}^{n} \omega_i \, \omega_j \, \sigma_{ij} \quad (2)$$

Subject to:

$$E(R_P) = \sum_{i=1}^{n} \omega_i \, \mu_i = K$$

$$\sum_{i=1}^{n} \omega_i = 1$$

$$\omega_i \geq 0 \; , \quad i = 1, \dots, n$$

With n financial assets it is possible to build a multitude of portfolios, by varying the weighting ω_i of the assets composing the portfolio [22]. The set of possible combinations of Expected Return and Risk for that limited number of assets (i.e., the portfolios that can be built) is called Feasible Portfolio Set, which represents the locus, within the return-risk space, of all admissible solutions to the Markowitz's optimization problem [16] [20].

2.3 Applications of Modern Portfolio Theory to Marketing

The research on the application of Modern Portfolio Theory within Marketing has been limited [23]. The theory has already been applied, nevertheless, to the formation of efficient portfolios of product lines, customer segments, retail formats and price promotions. Noteworthy are the works of [25], [23]), [24] and [26]. Cardozo and Smith [25] conducted the first formal study on the feasibility of application of Modern Portfolio Theory to Marketing decisions, particularly in terms of product portfolio. It was found that Markowitz's theory has a great potential as an analytical and planning tool for product decisions. Ryals [23] studied its application to customer or customer segment portfolios, using the Customer Lifetime Value as measure of return. In his study, Ryals states that the application of portfolio theory to Marketing has some limitations, such as (1) the possibility of the proposed solution causing interference with strategic issues of the company, (2) the complex (non-statistical[2]) correlation between clients and (3) the independence and freedom of clients in taking action (customers can choose to abandon a relationship, for instance).

Ryals [24] study the applicability of Markowitz's theory to Marketing portfolios, in general, on the premise that the allocation decisions of Marketing expense can be seen

[2] Relations that statistics, with its correlation coefficients, cannot explain or predict. Word-of-Mouth Marketing and the impact it has on clients is a fine example (Ryals, 2003).

as portfolio investment decisions. Nonetheless, the authors alert to, contrarily to financial instruments and portfolios, the investment in Marketing assets tends to affect their returns. In that sense, they propose some changes to Markowitz's framework before applying it to Marketing. According to these authors, the relationship between returns and Marketing expenses can be determined by statistical techniques to estimate the response of sales to Marketing stimuli. Knowing that a sales-response curve is typically S-shaped (meaning that reduced levels of Marketing expenditure do not have a significant impact on sales and that excessively high levels, beyond a certain amount, have a progressively decreasing effect over customer response), [24] proposed a modification to the original Modern Portfolio Theory framework, defining the portfolio's expected return on investment as a function of the investment on segment *i* and of its expected return.

Finally, [26] address the application of Modern Portfolio Theory to price promotions portfolios, equally emphasizing the limitations of applying Markowitz's theory to Marketing. Although the authors consider that the solution proposed by [24] on the response curve is an elegant solution, they argue that the introduction theoretical curves moves the study away from the pure empiricism that characterizes it. Moreover, [26]. underline the absence of a benchmark market or index with which to compare the performances of the portfolios, which compromises the use of the Sharpe Ratio. In response, the authors present an adaption to Sharpe Ratio's mathematical expression, eliminating from the numerator the term regarding the average risk-free interest rate, $E(R_f)$, leading to

$$\mathrm{SR}_P = \frac{E(R_P)}{\sigma_P} \tag{3}$$

The previous examples validate the hypothesis of using the mathematical concepts of Markowitz's [14]) framework in a variety of contexts and thus an application of Modern Portfolio Theory to the selection of Digital Marketing channels may now be proposed.

3 Proposed Framework

Just as in the previous examples, the application of Modern Portfolio Theory to the selection of Digital Marketing channels requires taking into consideration the main differences between the areas of Finance and Marketing, along with the difficulties in understanding and measuring Marketing's financial return. In this sense, some aspects of Markowitz's theory needed to be modified, namely the universe of *instruments* or *assets* to invest in, the measures of return to be used, the concepts of risk and diversification respecting a Digital Marketing environment, and finally the mathematical formulation of the optimization problem itself.

3.1 Assets

As instruments of the proposed framework, analogous to the financial assets in Harry Markowitz's original model, we shall have the different Digital Marketing channels –

websites, search engines (e.g., Google and Yahoo!), search ads, display ads, the e-mail, mobile devices and social media (such as forums, blogs, Facebook, Twitter, YouTube, LinkedIn, Pinterest, Tumblr or Instagram) – and tactics, which should be understood as specific ways of using some Digital Marketing channel to promote an advertising campaign. Consider Google Search and Google Ads, for instance; Google Search and Google Ads are two different tactics since they represent the use of Google online advertising on a *search* and on a *display* ad format, respectively. The Digital Marketing channels and tactics are hence the targets of analysis and selection within the proposed framework.

3.2 Returns

For measuring the return of our *assets*, as defined above, we shall use Marketing Metrics, using to that end the Web Analytics services. These metrics' suitability for the performance analysis will depend on the companies' objectives, which may be raising awareness, enhancing sales or building customer relationships (Chaffey *et al.*, 2006; Lindon *et al.*, 2009). Accordingly, the choice of the most appropriate return measure will be on the company, in line with its preferences. It may be the number of Clicks, the number of Impressions, the Clickthrough Rate, the conversion rates *Clicks into Leads* and *Leads into Sales*, the Cost per Click (CPC), the Cost per Mille (CPM), Cost per Action (CPA), among others.

Given the influence of Marketing expenditure (investment) over the returns, as reported by [24] and [26], it will be interesting to analyze, for each channel and each portfolio, the Marketing performance per amount invested. Thus, it is proposed as a measure of return of a channel or tactic *i*, in a past campaign *t*, the performance per currency unit invested:

$$R_i(t) = \frac{Marketing\ Performance_t}{Amount\ Invested_t} \tag{4}$$

In order to calculate the expected return of a channel or tactic *i* it is necessary to calculate the returns of all its historical records *t*, according to Equation 10, and then calculate the average of those returns, as shown in Equation 1. The portfolio's expected return will be then given by the weighted sum of the expected returns of the *n* channels (or tactics) with the respective weights ω_i, in line with Equation 1.

3.3 Risks

As in Markowitz's original model, the risk of a Digital Marketing channel or tactic will be expressed by the standard deviation of the return. However, in this context it will portray the risk inherent to the effectiveness of a channel or tactic[3] in delivering the intended message or in encouraging a specific behavior among customers and prospects. The ultimate goal of marketers should be to minimize this risk, opting for the most effective combination of channels or tactics.

---

[3] Or set of channels and tactics.

Consider, however, that in addition to the channel or tactic, this effectiveness may also depend on several other factors, such as (1) the quality of the advertising message's copy and design, (2) the adequacy of market segmentation, (3) the understanding of target segments' needs and interests, (4) the competition's offerings, (5) the target audience's predisposition to join and get involved with the content being launched and advertised, among others.

In this sense and in line with the definition of risk presented in the previous section, these risk factors might be categorized as pertaining to Market Risk or to Specific Risk. Risk factors related to the behavior of customers and to the competition affect all marketers and their advertising the same way and cannot be eliminated through diversification. Hence, from the conceptual viewpoint, these risk factors should be incorporated within the Market Risk. Conversely, following the same line of reasoning, those concerning the quality of the communication message's copy and design, the adequacy of target audience segmentation and the level of understanding of their needs and interests shall be considered as part of the Specific Risk, since they only affect a finite set of *assets*, such as an unattractive message or an advertising campaign supported by a poor market segmentation, for instance. This sort of risk factors, nevertheless, shall be taken as being constant and common to the whole universe of assets, as an assumption, not influencing the Marketing performance of any channel or tactic.

3.4 Diversification

In this framework, the diversification depicts the allocation of a limited budget to a set of channels or tactics to compose a Digital Marketing communications portfolio. It is assumed that the marketer has at his disposal a set of channels or tactics he can select, whose past returns, in terms of Marketing performance and associated investment, are known. Since the diversification of an investment allows to reduce the overall risk of a portfolio, for a given level of return, the apportioning of the Marketing expenditures across multiple channels or multiple tactics can, for a given level of return, minimize the overall risk of the Digital Marketing's channels portfolio, thus increasing its risk-adjusted return.

4 Proof of Concept and Validation

The proposed framework was empirically tested using data from a Portuguese Digital Marketing agency. This data concerns to three different Digital Marketing tactics used in a 2014's monthly online advertising campaign, for a period of six months, consisting of their monthly Marketing performance (according to digital Marketing Metrics such as impressions, generated leads and sales), and respective monthly investment. For the purpose of proof of concept, the analysis of portfolios was conducted having set as Marketing objective the maximization of brand awareness (notoriety). In line with that choice, the *number of impressions* was chosen as the metric for measuring the tactics' Marketing performance.

In its monthly campaigns, the digital marketing agency employed as Digital Marketing tactics Google Search, Google Display and Facebook Advertising. These tactics are associated with different returns and costs, which become known and accessible at the end of the month. Therefore, it was possible to obtain from the company's Web Analytics service the investment amounts and the resulting number of impressions, from January to June, for each tactic.

Identifying in the *data spreadsheet* the channels or tactics to be analyzed and inserting the inputs, it is possible to compute the portfolios of two and three channels that will best suit the marketers' objectives and interests, from the perspective of optimizing the risk-adjusted returns.

As a first step, the model computes the return of each tactic in each monthly campaign, by dividing the number of achieved impressions for the amount invested, according to Equation 10. Shortly after, these monthly returns of tactics are processed on the 2-channel and the *n*-channel sections using Equation 1, thus obtaining the expected return[4] (mean value) and risk (standard deviation) of each tactic[5] (Table 3).

In addition to calculating these tactics' expected returns and risks, covariance and correlation matrices are also automatically calculated, which allows to study and understand the combined behavior of each pair of tactics. The returns' covariance matrix is part of the results Summary Table for both 2 and *n* channels, while the correlations matrix is presented alone.

Table 1. Tactics' Expected Returns and Risk

Tactic	**Google Search**	**Google Display**	**Facebook**
Campaign	**Return (impressions/€)**	**Return (impressions/€)**	**Return (impressions/€)**
January	160,4	5 603,3	2 253,8
February	92,7	14 192,3	3 897,5
March	27,7	1 907,8	1 705,0
April	39,0	5 061,9	4 161,4
May	75,2	4 246,9	3 208,3
June	162,8	5 611,1	3 044,7
$E(R_i)$	**93,0**	**6 103,9**	**3 045,1**
Risk	**58,2**	**4 196,6**	**940,4**

The Excel model will then determine the Feasible Portfolio Set and the Efficient Frontier for the 2-channel and the *n*-channel scenarios, respectively. For the former scenario, it calculates the expected return and risk of the portfolios composed of varying investment weighting coefficients, $\omega_{\text{Google Search}}$ and $\omega_{\text{Google Display}}$, as previously explained, while for the latter it solves a series of different optimization problems,

[4] Hereby represented by $E(R_i)$.

[5] By definition, the 2-channel section presents the information alluding to the first two channels or tactics inserted in the data spreadsheet (in the shown example, these are Google Search and Google Display).

providing a new solution (efficient portfolio) every time a new value of desired return, *K*, is introduced. For each scenario, the model will calculate the Return-Risk Ratio of all the computed portfolios, identifying with two different colors the Minimum Variance Portfolio and the portfolio of maximum Return-Risk Ratio.

Afterwards, the Excel model uses this information to map in the space of Mean-Variance the Feasible Portfolio Set for the pair of tactics Google Search and Google Display, plotting the portfolios according to their Risk-Expected Return coordinates.

Table 2. Correlations Matriz for 3 Tactics

Tactic	Google Search	Google Display	Facebook
Google Search	1	0,240	-0,107
Google Display	0,240	1	0,580
Facebook	-0,107	0,580	1

Table 3. Summary of 3 Tactics

Tactic	Expected Return	Risk	Covariance Matrix		
Google Search	93,0	58,2	3 385	58 505	- 5861
Google Display	6 103,9	4 196,6	58 505	17 611 337	2 289 637
Facebook	3 045,1	940,4	-5 861	2 289 637	884 304

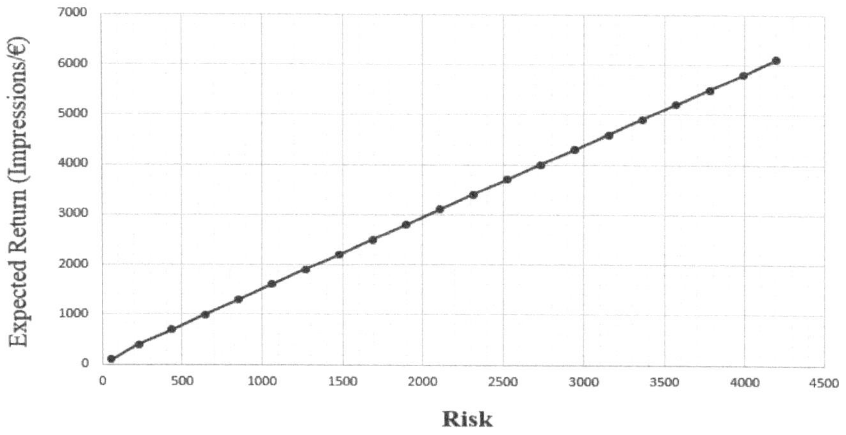

Fig. 1. Feasible Portfolio Set for 2 Tactics

Similarly to the analysis of portfolios of two tactics, the Excel model processes the resulting portfolios' risk and expected return and draws the Efficient Frontier in the space of Mean-Variance (Fig. 2):

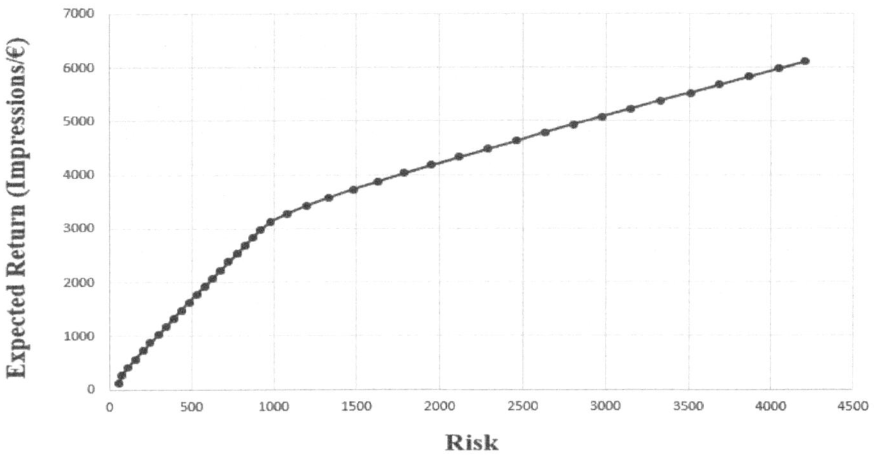

Fig. 2. Feasible Portfolio Set for 3 Tactics

Finally, setting in the 2-channel and in the *n*-channel sections the communication budget, the model suggests an amount to allocate to each of the analyzed tactics, depending on the marketer's degree of risk aversion. As predefined solutions, the model proposes the investment in the portfolio of highest risk-adjusted return (i.e., with highest Return-risk Ratio) or in the Minimum Variance Portfolio, also offering the user the chance to manually select any portfolio of the Efficient Frontier, thus taking advantage of marketers' deep market understanding and intuition.

5 Conclusions

This paper focused on applying Harry Markowitz's Modern Portfolio Theory to Digital Marketing investment decisions. The authors have adapted the original model and highlight the use of non-financial expected returns (e.g., risk premium and generated power) and the adaptation of the Sharpe Ratio, as it excludes the term referring to the expected return on risk-free assets. The model can be applied to the selection of channels or Digital Marketing tactics, using metrics such as return of investment performance indicators of Web Analytics by monetary unit invested (eg, number of impressions per Euro invested) and which includes the conversion of the Sharpe ratio in a simple ratio between return and risk of a portfolio, referred to as work-risk ratio return.

The proof of concept and validation model performed using the results of an online advertising of a Digital Marketing agency campaign, over a period of six months. It was found that the returns and risks of efficient portfolios varied uniformly (ie, higher expected return portfolios have higher associated risk and lower expected return portfolios have a lower risk), a result that is consistent with the assumptions established by Harry Markowitz (1952, 1959), the rational investor perspective. This reinforces the belief that Modern Portfolio Theory can be used in the rational selection of channels portfolios or tactics of Digital Marketing and is therefore considered as the most important result of the conducted proof of concept.

The proposed model in Excel enabled the calculation of Efficient Frontier to a scenario of two and three tactics, thus enabling marketers to optimize their spending Marketing by selecting a portfolio of Efficient Frontier that suits the level of risk they are willing to accept. However, it was found that, for the data used, that the correlation coefficients were partially or positive (on the order of 0.5) or close to zero, which implies, for the particular case study, of the ineffectiveness in diversifying the overall risk of the portfolio. The Variance Portfolio Minimum (efficient portfolio with lower risk) was, for both scenarios, the resulting portfolio allocation of the total budget available to the tactical Google Search. Thus, it was not substantiated by proof of concept that investment diversification leads to minimizing the overall risk of the portfolio. The performance of a new proof of concept with a larger sample could lead to more conclusive results.

References

1. Saxton, J.: New Media: The Growth of the Internet, Digital Television and Mobile Telephony and Implications for Not-for-Profit Marketing. International Journal of Nonprofit and Voluntary Sector Marketing 6(4), 347–363 (2001)
2. Constantinides, E.: The Marketing-Mix Revisited: Towards the 21st Century Marketing. Journal of Marketing Management 22(3-4), 407–438 (2006)
3. Lindon, D., Lendrevie, J., Rodrigues, J.V., Lévi, J., Dionísio, P.: Mercator XXI: Teoria e Prática do Marketing. Dom Quixote, Alfragide (2009)
4. Wind, J., Mahajan, V.: Digital Marketing. Symphonya. Emerging Issues in Management 1, 43–54 (2002), http://dx.doi.org/10.4468/2002.1.04wind.mahajan
5. Agarwal, R., Gupta, A.K., Kraut, R.: The Interplay Between Digital and Social Networks. Information Systems Research 19(3), 243–252 (2008)
6. Ryan, D., Jones, C.: Understanding Digital Marketing: Marketing Strategies for Engaging the Digital Generation. Kogan Page Limited, London (2009)
7. Chaffey, D., Ellis-Chadwick, F., Johnston, K., Mayer, R.: Internet Marketing: Strategy, Implementation and Practice, 3rd edn. Prentice Hall, Harlow (2006)
8. Kotler, P., Keller, K.L.: Marketing Management, 14th edn. Prentice Hall, Upper Saddler River (2012)
9. Kotler, P., Armstrong, G.: Principles of Marketing, 14th edn. Prentice Hall, Upper Saddler River (2012)
10. Eley, B., Tilley, S.: Online Marketing Inside Out. SitePoint Pty. Ltd., Victoria (2009)
11. Lindon, D., Lendrevie, J., Rodrigues, J.V., Lévi, J., Dionísio, P.: Mercator XXI: Teoria e Prática do Marketing. Dom Quixote, Alfragide (2009)
12. Dionísio, P., Rodrigues, J.V., Faria, H., Canhoto, R., Nunes, R.C.: b-Mercator: Blended Marketing, 2nd edn. Dom Quixote, Alfragide (2011)
13. Ascensão, C.P.: Google Marketing. Edições Sílabo, Lisboa (2011)
14. Markowitz, H.M.: Portfolio Selection. The Journal of Finance 7(1), 77–91 (1952)
15. Elton, E.J., Gruber, M.J.: Modern Portfolio Theory, 1950 to Date. Journal of Banking & Finance 21(11-12), 1743–1759 (1997)

16. Zhang, D., Lu, J.: Batch-Mode Computational Advertising based on Modern Portfolio Theory. In: Azzopardi, L., Kazai, G., Robertson, S., Rüger, S., Shokouhi, M., Song, D., Yilmaz, E. (eds.) ICTIR 2009. LNCS, vol. 5766, pp. 380–383. Springer, Heidelberg (2009)
17. Wu, M., Kong, D.-W., Xu, J.-P., Huang, N.-J.: On Interval Portfolio Selection Problem. Fuzzy Optimization and Decision Making 12(3), 289–304 (2013)
18. Paydar, Z., Qureshi, M.E.: Irrigation Water Management in Uncertain Conditions - Application of Modern Portfolio Theory. Agricultural Water Management 115(2), 47–54 (2012)
19. Markowitz, H.M.: Portfolio Selection: Efficient Diversification of Investments. John Wiley & Sons, Inc., New York (1959)
20. Elton, E.J., Gruber, M.J., Brown, S.J., Goetzmann, W.N.: Modern Portfolio Theory and Investment Analysis, 9th edn. John Wiley & Sons, Inc., New Jersey (2014)
21. Sharpe, W.F.: Risk, Market Sensitivity, and Diversification. Financial Analysts Journal, 84–88 (1995)
22. Sharpe, W.F., Alexander, G.J., Bailey, J.W.: Investments, 6th edn. Prentice Hall, Upper Saddle River (1998)
23. Ryals, L.: Making Customers Pay: Measuring and Managing Customer Risk and Returns. Journal of Strategic Marketing 11(3), 165–175 (2003)
24. Ryals, L., Dias, S., Berger, M.: Optimising Marketing Spend: Return Maximisation and Risk Minimisation in the Marketing Portfolio. Journal of Marketing Management 23(9-10), 991–1011 (2007)
25. Cardozo, R.N., Smith, D.K.: Applying Financial Portfolio Theory to Product Decisions: An Empirical Study. Journal of Marketing 47(2), 110–119 (1983)
26. Franco-Laverde, J., Littlewood, A., Ellis, C., Schraner, I., Varua, M.-E.: FMCG Portfolio Budget Allocation to Price Promotions using Modern Portfolio Theory (MPT). International Review of Business Research Papers 8(5), 16–30 (2012)

The Effect of Stockout Cause and Brand Equity on Consumer Preference in Online Retailing

Jun Ding[1], Qiang Lu[1], and Xianghua Chu[2,*]

[1] Harbin Institute of Technology Shenzhen Graduate School, Shenzhen, China
{haveding,qiang.lu.home}@gmail.com
[2] College of Management, Shenzhen University, Shenzhen, China
xianghua.chu@gmail.com

Abstract. Customers can browse the information of stockout items in online retailing, thus change their preferences for the stockout product. This study tests the hypothesis that brand equity moderates the effect of causes of stockout on consumer preference for the online stockout product. The results reveal that for high-equity branded products, consumers exhibit a slit decrease in preference for the product whether the product is out-of-stock due to heavy demand or short supply. By comparison, for low-equity branded products, consumers increase preference for the product if the product is out-of-stock due to excessive demand and decrease preference for the product if stockout cause is supply-based. These findings highlight the importance of taking the product's brand equity into consideration when stating different kinds of causes for stockout to consumers.

Keywords: Stockout cause, brand equity, consumer preference, online retailing.

1 Introduction

When on-shelf-availability (OSA) is seen from service-dominated logic perspective, only when consumers can select a product from the shelf can the retail supply chain create value [12]. Although high-tech supply chain management software and innovative inventory management theories are implemented today, Stock still remains a recurring problem for retailers and manufactures [1]. Therefore, research on stockout has been gaining much attention from academy and practitioners in recent years.

Stock would reduce profits because consumers would switch to other stores to purchase the product or drop the product [3,31]. Stockout causes negative emotion that negatively affects store image and decision satisfaction, decrease behavior intention [18]. Stockout also makes available alternatives seen more attractive, and accelerates consumer's purchase decision [16,19]. Although previous studies showed that frequent stockout decrease consumer preference for the stockout product if consumer's purchase interval is short [6], empirical study on the effect of stockout on consumer preference change is still lack.

* Corresponding author.

M. Gen et al. (eds.), *Industrial Engineering, Management Science and Applications 2015*,
Lecture Notes in Electrical Engineering 349, DOI: 10.1007/978-3-662-47200-2_63

With regard to causes for stockout, generally stockout happens when demand excels supply. In practice it is common to find products that are out-of-stock because vendor limits the supply, such as limited-quantity promotion [2] and limited-edition product. In line with previous studies, we classify stockout causes into supply-based and demand-based [20,21]. Supply-based stockout is caused by limited quantity that vendor wants to sell, while demand-based stockout is caused by excess quantity that consumers want to buy.

Moreover, consumer-based brand equity plays an important role in forming consumer preference [7]. Band equity is "the differential effect of brand knowledge on consumer response to the marketing of the brand" [17]. Brand equity positively affects consumer preference and purchase intention [9]. We discriminate products between high-equity branded products and low-equity branded products. Consumers that are more familiar with high-equity brand are more loyal to high-equity brand [4]. Generally speaking, consumers consider high-branded products to be of high quality [11] and are willing to pay more [13,25]. Therefore, we expect brand equity would affect consumer preference in stockout situation.

This study extends current understanding of the impact of stockout on consumer preference for the stockout product in online retailing. We combine the causes for stockout (i.e., stockout due to supply side vs. stockout due to demand side) and product's brand equity (i.e., high-equity brand vs. low-equity brand). We find that limited supply stockout would overall decrease consumer preference, and excess demand stockout would increase consumer preference only when the stockout product's brand equity is low.

This paper is organized as follows. First, we briefly discuss previous research concerning brand equity and causes for stockout on consumer preference, and present our hypotheses. Second, we continue with the description of research methodology and empirical results. Then, we end with directions for future research and a discussion of managerial implication.

2 Conceptual Foundations

2.1 Effect of Stockout on Consumer Preference

Stockout have mixed effects on consumer preference for the product. On one hand, stockout situation, which signifies the scarcity of the product, would increase consumer preference [2,26]. Consumers would engage in "bandwagon reasoning", using perceptions of relative demand to assess a product's worth, inferring that other consumers' demand implies the value [28]. Product can also be used to satisfy consumers' social needs [29]. For example, jewelries and expensive automobiles are used to demonstrate social status.

One the other hand, because consumers cannot buy the product immediately, the stockout situation doesn't meet consumers' expectation about the availability of the product. Consumers would be dissatisfied [18], especially when consumer wants to buy the product right away [15].

2.2 Supply-Based vs. Demand-Based Stockout

Stockout situation generally comes from consumers' demand exceeding vendor's supply. When considering the "scarcity effect" of stockout, the effect of demand scarcity is driven by a "social-proof mechanism," in which consumers rely on others' opinions as a clue to the value of a product [8]. Consumers also engage in "bandwagon reasoning," using perceptions of relative demand to assess a product's worth, inferring that other consumers' demand implies the value [28]. Behind this consideration is the notion that a product that has become scarce through high demand must be a good product, which may induce them to follow the lead of others. Supply-based scarcity doesn't supply information about quality of the product.

According to "locus of causality" theory, the locus of causality is regarded as external if one attributes a phenomenon to factors lie outside the individual, and internal if factor is located within oneself [24]. Thus, when consumers are informed that the stockout cause is limited supply, consumer would be likely to attribute the stockout situation for external cause. Thus the negative emotion stirred from stockout situation would decrease consumer preference for the product supplied by this supplier. When consumers are informed that the stockout cause is excess demand, consumers would attribute the cause for stockout to other consumers. Therefore, supply-based stockout decrease consumer preference, while demand-based stockout would increase consumer preference.

H1. Preference for the stockout product is higher when consumers are told that the stockout cause is excessive demand than limited supply.

2.3 Effect of Brand Equity

From consumer perspective, brand equity is conceptualized and measured base on each consumer, and denotes the added value endowed by the brand to the product from a cognitive psychology approach [22]. Some brands are with high equity while other brands are with relatively low equity [5]. People are familiar with and can easily recognize high-equity brands, like Coca-Cola and Pepsi, in a store or on the street [4]. When they are considering which to buy, they would be more likely to take those high-equity branded product into consideration. Thus, when high-branded product is out-of-stock, consumers would be more dissatisfied.

Consumers link high quality with high-equity branded products and expect to get better service [27]. The reality of stockout situation contradicts with consumer's expectation, and cause negative emotions [23]. Low-equity branded products do not offer as many benefits as high-equity brands, and are brought mainly for their low price. The stockout of low-equity branded products would not cause as much negative emotion as high-branded product.

Furthermore, based on consumers' experience, high-equity brands, such as Channel and Rolex, usually issue limited-edition product to enhance the uniqueness of the product, while low-equity brands seldom do. Supply-based cause for stockout of high-equity brand products seems more reasonable.

When stating the demand side stockout cause, on one side, reinsures consumers of the popularity of stockout product, which would increase consumer preference. On the other side, stockout of high-equity brand products would give rise to more negative emotion to consumer than low-equity brand.

H2. When cause for stockout is supply-based, the decrease of consumer preference is higher for high-equity branded product than low-equity branded product.

H3. When cause for stockout is demand-based, the increase of consumer preference is higher for low-equity branded product than high-equity branded product.

3 Study

3.1 Method

A 3×2 between-subjects design comprised three stockout conditions and two product types. The former were: absent, meaning that the product is in stock; present, demand-based; and present, supply-based. The latter were high-equity branded product and low-equity branded product.

Stockout causes were manipulated experimentally by means scenarios given to the participants, the same as former study [21]. We stated retailer's explanation for supply-based stockout as "Because this product is a limited edition, supply from vendor is limited", and demand-based stockout as "Because this product is very popular, consumers' demand exceed stock". We pre-tested the message on 23 respondents for various kinds of information to make sure that the manipulation was effective in attributing the stockout cause. All participants could correctly identify the stockout cause.

To select the brand equity products of interest, we created a shortlist of ten brands of detergent in China. Then 10 food experts, including four managers and 6 academics, classified those products as high-equity brand or low-equity brand. On the basis the evaluation, we selected one high-equity brand (Blue Moon) and one low-equity brand (Red Ross). Blue Moon is often seen on national TV for its advertise slogan "the leading band of detergent in China", while Red Ross is just a local brand that's never been seem on national TV.

We pre-tested the consumer-based brand equity by using scales derivate from [30]. Respondents' preference for "Blue Moon" brand is higher than "Red rose" brand ($M_{Bluemoon}$ = 3.90, $M_{Redrose}$ = 2.33,F (1,47) =17.8, p<.001).

In all conditions, participants reported their preference for the product in terms of its "attractiveness", in the aspects of bad/good, not nice/nice, and unlikeable/likeable, three seven-point scales adopted from Zhang and Buda (1999), and anchored by 1 = very low and 7 = very high (Cronbach's alpha = .877). This scale was also used in previous studies about consumer preference [21].

Finally, participants responded to manipulation check measures. None of respondents correctly guessed what the study was about, because of the use of between-subject design.

3.2 Participants

Participants were recruited from Internet through the hottest witkey website in China, www.zhubajie.com, on which we posted the recruiting information and a link to this study. This allowed for a mix sample of gender, age, employment, and geographical distribution. One IP address can only participate in the experiment once and all questions must be answered before submission.

Prior to the experiment, participants were asked whether they purchased detergent online and whether they encountered a stockout situation, how long ago did they buy detergent. Respondents that did not buy detergent online, or did not encounter a stockout, or did not buy detergent in six months were screened out. Participants were randomly assigned to one of the experimental scenarios.

3.3 Procedure

The written scenarios instructed participants to imagine being surfing an online retail website to buy detergent. Then participants were shown with the picture and description of the detergent to simulate the way that products are presented on the Internet. In the stockout condition, they were also told that the product they want to buy is out-of-stock due to limited supply or heavy demand according to which group participants were assigned to. After reading the scenario, participants answered questions concerning their preference for the product and manipulation check. At last, they provided information of gender, age and education background for demographic description. In order to increase participants' involvement to the experiment, we promised to choose two participants as lucky men, and each could receive the corresponding product as a prize.

In two weeks, 194 participants took part in the experiment. 46.4% participants were men; 53.4% participants were women. 31.3% participants were between 18 and 24. 47.4% participants were between 25 and 30. 21.3% participants were older than 30. 8% participants' education background was senior high school or technical secondary school. 37% participants' education background was junior college. 55% participants' education background were college and above.

3.4 Results

To verify the manipulation of stockout cause had been successful, participants were asked to identify whether the stockout occurrence they encountered was caused by limited supply or heavy demand. The outcome showed that all participants in supply-based or demand-based stockout condition correctly identified the type of stockout. We pre-tested the brand equity instead of asking participants to access consumer-based brand equity in the main study to prevent participants form scrutinizing the experiment's purpose.

A 2×2 ANOVA was performed on consumer preference with experimental group as independent variable. The main effects of both the brand equity (F (1, 194) = 31.175, p<.001, η^2= .142) and stockout cause (F (2, 194) = 3.517, p<.005, η^2 =.036) were significant. Participants in the excessive demand stockout group exhibited high preference for the stockout product that those in the limited supply group. H1 is supported. As expected, the interaction effect between cause for stockout and brand equity was significant (F (2, 194) = 2.171, p<.1, η^2=.026). Fig. 1 shows the mean of consumer preference in each condition.

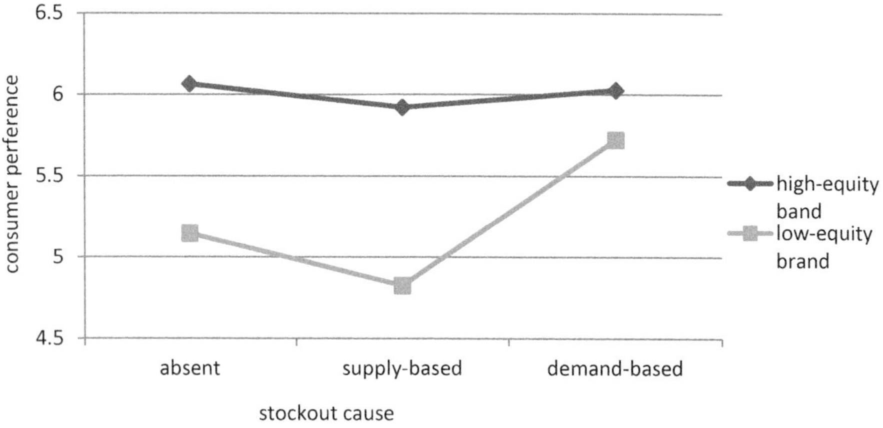

Fig. 1. Interaction Effect: Stockout Cause × Brand Equity

We then calculated the difference of consumer preference between the situations of stockout and absent of stockout. Table 2 showed the mean of consumer preference in each condition and mean difference between out-of-stock condition and corresponding in stock condition. Contrast analysis found that brand equity had no significant effect on consumer preference change when supply-based stockout cause was exhibited (M_{high}=-.0945, M_{low}=-.3485, F (1, 32) = 4.6, p >.1). H2 was not supported. When demand-based stockout cause was exhibited, the difference of increase of consumer preference between high-equity brand and low-equity brand significant (M_{high} = -.0529, M_{low} = .5707, F (1, 57) = .206, p <.05). H3 was supported.

The findings supported the proposition that the effect of out-of-stock situations on consumer preferences is moderated by consumer-based brand equity. When participants were told that the stockout cause come from the limited supply, they did not exhibit significant difference of decrease preference between high-equity branded products or low-equity branded products. On the other hand, when participants were told that the stockout cause came from excess demand, the increase preference of those shown with the low-equity brand product was higher than those presented with high-equity branded product.

Table 1. Group Means and Comparisons between Experimental Conditions and Control groups

Stockout cause	Brand type	Mean(I)	Mean of control group(J)	Mean difference(I-J)	p
Supply-based	Low-equity	4.8254	5.1473	-.3219	.367
	High-equity	5.9200	6.0652	-.1452	
Demand-based	Low-equity	5.6970	5.1473	.5497	.002
	High-equity	6.0270	6.0652	-.0382	

4 General Discussion

Instead of investigating the effect of stockout cause on consumer preference for available product in brick-and-motor retailing, we exam the effect of stockout cause on consumer preference for stockout product in online retailing. Empirical results show that consumers would generally decrease preference for stockout product of high-equity brand; no matter the stockout cause is supply-based or demand-based. However, consumers would decrease preference for stockout product of low-equity brand if the stockout cause is limited supply, and increase preference for the stockout product of low-equity brand if the stockout cause is excessive demand.

Our results show that in the stockout situation, brand equity and stockout cause would interact to influence consumer preference for the product. On one hand, consumers would decrease preference for the stockout product if they are told that the stockout cause is limited supply. The magnitude of decrease is not significantly different between high-equity brand products and low-equity brand products. On the other hand, consumers would increase preference for the low-equity branded stockout product if they are told that the stockout cause is excessive demand. Previous research just suggested that using the "extreme popular" statement can effectively retain the order [3]. This study shows only when the brand equity of the stockout product is low the positive effect of excess demand stockout cause statement on consumer preference would work.

5 Limitations and Future Research

Although the implications of our findings are discussed, these must be interpreted by taking their limitations into consideration, which would suggest useful directions for future research. First, the product we use in the experiment is detergent, which is inherent a functional product, and not used for conspicuous consumption. Preference for stockout product of categories that are more likely to issue limited edition, such as commemorative stamp and famous fashion designer's product, would be different. Thus, future research could expand the analysis by take product category into consideration to determine the most effective way to explain the stockout cause.

Second, even retailer or manufacture has explained the stockout cause; consumers would reason the real stockout cause. Such as when they can compare the stockout product with available products and find that the stockout product is inferior to available products, while explanation for stockout is excess demand, the explanation is contradicted with consumer reasoning. Future research could expand the analysis to see the effect of contradiction between explanation for stockout and consumer reasoning.

Third, as we've explained in prior section, cause stockout would also signify the scarcity of the product. The different influence of stockout and scarcity information on consumer preference still needs to be clarified, such as the magnitude of perceived scarcity and consumer's different attitude toward stockout and scarcity information.

Finally, stockout happens more frequently to promoted product [10], and the topic of promotional on shelf availability is still immature [14]. Future research also could investigate the effect stockout of a limited-time or limited-quantity promotion on consumer preference. For example, if consumer misses the limited time promotion or limited quantity promotion, would s/he blame self for not take chance of the opportunity or the supplier for limited time or quantity?

6 Managerial Implications

Retailers and manufactures have to face the problem of stockout, especially for high-equity brand providers. They must understand that it's virtually impossible and uneconomical to never run out of stock of any product [1].

Our empirical findings provide suggestions to manufactures and retailers about how to manage stockout. On one hand, stating the supply-based stockout cause is risky, even it's effective to increase consumer preference in some specific occasions. The demand-based for stockout cause is a relatively effective way to increase consumer preference for the stockout product. One the other hand, manufactures and retailers should access the relative position of brand equity in the product category, in order to increase consumer preference for the product when stockout happens.

For high-equity brand provider, the sure way to deal with stockout is to insure that they do not go out-of-stock. For low-equity brand provider, it would be unwise to tell consumer that the stockout is caused by limited supply, and must be more cautious to issue limited edition products. For when stockout happens, consumers would significantly decrease preference the low-equity brand product. The preferable way for low-equity brand provider to explain the stockout cause is to state the excessive demand from other consumers, which would reinsure consumers' popularity of the stockout product.

Although we proved empirically that it's better to inform consumers the excessive demand of stockout product,, but it would be unethical when retailer or manufacture explain the stockout cause as demand-based while the real cause is supply chain inefficiency or limited supply.

Acknowledgments. We acknowledge the financial support of the National Natural Science Foundation of China (NSFC). We also acknowledge the support from Shenzhen Key Laboratory of Urban Planning and Decision Making of Shenzhen Graduate School, Harbin Institute of Technology.

References

1. Aastrup, J., Kotzab, H.: Forty Years of Out-Of-Stock Research–and Shelves are Still Empty. The International Review of Retail, Distribution and Consumer Research 20(1), 147–164 (2010)
2. Aggarwal, P., Jun, S.Y., Huh, J.H.: Scarcity Messages. Journal of Advertising 40(3), 19–30 (2011)
3. Anderson, E.T., Fitzsimons, G.J., Simester, D.: Measuring and Mitigating the Costs of Stockouts. Management Science 52(11), 1751–1763 (2006)
4. Buil, I., de Chernatony, L., Martínez, E.: A Cross-National Validation of the Consumer-Based Brand Equity Scale. Journal of Product & Brand Management 17(6), 384–392 (2008)
5. Chandon, P., Wansink, B., Laurent, G.: A Benefit Congruency Framework of Sales Promotion Effectiveness. Journal of Marketing, 65–81 (2000)
6. Che, H., Chen, X., Chen, Y.: Investigating Effects of Out-Of-Stock On Consumer Stockkeeping Unit Choice. Journal of Marketing Research 49(4), 502–513 (2012)
7. Christodoulides, G., De Chernatony, L.: Consumer-Based Brand Equity Conceptualization and Measurement: A Literature Review. International Journal of Research in Marketing 52(1), 43–66 (2010)
8. Cialdini, R.B., Goldstein, N.J.: Social Influence: Compliance and Conformity. Annu. Rev. Psychol. 55, 591–621 (2004)
9. Cobb-Walgren, C.J., Ruble, C.A., Donthu, N.: Brand Equity, Brand Preference, and Purchase Intent. Journal of Advertising 24(3), 25–40 (1995)
10. Corsten, D., Gruen, T.: Desperately Seeking Shelf Availability: An Examination of the Extent, the Causes, and the Efforts to Address Retail Out-Of-Stocks. International Journal of Retail & Distribution Management 31(12), 605–617 (2003)
11. Dodds, W.B., Monroe, K.B., Grewal, D.: Effects of Price, Brand, and Store Information On Buyers' Product Evaluations. Journal of Marketing Research, 307–319 (1991)
12. Ehrenthal, J.C., Gruen, T.W., Hofstetter, J.S.: Value Attenuation and Retail Out-Of-Stocks–a Service-Dominant Logic Perspective. A Service-Dominant Logic View of Retail On-Shelf Availability, 179 (2014)
13. Erdem, T., Swait, J., Louviere, J.: The Impact of Brand Credibility On Consumer Price Sensitivity. International Journal of Research in Marketing 19(1), 1–19 (2002)
14. Ettouzani, Y., Yates, N., Mena, C.: Examining Retail On Shelf Availability: Promotional Impact and a Call for Research. International Journal of Physical Distribution & Logistics Management 42(3), 213–243 (2012)
15. Fitzsimons, G.J.: Consumer Response to Stockouts. Journal of Consumer Research 27(2), 249–266 (2000)
16. Ge, X., Messinger, P.R., Li, J.: Influence of Soldout Products On Consumer Choice. Journal of Retailing 85(3), 274–287 (2009)
17. Keller, K.L.: Conceptualizing, Measuring, and Managing Customer-Based Brand Equity. The Journal of Marketing, 1–22 (1993)
18. Kim, M., Lennon, S.J.: Consumer Response to Online Apparel Stockouts. Psychology & Marketing 28(2), 115–144 (2011)
19. Kramer, T., Carroll, R.: The Effect of Incidental Out-Of-Stock Options On Preferences. Marketing Letters 20(2), 197–208 (2009)
20. Ku, H.H., Kuo, C.C., Kuo, T.W.: The Effect of Scarcity On the Purchase Intentions of Prevention and Promotion Motivated Consumers. Psychology & Marketing 29(8), 541–548 (2012), doi:10.1002/mar.20541

21. Ku, H., Kuo, C., Fang, W., Yu, Y.: The Impact of Retail Out-Of-Stock Options On Preferences: The Role of Consumers' Desire for Assimilation Versus Differentiation. Marketing Letters, 1–14 (2013)
22. Leone, R.P., Rao, V.R., Keller, K.L., Luo, A.M., McAlister, L., Srivastava, R.: Linking Brand Equity to Customer Equity. Journal of Service Research 9(2), 125–138 (2006)
23. Mandler, G.: Mind and Body: Psychology of Emotion and Stress. WW Norton, New York (1984)
24. Meyer, J.P.: Causal Attribution for Success and Failure: A Multivariate Investigation of Dimensionality, Formation, and Consequences. Journal of Personality and Social Psychology 38(5), 704 (1980)
25. Netemeyer, R.G., Krishnan, B., Pullig, C., Wang, G., Yagci, M., Dean, D., Wirth, F.: Developing and Validating Measures of Facets of Customer-Based Brand Equity. Journal of Business Research 57(2), 209–224 (2004)
26. Parker, J.R., Lehmann, D.R.: When Shelf-Based Scarcity Impacts Consumer Preferences. Journal of Retailing 87(2), 142–155 (2011)
27. Rungtrakulchai, R.: The Relationship Between Price Deals. Perceived Quality, and Brand Equity for a High Involvement Product. AU Journal of Management 11(2) (2013)
28. Van Herpen, E., Pieters, R., Zeelenberg, M.: When Demand Accelerates Demand: Trailing the Bandwagon. Journal of Consumer Psychology 19(3), 302–312 (2009)
29. Vigneron, F., Johnson, L.W.: Measuring Perceptions of Brand Luxury. The Journal of Brand Management 11(6), 484–506 (2004)
30. Yoo, B., Donthu, N.: Developing and Validating a Multidimensional Consumer-Based Brand Equity Scale. Journal of Business Research 52(1), 1–14 (2001)
31. Zinn, W., Liu, P.C.: A Comparison of Actual and Intended Consumer Behavior in Response to Retail Stockouts. Journal of Business Logistics 29(2), 141–159 (2008)

Development of a Remote Controlled Mobile Robot for Toy Application Using RF Module in PIC Microcontroller

Rionel Belen Caldo[1,*], Donabel D. Abuan[1], and Elmer P. Dadios[2]

[1] Electronics and Communications Engineering Department
De La Salle University (DLSU), Manila, Philippines
[2] Manufacturing Engineering and Management Department
De La Salle University (DLSU), Manila, Philippines
rionel14_caldo@yahoo.com

Abstract. Designed, developed and presented in this paper is a remote controlled mobile robot for toy application using RF module in PIC microcontroller. The goal of this study is to provide a reliable way of communication over a Wi-Fi network for mobile robot using Radio Frequency (RF) wireless communication module. The proponents make use of model-based task allocation method. This effective task allocation method is made possible thru central commanding computer, which realizes reliable operation for identified tasks. The proponents built a mobile robot, with capability of responding to three simple movements: forward, turn left and turn right. The proponents make use of PIC microcontroller as its control unit and it was programmed using C++. The mobile robot will communicate with a central commanding server, which is to be controlled over RF wireless communication. Through a local area network, the computer will send the command movements to the mobile robot using RF module.

Keywords: Mobile robot, model-based task allocation method, PIC microcontroller, RF wireless communication.

1 Introduction

Undeniably, the introduction and application of robots had opened and widened horizons of advancement in varied fields. It allowed human beings to perform their tasks quickly, conveniently and safely. It then led to apprehending demand for robotics application (Y. Ota et.al, 2001). High ground adaptability is one important performance measure for the robot. In this criterion, the robot must be able to move on any uneven environment. Robot designers desired to have their robots able to move over much the same ground as humans are able to. It can be noted that robots need not only mobility, but also task-performing and supporting abilities for other tasks (Y. Ota et.al, 2001). In the field of multi-robot system, the system should assign

* Corresponding author.

M. Gen et al. (eds.), *Industrial Engineering, Management Science and Applications 2015*,
Lecture Notes in Electrical Engineering 349, DOI: 10.1007/978-3-662-47200-2_64

the task for each robot. There are two task allocation methods: negotiation-based and model-based method. In negotiation-based method, the system decides which robot to perform the task by negotiation (Y. Gao & Z. Luo, 2008). In the model-based method, instead of obvious negotiation, cooperation is realized by information communication. In addition, the bot has the task-choosing model. The input of the model is the sensor data, the information among robots, and the internal states of the robot. The task choosing decision will then be the output (Y. Gao & Z. Luo, 2008).

Recognition of the dynamic environment is required for cooperative motion by multiple mobile robots. Moreover, sensing and communication ability is considered crucial for the recognition from the practical viewpoint (S. Suzuki et.al, 1995). Suzuki, et.al in 1995 had already developed radio communication system, which enables each robotic agent to send messages to another specific agent by peer-to- peer communication, to agents within a certain group by groupcast, and to all the agents by broadcast. However, the radio communication system does not suit for local communication. In order to overcome these problems, they have developed an infra-red sensory system with local communication functionality, which enables not only detection of collisions against obstacles or other robots, but also local communication between robots (S. Suzuki et.al, 1995).

In this study, the proponents aim to provide a simple, yet convenient way of communication over a Wi-Fi network for mobile robot using Radio Frequency (RF) wireless communication module. The proponents propose an effective task allocation method thru central commanding computer to realize reliable operation for identified tasks. This study would likely apply to cooperative and performing robots. Most especially, due to its simplicity it could be most beneficial for toy robot applications.

2 Motion Control Strategy for Mobile Robot

Figure 1 shows the mobile robot's system configuration. The system block diagram is further divided into two major components namely: transmitter and receiver blocks. Each block contains four sections. The transmitter sends the control signal wirelessly and the receiver receives the instruction for execution. In the transmitter block, the RF module is connected to the microcontroller and is powered by 9V battery with capacitor as battery support. The main control unit is also connected to the voltage regulator. The 7805 regulator is used to regulate the output. Thus, maintaining constant voltage of 5V. On the other hand, the received block contains the L293D, which drives inductive loads in positive-supply application. The driver drives the left and right motors of the robot. It is powered by six pieces of 1.5 V in gradual control of the movement of the DC motors.

The proponents make use of three-stage control strategy for their mobile robot. This includes the human operator, the upper computer (commanding server) and the lower computer (mobile cube robots). The operator will remotely control the system's push buttons and the commanding server will send instruction to the mobile robot using Radio Frequency (RF) via Wi-Fi network.

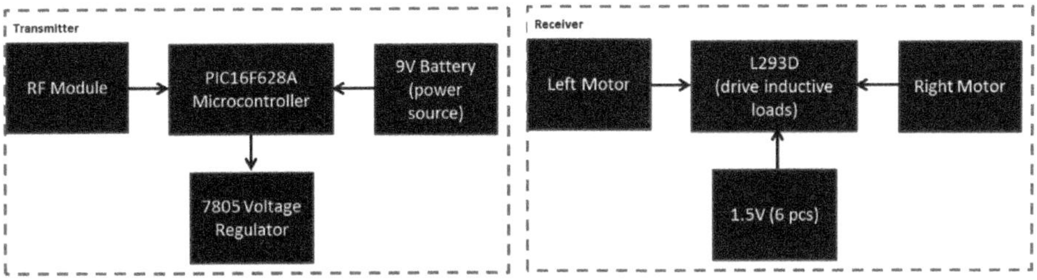

Fig. 1. System Configuration for Mobile Robot

Generally in mobile robotics, explicit communications between robots can only be made through: an infrared beam, or a radio-frequency transmission. The first one cannot allow passing through natural obstacles like walls, closed doors, and more generally opaque objects. So only the second one can be used to ensure a safe communication. The more efficient devices use a high frequency carrier with a bandwidth more or less large (N. Hutin et.al, 1998). In due course, the proponents considered to use RF module for sending and receiving of control signals. Since the study considers transmission and reception of control signals, the remote control with three simple movements will communicate and give instruction to the mobile robot. The robot agent is assumed of not missing any command, as two programs are developed for transmitting and receiving of control signals.

3 Design Considerations

In order to realize a mobile robot system, which can fulfil various tasks, flexibility in hardware and software is needed. A modular approach is then adopted. The design considerations section was divided into two (2). The hardware system block diagram and the software system block diagram as shown in *figures 2* and *3* respectively.

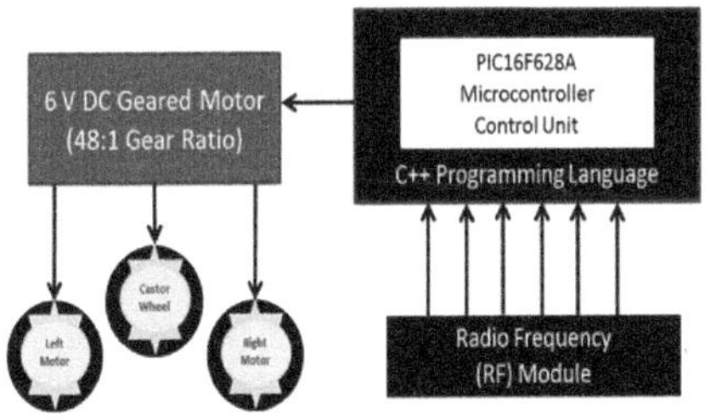

Fig. 2. Hardware System Block Diagram

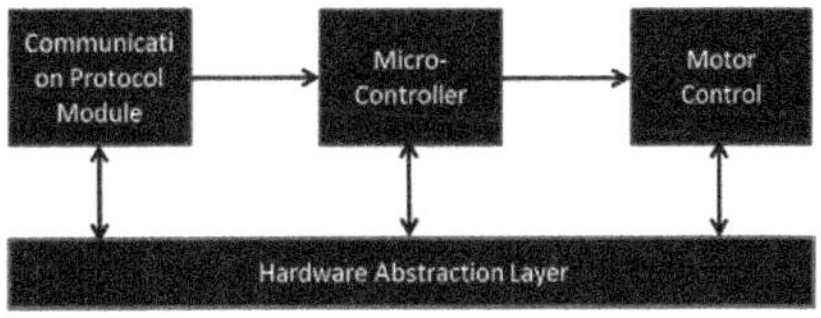

Fig. 3. Software System Block Diagram

3.1 Teleoperation Using Virtual Reality (VR) Technology

Implementing teleoperation using wireless technology to control robot remotely can significantly improve robotic system applications (N. Hutin et.al, 1998). It is then becoming a need to develop teleoperation systems, which allow full control of the robot in real environment. Sawaragi et al. investigated foundations needed in designing an interface system for robotic teleoperation. Hainsworth, on the other hand, presented a discussion of the requirements for user interfaces for teleoperation of mining vehicle systems. Zhong, et al. had adopted the virtual reality (VR) technology. Accordingly, VR technology is often used in lieu of establishing human-machine interface of teleoperation. In this technology, the operators manipulate robots on the spot by using joysticks, mouse or keyboards directly (G. Zhong et.al, 2012). In most of the studies concerning teleoperation, they regard the upper computer as a host and the lower computer as a slave. The proponents adopted this usual representation for VR teleoperation (N. Hutin et.al, 1998). As shown in *Figure 4*, the upper computer deals with the algorithm, decisions, and commands, while lower computer completes specific actions and sends feedback information.

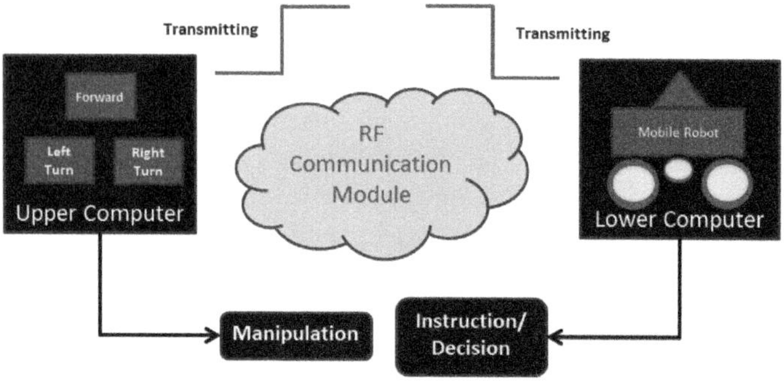

Fig. 4. Proposed teleoperation via RF module

4 Hardware Model

The hardware part of the design is consisted of RF module embedded in PIC microcontroller, actuators, and physical structure. All of these components had to work together to give the mobile robot the ability to sense its environment and take appropriate actions. The physical structure of the mobile robot is relatively inexpensive, requires minimal tools, and is easy to build. The PIC microcontroller and the quadruple high-current half-H drivers give the proponents much of the hardware; the motors, the driver and the control circuitry, in two cheap packages. The proponents make use of PIC16F628A. It is a powerful microcontroller, which is capable of executing instructions at 200 nanoseconds. It is low-cost of high performance, CMOS, fully-static, 8-bit and 18-pin microcontroller. It is easy to program, requiring only 35 single word instructions. The proponents make use of this microcontroller to transmit control signals to the mobile robot (Web-1). In receiving the control signals sent by PIC16F28A microcontroller, the proponents make use of L293D. It is designed to provide bi-directional drive currents of up to 600-mA at voltages from 4.5 V to 36 V. In controlling the left and the right motors of the mobile robot, the proponents make use of this driver, considering that it was designed to drive inductive load, in our case, DC motors (Web-2). The robot's design is consisted of two DC motors with shivel wheel, capable of rotating 90 degrees. The mobile robot is made capable of moving vertically and horizontally on a plane. In addition, the mobile robot follows command to move from a central module or controlled button. The communication technology to be used is RF.

5 Robot Implementation

5.1 Control Mechanisms of the Mobile Robot

The movements of the robot include forward, turn left and turn right. The move space of the robot is like a grid (in terms of one grid unit forward). The wheels are fixed to its axle and not steerable. It has castor wheel in front. In order to make the robot turn left or turn right, each wheel has its own DC motor, specifically a 6V geared DC motor. To make this happen, each wheel will be attached to its own controlled bidirectional motor. For forward movement, the same control signal will be sent to both wheels. The robot is designed to be controlled over RF wireless communication. When moving forward, both wheels are driven by their corresponding DC motors. When turning left, only the right wheel is driven. The left wheel stays idle and serves as a pivot. This is the same when turning right: only the left wheel is driven and the right serves as the pivot. The proponents make use of 6V DC geared motor and Pulse Width Modulation (PWM) to reduce the average DC to slowly drive the motors. Through the use of UHF RF module, the commanding server can easily enable the start and the stop option of the motors wirelessly.

5.2 Schematics

As shown in Figure 5, the transmit side of the RF module is connected to the pin 7 and the receive side of the RF module is connected to the pin 8 of the PIC microcontroller. On the other hand, the left and right motors of the mobile robot are connected to L293D IC as shown in Figure 6. The PIC microcontroller and the L293D IC were the two most important components of the system. In providing constant output voltage of 5V, the proponents make use of 7805 IC voltage regulator. Design applications of surge capacitors are taken into consideration as well in operating under severe stringent power system conditions. It prevents the equipment from damage due to voltage surges (Web-3).

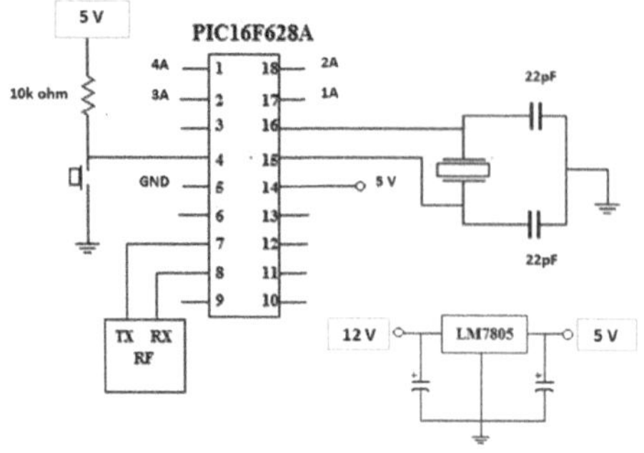

Fig. 5. Schematic diagram of data transmission of the mobile robot

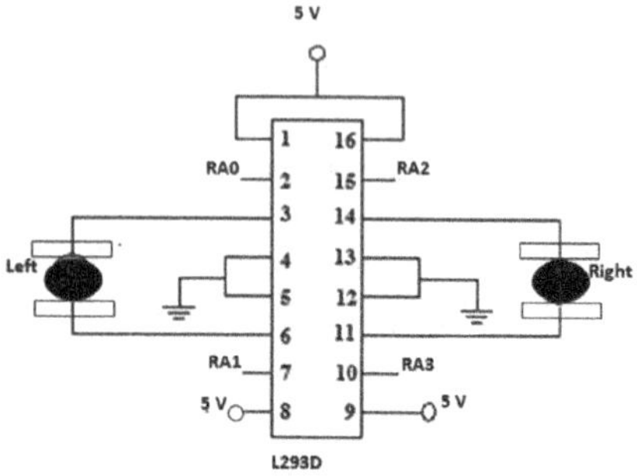

Fig. 6. Schematic diagram of data transmission of the mobile robot

5.3 Prototype

The complete mobile robot prototyping is exhibited in Figure 7. The transmitting and receiving side of the mobile robot (including the vital components within each module) is illustrated. By simply pressing the command button, assigned to the instruction, the mobile cube robot is controlled. There will only be three instructions known by the robot: Move Forward, Turn Left and Turn Right. The software side of the mobile robot requires two separate programs. The proponents make use of C++ language for transmission and reception of instructions from the remote control to the mobile robot.

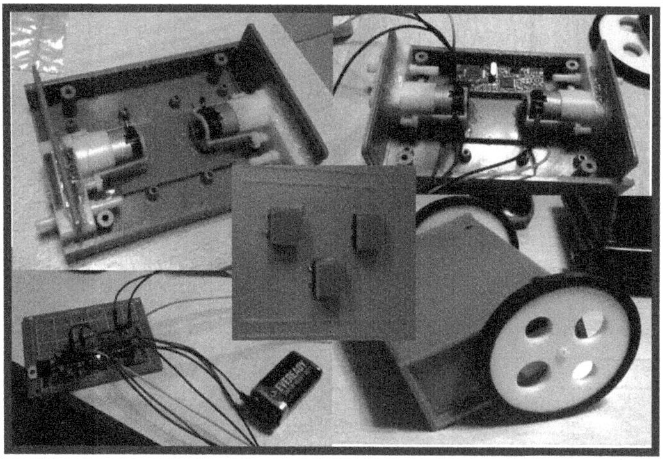

Fig. 7. General view of the mobile robot

6 Conclusions

In this paper, a mobile robot was designed and developed. The robot has an on-board PIC microcontroller as its control unit with built-in RF module, and it was programmed using C++. The proponents make use of model-based task allocation method. The robot control makes use of three-control strategy: human operator, upper computer (commanding server) and lower computer (mobile cube bots). The proponents have shown a Virtual Reality (VR) Teleoperation communication strategy, which allows information to be exchanged between the mobile robot and its environment. Physically, this communication is based on the utilization of RF wireless communication. The mobile robot was able to communicate with a central commanding server, which is controlled using C++. Through a local area network, the computer sent the command movements (forward, turn left and turn right) to the mobile robot using RF module.

7 Recommendations

Considering its distance constraints, it is recommended to put an antenna wire to sense the signal significantly longer. The proponents failed to include stepper motor in their mobile due to space constraints, considering they have to observe and limit themselves on the robot dimensions. In order to make the movement of the robot finer and smoother, it is recommended to put stepper motor. Also, it would also be best if a Graphical User Interface (GUI) will be developed to effectively control the movements of the mobile robot. Future researchers can further enhance the mobile robot by adding more features and instructions. Most importantly, other communication protocols might be tried and considered as well.

Acknowledgments. The authors gratefully acknowledge Jesus Christ, who is the source of knowledge and wisdom in the design and development of the mobile robot. They would like to give special commendation to Engr. Jerick Lee for his technical assistance. They would like to thank De La Salle University – Manila (DLSU) for giving scholarship and financial assistance for their doctorate studies. Also, the Research and Statistics Center of Lyceum of the Philippines University – Laguna (LPU-L) is gratefully acknowledged for funding the research publication and dissemination.

References

Zhong, G., Kobayashi, Y., Hoshino, Y., Emaru, T.: Intuitive Teleoperation of Nonholonomic Mobile Robot with a Manipulated based on Virtual Reality and Wi-Fi

Hutin, N., Pegard, C., Brassart, E.: A Communication Strategy for Cooperative Robots. In: Proceedings of the 1998 International Conference on Intelligent Robots and Systems (1998)

Suzuki, S., Asama, H., Uegaki, A., Kotosaka, S., Fujita, T., Matsumoto, A., Kaetsu, H., Endo, I.: An Infra-Red Sensory System with Local Communication for Cooperative Multiple Mobile Robots. In: 1995 IEEE (1995)

Gao, Y., Luo, Z.: Dynamic Task Allocation Method Based on Immune System for Cooperative Robots. In: Proceedings of the 2008 Seventh World Congress on Intelligent Control and Automation (2008)
Ota, Y., Yoneda, K., Muramatsu, Y., Hirose, S.: Development of Walking and Task Performing Robot with Bipedal Configuration. In: Proceedings of the 2001 International Conference on Intelligent Robots and Systems (2001)

Web sites:

Web-1, `http://support2.microsoft.com/kb/136403` (consulted December 5, 2014)
Web-2, `http://web.mit.edu/6.115/www/document/16f628.pdf` (consulted December 5, 2014)
Web-3, `http://www.ti.com/lit/ds/symlink/l293d.pdf` (consulted December 5, 2014)

Generalized Space Fourier Transform Method for the Analysis of Electrical Machines

Ankita Dwivedi*, S.K. Singh, and R.K. Srivastava

Department of Electrical Engineering,
Indian Institute of Technology (BHU)
Varanasi, India-221005
ankitadwd1808@gmail.com

Abstract. Spatial Fourier Transform theory is used to evaluate the performance characteristics of electrical machines. Yamamura's Space Fourier Transform theory applicable to linear induction motor is adapted for predicting the performance characteristics of different electrical machines.

Nomenclature

A= Magnetic Vector Potential (A/m)
D = Diameter of stator bore of original machine (m)
H_c = Coercivity of magnets (A/m)
I_S = Stator current (A)
J_S =Peak stator equivalent linear current density (A/m)
J_R = Peak rotor equivalent linear current density (A/m)
N = Number of turns per coil
T=Torque (Nm)
f= Frequency of stator supply (Hz)
g_c = Corrected air clearance using Carter's coefficient
h_C =Thickness of Magnets (m)
k_{w1}= Winding factor
l =Subscript layer number
m= Number of phases
n = Even number > 100
p= Number of Poles
v = Peripheral Linear velocity of rotor (m/sec)
w_t, w_{tt}, w_o, w_s, w_c = width of slot-teeth, tooth-tip, slot-openings and conductor respectively
μ_i = permeability of iron (H/m)
μ_l = Permeability of respective layer (H/m)
σ_l=Conductivity of respective layer (S/m)
τ= Pole pitch of original motor (m)
φ =phase difference between the two current densities
β =Fourier index of any layer
ω= Angular frequency (rad/sec)

---

* Corresponding author.

M. Gen et al. (eds.), *Industrial Engineering, Management Science and Applications 2015*,
Lecture Notes in Electrical Engineering 349, DOI: 10.1007/978-3-662-47200-2_65

1 Introduction

Analytical method provides a better insight of the effects of design parameters on the performance, design and optimization of an electrical machine. These methods are well known for their accuracy and lesser computational time as that of numerical methods (Finite Difference Method (FDM), Finite / Boundary Element Method (FEM / BEM)) and lumped parameter methods (equivalent circuit approach and generalized machine theory). Analytical techniques based on the solution of Maxwell's equations for predicting the performance characteristics of linear induction motor [1], induction motor [2], linear permanent magnet (PM) motor [3, 4], etc. have been reported in the literature. The determination of thrust force, torque, back-EMF and winding inductances in closed form is feasible. Fourier series expansion is also a frequently used method for determining the performance of electrical machines. Although, Fourier series approach provides accurate information of field distribution in the electrical machine, however the computational time is expected to be dependent on the speed of operation. In contrast to Fourier series method, in case of Fourier transform method the computational time is independent of the speed of operation.

The aim of this paper is to adapt and use the Yamamura's Space Fourier transform theory for a general solution of electromagnetic field equations of different electrical machines and its extension in determining the performance characteristics of Permanent Magnet Brushless AC Motor.

2 Space Fourier Transform Technique

Space Fourier transform theory has been reported widely in literature for determining the performance as well as losses due to longitudinal–edge effects in linear induction motor, transverse and longitudinal entry and exit end effects [1], MHD generators [5, 6] and eddy-current brake [7]. Unlike other method of analysis, the method uses the physical parameters of the machine for determining the magnetic field distribution and performance of special electrical machines.

For simplifying the analysis, an analytical layer model of the machine is developed, where different regions of the machine are represented by layers having homogeneous properties [8, 9]. This analytical model can be of two types- isotropic and anisotropic. Isotropic model assumes that the stator and rotor slotted regions are smooth and windings are present as a current sheet on the stator core. Such types of machine representation [1] is valid only for Linear Induction Motor having larger *entrefer* (stator iron to rotor iron distance), where the magnetic saturation in both the members of machines are not likely to occur. Most of the rotary machines have lesser *entrefer* as that of LIM, as such when isotropic model is applied; it gives error in the results. While in anisotropic model [9], the magnetic saturation in slotted regions of the stator and rotor are considered as separate layers having anisotropic material properties. The Residue theorem approach has an advantage of providing the closed form solutions. While in Parseval's theorem, the numerical integration in Fourier space is required.

2.1 Basic Maxwell's Equations

The general governing field equations for all machines can be given as,

$$\nabla^2 A = -\mu_0 \left(J + \nabla \times M + J_t + J_v \right) \tag{1}$$

where, J is the source current density of current carrying conductors of primary replaced by linear current density J_p, J_t is transformer induced eddy current density in moving conducting media, J_v is speed induced current density in moving conducting media and $\nabla \times M$ is the MMF due to magnets- replaced by linear current density J_s. For conventional machines, like induction motor and eddy current brake having no permanent magnet, $J_S = \nabla \times M$ is assumed to be absent. For eddy current brake excited by DC the transformer eddy current density J_t is absent. The Table 1 shows the use of governing field equation for different moving conducing media and for different angular frequency ω.

Table 1. Current densities in different machines

Current densities	Permanent Magnet $J_s = \nabla \times M$	Transformer induced $J_t = j\omega\mu\sigma A$	Speed Induced $J_v = \mu\sigma v \frac{\partial}{\partial x} A$
PM Machines	Yes	Yes	Yes
Synchronous Machines	Yes	Yes	Yes
Induction Machines– Linear , rotary, axial flux, tubular, circular etc.	$J_s = 0$	Yes	Yes
DC Dynamic braking of Induction Machines	$J_s = 0$	Yes	Yes
Eddy Current brake using DC foe high speed	$J_s = 0$	$\omega = 0$	Yes
Low frequency ECB for low speed	$J_s = 0$	Yes	Yes

The governing field equations for air gap or *'entrefer'*,

$$\nabla^2 . \mathbf{A} = 0 \tag{2}$$

and

$$\nabla . \mathbf{A} = 0 \tag{3}$$

The magnetic vector potential (MVP) is,

$$\vec{A} = A_z \vec{a}_z \tag{4}$$

and the flux densities for 2-D are,

$$B_x = \frac{\partial A_z}{\partial y} \qquad\qquad B_y = -\frac{\partial A_z}{\partial x}$$

2.2 Assumptions

The general assumptions made are:

i. The current sheet is present over a length of x=0 to x=L, where L=peripheral length of stator bore.
ii. The flux density has only radial and tangential component, thus magnetic vector potential has only z-component, as given in equation (4).
iii. Permeability of stator yoke is infinity, neglecting saturation in the yoke. The proposed method considers saturation in the stator core with anisotropic model, which is valid for machine having less *entrefer*.

2.3 Boundary Conditions

All field components are zero at z = ±∞. For general interface between limiting boundary and homogeneous layer, Neumann type boundary conditions applied are,

$$H_{x\bar{l}}(\xi) - H_{x\overline{l-1}}(\xi) = J(\xi) \tag{5}$$

Here, J(ξ) is the Fourier transform of linear current sheet present at any interface. At the interface between two layers,

$$A_l(\xi) - A_{l-1}(\xi) = 0 \tag{6}$$

Magnetic Vector Potential (MVP) has z-component only and is denoted as A or A_z. The Fourier transform of MVP $A_z(x, y)$ is $A_z(\xi, y)$ and is given by,

$$A(\xi, y) = \int_{-\infty}^{\infty} A_z(x, y) e^{-j\xi x} dx \tag{7}$$

3 General Analytical Model of Fourier Transform

The Kron's primitive machines based on lumped parameters approach is a known generalized machine theory applicable to most of the rotary machines. The method had been applied for prediction of performance characteristics of LIM by pole-by-pole approach [10]. The Space Fourier transform applicable to LIM is being adapted for analyzing different rotary machines. A generalized analytical model applicable for all machines is shown in Fig 1. It has two current sheets- initially representing generalized windings on both the members. Each source of magnetic flux source is replaced by an equivalent linear current sheet. The general Fourier index for 2D representing a particular layer takes the form [4],

$$\gamma_l = \sqrt{\{\xi^2 + j\omega\mu_l\sigma_l + j\mu_l\sigma_l v_l\xi\}}$$

In all the cases, the stator consists of primary windings as the source of excitation. The stator windings can be replaced by an equivalent linear current sheet [4] and its Fourier Transform is given by,

$$j_1(\xi) = j\left[e^{-j(k_1+\xi)L} - 1\right]\frac{J_1}{\xi + k_1} \tag{8}$$

where, J_1 for windings excited by a balanced AC source,

$$J_1 = \frac{m2\sqrt{2}k_w NI_S}{\tau p} \tag{9}$$

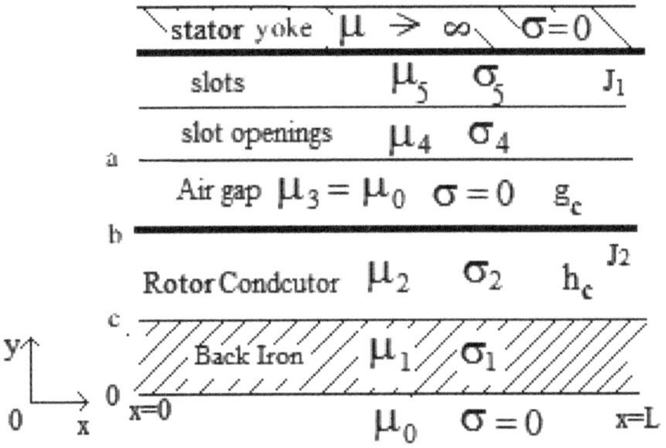

Fig. 1. A general anisotropic layer model with dual current sheets

and for windings excited by DC source,

$$J_1 = \frac{2NI_1}{L} \tag{10}$$

The rotor excitation in the form of alternate permanent magnet layers of BLDC / AC motor and for field winding of synchronous machines are represented by secondary linear current density.

$$j_2(\xi) = j\left[e^{-j(k_2+\xi)L} - 1\right]\frac{e^{\pm j0.5k_2\tau}\mathbf{J}_2}{\xi + k_2} \tag{11}$$

where, J_2 is the peak linear current density whose value depends on the type of field source and k_1 and k_2 represents the pole pitch of stator and rotor respectively.

The theory can be extended to phase by phase numerical calculations if $k_1 \neq k_2$. Further, it is evident that in case of PMBL motors, when space angle between the crusts of stator and rotor linear current densities is either φ=0° or φ=180°, the rotor permanent magnets will be aligned with the crust of stator rotating magnetic field, giving maximum normal force and no tangential force. The primary coils of stator are switched ON and OFF in synchronism with the rotor instantaneous position. The PM rotor follows the rotating magnetic field created by the stator winding currents. The magnetic rotor under steady state conditions runs at exact synchronous speed. It has been shown [4] that when rotor speed is equal to synchronous speed $v=v_s=2\pi f$, the displacement angle is φ=±90° [1], or in other words the phase difference between two current sheets is $\varphi = \pm j0.5k_2\tau_2$. Here ± sign indicates the direction of rotation.

In conducting moving media having finite conductivity,

$$\nabla^2 \mathbf{A}_\ell = j\omega\mu_\ell\sigma_\ell\mathbf{A}_\ell + v\mu_\ell\sigma_\ell\left(\frac{\partial \mathbf{A}_\ell}{\partial x}\right) \tag{12}$$

The rotor source of excitation is different for different electrical machines. It may be squirrel cage or wound rotor windings in case of induction motors, which can be replaced by a homogenized layer. Representation of PMs as equivalent current sheet is already explained [3]. Fig. 1 shows the general anisotropic analytical layer model with two current sheets. Slots and slot opening regions are neglected for isotropic model, while in anisotropic model, the material properties are homogenized to equivalent material properties. The anisotropic magnetic permeability in the region of slot openings and slot is adapted as [5],

$$\mu_{4x} = \frac{\mu_0\mu_1(\tau_s + \tau_w)}{\mu_1\tau_s + \mu_0\tau_w} \tag{13}$$

$$\mu_{4y} = \frac{\mu_1\tau_w - \mu_0\tau_s}{\tau_s + \tau_w} \tag{14}$$

$$\mu_{5x} = \frac{\mu_0\mu_1(\tau_c + \tau_t)}{\mu_1\tau_c + \mu_0\tau_t} \tag{15}$$

$$\mu_{5y} = \frac{\mu_1\tau_t - \mu_0\tau_c}{\tau_c + \tau_t} \tag{16}$$

thus in slot openings,

$$\mu_4 = \sqrt{\mu_{4x}^2 + \mu_{4y}^2}$$

and in slot regions,

$$\mu_5 = \sqrt{\mu_{5x}^2 + \mu_{5y}^2}$$

MVP at the surface of primary for a double-sided machine [1] may take a form,

$$A_{3z}(\xi,a) = \frac{j\mu_3 J_1}{2\pi} \int_{-\infty}^{\infty} \frac{\left[e^{-j(\xi+k)L} - 1\right] G(\xi,a)}{(\xi+k) H(\xi)} e^{j\xi x} d\xi \tag{17}$$

For single-sided excitation (single stator), MVP takes the form [1],

$$A_{3z}(\xi,a) = \frac{\mu_3 J_1}{2\pi\eta} \int_{-\infty}^{\infty} \frac{\left[e^{-j(\xi+k)L} - 1\right] W(\xi,a)}{\xi(\xi+k) K(\xi)} e^{j\xi x} d\xi \tag{18}$$

The characteristics equations for a generalized model taking saturation into account can be derived using governing field equation and homogenized material properties of different layers.

4 Magnetic Field and Torque Calculation

According to Lorentz force equation $\mathbf{F} = \mathbf{J} \times \mathbf{B}$. The normal or radial component of magnetic flux density produces thrust in linear motors and torque in rotary motors respectively. The flux density takes the form [4],

$$B_{3y}(\xi,a) = \nabla \times A_{3z} = -j\xi A_{3z}(\xi,a) \tag{19}$$

In case of linear model, thrust per unit width can be evaluated as,

$$F = \frac{1}{2}\int_0^L \mathrm{Re}\left(j_1^*(x) B_y(x)\right) dx \quad \text{(N/m)}$$

For rotary cylindrical machines, the end-effects are absent and current sheets are present only in the length 0 < x < L, as such torque developed takes the form [4] of,

$$T = 0.5Dp \int_0^L Re\left[j_P^*(x) B_{3y}(x)\right] dx \qquad \text{(Nm/m)} \tag{20}$$

4.1 Induction Motor

For squirrel cage and sheet rotor induction motor, secondary current sheet J_2 is assumed to be zero and rotor conductors are replaced by a non-magnetic conducting sheet having homogenized material properties while the transformer induced eddy current and speed induced eddy currents both are present. The primary current flows only in z-direction and the non-magnetic conducting sheet of finite thickness is travelling with a velocity v in x-direction.

The reported model was used for Single sided LIM, Double sided SLIM [1] and its application was extended to drag plate SLIM [11] and sheet rotor induction motor [12]. The model is useful for predicting torque versus slip characteristics for constant current excitation for regions of counter current braking (slip 2.0 to 1.0), motoring region (slip 1.0 to 0.0), and generating region (slip 0.0 to -1.0). It is also particularly useful for predicting the DC dynamic braking characteristics and low frequency-low speed braking characteristics. Fig 2. depicts the torque-speed characteristics of an induction motor when excited by AC or DC.

The generalized Space Fourier Transform model developed by Yamamura [1] has been found useful for predicting the performance characteristics of cylindrical induction motor , Tubular induction motor, axial flux induction motor and circular induction motor with either sheet secondary, squirrel cage secondary and solid rotor induction motor. The theory can be easily extended for synchronous motor, slip ring induction motor, PM motor if J_2 is considered to be present.

4.2 Permanent Magnet Brushless AC Motor

The PMBLAC motors has two sources of excitation namely; current in the stator windings and permanent magnets in the rotor. The two excitation sources are replaced by their equivalent current sheets. The magnetic vector potential (MVP) of each layer has to be determined separately utilizing the given boundary conditions. Since, the model has two conducting layers presenting primary windings and PM layer, the Fourier Transform of two current densities are given by equation (8) and (11).In order to establish the theory of space Fourier transform applicable to PMBLAC motor, it is assumed that,

$$k_1 = k_2 = k$$

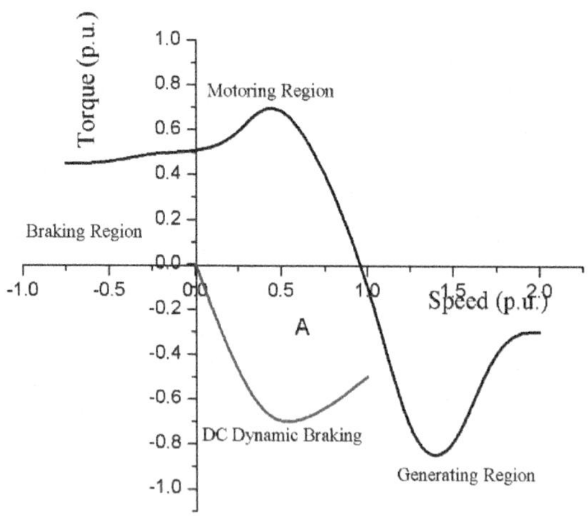

Fig. 2. Typical Constant current T-speed characteristics of linear, rotary, axial flux, tubular and circular induction motors

The MVP in the back iron supporting PMs, PM layer and in air gap is A_{1z}, A_{2z} and A_{3z} respectively. The torque is produced in the air gap by the interaction of two current sheets and thus, MVP of third layer at y=a can be obtained in the form,

$$A_{3z}(\xi,a)=\frac{\mu_3}{\xi}\left[\frac{W(\xi,a)}{K(\xi)}\right]j_P(\xi) \tag{21}$$

or

$$A_{3z}(x,a)=\frac{\mu_3}{2\pi}J_P\int_{-\infty}^{\infty}\left(e^{-j(\xi+k)L}-1\right)\frac{W(\xi,a)}{\xi(\xi+k)K(\xi)}e^{j\xi x}d\xi \tag{22}$$

The expressions $W(\xi,a)$ and $K(\xi)$ for any machine, can be derived using separation of variable. When integration of equation (22) is carried out using Residue Theorem, one of the poles will be,

$$\xi=-k \tag{23}$$

In case of rotary PMBLAC motor excited by an alternating current, the pole $\xi=-k$ is responsible for the generation of main torque in the machine. The model may be extended to numerically compute the thrust of a linear PMBL motor [6], wherein both the longitudinal end effects similar to that of linear induction motor are expected to be present which requires the computation of remaining poles of the characteristics equation,

$$K(\xi)=0 \tag{24}$$

This may result in large number of poles in second and third quadrant. However, for an ideal rotary PMBL motor such studies are ignored. The pole near origin on negative real axis in the complex-plane is a significant pole representing the pole pitch of primary stator winding. Fig 3. shows a typical torque-speed characteristics for different types of PMBLAC motors. The details of the analysis of BLAC motor using Parseval's approach is reported in [13].

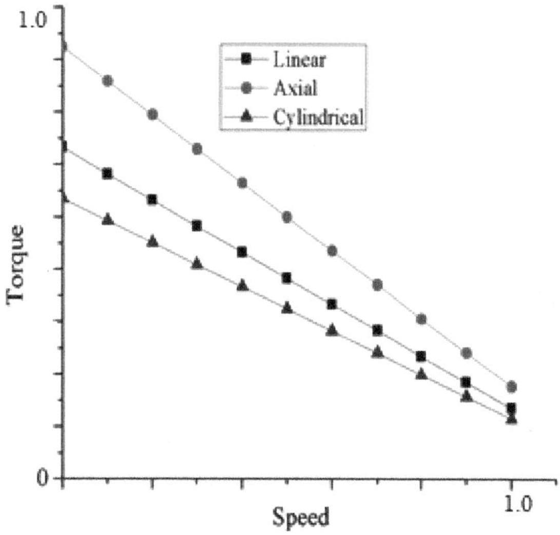

Fig. 3. Typical torque-speed characteristics for different shapes of PMBLAC motor

4.3 Eddy Current Brake

Eddy current brakes (ECB) are used to slow down a moving object. The braking torque produced can be used to load a smaller motor under test. Similar to induction motor, an ECB can take form as cylindrical, linear, axial, tubular and circular. The braking characteristics of all the topologies can be predicted with accuracy using the proposed adaptation of Space Fourier Transform. The braking characteristics takes the form of DC dynamic braking of induction motor as shown in Fig 2.

Parseval's method is found to be more suitable than Residue theorem, since no extra effort to separately compute the end-effect is required. The residue approach may also be extended to give pole-by-pole determination. This may probably resolve the possible errors caused by representation of finite secondary current sheet. Both the methods have been found to give appreciably good validation in many cases.

5 Conclusions

The paper presents the generalized Space Fourier Transform method for the analysis of electrical machines. The model is applicable to all types of electrical machines; linear, cylindrical, axial, tubular and circular. The extension of this method to PMBL motors using dual current sheet model is a new approach simplifying the determination of electromagnetic field distribution and performance of these motors. When pole-by-pole Residue approach is preferred for performance evaluation, the

CPU time is expected to be lesser than that of Parseval's approach and also that of Fourier series method. Space Fourier Transform technique is lesser known technique applied for the performance determination of electrical machines. The paper generalizes the method for different electrical machines. Although the model predicts the constant current determination of torque-speed, it can be easily extended for constant voltage analysis in conjunction with lumped parameter approach.

References

1. Yamamura, S., Ito, H., Ishulawa, Y.: Theories of Linear, Induction Motor and Compensated Linear Induction Motor. IEEE Trans. Power Apparatus and Syst. PAS-91(4), 1700–1710 (1972)
2. Sudhoff, S.D., Kuhn, B.T., Corzine, K.A., Branecky, B.T.: Magnetic Equivalent Circuit Modeling of Induction Motors. IEEE Trans. on Energy conv. 22(2), 259–270 (2007)
3. Deng, Z., Boldea, I., Nasar, S.A.: Forces and Parameters of Permanent Magnet Linear Svnchronous Machines. IEEE Trans. on Magn. 23(1), 305–309 (1987)
4. Isfahani, A.H.: Analytical Framework for Thrust Enhancement in Permanent-Magnet (PM) Linear Synchronous Motors With Segmented PM Poles. IEEE Trans. on Magn. 46(4), 1116–1122 (2010)
5. Fanucci, J.B., Kijewski, L.J., Ness, M.J.E.: Fringing Effects in an AC MHD Generator. In: Mather, N.W., Sutton, G.W. (eds.) Engineering Aspects of Magnetohydrodynamics, pp. 329–343. Gordon and Breach Science Publishers, Inc, N.Y (1963)
6. Sudan, R.N.: Interaction of a conducting Fluid Stream with a Travelling wave of Magnetic Field of Finite Extension. J. Appl. Phys. 34, 641–650 (1963)
7. Srivastava, R.K., Kumar, S.: An alternative approach for calculation of braking force of an eddy-current brake. IEEE Trans. Magn. 45(1), 150–154 (2009)
8. Mishkin, E.: Theory of The Squirrel-Cage Induction Machine Derived Directly From Maxwell's Field Equations. Quart. Jcrarn. Mecta. and Applied Math. VII(4) (1954)
9. Williamson, S.: The anisotropic layer theory of induction machines and induction devices. IMA J. Appl. Math. 17(1), 69–84 (1976)
10. Nondahl, T.A., Novotny, D.W.: Three-phase pole-by-pole model of linear induction machine. IEE Pros. B Elect. Power Appl. 127(2), 67–82 (1980)
11. Srivastva, R.K., Mahendra, S.N.: Drag Plate-Single-Sided Linear Induction Motor. In: IEEE Int. Conf. on Industrial Technology, Goa (India), vol. 2, pp. 699–702 (2000)
12. Dwivedi, A., Singh, S.K., Srivastava, R.K.: Equivalence between squirrel cage and sheet rotor induction motor, IE(I): Series B. Springer Press (in Press),
13. Dwivedi, A., Singh, S.K., Srivastava, R.K.: Analysis of Permanent Magnet Brushless AC Motor sing Two-Dimensional Fourier Transform-Parseval's Theorem Approach. In: Applied Electromagnetics International Conf (APPEIC), Bandung (Indonesia), December 16-18 (2014)

Product Attribute Analysis for Customer Involvement in Value Creation

Risdiyono

Mechanical Engineering Department, Faculty of Industrial Technology,
Islamic University of Indonesia, Yogyakarta 55584, Indonesia
risdiyono@uii.ac.id

A concept of design by customer (DBC) recently introduced by researchers aims to increase customer satisfaction by opening maximum possible channel for customers to involve in value creation. A comprehensive analysis is needed before applying the concept in order to maximize the opportunity for success. Proposed in this paper is product attribute analysis (PAA), a systematic method to determine the optimum level for customers to involve in defining product specifications so that their fitness to individual customer preference can be achieved without any problem in production process. The analysis starts with identifying the product attributes that customers may focus on and then finding out the corresponding building elements that contribute to the quality of the attributes. A progressive PAA evaluation table is utilized to conduct stage by stage assessments. Based on the table, the flexibility level of the attributes can be identified and the position of CIDPs can be assigned.

Keywords: product attribute analysis, design by customer, multi-channel for customer involvement, customization.

1 Introduction

Customer satisfaction has been identified as the most important factor determines the long-term success of a company. It can be achieved when customers perceive that the value of products and/or services they received is as (or exceed) what they expected. In order to create a premium value for customers, manufacturers should focus on the total customer experience, related to (1) superior solution to the need, (2) fair price, (3) treatment with respect, (4) emotional connection and (5) convenience (Berry 2002).

While the last three deal with the quality of services (customer experience in transaction processes), the first two are more related to the quality of product (customer experience in using the product). Focus on superior solution to the customer need may sound simple but in fact it is very complicated. Manufacturer must understand what people need and how to fulfill the need better than competitors. A key challenge will be how to ensure that the need manufacturer perceived is the real need customers demanded because it will differentiate between a successful product and a fail one. It is important to note that understanding customer need is not a trivial task especially in this era of personalization where the extent of market-of-one has been foreseen as a prospective driving force for the next transformation of global economy (Pine 2009).

M. Gen et al. (eds.), *Industrial Engineering, Management Science and Applications 2015*,
Lecture Notes in Electrical Engineering 349, DOI: 10.1007/978-3-662-47200-2_66

It is a nature that customers usually purchase products (goods or services) for a reason; they have a problem (a need) and expect manufacturer or service provider to come up with a solution (to fulfill their need). Their utmost satisfaction will be achieved when they get exactly what they want without compromising any requirement. As each customer may have her/his own preference, a rigid predictive-based product specification will be difficult to compete. As an alternative, a flexible responsive-based one should be adopted. A new concept of design by customer (DBC) to involve customers in value creation has been introduced as a way to achieve this purpose (Risdiyono and Koomsap 2011).

DBC is a new concept to satisfy every individual of customers by letting them to flexibly involve in value creation (defining product of their personal requirements) at any stage of value chain (by opening maximum possible channels). It does not mean that customers are given free hand to design in a blank space. Instead, they are guided to define the fittest alternative that meet the cost, schedule and the product requirements through the capabilities of a company. Depended mainly on the type of product and customer preference, the degree of customer involvement will vary from passive involvement (e.g. only giving voice of customer and let manufacturer to provide ready-made products) to active involvement (e.g. skinning personalization, mix and match customization, modification or even real designing process from scratch) before purchase (Fig. 1).

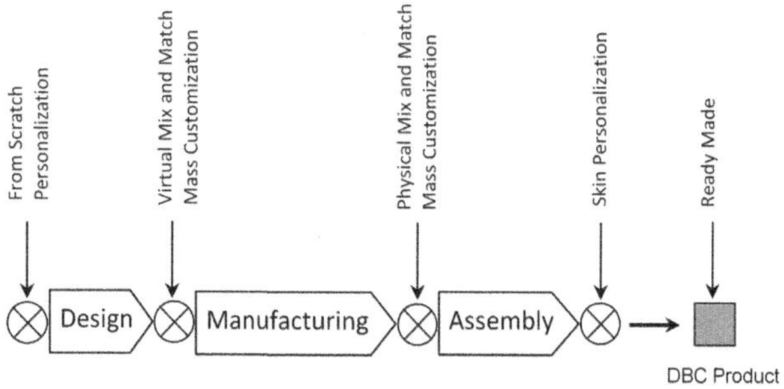

Fig. 1. Points of customer involvement in DBC system

When many channels are opened up for customers to flexibly involve in value creation, a stable system is needed to accommodate wide range of customer preferences. The system is required to provide optimum satisfaction to all customers and at the same time give maximum benefit for manufacturers. It is a big challenge for manufacturers on how to balance between individual customers' personal requirement, response time and cost as the position of point of customer involvement may not only influence the flexibility level but also cost and response time. If it is closer to the customer, lower cost and shorter response time can be achieved. Accordingly, if it is placed at the beginning of the production process, it could be assumed that higher cost and longer response time would be necessary. The problem will be more complicated when the product consists of numerous parts in a complex assembly system. Hence, it is recommended for manufacturers to comprehensively analyze their products in order to set in which level customers can involve in defining product specifications.

This paper proposes a systematic method namely product attribute analysis (PAA) to determine the product flexibility level in order to maximize the opportunities for success in design by customer concept realization.

2 Product Attribute vs. Building Element

Flexibility is considered as one of important aspects that customers consider in making buying decision. Anderson (2006) suggested giving customers flexibility in prices, service and delivery in order to increase market share, while the concept of DBC extends the flexibility to product specifications by enabling individual customers to involve in value creation through multi channels provided. The degrees of product specifications' flexibility have to be set appropriately in order to obtain clear competitive advantage for a manufacturing enterprise.

In general, product specification consists of product attribute and its assigned value. For illustration, color is a product attribute, and yellow, green, and red are the examples of assigned values. Based on market research, the value of each attribute will be set to meet the need of customers. In order to fit to individual customer requirements, the DBC system suggests that the product specifications should be adjustable. This means, customers should be involved in setting the value of attributes based on their preferences. Identifying which attributes that can be adjusted by customers and how to adjust are the main issues for manufactures to deal with.

Product attribute plays a very significant role in determining the success of a product in the market competition. In this research, product attribute is defined as the property or feature of a product that is thought to appeal to customers. It comprises tangible and intangible features that can be observed by its qualitative and/or quantitative entity. Since the main consideration on cost and schedule of component or parts are analyzed in the design phase (Jiao and Tseng 2000), it is reasonable to approach the PAA from the perspective of design, particularly the early stage of design. It is worthy to note that all products will always contain of three design dimensions i.e. (1) utilitarian design, which focuses on the practical benefits a product may provide, (2) kinesthetic design (ergonomics), which emphasizes how a user physically interacts with the product and (3) visual design which is driven by form, color, size and the desire to communicate value to consumers without necessarily interacting with the product (Noble and Kumar, 2008), of course with different level of content (Risdiyono and Koomsap, 2010b). For analysis purpose, product attribute can be simply derived from these design contents.

As illustrated in Fig. 2, a product can be considered as the assembly of building elements, i.e. part, sub-assembly, module, etc. that each individual has its own attributes. In general, the attributes of a part which is the simplest building element of a product consist of shape, size, color and material. In turn, the combination of the attributes of the building elements will form general product specification as it is a combinational result of overall product attributes. Therefore, the relationship between building elements and product attributes is very useful for establishing a framework for product attribute analysis as the modification of some building element properties may result in the change of some product attributes.

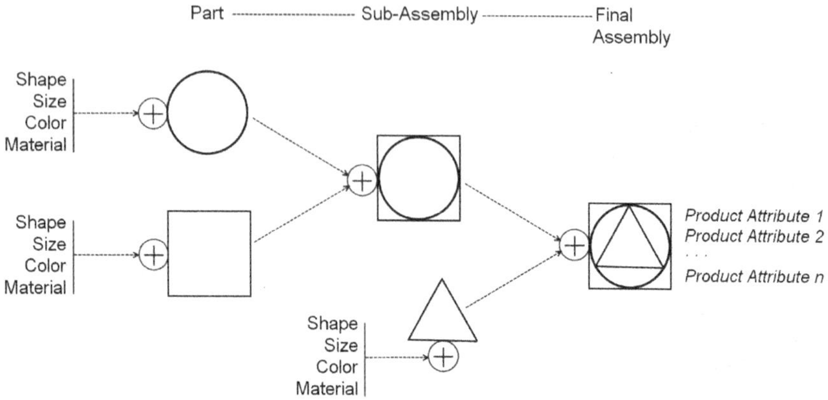

Fig. 2. The establishment of design contents of a product

As an illustration, let's consider ease of reading, one product attribute of a clock. Fig 3 shows two different wall clocks; the first one is an easy to read regular clock widely available in the market, while the second one is a unique but not easy to read analog binary clock designed by Liekens (2006). In this case, ease of reading is resulted from the combination of dials' attributes (e.g. shape, color, size), hands' attributes (e.g. shape, color, size) and background's attributes (e.g. motif, color). Hence, by altering the attributes of building elements, the overall product attribute will respectively be modified. It is up to customers decision to set their fittest attributes in case there are interrelationship conflicts among attributes (e.g. ease of reading might reduce the uniqueness), considering that customers know best about what they are looking for.

Therefore in the Design by Customer concept, customers are enabled to involve in specifying product specification of their preferences by altering the attributes of its building elements. The position of customer involvement decoupling point for every building element may vary depended upon its flexibility. When customers are allowed to modify shape, size, color or material of a building element, then individual personalization is provided. If customers are only allowed to choose from available variety of building element, a mix and match mass customization is offered (Fig. 4).

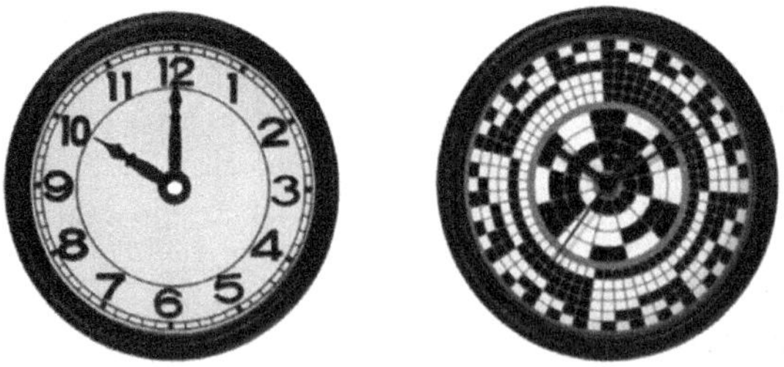

Fig. 3. Example of ease of reading attribute of wall clock

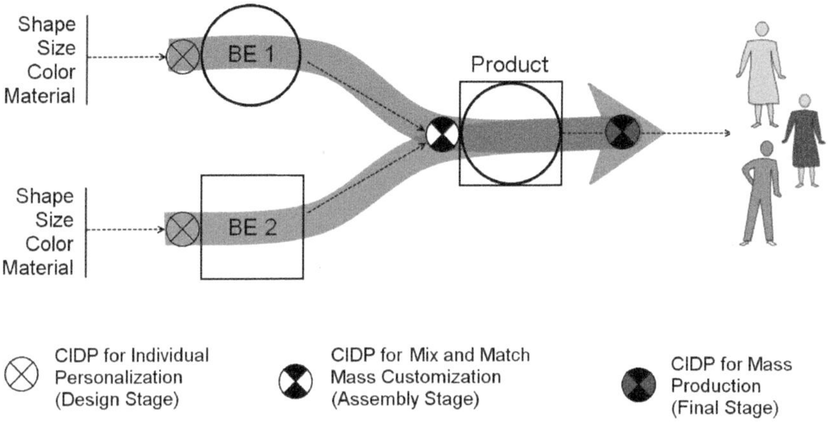

Fig. 4. CIDPs for Design by Customer concept

3 Factors Affecting Flexibility Level of an Attribute

In order to set the positions of the customer involvement decoupling point (CIDP) for an attribute, the type of the attribute whether it is a fix attribute, a mass customized attribute or an individual personalized attribute must be identified from its flexibility level. Three main factors for identifying the attribute flexibility include (1) customer interest, (2) manufacturing issue and (3) engineering and super system issues.

3.1 Customer Interest

One of the important issues manufacturers should consider before offering customer a product or service is the existence of customer interest (demand). Hence, before offering involvement to customers, evaluating the customer interest is a must. It can be done by analyzing the nature of customer preference, doing market survey, observing customer behavior after purchase (whether there is evidence that some customers try to make modification on a product attributes) or just using expert justification. The nature of customer preference towards each attribute is important to be analyzed whether it is homogeneous or heterogeneous. This analysis is recommended in order to reduce the complexity due to the number of attributes to be investigated. In case of battery, for example, the weight of battery, its life time and charging time are considered as attributes that all customers have similar preferences; they would like to have a light battery with life time as long as possible and charging time as quick as possible. How far manufacturer can come up with these ultimate product specifications will determine the success of the product. Hence these homogeneous attributes can be optimized by manufacturer without a need to involve customer in the value chain.

As a comparison, there is a variation in customer preference on shirt's attributes of size and color, which is quite difficult for manufacturer to define fixed value for these heterogeneous attributes to satisfy the need of all customers. In addition, the ease of customer involvement should also be considered. Customers should not have difficulties to involve either by mix and match design, parametric design or from scratch design.

3.2 Manufacturing Issues

Manufacturing issues related to the possibility to alter attributes of individual building elements in reasonable price and acceptable response time become a question needed to be addressed after there is evident that customers are interested to involve in value creation. The increase in overall production cost due to customer involvement should be reasonable (e.g. based on study by Goldsmith and Freiden (2004), it should not over 25%) while the total time from customer order to delivery must also be acceptable. In addition, the interdependent effect (i.e. the effect due to a change of one building element on other building elements) should also carefully be investigated as it will determine the complexity of manufacturing process.

3.3 Engineering and Super System Constrains

Some product attributes are sensitive to engineering constrains (i.e. constrain related to engineering issue, e.g. safety, stability, etc.) and super system constrains (i.e. constrain related to the bigger system outside, e.g. environment, laws, ethics, corporate image, etc.). Manufacturer, for example, should consider the possibility of 'bad design' created by customer that might relate to safety and gives poor effects on the image of the corporate.

4 Analysis of Product Attribute

Product attribute analysis (PAA) developed in this research is a tool to help manufacturers in decision making process to set the flexibility levels of product attributes. A product attribute is considered flexible if customers are allowed to make adjustment to their preference by modifying its related building elements. The degree of flexibility is depended upon the position of CIDP for every building element.

4.1 Building Element – Attributes Relationship

As mentioned earlier, there will always be a relationship between building elements and product attributes. Beside the strength of the relationship may vary for each, it is also possible that some of building elements might contribute to several attributes (e.g. the wheel of a bike does not only relate to attribute of speed, but also comfortability, ease of use, safety and beauty), while some might contribute only to a single attribute (e.g. paint and stripe might only relate to the attribute of beauty). On the other hand, some of attributes may depend upon several building elements while others only upon a single building element (Fig. 5). Therefore, identifying all these relationships is important, especially for assigning the weights of building elements' contribution on the product attribute. These weights are needed to calculate the flexibility level of the product. The assignment of a building element's contribution weight should be made based on the number of related attributes and their important level.

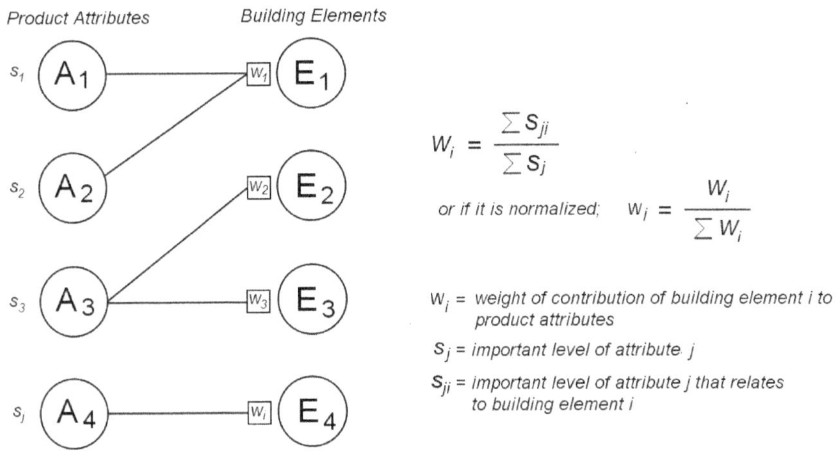

Fig. 5. Attributes – building elements relationship

4.2 Setting the Conditions for Attribute Flexibility Level

As the main objective of PAA is to optimize the points of customer involvement or CIDP in the production line, hence the analysis starts with assessing the flexibility of every building element's attributes. Customer involvement in individual personalization –IP– (i.e. from scratch and parametric design) and mass customization –MC– (i.e. mix and match) are investigated. The analysis consists of three stage assessments including (1) assessment on customer interest, (2) assessment on manufacturability, and (3) assessment on engineering and super system constrains manageability. The important items evaluated for each assessment are presented in table 1. In order to keep PAA simple but powerful enough to be used, a Go/No go analysis that usually be used in statistical process control has been adopted. In general, Go/No go analysis refers to a principle of pass/fail test (i.e. only using two conditions; o or 1). Hence evaluation involving numerical score should be converted to pass/fail score based on cut-offs defined by an evaluation standard provided. Considering all aforementioned factors affecting flexibility level, the PAA concept can be generated.

Table 1. Evaluated items for assessments

Assessment		Evaluated items		Go Condition for MC and IP
Customer Interest (CI)		*"Is there any demand and how easy for customers to involve?"*		1 ∩ (2 ∪ 3 ∪ 4) ∩ (5) = 1
		(1)	There is a variation of customers preference on the attribute	
		(2)	Market survey shows that customers are interested to involve in modifying the attribute	
		(3)	There is an evidence that some customers try to modify the attribute to fit their requirements	
		(4)	Expert says that there is a market for the offered attribute flexibility	
		(5)	It is possible for customers to easily involve in modifying the parameter either by using mix & match, parametric design or from scratch design	
Manufacturer Interest (MI)	Manufacturability (M)	*"Can we offer customer involvement in value chain in reasonable cost & acceptable response time?"*		6 ∩ 7 ∩ 8 = 1
		(6)	The overall production cost increase due to customer involvement is reasonable (e.g. based on study, maximum 25%)	
		(7)	The total time from customer order to delivery needed due to customer involvement is acceptable	
		(8)	There is no interdependent effect among related Building Elements that we cannot handle	
	Engineering & super system constrains manageability (E)	*"Is there any constrain related to engineering issue or super system?"*		9 ∩ 10 = 1
		(9)	Issues related to engineering (e.g. safety, stability) that may happen due to the customer involvement are manageable	
		(10)	Issues related to the bigger system outside (e.g. law, standard, ethic, corporate image) that may happen due to customer involvement are manageable	

4.3 Progressive Analysis for PAA

There are 3 factors assessed in this Go/No go flexibility analysis; customer interest, manufacturability and engineering & super system constrains manageability. Mass production can be offered regardless the condition of all these factors, while mix & match mass customization and individual personalization can be offered only if the conditions are passed. Therefore, a progressive analysis (or stage by stage assessment) can be applied; started from analyzing the flexibility of main building elements (i.e. the most downstream position of CIDPs, and then, when necessary, followed by the next upstream CIDPs. The analysis should stop when the CIDP of building element does not pass the conditions or when it has come to the simplest building element of a product (shape, size, color and material). Fig. 6 illustrates the flow of the analysis.

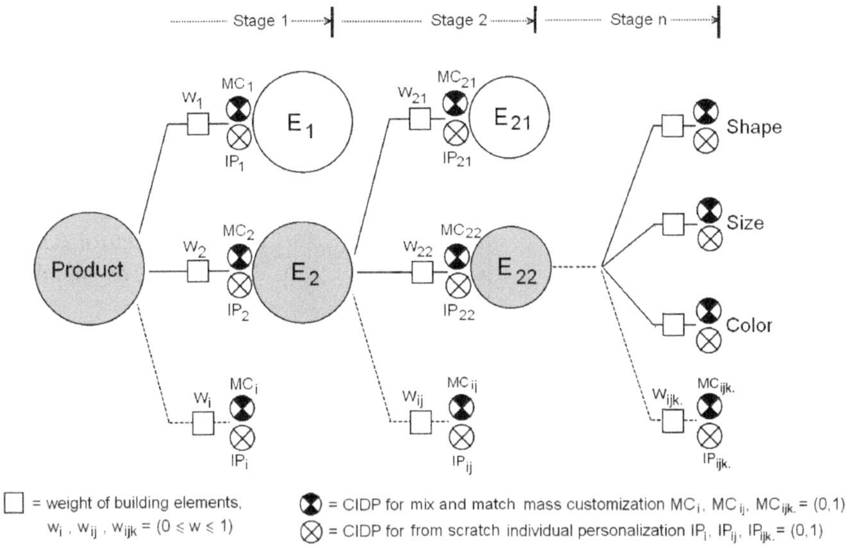

Fig. 6. The flow of progressive product attributes analysis

To calculate product flexibility, let FI_{ip} represents the flexibility index for from scratch individual personalization, and FI_{mc} represents the flexibility index for mix and match mass customization, then:

$$FI_{ip} = \sum_{i=1}^{m} \left((w_i \times IP_i) \times \sum_{j=1}^{n} \left((w_{ij} \times IP_{ij}) \times \sum_{k=1}^{p} (w_{ijk} \times IP_{ijk}) \times \sum \dots \right) \right)$$

$$FI_{mc} = \sum_{i=1}^{m} \left((w_i \times MC_i) \times \sum_{j=1}^{n} \left((w_{ij} \times MC_{ij}) \times \sum_{k=1}^{p} (w_{ijk} \times MC_{ijk}) \times \sum \dots \right) \right)$$

Hence, the overall flexibility index of the product is:

$$FI = \frac{FI_{ip} + FI_{mc}}{2}$$

When these indices are close to 1, the product specification is considered very flexible to allow customers to involve in defining product specifications. Contrarily, it they are close to 0, the product specification is very rigid and difficult to be adjusted by customers before purchase. Based on this analysis, the flexibility level of every building element could be assigned and the multi CIDP strategy could be applied accordingly.

5 Conclusion

A systematic method to analyze the product attributes flexibility level to support design by customer concept realization has been introduced in this paper. The concept of attribute flexibility aims on the fitness optimization of product specification to individual customer preference. Product attribute analysis is a tool to optimize the point of customer involvement in value creation by setting the degree of product attribute flexibility. It starts with identifying the product attributes that customers may focus on and then finding out the corresponding building elements that contribute to the quality of the attributes. The relationship between building elements and product attributes is evaluated based on the number of related attributes and their important level in order to get the weight of building elements' contribution to product attributes formation. However, the practical example of this method is still needed in order to see its applicability.

References

1. Anderson, C.: The Long Tail: Why the Future of Business is Selling Less of More. Hyperion, New York (2006)
2. Berry, L.L.: The Old Pillars of New Retailing. Harvard Business Review on Customer Relationship Management. HBS Press, Boston (2002)
3. Goldsmith, R.E., Freiden, J.B.: Have it your way: customer attitudes toward personalized marketing. Marketing Intelligence & Planning 22(2), 228–239 (2004)
4. Jiao, J., Tseng, M.M.: Fundamentals of product family architecture. Integral Manufacturing System 11(7), 469–483 (2000)
5. Lieken, A.: How to convert an analog wall clock into a binary clock (2006), `http://anthony.liekens.net/index.php/Misc/AnalogBinaryClock` (retrieved February 25, 2012)
6. Noble, C., Kumar, M.: Using product design strategically to create deeper consumer connections. Business Horizons 51(5), 441–450 (2008)
7. Pine, B.J.: The Future of Mass Customization. In: Special Talk in Mass Customization and Personalization Conference (MCPC 2009), Helsinki, Finland (2009)
8. Risdiyono, Koomsap, P.: Design by customer: concept and applications. Journal of Intelligent Manufacturing (2011), doi:10.1007/s10845-011-0587-4
9. Risdiyono, Koomsap, P.: Design by Customer: A Management of Flexibilities. In: Design Creativity 2010, pp. 323–328. Springer, London (2011)

Influence of Gender of Customers on Service Quality

S. Valli Devasena

Department of Commerce,
Mother Teresa Women's University,
Kodaikanal-624 102, Dindigul District, Tamil Nadu, India

Abstract. The customer choice and awareness have been increasing tremendously during this decade due to more open economy, the advent of information technology and media revolution, besides hectic competition for resources by banks. As markets have become increasingly competitive, customers can now immediately go elsewhere if they do not get what they want. Continuous improvement, gaining the competitive edge, increased market share, higher profits-none of these things is possible unless businesses can find new ways of maintaining the loyalty of existing customers. It takes only a few incidents and direct experiences for the knowledgeable customers to form an opinion about the quality of the services and the quality of the product offered. Hence, "customer service is not being viewed as just a business strategy but should become a corporate mission."

1 Introduction

"By entering into your premises, the customer is giving you opportunity to serve him, but you are not doing a favour by serving him" - Mahatma Gandhi.

The liberalisation and globalisation of Indian economy took place almost a decade ago. Ever since, the focus point in any service organisation has been "customer service", more so in the banking industry. The phrases such as "customer is the king in our business", "service to customer is service to God" are no more a myth but have turned out to be a reality. Customer service is the base for business expansion because of the stiff competition prevalent in the banking industry. With the advent of new private banks, the concept of "customer service" has become an important and pivotal issue in banks, whether it is in the public sector, private sector, co-operative sector and so on. The survival of banking business is dependent on customer services.

1.1 Customer Service-Significance

"The Banking sector industry is considered a service oriented industry. It has to render manifold services to the people who visit the banks. Customer service refers essentially to counter level inter face or through other modes with the customers." [1]

The issue of proper customer service is central to all business operations. That is why management experts have for long, considered customer service as an integral part of the growth strategy of their businesses.

M. Gen et al. (eds.), *Industrial Engineering, Management Science and Applications 2015*,
Lecture Notes in Electrical Engineering 349, DOI: 10.1007/978-3-662-47200-2_67

'The leader of an organisation should live and breathe customers' needs and should communicate this across the organisation."[2]
Hence, the major component of customer service is related to the involvement and commitment of the staff rendering such service. Since the marketing of financial service offered by bank very much depends on the quality of customer service and the satisfaction that customers derive from the services they receive the important criterion to judge the benchmark of a bank is customer satisfaction in terms of good service.

1.2 Banker and Customer Relationship

Today the relationship between the banker and customer has come under sharp focus both at the banker's as well as at the customer's ends. Many customers are expecting better service. The dominant questions which are bothering the minds of bank management today are how to improve customer service and competitive advantage. The products are almost the same; however, the battleground is service.

Table 1. Paradigm Shift – Scenario in India

Before 1991	*After1991*
Seller's market	Buyer's market
Protected market	Open market
Not many global brands	Increase in number of global brands
Friendly competition	Cut-throat competition
Patient customers	Demanding customers
Limited choice for customers	Increasing choice for customers
Limited role of service	Increased role of service
Speed @ will	Turbo speed
Fundamental standalone system	Enterprise system
IT-competitive advantage	IT-Enabler
Gaining new customers	Retaining existing customers
Monologue	Dialogue
Transaction	Relationship

Source: IBA Bulletin, August 2004, p-6.

2 Statement of Problem

Customers' preferences keep on changing at a rapid speed and their demands are turned insatiable. In order to cater to the changing preferences, bankers are bound to provide the services suitable to their needs to survive in the competition. Hence an attempt is made to analyse influence of gender of customers on service quality by Chi Square test

3 Review of Literature

Kamath[3] in his thesis entitled "Marketing of Bank Service with Special Reference to the Branches in Bombay City of Syndicate Bank" has concluded that quick and better services mattered in attracting and retaining a bank customer.

R.P Goyal[4] in his article "Customer Service in Banks" has underlined the importance of improving customer service in banks and suggested that it could be achieved by motivating and orienting the staff, simple systems and procedures and specific schemes to suit customer needs.

H.K. Bedbak[5] in his study entitled "Institutional Financing for Priority Sectors – An Analysis of Delay and Attitude," has analysed the delays in sanctioning of loans and attitude of institutional agencies towards the customers as borrowers.

Manjit Singh[6] in his project "A Study of the Impact of Bank Lendings on Weaker Sections – A Case Study of Agricultural Development Branch of State Bank of India, Moga" has reviewed the standard of living of beneficiaries and non beneficiaries in Moga of Madhya Pradesh.

R. Neelamegam[7] in his research study "Institutional Financing to Small Scale Industries" has reviewed the various types of institutional financing facilities available to small scale industrial units in Tamil Nadu.

Ranade[8] in his study entitled "Marketing of Deposit and Allied Service to Non-resident" customers concluded that guide service is the major factor influencing an NRI in the selection of a bank.

Eugene W. Anderson, Daes and Furness and Donald R. Lehmann[9] discussed the links between quality, expectations, customer satisfaction and profitability. The findings state that when quality and expectations increase, there is a positive impact on customer satisfaction and in turn, profitability.

4 Objectives of the Study

1 to analyse the influence of Gender status of customers on service quality in banks.
2 to offer suggestions for the improvement of customer services in State Bank of India

5 Scope of the Study

This study covers the customer services rendered by State Bank of India in Madurai city. As the study is an empirical study to identify the attitude of the customers towards the services rendered by the banker, the study has been focused towards customers who are the recipient of services and bank employees who are the agencies of delivery of services. As such, it has been projected from the point of view of bank employees and from the point of view of bank customers. It is analysed with reference to customers and employees attitude. The State Bank of India in Madurai city consists of 13 branches. The study was undertaken on the customers and Bank employees of 13 branches only.

6 Methodology

The present study is an empirical one based on survey method. Data were collected from both primary and secondary sources. The primary data were collected from banks' customers and bank employees by means of interview schedule and questionnaire.

7 Sampling Design

The study aims at analysing the attitude of customers of State Bank of India in Madurai city branches with regard its services.

242 bank employees were supplied with the questionnaire in 13 branches of State Bank of India in Madurai city. But only 240 respondents filled in the questionnaire. Out of these 240 respondents, 60 respondents were bank officials (Bank Managers and Officers). All of them have responded. The remaining was clerical which amounting to 60 per cent of the clerical staff selected on proportionate random sampling method. Indeed, the questionnaires were given through branch managers of the said 13 branches to the clerical staff and they got back researcher the questionnaire filled in by the clerical staff of the respective branches.

8 Geographical Area of the Study

The study covers the whole area of Madurai city only where the branches of the State Bank of India are situated. They are Amman Sannadhi Branch, Arasaradi Branch, Commercial Tax Complex, Madurai Agricultural Development Bank Branch, Madurai city Branch, Pasumalai Branch, Personal Banking Branch, Tallakulam Branch, Vinayaganagar Branch, West Tower Branch, Railway Station Branch and Madurai Main Branch.

9 Analysis of the Study

An attempt has been made to analyse the influence of Gender status on service quality of banks by Chi Square test.

9.1 Gender Status of the Respondents and their Opinion Level

An attempt has been made to analyse whether the gender status of the sample customers has a direct impact on the opinion level.

Table 2. Gender and Opinion Level of Sample Customers

Sl. No.	*Gender*	*Opinion Level*			*Total*
		Low	*Medium*	*High*	
1.	Male	55	279	95	429
2.	Female	37	145	39	221
	Total	92	424	134	650

Source: Computed from Primary Data.

From Table 2, it is inferred that among the 92 sample customers who have low-level opinion of banking services, 55 customers are male and 37 customers are female. Among medium opinion level customers, 279 are male and 145 are female.

Among high opinion level customers 95 and 39 are male and female respectively.

In order to test whether there is any relationship between the gender and opinion level of sample customers, chi-square test was applied.

9.2 II Null Hypothesis

There is no significant relationship between the gender of the respondents and their opinion level.

Table 3 shows the working of the chi-square test.

Table 3. Chi-square Test for Gender and Opinion Level of Sample Customers

Cell	*O*	*E*	*O-E*	*(O-E)2*	*(O-E)2/E*
R1 C1	55	60.72	-5.72	32.7184	0.538841
R1C2	279	279.84	-0.84	0.7056	0.002521
R1C3	95	88.44	6.56	43.0336	0.486585
R2C1	37	31.28	5.72	32.7184	1.045985
R2C2	145	144.16	0.84	0.7056	0.004895
R2C3	39	45.56	-6.56	43.0336	0.944548
Total					3.023374

Degrees of freedom = df = (c-1) (r – 1)
= (3 – 1) (2 – 1)
= 2
Calculated value of $\chi 2$ = 3.023374
Table value of χ 2 0.05 = 5.99

Since the calculated value is less than the Table value at five per cent level of significance, the null hypotheses is accepted. Hence, it is concluded that there is no significant relationship between the gender of the sample customers and their opinion level.

10 Suggestions

Recognition of service quality as a competitive weapon is relatively a recent phenomenon in the Indian Banking sector. Prior to the liberalisation era the banking sector in India was operating in a protected environment and was dominated by nationalised Banks. Banks at that time did not feel the need to pay attention to service quality issues and they assigned very low priority to identification and satisfaction of customer needs. Hence banks should concentrate more on tech savy products and services to have more customers and to retain their loyalty.

11 Conclusion

Customers vary in their expectations and attitudes and belong to wide socio-economic and cultural backgrounds. The gap between the expectations of customers and their fulfillment is the root cause of grievances which affects the image of the bank. To overcome this situation, there should be an effective monitoring mechanism and constant vigil over the services provided to customers. Since they have a wide choice of services and multiplicity of products they are more conscious of convenience and cost, safety and speed, respect and quality, courtesy and elegance. State Bank of India has to be very careful in responding to the needs of their customers in an intensely competitive and rapidly changing environment.

References

1. Bhaskara Rao, K.V.: Customer Service in Banks, Banking and Finance, p.8 (January 2006)
2. Sarkar, D.: TIPs for Customer Satisfaction, p.10. Beacon Books, New Delhi (1997)
3. Kamath. Marketing of Bank Service with Special Reference to the Branches in Bombay City of Syndicate Bank, BMP Ph.D., Thesis, Submitted to NIBM Pune (1979)
4. Goyal, R.P.: Improving Quality of Banking Service Commerce. Banking Supplement, p. 3, Bombay, July 30 (1983)
5. Bedbak, H.K.: Institutional Financing for Priority Sectors – An Analysis of Delay and Attitude. Commerce and Management 3(1), 20–30 (1983)

6. Singh, M.: A Study of the Impact of Bank Lending on Weaker Sections – A Case Study of Agricultural Development Branch of State Bank of India, Moga, Research Project Sponsored by Indian Council of Social Research, New Delhi (1983)
7. Neelamegam, R.: Institutional Financing to Small Scale Industries., Ph.D., Thesis Submitted to Madurai Kamaraj University, Madurai (1983)
8. Ranade, M.P.: Marketing of Deposit and Allied Service to Non-Resident., BMP, Ph.D., Thesis Submitted to NIBM, Pune (1985)
9. Anderson, E.W., Daes, F., Lehmann, D.R.: Customer Satisfaction, Market Share and Profitability: Findings from Sweden. Journal of Marketing 58, 53–66 (1994)
10. Chidambaram, R.M.: Promotional Mix for Bank Marketing. IBA Bulletin 16(3), 24–26 (1991)
11. Bhattacharyya, B.N.: Marketing Management and Innovators in the Light of Liberalisation. Prajnan 20(5), 419–425 (1991)
12. Zeithaml, V.A., Bitner, M.J.: Building Customer Relationship through Segmentation and Retention Strategies, Service Marketing, pp.169-199. McGraw Hill (1996)
13. Sadare, A.M.: Competency Building for Improving Customer Service in Banks. Banking and Finance 17(8), 19–21 (2004)
14. Financial Express (March 8, 2004)
15. Financial Express (May 29, 2004)
16. Krachenberg, A.R.: The Marketing of Services with Particular Reference to Commercial Bank Services, Ph.D., Thesis Submitted to University of Michigan (1963)
17. Marjorie, P.: Marketing and Banking – An Evaluation of Marketing Program Development in Selected Commercial Banks, Ph.D. Thesis Submitted to Indiana University (1965)
18. Dale, R.J.: Commercial Bank Selection: A Factor Analysis of the Determinants of Consumer Choice, Ph.D. Thesis University of Arkansas (1972)
19. Fitts, R.L.: Predictors of Usage Rates for Selected Commercial Bank Service - A Cross Sectional Approach Utilizing Socio-economic, Demographic and Attitudinal Variables, Ph.D., Thesis Submitted to University of Alabama (1975)
20. Goyal, R.P.: Improving Quality of Banking Service Commerce, Banking

The Service Quality of Indonesia's Logistics Service Provider in Preparation for ASEAN Economic Community

Mahendrawathi ER[1], Thananya Wasusri[2,*], Hanim Maria Astuti[1], and Anisah Herdiyanti[1]

[1] Information Systems Department, Institut Teknologi Sepuluh Nopember, Surabaya, Indonesia, 6011

[2] Graduate School of Management and Innovation, King Mongkut University of Technology Thonburi, Bangmod, Tungkru, Bangkok, 10140 Thailand

{mahendra_w, anisah}@is.its.ac.id, thananya.was@kmutt.ac.th, hanim@its-sby.edu

Abstract. Logistics Service Providers (LSPs) in Indonesia will play an increasingly significant part in the ASEAN Economic Community (AEC). To prepare for the single market place at the end of 2015, Indonesia LSPs need to understand their existing situation and how they can improve their performance. The study follows a prior research on Thai LSPs and thus the same questionnaire was distributed to Indonesia LSPs and users. The SERVQUAL method was utilized to identify the gaps between the LSPs and users. It was found that neither Indonesia LSPs nor Thai LSPs can perform their service to meet with their customers' expectation over the service delivered. The LSPs must improve their performance to satisfy their customers, which will directly drive product flows and eventually help them gain benefits from AEC.

Keywords: Logistic Service Provider (LSP), ASEAN Economic Community (AEC), Service Quality.

1 Introduction

Logistics is an important aspect in today's business environment. Logistics costs contribute a big part of the total product costs and therefore the amount spent on logistics becomes a significant factor of firm's competitiveness [1]. A study by Frost and Sullivan (2012) suggests that the total transportation, storage and courier in Indonesia is IDR 235.70 Trillion while the potential market for the outsourced and contract logistics service providers hidden in various sectors including manufacturing, agriculture, construction are up to IDR 997.22 Trillion in 2012. Furthermore, the size of transportation and logistics market is expected to grow 14.7 percent between 2011 and 2016.

The establishment of ASEAN Economic Community (AEC) that envisions a single market and single production base within ASEAN countries in 2015 is expected to

* Corresponding author.

M. Gen et al. (eds.), *Industrial Engineering, Management Science and Applications 2015*, Lecture Notes in Electrical Engineering 349, DOI: 10.1007/978-3-662-47200-2_68

open up ample opportunities for Logistic Service Providers (LSPs) in the Southeast Asia countries, yet it also brings major challenges. Indonesia's logistics position is lagging behind many other countries in ASEAN. Indonesia is ranked number 53 in the Logistics Performance Index 2014 published by the World Bank [15] behind Singapore (ranked 5), Malaysia (ranked 25), Thailand (ranked 35) and also Vietnam (rank 48) in various logistics aspects as summarized in table 1.

Table 1. Logistics Performance Index (LPI) of ASEAN Countries in 2014[15]

Country	Customs	Infrastructure	International shipments	Logistics quality and competence	Tracking and tracing	Timeliness
Singapore	4.01	4.28	3.70	3.97	3.90	4.25
Malaysia	3.37	3.56	3.64	3.47	3.58	3.92
Thailand	3.21	3.40	3.30	3.29	3.45	3.96
Vietnam	2.81	3.11	3.22	3.09	3.19	3.49
Indonesia	2.87	2.92	2.87	3.21	3.11	3.53
Philippines	3.00	2.60	3.33	2.93	3.00	3.07
Cambodia	2.67	2.58	2.83	2.67	2.92	2.75
Lao PDR	2.45	2.21	2.50	2.31	2.20	2.65
Myanmar	1.97	2.14	2.14	2.07	2.36	2.83

These situations give raise the need to understand the Indonesia's LSP condition and how they can improve their performance. Unlike manufacturing companies where product quality is more tangible and therefore easily measured, the quality of services provided by service providers are more difficult to measure and rely heavily on the perceptions of the customers. Service Quality (SERVQUAL) is a well-known technique that has been used to investigate the performance of service enterprises using the perspectives of the users.

There is a rich body of literature where SERVQUAL method was employed in various industries [9],[11],[3],[2]. Banomyong and Supatn (2011) specifically explored the shippers in Thailand to understand quality attributes that are important in choosing LSP companies. However, the literature on Indonesia LSPs is limited if not missing from the literature to date. The objective of this paper is to investigate the existing gaps between Indonesia LSPs and users using SERVQUAL, learn from similar study on Thai LSPs, and propose strategies to bridge the gap and improve the Indonesia's LSP performance in preparation for the ASEAN Economic Community 2015.

The paper is organised as follows. Section 2 describes literature that explore various aspects of LSP and SERVQUAL method; while Section 3 explains the methodology used in this study. Results obtained from the survey are described in Section 4, and followed by a brief review on findings from similar study conducted in Thailand (Section 5). Discussion and comparison with Thai's study are given in Section 6 along with practical implications for the Indonesian government and LSP.

2 Literature Review

The LSPs have a prominent role in the integration of supply chain, e.g. improving services quality to customer, and impacting overall business performance [12]. Larsen

(2000) in Seth et al. (2006) defines the role of an LSP as a hub of logistic services that can facilitate partnership, and vendor agreements. It helps a company to focusing on their core business process, while the logistic systems are being outsourced to LSPs. They provide a range of services from land transportation to water and air transportation via, e.g. truck, train, ship, and airplane. It also provides freight services for special products: bulk, containers, dangerous chemicals, frozen goods, and also within a specific time frame depending on the cost and services to provide, e.g. just in time, one day service. Other services may include warehouse facilities, operational services (bar coding, radio frequency), value added services (cross-docking, consolidation, pick and pack), and customs clearance.

There is a rich body of literature that address different aspects of logistics service provider including service quality measurement and investigating the relationships between service quality, customer's satisfaction and customer's loyalty. One of the most prominent methods to measure service quality is SERVQUAL method. SERVQUAL is defined service quality as the difference between customers' expectation and perceptions of services delivered by service enterprises[10]. On the one hand, Lewis and Booms (1983) also identified service quality as a measure to evaluate the service level delivered versus customer expectations; Cronin and Taylor (1992) viewed service quality derive from perception of performance alone. Although there are differences on service quality definition given, but the researchers share common viewpoints. First, service quality is an attitude and it is distinct from customer satisfaction. Second, perceptions of performance need to be measured and finally, the number and definition of dimensions depends on the service aspects.

Parasuraman et al. (1985) identified the service quality performance into 5 (five) Gaps. GAP1 is Customers' expectations versus management perceptions on customers' expectations. The result of GAP1 shows whether or not the service enterprises could understand the customers' expectations. GAP2 is management perceptions on customers' expectations versus service specifications identified by the service enterprises. This gap can identify the commitment to service quality. Although the management may understand customers' expectations very well, inadequate task standardization and absence of goal setting can occur. GAP3 is service specifications versus service delivery. The result of GAP3 shows the ability to deliver the service according the service specification. GAP4 measures service delivery versus external communication. Statements or advertisements made by the service enterprises can influence customers' expectations. This GAP will show the customers' expectations cannot be fulfilled at the time of the service delivered. GAP5 is the discrepancy between customer expectations and their perceptions of the service delivered. This gap easily shows the differences between customers' expectations and their perceptions.

Meanwhile, SERVQUAL has been utilized to measure service quality. There are five generic dimensions or factors to investigate. Van Iwaarden et al. (2003) pointed the five dimensions following these: 1) *Tangibility* that is physical facilities, equipment and appearance of personnel; 2) *Reliability* that is ability to perform the promised service dependably and accurately; 3) *Responsiveness* that is willingness to help customers and provide prompt service; 4) *Assurance* that is competence, courtesy, credibility and security to create trust and confidence on customers; and 5) *Empathy* that is caring and individualized attention that the firm provides to its customers.

Several authors have specifically investigated the issue related to service quality of Logistics Service Provider or Third Party Logistics (3PL). Seth et al (2006) stated that the SERVQUAL instrument provide a useful framework to assess service quality in 3PL relationships. Other authors also applied SERVQUAL for different type of 3PL companies including company that distribute consumer goods [9] and sea transport services[3].

Banomyong & Supatn (2011) conducted a survey on shipper's in Thailand to identify key freight logistics service quality attributes and their respective impacts on shipper's decision making process. They identified 24 freight logistics service quality and classified them into six attributes, while they also employed SERVQUAL model. The authors found that freight service quality significantly impacted shipper's selection and the service attributes that influenced how shippers selected 3PL are: accuracy of documents, EDI & e-commerce services, CRM, customer care, updated freight rates, consolidation provision. While research has been done in the context of Thai LSP, Indonesia is still absence from the literature to date. This paper aims at bridging this gap by conducting a study to assess the service quality of Indonesia's LSP.

3 Methodology

Based on the literature on SERVQUAL, two sets of questionnaires are administered: 1) questionnaire for the LSP, and 2) questionnaire for the LSP users. Both questionnaires consist of two main parts. The first part covered the background of the companies. In the second part of the user's questionnaire, users are asked their perceived and expectations of their LSP performance on five service quality i.e. empathy, assurance, responsiveness, reliability and tangibility. The questionnaire for LSP asked them to rate their performance and their perceptions of the customer's expectations on the five service quality dimensions. Likert scale of 1 – 5 are used to rate the LSP quality of service, with 1 being the lowest and 5 is the highest. The two set of questions were then reviewed by Thai experts with the support by Thailand Research Fund and National Research Council of Thailand.

In this study, data are collected from Indonesia LSPs and users. The LSP are chosen based on the database of ALI (Indonesia Logistics Association) and the questionnaire is distributed over the mailing lists. In addition, because some of the LSP in Indonesia are not a member of ALI, a search was done through Yellow Pages– a telephone directory of businesses. The first criterion in selecting the users is that a company must use Logistics Service Provider for some part of their business. The companies' names are obtained from the Indonesia Production and Operations Management (IPOM) memberships. The questionnaire is posted through the association's mailing lists. The LSP users are companies using the services of LSP. The companies are located in different regions in Indonesia of which have different types of services. For Thailand result, the data were obtained from the work of Tansakul et al. (2013a, 2013b).

4 Results of Indonesia

4.1 Descriptive of Respondents

Results are obtained from 32 LSPs located mostly in East Java, Jakarta, and Bali while 16 questionnaires are returned from the users. As shown in Figure 1, these companies offer different types of logistics services ranging from various transportation warehouse, custom, consulting and value added services such as packaging and labeling. In terms of revenue, most of LSP companies have revenue between $0.2 million and $1 million and six companies have a revenue over $1 million. The rest of the companies (12 companies) do not share their annual revenue information.

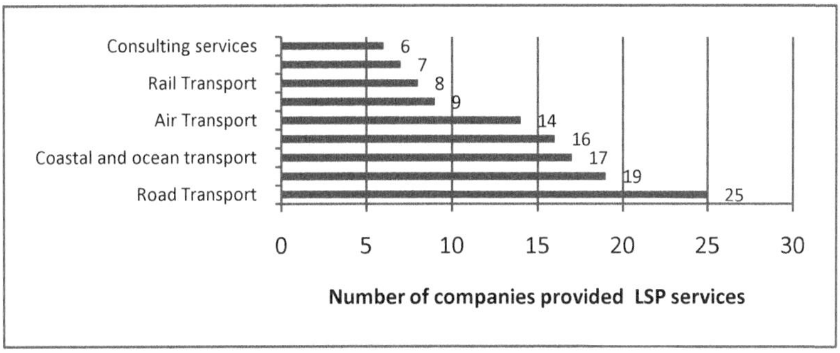

Fig. 1. Types of services offered by Indonesia LSPs [13]

4.2 Gaps' Results

Findings from both questionnaires are presented in Table 2 [14]. The findings are then analyzed to calculate the expectation gaps, perception gaps and service gaps by users. The rest of this part will discuss about these gaps.

Table 2. The gap score among Indonesia LSP and LSP users derived from SERVQUAL[14]

Dimensions	Expectation (E) Gaps		Perception (P) Gaps		Service Gaps	
	E-LSP	E-Users	P-LSP	P-Users	E-Users	P-Users
Empathy	4.43	4.26	3.85	3.11	4.26	3.11
Assurance	4.33	4.30	3.74	3.30	4.30	3.30
Responsiveness	4.49	4.45	3.85	3.16	4.45	3.16
Reliability	4.42	4.44	3.98	3.27	4.44	3.27
Tangibility	4.25	4.30	3.67	3.16	4.30	3.16

Expectation Gaps – Expectation (E) gap's column in Table 2 shows the expectation gaps between Indonesia LSPs and users. The gaps between LSP expectations and user expectations are relatively narrow. From these gaps, we argue that the

Indonesia's LSP can understand the user's expectations relatively well, particularly in terms of assurance, responsiveness and reliability. The result also shows that the user's expectation in terms of *tangibility* is higher than those of the LSP's expectation. In contrast, the LSPs overestimate user's expectation in terms of *empathy*.

Perception Gaps – According to perception (P) gap's column in Table 2, the Indonesia LSPs tend to overrate their service quality against the users perceptions. The average LSP perception on the five dimensions is 3.82 while the corresponding value from the users is only 3.20, which creates a 0.618 gaps.

LSP perceived the highest and the lowest quality are in terms of *reliability* and *tangibility*, respectively. However, the users thought that LSP score highest in terms of *assurance* and lowest in terms of *empathy*. The findings in the *empathy* dimension deserve further discussion. While the user's rate LSP's quality in terms *empathy* as the lowest of all the five dimensions, the LSP perceived their performance in that dimension relatively high. This leaves a big gap in this dimension. Referring to expectation gaps, the LSP realized that it is an important dimension for the user, which is shown in a high LSP expectation of *empathy*, even higher than the average user.

Big gaps were also found in terms of *reliability* and *responsiveness* which means that the customer perceived the LSP service in these two dimensions much less than what the LSP perceived they have provide to their customers. Certainly this recalls for improvement in the LSP side.

Service Gaps by Users – Service gap by users is related to the discrepancy between user expectations and user perceptions of the service delivered. As shown in Table 2, in all five dimensions there are an average of 1 point gap between what the user expected and what they perceived as being delivered by the LSP. The biggest gap according to the user is in terms of *responsiveness*. On the other hand, the narrowest gap is in terms of *assurance*. The situation raises awareness for the LSP to improve their responsiveness because customer's expectation is the highest in this dimension.

From the study, we also learned that Information Technology (IT) is the measure that has the biggest gaps in the *tangibility* dimension. Thus, LSP should continuously improve IT availability to support their services. Users' expectation and perception on the availability of IT are lower than the other tangibility measures. Timeliness in *reliability* dimension should be taken into account followed by safety working environment and basic service sufficiency. As the biggest gap lies in the *responsiveness* dimension, the priority should be given into these measures: quick response on customer complaint and providing information as a pre-transaction service for customer. To satisfy user in terms of *assurance* dimension, the LSP is expected to provide training to improve logistics staff's capability. In addition, flexibility service giving to users is the key to satisfy customer in *empathy* dimension.

5 Lesson Learned from Thai's LSP

Having an understanding toward the existing condition of Indonesia LSP and LSP users as explained in the previous section, this section provides result from a similar study conducted by Tansakul et al. (2013a, 2013b) on Thai LSP and LSP users. The study obtained the results from 67 Thai LSPs. About 51% of the respondents are located in Central of Thailand and the others are spread around Thailand. For the users, 92 questionnaires are returned. The respondents are mostly small and medium sizes with having annual revenue no more than $200 million. The services offered are road transport, customs clearance and warehousing. The rest of this section will discuss about the three gaps i.e. expectation gaps, perception gaps and service gaps by users from the work of Tansakul et al. (2013a, 2013b) are summed up (in Table 3) and analyzed.

Table 3. The gap score among Thai LSP and LSP users derived from SERVQUAL ([13,14])

Dimensions	Expectation (E) Gaps		Perception (P) Gaps		Service Gaps	
	E-LSP	E-Users	P-LSP	P-Users	E-Users	P-Users
Empathy	4.26	4.01	4.27	3.49	4.01	3.49
Assurance	4.19	4.08	4.18	3.62	4.08	3.62
Responsiveness	4.33	4.04	4.21	3.56	4.04	3.56
Reliability	4.34	4.15	4.24	3.64	4.15	3.64
Tangibility	4.22	4.03	4.15	3.55	4.03	3.55

Expectation Gaps – According to Table 3, it can be interpreted that Thai LSPs could understand their customers and they even overestimate customer expectation in every quality service dimension. Some significant gaps occurred in availability of vehicle and equipment and staff sufficiency, but it occurred by the over expectation of the LSP and it was found in other dimensions as well.

Perception Gaps – In term of perception gaps between Thai LSPs and their users, it can be said that the average scores considered by the users are moderate as the scores are around 3. On the other hand, Thai LSPs determine themselves very positively seen from their average scores are about 4. In other words, the average LSPs' perception on five dimensions is 4.21 while the average users' perception is only 3.57 and it is a 0.64 gap.

Thai LSPs considered that *tangibility* is the weakest performance and they rated the score at 4.15. It is similar to the users' perception that the *tangibility* is quite low comparing to the others, but the users' score is low at 3.55. The highest performance rated by the users is *reliability* performance. It was rated at 3.64 by the users while Thai LSPs considered themselves at 4.24. On the other hand, Thai LSPs perceived that they could do their best on the *empathy* dimension and the score was rated at 4.27, but the users perceived the *empathy* capability at 3.49 and it is the lowest score for the users.

Service Gap by Users - For the service gap by Thai users, the smallest gap is *assurance* dimension while the highest gap is *empathy*. It means that Thai LSPs can convey its assurance services to meet with the customers' expectation better than the other dimensions. Although, the perception of *reliability* is the highest, but its expectation is also the highest.

6 Discussion

6.1 Logistics Service Provider Gaps

Results from the two studies show that the users in Indonesia and Thailand have different expectations and perceptions. Overall, Indonesia users have higher expectations in all of five dimensions compared to Thai users. It can be seen that expectations of Indonesia and Thailand users are differently ranked. The quality service dimensions for Indonesia users are ranked as *responsiveness, reliability, tangibility, assurance*, and *empathy* consecutively. On the other hand, those of Thailand users are *reliability, assurance, responsiveness, tangibility,* and *empathy* consecutively. The results from the study also depicts that both Indonesia and Thailand LSPs could understand their customers' expectation pretty well. However, there are still gaps between LSP and users perceptions in the two countries. Bigger gaps are found between Indonesia LSPs' perception and users' perception. Indonesia LSPs have an average of 1 point gap while the average gap of Thai LSPs is less than 1 in every dimension.

These findings emphasis greater challenge faced by the Indonesia LSP. Firstly, the Indonesia customers have higher expectations and they perceived Indonesia LSP performance much less. Indonesia LSP must translate their customers' expectation into their standard working procedures. Furthermore, they should be able to correctly evaluate their performance related to their customers' perception.

Another important finding is that the users in each country rank the five dimensions differently. The Indonesia users rank for the five service dimensions are *responsiveness, reliability, tangibility, assurance*, and *empathy* consecutively. On the other hand, ranking from Thailand users are *reliability, assurance, responsiveness, tangibility,* and *empathy,* consecutively. In order to compete in the AEC era, Indonesia LSP must also be able to accommodate the customers' needs and requirements, distinguish different customers and tailor made their services.

In order to reduce service quality gaps, Indonesia and Thailand LSPs should improve themselves starting from having business strategies to comply with customer's expectation. Human resources should be developed and evaluated systematically and it is very important to incubate them with service mind. The use of IT to enhance their services and the information sharing with customers and suppliers are very essential. It is necessary to be available to service customer, the preventive maintenance of the vehicles and equipment should then be maintained. Work procedures should be standardized and to ensure service consistency among workers.

It can be said that the two countries must improve themselves in three aspects that are people, process and technology. Indonesia and Thailand logistics human resources must be developed in terms of the foundation of logistics management and other needed skills such as communication, critical thinking and service mind. Process must be reengineered focusing on value added perspective to achieve customer's satisfaction with time and cost reduction. Technology is the main point that both Indonesia and Thailand customers require. LSP must provide them with sufficient and efficient IT to plan, manage and monitor the logistics of products or packages.

6.2 Logistics Policy

In 2012 the Indonesia government have launched the National Logistics Systems policy to reach the 2020 vision of "Locally Integrated, Globally Connected for National Competitiveness and Social Welfare". The policy is based on supply chain management to reduce logistics cost through ICT implementation. The approach is on utilization of six key logistics drivers: key commodities, logistics infrastructure, service player and provider, logistics human resources, ICT and regulation harmonization. The second national logistics policy of Thailand has been launched and the period of policy covers from 2013 until 2017. The theme of the policy is to facilitate trades and supply chains for competitiveness. The policy can be classified into three main missions that are: 1) to strengthen Thai entrepreneurs to benefit from supply chain enhancement; 2) to enhance trade facilitation; and 3) capacity building and policy driving factors. Overall, the policies have the same direction and emphasis. However, Thai government have provided clearer actions. For example, according to the logistics policy, Thai LSPs will be supported to get connection between LSPs. Furthermore logistics professional standards for four positions including warehouse operator, inventory operator, forklift driver and road transport operator have just been launched. The professional standards can raise the effectiveness and efficiency of logistics activities. Consequently the service quality dimensions could be improved. The Indonesia government must fasten the implementation of the national logistics policy to prepare the LSP and other players in facing AEC.

7 Conclusion

This paper present findings from a survey conducted to measure the service quality of Logistics Service Providers in Indonesia in five dimensions including *empathy, assurance, responsiveness, reliability* and *tangibility*. Thai results of the similar study from prior research were also discussed. Gap analysis based on SERVQUAL methods showed that the LSP can predict their user's expectation relatively well, shown from narrow gaps between User's and LSP's expectations in all five dimensions. The biggest gap should be firstly bridge and these findings can help the LSP focus their effort to improve their performance. It is found that most of Indonesia and Thai LSPs are small and medium. They cannot cope with international logistics companies. The national policy should be implemented to support LSP especially small and medium sizes to improve their capabilities in facing the AEC.

Acknowledgement. The work presented in this paper is supported by International Collaboration Research Grant from Institute of Research and Public Services, Institut Teknologi Sepuluh Nopember. The authors also thanks Thai Research Fund for the support provided in the preliminary study.

References

1. Banomyong, R.: Thailand logistics costs/GDP. Thammasat Logistics Research Papers 3(1), 1–2 (2007)
2. Banomyong, R., Supatn, N.: Selecting Logistics Providers in Thailand: a Shipper's Perspective. European Journal of Marketing 45(3), 419–437 (2011)
3. Chen, K., Chang, C., Lai, C.: Service quality gaps of business customers in the shipping industry. Transportation Research Part E 45, 222–237 (2009)
4. Cronin, J.J., Taylor, S.A.: Measuring service quality: a Re-examination and extension. Journal of Marketing 56, 55–68 (1992)
5. Bookbinder, J.H., Tan, C.S.: Comparison of Asian and European Logistics Systems. Int. J. Physical Distribution and Logistics Management. 33(1), 36–58 (2003)
6. Frost, Sullivan: Indonesia Logistics Industry Outlook: 2012. An Executive Insight. Frost and Sullivan (2012)
7. van Iwaarden, J., van der Wiele, T., Ball, L., Millen, R.: Applying SERVQUAL to websites: An exploratory study. Int. J. of Quality & Reliability Management 20(8), 919–935 (2003)
8. Lewis, R.C., Bernard, H.B.: The Marketing Aspects of Service Quality. In: Shostack, L.B., Upah, G. (eds.) Emerging Perspectives on Services Marketing, pp. 99–107. American Marketing, Chicago (1983)
9. Neo, H.Y., Xie, M., Tsui, K.L.: Service quality analysis: Case study of 3PL company. Int. J. of Logistics Systems and Management 1, 64–80 (2004)
10. Parasuraman, A., Zeithaml, V.A., Berry, L.L.: A conceptual model of service quality and its implications for future research. J. of Marketing 49(3), 41–50 (1985)
11. Seth, T., Deshmukh, D.K., Vrat, P.: A Conceptual Model for Quality of Service in the Supply Chain. Int. J. of Physical Distribution & Logistic Management (2006)
12. Sumantri, Y., Lau, K.S.: The Current Status of Logistics Performance Drivers in Indonesia: An Emphasis on Potential Contributions of Logistics Service Providers (LSPs). J. of APIBTMS (2011)
13. Tansakul, C., Buddhakulsomsiri, J., Wasusri, T., Chaiwat, P., Kritjaoren, T.: A Study on Service Quality of Thai Logistics Service Providers. In: Proceeding of the 8th International Congress on Logistics and SCM systems (ICLS 2013), Tokyo, Japan (2013a)
14. Tansakul, C., Buddhakulsomsiri, J., Wasusri, T., Chaiwat, P., Kritjaoren, T.: A Gap Analysis in Service Quality of Thai Logistics Service Providers. In: Proceedings of the 4th International Conference on Engineering, Project, and Production Management (EPPM 2013), Thailand, Bangkok (2013b)
15. World Bank.: The Logistics Performance Index and Its Indicator. Washington. USA (2014)

Effects on Physical and Mechanical Properties of Thermochemical Treated Kenaf (*Hibiscus Cannabinus*) Fibres Composite Board

Mohamad Nurul Azman Mohammad Taib,
Mohd Ariff Jamaludin, and Masitah Abu Kassim

Faculty of Applied Sciences, University Teknologi Mara,
40450 Shah Alam, Selangor, Malaysia
aman_peace04@yahoo.com

Abstract. In this study, physical and mechanical properties of high density kenaf (*hibiscus cannabinus*) fibreboard (HDKF) were evaluated. Thermochemical treated kenaf fibres were treated with three different temperatures that were 100ºC, 130ºC and 150ºC in laboratory autoclave. The physical properties (thickness swelling and water absorption) and mechanical properties (static bending and internal bonding tests) were tested according to British Standards. The results showed that the thermochemical treatment was effective to reduce thickness swelling (TS) and water absorption (WA). Since the static bending test modulus of rupture (MOR) was reduced, whereas modulus of elasticity (MOE) was increased up to 130°C of treatment than started to drop when treated with 150°C of temperature. For internal bonding test results showed some improvement when treated with thermochemical treatment.

Keywords: kenaf, fibreboard, mechanical properties, morphological, thermochemical.

1 Introduction

Modification of lignocellulosic materials has been done by researchers to enhance the properties of the materials in terms of physical and mechanical properties [1],[2],[3]. Thermochemical treatment offers some potential to modify the wood or lignocellulosic material properties. This modification is related to overcome dimensional stability problem and increase in strength properties [4],[5]. Conventional treatment using sodium hydroxide has been known for its tendency to modify or give significant improvement on the mechanical properties and reduce the moisture absorption. It has been known that introduce the heat treatment on the wood or lignocellulosic material can enhance the resistance to moisture absorption and increase the durability of the material [1],[6],[7],[8],[9]. The modification using alkali treatment occurs when the hydrogen bonding in the network structure is disrupted, thereby increasing surface roughness [10]. Alkali treatment removes certain chemical component such as hemicelluloses, lignin, wax and oil that covering the external surface of the fibre cell

M. Gen et al. (eds.), *Industrial Engineering, Management Science and Applications 2015*,
Lecture Notes in Electrical Engineering 349, DOI: 10.1007/978-3-662-47200-2_69

wall. A chemical reaction occurs with depolymerization of cellulose and exposes the short length crystallites [10].

Kenaf or scientific name *Hibiscus cannabinus* has been known as a material that have good fibre properties that have similar characteristics as a Jute fibre. Kenaf is a warm seasonal plant and traditionally has been utilized as a sack and rope. Kenaf has two parts, outer part that called bast and inner part that called core [11]. Usually researchers utilize the bast part as a fibre because bast fibre has more roughly characteristic than core fibre.

Fibreboard consumption exceeds 618million m^2 with approximately 25% of that is medium density fibreboard (MDF). Usually binder that used for manufacturing commercial fibreboard is urea formaldehyde or UF resin that used for non-structural engineered wood market [12],[13]. For structural used the binder must have several criteria such as resistance to moisture sorption, structural bonding performance and durability [13],[14]. Some of common resin that used for structural product are phenol formaldehyde (PF) resin, melamine urea formaldehyde (MUF) and cyanate resin [15]. This resin has an ability to resist moisture and better bonding properties when compared with urea formaldehyde resin.

2 Materials and Methods

The HDKF panels were manufactured from kenaf fibres that were obtained from Malaysia Agricultural Research and Development Institute (MARDI). The fibres were further processed in Bio-composite Laboratory at Universiti Teknologi MARA Malaysia (UiTM), Shah Alam. The fibres were crushed using crusher machine to obtain fine fibres. Then, the fine fibres was sieved to get uniform fibres. The fine fibres were oven dried in oven until constant moisture of 5% was achieved. A liquid PF resin with a solid content of 40% was sprayed and mixed with fine fibres in a blender, Then the mixed resin and fibres was formed into a mat and cold pressed, then hot pressed to produce untreated and thermochemical treated HDKF. For thermochemical treated kenaf fibres HDKF, the fibres were processed in an autoclave prior to blend with PF. The fibres was boiled in autoclave with 3% concentration of Sodium Hydroxide at three different temperatures that were 100°C, 130°C, 150°C for 30 minutes. After autoclaving, the fibres were dried in a ventilated oven for 7 hours at a temperature of 103°C. The fibres were first dried to 5% MC and then blended with PF, formed into a mat, cold pressed and finally hot pressed. After hot pressing, the HDKF were left to condition in ambient laboratory condition. After that, the boards were trimmed to a final dimension of 30 cm X 30 cm X 1.4 cm. The target density was 0.9 ± 0.01 g/cm^3. The boards were cut into test samples of the standard size for testing according to the British Standard for physical properties (BS EN 317) and mechanical properties (BS EN 310 and BS EN 319) determination. The dimension of for water absorption and thickness swelling tests and the internal bonding (IB) test sample was 5.0cm X 5.0cm X 1.4 cm as shown in Fig. 1. The static bending test sample was 28cm X 5.0 cm X 1.4 cm in dimension as illustrated in Fig. 2. All the test samples were conditioned to constant mass in a humidity chamber with a mean relative humidity of 65% and a temperature of 20°C.

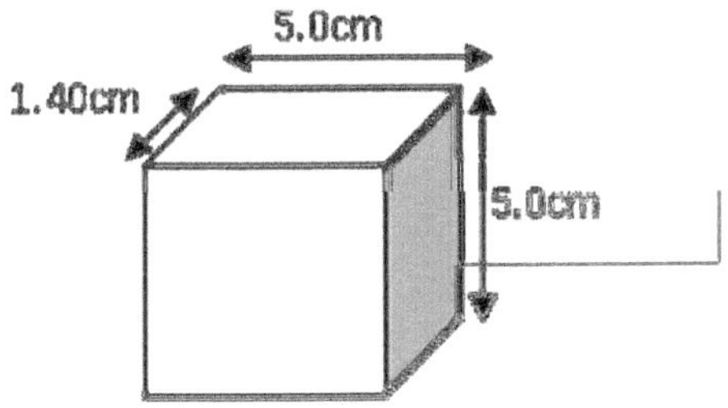

Fig. 1. Internal bonding test sample

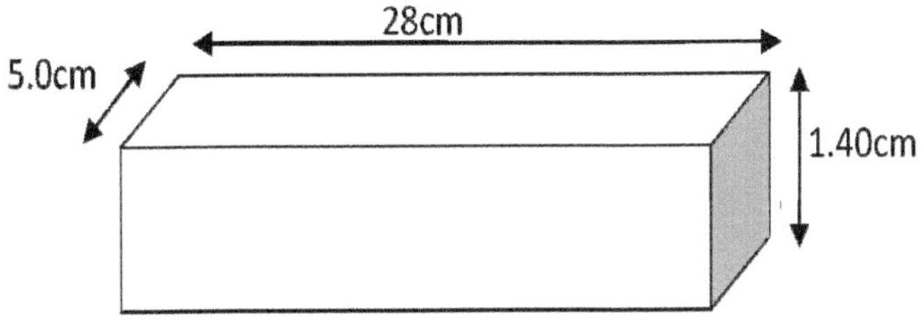

Fig. 2. Static bending test sample

3 Results and Discussions

Twelve samples have been used to determine the modulus of elasticity (MOE) and modulus of rupture (MOR) of HDKF made from thermochemically treated fibres and HDKF made from untreated fibres panels. Fig. 3 and Fig. 4 shows the MOE and MOR values. Based on multiple comparison test using Tukey's test all Types were significantly different except for Type 4 that was not significantly different from each other. The not significantly different of Type 4 was contributed by the effect of the thermochemical treatment on the fibres that degraded the structure and hydrolysis of the hemicelluloses too critical at the temperature 150°C thus reducing the MOE value. Similar study by [14] concluded that EFB that was boiled with NaOH would produced panels of lower mechanical properties compared to boards with untreated EFB fibres.

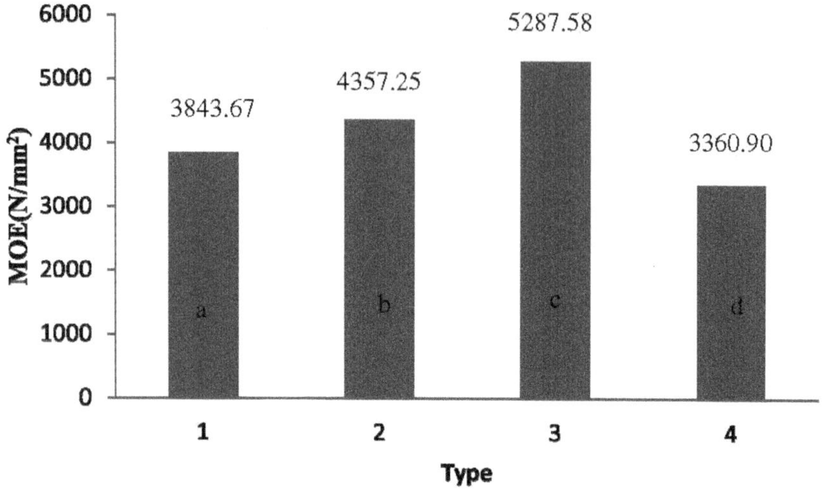

Fig. 3. Modulus of Elasticity of kenaf high density fibreboard Note: Group with the same letter indicates there is no significant different at $p < 0.05$

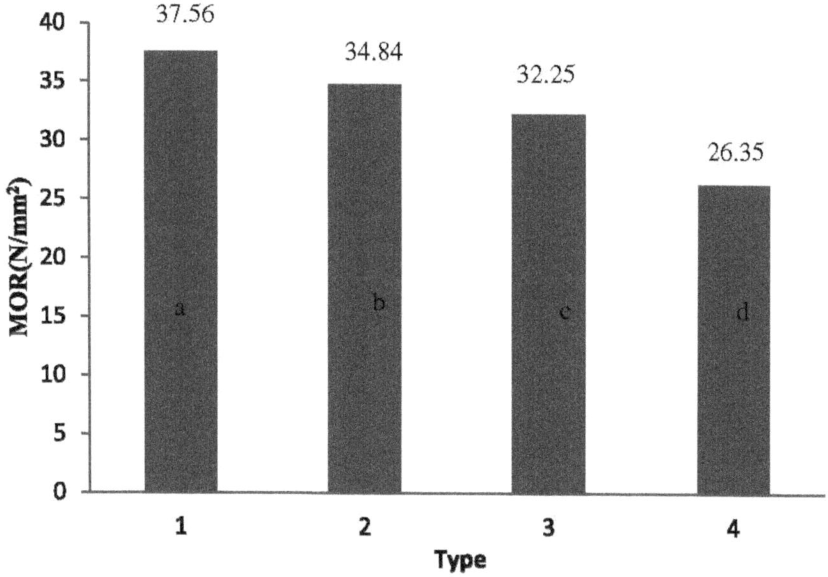

Fig. 4. Modulus of Rupture of kenaf high density fibreboard Note: Group with the same letter indicates there is no significant different at $p < 0.05$

Fig. 4 showed the MOR of test samples made from untreated fibre versus test samples made from thermochemically treated kenaf fibres. Based on multiple comparison test all the different Types of fibreboards were significantly different. The lower MOE and MOR were probably due to the softening and degradation of the fibres by the thermochemical treatment, thus lowering the fibres stiffness [16]. The degradation of fibres in terms of chemical component mainly hemicelluloses. A study by [17] stated that the reduction of mechanical properties was also contributed by the formation of soluble acidic chemicals such as formic acid and acetic acid from the degradation of hemicellulose. When these acids react with cellulose, the acids would depolymerised the cellulose chain by breaking down the long chain of cellulose in the form of crystalline structure to shorter chain resulting in lower strength of the fibre [18]. The polymerization by the acids and degradation by chemical treatments of these cellulose fibres can affect the MOR and MOE values [19]. The highest MOE value of Type 3 may be attributed by improve a mechanical interlocking and chemical bonding between the resin and fibres thus resulted in higher MOE values [20]. Previous study by [20] on the effect of alkalization on kenaf and hemp fibre composite also show higher flexural modulus of the test samples with treated fibres than the test samples with untreated fibres. Another research by [21] that conducted study on kenaf at elevated temperatures using alkali treatment also concluded that the kenaf have high MOE value due to remove of impurities (e.g. wax and pectin) on the fibre surface and make the fibre surface cleaner compared with untreated fibre surface thus, resulting in better interface bonding. Another reason for the highest MOE value of Type 3 test samples as compared to Type 1 test samples may be attributed to deep penetration of the resin into voids that developed (e.g. hemicelluloses, celluloses, wax and pectin) within the fibres after treatment. The deep penetration into the voids contributed to better binding between fibres and resin thus resulted in higher mechanical properties. The lowest result for Type 4 in Fig. 3 may attribute by the weaken of fibres after treated with thermochemical treatment at 150°C.

Result in Fig. 3 also showed that the increase of graph pattern in MOE can be achieved using thermochemical treatment up until 130°C (Type 3) at 150°C, the MOE decrease (Type 4). Further increase in temperature higher than 130°C would decrease the MOE value to below the value of the panels with untreated fibres based on the graph pattern in Fig. 3. This trend can be related of removing of chemical component (e.g. hemicelluloses and celluloses) at 150°C that is critical to the strength of the material.

The result in Fig. 4 showed lower MOR values for all the test samples made from thermochemical treated kenaf fibres when compared with the untreated samples. The lower of the MOR values is related to the reduction or degradation of the hemicelluloses and celluloses component that give rigidity to the kenaf fibres cell wall. This degradation could contribute to the lower strength properties of the fibres. Another reason is during alkali treatment at elevated temperatures, the potential for hydrolysis is quite high. This hydrolysis induced the destroying of the crystalline regions and increases the amorphous regions [22]. As a result, the cellulose becomes more flexible than before and tends to agglomerate and form lumps. Upon hot pressing the lumps would be pressed into much dense spots relative to the rest of the board. Such non-uniformity creates additional internal stress, particularly in bending which consequently lowers the MOR values of the boards. Previous study by [21] and [23] stated that the elasticity of the kenaf fibres and jute fibres increased when treated with alkaline solution but the strength of the fibres was reduced.

Fig. 5 includes the internal bonding of test samples made from untreated kenaf fibres versus test samples made from thermochemically treated kenaf fibres. Based on multiple comparison test using Tukey's for internal bonding all fibreboard types were significantly different except for Type 3 compared to Type 4. Type 4 was lowest when compared with others sample that made from thermochemical treated kenaf fibres, this was due to the extremely damaged fibres when treated at 150ºC. Using the thermochemical treatment at higher temperature made severe to fibre damage and make the bonding between fibres and resin worse. Further increase in the temperature would not increase the fibres-to-fibres bonding significantly.

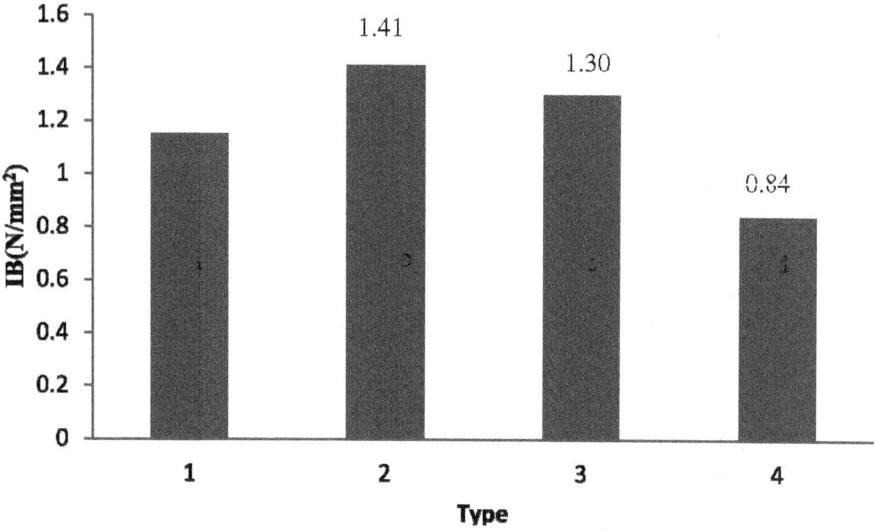

Fig. 5. Internal bonding of kenaf high density fibreboard Note: Group with the same letter indicates there is no significant different at $p < 0.05$

For thermochemically treated kenaf fibres shows that the IB was higher at 100°C (Type 2) as compared to similar samples subjected to 150ºC temperature. The higher of IB for Type 2 was related to modification of chemical component (e.g. hemicelluloses and cellulose) by sodium hydroxide that made the fibres shorter. The NaOH reacts with the hydroxyl groups on hemicelluloses in the fibres and leads to the destruction of the cellular structure therefore the fibre splits into filaments. This phenomenon is called the fibrillations, which breaks the untreated fibre bundle down into smaller fibrils by the dissolution of the hemicellulose. Fibrillation would increase the effectiveness of surface area for resin contact. The alkali treatment reduced the fibre diameter and increased the aspect ratio of the fibre itself. The short fibres improve the interfacial adhesion between the fibres [24].The lowest value for Type 4 in Fig. 5 was believed due to the extreme elevated temperature. The highest acceptable temperature for thermochemical treatment was 130ºC and over that temperature the fibres become unacceptably weakened because of excessive loss of the chemical component (e.g. hemicellulose and cellulose) in the fibres. Previous study by [24] also stated that the internal bonding strength is also dependent on the treatment and the size of the fibre used. Alkali treatment improves the fibre-resin adhesion because of the removal of natural and artificial impurities. The result indicates that treatment to a certain extent has been effective in enhancing the IB of the high density fibreboard (HDKF).

4 Conclusion

In terms of mechanical properties for bending test, panels made from thermochemically treated kenaf fibres have significant effects when compared with untreated kenaf fibres samples. Higher MOE values related with better binding between resin and fibres. The reduction pattern graph of the MOR values was related with the degradation of chemical component in the fibres especially the hemicelluloses that were degraded at elevated temperatures.

For internal bonding strength of test samples made from untreated kenaf fibres was lower when compared to the test samples made from thermochemically treated kenaf fibres at 100°C and 130°C except 150°C. Treatment with sodium hydroxide would increase the internal bonding of the samples but limited only at 130°C.

Aknowledgements. The researchers would like to thank the Universiti Teknologi MARA Malaysia, (UiTM) for providing facilities that has made this work possible.This work was financially supported by the Fundamental Research Grant Scheme (FRGS) Research Grant (600-RMI/FRGS 5/3/ (17/2013)).

References

1. Ayrilmis, N., Jarusombuti, S., Fueangvivat, V., Bauchongkol, P.: Effect of Thermal Treatment of Rubberwood Fibres On Physical and Mechanical Properties of Medium Density Fibreboard. J. of Tropical Forest 23, 10–16 (2011)
2. Bhuiyan, M.T.R., Hirai, N.: Study of Crystalline Behaviour of Heat –Treated Wood Cellulose During Treatments In Water. J. Wood Sci. 51, 42–47 (2005)
3. Hakkou, M., Petrissans, M., Zoulalian, A., Geradin, P.: Investigation of Wood Wettability Changes During Heat Treatment on the Basis of Chemical Analysis. Polymer Degradation and Stability 89 (2005)
4. NorulIzani, M.A., Paridah, M.T., Astimar, A.A., MohdNor, M.Y., Anwar, U.M.K.: Mechanical and Dimensional Stability Properties of Medium Density Fibreboard Produced From Treated Oil Palm Empty Fruit Bunch. Journal of Applied Science 12, 561–567 (2012)
5. Munawar, S.S., Umemura, K., Tanaka, F., Kawai, S.: Effects of Alkali, Mild Steam and Chitosan Treatments on the Properties of Pineapple, Ramie and Sanseveriafiber Bundles. J. Wood Sci. 54, 28–35 (2007)
6. Ayrilmis, N., Winandy, J.E.: Effects of Post Heat Treatment on Surfaces Characteristics and Adhesive Bonding Performance of Medium Density Fibreboard. Materials and Manufacturing Process 24, 594–599 (2009)

7. Tjeerdsma, B.F., Militz, H.: Chemical Changes In Hydrothermal Treated Wood; FTIR Analysis of Combined Hydrothermal and Dry Heat Treated Wood. HolzalsRoh-und werkstoff 63, 102–111 (2005)
8. Tjeerdsma, B.F., Boonstra, M., Militz, H.: Thermal Modification of Non-Durable Wood Species.Improved Wood Properties of Thermally Treated Wood. International Research Group on Wood Preservation, Document No. IRG/WP 98-40124 (1998)
9. Lawther, J.M., Sun, R., Banks, W.B.: Effect of Steam Treatmenr On the Mechanical Composition of Wheat Straw. Holzforschung 50, 365–371 (1996)
10. Akil, H., Omar, M., Mazuki, A., Safiee, S., Ishak, Z., Abu Bakar, A.: Kenaf Fiber Reinforced Composites: A Review. Materials and Design 32(8), 4107–4121 (2010)
11. Aisyah, H.A., Paridah, M.T., Sahri, M.H., Anwar, U.M.K., Astimar, A.A.: Properties of Medium Density Fibreboard (MDF) From Kenaf (2010), (Hibiscus Cannabinus L.) Core As Function of Refining Conditions. Composites: Part B (2012) (article in press)
12. Nemli, G., Aydin, I., Zekovic, E.: Evaluation On Some Properties of Particleboard As Function of Manufacturing Parameters. J. Mater. Des. 28, 1169–1176 (2007)
13. Winandy, J.E., Krzysik, A.: The Effects of Thermal Modification During Hot Pressing of MDF Composites. Part 1. Relative Effects And Benefits of Thermal Exposure. Wood Fibre and Science 39, 450–461 (2007)
14. Hunt, J.F., Vick, C.: Strength and Processing Properties of Wet-Formed Hardboards and Recycled Corrugated Containers and Commercial Hardboard Fibers. For. Prod. J. 49, 588–592 (1999)
15. Malonney, T.M.: Modern Particleboard and Dry-Process Fibreboard Manufacturing. Miller Freeman Publications San Francisco, USA (1993)
16. Jiang, J., Lu, J., Huang, R., Li, X.: Effects of Time and Temperature On the Viscoelastic Properties of Chinese Fir Wood. Drying Technology 27, 1229–1234 (2009)
17. Garrote, G., Dominique, H., Parajo, J.C.: Study On The Deacetylation of Hemicelluloses During The Hydrothermal Processing of Eucalyptus Wood. HolzalsRoh-und werkstoff 59, 53–59 (2001)
18. Hon, D.N.S., Shirashi, N.: Wood and Cellulosic Chemistry, 2nd edn, p. 914. Marcel Dekker, New York (2001)
19. Rowell, R.: Hand book of Wood Chemistry and Wood Composites. CRC Press, New York (2005)
20. Aziz, S.H., Ansell, M.P.: The Effect of Alkalization and Fibre Alignment On The Mechanical and Thermal Properties of Kenaf And Hemp Bast Fibre Composites: Part 1-Polyester Resin Matrix. Composites Science and Technology 64, 1219–1230 (2004)
21. Edeerozey, A.M.M., Hazizan, M.A., Azhar, A.B., Zainal, A.M.I.: Chemical Modification On Kenaf. Materials Letters 61, 2023–2025 (2007)
22. Browning, B.L.: Methods of Wood Chemistry, vol. 2, p. 387, 882 pages. Interscience Publishers, New York (1967)
23. Ishiaku, U.S., Yang, Y., Leong, Y.W., Hamada, H., Samba, T., Kitagawa, K.: Effects of Fiber Content and Alkali Treatment On The Mechanical and Morphological Properties of Poly(Lactic Acid)/Poly(Caprolactone) Blend Jute Fiber-Filled Biodegradable Composites. J. Biobased Material and Bioenergy. 1, 78–86 (2007)
24. Wong, E.D., Rzaliand, K., Yahya, S.S.N.: Manufacture of MDF from 5-year old Havea brasiliensis. Paper presented at the Bio-Based Composites. Proceedings of the 5th Pacific Rim Symposium, Canberra, Australia (2000)

Design Process Using Lean Six Sigma to Reduce the Receiving Discrepancy Report of ACE Logistics

Jervie Bersamin, Roselyn Drio, Ariane Lanel Lacibal, Camille Manalastas, Sheily Mendoza, Ghil Michael Danico Orallo, and Carl Timmothy Tan

University of Perpetual Help System DALTA, Las Piñas, Philippines

Abstract. ACE Logistics, Inc. operates cross-dock procedures where products from different suppliers are distributed directly to the customers. The study will focus on the dispatch process. It aims to reduce the cases of Receiving Discrepancy Reports (RDR) that induce charges, to analyze the present system, to identify the causes of RDR and to develop a proposed system for the company. It was carried out using the DMAIC (Define-Measure-Analyze-Improve-Control) model of Lean Six Sigma. The DMAIC Model identified that the factors that causes RDR are facility, equipment, process and man. The proposed system, which includes the improvement of the facilities layout and the cross dock operation process, could be used in order to reduce the cases of RDR.

Keywords: cross-dock, dispatch process, Receiving Discrepancy Reports (RDR), DMAIC (Define-Measure-Analyze-Improve-Control), Lean Six Sigma.

1 Introduction

1.1 Background of the Study

Logistics companies undergo complicated operations, and one those are the rejects or errors in both operation and outcomes. ACE Logistics, Inc. operates cross-dock procedures where products from suppliers are distributed directly to the customers. The study aims to reduce the number of occurrences of Receiving Discrepancy Reports (RDR) that cause unnecessary charges in the company through the application of Lean Six Sigma. The result of this study would help ACE Logistics regarding their problem with Receiving Discrepancy Reports (RDR) that causes them charges. It would likewise allow them to identify more improvements and enable their employees to be more efficient and more productive. The research could serve as a guide for the students and researchers conducting studies about Lean Six Sigma Logistics. The scope of the study is the dispatching process of ACE Logistics. It was focused on determining causes of Receiving Discrepancy Reports (RDR) in the company through the application of Lean Six Sigma. It does not include the prototype of the recommended equipment.

M. Gen et al. (eds.), *Industrial Engineering, Management Science and Applications 2015*, Lecture Notes in Electrical Engineering 349, DOI: 10.1007/978-3-662-47200-2_70

1.2 Gap and Missing Information

The cases of Receiving Discrepancy Reports (RDR) have been a problem for ACE Logistics. The reduction of the arising problems could be assured with the use of Lean Six Sigma. The researchers would continue researching and improving this system which could benefit the study and the company as well.

1.3 Objectives of the Study

The general objective of the study is to reduce the cases of Receiving Discrepancy Reports (RDR) that induce charges in ACE Logistics. The specific objectives of the research, which are to be dealt with through the use of Lean Six Sigma, are: to analyze the present system of ACE Logistics, to identify the causes of RDR in the company and to develop a proposed system for ACE Logistics.

2 Review of Related Literature

Six Sigma aims to understand and eliminate the negative effects of variations in processes. DMAIC (Define-Measure-Analyze-Improve-Control) is a step-by-step method, to understand and improve on organizational challenges [1]. Lean concepts are rooted in the Toyota Production System (TPS) [1]. It is a systemic method for the elimination of waste [2]. Excess inventory is at the top of the list of known wastes [1]. Lean Six Sigma improves process performance by combining Lean and Six Sigma to eliminate wastes [3]. Lean Six Sigma amplifies the strengths and minimizes the weaknesses of both approaches when used alone [4]. Logistics is the procurement, maintenance, and transportation of materials, facilities and personnel [5]. It involves the management of inventory [1]. Lean Six Sigma Logistics is the elimination of wastes through disciplined efforts to understand and reduce variation, while increasing speed and flow in the supply chain. Both Lean and Six Sigma provide distinctive disciplines and tools to logistics. Using these two will allow an organization to identify and deal with wastes and inefficiencies [1].

3 Methodology

Data gathering was made through observations, interviews, and through the reports provided by the company. The present system was analyzed through the application of Lean Six Sigma. It was carried out using the Define-Measure-Analyze-Improve-Control or DMAIC model. The problem was defined using value stream mapping which provides an overview of the entire process. Ishikawa Diagram was used to measure the process performance. Pareto chart was used to analyze the frequency of the problems and its causes. The proposed layout of the cross-dock warehouse was presented in order to attain improvement in the company. The Control Chart was applied to control the result from improved facility layout.

Table 1. Tabulated Objectives vs. General Procedural Steps

Objectives	Procedure
➢ To analyze the present system of ACE logistics	1. Data gathering through observations, interviews, and the reports provided by the company 2. Use of the DMAIC (Define-Measure-Analyze-Improve-Control) Model
➢ To identify the causes of Receiving Discrepancy Reports (RDR) in the company.	a. Define: Value Stream Mapping b. Measure: Ishikawa Diagram c. Analyze: Pareto Chart
➢ To develop a proposed system for ACE Logistics	d. Improve: Facilities Layout Improvement, Dispatch Process Improvement e. Control: Control Chart

4 Results and Discussion

4.1 Define

Figure 1 shows the value stream map of the cross-dock operations of ACE Logistics. The receiving process includes the docking of all cases that arrived, checking of Purchase Orders (P.O.) and scanning of the received cases. The cases are placed on the designated places inside the warehouse in the staging process. Pulling and pushing of the cases using materials handling equipment occurs in the traffic process. In the dispatch process, the cases are checked manually based on the given P.O., and then loaded on the designated trucks.

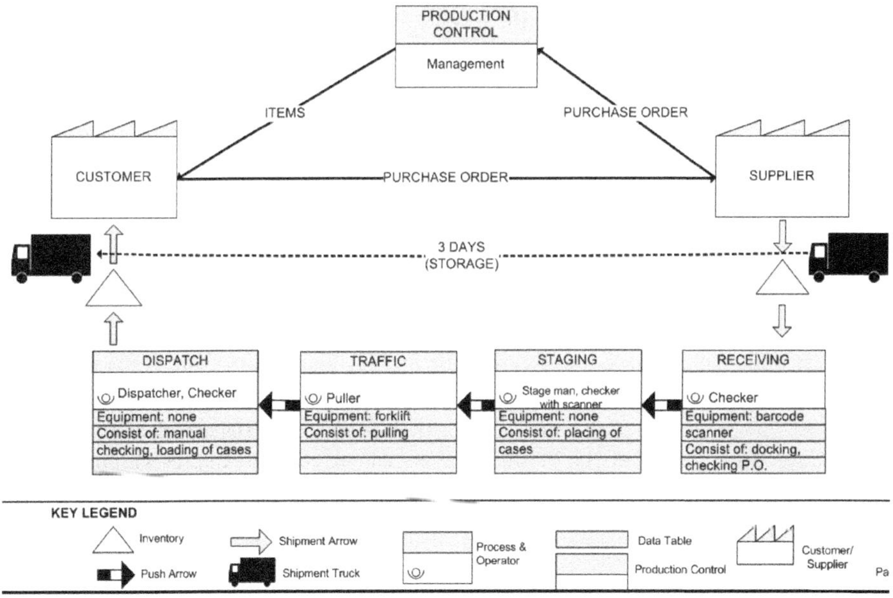

Fig. 1. Value stream map of the cross-dock operations of ACE Logistics

4.2 Measure

The Ishikawa Diagram was used as a tool to measure and to narrow down the scope into categories for subsequent analysis. The causes of rejects for the month of September, which has the most Receiving Discrepancy Reports (RDR), are insufficient aisle space, unorganized storage locations, lack of equipment, improper processes and damage caused by workers during loading. These had been narrowed and weighed into subcategories which are facility, equipment, process, and man. Figure 2 illustrates the Ishikawa Diagram of the causes of RDR for September.

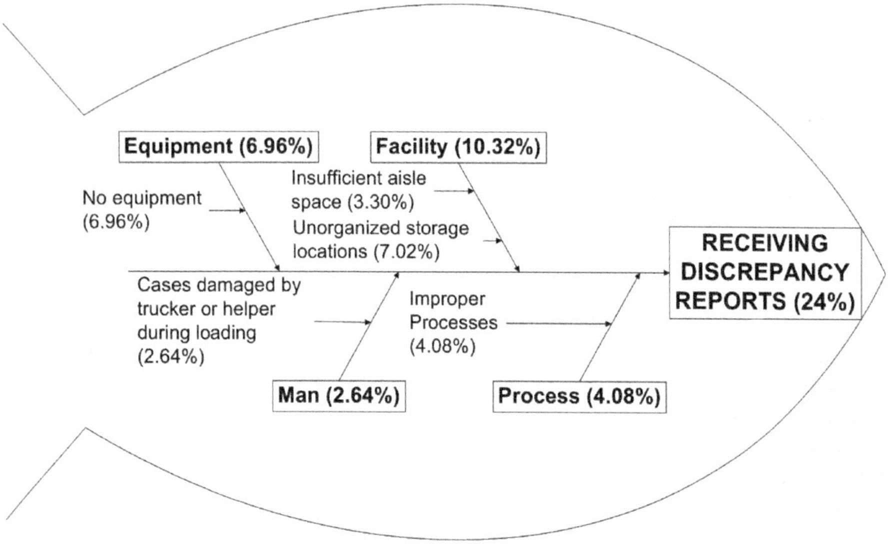

Fig. 2. Ishikawa diagram of the causes of RDR

4.3 Analyze

The Pareto Chart highlighted the major causes of Receiving Discrepancy Reports (RDR). It showed that the category that is most vital for improvement is the facility followed by equipment, process and man. The results were based from the measurements obtained in the Ishikawa Diagram. Figure 7 presents the Pareto Analysis of the causes of RDR in ACE Logistics.

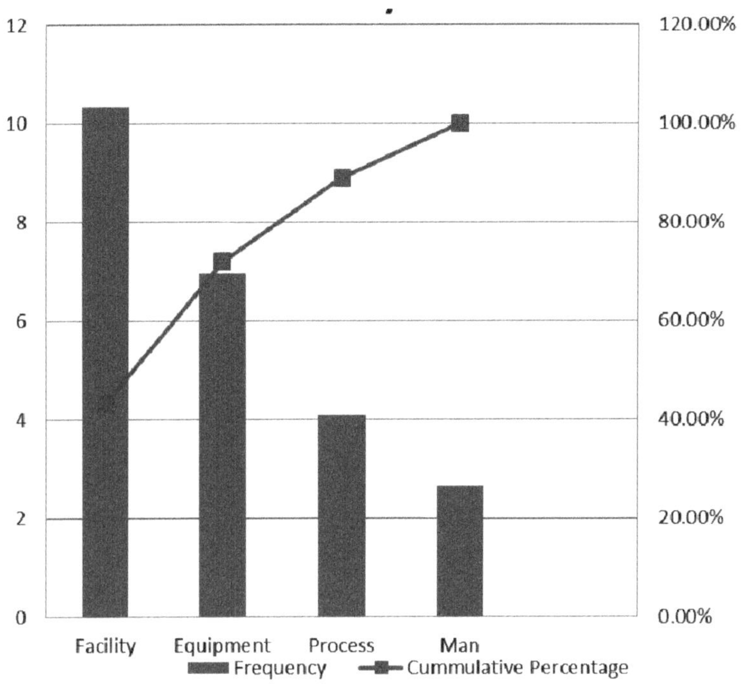

Fig. 3. Pareto chart of the factors that causes RDR

4.4 Improve

Facilities Layout Improvement. Figure 4 illustrates the present layout of the cross-dock warehouse which is subject for improvement. Figure 5 shows the proposed layout of the cross dock warehouse.

Figure 6 shows the results from the application of the proposed layout, which started in the month of October. The graph had shown the increasing percentage of RDR from July to September. The percentage of RDR started to decrease from the month of November. 18% of Receiving Discrepancy Reports (RDR) was reduced. 8% and 10% of RDR are reduced in November and December respectively, as shown in Figure 7.

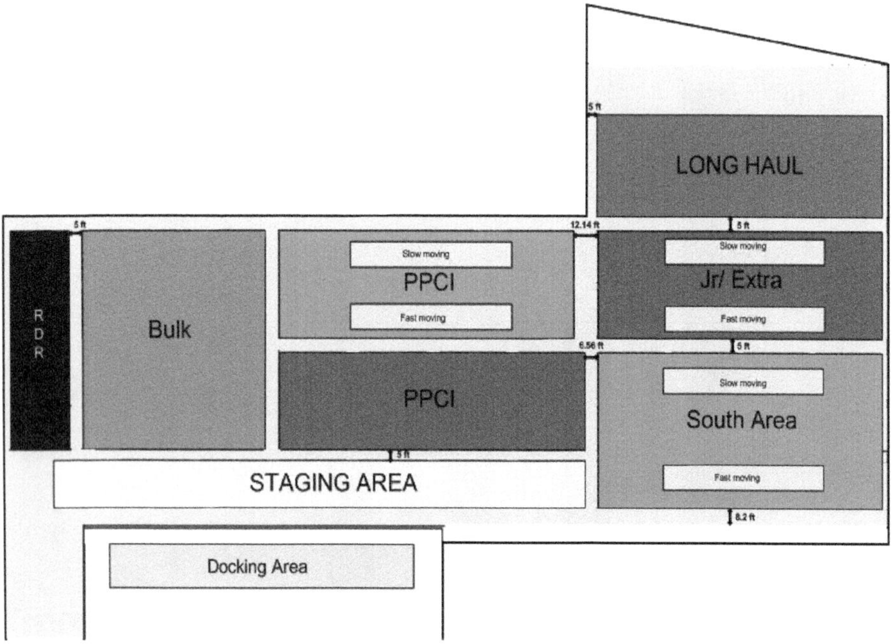

Fig. 4. Present layout

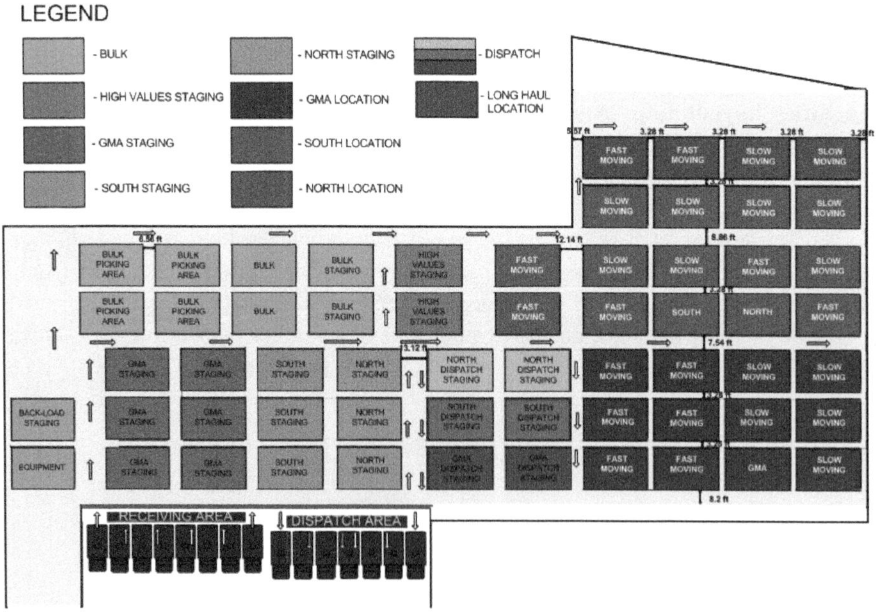

Fig. 5. Proposed layout

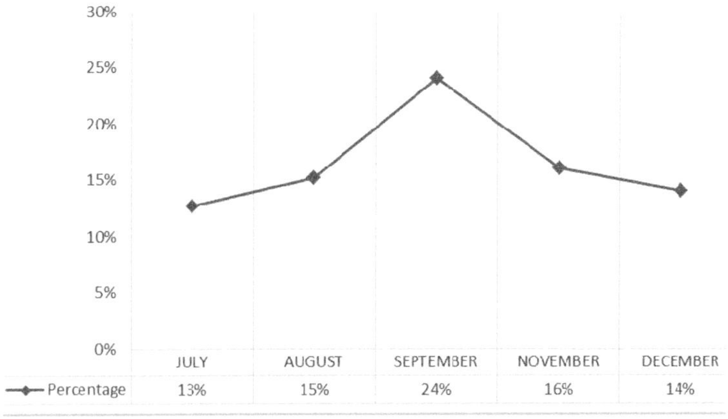

Fig. 6. Results based from the application of the proposed layout of cross dock warehouse

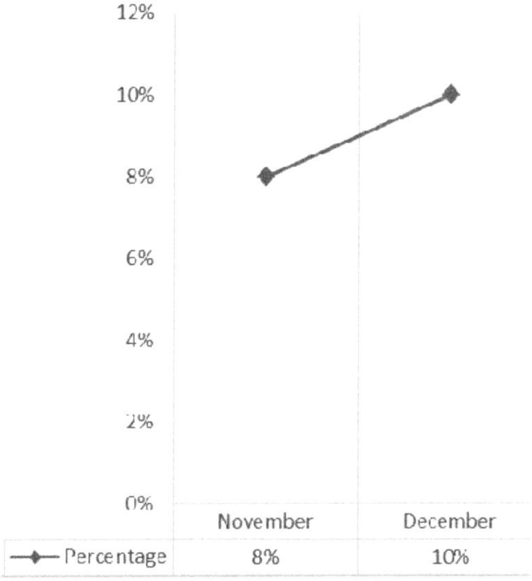

Fig. 7. Reduced percentages

Process Improvement. The flow charts are used to illustrate the sequence of operations required in the present and proposed system by which potential weaknesses and non-value added steps are made visual. Figure 8 shows flow chart of the present dispatch system of ACE Logistics. The proposed flow chart is shown in Figure 9.

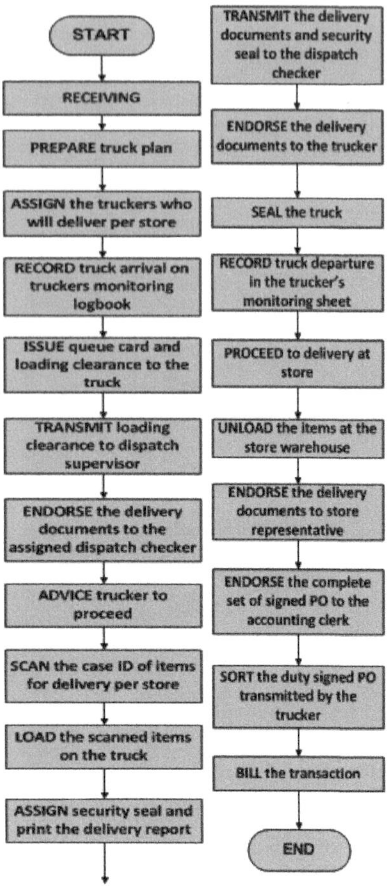

Fig. 8. Present Dispatch Process of ACE Logistics

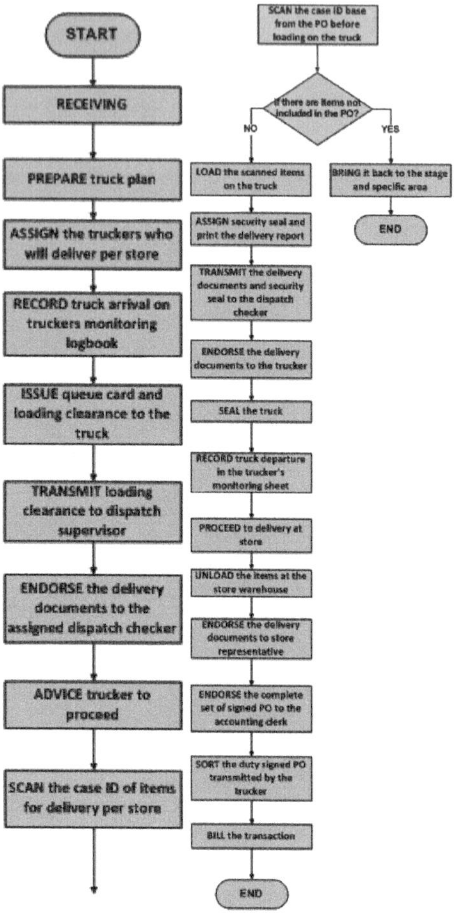

Fig. 9. Proposed Dispatch System of ACE Logistics

4.5 Control

The control chart determined the stability of the process performance. Figure 10 illustrates the present control chart, with 8.75 as the Upper Control Limit (UCL), 3.30 as the Central Limit (CL), and 0 as the Lower Control Limit (LCL). Figure 11 shows the proposed control chart with 25.55, 14.23 and 2.91 as the UCL, CL and LCL respectively. The 4 points that failed the tests for special causes in the present control chart had been reduced to 2, as shown in Table 1.

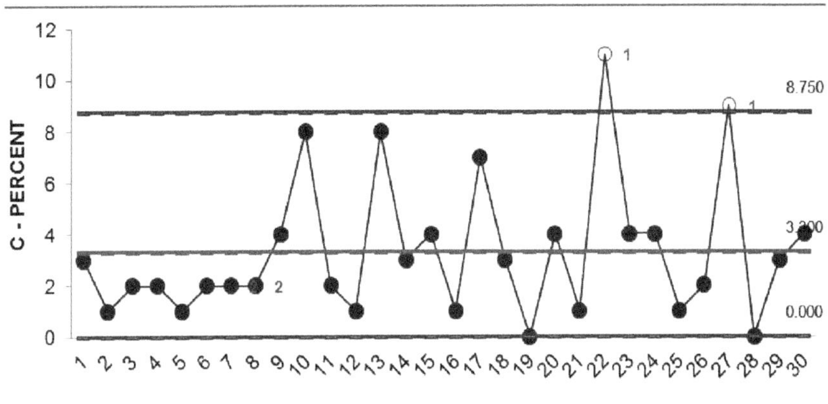

Fig. 10. Present control chart

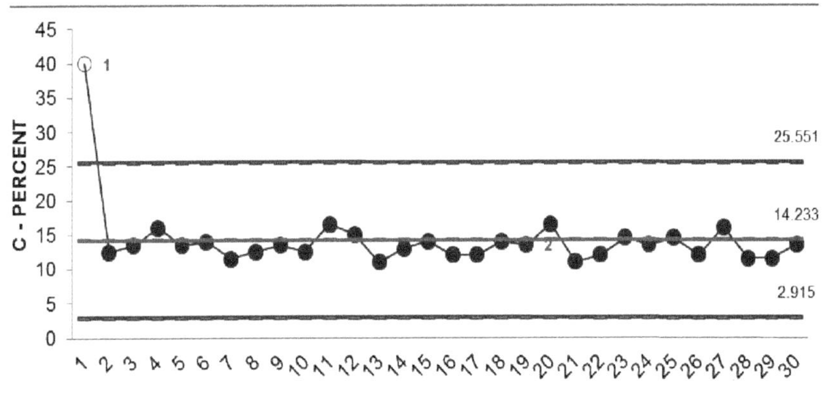

Fig. 11. Proposed Control Chart

Table 2. Number of points failing the test

Present Control Chart			Proposed Control Chart		
Observation No.	Test 1: 1 point more than 3 Stdev from CL	Test 2: 7 points in a row on same side of CL	Observation No.	Test 1: 1 point more than 3 Stdev from CL	Test 2: 7 points in a row on same side of CL
7		X	1	X	
8		X	19		X
10	X				
27	X				

5 Conclusions

The present system of ACE Logistics was analyzed through the Define phase of the DMAIC (Define-Measure-Analyze-Improve-Control) Model. The major causes of Receiving Discrepancy Reports (RDR) in the company were identified through the Measure and Analyze phase of the DMAIC Model. The sequences of the percentages are highest to lowest, from Facility, Equipment, Process and Man respectively. The proposed system, which includes the improvement of the facilities layout, in the Improve Phase, could be used in order to reduce the RDR. A total of 18% of RDR were reduced from the start of application. November and December have reduced 8% and 10% of RDR respectively. The net profit from July to September is Php 6,111,538.68, and Php 23,362,472.73 from November to December. There had been a great increase in profit because of the reduced cases of RDR and because November and December fall under the peak seasons.

6 Recommendation

The researchers recommend the application of a mobile barcode scanner in the dispatch process of ACE Logistics because of the following reasons: it could combat two of the causes of Receiving Discrepancy Reports (RDR), which are the lack of equipment and human errors; it is faster, more reliable and more time-saving than manual operations; compared to wired scanners, it allows the users to freely move the device towards the items at a wider range.

References

1. Goldsby, T., Martichenko, R.: Lean Six Sigma Logistics Strategic Development to Operational Success. J. Ross Publishing, Boca Raton (2005)
2. Lean Manufacturing, http://en.wikipedia.org/wiki/Lean_manufacturing
3. Lean Six Sigma, http://en.wikipedia.org/wiki/Lean_Six_Sigma
4. New to Lean Six Sigma? https://www.moresteam.com/new-to-lean-six-sigma.cfm
5. logistics, Merriam-Webster.com (2011), http://www.merriam-webster.com (January 17, 2015)

A Study on Carbon Emission Effects of Foreign Direct Investment in Secondary Industry of Shandong Province

Bin Xiong and Meng-jiao Wang*

Faculty of Management and Economics, Kunming University of Science and Technology, Yunnan, China
binxiong@gmail.com, 15969526351@163.com

Abstract. With China's rapid economic development in recent years, the scale of foreign direct investment has been expanding.In the meantime, environmental issues have been receiving increasing attention, since the concept of a low carbon economy is prevalent in the world. Secondary Industry is the major energy consumers and the emitted carbon dioxide has a significant negative influence on the environment.This paper discusses the background of low carbon economy and conducts an empirical study of the carbon dioxide emission effect of foreign direct investment in the Secondary Industryin Shandong Province. The conclusion is that foreign direct investment leads to more carbon dioxide emissions, which is not conducive to the development of green productivity and a low carbon economy in Shandong Province.

Keywords: Foreign Direct Investment, Low Carbon Economy, Secondary Industry, Carbon Dioxide Emission, Green Productivity.

1 Introduction

Recent years have witnessed prominent contradictions between economic development and resources and environment. China has become one of the most polluted countries in the world as a result of its rapid economic development. Consequently, much attention has been paid to such environmental problems as global warming resulting from carbon dioxide emissions. In order to deal with pollution, the concept of a low carbon economy has been widely agreed and promoted.

Shandong Province is one of the regions that attracts most foreign direct investment (hereafter referred to as FDI) in China. With the acceleration of industrialization, problems like exhaustion of resources, pollution and global warming are becoming increasingly prominent. As a major field of attracting FDI and a huge energy consumer, the Secondary Industry is the major source of carbon emissions. This paper tries to study the relevance between FDI in secondary industry and carbon dioxide emissions.

* Corresponding author.

M. Gen et al. (eds.), *Industrial Engineering, Management Science and Applications 2015*,
Lecture Notes in Electrical Engineering 349, DOI: 10.1007/978-3-662-47200-2_71

2 Current Situation of FDI Carbon Emission Effect

The research results of the carbon emission effect of FDI are divided into three categories. The first is the increase in carbon dioxide emissions from FDI (Sha Wenbin et al., 2006), the second is the decrease of carbon emissions from FDI (Wang Zhengming et al., 2013; Yao Yi et al., 2013; Song Deyong et al., 2011) and the third is the uncertain effect of FDI on carbon emissions(Li Zihao et al, 2010, 2011; Pendo Kivyiro et al., 2014). In terms of the Secondary Industry, research shows that FDI has increased the carbon emissions in secondary industries in China, as the total effects are negative (Dai Dier & Li Zihao, 2011). In contrast, the horizontal spillover effects of FDI, Forward Linkage Spillover Effects and Backward Linkage Spillover Effects have positive influence on the reduction of carbon emission intensity. Among these effects, forward linkage spillover effects are relatively obvious (Bi Kexin & Yang Chaojun, 2012). Besides, Shenggang Ren et al. (2014) studied the relationship among international trade of China's secondary industries, FDI and carbon emissions, the result of which is the introduction of FDI further increased carbon dioxide emissions.

To sum up, the research of Chinese and overseas scholars has not reached complete agreement about the carbon emission effect of FDI, and there are fewer studies concerning the carbon emission effect of FDI in the Secondary Industry or in different provinces. The research in this area is of great significance to the healthy development of foreign investment as well as the coordinated development of economy and environment.

3 The Carbon Emission Influence Mechanism of FDI in Secondary Industries

Research has shown that FDI can exert an influence on the environment through scale effects, structural effects, and technique and regulation effects.

3.1 Scale Effects

Scale effects refers to the influence of scale change caused by the incoming of FDI on the environment. The scale change of FDI may give play to scale effects through influencing overall economic scale. Serving to show the pollution caused by economic development,the Environmental Kuznets Curve vividly demonstrates the influence of scale effects. The curve is an inverse-U shape, which indicates that with the scale expansion during economic development, the environment will get worse at first and then will gradually improve.

Generally speaking, the incoming FDI will promote economic growth. It leads to a continuous increasing need for key elements and resources such as coal and oil. This, in turn, will cause more carbon dioxide emissions, which exerts negative influence on the sustainable development of the environment and economy. According to the Environmental Kuznets Curve, when the economic scale expands to a certain point, the government will have more capital, better management ability, and more awareness of environmental protection. They will enforce a series of regulations related to environment, which lead to the decrease of the degree of pollution and carbon dioxide emissions.

3.2 Structural Effects

Structural effects is the influence of the industrial structure of FDI on carbon emissions. The economy will enjoy sustainable development in developed countries when it grows to a certain extent. According to Vernon's "theory of product life cycle", developed countries will transfer pollution-intensive industries which have lost their domestic competitiveness to developing countries through foreign investment, since developing countries are more labor intensive and low in environmental cost. As a result, this exerts a negative influence on the environment of developing countries.

However, the influence of structural effects of FDI is also determined by the economic development level of the host countries. If the economic development of the host countries relies on heavy industries and FDI concentrates on resource-consuming, highly pollutive industries, there is no doubt that the environment will be destroyed and carbon dioxide emissions in the host country will go up. If the host country has a developed economy and advanced technology, FDI will flow to the tertiary industries. This will not do harm to the environment while promoting economic growth.

3.3 Technique and Regulation Effects

Technique and regulation effects is the influence brought about by the technique expansion of FDI. The incoming FDI from developed countries can provide financial support for the research and development of environmental protection technology. The introduction of advanced environmental-friendly technology, equipment and high quality talents facilitates the transformation of the production mode, the improvement of energy utility efficiency, the development of green productivity and the reduction of carbon emissions. The technology spillover effect of FDI and personnel flow is helpful to the improvement of the local enterprises' productivity, the transformation of production mode, economic growth and sustainable development. Moreover, because of the strong environmental protection awareness and strict environment management policies in developed countries, foreign enterprises will normally abide by the local environmental regulations more strictly than local enterprises. To some extent, this can play an exemplary role for local enterprises and is of benefit to the improvement of the environmental protection awareness of the local enterprises and the promotion of low carbon economy.

Regulation effects are closely related to technique effects. According to the Porter Hypothesis, strong regulation pressure and strict environmental protection standard will force and stimulate enterprises to devote to technique innovation to curb the pollutant emissions. Therefore, environmental regulation plays an important role in controlling local pollution caused by FDI. In places with weak environmental regulation, FDI may maximize its profit by employing low-cost techniques in production, which will increase pollution and carbon dioxide emissions, and will hinder the development of low carbon economy.

4 Current Situation of Carbon Dioxide Emissions and FDI in Shandong Province

4.1 Current Situation of Carbon Emissions in Shandong Province

According to the calculation guide of carbon emissions promulgated by the United Nations Intergovernmental Panel on Climate Change (hereafter referred as IPCC), the following formula is employed to calculate the carbon emissions in Shandong Province over the years:

$$A = \sum_{i=1}^{n} B_i \times C_i \qquad \text{Formula(1)}$$

In Formula (1), A is carbon emission; i is the type of energies; B is the consumption of energy i calculated according to standard coal; C is the coefficient of carbon emissions of energy i; coefficient of carbon emissions is derived from *the Guidelines for National Greenhouse Gas Inventories* by IPCC. The major energy consumption in Shandong Province is raw coal and crude oil, the coefficient of carbon emissions of which is: 0.7559(10^4t/10^4t)for raw coal and 0.5857(10^4t/10^4t) for crude oil. The amount of carbon dioxide emissions is the multiplication of the amount of carbon emissions by the coefficient 3.67.

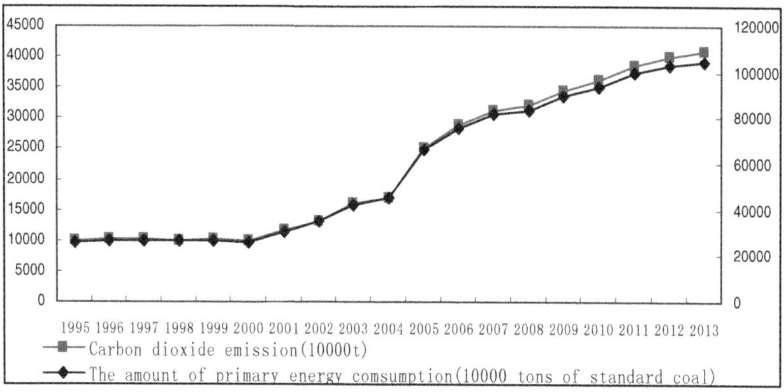

Fig. 1. Trend of Energy Consumption and Carbon Dioxide Emissions in Shandong Province during 1998-2013

According to the above-mentioned method, the trend of energy consumption and carbon dioxide emissions in Shandong Province during 1998- 2012 is calculated and sorted (as shown in Figure 1). The solid line indicates the amount of carbon dioxide emissions and the dotted line indicates the amount of primary energy consumption. From 1998 to 2013, the carbon dioxide emissions in Shandong Province is gradually increasing. The trend of energy consumption and carbon dioxide emissions is similar. Since different energies correlate with different coefficients of carbon dioxide, the similarity of energy consumption and carbon dioxide emissions suggests that the structure of energy consumption in Shandong Province is not varied and has not experienced much change over the years.

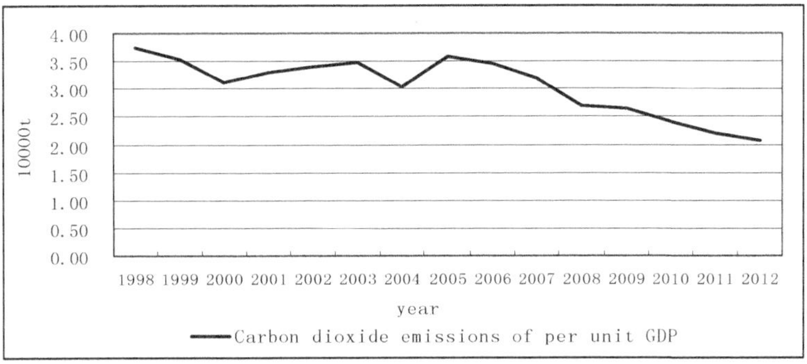

Fig. 2. Trend of the Amount of Carbon Dioxide Emissions per unit GDP (10000 ton/ 100 million Yuan) in Shandong Province

Figure 2 shows the trend of the amount of carbon dioxide emissions per unit GDP in Shandong Province during 1998-2012. The amount of carbon dioxide emissions per unit GDP is the ratio of the total amount of carbon dioxide emissions to local GDP. This index shows the differences between the speed of economic growth and that of carbon dioxide emission increase. In 2012, Shandong Provincial government promulgated work programs for controlling greenhouse gas emissions in "the twelfth Five-year Plan", which sets the goal that the amount of carbon dioxide emissions per unit GDP in Shandong Province must have a 18% decrease by 2015 in comparison with that in 2010 and a carbon emissions trading market will be established gradually.

According to the situation in Shandong Province, the amount of carbon dioxide emissions per unit GDP in Shandong Province presents a gradual decreasing trend. However, the continual increase of regional GDP is still based on the massive consumption of fossil energy. Therefore, the total amount of carbon dioxide emissions is still gradually rising with year.

4.2 The Industrial Structure of FDI in Shandong Province

As is shown in Figure 3, the secondary industry accounts for the largest share of actually used FDI in Shandong Province. The cumulative proportion of actually used FDI in the Secondary Industry was 73.35% by 2013. The majority of the investment goes to manufacturing and mining industries. In 2013, FDI in manufacturing industries was as high as 84.63%.

In recent years, the percentage of FDI in the Secondary Industry in Shandong has been decreasing whereas the tertiary industry is increasing. The industrial structure of FDI seems to more environment-friendly than before. However, the situation of FDI in the Secondary Industry will not alter in the short term.

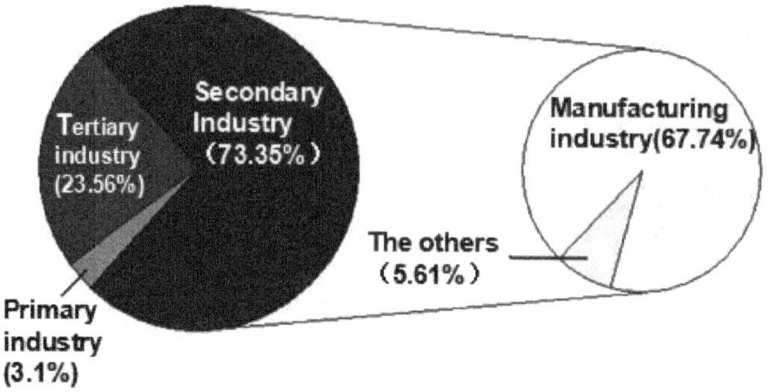

Fig. 3. The Cumulative Proportion of FDI in the First, Secondary and Tertiary Industries in Shandong Province in 2013

4.3 The Source Countries (Regions) of FDI in Shandong Province

The biggest FDI source country (region) in Shandong Province is Hong Kong, which accounts for 56.73%. The second is the Republic of Korea, which amounts to 8.54%. The third is British Virgin Islands, which has a proportion of 5.32%. The fourth is the Association of Southeast Asian Nations, which represents 5.05%. There difference of FDI between the British Virgin Islands and the Association of Southeast Asian Nations is very small. It can be seen that the source countries (regions) of FDI in Shandong Province are highly concentrated. Its structure and changing trend is not obvious. The FDI is primarily from developing countries while that from the developed countries in Europe and America accounts for a small proportion.

5 The Analysis of Carbon Emission Effect of FDI in the Secondary Industry in Shandong Province

5.1 The Analysis of the Effect of Actually Used FDI in the Secondary Industry on Carbon Dioxide Emissions

Statistics Selection

In order to study the relationship between FDI in the Secondary Industry and carbon emission in Shandong Province, the amount of industrial FDI in Shandong Province is used to reflect the actually used FDI, and the amount of the carbon dioxide emissions in Shandong Province is used to indicate carbon emission index. The time range of this study is from 2001 to 2013. The relevant statistics are from the *Shandong Statistic Year Book*.

Model Establishment

Emissions are established as follows:

$$LnE_t=c+\alpha LnGDPPC_t+\beta LnIPC_t+\mu_t \qquad \text{Model(1)}$$

In Model(1), E_t is used to indicate carbon dioxide emissions, the unit of which is based on 10000 tons; $GDPPC_t$ is used to show the per capita GDP in Shandong Province, the unit of which is Yuan; IPC_t is used to represent the per capita FDI in the Secondary Industry in Shandong Province. $GDPPC_t$ (in logs) and IPC_t (in logs) are independent variables and E_t (in logs) is a dependent variable. By ADF-Test and Co-integration test, the following regression model is reached:

$$LnE_t=2.4571+0.7225LnGDPPC_t+0.2122LnIPC_t$$
$$(0.0000)\ (0.0881)$$

Empirical Results
From empirical results, the increase of both per capita GDP and per capita FDI in the Secondary Industry in Shandong Province lead to more local carbon dioxide emissions and have a negative influence on the environment. However, carbon dioxide emissions from the actually used FDI in the Secondary Industry in Shandong Province are less than those from the per capita GDP.

5.2 The Correlation Analysis between Total Industrial Output Value from Foreign Enterprises and Carbon Dioxide Emissions

Statistics Selection
The total industrial output value of foreign-invested enterprises in Shandong is used to reflect the actual investment situation. The amount of carbon dioxide emissions in Shandong Province is used to show the carbon emission index. The time range is from 1993 to 2013. The relevant statistics are from the *Shandong Statistic Year Book.*

Correlation Analysis of Variables
By using SPSS, a correlation test is made through analyzing the amount of carbon dioxide emissions in Shandong Province and the total industrial output value of foreign-invested enterprises. The result is as follows:

Table 1. Correlation Analysis of the Amount of Carbon dioxide emissions and Total Industrial Output Value of Foreign-invested Enterprises

		E_t	WZ_t
E_t	Pearson Correlation	1	.977**
	Sig. (2-tailed)		.000
WZ_t	Pearson Correlation	.977**	1
	Sig. (2-tailed)	.000	

Table 1 reveals that the correlation coefficient is high between the amount of carbon dioxide emissions and the total industrial output value of foreign-invested enterprises. The two tailed p-value is 0.000. The analysis shows that the calculation model can be established.

Model Establishment

To study the correlation between the total industrial output value of FDI and carbon dioxide emissions, the index of the total industrial output value of foreign-invested enterprises and the index of the amount of regional carbon emissions over the years are selected. Environmental Kuznets Curve is simulated to explore whether the two indexes have a non-linear relationship similar to Environmental Kuznets Curve. With the amount of carbon dioxide emissions in Shandong Province during 1995-2013 and the total industrial output value of foreign-invested enterprises as dependent variables, the second model is established as follows:

$$E_t = c + \alpha WZ_t + \beta WZ_t^2 + \mu_t \text{Model} \qquad \text{Model (2)}$$

In Model (2), E_t is used to show carbon dioxide emissions, WZ_t is used to indicate the total industrial output value of foreign-invested enterprises in Shandong Province, WZ_t^2 is used to represent the square of WZ_t. α and β suggests the extent of the influence of different factors on carbon dioxide emissions. The WZ_t and WZ_t^2 factors are independent variables and E_t is a dependent variable. By ADF-Test and Co-integration test, the following regression model is arrived at by using STATA:

$$E_t = 8.2022 \times WZ_t - 0.000205 \times WZ_t^2 + 20946.46$$
$$(16.22) \qquad (-7.04)$$

The regression results show that the coefficient of Variable WZ_t is positive while that of Variable WZ_t^2 is negative. Therefore, the relationship between WZ_t and E_t is a downward parabola. The figure is as follows:

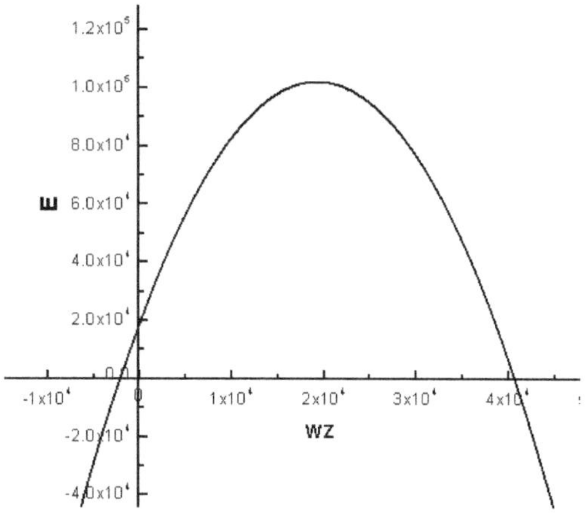

Empirical Results

It can be seen that carbon dioxide emissions will go up at first and then will gradually fall with the increase of the total industrial output value of foreign-invested enterprises. But according to the current situation in Shandong Province, the carbon emissions have not yet come to their turning point and are still on the left side of the curve. The current conclusion is that more total industrial output value of foreign-invested enterprises will lead to more carbon dioxide emissions.

6 Conclusions and Suggestions

According to empirical analysis, FDI in the Secondary Industry in Shandong Province leads to more carbon dioxide emissions. The more the total industrial output value of foreign-invested enterprises, the higher the amount of carbon dioxide emissions in the region. This shows that foreign-invested enterprises in the Secondary industry have not achieved green production. They are conducive to the increase in carbon dioxide emissions and harmful for the sustainable development of economy and environment. This has once again proved that FDI will intensify the industrial carbon emissions in Shandong Province. The negative effect has overshadowed its positive effect. Based on this conclusion, this paper is to put forward the following suggestions:

Firstly, it is necessary to expand the scale of attracting investment and to improve the quality of investment. More FDI from developed countries should be attracted and the investment environment must be favorable to low-carbon industries.

Secondly, the introduction of pollution-intensive industries must be strictly controlled. On the contrary, high energy consuming industries must be encouraged to use clean energy technology, improve energy utility efficiency, pay attention to the upgrading of industrial structure, promote the positive effect of green production and facilitate the transformation of patterns of provincial economic growth.

Thirdly, in the process of attracting investment, technology should be regarded as the lead and the concept of low carbon economy should be adopted. The exemplary role and competition effect of FDI industries should be used effectively. While obtaining carbon emission reduction technology from FDI industries, our own research and development capability must be strengthened. By studying foreign technology, green production must be developed to improve energy utility efficiency and to reduce pollution.

Fourthly, the function of environmental regulation must taken into consideration. Attention must be given to the advanced experience of foreign environmental regulation. In addition, proper environment-monitoring systems should be established. From enterprises' perspective, foreign enterprises should take the initiative to bring environmental objectives into their management strategies, self-regulate their behaviors concerning the environment and undertake the responsibilities of promoting the sustainable development of the economy and environment.

References

1. Wenbing, S., Tao, S.: The Environmental Effect of Foreign Direct Investment—An Empirical Study Based on China's Provincial Panel Statistics. World Economy Study (6) (2006)
2. Zhengming, W., Guimei, W.: Dynamic Carbon Emission Effect of International Trade and Investment. China Population Resources and Environment 23(5) (2013)
3. Yi., Y., Qin, N.: The Influence of Foreign Direct Investment on carbon intensity—An Empirical Study Based on China's Provincial Panel Statistics. Journal of Applied Statistics and Management 32(1) (2013)

4. Deyong, S., Yanchun, Y.: Foreign Direct Investment and the Carbon Emissions in China. China Population·Resources and Environment 21(01) (2011)
5. Zihao, L., Huihuang, L.: The Environmental Effect of Foreign Direct Investment — AnEmpirical Study Based on the Statistics of 36 Industries in China. Systems Engineering 28(11) (2010)
6. Zihao, L., Dier, D.: Foreign Direct Investment and the Carbon Dioxide emissions in China—An Empirical Study Based on Provincial Experience. Inquiry into Economic Issues (09) (2011)
7. Kivyiro, P., Arminen, H.: Carbon dioxide emissions, energy consumption, economic growth,and foreign direct investment: Causality analysis for Sub-Saharan Africa. Energy 74, 595–606 (2014)
8. Dier, D., Zihao, L.: The Carbon Emission Effect of Foreign Direct Investment—A Study Based on the Industrial Statistics in China. International Economics and Trade Research (05) (2011)
9. Kexin, B., Chaojun, Y.: The Influence of the Spillover Effect of Foreign Direct Investment on the Intensity of Industrial Carbon Emissions in China. Economic Management Journal (08) (2012)
10. Ren, S., Yuan, B., Ma, X.: International Trade, FDI (foreign direct investment) and Embodied Carbon Emissions: A Case Study of Chinas Industrial Sectors. China Economic Review (28), 123–134 (2014)

An Evaluation Performance of Log Periodic Dipole Antenna Based on the Parameter of Flux Density of the Solar Radio Burst Event

Z.S. Hamidi[1,*] and N.N.M. Shariff[2]

[1] School of Physics and Material Sciences, Faculty of Sciences, MARA University of Technology, 40450, Shah Alam, Selangor, Malaysia
[2] Academy of Contemporary Islamic Studies (ACIS), MARA University of Technology, 40450, Shah Alam, Selangor, Malaysia
zetysh@salam.uitm.edu.my, nnmsza@salam.uitm.edu.my

Abstract. In this work, an evaluation of the flux density of the solar radio burst is analyzed. We used a nineteenth (19) elements of rod aluminums' type as a conductor with of different sizes is being prepared to construct a log periodic dipole antenna (LPDA) from 45 - 870 MHz. The testing was carried out at the National Space Agency (PAN), Sg. Lang, Banting Selangor by connecting to the *Compound Low Cost Low Frequency Spectroscopy Transportable Observatory* (CALLISTO) spectrometer. We choose the input impedance, R_0 = 50 ohm for this LPDA antenna. From the analysis, the gain of the antenna is 9.3 dB. This antenna possibly captures a signal that covers about 0.08 m2 area of the Sun. It was found that the temperature of the burst that detected at the feed point of the antenna is 32 K. However, the signal becomes decrease to 28.75K while by the CALLISTO spectrometer as a receiver. We found that the isotropic source spectral power is 1576 W/Hz. Since the burst level above the background sky is 0.41 dB, the flux density of the burst is 5.5 x 10^{-21} W/m^2/Hz. We conclude that this antenna is suitable for to observe the Sun activities at low frequency region.

1 Introduction

The Log-Periodic Dipole Antenna (LPDA), first built in 1958 by Du Hamel and Dwight Isbell, an undergraduate researcher in the ECE Antenna Laboratory and the prepared by Carrel [1,2]. Since that, this type of antenna has been used widely and become as one of the essential parts in the recent explosion in information technology and wireless communications. The signal of electromagnetic radiation, in this case, a radio signal can be detected with a special arrangement of conductors. Based on principles of physics, the energy characteristics vary considerably below the lower frequency limit of the antenna. Simple design and effective detector are the most factors why this type is still relevant. Moreover, this antenna is also easy to construct and suitable to monitor the solar activities. The radiation characteristics vary considerably below the lower frequency limit of the antenna.

* Corresponding author.

M. Gen et al. (eds.), *Industrial Engineering, Management Science and Applications 2015*, Lecture Notes in Electrical Engineering 349, DOI: 10.1007/978-3-662-47200-2_72

In principle, the LPDA is a coplanar linear array of unequal and unequally spaced parallel linear dipoles fed by a twisted balanced transmission line. It contains of small, closely spaced half-wave dipoles. These dipoles are connected to the source using a twin transmission line in such a way that the phase is reversed at each connection relative to the adjacent elements. Each dipole is effective over a narrow band of frequencies determined by its length.

LPDA also been used to monitor solar activities of the sun based on the flux density of the burst in radio astronomy. During the explosion, these active regions relatively have strong magnetic fields with denser and hotter compared with their surroundings and enhanced radio burst typically from 10-100 minutes. This flux density is a main parameter to study the mechanisms of radio emission at specific wavelengths associated with physical condition in the source environment. The antenna can be used independently or by connecting to the dish of radio telescope with a tracking system. This detection is formed as a solar radio burst type.

There are five (5) solar radio burst designed as 1-V and each type is associated with different phenomena [3]. Solar radio burst type I is associated with a solar storm and an indicator of the initial eruption of the active region of the sun while type II is normally associated with Coronal Mass Ejections (CMEs) events. This phenomenon is one of the interesting phenomena that have been monitored since 1970s. Type II can be divided into two sub-types: (i) a hearing bones and (ii) harmonic structure. Type III and V occurred during solar flares. Meanwhile, type III is very dominant and could be observed frequently by singular, in groups, or storms compare with other burst. It can be identified as a drifting burst which is drifting from high to low frequencies. However type V is quiet rare and it has a smooth but short-lived continuum characteristics. The beginning of the new active region can be characterized by the type IV burst with a broad continuum and a fine structure [4].

There are some basic limitations that might affect the performance of the antenna. In order to get a better picture of the performance of LPDA, one of the methods that can be done is by determining the power flux density of the burst. This is important to distinguish the performance of LPDA during solar events occurred. There might be other factors that will affect the performance of LPDA such as interference from other nearest sources. Others factors such as the radiation pattern, polarization, operation of the frequency band, gain, input impedance and efficiency of an antenna which indicates the power or field strength radiated in any direction relative to that in the direction of maximum radiation also should be considered.

2 Methodologies

It should be noted that there are some basic parameters that might affect the performance of the antenna. In order to get a better image of the performance of LPDA, one of the approaches that can be done is by determining the power flux density of the burst. It is believed that temperature of solar radio burst is one the main parameters that very significant to study in details the structure and dynamism of burst during the high activity of the Sun. This is important to distinguish the performance of LPDA during solar events occurred. There might be other factors that will affect the performance of LPDA such as interference from other nearest sources.

We have constructed the LPDA at the Radio Cosmology Laboratory Physics Department, University of Malaya. One should consider which the best material that practical is and be able to conduct a current. Here, an aluminum type is being chosen seems it can be used for outdoor purpose with PVC insulator between the booms. It is tough, and does not rust as it has a protective oxide coating which quickly forms on freshly exposed aluminum. Beside inexpensive, it is easy to handle as it is light in weight. We cut it in different sizes based on frequency. It should be noted that the dimension of the antenna is based on the range of the frequency that we decided to observe. The maximum size of the element is 1.11 meters and the minimum size is 0.02 meters. Figure.1 shows the preparation of the element of the antenna.

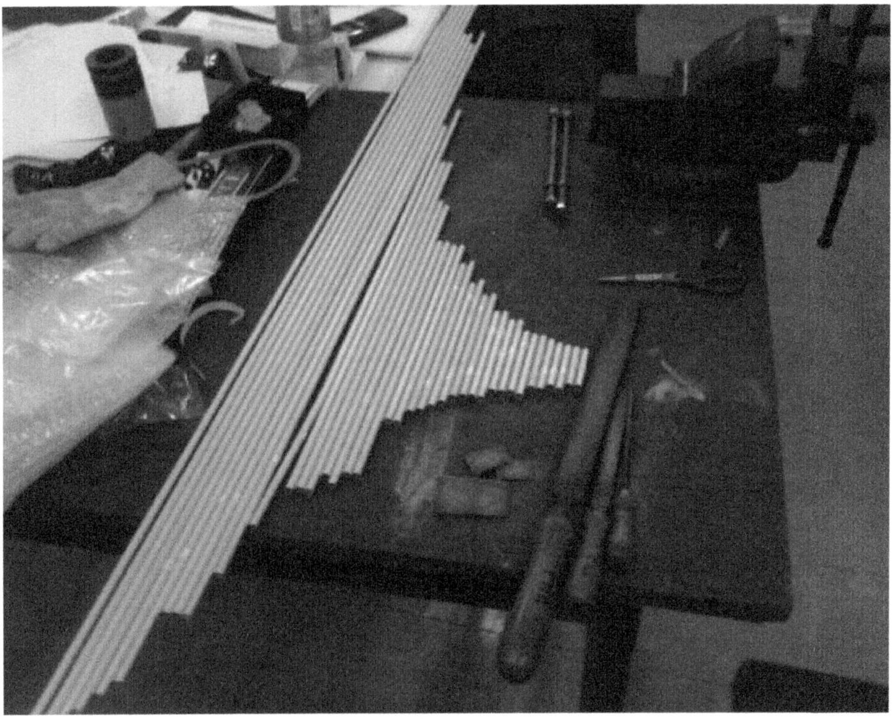

Fig. 1. Preparation of construction of LPDA

In order to obtain more gain than with a single log-periodic array, two arrays may be stacked. The diameter of these rods and tubes were chosen so as to incorporate the compactness as well as the ease of assembly for the antenna elements. Since it is dealing with transverse waves, all aerials are polarized. As this antenna is not automatically tracking yet, we decided to make sure that the directivity of an antenna is concentrated to the point of zenith direction. This is because the radiation of the Sun is higher during this point and the LPDA is more stable in this direction. A solar radio spectrometer CALLISTO was installed in February 2012 at the National Space Centre; Selangor, Malaysia. Fig. 2 shows the assembly of the LPDA at the National Space Center, Selangor Malaysia.

Fig. 2. Assemblies of LPDA at the National Space Centre, Selangor, Malaysia

3 Results and Analysis

The input impedance, R_0 =50 Ω is chosen for this LPD antenna [5]. For an effective transfer of energy, the impedance of the antenna and the impedance of the transmission cable connecting them must be the same and coordinated.

An alternating current element occurred to be a basic source of radiation according to the requirement of the dimension of LPDA. Based on theory, the performance of LPDA is a function of the antenna parameters τ and α [6]. The input impedance will increase toward the value of the unloaded feeder impedance as the value of τ is decreased. In particular, the input impedance depends on τ and α. The chosen tau (τ) and sigma (σ) give in the subtended angle of 3.43 degrees. In this case, it can be considered as a broadband type with a wide band of 825 MHz within 5.5 meters. As a result, the longest dipole in the array will be 3.96 m long [7]. It should be noted that the longer the boom the better the gain. However, due to the mechanical factors, it is not easy to construct a longer LPDA. The weight and dimension of the elements will make the antenna not become unstable. Unless, the material can be changed by using a copper, then it would be improved.

As indicated in Fig. 1, this LPDA antenna will have 19 elements. This antenna is connected CALLISTO (Compound Astronomical Low Cost Low Frequency Instrument

Spectroscopy Transportable Observatory) spectrometer in the operation room. This international network solar radio spectrometer is designed by the Institute of Astrophysics, ETH Zurich, Switzerland under the International Space Weather Initiative (ISWI) project in order to monitor solar activities in radio region within 24 hours. This spectrometer e-C07 having a detector sensitivity of 25mV/dB including control cables and radio frequency adapters was supplied by ETH Zurich. It consist a channel with 62.5 kHz resolution, while the radiometric bandwidth is about 300 kHz. Currently, there are 30 sites all over the world that used the same system. This collaboration is an alternative project in order to support a new country that is interested to study solar astrophysics.

At the receiving antenna, the electromagnetic energy is collected and converted into electrical energy and fed to the receiver. Only a tiny fraction of power is reflected back to the spectrum analyzer indicating that most of the power has been radiated by the antenna. Details values of the dimensions of the design of the LPDA are illustrated in Table I. We choose a solar burst type III that has been observed on 9th March 2012 (about 3:32 UT till 5:00 UT) for our case of study. The burst duration occur within 88 minutes. We focused on the duration starting from the pre-stage of solar flare till the post stage of the flare.

Table 1. The physical dimension of LPDA

No of elements	L(m)	R (m)	D(m)
1	3.96533329	5.551466963	1.110293326
2	3.17226668	4.441173636	0.888234674
3	2.53781338	3.552938962	0.710587750
4	2.03025073	2.842351212	0.568470208
5	1.62420061	2.273881003	0.454776173
6	1.29936050	1.819104830	0.363820944
7	1.03948842	1.455283885	0.291056759
8	0.83159075	1.164227125	0..232845411
9	0.66527261	1.164227125	0.186276331
10	0.53221809	0.745105382	0.149021067
11	0.42577448	0.596084315	0.119216855
12	0.34061959	0.476867459	0.095373486
13	0.27249567	0.381493973	0.076298790
14	0.21799654	0.305195182	0.061039032
15	0.17439723	0.244156150	0.048831227
16	0.13951779	0.195324922	0.039064982
17	0.11161423	0.156259940	0.031251986
18	0.08929139	0.125007954	0.025001589
19	0.07143311	0.100006365	0.020001271

Once the LPDA is successfully constructed, we test the antenna by evaluating the power flux density of solar burst. Therefore, we chose one of a good data. Solar radio burst type III that have been detected during 9th March 2012 associated with a large solar flare, type M6.3 has been chosen in this case. Based on calculation the temperature of the solar burst is 328.9K. It was found that the background sky temperature of the receiver of the antenna is 300.15 K. Figure 3 shows the solar burst (bright color) and sky background (dark color) from the observations.

In order to calculate the Signal to Noise Ratio (SNR), the Power Flux Density (PFD) will be measured. However, it is necessary to calculate the effective area of the antenna, A_{eff} at the first stage. In this case, the symbol G represents gain and G (dB) is a gain in dB units. Since the solar burst data is at frequency, f =240 MHz, this value will be selected and c is the speed of light in m/s.

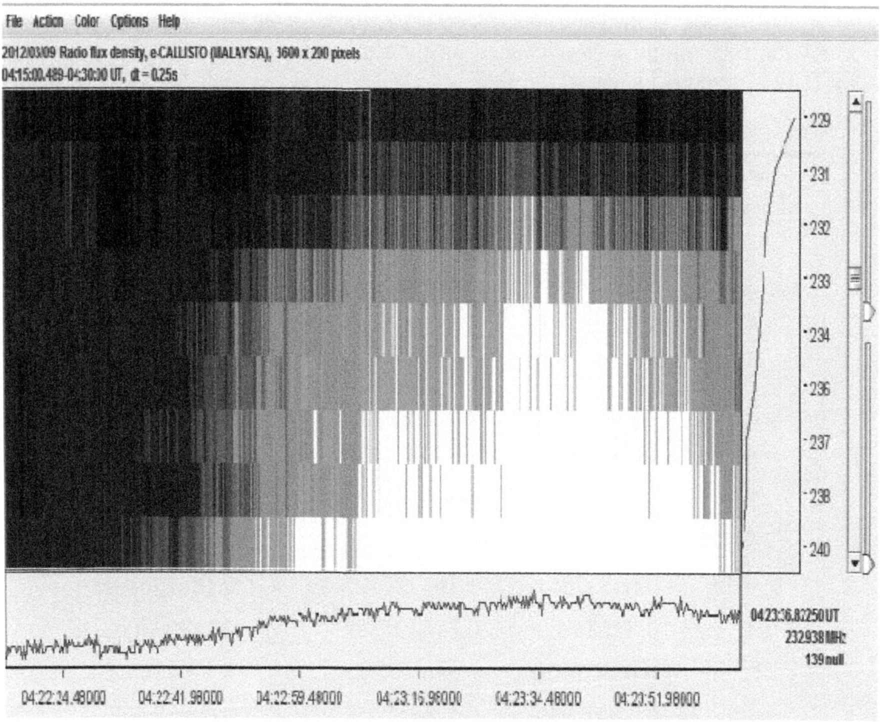

Fig. 3. The cold (blue) and hot (yellow) section of solar radio burst

In analysis process, we used a flux density and spectral power of Jovian and solar burst software to determine the other important parameters. One big challenge of this range is the population of different interference source that might affect the solar burst data. Yet it is still can be eliminated during the analysis process. Detailed analysis of LPDA and solar burst parameter is illustrated in Figure 4.

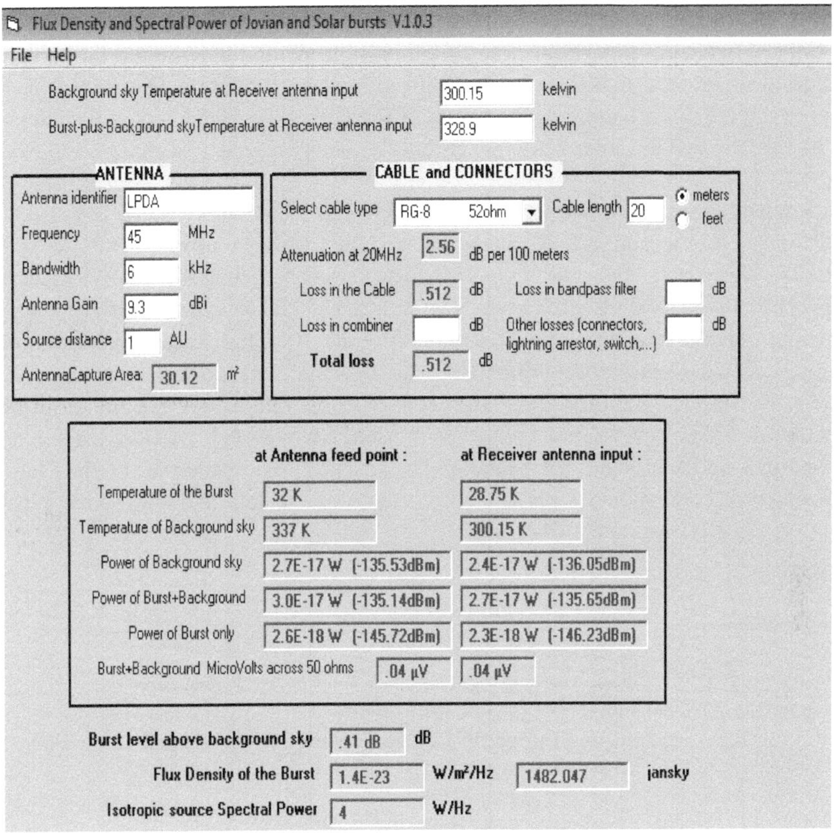

Fig. 4. The flux density software of solar burst

Based on the results, it was found that the gain of the antenna is 9.3 dB. This antenna potentially captures a signal that covers about 0.08 m^2 area of the Sun. By using 20 meters long of RG8 cable (52 Ω), there will be about 2.56dB of attenuation for every 20 meters. On average, the signal will lose about 0.512 dB per 100 meters.

In addition, the temperature of the burst that detected at the feedpoint of the antenna is 32K. However, The signal becomes decrease to 28.75 K while by the CALLISTO spectrometer as a receiver. It was found that the isotropic source spectral power is 1576 W/Hz. Since the burst level above the background sky is 0.41 dB , the flux density of the burst is 5.5 x 10^{-21} W/m^2/Hz.

4 Conclusions

While the main goal of wideband imaging is to obtain a high dynamic-range continuum solar burst profile, the reconstructed spectral structure can also be a useful astrophysical measurement. The flux density of the solar burst has been evaluated. Hence, this parameter is important seems the temperatures are different but the flux density, that is, the power landing on a square meter per hertz of bandwidth will be the same. Therefore, this is one way of calibration method that should be doing as a routine to make sure that the antenna is in a good performance. Our next task is to observe the radiation pattern of the signal of the LPDA. In conclusion, this antenna is suitable to observe the solar activities at low frequency region.

Acknowledgement. We are grateful to CALLISTO network; STEREO, LASCO, SDO/AIA, NOAA and SWPC make their data available online. This work was partially supported by the 600-RMI/FRGS 5/3 (135/2014), 600-RMI/RACE 16/6/2(4/2014and 600- RMI/RAGS 5/3 (121/2014) UiTM grants and Kementerian Pengajian Tinggi Malaysia. Special thanks to the National Space Agency and the National Space Centre for giving us a site to setup this project and support this project. Solar burst monitoring is a project of cooperation between the Institute of Astronomy, ETH Zurich, and FHNW Windisch, Switzerland, MARA University of Technology and University of Malaya. The research has made use of the National Space Centre Facility and a part of an initiative of the International Space Weather Initiative (ISWI) program.

References

[1] DuHamel, R.H., IsBell, D.E.: Broadband Logarithmically Periodic Antenna Structures. IRE National Convention Record, 119-128 (1957)

[2] Carrel, R.L.: Analysis and Design of the Log-Periodic Dipole Antenna, Elec. Eng. Dept. University of Illinois (1961)

[3] Kundu, M.R.: Solar Radio Astronomy. John Wiley (1965)

[4] Wild, J.P., Smerd, S.F., Weiss, A.A.: Solar Burst. Ann. Rev. Astron. Astrophysics 1, 291–366 (1963)

[5] Hamidi, Z.S., Anim, N.M., Hakimi, N.N.S., Hamzan, N., Mokhtar, A., Syukri, N., Rohizat, S., Sukma, I., Ibrahim, Z.A., Abidin, Z.Z., Shariff, N.N.M., Monstein, C.: Application of Log Periodic Dipole Antenna (LPDA) in Monitoring Solar Burst at Low Region Frequencies Region. International Journal of Fundamental Physical Sciences 2, 4 (2012)

[6] Smith, C.E.: Log Periodic Antenna Design Handbook, 1st edn., Ohio (1966)

[7] Hamidi, Z., Chumiran, S., Mohamad, A., Shariff, N., Ibrahim, Z., Radzin, N., Hamzan, N., Anim, N., Alias, A.: Effective temperature of the sun based on log periodic dipole antenna performance in the range from 45 Mhz to 870 Mhz. American Journal of Modern Physics 2, 4 (2013)

Attitude and Opinion of Bicycle-Helmet Signal

Pattama Longani[1,*], Orawit Thinnukool[1], and Anusorn Yodjaiphet[2]

[1] College of Arts, Media and Technology, Chiang Mai University, Chiang Mai, Thailand, 50200

[2] Department of Electrical Engineering, Faculty of Engineering, Chiang Mai University, Thailand, 50200

{pattama.longani,anusorn.rmutl}@gmail.com, orawit.t@cmu.ac.th

Abstract. This paper aims to studies the opinion, attitude and compliance of cyclists who ride a bicycle in Chiang Mai city, Thailand, with a Bicycle-Helmet Signal (BHS). The purpose of BHS is to help decreasing accidents for cyclists. The device automatically shows turning signal when cyclists lean their head to the left or right-hand side. The findings are based on a sample size of 30 which were taken from the local cyclists. Two groups have been separated to test the confidence, the group of cyclists using BHS and the other cyclists who use the same road with the first group. The results of five sections of questionnaires have been analyzed and show the quality of BHS. The details of opinions, attitudes and compliance are valuable for the design of BHS which provides safety on the road.

1 Introduction

Nowadays, bicycles are popular and the majority of people ride a bike. People also use the bicycle for recreational activities and that encourage people to use bicycles [1][2]. For example in US, an estimated 44.3 people younger than 21 years of age always ride a bicycle for more than five hours per week. However, reports of traffic accidents show that 300 children die and about 430,000 are injured annually [3]. Also, an annual police report in France reported road accidents caused an increase in the death proportion.

Bicycle helmets are used to protect the head from accidents; it reduces the risk of brain injury [4]. When using the helmet, 95% of people survive an accident [5][6]. Thus, the bicycle helmet significantly reduces the risk of injury and death from an accident. In bicycle helmet hardware nowadays, the designer has created several types that support safety features without the material for head cover [7]. The helmet attempts to protect the user's head by absorbing mechanical energy and protecting against penetration. Because of its benefits, many cyclists wear bicycle helmets.

There are many light products in the cycling market. Mostly are used only for giving the brightness. Cyclists also usually use their own hand for showing signals. However, it hard for other rider or driver to see the signal easily, especially at night.

* Corresponding author.

M. Gen et al. (eds.), *Industrial Engineering, Management Science and Applications 2015*, Lecture Notes in Electrical Engineering 349, DOI: 10.1007/978-3-662-47200-2_73

As a result, it becomes a lot of bike accidents. Some company have invented some light signal products to decrease the accident. For example;

- LED REMOTE TURNING SIGNAL BACKPACK which can trigger the LED lights on products by a wireless remote on the handlebar, http://www.sznicety.com
- ZACKEES TURN SIGNAL GLOVES, LED turn signal lights are built directly into the bike glove. There are sensors on the inside of both thumbs and the outside of both pointer fingers. Touch the two together and the LED will flash, https://zackees.com/,
- BIKE ZONE by Frank Guo, Hung Wang, and Sturt Morrow. It indicates a left or right turn by shining animated symbols that look like arrows on the ground, http://www.bikocity.com/lighted-bike-safety-zone-created-with-bike-light-concept/
- UON BICYCLE LIGHTS plug the light at the end plugs, Tap the top bottom to activate the signal light, https://www.facebook.com/UonLights
- BICYGNALS LED BICYCLE LIGHTS both front and rear signals are synchronized working by using wireless technology. sync. The light are controlled by flick of the thumb, http://www.geekalerts.com/bicygnals-led-bicycle-lights/

However, no company have invented the signal light controlled by leaning head to show turning signal before. Thus, we designed a circuit of light control with light materials that can be plug with a bicycle helmet that has some holes at the back. It helps cyclists show turning direction signal without unhand. Our Bicycle-Helmet Signal (BHS) has been designed as an alternative device to decrease accidents. We also survey sample test groups of local cyclist to ascertain the attitude, opinions and compliance regarding BHS. The comments, attitude, opinions and compliance have been taken into account in the questionnaire.

We will talk about BHS' design in Section 2. Methodology to find opinion, attitude, and compliance to BHS are stated in Section 3. The result from 30 respondents with five sections of questionnaires are presented in Section 4. Conclusions and future studies are remarked in Section 5.

2 Bicycle-Helmet Signal

The Bicycle-Helmet Signal (BHS) is a circuit uses a series of ten high-flux LED lights with the control of Gyroscope and Accelerometer to track neck movement of the cyclist. The circuit is kept in a rectangular box to protect the circuit from wet. It has been designed to show the signal in the case of turn left and turn right while the cyclist leans his head to the left or right, respectively. If he lean his head to the left hand side, the LED light will be twinkling to show the signal to turn left, likewise when you want to turn right. BHS provides different "beep" sound signal at the first time that the light twinkling, so the cyclist will know that the signal is work properly. The light will automatically off after a specific duration of time. The device can be removed from the bicycle helmet in order to insert a new battery (AAA) 1.5 watts or charge it directly with lituim-ion 3.8 volts 1000 mAh. The designing of bicycle-helmet signal device has been shown in Figure 1.

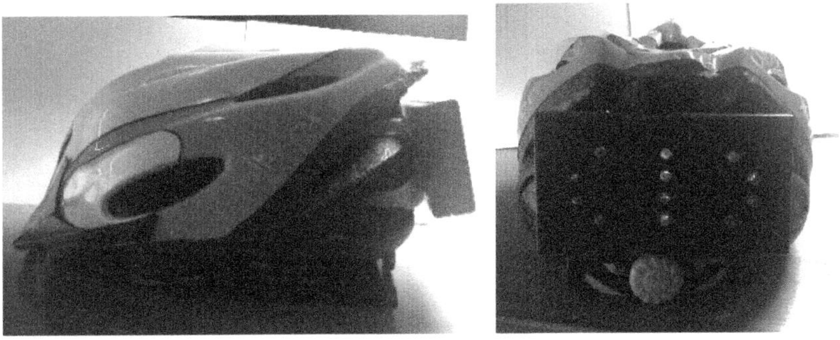

Fig. 1. Bicycle-Helmet Signal device

We have tested BHS by using local cyclists, include tourist police, as shown in Figure 2. All of them will experience both the user of BHS and the other cyclist who ride together with the one who uses BHS.

Fig. 2. Pilot test for BHS

3 Methodology

3.1 Interviewees

We gave the questionnaire to cyclists who ride a bicycle in Chiang Mai city, Thailand. Thirty anonymous have to use the BHS while riding before answering the questions.

3.2 Questionnaire

Participation in the survey was anonymous and voluntary, four sections with twenty closed questions were asked. Section 1 cover the attitude of the respondents to the traffic laws and wearing helmet while go cycling. Section 2 cover the opinion and attitude about BHS materials and its appearance. Section 3 cover the opinion and attitude about usability after using the BHS. Section 4 cover the confidence for the BHS can help decreasing accidents on the road when the respondent is the co-road user with the one who use BHS. Also, we have a special section for cyclists who wear cycling shoes with cleat. Questions cover opinion about tentative accident that may occur while using the cleat with BHS.

3.3 Statistics Analysis

Information is presented as the percentage of the respondents, opinions and attitudes of the questions in each part. Statistical analysis of the data from the questionnaire were performed using R program for Windows (R development core team, 2008). In the statistical analysis, descriptive statistics were used for analysis. We also used Pearson's chi-square for comparison, of which the statistical significance was defined as $p < 0.05$.

4 Results

In Based on the sample size of 30 measurements taken from the local cyclists who ride a bicycle in Chiang Mai city. The bar chart of five sections covering the attitudes, opinions and compliance for BHS are shown as the percentages with five rating scales. Let the meaning of the bar chart are stated in Figure 3. The results of the questionnaire have been shown, five sections of the attitude and opinion are shown below:

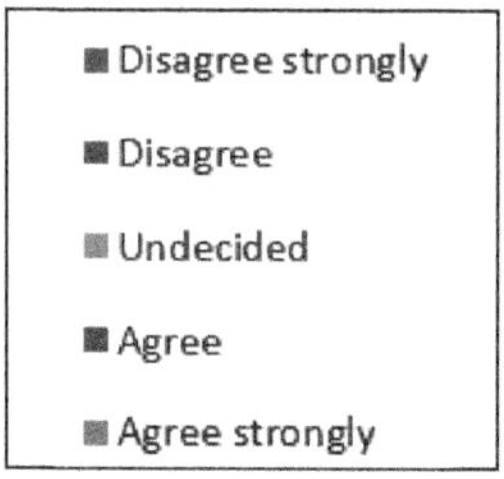

Fig. 3. The chart meaning

1. The attitude to the traffic laws and wearing helmet while go cycling are shown in Figure 4. The percentages are high showing that more than 70 % agree strongly with the questions. Our respondents will to obey the law and wearing helmets.

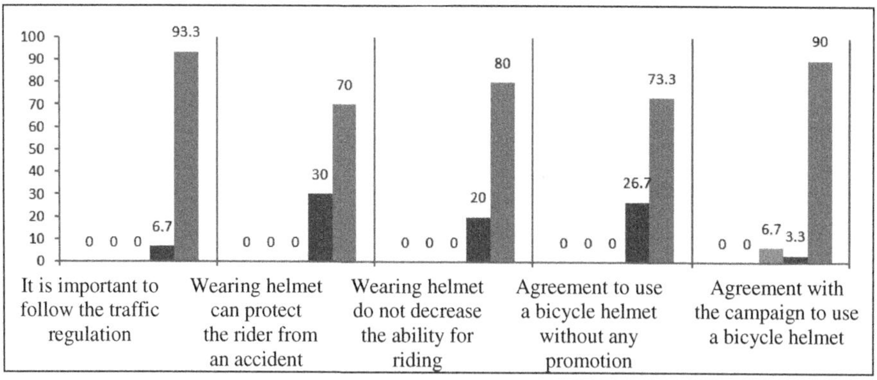

Fig. 4. The attitude of the respondents

2. Compliance for BHS materials and its appearance are shown in Figure 5. The size, material, weight, and the brightness of the light are considered. Respondents concerns about the brightness of the helmet lights and arrangement of the LED to convey meaning.

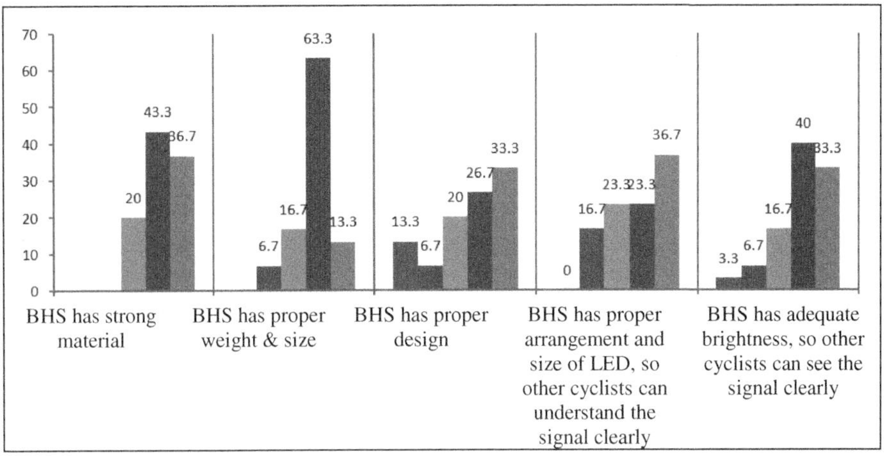

Fig. 5. The attitude of BHS design

3. Opinion and attitude about usability of BHS after using are shown in Figure 6. The results show the majority of the respondents believe in the usability of BHS. However, just a few cyclists disagree with the response of BHS and think that it is more dangerous when using BHS.

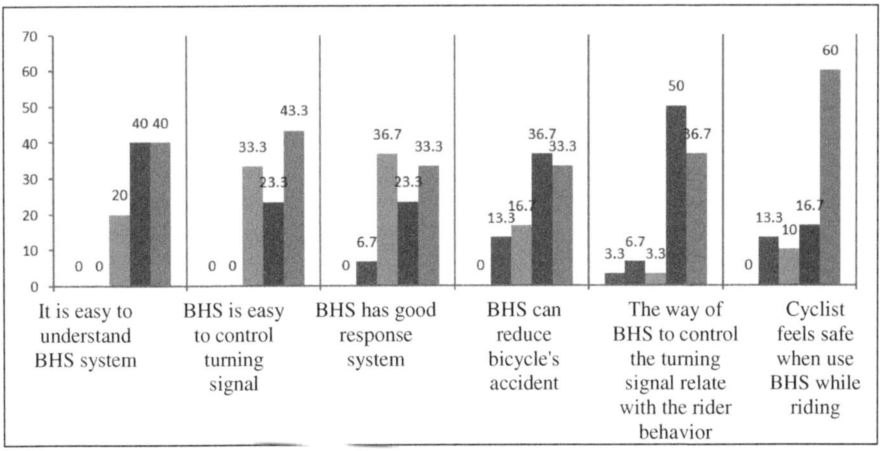

Fig. 6. The attitude of BHS usability

4. Confident that BHS can help decreasing accidents on the road, other cyclists opinion, are shown in Figure 7. The respondents suggest about the visibility of BHS, especially at the daytime. Many people still do not understand the meaning of the signal, so they do not feel safe while they are confusing.

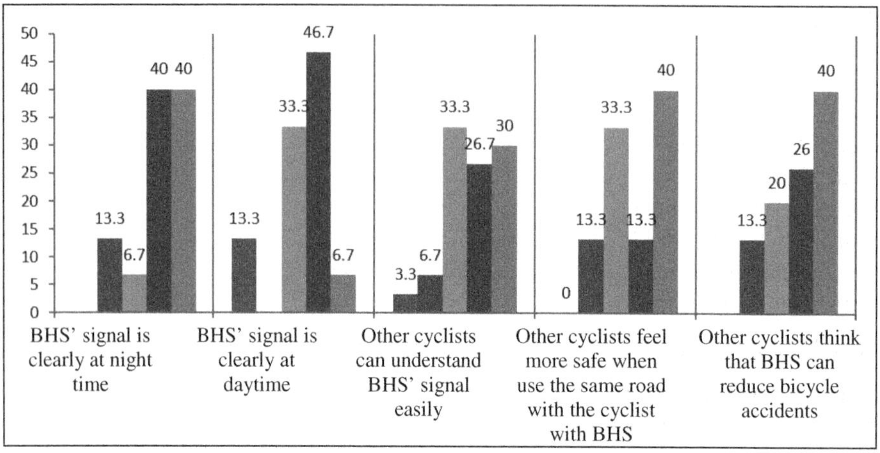

Fig. 7. The attitude of BHS signal

5. Opinion and attitude from the cyclists who wear shoes with cleat while using BHS are considered are shown in Figure 8. The respondents agree that they can control BHS while using cleats.

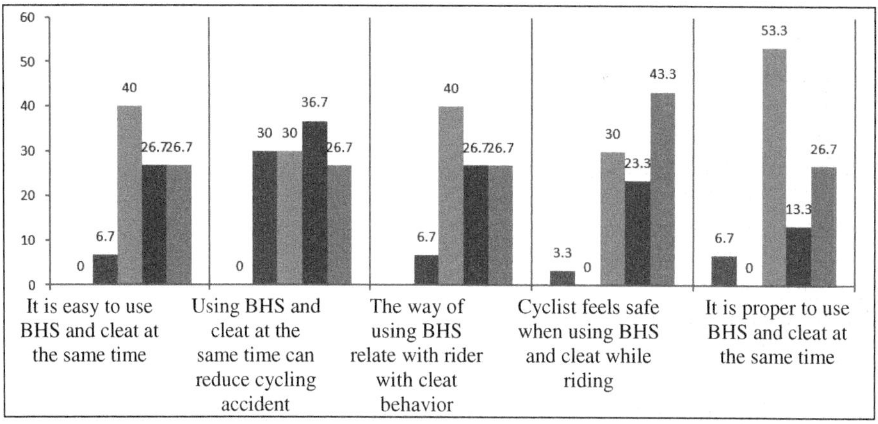

Fig. 8. The attitude of BHS design

We also have some comments from the respondent as follows. Some of the opinions suggest that the lightness and size are the important points that the designer needs to improve. The designer needs to reduce the shape of the helmet and improve the sensitivity of the light signal while turning.

Table 1 shows the statistical analysis for two groups of samples. First one is the group of cyclists who using the BHS while they are riding. Second is the other cyclists who ride the bicycle together with the cyclist who using BHS. We analyse each part of the questionnaire and the results show that the averages opinion, attitude and compliance between the two groups of are not different.

Table 1. Testing the opinion, attitude and compliance of the two groups of sample by using t-test (Independent Sample Test)

		F	Sig.	t	df	Sig. (2 tailed)
Opinion and attitude for BHS.	Equal variances assumed	.000	1.000	.000	28	1.00
	Equal variances not assumed			.000	28	1.00
Compliance with BHS' design	Equal variances assumed	.163	.689	.241	28	.812
	Equal variances not assumed			.241	27.95	.812
Opinion and attitude after use BHS.	Equal variances assumed	.000	1.000	.000	28	1.00
	Equal variances not assumed			.000	28.00	1.00
Confident for BHS. (other cyclists)	Equal variances assumed	.120	.732	.174	28	.863
	Equal variances not assumed			.174	27.98	.863
Attitude for using cleats and BHS at the same time.	Equal variances assumed	.000	1.000	.000	28	1.00
	Equal variances not assumed			.000	28.00	1.00

5 Conclusions

The questionnaire can tell us about the efficiency of BHS that is designed for help reducing accidents to cyclists. Opinions, attitudes and compliance of local cyclists are stated. The interesting comments about the material, the sensitivity of signal, the size and the brightness of the LED are indicated below:

1. Material: cyclists want to decrease the weight of BHS and also wish to use a nice design helmet.
2. Sensitivity of signal: cyclists want to adjust the sensitivity of signal when they want to control device to turning the direction. Because cyclists ride a bike in different speed, they want BHS to response all of their styles.

Delayed response may cause wrong communication to other riders and drivers on the road.

3. Size of LED: cyclists want to increase the size of the LED to make more understandable and increase the visibility to others who drive and ride on the same road.

Although, this project had a small sample, it also measured opinion, attitude and compliance for the first step of the design. To improve the effectiveness in the future, all responses from the questionnaire will facilitate improvement for the next generation BHS. We have some suggestion for the improvement as follows:

1. Design of a gadget has a lot of influence on the cyclist determination to use or not use.
2. Cyclists always ride in the different place with different climate, water resistant is important.
3. Because cyclists always ride at various speed and have different physical, a gadget should be adaptable.
4. Switching and Bluetooth are an alternative way to control the signal light while the cyclist's hand is still on the handlebar.

Acknowledgements. The authors would like to thanks Thai Health Promotion Foundation (Thai Health) for supporting fund to do the research, the supporting opportunity from College of Arts, Media and Technology, Chiang Mai University. We also extend our grateful to polices and cyclists in Chiang Mai, Thailand

References

1. Marcia, D.L.: The Bicvcle: Vehicle for a shall Planet. World Watch Paper 90, 7–8 (1989)
2. Stefanie, U., Daniel, M., Doris, K., David, A.D.: Bicycle helmet use and non-use – recently published research. Journal of Occupational Medicine and Toxicology 7, 9 (2012)
3. Richard, A.S., Jefrey, J.S.: Measuring community bicycle helmet use among children. Public Health Report 116, 113–116 (2001)
4. Carla, S.J., Klaus, Z., Uta, S., Andreas, W.U.: Attitude and opinion of neurosurgeons concerning protective bicycle-helmet use. Journal of Neurotrauma 27, 871–875 (2010)
5. James, G.: Cycle Helmets: The impacts of compulsory cycle helmet legislation on cyclist fatalities and premature deaths in the UK. Central Bedfordshire Council, 2-3 (2012)
6. Liu, B.C., Ivers, R., Norton, R., Boufous, S., Blows, S., Lo, S.K.: Helmets for preventing injury in motorcycle riders. The Cochrane Library 1, 4–7 (2012)
7. Cyclehelmets.org. Head up. Cycle (June/July, 2005)

Real Time Customer Satisfaction Index

Afizan Azman, Luwe Cheng Wong, Mohd Fikri Azli, Siti Zainab, Kirbana Jai Raman, and Sumendra Yogarayan

Faculty of Information Science and Technology (FIST), Multimedia University (MMU), Melaka, Malaysia
{afizan.azman,mfikriazli.abdullah,sitizainab.ibrahim}@mmu.edu.my, {luwechengwong,jpk_kirbz}@hotmail.com, mastersumen@gmail.com

Abstract. A Real Time and Automatic Customer Satisfaction Index is an application-based facial expression recognition system which is used to capture a person facial expression. The main idea of this system is to capture customer's facial expressions while using a product or services and act as an assessment tool to evaluate customer satisfaction with subjective evaluation (questionnaire) to understand customer's facial expressions. Hence, the system will enable product or services seller to get customer feedback immediately and it can help product or services seller to save time because the ordinary way to get customer feedback about a product or services is by doing a survey or interview.

1 Introduction

Human communication is a combination of both verbal and nonverbal interactions. Through facial expressions, body gestures and other non-verbal cues, a human can communicate with others which is especially true in the communications of emotions. In fact, studies have shown that a staggering 93% of affective communication takes place either non-verbally or paralinguistically (Mehrabian, 1971) through facial expressions, gestures, or vocal inflections. Computer systems open up new horizons for emotion detection by recognizing and detecting nonverbal cues via automated devices. Although these systems are still far from achieving the capacity of human perception, they are able to classify and assess user emotions through predetermined mathematical models with limited human intervention. The purpose of this project is to develop an application-based system known as "A Real Time and Automatic Customer Satisfaction Index" that is used to describe customer response while using a service or product in a real time using physiological measurement. It is a system that is able to recognize a person facial expression like neutral, happy, anger, fear, surprise and disgust.

2 Implementation and Testing

The proposed idea displays six universal expressions such as neutral, happy, sad, anger, disgust and surprise. It has functions of selecting or creating user, capturing

M. Gen et al. (eds.), *Industrial Engineering, Management Science and Applications 2015*,
Lecture Notes in Electrical Engineering 349, DOI: 10.1007/978-3-662-47200-2_74

photo and creating new expression class. After these three steps, it will be followed up by expression recognition. User will be able to abort the process if they intend to end it.

2.1 Design

As shown in the figure above, user must follow the process correctly in order to make sure the system does not abort. First of all, if user trying to access expressions recognition, the user will need to create new user then take photos else system abort if they pressed 'esc' key at this time the system will abort. After that, the user will also need to create new expressions class but also if the user pressed 'esc' key at this process, the system will abort. Last but not least, after create new expressions class completed, the user can now proceed to choose video format for capture and then expressions recognition. Finally, for the existing user part, the user will just need to select the previously created expressions file to proceed to expressions recognition

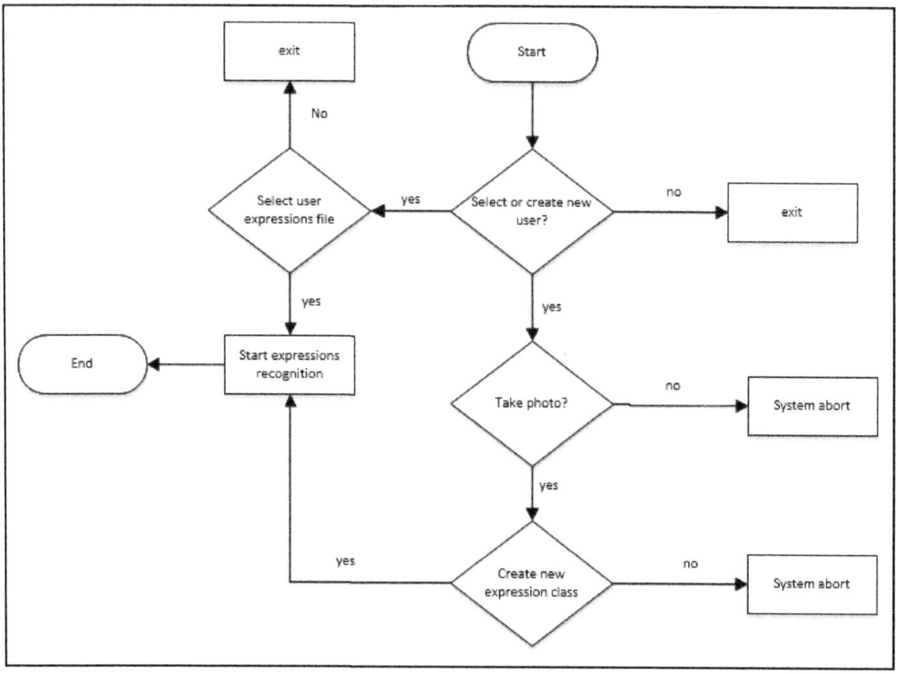

Fig. 1. Flowchart Diagram of Facial Expression Recognition System

2.2 Proposed Solution

K-Star is an instance-based classifier that is the class of a test instance is based upon the class of those training instances similar to it, as determined by some similarity function. Random Forest is a class for constructing a forest of random trees which means that is a group of un-pruned classification made from the random selection of samples of the training data. The dataset are trained and tested using 10-fold cross validation (Ghosal, Tikmani & Gupta, 2009) which means divide the dataset into ten parts (these are known as "folds"), nine of the parts are used for training and the last part used for testing. Then, with the same division, take another nine parts for training and the held-out part for testing. The whole things are repeated ten times, using a different segment for testing each time. Therefore, the accuracy results obtained as shown below can be used to estimate of prediction error.

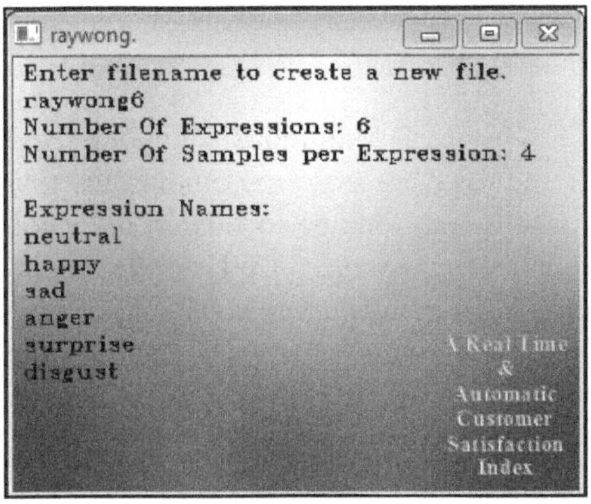

Fig. 2. Window of information required to create new user

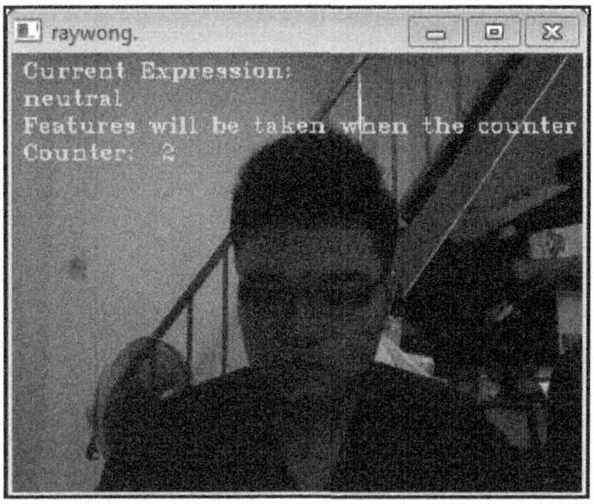

Fig. 3. Window to take neutral expression features

Fig. 4. Window to take happy expression features

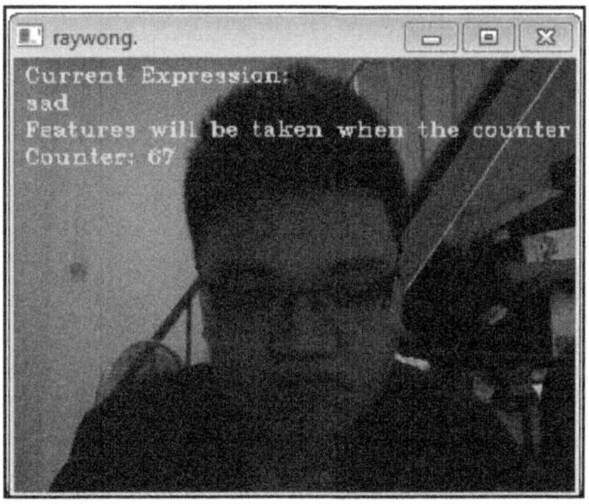

Fig. 5. Window to take sad expression features

Fig. 6. Window to take anger expression features

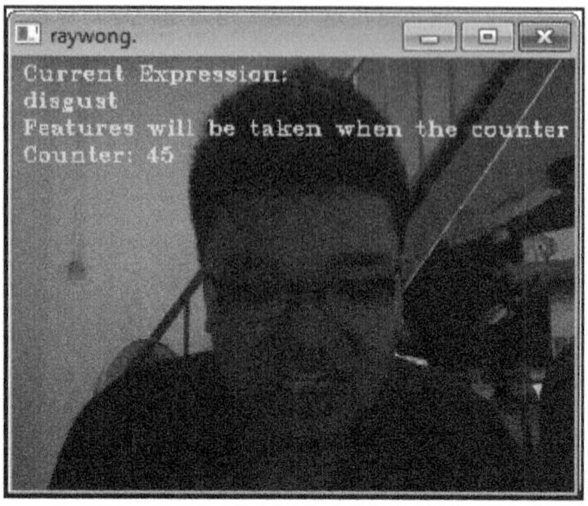

Fig. 7. Window to take disgust expression features

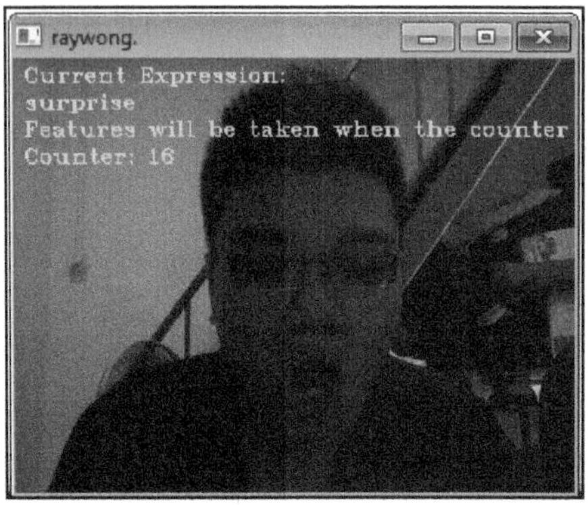

Fig. 8. Window to take surprise expression features

Fig. 9. Window for recognition

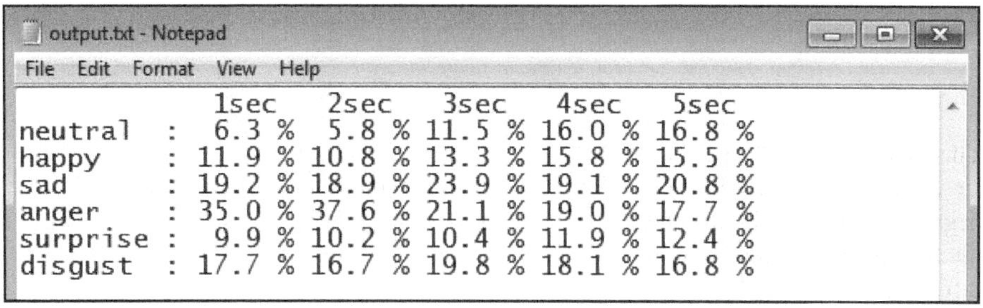

	1sec	2sec	3sec	4sec	5sec
neutral :	6.3 %	5.8 %	11.5 %	16.0 %	16.8 %
happy :	11.9 %	10.8 %	13.3 %	15.8 %	15.5 %
sad :	19.2 %	18.9 %	23.9 %	19.1 %	20.8 %
anger :	35.0 %	37.6 %	21.1 %	19.0 %	17.7 %
surprise :	9.9 %	10.2 %	10.4 %	11.9 %	12.4 %
disgust :	17.7 %	16.7 %	19.8 %	18.1 %	16.8 %

Fig. 10. Text file generated

A total of 19 people have been tested using the system, here is the outcome with the value of 1 is accurate while 0 is not accurate.

Table 1. Average Accuracy Results of K-Star Classifications Algorithm

Class	TP rate	FP rate	Precision	Recall	F-measure
Neutral	0.842	0.158	0.516	0.842	0.64
Happy	1	0.	1	1	1
Sad	0.737	0.032	0.824	0.737	0.778
Anger	0.632	0.084	0.6	0.632	0.615
Surprise	1	1	1	1	1
Disgust	0.368	0.011	0.875	0.368	0.519
Weighted Average	0.763	0.047	0.802	0.763	0.759

The average accuracy results of K-Star Classification Algorithm for happy and surprise is the best whereas neutral, sad and anger is slightly accurate.

Table 2. Average Accuracy Results of Random Forest Classifications Algorithm

Class	TP rate	FP rate	Precision	Recall	F-measure
Neutral	0.474	0.137	0.409	0.474	0.439
Happy	1	0	1	1	1
Sad	0.632	0.063	0.667	0.632	0.649
Anger	0.474	0.105	0.474	0.474	0.474
Surprise	1	1	1	1	1
Disgust	0.526	0.074	0.588	0.526	0.556
Weighted Average	0.684	0.063	0.69	0.684	0.686

The average accuracy results of Random Forest Classification Algorithm for happy and surprise is the more accurate compared to neutral, sad, anger and disgust.

Table 3. Confusion Matrix of K-Star

Neutral	Happy	Sad	Anger	Surprise	Disgust
16	0	2	1	0	0
0	19	0	0	0	0
5	0	14	0	0	0
5	0	1	12	0	1
0	0	0	0	19	0
5	0	0	7	0	7

Confusion matrix is commonly named contingency table. In this case of K-Star, only happy and surprised is accurately classified meanwhile there is few classes that is misclassified such as neutral, sad, anger and disgust.

Table 4. Confusion Matrix of Random Forest

Neutral	Happy	Sad	Anger	Surprise	Disgust
9	0	5	0	0	2
0	19	0	9	0	0
6	0	12	1	0	0
4	0	1	9	0	5
0	0	0	0	19	0
3	0	0	0	0	10

For confusion matrix of Random Forest, it is also seen that only happy and surprise is accurately classified meanwhile there is few classes that is misclassified such as neutral, sad, anger and disgust.

2.3 Validation

This section shows the accuracy results of using another dataset (Myunghoon Suk and Balakrishnan Prabhakaran, 2014) from another facial expressions recognition system. The facial expressions recognition system is based on mobile six different expressions which is anger, disgust, fear, happiness, sadness and surprise. The results obtained by using the same classifications algorithms, same testing and training methodology which are 10-fold cross validation and the same amount of instances in a dataset.

Table 5. The accuracy results of K-Star classifications algorithm

Class	TP rate	FP rate	Precision	Recall	F-measure
Anger	0.737	0.084	0.636	0.737	0.683
Disgust	0.789	0	1	0.789	0.882
Fear	0.737	0.021	0.875	0.737	0.8
Happiness	0.789	0	1	0.789	0.882
Sadness	0.579	0.126	0.478	0.579	0.524
Surprise	0.579	0.126	0.478	0.579	0.524
Weighted Average	0.702	0.06	0.745	0.702	0.716

Table 6. The accuracy results of Random Forest classifications algorithm

Class	TP rate	FP rate	Precision	Recall	F-measure
Anger	0.737	0.032	0.824	0.737	0.778
Disgust	0.789	0.042	0.789	0.789	0.789
Fear	0.737	0.042	0.778	0.737	0.757
Happiness	0.842	0	1	0.842	0.914
Sadness	0.526	0.063	0.625	0.526	0.571
Surprise	0.789	0.137	0.536	0.789	0.638
Weighted Average	0.737	0.053	0.759	0.737	0.741

Table 7. Confusion Matrix of Random-Forest

Anger	Disgust	Fear	Happiness	Sadness	Surprise
14	0	1	0	3	1
0	15	1	0	1	2
1	1	14	0	0	3
0	0	1	16	0	2
1	2	1	0	10	5
1	1		0	2	15

Table 8. Confusion Matrix of K-Star

Anger	Disgust	Fear	Happiness	Sadness	Surprise
14	0	0	0	4	1
0	15	0	0	2	2
0	0	14	0	2	3
1	0	2	15	0	1
3	0	0	0	11	5
4	0	0	0	4	15

As compared with Table 2.1 and Table 2.2 weighted average, our system algorithm Active Shape Model perform more or less equal to Cohn-Kanada (CK+) database ((P. Lucey, et al., 2010) system. Furthermore, another reason their system perform equalize because their uses Active Shape Model and Active Shape Model with Stasm (STASM) (S. Milborrow and F. Nicolls, 2008) which our system used Active Shape Model and Haar Features (Lili Lin, Xiangbiao Li & Wenhui Zhou, 2012).

2.4 Evaluation

Table 9. Evaluation of System

Description	Input	Output	Test
Select/Create new user window	Select user	Expressions class file window	Successful
	Create new user	Enter name	Successful
Expressions class file window	Select existing expression class file	Start recognition	Successful
	Create new expressions class file	Information required window	Successful
Create new expressions file window	Enter filename, number of expressions, number of samples per expressions, expressions name	Start recognition	Successful
Create new user window	Enter name	Take photos window	Successful
Take photos window	One photo	Error	Successful
	Two photos	Create new expressions file window	Successful
	Two or more photos	Create new expressions file window	Successful

3 Conclusion

Facial expressions recognition system is a system that allows users to capture or recognize a person facial expression. The main idea of this system is to capture customer's facial expressions while using a product or services. When the time the customer's expressions is captured at the same time the expressions percentage is also saved in a text file which consists of percentage of expressions of every seconds. Furthermore, a video are also recorded to AVI format video file when 'esc' key is pressed. The purpose of the text file and video file is used for analysis afterwards. In addition, a questionnaire is also used to understand customer's facial expressions.

The main reason of product seller to choose to use the system is because the product seller doesn't have to do survey/interviews to get customer feedback of the product and the process itself is time consuming. With this system, it can save a lot of time and that time can be used for other important things.

Last but not least, there are several limitations on this system such as the recognition or recording process is delaying/slow because when there are more things run on the camera, the more delay it is. The camera itself can run 30 frames per seconds, but after coding and things displayed on the camera, it left only few frame per seconds. Furthermore, when come to expressions recognition, the results may be vary due to some external reasons such as wearing a spectacle, with beards. In addition, the lighting condition while recognition is also an important issue because the results may become vary when recognizing in a dark lighting condition and when in a darker lighting condition the recognition process will become delay/slow. Lastly, another limitation of the system is the recognition accuracy is not stable and low sometimes.

3.1 Future Enhancement

In the future, in collaborations with other parties, an automatic process of recognition may be achieved with aid further study and research. Besides, security level need to be increase and more functions or ability need to be add on to make the system more useable. Last but not least, another enhancement would be the existing camera that can be replace to a better quality camera that is able to run 60 frames per seconds in order to solve the delay of the screen. Lastly, improvement on the accuracy of the recognition rate by at least thirty percent or even higher so that the system will be more valuable.

References

1. Ghosal, Tikmani, Gupta: Face Classification Using Gabor Wavelets and Random (2009)
2. Mehrabian, A.: Silent Messages (1st ed.). Wadsworth, Belmont (1971) ISBN 0-534-00910-7
3. Milborrow, S., Nicolls, F.: Locating facial features with an extended active shape model. In: Proceedings of the 10th European Conference on Computer Vision, pp. 504–513 (2008)
4. Lin, L., Li, X., Zhou, W.: Face Detection Algorithm Based on Haar and Local Binary Pattern Features. Journal of Convergence Information Technology 7(19), 602–610 (2012)
5. Suk, M., Prabhakaran, B.: Real-time Mobile Facial Expression Recognition System – A Case Study (2014)
6. Lucey, P., Cohn, J., Kanade, T., Saragih, J., Ambadar, Z., Matthews, I.: The extended cohn-kanade dataset (ck+): A complete dataset for action unit and emotion-specified expression. In: 2010 IEEE Computer Society Conference on Computer Vision and Pattern Recognition Workshops (CVPRW), pp. 94–101 (June 2010)

FLC-Based Indoor Air Quality Assessment for ASHRAE Standard Conformance

Rionel Belen Caldo*

College of Engineering and Computer Studies
Lyceum of the Philippines University – Laguna (LPU-L), Laguna, Philippines
Electronics and Communications Engineering
De La Salle University (DLSU), Manila, Philippines
rionel14_caldo@yahoo.com

Abstract. Designed and presented in this paper is Indoor Air Quality Assessment using fuzzy logic technology. Triangular membership functions for classified Air and Gaseous parametric categories are to be constructed in Matlab Fuzzy Logic Toolbox using ASHRAE, EPA and WHO standards. Air parameters include Temperature (T) and Relative Humidity (RH) while Gaseous parameters include Carbon Monoxide (CO) and Carbon Dioxide (CO_2). The parameters are to be classified as Very Good (VG), Good (G), Fair (F), Poor (P) and Very Poor (VP). The Sugeno-style of inference systems is utilized in this proposed to describe the fuzziness of assessing indoor air quality. There were 131 fuzzy rules formulated. The output parameter is classified as Highly Acceptable (HA), Acceptable (A), Just Acceptable (JA), Not Acceptable (NA), and Highly Not Acceptable (HNA). The fuzzy-based system is simulated and the results were further verified using the weighted means of 90 generated data samples.

Keywords: Indoor air quality, fuzzy logic, ASHRAE, Matlab toolbox.

1 Introduction

According to (Chih-Hung et al., 2014), in day time, most of us stay indoors for about 80% to 90%. Americans spend 90% of their time indoors according to Centers for Disease Control and Prevention (CDC) [2]. On the average, office workers stayed 40 hours per week in office buildings according to OSHA 3430-04 in 2011. Considering that we stayed indoors most of the time, we should be health conscious and it is a must for all of us to consider long-term health. Thus, research on indoor air quality assessment is deemed important and must be given emphasis. Indoor Air Quality (IAQ) must be assessed, as it affects the quality and efficiency of man's work. Experts believed as well that more people suffer from the effects of indoor air pollution than from the effects of outdoor air pollution. According to U.S. Department of Labor, a major concern to business, school, building and worker is IAQ, as it

* Corresponding author.

M. Gen et al. (eds.), *Industrial Engineering, Management Science and Applications 2015*,
Lecture Notes in Electrical Engineering 349, DOI: 10.1007/978-3-662-47200-2_75

impacts the health, comfort, well-being, and productivity of the building occupants. Occupational Safety and Health Administration (OSHA) recognizes the need of addressing poor IAQ concerns. Because of energy conservation measures provided for buildings, health effects related to IAQ have increased. The use of Heating, Ventilation, and Air Conditioning (HVAC) (HVAC) systems had increased indoor pollutants concentrations (Wei Cai et al., 2010). Because of this, many research enthusiasts paid attention to IAQ studies. In different countries, indoor environment studies were conducted, which focuses on office or residential buildings (Wei Cai et al., 2010). A new trend in modern architecture is intelligent building. Key features of this intelligent building include intelligent systems and environmental monitoring. One of the most important factors, which influence the performance of intelligent building, is environmental monitoring (Wang-Kun Chen, 2013). The most critical and difficult part in intelligent systems for controlling indoor environment lie not only on environmental standards, but primarily in the feelings and perceptions of people. Therefore, IAQ must consider the uncertainty and ambiguity derived from individual preferences (Wang-Kun Chen, 2013).

For a fact, Proportional Integral Derivatives (PIDs) could provide reasonable solutions for this research proposed. However, the use of PIDs could bring inconvenience in controlling the uncertainty of the dynamics of HVAC systems. This could be characterized easily using linguistic fuzzy rules. Fuzzy Logic Controllers (FLCs) appear as a viable alternative solution to traditional controllers, since it do not require mathematical model and FLCs are prepared to handle different criteria for dynamics of HVAC system according to the knowledge of a human expert (Solana, 2013). In this study, the proponent will use the concepts and principles of fuzzy logic in the simulation of identified air and gaseous parameters for the assessment of indoor air quality. The factors to be considered for classifying indoor air quality include the temperature and relative humidity for air parameters, and carbon monoxide and carbon dioxide for gaseous parameters respectively. The parameters will be categorized as Very Good, Good, Fair, Poor and Very Poor. The indoor air quality is being classified on the basis of its acceptation. Specifically, the output can be classified as Highly Acceptable, Acceptable, Just Acceptable, Not Acceptable, and Highly Not Acceptable. The proponent preferred to use the Sugeno-style of fuzzy inference system. In this study, the simplest type of membership function, which is the triangular membership function, will be used for its input and output parameter constructions. The proponent would verify the results using Matlab Fuzzy Logic Toolbox. This study will be simulated purely mathematical. This study excludes the use of sensors for data log, feedback and control mechanism, quality control and information management.

2 Methodology

As shown in **Figure 1**, air and gaseous input parameters are sampled and normalized. The normalized data inputs are to be fuzzified by the predefined input fuzzy sets. The fuzzified data will be triggered by the formulated fuzzy rule table.

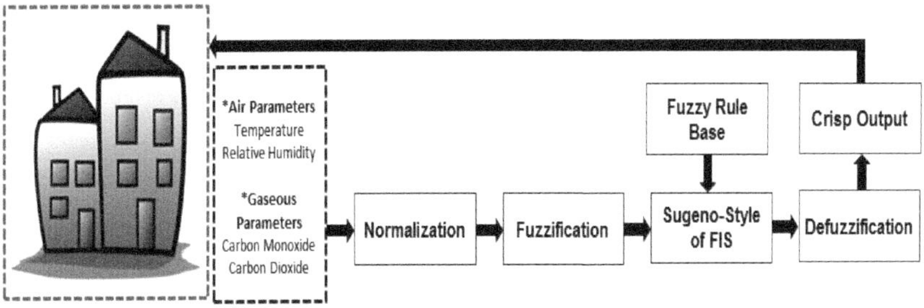

Fig. 1. Configuration of FLC-based indoor air quality assessment

Finally, the inferred data will be defuzzified to give crisp outputs of 1-Highly Not Acceptable, 2-Not Acceptable, 3-Jut Acceptable, 4-Acceptable and 5-Highly Acceptable. This assesses the IAQ, which later on can be improved in meeting IAQ standards and regulations. In this research, the proponent makes use of experimental study. In order for the proponent to effectively construct the membership function for input and output parameters, different standards and guidelines were used. The tolerance and specification limits set and maintained by ASHRAE, Environmental Protection Agency (EPA) and World Health Organization (WHO) were used for this process. Based on these international standards, the proponent identified the minimum and maximum possible range. The proponent generated 90 data for air and gaseous parameters using Excel VBA macro program. The specification limits and the generated data were used for categorizing the input parameters into "very good", "good", "fair", "poor" and "very poor". The membership functions for temperature, relative humidity, carbon monoxide and carbon dioxide were then constructed. Since the proponent makes use of Sugeno style of fuzzy inference system, it will only need him to set constants for linguistic class. The output parameter constants used are: **1** for Highly Not Acceptable (HNA), **2** for Not Acceptable (NA), **3** for Just Acceptable (JA), **4** for Acceptable (A), and **5** for Highly Acceptable (HA). The challenging part of fuzzy logic is forming its fuzzy rule. In building Fuzzy Associative Memory (FAM) matrix, perception of qualified experts about the linguistic description of indoor air quality was obtained. Also, the proponent had used the tolerance limits set by international standards. For simplicity of the ruling, the proponent assumed that all parameters are equal in terms of criticality. The method used for such data qualification and treatment is averaging technique and trial and error method. The proponent will make use of Matlab Fuzzy Logic Toolbox for simulation purposes.

The hierarchical structure for indoor air quality assessment includes two major blocks. First two blocks will be classified as very good, good, fair, poor and very poor. The last hierarchical level (indoor air quality assessment) classifies the indoor air quality as highly not acceptable, not acceptable, just acceptable, acceptable and highly acceptable. Listed below are sample rules used in this proposed for different hierarchical levels of structure: If *Temperature* is <very good> and *Relative Humidity* is <very good>; **Level 1:** *Then air parameters block is classified as* **<very good>** If *Carbon Monoxide* is <very good> and *Carbon Dioxide* is <very good>; **Level 2:** *Then gaseous parameters block is classified as* **<very good>** If *Air Parameters block* is

<very good> and *Gaseous Parameters block* is <very good>; **Level 3:** Then the last hierarchical level for Indoor Air Quality Assessment is classified as **< Highly Acceptable HA>**.

3 Membership Function Construction

In constructing the membership functions for the identified critical input parameters, the proponent makes use of the definition and description of air and gaseous quality parameter specifications used and adopted by AHSRAE stipulated in Indoor Air Quality Handbook (TSI Incorporated, 2013). Considering that OSHA does not have a general IAQ standard, as it provides only guidelines in addressing the most common workplace complaints about IAQ, the proponent preferred to use other notable standards. In this study, the proponent will make use of ASHRAE standard specifically the addendum to Thermal Environmental Conditions for Human Occupancy (stipulated in the Indoor Air Quality handbook standards and guidelines), EPA and WHO. Specifically, ASHRAE Standard 55-2010 and ISO 7730 will be used for T and RH parameters. ASHRAE Standard 62.1-2010 will be used for CO_2. On the other hand, for CO, EPA, ASHRAE and WHO standards will be used for 8 hr. TWA measure. By these standards, the proponent will build its FAM matrix and it will be verified using the generated 90 samples. These data samples will be clustered into three (3) shifts, as the proponent considers three measures. These measurements include data collection for shift A (6:00am to 2:00pm), shift B (2:00pm to 10:00pm) and shift C (10:00pm to 6:00am). The proponent will compute for the weighted mean of three shifts on a daily basis. For simulation purposes, the computed 30 weighted means will be plugged into the fuzzy system. In giving finer results, the proponent will use trial and error method.

4 Design Considerations

The indoor air quality system requires four (4) inputs from the user/operator. Air parameters include Temperature, which is measured in **°C** and Relative Humidity in terms of **%**. Gaseous parameters block comprised of Carbon Monoxide and Carbon Dioxide. Both parameters are measured in **ppm** unit. Considering that these parameters have different range of values, the inputs of the user will have to be normalized in fitting to its fuzzy scale of **0 to 1**. The normalized inputs will be fuzzified. Using FAM matrices, the rules trigger the output and it will be defuzzified using centroid of area for Sugeno style of inference system. The obtained crisp output will be used for linguistic classification in determining the acceptation of indoor air quality.

4.1 Fuzzy Associative Memory (FAM) Matrices for Indoor Air Quality Assessment

The FAM matric for air parameters block requires two inputs and classified into nine linguistic classes: VG - Very Good (8.5 to 9.5), G1 - Good1 (7.5 to 8.5), G2 - Good2 (6.5

to 7.5), F1 – Fair1 (5.5 to 6.5), F2 - Fair2 (4.5 to 5.5), P1 - Poor1 (3.5 to 4.5), P2 - Poor2 (2.5 to 3.5), VP1 - Very Poor1 (1.5 to 2.5) and VP2 - Very Poor2 (0.5 to 1.5). Gaseous parameters block, on the other hand, requires two inputs and classified into five linguistic classes: VG - Very Good (4.5 to 5.5), G - Good (3.5 to 4.5), F - Fair (2.5 to 3.5), P - Poor (1.5 to 2.5) and VP - Very Poor (0.5 to 1.5). Finally, Indoor air quality block requires two inputs and classified into five linguistic classes: HA - Highly Acceptable (4.5 to 5.5), A - Acceptable (3.5 to 4.5), JA - Just Acceptable (2.5 to 3.5), NA - Not Acceptable (1.5 to 2.5) and HNA - Highly Not Acceptable (0.5 to 1.5).

Considering that there were two input values and nine classifications for air parameters and two input values and five classifications for gaseous parameters, a total of **81** fuzzy rules were formulated for air parameters and **25** fuzzy rules were formulated for air parameters. For the final block, input parameters include the classified blocks of air parameters and gaseous parameters. There were two inputs and five classifications, which would give **25** fuzzy rule bases. Combining the three levels of assessment, a total of **131** fuzzy rules were formulated. This rule base formation is more efficient compared to **625** rules, which could be generated when we processed simultaneously the four input parameters and classify them into five distinctions. The linguistic variables were converted into its corresponding numerical values. The defuzzified output is interpreted in linguistic class for indoor air quality assessment.

4.2 FIS of Air Parameters

Based from ASHRAE standard 55-2010 and ISO 7730, we knew that the best temperature at summer is ranging from 23 to 28 °C and the best temperature at winter is ranging from 20 to 25.5°C. Temperature block defined nine membership functions as shown in Figure 2.

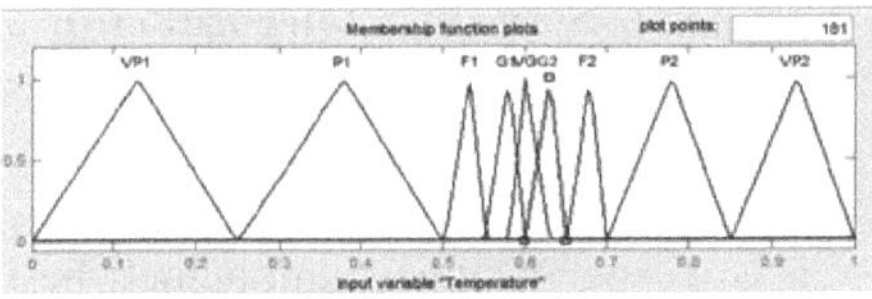

Fig. 2. Temperature (input)

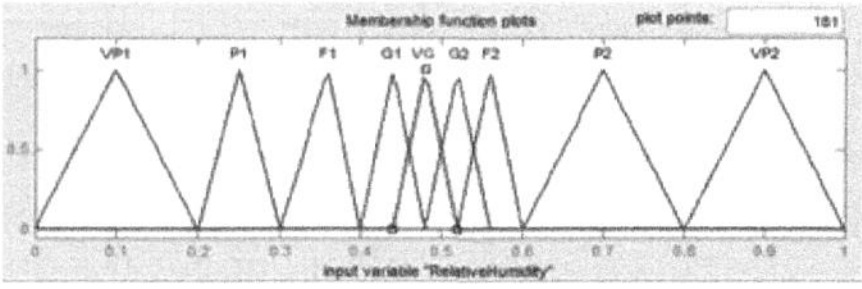

Fig. 3. Relative Humidity (input)

**(T-VG)*: concentration range of T block is 23 to 25 °C. **(T-G)*: concentration range of T block is 22 to 24 °C for T-G1 and 24 to 26 °C for T-G2. **(T-F)*: concentration range of T block is 20 to 22 °C for T-F1 and 26 to 28 °C for T-F2. **(T-P)*: concentration range of T block is 10 to 20 °C for T-P1 and 28 to 34 °C for T-P2. **(T-VP)*: concentration range of T block is 0 to 10 °C for T-VP1; 34 to 40 °C for T-VP2.

Based from ASHRAE standard 55-2010 and ISO 7730, we knew that the best relative humidity ranges from 30 to 65 %. Relative Humidity block defined nine membership functions as shown in Figure 3.

(RH-VG): concentration range of RH block is 44 to 52 %. *(RH-G)*: concentration range of RH block is 40 to 48 % for RH-G1 and 48 to 56 % for RH-G2. *(RH-F)*: concentration range of RH block is 30 to 40 % for RH-F1 and 52 to 60 % for RH-F2. *(RH-P)*: concentration range of RH block is 20 to 30% for RH-P1 and 60 to 80% for RH-P2. *(RH-VP)*: concentration range of RH block is 0 to 20 % for RH-VP1 and 80 to 100 % for RH-VP2.

4.3 FIS of Gaseous Parameters

Based from ASHRAE, WHO and EPA standards, we knew that the standard value of concentration of CO is 9 ppm for 8 hr. TWA; therefore any value below 9 ppm is normal and values higher than 9 ppm is considered to be Out-of-Control (OOC) or Out-of-Specifications (OOS). Carbon Monoxide block defined five membership functions as shown in Figure 4.

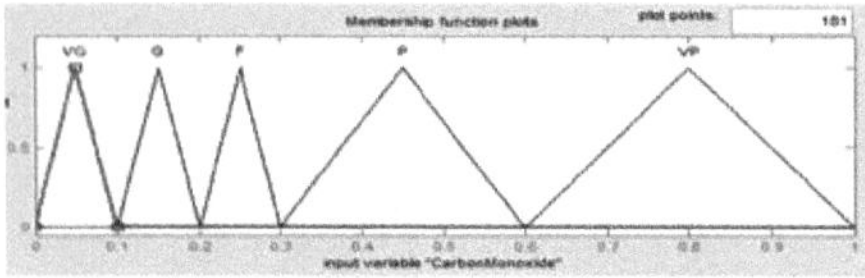

Fig. 4. Carbon Monoxide (input)

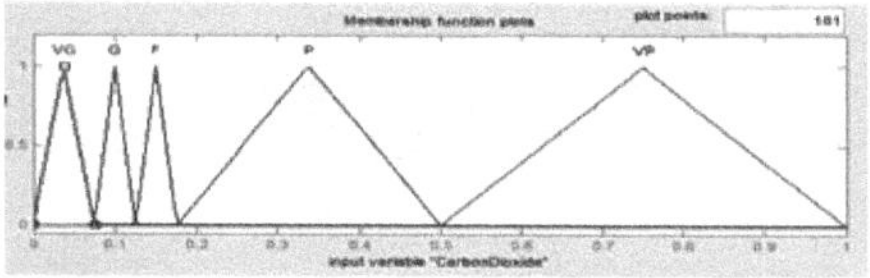

Fig. 5. Carbon Dioxide (input)

(CO-VG): concentration range of CO block is 0 to 3 ppm. *(CO-G)*: concentration range of CO block is 3 to 6 ppm. *(CO-F)*: concentration range of CO block is 6 to 9 ppm. *(CO-P)*: concentration range of CO block is 9 to 18 ppm. *(CO-VP)*: concentration range of CO block is 18 to 30 ppm.

Based from ASHRAE standard 62.1-2010, we knew that the standard value of the concentration of CO_2 is 700 ppm; therefore any value below 700 ppm is normal and values higher than 700 ppm is considered to be OOC or OOS. Carbon Dioxide block defined five membership functions as shown in Figure 5.

(CO_2-VG): concentration range of CO_2 block is 0 to 300 ppm. *(CO_2-G)*: concentration range of CO_2 block is 300 to 500 ppm. *(CO_2-F)*: concentration range of CO_2 block is 500 to 700 ppm. *(CO_2-P)*: concentration range of CO_2 block is 700 to 2000 ppm. *(CO_2-VP)*: concentration range of CO_2 block is 2000 to 4000 ppm.

5 Experiments and Analysis of Results

In simulating the FLC-based indoor air quality assessment system, the proponent conducted several tests for different values of input ranges on the basis of generated samples. For each parameter, 90 data were generated for tests 1 to 3 using Excel VBA macro program. The proponent computed for the weighted average of the generated data and the crisp outputs for each FIS were obtained as shown in Table 1.

Table 1. Simulation of FLC-based IAQ assessment based on data generation

Tests	Generated Data Samples: Temperature (T)	Relative Humidity (RH)	Carbon Monoxide (CO)	Carbon Dioxide (CO2)	Generated Data Samples (Normalized): Temperature (T)	Relative Humidity (RH)	Carbon Monoxide (CO)	Carbon Dioxide (CO2)	Crisp Output (Matlab Fuzzy Logic Toolbox): Air Parameters	Gaseous Parameters	Indoor Air Quality	Linguistic Assessment
1	20.85	56.59	8.14	841	0.52	0.57	0.27	0.23	3	3	3	Just Acceptable
2	34.40	80.80	17.54	1336	0.86	0.81	0.58	0.33	1	2	2	Not Acceptable
3	32.45	54.36	11.69	2350	0.81	0.54	0.39	0.59	2.5	2	3	Just Acceptable
4	24.10	63.83	8.40	1228	0.60	0.64	0.28	0.31	3	3	3	Just Acceptable
5	26.40	38.68	13.89	2582	0.66	0.39	0.46	0.65	3	2	3	Just Acceptable
6	23.64	56.49	11.14	2637	0.59	0.56	0.37	0.66	4	2	3	Just Acceptable
7	18.01	53.06	16.36	2542	0.45	0.53	0.55	0.64	3	2	3	Just Acceptable
8	27.71	48.76	26.00	3213	0.69	0.49	0.87	0.80	3.75	1	3	Just Acceptable
9	11.00	32.50	17.21	2323	0.28	0.33	0.57	0.58	3	2	3	Just Acceptable
10	11.48	54.14	15.59	2560	0.29	0.54	0.52	0.64	3	2	3	Just Acceptable
11	23.16	47.74	27.39	1006	0.58	0.48	0.91	0.25	5	2	4	Acceptable
12	10.40	61.40	16.80	1611	0.26	0.61	0.56	0.40	2	2	2	Not Acceptable
13	16.74	36.84	17.27	2383	0.42	0.37	0.58	0.60	3	2	3	Just Acceptable
14	16.87	50.34	10.36	2394	0.42	0.50	0.35	0.60	3.5	2	3	Just Acceptable
15	15.81	56.50	15.66	2462	0.40	0.56	0.52	0.62	3	2	3	Just Acceptable
16	15.16	66.74	21.69	2037	0.38	0.67	0.72	0.51	2	1	2	Not Acceptable
17	19.43	43.64	10.94	1061	0.49	0.44	0.36	0.27	3	2	3	Just Acceptable
18	13.85	42.14	11.78	2160	0.35	0.42	0.39	0.54	3	2	3	Just Acceptable
19	19.69	36.19	11.60	2094	0.49	0.36	0.39	0.52	3	2	3	Just Acceptable
20	15.76	54.12	11.40	2025	0.39	0.54	0.38	0.51	3	2	3	Just Acceptable
21	17.62	61.40	10.76	1529	0.44	0.61	0.36	0.38	2	2	2	Not Acceptable
22	29.31	46.42	19.09	1926	0.73	0.46	0.64	0.48	3	2	3	Just Acceptable
23	20.40	51.40	16.88	1620	0.51	0.51	0.56	0.40	4	2	3	Just Acceptable
24	18.38	53.66	10.78	1300	0.46	0.54	0.36	0.32	3	2	3	Just Acceptable
25	22.84	24.54	10.44	1729	0.57	0.25	0.35	0.43	3	2	3	Just Acceptable
26	24.44	24.47	19.40	1450	0.61	0.24	0.65	0.36	3.67	2	3	Just Acceptable
27	12.19	30.70	23.61	2346	0.30	0.31	0.79	0.59	3	1	2	Not Acceptable
28	25.45	62.69	17.37	2232	0.64	0.63	0.58	0.56	3	2	3	Just Acceptable
29	20.27	41.79	7.81	1268	0.51	0.42	0.26	0.32	4	3	4	Acceptable
30	12.87	31.42	11.98	2260	0.32	0.31	0.40	0.54	3	2	3	Just Acceptable

Table 2. Simulation of FLC-based IAQ assessment for possible crisp output values

Trials	Indoor Air Quality Parameters	Input Values	Input Values (Normalized)	Crisp Output (Air and Gaseous Parameters)	Air and Gaseous Parameters (Normalized)	Crisp Output (Matlab Fuzzy Logic Toolbox)	Linguistic Classification	True Error
1	Temperature (T)	24	0.6	5	0.91	5	High Acceptable (HA)	0
	Relative Humidity (RH)	44	0.44					
	Carbon Monoxide (CO)	1	0.03	5	0.91			
	Carbon Dioxide (CO2)	200	0.05					
2	Temperature (T)	25	0.68	4.6	0.84	5	Highly Acceptable (HA)	0
	Relative Humidity (RH)	46.4	0.464					
	Carbon Monoxide (CO)	2.5	0.08	5	0.91			
	Carbon Dioxide (CO2)	250	0.06					
3	Temperature (T)	23	0.58	5	0.91	5	Highly Acceptable (HA)	0
	Relative Humidity (RH)	48	0.48					
	Carbon Monoxide (CO)	5	0.17	4	0.73			
	Carbon Dioxide (CO2)	450	0.11					
4	Temperature (T)	27	0.68	3	0.55	4	Acceptable (A)	0
	Relative Humidity (RH)	54	0.56					
	Carbon Monoxide (CO)	4	0.13	4	0.73			
	Carbon Dioxide (CO2)	425	0.11					
5	Temperature (T)	22	0.55	3.75	0.68	4	Acceptable (A)	0
	Relative Humidity (RH)	53	0.53					
	Carbon Monoxide (CO)	8	0.27	3	0.55			
	Carbon Dioxide (CO2)	550	0.14					
6	Temperature (T)	21	0.52	3	0.55	3	Just Acceptable (JA)	0
	Relative Humidity (RH)	31	0.31					
	Carbon Monoxide (CO)	8.5	0.28	3	0.55			
	Carbon Dioxide (CO2)	650	0.16					
7	Temperature (T)	15	0.38	2	0.36	2	Not Acceptable (NA)	0
	Relative Humidity (RH)	25	0.25					
	Carbon Monoxide (CO)	14.5	0.48	2	0.36			
	Carbon Dioxide (CO2)	1750	0.44					
8	Temperature (T)	29	0.73	2	0.36	2	Not Acceptable (NA)	0
	Relative Humidity (RH)	61	0.61					
	Carbon Monoxide (CO)	17	0.57	2	0.36			
	Carbon Dioxide (CO2)	800	0.2					
9	Temperature (T)	5	0.13	1	0.18	1	Highly Not Acceptable (HNA)	0
	Relative Humidity (RH)	11	0.11					
	Carbon Monoxide (CO)	26	0.87	1	0.18			
	Carbon Dioxide (CO2)	3500	0.88					
10	Temperature (T)	38	0.96	1	0.18	1	Highly Not Acceptable (HNA)	0
	Relative Humidity (RH)	92	0.92					
	Carbon Monoxide (CO)	19	0.63	1	0.18			
	Carbon Dioxide (CO2)	2100	0.53					

To further verify the accuracy of the formulated rules, the proponent conducted and intentional 10 trials as shown in Table 2 to accommodate possible crisp outputs of Highly Acceptable (HA), Acceptable (A), Just Acceptable (JA), Not Acceptable (NA) and Highly Not Acceptable (HNA) and it was verified that the fuzzy based system is working properly in accordance to programmer-defined rules.

6 Conclusion

In this paper, the proponent makes use of fuzzy logic in classifying indoor air quality. The paper presents a technique employing fuzzy logic for automatic classifying of indoor air quality system. It presents a cheaper approach to the classification of indoor air quality, which will also apply most likely to the classification of any other related applications. There were three FIS created: Air parameters, Gaseous parameters and Indoor Air Quality assessments. There were four input fuzzy sets and these were normalized and fuzzified using Sugeno-style of FIS. There were 131 fuzzy rules established. The basis for the construction of the membership function were the identified experts and standards set and maintained by ASHRAE standard 55-2010, ASHRAE standard 62.1-2010, EPA and WHO. The FLC-based indoor air quality assessment was simulated using Matlab Fuzzy Logic Toolbox and the data being generated using Excel VBA macro program. The results obtained can be used for possible improvement of indoor air quality in meeting global standards.

Acknowledgments. The author acknowledges Jesus Christ, who is the source of wisdom and knowledge in the study of fuzzy-based system for indoor air quality classification. The author is thankful to Dr. Elmer P. Dadios for giving insights and sharing his expertise on fuzzy logic concepts. He would like to thank De La Salle University for giving financial assistance for his doctorate studies. Also, the Research and Statistics Center of Lyceum of the Philippines University – Laguna (LPU-L) is gratefully acknowledged for funding the research publication and dissemination.

References

Wu, C.-H., Ma, L.-S., Chen, C.-H., Liu, Y.-W.: A Design of Fuzzy Logic Controller for Conforming to the Regulations of Indoor Air Quality and Thermal Comfort. In: Proceedings of the 4th Annual IEEE International Conference on Cyber Technology in Automation, Control and Intelligent Systems, Hong Kong, China, June 4-7 (2014)

Molina-Solana, M., Ros, M., Delgado, M.: Unifying Fuzzy Controller for Indoor Environment Quality. In: Proceedings of the 2013 International Joint Conference on IFSA World Congress and NAFIPS Annual Meeting, IFSA/NAFIPS (2013)

Occupational Safety and Health Administration U.S. Department of Labor (OSHA 3430-04), Indoor Air Quality in Commercial and Institutional Buildings (2011)

TSI Incorporated, Indoor Air Quality Handbook: A Practical Guide to Indoor Air Quality Investigations (2013)

Chen, W.-K.: Study on Fuzzy Determination of Indoor Environmental Quality in the Intelligent Building. In: Proceedings of the 2013 International Conference on Machine Learning and Cybernetics, Tianjin, July 14-17 (2013)

Cai, W., Wang, D., Wen, X., Yu, X.: Indoor Air Quality Assessment in an Art Gallery with an HVAC System. In: Proceedings of the 4th International Conference on Bioinformatics and Biomedical Engineering, iCBBE (2010)

Artificial Intelligent System to Stop Bots from Playing Online Games

Gagandeep Singh[1], Pooja Choudhary[2], and Vikrant Sharma[2]

[1] Punjab Institute of Technology, Kapurthala, India
[2] CT Institute of Engineering Management and Technology, Jalandhar, India
{gagandeepct,vikrant.sharma80}@yahoo.co.uk,
poojachoudhary80@gmail.com

Abstract. With the wide spread popularity of internet in the field of entertainment industry, online games emerged as one of the biggest sources of entertainment. Many players all over the globe play online and compete against each other. But unfortunately not all players playing online games are honest. Some players use unfair means to have advantage over the others, the use of bot is the one among them. In this paper we will purpose a new Artificial Intelligent System which makes use of both hardware and software to identify the bots and to stop them from playing the game. Later in the paper we discussed how this technique is better than the existing techniques and what more should be done in order to stop bots from playing online games.

Keywords: Bots, AIS, Gold farming

1 Introduction

Video games had been a big source of entertainment from the past few decades. With the wide spread popularity of internet to every household around the globe, it played a massive role in the entertainment industry. Internet joined people from all over the world to play games across the globe and compete against one another. Today billions of players from all over the globe enjoy playing online games of their choice. The earlier source of entertainment is now also an opportunity of making money. In order to buy these products online the players used to play the online games for a long time which increased the popularity of the game. For some people who were seeking this as an opportunity made it as their business. They employed people to play online games for them and make more and more game currency, *commonly called gold farming* and with this currency they buy products online and sell these products at cheaper rate to the other players in exchange of real money. In 2009, game played on social network such as Facebook, games that primarily derive revenue from the sale of virtual goods, brought in 1 billion USD, and increased to 1.6 billion in 2010. Worldwide, 7.3 billion USD was made from virtual goods that same year [1]. Blizzard Entertainment`s World of Warcraft (WoW)- surpassed 11.5 million subscribers in December 2008, making in an estimated US $150 million subscription fee per month [2].

M. Gen et al. (eds.), *Industrial Engineering, Management Science and Applications 2015*,
Lecture Notes in Electrical Engineering 349, DOI: 10.1007/978-3-662-47200-2_76

Unfortunately, not everyone plays the game fairly, and this has caused nuisance and irritation to many honest players in the community. As a result many have either stopped playing or turned into cheaters themselves [3], and in turn have hindered the growth of the industry. Use of bots is one among many unfair way to cheat. The word "Bot" is derived from last three words of "Robot". A robot is the machine which reduces human effort by doing work on behalf of human. Similarly a "bot" is an artificially intelligent program that controls the player playing online game. It reads the game environment from the memory map of the game and then based on the analysis of the environment it takes the action accordingly.

The use of bots causes a huge loss to the gaming companies. The player who uses bot can run the program for a long time which human player cannot. Hence the player using the bot can buy all the expensive items in the game. With more players using bots to buy the expensive goods in the game, the price of that item increases as a result of which the honest player playing the game loses the interest in the game. As a result of this they switch to some other game and hence the popularity of the game is affected.

2 Related Work

The existing techniques mostly focus on the use of software methods. Some of the important approaches are discussed as follow:

Human Interactive Proofs (HIP) [4] increasingly present in nowadays computer systems. The idea behind the concept is to make a certain system able to identify whether it is dealing with a human being or an automatic computational procedure, or bot. A CAPTCHA, Completely Automatic Public Turing test to tell Computers and Humans Apart, is a class of HIPs whose purpose is to distinguish humans from computers. CAPTCHAs have the attractive property that they do not have to be as computationally intractable as traditional cryptographic problems if an adversary can hire a human to take the test more cheaply than it would cost to break the test computationally, the test is secure enough. CAPTCHAs have the attractive property that they do not have to be as computationally intractable as traditional cryptographic problems if an adversary can hire a human to take the test more cheaply than it would cost to break the test computationally, the test is secure enough [5].

Fig. 1. CAPTCHA

In Human Observation Proofs (HOPs) [6] we differentiate bots from humans from their behaviour. HOPs differentiate bots from human players by passively monitoring input actions that are difficult for current bots to perform in a human-like manner. While playing the game humans are more adventitious and they tend to explore the environment in the game. In case of bots they read the memory map of the game and hence they act accordingly. HOPs provide continuous monitoring throughout a session and they are non-interactive, i.e., no test is presented to a player, making HOPs completely non-obtrusive [7]. Basically bots are detected by HOP ether at server end by analysing the set of inputs received and checking behaviour of the player, or by polling done by players playing the game by which the players playing the game. Some of the HOPs are illustrated in fig below:

Action	Definition
Keystroke	The press and release of a key.
Point	A series of continuous mouse cursor position changes with no mouse button pressed; the time-stamps for each pair of cursor position changes are no more than 0.4 seconds apart.
Pause	A period of 0.4 seconds or longer with no actions.
Click	The press and release of a mouse button; the cursor travels no more than 10 pixels between the press and release.
Point-and-Click	A point followed by a click within 0.4 seconds.
Drag-and-Drop	The press and release of a mouse button; the cursor travels more than 10 pixels between the press and release.

Fig. 2. Human Observation Proofs

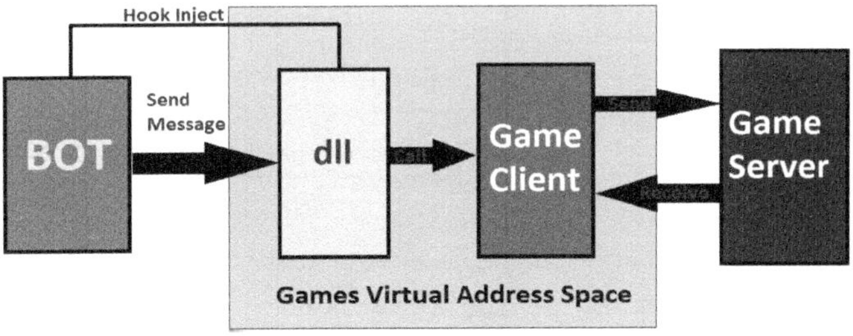

Fig. 3. Shows How bots are injected in .dll files of game client

Wu Chun et al. [9] in their paper "The Study of Bot Technology for Online Games" discussed about Direct Link Library or .dll files are almost used by all the API functions. DLL file cannot be run independently and be loaded and is called by a process. When a game is running, bots is loaded into the .dll file by using a hook. During the process of game play the game process keeps on sending the information to the .dll file which checks the information and performs the operations according to the action which is written in the bots program.

There are two types of DLL injection techniques:-

1. ***Dynamic dll injection technique:*** It is a technology which injects the DLL code into the running process.
2. ***Remote dll injection technique:*** Remote DLL injection technique is creating a remote thread in another process, which is aim to enter into the invaded process memory address space. In the process, we can create a new thread in another remote process using the function reatRemoteThread, and the new thread is created to share the remote process address space.

Hook Technology: Hook is a platform of Windows message handling mechanism, on which application can set the message for appointed window; moreover the appointed window can be created by other processes. When message arrived, it was dealt with before the aim window of processing function.

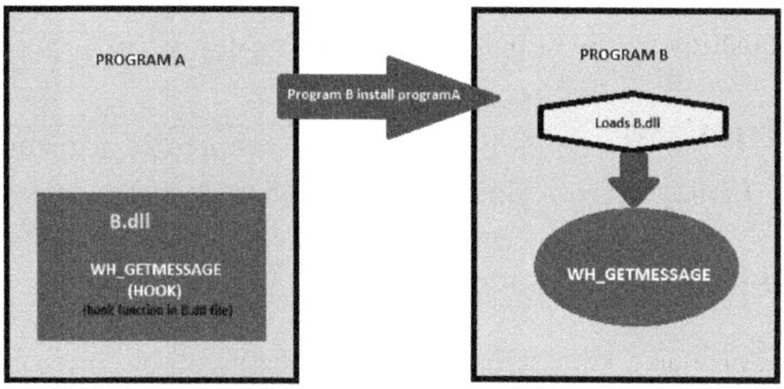

Fig. 4. Hook Technology

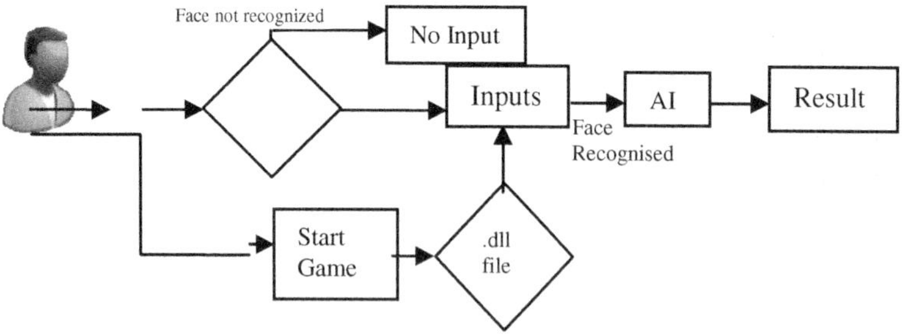

Fig. 5. Artificial Intelligent System

3 Methodology

In the existing approaches so far, the researchers mostly made use of software methods to stop bots from playing online games. In the study we made so far, we found that only software methods are not that efficient to stop bots from playing online games while the hardware methods used so far are efficient but expensive and difficult to implement. Some of the hardware methods also need modification in gaming hardware devices which is not feasible for present games. In our purposed artificially intelligent system we made use of both hardware and software devices to stop bots from playing online games. In this approach we used webcam as a hardware device, which is commonly used device these days. While Artificial Intelligent System will analyze the input and separate bots from humans.

When the user starts the game, the .dll file scanner will start scanning all the .dll files received from the user while loading the game into the server. This test is performed because sometimes the programmer hook a .exe inside a .dll file which will automatically takes the input from the linked file and takes action accordingly. The .dll file scanner will scan all files that are being loaded into the system and compare those with existing files and block any attachments or links that are found in file. At the same time the Face Reorganisation System (FRS) will check if the person playing the game is really sitting in front of computer or not. The input from the user will only be taken if the system will recognise human face in the cam while playing the game else the input will stop.

Once the input is taken from the user, on next step we introduced an Artificial Intelligent System, the function of this system is to block if there is still any bot which is found in the system. For doing so we will take parameters like character behaviour, sequence of keys pressed, time of target and idle time of player.

Character behaviour: Although the bots acts on the behalf of humans but yet but even the best bots shows different behaviour. This difference in behaviour separates bots from the humans. We can study this behaviour of bots at server side and further we can train an AIS system which can recognize this odd behaviour and when the server will recognize same odd behaviour from the user, our AIS will recognize this odd behaviour and block the user.

Sequence of Keys: When the user play a game online, the actions of the character is controlled by user by pressing the keys. This key sequence is received at the server side and the player moves accordingly. In case of human players this sequence keeps on changing. While in case of bots this sequence repeat after certain interval. For this we will train our AIS in such a way that it will keep check on the sequence of keys it receive, it will record this sequence and compare this with the new sequence received. In case of bots the sequence received will be same while in case of human this sequence will change every time.

Time of Target: Some bot users use Aimbot which is an artificially aiming bot to aim at enemies while some bots detects the enemies from the memory map of the game. To stop them our artificial intelligent system will be trained in such a way that it will record the time taken to shoot the enemy. Our system will trace the time taken to shoot the enemy and our system will calculate the average time taken to kill the enemies. If the average time taken to kill the enemies is too small as compared to rest then our AIS will analyze that the player is using Aimbot and it will block the user from playing ahead.

Idle Time of player: It is the time interval for which the player will remain inactive during the course of game play. Human players are adventurous and like to explore environment so they are continuously active in the game while the bots have the tendency to wait for the enemy and shoot. Hence bots can remain inactive for a long time while the humans can`t. The bots could be differentiated from humans in case they very long inactive time. Our AIS will calculate this idle time and will block all users who remain idle for a long time.

Based on the above scenarios our AIS will block all those users who will try to seek unfair advantage in the game through the use of bots.

4 Benefits

The so far existing approaches are mostly software based. But from our study we found that use of software`s alone is not enough to stop bots from playing online games. So we designed an algorithm that made use of both hardware and software system to stop bots from playing online games. We used camera to make sure if the player playing the game is actually in front or not. The AIS we used ahead makes sure that the input received is from human only and not from a program. The use of both camera and AIS makes our system much more secured and safe, and will reduce the attack from bots up to a great extinct.

5 Conclusion

In this paper we discussed about various existing approaches that were used to stop bots from playing online games. Since these approaches were mostly software dependent so they were not able to stop bots completely from playing online games. In our approach we made use of both hardware and software in order to stop the bots from playing online games. The use of both camera and AIS makes our system much more secured and safe, and will reduce the attack from bots up to a great extinct. But still our technique is a little complex and expensive, moreover this technique will be tough to apply on existing games.

References

Wikipedia (June 2013), http://www.en.wikipedia.org/wiki/virtual_goods

Mitterhofer, S., Kruegel, C., Kirda, E., Platzer, C.: Server-Side bot detection in massively multiplayer online games. IEEE Security & Privacy 7(3), 29–36 (2009)

Pritchard, M.: How to hurt the hackers the scoop. http://www.gamasutra.com/view/feature/3149/how to hurt the hackers the scoop .php (2000)
Gusfield, D.: Algorithms on Strings, Trees, and Sequences: Computer Science and Computational Biology. Cambridge Univ. Press (1997)
Heeks, R.: Current Analysis and Future Research Agenda on 'Gold Farming': RealWorld Production in Developing Countries for the Virtual Economies of Online Games. Working Paper Series, vol. 32 (2008)
Gianvecchio, S.: Detecting Bots in Online Games using Human Observational Proofs. IEEE (2008)
Gianvecchio, S., Wu, Z., Xie, M., Wang, H.: Battle of botcraft: fighting bots in online games with human observational proofs. In: Proceedings of the 16th ACM Conference on Computer and Communications Security, pp. 256–268. ACM (2009)
Gaspareto, O.B., Barone, D., Schneider, A.M.: Neural networks applied to speed cheating detection in online computer games. In: Fourth International Conference on Natural Computation, ICNC 2008, vol. 4, IEEE (2008)
Chun, W., Guo-Hun, Z., Yong-Hua, W., Rong, X.: The Study of Bot Technology for Online Games. In: 3rd International Conference on Genetic and Evolutionary Computing, WGEC 2009, pp. 654–656. IEEE (2009)
Yu, S.-Y., Hammerla, N., Yan, J., Andras, P.: A statistical aimbot detection method for online FPS games. In: The 2012 International Joint Conference on Neural Networks (IJCNN), pp. 1–8. IEEE (2012)

Multi-lane Detection Based on Original Omni-Directional Images*

Chuanxiang Li, Bin Dai, and Tao Wu

College of Mechatronic Engineering and Automation,
National University of Defense Technology, Changsha, P.R. China, 410073
daibin@nudt.edu.cn

Abstract. A lane detection method is presented based on original omni-directional images. The parameterized representation of curves in panoramic images is provided by analyzing the projection model of omni-directional multi-camera system. Those lines of lane markings in image can be described by the parameter of lane markings in world coordinate system. The results of line fitting into world coordinate system are directly used to fit lane model for both parallel lane and non-parallel lanes. The performance of feature extractor and effectiveness of proposed lane detection method is verified on real world data.

Keywords: Intelligent vehicle, lane detection, Omni-directional image, projection model, non-parallel lane, curve fitting.

1 Introduction

Lane detection is an important part of intelligent vehicles. Lanes provide vital information of road and enhance ego-vehicle's localization capability [1].There are plentiful vision-based lane detection methods reported in recent works like [2] and [3]. However, most of the exiting works use the conventional cameras that have a relative small field of view comparing the omni-directional camera. In this paper, a lane detection method is proposed based on omni-directional multi-camera system, which can acquire information from more than 80% of the full 360 degree sphere.

According to the projection model of omni-directional camera, either straight lines or curves on the road plane are projected as complex curves in panoramic image. Because of the panoramic image with distort, the previous feature extraction methods and lane model fitting methods, which perform efficient on conventional camera images, are not stable.

There are also some methods for lane detection using omni-directional camera [4, 5, 6, 7]. Those methods operate on bird's-eye view images, which are transformed for the original panoramic images with the knowledge of the camera's intrinsic and extrinsic parameters. In the bird's-eye images, the width of lane markings and lane

* This work was supported by National Natural Science Foundation of China: 61075043 and 61375050.

M. Gen et al. (eds.), *Industrial Engineering, Management Science and Applications 2015*,
Lecture Notes in Electrical Engineering 349, DOI: 10.1007/978-3-662-47200-2_77

are approximately constant. It is an advantageous property that facilitates feature extraction and lane model fitting. But the transformation is associated with computational cost and loss of information. In this paper, our algorithm operates on original panoramic images. A new feature extractor is proposed mainly based on the photometric features of lane markings.

With the feature maps, the most important question is to fit the curves in panoramic image. Many algorithms for extracting curves from panoramic images are shown [8, 9, 10]. The results of curves fitting of those methods are the parameters of curves in panoramic images or three-dimensional spherical coordinate system. With the intelligent vehicle application, the acquired curves need to be transformed to world coordinate system for lane model fitting. In this paper, analyzing the projection model of omni-directional multi-camera system, the curves are described by the parameter of line in world coordinate. The parameter of curves can be calculated using two points of curves and used to fit lane model directly.

The paper is organized as follows. We introduce the projection mode of omni-directional multi-camera system and give the parameterized representation of curves in panoramic image in section 2. In the section 3, lane detection algorithm based on panoramic images is proposed. We provide experimental results of feature extraction and lane detection on real world data in the section 4. Eventually, we conclude this paper in section 5.

2 Lines in Panoramic Image

Our Ladybug3 camera is composed of six synchronized cameras (five in horizontal ring and one on top), which is 1624×1232 pixels. Six raw images are mapped onto a sphere of fixed radius and acquired the three-dimensional mapping coordinates to the sphere. The three-dimensional spherical coordinate system of camera, the spherical coordinate of panoramic image, and the image axis are shown in Fig. 1. There are several projection methods for transforming the image on sphere to the two-dimensional image: radial projection, cylindrical projection, dome projection and cubic projection. Radial projection is one of the most popular projection methods, and the output image is easy to use. The panoramic images from radial projection model are usually two-dimensional so that it can be displayed and processed easily.

The panoramic images have two spherical θ coordinate for horizontal, and φ for vertical. The projection equation is following:

$$\begin{cases} \theta = \operatorname{atan} 2(Y, X) \\ \varphi = \arccos\left(\dfrac{Z}{\sqrt{X^2+Y^2+Z^2}}\right) \end{cases} \tag{1}$$

where (X,Y,Z) is a point in the spherical coordinate system. (θ,φ) are the spherical coordinate of panoramic image as shown in Fig. 1. The range of value of θ is $[-\pi,\pi]$ and the range of value of φ is $[0,\pi]$.

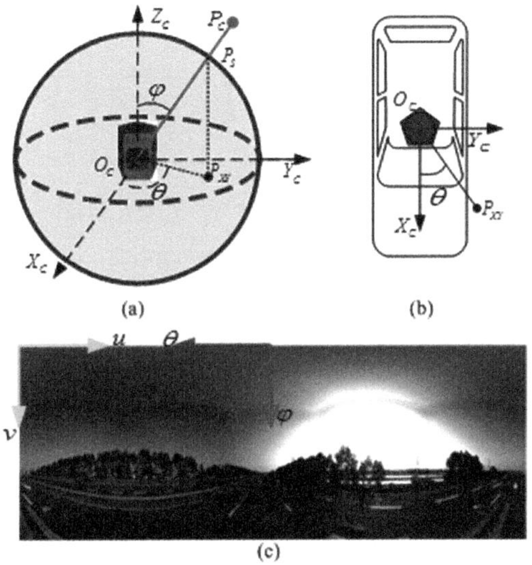

Fig. 1. Definition of coordinate system. (a) Three-dimensional spherical coordinate system; (b) Bird's-eye view of spherical coordinate system; (c) Spherical coordinate of panoramic image and the image axis

The image axes (u,v) to spherical coordinates (θ,φ) of panoramic image are defined respectively as following:

$$\begin{cases} u = \dfrac{\pi - \theta}{2\pi} \bullet nCols \\ v = \dfrac{\varphi}{\pi} \bullet nRows \end{cases} \tag{2}$$

where $nCols \times nRows$ is the size of the panoramic images. Rearranging the equation (1) and (2), we get the projective relation between the point (X,Y,Z) in the spherical coordinate system and (u,v) in image coordinate system:

$$\begin{cases} u = \dfrac{\pi - \operatorname{atan2}(Y,X)}{2\pi} \bullet nCols \\ v = \dfrac{\arccos(\dfrac{Z}{\sqrt{X^2+Y^2+Z^2}})}{\pi} \bullet nRows \end{cases} \tag{3}$$

Line in the camera coordinate frame can be represented as:

$$\frac{X - X_1}{X_2 - X_1} = \frac{Y - Y_1}{Y_2 - Y_1} = \frac{Z - Z_1}{Z_2 - Z_1} \tag{4}$$

where (X_1,Y_1,Z_1) and (X_2,Y_2,Z_2) are two points on the line. The line segment of lane markings or boundaries of lane are on road plane. The parameter Z of those line segments is a constant, which is determined by the height of camera. The line segments on road plane can be simplified as:

$$\begin{aligned} Y &= aX + b \\ Z &= -Z_0 \end{aligned} \tag{5}$$

where (a,b) are the parameters of line and Z_0 is the height of camera system. The straight lines on road plane are projected in panoramic image as following:

$$\left(\frac{-b}{a + \tan\left(\frac{2\pi u}{nCols}\right)} + \frac{ab}{a^2+1} \right)^2 + \left(\frac{b}{a^2+1} \right)^2 + \frac{(-Z_0)^2}{a^2+1} = \frac{(-Z_0)^2}{a^2+1} \bullet \frac{1}{\left(\cos\left(\frac{2\pi v}{nRows} \right) \right)^2} \tag{6}$$

From above equations, the curves in image coordinate system can be represented by the parameters of straight line in camera coordinate system.

3 Multi-lane Detection Algorithms

In this section, we detail the lane detection method based on original panoramic images. Firstly, the features of lane markings are extracted using a novel extractor from original panoramic images. With the feature map, the parameters of line segments of lane markings or lane boundaries are then calculated by two points on the lines. Lastly, non-parallel and parallel lanes are fitted in uniform lane model.

3.1 Feature Extraction in Panoramic Image

In radial projection model, the pixels of top and bottom image are stretched out. The straight lines and curves of lane markings are stretched as curves in panoramic images. Therefore, most of lane markings feature extractors introduced in [11], which perform efficient on conventional camera images, are disable.

A feature extractor is designed for panoramic images in this paper. For each pixel (x, y) in the panoramic image, horizontal and vertical intensity averages are computed as follow:

$$\overline{I}_{horizonal}(x, y) = \frac{1}{8S_M + 1} \sum_{y_h = y - 4S_M}^{y_h = y + 4S_M} I(x, y_h) \tag{7}$$

$$\overline{I}_{vertical}(x, y) = \frac{1}{8S_M + 1} \sum_{x_v = x - 4S_M}^{x_v = x + 4S_M} I(x_v, y) \tag{8}$$

where S_M implies the maximal width of lane markings, which is an approximate linear function of the vertical image coordinate. Given the threshold T, if the intensity of pixel $I(x, y)$ is higher than either $T + \overline{I}_{horizonal}(x, y)$ or $T + \overline{I}_{vertical}(x, y)$, the pixel is selected as candidate lane markings feature. Then the set of connected pixels wider than the minimal width of lane markings S_m are considered as lane markings.

3.2 Curve Fitting

According to the equation (6), each curve $c_i (i = 1,2,...,n)$ in panoramic images space can be represented by the parameter a_i and b_i, which are the parameters of straight line in camera coordinate system. The parameters of each curve segments in image space can be calculated using two points on this curve. Let $P_1 = (u_1, v_1)$ and $P_2 = (u_2, v_2)$ be the two points of curve. Four pairs of a_i and b_i can be calculated with points P_1 and P_2 according to equation (6), but only one pair of parameter is real truth. In order to determine the truth of parameter of curve, four virtual curves in panoramic image are generated using the four pair of parameters. Those virtual curves are then compared with real curve and the real truth can be determined.

Another important thing worth mentioning is the accuracy of parameter a_i and b_i of curve c_i depends on the location of P_l and P_2. As shown in Fig. 2 that the color curves are calculated using corresponding color endpoints respectively. The red and blue endpoints are on edge of lane marking and green endpoints are on the center line of lane marking. It can be seen from the bottom of Fig. 2 that the green curve coincide with all curve segments belonging to the discontinuously lane markings. However, the red curve and blue curve are not the best results. Compared with edge of lane marking, the centre line of lane marking is benefit to get accurate parameter of lane marking. In this way, the endpoints of centre line of lane markings are chose to calculate the parameter of curves.

Fig. 2. Curves Fitting. Color points are the pair of endpoint of lane marking and the color curves are fitted using corresponding endpoints

3.3 Lane Model Fitting

Lane model is the assumption about the lane structure in the real world environments. In this paper, lane model is defined as straight lane model, and each lane has approximately lane markings or boundaries. Considering the splitting or merging of lane markings, a single lane may not parallel to others on the road plane. Each single lane $L_j (j = 1, 2, ..., N)$ is described by three parameters:

$$L_j = \{A_j, B_j, w_j\} \tag{9}$$

where A_j and B_j are the parameter of centre line of lane L_j , w_j denotes the width of lane.

In urban environments, non-parallel lanes must be considered at intersections, in splitting lane or merging lane. The parameters A_j of parallel lanes are approximately equal in world coordinate system. However, there are specific differences of A_j between non-parallel lanes. With the parameters A_j of candidate curves, several cluster centers $C_k (k = 1, 2, ..., C_M)$ are generated using K-means clustering algorithms. Lane model is fitted as shown in Algorithms 1.

Algorithm 1 Lane Model Fitting

Input: The curves $c_i = (a_i, b_i)(i = 1, 2, ..., n)$

$r \leftarrow 1$
For $k \leftarrow 1$ to C_M
  $r_l \leftarrow 0$
  For $i_x \leftarrow 1$ to n
    For $i_y \leftarrow 1$ to n
      If $dist(c_{i_x}, c_{i_y}, C_k) < \phi$ and $|width(c_{i_x}, c_{i_y}) - W| < \gamma$
        $r_l \leftarrow r_l + 1$
        $A_{r_l} \leftarrow a_m$
        $B_{r_l} \leftarrow b_m$
        $w_{r_l} \leftarrow width(c_{i_x}, c_{i_y})$
      End If
    End For
  End For
  If $r_l > 0$
    $r \leftarrow r + 1$
  End If
End For
Return $\{A_{r_l}, B_{r_l}, w_{r_l}\}$

An array of parallel lanes have common the direction of road. Non-parallel lanes mean multiple directions of road on the road plane. In Algorithm 1, r is the number of direction of road on the road plane and r_l denote the number of parallel lane sharing the common direction. c_{i_x} and c_{i_y} are two curves of all candidate curves. Let L_{r_l} contains two boundaries c_{i_x} and c_{i_y}, and $width(c_{i_x}, c_{i_y})$ is the width of lane L_{r_l}. W is the defined width of lane. a_m and b_m are the parameter of centre line of lane L_{r_l}. The function $dist(c_{i_x}, c_{i_y}, C_k)$ gives the average distance between cluster center C_k and the direction of those two curves. When the distance is lower than threshold ϕ and the width of lane fall in $[W-\gamma, W+\gamma]$, the lane will be treated as reasonable result.

4 Experimental Results

In this section, experimental results of the proposed lane detection system on real world data are presented. Firstly, the performance of the proposed feature extractor is acquired by comparing features maps with the ground truth images. In order to evaluate the performance of feature extractor on original panoramic images and on bird's-eye view images, 599 road images are selected to construct the datasets. The ground truths of lane markings in all those images are labeled manually. And the frames are selected from variable scenes. The bird's-eye view images and ground truth of those images are generated using the same extrinsic parameters of camera. And the performance of proposed feature extractor on flat-plane images is provided.

The original panoramic images are of 2048×1024 pixels, and the corresponding bird's-eye view images are 301×501 pixels. The average number and percentage of pixel of lane markings in original image and bird's-eye view are shown in Table 1. According to the table, the loss of lane markings feature is accounting for 60.76 % during transforming the original panoramic images to bird's-eye view images. The average percentage of lane markings in bird's-eye view image is higher, since the bird's-eye view image just consists of road plane scene.

Table 1. Pixel of lane markings in original image and bird's-eye view image

	Original Image	Bird's-eye view Image
Number	10158.17	3985.49
Percentage	0.0048	0.0264

Receiver Operating Characteristic curves (ROC) and Dice Similarity Coefficient curves (DSC) are chosen as evaluation metric to quantify the performance of feature extractor. The area under ROC curves corresponds with the performance of feature extractors. The optimal value of the threshold and the best possible performance of algorithm can be described by the maximum value of DSC curves.

For the same feature extractors on original images and bird's-eye view images, the ROC curves and DSC curves on our dataset are shown in Fig. 3 and Fig. 4. It can be seen from the area under ROC curves and the maximal of the DSC curves, the proposed feature extractor performs much better on original panoramic image. At the same time, the width of DSC curve peak demonstrates the robustness of feature extractor on original image is better than it on bird's-eye view images. In term of feature extraction from panoramic image, extractor operated on original image performs better than on bird's-eye view image. One reason for the performance of the feature extractor decline on bird's-eye view images is information loss during the transformation. In addition, one may attribute to bright stripe noise, which is introduced by transformation.

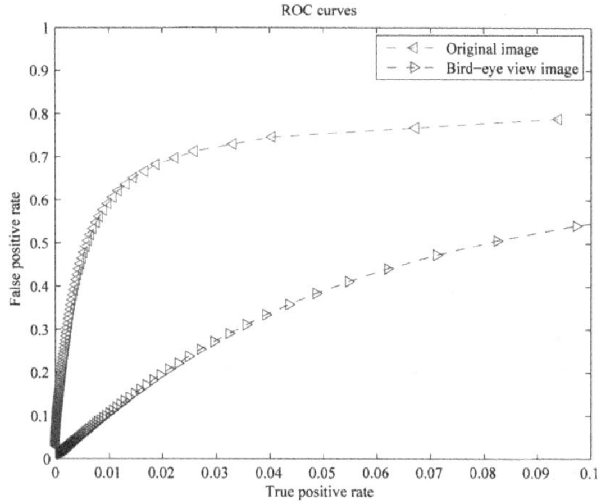

Fig. 3. ROC curves of the same feature extractor on original image and bird's-eye view image

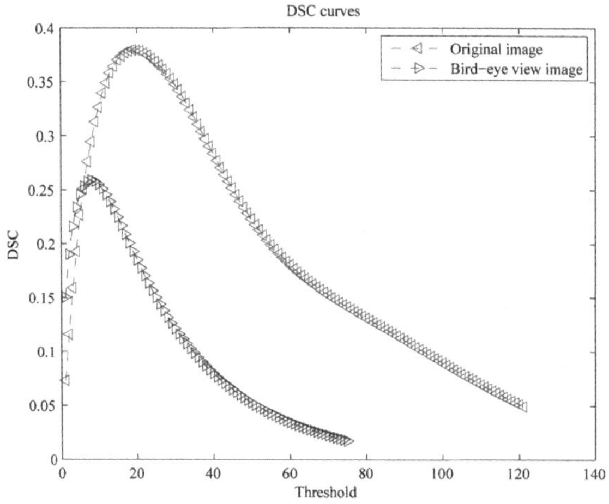

Fig. 4. DSC curves of the same feature extractor on original image and bird's-eye view image

Some qualitative results for proposed lane detection method are given in Fig. 5. Those results represent different road scenes: change lane, entering tunnel, rain, splitting lane, merging lane, shadow.

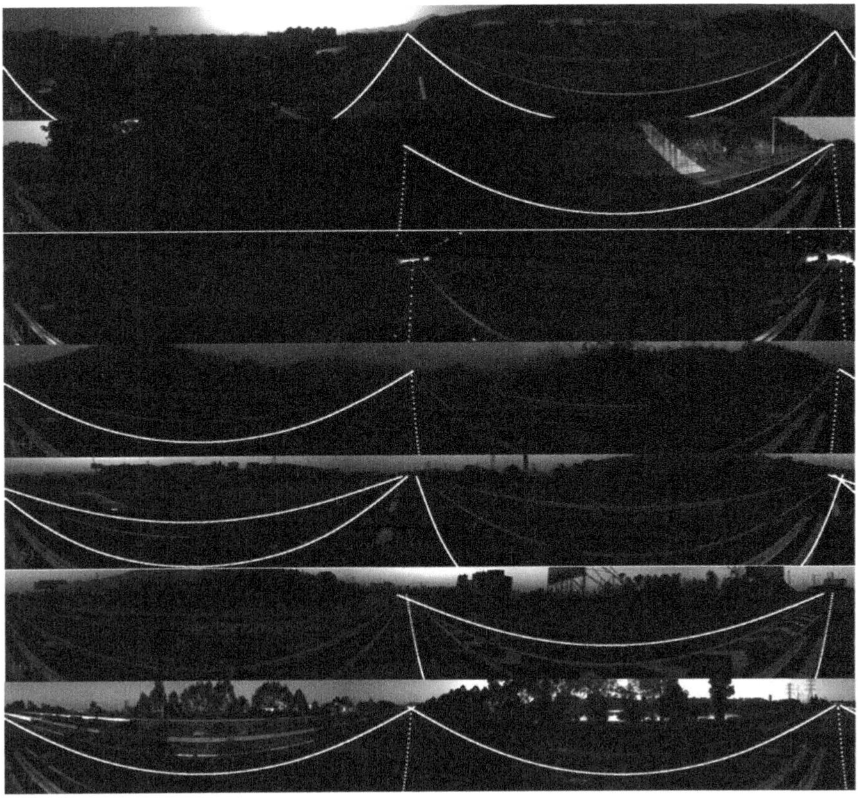

Fig. 5. Some results of lane detection. The white lines denote the centre line of lane. (1) Change lane; (2) entering tunnel; (3) in tunnel; (4) rain; (5) splitting lane; (6) merging lane; (7) shadow of tree

5 Conclusion

In this paper, multi-lane detection method based on omni-directional image was proposed. Instead of transforming the panoramic images to bird's-eye view images for lane detection, our algorithms detect lane from the original omni-directional images. According to the projection model of omni-directional multi-camera system, the curve can be represented by the parameters of line in world coordinate system. With the curve segments in world coordinate system, lane model is fitted to get the parameter of lanes. And the number of direction of road on the road plane and the number of lane of each lane direction can be calculated during lane model fitting. The proposed lane detection method was test on our real car experiments. Experimental results show that the performance of the same feature extractor operating on original images is better than its' on bird's-eye view images. And Real data tests demonstrate the effectiveness of the proposed lane detection method, especially in adverse light condition.

Future works will be focus on the development of a comprehensive lane model for either straight or curve road. In addition, the lanes markings feature extractors operating on panoramic images still need further investigation.

References

1. Tao, Z., Bonnifait, P., Frémont, V., Guzman, J.I.: Lane marking aided vehicle localization. In: 16th International IEEE Conference on Intelligent Transportation Systems, Netherlands, (2013)
2. Yenikaya, S., Yenikaya, G., DÜVen, E.: Keeping the vehicle on the road - a survey on on-road lane detection system. ACM Computing Surveys 46(1), 1–43 (2013)
3. Hillel, A.B., Lerner, R., Levi, D., Raz, G.: Recent progress in road and lane detection: a survey. Machine Vision and Applications 25(3), 727–745 (2014)
4. Amemiya, M., Ishikawa, K., Kobayashi, K., Watanabe, K.: Lane detection for intelligent vehicle employing omni-directional camera. In: SICE Annu. Conf., pp. 2166–2170 (2004)
5. Ishikawa, K., Odagiri, K., Kobayashi, K., Watanabe, K.: Lane detection by using omni-directional camera for outdoor terrains. In: SICE Annu. Conf., Osaka, Japan (2002)
6. Cheng, S.Y., Trivedi, M.M.: Lane tracking with omni-directional cameras: algorithms and evaluation. EURASIP J. on Embedded Systems 2007(46972), 1–8 (2007)
7. Li, C.X., Dai, B., Wu, T., Nie, Y.M.: Multi-lane detection in urban driving environments employing omni-directional camera. In: Proc. of the IEEE Int. Conf. on Information and Automation, pp. 284–289 (2014)
8. Mei, C., Malis, E.: Fast central catadioptric line extraction, estimation,tracking and structure from motion. In: International Conference on Intelligent Robots and Systems, pp. 4774–4779 (2006)
9. Bazin, J.-C., Demonceaux, C., Vasseur, P., Kweon, I.: Rotation estimation and vanishing point extraction by omnidirectional vision in urban environment. International Journal of Robotics Research 31(1), 63–81 (2012)
10. Barreto, J.P., Araujo, H.: Fitting conics to paracatadioptric projections of lines. Computer Vision and Image Understanding 101(3), 151–165 (2006)
11. Veit, T., Tarel, J.-P., Nicolle, P., Charbonnier, P.: Evaluation of roadmarking feature extraction. In: Proc. of the 11th Int. IEEE Conf. on Intelligent Transportation Systems, pp. 174–181 (2008)

A Framework for Text Classification Using Intuitionistic Fuzzy Sets

Peerasak Intarapaiboon

Department of Mathematics and Statistics, Faculty of Science and Technology, Thammasat University, Thailand
peerasak@mathstat.sci.tu.ac.th

Abstract. Due to massively increasing of web pages and online documents, one of crucial processes to handle those documents is automatic (or at least semi-automatic) text classification. Based on the concept of intuitionistic fuzzy set (IFS), a framework for text classification is presented. In the framework, we introduce statistical methods to represent each document as an IFS and to learn a pattern of each document class. Then, a similarity measure for IFSs is applied in order to assign the most relevant class to a new document. The proposed framework with various similarity measures for IFSs is evaluated by benchmark datasets. The experimental results show that our framework yields satisfactory results.

Keywords: Intuitionistic fuzzy sets, similarity measures, classification problems.

1 Introduction

Due to exponential growth of web pages and online documents, text classification or text categorization (TC), which involves assigning a textual document to one of predefined classes, attracts more attention from researchers. Since this task can be seen as a classification problem in a machine learning point of view, several frameworks using varieties of classification techniques have been proposed, ranging from naive bayes to support vector machines. Reviews of recent research on TC using those classification techniques can be found in several articles, e.g. [1], [2], [3], [4], [5].

Recently, intuitionistic fuzzy set (IFS) [6] has been much explored in both theory and application. Differing from representation of a fuzzy set (FS) [7], an IFS considers both the membership and non-membership degrees of elements with respect to such a set. IFS is therefore more flexible to handle the uncertainty than FS. Measuring similarity and distance between IFSs is one of most research areas to which many researchers have paid their focus. Dengfeng and Chuntian [8] gave the definition of similarity measures between IFSs and proposed measures based on high membership and low membership functions. Liang and Shi [9] showed some counter-intuitive cases obtained from the measures proposed by Dengfeng and Chuntian [8] and then presented several similarity measures to overcome those cases. Later on, Mitchell [10]

M. Gen et al. (eds.), *Industrial Engineering, Management Science and Applications 2015*,
Lecture Notes in Electrical Engineering 349, DOI: 10.1007/978-3-662-47200-2_78

claimed that the rationale behind unreasonable cases was the weakness of the definition for similarity measures. Thus, a stronger definition for grading similarity degree between IFSs, which has been widely accepted in this research area, was defined. Moreover, some measures based on a statistical viewpoint were proposed. Hung and Yang [11] adopted the Hausdorff distance for developing several similarity measures. Xu [12] introduced the concepts of positive and negative ideal IFS and extended some similarity measures by assigning weights. The proposed measures were applied to solve multi-attribute decision making problems. Khatibi and Montazer [13] conducted experiments for bacterial classification using similarity measures for FSs and IFSs. The results indicated that each measure for IFSs outperformed that for FSs. In [14], cosine similarity measures for IFSs were proposed and applied to a small medical diagnosis problem; then, these measures were modified to satisfy the similarity definition by Hwang et al. [15]. Reviews of similarity measures for IFSs were presented in [16], [17].

By the success of research in IFS, especially similarity measurement, it is anticipated that IFS technologies will contribute to TC. Hence, this work presents a framework aimed to connect between the world of TC and that of IFS. In the framework, a method for representing textual documents in terms of IFSs is introduced. To assign an appropriate class to a document, a similarity measure for IFSs plays an important role. For this point, various similarity measures are employed to evaluate their performance.

2 Intuitionistic Fuzzy Sets and their Similarity Measures

In this section, some basic concepts for IFSs and their similarity measures are presented. Let $X = \{x_1, x_2, \dots, x_h\}$ be a discrete universe of discourse and *IFS(X)* be the class of all IFSs of X. Atanassov [6] defined an IFS A in *IFS(X)* as follows:

$$A = \{\langle x, \mu_A(x), \upsilon_A(x)\rangle | x \in X\} \quad (1)$$

which is characterized by a membership function $\mu_A(x)$ and a non-membership function $\upsilon_A(x)$. The two functions are defined as:

$$\mu_A: X \longrightarrow [0,1], \quad (2)$$

$$\upsilon_A: X \longrightarrow [0,1], \quad (3)$$

such that

$$0 \leq \mu_A(x) + \upsilon_B(x) \leq 1, \ \ \forall x \in X. \quad (4)$$

In the IFS theory, the hesitancy degree of x belonging to A is also defined by:

$$\pi(x) = 1 - \mu_A(x) - \upsilon_B(x). \quad (5)$$

Some of relations on IFSs related with this paper are listed below:

$$A \subseteq B \text{ iff } \mu_A(x) \leq \mu_B(x) \text{ and } \upsilon_A(x) \geq \upsilon_B(x) \quad (6)$$

$$A = B \text{ iff } A \subseteq B \text{ and } B \subseteq A \tag{7}$$

Definition 1. [10] *A similarity measure S for IFS(X) is a real function*

$$S: IFS(X) \times IFS(X) \longrightarrow [0,1],$$

which satisfies the following properties:

P1: $0 \le S(A,B) \le 1,\ \forall A,B \in IFS(X)$
P2: $S(A,B) = S(B,A),\ \forall A,B \in IFS(X)$
P3: $S(A,B) = 1$ iff $A = B$
P4: If $A,B,C \in IFS(X)$ and $A \subseteq B \subseteq C$, then $S(A,C) \le S(A,B)$ and $S(A,C) \le S(B,C)$

Assuming $A = \{\langle x_i, \mu_A(x_i), \upsilon_A(x_i)\rangle | x_i \in X\}$ and $B = \{\langle x_i, \mu_B(x_i), \upsilon_B(x_i)\rangle | x_i \in X\}$ in *IFS(X)*, Table 1 highlights some similarity measures between IFSs. These measures can be classified into three major groups: non distance-based measurement, distance-based measurement, and hybrid measurement. The first group consists of C_{IFS}, S_{IFS}, and S_Y. The measure C_{IFS} calculates the average of inner products of normalized vectors generated from *A* and *B* using membership and non-membership degrees. The measure S_{IFS} calculates the average of the similarity degrees obtained by applying C_{IFS} into three pairs of vectors generated from *A* and *B*. S_Y is defined in terms of membership degree, non-membership degree, hesitancy degree, and the maximum values of membership. The second group contains the others except S_F^m. The measures in this group are derived from the l_p distance measure using various pieces of information. For example, the median values of membership intervals are applied in S_d^p, while the hesitancy degrees are in S_Z. For the last group containing S_F^m, a similarity degree is obtained by averaging similarity degrees calculated by a distance-based measure and a set-theoretic measure. As a new source of information, the convex combination of the endpoints of the interval restricting the membership degree of an IFS, e.g., $\chi_j(A(x_i))$, is applied to the distance-based measure.

3 An IFS-Based Framework for Text Classification

3.1 IFS-Based Document Representation

In the preprocessing step, a document is represented in terms of Bag of Word (Bow). The BoW is used to form a vector for representing a document using the frequency of each word determined as a relevant feature for classification. Such words can be selected by some feature selection techniques. Assume h words, i.e., $w_1, w_2, \dots, w_h$, are selected for this representation. Document d_i can be represented by

$$d_i = (n_{i,1}, n_{i,2}, \dots, n_{i,h}),$$

where $n_{i,j}$ denotes the number of occurrences of w_j in d_i, $j = 1,2,\dots,h$.

Table 1. Some similarity measures between IFSs

Ref.	Expression
[8]	$S_d^p(A,B) = 1 - \frac{1}{\sqrt[p]{h}} \sqrt[p]{\sum_{i=1}^{h} \lvert\varphi_A(i) - \varphi_B(i)\rvert^p}$ where $\varphi_k(i) = (\mu_k(x_i) + 1 - \nu_k(x_i))/2$, $k = \{A,B\}$, and $p = 1,2,3,\ldots$
[9]	$S_e^p(A,B) = 1 - \frac{1}{\sqrt[p]{h}} \sqrt[p]{\sum_{i=1}^{h} \left(\frac{\lvert\mu_A(x_i)-\mu_B(x_i)\rvert + \lvert\nu_A(x_i)-\nu_B(x_i)\rvert}{2}\right)^p}$ where $p = 1,2,3,\ldots$
[10]	$S_{mod,p}(A,B) = \frac{1}{2}\left(\rho_\mu(A,B) + \rho_f(A,B)\right)$ where $\rho_\mu(A,B) = S_d^p(\mu_A(x_i), \mu_B(x_i))$ and $\rho_f(A,B) = S_d^p(1-\nu_A(x_i), 1-\nu_B(x_i))$
[12]	$S_Z(A,B) = 1 - \left[\frac{1}{2h}\sum_{i=1}^{h}((\mu_A(x_i)-\mu_B(x_i))^p + (\nu_A(x_i)-\nu_B(x_i))^p\right.$ $\left. +(\pi_A(x_i)-\pi_B(x_i))^p)\right]^{\frac{1}{p}}$ where $p = 1,2,3,\ldots$
[14]	$C_{IFS}(A,B) = \frac{1}{h}\sum_{i=1}^{h} \frac{\mu_A(x_i)\mu_B(x_i)+\nu_A(x_i)\nu_B(x_i)}{\sqrt{\mu_A^2(x_i)+\nu_A^2(x_i)}\sqrt{\mu_B^2(x_i)+\nu_B^2(x_i)}}$
[15]	$S_{IFS}(A,B) = \frac{1}{3}\left(C_{IFS}(A,B) + C_{IFS}^{*}(A,B) + C_{IFS}^{**}(A,B)\right)$ where $C_{IFS}^{*}(A,B) = \frac{1}{h}\sum_{i=1}^{h} \frac{\varphi_A(x_i)\varphi_B(x_i)+\nu_A(x_i)\nu_B(x_i)}{\sqrt{\varphi_A^2(x_i)+\nu_A^2(x_i)}\sqrt{\varphi_B^2(x_i)+\nu_B^2(x_i)}}$, with $\varphi_k(x_i) = \frac{1+\mu_k(x_i)-\nu_k(x_i)}{2}$, $k = A, B$ and $C_{IFS}^{**}(A,B) = \frac{1}{h}\sum_{i=1}^{h} \frac{(1-\mu_A(x_i))(1-\mu_B(x_i))+(1-\nu_A(x_i))(1-\nu_B(x_i))}{\sqrt{(1-\mu_A(x_i))^2+(1-\nu_A(x_i))^2}\sqrt{(1-\mu_B(x_i))^2+(1-\nu_B(x_i))^2}}$
[18]	$S_{new,p}(A,B) = 1 - \sqrt[p]{\sum_{i=1}^{h} w_i\left(\lvert\mu_A(x_i)-\mu_B(x_i)\rvert\right)^p} - \sqrt[p]{\sum_{i=1}^{h} w_i\left(\lvert\nu_A(x_i)-\nu_B(x_i)\rvert\right)^p}$ with $w_i \geq 0$ and $\sum_{i=1}^{h} w_i = 1$ and $p = 1,2,3,\ldots$
[19]	$S_{\Upsilon} = \frac{1}{2h}\sum_{i=1}^{h}\left(\sqrt{\mu_A(x_i)\mu_B(x_i)} + 2\sqrt{\nu_A(x_i)\nu_B(x_i)}\right.$ $\left. + \sqrt{\pi_A(x_i)\pi_B(x_i)} + \sqrt{(1-\nu_A(x_i))(1-\nu_B(x_i))}\right)$
[20]	$S_F^m = \frac{1}{2}\left(S^d(A,B) + S^{mix}(A,B)\right)$, where $S^d(A,B) = 1 - \sqrt{\frac{1}{h}\sum_{i=1}^{h}\left(\frac{1}{m+1}\sum_{j=0}^{m}\left[\chi_j(A(x_i)) - \chi_j(B(x_i))\right]^2\right)}$, $S^{mix}(A,B) = \frac{\sum_{i=1}^{h}(min\{\mu_A(x_i),\mu_B(x_i)\}+min\{1-\nu_A(x_i),1-\nu_B(x_i)\})}{\sum_{i=1}^{h}(max\{\mu_A(x_i),\mu_B(x_i)\}+max\{1-\nu_A(x_i),1-\nu_B(x_i)\})}$, $\chi_j(k(x_i)) = (1-\frac{j}{m})\mu_k(x_i) + \frac{j}{m}(1-\nu_k(x_i))$, $j = 0,1,2,\ldots,m$, $m = 1,2,3,\ldots$, and $k = \{A,B\}$

Next, the obtained vectors are converted in an IFS form. We begin the transformation with defining the universe $X = \{\text{LargeNumber}_1, \ldots, \text{LargeNumber}_h\}$.[1] Every value $n_{i,j}$ in the vector-based representation of the document i is then converted in terms of the three degrees of LargeNumber_j as the following steps:

1. $n_{i,j}$ is normalized by standard deviation, i.e.,

$$z_{i,j} = \frac{n_{i,j} - \bar{X}_j}{s_j} \tag{8}$$

[1] Each element in X is aimed at indicating a level of the frequency of a selected word in a particular document.

where $\bar{X}_j$ and s_j are the mean and the standard deviation, respectively, of the feature word w_j.

2. Denoted by $\mu_{i,j}$ and $\nu_{i,j}$, membership and non-membership degrees, respectively, of the document i with respect to LargeNumber_j are determined by weighted sigmoid functions:

$$\mu_{i,j} = r_j \frac{1}{1+e^{-z_{i,j}}}, \tag{9}$$

$$\nu_{i,j} = r_j^* \frac{1}{1+e^{z_{i,j}}}, \tag{10}$$

where $r_j, r_j^* \in (0,1]$ are weights for LargeNumber_j.

3. Denoted by $\pi_{i,j}$, the hesitancy degree of the document i with respect to LargeNumber_j is calculated by (5), i.e.,

$$\pi_{i,j} = 1 - \mu_{i,j} - \nu_{i,j}.$$

3.2 Pattern Learning

This section presents a procedure for learning patterns of document classes in terms of IFSs. Assume that there are l pre-defined classes referred to as $C_1, C_2, \dots, C_l$; m training documents as $d_1, d_2, \dots, d_m$; and h word features as $w_1, w_2, \dots, w_h$. A pattern for C_k, denoted by P_k, is defined by

$$P_k = \left\{\langle \text{LargeNumber}_j, \bar{\mu}_{kj}, \bar{\upsilon}_{kj}\rangle\right\}_{j=1}^{h},$$

where $\bar{\mu}_{kj}$ and $\bar{\upsilon}_{kj}$ are the average values of membership and nonmembership, respectively, of the word feature w_j observed from d_i belonging to C_k. More precisely,

$$\bar{\mu}_{kj} = \frac{\sum_{i=1}^{m}(\mu_{ij}\chi_k(d_i))}{\sum_{i=1}^{m}\chi_k(d_i)} \quad \text{and} \quad \bar{\upsilon}_{kj} = \frac{\sum_{i=1}^{m}(\nu_{ij}\chi_k(d_i))}{\sum_{i=1}^{m}\chi_k(d_i)},$$

where

$$\chi_k(d_i) = \begin{cases} 1, & d_i \in C_k \\ 0, & \text{otherwise} \end{cases}.$$

3.3 Classification

To classify a new document d_t, the document is represented as an IFS, by using the same parameter values as in the training step. Intuitively, d_t is grouped into class C' such that its pattern is closest to d_t. More formally,

$$C' = \text{argmax}_{C_k}\{Sim(P_k, d_t)\},$$

where *Sim* is a similarity measure and P_k is an IFS-based pattern of class C_k.

Table 2. Data set characteristics

Dataset	No. of articles	No. of words	Class distribution
BBC	2225	9635	474:374:434:506:437
BBCSport	737	4613	101:124:265:147:100

4 Datasets and Experimental Settings

4.1 Datasets

In this paper, two news datasets,[2] namely BBC and BBCSport, were used for experiments. BBC [21] consists from 2225 news articles of 5 topical areas, i.e., business, entertainment, politics, sport, and technology. With basic preprocessing steps including stemming, stop-word removal, low term frequency filtering, 9635 words are obtained for representing each article. On the other hand, BBCSport [21] contains 737 sports news articles of 5 areas, i.e., athletics, cricket, football, rugby, and tennis. Each article is represented by the frequency of 4613 words obtained from the same processes applied to BBC. Table 2 summarizes important characteristics of each dataset.

4.2 Experimental Schema and Parameter Setting

As seen in Table 2, the feature-space dimensions (No. of words) of both datasets are very high. To reduce such dimensions, one effective measure, i.e., *information gain (IG)* was used for evaluating feature relevance. Top 20% and 30% of features ranked by their IG values for BBC and top 50% and 80% of those for BBCSport were selected.

To retain the independence of the data in use, the 5-fold cross-validation was adopted for evaluating the proposed framework. Each dataset was divided roughly equally into five disjoint subsets, one of which contained approximately 20 percent of each news group. Every subset was once used as test data and four times as training data.

In our experiment, we introduce a strategy for setting the weights r_j and r_j^* in (9) and (10). The strategy sets the two weights based on statistical characteristics of feature words by the formula below:

$$r_j = r_j^* = \left|\frac{1-s_j}{1+s_j}\right|, \tag{11}$$

where s_j is the standard deviation of the number of occurrences of w_j in training documents. In this paper, all IFS similarity measures listed in Table 1 were used when parameter p's of $S_d^p, S_e^p, S_{mod,p}, S_Z$, and $S_{new,p}$ were assigned by 2; m of S_F^m by 6; and each w_i of $S_{new,p}$ was equally set to 1/(No. of features).

[2] Available at `http://mlg.ucd.ie/datasets`.

4.3 Evaluation Metric

To evaluate the performance of classification with respect to class C_i, we used recall, precision, and F-measure denoted by R_i, P_i, and F_i, respectively. They are defined as:

$$R_i = \frac{TP_i}{TP_i+FN_i}, \quad P_i = \frac{TP_i}{TP_i+FP_i}, \quad F_i = \frac{2\times R_i\times P_i}{R_i+P_i},$$

where TP_i (true positive) is the number of documents correctly classified to C_i; FN_i (false negative) is the number of documents in C_i classified to C_j; and FP_i (false positive) is the number of documents in C_j classified to C_i with $C_i \neq C_j$.

Since every sub-dataset for the 5-fold cross-validation contains 5 classes and the numbers of documents belonging in the classes are not equal, all experiments are multiclass imbalance scenarios. To show average performance over all classes, mean F-measure (MFM) and macro average geometric (MAvG), which are widely used in this situation were applied. These two evaluation metrics, defined in [22], are as follows:

$$\text{MFM} = \frac{\sum_{i=1}^{l} F_i}{l}, \quad \text{MAvG} = \sqrt[l]{\prod_{i=1}^{l} R_i},$$

when l is the number of predefined classes in the dataset.

4.4 Baselines

The proposed framework was compared with classification by four models, i.e., Decision Tree (DT) using C4.5, Naive Bayes (NB), *k*-Nearest Neighbor (kNN), and Support Vector Machine (SVM) based on the RBF kernel. The Weka machine learning suite[3] was employed for classifier learning and evaluation, using its default parameters. As observed during the learning process, 3-NN performed slightly better than 1-NN, 5-NN, and 7-NN on our training data sets, and was chosen as a representative of *k*-NN.

5 Experimental Results and Discussion

Using the similarity measures depicted in Table 1 and the four classifiers, Tables 3 and 4 give the evaluation results on BBC and BBCSport respectively. In each table, the first column presents the percentage of selected features. The next one expresses a similarity measure or a classification model used in experiment. The last two columns show experimental results in terms of the average of MFM (in percent) and MAvG (in percent) along with standard deviation (SD) from the 5-fold cross-validation. Note that, on the BBC dataset, even the memory was extended, Weka ran out of memory in the training process of SVM when 30% of features were selected.[4] Hence, there is no report of the case in Table 3.

---

[3] Available at `http://www.cs.waikato.ac.nz/ml/weka`.

[4] The experiments were all performed on a standard PC with Intel Core i7 processor (3.40 GHz, 8.00 GB RAM), running Windows-7 operating system.

Table 3. Evaluation results on BBC

N% of features	Measure/ Classifier	MFM±SD	MAvG±SD
20	S_d^p	94.13±1.27E-02$^{[\uparrow,\uparrow,-,-]}$	93.80±1.44E-02$^{[\uparrow,\uparrow,-,-]}$
	S_e^p	94.13±1.27E-02$^{[\uparrow,\uparrow,-,-]}$	93.80±1.44E-02$^{[\uparrow,\uparrow,-,-]}$
	$S_{mod,p}$	94.13±1.27E-02$^{[\uparrow,\uparrow,-,-]}$	93.80±1.44E-02$^{[\uparrow,\uparrow,-,-]}$
	S_Z	94.13±1.27E-02$^{[\uparrow,\uparrow,-,-]}$	93.80±1.44E-02$^{[\uparrow,\uparrow,-,-]}$
	C_{IFS}	92.81±1.65E-02$^{[\uparrow,\uparrow,-,-]}$	92.31±1.88E-02$^{[\uparrow,\uparrow,-,-]}$
	S_{IFS}	92.81±1.65E-02$^{[\uparrow,\uparrow,-,-]}$	92.31±1.88E-02$^{[\uparrow,\uparrow,-,-]}$
	$S_{New,p}$	94.13±1.27E-02$^{[\uparrow,\uparrow,-,-]}$	93.80±1.44E-02$^{[\uparrow,\uparrow,-,-]}$
	S_Y	94.34±1.14E-02$^{[\uparrow,\uparrow,-,-]}$	94.05±1.32E-02$^{[\uparrow,\uparrow,-,-]}$
	S_F^m	80.28±2.01E-02$^{[\uparrow,\uparrow,\downarrow,\downarrow]}$	77.26±2.79E-02$^{[\uparrow,\uparrow,\downarrow,\downarrow]}$
	DT	69.84±2.63E-02	69.35±2.77E-02
	*k*NN	60.15±3.83E-02	53.16±3.70E-02
	NB	93.85±8.43E-03	93.94±8.89E-03
	SVM	94.41±1.48E-02	94.48±1.54E-02
30	S_d^p	94.18±1.24E-02$^{[\uparrow,\uparrow,-]}$	93.85±1.43E-02$^{[\uparrow,\uparrow,-]}$
	S_e^p	94.18±1.24E-02$^{[\uparrow,\uparrow,-]}$	93.85±1.43E-02$^{[\uparrow,\uparrow,-]}$
	$S_{mod,p}$	94.18±1.24E-02$^{[\uparrow,\uparrow,-]}$	93.85±1.43E-02$^{[\uparrow,\uparrow,-]}$
	S_Z	94.18±1.24E-02$^{[\uparrow,\uparrow,-]}$	93.85±1.43E-02$^{[\uparrow,\uparrow,-]}$
	C_{IFS}	92.91±1.66E-02$^{[\uparrow,\uparrow,-]}$	92.42±1.86E-02$^{[\uparrow,\uparrow,-]}$
	S_{IFS}	92.91±1.66E-02$^{[\uparrow,\uparrow,-]}$	92.42±1.86E-02$^{[\uparrow,\uparrow,-]}$
	$S_{New,p}$	94.18±1.24E-02$^{[\uparrow,\uparrow,-]}$	93.85±1.43E-02$^{[\uparrow,\uparrow,-]}$
	S_Y	94.39±1.13E-02$^{[\uparrow,\uparrow,-]}$	94.08±1.31E-02$^{[\uparrow,\uparrow,-]}$
	S_F^m	82.37±1.92E-02$^{[\uparrow,\uparrow,\downarrow]}$	79.79±2.45E-02$^{[\uparrow,\uparrow,\downarrow]}$
	DT	66.45±3.03E-02	65.72±3.31E-02
	*k*NN	49.75±3.21E-02	41.23±3.41E-02
	NB	94.09±8.17E-03	94.19±8.69E-03

Table 3 indicates that, regardless of similarity measures, our framework yields much higher performance than DT and *k*NN. Concretely, using 30% of features, the maximum gaps of MFM and MAvG are about 45% and 50% (i.e., 94.39% of MFM from S_Y and 49.75% from *k*NN; and 94.08% of MAvG from S_Y and 41.23% of from kNN). Comparing with NB, all distance-based measures and S_Y yield slightly better MFM, but lower MAvG. In terms of both MFM and MAvG, the cosine-based measures, i.e., C_{IFS} and S_{IFS}, produce a few lower performance, while S_F^m does apparently less accuracy. Comparing with SVM, distance based and non-distance based measures show comparable results, while S_F^m gives much lower. Likewise, Table 4 reveals that our framework with every measures clearly outperforms DT and *k*NN. Using measures in both distance and non-distance groups, accuracy is comparable to NB and higher than SVM. The performance of S_F^m is lower than that of NB and is in the same level of SVM.

To gain insight this comparison, a two-tailed paired t-test at 5% for each metric was performed. The comparison result depicts as $\uparrow$,$-$, or $\downarrow$ meaning that the measure is statistically superior, equal, or inferior to the classifier. The order of classifiers to be compared is DT, *k*NN, NB, and SVM. For instance, in Table 4, the comparison result of MAvG obtained from C_{IFS} to that from the classifiers reported in the fifth row is $[\uparrow,\uparrow,\downarrow,-]$. It means that C_{IFS} is significantly better than DT and kNN, worse than NB, and equal to SVM.

Table 4. Evaluation results on BBCSport

N% of features	Measure/ Classifier	MFM±SD	MAvG±SD
50	S_d^p	94.78±8.64E-03[↑,↑,−,↑]	93.54±1.18E-02[↑,↑,−,−]
	S_e^p	94.78±8.64E-03[↑,↑,−,↑]	93.54±1.18E-02[↑,↑,−,−]
	$S_{mod,p}$	94.78±8.64E-03[↑,↑,−,↑]	93.54±1.18E-02[↑,↑,−,−]
	S_Z	94.78±8.64E-03[↑,↑,−,↑]	93.54±1.18E-02[↑,↑,−,−]
	C_{IFS}	93.37±1.41E-02[↑,↑,−,↑]	91.45±1.85E-02[↑,↑,↓,−]
	S_{IFS}	93.37±1.41E-02[↑,↑,−,↑]	91.45±1.85E-02[↑,↑,↓,−]
	$S_{New,p}$	94.78±8.64E-03[↑,↑,−,↑]	93.54±1.18E-02[↑,↑,−,−]
	S_Y	94.48±8.96E-03[↑,↑,−,↑]	93.47±1.27E-02[↑,↑,−,−]
	S_m^F	89.13±1.94E-02[↑,↑,↓,−]	85.80±2.36E-02[↑,↑,↓,−]
	DT	69.89±3.86E-02	68.92±4.16E-02
	kNN	42.17±9.44E-02	30.18±8.91E-02
	NB	95.86±1.42E-02	96.26±1.66E-02
	SVM	89.44±3.98E-02	90.91±3.52E-02
80	S_d^p	94.68±6.59E-03[↑,↑,−,−]	93.49±1.10E-02[↑,↑,−,−]
	S_e^p	94.68±6.59E-03[↑,↑,−,−]	93.49±1.10E-02[↑,↑,−,−]
	$S_{mod,p}$	94.68±6.59E-03[↑,↑,−,−]	93.49±1.10E-02[↑,↑,−,−]
	S_Z	94.68±6.59E-03[↑,↑,−,−]	93.49±1.10E-02[↑,↑,−,−]
	C_{IFS}	93.61±5.30E-03[↑,↑,−,−]	91.80±1.00E-02[↑,↑,↓,−]
	S_{IFS}	93.61±5.30E-03[↑,↑,−,−]	91.80±1.00E-02[↑,↑,↓,−]
	$S_{New,p}$	94.68±6.59E-03[↑,↑,−,−]	93.49±1.10E-02[↑,↑,−,−]
	S_Y	94.71±1.05E-02[↑,↑,−,−]	93.72±1.46E-02[↑,↑,−,−]
	S_m^F	90.13±2.17E-02[↑,↑,↓,−]	88.85±2.42E-02[↑,↑,↓,−]
	DT	66.60±5.25E-02	65.50±5.00E-02
	kNN	40.36±2.43E-02	28.58±1.55E-02
	NB	95.50±1.72E-02	95.80±1.74E-02
	SVM	90.10±3.77E-02	91.20±4.28E-02

6 Conclusions

This paper presents an IFS-based framework for text classification. With basic statistical concepts, documents can be represented in terms of IFSs, and patterns of document classes can be constructed. Experiments with real datasets show that our framework with some existing similarity measures for IFSs yields the satisfactory results comparing with the results from the four traditional classifiers i.e., Decision tree, Naive bayes, k-NN, and SVM. As a future work, we will explore other techniques to represent documents in terms of IFSs and to learn class patterns.

Acknowledgement. This research is supported by Thammasat University Research Fund under the TU Research Scholar, Contract No. 1/8/2557.

References

1. Sebastiani, F.: Machine learning in automated text categorization. ACM Comput. Surv. 34, 1–47 (2002)
2. Ikonomakis, M., Kotsiantis, S., Tampakas, V.: Text Classification using Machine Learning Techniques. WSEAS transactions on Computers 8, 966–974 (2005)
3. Baharudin, B., Lee, L.H., Khan, K.: A Review of Machine Learning Algorithms for Text-Documents Classification. Journal of Advances in Information Technology 1, 4–20 (2010)
4. Dalal, M.K., Zaveri, M.A.: Automatic Text Classification: A Technical Review. International Journal of Computer and Applications 28, 37–40 (2011)
5. Dasari, D.B., Rao, V.G.: Text Categorization and Machine Learning Methods: Current State of The Art. Global Journal of Computer Science and Technology Software & Data Engineering 12, 37–46 (2012)
6. Atanassov, K.: Intuitionistic Fuzzy Sets. Fuzzy Set Syst 20, 87–96 (1986)
7. Zadeh, L.A.: Fuzzy Sets. Information Control 8, 338–353 (1965)
8. Dengfeng, L., Chuntian, C.: New Similarity Measures of Intuitionistic Fuzzy Sets and Application to Pattern Recognition. Pattern Recogn. Lett. 23, 221–225 (2002)
9. Liang, Z., Shi, P.: Similarity Measures on Intuitionistic Fuzzy Sets. Pattern Recogn. Lett. 24, 2687–2693 (2003)
10. Mitchell, H.B.: On the Dengfeng-Chuntian Similarity Measure and Its Application to Pattern Recognition. Pattern Recogn. Lett. 24, 3101–3104 (2003)
11. Hung, W.-L., Yang, M.-S.: Similarity Measures of Intuitionistic Fuzzy Sets Based on Hausdorff Distance. Pattern Recogn. Lett. 25, 1603–1611 (2004)
12. Xu, Z.: Some Similarity Measures of Intuitionistic Fuzzy Sets and Their Applications to Multiple Attribute Decision Making. Fuzzy Optim. Decis. Making. 6, 109–121 (2007)
13. Khatibi, V., Montazer, G.A.: Intuitionistic Fuzzy Set VS. Fuzzy Set Application in Medical Pattern Recognition. Artif. Intell. Med. 47, 43–52 (2009)
14. Ye, J.: Cosine Similarity Measures for Intuitionistic Fuzzy Sets and Their Applications. Math. Comput. Model. 53, 91–97 (2011)
15. Hwang, C.-M., Yang, M.-S.: Modified Cosine Similarity Measure between Intuitionistic Fuzzy Sets. In: Lei, J., Wang, F.L., Deng, H., Miao, D. (eds.) AICI 2012. LNCS, vol. 7530, pp. 285–293. Springer, Heidelberg (2012)
16. Li, Y., Olson, D.L., Qin, Z.: Similarity Measures between Intuitionistic Fuzzy (Vague) Sets: A Comparative Analysis. Pattern Recogn. Lett. 28, 278–285 (2007)
17. Papakostas, G.A., Hatzimichailidis, A.G., Kaburlasos, V.G.: Distance and similarity measures between intuitionistic fuzzy sets: A comparative analysis from a pattern recognition point of view. Pattern Recogn. Lett. 34, 1609–1622 (2013)
18. Julian, P., Hung, K.-C., Lin, S.-J.: On the Mitchell similarity measure and its application to pattern recognition. Pattern Recogn. Lett. 33, 1219–1223 (2012)
19. Song, Y., Wang, X., Lei, L.: Xue. A.: A New Similarity Measure between Intuitionistic Fuzzy Sets and Its Application to Pattern Recognition. Abstr. Appl. Anal, 1-11 (2014)
20. Farhadinia, B.: An Efficient Similarity Measure for Intuitionistic Fuzzy Sets. Soft Comput. 18, 85–94 (2014)
21. Greene, D., Cunningham, P.: Producing Accurate Interpretable Clusters from High-Dimensional Data. In: Jorge, A.M., Torgo, L., Brazdil, P.B., Camacho, R., Gama, J. (eds.) PKDD 2005. LNCS (LNAI), vol. 3721, pp. 486–494. Springer, Heidelberg (2005)
22. Sun, Y., Kamel, M.S., Wang, Y.: Boosting for Learning Multiple Classes with Imbalanced Class Distribution. In: The 6th International conference of data mining (ICML 2006), pp. 592-602 (2006)

An Adaptive Incremental Fuzzy TSK Controller Combined with Evolutionary Optimization

Niusha Shafiabady[1], Rajprasad K. Rajkumar[1], Dino Isa[1], J. Michael Menke[2], and M.A. Nima Vakilian[2]

[1] University of Nottingham, Malaysia Campus
[2] IMU, Kuala Lumpur, Malaysia
{Niusha.shafiabady,Rajprasad.Rajkumar, Dino.isa}@nottingham.edu.my, MichaelMenke@imu.edu.my, Nima_vakilian@yahoo.co.uk

Abstract. In this paper fully informed particle swarm optimization (FIPSO) is used as a tool to create an incremental method for design and optimization of Takagi-Sugeno-Kang (TSK) fuzzy controllers. The controller is refined as time goes on, starting from a single controlling rule and evolving as FIPSO leads it to optimize itself. The controller is adaptive and can adjust itself to changes in the parameters of the system as it updates its rule base and parameters. The proposed scheme was applied to the inverted wedge that is a planar robot with two degrees of freedom and a single control input. The inverted wedge is an under actuated system known to be a good test bed for the development of unconventional advanced control techniques. The adaptive fuzzy TSK controller achieved the control objective in a short period of time which shows the efficiency of the proposed method.

Keywords: Fully Informed Particle Swarm Optimization, Incremental evolutionary design, Fuzzy controller and TSK model.

1 Introduction

Using conventional techniques, designing controllers for plants with nonlinear dynamics and modeling uncertainties can be quite challenging. Designing a good controller for the planar robot called inverted wedge that is a nonlinear under actuated initially unstable system is quite a challenging task. By making use of heuristics, intelligent control techniques potentially can greatly simplify the synthesis of a controller for such plants. Although the analysis of intelligent techniques such as fuzzy systems, neural networks, and evolutionary programming, has in many cases been widely studied and show promising results, these techniques are used relatively infrequently in industry. This is possibly because of the lack of experimental implementation and analysis of these techniques. The concept of fuzzy logic and artificial neural network for control problem has been grown into a popular research topic in recent years [1, 3]. The reason is that the classical control theory usually requires a mathematical model for designing the controller. The inaccuracy of mathematical modeling of the plants usually degrades the performance of the controller, especially for nonlinear and complex control problems [4].

M. Gen et al. (eds.), *Industrial Engineering, Management Science and Applications 2015*, Lecture Notes in Electrical Engineering 349, DOI: 10.1007/978-3-662-47200-2_79

On the contrary, the fuzzy logic controller and the artificial neural network controller, they offer a key advantage over traditional adaptive control systems. That is, they do not require mathematical models of the plants. The traditional neural networks can learn from data and feedback, but the meaning associated with each neuron and each weight in the network is not easily understood. Alternatively, the fuzzy logical models are easy to appreciate, because it uses linguistic terms and the structure of if-then rules. However, as compared with the neural networks, learning ability is lack of fuzzy logic. Evolutionary algorithms have been used for tuning and optimization of fuzzy systems in general and fuzzy controllers in particular [2-5]. Fuzzy system design is concerned with defining the membership functions and fuzzy rules. The design process is accompanied by an evolutionary algorithm like FIPSO to tune the parameters of the knowledge base. Augmentation of the fuzzy rule base [2] has been used for accuracy improvement [5-7]. In this paper a TSK fuzzy control system has been applied to a nonlinear plant. The peculiarity of the method is the incremental design and optimization of first order TSK fuzzy controller by means of an efficient and powerful evolutionary algorithm. The controller starts with a single control law and refines itself by the help of FIPSO that helps the controller to refine itself. Structural augmentation is intertwined with the evolutionary adaptation of the additional parameters with the objective to improve the control performance. The key idea of this divide-and-conquer approach is to apply the concept of gradual refinement of the fuzzy model to the incremental design of TSK fuzzy controllers for nonlinear systems. An evolution strategy is used as a tool for incremental optimization of the TSK fuzzy controller. Different controllers have been designed to control the same system in [1] and the proposed controller has been able to achieve a better result than the previous works. This paper is organized as follows. Section 2 explains the control technique. Section 3 describes the inverted wedge that is the system to be controlled and section 4 gives a brief explanation on FIPSO. Section 5 explains the fuzzy controller. Section 6 describes the evolutionary augmentation of fuzzy controller and section 7 explains the results. Finally the conclusion is given in section 8.

2 The Control Technique

A TSK fuzzy control system for a nonlinear plant is composed of a set of linear models at different operating points. The activation rate of each rule in its antecedent part decides the weight that is given to that specific rule. The stability of this control technique has been discussed in the previous works [2, 7].

In general, a linear system's equation is defined as follows.

$$\dot{x} = Ax + Bu \tag{1}$$

A TSK fuzzy control problem can be modeled as a set of optimal control problems if a set of linear models with different operating points are assumed and an optimal controller is designed for each operating point. The optimal control law minimizes the following cost function.

$$J = \int_{=0}^{\infty} x(t)Qx(t) + u(t)Ru(t) \tag{2}$$

The input u is given by a state feedback control law mentioned below.

$$u = -R^{-1}B^{T}Xx \tag{3}$$

Where X is the Riccati equation's solution that is given below.

$$-XBR^{-1}B^{T}X + X^{T}A + A^{T}X + Q = -\dot{X} \tag{4}$$

By assuming that X is constant Riccati equation equals with zero and is called algebraic Riccati equation. The optimal control law implemented in a TSK fuzzy controller system is given in general by this rule. *If Inp is A then* $u = k_0 + \sum_i k_i x_i$

Here the gain factors k_i are decided by (4). The augmentation methods and evolutionary algorithms are applied to update the rules and find the gain factors while the controller evolves.

3 The Inverted Wedge

The inverted wedge [1], shown in Fig. 1, is an under actuated system as it has two degrees of freedom and one control input. It can be modeled as a nonlinear system in the form of Eq. (6).

$$\dot{x}(t) = f(x(t), u(t)) \tag{5}$$

Here x is the state vector of the system. It consists of a motor and a cart that can be moved on the bar by the motor to keep the wedge balanced. The states of the system are the position and the velocity of the cart ($\gamma(t)$ and $\dot{\gamma}(t)$), the angular position and velocity of the wedge ($\alpha(t)$ and $\dot{\alpha}(t)$), and the motor current $I(t)$. The aim is to keep $\alpha = 0$ that guarantees the wedge's balance. The system's dynamic equations are [1] mentioned below.

$$\begin{aligned}
\ddot{\gamma} &= \frac{1}{m}\left(\frac{KI}{r} - K_f\dot{\gamma} - mg\sin\alpha\right),\\
\ddot{\alpha} &= \frac{1}{J_w}(Mgl\sin\alpha - mg(-d\sin\alpha + \gamma\cos\alpha)\\
&\quad - m(d^2\dot{\alpha}\quad d\dot{\gamma} + \gamma^2\dot{\alpha}),\\
\dot{I} &= \frac{1}{L}\left(u(t) - RI - K_b\frac{\dot{\gamma}}{r}\right)
\end{aligned} \tag{6}$$

Using Eq. (7) the state vector defined in (6) can be written as

$$\dot{x} = \begin{bmatrix} x_2 \\ \frac{1}{m}(\frac{K}{r}x_5 - K_f x_2 - mg\sin x_3) \\ x_4 \\ \frac{1}{J_w}((MgL - mg)\sin x_3 - mgx_1\cos x_3 \\ + mdx_2 - m(d^2 + x_1)x_4) \\ \frac{1}{L}(u(t) - Rx_5 - \frac{K_b}{r}x_2) \end{bmatrix} \quad (7)$$

Where the state vector is $x = [x_1, x_2, x_3, x_4, x_5]^T = [\gamma, \dot{\gamma}, \alpha, \dot{\alpha}, I]^T$ and the plant's parameters are given in Table 1.

Table 1. Summary of plant parameters and variables

Plant Parameters		
J_w	1.4029 kgm^2	Inertia of wedge
M	4.86 kg	Mass of wedge
M	1.25 kg	Mass of cart
L	0.14m	Distance between pivot and COG of wedge
D	0.17 m	Distance between pivot and sliding surface
R	0.019 m	Radius of gear
K_f	20.0 N/m/s	Friction coefficient
R	1.9 Ω	Motor resistance
L	0.004 H	Motor inductance
K	14.8 oz-in/A	Motor torque constant
K_b	11.0 V/kRPM	Back EMF constant

The aim of the optimal control is to find a control law that minimizes the cost function. The cost function is composed of three terms. The first term J_0 is defined as follows.

$$J_0 = (t_{final} - t_{end}) \quad (8)$$

Where t_{end} is the time that the controller has been able to achieve its goal and t_{final} is the final control time that is assumed to be 10 seconds in this study as the system has to be stabilized in maximum 10 seconds. The second term of the cost function is J_1 that is computed as follows.

$$J_1 = \frac{1}{t_{end}} \int_0^{t_{end}} (\alpha^2 + \frac{\gamma^2}{\beta}) dt \tag{9}$$

Where α and γ indicate the angle to be controlled and the cart position and $\beta = 10$ is used as a weighting factor to indicate that the angle is of much more importance than the position regarding the control goal.

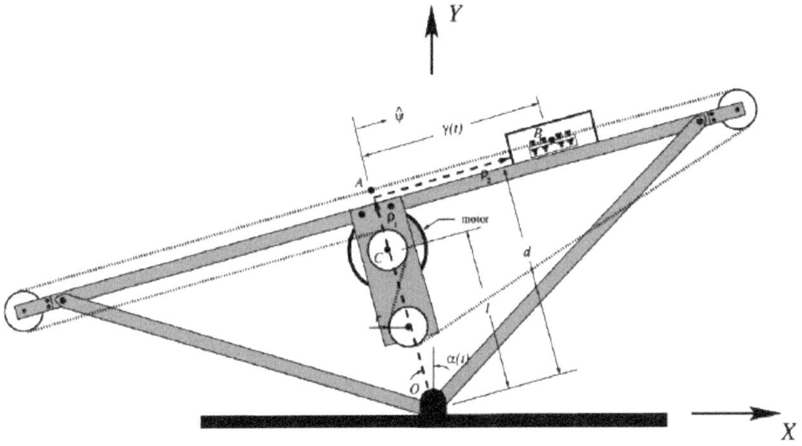

Fig. 1. The inverted wedge [3]

The third term is related to control signal that is weighted by matrix R in LQR design, which should not exceed the limit u_{max} here. It is defined as follows.

$$J_2 = \frac{1}{t_{end}} \int_0^{t_{end}} \frac{(|u| - u_{max})^2}{\tau} \tag{10}$$

In Eq. (11) $\tau = 10$ and is again the weighting factor to indicate that the input is less important in control. Finally the cost function is computed as mentioned here.

$$J = J_0 + J_1 + J_2 \tag{11}$$

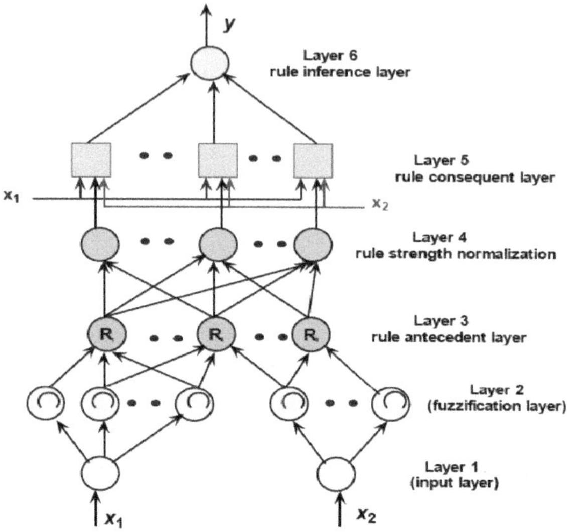

Fig. 2. TSK fuzzy controllers structure

Any other control aims can be added to this cost function if needed. This cost function plays the role of the fitness function that is used for optimizing the gain factor population by applying fully informed particle swarm optimization.

4 Fully Informed Particle Swarm Optimization

The canonical particle swarm algorithm is a new approach to optimization, drawing inspiration from group behavior and establishment of social norms. It is gaining popularity because of its being easy to use and its speed of convergence. This method keeps the individuals fully informed that means unlike simple PSO, every single member of the population's experience is taken into account.

$$\begin{aligned}&\vec{v}_i(t) = \phi_1 \vec{v}_i(t-1) + \vec{r}_1 \vec{c}_1 \otimes \\&(\vec{x}_{personalbest} - \vec{x}_i(t)) + \vec{r}_2 \vec{c}_2 \otimes (\vec{x}_{globalbest} - \vec{x}_i(t)) \\&\vec{x}_i(t) = \vec{v}_i(t) + \vec{x}_i(t-1) \\&\phi_1 = 1 - 0.5 \times \frac{1-t}{1-t_{\max}}\end{aligned} \tag{12}$$

As it is clear in Eq. (13), the experience of every member of the population is taken into consideration and this method has a better performance than the simple PSO [8].

5 The Fuzzy Controller

The structure of the TSK fuzzy controller is shown in Fig. 2. The first layer is the input layer which transfers the data to the second layer with no computation. The second layer is the fuzzification layer that computes the degree to which an input value belongs to a fuzzy set. The third layer computes the antecedent part of a rule by applying a t-norm to the outputs of the second layer according to the definition of each rule. The fourth layer normalizes the strength of each rule using the weights as mentioned here.

$$\overline{w}_i = \frac{w_i}{\sum_i w_i}, i = 1,2,... \tag{13}$$

The fifth layer computes the rules' consequents. In first order TSK fuzzy systems, the weighted consequents of the rules are computed as mentioned below.

$$\overline{w}_i f_i = \overline{w}_i (a_i x_1 + b_i x_2 + ... + q_i) \tag{14}$$

Here the number of the coefficients (n) depends on the number of the input variables (m) that are fed to the fuzzy system where $n=m+1$. Finally the sixth layer is the rule inference system that computes the overall output as follows.

$$overalloutput = \sum_i \overline{w}_i f_i \tag{15}$$

For the system defined here f_i for the TSK controller can be defined as follows.

$$f_i = k_0 + k_1\alpha + k_2\dot{\alpha} + k_3\gamma + k_4\dot{\gamma} \tag{16}$$

Where k_i, $i=0,1,...,4$ are the parameters to be trained by FIPSO.

6 Evolutionary Augmentation of Fuzzy Controllers

The augmentation is performed to incrementally evolve the controller. Each refinement step consists of an augmentation in which a single fuzzy set is split into multiple membership functions. The rules with antecedents that refer to the split fuzzy sets are duplicated. The augmentation is illustrated in Fig. 3 for two state space variables α and $\dot{\alpha}$. The gain factors are assumed to be $(k_0, k_\alpha, k_{\dot{\alpha}}, k_\gamma, k_{\dot{\gamma}})$ and each fuzzy rule is in the form mentioned here.

R: If α is A and $\dot{\alpha}$ is B then $u = k_0 + \alpha \times k_\alpha + \dot{\alpha} \times k_{\dot{\alpha}} + \gamma \times k_\gamma + \dot{\gamma} \times k_{\dot{\gamma}}$ (17)

In each phase the fuzzy sets are spilt and the rules evolve until the desired outcome is achieved. The gain factors are initialized randomly for the first generation and eventually they evolve to construct the conclusion part's parameters using FIPSO.

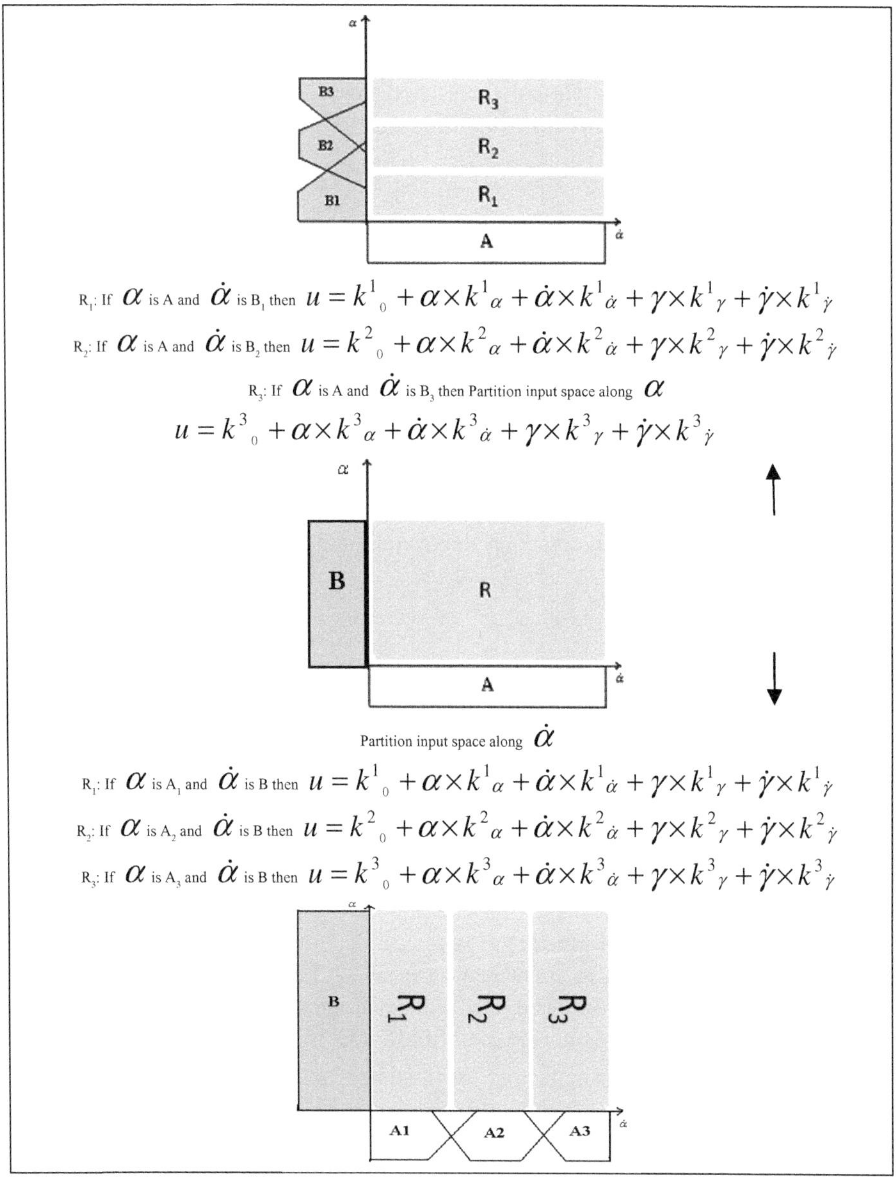

Fig. 3. Partition of state space for fuzzy controller

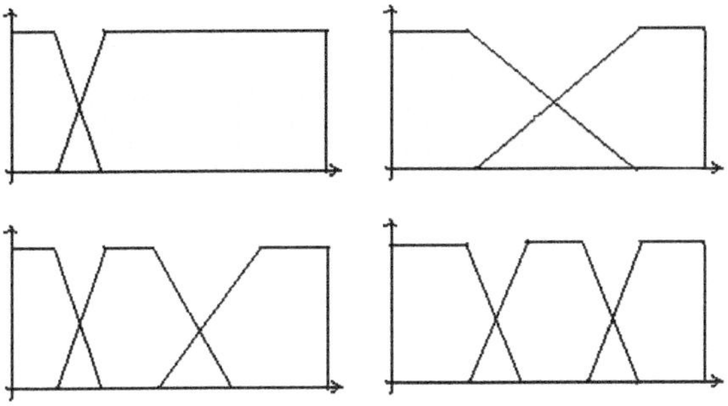

Fig. 4. (Left) Insertion of a fuzzy set in the core, (Right) Insertion of a fuzzy set in between two fuzzy sets

Two operations can be used for splitting the fuzzy sets, insertion of a fuzzy set in the core and insertion of a fuzzy set between two fuzzy sets as shown in fig. 4. The insertion shown in fig. 4 (left) changes the cores but the rules' conclusions remain intact and are inherited to new born rules created by the new set. The insertion of a fuzzy set between two adjacent fuzzy sets as shown in fig. 4 (right) leaves the cores of the initial fuzzy sets intact and a new set is inserted between them. A combination of these two methods can be used for splitting the antecedent parts of the rules. After performing the split the optimal gain factors are to be computed by FIPSO. This evolution strategy diverges into multiple subpopulations and the population with the best control performance according to the cost function (12) survives and is chosen as the best solution and the other members of the population are discarded.

7 Results

The controller is implemented in MATLAB and the results given in fig. 5 and fig. 6 show that it has been capable of controlling the position of the cart and the position of the wedge in a short time. Fig. 7 and fig. 8 show the cart and the wedge position controlled starting from a different initial position.

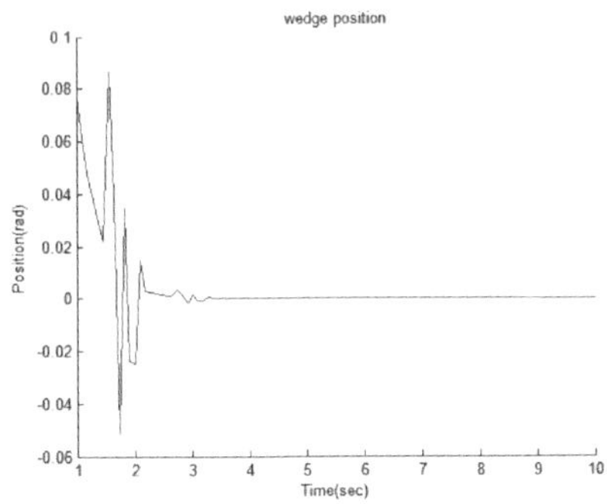

Fig. 5. The result of wedge angle

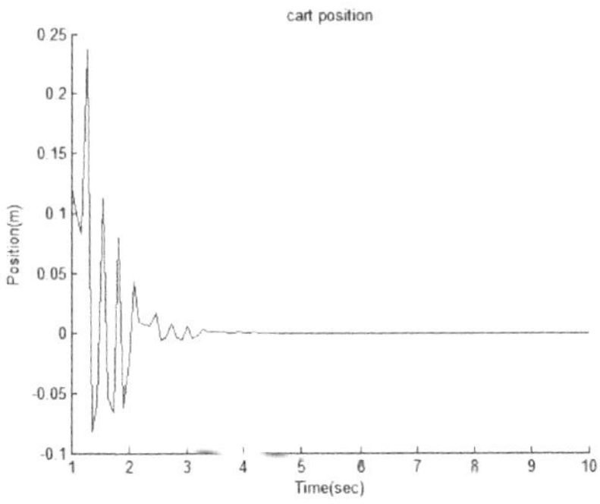

Fig. 6. The result of cart position

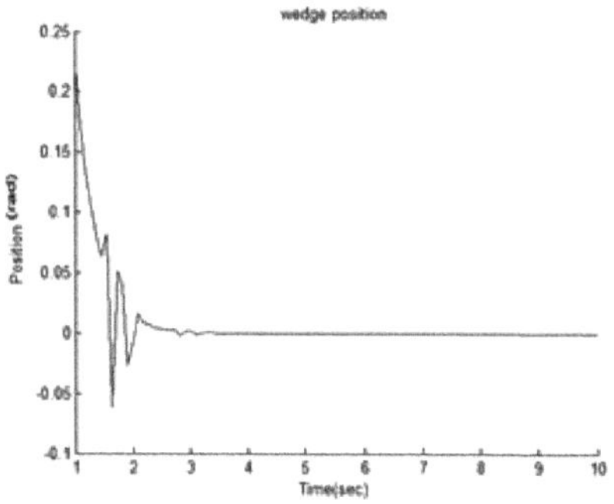

Fig. 7. The result of wedge angle

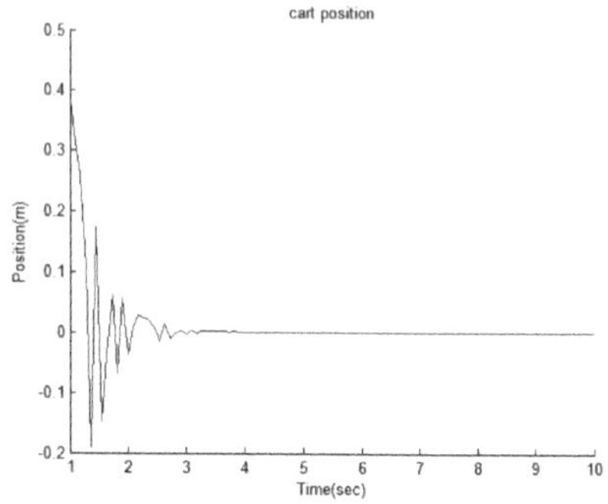

Fig. 8. The result of cart position

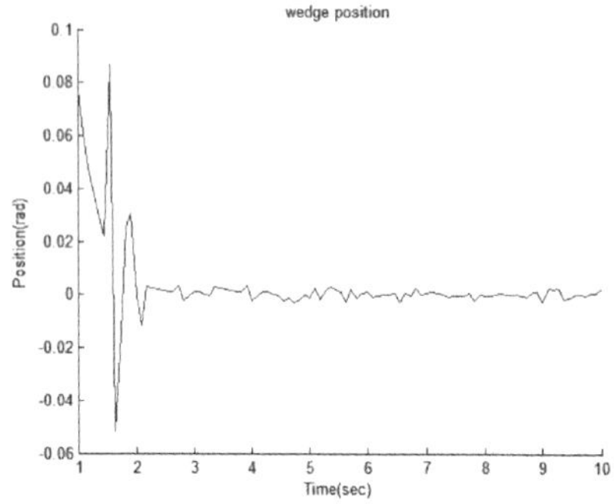

Fig. 9. The result of wedge angle after adding noise

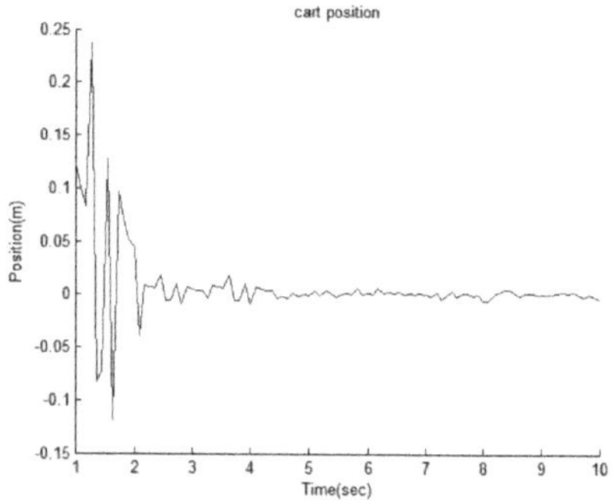

Fig. 10. The result of cart position after adding noise

A Gaussian noise n that is zero-mean is added to the wedge's angle α and its results are shown in fig. 9 and fig. 10. This noise works as an external force that shakes the wedge continuously. The results show that the controller has been able to reach the control objective in presence of noise too.

Fig. 11 and fig. 12 show the best results achieved by the controller designed in [1] for the same system.

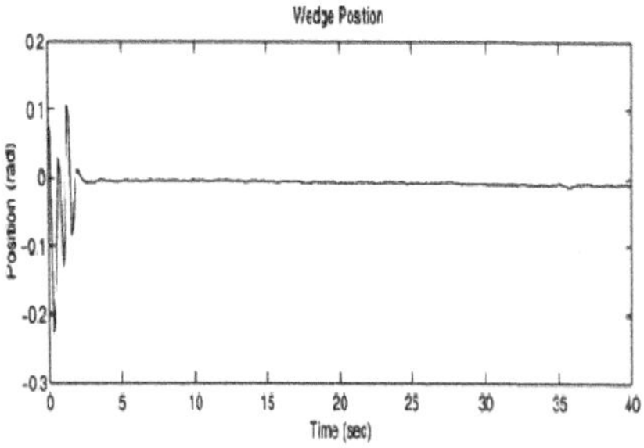

Fig. 11. The result of wedge angle

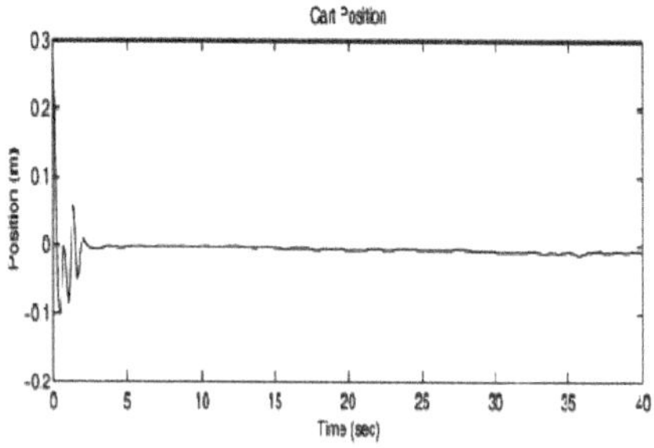

Fig. 12. The result of cart position

As it was mentioned earlier the aim of controlling the wedge is keeping $\alpha = 0$ and this aim is achieved with less fluctuation in the wedge's angle by the controller used here in comparison to previous work done on this system [1]. As the results show the controller has been able to perform its task in less than 4 seconds and achieved the goal well.

8 Conclusion

We propose an adaptive fuzzy TSK control scheme that evolves through time based on changes in the system parameters. The controller augments the rule base. Augmenting the rule base is an appropriate method for designing the incremental fuzzy controllers as it does not require expert knowledge of the controlled system except for the cost function that is to be optimized and this knowledge is easy to gain. Using FIPSO to determine the consequent gains by defining a goal-based cost function is also efficient for this purpose. This control scheme can be assumed to be an adaptive control method as it evolves through time and the changes in parameters will be observed by the controller. Applying the controller to an under actuated system has shown a good performance in comparison with other controlling methods applied to the same system. The incremental fuzzy controller design can be used as a good control solution for many problems. Other optimization techniques like simulated annealing can be applied here to find the parameters. TS fuzzy system can be also used for the control purpose.

References

1. Moore, M.L., Muscchio, J.T., Passino, K.M.: Genetic Adaptive Control for an Inverted Wedge: Experiments and Comparative Analyses. Engineering Applications of Artificial Inteligence (2001)
2. Hoffman, F., Schauten, D., Holeman, S.: Incremental Evolutionary Design of TSK Fuzzy Controllers. IEEE transactions on Fuzzy Systems 15(4) (2007)
3. Porter, L.L., Passino, K.M.: Genetic Adaptive and Supervisory Control. Journal of Intelligent Control and Systems 2(1) (1998)
4. Hoffman, F., Nelles, O.: Genetic Programming for model selection of TSK Fuzzy Systems. J. Inf. Sci. 136, 7–28 (2001)
5. Angelov, P., Buswell, R.: Identification of Evolving Fuzzy Rule-based Models. IEEE Transaction on Fuzzy Systems 10(5), 667–677 (2002)
6. Lam, H.K., Leung, F.H., Tam, P.K.: Design and Stability Analysis of Fuzzy Model-based Nonlinear Controllers for Nonlinear Systems Using Genetic Algorithm. IEEE Trans. Syst., Man Cybern, Part B 33(2), 250–257 (2003)
7. Roubos, J., Babuska, R., Bruijn, P., Verbruggen, V.: Predictive Control by Local Linearization of Takagi-Sugeno Fuzzy Model. In: Proc. FUZZ-IEEE 1998, pp. 37–42 (1998)
8. Shafiabady, N., Nekoui, M.A.: Optimal Evolutionary-Based Deployment of Mobile Sensor Networks. In: Deng, H., Wang, L., Wang, F.L., Lei, J. (eds.) AICI 2009. LNCS, vol. 5855, pp. 563–567. Springer, Heidelberg (2009)

9. Shir, O.M., Back, T.: Dynamic Niching in Evolution Strategies with Covariance Matrix Adaptation. In: IEEE CES (2005)
10. Wang, L., Maciejewsi, A.A., Siegel, H.J., Roychowdhury, V.P., Eldridge, B.D.: A Study of Five Parallel Approaches to a Genetic Algorithm for the Traveling Salesman Problem. Intelligent Automation and Soft Computing 11(4), 217–234 (2005)
11. Engelbrecht, A.P.: Computational Intelligence: an introduction, 2nd ed., John Wiley Press (2007)
12. Shafiabady, N., Teshnehlab, M., Nekoui, M.A.: Some Applications of S.T (Shafiabady-Teshnehlab) Evolutionary Optimization Algorithm. International Journal of Innovative Computing, Information and Control 8(2) (February 2012)
13. Ho, W.H., Chou, J.H.: Design of optimal controllers for Takagi-Sugeno fuzzy-model-based systems. IEEE Transactions on Systems, Man and Cybernetics A 37, 329–339 (2007)
14. Shim, H., Jo, N.H.: Determination of stability with respect to positive orthant for a class of positive nonlinear systems. IEEE Trans. on Automatic Control 53(5), 1329–1334 (2008)
15. Fornasini, E., Valcher, M.E.: Linear Copositive Lyapunov Functions for Continuous-Time Positive Switched Systems. IEEE Trans. on Automatic Control 55(8), 1933–1937 (2010)
16. Benzaouia, A., Hmamed, A., El Hajjaji, A.: Stabilization of controlled positive discrete-time T-S fuzzy systems by state feedback control. International Journal of Adaptive Control & Signal Processing (2010), `http://onlinelibrary.wiley.com/`
17. Cococcioni, M., Lazzerini, B., Marcelloni, F.: On reducing computational overhead in multi-objective genetic Takagi-Sugeno fuzzy systems". Applied Soft Computing 11(1), 675–688 (2011)

Ground Grid Integrity Testing Using Matlab Fuzzy Logic Toolbox

Bryan M. Dimayuga[1], Kevin Martin E. Jaron[1], Alexander T. Montero[1], Mark Kenneth Z. Peros[1], and Rionel Belen Caldo[2,*]

[1] Electrical Engineering Department,
Lyceum of the Philippines University-Laguna (LPU-L), Laguna, Philippines
[2] Electronics and Computer Engineering Department,
Lyceum of the Philippines University-Laguna (LPU-L), Laguna, Philippines
rionel14_caldo@yahoo.com

Abstract. In this study, the proponents make use of the concepts of fuzzy logic in simulating the ground grid integrity test of LPU-L SHL bldg. as Highly Acceptable (HA), Considerably Acceptable (CA), Just Acceptable (JA), Poor (P) and Critical (C). The input parameters include grounding conductor (conductivity), earth resistivity (Ω-°C) and grounding electrode (Ω -mm). The parameters in the model are acquired through the use of standards prescribed by the Philippine Electrical Code (PEC) and the Institute of Electrical and Electronics Engineering (IEEE). This study aims to provide a mathematical model to assess the integrity of the grounding system of the SHL bldg., design a fuzzy-based system, simulate and verify the effectiveness of the results. The proponents preferred to use the triangular membership functions and Sugeno-style of fuzzy inference systems.

Keywords: fuzzy logic, ground grid integrity, Sugeno-style, Matlab Fuzzy Logic Toolbox.

1 Introduction

The effectiveness of Electrical system design and wiring of an institutional building depends on the reliability of its components. This includes proper functions of protective devices (such as fuses, circuit breakers and contactors for motors), proper insulation of conductors and also its grounding system. There are drawbacks and issues in electrical safety. This includes but not limited to electrical ground faults, short circuit currents, lightning and other transients often do occur in an institutional building. In this regard, issues of electrical safety when servicing electrical equipment has acquired growing importance. By establishing the new principles and methods of protection, taking into account advances in science and practice of electrical safety are only some of the ways to improve electrical safety conditions (Switzer, 1999).

A properly designed, installed and maintained grounding system is very important for a safe and effective electrical system in an institution. The most important reason

* Corresponding author.

M. Gen et al. (eds.), *Industrial Engineering, Management Science and Applications 2015*,
Lecture Notes in Electrical Engineering 349, DOI: 10.1007/978-3-662-47200-2_80

for effective grounding is to protect people. Second, is to include protection of structures and equipment for unintentional contact with energized lines. This also ensures the maximum safety for electrical system faults (Switzer, 1999).

It is important to keep in mind that the requirements contained in the Institute of Electronics and Electrical Engineers (IEEE) codes or any codes that can be used as a standard for electrical system design constitute minimum electrical installation requirements. These minimum requirements cannot ensure that the equipment will perform satisfactorily. For this reason, electrical practioners often require additional grounding components. One of these consists of a copper conductor that is directly connected to earth and installed in the perimeter of the building. The steel building columns and some non – current carrying metallic frame of electrical equipment or some electrical part of the system are connected to this copper conductor to complete the grounding system (Surbrook and Althouse, 2008).

There are many factors in determining the overall integrity of the grounding system. The voltage drop, resistance and the continuity and the earth resistance can significantly impact the overall resistance of the grounding system. The moisture content, mineral content, soil type, soil contaminants and any other related factors determine the overall resistivity of the earth. To properly design a grounding system, the earth resistivity must be measured and also must be in a good condition to establish a low resistive grounding (Switzer, 1999).

The testing and evaluation of the integrity of the grounding system to determine its actual condition is the first step in the process to correct problems. The study is focused on creation of a new approach towards establishing condition monitoring for grounding integrity. Considerable benefits such as time and labor reduction for the grounding devices investigation with increase of accuracy of failures location can be achieved by using this proposed technique.

In this study, the proponents will use the concepts and principles of fuzzy logic in simulation of the ground grid integrity test. The factors and parameters to be considered for classifying the integrity of the grounding system include the grounding conductor (conductivity), earth resistance (Ω-^{o}C) and the grounding electrode (Ω -mm).

The parameters will be categorized as Very Good (VG), Good (G), Satisfactory (S), Poor (P) and Critical (C). The proponents will use triangular membership functions for its input and output parameters and it would employ the Sugeno style of fuzzy inference system. The proponents would verify the results using Matlab Fuzzy Logic Toolbox and it will be compared to derived formulas in Excel. This study will be simulated purely mathematical.

2 Fuzzy Rule Based System

There are fuzzy rules constructed to assess the integrity of the grounding system such as Highly Acceptable (HA), Considerably Acceptable (CA), Just Acceptable (JA), Poor (P) and Critical (C).

A hierarchical structure was constructed for the simulation of the grounding system, *Refer to figure 1.* The second level characterizes the grounding conductor integrity, earth resistance and the grounding electrode to obtain an acceptable grounding system for monitoring and surveillance purposes. The last hierarchical level characterizes the integrity of the grounding system. The following are the sample rules stored at three different hierarchical levels of structure:

If the Voltage Drop is<good> and the Resistance is<good> the Watts – Loss is<satisfactory> and the Continuity is<very good>
Then the Grounding Conductor is<good>

If the Soil Resistance is<poor>, the Soil temperature is<very good> and the moisture content is<very good>
Then the Earth Resistance is<satisfactory>

If the Depth is<poor>, the Electrode Resistance is<critical> and the spacing is<very good>
Then the Grounding Electrode is<poor>

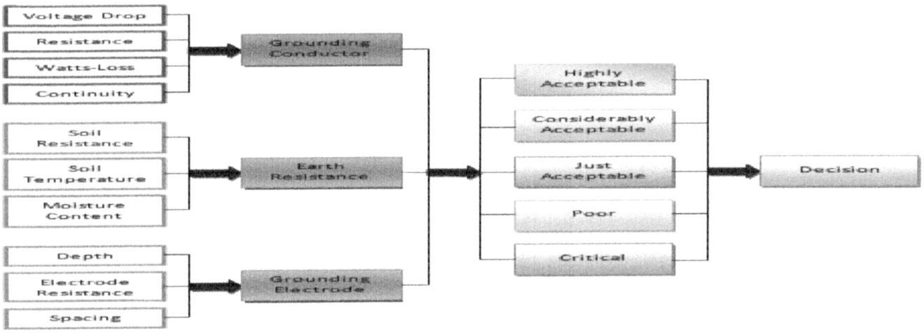

Fig. 1. Hierarchical Structure for the simulation of the integrity of the grounding system

The proponents constructed a precedence graph that can be used in analyzing the sequence of the tasks performed by the operators in line with the simulation of the integrity of the grounding system. Figure 2 shows a precedence graph with three major tasks and ten minor tasks to assess the integrity of the grounding system. Table 1 represents the standards for the simulation of the assessment for the integrity of the grounding system.

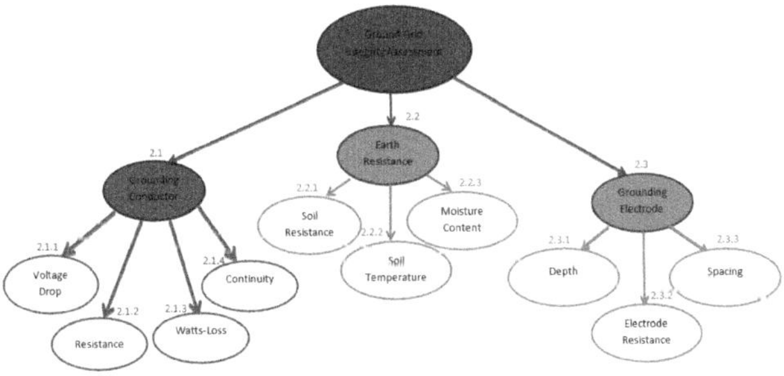

Fig. 2. Precedence for the simulation of the assessment for the integrity of the grounding system

Table 1. Standards used for the simulation of the assessment for the integrity of the grounding system

GROUNDING CONDUCTOR (CONDUCTIVITY)					
ADJECTIVAL RATING	VERY GOOD	GOOD	SATISFACTORY	POOR	CRITICAL
Voltage Drop (0.03 Volts / Feet)	0 to 0.01	0.01 to 0.02	0.02 to 0.04	0.04 to 0.06	0.06 to 1.0
Resistance (0.955mΩ - 2.626mΩ per Feet for Copper conductors)	0 to 0.955mΩ	0.955 mΩ to 1.3728 mΩ	1.3728 mΩ to 1.7905 mΩ	2.208 mΩ to 2.626 mΩ	>2.626 mΩ
Watts – Loss (342.726mW - 942.408mW per Feet)	0 to 324.726	342.726 to 492.6465	492.6465 to 642.567	642.567 to 792.4875	>792.4875
Continuity	Short	----	-----	-----	Open

EARTH RESISTANCE (OHM – °C)					
ADJECTIVAL RATING	VERY GOOD	GOOD	SATISFACTORY	POOR	CRITICAL
Soil resistance (Ω)	0 to 1.25	1.25 to 2.5	2.5 to 3.75	3.75 to 5	>5
Soil Temperature	HIGH TEMP.	MODERATE TEMP.	LOW TEMP.	VERY LOW TEMP.	BELOW FREEZING POINT
Moisture Content	80% - 100%	60% - 80%	40% - 60%	20% - 40%	0% - 20%
GROUNDING ELECTRODE(Rod and Pipe Electrodes) (OHM - MM)					
ADJECTIVAL RATING	VERY GOOD	GOOD	SATISFACTORY	POOR	CRITICAL
Depth (mm, Rod and Pipe Electrodes)	>750	562.5 to 750	375 to 562.5	187.5 to 375	0 to 187.5
Electrode Resistance	0 to 6.25	6.25 to 12.5	12.5 to 18.75	18.75 to 25	>25
Spacing (mm)	>1800	1350 to 1800	900 to 1350	450 to 900	0 to 450

3 Matlab Fuzzy Logic Toolbox

The linguistic variables are commonly used instead of numerical variables in fuzzy logic system. Fuzzification is the process of converting a numerical variable (real number or crisp variables) into a linguistic variable (fuzzy number or fuzzy variable). The perception, experience and the general knowledge of the system behavior serve as the derivation that will act as the control rules that relate the fuzzy output to the fuzzy inputs. In this study, the proponents make use of the averaging technique in deriving its membership functions. The rule table for the designed fuzzy logic system for ground grid integrity assessment is given in Table 2.

Table 2. Fuzzy Associative Memory (FAM) Matrix for Ground Grid Integrity Assessment

Count	Weight	Grounding Conductor	Earth resistivity	Grounding Electrode	Ground Grid Integrity (Classified Value)	Ground Grid integrity (Linguistic Class)
0	w1	5	5	5	5.00	HA
1	w2	5	5	4	4.68	CA
2	w3	5	5	3	4.36	CA
3	w4	5	5	2	4.05	CA
4	w5	5	5	1	3.73	JA
5	w6	5	4	5	4.77	CA
6	w7	5	4	4	4.45	CA
7	w8	5	4	3	4.14	CA
8	w9	5	4	2	3.82	JA
9	w10	5	4	1	3.50	JA
10	w11	5	3	5	4.55	CA

From the combination of the input parameters such as grounding conductor, earth resistance and grounding electrode, 125 fuzzy rule bases were able to formulate. The triangular figures of the associated function of this arrangement presume that for any particular input there is only one dominant fuzzy subset. The linguistic variables are converted into a numerical variable.

Creating, editing and observing the fuzzy inference system makes use of five primary Graphical User Interfaces (GUIs). It comprise of Fuzzy Inference System (FIS) Editor, Membership Function Editor, Rule Editor, Rule Viewer and Surface Viewer. If changes were made to the FIS of one of the toolbox, the effect can be seen in other GUIs since it is dynamically connected with each other. In addition to these five primary GUIs, the toolbox includes the graphical ANFIS Editor GUI, which is used for building and analyzing Sugeno-types adaptive neural fuzzy inference systems (Caldo, 2013).

The method used in this study for Matlab Fuzzy Logic Toolbox simulation is the Sugeno or Takagi-Sugeno-Kang of fuzzy inference that was introduced in 1985 and it is similar to Mamdani method in many respects. The first two parts of the fuzzy inference process, fuzzifying the inputs and applying the fuzzy operator are exactly the same. Sugeno output membership functions are either linear or constant unlike the Mamdami (Caldo, 2013). In this paper, the proponents think about the use of constants as output membership functions

3.1 Derivation of Input to the Fuzzy Sets

In this paper, the proponents derived inputs for the fuzzy sets Grounding Conductor, Earth Resistance and Grounding Electrode. The grounding conductor has a factor of 45.45%. They have used 20%, 30%, 20% and 30% factors for Voltage drop, Resistance, Watts – loss and Continuity respectively. The Resistance and Continuity have higher percent contribution considering that they are critical parameters of the grounding system. The Earth Resistance has a factor of 22.73%. They have used 40%, 40% and 20% factors for Soil Resistance, Soil Temperature and Moisture Content respectively. Finally, the Grounding Electrode has a factor of 31.82%. They have used 28.57%, 42.86% and 28.57% factor for Depth, Electrode Resistance and Spacing. Each factor being derived was based from the importance of each parameter in a grounding system as noted by PEC and IEEE standards.

4 Data and Results

In presenting the data and results of the study, the proponents have used the rule editor (Figure 3) and rule viewer (Figure 4) for ground grid integrity testing using Matlab fuzzy logic toolbox. It is where the FAM matrix of 125 rules is plugged in. The proponents conducted 10 tests to determine the reliability of the fuzzy system for each linguistic classification. Table 3 shows the simulation results, which classifies the integrity of the grounding system as Highy Acceptable, Considerably Acceptable, Just Acceptable, Poor or Critical. Based from the results obtained, it could be analyzed that there is a perfect correlation between fuzzy system for, PEC and IEEE standards for ground grid integrity test as shown in Table 4.

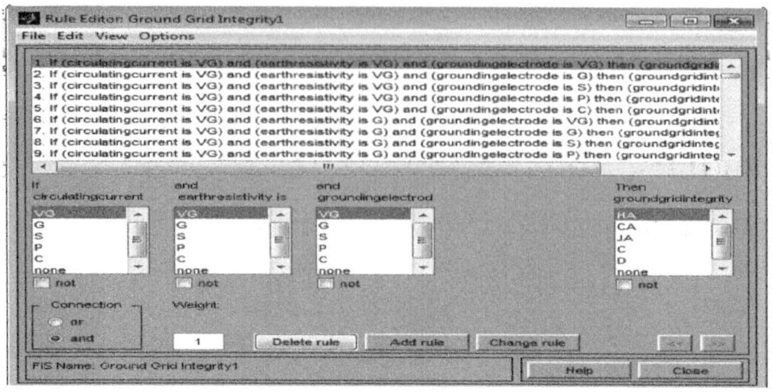

Fig. 3. Rule Editor for Ground Grid Integrity Assessment

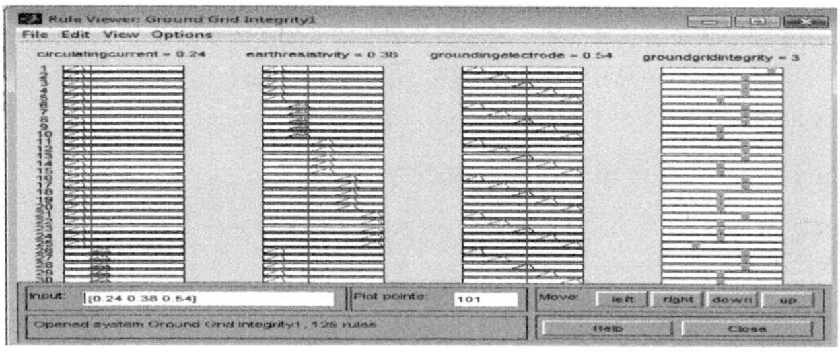

Fig. 4. Rule Viewer for Ground Grid Integrity Assessment

Table 3. Testing Results using MatLab Fuzzy Logic Toolbox

Trials	Board Predictive Assessment Input Parameters		Input Values (Normalized)	Crisp Output (Matlab Fuzzy Logic Toolbox)	Linguistic Classification
1	Ground Grid Integrity	Grounding Conductor	0.24	3	Just Acceptable
		Earth Resistivity	0.38		
		Grounding Electrode	0.54		
2	Ground Grid Integrity	Grounding Conductor	0.36	4	Considerably Acceptable
		Earth Resistivity	0.54		
		Grounding Electrode	0.19		
3	Ground Grid Integrity	Grounding Conductor	0.11	3	Just Acceptable
		Earth Resistivity	0.78		
		Grounding Electrode	0.88		
4	Ground Grid Integrity	Grounding Conductor	0.57	2	Poor
		Earth Resistivity	0.66		
		Grounding Electrode	0.92		
5	Ground Grid Integrity	Grounding Conductor	0.16	4	Considerably Acceptable
		Earth Resistivity	0.19		
		Grounding Electrode	0.35		
6	Ground Grid Integrity	Grounding Conductor	0.87	1	Critical
		Earth Resistivity	0.72		
		Grounding Electrode	0.69		
7	Ground Grid Integrity	Grounding Conductor	0.52	3	Just Acceptable
		Earth Resistivity	0.08		
		Grounding Electrode	0.75		
8	Ground Grid Integrity	Grounding Conductor	0.17	5	Highly Acceptable
		Earth Resistivity	0.19		
		Grounding Electrode	0.19		
9	Ground Grid Integrity	Grounding Conductor	0.38	3	Just Acceptable
		Earth Resistivity	0.78		
		Grounding Electrode	0.27		
10	Ground Grid Integrity	Grounding Conductor	0.32	2	Poor
		Earth Resistivity	0.76		
		Grounding Electrode	0.67		

Table 4. Verification of Fuzzy based results with PEC/IEEE Standards

Trials	Parameters		Actual Input Values	PEC standards	IEEE standards	PEC/IEEE standards	Linguistic classification (fuzzy system for ground)		Linguistic classification (fuzzy system for ground)		%	
1	Grounding Conductor	Voltage Drop	0.011 V/ft	N/A	0.03 V/ft	4 - Good	VG	Good	HA	CA	20	45.45
		Resistance	1.22mΩ/ft	955.34 μΩ/ft - 2.626mΩ/ft	N/A	4 - Good					30	
		Watts-Loss	353mW/ft	342mW/ft - 942.08 mW/ft	N/A	4 - Good					20	
		Continuity	short	short	short	5 - Very Good					30	
	Earth Resistance	Soil Resistance	3.88Ω	N/A	5Ω	3 - Satisfactory	S		JA		40	22.73
		Soil Temperature	M	N/A	N/A	4 - Good					40	
		Moisture Content	38%	N/A	N/A	2 - Poor					20	
	Grounding Electrode	Depth	369mm	750 mm	N/A	3 - Satisfactory	G		CA		28.57	31.82
		Electrode Resistance	7.42Ω	25Ω	N/A	4 - Good					42.86	
		Spacing	1850mm	1800 mm	N/A	5 - Very Good					28.57	
2	Grounding Conductor	Voltage Drop	0.053 V/ft	N/A	0.03 V/ft	2 - Poor	VG	Good	HA	CA	20	45.45
		Resistance	0.983mΩ/ft	955.34 μΩ/ft	N/A	4 - Good					30	
		Watts-Loss	236mW/ft	942.08 mW/ft	N/A	5 - Very Good					20	
		Continuity	short	short	short	5 - Very Good					30	
	Earth Resistance	Soil Resistance	0.9Ω	N/A	5Ω	5 - Very Good	G		CA		40	22.73
		Soil Temperature	M	N/A	N/A	4 - Good					40	
		Moisture Content	8%	N/A	N/A	1 - Critical					20	
	Grounding Electrode	Depth	203mm	750 mm	N/A	2 - Poor	S		JA		28.57	31.82
		Electrode Resistance	18.6Ω	25Ω	N/A	3 - Satisfactory					42.86	
		Spacing	1389mm	1800 mm	N/A	4 - Good					28.57	

5 Conclusions

This paper introduces a tangible tool, which could be possibly used in determining the effectiveness of the integrity of the grounding system. The proposed method was implemented systematically using Matlab Fuzzy Logic Toolbox and it showed that fuzzy-based system for ground grid integrity test is simple, available, reliable and effective. The constructed fuzzy logic algorithm was verified experimentally through successful tests.

The proponents were able to establish a distinctive approach towards unsophisticated way of assessing the integrity of the grounding system, which would apply to low voltage dc application. It presents a cheaper and quicker method, which will also apply most likely to the improvement, development and maintenance of an effective and reliable grounding system for institutional buildings.

Acknowledgement. The authors acknowledge Engr. Jose Lumbera, OIC of High Voltage Equipment Testing and Diagnostics Department of National Grid Corporation of the Philippines for giving them an ample time to share his knowledge and expertise regarding the ground grid integrity testing. Also the authors would like to thank LPU-L for providing all the financial needs in pursuing this study.

References

Theraja, B.L., Theraja, A.K.: ABC of Electrical Engineering. S. Chand and Company LTD (2012)

Robbins, A.H., Miller, W.C.: Circuit Analysis: Theory and Practice, 3rd edn. Thomson/Delmar Learning (2003)

Philippine Electrical Code, Institute of Integrated Electrical Engineers of the Philippines (2009)

Caldo, R.B.: Fuzzy Logic Derivation and Simulation of a Three-Variable Solar Water Heater Using Matlab Fuzzy Logic Toolbox. In: Proceedings of the November 2013 IEEE International Conference
Keith Switzer, W.: Practical Guide to Electrical Grounding (1999)
Surbrook, T.C., Althouse, J.R.: Interpreting the National Electrical Code 8th Edition (2008)
81 – 2012 – IEEE Guide for Measuring Earth Resistivity, Ground Impedance, and Earth Surface Potentials of a Grounding System

Filtering as a Tool of Diversity in Ensemble of Classifiers

Eva Volna, Martin Kotyrba, and Vaclav Kocian

Department of Informatics and Computers, University of Ostrava, Ostrava, Czech Republic
{eva.volna,martin.kotyrba,vaclav.kocian}@osu.cz

Abstract. This paper discusses possibilities of using ensembles of neural-networks-based classifiers in pattern recognition and classification. Attention is paid to systems that minimize demands on data preprocessing. Minimizing of requirements for preprocessing leads automatically to systems that are able to sufficiently classify the submitted data into predefined classes without knowledge of details of their significance. In our experiment, we try to increase diversity of classifiers by various filtering methods. The methods proposed in this paper come out from a technique called boosting, which is based on the principle of combining a large number of so-called weak classifiers into a strong classifier. All proposed improvements are experimentally verified.

Keywords: filtering, ensemble of classifiers, diversity, neural networks.

1 Introduction

The goal of ensemble learning methods is to construct a collection (an ensemble) of individual classifiers that are diverse and yet accurate. If this can be achieved, then highly accurate classification decisions can be obtained by voting the decisions of the individual classifiers in the ensemble. The two most popular methods for creating ensembles are boosting [8] and bagging [1]. Boosting is reported to give better results than bagging. Both of them modify set of training examples to achieve diversity of weak learners in ensemble. As alternative methods we can mention randomization based on a random modification of base decision algorithm [3] or Bayesian model averaging which can even outperform boosting in some cases [2]. There have been examined several methods to improve performance of boosting [5].

We have provided an experimental study which was aimed to explore the possibility of using neural network as the base algorithms for so called weak classifiers. We were interested in the possibility of using neural networks in place of commonly used decision trees. Our analysis is based on the idea that it is more efficient to create a number of imperfectly adapted networks, which are small in their topology than one perfectly adapted a sophisticated network. In the experimental work, each neural network was adapted by only one pass through the training set (a sloppy neural networks adaptation). We have proposed this method of learning on the basis of the property of neural networks which have been noticed during another work [6].

M. Gen et al. (eds.), *Industrial Engineering, Management Science and Applications 2015*,
Lecture Notes in Electrical Engineering 349, DOI: 10.1007/978-3-662-47200-2_81

2 Diversity of Classifiers

The success of an ensemble system - that is, its ability to correct the errors of some of its members - rests squarely on the diversity of the classifiers that make up the ensemble. If all classifiers provided the same output, correcting a possible mistake would not be possible. Therefore, individual classifiers in an ensemble system need to make different errors on different instances. If each classifier makes different errors, then a strategic combination of these classifiers can reduce the total error. Specifically, an ensemble system needs classifiers whose decision boundaries are adequately different from those of others. Such a set of classifiers is said to be diverse.

Classifier diversity can be achieved in several ways.

- To use different training datasets to train individual classifiers, where training data subsets are drawn randomly, usually with replacement, from the entire training data. To ensure that individual boundaries are adequately different, despite using substantially similar training data, weaker or more unstable classifiers are used as base models, since they can generate sufficiently different decision boundaries even for small perturbations in their training parameters.
- To use different training parameters for different classifiers. For example, a series of multilayer perceptron (MLP) neural networks can be trained by using different weight initializations, number of layers / units, error goals, etc. Adjusting such parameters allows one to control the instability of the individual classifiers, and hence contribute to their diversity.
- To use entirely different type of classifiers, such MLPs, decision trees, nearest neighbour classifiers, and support vector machines can also be combined for added diversity.
- To use different features, or different subsets of existing features. In fact, generating different classifiers using random feature subsets is known as the random subspace method [4]. Just the approach we have used in our work.

Each individual classifier in an ensemble system allows to generate different decision boundaries. If proper diversity is achieved, a different error is made by each classifier, strategic combination of which can then reduce the total error. Figure 1 graphically illustrates this concept, where each classifier - trained on a different subset of the available training data - makes different errors (shown as instances with dark borders), but the combination of the three classifiers provides the best decision boundary.

3 Sloppy Neural Networks Adaptation

At first, we have proposed a method of learning in which the neural networks were not fully, nor well adapted. We have proposed this method of learning on the basis of a property of neural networks, which we have noticed during experiments with time series data [10]. A very major part of neural networks' adaptation is performed during the first pass. We have used neural networks as generators of weak classifiers only, i.e. such classifiers which are slightly better than a random variable with uniform distribution. For weak classifiers, their diversity is more important than their accuracy. Therefore, it has seemed appropriate to use a greedy way in order to propose classifiers. This approach uses only the power of the neural network adaptation rule in the early stages of its work and thus time is not lost due to a full adaptation of the classifier.

We performed a simple experiment that has shown a typical evolution of a neural network error during adaptation. During the experimental study, we used two different types of networks in three configurations, see Table 1. We tested 1000 instances of each configuration. To improve the conclusiveness of the test, we used techniques to increase the diversity of classifiers, which are described below.

In Table 1 we have used the following nomenclature:

- x – output of neuron in lower layer,
- t – required (expected) output value,
- y_{in} – input of neuron ,
- y_{out} – output of neuron ,
- α – learning parameter – by this parameter an adaptation speed can be adjusted,
- φ – formula for calculating a neuron output value (activation function) $y_{out} = \varphi(y_{in})$,
- Δw – formula for calculating a change of a weight value.

Table 1. Neural networks used in the sloppy adaptation test

	Hebb	**BP_5**	**BP_20**
Values	Bipolar	Bipolar	Bipolar
Hidden	0	5	20
φ	Identity	$\frac{2}{1+\exp(-y_{in})}-1$	
Δw	$x \cdot t$	$\alpha x(t-y_0)\frac{1}{2}(1+y_{out})(1-y_{out})$	

The acquired data is plotted in graphs, see Figures 1-2. They provide values from the first 50 cycles of each neural network adaptation only. We have monitored minimal, maximal and average error in each cycle. The experimental study demonstrated that neural networks could be suitably used in the way which we proposed. From Figures 1-2, we can see that the main deal of the adaptation work is done during the first epoch of learn. This means that the neural network is adapted through one cycle only.

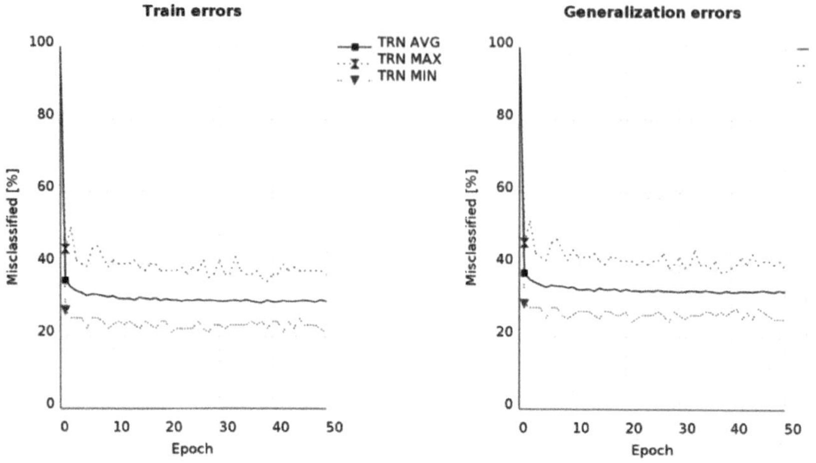

Fig. 1. Train and generalization error development in the first 50 learning cycles. Statistics from 1000 of Hebbian networks. The first iteration is marked.

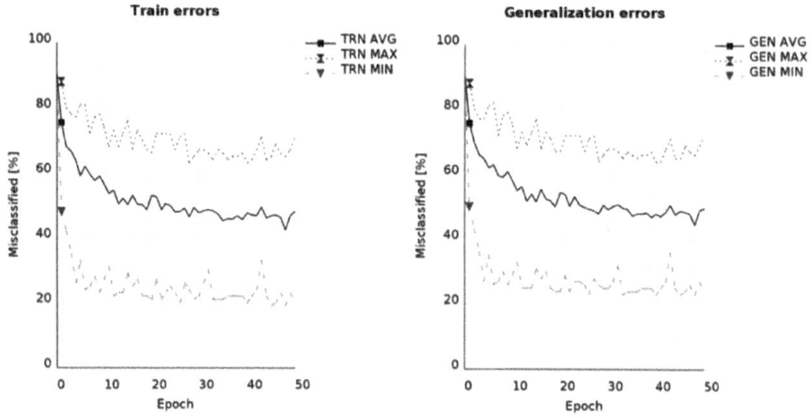

Fig. 2. Train and generalization error development in the first 50 learning cycles. Statistics from 1000 of Back propagation networks with 5 neurons in the hidden layer. The first iteration is marked.

Graph analysis Figure 1 and Figure 2 is the following:

- Almost each tested classifier has its error significantly smaller than a random classifier after the first iteration. Exceptions were related to BP classifier only.
- In the next iterations, a remarkable improvement in performance of a classifier occurs for BP classifier only.
- Some instances of BP classifier achieve success of around 80% in the first iteration.

- 'Constricted' form of Hebb graph implies that the diversity of the linear classifier is less than at multilayer BP networks. In fact, the diversity of Hebb classifier in the basic configuration equals zero because the adaptation rule has no random parameters. In the graph in Figure 2, we can see diversity obtained exclusively by filtering which is described further in the text.

In the first cycle, the 'exploitation' of classifier is unambiguously the highest. Back propagation network with 20 hidden neurons has achieved about 75% successfulness in the first cycle, about 82% in another 49 cycles. This manner allowed to generate more classifiers at the same time.

Sloppy adaptation works with all neural network based classifiers. The sloppy adaptation makes it possible to utilize the sophisticated Back propagation neural networks in the ensembles of classifiers. The method does not affect the classifier's algorithm. It only exploits an observed property of the Back propagation neural network.

4 Test of Performance and Diversity

The experiment has been conducted over the MNIST database of handwritten digits [7], which includes a training set of 60,000 examples, and a test set of 10,000 examples. MNIST database is commonly used for training various image processing systems. MNIST is also widely used for training and testing in the field of machine learning and pattern recognition methods on real-world data while spending minimal efforts on preprocessing and formatting.

We have used a total of five types of neural networks in the study as well as AdaBoost [5] that is a method for improving the accuracy of a learning algorithm (a.k.a. base learner). This is achieved by iteratively calling the base learner on re-sampled training data, and by combining the so-produced hypothesis functions together to form an ensemble classifiers. Each ensemble always used a specific base of algorithms. A base of algorithms is a set of neural networks whose instances form the ensemble.

The classifier is an alias for an instance of a neural network. We used neural networks only, no other types of classifiers. In total, we used five different types of neural networks, four single-layer and one multi-layer. Each classifier was created separately and adapted by only one pass through the training set. We have used codes N1-N4 for single-layer networks, and N5 for a two-layer network. The proposed ensembles of neural-networks-based classifiers are basically a set of m classifiers. All the *m* classifiers work with the same set of *n* inputs. Each of the m classifiers tries to learn to recognize objects of one class in the input patterns of size *n*.

All the neural networks used the *winner-takes-all strategy* for output neurons $(Y_1,\ldots,Y_m)$ when they worked in the active mode [9]. So only one output neuron with the highest y_{out} value could be active. The Y_i is considered the winner if and only if $\forall j: y_j < y_i \vee (y_j = y_i \wedge i < j)$, i.e. the winner is the neuron with the highest output value y_i. In the case that more neurons have the same output value, the winner is considered the first one in the order. All these neural networks used the algorithm for an elimination of irrelevant inputs as proposed in [6].

For purpose of our experiment we have defined an ensemble as a group of 100 classifiers generated over the same set of algorithms with the same configuration of the generator. A total of six sets of ensembles were created. Each set of ensembles has been generated over another subset of available algorithms. One set of ensembles has been made over all available algorithms, the other five sets always used only one of the algorithms. We have used six bases of algorithms in total: N1 represents Adaline, N2 represents delta rule, N3 represents Hebbian network, N4 represents perceptron, N5 represents Back Propagation network and the sixth base N1-N5 represents all ensembles contain 20 specific instances of a specific type. Adaline did not perform well with the basic learning rule $\alpha x(t - y_{in})$. We assume, that the cause lays in the relatively big number of patterns and inputs and therefore possibly the big value of $(t - y_{in})$. That is, why we have normalized value of $(t - y_{in})$ by the sigmoid function. Classifiers were generated as instances of N1-N5 algorithms during the experiment. The accuracy of each generated classifier was verified on both the test and the training set. Twelve different configurations have been tested within each set. Each configuration was tested 50 times in every set of ensembles. Therefore there has been created and tested $6 \cdot 12 \cdot 50 \cdot 100 = 360000$ different instances of neural networks, 'packaged' into 3600 ensembles during the experimental study. Figure 3 shows the implementation procedure of the experiment.

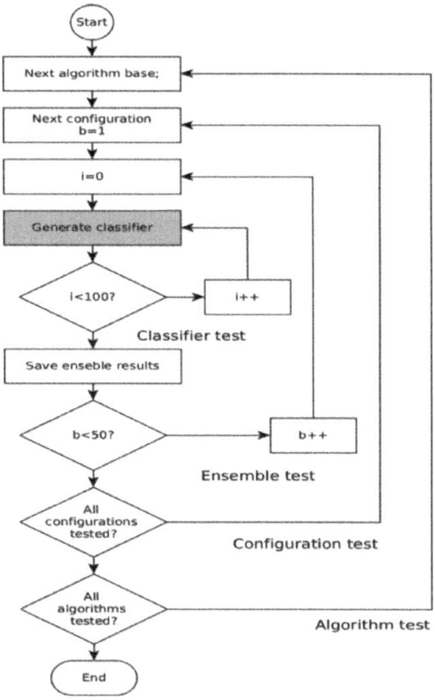

Fig. 3. Creating and testing the ensembles of the neural network classifiers

The experiment was conducted over the data from the MNIST database [7], as was mentioned above. Therefore the training contains 60000 patterns and the testing contains 10000 patterns. Patterns are stored in the database as a 28x28 pixel images with

256 grayscale. As the examples represent digits, it is obvious, that they can be divided into 10 classes. Therefore all the tested neural networks had a total of 10 output neurons. The highest number of input neurons was 28 x 28 = 784. Actual number depended on the current teaching strategy (see description of proposed filters below). The number of neurons in the hidden layer of the back propagation network was determined experimentally.

5 Filtering as a Method of Classifiers Diversity Enhancing

We have proposed several approaches to improve performance of boosting [5]. In our experiment, we try to increase diversity of classifiers by various filtering methods. The main idea of the input filter is, that classifier 'sees' only a part of the input. It forces the classifier to focus its attention only on certain features (parts of pattern). It should increase the diversity of individual classifiers generated. Input filter is represented by a bitmap ***F*** of the same size as the original pattern. The classifier 'sees' the input pattern 'through' matrix ***F*** while only bits of ***F***, which have value of *1* are 'transparent'. The blank filter is represented by a matrix whose pixels are all set to value of *1*. Topology of classifier always reflects the current filter in the sense that the number of input neurons is equal to the number of bits with value of *1* in the bitmap ***F***, i.e. the number of pixels that are visible to classifier. It implies that the topology of classifiers when using a non-blank filter is smaller and less demanding in terms of memory space and CPU time. However, it is also clear that the classification of filtered patterns is less accurate because the classifier cannot see the whole pattern, but only a part of it. Figure 4 shows the example of application of different filters to the pattern representing the number '8'.

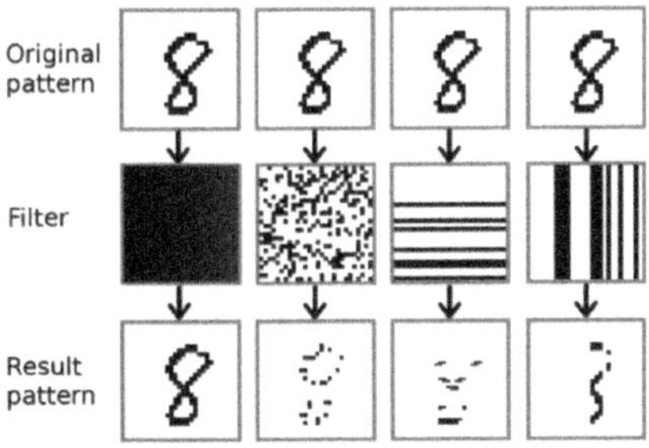

Fig. 4. Example of use of different filters on the input pattern with the image number '8' In the top row we can see four different filter matrices, in the bottom row there are the results of the filtration (what the classifier can see)

We have used the diversity of classifiers in the ensembles as the criterion for judging the success of algorithms and configurations. Moreover, we have focused mainly on performance at the test set of patterns. We have expressed the diversity as the reciprocal of the count of patterns, which were not correctly classified by any of the classifiers in the ensemble. The smaller number of unrecognized patterns means the more successful ensemble as well as the more successful configuration and base of algorithms. If the diversity on the testing set was remarkably worse than the diversity on the training set, we have experienced over-fitting.

In the subsequent text we will use the following nomenclature:

- TMin/TMax/TAvg – minimum/maximum/average percentage of unrecognized patterns in the training set
- GMin/GMax/GAvg – minimum/maximum/average percentage of unrecognized patterns in the test set
- In Figure 5 we can see that the best result was produced by the Back propagation network. Although its minimum reached error is slightly worse than the one of delta rule or the mixed algorithm, its average performance is the best. It also shares the best G/T value with the perceptron, which is 1,42. On the other side, the Hebb is the worst it gives the worst, performance on both the average error and the G/T value, which is 1,58. G/T means GAvg/ TAvg error ratio, which represents generalization capabilities of ensembles by used base algorithm. The higher value indicates the higher overfitting. So the smaller number means the higher quality of the classifier. The values of G/T >1 indicates the typical behaviour when an ensemble performs better on the train set then on the test one. A value of 1 would indicate that the ensemble performed same on both the train and the test set. The value of G/T <1 would indicate really 'intelligent' and unlike behaviour - the better performance on the test set than on the train one.

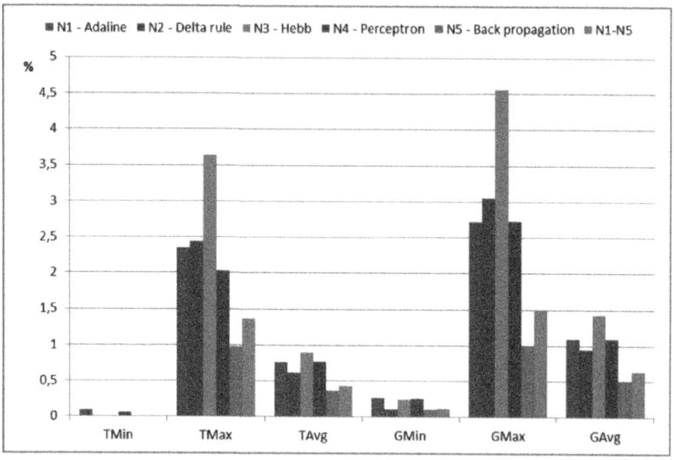

Fig. 5. Comparison of quality of ensembles by used base algorithm [%]

In Figure 6 we can investigate the influence of different filters on the ensembles behaviour. It is clear from the values in the table, that the filtering is the right way to go through. The filtering method put the ensemble's performance forward in both the average error and the generalization capabilities. We can also see that the streak filter performs significantly better than the random one. Its G/T value is 1,31. The effectiveness of a quality of *ensembles with and without filtering* was verified experimentally.

Each calculation was run 50 times, to achieve statistically significant results. There were recorded values of the error function E (1).¨

$$E = \sum_{l=1}^{q} E_l(\mathbf{w}) \tag{1}$$

$E_l(w)$ is the squared error for a particular training pattern l ($l = 1, \ldots, q$), where q is the number of patterns (2). t_k represents the target for pattern l on node k, whilst y_k represents the actual output at that node. Y represents a set of output units.

$$E_l(\mathbf{w}) = \frac{1}{2}\sum_{k \in Y}(y_k - t_k)^2 \tag{2}$$

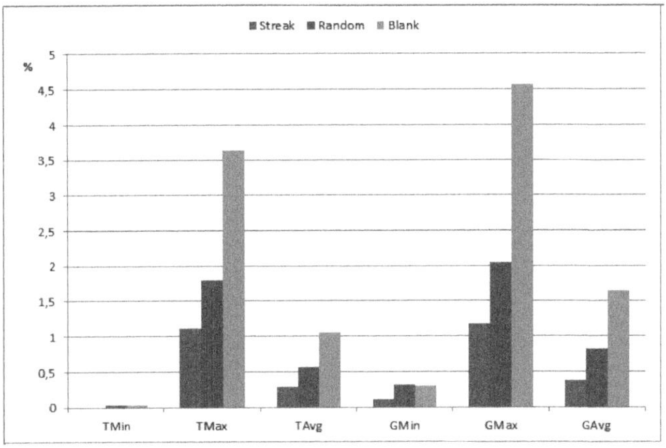

Fig. 6. Comparison of quality of ensembles by filtering. The filtering method put the ensembles' performance forward in both the average error and the generalization capabilities.

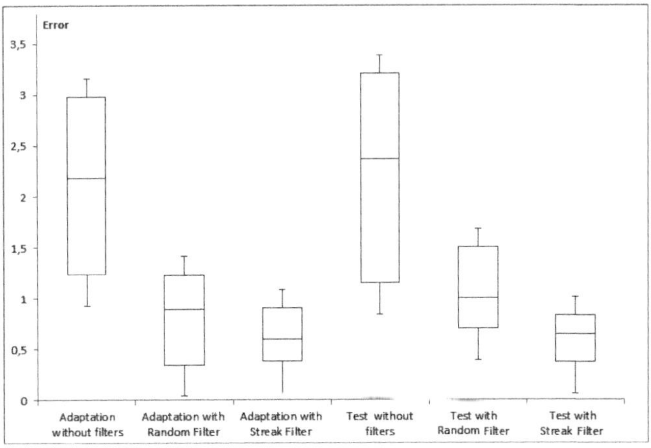

Fig. 7. Results representing processes of adaptation and testing belonging to ensembles with and without various filtering methods

6 Conclusion

We have verified the statistical significance of differences between the results with and without various filtering methods during both phases of adaptation and testing. The obtained values were compared by Two tailed T-test with a significance level of α=0.05. Differences between *quality of ensembles with and without filtering* (ensembles are represented by their error function *E*) is statistically significant, therefore we can reject the null hypothesis H_o ($\mu_{with\_filtering} = \mu_{without\_filtering}$). Results representing processes of adaptation and testing belonging to ensembles with and without filtering are shown graphically in Figure 7. The box-plots represent values of error function *E* of adaptation as well as generalization with and without various filtering methods, when every calculation was run 50 times. The median is indicated by the horizontal line that lies inside the box dividing it into two parts.

It was experimentally tested that Input filters increase the diversity of the neural networks. As the method reduces the size of input vector, it also increases the speed of the classifier in both the learning and the classifying. During experiments the method reduced the size of the neural networks by 0-80% depending on used input filter (Fig.4). It is obvious, that filters can reduce accuracy of the classifier, but this fact is not vital as the classifiers are used in the boosted ensembles.

Acknowledgments. The research described here has been financially supported by University of Ostrava grant SGS16/PRF/2014. Any opinions, findings and conclusions or recommendations expressed in this material are those of the authors and do not necessarily reflect the views of the sponsors.

References

1. Breiman, L.: Bagging predictors. In: Machine Learning, pp. 123-140 (1996)
2. Davidson, I., Fan, W.: When efficient model averaging out-performs boosting and bagging. In: Fürnkranz, J., Scheffer, T., Spiliopoulou, M. (eds.) PKDD 2006. LNCS (LNAI), vol. 4213, pp. 478–486. Springer, Heidelberg (2006)
3. Dietterich, T.G., Fisher, D.: An experimental comparison of three methods for constructing ensembles of decision trees. In: Bagging, boosting, and randomization. Machine Learning, pp. 139–157 (2000)
4. Ho, T.K.: Random subspace method for constructing decision forests. IEEE Transactions on Pattern Analysis and Machine Intelligence, 832–844 (1998)
5. Iwakura, T., Okamoto, S., Asakawa, K.: An adaboost using a weak-learner generating several weak hypotheses for large training data of natural language processing. IEEJ Transactions on Electronics, Information and Systems 130, 83–91 (2010)
6. Kocian, V., Volná, E.: Ensembles of neural-networks-based classifiers. In: Proceedings of the 18th International Conference on Soft Computing, Mendel 2012, Brno, pp. 256–261 (2012)

7. LeCun, Y., Cortes, C., Burges, C.: The MNIST Database (retrieved march 2014), http://yann.lecun.com/exdb/mnist/
8. Schapire, R.E.: A brief introduction to boosting. In: Proceedings of IJCAI 1999, pp. 1401–1406. Morgan Kaufmann Publishers Inc, San Francisco (1999)
9. Volna, E., Kocian, V., Kotyrba, M.: Boosting of Neural Networks over MNIST Data. In: Proceedings of NCTA 2014, pp. 256–263. Scitepress, Portugal (2014)
10. Volná, E., Janošek, M., Kocian, V., Kotyrba, M.: Smart time seris prediction. In: Proceedings of SOCO 2013, pp. 211–220. Springer, Heidelberg (2013)

Mining High Utility Patterns in Different Time Periods

Show-Jane Yen and Yue-Shi Lee

Department of Computer Science and Information Engineering,
Ming Chuan University, Taoyuan County, Taiwan
{sjyen,leeys}@mail.mcu.edu.tw

Abstract. Mining *frequent itemset* generates frequently purchased itemsets. However, a frequent itemset may not be the itemset with high value. *Mining high utility itemset* considers both of the profits and purchased quantities for the items, which is to find the itemsets with high utility for the business. However, mining high utility itemsets from the whole database may lose the high utility itemsets in a specific time period. Mining *high utility itemsets in different time periods* considers when the itemsets will be high utility, which can provide business manager to promote high utility itemsets in a specific time period.

In this paper, we propose an approach for mining high utility itemsets in different time periods. We first partition the database according to the user-specified basic time periods, and then combine the transactions in the continuous basic periods to find high utility itemsets in the continuous periods. Besides, we remain the itemsets which are not high utility in the previous periods, but may be high utility after combining the previous periods with the current basic period, such that the high utility itemsets in continuous basic periods will not be lost.

Keywords: Data mining, Knowledge discovery, High utility itemset, Time period.

1 Introduction

In this section, we first introduce some preliminaries for mining high utility itemsets [7]. Let $I=\{i_1, i_2, ..., i_m\}$ be the set of all the items. An itemset X is a subset of I and the *length* of X is the number of items contained in X. An itemset with length *k* is called a *k-itemset*. A transaction database $D=\{T_1, T_2, ..., T_n\}$ contains a set of *transactions* and each transaction has a unique transaction identifier (TID). Each transaction contains the items purchased in this transaction and their purchased quantities. The purchased quantity of item i_p in a transaction T_q is denoted as $o(i_p,T_q)$. The *utility of item* i_p *in* T_q is $u(i_p, T_q)=o(i_p, T_q)\times s(i_p)$, in which $s(i_p)$ is the profit of item i_p. The utility of an itemset X in T_q is the sum of the utilities of the items contained in $X \subseteq T_q$, which is shown in expression (1). If $X \not\subset T_q$, $u(X, T_q)=0$. The *utility of an itemset X* in D is the sum of the utilities of X in all the transactions containing X, which is shown in expression (2). An itemset X is a *high utility itemset* if the utility of

M. Gen et al. (eds.), *Industrial Engineering, Management Science and Applications 2015*,
Lecture Notes in Electrical Engineering 349, DOI: 10.1007/978-3-662-47200-2_82

X in D is no less than *minimum utility* (*MU*), which can be computed by multiplying the total utility of the database and the user-specified *utility threshold.*

$$u(X,T_q) = \sum_{i_p \in X \subseteq T_q} u(i_p, T_q) \quad (1)$$

$$u(X) = \sum_{X \subseteq T_q \in D} u(X, T_q) \quad (2)$$

Table 1. Transaction database

Item / TID	A	B	C	D	E	Item / TID	A	B	C	D	E
1	3	2	18	0	1	14	0	2	12	0	2
2	0	6	0	1	1	15	3	0	0	0	1
3	2	0	1	1	1	16	2	1	5	0	0
4	1	2	0	1	1	17	0	2	2	0	1
5	0	0	4	0	2	18	1	1	0	0	1
6	1	0	0	1	0	19	0	2	2	1	2
7	0	10	0	1	1	20	0	0	10	0	2
8	3	0	25	3	1	21	0	5	4	0	2
9	1	1	0	0	0	22	1	3	0	0	1
10	0	6	2	2	1	23	0	1	5	2	0
11	1	0	0	0	1	24	0	1	5	2	0
12	0	2	25	0	0	25	3	0	10	0	2
13	1	1	0	0	0						

Table 2. Profit Table

Item	A	B	C	D	E
Profit	3	10	1	6	5

For example, Table 1 is a transaction database, in which each number in a (TID, Item) entry represents the purchased quantity for an item in a transaction. Table 2 is the profit table which records the profit for each item in Table 1. Suppose the utility threshold is 60%. Because the total utility of the database is 141, the minimum utility MU is 141x60%=84.6. The utility of itemset {B,D} in Table 1 is u({B,D})= (1×10+1×6) + (6×10+2×6) + (3×10+1×6) =124 ≥ 84.6. Therefore, the itemset {B,D} is a high utility itemset.

$$TU(T_q) = \sum_{i_p \in T_q} u(i_p, T_q) \quad (3)$$

$$TWU(X) = \sum_{X \subseteq T_q \in D} tu(T_q) \quad (4)$$

For mining frequent itemset [1], all the subsets of a frequent itemset are frequent, that is, there is a downward closure property for frequent itemsets. However, the property is not available for high utility itemsets, since a subset of a high utility itemset may not be a high utility itemset.

Therefore, Liu et al. [7] proposed a *Two-Phase algorithm* for mining high utility itemsets. They defined *transaction utility* TU and *transaction weighted utility* TWU for an itemset X, which are shown in expressions (3) and (4), respectively.

If the TWU for an itemset is no less than MU, then the itemset is a *high transaction weighted utility itemset (HTWUI)*. According to expression (4), the TWU for an itemset X must be greater than or equal to the utility of X in D. Therefore, if X is a high utility itemset, then X is a HTWUI. All the subsets of a HTWUI are also HTWUIs. Therefore, there is a downward closure property for HTWUIs. The first phase for the Two-Phase algorithm [7] is to find all the HTWUIs which are called *candidate high utility itemsets* by applying Apriori algorithm[1]. Two-Phase algorithm scans the database again to compute the utilities of all the candidate high utility itemsets and find high utility itemsets in the second phase.

However, in the first phase, Two-Phase algorithm needs to repeatedly scan the database and search a large number of candidate HTWUIs to generate candidate high utility itemsets. If the minimum utility is small, a huge number of candidate high utility itemsets will be generated. In the second phase, Two-Phase algorithm needs to scan the large database again and search a huge number of candidate high utility itemsets, which would significantly degrade the mining performance. In order to reduce the number of database scans, Ahmed et al. proposed HUC-Prune algorithm [2], which applies FP-Growth algorithm [6] to generate candidate high utility itemsets and then scans the database and searches the candidates to find the high utility itemsets. Although HUC-Prune only scans database three times, it still needs to generate candidate high utility itemsets and scan the whole database to find the high utility itemsets.

For the above approaches, the high utility itemsets are discovered from the whole database. Therefore, we still cannot know when the itemsets should be promoted to gain the high profits. For example, because the itemset {outerwear, long socks} would be bought frequently in winter, but this itemset is rarely purchased in the other seasons. Therefore, an itmeset may not be high utility in the whole year, but it can be high utility in winter. If the whole transaction database can be partitioned by seasons, then the high utility itemsets can be discovered in each season.

Some researches considered the on shelf time periods of products. In 2007, Huang, et al. [3] proposed the algorithm *Twain* to find the frequent itemsets at the longest common continuous on shelf periods of time, since they think that there will be different on shelf time periods for different items. However, this approach will lose some frequent itemsets if the itemsets are not frequent in the first time period, since the supports of these itemsets cannot be accumulated to the next time periods. In 2011, Lan and et al. [5] think that the utilities of the itemsets should be computed at the on shelf time periods of the itemsets. They specified the on shelf time periods for each item, and proposed an algorithm to find the high utility itemset in which the items have common on shelf time periods. However, it is unreasonable to specify the fixed time periods and assign the on shelf time periods for each item according to the fixed time periods. Moreover, this approach still cannot identify which itemsets will be high utility in which time periods.

In this paper, we propose an algorithm for mining high utility itemsets in different time periods. We first partition the database according to the user-specified basic time periods, and find the high utility itemsets and the *potential high utility itemsets* from each partition of the database with the time period. The utility of a potential high utility itemset is near to the minimum utility but not satisfy the minimum utility, which is used to decrease the information loss in the continuous time periods. After discovering all the high utility itemsets and the potential high utility itemsets for each basic time period, these itemsets in the continuous time periods would be combined to find the high utility itemsets or the potential high utility itemsets in the continuous time periods.

2 Our Approach

In this section, we describe our algorithm for mining Continuous Temporal High Utility Itemsets (CTHUI). CTHUI first scans the database once to compute the total utility in the database and the TWU of the items in each time period and in the whole database. For each transaction, CTHUI sorts the items by their TWU in descending order, and removes the items whose TWU is less than *minimum potential utility* which is the product of the *user-specified potential utility threshold* and the total utility. An itemset is a potential high utility itemset if its utility is less than the minimum utility but no less than the minimum potential utility.

Table 3. The utility for each item in each time period

Item	Total	P_1	P_2	P_3	P_4	P_5
E	775	189	247	64	103	172
B	734	157	203	100	104	170
C	574	84	136	87	106	161
D	485	123	256	0	38	68
A	310	104	79	21	39	67
Total utility in each period		189	269	122	124	199
Minimum utility		56.7	80.7	36.6	37.2	59.7
Minimum potential utility		37.8	53.8	24.4	24.8	39.8

For example, the transaction database in Table 1 is partitioned into 5 basic time periods: P_1 contains TIDs 1, 2, 3, 4 and 5; P_2 contains TIDs 6, 7, 8, 9 and 10; P_3 contains TIDs 11, 12, 13, 14 and 15; P_4 contains TIDs 16, 17, 18, 19 and 20; P_5 contains TIDs 21, 22, 23, 24 and 25. The utility threshold and the potential utility threshold are set to be 30% and 20%, respectively. For the first transaction TID 1, the transaction utility $TU_1 = 3\times3 + 2\times10+18\times1+1\times5 = 52$. Because items A, B, C and E are contained in the transaction TID 1, TWU(A) = TWU(B) = TWU(C) = TWU(E) = 52 so far. Similarly,

$TU_2 = 6 \times 10 + 1 \times 6 + 1 \times 5 = 71$ and TWU(B) = TWU(E) = 52 + 71 = 123 after scanning the previous two transactions. After scanning the five transactions in the first period, the total utility of the first period is $PTTU_1 = TU_1 + TU_2 + TU_3 + TU_4 + TU_5 = 189$. After scanning the transactions in all the periods, the total utility of each period and the TWU of each items in each period can be obtained, which are shown in Table 3. From Table 3, we can see that the minimum potential utility is 122×20% = 24.4 in the third period, so items A and D can be removed in the third period.

For each time period, CTHUI finds the (potential) high utility itemsets from the transactions in the period, and process them from the first period. For each (potential) high utility itemset in the current period, CTHUI accumulates the utility of an itemset in the previous period to the utility of the itemset in the current period. For an itemset, there are four cases will occur:

Case 1: The itemset is high utility in the current period P_j, but potential high utility in the last continuous periods $P_i, \ldots, P_{j-1}$ ($j > i \geqq 1$). If the total utility of the itemsets in periods $P_i, \ldots, P_{j-1}, P_j$ is no less than the minimum utility, then the itemset is high utility in periods $P_i, \ldots, P_{j-1}, P_j$. Otherwise, if the total utility is no less than the minimum potential utility, then the itemset is a high utility itemset in period P_j, and potential high utility in the continuous periods $P_i, \ldots, P_{j-1}$.

Case 2: The itemset is high utility in both the current period P_j and the last continuous periods $P_i, \ldots, P_{j-1}$ ($j > i \geqq 1$). In this case, the itemset is high utility in the continuous periods $P_i, \ldots, P_{j-1}, P_j$.

Case 3: The itemset is potential high utility in the current period P_j, and high utility in the last continuous periods $P_i, \ldots, P_{j-1}$ ($j > i \geqq 1$). If the total utility of the itemsets in periods $P_i, \ldots, P_{j-1}, P_j$ is no less than the minimum utility, then the itemset is high utility in periods $P_i, \ldots, P_{j-1}, P_j$. Otherwise, if the total utility is no less than the minimum potential utility, then the itemset is a potential high utility itemset in period P_j, and high utility in the continuous periods $P_i, \ldots, P_{j-1}$.

Case 4: The itemset is a potential high utility itemset in both the current period P_j and the last continuous periods $P_i, \ldots, P_{j-1}$ ($j > i \geqq 1$). In this case, the itemset is a potential high utility itemset in the continuous periods $P_i, \ldots, P_{j-1}, P_j$.

CTHUI sequentially scans the transactions in each time period to construct *Utility-tree* (*Utree*) which is similar to FP-tree structure [6]. From the Utree structure, high utility itemsets and potential high utility itemsets can be generated. An Utree consists of a null root, a set of nodes and a *header table*. Each node, except the root node, in the Utree consists of *item-name*, *CID*, *utility* and *item-link*. The item-name registers which item this node represents, CID and utility register the transaction ID and the utility of the item in this transaction, and item-link links to the next node in the Utree carrying the same item or null if there is none. There is an item-link structure for each HTWU item. Each entry in the header table consists of two fields: *item-name* and *head of item-link* which points to the first node in the Utree carrying the same item-name.

The construction of Utree algorithm is described as follows: First, a null root node is created. For each transaction t in the database D, the HTWU items in the transaction t are sorted by their TWU in descending order and the non-HTWU items in t are removed. Let n_0 is a root node. For the sorted transaction T = $\{i_1, i_2, \ldots, i_m\}$, the TID T and the utility of each item in T are recorded in the node with this item and a new node with item i_{k+1} is created as a child of the node with item i_k ($\forall\ k, q \le k \le m-1$) for the path $P = <n_0, n_1, n_2, \ldots n_q, \ldots n_r>$ ($r \ge 1$) if the item of node n_j is i_j ($\forall\ j, 1 \le j \le q$) and the item of node n_{q+1} is not i_{q+1}. We use the above example to illustrate our algorithm CTHUI.

From Table 3, we can see that the order of the HTWU items is E, B, C and A which are ordered by their TWU in descending order. For the first transaction TID 1, a node E is created as child of the root node, and the TID 1 and utility u(E, TID 1) = 1x5 = 5 are recoded in node E. The link of item E in the header table is linked to the node E. For the next ordered item B in TID 1, a node B is created as the child of node E because there is no child node B for node E. the utility u(B, TID 1) = 2 × 10 = 20 and TID 1 are recoded in node B. After scanning the first transaction in Table 1, the Utree is shown in Figure 1.

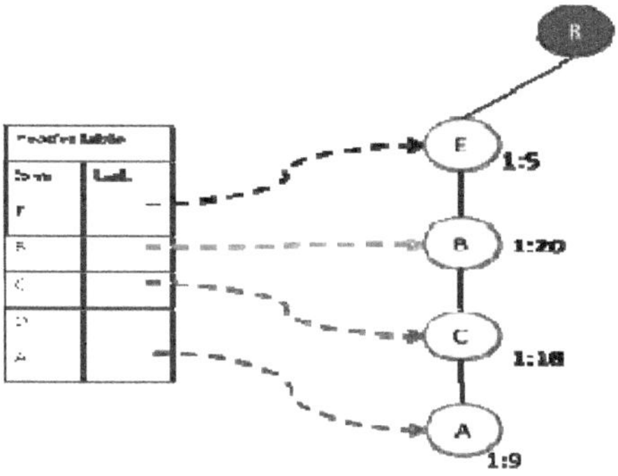

Fig. 1. The Utree structure after processing the first transaction in Table 1

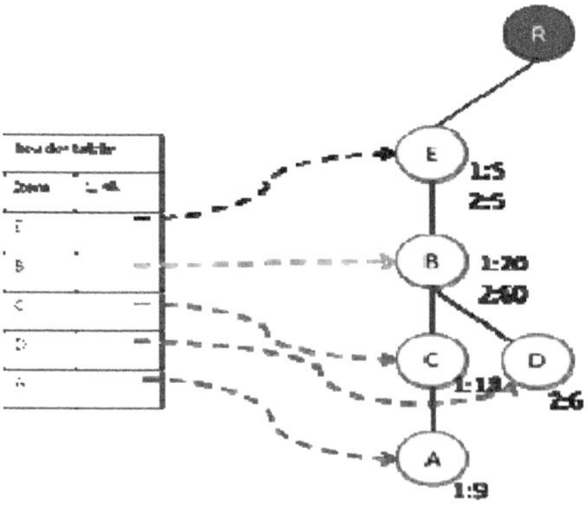

Fig. 2. The Utree structure after processing the second transaction in Table 1

For the second transaction TID 2 in Table 1, the ordered items are E, B and D. Because node E has been the child of the root node in Figure 1, the utility u(E, TID 2) = 1 × 5 = 5 and TID 2 are recoded in node E. Similarly, the utility u(B, TID 2) = 6 × 10 = 60 and TID 2 are recoded in node B. Because there is no child node D for node B, a node D is created as the child node of node B, and utility u(D, TID 2) = 1 × 6 = 6 and TID 2 is recorded in node D. After scanning the second transaction TID 2, the Utree structure is shown in Figure 2. After processing all the transactions in the first period P_1, we can obtain the Utree structure which is shown in Figure 3.

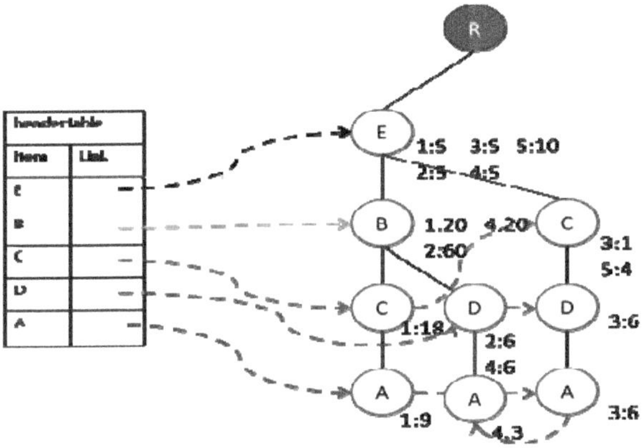

Fig. 3. The Utree structure after processing all the transactions in period P_1

Table 4. High utility itemsets in the first time period P_1

HUI	Utility	Periods
ABE	62	[P_1, P_1]
DBE	102	[P_1, P_1]
DB	92	[P_1, P_1]
BE	115	[P_1, P_1]

Table 5. Potential high utility itemsets in the first time period P_1

PHUI	Utility	Periods
ACBE	52	[P_1, P_1]
ACB	47	[P_1, P_1]
ACE	44	[P_1, P_1]
AB	52	[P_1, P_1]
CBE	43	[P_1, P_1]
CB	38	[P_1, P_1]
CE	43	[P_1, P_1]

Table 6. High utility itemsets in different time periods after processing P1 and P2

HUI	Utility	Periods
ABE	62	$[P_1, P_1]$
DBE	290	$[P_1, P_2]$
DB	270	$[P_1, P_2]$
BE	285	$[P_1, P_2]$

For each item x in the header table of Utree, our algorithm CTHUI retrieves each path from the node x to the child of the root via the item-links of item x. For each retrieved path $a_1 \rightarrow a_2 \rightarrow \ldots \rightarrow a_n \rightarrow x$, CTHUI generates all the combinations of the items in the path, such as $\{a_1\}$, $\{a_2\}$, …, $\{a_n\}$, $\{a_1a_2\}$, …,$\{a_{n-1}a_n\}$, $\{a_1a_2a_3\}$, … and $\{a_1a_{2...}a_n\}$, and accumulates the utilities of the nodes for each combination and node x with the same TID. For each combination, the item x is implicitly attached to form a candidate itemset, such as $\{a_1x\}$, $\{a_2x\}$, … $\{a_nx\}$, $\{a_1a_2x\}$, …, $\{a_{n-1}a_nx\}$, $\{a_1a_2a_3x\}$, … and $\{a_1a_{2...}a_nx\}$. For example, the first retrieved path for item A in Figure 3 is E→B→C→A, in which 7 itemsets {EA}, {BA}, {CA}, {EBA}, {ECA}, {BCA} and {EBCA} can be generated. For the utility of itemset {EA}, because the common TID for node E and node A in the path E→B→C→A is 1, the utility of {EA} in this path is 5 + 9 = 14. The utility of itemset {EBA} can be computed from the two paths E→B→C→A and E→B→D→A for item A, which is (5 + 20 + 9) + (5 + 20 + 3) = 62. By using this way, the utilities for all the itemsets containing item A can be obtained. After processing each items in the header table, the high utility itemsets and potential high utility itemsets in the first time period are shown in Table 4 and Table 5, respectively.

Table 7. Potential high utility itemsets after processing P1 and P2

PHUI	Utility	Periods:Utility
CBE	110	$[P_1, P_2]$
CB	100	$[P_1, P_2]$
ADCE	57	$[P_2, P_2]$
DCE	67	$[P_2, P_2]$
DC	57	$[P_2, P_2]$
DCBE	79	$[P_2, P_2]$
DCB	74	$[P_2, P_2]$

By using the same method, the high utility itemsets and potential high utility itemsets in the second time period P_2 can be obtained. CTHUI checks if these itemsets satisfy the above four cases. For example, the utility of itemset {DBE} in P_2 is 188, which is greater than the minimum utility in P_2. Therefore, itemset {DBE satiafies case 1, which is a high utility itemset in the continuous time periods $[P_1, P_2]$, and u({DBE}) = 102 + 188 = 290 in $[P_1, P_2]$. The utility of itemset {CBE} in P_2 is 67,

Table 8. High utility itemsets in different continuous time periods for Table 1

THUI	Utility	Periods
ABE	62	$[P_1, P_1]$
DBE	290	$[P_1, P_2]$
DB	270	$[P_1, P_2]$
BE	505	$[P_1, P_5]$
CB	325	$[P_1, P_5]$
CBE	310	$[P_1, P_5]$
DCBE	38	$[P_4, P_4]$
CE	98	$[P_4, P_5]$

which is greater than minimum potential utility but less than minimum utility. Therefore, itemset {CBE} satisfies case 4, which is a potential high utility itemset in the continuous time periods $[P_1, P_2]$. After processing all the generated itemsets in P_2, the high utility itemsets and potential high utility itemsets in different time periods are shown in Table 6 and Table 7, respectively.

For the high utility itemsets and potential high utility itemsets in $[P_i, P_k]$ $(1 \leqq i \leqq k \leqq j)$, if $k \neq j$, then the itemset needs to be removed before processing the itmesets in the (j+1)th time period, since the time periods of the itemset cannot continue to the next time period. After processing all the high utility itmesets and potential high utility itemsets in the five time periods of the transaction database in Table 1, the high utility itemsets in different continuous time periods are shown in Table 8.

3 Experimental Results

For the performance evaluation, because there is no research on mining high utility itemsets in different time periods, we compare our algorithm CTHUI with THUI-Mine [4] for mining high utility itemsets in a data stream. The synthetic dataset T5I4D100K is generated based on [7], in which T is the average number of the items in a transaction, I is average number of the items in a potential frequent itemsets, and D is the number of the transactions in the database. The synthetic database is equally partitioned into 10 partitions, and the potential utility threshold is less than utility threshold 10%. All our experiments were conducted on Intel® CORE (TM)2 Quad CPU Q9400 @2.66GHz, 4B memory, using C# Programming Language and running on Microsoft windows 7 environment.

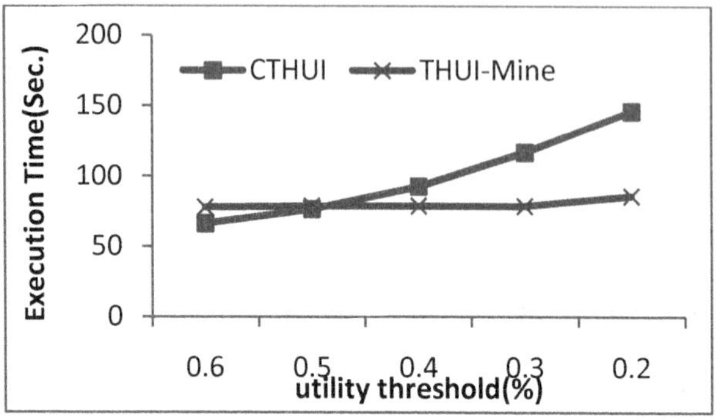

Fig. 4. The execution times for algorithms CTHUI and THUI-Mine

Figure 4 and Figure 5 show the execution times and memory usages for the two algorithms. The number of high utility itemsets generated by CTHUI is larger than that of THUI-Mine, because CTHUI algorithm needs to generate high utility itemsets in different time periods, but THUI-Mine only generates the high utility itmesets in the current dataset. Besides, CTHUI needs to keep all the high utility itemsets and potential high utility itemsets in each time periods. Therefore, CTHUI takes more time and more memory space than THUI-Mine.

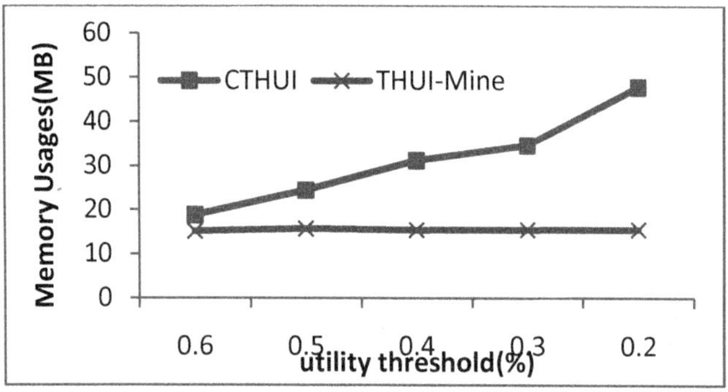

Fig. 5. The memory usages for algorithms CTHUI and THUI-Mine

4 Conclusions

Mining high utility itemsets from the whole database will lose the high utility itemsets in a specific time period. Therefore, we cannot know when the itemsets should be promoted to gain the high profits. In this paper, we propose an algorithm for mining high utility in different time periods which can provide the information about what time the high utility itemsets should be promoted. Although our algorithm takes more execution time and memory space than THUI-Mine for mining high utility itmesets in a data stream, our algorithm can get more useful information than that of THUI-Mine.

References

1. Agrawal, R., Srikant, R.: Fast Algorithms for Mining Association Rules. In: Proc. of 20th International Conference on Very Large Databases, Santiago, Chile, pp. 487–499 (September 1994)
2. Ahmed, C.F., Tanbeer, S.K., Jeong, B.-S., Lee, Y.-K.: "An Efficient Candidate Pruning Technique for High Utility Pattern Mining. In: Proc. of the 13th Pacific-Asia Conference on Knowledge Discovery and Data Mining, pp. 749–756 (April 2009)
3. Chen, M.S., Dai, B.-R., Huang, J.W.: Twain: Two-end association miner with precise frequent exhibition periods. ACM Transactions on Knowledge Discovery from Data 1(2), Article 8 (2007)
4. Chu, C.J., Liang, T., Tseng, V.S.: An efficient algorithm for mining temporal high utility itemsets from data streams. Journal of System Software 81(7), 1105–1117 (2008)
5. Hong, T.P., Lan, G.C., Tseng, V.S.: Discovery of high utility itemsets from on-shelf time periods of products. Expert Systems with Application 38(5), 5851–5857 (2011)
6. Han, J., Pei, J., Yin, Y.: Mining Frequent Patterns without Candidate Generation. In: Proc. of Int. Conf. on Management of Data (ACM SIGMOD), pp. 1–12 (May 2000)
7. Liu, Y., Liao, W.-K., Choudhary, A.: A Fast High Utility Itemsets Mining Algorithm. In: Proc. of the 1st Int. Workshop on Utility-Based Data Mining, pp. 90–99 (August 2005)

Content Based Image Retrieval Using Fuzzy Texton and Shearlet Transform

Sudhakar Putheti[1], P.R. Krishna Prasad[2], and Srinivasa Reddy Edara[1]

[1] Computer Science and Engineering, Acharya Nagarjuna University, Guntur, A.P., India
[2] Computer Science and Engineering, JNTU Kakinada, Kakinada, A.P., India
sudhakarp0101@gmail.com, {krishna_palli,edara_67}@yahoo.com

Abstract. This paper presents a novel and an efficient content based image retrieval technique based on content color, texture and shape, which are primary low level features to describe image. The proposed method uses fuzzy texton method to extract texture features and discrete shearlet transform for shape features in addition to color features. Fuzzy Texton is a way to describe the texture properties in image analysis. Shearlet transforms can analyze signals defined not only at various orientations but also at multiple scales. The test results emerged more efficient than all the predecessor peer techniques like Texton Co-occurrence Matrix (TCM), Texton Multi Histogram (TMH) and Texels of Fuzzy Texton (TT).

Keywords: Fuzzy Texton, CBIR, HSV, Discrete Shearlet Transform.

1 Introduction

With an explosive growth of digital image collections, content based image retrieval (CBIR) has been emerged as one of the most active problems in computer vision as well as multimedia applications. The target of content-based image retrieval (CBIR) [1] is to retrieve images relevant to a query of a user, which can be expressed by example. In CBIR, an image is described by automatically extracted low-level visual features, such as color, texture and shape [2-4]. When a user submits one or more query images as examples, a criterion based on this image description ranks the images of an image database according to their similarity with the examples of the query and, finally, the most similar are returned to the Digital image retrieval systems. Since 1990's, Content Based Image Retrieval (CBIR) has attracted great research attention [5-6]. Early research was focused on finding the best representation for image features. The current work primarily focuses on using Fuzzy Textons and Shearlet Transformation methods for CBIR.

Texton proposed by Julesz conceptually more than 20 years ago [7, 8] is one of the useful texture analysis concept central to the CBIR. Though texton is defined as a set of blobs or emergent patterns sharing a common property all over the image, a

M. Gen et al. (eds.), *Industrial Engineering, Management Science and Applications 2015*,
Lecture Notes in Electrical Engineering 349, DOI: 10.1007/978-3-662-47200-2_83

thorough exploration of the texton properties still remains a challenge. As part of the exploration, shape one of the discriminating features is considered for the extraction accurate information of the image [9]. This paper presents wavelet decomposition method to extract shape feature. Traditional methods such as wavelet and Gabor transforms are not capable of efficiently capturing directional features from data due to their limited directional sensitivity.

On the other hand, Shearlet transforms can analyze signals defined not only at various orientations but also at multiple scales; they provide efficient mathematical and computational methods to address singularities such as image edges [10]. During the retrieval, the similarity between two images is calculated by summing the distances of individual low-level features with fixed weights. This paper attempts to analyze the texture as well as the similarities between the images to efficiently extract the images.

The rest of the paper is organized as follows. Section 2, research efforts on fuzzy texton and shear wavelet transform are reviewed. In Section 3, the proposed image retrieval method is discussed. In Section 4, Experiments, performance evaluations, and discussions are given. Finally, Section 5 concludes our work.

2 Existing Methods

For image retrieval in CBIR various algorithms have been designed to extract the color and texture and shape features. There are some algorithms that combine both color and texture features, such as the integrative co-occurrence matrix [11], texton co-occurrences matrix [12], multi-texton histogram (MTH) [13], color edge co-occurrence histogram (CECH) [14], color auto-correlograms [15], etc. In recent years, many methods for shape representation and recognition have been proposed. An advanced review of shape representation techniques can be found in [16, 17]. One can distinguish between contour-oriented and region-oriented shape description techniques. The first method describes the object outlines and the second describes the object areas. Suitable contour and region primitives are first used in order to extract the relevant objects from a given image (e.g., Freeman codes [18], polygon lines [19], and pixels or image squares [20]). Based on this data, shape description can be performed by applying a feature vector in order to represent the given objects. In this paper, we use contour polygons and apply feature vectors that employ new multi-scale transform called the Shearlet transform.

2.1 RGB Vs HSV Color Models

The disadvantage with RGB model is the change in its behavior when the illumination in an image is altered. This is overcome by using HSV model. HSV defines a color by its hue, saturation and value. HSV describes colors as humans perceive them. This perceptual relevance might make it a better choice than RGB for image retrieval. HSV has an advantage as hue is invariant under the orientation of the object with respect to illumination and camera direction. Hence, it is used in the proposed method.

2.2 Fuzzy Texton

Fuzzy Texton describes the texture property better than the Texton since Fuzzy texton uses the fuzzy texture unit (FTU) instead of texture unit (TU). By including fuzzy concept to the texton all ranges of the values including '2' in the FTU will be generated. It generates 5^8 distinct textures instead of $4^{8.}$ It covers total spectrum which is not possible when only texton is used. The main idea of the proposed method (TFT) is to obtain Texton images by applying the 12 types of 3 x3 texton templates [21].

2.3 Discrete Shearlet Transform

This paper is based on a new multiscale transform called the shearlet transform. It is a multidimensional version of the traditional wavelet transform, and is designed to address anisotropic and directional information at various scales. Indeed, the traditional wavelet approach, which is based on isotropic dilations, has a very limited capability to account for the geometry of multidimensional functions. In contrast, the analyzing functions associated with the shearlet transform are highly anisotropic, and, unlike traditional wavelets, are defined at various scales, locations and orientations. As a consequence, this transform provides an optimally efficient representation of images with edges [22]. We refer to G. Easley et. al for more details about the comparison of shearlets and other orientable multiscale transforms[23].

3 Proposed Method

An efficient image retrieval technique is required to improve the success rate of CBIR. The block diagram of the proposed CBIR system is shown in Fig. 1. It consists of two phases: database building (off-line) and query processing (on-line) phase. The following are the steps in each phase:

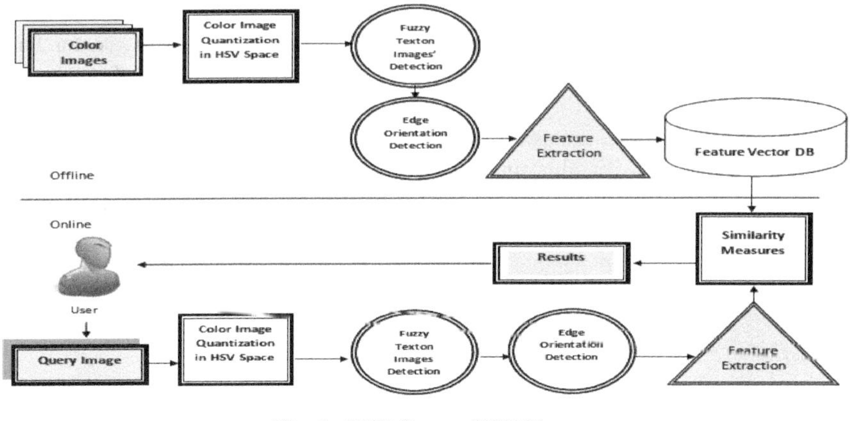

Fig. 1. CBIR System HSV Planes

3.1 Decomposition of Color Image into HSV Planes

The input true color image is decomposed into H, S, and V color planes as shown in Figure 2. One of the advantages of HSV model is it is insensitive to changes in illumination. Each color plane has a value ranging from 0 to 255 giving a total of 16777216 different colors. The dimensionality of feature vector depends on the color quantization levels. Hence HSV channels are uniformly quantized into 4 levels so that 64 colors are obtained [21].

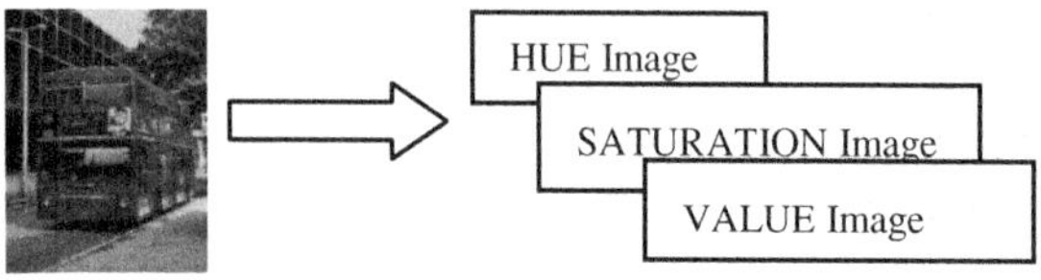

Fig. 2. HSV Planes

3.2 Color Quantization and Extraction of FuzzyTexton Images

As per Julesz description a texton is a pattern which is shared by an image as a common property. Textures are formed only if the adjacent elements lie within the neighborhood. The critical distances that depend on the texture element size, between texture elements are used to incline the texton. Texture can be decomposed into elementary units, viz the Texton classes of colors, elongated blobs of specific width, orientation, aspect ratios and terminators of elongated blobs. If the elongated elements are not jittered in orientation, texture gradients increase at boundaries. Thus a small sub image of size 3 x 3 is used to obtain texton gradient. Here we have proposed 12 textons of 3 x 3 grids. The computational complexity for using the overlapped components of 12 textons is also less to obtain final texton image. The texture property determination depends on the values of Texture Unit of an image plane [21]. The basic unit of the method is defined by a central pixel and its eight neighbors, forming a 3x3 pixel square. This minimal square image has the local texture information of the central pixel in all the directions.

In a digital image the main goal is to extract the local texture information of a neighborhood of pixels. In our case the size of the neighborhood is 3x3 pixels. This pattern of the image, consisting 9 pixels, is denoted by a set $V=\{V_0, V_1, V_2 \ldots V_8\}$, where V_0 represents the intensity value of the central pixel and V_i $(1 \leq i \leq 8)$ the intensity value of each neighboring pixel. The smallest complete unit which best characterizes the local texture aspect of a given pixel and its neighborhoods, in all eight directions of a square raster, is Texture Unit (TU) that is defined, by $TU=\{E_1, E_2, \ldots, E_8\}$, where:

$$E_i = \begin{cases} 0 & \text{if } V_i < V_0 \text{ and } V_i < p \\ 1 & \text{if } V_i < V_0 \text{ and } V_i > p \\ 2 & \text{if } V_i = V_0;\ 1 \leq i \leq 8 \\ 3 & \text{if } V_i > V_0 \text{ and } V_i < q \\ 4 & \text{if } V_i > V_0 \text{ and } V_i > q \end{cases} \tag{1}$$

here p and q are user defined values and each element E_i occupies the same position as pixel *i*. An example is shown in Figure 3.

121	190	67
225	180	252
180	170	201

(a)

1	3	0
4		4
2	1	3

(b)

a	b	c
h		d
g	f	e

(c)

Fig. 3. (a) Hue levels of an image part.(b) Texture Unit to the central pixel.(c) Texture Unit Ordering.

As to each element of the TU can be assigned one of the possible values 0, 1,2,3 or 4, the total number of Texture Units is $5^8 = 16777216$. These Units can be labeled and ordered in different ways; here we will label each TU as a 5-base number, named Texture Unit Number, N_{TU} according to next formula:

$$N_{TU} = \sum_{i=1}^{8} E_i \cdot 5^{\frac{i-1}{2}} \tag{2}$$

Where i is the position of the Texture Unit box, and E_i is the value of the box (0, 1, 2, 3 or 4). Moreover, the 8 elements can be ordered differently. If they are ordered clockwise, as shown in Figure 1(c), the first element can take eight possible positions, from the top left a to the middle left h, and then the 16777216 texture units can be labeled by the above formula under eight different ordering ways (from a to h).

A more detailed study of texture unit indicates that the absent TU's involve two's in their texture unit. This is the case when neighbors and central pixels have the same values. If there is a lack of two's then TU will take only 0,1,3 and 4 which means that the possible real number of different textures are 4^8 instead of 5^8, that is 65536 and 390625. Thus the power of the method is misused and it impacts the texture Unit number also. To overcome this, fuzzy texture is used in the proposed method.

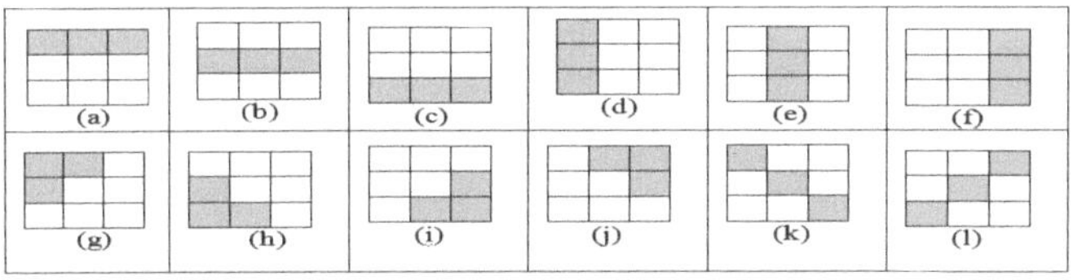

Fig. 4. 12 types of 3 x 3 texton templates

The Fuzzy Texture unit boxes (FTUB) and Fuzzy Texture Unit Numbers (FTUN) are employed in the method by Aina Barceló. et. al [24].

In this proposed work, FTUB and FTUN are used during the period of quantifying the texton images. Then, the resultant textons are called as fuzzy texton images. As part of the detection process of fuzzy texton, the 3 x 3 grids can detect the textons in all directions and also corners of the textures. If three pixels are highlighted and have the same value then the grid will form a fuzzy texton as shown in Figure 5.

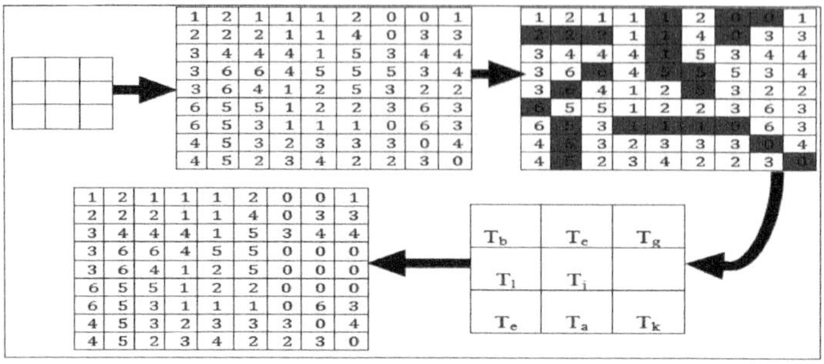

Fig. 5. Fuzzy Texton Detection Process

The main idea of the proposed method (TFT) is, Texton images are obtained by applying the 12 types of 3 x3 texton templates [21] as shown in figure 4 on HSV planes as shown in figure 3.

3.3 Edge Orientation Using Discrete Shearlet Transform

An algorithm for edge detection based on shearlets was introduced in [25, 26], where a discrete shearlet transform was described with properties designed for this task. In fact, the discrete shearlet transform which was presented for image denoising, produces large sidelobes around prominent edges which interfere with the detection of the edge location. In contrast, the special discrete shearlet transform introduced in [25, 26] is not affected by this issue since the analysis filters are chosen to be consistent with the theoretical results in [27, 28], which require the shearlet generating function to satisfy specific symmetry properties in the Fourier domain.

The first step of the shearlet edge detector algorithm consists of selecting the edge point candidates of a digital image u[m1,m2]. They are identified as those points $(\overline{m_1}, \overline{m_2})$ which, at fine scales j, are local maxima of the function.

$$M_j u[m_1, m_2]^2 = \sum_l \left(SHu[j, l, m_1, m_2]\ \right)^2 \tag{5}$$

Here $SHu[j, l, m_1, m_2]$ denotes the discrete shearlet transform. According to the properties of the continuous shearlet transform summarized above, we expect that, if $(\overline{m_1}, \overline{m_2})$ is an edge point, the discrete shearlet transform of u will behave as

$$|SHu[j, l, \overline{m_1}, \overline{m_2}]| \sim C2^{-\beta j} \tag{6}$$

where $\beta \geq 0$. If, however, $\beta < 0$ (in which case the size of |SHu| increases at finer scales), then $(\overline{m_1}, \overline{m_2})$ will be recognized as a spike singularity and the point will be classified as noise. Using this procedure, edge point candidates for each of the oriented components are found by identifying the points for which $\beta \geq 0$. Next, a non-maximal suppression routine is applied to these points to trace along the edge in the edge direction and suppress any pixel value that is not considered to be an edge. Using this routine, at each edge point candidate, the magnitude of the shearlet transform is compared with the values of its neighbours along the gradient direction (this is obtained from the orientation map of the shearlet decomposition). If the magnitude is smaller, the point is discarded; if it is the largest, it is retained. Extensive numerical experiments have shown that the shearlet edge detector is very competitive against other edge detectors, and its performance is very robust in the presence of noise. A shearlet edge detector is compared against the wavelet edge detector and the Sobel and Prewitt edge detectors. The performance of the edge detectors is assessed using the Pratt's Figure of Merit, which is a fidelity function ranging from 0 to 1, where 1 is a perfect edge detector. This is defined as

$$FOM = \frac{1}{max(N_e, N_d)} \sum_{k=1}^{N_d} \frac{1}{1+\alpha d(k)^2} \tag{7}$$

where N_e is the number of actual edge points, N_d is the number of detected edge points, d(k) is the distance from the k-th actual edge point to the detected edge point and α is a scaling constant typically set to 1/9. The numerical test reported in the figures show that the shearlet edge detector consistently yields the best value for FOM.

The continuous shearlet transform has the ability to precisely characterize the geometry of the edges. These properties directly lead to an effective algorithm for the estimation of the edge orientation, which was originally introduced in [36]. Specifically, by taking advantage of the parameter associated with the orientation variable in the shearlet transform, the edge orientations of an image u, can be estimated by searching for the value of the shearing variable s which maximizes $SHu(a,s,p)$ at an edge point p, when a is sufficiently small. Discretely, this is obtained by fixing a sufficiently fine scales (i.e., $a = 2^{-2j}$ sufficiently "small") and computing the index $\tilde{l}$ which maximizes the magnitude of the discrete shearlet transform $SHu[j,l,m]$ as

$$\tilde{l}(j,m) = argmax_l |SHu[j,l,m]| \tag{8}$$

Once this is found, the corresponding angle of orientation $\theta_{\tilde{l}}(j,m)$ associated with the index $\tilde{l}(j,m)$can be easily computed. The sensitivity of the shearlet transform to the edge orientation is useful for the extraction of landmarks, another imaging application, which is important in problems of classification and retrieval. To illustrate the general principle, consider large smooth regions separated by piecewise smooth curves. The junction point A, where three edges intersect, is certainly the most prominent object in the image, and this can be easily identified by looking at values of the shearlet transform. In fact, if one examines the discrete shearlet transform $SHu[j_0,l,m_0]$, at a fixed (fine) scale j_0 and locations m_0, as a function of the shearing parameter l, the plot immediately identifies the local geometric properties of the image. Specifically, one can recognize the following four classes of points inside the image. At the junction point $k_0 = A$, the function $|SHu[j_0,l,m_0]|$ exhibits three peaks corresponding to the orientations of the three edge segments converging into A; at the point $m_0 = B$, located on a smooth edge, $|SHu[j_0,l,m_0]|$ has a single peak; at a point $m_0 = D$, inside a smooth region, $|SHu[j_0,l,m_0]|$ is essentially flat; finally, at a point $m_0 = C$ "close" to an edge, $|SHu[j_0,l,m_0]|$ exhibit two peaks, but they are much smaller in amplitude than those for the points A and B.

A similar behavior was observed, as expected, for more general images, even in the presence of noise. Based on these observations, a simple and effective algorithm for classifying smooth regions, edges, corners and junction points of an image was proposed and validated in [26].

3.4 Extraction of Texels

After the extraction of fuzzy texton images need to extract the texels from them. The local properties used to extract the feature vectors used here come under two categories; color, texture and shape. Some of the important features of texture properties are Local Homogeneity, Cluster Shade and Cluster Prominence.

i) Local homogeneity: $\sum_{i,j=0}^{n} \frac{1}{1+(i-j)^2} c(i,j)$ (9)

ii) Cluster shade: $\sum_{i,j=0}^{n} \left(i - M_x + j - M_y\right)^3 c(i,j)$ (10)

iii) Cluster Prominence: $\sum_{i,j=0}^{n} (i - M_x + j - M_y)^4 c(i,j)$ (11)

where $M_x = \sum_{i,j=0}^{n} iC(i,j)$ (12)

and $M_y = \sum_{i,j=0}^{n} jC(i,j)$ (13)

There are three important properties regarding color information. They are Color Expectancy, Color Variance and Skewness.

i) Color Expectancy: $E_i = \frac{1}{N} \sum_{j=1}^{N} C_{ij}$ (14)

ii) Color Variance: $\delta_i = (\frac{1}{N} \sum_{j=1}^{N} (C_{ij} E_i)^2)^{\frac{1}{2}}$ (15)

iii) Skewness: $\sigma_i = (\frac{1}{N} \sum_{j=1}^{N} (C_{ij} E_i)^3)^{\frac{1}{3}}$ (16)

4 Results and Discussion

The results demonstrate the performance of the method over Corel database. We compared our results with two other methods: TT [29] and TFT [21]. Both are based on edge features without image segmentation, and TFT is the origin of our method. HSV channels are of a true color image. Figure 6 shows the true color image.

(a) (b) (c) (d)

Fig. 6. (a) Original image (b) Hue image (c) Saturation image (d) Value image

4.1 Distance Measure

We can use different distance metrics for matching such as an N-dimensional feature vector F = [F1, F2...FN]. It is extracted from every image of database and stored in database. Let Q = [Q1, Q2, Q3,…,QN] be the feature vector of query image. A simple distance measure [30] whose time complexity is very less when compared with others like Euclidean (no square or square root operations) when we consider large databases, is given by

$$F, Q) = \sum_{i=0}^{N} \frac{|F_i - Q_i|}{1 + F_i + Q_i} \tag{17}$$

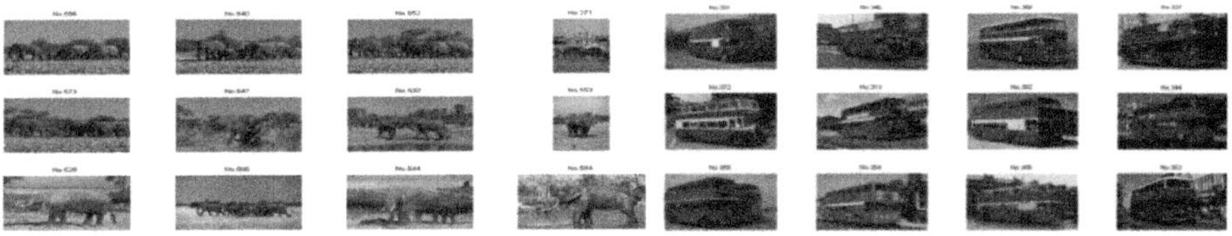

(a) (b)

Fig. 7. (a) 11 out of 12 images (b) 12 out of 12 images

The standard Dataset SIMPLIcity which contains 1000 images of 10 categories with each categeriy consists of 100 images is used to test the proposed method. As part of that, Figure 7(a) shows retreival of the resultant image as 11 out of 12 images query image, elephant image. Similarly Figure 7(b) shows retrieval of the resultant images for the buses.

4.2 Performance Measure

Most common measurements are used to evaluate the performance of image retrieval methods are Precision, Recall and Accuracy curves [31].

$$\text{Precision is given by } P(N) = \frac{I_N}{N} \tag{18}$$

$$\text{Recall is given by } R(M) = \frac{I_M}{M} \tag{19}$$

$$\text{Accuracy } \quad A(N) = \frac{(P(N) + R(N))}{2} \tag{20}$$

Where I_N is the number of images retrieved in the top N positions that are similar to the query image and M is the total number of images in the database similar to the query image. Table 1 shows the comparison results of Proposed Method with Texels of Fuzzy Textons (TFT)[21] and TT[29].

Table 1. Comparision of Proposed Method with TT and TFT

Data Bases	Performance	Method		
		TT	TFT	Proposed
SIMPLIcity	Precision (%)	72.76	74.63	75.87
	Recall (%)	8.86	9.06	9.47
Corel – 5000	Precision (%)	58.43	60.32	61.89
	Recall (%)	6.83	7.48	8.03
Corel–10000	Precision (%)	43.56	44.93	46.37
	Recall (%)	5.77	5.93	6.22

References

1. Datta, R., Li, J., Wang, J.Z.: Content-Based Image Retrieval: Approaches and Trends of the New Age. Multimedia Information Retrieval, 253–262 (2005)
2. Ma, W.Y., Zhang, H.J.: Content-Based Image Indexing and Retrieval. In: Handbook of Multimedia Computing, CRC Press, Boca Raton (1999)
3. Ionescu, B., Lambert, P., Coquin, D., Buzuloiu, V.: Color based content retrieval of animation movies: a study. In: Proceedings of the International Workshop on Content-Based Multimedia Indexing (CBMI 2007), Talence, France, pp. 295–302 (June 2007)
4. Rui, Y., Huang, T.S., Chang, S.-F.: Image retrieval: current techniques, promising directions, and open issues. Journal of Visual Communication and Image Representation 10(1), 39–62 (1999)
5. Jing, F., Li, M., Zhang, H.-J., Zhang, B.: Relevance Feedback in Region-Based Image Retrieval. IEEE Trans. Circuits Syst. Video Technol. 14(5), 672–681 (2004)
6. Tong, S., Chang, E.: Support Vector Machine Active Learning for Image Retrieval. ACM Multimedia, 107–118 (2001)
7. Mao, J., Jain, A.: Texture classification and segmentation using multi-resolution simultaneous autoregressive models. Pattern Recognition 25(2), 173–188 (1992)
8. Liu, F., Picard, R.: Periodicity, directionality, and randomness: wold features for image modeling and retrieval. IEEE Trans. Pattern Anal. Mach. Intell. 18(7), 722–733 (1996)
9. Hiremath, P.S., Pujari, J.: International Journal of Image Processing 2(1)
10. Marakakis, A., Galatsanos, N., Likas, A., Stafylopatis, A.: Probabilistic Relevance Feedback Approach for Content-Based Image Retrieval Based on Gaussian Mixture Models. IET Image Process 3(1), 10–25 (2009)
11. Palm, C.: Color texture classification by integrative co-occurrence matrices. Pattern Recognition 37(5), 965–976 (2004)
12. Liu, G.-H., Yang, J.-Y.: Image retrieval based on the texton co-occurrence matrix. Pattern Recognition 41(12), 3521–3527 (2008)
13. Liu, G.-H., Zhang, L., et al.: Image retrieval based on multi-texton histogram. Pattern Recognition 43(7), 2380–2389 (2010)
14. Luo, J., Crandall, D.: Color object detection using spatial-color joint probability functions. IEEE Transactions on Image Processing 15(6), 1443–1453 (2006)
15. Huang, J., Kumar, S.R., Mitra, M., et al.: Image indexing using color correlograms. In: IEEE Conference on Computer Vision and Pattern Recognition, pp. 762–768 (1997)

16. Zhang, D., Lu, G.: Review of shape representation and description techniques. Pattern Recognition 37(1), 1–19 (2004), View at Publisher · View at Google Scholar · View at Scopus
17. Pitas, I.: Digital Image Processing, Algorithms and Application. John Wiley & Sons, New York (2000)
18. Freeman, H.: Techniques for the digital computer analysis of chain-encoded arbitrary plane curves. In: Proceedings of National Electronics Conference, vol. 17, pp. 421–432 (1961)
19. Nabout, A.: Modular Concept and Method For Knowledge Based Recognition of Complex Objects in CAQ Applications, Series 20(92). VDI Publisher (1993)
20. Chang, Y.L., Li, X.: Adaptive image region-growing. IEEE Transactions on Image Processing 3(6), 868–872 (1994), View at Publisher · View at Google Scholar · View at Scopus
21. Putheti, S., Edara, S.R., Edara, S.: CBIR using Texels of colour Fuzzy Textons. In: 2012 12th International Conference on Hybrid Intelligent Systems (HIS), pp. 461–467 (2012)
22. Guo, K., Labate, D.: Optimally sparse multidimensional representation using shearlets. SIAM J. Math. Anal. 9, 298–318 (2007)
23. Easley, G., Labate, D.: Sparse Directional Image Representations using the Discrete Shearlet Transform. To appear in Appl. Comput. Harmon. Anal. (2008)
24. Barceló, A., Montseny, E., Sobrevilla, P.: On Fuzzy Texture Spectrum for Natural Microtextures Characterization, EUSFLAT - LFA (2005)
25. Yi, S., Labate, D., Easley, G.R., Krim, H.: Edge detection and processing using shearlets. In: Proc. IEEE Int. Conference on Image Proc., San Diego, October 12-15 (2008)
26. Yi, S., Labate, D., Easley, G.R., Krim, H.: A Shearlet approach to edge analysis and detection. IEEE Trans. Image Proc. 18(5), 929–941 (2009)
27. Guo, K., Labate, D.: Characterization and analysis of edges using the continuous shearlet transform. SIAM J. Imaging Sciences 2, 959–986 (2009)
28. Guo, K., Labate, D., Lim, W.: Edge analysis and identification using the continuous shearlet transform. Appl. Comput. Harmon. Anal. 27, 24–46 (2009)
29. Putheti, S., Edara, S.R., Edara, S.A.: CBIR using texels of RGB colour textons. In: 2011 World Congress on Information and Communication Technologies (WICT), pp. 1020–1025 (2011)
30. Russel, R., Sinha, P.: Perceptually absed comparision fo Image Similarity Metrics. MIT AI Memo 001-014 (2001)
31. Liu, G.-H., Yang, J.–Y.: Image retrieval based on multi texton histogram. Int. J. Pattern Recognition 43 (2010)

Assessing Lean Implementation

Timo Schröders[1] and Virgílio Cruz-Machado[2]

[1] Ruhr University Bochum, Faculty of Mechanical Engineering, Germany
[2] UNIDEMI (Department of Mechanical and Industrial Engineering), Faculdade de Ciências e Tecnologia, Universidade Nova de Lisboa, Portugal
vcm@fct.unl.pt

Abstract. The lean management philosophy is applied to improve the productivity and customer satisfaction of an organization. Though lean implementations often fail or are not sustainable. Based on this, an assessment tool is developed. It consists of 24 criteria that are divided in the 4 categories: leadership, culture, knowledge, and process. The category culture is subdivided in: improvement, empowering, and partnering. The rating of each criterion works by interviews on the management and supervisor level, and surveys on all levels of the company. To improve the score of low rated criteria, different countermeasures for each criterion are named. By gathering the knowledge of different research and case studies, this tool can help companies to implement lean in a sustainable way.

Keywords: Lean management, lean implementation, performance assessment.

1 Introduction

In the past decades, lean as a process management philosophy has gained popularity in the manufacturing sector, and became a standard in large scale enterprises. During the last years, the implementation of lean increased also in small and medium-sized companies (SMEs), and the philosophy was adopted by other sectors like service, administration, and research. The goal of lean is to satisfy the customer through on time delivery and high quality products by focusing on identifying and eliminating waste throughout a product's entire value stream (including the company's supply chain network) (Bhasin, 2011; Scherrer-Rathje et al., 2009). Womack and Jones (1996) define waste as any human activity that absorbs resources but creates no value. Ohno (1988) identified seven types of waste: defect, overproduction, transportation, unnecessary motion, waiting, inventory, and inappropriate processing. Later, Liker (2004) added an eighth type of waste: unused employee creativity. Environmental waste is considered the ninth.

The benefits of a lean state in a company are: shorter cycle times, shorter lead times, lower work in progress, faster response time, lower cost, greater production flexibility, higher quality, better customer service, higher revenue, higher throughput and increased profit (Bhasin, 2011).

M. Gen et al. (eds.), *Industrial Engineering, Management Science and Applications 2015*, Lecture Notes in Electrical Engineering 349, DOI: 10.1007/978-3-662-47200-2_84

Though lean is being studied extensively for more than 30 years, companies still face difficulties in the implementation of lean in a sustainable way. The question why lean implementations fail is discussed in literature (e.g. Scherrer-Rathje et al., 2009; Liker and Rother, 2013; Miina, 2012). This paper focuses on finding a solution for these problems through developing an assessment tool. This tool includes countermeasures based on literature and the combination of different business models, techniques and award systems.

2 Assessment Tool

The objective of this paper is to develop an assessment tool that helps a company to implement lean in a sustainable way. This tool is based on the reviewed success factors, business models, and further literature. The tool consists of 24 criteria which are divided in 4 categories: leadership, culture, knowledge, and process. Culture is subdivided in the subcategories: improvement, empowering, and partnering. Each criterion can be rated on a scale from 1 to 10 (10 being the highest possible score). This rating works by interviews on the management and supervisor level and surveys on all levels of the company. External experts can be consulted for an objective third-party feedback and countermeasures may be applied if its rating is low.

2.1 Leadership

The category leadership includes 6 criteria:

#1 Business plans providing resources, encouragement, and time for improvement: For a successful and sustainable lean approach, it is essential that the company has clear visions and an elaborated long-term strategy. Beside the goal of every step, it has to be clear how to reach it. Resources (human, financial, and material), encouragement of employees, and enough time for improvement are important factors in the strategy. To achieve a high rating in this criterion, managers have to analyze the whole company with it stakeholders, strategic objectives, business processes, and methods. The score can be increased by the application of *hoshin kanri*, a method devised to capture and cement strategic goals. By developing a lean house, the company can interpret their individual philosophy of lean in an own production system model. This concept is based on the theory that every company has its own understanding of lean and should focus on its own production philosophy. A detailed description of the lean house is available in literature (Miina, 2012). Another approach is the implementation of lean accounting where the focus lies on cost reduction, while standard accounting focuses on cost maintenance (Schlichting, 2009).

#2 Visible management commitment, participation, and financial support: Visible and active management commitment is crucial for the success of a lean implementation. A lack of commitment may lead to a set of other issues like limited access to resources, lengthy decision-making processes, and communication breakdowns (Scherrer-Rathje et al., 2009). Also, employees may not realize the importance of lean, and participating sufficiently. The importance of financial support is evident, since every change, especially an extensive one like a lean approach, is related to investments that will take some time to pay off. By weekly walks on the floor level (frequency can be reduced after lean is implemented), including conversation with the employees, the management can show their commitment and enhance the internal communication. The participation in strategic meetings is crucial to obtain information about all processes in the company and to show the importance of the approach. To ensure enough financial resources, the funding of the lean approach has to be appropriate and included in the business plan.

#3 Communication of the company's visions and values: It is important that every employee of the company understands its visions and values, and identifies himself with them. It leads to an higher motivation, and changes in a process can be understood and accepted easier. In the absence of this understanding, workers could fall back to their old ways after changes, and the lean approach will not be sustainable. This can be counteracted by regular info and team meetings in small groups, including all employees. Furthermore, the distribution or installation of info booklets and posters can improve the communication of a company's ideals.

#4 Communication and measurement of improvement: The communication of lean benefits is important to motivate the employees and indorse the management in the lean approach. It is a crucial factor to convince everyone in the company that lean is the right choice, and to make sure that employees keep the new processes and do not fall back to old patterns. One way to measure and communicate the improvement trough lean is to implement a pilot project in a small area of the company. The benefits can be shown very quickly, facilitating the implementation of lean in other areas (Scherrer-Rathje et al., 2009). Another way to communicate the improvement, is to measure it by data in regular reviews (e.g. productivity of workspaces before and after the implementation, cost reduction, etc.) and present it at team meetings or on posters in the company. E-mail newsletters that communicate the improvement can also be applied.

#5 Philosophy that encourages and recognizes innovations and improvement: An important aspect of lean is that it includes all employees. Everyone should try to improve the process they are dealing with, since they might have the best impression of its weaknesses and failures (Dombrowski and Mielke, 2013). An improvement culture is crucial to make sure the employees feel free to develop ideas and innovations. It is based on a different understanding of failure: every failure shows possibilities for improvement and learning. The goal is to find the root cause of the failure and to make sure, it will not occur again. Furthermore, the employees need support from the management in order to maintain improvement activities at all levels and in all processes. Supervisors have to pick up the ideas and help to apply them in practice.

#6 Employee involvement in lean approach: The involvement of employees is a crucial factor for the success of a lean implementation. A lean conversion cannot be implemented in a top-down manner and managers have to recognize the knowledge of their employees, and accept their input in problem solution processes (Schlichting, 2009). It keeps the workers motivated and enhances their contribution to the lean approach. Improvement will only be sustained, if the operators understand the reasons behind it and this can only be achieved by involving them early in the process. To improve the employee involvement, information about the lean conversion has to be made visual to the workers. Trough so called visual management, the operators will be enabled to identify and understand problems on the floor level and therefore feel the need to improve. It includes info boards with daily data, which help the employees understand the situation on the shop floor, and production boards, which shows the productivity per cell/unit and make the production numbers transparent to everybody. Additionally, kaizen events can be arranged. In these events, the operators, managers, and owners of a process come together, map the existing process, and improve it with buy-in from all parties related to the process. Trough job rotation, operators will be promoted to be "cross-trained" to work in more than one cell/process.

2.2 Culture

The category culture is divided in the subcategories: improvement, empowering, and partnering. It includes 9 criteria in total:

2.2.1 Improvement

#7 Continuous improvement: The concept of continuous improvement is crucial to keep a lean conversion sustainable and is mentioned as being the most important step in a lean journey by Miina (2012). It takes place in cycles, where an organization never stops to question its processes. A basic part of continuous improvement is to make the need for improvement visible, so it will be made with full conviction, to ensure it will take its full effect and be sustainable. Kaizen can be used to systematically detect and eliminate waste according to the Plan-Do-Check-Act-cycle (PDCA). There are different Kaizen activities, for example, value stream design, continuous improvement process workshops, periodic meetings, process reengineering, 5xWhy, or daily short stand-up meetings. By measuring quality and non-quality cost systematically, needs for improvement can be illustrated and eliminated by using the PDCA. The use of benchmarking techniques helps to establish new improvement standards. By comparing processes, standards, or even organizations, improvement benefits can be made visible and they can be adopted to create new improvement standards (Miina, 2012).

#8 Self-assessment: As mentioned before, a company has to improve continuously to achieve a lean state. Self-assessments and reviewing all processes in the organization is crucial. Only by doing it regularly, mistakes or non-effective processes can be eliminated and the quality of the work kept high. The implementation of self-controlling interdisciplinary teams gives a wide view from different perspectives on a process, and ensures that mistakes are found and handled in a proper way (Scherrer-Rathje et al., 2009).

#9 Long term orientation: The management has to think in long term and devolve this type of thinking to the whole company. A lean approach does not always pay off financially in a short time, because every change is carried by investments. The productivity can also decrease on the short run, since the employees have to get used to the new standards. But, if an organization manages to keep lean sustainable and does not stop to believe in it, it will benefit from it on the long run. This has to be included in calculation and business plans, and everyone in the company has to be aware of it.

2.2.2 Empowering

#10 Suggestion system: To encourage the employees to be an active element in the lean conversion, and present suggestions, it has to be sure that these suggestions will be heard. If they are, the processes will be improved and the motivation of the workers will be higher, since they will feel as an important part of the company. Therefore, a suggestion system has to be established. This can be achieved by regular surveys, regarding improvement of processes, and meetings, where every employee can present his ideas and suggestions to his superior. Furthermore, there should be suggestion forms available at every workstation and a mailbox that is cleared and suggestions taken in consideration at least once a week.

#11 Recognition and reward system: Beside suggestions, good work has to be reconditioned and rewarded. In this way the employees keep their motivation and the productivity can be improved. The first step is to assess and compare the work processes, for example, by shift or workstation. The results should be presented on posters at the shop floor, to make them visible for everyone, so as to increase the motivation of each employee. A bonus system should be installed to reward good work. The employees of the most productive shift of a month/year can be, for example, invited to a diner or an extracurricular activity. Or each worker of the shift receives a bonus (Losonci et al., 2011). Through this, a positive competition will be initiated, teambuilding will be supported, and the productivity of each shift or workplace will increase.

#12 Formal teams (with autonomy): To ensure a fast decision making process, formal and interdisciplinary teams with autonomy to decide (without the need to ask a superior) should be installed. By regular meetings of these interdisciplinary teams, the problems in processes can be assessed with different points of view. Afterwards, the best solution can be found and directly implemented, without waiting days or weeks for the permission of a superior or the management.

#13 Employees at all levels meet with customers: Lean focuses on the customer. To enable this, a company has to know the customer and their needs in detail. Beside the management, which is usually in contact with the customers, every employee should meet the customers if possible. Events to connect the staff with customers can be arranged or different employees can be sent to fairs. This ensures that the workers know the needs of the customers and can use this knowledge to improve processes.

2.2.3 Partnering

#14 Company and suppliers/customers focus on improvement: To achieve a high productivity, a company has to work together with its suppliers and customers. As mentioned before, the needs of the customer have to be known by the company. In the same way the supplier needs to know the needs of the company. By cooperation, focused on improvement, the efficiency can be increased on both or all three sides. Regular meetings of interdisciplinary teams (of both or all three parties) ensure the exchange of knowledge and increase communication. Even special topics and problems can be discussed since the teams are interdisciplinary and know the subject matter. Based on these meetings, quality agreements should be established to ensure that the outcomes will be sustainable.

#15 Community, in terms of cooperative improvements: Knowledge is one of the most important factors to achieve improvement. By sharing information, practices, and experience with other companies that have similar processes, problems and mistakes that occur on the way to improvement can be avoided. Therefore, it is helpful to build up a community, for example, with competitors or companies in comparable sectors and exchange information and knowledge with them. To find the best method, the different practices and their results should be assessed, compared, and discussed.

2.3 Knowledge

The category knowledge includes 4 criteria:

#16 Skill of management: The management and its skill is crucial for a lean approach. It has to lead and motivate all employees. Dombrowski and Mielke (2014) mention lean leadership as a way to make a lean implementation sustainable. For achieving this, the managers have to develop themselves over years and have a deep knowledge about the whole company with all of its processes. To get the skills necessary for becoming a good leader, managers can participate in workshops and trainings. Furthermore, self-improvement can be achieved by self-reflection and literature study.

#17 Knowledge of and experiences with lean practices: A deep knowledge about lean and experiences with lean practices on the management level are obligatory for a sustainable lean approach. It is necessary to make the right decisions at the right time, and choose the suitable tools. If the knowledge or experiences is not given, external experts can be hired to assist during the lean approach. Taking into consideration that this help is temporary, managers have to participate in workshops and trainings to improve in this field. Additionally, lean literature should be studied and applied in consultation with the external experts.

#18 Skill of workforce / #19 Training and improvement: Beside the management, also all employees of the company need to have the necessary skills to implement lean and apply the lean tools. The skill of workforce is another crucial factor, beside their motivation and understanding of the lean effort and its benefits. Even if the workers want to apply the changes of the lean program, it will not be sustainable if they do not have the necessary knowledge and capabilities. The employees are expected to conduct the lean program, for example, by the use of problem-solving skills, which requires long-term employee development. This can only be achieved through daily development by coaching, and as far as possible, every employee should be developed at its individual level. These trainings should take place in short cycles based on PDCA.

Workers should be encouraged to learn from each other. This can be accomplished by monthly communication meetings with workers from the same workplace or overlapping shifts, where workers can exchange their knowledge and practical experiences at a workstation. The development of employees is a never ending process, and the trainings have to be continuous as well. Knowledge and skills should be regularly reviewed to assess improvement and identify the training topic in need. A small leader-to-employee ratio ensures that every worker will be developed in a proper manner, because the leaders can pay attention to each single employee. Dombrowski and Mielke (2014) recommend a ratio of 1:5 at operational level, while it can be up to 1:10 at higher levels.

2.4 Process

The category process includes 5 criteria:

#20 Systematic identification and elimination of all types of waste: As mentioned before, the nine types of waste are: over production, unnecessary stock, inefficient transportation, unnecessary motion, waiting times, rejects & defects, inappropriate processing. The basic idea behind lean is to eliminate all of them. Therefore, they have to be identified systematically in a process. All work methods and organizational processes have to be comprehensively documented and verified (Bou-Llusar et al., 2009). Quality manuals and organizational processes have to be periodically revised, to keep the processes at its highest level. If a waste is identified, there have to be changes made in the process. These changes can be suggested by employees, found in literature, or developed in special meetings. After these changes are implemented by teaching the employees and placement of manuals at the workstations, they have to be made sustainable and reviewed regularly. A system of indicators should be installed to check if unwanted changes are made in a process.

#21 Standardization of processes: The standardization of processes is one of the basics of lean and represents control of the process, constructing the process as simply as possible (Duarte and Cruz-Machado, 2013). It reduces improvisation or ineffective actions, makes work easier, and provides security. Dombrowski and Zahn (2011) say that "standardization in the context of working standards means that specific methods (e.g. construction methods) have to be standardized, coordinated with suppliers, and be state of the art". To define standards, the best possible process has to be found. This can be made in meetings with workers, since they know their workplace and the process best. The implementation of new standards can be conducted by training the affected employees, and installation of posters of the standardized process to ensure it stays in the mind of the workers. Moreover, there should be reviews to check if the standardized process is applied.

#22 Cellular process arrangement and improvement: Cellular manufacturing is a workplace design model, which takes advantage of the similarity between parts or processes. Similar machines are placed close together or machines are grouped together, according to the similarity of the parts produced. Processes are more robust to machine breakdowns, the material flow improves (reduce of transport, waiting and process times), and lead times are reduced (Bhasin and Burcher, 2006).

#23 Just-in-time (JIT): JIT is one of the two pillars of the Toyota Production System (TPS), which is the system lean management is based on. It reduces the inner waste of resources with the smallest investment achieving the biggest output. Its main objective is to produce the required number at the moment the customer orders it (customer orientation). According to R. Haak (2006) JIT means the parts needed to reach the next processes step at the time they are needed in a flow process. Furthermore, it is mentioned that a company which established this flow throughout can approach zero inventory. Another method is simultaneous engineering (SE), where fixed contact persons in every department are responsible to ensure the parallelization of tasks. Detailed information about the different methods can be found in literature (Dombrowski and Zahn, 2011). Quick changeover is another concept to reduce downtime. It enables a company to produce small volumes of a large variety of products while still maintaining the advantages of mass production.

#24 Automation (Jidoka) and multiprocess handling: The other pillar of the TPS is Jidoka, which can be translated as automation. It includes the concept of automation and of autonomous monitoring for defects and elimination of their causes. Every process has to be analyzed on opportunities for automation, and if it is possible and useful, the process should be automatized. Since a defect can cause an immediate standstill of the machines in automated processes, the production workers have to be in a position to find the defect as soon as possible and fix it to minimize the production down time (in-station quality control) (Haak, 2006). The so called Andon line, allows employees to call for help if they notice a problem and even stop the production line if the problem cannot be solved directly (Schlichting, 2009). Multiprocess handling means that one operator works across several machines in one process, rather than have different operators. Since the product will flow with the operator, the workflow will be constant.

3 Discussion and Conclusion

The purpose of this paper is to develop an assessment tool that helps an organization to implement lean in a sustainable way. Initially, the success factors of a lean approach have been identified and different business models have been reviewed. The research has shown that there are 4 main categories of crucial factors (leadership, culture, knowledge and process), including 24 criteria in total. Based on the literature, different countermeasures were chosen for each criterion. A company can rate the criteria to reveal its needs for improvement, and react immediately by applying the respective countermeasures. This tool gathers a lot of knowledge and information, by combining various findings of different research and case studies. It helps organizations to implement lean, but the management has to be aware, that every lean approach is individual. The conditions and environment change from company to company, thus the proper use of the proposed countermeasures.

The limitations of this developed tool are related to the lack of experience. Though the tool is partly based on case studies and assessments, it is still a theoretical approach and it has not been verified in practice. Further research is necessary to analyze the impact of this tool on lean implementations. Additionally, focus could be placed on the rating, and a model to measure the value of a criterion could be developed.

References

Bhasin, S.: Improving performance through Lean. International Journal of Management Science and Engineering Management 6(1), 23–36 (2011)

Bhasin, S., Burcher, P.: Lean viewed as a philosophy. Journal of Manufacturing Technology Management 17(1), 56–72 (2006)

Bou-Llusar, J.C., Escrig-Tena, A.B., Roca-Puig, V., Beltrán-Martín, I.: An empirical assessment of the EFQM excellence model: evaluation as a TQM framework relative to the MBNQA model. Journal of Operations Management 27(1), 1–22 (2009)

Dombrowski, U., Mielke, T.: Lean Leadership–Fundamental Principles and their Application. Procedia CIRP 7, 569–574 (2013)

Dombrowski, U., Mielke, T.: Lean Leadership–15 Rules for a Sustainable Lean Implementation. Procedia CIRP 17, 565–570 (2014)

Dombrowski, U., Zahn, T.: Design of a lean development framework. In: 2011 IEEE International Conference on Industrial Engineering and Engineering Management (IEEM), pp. 1917–1921 (2011)

Duarte, S., Cruz-Machado, V.: Modelling lean and green: a review from business models. International Journal of Lean Six Sigma 4(3), 228–250 (2013)

Haak, R.: Implementing Process Innovation—The Case of the Toyota Production System. In: Management of Technology and Innovation in Japan, pp. 185–203. Springer (2006)

Liker, J., Rother, M.: Why lean programs fail, Lean Enterprise Institute (2013)

Liker, J.K.: The Toyota Way: 14 Management Principles from The World's Greatest Manufacturer. McGraw-Hill (2004)

Losonci, D., Demeter, K., Jenei, I.: Factors influencing employee perceptions in lean transformations. International Journal of Production Economics 131(1), 30–43 (2011)

Miina, A.: Lean Problem: Why Companies Fail with Lean Implementation? Management 2(5), 232–250 (2012)

Ohno, T.: Toyota Production System: Beyond large scale production. Productivity Press, Portland (1988)

Scherrer-Rathje, M., Boyle, T.A., Deflorin, P.: Lean, take two! - Reflections from the second attempt at lean implementation. Business Horizons 52, 79–88 (2009)

Schlichting, D.I.B.C.: Sustaining Lean Improvements (Doctoral dissertation, Worcester Polytechnic Institute) (2009)

Womack, J.P., Jones, D.T.: Lean thinking: banish waste and create wealth in your corporation, Simon and Schuster (1996)

Concealing of $Al_2(SO_4)_3$ Stain by Spray Coating Process

Chachsanun Srisoy* and Suksan Prombanpong

Department of Production Engineering, Faculty of Engineering,
King Mongkut's University of Technology Thonburi (KMUTT), Bangkok, Thailand
chachsanun@hotmail.com, suksan.pro@gmail.com

Abstract. In hard anodizing process which requires an aluminum part soaked in the electrolysis solution in order to form anodic (aluminum oxide) film on the aluminum part so that it will be harder and durable for work. Although, this part will be drained afterwards by ionize water to eliminate impurities from it, aluminum sulfate stain may be traceable from the part. It is especially for cookware products that require clear coating on an anodized part for aesthetic purpose. It can be imagined that the trace mark from aluminum sulfate can scrap this part or it has to be reworked. However, in coating process, there are three coating steps named prime, middle, top that may aid to conceal the mark as long as the required coating thickness is within specification. However, only middle level is concerned in this work. Therefore, the objective of this paper is two folds. One is to investigate the relationship of spray coating parameters to dry film thickness (DFT) and the other is to determine DFT which is able to conceal the abovementioned trace on the hard anodizing part.

Keywords: $Al_2(SO_4)_3$ stain, Hard anodizing, Spray coating.

1 Introduction

Hard anodizing is an electrochemical process for surface hardening on an aluminum part where an anodic substrate is built up and aluminum surface is protected and hardened. The process is quite simple where an aluminum work part is an anode and the aluminum bar is normally used as cathode. Then the aluminum parts are submerged in sulfuric acid electrolyte. The direct current will be applied to the bath tank and aluminum oxide (Al_2O_3) will be generated on an aluminum surface. For every two microns of anodic (oxide) film produced one micron of the base metal is consumed in the reaction, and pro-rata. The anodic film is transparent and therefore the type of pretreatment given to the aluminum can influence the appearance of the final product before anodizing [1,2,3,4,5]. Due to the fact that the chemical reaction between sulfuric acid and aluminum create aluminum oxide and aluminum sulfate where aluminum oxide built up at the aluminum substrate becomes anodic film. However, sulfuric acid and aluminum sulfate solution still wet the aluminum part;

* Corresponding author.

M. Gen et al. (eds.), *Industrial Engineering, Management Science and Applications 2015*,
Lecture Notes in Electrical Engineering 349, DOI: 10.1007/978-3-662-47200-2_85

consequently, it will be neutralized by sodium hydroxide solution and ionized water before transferring to oven for quick dry. Although the aluminum part is dried by heat wave at a short period of time, liquid can be trapped in some area and cause trace mark on the part as shown in Fig. 1. Normally, an anodized cookware will be clear coated or colored spray coated for good appearance. The trace mark may be eminently seen by eyes and will be scraped if it is clear coated otherwise it could be reworked by removing hard anodized film and reprocessing as described earlier. Therefore, the main objective of this paper is two folds. The first one is to investigate the relationship of spray coating parameters to dry film thickness (DFT) and the second one but more important is to determine DFT which is able to conceal the abovementioned trace on the hard anodizing part. The key research issue is that at what DFT should be used to reduce an adverse effect of trace mark. It should be noted that if coating thickness or DFT is too thin, it might not be able to conceal the trace mark. On the other hand, if coating thickness (DFT) is too thick, it may be able to conceal the trace mark but other defects such as peeling or droplet may be occurred and it is not desirable.

Fig. 1. Trace mark appears on a part

1.1 The Relationship between DFT and Trace Mark

DFT is dry film thickness which is used to measure coating thickness. Typically, the shape of spray area is rather an ellipse as shown in Fig. 2 and it can even be distorted under the non-perpendicular setups of spray gun to the target surface as depicted in Fig. 3 In addition, the spray coverage increases with the span angles of spray gun and the stand-off distance between the spray gun and target surface. In spray coating process, the spray area (A) can be geometrically expressed as [6]:

$$A = \frac{\pi * Z^2 \tan\alpha \tan\beta}{\cos\theta} \qquad (1)$$

where Z, α, β and θ are the spray gun standoff-distance, span angle respected to the major axis, span angle respecting to the minor axis, and the inclination angle of spray gun, respectively.

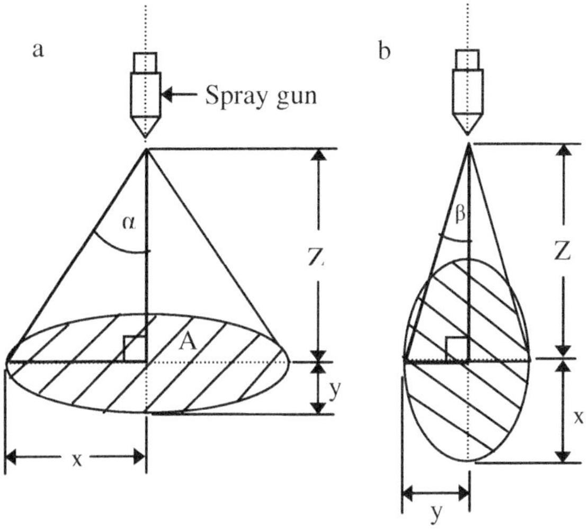

Fig. 2. Span angles and standoff-distance in the spraying process: (a) front view; (b) side view

By taking the assumptions that there is no drag force during the spraying process and the distribution of atomization across the spray is uniform. It can be implied that the coating layer thickness is uniform.

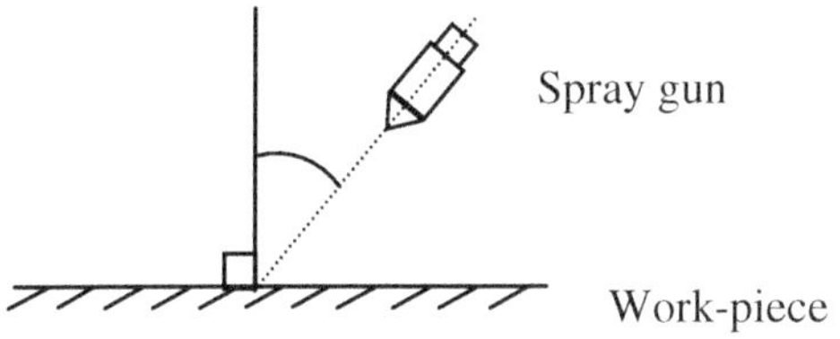

Fig. 3. Inclination angle of spray gun

Moreover, by neglecting the loss of spray and taking the conservation of mass into account, the coating layer thickness can be estimated when the volume of coating material is given. However, this thickness is according to the coating material in liquid status. After water or other liquid solutions is dried, the coating material in solid form is remained as a coating layer, whose thickness is normally thinner than the wet deposition. The dry film thickness is of interest in many painting and coating processes, needing to be well justified. Based on the assumptions given above and the spray area expressed in (1), the dry film thickness (T_{DFT}) can be parametrically calculated by:

$$T_{DFT} = k * \frac{m * \cos\theta}{\rho_{cm} * \pi * Z^2 * \tan\alpha \tan\beta} \quad (2)$$

Where m, ρ_{cm} and k are weight of coating material in liquid solution, coating material density, and coefficient of spray coating process respectively.

A few number of literatures conducted research in spray coating process. It can be summarized that important variables in spray coating are nozzle size, spray time, air pressure and spray distance [7,8,9,10]. DFT is a measurement of coating thickness which is related to an ability of concealing the trace mark. Thus, in this study for each experiment DFT will be measured and also the appearance of the part to check whether the trace mark can be concealed or not.

2 The Experimental Design

The experimental design technique is widely used when concerning of multi-variables problem and the experiment should be performed to obtain the solution [11]. In the coating process, there are 3 coating levels named prime, middle, and top. Note that thickness at middle level plays the most crucial role to the result. This is due to the fact that the purpose of prime level coating is to increase roughness to the hard anodized part for better spray result at middle level so the prime level thickness is not too critical. Likewise, the top level is a clear coating which is to increase shiny appearance. As mentioned earlier the key parameters of spray gun are nozzle size, spray time, and air pressure. Thus, to perform an experiment all of these variables must be designed and they are tabulated in Table 1. Note that each variable is defined maximum and minimum value as an example minimum spray time is 1.5 whereas maximum spray time is 2.0 s. The 2k factorial design is used in the experimental design; therefore, eight experiments will be conducted for 5 replications. Thus a total of 40 sets of experiment will be conducted. Note that each experiment will be randomly conducted due to statistical practice.

Table 1. Level of process parameters

Factors	Min	Max
Nozzle size (round)	0.5	1
Spray time (s)	1.5	2.0
Air pressure (bar)	2	3

3 Results

The experimental data will be processed by MINITAB software program for variance analysis and will be described next.

3.1 Statistical Analysis

In order to obtain the optimal solution, the experimental design must be performed. This involves an experiment on a set of data. In this case, 40 set of experiments were conducted. Then all data must be analyzed and verified to ensure no bias occurring during the experiment. Thus, the normal probability test must be performed. It is found that the p-value obtained from the analysis was 0.118, indicating that the data was in the normal distribution. Then, the equal variance test was also conducted to justify whether all test conditions share the same variance. The p-value obtained from this analysis was 0.298, showing that there is no deviation in variance of each test. The p-value of random test is 0.601 as shown in Table 2.After the data were verified through normal distribution and equal variance test, the analysis of variance (ANOVA) was subsequently conducted to determine the effects of these three parameters on the number of defective part. The criteria used to determine the significant factors are the p-value. If the p-value is less than 0.05, this corresponding value is considered significant to the response value. The obtained results of p-value in the experiment are tabulated in Table 3.

Table 2. P-value of model adequacy

Model Adequacy Checking	P-value
Normal Distribution	0.118
Equal variance	0.298
Randomization	0.601

Table 3. Result of ANOVA

Source	DF	Seq SS	Adj MS	F	P-value
Nozzle Size	1	1537.60	1537.60	896.23	0.000
Spray Time	1	172.23	172.23	100.39	0.000
Air Pressure	1	291.60	291.60	169.97	0.000
Nozzle Size*Spray Time	1	75.63	75.63	44.08	0.000
Nozzle Size*Air Pressure	1	52.90	52.90	30.83	0.000
Spray Time*Air Pressure	1	21.03	21.03	12.26	0.001
Nozzle Size*Spray Time* Air Pressure	1	38.03	38.03	22.16	0.000
Error	32	54.90	1.72		
Total	39	2243.90			
S = 1.30982, R-Sq = 97.55%, R-Sq (pred) = 96.18%, R-Sq (adj) = 97.02%					

3.2 Effect of Spray Conditions

The result from Table 3 can be concluded that all 3 parameters i.e. nozzle size, spray time and air pressure are main effect to the ability to conceal the trace mark. The optimal condition is that nozzle size at 0.8 turning, spray time at 1.50 s, and air pressure at 2 bar respectively. In addition, the regression model can be obtained as shown in (3)

$$DFT = 83.7-127.2x_1-52.2x_2-29.3x_3+100x_1x_2+45.4x_1x_3+17.6x_2x_3-31.2x_1x_2x_3 \quad (3)$$

Where x_1 is nozzle size, x_2 is spray time and x_3 is air pressure.

The main effect plot of nozzle size, spray time, air pressure to DFT is shown in Fig. 4-6 respectively. It can be seen from Fig. 4 and 5 that an increase of variables, DFT is increasing. On the contrary, if air pressure is increasing, DFT is decreasing.

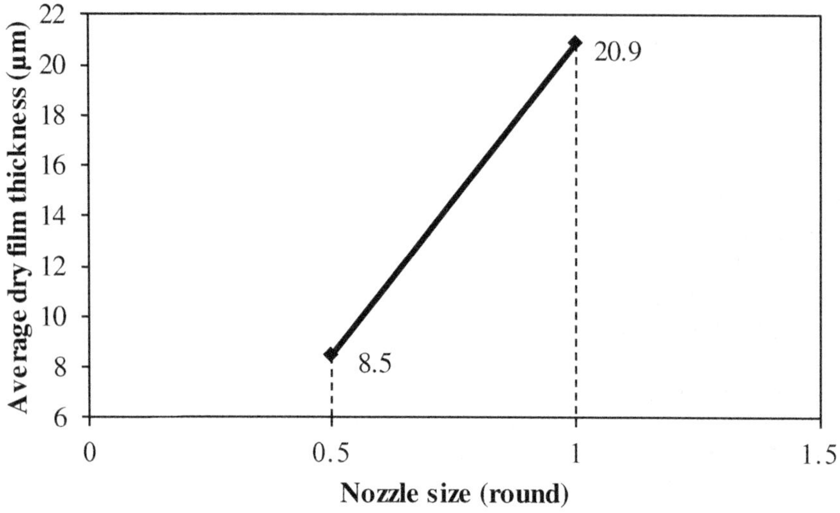

Fig. 4. Main effect plot of nozzle size

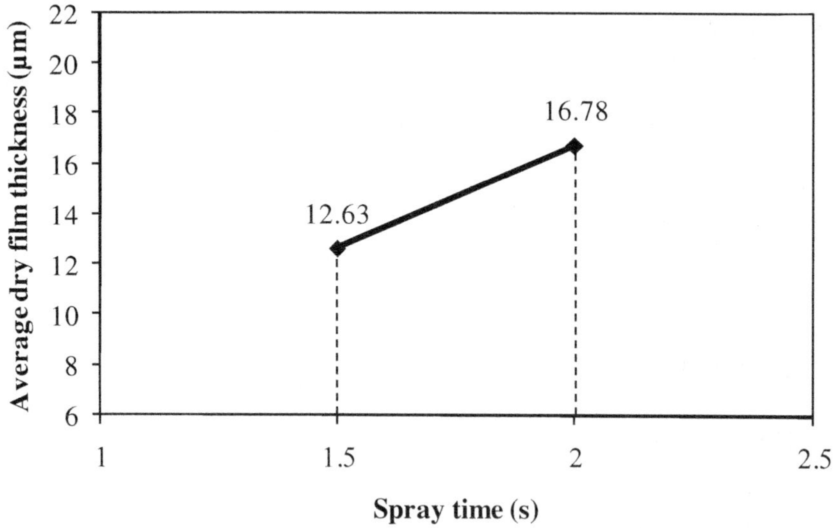

Fig. 5. Main effect plot of spray time

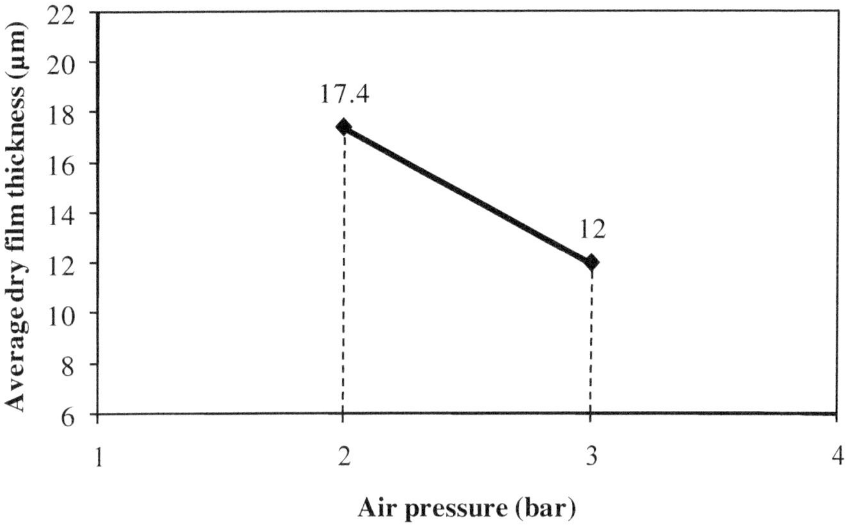

Fig. 6. Main effect plot of air pressure

4 Discussion and Conclusion

The problem of aluminum sulfate trace mark can be subsided during spray coating by increasing coating thickness so that coating material can conceal the mark. However, if DFT is too thick, other defects can occurs. Therefore, an experimental design is implemented to see the relationship between DFT and spray coating parameters i.e. nozzle size, spray time, and air pressure. It can be concluded that at an optimum conditions of DFT at 15.6 µm at middle coating level provides the most favorable result.

References

1. Samon, H.: Preparation and Evaluation of Hard Anodized Coating on Aluminium, King mongkuts Univesity of Technology Thonburi, Thailand (1994)
2. Parinya, S.: Aluminum Color-Coating Machine by Anodize Methodology. Rajamangala University of Thechnology Krungthep, Thailand (2010)
3. Trompette, J.L., Arurault, L., Fontorbes, S., Massot, L.: Influence of the Anion Specificity on the the Electrochemical Corrosion of Anodized Aluminum Substrates. Electrochimica Acta 55, 2901–2910 (2010)
4. Chung, C.K., Liao, M.W., Chang, H.C., Lee, C.T.: Effects of Temperature and Voltage Mode on Nanoporous Anodizc Aluminum Oxide Files by on-Step Anodization. Thin Solid Films 520, 1554–1558 (2011)
5. Han, X.Y., Shen, W.Z.: Mproved Two-Step Anodization Technique for Ordered Porous Anodic Aluminum Membranes. Journal of Electroanalytical Chemistry 655, 56–64 (2011)

6. Luangkularb, S., Prombanpong, S.: Material Consumption and Dry Film Thickness in Spray Coating Process. In: Proceedings of the 47th CIRP Conference on Manufacturing Systems, vol. 17, pp. 789–794 (2014)
7. From, P., Gunnar, J., Gravdahl, J.: Optimal paint gun orientation in spray paint applications-experimental results. Proc. IEEE Transaction on Automation Science and Engineering 8, 438–442 (2011)
8. Kout, A., Muller, H.: Parameter optimization for spray coating. Proc. of the Advances in Engineering Software 40(10), 1078–1086 (2009)
9. Sheng, W., Chen, H., Xi, N., Chen, Y.: Tool path planning for compound surfaces in spray forming processes. Proc. IEEE Transaction on Automation Science and Engineering 2, 240–249 (2005)
10. Youdong, C., Xiaodong, L.U., Hongxing, W.: Trajectory planning based on latitude for painting on spherical surface. In: Proc. IEEE Industrial informatics (INDIN), pp. 431–435 (2012)
11. Somgumnerd, J., Tangwarodomnukun, V., Prombanpong, S.: Effects of Flap Wheel Grinding Parameters on Surface Roughness for Stainless Steel. The Applied Mechanics and Materials, 548–549, 506-509 (2014)

Developing Interfaces Based on Services to the Cloud Manufacturing: Plug and Produce

Eduardo Cardoso Moraes[1,*], Herman Augusto Lepikson[2], and Armando Walter Colombo[3]

[1] Federal Institute of Alagoas-IFAL, Federal University of Bahia - PEI/UFBA
Maceió, Alagoas, Brazil/Salvador, Bahia, Brazil
[2] CTAI/PEI, Federal University of Bahia-UFBA
Salvador, Bahia, Brazil
[3] University of Applied Sciences Emden/Leer, i2ar
Emden, Germany
eduardo.moraes@ifal.edu.br, herman@ufba.br,
armando.colombo@hs-emdenleer.de

Abstract. In the last decade the industrial automation had been making important changes integrating Internet and other emergent technologies, concepts, methods and tools, as cloud computing and Service Oriented Architecture(SOA) approaches. As a first consequence, the industrial automation context is also increasing in complexity, presenting new challenges for engineers and managers. Examples include dealing with autonomous cyber physical systems, evolvable heterogeneous structures that do not appear completely formed and where functions and goal are added, removed or modified along the life cycle, as well as handling emergent properties and behaviours of entire systems, e.g. resulting from integrating new systems being not localized to any single system component. [1]. The objective of this paper is to make a review in industrial automation background, analyze the technological trends and challenges and propose a flexible and autoconfigurable service-oriented model, making up basis for the cloud manufacturing. In this model the entities are loosely coupled, due the fact that a service's functionality communication is showed through its interfaces.

Keywords: SOA, CPS, cloud manufacturing.

1 Introduction

Nowadays enterprises face a more and more dynamic market. In order to be competitive and be profitable they are encouraged to respond to market changes as fast as possible. Globalization, high variability, mass production, cost reduction pressure, reduced product life-time, creates the need of permanently changing business goals. It becomes vital specially because it increases competitive pressures, hence enterprises

* Corresponding author.

M. Gen et al. (eds.), *Industrial Engineering, Management Science and Applications 2015*,
Lecture Notes in Electrical Engineering 349, DOI: 10.1007/978-3-662-47200-2_86

face the necessity to quickly capitalize opportunities when they arise. The main challenge in doing so is the act of transformation of already established organizations, processes, infrastructure and systems into a dynamic business model.

The modern companies need to be agile and dynamically support decision making processes at several levels. To elaborate this scenario, information is a critical need to be available at the right point in a timely manner at several levels. Especially with intra and inter collaboration in mind, providing fine grained info when it is needed and in the right form is a challenging task.

In the last decade the industrial automation had been making important changes integrating Internet and other emergent technologies, concepts, methods and tools, as cloud computing and Service Oriented Architecture(SOA) approaches. As a first consequence, the industrial automation context is also increasing in complexity, presenting new challenges for engineers and managers. Examples include dealing with autonomous cyber physical systems, evolvable heterogeneous structures that do not appear completely formed and where functions and goal are added, removed or modified along the life cycle, as well as handling emergent properties and behaviours of entire systems, e.g. resulting from integrating new systems being not localized to any single system component. [1]. Moreover, by using advanced data analytics, networked machines will be able to perform more efficiently, collaboratively, autonomous and resiliently. Such trend is transforming manufacturing industry to the next generation, namely Industry 4.0[2].

The objective of this paper is to make a review in industrial automation background, analyze the technological trends and challenges and propose a flexible and autoconfigurable service-oriented model, making up basis for the cloud manufacturing. In this model the entities are loosely coupled, due the fact that a service's functionality communication is showed through its interfaces[3].

1.1 Structure of the Paper

The paper is structured as follows. Chapter 2 presents a brief overview of the state-of–the-art in the area of industrial automation. Chapter 3 discuss about new trends and approaches of the industry 4.0. Chapter 4 introduces a novel, flexible and autoconfigurable service-oriented model and details the major characteristics. The final chapter contains the conclusions and points out future challenges.

2 Industrial Automation Background

As discussed last chapter agility and flexibility are fundamental requirements for modern industrial automation enterprises in order to reach customers' expectations and challenges that come by globalization, environmental and working conditions regulations, improved standards for quality and fast technological mutation. A modern agile organization requires a deep integration and straight alignment between high level processes, for example complex supply chain management, storage and production management. If a lower level (e.g. devices level behavior) environment cannot accomplish the desired agility goals, the overall system will be incapable of delivering the expected agile performance.

The industrial automation field should be able to overcame know issues as explained by[4] :

- Long time for system design, commissioning and setup.
- Complex and time-consuming reengineering phase to trail requirements variations.
- Inflexible centralized/hierarchical implementations.
- Scalability involves exponential complexity.
- No default fault-tolerance or redundancy.
- Incompatibility between different vendors equipment and legacy systems.
- Shop floor isolated from higher level business Information and Communications Technologies (ICT).

Normally, today's manufacturing systems are structured in a 5-level hierarchical model. This hierarchical model is explicated in ISA-95, it is an international standard from the International Society of Automation for developing an automated interface between enterprise and control systems. The standard defines hierarchical models defining data flows, objects, operation activity models, attributes to improve the exchange of information between different systems. The limitation is this standard is abstract and every integrator can implement in a different manner. This model is structured in UML models, which are the basis for the development of standard interfaces between ERP and MES systems.

SCADA (supervisory control and data acquisition) and Programmable logic controller (PLC) systems are in the heart of the modern industrial infrastructure[5]. SCADA controls and monitor lower levels in a centralized way and spread out data. Most control actions are performed automatically by RTUs(Remote Terminal Units) or by PLCs. Host control functions are usually restricted to basic overriding or supervisory level intervention. The rapid changes in the networked embedded systems and the way industrial applications are designed and implemented, call for a shift in the architectural paradigm. Next generation SCADA and DCS systems will be able to deal with intra and inter collaboration with the shop-floor devices as well as heterogeneous networks and enterprise applications. Cloud driven by (web) service based interactions will enable stronger coupling of real-world and the business side, leading to a new generation of monitoring and control applications and services witnessed as the integration of large-scale systems of systems that are constantly growing to address new user needs.

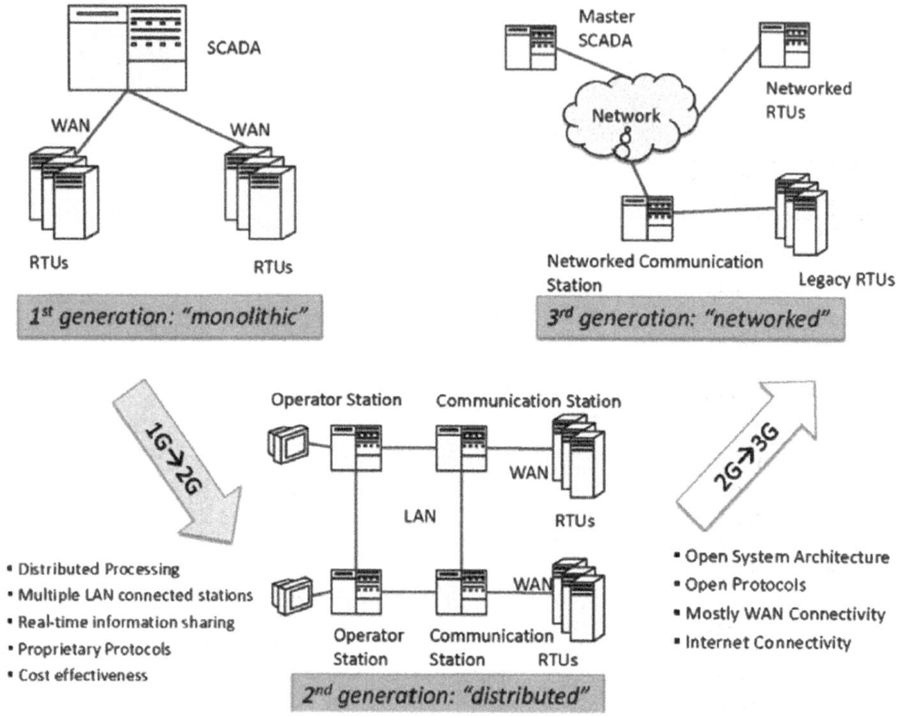

Fig. 1. Generations of the PLC/SCADA systems [5]

Figure 1 shows the evolution of PLC and SCADA systems, from centralized and using proprietary protocols, in a static scenario to meet at the second generation a distributed processing and data sharing, multiple LANs, computer operation stations, and we are on the way to use open architectures and protocols, most of the had been benefited with the high speed and low price of Ethernet networking, moving towards a wide internet connectivity, using service oriented approach and Cloud manufacturing. This evolvable context depends of constant effort towards laying down the fundamentals of structuring complex production systems, defining models and exchange formats of information, most of them follow the ISA 95 ISA 95 / IEC 62264 standard representing different aspects of equipment used or to define suitable protocols for horizontal or vertical communication. There are various standards addressing that area on global or specific level, Colombo[1] and Cândido [6] provide a good overview about these efforts.

Networked manufacturing is to realize information integration and resource sharing inside and outside the manufacturing enterprise based on the advanced network techniques.

Nowadays the shop floor is still isolated from higher level business Information and Communications Technologies (ICT), but the need of collaborative manufacturing of products, and improvement production is a trend of enterprises. One of the biggest current research challenges in networked manufacturing is to implement the information integration and sharing of the heterogeneous and distributed application systems existed in the manufacturing enterprise.

From a technical perspective, different applications are written in different programming languages, running on different platforms, and communicating through different networks. From a business perspective, they provide different services and have to meet different integration needs[7]. Therefore, a fundamental challenge in the networked manufacturing of real-time information analysis and integration and sharing among different application systems, and how to construct an information integration service platform for reach the strategic level and providing on demand changes in a short time space. For this happen the amount of data is generated by real world devices needs to be integrated, processed within a specific context and communicated on-time, traditional approaches aiming at the efficient data inclusion in enterprise services need to be changed[5].

3 Technology Trends for Industry 4.0

The advances in Information technology(IT), physical components and network communication led to a new generation of embedded devices with increasing intelligence and networking capabilities, providing new possibilities of distributed process, integration and autonomy[8].

In this context new elements will be connected and the next step is illustrated as Internet of things(IoT) were the Cyber Phisycal Systems(CPS) resources will connect each other in a natural extension of the networking paradigm into control devices by allowing the different networks to join and form a homogenous networking. The fundament of Cyber-Physical Systems and Internet of Things are Service oriented architecture(SOA), cloud computing, System of system integration, embedded resources, etc. In this article we don´t desire to exhaust these emerging and important technologies, but give an overview of the trends and the possible impacts of them in manufacturing.

3.1 CPS

A cyber physical system (CPS) is a new concept that connects the virtual world with the physical world. There are no consensus definition, but The National Science Foundation (NSF) CPS Summit defines CPS as a physical and engineered system whose operations are monitored, coordinated, controlled and integrated by a computing and communication core[9].

A CPS has two parallel networks to control, namely a physical network of interconnected components of the infrastructure and a cyber network comprised of intelligent controllers and is the result from multiple advances such as embedded systems and sensor networks the communication links among them. By integrating the "triple c" capabilities (computing, communication, control) and incorporating research, CPS integrates these networks dependably, safely, securely, efficiently and in real time . It is achieved by using a set of sensors, actuators , RTUs, PLCs and communication devices (routers, GSM modems), as shown in Fig. 2[2].

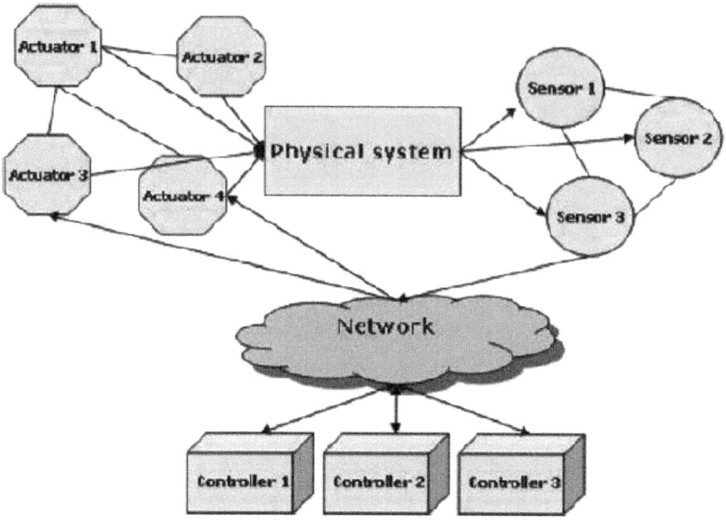

Fig. 2. Architecture of CPS [9]

By embedding more intelligence is each device, the overall system agility is expected to improve by having more autonomous, intelligent and self-contained devices. Also, the level of abstraction and complexity of the exchanged information will rise and devices low-level intricacies will be circumvented. Devices easier to setup, manage, monitor and diagnose are a key-factor during reengineering or down-time phases by saving a considerable amount of integration time that subsequently conditions production capacity.

The connection between shop-floor devices and the enterprises services is fundamental to create more sophisticated high level services and to support more reliable decision making and feedback.

3.2 Service Oriented Architecture (SOA) Input in Industrial Automation

SOA is an approach originated in IT field, it tries to align the IT goals with the world of business. Many times projects failed due not understand deeper the stakeholds expectations and not align the technical view with the business strategy. To minimize this pain the Service oriented architecture is a proposal based on the concept of software service. A software service is performed by a software program. It produces effects that have value to the people or organizations that are its consumers. The approach is based on some paradigms as low coupling, distributed processing, reuse, components, messaging, Internet based protocols, a complete overview can be found at [10].

The manufacturing world using this approach to create services, that can be associated and used in a cross-layer way. The hierarchical 5-level structure as described will became more flexible and the survival enterprises will need to adjust their process in a lower time to deal with the newcomer context. To reach this aim is vital the higher integration in a horizontal e vertical level. Enterprise applications will be able to connect directly if needed to devices, without the use of proprietary drivers, while non-web-service enabled devices can still be attached and their functionality wrapped by service mediators or at middleware layer Peer to peer communication among the devices is already pushing SOA concepts down to device layer and create new opportunities for functionality discovery and collaboration.

Networked embedded systems have become more powerful with respect to computing power, memory, and communication; therefore they are starting to be built with the goal to offer their functionality as one or more services for consumption by other devices or services. Due to these advances we are slowly witnessing a paradigm shift where devices can offer more advanced access to their functionality and even host and execute business intelligence, therefore effectively providing the building blocks for expansion of service-oriented architecture concepts down to their layer. As such, event based information can be read, processed on-device and in-network , without the need of storage in intermediate databases and processing by third parties, and eventually be delivered to the corresponding business processes. This capability provides new ground for approaches that can be more dynamic and highly sophisticated, and that can take advantage of the context specifics available[5].

Web services are suitable and capable of running natively on embedded devices, providing an interoperability layer and easy coupling with other components in highly heterogeneous shop-floors. Device Profile for Web Services (DPWS) and OPC UA are emerging technologies for realizing web service enabled controllers and devices. Several projects such as SIRENA (www.sirena-itea.org), SODA (www.soda-itea.org), SOCRADES (www.socrades.eu) and IMC-AESOP (http://www.imc-aesop.eu/) have experimented with SOA-ready industrial automation devices and their integration on industrial applications.

4 Developing Interfaces to the Cloud Manufacturing

Previous chapter described the advances from IT world that are impacting on enterprises worldwide. Cyber physical Systems (CPS), Service Oriented Architecture(SOA), Internet of things and cloud computing have been identified as main business technology trends that will reshape enterprises worldwide. The manufacturing industry is going through transformations enabled by IT and these smart technologies. Cloud computing is one of main smart technologies and derived by this trend recently was proposed the concept of cloud manufacturing.

The figure 3 describes the different models, hierarchy and technologies for the industrial automation. It is clear the plurality of technologies and standards from shop floor from to the ERP level. The differences are not only information model but into the nature of the subjects. For example the way that one shop floor device captures data is cyclical and the systems in an upper level work with event data or information extracted from a databases.

To integrate these complexity, the IMC-AESOP project proposed the Complex event Processing(CEP), that is a mediator between the devices and the upper levels.

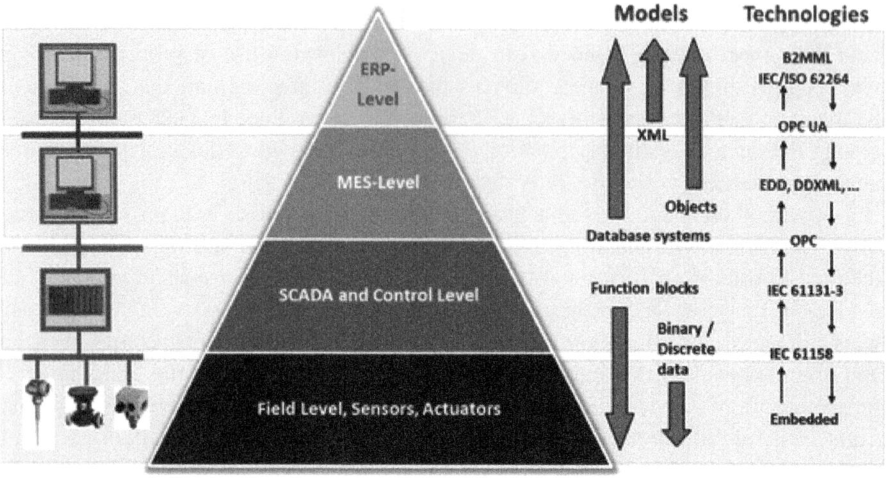

Fig. 3. Levels, models and technologies industrial automation [1]

This mediator received the data from the devices and manage/monitor/control this data according to upper levels definitions, like MES. This work desires to go on the results from the IMC-AESOP and define interfaces from the shop floor level to improve reconfiguration of production devices connected and components that have to be automatically integrated into the existing communication network. This applies to new applications or the old infrastructure. Particularly on the shop floor level the transmission of the process relevant data is time critical and requires a real-time capable communication network. Today it is usually realized with field bus technology. But recently industrial Ethernet technology is substituting the fieldbuses on the shop floor for Ethernet-based networks and with this changes will be possible to avail the standards from IT world using interfaces. Interface here is describe as a service, with schemas(metadata) describing the rules and objects. Service-oriented applications provide an integrated communications infrastructure, down from the lowest levels of the manufacturing device hierarchy up into the manufacturing enterprise's higher level business process management systems. It provides the necessary support for better decision-making, as well as highlights the inherent complexity of activity.

Figure 4 describes how is modeled these interfaces and how it can interact in a web-based and web-service oriented platform. The advance of this approach is that using interfaces the provider and the client uses the Universal Description, Discovery, and Integration (UDDI) protocol to discover the available services using a XML file with a pre-defined structure. It is built on the Simple Object Access Protocol (SOAP) data communication standard, UDDI creates a global, platform-independent, open architecture space that will benefit business.

Plug and produce systems should adapt automatically to new scenarios such as new product variants or to new production modules. To develop such a plug and produce capability, a different engineering chain is needed: Automation engineers will specify production aims, for example sub components or products, instead of automation algorithms. To reach reconfiguration and the plug and produce goal, these interfaces should provide methods for automatic integration of devices in an IP based network without configuration and setup by means of specifying services for discovery, description and control of devices on the device level.

The figure 4 describes our proposal, it is based on three essential requirements:

- Identification of signals: The data processing to the variables of the control software has to proceed without manual interventions. A solution to this problem could be the definition of the semantic of all signals in a machine-readable.
- Conventional operations the IO-Controller get its configuration data from the engineering tool, we propose to use an orchestration unit to do that task.
- Autoconfiguration: Every time when the devices are first connected in the nework they get in touch with the orchestration, and the orchestration should have previously information about the device. The devices Plug-and-Produce ready requires the ability to increase and remove devices from the network without manual configuration of them.
- Real-time communication: In the automation industry networks on the shop floor based on SOA RTEs are in development. RTEs need to be configured during the design of the system in an engineering tool.

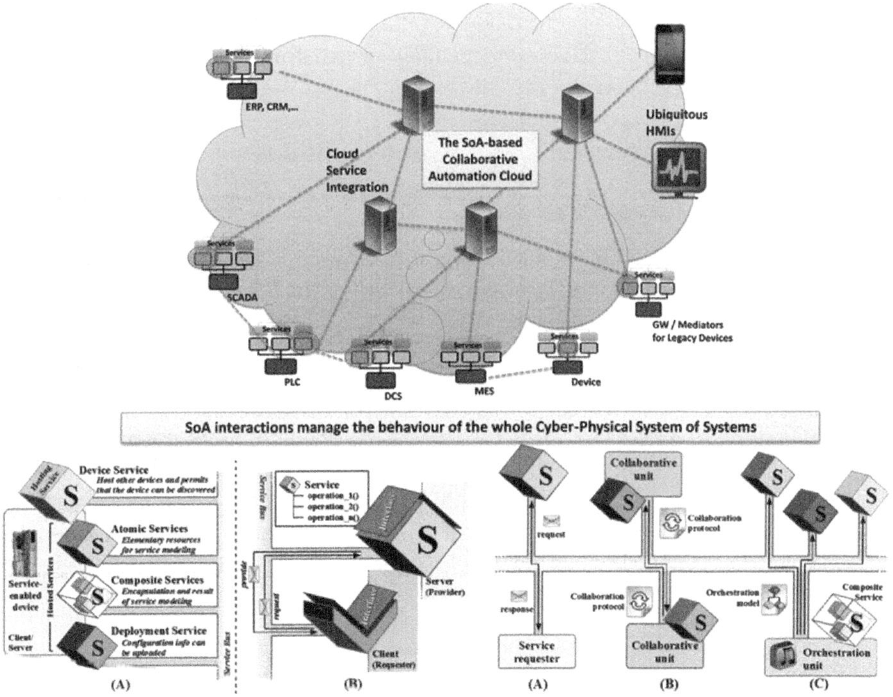

Fig. 4. Interfaces for cloud manufacturing [1]

5 Conclusions and Future Directions

As described in the paper the next revolution of industrial automation will be based in Smart devices, open protocols, Internet connectivity with higher interaction between devices resulting in combined systems that will have to handle these complexity, which the individual components alone would be unable to do, and to with a virtualization of the physical devices being distributed geographically and with a evolvable behavior.

Nowadays in shop floor we have resource constraint devices and in a network with a huge amount of devices this might congest the network and overload. Integrating CPS with services, production and logistics in the current industrial practices, it would change today's factories into an Industry 4.0 with significant economic potential. For instance, a joint report by the Fraunhofer Institute and the industry association Bitkom said that German gross value can be boosted by a cumulative 267 billion euros by 2025 after introducing Industry 4.0[11].

Moving from production-oriented manufacturing to service-oriented manufacturing, cloud manufacturing seems to be a natural next step. The manufacturing resources are distributed will be encapsulated into cloud services and managed in a centralized or decentralized way. Clients can use cloud services according to their requirements.

Cloud manufacturing can offer services ranging from product management manufacturing, testing, design and all other stages of a product life cycle.

SOA as well as the described technologies such as cloud-based approaches may act as glue for a paradigm revolution, in few years we will testimony big changes in manufacturing area.

Acknowledgments. The authors would like to acknowledge Brazilian agency CAPES (Coordenação de Aperfeiçoamento de Pessoal de Nível Superior) for the support, PDSE scholarship number 1849/2014.

References

1. Colombo, A.W., Karnouskos, S., Bangemann, T.: A system of systems view on collaborative industrial automation. In: 2013 IEEE International Conference on Industrial Technology (ICIT), pp. 1968–1975 (2013)
2. Lee, J., Bagheri, B., Kao, H.-A.: A Cyber-Physical Systems architecture for Industry 4.0-based manufacturing systems. Manufacturing Letters 3, 18–23 (2015)
3. Cardoso Moraes, E., Augusto Lepikson, H.: Service-Oriented Framework for Oil Fields Automation. Applied Mechanics and Materials 496, 1438–1441 (2014)
4. Cândido, G., Colombo, A.W., Barata, J., Jammes, F.: Service-oriented infrastructure to support the deployment of evolvable production systems. IEEE Transactions on Industrial Informatics 7, 759–767 (2011)
5. Karnouskos, S., Colombo, A.W.: Architecting the next generation of ser-vice-based SCADA/DCS system of systems. In: IECON 2011-37th Annual Conference on IEEE Industrial Electronics Society, pp. 359–364 (2011)

6. Cândido, G., Barata, J., Colombo, A.W.: Service-oriented infrastructure at device level to implement agile factories. In: 2012 IEEE International Conference on Systems, Man, and Cybernetics (SMC), pp. 1171–1176 (2012)
7. Yang, J., Li, J., Deng, X., Xu, K., Zhang, H.: A web services-based approach to develop a networked information integration service platform for gear enterprise. Journal of Intelligent Manufacturing 23, 1721–1732 (2012)
8. Moraes, E.C., Lepikson, H.A.: Potential Impact for Energy Consumption with Service Oriented Architecture. In: Advanced Materials Research, pp. 1925–1929 (2014)
9. Parvin, S., Hussain, F.K., Hussain, O.K., Thein, T., Park, J.S.: Multi-cyber framework for availability enhancement of cyber physical systems. Computing 95, 927–948 (2013)
10. Xu, L.D.: Enterprise Systems: State-Of-The-Art And Future Trends. IEEE Transactions on Industrial Informatics 7 (2011)
11. Lee, J., Lapira, E.: Predictive factories: the next transformation. Manuf. Leadership J (2013)

Optimization of Teflon Spraying Process for Non-stick Coating Application

Oraphan Poonkwan*, Viboon Tangwarodomnukun, and Suksan Prombanpong

Department of Production Engineering, Faculty of Engineering,
King Mongkut's University of Technology Thonburi (KMUTT), Bangkok, 10140, Thailand
{oraphan.pk,suksan.pro}@gmail.com, viboon.tan@kmutt.ac.th

Abstract. Spray coating process is a simple technique to develop a thin and uniform coating layer on a substrate, where the atomized coating material is transferred to the target surface for deposition by the assist of pressurized air flow. However, the utilization of coating material and deposition rate associated with this technique can be extremely low under the improper setup of spraying conditions. The process study and optimization of spray coating were experimentally examined in this paper to provide an insight into the process. Teflon was employed as the coating material due to its renowned on the non-stick feature. Air pressure, spray time, nozzle size and spray distance were optimized with regard to the material consumption and coating layer thickness. As per the experimental results, the small nozzle size of spray gun should be applied together with the low air pressure, low spray time and short spray distance from the target to minimize the Teflon consumption and to accurately control the dry film thickness obtained. The predictive models for these two responses were also presented in this research.

Keywords: Spray coating, Material consumption, Dry film thickness.

1 Introduction

Spray coating process has played a significant role in forming a thin protective film on the substrate to enhance some surface properties and/or aesthetic feature. The process has been normally implemented for coating Teflon on a metal substrate for the purpose of non-stick function. This is mostly recognized in the cookware manufacturing and surface coating of some specific machine tools in order to minimize the use of lubricant. The coating material in liquid status is pressurized and fed into an air spray gun where the material is mixed with pressurized air, atomized and then flew to deposit on the target surface. The coating layer is gradually developed on the target due to the surface tension and adhesion force between the coating material and substrate surface. Despite maintaining a distance between the spray gun and target surface, some of the atomized material is spread away from the jet mainstream facing the target surface. This is attributed to the air entrainment and turbulent flow occurring

* Corresponding author.

M. Gen et al. (eds.), *Industrial Engineering, Management Science and Applications 2015*,
Lecture Notes in Electrical Engineering 349, DOI: 10.1007/978-3-662-47200-2_87

around the jet. As such, the coating material consumption and the control of film characteristics could be the important concerns that critically affect the processing time and costs.

Luangkulab et al [1], Kout and Muller [2] and Winnichi et al [3] performed the optimization of spray coating process in order to minimize the coating material consumption while maintain the coating layer thickness conforming to the specified value. The full-factorial experimental design is generally used to reveal the effect of spray coating parameters on the process performance. This experimental design can also be found in the other studies investigating the similar coating processes; i.e. plasma spray, thermal spray, high velocity oxy-fuel methods [4-7]. In addition, Luangkulab suggested that the main parameters for spray coating process are air pressure, spray time, nozzle size of spray gun and spray distance [1]. These factors can affect the coating material consumption and dry film thickness (DFT) obtained. Furthermore, the spray gun position can also induce the coating performance as reported in [8-10]. However, there is very little discussion on the effect of spray gun position on the material consumption and control of coating layer thickness in the Teflon coating process. Hence, the optimization of such aspects was to be examined in this research, where the implication of this work could bring a guideline for process setup and manufacturing controls in an effective manner.

2 Materials and Method

Teflon was selected as the coating material used in this study, where it was sprayed on an aluminum plate by the fan-pattern air spray gun. The spray gun was positioned at 29.5 cm above the target surface with the tilt angle of 45° respecting to the work surface as shown in Fig. 1. With this setup, the coating layer obtained was likely uniform and provided the smooth surface texture of coating layer. Four process parameters were considered, i.e. air pressure, spray time, nozzle size and spray distance, and

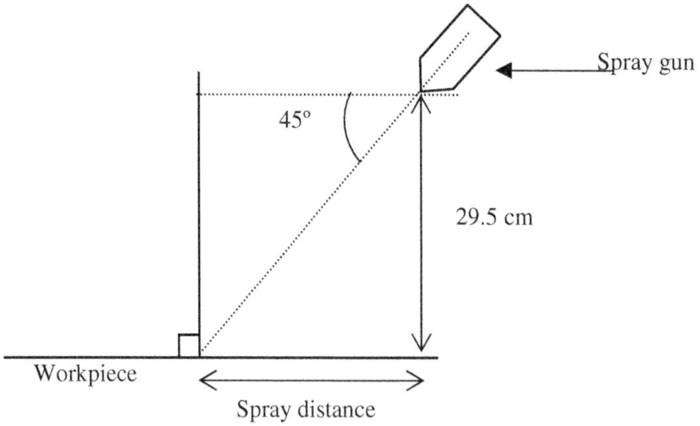

Fig. 1. Positions and angle of spray gun respecting to the workpiece surface

each of which had two levels to be tested as given in Table 1. Hence, the experiments were performed with regard to the 2^4-factorial design. The material consumption and dry film thickness were measured with the replication of 2 for each condition. The results were analyzed by the analysis of variance in order to statistically determine the degree of significance.

Table 1. Process parameters used in this study

Factors	Min	Max
Air pressure (bar)	2.5	3.5
Spray time (s)	1.8	2.0
Nozzle size (degree)	675	1170
Spray distance (cm)	22	24

3 Results and Discussions

3.1 Statistical Analysis

The normality, constant variance and randomization tests were performed to statistically justify the adequacy of experimental results, and all of them were in the 95% confidence interval. The analysis of variance (ANOVA) was used to determine the significant level of the process parameters on the material consumption and dry film thickness. As given in Table 2, the spray time, nozzle size and their interaction were the factors that can induce the significant change to the coating material consumption. In addition, the dry film thickness was found to be affected by all parameters including some interactions as presented in Table 2.

Table 2. Analysis of variance (ANOVA) results for material consumption and dry film thickness

Factors	P-value for material consumption	P-value for dry film thickness
Air pressure (bar)	0.599	0.000
Spray time (s)	0.000	0.002
Nozzle size (degree)	0.000	0.000
Spray distance (cm)	0.112	0.006
Air Pressure*Spray Time	0.227	0.152
Air Pressure*Nozzle Size	0.355	0.685
Air Pressure*Spray Distance	0.606	0.000
Spray Time*Nozzle Size	0.001	0.037
Spray Time*Spray Distance	0.748	0.036
Nozzle Size*Spray Distance	0.769	0.000

3.2 Effects of Spray Conditions

The effects of air pressure on the average material consumption and dry film thickness are shown in Fig. 2, indicating that the increased air pressure decreases the dry film thickness while it has no effect on the coating material consumption. Since the increase in air pressure increases the mixture proportion of air to the coating material per a spray shot, the amount of material consumed is basically no change as depicted in Fig. 2(a). However, a higher air pressure can introduce a finer material droplet which normally has a lower momentum than the low pressure or less portion of air mixture setup. Regarding the air entrainment and drag force at the nozzle exit, the fine material particles are rather blowed randomly at a certain distance from the nozzle tip, thus decreasing the amount of droplet deposition so as the dry film thickness obtained.

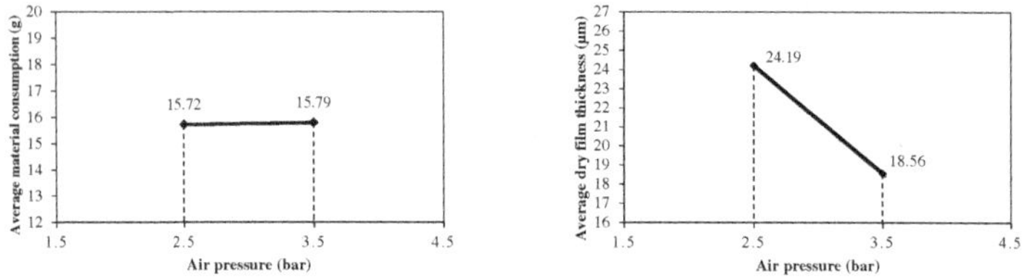

(a) (b)

Fig. 2. Effect of air pressure on: (a) average material consumption; (b) average dry film thickness

The effects of spray time are shown in Fig. 3, where the increase in this factor increases the material consumption and coating layer thickness. These findings are also similar to the effects of nozzle size (Fig. 4) in that a greater amount of material passing through the nozzle can cause a thicker deposition on the target surface. However, the use of large nozzle size can result in the large size of droplet which could affect the uniformity and texture of the coating layer. Hence, it can be noted that a trade-off between these two factors is to be considered for yielding the best coating performance.

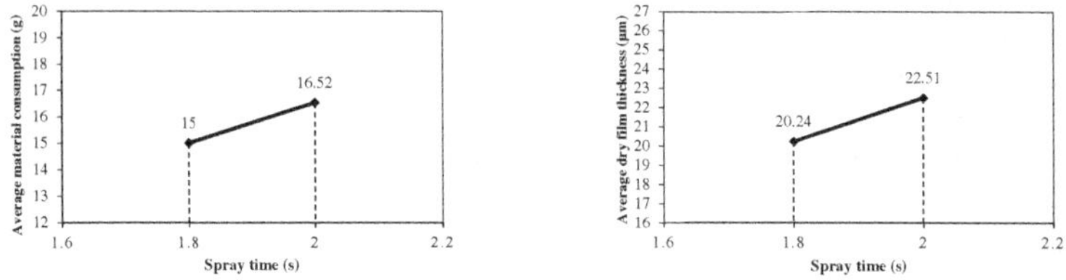

(a) (b)

Fig. 3. Effect of spray time on: (a) average material consumption; (b) average dry film thickness

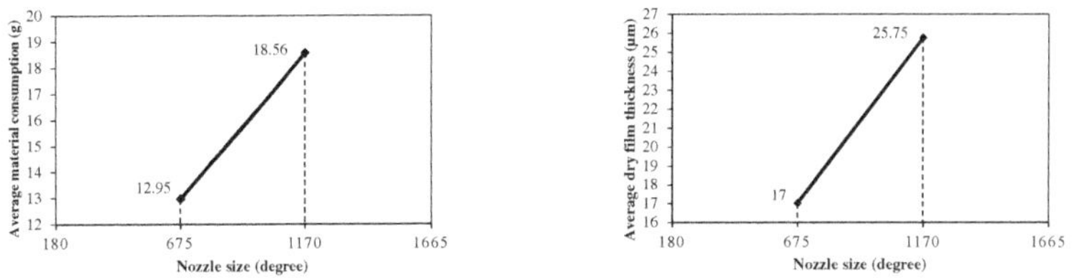

(a) (b)

Fig. 4. Effect of nozzle size on: (a) average material consumption; (b) average dry film thickness

Spray distance was found to be marginal on the material consumption aspect as shown in Fig. 5(a), while a slight change on the dry film thickness can be seen under the different distances (Fig. 5(b)). Using a farther spray distance (24 cm in this study), the spray droplet has to travel longer in air, where its momentum tends to decrease due to the drag. This thus results in the attenuated deposition rate so as the thickness of coating layer. Though a shorter distance can provide a thicker film, the spray coverage area becomes smaller, hence requiring a longer processing time to complete the whole workpiece.

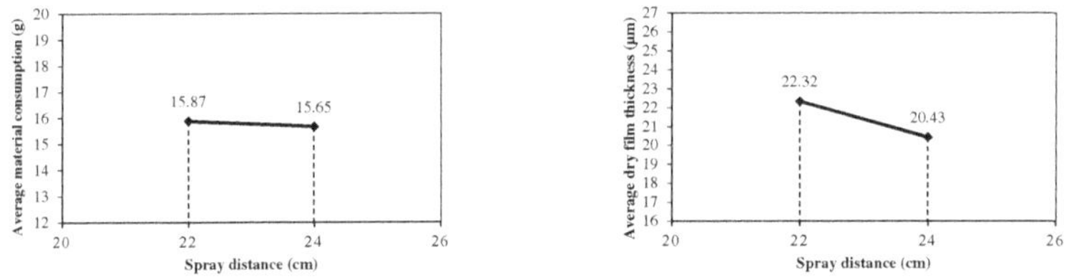

(a) (b)

Fig. 5. Effect of spray distance on: (a) average material consumption; (b) average dry film thickness

3.3 Process Optimization and Modeling

The process optimization was performed to minimize the coating material consumption and also yield the dry film thickness ranging in between 18-23 μm. The optimum settings were as follow: air pressure of 2.5 bar, spray time of 1.8 s, spray nozzle size of 676 degrees and spray distance of 22 cm. By using this condition, the calculated material consumption (MC) and dry film thickness (DFT) were 12.4653 g and 18.9158 μm, while those obtained from the confirmation test were 12.55 g and 18.94 μm. The 1-sample t-test was performed to assess the comparison, and the result showed that the responses obtained from the prediction and experiment were statistically identical. The formed aluminum plates for cooking pan that present the raw aluminum surface and Teflon-coated surface are shown in Fig. 6. With the use of regression analysis, the predictive models for the material consumption and dry film thinness can be expressed by:

$$DFT = 29.1259 - 67.29x_1 + 11.3344x_2 + 0.210667x_3 - 1.2479x_4 + 2.68063x_1x_4 - 0.00839015x_3x_4 \quad (1)$$

$$MC = 9.43693 - 2.17557x_2 - 0.00879545x_3 + 0.0105934x_2x_3 \quad (2)$$

where x_1, x_2, x_3 and x_4 are air pressure, spray time, nozzle size and spray distance, respectively. As the models were presented as a function of the statistically significant parameters, they could induce an accurate prediction for the material consumption and film thickness in the Teflon spray coating of aluminum.

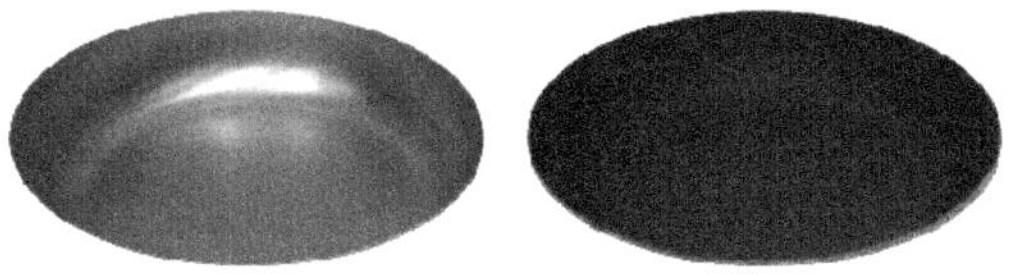

(a) (b)

Fig. 6. Formed aluminum plate: (a) before Teflon coating; (b) after Teflon coating

4 Conclusions

This study presented the effects of air pressure, spray time, nozzle size and spray distance on the coating material consumption and dry film thickness in the Teflon coating of aluminum plate. By using the small nozzle size, low air pressure, low spray time and short spray distance, the material consumption can be minimized and the thickness of coating layer can be conformed to the requirement. The optimum condition and process models suggested in this paper can be used for process planning and control in order to gain a better performance of Teflon coating process.

References

1. Luangkularb, S., Prombanpong, S.: Material Consumption and Dry Film Thickness in Spray Coating Process. In: Proceedings of the 47th CIRP Conference on Manufacturing Systems, vol. 17, pp. 789–794 (2014)
2. Kout, A., Muller, H.: Parameter Optimization for Spray Coating. Advances in Engineering Software 40, 1078–1086 (2009)
3. Winnicki, M., Małachowska, A., Ambroziak, A.: Taguchi Optimization of the Thickness of a Coating Deposited by LPCS. Archives of Civil and Mechanical Engineering 14, 561–568 (2014)
4. Yusoff, N.H.N., Ghazali, M.J., Isa, M.C., Daud, A.R., Muchtar, A., Forghani, S.M.: Optimization of Plasma Spray Parameters on The Mechanical Properties. Materials and Design 39, 504–508 (2012)
5. Murugan, K., Ragupathy, A., Balasubramanian, V., Sridhar, K.: Optimizing HVOF Spray Process Parameters to Attain Minimum Porosity and Maximum Hardness in WC–10Co–4Cr Coatings. Surface and Coating Technology 247, 90–102 (2014)

6. Pierlot, C., Pawlowski, L., Bigan, M., Chagnon, P.: Design of Experiments in Thermal Spraying: A Review. Surface and Coatings Technology 202, 4483–4490 (2008)
7. Forghani, S.M., Ghazali, M.J., Muchtar, A., Daud, A.R., Yusoff, N.H.N., Azhari, C.H.: Effects of Plasma Spray Parameters on TiO2-coated Mild Steel using Design of Experiment (DOE) Approach. Ceramics International 39, 3121–3127 (2013)
8. Hong, S., Wu, Y., Wang, B., Zheng, Y., Gao, W., Li, G.: High-Velocity Oxygen-Fuel Spray Parameter Optimization of Nanostructured WC–10Co–4Cr Coatings and Sliding wear Behavior of the Optimized Coating. Materials and Design 55, 286–291 (2014)
9. Kang, J.J., Xu, B.S., Wang, H.D., Wang, C.B.: Influence of Spraying Parameters on the Microstruc-ture and Properties of Plasma-Sprayed Al2O3/40%TiO2 Coating. Proceedings of Heat Treatment and Surface Engineering 50, 169–176 (2013)
10. Song, E.P., Ahn, J., Lee, S., Kim, N.J.: Effects of Critical Plasma Spray Parameter and Spray Distance on Wear Resistance of Al2O3–8 wt.%TiO2 Coatings Plasma-Sprayed with Nanopowders. Surface and Coatings Technology 202, 3625–3632 (2008)

Multidimensional Process Analytical System for Manufacturing Management

Poonphon Suesaowaluk

Graduate School of eLearning in Information Communication Technology
Assumption University, Bangkok Thailand
poonphonssw@au.edu

Abstract. Industrial or manufacturing management is a complex task that requires prior information from several entries to support in order to make the right decision with efficient solution and obtain the optimum profit. Business intelligent technology approach allows business organization to utilize information in a form of heterogeneous view and benefits several objectives for uses. This work employs business intelligent concepts to develop simplified real time process analysis systems for supportive manufacturing management. With many desirable benefits provided, multiple objectives of analysis derive from the single database. The system trial in homogenous or mass productions results in displaying the data analysis and enabling managers to view the multidimensional information based on a variety of perspectives such as a process improvement, inventory control, material management, quality assurance and eventually governance management in manufacturing. Furthermore, this work can be applicable to different configuration products.

Keywords: Business Intelligent, Manufacturing Management, Material Resource Planning, Multidimensional Data Analysis, Real Time System.

1 Introduction

Since 1960s, the growth of world economy has influenced the industries that produce the goods as appliance to serve the needs of human livelihood. The evolution of industries had been developed as mass production simultaneously with technology. The evolution of information technology has driven the business growth and becomes the important tool to efficiently make the right decision with a better performance to reach the successful goal. Most wise managers of business organizations have been searching an appropriate tool to analyze information and data based on the business direction as well as enhancing the profitability to the organization. Presently, information systems have been developed in various functions with multi-purposes using multidimensional displaying called Business Intelligence or BI [1]. BI concept spreads to industry or manufacturing in various strategies such as data mining and multi-dimensional data analysis.

M. Gen et al. (eds.), *Industrial Engineering, Management Science and Applications 2015*,
Lecture Notes in Electrical Engineering 349, DOI: 10.1007/978-3-662-47200-2_88

Multidimensional data analysis has been implemented to several automation manufacturing industries in order to produce mass volume of production. To manage the mass production or manufacturing, it is necessary to look up for the information from cooperative departments that work to produce the goods. The common manufacturing has implemented several information systems such as financial, logistic, procurement, store, warehouse, and inventory. These departments mostly store the data concurrently for individual propose separately, so it causes redundancy of database. In fact, they may be reciprocal. The corresponding among in-process information is normally used for quality and process improvement. For example, a defective unit that is used for process analysis and quality control can be simultaneously used for material planning, and inventory control. It is possible to utilize and manipulate the relationship information among in-process data for multi-proposes of quality and material management concurrently by applying BI concepts as a multidimensional data analysis.

2 Related Works

Material control and planning have been a complicated problem for mass production and assembly manufacturing due to the large volume of goods consumed in a various large amount of raw material and components of those units. The explicit examples include assembly process, the incorporate parts, components among the product type, which are composed of material (acronym as m) m1, m2, m3,.., mn. These materials have a unique identified part number whereas some materials may be shared with different products such as Product 1 (acronym as P). P1 may be assembled with material m1, m2, m3 while the product P2 may be assembled with m1, m3, m4. For instance, the hard disk drive (HDD) is actually built for many models with different performances or amount of capability storage and space. The sharable parts such as wires and suspensions are utilized for more than one models with different assigned part-number and controlled by configuration. This causes complexity and difficulty to track the actual materials and inventories.

Moreover, traditional modern manufacturing and the process system are mostly implemented in a real time manner in order to specify the status of the production or process performance. There are several works introduced the mechanism dealing with Manufacturing Resource Planning, MRP, such as One Kind of Production, OKP. In fact, OKP was an integrated production platform to improve production efficiently and profitability [2]. An Economy Manufacturing Enterprise System [3] was designed an application systems to assist management and monitor the overall process performance to all products and processes. The system was implemented for interconnected data display supporting process improvement.

3 Multidimensional Manufacturing Processing System

This paper presents the proposed system to support manufacturing management in a multidimensional aspect of information such as the material planning system, production planning, process and quality improvement. This is an extension module to the previous work of the Enterprise Manufacturing System. Multidimensional Manufacturing Process System was designed with low investment in order to provide the report of process performance by manipulating information recorded during the actual process in the production line of plastic molding. This is created by measurement of product unit denoted by controlling specification requirement under disciplinary agreement.

3.1 Database Implementation

Generally, implementation of relational database system is robust for data manipulation. Especially, it is always used in business intelligence [4], which selects this strength and applies it into the initiate four major modules database. The first module is recording goods items with details about product information such as product name, part number, type of material, and unit. The second module is manipulated based on routing records information of product, process, testing machine, and operators working at those processes. The third module is defective manipulation which record product lot number, process code, time in and time out of each product at those processes. The last module is declaring the details of defectives. The flows of data and attribute of database described in Figure 1 and 2 respectively.

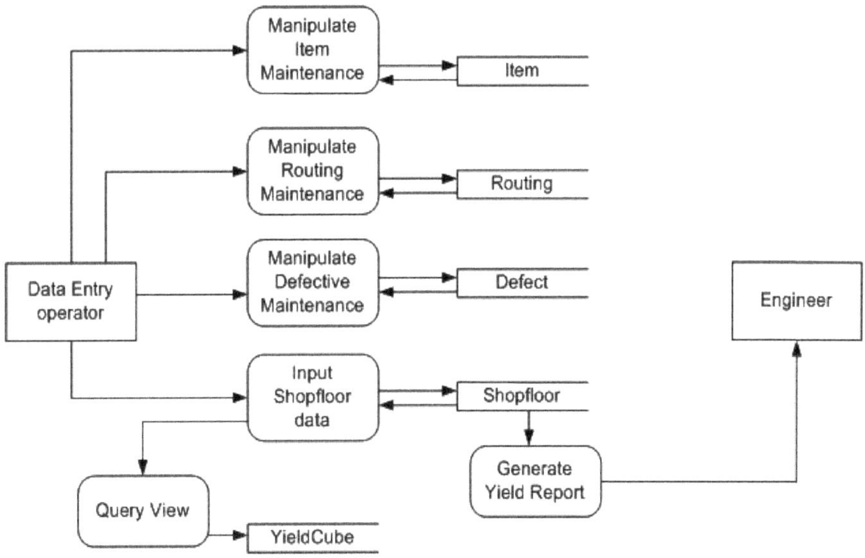

Fig. 1. Context diagram of database and its flows

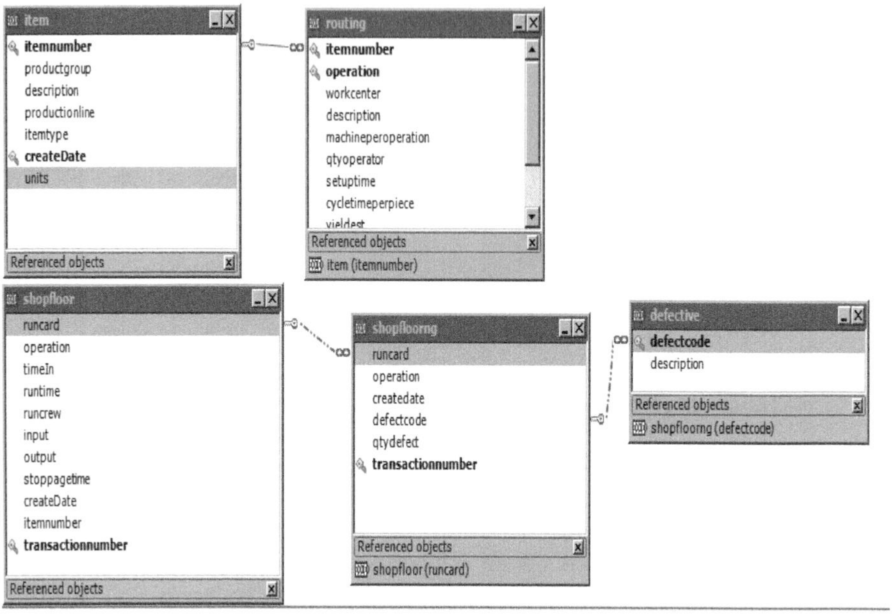

Fig. 2. Database attributes and its relationship

3.2 Development Tools

The system was developed by the open source tools of Java programming and the form of web base Client-Server architecture [5,6] of tree layers: interface, data manipulation algorithm and database. The primary data will be entered by an operator who is in-charge in production line while performing measurement unit at each process. The data input will be stored in four types as described.

The example of system and database implementation is described as follows:

```
Package  Manufacturing;
import java.io. IOException;
import java.util.Vector;
import  javax.servlet.ServletException;
import  javax.servlet.http.HttpServlet;
import  javax.servlet.HttpServletRequest;
import  javax.servlet.HttpServletResponse;
import org.apache.log4j.Logger;
import utility.UtilityFunction;

import constants.GeneralConstans;
/** Servlet implementation class ShopfloorAddNG*/
Public class ShopfloorAddNG extends HttpServlet {
                ......
}
```

The example implementation of shopfloor database;

```
Package  Manufacturing;
import java.sql.Timestamp;

public class RuncardModel  {
    private String runcard;
    private String itemnumber;
    private Timestamp createdate;
    private double qtyinput;
    private String runcardtype;
    private String status;
    private String workorder;
  public String gerRuncard () {
       return runcard;
  }
....
}
```

The modules of Generate Yield Report and Query View would help obtain the different views as desired types of information displaying. The linkage association of the database is made as an array shown in Figure 3. The array of data could be demonstrated in dimension of specifying date, product, process, defective types and quantities, product part number, input. The array condition would contribute the multidimensional information such as index of specific period of time, product, process, defective both for individual product and percentage fraction to the overall at a certain desired dimension.

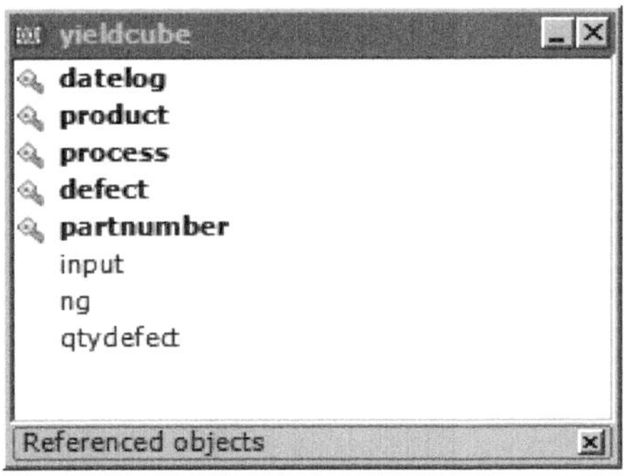

Fig. 3. The key array for report displaying

The proposed system has been developed and interfaced with the real-time process measurement system. This work is initiated to serve the factories and used to control the process by measure the product can build in real-time manner. The system has been recorded for the measurement products of the manufacturing whereas the process has been performed to build the products in the production line. The information regarding the process has been recorded to the database for multi-purpose analysis. During the measurement process, the rejected units would be recorded with defective, including the type of defective, input, and output. It would be applied to all process since the first operation throughout the last operation that delivers the finished goods for shipping to customers.

3.3 System Verification and Validation

The validation of system has been made for the trial run by simulating the data to the system and then generating the report with details of input, defective and finally the yield of that process. The multi-dimensional data can be displayed in various views of process performance and performs analysis resulting in heterogeneity aspects such as reporting the actual type of defect for defective analysis, performance of individual measurement machines, based on process or product type, operators, and specific period of time.

The outcome of this system could be beneficial to both process analysis and the material management. The example report is demonstrated in Table 1 indicating that a situation specifies a given period of time throughout the process; this could be noticeable that any model or product types contains amount of input, defective and factual percentage of the overall defective.

Table 1. Example report with specify the overall of the products, quantity input, defective

datelog	(All)		
process	(All)		
	Values		
Row Labels	Sum of input	Sum of qtydefect	Sum of %defective
Model46X4	9132	0	0.00%
Model47X4	30949	0	0.00%
Model48X4	109347	338	0.31%
Model50X4	19200	0	0.00%
Model51X4	131358	130	0.10%
Model66X4	26258	26	0.10%
Model84X4	135866	147	0.11%
Model85X4	101777	247	0.24%
Model86X4	110617	99	0.09%
Model87X4	75786	65	0.09%
Model88X4	58890	80	0.14%
Model89X4	53288	125	0.23%
Model90X4	2368	0	0.00%
Grand Total	864836	1257	0.15%

To obtain all of the rejected parts, calculation is denoted by equation (1), where all rejected parts show a combination of all processes (abbreviated as Pri, i=1,2,3,.., n and all products (Pdj, j=1,2,3,…k).

$$\boldsymbol{All\,rejected\,parts} = \sum_{i=1}^{n} \boldsymbol{Pri} \sum_{j=1}^{k} \boldsymbol{Pdj} \tag{1}$$

4 Conclusion

The result of the system contributes to support manufacturing management considered based on the relationship of information among process analysis information from a single database, manipulated for material planning, inventory control, process improvement, minimize material in stock, and quality assurance in order to reach the optimum quality goals. It is a low cost using single database for multi-purposes with many benefits as governance managerial in conclusion.

This system is effective to homogeneity product pattern such as the lot of plastic molding unit which has the same identity. However, this concept could be developed for different configurations of assembly products which are composed of different piece parts.

References

1. Laudon, K.C., Laudon, J.P.: Management information systems in managing the digital firm, 12th edn. Pearson Education, Harlow (2012)
2. Zhao, W., Zhang, H., He, Z., Tan, R., Jiang, P.: Improved MRP System of OKP Enterprise Based on Product Platform. In: Management of Innovation and Technology, ICMIT, pp. 1240–1245. IEEE Press (2008)
3. Suesaowaluk, P.: An Economy Manufacturing Enterprise System. In: International Conference on Information Science, Electronics and Electrical Engineering (ISEEE 2014), pp. 609–619. IEEE Press, Japan (2014)
4. Hoffer, J.A., Mary, S., Heikki, T.: Modern Database Management, 10th edn. Prentice-Hall, Upper Saddle River (2011)
5. Tanenbaum, A.S., Steen, M.V.: Distributed Systems in Principles and Paradigms, 2nd edn. Pearson International Edition, Prentice Hall, Vrije University, Amsterdam (2007)
6. Kendall, K.E., Kendall, J.E.: Systems Analysis and Design, 8th edn. Pearson Education, Harlow (2011)

The Design of Machine Cluster for Loading and Unloading Slider in the Hard Disk Drive Manufacturing

Amarin Wongsetti* and Suksan Prombanpong

Department of Production Engineering, Faculty of Engineering, King Mongkut's University of Techonology Thonburi (KMUTT), Bangkok, Thailand
amarin.w@mail.kmutt.ac.th, suksan.pro@gmail.com

Abstract. This paper is aimed at demonstrating a design of a loading and unloading station cell for transporting the Slider, one of the critical components in a hard disk drive, which is attached at the end of the head gimbal arm, for electronic inspection and verification before transferring to the Head Gimbal Assembly (HGA) line. This is due to the fact that the Slider must be defect free so that a hard disk drive can perfectly perform its function as a data storage device. Therefore, each Slider must be 100% quality checked and identified as a conforming or nonconforming product and it must be done by an automated testing machine. The required yearly demand of a Slider for testing is around 160 million pieces. Therefore, a design to meet this criterion is a must and must be cost-effectiveness. A machine cluster concept is proposed to design load/unload station configuration. It is found that the proposed design is consisted of four cells which is composed of 44 automated load/unload machines with only one worker to perform feeding service to these cells.

1 Introduction

A Slider used in the hard drive Head Gimbal Assemblies is of important part in the hard disk drive manufacturing industry. Its main function is to slide up and down between plates in a hard disk drive. This Slider is made of wafer which is trimmed in shape and then assembled in Head Gimbal Assemblies (HGAs). Then the next step is a Head Stack Assembly (HSA) where the HGAs are stacked and assembled together and then move to the final assembly line where all components are assembled in the frame of a hard disk drive. Fig. 1 shows operation flow of a hard disk drive manufacturing process. Production inspection of hard drive HGAs requires the accurate measurement of a number of critical factors in order to meet tight tolerances. These crucial factors include pitch and roll angles, Slider positioning, and Slider orientation/alignment issues. Due to these complex assembly factors, an HGA inspection platform must combine highly accurate and repeatable measurement capabilities with versatile lighting, multiple magnification options, and advanced programming functionality.

* Corresponding author.

M. Gen et al. (eds.), *Industrial Engineering, Management Science and Applications 2015*,
Lecture Notes in Electrical Engineering 349, DOI: 10.1007/978-3-662-47200-2_89

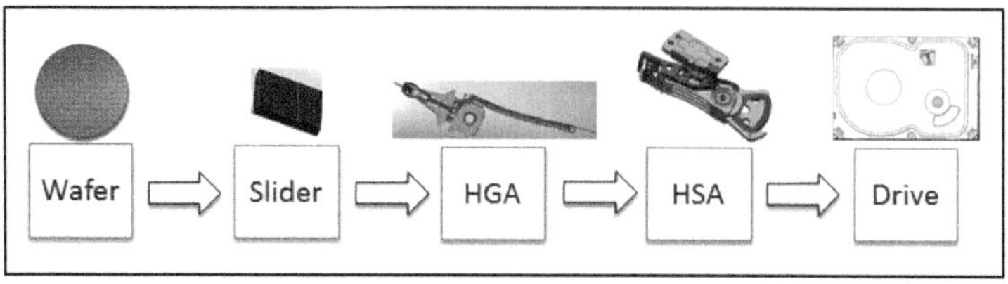

Fig. 1. The operation flowchart of hard disk drive manufacturing

A HGA manufacturing process starts with FOLA (Front of Line Automation) for assembling Slider into HGA, then cleaning by deionization water, LSA (Laser Static Attitude Adjust Machine) for adjustment HGA Slider pitch and GL (Gram-Load) for measuring the flexibility of HGA's flexure. Finally, testing is performed in HGA ET (Head Gimbal Assembly Electrical Testing) station where the good HGA will be accepted and delivered to HSA process; otherwise, it will be rejected and most of the defective HGAs are due to Slider malfunction. The flow process can be depicted in Fig. 2(a). This is due to the fact that a Slider plays an important function in a HGA; therefore, an incoming Slider requires a tight inspection within the process as shown in Fig.2 (b). A Slider must be inspected by electrical testing (ET) and this can be accomplished by load/unload system module.

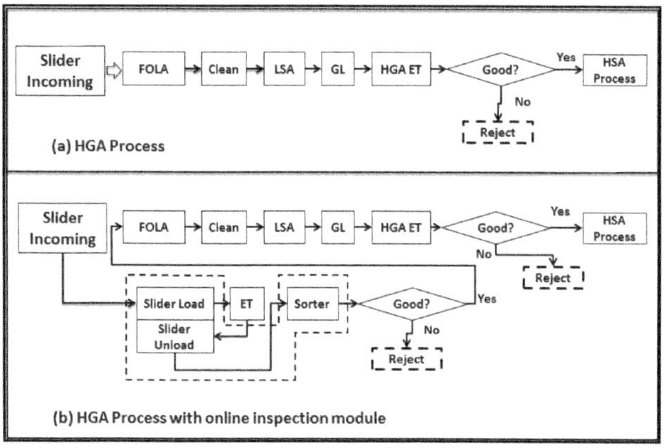

Fig. 2. A flow process chart of (a) HGA Process. (b) HGA process with online inspection module.

In this module, a Slider will be loaded and transferred to ET for testing and subsequently unloaded, and sorted to separate defective Sliders. Only defect free Slider will be further processed to FOLA. More details of the load/unload system module will be discussed in section 2.1. Thus the load/unload system module must be designed and the prototype was built and tested. This module can accomplish the task at a speed of 475 pieces per hour. However, the required yearly production demand is around 162,837,486 pieces which is equivalent to 21,245 pieces per hour. Thus, it is obvious that a single load/unload system module cannot satisfy the required production and multiple ones are inevitable and must be designed with cost effective as much as possible which is the main purpose of this paper.

Single manufacturing cells (SMC) in form of machine cluster concept will be applied to determine a number of required machines and optimal number of workers to serve the clusters [1]. A number of related literatures in this field are investigated and will be briefly described. Prombanpong and Seenpipat presented a worker assignment design for machine cluster in the manufacturing cell [2]. The optimum number of workers is determined to service a machine cluster. Prombanpong et al. applied single manufacturing concept for a determination of number of AGVs (Automated Guided Vehicle) in a flexible manufacturing cell and verified a result by a simulation PROMODEL software program [3]. Hu and Zhou utilized mathematical model to optimize machine configuration and job scheduling strategy in semiconductor industry [4]. Particle Swarm Optimization (PSO) technique is also proposed to utilize in machine cell layout and design [5, 6]. Tariq and Ghafoor applied hybrid genetic algorithm in single machine cells design [7]. Nguyen and Takakuwa applied simulation technique for manufacturing line design in automobile industry [8]. Pierreva and Plaquin applied Group Machines into Manufacturing Cells for improving material flow and utilized the Genetic Algorithms (GA) for determining the optimal solution [9]. Xing and Marwala applied swarm intelligence on the intra-cell machine re-layout (ICMRL) problem for finding a better solution [10]. Lozano and Gimnez proposed Tabu search algorithm for searching the optimal number of machine per cell on the cellular manufacturing design process [11].

2 Load/Unload System Cell Design

As mentioned earlier a single load/unload system module which its production rate is 475 pieces per hour will not be able to satisfy the required production of 21,245 pieces per hour. Thus, a multiple number of load/unload systems must be designed under various constraints such as machine availability, utilization, and limited space. Machine availability is a measurement of machine readiness for work. If a machine is malfunction and needs a repair, a percentage of machine availability will be less than 100 and lower productivity is ensued. Thus, the machine availability factor must be taken into account in design process. Likewise, machine utilization is another performance measurement in terms of obtained output compared with ideal one. If machine has no idle time, it should be able to produce output according to the design calculation. However, in practice there may be some problems i.e. short of incoming work parts, wait for worker to attend the machine, or quality problem occurring during production and cause machine idle. As a result, a lower productivity can be expected and therefore, machine utilization must also be included in the design process.

2.1 Load/Unload System Module

The prototype of load/unload system was designed and it is composed of 3 main systems named system 1 to system 3. The system 1 is basically designed to load and unload Bola tray which is used to position a Carrier where a Slider is attached to during electrical testing. A figure of Bola tray, Carrier, Slider, and Slider tray is shown in Fig.3 (a),(b),(c) and (d) respectively. The system is designed for Carrier replacement at the end of Carrier life which has been used for 3000 running cycles.

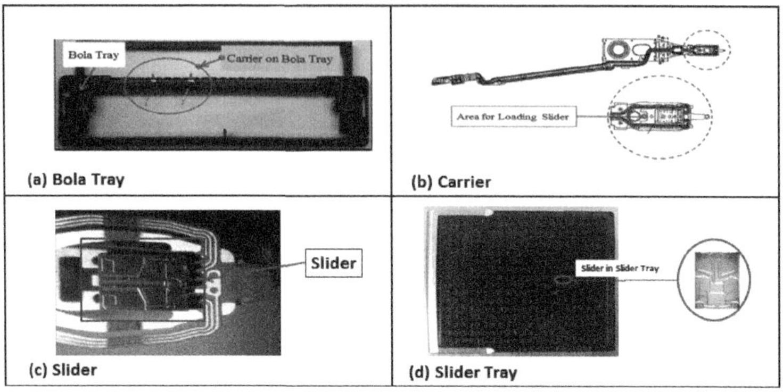

Fig. 3. (a) Bola tray (b) Carrier (c) Slider (d) Slider tray

This system ensures proper functions of the Carrier. The system 3 is dispensing area for loading and unloading of a Slider tray. The load/unload system module is 1.22 m. in length and 0.92 m. in width as shown in Fig. 4.

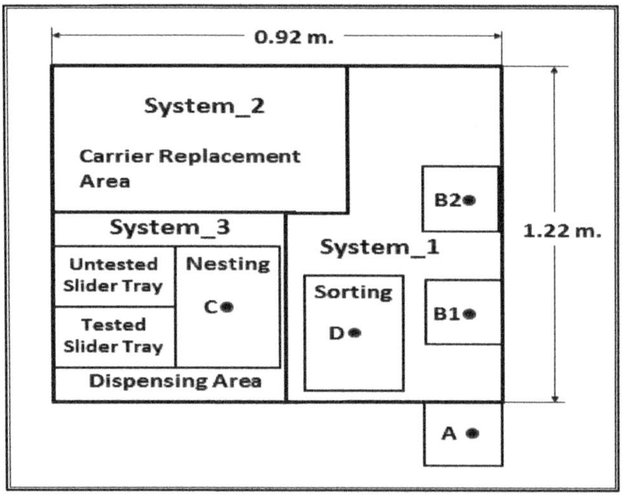

Fig. 4. The load/unload Slider system module

The working mechanism of the load/unload Slider module in Fig. 4 can be explained as the following. The mechanism begins with the Slider which has already been processed from ET and it is mounted to the Carrier which is positioned in Bola tray will be fed into the stacker (A). The stacker will transfer the mentioned Bola trays to dock 1 (B1) and dock 2 (B2) located in the system 1. Then, each Carrier will be transferred to nesting position (C) for dismounting the Slider and will be sorted in the sorting area (D). At the same time the untested Slider from a Slider tray in a dispensing area will be mounted to the Carrier already positioned in the nesting area (C) and this Carrier will be ready for returning to ET for testing.

2.2 Cell Layout Design

In the cell design process, the most important factor is that the cell must be capable of transferring enough Sliders for electrical testing which is at the production rate of 21, 245 pieces per hour. Since the available space is around 340 square meter or 20 m. by 17 m. Thus, it is plausible to configure as inline layout where a series of load/unload system modules are consecutively positioned next to each other to form a cell of m machines and other cells can be created to form n cells as shown in Fig. 5.

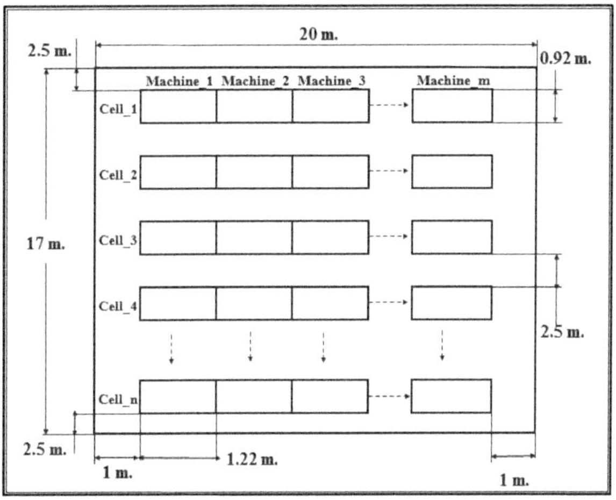

Fig. 5. Cell configuration as an inline layout

Each module will be linked by a roller conveyor where a Bola tray can be transported along the cell. Forming as a cell and linked by a conveyor will substantially reduce a number of stacker to only two stackers per cell i.e. feeding and receiving as shown in Fig. 6. This will be cost effectiveness since it is not necessary to install each stacker for each load/unload system module. However, an intelligent system for traffic control must be installed to prevent a collision. Since the cycle time of each system module is constant and identical, the traffic strategy is designed in such a way that loading and unloading of Bola trays to and from the docking area of all modules will be simultaneously performed. Therefore, all Bola trays on the conveyor will concurrently move and stop. As a consequent traffic pattern becomes much easier to design and implement.

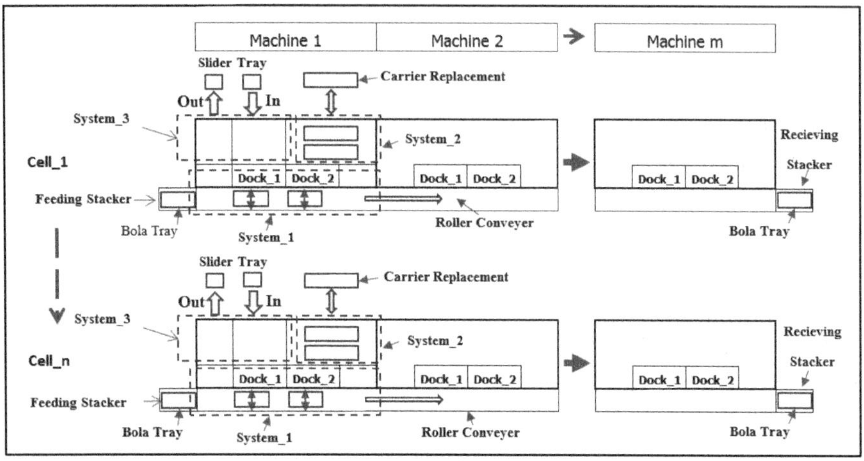

Fig. 6. Cell design for load/unload system module

2.3 Number of Load/Unload System Modules in a Cell

A determination of number of system modules in a cell can be determined from (1).

$$\text{Total number of modules} = \frac{WL_{total}}{AT} \tag{1}$$

Where WL_{total} is a total workload and AT is available time to accomplish the work. Hence WL_{total} can be calculated from (2).

$$WL_{total} = WL_s + WL_{st} + WL_{cr} \tag{2}$$

Where WL_s is scheduled workload, WL_{st} is setup workload and WL_{cr} is Carrier replacement workload. In this case, scheduled workload, setup time and Carrier replacement time in a year is 325,676.07, 60.83, 150.77 hours in a year. Substitute in (2) we obtain:

$$WL_{total} = 325,676.07 + 60.83 + 150.77 = 325,887.67$$

AT is available time per year which can be calculated from (3).

$$AT = T_w \times A \times U \tag{3}$$

Where T_w is working time in a year, A is machine availability and U is utilization. In this case working time (T_w) in a year is 7,665; A is 0.9903 and U is 0.9997. Substitute in (3) AT is equal to 7,588.37 hours per year. Thus substitute a total number of modules are 42.9 modules. Since the available space is 340 square meter which is 20 by 17 meter. The module is 1.22 meter long and margin between modules is 2.5 meter; therefore, we can comfortably design 11 modules in a cell. As a result, we need a total of 4 cells as shown in Fig. 7.

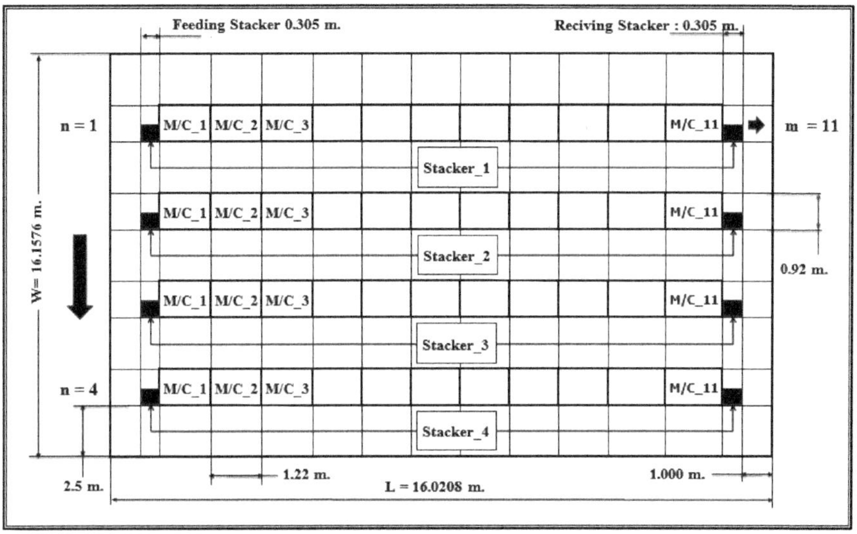

Fig. 7. The final layout machine cluster in the cell layout

2.4 Determination of Optimum Number of Workers

In one cycle a worker is required to attend the designed cell by loading Bola trays to the feeding stacker at the beginning of the cell, loading Slider tray of each machine in the cell and unloading Bola trays from the receiving stacker at the end of the cell. If the system module cycle time is longer than that of worker service time, the worker is able to attend another module and perform the repetitive task as mentioned. However, the bottom line is that the system module must not experience idle time; otherwise the production rate will be lowered than expected which is not a desirable situation. Therefore, it is possible to allow some idle time to a worker. Thus it is important to determine a number of system modules that one worker can attend to and this can accomplish from (4) as the following.

$$N = \text{Maximuminterger} \leq \frac{(T_m + T_s)}{(T_s + T_r)} \tag{4}$$

Where N= number of modules that one worker can attend to, T_m is machine or module cycle time, T_s is worker service time and T_r is worker repositioning time. In this this case Tm, T_r and T_s is 1,344, 1.511, 6.872 seconds respectively. Thus, substitute these parameters in (4)

$$N = \text{Maximum interger} \leq \frac{(1344 + 6.872)}{(6.872 + 1,511)}$$

$$N \leq 161.14$$

Thus, N is equal to 161 modules. Meaning that only one worker is enough to attend 161 modules; therefore, this designed cell cluster which is accommodated only 44 modules can be attended by only one worker. Thus, in this case a worker will work around 6.15 minutes and enjoys idle time around 975.15 seconds or it is equivalent to 16 minutes before coming back to repeat the task. As a result, a utility worker in the production line can be assigned to perform this task and can do other tasks during free time. The walking route of a worker during serving cluster is shown in Fig. 8. The worker will start from the first module in cell 1 and follow the path until the last module in cell 4 is served.

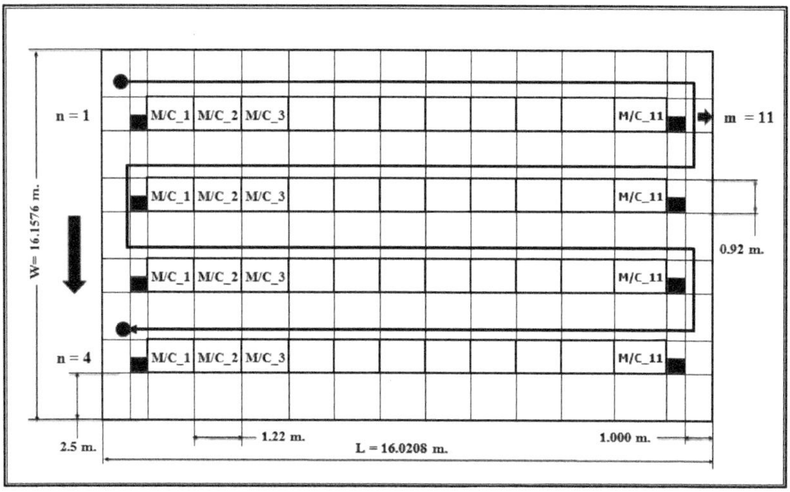

Fig. 8. Path of service worker in the cluster

3 Conclusion and Discussion

In the HGA assembly for hard disk drive manufacturing industry, The Slider plays an important role to the functional of a hard drive. Thus it is important to inspect the Slider before assembly. Thus the automated load/unload system is designed for launching the Slider for electrical testing. This module requires mount and dismount the Slider to the Carrier and the prototype module is designed. It requires around 43 modules to satisfy annual production demand. Thus, a single manufacturing cell in form of machine cluster concept is implemented. It is found that with the available space one cell can accommodate up to 11 modules and therefore 4 cells are designed. In order to reduce a number of stackers which is required to dispense a Bola tray, a roller conveyor is used to link all modules within a cell. This is due to the fact that each module still requires a worker to serve the Slider tray at each cycle and it is found that one worker is enough to support up to 116 modules.

The single manufacturing cell concept is useful for determination of required work cell to satisfy production demand. It is important to include machine availability, defect rate, utilization in the calculation since these factors has an adverse effect to the production. Once a number of machines were determined, a machine cluster concept should be attempted to reduce labor cost or perhaps any common devices can reduce hardware installation cost. It should be noted that the equation (4) listed in this paper is very useful to calculate a number of machines attended by one worker. Although it looks quite simple, only a few people recognize and hardly use it.

This project is currently under implementation with a speculated budget of 10 million US dollars.

References

1. Groover, M.P.: Automation, Production Systems, and Computer Integrated Manufacturing, 3rd edn., pp. 394–395. Prentice Hall International, London (2008)
2. Prombanpong, S., Seenpipat, W.: A Worker Assignment for Machine Custer in the Manufacturing Cell. In: Proceedings of the 2011 IEEE Industrial Engineering and Engineering Management, pp. 352–356 (2011)
3. Prombanpong, S., Kiattiphatthananukul, W., Songsanan, A., Sukin, A.: The Design of an AGV in the Manufacturing Cell. In: Proceedings of the 2012 IEEE Industrial Engineering and Engineering Management, pp. 1006–1009 (2012)
4. Hu, Z., Yan, L., Zhou, W.: Optimized Machine Configuring and Job scheduling Strategy for High Efficiency Semiconductor Manufacturing. In: Proceedings of Computational Intelligence and Software Engineering (CISE), Wuhan, pp. 1–4 (2009)
5. Yamada, Y., Ookoudo, K., Komura, Y.: Layout Optimization of Manufacturing Cells and Allocation Optimization of Transport Robots in Reconfigurable Manufacturing Systems using Particle Swarm Optimization. Int. J. of Intelligent Robots and Systems (IROS) 2, 2049–2054 (2003)
6. Durán, O., Rodriguez, N., Consalter, L.A.: A PSO-Based Clustering Algorithm for Manufacturing Cell Design. In: Proceedings of Institute for Computer Sciences, Social-Informatics and Telecommunications Engineering (ICST), Brussels, p. 53 (2008)
7. Tariq, A., Hussain, I., Ghafoor, A.: Consideration of single machine cells in designing Cellular Manufacturing System using a Hybrid Genetic Algorithm. In: Emerging Technologies (ICET), Islamabad, pp. 6–10 (2007)
8. Nguyen, M.D., Takakuwa, S.: Emergence of simulations for manufacturing line designs in Japanese automobile manufacturing plants. In: Simulation Conference, WSC, Austin, pp. 1847–1855 (2008)
9. Pierreval, H., Plaquin, M.F.: A Genetic Algorithm Approach to Group Machines into Manufacturing Cells. In: Computer Integrated Manufacturing and Automation Technology, New York, pp. 267–271 (1994)
10. Xing, B., Gao, J., Nelwamondo, F.V., Battle, K., Marwara, T.: Swarm Intelligence in Cellular Manufacturing: Application of Hybrid Ant System to Manufacturing Cell Machine Layout Problem. In: World Automation Congress (WAC), Kobe, pp. 1–6 (2010)
11. Lozano, S., Adenso, B., Salinas, I., Gimnez, L.: A One-Step Tabu Search Algorithm for Manufacturing Cell Design. Journal of the Operational Research Society 50, 509–516 (1999)

A Study of Downloading Game Applications

Long-Sheng Chen* and Chen-Wei Yen

Department of Information Management,
Chaoyang University of Technology, Taichung, Taiwan
{lschen,s10014628}@cyut.edu.tw

Abstract. With the growth of communications services and mobile devices, it leads to the popularity of mobile operating systems such as Android and iOS, and bring the market growth opportunities for the software industry. Among lots kinds of applications (APP), game APPs have attracted much attention. According to related research, it estimates that in 2015 the market value of game APPs will reach to 20% of the overall game market. In the fierce competition, how to design products for satisfying the needs of users, and attract users to download have become one of important issues for developers of game APPs. Therefore, the main purpose of this study is to define the potential factors that influence users to download game APP based on the technology acceptance model (TAM) and innovation diffusion theory (IDT). Then, two feature selection methods, Support Vector Machine-Recursive Feature Elimination (SVM-RFE) and Information Gain (IG) will be utilized to identify the key factors of downloading game APPs. Finally, a comparison between different groups will be provided. The results can be used as a reference for game developers and researchers to design game APPs.

Keywords: Game APP, Diffusion of innovation theory, Technology acceptance model, Feature selection, SVM-RFE.

1 Introduction

As the number of users of smart phones, more and more people are gradually driving the use of software applications (APP) in mobile operating environments [12]. Among lots kinds of APPs, game Apps have attracted much attention since 2007 when iPhone come to the market. Handheld devices combine multi-functionality (touch screen, dynamic sensors, positioning systems, enhanced display, access high amount, and so on) with the wireless network, and change the types of Internet games. Many innovations are emerging, such as APP stores, online multiplayer games, multimedia entertainment combined with social media, games combined with motion sensors [10]. A new business model evolved into the mobile phone industry, for example, two popular game vendors, Greystripe and GluMobile, offer free games, and add advertising in the game to increase your click-through-rate (CTR). Some companies make game APPs combined with social media, such as Facebook [9].

* Corresponding author.

M. Gen et al. (eds.), *Industrial Engineering, Management Science and Applications 2015*,
Lecture Notes in Electrical Engineering 349, DOI: 10.1007/978-3-662-47200-2_90

In order to attract even more customers to use, App developers made a number of promotion ways, for example, providing free or trial App to customers; prepayment could obtain more services; providing a complete free App to customers, then through in-app purchases or click on mobile ads, customers can get extra content and services [12]. Such ways are also expanded to gaming revenue model which contains thre major models, in-APP purchases for free APP, paid APP, in-APP purchases for paid APP. According to eMarketer survey shows that the in-APP purchases for free APP will become main stream [8]. Nevertheless, free game APPs still differ from paid game APPs, mainly due to the paid games invested by the well-known game maker (EA, Gameloft, CAPCOM, etc.), so that paid game APPs have good quality.

On the other hand, after the game APPs launched, in fact, the usage rate and popularity are not good. Past studies have also confirmed that both the map services and game APPs are not automatically used by mobile devices users [23]. Therefore, how to promote game APPs has become a critical issue for App developers. They need to consider additional factors which come from the voice of cusotmers. A survey in the United States pointed out that users' interests in mobile entertainment services, in addition to population statistics and personal privacy issues, are highly related to communications technology [19].

There are many researches focused on mobile entertainment in the past decade. But for the current development of mobile games, whether from social media or game design viewpoints, it still lacks a more comprehensive explanation [9]. Most of related works focus on qualitative studies, relative few papers used data mining approaches. For examples, Hsu and Lin [12] investigated the use of paid App user issues. They indicated that there are significant differences between actual users and potential users. Liu and Li [17] investigated the influence of the environment on the use of hedonic mobile service (mobile game). They found that use of the environment is an important prediction factor which directly or indirectly affect the perceived ease of use, perceived usefulness, hedonic perception, cognition focused sex, attitudes and behavioral intentions. And perceived usefulness, perceived hedonic and cognitive focus of all have a positive impact on the attitude of acceptance of mobile games.

Consequently, this study aims to define the potential factors that influence users to download game APP based on the technology acceptance model (TAM) [6] and innovation diffusion theory (IDT) [20]. Then, feature selection methods including Support Vector Machine-Recursive Feature Elimination (SVM-RFE) and Information Gain (IG) will be employed to select the important factors of downloading APPs for free game APPs and paid game APPs, respectively. Finally, a comparison between different groups will be provided. The results can be used as a reference for game developers and researchers to design game APPs.

2 Related Works

2.1 Applications

APPs greatly change the way people use computers and the experience of the past for the phone. Nowaday, mobile devices are becoming the core of communication and information needs, and APP brought to the application level is closely related to our life [4]. The APP store like a software trading market offers people instant download APPs [14]. Researchers wonder what kind of mobile service is the killer application that can drive mobile services widely [25], for examples, short message services (SMS) and DoCoMo's i-mode in Europe and Japan, mobile banking, m-commerce, location-based catalog and map services [15].

Many APP related researches applied TAM and IDT to find why internet users used APPs [24]. A study used TAM to explore the key factors of using mobile phone games are use context, perceived usefulness, cognitive concentration, attitude, and perceived enjoyment [17]. Relative few works focus on paid APPs. They found that confirmation of the perceived value and satisfaction are positively shut; value for money, evaluation, and application of alternative free App is a direct impact on the user purchase intent. After comparing the actual users and potential users, they found the actual users purchase intent by value for money, satisfaction, and free substitution App influence; potential users purchase intent by value for money, and App evaluate the impact of social value [12] .

2.2 Diffusion of Innovations and Technology Acceptance Model

In this work, we employ IDT and TAM to assist us to define factors of game APPs. IDT developed by Rogers[20], he pointed out that "diffusion" means the process which one innovation spreads to every members of social systems through communication channels. With time going, this innovation will be gradually accepted.

Innovation diffusion process must include four elements: Innovation, Communication Channel, Social System, and Time, where time is an important factor in the innovation diffusion process. Rogers used time as a basis to divide innovation adopters into innovators, early adopters, early majority, late majority and laggards. In addition, other factors that affect the adoption rate of innovations include the type of innovation and decision-making, communication channels, social systems, and efforts of spokesman.

TAM [6], simplified from Theory of Reasoned Action (TRA), is now commonly used to investigate user acceptance of information technology/services. TAM model consists of two major cognitive factors, namely "perceived usefulness" and "perceived ease of use" [13]. TAM is widely used to explore the use of information technology factors. But, in order to pursue common and simple, TAM was adapted to the most innovative environment using any technology. It contains only the perceived usefulness and perceived ease of use. To explore new and innovative uses, only two determinants is too simple and may lack behavioral richer interpretation. Related research integrated "the channels of communication" of IDT into TAM to enhance the disavantage of TAM in lacking the concept of social influence. And, it confirmed that the social influence is likely an important factor to affect people decide to use mobile services.

2.3 Feature Selection

The main purpose of feature selection is to increase the efficiency of learning tasks and improve the learning performance by excluding irrelevant features [7]. Based on the labels of training set, feature selection algorithm mainly divided into supervised, unsupervised and semi-supervised groups. In the group of supervised feature selection, it contains filters, wrappers, and embedded methods [21]. In this study, we use SVM-RFE and IG. SVM-RFE is a wrapper approach used in two-class circumstances. It was demonstrated that the features selected by SVM-RFE yielded better classification performance than the other methods [11]. SVM-RFE uses the weights of SVM to rank the feature for their removal. In the work of [26], SVM-RFE has successfully applied to identify product features.

IG is another used feature selection method in this work. IG is one of the most common attribute evaluation methods [27]. This univariate filter provides an ordered ranking of all the features and then a threshold is required. IG can used to calculate the degree of importance of each factor. IG has also been used in many non-medical fields, such as learning strategies, design contests, forecasting market [1].

3 Employed Approach

The implemental procedure contains 7 major steps in this study. The concise steps can be found as follows.

Step 1: Define factors of game APP
Based on innovation diffusion theory (IDT), the technology acceptance model (TAM), and some related literature, we attempt to define potential factors of game APPs for downloading free or paid game Apps.

Step 2: Design questionnaire
Based on the defined factors in step 1, we can develop a questionnaire to measure the importance level for downloading free or paid game APPs. Briefly speaking, this questionnaire contains three parts.

Part I: Basic information of respondents
Part II: The question items of different defined factors to measure the importance levels for downloading free or paid game Apps.
Part III: The probability of downloading free or paid game Apps.

Step 3: Pre-test questionnaire
This step is mainly to avoid respondents misunderstand questionnaire. Before issuing a formal questionnaire, we use one-by-one interview to pretest the designed questionnaire.

Step 4: Data collection
In this study, a questionnaire survey is implemented from the internet. Those who have used APPs will be collected for further analysis.

Step 5: Data pre-process
To explore the users' download intensions, the five-point Likert scale responses of question items in Part III will be re-organized, since support vector machine recursive feature elimination (SVM-RFE) and IG are designed for binary classification. The responses 1 and 2 will represent that customers will not download game APPS if their selected important factors in Part II will be fulfilled; 3, 4 and 5 represent that customers will download game APPS if their selected important factors in Part II will be fulfilled.

Step 6: Implement feature selection

Step 6.1: SVM-RFE
In this work, we use SVM-RFE. The algorithm of SVM-RFE can be found as follows.
Inputs:
Training examples

$$X_0 = [X_1, X_2, \dots, X_k, \dots, X_l]^T \tag{1}$$

Class labels

$$y = [y_1, y_2, \dots, y_k, \dots, y_l]^T \tag{2}$$

Initialize:
Subset of surviving features

$$s = [1, 2, \dots, n] \tag{3}$$

Feature ranked list

$$r = [] \tag{4}$$

Repeat until

$$s = [] \tag{5}$$

Restrict training examples to good feature indices

$$X = X_0(:, s) \tag{6}$$

Train the classifier

$$\alpha = SVM - train(X, y) \tag{7}$$

Compute the weight vector of dimension length(s)

$$W = \sum_k \alpha_k y_k x_k \tag{8}$$

Compute the ranking criteria

$$c_i = (w_i)^2, \text{ for all } i \tag{9}$$

Find the feature with smallest ranking criterion

$$f = \arg\min(c) \tag{10}$$

Update feature ranked list

$$r = [s(f), r] \tag{11}$$

Eliminate the feature with smallest ranking criterion

$$s = s(1 : f-1, f+1; length(s)) \tag{12}$$

Output:

Feature ranked list

$$r \tag{13}$$

Step 6.2: Information Gain

Information gain (IG) is the easiest and fastest way of select optimal attribute subset. The definition of IG is based on entropy. Assume we have examples $N = \{1, 2, \cdots, n\}$, attributes $D = \{1, 2, \cdots, d\}$, and class $K = \{1, 2, \cdots, k\}$, from equations (14)~(16), we can define IG.

$$Entropy(N) = \sum_{i=1}^{K} P_i \log_k \left(\frac{1}{P_i}\right) = -\sum_{i=1}^{k} P_i \log_k P_i \tag{14}$$

$$Entropy(D_j) = \sum_{i=1}^{|D_j|} \frac{D_{ji}}{N} \times Entropy(D_{ji}) \tag{15}$$

$$IG(D_j) = Entropy(N) - Entropy(D_j) \tag{16}$$

Step 7: Draw conclusions

Through analysis of the results of step 6.1 and 6.2, we will identify important factors for different groups for downloading free or paid game Apps.

4 Results

From available literatures, we define 14 potential factors of game APPs. Table 1 summarizes these factors, their sources and supports. A total of 135 questionnaires were collected. After removing invalid questionnaires, 130 valid questionnaires remain for further analysis. These collected examples will be divided into (1) leading group (innovators, early adopters), (2) majority group (early majority, late majority), and (3) laggard group (laggards), for free and paid game APPs, respectively. Table 2 lists the basic statistics for these three groups.

In this study, 5-fold cross-validation experiment has been employed. Those factors whose values are greater than the average will be picked up as an important factor. Results of SVM-RFE and IG are given in Tables 3 and 4.

Table 1. The defined potential factors of game APPs

Source	No.	Factors	Supports	Source	No.	Factors	Supports
TAM	A1	Perceived usefulness	[5, 13, 17, 22]	IDT	A8	Image	[13, 18]
TAM	A2	Perceived ease of use	[5, 13]	IDT	A9	Mass media	[3, 20]
IDT	A3	Compatibility	[13, 20]	IDT	A10	Interpersonal channels	[3, 20]
IDT	A4	Visibility	[13, 18]	Others	A11	Perceived enjoyment	[17]
Others	A5	Voluntariness	[13, 18]	Others	A12	Cognitive concentration	[17]
IDT	A6	Result demonstrability	[13, 18]	Others	A13	Perceived risk	[2, 16]
IDT	A7	Trialability	[13, 20]	Others	A14	Use context	[17]

Table 2. Statistics of collected samples for different groups

Group / Item	Leading group	Majority group	Laggard group
Gender	Male: 72% Female: 28%	Male: 54% Female: 46%	Male: 45% Female: 55%
Age	18~30 years old (100%)	18~30 years old (96%) 31~40 years old (1%) >40 years old~ (3%)	18~30 years old (64%) 31~40 years old (36%)
Income per month	<5K NTD(36%) 5K~10K NTD(26%) 10K~20K NTD(10%) 20K~50K NTD(28%)	<5K NTD(43%) 5K~10K NTD(20%) 10K~20K NTD(13%) 20K~50K NTD(23%) >50K NTD(3%)	<5K NTD(9%) 5K~10K NTD(27%) 10K~20K NTD(18%) 20K~50K NTD(45%)
Education	Undergraduate (64%) Master & above (36%)	Undergraduate (76%) Master & above (24%)	Undergraduate (64%) Master & above (36%)
Game information	Keyword Ads (13%) Shared by friends (28%) Magazines (5%) Social media marketing (44%) Mobile Ads (3%) Others (8%)	Keyword Ads (6%) Shared by friends (34%) Magazines (1%) Social media marketing (48%) Mobile Ads (6%) Others (5%)	Shared by friends (27%) Social media marketing (55%) Mobile Ads (9%) Others (9%)
Game usage time	<3hrs (82%) 4~6hrs (10%) 7~9hrs and above (8%)	<3hrs (69%) 4~6hrs (28%) 7~9hrs and above (4%)	<3hrs (91%) 4~6hrs (9%)
Paid downloads	Ever (15%) Never (85%)	Ever (10%) Never (90%)	Ever (9%) Never (91%)
The payment amount of paid downloads	<1.99USD (17%) 1.99~2.99USD (33%) >2.99USD (50%)	<1.99USD (63%) 1.99~2.99USD (25%) >2.99USD (13%)	<1.99USD (100%)
In-APP purchase	Ever (49%) Never (51%)	Ever (36%) Never (64%)	Ever (9%) Never (91%)
The amount of in-APP purchase	<1.99USD (42%) 1.99~2.99USD (21%) >2.99USD (37%)	<1.99USD (28%) 1.99~2.99USD (28%) >2.99USD (45%)	<1.99USD (100%)

Table 3 shows the results of important factors for free downloads. We can find the important factors for both SVM-RFE and IG are Perceived usefulness (A1), Perceived ease of use (A2), Compatibility (A3), Visibility (A4), Voluntariness (A5), Result demonstrability (A6), Trialability (A7), Image (A8), and Use context (A14) for whole examples. In leading group, the important factors are Perceived ease of use (A2), Visibility (A4), Voluntariness (A5), Result demonstrability (A6), Image (A8), Interpersonal channels (A10), and Use context (A14). In majority group, the crucial factors are Perceived usefulness (A1), Perceived ease of use (A2), Compatibility (A3), Voluntariness (A5), Result demonstrability (A6), Trialability (A7), Image (A8), Use context (A14). In laggard group, the important factors are Perceived usefulness (A1), Perceived ease of use (A2), Compatibility (A3), Voluntariness (A5), Result demonstrability (A6), Trialability (A7), Image (A8), and Use context (A14).

Table 3. Summary of feature seleciton results for different groups (FREE downloads)

Factors	Leading group		Majority group		Laggard group	
	SVM-RFE	IG	SVM-RFE	IG	SVM-RFE	IG
A1	✓	✓		✓	✓	✓
A2	✓	✓	✓	✓	✓	✓
A3	✓	✓	✓		✓	✓
A4	✓	✓	✓	✓		✓
A5	✓	✓	✓	✓	✓	✓
A6	✓	✓	✓	✓	✓	✓
A7	✓	✓		✓	✓	✓
A8	✓	✓	✓	✓	✓	✓
A9	✓		✓		✓	
A10	✓		✓	✓	✓	
A11		✓	✓		✓	
A12			✓			
A13	✓		✓		✓	
A14	✓	✓	✓	✓	✓	✓

Table 4. Summary of feature seleciton results for different groups (PAID downloads)

Factors	Leading group		Majority group		Laggard group	
	SVM-RFE	IG	SVM-RFE	IG	SVM-RFE	IG
A1	✓	✓		✓	✓	
A2	✓	✓	✓	✓	✓	✓
A3	✓		✓		✓	
A4	✓	✓	✓	✓		✓
A5	✓	✓	✓	✓	✓	✓
A6	✓	✓	✓	✓	✓	✓
A7	✓	✓		✓	✓	✓
A8	✓		✓	✓	✓	✓
A9	✓		✓		✓	
A10	✓	✓	✓	✓	✓	✓
A11		✓	✓		✓	
A12			✓			
A13	✓		✓		✓	
A14	✓	✓	✓	✓	✓	✓

Table 4 shows the results of important factors for paid downloads. For whole examples, the selected important factors are Perceived usefulness (A1), Perceived ease of use (A2), Visibility (A4), Voluntariness (A5), Result demonstrability (A6), Trialability (A7) , Interpersonal channels (A10), and Use context (A14). In leading group, the important factors are Perceived ease of use (A2), Visibility (A4), Voluntariness (A5), Result demonstrability (A6), Image (A8), Interpersonal channels (A10), and Use context (A14). In majority group, the selected important factors are Perceived ease of use (A2), Voluntariness (A5), Result demonstrability (A6), Trialability (A7), Image (A8), Interpersonal channels (A10), and Use context (A14). In laggard group, the important factors are Perceived ease of use (A2), Voluntariness (A5), Result demonstrability (A6), Trialability (A7) , Image (A8), Interpersonal channels (A10), and Use context (A14).

5 Conclusions

The purpose of this study is to confirm the key factors of downloading game Apps. According to the available literatures, 14 potential factors have been defined. SVM-REF and IG feature selection methods have been employed to select the important factors. The results indicated that no matter whether free or paid games App, Perceived ease of use (A2), Voluntariness (A5), Result demonstrability (A6), Use context (A14) are important factors. The discovered results can provide game APP developers to design products in the future.

Regarding the potential directions of future works, additional feature selection methods could be used to evaluate their effectiveness. In addition, different types of APPs might be considered as research topics. These ways might double confirm the results obtained from this study.

Acknowledgement. This work was financially supported in part by Ministry of Science and Technology of Taiwan (Grant No. MOST 101-2628-E-324-004-MY3). Authors express our deeply appreciation to the funding.

References

1. Evangelou, E., et al.: Perceived information gain from randomized trials correlates with publication in high–impact factor journals. Journal of Clinical Epidemiology 65(12), 1274–1281 (2012)
2. Al-Jabri, I.M., Sohail, M.S.: Mobile banking adoption: application of diffusion of innovation theory. Journal of Electronic Commerce Research 13(4), 379–391 (2012)
3. Bhattacherjee, A.: Acceptance of e-commerce services: the case of electronic brokerages. IEEE Transactions on Systems, Man and Cybernetics, Part A: Systems and Humans 30(4), 411–420 (2000)
4. Butler, M.: Android: Changing the mobile landscape. IEEE Pervasive Computing 10(1), 4–7 (2011)

5. Davis, F.D.: Perceived usefulness, perceived ease of use, and user acceptance of information technology. MIS Quarterly, 319–340 (1989)
6. Davis, F.D., Bagozzi, R.P., Warshaw, P.R.: User acceptance of computer technology: a comparison of two theoretical models. Management Science 35(8), 982–1003 (1989)
7. ElAlami, M.E.: A filter model for feature subset selection based on genetic algorithm. Knowledge-Based Systems 22(5), 356–362 (2009)
8. eMarketer (2013), http://www.emarketer.com/Article/In-App-Purchases-Take-Over-App-Revenues/1010491
9. Feijoo, C., et al.: Mobile gaming: Industry challenges and policy implications. Telecommunications Policy 36(3), 212–221 (2012)
10. Fenez, M.: Global Entertainment and Media Outlook: 2009-2013. PricewaterhouseCoopers, New York (2009)
11. Guyon, I., et al.: Gene selection for cancer classification using support vector machines. Machine Learning 46(1-3), 389–422 (2002)
12. Hsu, C.-L., Lin, J.C.-C.: What Drives Purchase Intention for Paid Mobile Apps?–An Expectation Confirmation Model with Perceived Value. Electronic Commerce Research and Applications (2014)
13. Hsu, C.-L., Lu, H.-P., Hsu, H.-H.: Adoption of the mobile Internet: An empirical study of multimedia message service (MMS). Omega 35(6), 715–726 (2007)
14. Jansen, S., Bloemendal, E.: Defining App Stores: The Role of Curated Marketplaces in Software Ecosystems. In: Herzwurm, G., Margaria, T. (eds.) ICSOB 2013. LNBIP, vol. 150, pp. 195–206. Springer, Heidelberg (2013)
15. Kivimaki, A., Fomin, V.: What makes a killer application for the cellular telephony services? In: 2001 2nd IEEE Conference on Standardization and Innovation in Information Technology. IEEE (2001)
16. Kleijnen, M., De Ruyter, K., Wetzels, M.: Consumer adoption of wireless services: discovering the rules, while playing the game. Journal of Interactive Marketing 18(2), 51–61 (2004)
17. Liu, Y., Li, H.: Exploring the impact of use context on mobile hedonic services adoption: An empirical study on mobile gaming in China. Computers in Human Behavior 27(2), 890–898 (2011)
18. Moore, G.C., Benbasat, I.: Development of an instrument to measure the perceptions of adopting an information technology innovation. Information Systems Research 2(3), 192–222 (1991)
19. Rice, R.E., Katz, J.E.: Assessing new cell phone text and video services. Telecommunications Policy 32(7), 455–467 (2008)
20. Rogers, E.M.: Diffusion of innovations. Simon and Schuster (2010)
21. Tang, J., Alelyani, S., Liu, H.: Feature Selection for Classification: A Review. Data Classification: Algorithms and Applications. CRC Press (2013)
22. Tornatzky, L.G., Klein, K.J.: Innovation characteristics and innovation adoption-implementation: A meta-analysis of findings. IEEE Transactions on Engineering Management 29(1), 28–45 (1982)
23. Verkasalo, H., et al.: Analysis of mobile Internet, map application and game adoption among smartphone users. Proceedings of the paper presented at the 8th Global Mobility Roundtable (GMR 2009), Cairo, Egypt (2009)

24. Verkasalo, H., et al.: Analysis of users and non-users of smartphone applications. Telematics and Informatics 27(3), 242–255 (2010)
25. West, J., Mace, M.: Browsing as the killer app: Explaining the rapid success of Apple's iPhone. Telecommunications Policy 34(5), 270–286 (2010)
26. Yang, C.-C.: A classification-based Kansei engineering system for modeling consumers' affective responses and analyzing product form features. Expert Systems with Applications 38(9), 11382–11393 (2011)
27. Zhang, H., Ren, Y.-G., Yang, X.: Research on Text Feature Selection Algorithm Based on Information Gain and Feature Relation Tree. In: 2013 10th Web Information System and Application Conference (WISA). IEEE (2013)

Explicating the Trends of China's Logistics Services for Electronic Commerce

Mu-Chen Chen[1], Pei-Ju Wu[2], and Wei-Hua Xiong[3]

[1] Department of Transportation and Logistics Management, National Chiao Tung University
4F, No. 118, Sec. 1, Chung-Hsiao W. Road, Taipei, Taiwan
[2] Department of Transportation Technology and Management, Feng Chia University
No.100, Wenhwa Road, Seatwen, Taichung, Taiwan
[3] Master of Logistics Engineering Center, University of Science and Technology of China
No. 96, JinZhai Road, Baohe District, Hefei, Anhui, P.R. China
ittchen@mail.nctu.edu.tw, wupj@fcu.edu.tw,
xwh52041@mail.ustc.edu.cn

Abstract. The surge e-commerce demand stimulates the corresponding huge logistics requirements in China. However, most managers have little knowledge about how to well manage e-commerce logistics. This study systematically investigates the logistics system of two representative e-commerce platforms (i.e., Big Taobao and JD.COM) in China. Furthermore, the SWOT analysis was conducted to explore the logistics system of Big Taobao and JD.COM, respectively. Finally, the crucial operations strategies were devised based on the results of SWOT analysis.

Keywords: Electronic commerce, E-commerce logistics, SWOT analysis, Operations strategies.

1 Introduction

In 2000-2002, the Chinese electronic commerce (EC) industry has experienced a period of freezing and adjustment, and around four years later, it became to get recovery and warmer [1]. In 2010, the EC industry in China had a rapid development, and it became the world's largest Internet market [2]. Until the year of 2010, China's EC has matured, whether from internal or external environmental conditions, and it showed explosive growth. Especially, the large EC platform dominated the Chinese EC industry. For instance, the Business-to-Consumer (B2C) model such as JD.COM and Tmall.COM, and the Consumer-to-Consumer (C2C) model such as Taobao.COM have been the large EC platforms. In December 2013, China's online shopping users reached 302 million [3], and the market size of year 2013 was over 2 trillion RMB [4]. In the second quarter of 2014, the sales of China's online retail market reached 1.0856 trillion RMB, accounting for 8.7 percent of total retail sales [5]. Due to the "11.11" Shopping Festival, this proportion has been continuing to expand in 2014.

M. Gen et al. (eds.), *Industrial Engineering, Management Science and Applications 2015*,
Lecture Notes in Electrical Engineering 349, DOI: 10.1007/978-3-662-47200-2_91

E-commerce together with the Internet has become an important tool in logistics and supply chain management [6]. With the rapid development of China's EC industry, the logistics sector for EC in China also gained rapid growth. In the second quarter of 2012, the companies of domestic online shopping courier scaled up to 46.83 billion RMB in revenue, growth in the percentage of 39.8% [7]. In 2013, China's enterprises of express delivery services totaled 9.2 billion transactions, and the market size rose to second place in the world with a growth rate up to 61.6% [8]. Nowadays, China's EC logistics model can be divided into three categories: self-built logistics system model, third-party logistics model and the hybrid of self-built logistics system model and third-party logistics model. The famous EC company using the self-built logistics system model is JD.COM; the hybrid model, Alibaba Group's Big Taobao representing Taobao and Tmall.COM. Due to the cost, the small- and medium-sized EC companies generally use the third-party logistics model.

E-commerce and logistics are always of mutual restraint and mutual development; however, in the past, EC companies have not paid attention on logistics leading to the relative lack of e-logistics development. Such a situation restricts the development of EC. This study aims to explicate the trends of logistics service in China by investigating the current status and future development of EC in China. Two big EC platforms, Big Taobao and JD.COM are used as the subjects of investigation in this study. Furthermore, the SWOT analysis are conducted to better understand e-commerce logistics in China and to develop the operations strategies for firms to implement e-commerce logistics in China.

2 China's Electronic Commerce

2.1 The Status Quo of China's EC Industry

Currently, the future development of EC in China generally would like to be optimistic, and it has brought not only new challenges, but also opportunities to logistics management [9]. From the analysis of the external environment, such as the consumption level, the Internet penetration continues to increase. According to the internal factors such as the regulation of recent online shopping market, consumers have accepted this transaction pattern. Moreover, not only the online shoppers in China's first-tier cities have been decreasing, but also the online shopping is gradually extending to the third- or forth-tier cities. In the past five years, China's EC sector has a dramatic growth in terms of sales, number of online shoppers, number of enterprises, number of direct employees, etc. In addition, EC sales have grown at a fast pace in recent years [10]. In 2011, the growth rate of revenue of the EC in China is 70.2 %, in 2012, 66.2 %. In 2013, it still maintains a high growth rate of 42.4 % [11]. In 2010, the China's EC market is 4.5 trillion RMB, and by 2013, it surges to 10.2 trillion RMZ. For the number of online shoppers, in 2010, it is 158 million, and it reaches 312 million in 2013. The numbers of Chinese B2C and C2C enterprises surges to 29,303 in 2013 from 9,962 in 2010, and the number of direct employees reaches 235 million in 2013 [3].

Two essential players in China's EC industry are Alibaba Group's Big Taobao and JD.COM. Both their market size and business models have the typical significance in China. Accordingly, this study investigates the current status of these two big EC players and their needs in the field of logistics services to better understand how to implement EC logistics in China

2.2 The Status Quo and Business Model of Big Taobao

The Alibaba Group which was founded in 1999 in Hangzhou, Zhejiang Province, China, after years of development, has already firmly been at the leading position in China's EC industry. Its core business units such as Tmall.COM in charge of B2C, Taobao in charge of C2C, and the third-party online payment platform, Alipay, all occupy the largest market share in China. In the field of logistics, Alibaba Group is also actively deploying its CAINIAO Network. The Alibaba Group has gradually formed an open system including EC, logistics, finance and other EC ecosystems.

From the EC business unit, Taobao is the comprehensive online shopping platform in charge of C2C. Its main commodity categories covers more than ten categories of clothing & apparel, hats & bags, outdoor & sports, digital phone, home appliances, baby products, and so on. As early as in 2010, Taobao's registered users reaches 370 million, and the online products are 800 million pieces [12]. In the second quarter of 2014, Taobao's deals reach 50.37 % of China's online retail market, and it became the largest EC platform in China [13].

Tmall.COM is a comprehensive B2C platform, which is different from Taobao's C2C business model. Nonetheless, Tmall.COM's business model is based on the platform of Taobao, in which individuals can deal directly with others on the Internet. Specifically, companies can use Tmall.COM to directly link to consumers to conduct online transactions. Furthermore, sellers and buyers can use the services provided by Taobao for free in the transaction process. Even in China B2C market, Tmall.COM is also the largest player, which accounts for 57.4 %, whereas the second largest player, JD.COM, only accounts for 21.1 % of the market. The other players (e.g., Suning, Amazon) only occupy a small amount of market in China, and it's difficult for them to compete with these two giants [14]. Regarding the profit model, Taobao is quite different to the foreign EC platforms such as eBay. For the sources of profit of eBay, the advertising occupies only 5% of total revenue, and most of the profit is generated in the course of trade [15]. In contrast, Taobao's main source of profit is from the advertising [16]. In 2013, its advertising revenue accounts for 66.46% of total revenue [13].

2.3 The Status Quo and Business Model of JD.COM

JD.COM, in its earlier stage, executes all the functions by itself. Recently, after the adjustment of strategy, the online shopping platform and logistics have been outsourced to the third-party service providers. The outsourcing strategy makes JD.COM have amazing growth in recent years. In 2011, 2012 and 2013, the revenue growth rates of JD.COM's are 146.2 %, 95.8 %, and 70.9 %, respectively. JD.COM possesses the much higher growth rate than the average of EC industry in China.

Nowadays, JD.COM is China's second largest B2C platform, it ranks only after Tmall.COM. For the main products sold, JD.COM and Tmall.COM have a high level of overlapping. The overall market share and number of SKUs of JD.COM are also behind Tmall.COM; however, it is worth noting that JD.COM dominates the B2C market of 3C products with the largest market share. JD.COM relies on its internal resources, brand protection and self-built logistics.

3 China's Logistics Services for Electronic Commerce

Alibaba Group's Big Taobao and JD.COM are the most representative of the two companies in the China e-commerce industry. Meanwhile, these two e-commerce enterprise's logistics models can be viewed as the essential mainstream of the e-commerce logistics models in China. Hence, the e-commerce logistics of Big Taobao and JD.COM are analyzed as follows.

3.1 The Big Taobao's E-commerce Logistics Model

The Big Taobao's e-commerce logistics model used to be criticized by its consumers involving the low efficiency and the inferior quality of logistics service. The main reason of its weak logistics service is that the Big Taobao separately employed STO Express, YTO Express, ZTO Express, BEST express and YUNDA express as its main third party logistics providers [17], which lacks the seamless integration.

Nevertheless, the Big Taobao got things turned around. In 2013 the Alibaba Group coordinated YINTAI Group, FORCHN HOLDINGS Group, FOSUN Group, SF express, STO Express, YTO Express, ZTO Express and YUNDA express to set up the CAINIAO Network Technology Co., Ltd., initiating the first step in self-built logistics of the Alibaba Group [17]. For example, in 2014 the "11.11" Shopping Festival began to participate in the operation of its e-commerce logistics platform, which indicates that their self-built logistics system has taken molding beginning to be effective.

The CAINIAO Network is not just a logistics company, but is a Chinese e-commerce logistics network platform. The CAINIAO Network is implemented through "cloud net - ground net" logistics network model. Herein, ground net means the deployment of e-commerce logistics facility network. Specifically, the CAINIAO Network aims to integrate the whole China's e-commerce logistics resources to optimize resource allocation. On the other hand, cloud net means the data analytics of logistics data. That is, the CAINIAO Network seeks to let logistics companies share their data within the platform and to conduct data analysis for creating valuable information such as logistics demand forecasting.

3.2 The JD.COM's E-commerce Logistics Model

JD.COM is not just a pure e-commerce company, but it is a company to provide total supply chain solutions. Particularly, JD.COM effectively integrates suppliers and end sellers through JD.COM platform.

Currently, JD.COM has been constructed covering 1000 counties self-built logistics system consists of five logistics centers and nearly 1400 distribution sites, using its self-built logistics for distribution of orders accounted for more than 85% of all orders [18]. JD.COM also launched a campus business offices, subway pickup point and communities pick up points to reflects the importance of the last mile logistics [19].The self-built logistics system of JD.COM is the most extensive coverage and it also has the most complete logistics facilities in China's EC industry. JD.COM self-built logistics system framework is divided into three modules: the layout logistics warehousing; transportation and distribution system; logistics network system. The following is a detailed analysis of its three modules:

(1) The layout logistics warehousing: JD.COM self-built logistics system established five primary logistics center in Beijing, Shanghai, Guangzhou, Chengdu, Wuhan, occupied the north, east, south, southwest and central China, respectively, basically covering the major urban centers of China [20]. To the end of 2012, JD.COM self-built warehousing area has more than 1 million square meters [21, 22]. At the same time, JD.COM spent a huge cost of construction of "Asia No. 1" modern logistics center, which is China's largest and most advanced EC logistics center. "Asia No.1" is located at Shanghai Jiading, constructed in two phases, planning construction area of 200,000 square meters. It puts into operation a positioning for medium-sized merchandise warehouses, with a total construction area of about 100,000 square meters, divided into four regions- dimensional reservoir area, multi-loft picking area, production work area, and shipping sortation area. Its powerful capabilities of order processing can support JD.COM's annual growth in logistics orders, and achieve the efficiency of logistics services and quality.
(2) Transportation and distribution system: There are nearly 20,000 staff in JD.COM self-built logistics system, more than 1,000 vehicles owned [23], and considering air freight qualifications to prepare for the coming aircraft within JD.COM self-built logistics system. Furthermore, JD.COM has a unique "warehouse + matching" mode, which is a real-time replenishment system to analyze the sales demand for timely replenishment based on the actual situation.
(3) Logistics network system: JD.COM has completed the upgrade of the logistics system named "Qinglong" in 2013. The "Qinglong" system not only supports JD.COM's internal logistics services, but also supported to undertake external logistics business, to effectively expand the service types and scope of JD.COM self-built logistics such as on-site reverse return services to consumers or third-party companies.

4 SWOT: CAINIAO Network and JD.COM Self-Built Logistics System

CAINIAO Network coordinates various resources involving physical facilities and operations data to fulfill the optimal Chinese e-commerce logistics network platform. On the other hand, JD.COM develops its own logistics system to offer total logistics solutions. This study conducted the SWOT analyses for the logistics system of

CAINIAO Network (Table 1) and JD.COM (Table 2), respectively. Furthermore, the corresponding operations strategies were developed based on the analytical results of the SWOT analyses.

Table 1. The SWOT matrix of the CAINIAO network

Strengths (S)	**Weaknesses (W)**
1. Sufficient logistics orders Taobao and Tmall.COM represents the largest C2C and B2C market, respectively. These two platforms have plenty of orders and belongs to the Alibaba Group, which provide enough support for the CAINIAO network. 2. Abundant financial support Logistics system needs huge funding to build up. The Alibaba Group have plentiful financial resources which can support the CAINIAO network. 3. Powerful technical support Alibaba Group has its own research institute with a number of outstanding technical team, which can solve technical problems for the CAINIAO network.	1. Absence of logistics experience The Alibaba Group has no prior experience in self-built logistics, and logistics is not its core abilities. For example, the lack of professional logistics personnel in its talent pool. Hence, the CAINIAO network is more likely to suffer logistics issues without gaining enough professional logistics support from the Alibaba Group. 2. Running late for logistics system CAINIAO Network started relatively late. It needs time to establish a complete logistics system and is more likely to lose competitive advantage.
Opportunities (O)	**Threats (T)**
1. Brilliant EC and logistics environment in China The surge C2C and B2C demands in China stimulate the rapid development of Chinese e-commerce. Moreover, the Chinese opening and reforming policy provides another opportunity for the evolution of the CAINIAO network. 2. In need of integrating logistics resources Numerous small and medium-sized logistics providers strive to compete with each other, which squanders logistics resources. Correspondingly, the CAINIAO Network can consolidate separate logistics resources and create total optimization for China's EC logistics.	1. Competing self-built e-commerce logistics More than one self-built logistics system would grab the market share of e-commerce logistics. For example, the second largest B2C platform, JD.COM, has its self-built logistics system. JD.COM's e-commerce logistics would endanger the development of CAINIAO Network. 2. Rivaling third-party logistics providers The CAINIAO Network and third-party logistics providers have the same interest which is to integrate logistics resources. Third-party logistics providers may oppose the formation of the CAINIAO Network.

(1) S-O strategy: Brilliant e-commerce development in China triggers more and more logistics providers engaging into e-commerce logistics. Nonetheless, logistics providers fight each other to grab e-commerce orders and eventually waste logistics resources. Correspondingly, the powerful e-commerce company, the Alibaba Group, has capability to coordinate logistics providers to deal with e-commerce orders. Specifically, the CAINIAO network generated by the Alibaba Group offers a platform to consolidate not only the e-commerce orders (demands) but also logistics (supplies). By doing so, the CAINIAO network not only catches e-commerce market but also occupies logistics resources. Furthermore, rich e-commerce demand and supply data can enhance the logistics abilities of the CAINIAO network such as logistics demand forecasting, logistics scheduling, etc.

(2) W-O strategy: Even though the CAINIAO network does not develop in the early stage and lack of rich logistics experience, the current logistics market have various logistics resources for the CAINIAO network to dig out. In particularity, the CAINIAO network can make use of the existing logistics resources to supplement its logistics weaknesses. For instance, the CAINIAO network can recruit experienced logistics professionals to implement its e-commerce logistics.

(3) S-T strategy: The CAINIAO network confronts competing self-built e-commerce logistics and third-party logistics providers to get e-commerce logistics orders. Based on the substantial e-commerce logistics orders from the Alibaba Group, the CAINIAO network can keep having sufficient e-commerce logistics orders. Furthermore, the CAINIAO network should take advantage of its rich financial and technical resources to devise comparative logistics services to attract more logistics orders.

(4) T-W Strategy: The CAINIAO Network should minimize the impact of weak logistics experience and avoid directly battle logistics rivals. Specifically, the CAINIAO Network should attract logistics providers to join its network (from rivals to partners). Moreover, the CAINIAO Network can target specific logistics services or create unique logistics services (niches) before it absorb sufficient logistics knowledge and possess rich logistics experience.

(1) S-O strategy: JD.COM can gather various logistic providers according to its diverse logistics services. Moreover, integrating various logistic providers can enhance JD.COM capabilities for total logistics services. Besides e-commerce logistics orders, JD.COM can entice comprehensive logistics demands.

(2) W-O strategy: JD.COM should yield an appealing plan (e.g., revenue sharing) to attract logistic providers taking part in JD.COM's logistics system. By doing so, JD.COM can not only diminish its financial burden (e.g., utilizing partners' logistics facilities) but also have various experts dealing with varying issues.

(3) S-T strategy: JD.COM can takes its well-developed logistics system as a holistic logistics platform to cooperate (not flight) with other self-built e-commerce logistics and third-party logistics providers.

Table 2. The SWOT matrix of the JD.COM

Strengths (S)	**Weaknesses (W)**
1. Mature logistics system JD.COM established its logistics system in the early stage compared with other self-built e-commence logistics system. Hence, JD.COM can offer comprehensive logistics services and have rich logistics experience. 2. Superior logistics services JD.COM self-built logistics system offers total supply chain services for its customers. Therefore, JD.COM has better ability to control the logistics quality and retain the loyalty of its customers.	1. Financial burden JD.COM has incurred enormous costs to conduct its logistics system. 2. Broad logistics services JD.COM aims to involve various logistics services, which may disperse its limited resources and need diverse logistics professionals.
Opportunities (O)	**Threats (T)**
1. Vast logistics demand In addition to e-commence logistics demand in China, other logistics requirements also can be serviced by the JD.COM with total logistics solutions. 2. Unconsolidated logistics providers Abundant logistics providers exist in China logistics market. JD.COM's logistics system can integrate individual logistics resources.	1. Threatening self-built e-commerce logistics More and more e-commerce companies have conducted their own logistics services such as the CAINIAO network. 2. Pressurizing third-party logistics providers Third-party logistics providers aim to offer logistics services for e-commerce logistics, which conflicts JD.COM's interests.

(4) T-W Strategy: JD.COM should identify unprofitable logistics services and surrender them to minimize financial burden and wide logistics services as well as to avoid competitors' threats.

5 Conclusions

The surge e-commerce demands triggers the corresponding huge logistics requirements in China. Nonetheless, most managers has little knowledge to manage e-commerce logistics. This study explores the logistics system of two representative e-commerce platform (i.e., Big Taobao and JD.COM) in China. Moreover, the SWOT analysis was conducted to explore the logistics system of Big Taobao and JD.COM, respectively.

Based on the analytical results of the SWOT analysis, the essential e-commerce logistics strategies can be summarized as follows. First, an e-commerce logistics platform should be created to integrate varying e-commerce orders and various logistics providers (especially for numerous small and medium-sized logistics providers in China). Second, a knowledge-based system should be developed to gather e-commerce logistics experience. Then, e-commerce logistics analyses can be conducted to generate logistics operations such as demand forecasting and logistics scheduling. Finally, e-commerce logistics training programming should be devised to ensure sufficient professions for tackling e-commerce logistics issues.

Acknowledgments. This work is partially supported by Ministry of Science and Technology, Taiwan, R.O.C. under grant MOST 103-2410-H-009-029-MY3.

References

1. China_Electronic_Commerce_Research_Center. 1997-2009: Chinese electronic commerce twelve years investigation report (2009), http://B2B.toocle.com
2. Clemes, M.D., Gan, C., Zhang, J.: An empirical analysis of online shopping adoption in Beijing, China. Journal of Retailing and Consumer Services 21, 364–375 (2014)
3. China_Internet_Network_Information_Center. 2013 China's online shopping market research report (2014), http://www.cnnic.net.cn/hlwfzyj/hlwxzbg/201409/P020140901332431510284.pdf
4. The_Boston_Consulting_Group. The global electronic commerce new leader: Master China network shop-ping new ecological (2011), http://www.bcg.com.cn/export/sites/default/cn/files/publications/reports_pdf/BCG_The_Worlds_Next_E-Commerce_Superpower_Nov_2011_CHN_Updated.pdf
5. China_Electronic_Commerce_Research_Center. Chinese online retail market data monitoring report (2014), http://www.100ec.cn/zt/upload_data/down/2014shwlls.pdf
6. Mi, J., Ng, J.: Barriers to e-commerce policy in logistics: an exploratory study of the Pearl River Delta, China. International Journal of Logistics: Research and Applications 12, 195–212 (2009)
7. China_Electronic_Commerce_Research_Center. 2008-2013 China's online shopping courier companies' revenue scale (2012), http://www.100ec.cn/
8. Deloitte_Touche_Tohmatsu, Development& Research Center of the State Post Bureau. China's Express Delivery Industry Development Report (2014) (in Chinese)
9. Gunasekaran, A., Ngai, E.W.T., Cheng, T.C.E.: Developing an E-logistics System: A case study. International Journal of Logistics: Research & Applications 10, 333–349 (2007)
10. Huang, X., Dai, X., Liang, W.: BulaPay: a novel web service based third-party payment system for e-commerce. Electronic Commerce Research 14, 611–633 (2014)
11. Huxiu.COM. JD recent five-year average sales growth of 210 % (2012), http://www.huxiu.com/article/1873/1.html
12. Alibaba. Taobao Online Shopping Data 2010 (2011) (in Chinese)
13. Enfodesk. Enterprise Analysis Report - Alibaba (2014), http://www.enfodesk.com/SMinisite/newinfo/reportdetail-id-416990.html
14. China_Electronic_Commerce_Research_Center. 2014 Chinese electronic commerce market data (2014), http://www.100ec.cn/zt/upload_data/down/2014shch.pdf

15. Huang, J.-J.: Key Competitiveness Model of C to C E-Commerce Profit Model-Taobao and eBay China Competitive Strategies Analyzing: Tongji University (2008)
16. Xuan, L., Zeng, X.-Q.: "Taobao" profit model analysis. Inquiry into Economic Issues 1 (2012) (in Chinese)
17. Lu, X.-T.: What is CAINIAO network (2013), http://tech.sina.com.cn/i/csj/blog/2013-05-29/1404/1669181277/637daf5d01018npl.shtml
18. China Electronic Commerce Research Center. JD self logistics and distribution orders have accounted for 85 % of all orders (2013), http://b2b.toocle.com/detail-6096388.html
19. Ramanathan, R.: E-commerce success criteria: determining which criteria count most. Electronic Commerce Research 10, 191–208 (2010)
20. Wei, G., Kui, Y.: Analysis of JD mall logistics distribution model. Technology and Market 2 (2013) (in Chinese)
21. Yan, H.S.: Jingdong open the logistics platform, Liu Qiang Dong pick upgrade "Cats & Dogs". IT Time Weekly 17 (2013) (in Chinese)
22. Xu, Y.Z.: Do you have "JD"– B2C logistics era. Programmable Controller & Factory Automation 6 (2011) (in Chinese)
23. Qing, Y.S.: On Jingdong self built logistics system. Heilongjiang Science and Technology Information 30 (2013) (in Chinese)

Measuring Customer Relationship Marketing Outcomes in the Greek Banking Sector

Ilias Santouridis* and Melania Stoumbou

Department of Accounting and Finance, TEI of Thessaly, 41110 Larissa, Greece
isant@teilar.gr, stouboumel@yahoo.gr

Abstract. Banks were among the pioneers in developing and implementing relationship marketing strategies, since they were between the first recipients of the radical changes that took place in global economy in recent years. This study is part of an ongoing research project, which among others, aims to develop an instrument to measure the outcome of the implementation of relationship marketing strategies. Field research was conducted among customers of Greek banks. Principal component analysis was then performed in order to identify latent factors among the questionnaire items. The results obtained were satisfactory, since 7 factors were extracted, which were labeled as relationship value, crisis management, personalized sales, customer focus, flexibility, communication quality and reliability. Moreover, regression analysis verified the significant explanatory power of the factors emerged over customer loyalty.

Keywords: relationship marketing, CRM, banking, performance monitoring.

1 Introduction

Contemporary economy, which is greatly influenced by globalization, increasing competition and advances in communication and information technology, force companies to gradually depart from traditional marketing principles and adopt a customer-centric approach by focusing on managing customer relationships [1]. As a result of this, the focus of differentiation has been shifted from products and services towards customers [2].

Relationship marketing (RM) is an enterprise wide approach having the target to offer answers to the above considerations by aiming to understand and influence customer behavior in order to improve customer acquisition and retention as well as profitability [3]. RM promotes a holistic approach by introducing the notion of customer life-cycle, including stages such as acquisition, retention, expansion and recovery, and promoting the recognition and management of a wide range of company relationships (e.g. internal and external customers, suppliers, competitors, companies in other market segments).

---

* Corresponding author.

M. Gen et al. (eds.), *Industrial Engineering, Management Science and Applications 2015*,
Lecture Notes in Electrical Engineering 349, DOI: 10.1007/978-3-662-47200-2_92

RM strategies rely heavily on gathering and processing customer data in order to yield information that will help companies to develop and implement customer-centric strategies. As a result of this, information technology plays a pivotal role as an enabler of successful RM strategies implementation. Customer Relationship Management (CRM) software systems are being widely used since the late '90s as information management enabling relationship marketing solutions [4]. However, as early efforts for the adoption of such systems have shown, CRM implementations can pose significant challenges. Digman [5] emphasizes that "*the majority of CRM initiatives fail to meet expectations*". One of the most important lessons learned from early adopters is that there is a need to closely monitor the performance of RM strategy implementations, which utilize CRM systems and tend to be expensive.

Banks were among the service industry pioneers in RM strategies adoption, since they were between the first recipients of the radical changes that took place in global economy during the past few decades [6]. They operate in a highly competitive environment, where the acquisition of new customers is often proved much more expensive than retaining and expanding the ones they already have.

This study is part of an ongoing research project aiming to identify and understand the emerging roles of accounting and finance professionals in the contemporary digital economy. Gummesson [7] acknowledged that a company has to incorporate relationship aspects not only in its marketing plan but also in its corporate business plan. This led him to the conclusion that accounting and financial metrics must also include the company's return on relation relationships (ROR), which has been defined as "*the long-term net financial outcome caused by the establishment and maintenance of an organization's network of relationships*". Taking this stance, this study presents an effort to develop an instrument measuring the outcome of the implementation of RM strategies in the banking sector in Greece, by investigating how customers experience this approach. The utilization of such an instrument would provide banks with a valuable tool that facilitates the monitoring of the performance of their RM strategies.

2 Background

2.1 Relationship Marketing

Interest in RM began to grow in the '90s. Research published at that time (e.g. [8]) revealed the great impact on profitability that can be achieved even by small increases in customer retention rates. This made the marketing community more conscious of the necessity to manage customer relationships in the long term. Today it is widely accepted that customers are at the core of business and therefore a company's success depends heavily on effectively managing relationships with them.

According to Egan [9], despite the fact that considerable research and practice took place during the past few decades, RM may still be regarded as an 'umbrella philosophy' with plenty of variations and definitions rather than a unified concept with commonly accepted objectives and strategies. Perhaps the most comprehensive definition of RM was given by Gronroos [10], who defined RM as the marketing approach aiming to "*identify and establish, maintain and enhance and, when necessary, terminate relationships with customers and other stakeholders, at a profit so that the objectives of all parties involved are met; and this is done by mutual exchange of fulfilment and promises*". This definition implies that since each customer must be treated individually, the company must be in the position to recognize the relationship stage each customer is in and treat them in a different manner. The introduction of various consecutive relationship stages in RM thinking lead to the formation of the customer life-cycle concept. Various relevant models have been reported in literature, examples of which are the awareness, exploration, expansion, commitments and dissolution stages proposed by Dwyer et al [11] and the suspect, prospect, first-time customer, repeat customer, client, advocate, member and partner stages reported by Kotler [12].

One of the vital questions raised regarding the implementation of an RM strategy is how to measure its performance in order to monitor its effectiveness. Wang et al. [13] pointed out that a widely accepted way to measure RM performance is in terms of customer behaviors "*since they are the underlying sources of value of current customers of a firm and have the potential to increase the future revenue streams associated with them and those of prospective customers*". Moreover, since one of the main RM objectives is to assure steady levels of revenues and maximization of customer lifetime value, customer behaviors, which could bring revenues to the firm, are believed to be strategically important [14, 15].

2.2 Relationship Marketing in Banks

Andersen et al. [16] claim that the trust and commitment, which can be developed between banks and their customers, are considered to be the most powerful competitive advantage for banks, a fact that explains to a certain degree why financial institutions have moved from transaction-based banking to a relational approach. Their research showed that the annoyance of bank customers is increased from product focused strategies, since they expect a more advisory and caring treatment. Therefore, a customer relational approach in the highly competitive banking industry, where products and services tend to be more and more standardized, offers a differentiating factor. The improvements that can be achieved by banks adopting an RM approach can be summarized as follows [17]:

- greater customer satisfaction due to better service offering,
- greater business coherence, since the bank defines objectives directly linked to customers,
- lower costs,
- increased service effectiveness resulting from the acquisition of complete and correct customer information,
- new business opportunities because of the establishment of long-term relationships with customers,

- more accurate and successful customer segmentation, which can distinguish groups of profitable and less profitable customers and thus create business plans for each case,
- acquisition of significant sales and marketing information reflecting customers' needs, thus helping to fulfil their expectations.

2.3 Customer Relationship Management Systems

Rababah et al. [18] described CRM as "*the building of a customer-oriented culture by which a strategy is created for acquiring, enhancing the profitability of, and retaining customers, that is enabled by an IT application, for achieving mutual benefits for both the organization and the customers*". Therefore, CRM can be considered as an IT system, which is an RM strategy enabler [4].

Business processes, which a CRM system can support, can be classified into three groups, namely delivery processes (activities associated with sales, service support, marketing promotions and customer complaints), support processes (activities oriented towards what the market needs with the intention to develop loyal customers) and analysis processes (activities entailing analysis of the data acquired by processes from the other two groups and the creation of value leading to service innovation) [19]. Especially in the banking sector, the utilization of data relevant to customers' needs and behaviour, can lead to the identification of the most important customers, support the development of relationships with prospective customers and estimate not only the generated revenues but also any future possible opportunities for investments [20].

3 Research Methodology

The research methodology was comprised of the following 4 stages:

1. Development of a questionnaire, aiming to capture the outcome of the implementation of RM strategies, by investigating how customers experience this approach.
2. Conduct of an interviewer-administered field research, using the questionnaire developed in stage 1.
3. Conduct of Principal Components Analysis (PCA) on the survey results, in order to identify latent factors within the questionnaire.
4. Conduct of multiple regression analysis, using the latent factors emerged as independent variables and customer loyalty as the dependent. This will examine the degree to which the emerging factor structure explains the variance of a core RM target, such as customer loyalty.

The questionnaire utilized was developed by adopting the scale proposed by Abratt and Russell [21]. Their objective was to establish whether RM is appropriate for marketing financial services to private banking clients in South Africa, by measuring RM outcomes from the customers' perspective. They initially designed a 53-item instrument based on literature review and in-depth interviews with bank manages and academics. This was used to conduct a survey on both private and non-private banking customers. The statistical analyses of the survey data lead to the proposal of a 43-item scale. These items were allocated to 3 distinct customer life-cycle stages, namely starting a relationship, early stages of a relationship and once a relationship has been established. Moreover, they also denoted a pool of items, parallel to all the aforementioned life-cycle stages, as general relationship marketing criteria.

The rationale behind the choice of this particular scale lies on the fact that it measures RM outcomes from all basic customer life-cycle stages and therefore it can provide a solid basis as a starting point.

After discussion with 3 managers of bank branches, it was decided to exclude 13 items of the initial 43 of the Abratt and Russell [21] scale, as they were deemed to focus exclusively on private banking and did not fit the aim of the present research. The resulting 30-itemed scale was then pilot-tested with 10 bank customers. The results of the pilot test proved to be very satisfactory, since all the respondents found the questionnaire items understandable, while no blurring questions that would have to be rephrased were found.

The survey target sample consisted of customers of the four systemic Greek banks, all of which employ CRM technology to manage customer relationships. All respondents were interviewed in Larissa, a city with intense economic, political and social life, located in central Greece. They were approached at local branches, while waiting to be served. It must be noted that since the majority of bank customers have accounts in more than 1 bank, the respondents were asked to answer the questionnaire having in mind the bank they were using more often for their transactions. The resulting sample comprised of 289 valid questionnaires. The demographic analysis of the sample showed that it was largely representative of the general population, since it included customers having various ages, educational levels and incomes.

4 Data Analysis and Results

PCA was conducted to identify latent factors within the questionnaire items. Seven factors with eigenvalues greater than one [22] were extracted from the data, as shown in Table 1. These principal components accounted for approximately 64% of the total variation. Preceding PCA, the Bartlett sphericity and Kaiser–Meyer–Olkin (KMO) tests verified the appropriateness of the sample.

A cut-off of 0.50 was used for item scale selection and normalized varimax rotation was adopted to bring about simple and interpretable structure. Six items did not load on any of the 7 emerged factors (<0.50) and were excluded from the final model. Following an inspection of the items' loadings on each factor, the 7 distinct principal components identified were named as reliability (F1 in Table 1), relationship value (F2), personalized sales (F3), crisis management (F4), flexibility (F5), customer focus (F6) and communication quality (F7).

Table 1. Principal components analysis

	Item	F1	F2	F3	F4	F5	F6	F6
1	The prices of the offered services and products are important	-	0.732	-	-	-	-	-
2	I want a better relationship with the bank	-	0.765	-	-	-	-	-
3	The quality of services affects the relationship with the bank	-	0.762	-	-	-	-	-
4	Client referred to client	-	0.709	-	-	-	-	-
5	The bank is available in crisis	-	-	-	0.725	-	-	-
6	I am confident in advice from bank	-	-	-	0.631	-	-	-
7	The bank can be counted on to do what is right	-	-	-	0.684	-	-	-
8	The bank provides me with value – enhancing services	-	-	0.533	-	-	-	-
9	The bank knows what is important for me	-	-	0.602	-	-	-	-
10	The bank provides me with cross-sale products	-	-	0.724	-	-	-	
11	The bank offers me a wide range of products and services	-	-	0.551	-	-		-
12	The bank focuses on individualized client services	-	-	-	-	-	0.627	-
13	The bank focuses to the client	-	-	-	-	-	0.679	-
14	The bank has skills to manage client's account	-	-	-	-	0.725	-	-
15	The bank is flexible	-	-	-	-	0.680	-	-
16	The bank responds quickly to any possible errors	0.656	-	-	-	-	-	-

Table 1. (*continued*)

17	The bank is reliable	0.799	-	-	-	-	-	-
18	There is mutual trust between me and the bank	0.704	-	-	-	-	-	-
19	The bank is consistent with the delivery of service	0.684	-	-	-	-	-	-
20	The bank is responsive to my requests	0.646	-	-	-	-	-	-
21	The clients' expectations from the bank are met	0.675	-	-	-	-	-	-
22	The bank delivers the services it promises	0.669	-	-	-	-	-	-
23	There is open communication between me and the bank	-	-	-	-	-	-	0.759
24	There is frequent interaction between me and the bank	-	-	-	-	-	-	0.827
	Eigenvalues	*9.075*	*2.203*	*1.497*	*1.258*	*1.148*	*1.055*	*1.004*
	Percent of total variation	*33.609*	*8.161*	*5.543*	*4.659*	*4.253*	*3.908*	*3.720*
	Cumulative Percent of total variation	*33.609*	*41.770*	*47.313*	*51.972*	*56.225*	*60.133*	*63.853*

Inter-item analysis was then used (Table 2) to verify the scale for internal consistency and reliability [23]. More specifically, Cronbach's coefficient alpha was calculated for each emerged factor. The values for 4 factors (reliability, relationship value, personalized sales and communication quality) ranged between 0.708 and 0.894. Therefore, all these 4 factors exhibited values well over the minimum acceptable reliability level of 0.7 [23]. On the other hand, the alpha values for crisis management, flexibility and customer focus were 0.659, 0.548 and 0.549 respectively. Up to a certain extent, these relatively low values can be attributed to the fact that these factors comprise only 2 or 3 items.

Table 2. Internal reliability analysis

Dimensions	Number of items	Cronbach's alpha
Reliability	7	0.894
Relationship value	4	0.780
Personalized sales	4	0.727
Crisis management	3	0.659
Flexibility	2	0.548
Customer Focus	2	0.549
Communication quality	2	0.708

Multiple regression analysis was the conducted (Table 3) in order to determine the extent to which the emerged RM outcomes factor structure explains the variance of customer loyalty and determine which factors pose significant influence. It was opted to exclude flexibility, crisis management and customer focus from the model, since they exhibited low Cronbach alpha values.

Table 3. Regression analysis

Independent Variables	Coefficient Beta	Sig.
(Constant)		.541
Sex	-.023	.613
Age	.039	.405
Educational Level	-.026	.567
Income	-.096	.057
Relationship value	.228	.000***
Personalized sales	.093	.082
Reliability	.442	.000***
Communication quality	.189	.000***

Dependent variable: Customer loyalty, **Adjusted R^2**=0.507.

5 Conclusions

This study presented work aiming to develop an instrument to measure the outcome of the implementation of RM strategies in the banking industry. The results obtained form part of the findings of an ongoing research project, which aims to identify the necessary competencies of accounting and finance professionals in the digital economy era.

The results acquired from PCA were very encouraging, since seven factors were extracted. Following an examination of the items' loadings on each factor, the seven distinct factors were labeled as relationship value, crisis management, personalized sales, customer focus, flexibility, communication quality and reliability. The Cronbach alpha reliability test showed that relationship value, personalized sales, communication quality and reliability had values well above the 0.7 limit. These results are considered as significant findings, since these factors represent typical RM outcomes as perceived by customers. However, the remaining 3 of the extracted factors, namely customer focus, flexibility and communication quality failed to pass the reliability test, since their values were below 0.7. This problem could be possibly rectified by enriching the questionnaire with items aiming to capture the respondents' answers on issues relevant to the 3 aforementioned factors.

PCA was followed by multiple regression analysis having customer loyalty as the dependent variable and the emerged factors that passed the 0.7 value Cronbach alpha test as independent. The results show that the emerged RM outcome factors exhibit a strong customer loyalty predictive power, since they explain 50.7% of its variance. Moreover, relationship value, reliability and communication quality were found to have a very significant positive effect on loyalty.

Acknowledgments. This research has been co-financed by the European Union (European Social Fund-ESF) and Greek national funds through the Operational Program "Education and Lifelong Learning" of the National Strategic Reference Framework (NSRF) – Research Funding Program: ARCHIMEDES III. Investing in knowledge society through the European Social Fund References.

References

1. Bose, R.: Customer relationship management: key components for IT success. Industrial Management & Data Systems 102(2), 89–97 (2002)
2. Teo, T.S.H., Devadoss, P., Pan, S.L.: Towards a holistic perspective of customer relationship management (CRM) implementation: A case study of the Housing and Development Board, Singapore. Decision Support Systems 42(3), 1613–1627 (2006)
3. Swift, R.S.: Accelerating Customer Relationship using CRM and Relationship Technologies. Prentice Hall PTR, Upper Saddle River (2001)
4. Ryals, L., Payne, A.F.T.: Customer relationship management in financial services: Towards information enabled relationship marketing. Journal of Strategic Marketing 9(1), 1–25 (2001)
5. Digman, L.: CRM: Dream or Nightmare? ZDNet (2002), `http://www.zdnet.com/news/crm-dream-or-nightmare/298017` (last accessed on January 29, 2015)
6. Durkin, M.G., Howcroft, B.: Relationship marketing in the banking sector. Marketing Intelligence and Planning 21(1), 61–71 (2003)
7. Gummesson, E.: Return on relationships (ROR): the value of relationship marketing and CRM in business-to-business contexts. Journal of Business and Industrial Marketing 19(2), 136–148 (2004)
8. Reichheld, F.F., Sasser Jr., W.E.: Zero defects comes to services. Harvard Business Review, 105–111 (September-October 1990)

9. Egan, J.: Relationship Marketing: Exploring Relational Strategies in Marketing, 3rd edn. Pearson Education, England (2008)
10. Gronroos, C.: From marketing mix to relationship marketing: towards a paradigm shift in marketing. Management Decisions 32(2), 4–20 (1994)
11. Dwyer, F.R., Schurr, P.H., Oh, S.: Developing byer-seller relationships. Journal of Marketing 51(2), 11–27 (1987)
12. Kotler, P.: Method for the millennium. Marketing Business, 26–27 (February 1997)
13. Wang, Y., Lo, H.P., Chi, R., Yang, Y.: An integrated framework for customer value and customer-relationship-management performance: a customer based perspective from China. Managing Service Quality 14(2/3), 169–182 (2004)
14. Bolton, R.N., Lemo, K.N., Verhoef, P.C.: The theoretical underpinnings of customer asset management: a framework and propositions for future research. ERIM Report Series Research in Management, ERS-2002-80-MKT, ERIM, Rotterdam, The Netherlands (2002)
15. Grant, A.W.H., Schlesinger, L.A.: Realize your customer's full profit potential. Harvard Business Review 73(5), 59–62 (1995)
16. Andersen, B., Henriksen, B., Aarseth, W.: Holistic performance management: An integrated framework. International Journal of Productivity and Performance Management 55(1), 61–78 (2006)
17. Bergeron, B.: Essentials of CRM: Customer Relationship Management for Executives. John Wiley & Sons (2001)
18. Rababah, K., Mohd, H., Ibrahim, H.: Customer Relationship Management, No. CRM) Processes from Theory to Practice: The Pre-implementation Plan of CRM System. International Journal of e-Education, e-Business, e-Management and e-Learning 1(1), 22–27 (2011)
19. Geib, M., Reichold, A., Kolbe, L., Brenner, W.: Architecture for Customer Relationship Management Approaches in Financial Services. In: Proceedings of the 38th International Conference on System Science, Hawaii, USA (2005)
20. Zineldin, M.: Quality and customer relationship management (CRM) as competitive strategy in the Swedish banking industry. The TQM Magazine 17(4), 329–344 (2005)
21. Abratt, R., Russell, J.: Relationship marketing in private banking in South Africa. International Journal of Bank Marketing 17(1), 5–19 (1999)
22. Kaizer, H.F.: The application of electronic computers to factor analysis. Educational and Psychological Measurement 20(1), 141–151 (1960)
23. Nunnally, J.C., Bernstein, I.H.: Psychometric Theory, 3rd edn. McGraw-Hill, New York (1994)

Analysis of Websites of Top Global Logistics Providers by a Trust Building Framework

Lachana Ramingwong and Sakgasit Ramingwong

Department of Computer Engineering, Faculty of Engineering, Chiang Mai University, Chiang Mai, Thailand
{lachana,sakgasit}@eng.cmu.ac.th

Abstract. Websites are one of the most critical mechanics which are used to establish transaction between consumers and businesses. Gaining trust from the customers can be challenging, especially for new entrepreneurs. This paper analyzes websites of top global logistics providers based on a trust building framework. The results reveal motivating factors that are valuable and critical for building trust on the websites. In this paper, common factors and imperfection affecting trust of the top global logistics websites are uncovered. The findings provide guidelines in building trust for any emerging logistics businesses, or other E-business.

Keywords: Trust, Logistics, Services, Websites, Information Technology, E-Business.

1 Introduction

Trust is a key factor which influences customers to purchase [1]. This is especially true in E-business where websites are the major gateway. Moreover, website trust becomes a crucial element that influences a success story or failure of businesses [2][3][4]. In order to encourage a purchase, businesses strive to build an informative, accessible and trustworthy website [5]. A small business with a high quality website may outsell a large business with an unreliable website [6].

It is not an overstatement to articulate that the websites of the top global logistics services have the highest quality among other websites. However, it is also undeniable that these businesses need high standard gateways to secure their creditability as well as encourage business transactions and craft consumer trust. Thus, it is important for competitors to gain a deeper understanding of trust elements of these portals. The comparative investigation and analysis could provide insights into the factors contributing to trust building.

Trust building framework could be used to investigate the elements of trust on websites. Based on the Unified Theory of Acceptance and Use of Technology (UTAUT/UTAUT2) [7] and E-Service Quality (E-S-QUAL and E-RecS-QUAL) [8], websites' electronic service quality can be measured. Therefore, a number of trust factors can be identified. Examples of these factors include performance, motivation, availability, privacy, compensation and responsiveness.

M. Gen et al. (eds.), *Industrial Engineering, Management Science and Applications 2015*,
Lecture Notes in Electrical Engineering 349, DOI: 10.1007/978-3-662-47200-2_93

This paper attempts to explore and analyze trust factors of the websites of top logistics providers based on a trust building framework [9]. The second section of this paper describes the background and details of the trust building framework. The third section explores the top logistics services and their websites. The analysis of these websites is displayed and discussed in the fourth section. Finally, the fifth section concludes this paper.

2 A Trust Building Framework

A trust building framework proposed in 2014 [9] was used in this study. A questionnaire based on Unified Theory of Acceptance and Use of Technology (UTAUT/UTAUT2) and E-Service Quality (E-S-QUAL and E-RecS-QUAL) was developed and participated by 473 users of the E-Business website. A factor analysis suggested that four main elements had a major connection to trust. These factors include fulfilment, efficiency, customer service and communication. Figure 1 illustrates this trust building framework. While several trust frameworks were developed [10][11][12][13], most of them emphasizes external trust factors rather than internal factors. Examples of external factors include technological solution, legal environment and market place. These factors are difficult to control and, thus, are not included in this study.

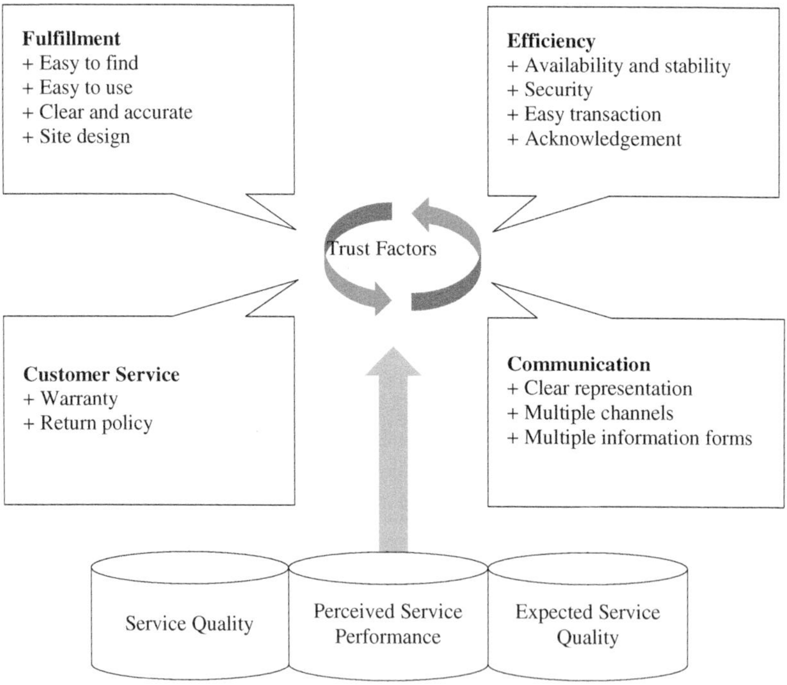

Fig. 1. A trust building framework

Fulfilment is the element that highlights the importance of usability of the website. Customers indicated that they need the website to be easily navigated, so that the products can be found effortlessly. The navigation menus should be highly accessible, efficient, and easy to understand. The design of the website should be clear and provide necessary information.

The second element, efficiency, involves the availability of the websites. This also includes protection of privacy and security systems. The transaction should be straightforward, logical, flexible and secured. In addition, a feedback system should be available to acknowledge the transactions.

Customer service, the third element, mainly deals with after purchase issues such warranty and return policy. The customers indicated that the warranty period should be clearly announced and the provider should accept return if products are dissatisfied. The history of claimed guarantee is also reported to help improving trust. People are likely to buy if warranty and return information are available.

The final trust element is the communication between stakeholders. Information on communication channels should be clear. Multiple channels of communication, such as email, phone, or other messenger services, are preferred. The customers also appreciated the websites that provide sufficient information on the products and services.

Efficient installation of these four elements would encourage the consumers to be confident in establishing business transaction with the E-business providers and, moreover, generate loyalty. Each element is considered to have the same level importance in this framework.

3 Top Logistics Services and Their Websites

Armstrong & Associates announced the top 50 global third party logistics providers [14]. Among the list, the top five providers were reported to achieve more than $10,000 million USD in 2013. This includes DHL Supply Chain & Global Forwarding, Kuehne + Nagel, DB Schenker Logistics, Nippon Express and C.H. Robinson Worldwide. This section investigates the homepage of these organizations. The screen shots from the default view of these websites were taken in two modes. The website views obtained from the two modes of access are referred as laptop view and mobile view, respectively. They are illustrated in the following sections. The screenshots from the first mode are created via Google Chrome browser from a laptop computer with a resolution of 3200x1800 pixels. The second mode's screenshots are generated from a vertical view from Safari browser on a mobile device with a resolution of 640x1136 pixels. The websites accessed in first mode are also tested for load time by Pingdom, an online testing service [15].

3.1 DHL Supply Chain and Global Forwarding

The homepage of DHL Supply Chain & Global Forwarding [16] provides almost all essential trust elements. Links to services and general company information are on the same main menu. Figure 2 shows that their main services, such as tracking and quoting, are clearly visible and can be quickly accessed through the right menu. The website load time takes less than 2 seconds. A secure login system for tracking purposes is available. The warranty and return policies are not accessible from the first page but can be found from the search menu. The communication option of this organization is excellently presented. The customers are given a number of contact options based on their purchased services and locations. The main disadvantage of DHL website is it is not fully responsive as can be seen from Figure 2(b) that the information presented in the vertical view is unfortunately less informative when being accessed from a smaller screen.

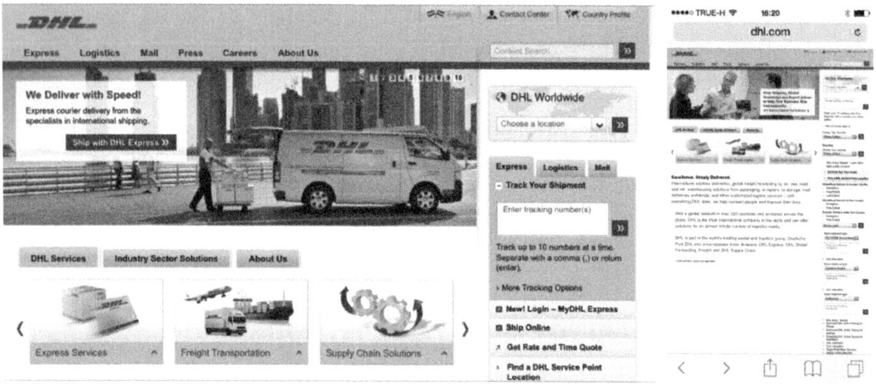

(a) (b)

Fig. 2. DHL Supply Chain & Global Forwarding homepage [16]
(a) laptop view (b) mobile view

3.2 Kuehne + Nagel

As can be seen in Figure 3, Kuehne + Nagel [17] features their tracking service and quotation requesting on the main and side menus. General information on the organization is represented on a separated and less emphasized menu. An option of secure login is also provided. The insurance policy is conveniently attached in all freight menus. The load time of this website is slightly more than 3 seconds. Multiple modes of communication can be retrieved via the clearly presented contact menu. Yet, similar to DHL website, the website of Kuehne + Nagel is not fully responsive. The customers might find that all contents are rather small when being accessed via a mobile device.

Fig. 3. Kuehne + Nagel homepage [17] (a) laptop view (b) mobile view

3.3 DB Schenker Logistics

DB Schenker Logistics website [18] has a main menu addressing both their services and corporate information. Information on freight services can be accessed via the first option of the main menu. A tracking option is slightly hidden in a tab on the right menu. Similar to the other providers, DB Schenker provides a secure login option as well as multiple channels of communication and feedbacks. The load time takes slightly lower than 3 seconds. However, information on insurance or warranty is not visible from the first page. The portal itself is not fully responsive and could be difficult to read in smaller screens.

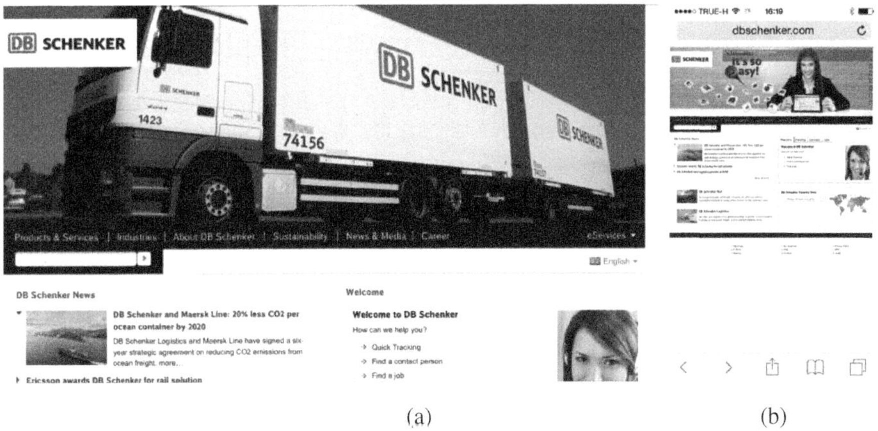

Fig. 4. DB Schenker Logistics homepage [18] (a) laptop view (b) mobile view

3.4 Nippon Express

The website of Nippon Express [19] is similar to previous organizations. The main menu represents corporate information as well as information on their services and solutions. A section for tracking shipment is clearly provided on the right of the portal. Nippon Express seems to highlight on information of their global operation since there are a number of options and tabs which links to regional offices. Several modes of communication as well as their terms and conditions are provided. Information on insurances and warranties can be inquired from the search menu. The website takes less than 3 seconds to load. However, similar to those of other top logistics companies, it is not fully responsive.

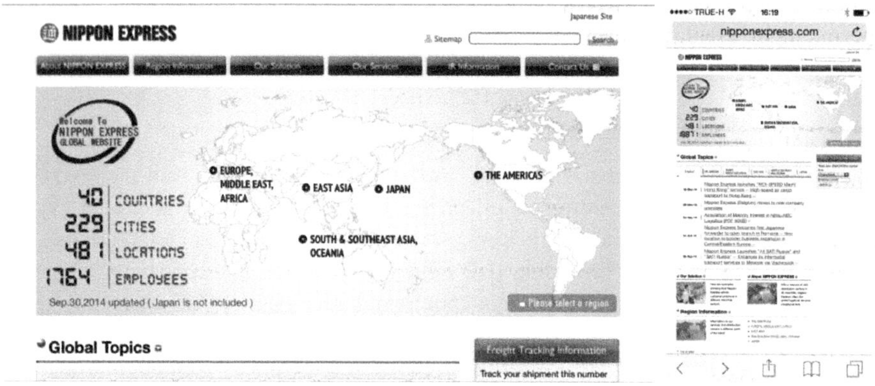

(a) (b)

Fig. 5. Nippon Express homepage [19] (a) laptop view (b) mobile view

3.5 C.H. Robinson Worldwide

The website of C.H. Robinson Worldwide [20] has a classic menu offering information on their services and company profiles on top and bottom of their website. Three outstanding menus on login system, request for quotation and office location are also provided. Various communication channels can be accessed via the office location option. It takes slightly longer than 1 second to load the entire portal. Terms of use, warranties and insurance options can be found from the search menu. Interestingly, C.H. Robinson Worldwide website is responsive when being viewed from a mobile device. Three special menus on logistics, about us and resources are highlighted on the vertical view of this website.

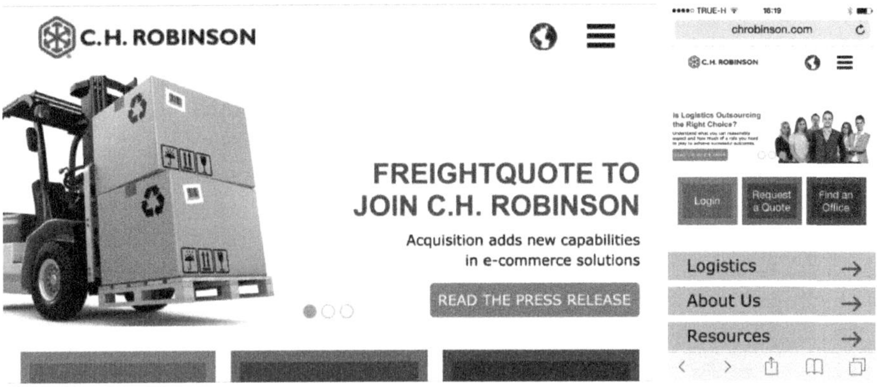

(a) (b)

Fig. 6. C.H. Robinson Worldwide homepage [20] (a) laptop view (b) mobile view

4 Analysis and Discussion

All trust elements and their scores from the top logistics providers' websites are summarized in Table 1. The average score reflects the overall quality of each trust element. The score is given on a Likert scale of 1 to 5. The score of 1 indicates very low satisfactory in that dimension, while 5 indicates very high satisfactory. Each trust elements is comprised of related aspects of quality service. Fulfillment element represents visual, navigation and precise production information aspects. Efficiency element represents availability, transaction processing and data security aspects of websites. Customer service element involves after purchase related policies. Communication element involves communicating product information and providing a wide variety of communication channels to consumers.

While all of the websites appear to be aesthetically attractive at first glance, their performances are remarkably different in several areas of trust elements. Visual design of websites and website usability [21] are suggested to have a positive effect on the trustworthiness of the business [22]. However, website usability of these websites is not of the same degree. This, therefore, suggests differences in E-service quality offered. Detailed analysis of each trust elements are discusses next.

Table 1. Trust elements of websites of the top logistics services

	Fulfilment	Efficiency	Customer Service	Communication	Average Score
DHL Supply Chain & Global Forwarding	4.38	4.28	2.50	3.40	**3.64**
Kuehne + Nagel	3.89	4.50	4.00	3.80	**3.86**
DB Schenker Logistics	3.89	3.86	2.50	4.20	**3.24**
Nippon Express	3.75	3.50	2.50	3.20	**3.24**
C.H. Robinson Worldwide	4.75	4.00	2.50	4.60	**3.96**

4.1 Fulfillment

The website is likely to be trusted if it fulfills consumers in their visual aspects along with easy access navigation and accurate production information. All top logistics providers' websites achieve a considerable high score on average for this category of trust elements. It is clear to see that their visual aspects play an important part. However, website navigation is also one of the most important factors that contribute to building trust. Whereas the majority of websites use drop down navigation menus for easy navigation and to speed up information seeking activity, some does not. As a result, the website without drop down navigation menus requires more clicks and consumers also need more time to look for information they want. Typically, if consumers can effortlessly find the information, they are likely to stay on the website. Some websites, i.e. DHL and C.H. Robinson, score higher than the rest due to the combination of visual aspects and easy access menus.

4.2 Efficiency

The websites are efficient when they are available by the time consumers want to use them. The websites should process transactions without difficulty and secure consumers' data. Four websites receive high score (> 4.00) in this category except one website. This is caused by inability of the website to provide a clear problem/error report. If the site quickly reports problem or error, consumers may be able to find the way or solve problem on their own. That signifies the importance of providing an appropriate assistance to websites' users when they need. Otherwise, it may result in frustration and a decision to leave the website.

Yet data security is important for protecting consumers' personalized data, accessibility of this feature on some websites are not noticeable. Three websites out of five openly reserve a space on their main page to make way for consumer authentication.

4.3 Customer Service

Customer service information such as warranty, insurance and claim is also essential factor for gaining trust from consumers. Only Kuehne + Nagel provides an easy access to this information by including it in a drop down menu of every service. Hence, consumers does not need to spend time looking for this information on other pages. For other websites, consumers have to get their hands on the search menu in order to find related results and perform sensemaking activities on their own to obtain insurance or claim information. The latter case is likely to be more difficult and time-consuming and, furthermore, could affect trust value.

4.4 Communication

All websites offer multiple communication channels. C.H. Robinson's website is the only one that explicitly shows contact information on the first page, while other websites need consumers to continue surfing the websites to find out. C.H. Robinson's website is outstanding in the way that it communicates very well with consumers. Although the homepage is longer than the screen size, the global navigation menu is readily accessible whether consumers are at the top or bottom of the page. This enables fast product inquiry.

5 Conclusion

This paper studied and analyzed the factors influencing trust of websites of the top five logistics providers. A framework for trust building of E-Commerce websites is used to provide structure for the analysis. Four trust elements; fulfillment, efficiency, communication and customer service, of these websites are analyzed in great details. Unsurprisingly, the average scores of them are quite high. The lowest score is 3.24, while the highest score is 3.96. The notable differences between the lowest score and the highest score the ability to fulfill and communicate with consumers. It can also be implied that the websites that communicate well are likely to fulfill their consumers better and be more efficient in helping their consumers complete the tasks. The application of this framework with websites of the top global logistics provider confirms that the framework can be generalized with other kind of websites. In a future work, identification of further trust elements should be studied in order to develop a more detailed guideline for building trust to websites.

References

1. Ganguly, B., et al.: The effects of website design on purchase intention in online shopping: the mediating role of trust and the moderating role of culture. International Journal of Electronic Business 8(4), 302–330 (2010)
2. Urban, G.L., Sultan, F., Quall, W.J.: Placing trust at the center of your Internet strategy. Sloan Management Review 42(1) (2012)
3. Srinivasan, S.: Role of trust in e-business success. Information Management & Computer Security 12(1), 66–72 (2004)
4. Dubelaar, C., Sohal, A., Savic, V.: Benefits, impediments and critical success factors in B2C E-business adoption. Technovation 25(11), 1251–1262 (2005)
5. Warrington, T.B., Abgrab, N.J., Caldwell, H.M.: Building trust to develop competitive advantage in e-business relationships. Competitiveness Review: An International Business Journal 10(2), 160–168 (2000)
6. Lee, Y., Kozar, K.A.: Investigating the effect of website quality on e-business success: An analytic hierarchy process (AHP) approach. Decision Support Systems 42(3), 1383–1401 (2006)
7. Venkatesh, V., Morris, M.G., Davis, G.B., Davis, F.D.: User acceptance of information technology: Toward a unified view. MIS Quarterly 27(3), 425–478 (2003)
8. Parasuraman, A., Zeithaml, V.A., Malhotra, A.: ES-QUAL: A multiple-item scale for assessing electronic service quality. Journal of Service Research 7(3), 213–233 (2005)
9. Panya, A., Ramingwong, L.: A Framework for Building Trust in Business to Consumer E-Commerce in Thailand. In: 2013 19th Asia-Pacific Conference on Communications (APCC), pp. 485–489. IEEE (2013)

10. Siau, K., Shen, Z.: A framework for Building customer trust in mobile commerc. Communications of the ACM 46(4), 91–94 (2003)
11. Mahadevan, B., Venkatesh, N.S.: Building On-line Trust for Business to Business E Commerce. In: IT Asia Millenium Conference, Bombay, India, pp. 29–30 (November 2000)
12. Paakki, M.K.: Framework for Consumer Related Trust Issues in E-Commerce. Frontiers of E-Business Research, 332–339 (2004)
13. JIS.: Research of Online Trust in C2C E-Commerce Based on Institutions. Communications in Information Science and Management Engineering (2011).
14. Supply Chain 247.: 2013 Top 50 Global & Domestic Third-Party Logistics Providers (2014), http://www.supplychain247.com/article/2013_top_50_global_top_30_domestic_3pls/seko_logistics
15. Pingdom, A.B.: Website Speed Test (2015), http://tools.pingdom.com/
16. DHL International GmbH.: DHL | Global | English (2015), http://www.dhl.com/
17. Kuehne + Nagel Inc.: Kuehne + Nagel: Home (2015), http://www.kn-portal.com/
18. Deutsche Bahn, A.G.: DB Schenker – Transportation & Logistics (2014), http://www.dbschenker.com/
19. Nippon Express (2015), http://www.nipponexpress.com/
20. Robinson, C.H.: Worldwide Inc. (2015), http://www.chrobinson.com/
21. Flavián, C., Guinalíu, M., Gurrea, R.: The role played by perceived usability, satisfaction and consumer trust on website loyalty. Information & Management 43(1), 1–14 (2006)
22. Karimov, F.P., et al.: The effect of website design dimensions on initial trust: A synthesis of the empirical literature. Journal of Electronic Commerce Research 12(4), 272–301 (2011)

An Empirical-Based Construction of the Multi-purpose Process Reference Model for Hospital Supply Chain

Wirachchaya Chanpuypetch and Duangpun Kritchanchai*

Department of Industrial Engineering, Faculty of Engineering,
Mahidol University, Nakhonpathom, 73170, Thailand
wirachchaya@gmail.com, duangpun.skr@mahidol.ac.th

Abstract. Nowadays, a lack of continuity and a lack of information sharing, often known as "fragmented system" have been found in Thai hospital supply chain. This issue is a critical problem in healthcare domain. To solve this situation, the fragmented pieces of business processes can be assembled by business process reengineering (BPR). A tool for understanding systems and determining whether changes are effective is needed as one foundation. Likewise, a standard view of collaboration among individuals and functions are required. Therefore, this study aims to construct the multi-purpose of hospital supply chain process reference model (H-PROM) in order to achieve BPR in healthcare. In order to develop the model, the empirical-based construction method is proposed. At the first two stage of the methodology, some results are presented in this paper. The hospital supply chain process reference framework is constructed based on the empirical foundation that acquired from hospital case studies.

1 Introduction

Within Thai service sector, the healthcare industry is growing quickly and now coming to be the medical hub of Asia. It is an important service industry that provides quality of care and the safest to patients [1]. Currently, a research study of supply chain management (SCM) in healthcare field is challenge and increasing interest in global. Aktaş, *et al.* [2] indicate that healthcare systems are involving a variety of factors that influence their efficiency. The factors may be uncertain, and the information incomplete. Additionally, healthcare provider organizations are high-reliability organization where managers make decisions in an extremely unpredictable and dynamic environment [2]. Thus, this context is more complex than other industries because of the impact on people's health requiring adequate and accurate medical supply according to the patient's needs. Nowadays, a series of research in healthcare have been distributed continuously. However, Dobrzykowski, *et al.* [1] significantly recognise that research on SCM in this context is beginning and needs a thorough understanding.

* Corresponding author.

M. Gen et al. (eds.), *Industrial Engineering, Management Science and Applications 2015*,
Lecture Notes in Electrical Engineering 349, DOI: 10.1007/978-3-662-47200-2_94

Hence, we first try to understand the AS-IS situation of healthcare supply chain in Thailand through the preliminary empirical investigation with the multiple-case study. As the analysis and outcomes, there is a varied different information systems involved in the hospital SCM of pharmaceutical products. A lack of continuity and a lack of information sharing, often known as "fragmented system" have been found. This occurrence as a critical problem in healthcare domain. Hammer [3] realises that it is a major source of inferior quality. Besides, delays and errors are also unavoidable on fragmented process structures [3]. To solve this situation, the fragmented pieces of business processes can be assembled by business process reengineering (BPR) [3].

BPR has been applied in healthcare over decades. However, there is no success story available about BPR implementation [4]. Patwardhan and Patwardhan [4] claim that healthcare setting has the same characteristics of industrial environment with much more complexity and a very uncertain process of care. Many healthcare processes are nonlinear and cannot be simplified into strings of quick sequential tasks [5]. McConnell [5] suggests that healthcare organisations are indeed unique in terms of the output they produce, but they are not necessarily unique in terms of the management processes employed. Notwithstanding, healthcare organisation is different in one crucial way. Preserving life and restoring health are concerned as the focal point [4, 5]. Patients need are focused rather than organisation need [4].

To achieve BPR in healthcare setting, we require a clear business need to raise awareness of BPR in this environment. BPR publications that have been conducted in healthcare environment were then reviewed. In the book, McConnell [5] confirms that hospital system is the most difficult to modify. Patwardhan and Patwardhan [4] and McConnell [5] define healthcare system is an interlinked, interdependent, and interdepartmental. Healthcare processes are interconnected to each other [6]. Structural and cultural change needs time to be developed and accepted in all levels in healthcare setting [4, 6]. Moreover, Khodambashi [6] indicates that resistance to change exist in healthcare. Healthcare professionals are not recognize for the change [4, 5]. Besides, the information necessary to manage the change process is also often extremely lacking in healthcare organisations [7]. MacPhee [7] realises that continuous change in complex healthcare environments for improving the quality of care delivery is a challenge.

Accordingly, Pringle, *et al.* [8] point out one cornerstone of healthcare improvement that is providing a tool for understanding systems and determining whether changes are effective. Moreover, a high degree of collaboration among individuals and functions are also often required [5]. Walters and Jones [9] suggest an effective business model that should be created for healthcare BPR. A holistic approach lead to acceptance to change needs to be viewed. Besides, as any improvement in this setting, Francis and Alley [10] indicate that everyone in organisation must make a commitment to patient as the primary motivator. They also emphasise that the change will be doomed to failure if the implementation is made without healthcare professionals support.

Therefore, this paper intends to propose the business model for healthcare BPR. Basically, BPR requires the end-to-end processes view of organisation as an essential ingredient [3]. Davenport [11] also underlines a process perspective as standard view can facilitate communications about how the business operates, enable smooth handoffs across process boundaries, and make possible comparative measure of performance within a company. However, standardised end-to-end process for hospital supply chain has still not been presented [12]. Davenport [11] claims that lack of standardised forms of representation and lack of models can solve by providing a reference model-based design recommendations. Thus, in this paper, the proposed business model is constructed in the form of process reference model of hospital supply chain (H-PROM).

This paper is organised as follows. In section 2, the overview of process reference model (PRM) is reviewed and described. Then the purposes of H-PROM and its components are determined. The section 3 provides an empirical-based construction method for construction the H-PROM. The framework of H-PROM is presented in section 4. Finally, conclusion and the further research are depicted.

2 Process Reference Model (PRM)

In BPR, the provided end-to-end process view as a blueprint or process template can increase the efficiency and effectiveness. This process template is typically called process reference model (PRM) [13]. Blecken [14] describes that PRM is a specific kind of reference model that focuses on the behavioral aspects of an organisation through the analysis of its business processes. It allows management to be much more confident that the changes desired in business process performance are the right changes, and that performance improvements can be predicted, achieved and measured [15].

In several contexts, many researchers develop the specified PRM with various purposes. Blecken [14] proposes PRM to present the standardisation for communication and collaboration in humanitarian supply chain. Thomas, *et al.* [16] provide PRM of event management to make recommendation for application system and organisation design. A set of process model by Verdouw, *et al.* [17] can be used to support the design of customised fruit supply chain. Moreover, it is a basis for information system implementation. Some PRM by Rabe, *et al.* [18] aims to help management to support strategic decision and handle cross-organisation processes.

Consequently, to construct H-PROM, the business needs that were introduced to archive BPR in healthcare setting are considered. An effective business model for understanding systems and leading to acceptance to change is needed. Thus, we suggest that H-PROM should provide with multi-purpose of the model including 1) to provide the standardisation for facilitating of interdepartmental communication and collaboration and 2) to act as a strategic decision support tool for leading to acceptance to change.

To develop the multi-purpose of H-PROM, we then review the relevant literature to understand a common structure of PRM. Reichert, *et al.* [19] construct PRM for master data management. They follow the ARIS conventions for designing the process reference architecture and each main process is detailed. Then the developed reference model is applied in real life context through case studies. Thomas, *et al.* [16] provide PRM of event management. First, they create the reference model framework by inductive procedure then the constructed framework is accomplished with a deductive action. The process modelling language, event-driven process chain (EPC), is integrated into their framework as the modelling of detail model. For supply chain design and configuration, Rabe, *et al.* [18] offer multi-purpose reference model for supply chain to handle the cross-organisation processes. Their framework includes the description of the as-is situation and the definition of a to-be model. Besides, the simulation is provided.

According to a review of previous researches, PRMs are composed with at least two components including process reference framework and documentation of the detailed model [16, 18, 19]. Furthermore, an application of the constructed PRMs is included such as simulation scenario [18], business process execution with modelling language technique [14], applying in real case studies [19].

Hence, H-PROM is composed of three main components in this study. To standardise, hospital supply chain process reference framework and the documentation of the detailed model are provided for facilitating communication and collaboration. In addition, the application of H-PROM should be also composed in order to overcome change resistance and handle organisation performance. The methodology of H-PROM construction is conducted in the following section.

3 Empirical-Based Construction Method of H-PROM

In PRM construction, the corresponding empirical reference should be linked to the element of a reference model. Scheer and Nüttgens [20] state that an empirical-based construction is the best practice, although a reference model can be developed theoretically or utilize document process know-how. However, a standard detailed description of an empirical-based construction method of PRM is still not provided. Based on a review and synthesis of the relevant literature, in this study, the provided methodology of H-PROM construction consists of four main stages which are illustrated in Figure 1 and described as follows.

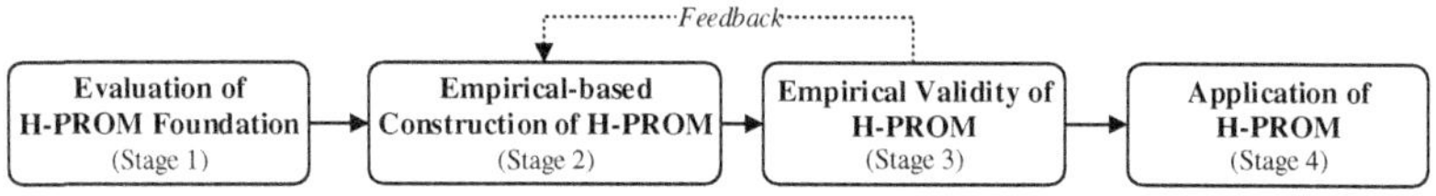

Fig. 1. The methodology of H-PROM construction

3.1 *Stage 1* - Evaluation of H-PROM Foundation

At the beginning stage, foundations for reference modeling are evaluated including SCM framework (namely: The Supply Chain Planning Matrix (SCP matrix), Global Supply Chain Framework (GSCF), the H-Retail framework, and Supply Chain Operation Reference (SCOR) model) and modeling language technique (namely: Integrated definition language (IDEF), Event-driven Process Chain (EPC), Unified Modelling Language Activity Diagram (UML-AD), and Business Process Modelling Notation (BPMN)). These foundations are selected based on an evaluation criteria that can be determined through the literature review. The alternatives of the evaluation can be guided by the purpose and scope of H-PROM construction [14]. The results of this stage are the foundation in the subsequent stages.

3.2 *Stage 2* - Empirical-Based Construction of H-PROM

The second stage, empirical foundation is acquired as a source of business process knowledge for PRM construction [14, 21]. First, based on the SCM framework selected in *Stage 1*, the semi-structured interview questions is conducted to:

- identify the hospital key supply chain business processes
- identify the hospital functional silos
- obtain the phenomenon or details of the hospital activities
- define the tasks at each function area and supply chain business process
- perceive the problems of hospital supply chain

The four cases of a large hospital providing tertiary care and teaching hospital where convenience to access in Thailand were selected for this study. For interviewees, we considered the criteria for selecting them such as accessible and available to give the relevant information. The managers at senior level who could provide information on the relevant hospital supply chain business processes are interviewed such as head of hospital pharmacy department, pharmacy purchasing manager, pharmacy warehouse manager, and pharmacy store manager. According to the collected empirical data, the hospital supply chain process reference framework can be constructed by modifying the selected SCM framework. Afterwards, the standard descriptions of each task in the constructed framework are annotated based on the generic principles of SCM. To build a robust model, the acquired empirical data will be grounded through the qualitative analysis method (grounded theory) to find the characteristics of hospital supply chain in Thailand. All reference process patterns are integrated into the constructed hospital process reference framework and illustrated with the modelling language technique. Consequently, the developed H-PROM can be presented in standardised form. Furthermore, the generic problems of hospital supply chain can be emerged at this stage. Then the appropriate solutions will be reviewed and analysed to propose the application of H-PROM in *Stage 4*.

3.3 *Stage 3* - Empirical Validity of H-PROM

Empirical validity should be comprised in a construction of PRM. The developed reference model can be validated and proof by surveying domain experts through questionnaire or online discussion forums [21]. The quality of a developed concrete PRM are evaluated by applying the checklists with a list of questions that guide reviewers [21]. These checklists can be extended with criteria for good reference model's, namely adaptability, understandability, accessibility within organisation, and the inclusion of key issues of specific domains [21]. Finally, the constructed PRM will be considered valid.

3.4 *Stage 4* - Application of H-PROM

The application of a constructed reference model is crucial in order to ensure that the PRM delivers efficient solutions to problems of the application domain [14]. Hence, at the final stage, some generic problem of hospital supply chain emerged from *Stage*

3 is considered to provide the solution. The application of H-PROM will be proposed in order to overcome change resistance and handle organisation performance.

4 Hospital Supply Chain Process Reference Model (H-PROM)

According to the methodology, one foundation of reference modelling is analysed based on the published literature to generate a backbone of the reference model in the first stage. Next stage, empirical foundation that was acquired from hospital case studies were used to frame the framework and construct H-PROM. In this section, we present some result of *Stage 1* and *Stage 2* as follows.

4.1 The Suitable SCM Framework

The appropriate SCM framework can be used to frame an effectively navigable directory for PRM construction. Thus, the four candidates of generic framework for SCM were reviewed in this study. According to the purposes of PRM construction, the three main criteria for selecting the frameworks are considered based on Moberg and Vitasek [22]. These criteria are the framework should 1) provide a structure for the presentation of organisation view, 2) support the creation of organisation end-to-end view across the borders of departments or organisation, and 3) provide a way that supports the standardisation of organisation end-to-end processes. The comparison of SCM frameworks is presented in Table 1.

Table 1. Comparison of SCM framework

Criteria	The Retail-H	SCP Matrix	GSCF	SCOR
Presentation of organisation view	Represent the retail information system and the business administrative functions.	Represent in the two dimensions including planning horizon and supply chain process.	Represent coordination of operation tasks and processes within and between organisations in the supply chain.	Identify the unique processes a supply chain requires to support the objective of fulfilling customer orders.
Creation of organisation end-to-end view across the borders of organisation	Be a structural to develop a reference model for the design of retail information systems.	Focus in the decomposition of decisions and structuring planning problems.	Focus on integration across all organisation functions.	Focus on interaction among a few key functions and cannot describe every activity including integrated cross-functional areas.
Standardisation of organisation end-to-end processes	Aim to enhance the orientation within the heap of information models applied in the retail sector.	Be a framework for categorizing SCM planning tasks.	Suggest the implementation of standard cross-functional business processes. The blueprint for managing the process can be defined.	Suggest the implementation of standard cross-functional business processes and help standardise the description of the supply chain architecture.

Based on these criteria, the GSCF will be used in the following to serve as a reference framework for H-PROM construction. The selected SCM framework can facilitate a cross-functional or cross-enterprise view of the organisation and its supply chain partners [23]. It can display a collaboration among individuals and functions [23] that is a need for hospital system improvement [5]. The GSCF can bring the constructed H-PROM to meet its purpose of providing the standardisation for facilitating of interdepartmental communication and collaboration. Moreover, the format of GSCF is easy to understand.

Nevertheless, the GSCF originates in manufacturing and therefore modification are necessary in applying the approach to the healthcare industry. The key supply chain business processes that cut across hospital organisation and functional silo will be defined. The operational tasks that reside inside a functional silo can be specified and described its set of activities.

4.2 The Constructed Hospital Supply Chain Process Reference Framework

In this section, the acquired empirical data is used to build the generic process framework as the backbone of hospital supply chain business processes. The data from the 4 cases of tertiary hospital in Thailand were collected through the semi-structure interview guideline. Consequently, the key hospital supply chain business processes and functional silos are defined and described through the set of questions (see the objectives' question in section 3.2) for framing the process reference framework. The hospital's functions across the key hospital supply chain processes can be obtained.

Hospital Supply Chain Business Processes: Horizontal Axis. In order to form the generic set of hospital supply chain processes, we present the eight key supply chain processes that are included in the framework with the descriptions by Lambert and Cooper [24] to interviewees. They were asked to identify the key supply chain business processes for pharmaceutical products management in hospital. As a result, the horizontal axis of the GSCF can be modified. All hospital cases reflect overall main hospital operation flows on the four key supply chain processes. These processes represent in the generic hospital supply chain process reference framework as follows.

- Demand management process is concerned with balancing the demand with the supply chain capabilities [24]. In hospital, demand is created by a physical doctor as the drugs needed to treat his/her patient. It needs to be forecasted and planned to facilitate fulfillment of supplies on a periodic basis. Sourcing options are also considered at the time of order receipt [24].
- Order fulfillment process provides for timely and accurate delivery of customer orders. This process includes all activities necessary to design a network and enable a firm to meet customer requests [24].
- Supplier relationship management process is represented that how an organisation interacts with its suppliers. This process is renamed from the procurement process [24] The development of purchasing function such as EDI is considered. Implementation of tracking process is also included.
- Return management process is the SCM process by which activities associated with returns, reverse logistics and avoidance are managed within the hospital and across key members of the supply chain.

Hospital Supply Chain Functions: Vertical Axis. In order to identify hospital functional silos, interviewees were asked to point out the main hospital functions or departments involved pharmaceutical products management. As a result, hospital pharmacy of all hospital cases relates with the four main locations including purchasing department, hospital central warehouse, pharmacy room, and ward. These departments provide their own functions to manage pharmaceutical products. Accordingly, the four main hospital's functions can be identified namely procurement, warehousing and distribution, replenishment, and prescribing and dispensing. Then the phenomenon of each hospital supply chain process across each hospital's function is described to define the hospital's tasks in the following subsection.

The Defined Tasks and Descriptions. Based on the GSCF framework, Croxton, *et al.* [25] state that the appropriate coordination must be placed across the various functions then the supply chain processes will be effective and efficiency. Thus, this part aims to define the tasks to place across each hospital's function and each supply chain process. According to the modified framework, the backbone of hospital supply chain business processes is provided. Interviewees were asked to describe their related activities at each junction of a functional area and supply chain processes. From all hospital cases, the core phenomenon are identically explained. Then the hospital's reference tasks can be defined. For example, the demand management process at function of prescribing and dispensing, physical doctor creates demand for prescription by determining the drugs needed to treat his/her patient. At wards, physical doctor make prescription and then send prescription to pharmacy department for dispensing drugs. This core phenomenon is defined as 'drug prescribing' task. Then all descriptions of core phenomenon and the defined reference tasks are presented in Table 2.

Table 2. List of hospital's reference tasks

Supply chain processes / Functions		Reference tasks and descriptions
H-Demand management	Procurement	*Sourcing*: Purchasing department negotiates contract with vendor that provides concession in purchase price and assign hospital agreement.
	Warehousing & distribution	*Hospital demand forecasting*: Hospital central warehouse produces a forecast for prediction future requirement. Forecasting data are distributed to determine the drug requirements at the stock keeping unit level. Then the amount of requisition is considered.
	Replenishment	*Internal demand planning*: A pharmacist at pharmacy room checks a stock keeping unit level. A forecast is used as an input to the demand planning. If lower than 70-80% of stock on hand, the requisition is determined with adding judgment approach based on experience.
	Prescribing & dispensing	*Drug prescribing*: A physical doctor creates demand for prescribing by determining the drugs needed to treat his/her patient. At wards, a prescription is made and sent to pharmacy room.

Table 2. (*continued*)

Supply chain processes / Functions		Reference tasks and descriptions
H-Order fulfillment	Procurement	*Requisition management*: Purchasing department receives order requisition from hospital central warehouse and commit order request.
	Warehousing & distribution	*Internal requisition management*: Hospital central warehouse receives an internal order requisition that be requested for replenishment. Then the hospital's pharmacy repository is searched for the requested drugs. The drugs are prepared for internal distribution. If stock lower than 70-80%, a pharmacist at warehouse makes a purchase request and send to purchasing department.
	Replenishment	*Requesting for replenishment*: A pharmacist at pharmacy room creates internal order requisition to request the ordering of the drug items. Then an internal requisition is sent to hospital central warehouse.
	Prescribing & dispensing	*Drug dispensing*: Hospital pharmacy receives a prescription. Then a pharmacist reviews drug order and pharmacy patient record. If a problem with a prescription is discovered, pharmacist must consult the doctor. The prescribed drugs are prepared for dispensing to patients. A pharmacist provides information to the patient and dispenses drug.
H-Supplier relationship management	Procurement	*Order placing*: Purchasing department searches supplier lists for the required drugs. A purchase order is created and approved then released to supplier.
	Warehousing & distribution	*Drug receiving and placing*: When a shipment arrives, a hospital committee inspects the ordered drugs then the accepted items are received. After data entry is completed, an inventory level is updated. The received drugs are stored and kept in good condition at the hospital central warehouse.
	Replenishment	*Internal drug receiving and placing*: When an internal order requisition arrives, a pharmacist inspects the requested drugs then the accepted items are received. For data entry, the information captured from supplier should be concerned. After data entry is completed, an inventory level at a storage location (pharmacy room) is updated. The received drugs are stored and kept in good condition at a pharmacy room.
	Prescribing & dispensing	*Drug prescribing and dispensing record keeping*: After drugs dispense, inventory on hand is updated. All list of drug orders are kept prescribing and dispensing record for each patient at least for five years. The information captured from the supplier or manufacturer should be also concerned.
H-Return management	Procurement	*Return policy management*: Hospital assigns the agreement for product return with vendor such as expired drug. The product return is coordinated by purchasing department. In a drug recall, a drug is withdrawn or removed from the pharmaceutical distribution chain. Hospital will receive recall information and document that initiated from supplier. Then the recall information is forwarded to hospital pharmacy.
	Warehousing & distribution	*Drug return management*: Hospital central warehouse checks the correctness of the returned drug items which are sent from a pharmacy room.
	Replenishment	*Reverse logistics*: Pharmacist at pharmacy room checks the returned drugs that unused or the expired drugs in repository. The returned drug form is made. Then the collected return drugs are sent to hospital central warehouse. For drug recalls, the recalled drugs in pharmacy repository are identified and then returned to hospital central warehouse.
	Prescribing & dispensing	*Drug return / Drug recall*: Unused drugs that are prepared for dispensing must be returned to the hospital pharmacy. For drug recall, hospital pharmacy identifies and contacts the patients who receive the recalled drug for returning to hospital.

Finally, the constructed hospital supply chain processes reference framework is illustrated in Figure 2.

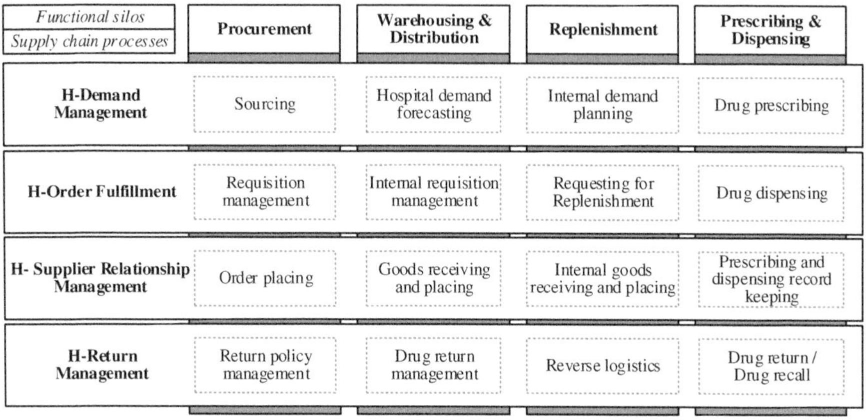

Fig. 2. Hospital supply chain process reference framework

5 Discussion and Conclusion

This study started by understanding the AS-IS situation of healthcare supply chain in Thailand through preliminary empirical investigation. Then the fragmented pieces of business processes have been found in this setting and BPR can solve this situation. In order to BPR in any context, the end-to-end process view of organisation is required and a process perspective as standard view can much more facilitate. Thus, PRM should be provided for hospital supply chain setting.

To develop H-PROM, business needs must be recognized for determining the purposes of the model. The four highlight issues can be found including healthcare professionals are key players, healthcare system focuses on patient and the ultimate goal is safety, healthcare is interdepartmental systems, and resistance to change exist. Accordingly, we suggest the constructed H-PROM with multi-purpose in this study. The H-PROM should 1) provide standardisation for facilitating of interdepartmental communication and collaboration and 2) act as a strategic decision support tool for leading to acceptance to change. Then the empirical-based construction method of H-PROM is provided in this paper.

According to the model construction methodology, firstly, we selected the suitable SCM framework based on the determined criteria. Then empirical foundation from hospital case studies was acquired based on the selected framework and theory of SCM to frame hospital supply chain process reference framework. Finally, descriptions of phenomenon were used to define the hospital's tasks and then the framework can be constructed. The some results of the first two stages can speed up all stakeholders to understand in an organisation relationship through standard view. Besides, the model could expand the boundary of supply chain operations management towards strategic decision support solution.

For further research, to build a robust model, all hospital reference process patterns must be grounded and integrated into the constructed hospital supply chain process reference framework. Moreover, the generic problems of hospital supply chain should be extracted to provide the appropriate solutions as the application of H-PROM.

References

1. Dobrzykowski, D., et al.: A Structured Analysis of Operations and Supply Chain Management Research in Healthcare (1982–2011). International Journal of Production Economics 147, Part B, 514–530 (2014)
2. Aktaş, E., et al.: A Decision Support System to Improve the Efficiency of Resource Allocation in Healthcare Management. Socio-Economic Planning Sciences 41(2), 130–146 (2007)
3. Hammer, M.: Reengineering Work: Don't Automate, Obliterate. Harvard Business Review, 104–112 (July-August 1990)
4. Patwardhan, A., Patwardhan, D.: Business Process Re-Engineering – Saviour or Just Another Fad? International Journal of Health Care Quality Assurance 21(3), 289–296 (2008)
5. McConnell, C.R.: Umiker's Management Skills for the New Health Care Supervisor. Jones & Bartlett Learning, Ontario (2010)
6. Khodambashi, S.: Business Process Re-Engineering Application in Healthcare in a Relation to Health Information Systems. Procedia Technology 9, 949–957 (2013)
7. MacPhee, M.: Strategies and Tools for Managing Change. J. Nurs. Adm. 37(9), 405–413 (2007)
8. Pringle, M., et al.: Measuring "Goodness". Individuals and Healthcare Systems 325, 704–707 (2002)
9. Walters, D., Jones, P.: Value and Value Chains in Healthcare: A Quality Management Perspective. The TQM Magazine 13(5), 319–335 (2001)
10. Francis, S.D., Alley, P.G.: A "Patient Focus Review" of Surgical Services. Business Process Re-engineering & Management Journal 2(1), 48–62 (1996)
11. Davenport, T.H.: The Coming Commoditization of Processes. Harvard Business Review 83(6), 100–108 (2005)
12. Kriegel, J., et al.: Advanced Services in Hospital Logistics in the German Health Service Sector. Logist. Res. 6(2-3), 47–56 (2013)
13. Fettke, P., Loos, P.: Reference Modeling for Business Systems Analysis. Idea Group Pub. (2007)
14. Blecken, A.: Supply Chain Process Modelling for Humanitarian Organizations. International Journal of Physical Distribution & Logistics Management 40(8), 675–692 (2010)
15. Stewart, G.: Supply-Chain Operations Reference Model (Scor): The First Cross-Industry Framework for Integrated Supply-Chain Management. Logistics Information Management 10(2), 62–67 (1997)
16. Thomas, O., et al.: Reference Model-Based Event Management. International Journal of Event Management Research 4(1), 38–57 (2008)
17. Verdouw, C.N., et al.: Process Modelling in Demand-Driven Supply Chains: A Reference Model for the Fruit Industry. Computers and Electronics in Agriculture 73(2), 174–187 (2010)

18. Rabe, M., et al.: Reference Models for Supply Chain Design and Configuration. In: The Winter Simulation Conference (WSC 2006), pp. 1143–1150 (2006)
19. Reichert, A., et al.: A Reference Process Model for Master Data Management. In: The 11th International Conference on Wirtschaftsinformatik (WI 2013), Leipzig, pp. 817–845 (2013)
20. Scheer, A.-W., Nüttgens, M.: Aris Architecture and Reference Models for Business Process Management. In: van der Aalst, W.M.P., Desel, J., Oberweis, A. (eds.) Business Process Management. LNCS, vol. 1806, pp. 376–389. Springer, Heidelberg (2000)
21. Galster, M., Avgeriou, P.: Empirically-Grounded Reference Architectures: A Proposal. In: The Joint ACM SIGSOFT Conference, Boulder, Colorado, USA, pp. 153–158 (2011)
22. Moberg, C.R., Vitasek, K.: Time to Remodel. 2008 Quarter 3 (2008), http://www.supplychainquarterly.com/print/scq200803scmmodel/ (cited July 22, 2014)
23. Lambert, D.M., et al.: An Evaluation of Process-Oriented Supply Chain Management Frameworks. J. Bus. Logist. 26(1), 25–51 (2005)
24. Lambert, D.M., Cooper, M.C.: Issues in Supply Chain Management - Don't Automate, Obliterate. Industrial Marketing Management 29(1), 65–83 (2000)
25. Croxton, K.L., et al.: The Supply Chain Management Processes. International Journal of Logistics Management 12(2), 13–36 (2001)

Smartphone Based Healthcare Platform and Challenges

Bofan Song, Bingwen Yu, Dan Zhu, Wei Jin, and Ying Mu*

State Key Laboratory of Industrial Control Technology, Zhejiang University, Hangzhou, China
muying@zju.edu.cn

Abstract. With the rapid penetration of mobile devices into society, potential of using smartphones in healthcare are very exciting. Smartphone based healthcare platforms bringing the advanced laboratory instruments to rural areas, develop-ing countries and in case of emergency. It is foreseeable that the ubiquitous smartphone based platform would generate tremendous data. We need to find better ways to manage and analyze them. Other challenges involving reliability, usability and security are also important issues to take into account. Mobile is the future of healthcare. The way of healthcare will be changed stupendously after mitigate the barriers and limitations.

Keywords: m-Health, healthcare, healthcare platform, smatphone, network, chanllenges.

1 Introduction

Mobile communication devices, associates with Internet and social media, present opportunities to enhance on-site disease detection, disease prevention and management by extending health service beyond the reach of traditional care-resource limited settings-an approach referred to as m-Health [1]. Nowadays, Smartphones are integrated devices of the state-of-art technologies such as fast CPUs, high-resolution high-definition digital cameras, user-friendly interfaces, precise GPS, gyro sensors and so on. The latest generation of smartphones is more likely to be viewed as hand-held computers rather than as phones. With the rapid development, an estimated six billion cellphone subscriptions exist worldwide with affordable price, and over 70% of the users are living in developing countries (www.itu.int/ict/statistics). The almost universal availability of cell phones means that most people are connected with data transmission capability as well as photography, positioning and computation, allowing the implementation of many applications. "Mobile computing platforms are beginning to make inroads into the laboratory-serious prospect or fairy tale? [2]"

As early as 1905, Eindhoven *et al.* transmitted ECG using telephone line. How-ever, the first telemedicine system with practical value appeared in the 1950s, which transmitted medical data through telephone lines. In the 1970s, the telephone-network has been used to transmit medical images [3]. At the end of 20_{th} century, modern microelectronics, communication technologies, and computer networks promoted the

* Corresponding author.

M. Gen et al. (eds.), *Industrial Engineering, Management Science and Applications 2015*,
Lecture Notes in Electrical Engineering 349, DOI: 10.1007/978-3-662-47200-2_95

development of m-Health. In recent years, cutting edge smartphones and rapid mobile networks make this field full of imagination. It is clear that the mobile device has tremendous potential to transform the traditional healthcare and clinical intervention in the community and developing countries.

Several previous studies have evaluated healthcare and public health interventions using mobile phones, especially in the healthcare data collection and collation. Mobile phones has been used in developing countries to support remote healthcare and telemedicine, including off-site medical diagnosis and HIV care treatment in rural areas which serve as information support apparatus [4, 5]. Smart phone with the feature of GPS (Global Positioning System) offer many opportunities to variety of appli-cations, including chronic disease sufferers assistance, epidemiology, public health surveillance and community data real-time collection. The main advantages of mobile phones in health and clinical practice involve continuous uninterrupted data stream, high-definition camera, portability and the capability through powerful computing power to multimedia software applications supporting. Mobile device is also an at-tractive tool for cost reduction and bring substantial economic benefits, which most countries are facing heavy pressure and burdens of excessive increase of medical expenditures [6].

2 Healthcare Platform Based on Smart Phone

The advances of smart phone cameras provide various opportunities of microscope applications based on the smart phone [7]. Microscopy is one of the most useful tools in medical field. Besides microscopy, these advanced imaging, optoelectronic units or electronic sensing technologies embedded in smart phones can also be utilized for various POC, telemedicine and mobile health related applications. Many interesting and valuable design have been reported recently, including cytometry, blood analysis, bacteria and viruses detection [8], infectious diseases diagnosis [9], chronic patient monitor, allergen sensing platform, electrocardiogram rhythms monitor, parasites detection based on electrochemical [10], human eye refractive errors estimation, label-free detection device, cataract detection and so on.

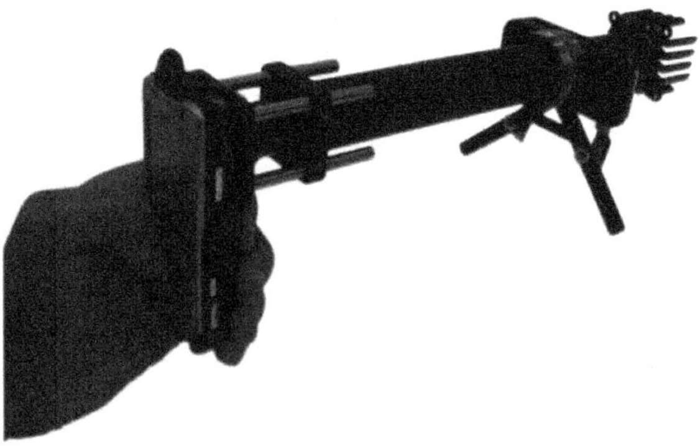

Fig. 1. Microscope designed by prof. Fletcher's Lab at the University of California Berkeley

3 Networks

Increasingly information and communication technologies are viewed as means for efficient and effective delivery of health care services considering the soaring medical budget. With the development of healthcare information highway, health resources can be utilized more efficiently, and the medical service would be improved, as well as the management [11]. The patient information and health data can be uploaded, recorded, and checked at anytime, anywhere as long as the information highway is accessible. The patient can be tested at home, field, in the case of emergency or rural area using the point-of-care test devices based on smart phone, and report the test results to the server immediately via mobile network. Through the health information highway, clinician can check the patient data from the hospital server, and make the diagnosis in real-time. The pathologist also can check the test results from the Lab server at once, and give his opinions and feedbacks on-line. The health information highway connects the detection equipment, clinician, pathologist and the patient worldwide. The health information highway also can complete the payment process by networking the insurance corporation. When the data becomes huge, statistical database and spatio-temporal analysis would be essential, and government depart-ments or organizations such as center of disease control (CDC) can check the data through the health information highway to prevent and control.

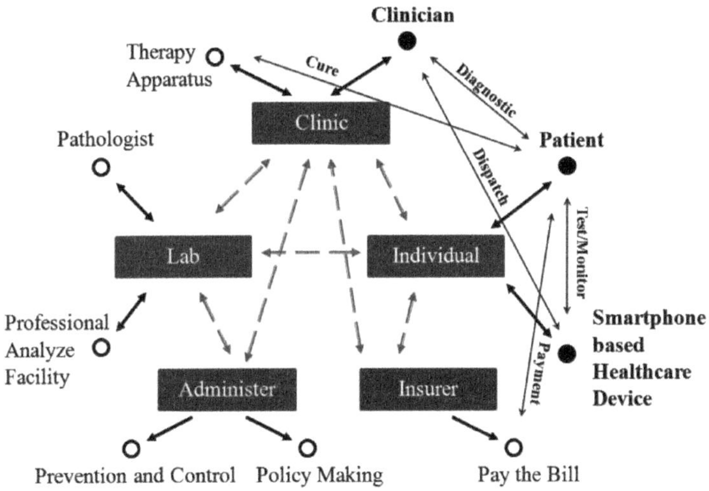

Fig. 2. Future healthcare networks

4 Big Data Problem

It is foreseeable that the ubiquitous smartphone based cost-effective detection devices would generate tremendous data, and need to be managed and analyzed [12]. Perhaps the biggest challenge of the future m-Health concerns the volume of information that smart mobile devices will produce. Investigators will have millions or perhaps billions of data points to read and analyze instead of hundreds. Potentially continuous flows of data from thousands of patients will come day and night from all around the world. Current technologies are not able to handle, store, retrieve or analyze such amount of masses of data from variety of sources and types. The latest state of technologies like machine learning, data mining and other modern analytic methods will be required to confront the mountains of data. In the future, we would be able to dynamically track diseases, different species or infectious outbreaks temporally and spatially, and would be able to better identify and investigate these spatio-temporal patterns at a much larger scale. If so, an important tool is providing, especially for epidemiology prevention and control. Such a platform could even essential to government for health-care issues which local and global data can be studied.

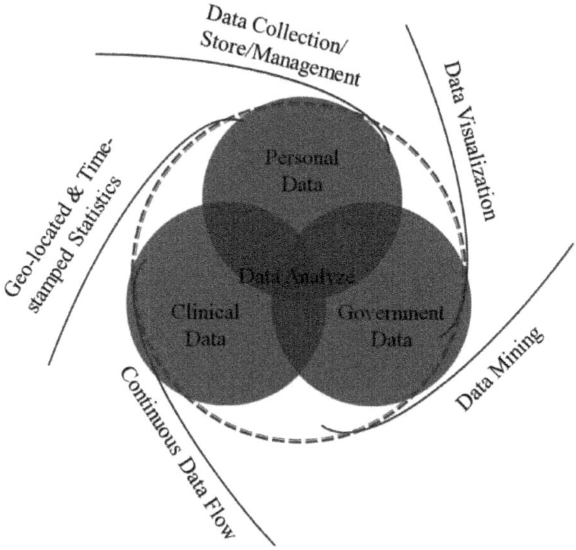

Fig. 3. Big data problem of future healthcare

5 Challenges

There are many challenges to the development of the smartphone based healthcare platforms, healthcare networks and services. Most importantly, the operation of the platforms must be seamless and autonomous. It is very possible that the patient might even be unconscious during times of medical emergency, which does not able to operate software or device. The target group users are usually not well-trained, does not have any familiarity with technology. Reliability is also an important issue for system and service, due to the site these platforms been used are usually far away from technical teams and professional personnel, the platform may give the user very negative sensation in the case of malfunctioning. Usability is also important for such platforms. How to implement cutting edge intelligent mechanisms or complex algorithm on old style mobile devices with limited resources is also an important issue to take into account.

Notwithstanding all of the benefits by using mobile devices and networks in healthcare and clinical practice, problems and limitations exist. Battery capacity and charging is one of the big issues of mobile devices and related platforms in case of emergency. In the future, the use of cloud computing resources can offset the limitation of mobile processing speeds and memory capability. However the transmission speed is an important issue. Mobile phones are smaller, more portable and less obtrusive. But factors such as loss or theft of devices may impact upon the security of healthcare records and confidential digital information. There are some security risks by using mobile services in healthcare. The security of patient data held on mobile devices is an important problem need to concern.

6 Conclusion

With the rapid proliferation and deep penetration of mobile devices into society, opportunities to exploit the potential of smartphones in healthcare are very exciting. Mobile health (m-Health) applications has grown rapidly, smartphones has been adopted by many clinicians and health workers in a diverse range of practices. Smart-phones are raising tools for patients to access health information, contact with their healthcare providers and participate in their own care actively.

The use of smartphones for the development of point-of-care platforms bringing the advanced laboratory instruments to rural areas, field conditions and resource-limited settings and in the case of emergency. This might provide more cost-effective and powerful analysis tools to developing countries, which improve their research labs and institutions and result in better research outcomes. Mobile phones will change the way that healthcare is conducted, especially in developing countries, and impacting the existing practices in medicine.

Go mobile is the natural progression of healthcare in the future, because smart-phones can offer a convenient solution to many applications. Smartphones are useful for clinicians to communicate advices, guidelines and updates to a distributed com-munity in an easy and cost effective manner. Clinicians also can use smartphones to update the latest medical techniques. Mobile phones are useful for monitoring, analyzing and diagnosing when clinicians are far away from patients. Smartphones are agile, handheld, easy to use and can be used on the move, therefore appropriate to the medical and health related professions. However, problems and limitations still existed by using smartphones and related platforms in healthcare. We need to face mountains of data and find better ways to handle, store, transmit and analyze. Other problems such as reliability or usability are also important for these devices, apps and platforms' development.

References

1. Deborah, E., Ida, S.: Open mHealth Architecture: An Engine for Health Care Innovation. Science 330, 759–760 (2010)
2. Editorial.: Mobile Computing Platforms such as iPhone are Beginning to Make Inroads in-to the Laboratory-Serious Prospect or Fairy Tale? Nat. Methods. 7, 87 (2010)
3. Lemeke, H.U.: Communication Networks for Medical Image Transmission. Trahlenther Onkol. 169, 512–520 (1993)
4. Blaya, J.A., Fraser, H.S., Holt, B.: E-health Technologies Show Promise in Developing Coutries. Health Affairs 29, 244–251 (2010)
5. Kaplan, W.: Can the Ubiquitous Power of Mobile Phones be Used to Improve Health Out-comes in Developing Coutries? Global Health 2, 9 (2006)
6. Setia, H., Glyn, L.: Managing Obesity through Mobile Phone Applications: a State-of-the-art Review from a User-centered Design Perspective. Pers. Ubiquit. Comput. 18, 2003–2023 (2014)
7. Saumyakanti, K., Michel, O.: Toward Single-Molecule Microscopy on a Smart Phone. ACS Nano 7, 8340–8343 (2013)
8. Zhu, H., Uzair, S., Aydogan, O.: Quantum dot Enabled Detection of Escherichia Coli Using a Cell-phone. Analyst 137, 2541–2544 (2012)
9. Cardoso, L.P.V., et al.: Development of a Quantitative Rapid Diagnostic Test for Multiba-cillary Leprosy Using Smart Phone Technology. BMC Infectious Diseases 13, 497 (2013)
10. Lillehoj, P.B., Huang, M.C., Truong, N., Ho, C.M.: Rapid Electrochemical Detection on a Mobile Phone. Lab. Chip. 13, 2950–2955 (2013)
11. Eramo, L.A.: Patient Portals: Express Lane on the Health Information Highway. J. AHIMA 83, 24–28 (2012)
12. Alan, R.W.: Big Data In Health: a New Era for Research and Patient Care. Health Affairs 33, 1110 (2014)

Operating Rooms Decision Optimization Integrating Surgery Planning and Nurse Rostering

Siyu Wang, Changyue Ma, and Wei Xiang*

Faculty of Mechanical Engineering and Mechanics, Ningbo University, P.R. China 315211
xiangwei@nbu.edu.cn

Abstract. Operating room (OR) scientific management calls for an efficient decision making to determine the elective surgeries date and assign qualified nurses to shifts over a certain period. Such two decisions, the surgery planning and nurse rostering, are classified as the mid-term decision. They have mutual influence to each other but are solved independently in the literature. A mid-term OR decision optimization integrating surgery planning and nurse rostering is described and an ACO approach is proposed to solve such multi-objective optimization problem. The test case from the 10-days actual hospital operation data is used to evaluate the proposed approach. The comparison analysis shows the advantage of the integrated decision in balancing the daily surgeries quantity and the nurses' workload.

Keywords: Mid-term decision, integration, Surgery planning, Nurse rostering, Ant colony optimization.

1 Introduction

Nowadays hospitals in China are facing a trend of increasing demand for surgeries due to the aging population. In turn, such increasing demand brings pressure in OR management to deliver quality surgery services to patients in time. In most china hospitals, there is no systematic surgery planning but surgeons from different specialties determine individual surgery date according to his timetable. Therefore, it is quite common that there are large variations in the number of daily surgeries. Most of the time OR nurses have to work overtime, however it is also the case that it may have nurses in some qualified level to be idle because of less surgeries to be performed. Such unbalanced utilization has been a difficult and a challenge problem in OR management. In this context, it is essential to provide an efficient decision making strategy for OR management.

Surgery planning and nurse rostering are two core decision-making in OR. Both of them can be classified as the mid term decision. The former is to determine the surgeries date and the latter one is to determine the shift plan for all nurses. For the OR management in real hospital, it is the surgeons from individual specialties who

* Corresponding author.

M. Gen et al. (eds.), *Industrial Engineering, Management Science and Applications 2015*,
Lecture Notes in Electrical Engineering 349, DOI: 10.1007/978-3-662-47200-2_96

determines when will the surgeries to be performed. Although such decision will take into account all patients in this specialty. However, it is not possible for decision maker to know surgeries from other specialties. Such information barrier leads to the large fluctuation in the numbers of surgeries in ORs each day. For nurse rostering, it is the head nurse from OR who determines the shift plan for all nurses over a period of time according to the labor policy, skill and qualification level and the nurse preferences. Usually it is a monthly roster or a weekly roster. Nowadays in China, it is common that the nurse rostering is still a kind of experience-based decision and done manually by the head nurse in OR.

Just like the real life OR management, researches on surgery planning and nurse rostering are separately conducted in operation research community. Several good review works can be found respectively in OR planning and scheduling [1-4], and nurse rostering [5-7]. Very few researches try to integrate these two decision problems. Some researches considered the integration in surgery scheduling and nurse rostering constraints [8-9]. However, the integration exists in short-term usually daily OR management. As pointed out by Litvak and Long [10], one common problem at OR management is the extreme variation in daily workload which makes staffing the ORs difficult at best. And He et al.[11] further demonstrated that incorporating surgery planning information like the exact number and type of cases for each specialty into staff scheduling can reduce total staffing costs by an average of 49% relative to those without using the case information. Those research works pointed out the importance of integrating mid-term OR decisions. Beliën and Demeulemeester [12] is the first and perhaps the only paper which developed an integrated nurse and surgery scheduling system for a weekly planning problem. The integration in their work was performed in sequence.

In view of the above problems commonly arise in real life hospital, this paper proposed an integrating decision-making which involves the surgery planning and the nurse rostering. Several real life constraints, e.g. the specialty, surgery's due date, the nurse's preference, etc, are further considered in such integrated decision model. An ant colony algorithm with a two-layer ant graph is proposed to solve the integrating decision model. Test case from a real life hospital in China is used to evaluate the proposed approach.

2 OR Decision Problem Description

2.1 Problem Description

It is assumed that there is a set of elective surgeries from a set of specialties to be performed in an operating suite with parallel multi-functional operating rooms. Medical staffs considered in OR in this work mainly include nurses. Nurses are regarded as heterogeneous with different qualifications, competences, shifts, roles (i.e. nurse scrub and circulator) and job titles (i.e. fulltime and contract). Other resources, like anesthetists, pre-operative holding units (PHU) beds, post anesthesia care unit (PACU) beds are considered as homogenous that can be assigned to deal with any surgeries. Each surgery is associated with surgery demand and resource demand. Surgery demand is to specify the particular surgery specialty (SS) and the estimated surgery duration, which can vary from surgeon to the next even for the same SS. Resource demand indicates all resources required for the complete stages (i.e. pre-operative stage, surgery stage, and post-operative stage) of a surgery. A surgery has to be performed by a team of medical staffs. Nurses, anesthetists, and surgeons have staff preferences in forming such team.

2.2 Objectives

The integrated surgery planning and OR nurse rostering is a mid-term decision to determine the surgery date of all elective surgeries and the working date/ shift of medical staffs within a period of time. It can be regarded as a multi-objective optimization (MO) problem with objectives from different perspectives. Since the large variation in daily surgeries is a serious problem in real life OR management, a balanced daily surgeries quantity is one of the important goal for OR decision. To match surgeries to qualified staffs also requires an appropriate workload allocation in staffs. A balanced nurses' workload has direct relationship with ORs' operation cost. Last, but not least is a satisfaction on staffs. Only when staffs' preferences are satisfied, a higher quality healthcare service can be provided to the patients. Therefore, the OR decision optimization problem in this work proposes three objectives: minimizing daily surgeries variation, minimizing staffs workload variation, and maximizing staffs preference satisfaction.

2.3 Constraints

In addition, several the real life constraints commonly found in hospital OR management should be taken into account as well to make the decision more practical in application. Those constraints are further classified as hard constraints and soft constraints.

1. Soft constraints on specialty-surgery schedule

Usually hospitals have an individual weekly specialty–surgery schedule. It suggests an allocation of surgery specialties to an exact day within a week. For example, specialty-gynecology is assigned to everyday from Monday to Friday, and specialty- orthopedic is only assigned to Wednesday. Therefore, such schedule should be taken into account when dealing with surgery planning. However, it is not a compulsory. The emergent surgery should always be arranged immediately no matter its belonging specialty is in the scheduled day or not. It is considered as a kind of "try to follow" soft constraint.

2. Hard constraints on nurse rostering

Some regulations on labor law and the compulsory requirement for surgery are classified as the hard constraints which must not be violated in decision making.

- Each nurse can only do one shift work a day at most ;
- Within one roster cycle, individual nurse's longest working hours cannot exceed the specified upper bound, and cannot be less than the lower bound as well;
- Surgeries demand in qualified staffs must be meet on that shift;
- Nurses can only do nursing work in surgeries equal to or lower than their qualified level;

3. Soft constraints on preferences

There also exist some staff preferences and shift preferences especially in nurse rostering. For example, some staffs feel comfortable in a team and can have a high efficiency working together. Some prefer to take day shift while some prefer to always take night shift. Those constraints are not compulsory either. Staffs will have higher satisfaction once their preferences are met.

3 ACO Approach for OR Decision Optimization Problem

Either the surgery planning or the nurse rostering problem is the complex combinatorial optimization problem. To integrate such two decision problems as one decision optimization further increases the computational complexity. Therefore, a meta-heuristic approach is preferred to quickly find a sub-optimal solution. An ACO approach is adopted to solve the above mid-term integrating decision optimization.

3.1 A Two-Layer Dynamic Ant Graph Model

Since the integration problem not only calls for determining surgery date for individual patients, but also have to settle nurse roster as well, the solution of such decision optimization problem should provide both information. It cannot be realized by the classical single ant graph model. A two-layer dynamic ant graph model represents both surgery nodes and resources nodes is proposed in this work as shown in Fig.1.

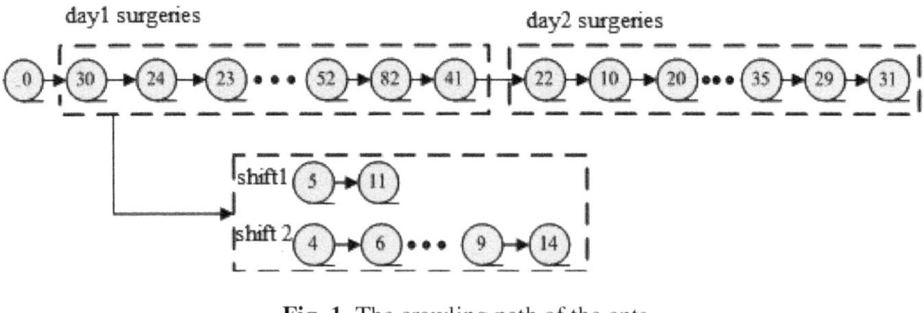

Fig. 1. The crawling path of the ants

Surgery layer is composed by all elective surgeries nodes and an additional dummy node as the start node. Surgeries nodes are further have to be grouped in a day set. The numbers of surgeries within one day is depended on the surgeries duration, the capability of ORs and the normal working hours. Therefore, the number of nodes in each day set is dynamic in each ant cycle. Nurse layer consists of all nurses to be assigned to a day with associated surgery nodes. Surgeries nodes determine the resources demanding, and further determine the nurses' nodes within a day set. The arc linked nodes in one day set has no meaning, but the nodes in which day set has a specific meaning and should be recorded as a pheromone.

3.2 The Heuristic Definition

Due to the two-layer ant graph structure, the ants sense differently when crawling over the surgery layer and nurse layer. The corresponding heuristics in two layers is defined.

1. the heuristics in surgery layer

Individual surgery has a "due" which means patient must be undergone the surgery before that date for his severity. When approaching to the due, a higher heuristics should be assigned so as to increase the possibility to be selected next.

$$\eta_{ij} = e^{-(Due_i - day_j)} \tag{1}$$

2. the heuristics in nurse layer

Within nurse layer, the ant choosing the next visited node depends on the nurses' qualification constraints, the nurses' shift preference and the nurses' staff preference. The node with higher matching on these three areas should be assigned with higher heuristics as shown in Equation (2). K is the feasible nurses set, D_j is selected nurses set on day j. I_j represents the surgeries set in day j, $shift\_preference_{kjm}$ represents the shift preference of nurse k in day j shift m. $qualification(k, I_j)$ returns a value to indicate the matching in qualification of nurse k for surgeries in I_j. $matching(k, D_j)$ returns a value indicating the nurse k's staff preference with those nurses already assigned to day j shift m.

$$\eta_{jk} = 1 / shift\_preference_{kjm} \cdot qualification(k, I_j) \cdot matching(k, D_{jm}) \tag{2}$$

3.3 The Algorithm Description on the Pareto Set-Base ACO for MO

The proposed mid-term integrated decision problem is a multi-objective optimization problem with three objectives. Moreover, the three objectives are conflicting. There may not exist a single solution that simultaneously minimizes each individual objective. The proposed ACO is further combined with the Pareto set for every ant cycle, then keep update such Pareto set along with the iterations. The final decision can be the one based on decision maker's preference in such Pareto optimal set. The detail algorithm is explained as follows:

Step1: Put m ants on arbitrary node and initialize pheromone trial

Step2: Construct a feasible solution set S by ants traversing the two-layer ant graph

Step3: Construct a Pareto set PS_{now}, and form an iteration feasible set $S_{iteration} = PS_{iteration} \cup PS_{now}$

Step 4: Construct a final iteration Pareto set $PS_{iteration}$ by set initial feasible solution set as the iteration feasible set $S = S_{iteration}$

Step 5: Update the pheromone value in two-layer ant graph.

Step 6: Repeat from Step 1~5 until the maximum iteration number is reached.

Step 7: Select a solution S* from a final Pareto set $PS_{iteration}$ according to decision maker's preference.

4 Case Study

The proposed ACO algorithm is implemented with MATLAB and is run on a PC running Windows XP with Intel Core5 @2.79GHz and 3GB of memory. The 10 days real operation data (from Oct, 22, 2013 to Nov, 04, 2013) in a typical medium scale hospital in China is collected and built as a test case to evaluate the proposed ACO approach. The test case involves 10 ORs, 6 PHU beds, 10 PACU beds, and staffed with totally 30 nurses and 27 ANs. All ORs are open during the day shift period (starting from 7:30 AM to 4:30 PM) of week day (Monday to Friday). Totally there are 427 elective surgeries from 23 specialties performed during 10 working days. Nurses are classified into 3 grades. Two shifts, day shift and night shift are involved in nurse rostering. The nurses'staffs preference and shift preference are not considered.

Table 1. The detail information on the actual schedule and optimal schedule by ACO

	day	1	2	3	4	5	6	7	8	9	10
surgeries	Actual	44	41	37	48	55	32	46	39	40	45
	ACO	47	36	42	40	46	47	36	47	50	30
Total time (h)	Actual	66.9	74.9	64.3	79.4	93.7	64.3	80.8	62.7	71.1	60.9
	ACO	72.9	72.5	73.9	73.0	73.2	71.4	72.4	73.9	72.4	63.6
nurses	Actual	20	21	21	21	22	22	20	21	19	20
	ACO	21	21	21	20	19	18	21	20	20	15

The detail information like the number of the daily surgeries, the total ORs time, and the number of nurses assigned of both the actual schedule from hospital and the optimal results generated by the proposed ACO approach are listed in Table 1. The comparison analysis can be found in Table 2. The first observation from Table 2 is the variation in daily surgeries of the ACO schedule is smaller than those in actual schedule. The maximum number is reduced from 55 to 50 and the minimum number is increased from 32 to 36. The balanced number of surgeries sometime cannot really reflect the equilibrium because the surgeries duration may be varied a lot. Such equilibrium advantage in ACO schedule can be further proofed by the smaller variation in total daily OR time. The maximum value reports a great reduction from 93.76 to 73.91 and the minimum value has a little increase. The last statistics measurement is the coefficient of variation in working time (CVWT) of resources. It is defined as the ratio of the standard deviation to mean and is used to evaluate the balance of resource utilization. The smaller value in CVWT means a more balanced utilization. From Table 2, both CVWT on surgeries and nurses have a smaller value by ACO compared with those by actual schedule in hospital.

Table 2. The comparison between actual schedule and optimal schedule

Approach	Number of daily surgeries		Total daily OR time (hr)		CVWT	
	Maximum	Minimum	Maximum	Minimum	Surgeries	Nurses
Actual	55	32	93.76	60.96	0.29	0.15
ACO	50	36	73.91	63.57	0.26	0.13

5 Conclusion

In real life OR management, the surgeons in specialties dominate the decision making in surgery planning and the head nurse is in charge of nurse rostering. Those two decisions are separately made by different parties. It may cause the common problem of a large variation in daily surgeries and daily nurses' workload. This paper aims to integrate such two separated mid-term OR decision problems and proposes an ACO approach to solve the integration decision optimization problem. A specific two-layer ant graph model is proposed and the corresponding heuristics in two layers are designed. Moreover, the Pareto set is combined to ACO algorithm as well to solve the multi-objective optimization problem with possible conflicting objectives. A test case from a real hospital is build by collecting a 10-days operation data and is used to evaluate the proposed ACO approach for the integrated decision optimization problem. The comparison analysis shows the advantages of the integration of the mid-term OR decision by the proposed ACO approach in reducing the variation of daily surgeries and nurses' workload. More test cases including the nurses' preferences will be the future work to comprehensively validate the proposed approach.

Acknowledgement. The Project is supported by Zhejiang Provincial Natural Science Foundation of China (LY12G01007), Ningbo Natural Science Foundation (2013A610109) and K.C.Wong Magna Fund in Ningbo University.

References

1. Cardoen, B., Demeulemeester, E., Beliën, J.: Operating room planning and scheduling: a literature review. Eur. J. Oper. Res. 201(3), 921–932 (2010)
2. Erdogan, S.A., Denton, B.T.: Surgery planning and scheduling. In: Cochran, J.J., Cox, L.A., Keskinocak, P., Kharoufeh, J.P., Smith, J.C. (eds.) Wiley Encyclopedia of Operations Research and Management Science. John Wiley & Sons (2010)
3. Guerriero, F., Guido, R.: Operational research in the management of the operating theatre: a survey. Health Care Manag. Sci. 14, 89–114 (2011)
4. May, J.H., Spangler, W.E., Strum, D.P., Vargas, L.G.: The Surgical Scheduling Problem: Current Researchand Future Opportunities. Prod. Oper. Manag. 20(3), 392–405 (2011)
5. Ernst, A., Jiang, H., Krishnamoorthy, M., Owens, B., Sier, D.: An annotated bibliography of personnel scheduling and rostering. Annals of Operations Research 127(1), 21–144 (2004)
6. Cheang, B., Li, H., Lim, A., Rodrigues, B.: Nurse rostering problems, a bibliographic survey. European Journal of Operational Research 151(3), 447–460 (2003)
7. Burke, E., De Causmaecker, P., Van Landeghem, H.: The state of the art of nurse rostering. Journal of Scheduling 7(6), 441–499 (2004)
8. Mobasher, A., Lim, G., Bard, J.F., Jordan, V.: Daily scheduling of nurses in operating suites. IIE Transactions on Healthcare Systems Engineering 1(4), 232–246 (2011)
9. Wei, X., Jiao, Y., Gino, L.: Modified ant colony algorithm for surgery scheduling under multi-resource constraints. Advances in Information Sciences and Service Sciences 5(9), 810–818 (2013)
10. Litvak, E., Long, M.C.: Cost and quality under managed care: Irreconcilable difference? The American Journal of Managed Care 6, 305–312 (2000)
11. He, P.B., Dexter, F., Macario, A., Zenios, S.: The timing of staffing decisions in hospital operating rooms: incorporating workload heterogeneity into the newsvendor problem. Manufacturing & Service Operations Management 14(1), 99–114 (2012)
12. Beliën, J., Demeulemeester, E.: A branch-and-price approach for integrating nurse and surgery scheduling. European Journal of Operational Research 189(3), 652–668 (2008)

Linking Hospital Supply Chain Processes and Performance to Identify Key Performance Indicator

Tuangyot Supeekit[1], Tuanjai Somboonwiwat[1], and Duangpun Kritchanchai[2]

[1] Department of Production Engineering,
King Mongkut's University of Technology Thonburi
Bangmod, Thungkru, Bangkok, Thailand
[2] Department of Industrial Engineering,
Mahidol University Salaya, Phuttamonthon, Nakhonpathoom, Thailand
tuangyot.sup@gmail.com, tuanjai.som@kmutt.ac.th,
duangpun.skr@mahidol.ac.th

Abstract. Previous studies of hospital supply chain performance measurement normally focus on either the performance of the hospital supply chain to be measured and the framework/methodology for performance measurement, or the performance of certain activities/processes in the chain for improvement. There is limited number of studies that explore the entire hospital supply chain performance concurrently with the processes in the hospital supply chain. This paper attempts to explore the performance of hospital supply chain and link them to the activities of the hospital supply chain in order to see the relationship of performance on the hospital supply chain activities to eventually identify the key performance indicator for the particular process. The result of the study shows that relationship of hospital supply chain performance, the business process and the link between the two can yield the key performance indicator for the considered process.

1 Introduction

The concept of supply chain has been implemented to manage hospital operations for many years. The hospital supply chain is implemented in different areas of the chain. To ensure the success of supply chain implementation requires performance measurement. Performance measurement is a management tool which is the process to evaluate the operation of the business or organization [1, 2]. Performance measurement of supply chain should be measured from supply chain perspective which can be investigated through business process of the supply chain to ensure coverage and completeness and to create efficiency to the supply chain as a whole [3]. The previous studies regarding hospital supply chain performance seem incomplete. Some studies present only the performance framework without considering the supply chain of hospital such as [4, 5]. Some studies aim the improve the particular process so the focus is place on certain activities in hospital supply chain without considering performance of hospital supply chain as a whole [6, 7]. Some of them, though covered, neglect the

M. Gen et al. (eds.), *Industrial Engineering, Management Science and Applications 2015*,
Lecture Notes in Electrical Engineering 349, DOI: 10.1007/978-3-662-47200-2_97

hospital supply chain structure arrangement such as [8, 9]. There is limited number of studies that explore the entire hospital supply chain performance concurrently with the processes in the hospital supply chain. This paper attempts to explore the performance of hospital supply chain and link them to the activities of the hospital supply chain in order to see the relationship of performance on the hospital supply chain activities to eventually identify the key performance indicator for the particular process.

The rest of the paper is organized into 5 sections. Section 2 presents the business processes in hospital supply chain. Performance measurement in hospital supply chain is presented in section 3. In section 4, attempt to link performance measurement to the hospital supply chain in order to identify the key performance indicator is presented. Then the section 5 presents conclusion and discussion for future research.

2 Business Process in Hospital Supply Chain

In order to study the business process in the hospital supply chain, the structure of the supply chain must first be explored. Supply chain of hospital can be considered more complex than other supply chains. The ultimate customer of the hospital supply chain is patients who visit the hospital seeking diagnosis and treatment [10]. To fulfill the requirement of patient require both services from medical professionals with flow along clinical care processes and products such as pharmaceutical and medical supplies which flow in the supporting processes [11, 12]. The hospital supply chain in this study can be illustrated in fig.1. To understand the operations of hospital supply chain for successful management of the supply chain, therefore, must explore all activities in both clinical care and supporting processes as follows.

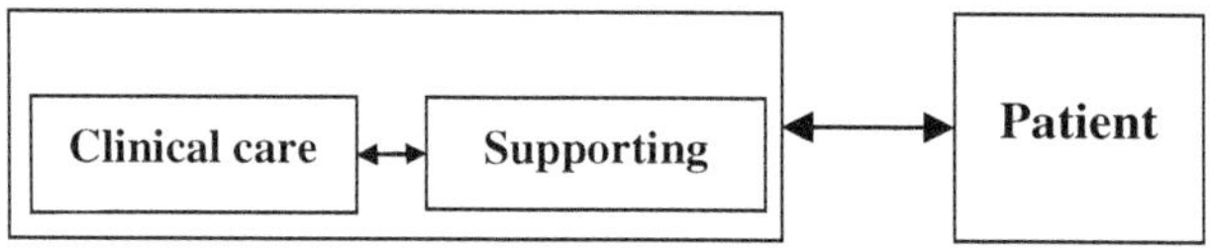

Fig. 1. Hospital Supply Chain

2.1 Clinical Care Processes

Clinical care processes provide medical service to patients. Health professional including doctors and nurses are the main resources for these processes. The medical tools and equipment are also required when delivering the care to patient. The clinical care processes are normally conducted at the clinics such as Internal Medicine, Surgery, Neurology and Pediatrics. According to [13-15], the clinical care processes include the following processes.

Admission. Upon arrival patient must go through these process so that they can be admitted to be cared and treated. These processes might include processes such as registration, triage, and inpatient admission.

Patient Investigation. This is the processes are normally performed by the nurses and doctors to discover the patient's requirements so that the care and treatment can be planned in advance.

Care Delivery. After the requirements are known, the treatment are planned. This process of care and treatment for patient would be carried on. This might also include continuity of care.

Discharge/Transfer. This process is normally the last process at a particular hospital. These processes occur when the patient is treated as planned (discharge) or when the patient's illness is beyond hospital's proficiency.

Demand/Capacity Management. This process is normally performed due to limited resources in the hospital to ensure that the waits and delays is lessened [16].

2.2 Supporting Processes

Supporting processes aid in the completion of clinical care processes. To diagnose or treat patient in patient care services requires medical materials such as pharmaceutical, medical supplies, and medical apparatus. These processes normally occur into two major units: medical supporting units [12, 17-19] such as radiology, laboratory and pharmacy; and non-medical supporting units [1, 7, 12] such as patient porting service, materials management, and sterilized services. The supporting processes involve the following processes.

Sourcing and Purchasing. These are the processes to procure or source medical materials from suppliers or supporting units when needed.

Order Management. The processes are performed by the supporting units to ensure that medical materials are arranged for the clinical care units in a timely manner.

Production or Preparation. These processes occur internally in the supporting units to prepare medical materials for clinical care units.

Transportation and Distribution. These are the processes that transfer the materials or patient to the point of care or where the materials or patients are required for curing, treating and others.

Stock Management. These processes are the processes to ensure the availability of medical materials while keeping the cost lowest possible.

The clinical care and supporting processes can be illustrated in fig. 2

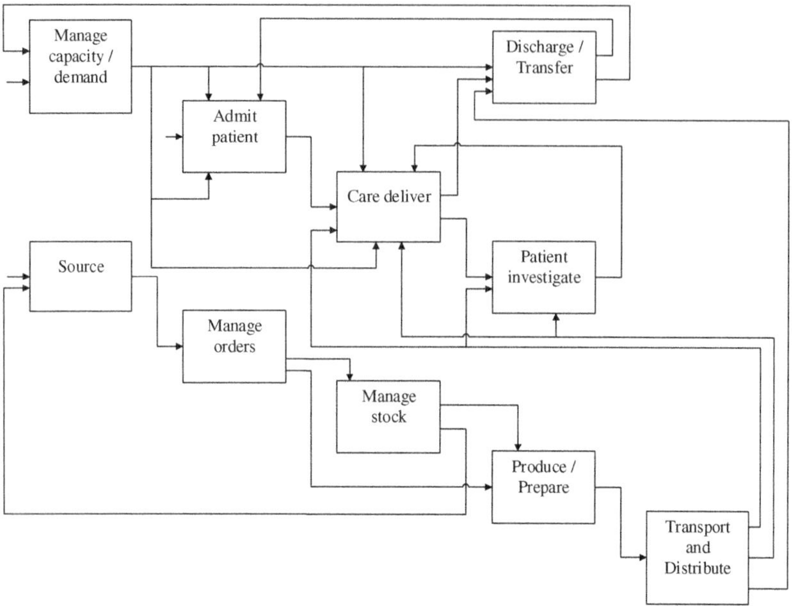

Fig. 2. Business Processes in Hospital

3 Performance in Hospital Supply Chain

Performance measurement is the analytical process which can provide information for improvement decision making. To suggest the improvement in a supply chain, the performance measurement must be specific to the particular supply chain to ensure overall supply chain efficiency [3, 20, 21]. Therefore, the performance measurement in the hospital supply chain must consider hospital supply chain efficiency in clinical care and supporting processes as well as the performance that would affect patients as per being specific to the hospital supply chain. Assertions from [22, 23] state that the patient safety should be taken into consideration to ensure the good outcome to patient. Therefore, to measure the supply chain as a system or to create the supply chain performance measurement system for the hospital supply chain should consider the supply chain efficiency and goal of patient safety altogether. Moreover, the performance measurement must possess a clear explanation of performance criteria and a relationship model between metrics [20, 24]. This section would explore the hospital supply chain performance criteria including supply chain efficiency and patient safety.

3.1 Supply Chain Efficiency

Supply chain efficiency is defined as the efficiency which takes into account multiple performance criteria related to the supply chain processes for the certain supply chain and the integration of supply chain members [8, 25-27]. The criteria which are normally asserted when considering supply chain efficiency include cost and time [8, 26, 27]. Reliability is also asserted to measure supply chain efficiency in supply chain context [26]. Moreover, productivity is also included [27] in measuring the supply chain efficiency since it refers to the efficient use of resources.

Therefore, the term supply chain efficiency in this study comprises of the following:

Cost. – Ability to manage costs effectively

Clinical care process cost (CPC). Operational costs, cost of resources, (Total) cost of treatment.

Supporting process cost (SPC). Processing cost, supply cost, cost of materials

Time. – Ability to manage time of process as well as referring to responsiveness

Clinical care process time (CPT). Response time, treatment time, turnaround time

Supporting process time (SPT). Cycle time, response time

Reliability. – Ability to arrange to process to ensure accuracy and respond needs.

Clinical care process reliability (CPR). Accuracy of services, timeliness of service, accessibility to service

Supporting process reliability (SPR). Service level, fill rate, forecasting accuracy

Productivity. – Ability to operate any processes with a certain number of resources

Clinical care process productivity (CPP). Patient throughput, professional productivity, resource utilization, resource availability

Supporting process productivity (SPP). Inventory turnover, resource availability, staff productivity

3.2 Patient Safety

The other performance of hospital supply chain is the performance that affects patients: patient safety. It is believed that patient safety can be considered the ultimate goal of hospital supply chain [23, 28]. Typical studies of patient safety are normally around mistakes and errors such as medication errors and wrong site surgery [29-31]. Errors are, however, not the only criteria in patient safety as asserted in [32-34]. They claim that patients would also be unsafe if they are not properly diagnosed or treated, or they do not receive treatment in a timely manner. Therefore, patient safety would include the following aspects

Free from Errors. – The processes or operations must contain no errors that might be harmful to patients

No Delay in Treatment. – The processes contains no delays in providing diagnosis and treatment to patients

Completeness of Treatment. – The processes should provide all the care, diagnosis and treatment they may require.

3.3 Relationship among the Supply Chain Efficiency and Patient Safety

In a complex system, all criteria are related, either directly or indirectly, making it difficult to define a specific aspect in isolation [35]. Articles [20, 24] also assert that there are relations in performance measurement. An alteration in one performance area could affect others. It is, therefore, useful to study relationships among performance criteria. Causal relationships also enable enterprises to ascertain the values of various activities performed by relating them to definite result. The causal relationships can be used to acquire the most important factors or criteria for decision making [36, 37]

The relationships among hospital supply chain performance area of supply chain efficiency in clinical care and supporting processes, and patient safety is studied by Supeekit et al. [38] applying DEMATEL methodology to identify the relationship. The result shows that the relationship among clinical care efficiency, supporting process efficiency, and patient safety exists. The relationship can be depicted in fig.3

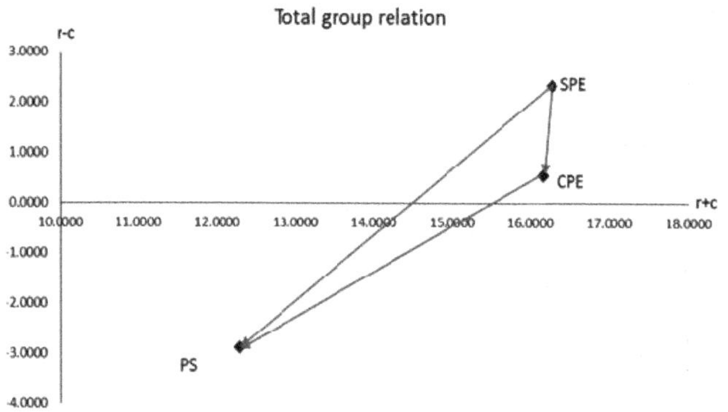

Fig. 3. Business Processes in Hospital

In fig.3, the results show that supporting process efficiency has influence on clinical care process efficiency and patient safety; clinical care process efficiency also affects patient safety. Apart from the result presented in fig.3, Supeekit et al. [38] also analyze the most important performance aspects in clinical care and supporting processes as shown in fig 4.

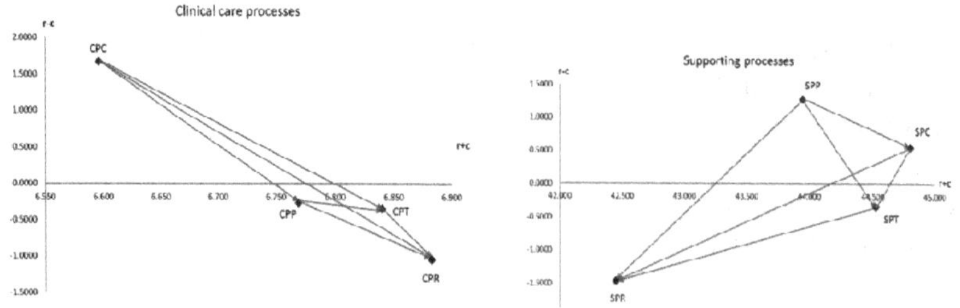

(a) (b)

Fig. 4. Relationship within clinical care efficiency (4a) and supporting efficiency (4b)

In clinical care process (fig. 4a), the most important performance, from DEMATEL analysis, is clinical care reliability because the clinical care directly affect patients. However, reliability of clinical care processes are affected by other performance criteria: clinical care time, clinical care productivity and clinical care cost. Looking at the most influence performance criteria in clinical care process is the clinical care cost. Improvement in clinical care cost would affect its time, productivity and reliability. Therefore, it is worth improving cost performance in clinical care processes. If cost is the issue that it is very limited, then consider to improve time, then productivity. Similarly in supporting processes, the most important performance is supporting process cost due to the high value of medical materials which need attention. The supporting process cost, which has effect on supporting process time and supporting process reliability, is affected from supporting process productivity. The supporting process productivity also affect supporting process time and supporting process productivity. Therefore, improvement in supporting process cost and productivity can both yield improvement in others.

The relationship study in [38] is studied from the performance perspective. However, to improve hospital supply chain, the processes and activities in the hospital supply chain need to be improved. This study attempt to map performance of the hospital supply chain to the business processes of hospital supply chain aiming that the performances are measured for improvement.

4 Linking Performance Measurement to Business Processes in Hospital Supply Chain

Previous sections present business processes in hospital supply chain and performance measurement separately. Though the performance which need improvement in the hospital supply chain is known, the information of the process to be improved remains unknown. Improvement of any supply chain normally can be achieved through an improvement of processes in the supply chain. It is, therefore, important to learn what process in the hospital supply chain needs to be measured for improvement.

According to [38], important performance to measure in the clinical care processes include only clinical care cost since it is shown that improvement in clinical care cost could yield improvement in three others: clinical care time, clinical care productivity and clinical care reliability. Similarly, the performance in supporting processes to be focus on is supporting process productivity. However, all the performance of hospital supply chain is mapped with the business process. Then the important performance is linked to the process later.

4.1 Mapping Performance Measurement to the Business Processes

This section maps all the performance measurement in section 3 to business processes in section 2 according to their activities in the business processes. The mapping is shown in table 1.

Table 1. Mapping performance to business processes.

Area	Business processes	Performance criteria			
		Cost	Time	Reliability	Productivity
Clinical care	Admit patient		Time to admit	Accessibility	
	Patient investigate	Cost of testing Cost of capital asset	Patient wait time	Accuracy of results	Availability of tool Productivity of lab staff
	Care deliver	Cost of care Cost of capital	Waiting time Response time to unplanned issue	Accuracy of diagnosis or treatment	Doctor availability Nurse productivity
	Discharge / transfer		Time to discharge / transfer	Accuracy of discharge process	
	Manage cap./demand		Patient turnaround time		Resource utilization Bed turnover
Supporting	Source / procure	Cost to issue orders Total cost of supplies	Sourcing response time Order lead time	Lead time variability Fill Rate Percentage of rush orders	
	Manage orders		Time spent to order Order fulfillment cycle time	Accuracy of order	
	Manage stock	Operating cost Inventory days of supply Value of buffer stock		Stock out at point of use Stock accuracy	Inventory turnover
	Produce / prepare	Operating cost	Production / preparation time	Pick and pack accuracy	Availability of tools and equipment Staff productivity
	Transport / distribute		Time for transportation Cycle time of transportation	On time delivery Response time to urgent request	Availability of porter

Table 1 shows common performance measures from literatures. It can be seen that not all the performance criteria that need to be measured since some performance criteria are not the key performance of the process. For example, the process of admitting patient is the process of decision making whether the coming patient can be admitted to be cured in the hospital as inpatient or can be arranged to see the doctor for diagnosis and treatment as outpatient. Performances that would be able to use as the key performance measures for admission process are time to admit which express the time to decide. The longer time admitting process could prolong patient treatment process and this might end up with jeopardize the safety of patients. The other performance measure of the admission process is the accessibility of patient which can be determined by the percentage of patients who are not admitted to the care processes. This is just the location that the patient admission record can be collected to expose inadequate capacity of clinical care processes. The other process performance measures can be similarly reflected.

4.2 Linking Important Performance Measurement to Business Processes: An Example

This section attempt to link the important performance measurement presented in section 3.3 to hospital supply chain business processes. From section 3.3, it is known that the supply chain efficiency in supporting processes affects supply chain efficiency in clinical care process. It is also known that the most critical performance criteria in supporting process efficiency is supporting process productivity. The unknown is what process that need to be taken into consideration.

Considering the hospital supply chain business processes depicted in fig. 2, the example to discuss in this paper is transportation and distribution process which influences clinical care process: care deliver, patient investigate, discharge/transfer.

Transportation and Distribution Process. The part of transportation and distribution processes can be illustrated in fig. 5.

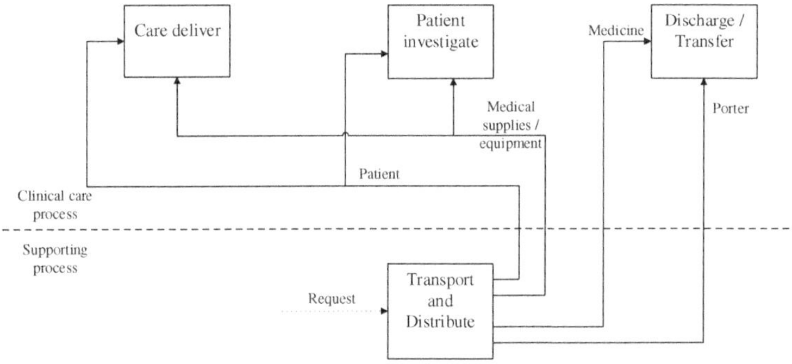

Fig. 5. Transportation and distribution process and its related processes

Transportation and distribution in hospitals involves moving patients, medicine, and other medical supplies to the point of use. Care delivery process which is operated by doctors requires medical supplies or apparatus. Distribution process ensures availability of apparatus at the point of use so that the doctor can use them to treat patients. Similarly, the patient investigation process also requires medical apparatus which is transported by transporters. If the transportation and distribution process is efficient, the departments where care and investigation are performed need not stock many medical supplies and the nurses or doctors can spend more time on patient care instead of stock counting. Moreover, care delivery and patient investigation need patient to be present at the location. If the patient is immobile, transportation service is required. The transportation process again must be efficient to ensure the presence of patients. For the discharge process, the patient requires medicine which is normally transported by porter. Again, if the transportation process is efficient, the time of discharge process may also be shorten. The study of supporting process efficiency informs that supporting process productivity is the most critical performance criteria. In the case of transportation and distribution process, productivity of transportation process; productivity of porters or carriers can be used as the Key Performance Indicator of transportation process.

5 Discussion and Conclusion

The method of linking business processes to the performance measurement as shown in section 4.2 can be applied to identify the key performance indicators. This method can be employed to other business processes to identify the key performance indicator in order to ensure the success of particular process without being confused with plenty of indicators. However, the mapping of business process and performance measure from literatures presented in section 4.1 is still incomplete. This map can further be completed by taking into consideration the business processes. For example, the discharge process contain no cost performance. If cost performance is required (as clinical care cost is the most critical performance), the study of performance measure incorporated with business process should be performed. The cost of discharging process depends on the resources used for discharging the patient. Time of discharging process occupy the nurse working hours which can then be interpreted into the cost of nurse who perform the discharge process. The other missing performance measure if required can be achieved by the same approach.

References

1. Kumar, A., Ozdamar, L., Ng, C.P.: Procurement performance measurement system in the health care industry. International Journal of Health Care Quality Assurance 18(2), 152–166 (2005)
2. Neely, A., et al.: Performance measurement system design: Should process based approaches be adopted? International Journal of Production Economics 46, 423–431 (1996)
3. Beamon, B.M.: Measuring supply chain performance. International Journal of Operations and Production Management 9(3), 275–292 (1999)
4. da Encarnação Filipe Amado, C.A., Pereira dos Santos, S., Challenges for performance assessment and improvement in primary health care. Health Policy 91(1), 43–56 (2009)
5. Purbey, S., Mukherjee, K., Bhar, C.: Performance measurement system for healthcare processes. International Journal of Productivity and Performance Management 56(3), 241–251 (2007)
6. Abu Bakar, A.H., et al.: Measuring supply chain performance among public hospital laboratories. International Journal of Productivity and Performance Management 59(1), 75–97 (2010)
7. Kumar, A., Ozdamar, L., Zhang, C.N.: Supply chain redesign in the healthcare industry of Singapore. Supply Chain Management: An International Journal 13(2), 95–103 (2008)
8. Aronovich, D., et al.: Measuring Supply Chain Performance: Guide to Key Performance Indicators for Public Health Managers, USAID and T.O. DELIVER PROJECT, Editors, p. 52. U.S. Agency for International Development, Arlington, Va (2010)
9. HSCMW group, H.S.C.M.W.G., Performance Measurement: a report, OntarioBuys, Editor. Treasury Board Office of the Ontario Ministry of Finance, Ontario (2009)

10. Dobrzykowski, D., Vonderembse, M.A.: Healthcare supply chain and IS strategies for improved outcomes. In: POMS 20th Annual Conference, Orlando, Florida, pp. 1–19 (2009)
11. Parnaby, J., Towill, D.R.: Engineering cellular organisation and operation for effective healthcare delivery supply chains. The International Journal of Logistics Management 20(1), 5–29 (2009)
12. Singh, M.: Chronic care driving a fundamental shift in health care supply chains, in MIT Research Report, MIT (2008)
13. Bhattacharjee, P., Ray, P.K.: Patient flow modelling and performance analysis of healthcare delivery processes in hospitals: A review and reflections. Computers & Industrial Engineering (2014)
14. Kujala, J., et al.: Time-based management of patient processes. Journal of Health Organization and Management 20(6), 512–524 (2006)
15. Rahimnia, F., Moghadasian, M.: Supply chain leagility in professional services: how to apply decoupling point concept in healthcare delivery system. Supply Chain Management: An International Journal 15(1), 80–91 (2010)
16. Kosnik, L.: Breakthrough Demand–Capacity Management Strategies to Improve Hospital Flow, Safety, and Satisfaction. In: Hall, R. (ed.) Patient Flow: Reducing Delay in Healthcare Delivery. Springer, New York (2013)
17. Swinehart, K.D., Smith, A.E.: Internal supply chain performance measurement A health care continuous improvement implementation. International Journal of Health Care Quality Assurance 18(7), 533–542 (2005)
18. Narayana, S.A., Kumar Pati, R., Vrat, P.: Managerial research on the pharmaceutical supply chain – A critical review and some insights for future directions. Journal of Purchasing and Supply Management 20(1), 18–40 (2014)
19. Mustaffa, N.H., Potter, A.: Healthcare supply chain management in Malaysia: a case study. Supply Chain Management: An International Journal 14(3), 234–243 (2009)
20. Cai, J., et al.: Improving supply chain performance management: A systematic approach to analyzing iterative KPI accomplishment. Decision Support Systems 46(2), 512–521 (2009)
21. Gunasekaran, A., Patel, C., McGaughey, R.E.: A framework for supply chain performance measurement. International Journal of Production Economics 87(3), 333–347 (2004)
22. Barnabè, F.: A system dynamics-based Balanced Scorecard to support strategic decision making. International Journal of Productivity and Performance Management 60(5), 446–473 (2011)
23. Battini, D., et al.: Healthcare supply chain modelling: a conceptual framework. In: POMS 20th Annual Conference, Orlando, Florida (2009)
24. Berrah, L., Clivillé, V.: Towards an aggregation performance measurement system model in a supply chain context. Computers in Industry 58(7), 709–719 (2007)
25. Shafiee, M., Hosseinzadeh Lotfi, F., Saleh, H.: Supply chain performance evaluation with data envelopment analysis and balanced scorecard approach. Applied Mathematical Modelling (2014)
26. Goedhals-Gerber, L.L.: The Measurement of Supply Chain Efficiency: Theoretical Considerations and Practical Criteria. In: Logistics Management Program. Stellenbosch University, Matieland (2010)
27. Pettersson, A.: Measurements of Efficiency in a Supply Chain, in Logistics Management Program. Luleå University of Technology, Stockholm (2008)
28. Babaie, K.: Promoting patient safety: One company's example. The Case Manager 17(6), 54–59 (2006)

using the verbal and non-verbal parameters [2,4]. Evaluating the worker in a production system is important to increase the efficiency, achieve high quality product and minimizing production lead time with a good sensibility [4].

Ushada *et al.* [4] have proposed a daily worker evaluation model for small medium-scale food production system using Kansei Engineering. The model consist of worker capacity assessment and worker performance evaluation sub-models. Kansei Engineering is an approach to analyze the worker sensibility by using comparison between verbal and non-verbal parameters [2,3]. Kansei Engineering is a method to extract the verbal and non-verbal parameters, to help human identifying the needs, preference or achieving satisfaction for work system/product [5]. Nagamachi [2] stated that Kansei is Japanese word means the human's psycological feeling as well as embracing physiological issues.

Profile of Mood States (POMS) has been applied to extract Total Mood Disturbance (TMD) as the verbal parameters in worker's physiological measurement [6,7,3]. Heart rate has been applied as the non-verbal parameters in worker's physiological measurement [8,4]. In this paper, Kansei's physiological measurement for the worker was evaluated as the response to workplace environmental parameters.

The research objectives are: 1) To evaluate the application of total mood disturbance and heart rate for Kansei's physiological measurement; 2) To analyze the Kansei's physiological measurement in small-medium sized enterprises. The research advantages is to support the development of Kansei Engineering-based Sensor for AgroiNdustry (KESAN) based on a daily worker evaluation model for small-medium sized enterprises. KESAN is expected to assists industrial management to challenge AEC by pursuing the appropriate worker assignment for shift schedulling and environmental set point for the workplace comfortability.

2 Case Studies and Methods

Four case studies of SME food production system in Special Region of Yogyakarta (DIY), Indonesia were evaluated. These SMEs produces indigenous food product of Bakpia, Cracker, Fish Chips and Tempe. Food SME was decided as the research object since its product popularity in Indonesian consumers. Besides, the SMEs support the tourism industry and contribute to valuable support for food sovereignity. The profiles of production system can be described as follows:

2.1 Case Studies

2.1.1 Bakpia's SME

Bakpia is a snack food which is produced from combination of baked flour and mung beans with addition of various indigenous taste. This kind of SME produced handgift product with the local wisdom and contributed to local tourism industry. The production systems consists of 5 (five) work stations as paste making, flouring, paste mixing, filling and baking. In this kind of production system, the Kansei's physiological measurement was possibly influenced by the worker skill. The workers are required to produce the bakpia in the uniformity by combing the skill and experience.

2.1.2 Cracker's SME

Cracker is a complementary food which is made from flour, stamped and fried. This kind of SME produce a chained product which is distributed to various restaurant and local food stall. The production system consists of 8 (Eight) work stations as spice making, basic paste mixing, mixing of cracker paste, squeezing, stamping, boiling, winding and frying. In this kind of production system, the Kansei's physiological measurement was possibly influenced by outdoor environment in drying process.

2.1.3 Fish Chips's SME

Fish chips is a snack and complementary food, made from baby fresh water fish, deep fried as chips. This kind of SME produce a local product with the added value for food sovereignity. The production systems consists of 9 (nine) work stations as cutting, filleting, cleaning, spicing, marinated process, flouring, first frying, second frying and spinning. In this kind of production system, the Kansei's physiological measurement was possibly influenced by quality and sanitation requirement. Requirement of quality and sanitation takes longer processing and cycle time.

2.1.4 Tempe's SME

Tempe is Indonesia popular complementary food which is produced from fermented soybean. The production systems consists of 10 (ten) works stations where 1 (one) worker responsible for 1 (one) work station. The work stations are threshing, washing, soaking, first boiling, crushing, peeling, second soaking, second boiling, fermentation and packaging. In this kind of production system, the Kansei's physiological measurement was possibly influenced by indoor workplace environment. The indoor environment was influenced by bioproduction systems which required the value-added biprocess such as boiling and fermentation.

2.2 Methods

2.2.1 Profile of Mood States

The Total Mood Disturbance (TMD) score were measured using the paper-based questionnaire of Profile of Mood States (POMS) [6]. Profile of Mood States (POMS) was selected as verbal parameter because it can measure the affective mood state fluctuation in a wide variety of populations [6]. POMS are intended to measure five identifiable negative moods with affective states of worker as Tension-Anxiety, Depression, Anger-Hostility, Fatigue, Confussion and two positive moods as Vigor and Friendliness. POMS are measureable using questionnaire of 65 words declaring the worker mood. Questionnaire result is analyzed using Total Mood Disturbance (TMD) score in equation 1 [6] :

$$TMD = TA + D + AH + F + C - V - FR \tag{1}$$

2.2.2 Heart Rate

Heart Rate (HR) was selected as physiological parameter due to its applicability, reliability, and provides good accuracy on manual measurement [8]. HR was measured using wrist blood pressure monitor during various days depend on the permission of SME's management. Heart rate was measured respectively in Bakpia's SME during 8 working days, Cracker's SME during 7 working days, Fish chips's SME during 8 working days and Tempe's SME during 6 working days.

3 Results and Discussions

Table 1 indicated that maximum TMD score of 8.3 confirmed second maximum heart rate of 14.1 in the worker B1. Worker B2 has lower standard deviation than B1 eventough the maximum % deviation is 19.8. It is indicated that the highest sensibility in Bakpia's SME is worker B1 of paste making. The highest sensibility was confirmed by the highest value of B1's workplace relative humidity in Fig.1b, while the temperature is relatively similar to other workplace in Fig. 1a

Table 1. Kansei's physiological measurement in Bakpia SME ($n = 8$)

Worker	TMD		%	Heart Rate		%
	Before	After	Deviation	Before	After	Deviation
B1	6.7 ± 1.0	7.3 ± 0.6	8.3	91.9 ± 5.5	104.9 ± 4.9	14.1
B2	6.7 ± 0.7	6.6 ± 0.2	-1.3	84.5 ± 6.3	101.3 ± 11.3	19.8
B3	6.7 ± 0.8	6.9 ± 0.7	2.1	84.8 ± 7.1	92.6 ± 4.3	9.3
B4	7.0 ± 0.4	6.8 ± 0.6	-2.8	88.6 ± 8.6	94.9 ± 10.0	7.1
B5	6.3 ± 1.3	6.4 ± 1.5	2.1	98.6 ± 13.5	104.0 ± 14.4	5.4
	Maximum		8.3	Maximum		19.8
	Minimum		-2.8	Minimum		5.4

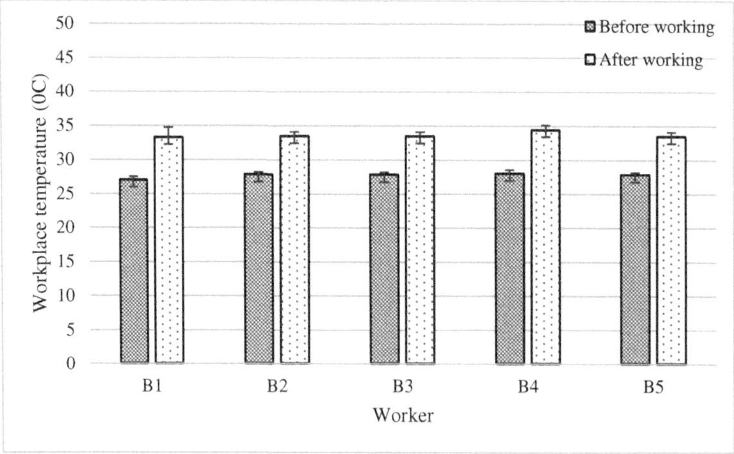

(a)

Fig. 1. Workplace environment in Bakpia's SME: (a) Temperature; (b) Relative humidity

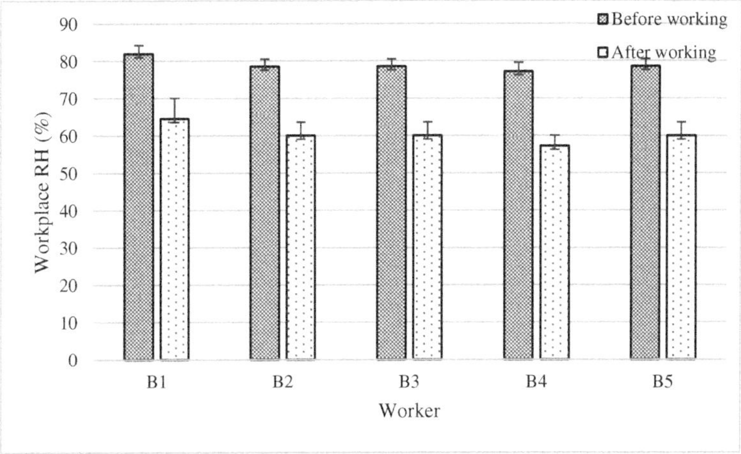

(b)

Fig. 1. (*continued*)

Table 2 indicated the entire positive value of % deviation of TMD score with the maximum value of 9.3 and minimum of 0.8. It can be concluded the mood of worker is very sensible after finishing the work. Table 2 indicated that worker C7 has the highest % deviation TMD score of 9.3 and the lowest % deviation heart rate of 3.5. It can be concluded that the TMD score has the opposite value to heart rate in Cracker's SME. Figs 2a and 2b concluded that the workplace temperature and relative humidity has less impact on sensibility of TMD and heart rate.

Table 3 indicated that Fish Chips's SME has the highest of maximum and minimum value of TMD score compared to others SME in Tables 1, 2 and 4. The maximum TMD score is indicated by worker of FC4 and FC5 and the minimum score by worker of FC1. Besides, fish chips's SME has the highest % deviation of heart rate compared to others SME in Tables 1, 2 and 4. Figs 3a and 3b concluded that the workplace temperature and relative humidity has less impact to sensibility of TMD and heart rate.

Table 2. Kansei's physiological measurement in Cracker SME (n = 7)

Worker	TMD		%	Heart Rate		%
	Before	After	Deviation	Before	After	Deviation
C1	6.2 ± 0.2	6.3 ± 0.2	2.2	78.7 ± 3.5	83.7 ± 7.4	6.4
C2	8.6 ± 1.1	9.6 ± 0.8	6.5	74.1 ± 10.2	77.7 ± 15.7	4.8
C3	6.4 ± 0.4	6.5 ± 0.4	1.5	82.0 ± 7.9	91.4 ± 9.3	11.5
C4	8.2 ± 0.4	8.3 ± 0.3	1.1	86.7 ± 6.6	100.0 ± 10.6	15.3
C5	8.6 ± 1.0	9.0 ± 0.5	4.6	82.6 ± 9.4	89.0 ± 11.2	7.8
C6	8.4 ± 0.5	8.9 ± 1.1	5.4	77.9 ± 6.5	89.6 ± 12.7	15.0
C7	9.0 ± 0.8	9.9 ± 0.6	9.3	86.4 ± 6.4	89.4 ± 6.4	3.5
C8	7.1 ± 0.9	7.2 ± 0.9	0.8	82.4 ± 9.7	95.1 ± 9.9	15.4
C9	8.2 ± 1.1	8.7 ± 0.8	6.9	77.4 ± 10.4	87.6 ± 10.0	13.1
C10	8.4 ± 0.5	8.9 ± 1.1	5.4	77.9 ± 89.6	104.0 ± 14.4	15.0
	Maximum		9.3	Maximum		15.4
	Minimum		0.8	Minimum		3.5

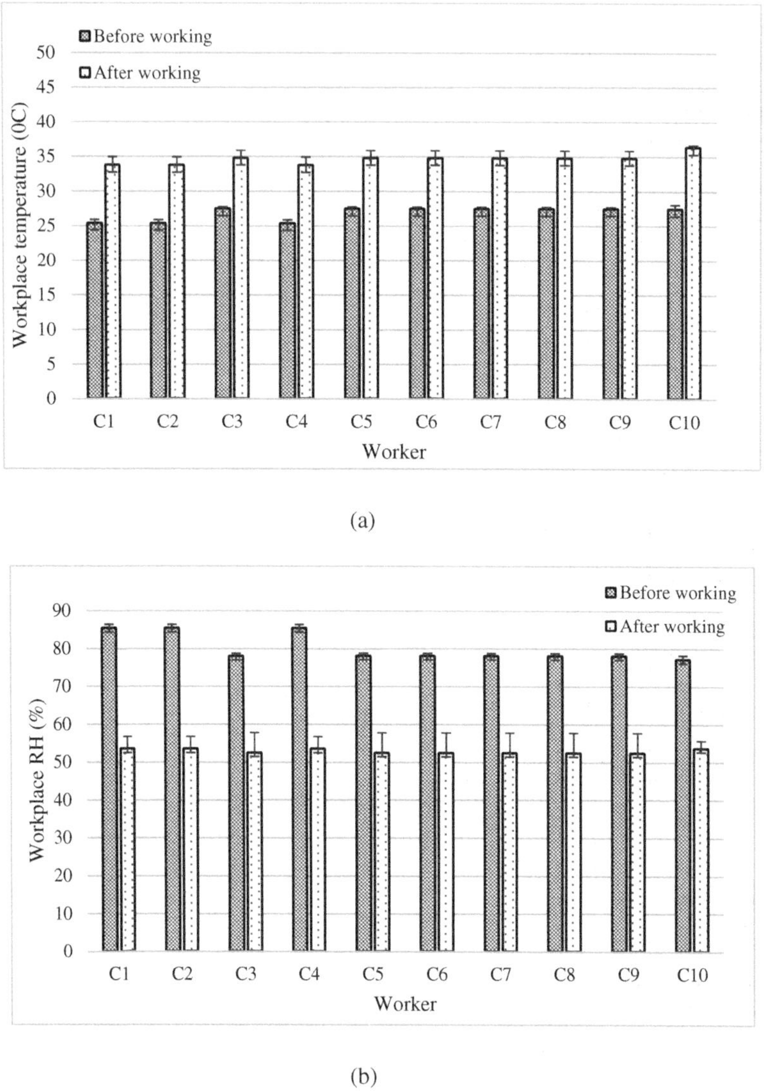

Fig. 2. Workplace environment in Cracker's SME: (a) Temperature; (b) Relative humidity

Table 3. Kansei's physiological measurement in Fish Chips SME (n = 8)

Worker	TMD		%	Heart Rate		%
	Before	After	Deviation	Before	After	Deviation
FC1	8.3 ± 0.5	6.7 ± 0.6	-19.3	85.1 ± 3.6	95.5 ± 4.9	12.2
FC2	7.8 ± 0.3	7.6 ± 0.5	1.8	78.8 ± 4.9	85.1 ± 6.4	8.1
FC3	6.7 ± 0.3	6.6 ± 0.3	-1.6	90.5 ± 4.9	97.6 ± 3.0	7.9
FC4	5.8 ± 0.5	6.5 ± 0.5	10.8	87.9 ± 4.2	93.8 ± 5.8	6.7
FC5	5.8 ± 0.5	6.5 ± 0.5	10.8	87.9 ± 4.2	93.8 ± 5.8	6.7
FC6	7.5 ± 0.6	7.3 ± 0.5	-1.5	82.6 ± 4.5	86.0 ± 12.0	4.1
FC7	8.1 ± 0.2	8.2 ± 0.3	2.2	79.5 ± 8.4	96.8 ± 2.6	21.7
FC8	8.1 ± 0.2	8.2 ± 0.3	2.2	79.5 ± 8.4	96.8 ± 2.6	21.7
FC9	7.5 ± 0.6	7.3 ± 0.5	-1.5	82.6 ± 4.5	86.0 ± 12.0	4.1
	Maximum		10.8	Maximum		21.7
	Minimum		-19.3	Minimum		4.1

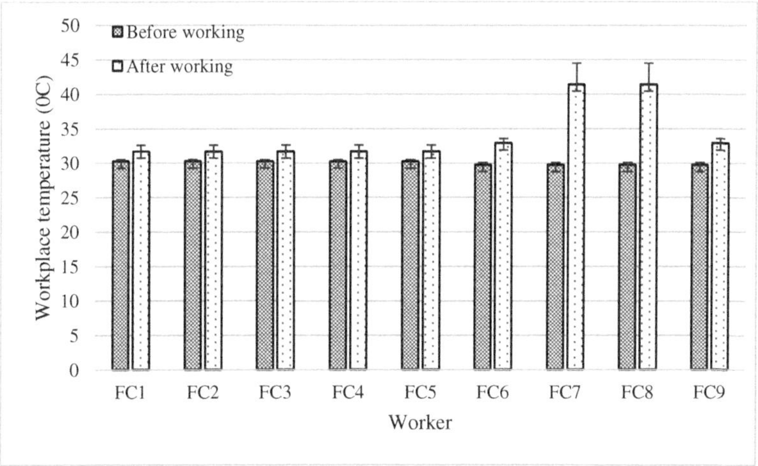

(a)

Fig. 3. Workplace environment in Fish Chips's SME: (a) Temperature; (b) Relative humidity

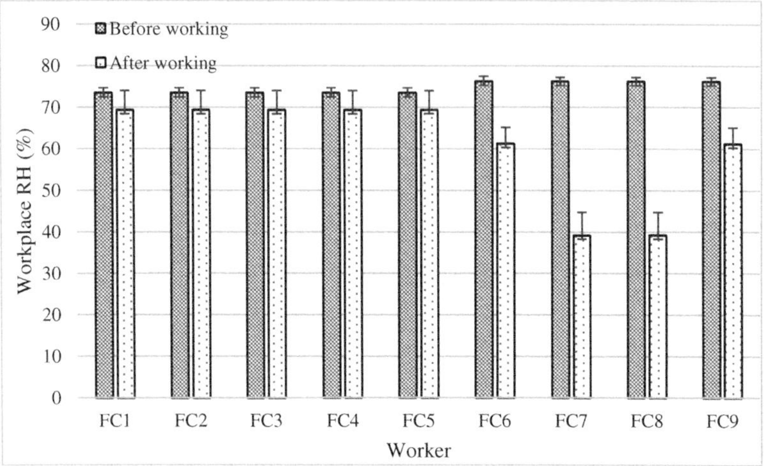

(b)

Fig. 3. (*continued*)

Table 4. Kansei's physiological measurement in Tempe SME (n = 6)

Worker	TMD		%	Heart Rate		%
	Before	After	Deviation	Before	After	Deviation
T1	8.5 ± 1.7	8.3 ± 1.4	-2.2	71.7 ± 1.8	77.7 ± 1.2	8.4
T2	8.2 ± 1.7	7.8 ± 0.8	-4.9	72.2 ± 1.2	81.2 ± 1.0	12.5
T3	7.9 ± 1.4	8.2 ± 1.4	3.1	73.5 ± 0.8	76.7 ± 1.4	4.3
T4	8.5 ± 0.7	8.3 ± 1.0	-1.9	74.2 ± 1.5	75.5 ± 1.2	1.8
T5	8.2 ± 1.2	7.7 ± 1.0	-5.8	77.3 ± 2.1	82.5 ± 1.4	6.7
T6	8.8 ± 0.4	8.7 ± 1.1	-1.0	76.7 ± 1.0	85.2 ± 0.8	11.1
T7	8.1 ± 0.8	8.2 ± 1.4	0.7	84.7 ± 1.0	86.3 ± 0.5	2.0
T8	8.6 ± 1.1	8.4 ± 0.5	-2.5	77.3 ± 1.2	82.5 ± 1.4	6.7
T9	9.0 ± 1.2	8.8 ± 1.9	-2.5	81.5 ± 3.1	84.7 ± 1.5	3.9
T10	7.9 ± 1.1	8.2 ± 1.0	3.7	77.2 ± 1.0	87.3 ± 1.0	13.2
	Maximum		3.7	Maximum		13.2
	Minimum		-5.8	Minimum		1.8

Table 4 indicated that the maximum TMD score of 3.7 confirmed the second maximum heart rate of 13.2 in the worker T10. It is indicated that the highest workload in Bakpia's SME is worker B1 of paste making. Table 4 indicated that Tempe SME has the lowest % deviation of maximum TMD compared to others SME. It is an interesting result since the workplace environmental parameters in Tempe's SME is the highest among others SME as shown in Figs 4a and 4b.

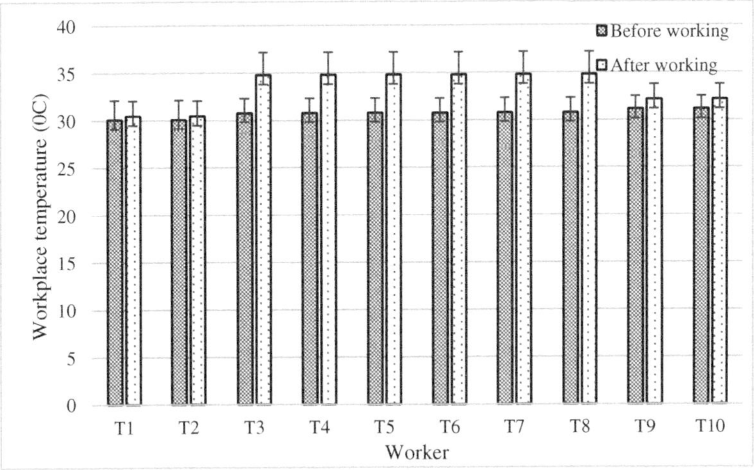

(a)

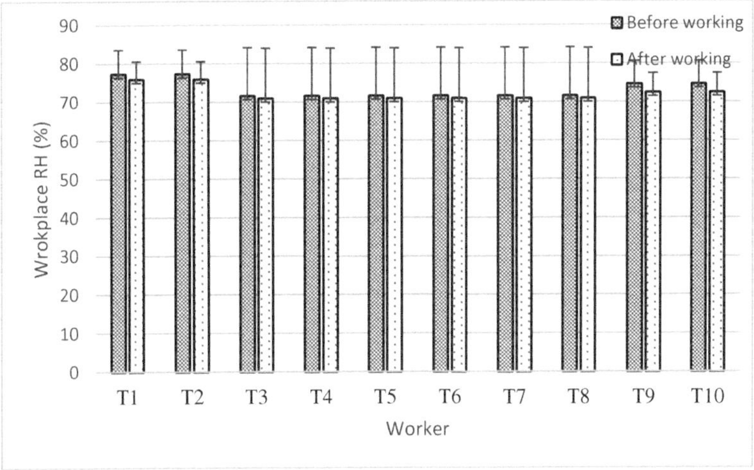

(b)

Fig. 4. Workplace environment in Tempe's SME: (a) Temperature; (b) Relative humidity

4 Conclussions

Kansei's physiological measurement were pursued in production systems of indigenous food industry in Special Region of Yogyakarta (DIY), Indonesia. The measurement was pursued in daily check-in before working and check-out after working. The research results indicated TMD and heart rate were sensible to measure physiological response to workplace environmental parameters. Workplace environment has greater impact to the sensibility of worker mood and heart rate in Bakpia SME and Tempe, while it has less impact in Cracker and Fish Chips industry.

Acknowledgments. We would like to express our sincere gratitute and acknowledge the financial support from Directorate General for Higher Education, Indonesian Ministry of Education and Culture for 2014 (No.: LPPM-UGM/1008/LIT/2014) and 2015 (No.: 310/LPPM/2015) International Collaboration Competitive Research Grant for International Publication-Universitas Gadjah Mada. We would like to express our sincere gratitute to industries of Bakpia, Crackers, Fish Chips and Tempe. Finally, we would like to acknowledge the support for data acquisition from Arum Baashitu, Reza Purnama Triendy, Dwita Rahmawati and Dzikri Rahadian Fudholi.

References

1. Anonym, ASEAN Economic Community Blueprint. ASEAN Secretariat, Jakarta (2008) ISBN 978-979-3496-77-1
2. Nagamachi, M.: Kansei Engineering as A Powerful Consumer-oriented Technology Development. App. Ergon. 33, 289–294 (2002)
3. Ushada, M., Mulyati, G.T., Guritno, A.D., Murase, H.: Combining Drum-Buffer-Rope algorithm and Kansei Engineering to Control Capacity Constrained Worker in A Bioproduction System. In: 2013 International Federation of Automatic Control (IFAC) Biorobotics Conference, Japan Paper No. 11 (2013)
4. Ushada, M., Okayama, T., Suyantohadi, A., Khuriyati, A., Murase, H.: Daily Worker Evaluation Model for Small Medium-scale Food Production System Using Kansei Engineering and Artificial Neural Network. Agriculture and Agricultural Science Procedia 3, 84–88 (2015)
5. Nagamachi, M.: Kansei Engineering: A New Ergonomic Consumer-oriented Technology for Product Development. Int. J. of Indust. Ergon. 15, 3–11 (1995)
6. McNair, D.M., Lorr, M., Droppleman, L.F.: Manual for the Profile of Mood States (Rev. edn.). Educational and Industrial Testing Service, San Diego (1992)
7. Ujita, Y., Murase, H.: Profile Of Mood States Software for Application of Baum Test Characteristic. In: Proceedings of the 3rd International Symposium on Machinery and Mechatronics For Agriculture and Biosystems Engineering (ISMAB), Seoul, Korea, November 23-25 (2006)
8. Smolander, J., Juuti, T., Kinnunen, M., Laine, K., Louhevaara, V., Mannikko, K., Rusko, H.: A New Heart Rate Variability-based Method for The Estimation of Oxygen Consumption without Individual Laboratory Callibration: Application Example on Postal Workers. App. Ergon. 39(3), 325–331 (2008)

Investigation of Customer and Technical Requirements for Designing an Ergonomics Notebook Soft Case Using Quality Function Deployment (QFD) Approach

Hilma Raimona Zadry* and Defri Arif Irfansyah

Department of Industrial Engineering, Faculty of Engineering, Universitas Andalas
Kampus Unand Limau Manis, Padang, West Sumatera, Indonesia
hilma@ft.unand.ac.id, defryirfans@yahoo.co.id

Abstract. Notebook experienced rapid growth following the human needs. This is evidenced by the increase of notebook sales every year. The problem of shortage and mismatch facilities and supporting infrastructure causes of notebook users force their capacity to use the notebook with non ergonomic position. A preliminary survey through interviews found that the notebook soft case not yet provide ease of mobilization, convenience of use, and protection of health to the users. Therefore it is required an alternatives design of notebook soft case which is ergonomics and multifunction. One of the methods used to design the notebook soft case is Quality Function Deployment (QFD). In this study, QFD is implemented to determine the customer and technical requirements. The results indicate that the material selection which has impact resistant and waterproof is very important to design an ergonomics and multifunctional notebook soft case.

1 Introduction

Since the mid-1970s, the development of technology, especially computer technology is increasing very rapidly along with the times. Advances in computer technology produce a very rapid development of the computer and its use is widespread [1]. Computers then evolved toward smaller size, from desktop into a notebook [2]. The use of notebook today is very rapidly compared to desktop. People prefer notebooks because its more flexible and fashionable [3].

The increasing quantity of laptop users is not comparable with the availability of facilities and infrastructure support in the public space. One of them is to use notebook on the lap in a long time. It can cause pain in the muscles, back, waist, neck, thighs, eye disorders and impaired concentration [4]. It also can cause skin damage because of the radiation [5]. Another thing to be a priority in the notebook uses is mobilization and protection. Notebooks require protection from external threats that can cause physical defects or damage to the notebook system. Based on the above reasons, it gives birth to a notebook soft case. Nowadays, notebook soft case becomes a

* Corresponding author.

M. Gen et al. (eds.), *Industrial Engineering, Management Science and Applications 2015*,
Lecture Notes in Electrical Engineering 349, DOI: 10.1007/978-3-662-47200-2_99

trend among notebook users. Soft case also takes on the role as a notebook bag that is simple, practical, lightweight, and easy to use fashionable [3].

Preliminary survey through interviews found that the notebook soft case still not provide ease of mobilization, protection, comfort and health assuredness for users when using a notebook. In addition, the existing notebook soft case cannot cover the lack of facilities and infrastructure availability to support the use of a notebook in a public space. Based on the above information, it can be concluded that the notebook soft case on the market today are still not satisfy the users. Therefore this study investigates the customer and technical requirements related to the notebook soft case which is the preliminary stages for designing ergonomics and multifunctional notebook soft case.

With regard to the design methods for new product development, Quality Function Deployment (QFD) is an important methodological approach to increase customer satisfaction and reduce the product costs and development cycle time. It is also an ultimate tool to increased time and resources saving throughout all stages – design to production planning [6]. QFD has been successfully applied by industries around the world [7-9]. Therefore, this study used QFD method to investigate the customer and technical requirements for designing ergonomics and multifunctional notebook soft case.

2 Methodology

2.1 Participants

There were 115 university students from various universities in Padang City; West Sumatera participated in the study. Forty participants were involved in the survey through interview, while 75 participants were involved in answering the survey questionnaires. The criteria of the participants are those who usually use notebook and the soft case in everyday activities. The participants were randomly selected by personal visit and they were full consent to participate. Data was collected from November to December 2014.

2.2 Data Collection

Determining the user needs is probably the most important step in the product planning process. Information about the customer needs and their priorities can be gained through various marketing methods, e.g. questionnaires, interviews, and brainstorming techniques [10]. This study collects data using survey methods, through interview and questionnaire. The interviews were conducted to know in general how customers respond to the presence of notebook soft case on the market today as well as the characteristics of the customers towards the desired notebook soft case in the future. The interview is a preliminary stage that serves as a reference in designing the research questionnaire. It is also a method to get an initial picture of consumer expectations for designing an ergonomics notebook soft case.

The survey questionnaire was developed based on the data from the initial interviews. It also refers to the dimensions of quality according to David A. Garvin [11]. The criteria used are performance, features, durability, conformance to standards, aesthetics as well as the price. The questionnaire aims to investigate the customer needs and desires related to the notebook soft case. It covers a range of topics including personal backgrounds and customer requirements to the design of notebook soft case. Scale used for the design of the questionnaire is a Likert rating scale with a range of 1-5 for the assessment is not important (1) to very important (5).

2.3 Data Analysis

The collected survey questionnaires were tested for validation and reliability. Then, the collected data were processed using QFD design through House of Quality (HoQ). HoQ consists of several activities supported by various tables and matrices. The basic idea is to translate customers' requirements into the appropriate technical requirements for each stage of product development and production. The procedures of QFD design in this study are divided into the following steps:

a. Determine the customer requirements and the customer important ratings
b. Determine the technical characteristics
c. Translating customer requirements into measurable technical requirements
d. Determine the relationship of each technical characteristics
e. Determine the priority of the technical characteristics

3 Results

3.1 Interview and Questionnaire

The process of identification of customer needs is the first step in designing a product. The customer requirements were obtained from the initial interview survey is presented in Table 1.

Table 1. List of Customer Requirements for Notebook Soft case derived from Interviews

No	Customer Requirements	Total
1	Safe and protected	26
2	Attractive design	7
3	Thin design	2
4	Waterproof	22
5	Lightweight	5
6	Availability of storage for goods	5
7	Availability of portable desk	3
8	Availability of external fan	1
9	Multifunction	4
10	Adjustable and multifunction straps	10
11	Thick foam on the straps	4
12	Affordable prices	1
13	Comfortable when used and carried	7

Based on those needs obtained from interview, a questionnaire was developed to gain reliable data about customer requirements with respect to customer satisfaction. It also refers to the five dimensions of Garvin quality namely performance, features, durability, conformance, aesthetics, and price. The lists of criteria related to the customer requirements in the questionnaire were presented in Table 2.

Table 2. The lists of criteria related to the customer requirements in the questionnaire

Quality Dimension	Criteria
Performance	Comfortable when used
	Comfortable when carried
	Safe when used
Features	Availability of back strap
	Availability of tote strap
	Adjustable straps
	Thick foam on the straps
	Availability of storage for goods
	Availability of raincoat
	Availability of external fan
	Availability of tilt adjusting
	Availability of portable desk
	Availability of wheels
	The closed soft case is semi-automatic
Durability	Strong material
	Material safe for health
	Material protects from impact
	Lightweight material
	Waterproof material
Conformance	In accordance with the notebook size
Aesthetics	Simple and minimalist
	Thin
	Colorful
Price	In accordance with the purchasing power
	In accordance with the quality offered
	In accordance with the perceived benefits

3.2 Quality Function Deployment (QFD)

The identification of the customer requirements was conducted by distributing questionnaires to 75 participants. The answers were weighted in order to obtain the measured value. Measurement scale used was a Likert scale with the scale from 1 (not important) to 5 (very important). The collected questionnaire data were tested for validity, reliability and adequacy. The results show that all data obtained were valid, reliable and sufficient for the next analysis. Table 3 shows the customer requirements and customer important ratings for designing multifunctional notebook soft case.

Table 3. Customer Important Ratings

No.	Customer Requirements	Customer Important Ratings
1	Material protects from impact	4.73
2	Waterproof material	4.64
3	In accordance with the purchasing power	4.61
4	Safe when used	4.56
5	Lightweight material	4.56
6	Comfortable when carried	4.52
7	Material safe for health	4.45
8	In accordance with the quality offered	4.45
9	Comfortable when used	4.43
10	In accordance with the perceived benefits	4.40
11	In accordance with the notebook size	4.35
12	Strong material	4.28
13	Availability of storage for goods	4.24
14	Adjustable straps	4.17
15	Availability of raincoat	4.11
16	Thick foam on the straps	4.03
17	Simple and minimalist	3.99
18	Availability of tote strap	3.92
19	Colorful	3.59
20	Availability of external fan	3.31
21	Availability of tilt adjusting	3.31
22	Thin	3.25
23	Availability of back strap	3.24
24	Availability of portable desk	3.11
25	The closed soft case is semi-automatic	2.76
26	Availability of wheels	2.04

The results from the questionnaire survey in Table 3 revealed information on the average weighting of ergonomic and multifunction demands of notebook soft case. The table shows that of the 26 items of questions posed to the participants, 16 items are considered very important by participants in a notebook soft case. According to the customers, material that can protect the notebook from impacts is the most important need to be met when designing the notebook soft case, followed by waterproof material, in accordance with the purchasing power, etc. This information was used directly in the QFD analysis as a basis for translating the customer requirements into technical characteristics expressed in technical terms.

Technical characteristics used to represent the voice of developers. Determination of the technical characteristics was done by consultation and discussion with experts in the respective products related to material and manufacturing. The technical characteristics used in the QFD analysis were ergonomics design, dimension, material strength, material type and quality, color variations, external fan design, notebook desk design, and strap design.

The next step is determination of the relationship between the different technical characteristics. It is determined, in order to identify particularly the inverse relationships. An inverse relationship between different product characteristics means that improving one property leads to a deterioration of the other. A strong positive relationship is illustrated with a '√√', a positive with a '√', no relationship with 'blank', a strong negative with a 'xx' and a negative with a 'x' [12]. By comparing different technical characteristics, their relationships are identified. For example, the relationship between the material strength with the material type and quality is a strong positive relationship. It means the better the material type and quality, the better the material strength. Relationship between the different technical characteristics is presented in Figure 1.

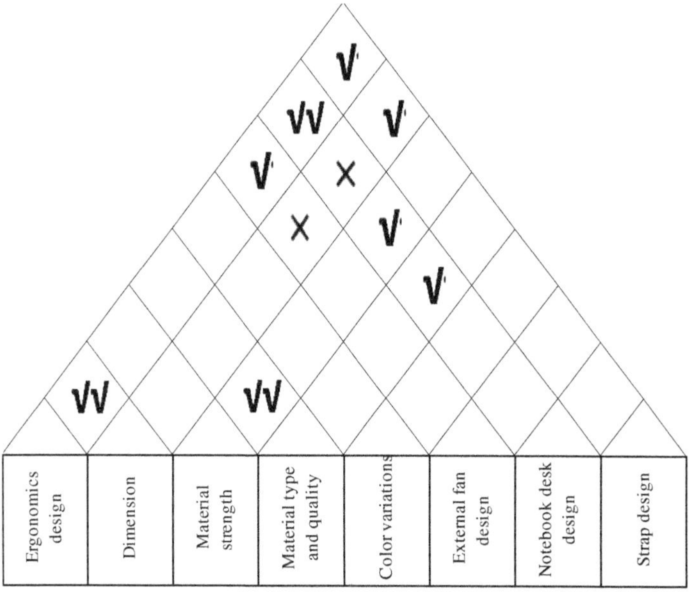

Fig. 1. Relationships between Different Technical Characteristics

The relationships between consumer requirements and technical characteristics have to be established in order to identify important product properties. This relationship is made in the form of a matrix by using the symbols according to Cohen 1995. Based on discussion with experts from material field and manufacturing, the team assigns a 'strong', 'medium', 'weak' or 'no' relationship value to each specific pairing. For this an appropriate scale (9–3–1) is applied, illustrated using symbols (•-o-Δ). This stage is one of the key elements of the QFD method because it is the one which permits the transition to be made between what the user wants and what the designer offers [13]. Figure 2 shows the relationship between customer requirements and technical characteristics for the design of notebook soft case.

An example of the relationship between the customer requirement and the technical characteristics is waterproof material with the material type and quality which has a strong relationship. That is, the better the selection of the material type and quality, the better the durability of product to the water. After determining the relationship between customer requirements and technical characteristics, the next step is to determine what technical characteristics are influencing the customer satisfaction. It shows which technical characteristics will be a priority in product development. Priority value is determined by multiplying the customer important ratings and numerical weighing of the relationship, then summing these together, each product characteristic is given an overall weighting.

	Technical Characteristics	Ergonomics design	Dimension	Material strength	Material type and quality	Colour variations	External fan design	Notebook desk design	Strap design
Customer Requirements	**Customer Important Ratings**								
Comfortable when used	4.43	•					•	•	
Comfortable when carried	4.52	•	•			o			o
Safe when used	4.56			•	•		o	•	
Availability of back strap	3.24					Δ			Δ
Availability of tote strap	3.92					Δ			o
Adjustable straps	4.17	•	o						o
Thick foam on the straps	4.03								Δ
Availability of storage for goods	4.24	o							
Availability of raincoat	4.11	Δ							
Availability of external fan	3.31	o	o				o		
Availability of tilt adjusting	3.31	o	o					Δ	
Availability of portable desk	3.11	o	o	o		Δ		o	
Availability of wheels	2.04	Δ							
The closed soft case is semi-automatic	2.76	Δ							
Strong material	4.28			•					
Material safe for health	4.45				•				
Material protects from impact	4.73			•	•		o	o	
Lightweight material	4.56				o		o	•	
Waterproof material	4.64			•	•				
In accordance with the notebook size	4.35		•				o	o	
Simple and minimalist	3.99		o			Δ			
Thin	3.25		o					o	
Colorful	3.59					o			
In accordance with the purchasing power	4.61			•	o	Δ			
In accordance with the quality offered	4.45				•				
In accordance with the perceived benefits	4.40				o		o	o	
Priority Values		**168.90**	**143.20**	**214.80**	**246.30**	**43.20**	**117.60**	**184.70**	**45.10**
% Priority		**14.51%**	**12.30%**	**18.46%**	**21.16%**	**3.71%**	**10.10%**	**15.87%**	**3.88%**

Fig. 2. Relationships between Customer Requirements and Technical Characteristics

4 Discussion

The survey shows that of the 26 items of questions posed to the respondents, 16 items are considered very important by respondents in a notebook soft case. Materials that can protect the notebook from impact become the highest priority of the customer requirements with a value of 4.73. It is because it relates directly to the notebook security from external threats. Proper material selection is certainly considered to protect the notebook from a crash that could result in damage and defects on the notebook. Notebook falls from height are things that might happen, thus, it is needed a good material to protect notebook from impact. Those kinds of materials commonly used to protect the notebook from impact are foam. Foam is a material that can dampen vibration on impact. The thicker the foam wind used, the better its function in protecting the notebook from impact.

Waterproof material are the second highest of the customer important ratings with a value of 4.64. Notebooks are vulnerable device be damaged if exposed to water. Assuredness notebook security in the mobilization is very important for customers. Selection of waterproof material is the perfect solution to maintain security of notebook from damage in unfavorable conditions such as rain. Waterproof material selection will surely make customers more comfortable and not have to worry to carry a notebook in the current rainy conditions. These materials such as cloth that is impermeable to water. It is a combination of the polyester and nylon.

Price conformity with the purchasing power was in the third place of customer important ratings with a value of 4.61. The design of notebook soft case was adjusted to the purchasing power of the people. Therefore, in the choice of materials and components of the design, it is considered to use affordable and qualified materials according to the specifications of customer requirements. Availability of wheels on the notebook soft case is in the last place of customer important ratings with a value of 2.04. Giving the wheels on the notebook soft case adversely affect the security of notebook such as knocks and falls. Notebook position should be in a fixed position during use.

Quality Function Deployment method aims to design more targeted products in accordance with customer desires. QFD translates customer requirements into the form of product attributes to meet the customer requirements. This study use QFD until HOQ phase 1. The next phase until product design will be done at the future research.

5 Conclusion

This research has used the Quality Function Deployment (QFD) method to investigate the customer and technical requirements in designing the ergonomics and multifunctional notebook soft case. The results indicate that the material selection which has impact resistant and waterproof is very important to design an ergonomics and multifunctional notebook soft case. For the future research, TRIZ method will be used to solve the contradictions between technical requirements continued with the development of QFD phase 2 to determine the design characteristics. Thus, the ergonomics and multifunctional notebook soft case will be designed in accordance with the conditions and the latest technology.

References

1. Zulkarnaian, A.: Peranan Laptop Support Dalam Mengurangi Kelelahan pada Pengguna Laptop. INSAN 14(2), 97–107 (2012)
2. Sudirman, I., Wahono, R.S.: Sejarah Komputer. Kuliah Pengantar Ilmu (2003), http://Komputer.com

3. Hendra, Octaviani, D.F.: Keluhan Kesehatan Akibat Penggunaan Laptop pada Mahasiswa FKM UI. Jurnal Universitas Indonesia, 1–19 (2012)
4. Thandung, D., Lintong, F., Supit, W.: Tingkat Radiasi Elektromagnetik Be-berapa Laptop dan Pengaruhnya Terhadap Keluhan Kesehatan. Jurnal e-Biomedik 1(2), 1–7 (2013)
5. Saptriyawati. Dampak Negatif Memangku Laptop, `https://saptriyawati.wordpress.com/2011/04/16/dampak-negatif-memangku-laptop/`
6. Shil, N.C., Ali, M.A., Paiker, N.R.: Robust Customer Satisfaction Model using QFD'. International Journal of Productivity and Quality Management 6(1), 112–136 (2010)
7. Bergman, B.K.B.: Quality from customer needs to customer satisfaction. McGraw-Hill, London (1994)
8. Geuma, Y.J., Kwak, R., Yongtae Park, Y.T.: Modularizing services: A modified HoQ approach. Computers & Industrial Engineering 62, 579–590 (2012)
9. Vezzetti, E., Moos, S., Kretli, S.: A product lifecycle management method-ology for supporting knowledge reuse in the consumer packed goods domain. Computer-Aided Design 43, 1902–1911 (2011)
10. Bergquist, K., Abeysekera, J.: Quality Function Deployment (QFD) - A means for developing usable products. Int. J. Ind. Ergon. 18, 269–275 (1996)
11. Garvin, D.A.: Managing quality: The strategic and competitive edge. Simon and Schuster (1988)
12. Cohen, L.: Quality Function Deployment: How to Make QFD Work for You. Addison-Wesley Publishing Company, Massachusetts (1995)
13. Marsot, J.: QFD: a methodological tool for integration of ergonomics at the de-sign stage. Appl. Ergon. 36(2), 185–192 (2005)

Demographic Characteristics in Correlation with Household Electricity Use

Lusi Susanti*, Prima Fithri, and Karin Bestarina

Department of Industrial Engineering, Andalas University, Padang, West Sumatra, Indonesia
{susantilusi,pfithri28,karinbestarina}@gmail.com

Abstract. Household has been identified by researchers as an important target group of energy conservation. User behavior in utilizing electricity becomes significant. Energy consumption in the residential sector offers an important opportunity for conserving resources. This study attempts to identify the role of the economic behaviour of resident households. In this paper, we analyse the extent to which the use of electricity is determined by the technical specifications of the dwelling as compared to the demographic characteristics of the residents. Analysis is based on a sample of 210 Padang homes and their occupants. Results show that electricity consumption varies more directly with household composition. The results provide a scientific evidence for stakeholders on the potential of controlling electricity consumption and designing energy policy in residential sector

Keywords: Electricity, Consumption, Energy Conservation, Household, Demography.

1 Introduction

INDONESIA currently is one of the fastest growing developing countries in Asia. Indonesian economic development encourages the improvement in all sectors such as industrial and commercial sector as well as the people's welfare. During the years 2001 to 2009 the GDP per capita continues to increase sharply, which the average was as much as 15% annually. In 2001 the GDP per capita amounted to US$ 748, and in 2009 became US$ 2,698 (Statistics Indonesia, 2010; CDIEMR, 2010). These situations are predicted to keep continued over coming years due to revival of the world economy.

However, with its rapid economic growth, the disparity between the energy consumption and production are becoming increasingly greater in recent years. Indonesia final energy consumption in the period 2000 to 2010 has jumped almost two-time, from 777.9 million BOE to be 1182.1 million BOE (BPPT, 2012). Indonesia energy consumption are predicted to increase steadily with the growth rate of 5% per year until 2030 if no energy saving efforts or implementation of policies related to energy conservation and energy efficiency are undertaken, or in other words still apply to business as usual (BAU). To avoid energy crisis and at the same time to maintain

* Corresponding author.

M. Gen et al. (eds.), *Industrial Engineering, Management Science and Applications 2015*,
Lecture Notes in Electrical Engineering 349, DOI: 10.1007/978-3-662-47200-2_100

energy sustainability, Indonesia government has set a long-term target to reduce energy intensity of at least 1% per year until 2025 (BPPT, 2012). In accordance with the national energy policy targets, it means the total energy consumption in 2025 has to decrease up to 50% with energy conservation scenario. It is therefore urgent for Indonesia to improve the efficiency of energy use and encourage energy conservation.

The electricity placed as the highest ranks fourth of energy consumed after fuels, biomass energy, and coal. Although the use of electricity continues to rise, according to the data from the Directorate General of Electricity Ministry of Energy and Mineral Resources in 2013, Indonesia's electrification ratio is still around 80.1%. Indonesian Government thrives for the electrification ratio reached until 100% in 2020 through a national energy policy. This target would have to be balanced with the ability to supply electricity to the customers by the Power Users of Government Services, here in after referred to PLN.

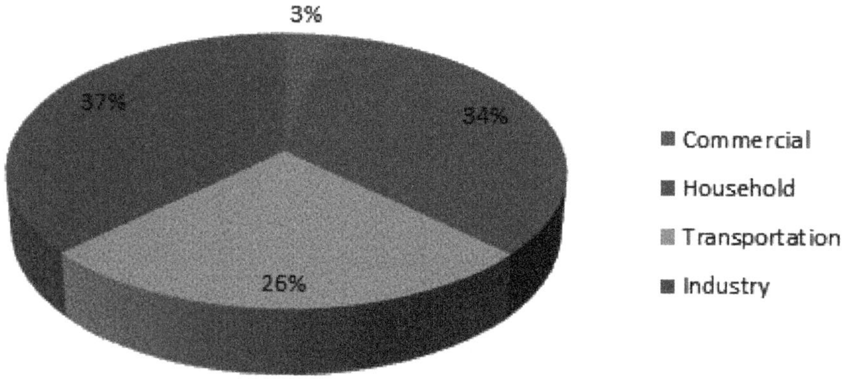

Fig. 1. Indonesia energy share by sector

Household has been identified by researchers as an important target group of energy conservation. To illustrate, households account for 25% of the total energy consumption in the US, 26% in Japan, 50% in Saudi Arabia (Saidur et. Al, 2007). In Indonesia, household is responsible for 34% of total energy consumption, as seen in Fig. 1 (Statistics Indonesia, 2010). Electricity still dominated the energy type used in residential sector. Research on energy-saving behavior of households in urban areas has been done before. The potential for energy saving in household electricity can be conducted through three approaches: the efficiency of household appliances, the energy saving campaign, and the government policies (Guerra-Santin and Visscher, 2009). In line with these studies, Sukarno (2013) explained that the lifestyle and the household appliances become the main factor of electrical energy consumption in the household sector. Wang et al., 2011; Li et al., 2009; Ouyang and Hokao, 2009; Zhou et al., 2008; Xiaohua and Zhemnin, 2005 also have surveys about household energy consumption in Asia, which there are several factors that affect households energy-saving behaviors.

Padang city with the population of 765,450 and the wide area of 694.96 km^2, being one of the cities with a rapid growth in infrastructure and business. The development of infrastructure and facilities are done continuously which followed by the increase in electricity consumption. Febrizal (2012), has been conducting research to predict a large escalation in the consumption of electrical energy in Padang city in 2012-2020.

In this paper, the author initiated an investigation on urban energy consumption in Padang city, West Sumatra Province, Indonesia. Previous investigation found that electricity consumption in households sector in Padang was responsible for 92% of total consumption in Padang compared to other sectors (Sukarno et.al, 2013). This trend seems to increase in the future due to increase of population and improvement of living standard in Padang. It is therefore, very urgent and rational to relieve the pressure of electricity shortage in household sector by further encouraging the occupants' behavior toward energy saving.

2 Research Methods

Understanding the key determinants of residential energy consumption is important for the design and implementation of effective policies to reduce energy consumption of the residential building sector. A set of structured interview protocol was prepared to analyse detailed micro-economic data on energy consumption, occupant characteristics and dwelling specifications for 210 households; all of them were the PLN's customers in Padang. We focus on homeowners, as we lack critical information on rental units, and the type of rental contract -- net or gross -- which may have a substantial influence on energy consumption. Collecting the data used door-to-door survey method. The investigation focused on electricity consumption. Electricity bills (in Rupiah) and demographic characteristics of each individual household, such as age of the household head, number of persons in the household, family composition, ethnicity and (for a subset of households) annual income of the home samples were collected. We compare the importance of these demographic characteristics with the energy consumption pattern of households Padang. Details of demographic characteristics of the samples are displayed in

Table 1. Demographic characteristics of respondents

No	Characteristics
1	Gender
2	Age
3	Housing area
4	Occupation
5	Number of persons in the household
6	Education level of family member
7	Monthly Income
8	Payment method of electricity bills
9	Power level
10	Type and number of home appliances used

According research conducted by Bachus, et al. (2011), energy consumption pattern in residential sector are mainly triggered by following factors:

Attitude
Most of the literature that discusses the energy savings in households suggests that there is a relation between attitudes and behavior. The influence of behavior is mostly on micro-personal level. Individual attitudes in using the energy are triggered by socio-economic factors. In this study, the attitudes which is investigated is related to the action in the peak load hours, the action of discharge the electrical equipment when not in use, and the action to encourage energy saving habits to the family members.

Household Income
It is believed that the most dominant factor of energy consumption in residential sector is household income. The theory of the Kuznets relationship (Figure 3) is pretty explained the correlation between household income and energy consumption.

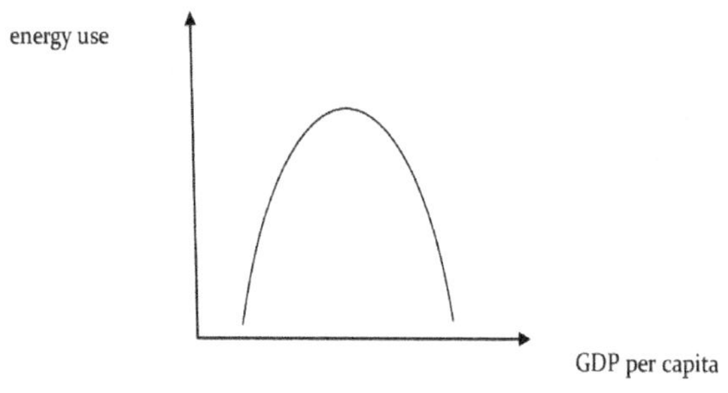

Fig. 2. Kuznets Curve

Prices of Energy
These factors may affect the patterns of electricity consumption in household followed by the number of family members, age, socio-economic situation of the family members, the number and types of household appliances. This research will be investigated the response of the family member to the household electricity tariffs and the actions taken when electricity rates fluctuate.

Knowledge
Most of the studies in energy use assume that public are aware of the amount and the type of energy used. In fact, most of the households do not really know what kind of energy they pay when using a number of tools to support their daily activities, due to the lack of information received. Basically, people need information in order to have knowledge about saving energy. Consistent information through social interaction within the community is able to trigger an effective energy-saving behavior. So that, it is important to study further the knowledge of households in using electricity.

Other Determinants
There are many other factors that influence energy-saving behavior of households, one of those factors is the size and design of the house. The design of the house is related to the energy enter and out of building, ventilation, lighting systems, as well as room condition.

Statistical analysis using Cross tabulation analysis is used to examine whether there is any correlation between two variables. In this study, cross tabulation analysis is intended to provide an understanding of the relationship between the demographic characteristics of households with electrical energy consumption patterns which decomposes on five indicators.

3 Results and Discussion

3.1 Description of Respondents

Characteristics of Respondents
Data of demography shows that the majority of respondents are female with a percentage of 63% and 37% men. Most of the respondents are people aged >50 years (40%), aged 41-50 years (40%), aged 31-40 years (15%), aged 21-30 years (3%), and aged <21 years (2%). Respondents are in various kinds of jobs such as, 23% are government/private employees, 21% are housewives, 19% are self-employee, 15% are traders, 10% are teachers, 6% are retired, 2% are laborers, 1% are farmers, and 2% are others.

Based on the education level of family members, 1% graduates doctor degree, 10% graduates master degree, 59% graduates bachelor/undergraduate degree, 24% graduates high school, 5% graduates junior high school graduates, and 1% graduates primary school. Based on income per month (in Indonesian Rupiah, 1 USD ≈ 12,000 IDR), 7% has income >10 million, 15% has income 5-10 million, 38% has income3-5 million, 24% has income 1.5-3 million, and 16% has income<1.5 million.

Based on the method of electricity payment, 89% using post-paid method and the remaining 11% have used a prepaid method. Based on the classification of the installed power, 47% registered as a customer with 900 VA power, 27% power 1300 VA, 2200 VA power12%, 12% and 450VApower, and as much as 2% was recorded using the power > 2200 VA.

Possession of Home Appliances
The use of electronic equipment can provide information on the implications of household electricity consumption. In addition, the type and amount of electronic equipment used reflects the attitude and behavior patterns of the households. It can be targeted to potential electrical energy savings. Based on the interview, it is known that most of electronic equipment used by households is used for cooking purposes. Each of the household has at least one unit of rice cooker, water dispenser, refrigerator, juice blender. The need for air conditioning equipment is quite large, considering the city of

Padang is in the tropical region crossed by the equator. A total of 61 among 210 homes have both AC and electric fan.

For lighting, 63 households still do not use energy-saving lamps. The majority argues that the price for energy saving lamps is too high while it does not necessarily have significant benefits in reducing electricity bills. Almost all of the households have supporting machines in doing their activities such as iron, machine washing. Television became the primary source of entertainment for the respondents, the total number of television households in the sample that is 280 units. For example, one of the respondents has at least 1 to 5 television units. Most respondents have started to use laptop than computer (see Table 2).

Table 2. Number of Electric Appliances Owned By Households

Appliance	Number of household based on the appliance possession			
	1 items	2 items	3 items	>3 items
Rice cooker	186	16	0	0
Water dispenser	153	2	0	0
Refrigerator	186	13	0	0
Juice blender	144	4	0	0
Electric stove	8	0	0	0
Handy mixer	87	0	0	0
Microwave	24	0	0	0
Coffee maker	6	0	0	0
Toaster	32	0	0	0
Food processor	2	0	0	0
Air conditioner	38	12	0	0
Electric fan	93	59	27	11
Television	141	41	16	2
Personal Computer	54	5	0	0
Laptop	68	53	15	7
DVD/CD Player	90	2	0	0
Video game	31	1	0	2
Handphone	26	33	33	85
Washing machine	142	5	0	0
Iron	191	9	0	0
Water pumps	106	3	0	0
Har dryer	14	1	0	0
Vacuum cleaner	27	1	0	0
CFL	61	20	7	59
Incandescent bulb	11	24	9	51

Households Characteristics and Electricity Use

Attitude

Household characteristics that influence the attitude of energy saving in the household are the income, the number of family members, and the building area of the house. The area of the house and the family income influence significantly. When the peak load, 39% of households turning off two or more lights while 11% of non-lethal electrical equipment while power is still adequate dominated by households with middle to upper incomes. The action of turn off the electronic equipment during the peak load was motivated by the hope of cheaper electricity bills. The results showed that most of the respondents (98%) suggested energy-saving habits to their family members.

Electricity Price

The household response to the energy price is influenced by the head of household's occupation, the level of education, and the income. The level of education and household income has significant influence. The households with the highest education level of family members is in high school, but has arrange of 3-5 million income considered that the electricity rates is cheap. The households with level of education is undergraduate degree with 1-3 million rupiah income per month also found the electricity rates is cheap. 51% of the households choose to reduce power consumption when the electricity rates increase. However, only 11% of households who choose to replace equipment with more energy efficient.

Household Income

Gender, housing size, jobs, education level, household income, electricity bill payment method, and customer class influence household spending on electricity and electronic equipment.20% of respondents said replacing electronic equipment with more energy efficient when getting more income. From the survey15% of households know not to use energy saving lamps (CFL). The level of education greatly affects the choice. Average household that does not use energy-saving light is the highest level of junior and senior high school education. They argued that the prices of the energy-saving lamps are more expensive.

Knowledge

Household knowledge about energy saving attitude is influenced by gender, age, house size, number of family members, income, utility bills payment method, and customer class. Age very significant influence on the probability of respondents informed electrical energy saving. Media is considered to be a potential for energy-efficient information dissemination is a television, chosen as much as 91% of respondents.

Other Determinants

Home building design influence on power consumption, especially lighting needs. 35% of households do not pay attention to the house design in an effort to save electricity. Mostly, because they do not have knowledge of it while other households have limited fund for the re-design of the house. Other factors include the priority of respondents in choosing electronic equipment. Fig 4 shows that 39% households purchase electric appliances because of the quality of the equipment. Only 3% households think that electric power in watt is important factor on choosing the electric appliances.

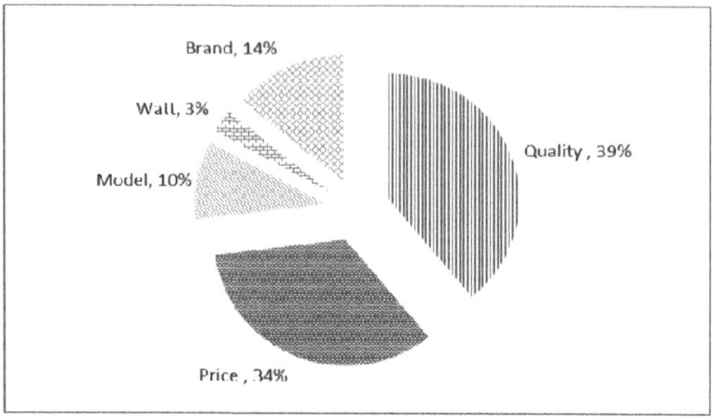

Fig. 3. Driving factor in purchasing electric appliances

3.2 Correlation between Demographic of Respondents and Electricity Use

A. Gender
Gender has been found not to be a significant factor of electricity consumption in household. There was no significant difference between male and female respondents toward household energy consumption pattern. However, female respondents seemed to be more aware of energy bills/prices as it would affect the household expense. For instance, female respondents tended to change the incandescent lamps with the energy saving lamps if they had additional income. This attitude was seemingly caused by the fact that the household paid less of energy bill if they used energy saving lamps.

B. Age
There is no significant difference in term of age toward energy saving behaviour and consumption pattern. As more than 90% of total respondents aged > 21 years old, they were considered already mature to manage household, especially in managing expense.

C. Housing area
Housing area affected the electricity consumption of household. The more spacious housing are, residents will have a tendency to pay more attention to the use of electrical equipment. This finding is coincident with that found by (Bachus dan Ootegam, 2011)

D. Occupation
Entrepreneurs and teachers/lecturers tended to invest more on energy saving equipment in order to reduce electricity consumption rather compared to respondents with other occupation. The two groups of occupation seemed to be more educated hence their perspective toward energy consumption behavior was also more long-term vision.

E. Number of Persons in the Household
The bigger the number of persons in household the better the household behaves toward energy efficient. This condition triggered by the fact that the more the number of family members, the tendency of saving electricity attitude will be more visible. Decision-making in the family will be more complicated than the individual, because

it involves two or more people Yuliati and Nurasrina (2012). The larger the family the more the difficulty of decision-making process because more people need to be considered and the more resources must be prepared.

F. Education level of family member
Education or more specific the knowledge of household member about importance of energy for our sustainable living played important role to encourage family member to be more aware of energy use. Education affects the attitude of saving electrical energy as it affects the person's decisions and consumption patterns Yuliati and Nurasrina (2012.

G. Monthly income
Household income related to the purchasing power of households either the ability to pay utility bills and household spending on electronic equipment. Respondents with high and stable income tended to consume more energy than others.

H. Payment method of electricity bills
There are two types of payment method for electricity bills – pre-paid and post-paid. In this study, there was no significant difference between the two methods on electricity consumption pattern in household. PLN actually encourages households to migrate from post-paid to prepaid method, but in the majority of people still thought that the prepaid method was much more expensive than postpaid causing still few households used prepaid payment method of electricity bills.

I. Power level
Class customer rates, income, and education are interrelated characteristics. Individuals with high incomes will put huge resources to meet the needs of the household. But most of individuals with good incomes are well educated person. Households that installed > 2200 KVA tended to loose in control of energy use in the house. However, as education also played role in changing behavior, high income people with better education and knowledge behaved more efficiently.

J. Type and number of home appliances used
Identification and analysis of household electronic equipment useful for understanding the characteristics of the electrical energy consumption in households. Electricity consumption in a household with other household varies, depending on the installed equipment and systems class customer rates.

4 Conclusion

This research observes the behavior patterns of electrical energy saving in households and seeks the potential of electricity saving through behavioral changing in this sector. Padang demography related to the number of samples participated in this study are classified into the social, economic, and use of technology aspects. The results of the interviews show that households tend to unconcern about their electricity saving habit because the energy prices are reasonably cheap so that in range of affordability of the households to pay, especially for households who have middle to high income. The households tend to consume more electricity as long they afford to pay the electricity bills. The households in Padang are not so well informed about energy efficient equipment, so it is not a priority in their effort to control power consumption. Households need to be more educated through sufficient information primarily from broadcast media to help them taking action in behaviorally change in electricity energy saving. These findings will provide a scientific evidence for stakeholders on the potential of controlling electricity consumption and designing energy policy by government in residential sector.

Acknowledgment. This research was financially supported by Directorate General for Higher Education with Scheme International Research Collaboration and Publication in year 2015. The authors also thank to households as samples in this study.

References

1. Statistics Indonesia, Badan Pusat Statistik (2010)
2. CDIEMR. Handbook of energy and economic statistic of Indonesia (2010)
3. BPPT. Annual Report BPPT 2012 (2012)
4. Saidur, R., Masjuki, H.H., Jamaluddin, M.Y.: An application of energy and energy analysis in residential sector of Malaysia. Energy Policy 35, 1050–1063 (2007)
5. Guerra-Santin, O., Itard, L., Visscher, H.: The effect of occupancy and building characteristics on energy use for space and water heating in Dutch residential stock. Energy and Buildings 41, 1223–1232 (2009)
6. Sukarno, I., Hiroshi, M., Susanti, L.: Energy Consumption Based on a Cohort Model in a Local City of Indonesia. In: 12th International Conference on Sustainable Energy Technologies, Hongkong (2013)
7. Wang, Z., et al.: Determinants and policy implications for household electricity- saving behavior: evidence from Beijing, China. Energy Policy 39, 3550–3557 (2011)
8. Li, G., Niu, S., Ma, L., Zhang, X.: Assesment of environmental and economic costs of rural household energy consumption in Loess Hilly Region, Gansu Province, China. Renewable Energy 34, 1438–1444 (2009)
9. Ouyang, J., Hokao, K.: Energy saving potential by improving occupants behavior in urban residential sector in Hangzhou City, China. Energy Policy 41, 711–720 (2009)
10. Zhou, Z., Wu, W., Chen, Q., Chen, S.: Study on sustainable development of rural household energy in northern China. Renewable and Sustainable Energy Reviews 12, 2227–2239 (2008)
11. Xiaohua, X., Zhenmin, F.: Study on affecting factors and standard of rural household energy consumption in China. Renewable and Sustainable Energy Reviews 9, 101–110 (2005)
12. Febrizal, Y.: Electricity Energy Needs Prediction of Padang City Until 2020. Electrical Department Journal of ITP, No. 1 (January 2012)
13. Bachus, K., Van Ootegam, L.: Determinants of Energy Saving Behaviour by Households. INESPO Project (2011)
14. Yuliati, L.N., Nurasrina, I.: Pesan, Kesadaran, dan Perilaku Hemat Listrik Rumah Tangga. Jur. Ilm. Kel. & Kons 5(1), 88–95 (2012) ISSN: 1907-6037 (in Indonesian)

Central Composite Design for the Experiments with Replicate Runs at Factorial and Axial Points

Haeil Ahn

Department of Industrial Engineering
Seokyeong University, Seoul, Korea 136-704
hiahn@skuniv.ac.kr

Abstract. In this paper, we have considered for study the central composite design (CCD) with a view to extending the related second-order response surface models. In the CCD, there are three types of design points, such as the factorial, axial and center points. We devised a new type of CCD for the experiments with replicate runs at such design points, not just at the design center. This type of experimental design, named as CCD-R, proves to be flexible in the sense that there exist a myriad of exactly orthogonal and nearly rotatable designs. It is also demonstrated that the proposed CCD-R is useful, especially when a number of replicate runs are desired in the experiment for one reason or another. Since the second-order generalized regression models provide a good framework of the method, an attempt is given to the problem of logistic regression analysis for categorical data based on the CCD-R.

Keywords: central composite design, second-order, logistic regression.

1 Introduction

When it comes to developing an improved process, some changes should be made on purpose to the input variables of a process, so that we can identify the reasons for changes in the responses and improve the operating conditions.

When a two-level full or fractional factorial experiment is conducted in an ongoing process, an experimenter might have to consider using the current operating condition as the design center for the design and employing quadratic terms in the model in order to test the linearity of the region of interest. A few center runs would suffice to detect the quadratic effects against the hypothesis of linearity over the region of exploration. However, there are situations where the parameters of quadratic terms should be separately estimated. In order to estimate the parameters of the quadratic terms, the second-order model is required. To estimate all the parameters in the second-order model, more independent runs at some more design points such as center and axial points are required.

A simple and effective solution for this problem is to augment the full (2^k) or fractional (2^{k-p}) factorial design with a center and $2k$ axial points, where k is the number of factors and p is the index of fraction.

M. Gen et al. (eds.), *Industrial Engineering, Management Science and Applications 2015*,
Lecture Notes in Electrical Engineering 349, DOI: 10.1007/978-3-662-47200-2_101

The orthogonal CCD was introduced by Box and Wilson (1951). Later, the rotatable CCD was proposed by Box and Hunter (1957). The concept of the orthogonality and the rotatability plays a key role in designing the experiment and determining the axial distance value α. The α values needed for the orthogonality are usually different from those for the rotatability, except some special cases.

In relation to the CCD, the slope-rotatability was studied by Hader and Park (1978), Park (1987), and Park and Kim (1992). Kim (2002) extended the original CCD concept to what they call CCD of second type or CCD2. In the CCD2, the positions of axial points are indicated by two axial distance values, say, α_1 and α_2. In the CCD2, the orthogonality and the rotatability of the design are achievable at the same time by employing double axial points. Kim and Ko (2004) discussed in some detail on the slope-rotatability of the CCD2. The statistical properties are further examined by Kim and Park (2006).

In practice, an experimenter is often encountered by some cases that a certain number of replicate runs are needed at factorial or axial points for one reason or another. In this study, the CCD with replicate runs, named as CCD-R, is considered for study. In the case of the CCD-R, it would be also desirable to achieve the orthogonality and even the rotatability, if possible, of the design. We have used a simple, but illustrative example to explain how to implement generalized and/or logistic regression analysis based on the CCD-R.

2 Central Composite Design

2.1 Orthogonal CCD

The orthogonal CCD has evolved from the use of two-level factorial designs in the general regression framework, where a potential concern is the assumption of linearity in the factor effects. We assume that the k factors are quantitative, and the levels of a factor are coded by -1, 0, or +1. A first-order model consists of constant, linear and interaction terms and takes the form of

$$y = \beta_0 + \sum_{i=1}^{k} \beta_i x_i + \sum_{i=1}^{k-1} \sum_{i<j=2}^{k} \beta_{ij} x_i x_j + \varepsilon \tag{1}$$

The model contains $1 + k + k(k-1)/2$ parameters, to estimate main effects and interaction effects. The CCD method consists of adding center point to the 2^k design. It is customary for an experimenter to add some center point runs to test the linearity of the model. These consist of n_C replicate runs at the design center. If the curvature around the design center is significant, the region of exploration probably is not linear any more. The design center might be very close to or near the local optimum. For this reason, the factorial design augmented with the design center can provide the framework of the second-order model. The second-order model includes the second order terms.

$$y=\beta_0+\sum_{i=1}^{k}\beta_i x_i+\sum_{i=1}^{k-1}\sum_{i<j=2}^{k}\beta_{ij}x_i x_j+\sum_{i=1}^{k}\beta_{ii}x_i^2+\varepsilon. \tag{2}$$

Since the model contains $1+2k+k(k-1)/2$ parameters, there must be at least this number of distinct design points. Without the axial runs, the variation due to the second-order terms cannot be orthogonally decomposed. The basic concepts of the CCD for $k=2$ and $k=3$ are depicted in Figure 1, where the bold dots are the design points.

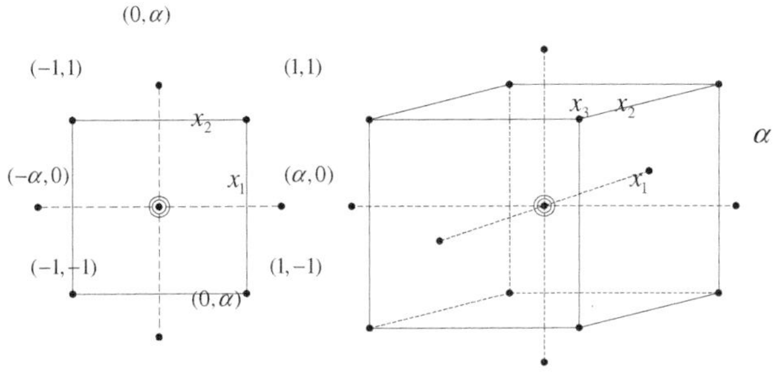

Fig. 1. CCD for $k=2$ and $k=3$

2.2 Exactly Orthogonal CCD-R

Under the premise that the some replicate runs are available at the factorial points and the axial points, the number of which are n_F and n_A, respectively. The centered square columns can be given as follows:

$$\begin{matrix} x_1^2-m & x_2^2-m & \cdots & x_k^2-m \end{matrix}$$

$$\left.\begin{bmatrix} 1-m & 1-m & \cdots & 1-m \\ 1-m & 1-m & \cdots & 1-m \\ 1-m & 1-m & \cdots & 1-m \\ \cdots & \cdots & \ddots & \vdots \\ 1-m & 1-m & \cdots & 1-m \end{bmatrix}\right\} F\times n_F$$

$$\left.\begin{bmatrix} \alpha^2-m & -m & \cdots & -m \\ \alpha^2-m & -m & \cdots & -m \\ -m & \alpha^2-m & \cdots & -m \\ -m & \alpha^2-m & \cdots & -m \\ \vdots & \vdots & \ddots & \vdots \\ -m & -m & \cdots & \alpha^2-m \\ -m & -m & \cdots & \alpha^2-m \end{bmatrix}\right\} 2k\times n_A \,. \tag{3}$$

$$\left.\begin{bmatrix} -m & -m & \cdots & -m \\ -m & -m & \cdots & -m \\ \vdots & \vdots & \ddots & \vdots \\ -m & -m & \cdots & -m \end{bmatrix}\right\} n_C$$

, where F is the number of factorial points and m defined as the average of a second-order column can be identified as follows:

$$m = \frac{F \cdot n_F + 2\alpha^2 \cdot n_A}{N} = \frac{F \cdot n_F + 2\alpha^2 \cdot n_A}{F \cdot n_F + 2k \cdot n_A + n_C}. \tag{4}$$

It can be shown that the centered square columns are orthogonal to the column of ones by proving the following condition.

$$\sum_{u=1}^{N} (x_{iu}^2 - m) = 0, \quad i = 1, 2, \cdots, k \quad . \tag{5}$$

It is also easy to see that the centered square columns are also orthogonal to the linear columns and the interaction columns. The necessary condition that the centered square columns are orthogonal to each other is the following:

$$\sum_{u=1}^{N} (x_{iu}^2 - m)(x_{ju}^2 - m) = 0 \ . \tag{6}$$

for $i = 1, \cdots, k-1$ and $j = 2, \cdots, k$ $(i < j)$. The orthogonal decomposition of a regression model boils down to the fact that $\mathbf{X'X}$ is a diagonal matrix, where $\mathbf{X}$ is the regression matrix of the model. The orthogonality condition of the centered square columns can be summarized as follows:

$$m^2 N = F n_F \tag{7}$$

Since $m = (Fn_F + 2\alpha^2 n_A)/N$, the α value can be determined as follows:

$$\alpha = \left(\frac{\sqrt{Fn_F(Fn_F + 2kn_A + n_C)} - Fn_F}{2n_A} \right)^{\frac{1}{2}} = \left(\frac{\sqrt{F\left(F + 2k \cdot n_A/n_F + n_C/n_F\right)} - F}{2 \cdot n_A/n_F} \right)^{\frac{1}{2}}. \tag{8}$$

The α values for various k and n_C / n_F are displayed in Table 1

Table 1. α for k and n_C / n_F where $n_A / n_F = 1$

n_C / n_F		0	1	2	3	4	5	6	7	8	9	10
k	2	0.910	1.000	1.078	1.147	1.210	1.267	1.320	1.369	1.414	1.457	1.498
	3	1.136	1.215	1.287	1.353	1.414	1.471	1.525	1.575	1.623	1.668	1.711
	4	1.341	1.414	1.483	1.547	1.607	1.664	1.719	1.771	1.820	1.868	1.914
	5	1.527	1.596	1.662	1.724	1.784	1.841	1.896	1.949	2.000	2.049	2.097
	6	1.694	1.761	1.824	1.885	1.943	2.000	2.055	2.108	2.159	2.209	2.257

It would be a better idea, if we could make the value of n_A/n_F positive integer valued, so that α values can reduce until it becomes close to or smaller than one.

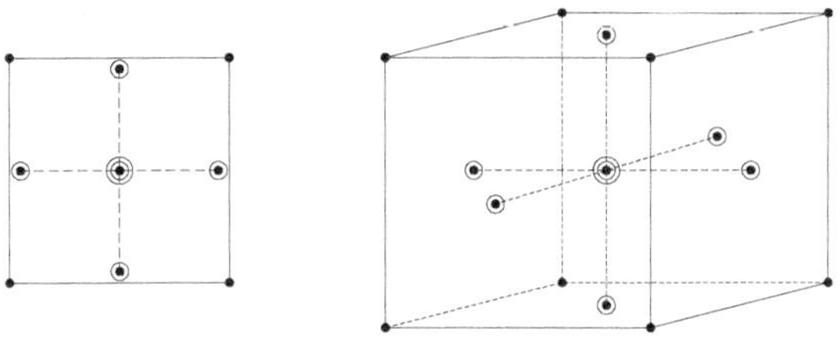

Fig. 2. CCD-R for $k=2$ and $k=3$

The axial distance value α gets larger as the number of replicate runs at the center point (n_C) increases, which is an unfavorable phenomenon for the experimenters. Fortunately, however, we can make the α value smaller by increasing the number of replicate runs at the axial points (n_A) as shown in Figure 2 CCD-R for $k=2$ and $k=3$.

We observed that CCD-R's were also available even without the design center.. Without the design center, it may not be a CCD in the real sense. It is what we call axial composite design with replicate runs (ACD-R). The ACD-R requires smaller α value as compared to CCD-R.

Nevertheless, we do not have to totally exclude the center point in the design. The coordinate of the design center corresponds to the current operating condition, under which plenty of data are available. By increasing the number of replicate runs at the current operating condition, we can improve the precision of the parameter estimation of the model.

For example, if $k=3$ and $n_F=2$, then $n_A=2$. In order to devise a CCD-R experiment, if we choose $n_C/n_F=4$, then $\alpha=1.414$ and $n_C=n_F \cdot n_C/n_F=8$. For an ACD-R experiment, if we let $n_C/n_F=0$, we get $\alpha=1.136$ and $n_C=0$.

Table 2. α for k and n_C/n_F where $n_A/n_F=2$

n_C/n_F		0	1	2	3	4	5	6	7	8	9	10
k	2	0.856	0.896	0.933	0.968	1.000	1.030	1.059	1.086	1.112	1.136	1.160
	3	1.078	1.114	1.147	1.179	1.210	1.239	1.267	1.294	1.320	1.345	1.369
	4	1.287	1.321	1.353	1.384	1.414	1.443	1.471	1.498	1.525	1.550	1.575
	5	1.483	1.515	1.547	1.577	1.607	1.636	1.664	1.692	1.719	1.745	1.771
	6	1.662	1.694	1.724	1.755	1.784	1.813	1.841	1.869	1.896	1.923	1.949

For another example, we can make use of Table 2. If $k=2$ and $n_F=1$, then $n_A=2$. For a CCD-R experiment, if we let $n_C/n_F=4$, then $\alpha=1.0$ and $n_C=n_F\cdot n_C/n_F=4$. For an ACD-R experiment, let $n_C/n_F=0$, then $\alpha=0.816$ and $n_C=n_F\cdot n_C/n_F=0$.

Table 3. α for k and n_C/n_F where $n_A/n_F=3$

n_C/n_F		0	1	2	3	4	5	6	7	8	9	10
k	2	0.816	0.841	0.865	0.887	0.908	0.928	0.947	0.965	0.983	1.000	1.016
	3	1.035	1.056	1.078	1.098	1.117	1.136	1.155	1.172	1.190	1.206	1.223
	4	1.245	1.266	1.286	1.306	1.325	1.344	1.362	1.380	1.397	1.414	1.431
	5	1.446	1.466	1.486	1.506	1.525	1.544	1.562	1.581	1.598	1.616	1.633
	6	1.633	1.653	1.673	1.693	1.712	1.731	1.750	1.768	1.786	1.804	1.821

In this manner, we can develop a series of tables and devise a variety of CCD-R's or ACD-R's using the axial distance value α given in Table 1, Table 2, Table 3 and so on

2.3 Exactly Orthogonal and Rotatable CCD-R

A rotatable CCD-R is subject to the same conditions as the rotatable CCD. The necessary and sufficient conditions for the rotatability are as follows:

1. All odd moments through order 4 are zero.
2. $\sum_{u=1}^{N} x_{iu}^4 = 3\sum_{u=1}^{N} x_{iu}^2 x_{ju}^2$

According to Kim (2002), the orthogonality condition for CCD2 is given by

$$\alpha_1^2+\alpha_2^2=\frac{\sqrt{F(F+4k+n_C)}-F}{2}. \tag{9}$$

The rotatability condition for CCD2 is derived as follows:

$$\alpha_1^4+\alpha_2^4=F. \tag{10}$$

In this case, $N=F+4k+n_C$. Converted to CCD2-R version, these two conditions should look as follows:

$$\alpha_1^2 + \alpha_2^2 = \frac{\sqrt{F(F + 4k \cdot n_A/n_F + n_C/n_F)} - F}{2 \cdot n_A/n_F} \tag{11}$$

$$\alpha_1^4 + \alpha_2^4 = F \cdot n_F/n_A \tag{12}$$

By solving these two equations simultaneously, we can obtain the exactly orthogonal and rotatable CCD2-R.

3 Illustrative Example

3.1 LED Manufacturing

A manufacturing company produces light-emitting diodes (LEDs) of different types and colors. An LED is encapsulated by a mold, the material of which is known as epoxy. The mold manufacturing is subject to a curing process. The LEDs are being produced by the batch of size 200. An LED is defective, unless the color of the light emitted through the mold of an LED belongs to a specific bandwidth of a color spectrum. An experimenter is seeking a method of decreasing the defective rate of LEDs of a specific color. There are three variables of interest, the amount of pigments applied to the mold material, curing time, and curing temperature. The experimenter is uncertain of the optimality of current curing process and the linearity assumption over the domain of interest. He decides to conduct a 2^3 design (with a single batch of each factorial run) augmented with four center points, the operating condition of which is the current operating condition. In addition, the experimenter decided to include the axial points.

3.2 Augmentation

The three factors are quantitative, and the levels of a factor are coded by -1, 0, or +1. The data are grouped as frequencies for each combination of factor levels. There are fifteen design points within the table, each of which corresponds to a certain operating condition.

In the case of this example, $k = 3$, $n_F = 200$, $n_A = 400$ and $n_C = 800$, hence $n_A/n_F = 2$ and $n_C/n_F = 4$. The axial distance value α depends on the existence of the design center. In the presence of the design center, $n_C/n_F = 4$ and hence it is desirable to set α to 1.210 for the orthogonality of the design, as can be seen from Table 2. As mentioned above, the adoption of the design center is optional. Without the design center, we had better set α value to 1.078.

Table 4. Augmented Design Matrix and Attribute Data

Design point	Factors, Interactions and Square Columns										No. of Bad Items	Sample Size
	1	x_1	x_2	x_3	x_1x_2	x_1x_3	x_2x_3	x_1^2	x_2^2	x_3^2		
1	1	-1	-1	-1	1	1	1	1	1	1	5	200
2	1	1	-1	-1	-1	-1	1	1	1	1	4	200
3	1	-1	1	-1	-1	1	-1	1	1	1	8	200
4	1	1	1	-1	1	-1	-1	1	1	1	9	200
5	1	-1	-1	1	1	-1	-1	1	1	1	19	200
6	1	1	-1	1	-1	1	-1	1	1	1	26	200
7	1	-1	1	1	-1	-1	1	1	1	1	21	200
8	1	1	1	1	1	1	1	1	1	1	20	200
9	1	$-\alpha$	0	0	0	0	0	α^2	0	0	14	400
10	1	α	0	0	0	0	0	α^2	0	0	12	400
11	1	0	$-\alpha$	0	0	0	0	0	α^2	0	10	400
12	1	0	α	0	0	0	0	0	α^2	0	16	400
13	1	0	0	$-\alpha$	0	0	0	0	0	α^2	11	400
14	1	0	0	α	0	0	0	0	0	α^2	13	400
Center	1	0	0	0	0	0	0	0	0	0	15	800

The actual regression matrix $\mathbf{X}$ is shown in Table 5.

Table 5. Regression Matrix

Design point	Factors, Interactions and Centered Square Columns										Data
	1	x_1	x_2	x_3	x_1x_2	x_1x_3	x_2x_3	x_1^2-m	x_2^2-m	x_3^2-m	No. of Batches
1	1	-1	-1	-1	1	1	1	0.4227	0.4227	0.4227	1
2	1	1	-1	-1	-1	-1	1	0.4227	0.4227	0.4227	1
3	1	-1	1	-1	-1	1	-1	0.4227	0.4227	0.4227	1
4	1	1	1	-1	1	-1	-1	0.4227	0.4227	0.4227	1
5	1	-1	-1	1	1	-1	-1	0.4227	0.4227	0.4227	1
6	1	1	-1	1	-1	1	-1	0.4227	0.4227	0.4227	1
7	1	-1	1	1	-1	-1	1	0.4227	0.4227	0.4227	1
8	1	1	1	1	1	1	1	0.4227	0.4227	0.4227	1
9	1	-1.21	0	0	0	0	0	0.8868	-0.5774	-0.5774	2
10	1	1.21	0	0	0	0	0	0.8868	-0.5774	-0.5774	2
11	1	0	-1.21	0	0	0	0	-0.5774	0.8868	-0.5774	2
12	1	0	1.21	0	0	0	0	-0.5774	0.8868	-0.5774	2
13	1	0	0	-1.21	0	0	0	-0.5774	-0.5774	0.8868	2
14	1	0	0	1.21	0	0	0	-0.5774	-0.5774	0.8868	2
Center	1	0	0	0	0	0	0	-0.5774	-0.5774	-0.5774	4
Weighted Total	24	0	0	0	0	0	0	0	0	0	18

All the columns are orthogonally decomposed and therefore the matrix $\mathbf{X}'\mathbf{X}$ reduces to a diagonal matrix shown in the following.

$$\mathbf{X}'\mathbf{X}=\begin{bmatrix} 24 & 0 & 0 & 0 & 0 & 0 & 0 & 0 & 0 & 0 \\ 0 & 13.86 & 0 & 0 & 0 & 0 & 0 & 0 & 0 & 0 \\ 0 & 0 & 13.86 & 0 & 0 & 0 & 0 & 0 & 0 & 0 \\ 0 & 0 & 0 & 13.86 & 0 & 0 & 0 & 0 & 0 & 0 \\ 0 & 0 & 0 & 0 & 8.0 & 0 & 0 & 0 & 0 & 0 \\ 0 & 0 & 0 & 0 & 0 & 8.0 & 0 & 0 & 0 & 0 \\ 0 & 0 & 0 & 0 & 0 & 0 & 8.0 & 0 & 0 & 0 \\ 0 & 0 & 0 & 0 & 0 & 0 & 0 & 8.57 & 0 & 0 \\ 0 & 0 & 0 & 0 & 0 & 0 & 0 & 0 & 8.57 & 0 \\ 0 & 0 & 0 & 0 & 0 & 0 & 0 & 0 & 0 & 8.57 \end{bmatrix} \tag{13}$$

3.3 Logistic Regression Model

The attribute data lend itself to the general or logistic regression model. We would like to test the linearity of the region of exploration. The logistic regression model is being dealt with in Myers and Montgomery (2002), Fleiss *et al.* (2013) and Agresti (2014). Actually, we can use the following centered model to fit into the data.

$$\text{logit}\left[\pi(\mathbf{x})\right]=\beta_0'+\sum_{i=1}^{3}\beta_i x_{iu}+\sum_{i=1}^{2}\sum_{i<j=2}^{3}\beta_{ij}x_{iu}x_{ju}+\sum_{i=1}^{3}\beta_{ii}(x_{iu}^2-m) \tag{14}$$

for $u=1,2,\ldots,N$. Strictly speaking, the variation due to error is not normally distributed in the case of logistic regression. But it is also known to be asymptotically normal when the sample size is large enough as pointed out in Fleiss *et al.* (2013).

The logistic regression algorithm of MINITAB is used to estimate the parameters and related confidence intervals of the model.

3.4 Estimation of Parameters

The Minitab output is displayed in Figure 3. As it turned out, the estimates for β_0', β_3, β_{11}, β_{22}, and β_{33} are significant at 95% significance level.

Prediction Variables	Coeff	SE	Z	P	OR	95% CI Lower	Upper
Const	-3.30600	0.0840052	-39.35	0.000			
x1	0.0143515	0.0903660	0.16	0.874	1.01	0.85	1.21
x2	0.132390	0.0916564	1.44	0.149	1.14	0.95	1.37
x3	0.499378	0.0963390	5.18	0.000	1.65	1.36	1.99
x1x2	-0.0552992	0.100646	-0.55	0.583	0.95	0.78	1.15
x1x3	0.0479896	0.106712	0.45	0.653	1.05	0.85	1.29
x2x3	-0.170805	0.107846	-1.58	0.113	0.84	0.68	1.04
x1^2	0.515763	0.134598	3.83	0.000	1.67	1.29	2.18
x2^2	0.507722	0.134980	3.76	0.000	1.66	1.28	2.16
x3^2	0.350694	0.140461	2.50	0.013	1.42	1.08	1.87

Fig. 3. Parameter Estimates for the Logistic Model

Moreover, we are quite sure of the fact that the design center or the current operating condition is near the local minimum, since the estimates of the second-order terms are all positive. The region of exploration may as well be scrutinized afterward to pinpoint the minimum point.

Since the variation due to each term is orthogonally decomposed, the variance-covariance matrix $\mathbf{(X'X)^{-1}}$ is also diagonal. The calculation of confidence interval at any point in the region of exploration can be conducted separately because the variation due to each term is orthogonally decomposed.

4 Conclusion and Further Study

We propose that an experimenter can devise a CCD-R experiment with smaller α values by increasing the number of replicate runs at axial points. The α value gets smaller as the number of replicate runs at the axial points (n_A) increases.

Although we can improve the precision of the parameter estimates of the model by increasing the number of replicate runs at the design center, we observed that an experimental design was also possible even without the design center. Without the design center, it is an axial composite design with replicate runs (ACD-R). The ACD-R requires smaller α value as compared to the CCD-R.

We have used a simple example to explain how to implement general and/or logistic regression analysis based on the CCD-R. We used the general and logistic regression program to examine and demonstrate the usefulness of the CCD-R in analyzing the general or logistic regression model.

If there is anyone who wishes to devise a CCD-R that enjoys the orthogonality and the rotatability at the same time, we recommend that the CCD2 can be extended to the CCD2 with replicate runs, say, CCD2-R or ACD2-R.

References

1. Agresti, A.: Categorical data analysis. John Wiley & Sons (2014)
2. Box, G.E.P., Hunter, J.S.: Multifactor Experimental Designs for Exploring Response Surfaces. The Annals of Mathematical Statistics 28, 195–241 (1957)
3. Box, G.E.P., Wilson, K.B.: On the Experimental Attainment of Optimum Conditions. Journal of the Royal Statistical Society, Series B 13, 1–45 (1951)
4. Fleiss, J.L., Levin, B., Paik, M.C.: Statistical methods for rates and proportions. John Wiley & Sons (2013)
5. Hader, R.J., Park, S.H.: Slope-rotatable central composite designs. Technometrics 20(1978), 413–417 (1978)
6. Kim, H.J.: Extended central composite designs with the axial points indicated by two numbers. The Korean Communications in Statistics 9, 595–605 (2002)
7. Kim, H.J., Ko, Y.M.: On slope rotatability of central composite designs of the second type. The Korean Communications in Statistics 11, 121–137 (2004)

8. Kim, H.J., Park, S.H.: Statistical Properties of Second type Central Composite Designs. The Korean Journal of Applied Statistics 19(2), 257–270 (2006)
9. Myers, R.H., Montgomery, D.C.: Response Surface Methodology – Process and Product Optimization Using Designed Experiments, 2nd edn. Wiley Series in Probability and Statistics (2002)
10. Park, S.H., Kim, H.J.: A measure of slope-rotatability for second order response surface experimental designs. Journal of Applied Statistics 19, 391–404 (1992)

A Software Trustworthiness Measure Based on the Decompositions of Trustworthy Attributes and Its Validation

Hongwei Tao[1], Yixiang Chen[2,*], and Jianmin Pang[1]

[1] State Key Laboratory of Mathematical Engineering and Advanced Computing, Zhengzhou, 45002, China
[2] MoE Engineering Center for Software/Hardware Co-Design Technology and Application, East China Normal University, Shanghai, 200062, China
yxchen@sei.ecnu.edu.cn

Abstract. There exist many software trustworthiness measures based on the decompositions of attributes. However, few of them are concerned with using more rigorous approaches to measure software trustworthiness and carrying out the theoretical validation of measures. Axiomatic approaches formalize the empirical understanding of a software attribute through the definition of a set of desirable software measure properties. It can provide precise and formal terms for the quantification of software attributes. We once used axiomatic approaches to measure software trustworthiness based on multi-dimensional attributes, in this paper, based on previous works, we apply axiomatic approaches to measure software trustworthiness based on the decompositions of attributes, propose the desirable properties of software trustworthiness measures based on the decompositions of attributes, present a software trustworthiness measure and validate this measure from the theory by proving that it complies with the set of properties. Compared with the popular software trustworthiness measures, this measure can evaluate software trustworthiness better in view of our properties.

Keywords: Trustworthy software, Software trustworthiness measure, Axiomatic approaches.

1 Introduction

With the increasing dependence on software, the software trustworthiness have received extensive attention in recent years, meanwhile, the measurement of software trustworthiness which is the quantification of software trustworthiness has been a hot topic for researchers [1]. The software trustworthiness can be characterized by many attributes [2-9], which are described as trustworthy attributes in this paper. Trustworthy attributes are normally at too high level to be measurable directly, hence they are further subdivided into sub-attributes. Many software trustworthiness measures based on the

* Corresponding author.

M. Gen et al. (eds.), *Industrial Engineering, Management Science and Applications 2015*,
Lecture Notes in Electrical Engineering 349, DOI: 10.1007/978-3-662-47200-2_102

decompositions of trustworthy attributes are proposed. Typical ones include evidence theory [4], weakness analysis [5], data mining [6], uncertain theory [7], Bayesian Networks [8] etc.. However, few researches focus on measuring software trustworthiness with more rigorous approaches and validating the measures from theory. Rigorous measurement of software trustworthiness can provide very important help in the evaluation and improvement of software trustworthiness. Theoretical validation is a required activity for using or defining measures that make sense and it is a necessary step for the empirical validation of measures [10]. Beyond their mathematical appearance, axiomatic approaches are actually ways to formalize common sense [11]. They have been applied to measure internal attributes, for example size, complexity, cohesion and so on [10-15]. In order to make the software trustworthiness measures more rigorous, we once used axiomatic approaches to measure software trustworthiness based on trustworthy attributes [16-17]. In this paper, we extend the previous works to apply axiomatic approaches to measure software trustworthiness based on decompositions of trustworthy attributes, propose the desirable measure properties from the view of decomposition of trustworthy attributes, give a software trustworthiness measure based on decomposition of trustworthy attributes and carry out theoretical validation of this model. Comparative study shows that this model is better from the point of the measure properties proposed by us.

The rest of this paper is organized as follows. The desirable measure properties from the point of decomposition of trustworthy attributes are described in section 2. We introduce a software trustworthiness measure based on decomposition of trustworthy attributes and carry out its theoretical validation in section 3. In section 4 we make a case study. The comparative study is given in section 5. The conclusions and future work come in the last section.

2 Properties of Software Trustworthiness Measures Based on the Decompositions of Trustworthy Attributes

Trustworthy attributes are separated into critical attributes and non-critical attributes. The trustworthy attributes that a software must have are called critical attributes and the other trustworthy attributes are referred as non-critical attributes [16]. With the same symbols as that given in [16], $y_1,\ldots,y_m$ are used to represent the degree of critical attributes and y_i $(m + 1 \le i \le m + s)$ are used to express the degree of non-critical attributes. Suppose there are n sub-attributes that comprise the trustworthy attributes and let their values separately be $x_1,\ldots,x_n$. Let T be a software trustworthiness measure function regarding $y_1,\ldots,y_{m+s}$ and y_i $(1 \le i \le m + s)$ be trustworthy attribute measure functions w. r. t. $x_1,\ldots,x_n$. The following are the desirable properties of software trustworthiness measures based on the decompositions of trustworthy attributes.

1) Monotonicity

$$\frac{\partial T}{\partial y_i} \ge 0, \frac{\partial y_i}{\partial x_k} \ge 0, 1 \le i \le m+s, 1 \le k \le n .$$

It means that the increase of a trustworthy attribute does not lead to software trustworthiness to decrease, and the increase of a sub-attribute does not lead to trustworthy attribute to decrease.

2) Acceleration

$$\frac{\partial^2 T}{\partial {y_i}^2} \le 0, \frac{\partial^2 y_i}{\partial {x_k}^2} \le 0, 1 \le i \le m+s, 1 \le k \le n .$$

It implies that the increase of the trustworthy attributes and sub-attributes lead to their utilization efficiency to decrease.

3) Sensitivity

$$0 \le \frac{\partial T}{\partial y_i}\frac{y_i}{T} = f_1(y_i, w_i),\ 0 \le \frac{\partial y_i}{\partial x_k}\frac{x_k}{y_i} = f_2(x_i, w_{ik}),\ 1 \le i \le m+s, 1 \le k \le n .$$

Where w_i is the weight of y_i and f_1 is a function regarding y_i and w_i, w_{ik} is the weight of x_{iw}, f_2 is a function regarding x_i and w_{ik}. Sensitivity is used to describe the percentage changes of software trustworthiness (or trustworthy attributes) caused by the percentage changes of trustworthy attributes (or sub-attributes). They should be positive and related to the corresponding attributes and their weights. The software trustworthiness should be more sensitive to the minimal critical attribute compared with its weight.

4) Substitutivity

(a). $(\exists c_1, c_2 \in \Re^+) c_1 \le \sigma_{ij} \le c_2$,where

$$\sigma_{ij} = \frac{d(y_i/y_j)}{d(-\frac{\partial T/\partial y_j}{\partial T/\partial y_i})} \times \frac{-\frac{\partial T/\partial y_j}{\partial T/\partial y_i}}{y_i/y_j}, 1 \le i, j \le m+s, i \ne j \tag{1}$$

are used to give expression of the difficulty of the substitution between the trustworthy attributes.

(b). $(\exists c_3, c_4 \in \Re^+) c_3 \le \sigma_{ikl} \le c_4$,where

$$\sigma_{ikl} = \frac{d(x_k/x_l)}{d(-\frac{\partial y_i/\partial x_l}{\partial y_i/\partial x_k})} \times \frac{-\frac{\partial y_i/\partial x_l}{\partial y_i/\partial x_k}}{x_k/x_l}, 1 \le i \le m+s, 1 \le k, l \le n, k \ne l \tag{2}$$

are used to express the difficulty of the substitution between the sub-attributes.

The smaller σ_{ij} is, the more difficult the substitution between the y_i and y_j is. Likewise, the smaller σ_{ikl} is, the more difficult the substitution between the x_k and x_l that comprise the trustworthy attributes y_i is. This property states that the trustworthy attributes and sub-attributes can substitute each other to some extent. Moreover, the substitutivity between critical attributes and non-critical attributes should be more difficult than either the substitutivity among critical attributes or that among non-critical attributes, and the substitutivity among sub-attributes should be more difficult than the substitutivity among trustworthy attributes.

5) Expectability

If $y_0 \le \min\{y_1,\ldots,y_{m+s}\}$, then $y_0 \le T \le \max\{y_1,\ldots,y_{m+s}\}$.

That property asserts that if all the trustworthy attributes achieve the user expectation, then the software trustworthiness should meet the user expectation too and be less than the maximum of all the trustworthy attributes.

3 A Software Trustworthiness Measure Based on the Decompositions of Trustworthy Attributes

In this section, we propose a software trustworthiness measurement based on the decompositions of trustworthy attributes and validate it by showing that it complies with properties set described in section 2. α and β are used to distinguish the contribution of critical attributes and non-critical attributes to the software trustworthiness, they are required to satisfy that $\alpha+\beta=1$ and $0\le\beta<0.5<\alpha\le 1$. All critical attributes are proportioned into $\alpha_1,\ldots,\alpha_m$ with $\sum_{i=1}^{m}\alpha_i=1$ and $0\le\alpha_i\le 1$. Similarly, we let β_j $(m+1\le j\le m+s)$ with $\sum_{j=m+1}^{m+s}\beta_j=1$ and $0\le\beta_j\le 1$ express the relative importance of the non-critical attributes. We put different weights w_{ik} $(1\le i\le m+s,\ 1\le k\le n)$ on sub-attributes that constitute the *ith* trustworthy attribute which follows $\sum_{k=1}^{n} w_{ik}=1$ and $0\le w_{ik}\le 1$. For simplicity, in the following of this paper we denote the i with $\min_{1\le i\le m}\{y_i\}$ by min, the i with $\max_{1\le i\le m}\{y_i\}$ by max, the i with $\min_{m+1\le i\le m+s}\{y_i\}$ by min′, the i with $\max_{m+1\le i\le m+s}\{y_i\}$ by max′ and set $y_{\min}=\min_{1\le i\le m}\{y_i\}$, $y_{\max}=\max_{1\le i\le m}\{y_i\}$, $y_{\min'}=\min_{m+1\le i\le m+s}\{y_i\}$, $y_{\max'}=\max_{m+1\le i\le m+s}\{y_i\}$.

Definition 1 (A software trustworthiness measurement based on the decompositions of trustworthy attributes M3).

$$\begin{cases} T=\{\alpha[(\frac{y_{\min}}{10})^{\varepsilon}\, y_1^{\alpha_1} y_2^{\alpha_2}\cdots y_m^{\alpha_m}]^{-\rho}+\beta[y_{m+1}^{\beta_{m+1}} y_{m+2}^{\beta_{m+2}}\cdots y_{m+s}^{\beta_{m+s}}]^{-\rho}\}^{-\frac{1}{\rho}} \\ y_i=(\sum_{k=1}^{n} w_{ik}x_k^{-\rho_i})^{-\frac{1}{\rho_i}},\ 1\le i\le m+s \end{cases}$$

where

1) *ϵ is used to control the effect of the minimum critical attribute on the software trustworthiness with* $0\le\epsilon\le\min\{1-\alpha_{min}, \frac{\ln y_0-\ln y_{\min}}{\ln y_{\min}-\ln 10}\}$, *of which y_0 is the user expectation;*
2) *$0<\rho$ is a parameter related to the substitutivity between critical and noncritical attributes, the bigger ρ is, the more difficult the substitutivity between critical and non-critical attributes is;*
3) *$0<\rho\le\rho_i$ $(1\le i\le m+s)$ are parameters that are associated with the*

substitutivity among the sub-attributes that constitute the ith trustworthy attribute, similarly, the bigger ρ_i is, the more difficult the substitutivity among the sub-attributes is;

4) $1 \le x_k \le 10$, $1 \le k \le n$.

For convenience, in the rest of this paper, we let

$$a_1 = [(\frac{y_{\min}}{10})^{\varepsilon} y_1^{\alpha_1} y_2^{\alpha_2} \cdots y_m^{\alpha_m}]^{-\rho}, b_1 = [y_{m+1}^{\beta_{m+1}} y_{m+2}^{\beta_{m+2}} \cdots y_{m+s}^{\beta_{m+s}}]^{-\rho}.$$

Claim 1. (a) $1 \le y_i \le 10$, $1 \le i \le m + s$.

(b) $1 \le T \le 10$ and T complies with expectability.

Proof. (a) Because $0 < \rho_i$ and $0 \le w_{ik}$ $(1 \le i \le m + s,\ 1 \le k \le n)$, then

$$\sum_{k=1}^{n} w_{ik} \{\max_{1 \le k \le n}\{x_k\}\}^{-\rho_i} \le \sum_{k=1}^{n} w_{ik} x_k^{-\rho_i} \le \sum_{k=1}^{n} w_{ik} \{\min_{1 \le k \le n}\{x_k\}\}^{-\rho_i}.$$

Substituting $\sum_{k=1}^{n} w_{ik} = 1$ and then raising to $-1/\rho_i$ power in the above inequality, it follows that $\min_{1 \le k \le n} \{x_k\} \le y_i \le \max_{1 \le k \le n} \{xk\}$. Because $1 \le \min_{1 \le k \le n} \{x_k\} \le \max_{1 \le k \le n} \{x_k\} \le 10$, then $1 \le y_i \le 10$.

(b) Since $0 \le \alpha_i \le 1$ $(1 \le i \le m)$, $\sum_{i=1}^{m} \alpha_i = 1$, then

$$\frac{y_{\min}^{1+\varepsilon}}{10^{\varepsilon}} \le (\frac{y_{\min}}{10})^{\varepsilon} y_1^{\alpha_1} y_2^{\alpha_2} \cdots y_m^{\alpha_m} \le y_{\max}.$$

Similarly, because $0 \le \beta_j \le 1, m + 1 \le j \le m + s$, $\sum_{j=m+1}^{m+s} \beta_j = 1$, therefore

$$y_{\min'} \le y_{m+1}^{\beta_{m+1}} y_{m+2}^{\beta_{m+2}} \cdots y_{m+s}^{\beta_{m+s}} \le y_{\max'}.$$

Because of $0 < \rho$ and $0 \le \alpha, \beta \le 1$, $\alpha + \beta = 1$ we obtain that

$$\min\{\frac{y_{\min}^{1+\varepsilon}}{10^{\varepsilon}}, y_{\min'}\} \le T \le \max\{y_{\max}, y_{\max'}\} = \max\{y_1, \cdots, y_{m+s}\}.$$

Notice that we have proved that $1 \le y_i \le 10$, $1 \le i \le m + s$, therefore $1 \le T \le 10$.

If $y_0 \le \min\{y_1,\ldots,y_{m+s}\}$, because $0 \le \epsilon \le \min\{1-\alpha_{min}, \frac{\ln y_0 - \ln y_{\min}}{\ln y_{\min} - \ln 10}\}$, it follows that

$$y_0 \le \min\{\frac{y_{\min}^{\varepsilon}}{10^{\varepsilon}}, y_{\min'}\} \le \max\{y_1, \cdots, y_{m+s}\},$$

then we can deserve that $y_0 \le T \le \max\{y_1,\ldots,y_{m+s}\}$. □

Claim 2. Monotonicity is satisfied by T.

Proof. By calculating the partial derivatives of T with respect to y_i $(1 \le i \le m + s)$, we can obtain

$$\frac{\partial T}{\partial y_i} = \begin{cases} \alpha\alpha_i T^{1+\rho} a_1 y_i^{-1} & i \neq \min, 1 \leq i \leq m \\ \alpha(\alpha_i + \varepsilon) T^{1+\rho} a_1 y_i^{-1} & i = \min \\ \beta\beta_i T^{1+\rho} b_1 y_i^{-1} & m+1 \leq i \leq m+s. \end{cases}$$

From the Claim 1, we know that $1 \leq y_i \leq 10(1 \leq i \leq m+s)$, $1 \leq T \leq 10$. And because for $1 \leq i \leq m, m+1 \leq j \leq m+s$, $0 \leq \alpha$, α_i, $\alpha_{min} + \epsilon \leq 1$, $0 \leq \beta$, $\beta_j \leq 1$. Then

$$\frac{\partial T}{\partial y_i} \geq 0, 1 \leq i \leq m+s.$$

Since for $1 \leq i \leq m+s$, $1 \leq k \leq n$, $1 \leq x_k \leq 10$ and $0 \leq w_{ik} \leq 1$, $\sum_{k=1}^{n} w_{ik} = 1$.Then we can derive

$$\frac{\partial y_i}{\partial x_k} = w_{ik} (\sum_{k=1}^{n} w_{ik} x_k^{-\rho_i})^{-\frac{1}{\rho_i}-1} x_k^{-\rho_i - 1} \geq 0.$$

Therefore, monotonicity holds for T. □

Claim 3. T complies with the acceleration.
Proof. By computing the second order partial derivatives of T with respect to $y_i (1 \leq i \leq m+s)$, we get that

$$\frac{\partial^2 T}{\partial y_i^2} = \begin{cases} \left[\alpha\alpha_i^2 (1+\rho)\alpha a_1 T^{\rho} - \alpha\alpha_i^2 \rho - \alpha\alpha_i\right] T^{1+\rho} a_1 y_i^{-2} & i \neq \min, 1 \leq i \leq m \\ \left[\alpha(\alpha_i + \varepsilon)^2 ((1+\rho)\alpha a_1 T^{\rho} - \rho) - \alpha(\alpha_i + \varepsilon)\right] T^{1+\rho} a_1 y_i^{-2} & i = \min \\ \left[\beta\beta_i^2 (1+\rho)\beta b_1 T^{\rho} - \beta\beta_i^2 \rho - \beta\beta_i\right] T^{1+\rho} b_1 y_i^{-2} & m+1 \leq i \leq m+s \end{cases}$$

Because $0 \leq \alpha_i \leq 1(1 \leq i \leq m)$, $0 \leq \beta_j \leq 1(m+1 \leq j \leq m+s)$ and $0 < \alpha a_1 T_\rho \leq 1$, $0 < \beta b_1 T_\rho \leq 1$, $\alpha_{min} + \epsilon \leq 1$. It follows that

$$\begin{cases} \alpha\alpha_i^2 (1+\rho)\alpha a_1 T^{\rho} - \alpha\alpha_i^2 \rho - \alpha\alpha_i \leq 0 & i \neq \min, 1 \leq i \leq m \\ \alpha(\alpha_i + \varepsilon)^2 (1+\rho)\alpha a_1 T^{\rho} - \alpha(\alpha_i + \varepsilon)^2 \rho - \alpha(\alpha_i + \varepsilon) \leq 0 & i = \min \\ \beta\beta_i^2 (1+\rho)\beta b_1 T^{\rho} - \beta\beta_i^2 \rho - \beta\beta_i \leq 0 & m+1 \leq i \leq m+s \end{cases}$$

Therefore, we have

$$\frac{\partial^2 T}{\partial y_i^2} \leq 0, 1 \leq i \leq m+s.$$

Likewise, for $1 \leq i \leq m+s$, $1 \leq k \leq n$, we have

$$\frac{\partial^2 y_i}{\partial x_k^2} = w_{ik} (1+\rho_i)(\sum_{k=1}^{n} w_{ik} x_k^{-\rho_i})^{-\frac{1}{\rho_i}-1} x_k^{-\rho_i - 2} (\frac{w_{ik} x_k^{-\rho_i}}{\sum_{k=1}^{n} w_{ik} x_k^{-\rho_i}} - 1) \leq 0.$$

In summary, T satisfies the acceleration. □

Claim 4. Sensitivity holds for T.
Proof. It is easy to prove that

$$0 \le \frac{\partial T}{\partial y_i}\frac{y_i}{T} = \begin{cases} \alpha\alpha_i a_1 T^{\rho} & i \ne \min, 1 \le i \le m \\ \alpha(\alpha_i + \varepsilon) a_1 T^{\rho} & i = \min \\ \beta\beta_i b_1 T^{\rho} > 0 & m+1 \le i \le m+s \end{cases}$$

Then we can obtain the conclusion that T is sensitive to all trustworthy attributes and the minimal critical attribute affects on the whole software trustworthiness more than other attributes by adding of ϵ relative to its importance.

Similarly, for the sensitivity of *ith* ($1 \le i \le m + s$) trustworthy attribute about the *kth* ($1 \le k \le n$) sub-attribute, we can obtain that

$$0 \le \frac{\partial y_i}{\partial x_k}\frac{x_k}{y_i} = \frac{w_{ik} x_k^{-\rho_i}}{\sum_{k=1}^{n} w_{ik} x_k^{-\rho_i}}.$$

In a word, sensitivity is satisfied by T. □

Claim 5. T satisfies substitutivity.
Proof. According to the equation (2), the substitutivity among the sub-attributes which compose of the *ith* ($1 \le i \le m + s$) trustworthy attribute can be determined as follow

$$\sigma_{ijk} = \frac{1}{1+\rho_i}, 1 \le i \le m+s, 1 \le k, l \le n, k \ne l$$

Likewise, according to the equation (1), for the substitutivity among critical attributes, we have $\sigma_{ij} = 1$, ($1 \le i, j \le m$, $i \ne j$), for the substitutivity among non-critical attributes $\sigma_{ij} = 1$, ($m + 1 \le i, j \le m + s$, $i \ne j$), and the substitutivity between critical and non-critical attributes can be derived as

$$\sigma_{ij} = \begin{cases} \dfrac{1 + \dfrac{y_i}{y_j}\dfrac{\partial T/\partial y_i}{\partial T/\partial y_j}}{(1+\rho\beta_i) + (1+\rho\alpha_j)\dfrac{y_i}{y_j}\dfrac{\partial T/\partial y_i}{\partial T/\partial y_j}} & j \ne \min, 1 \le j \le m, m+1 \le i \le m+s \\ \dfrac{1 + \dfrac{y_i}{y_j}\dfrac{\partial T/\partial y_i}{\partial T/\partial y_j}}{(1+\rho\beta_i) + (1+\rho\varepsilon+\rho\alpha_j)\dfrac{y_i}{y_j}\dfrac{\partial T/\partial y_i}{\partial T/\partial y_j}} & j = \min, m+1 \le i \le m+s. \end{cases}$$

It is easy to prove that for $j \ne min$, $1 \le j \le m$, $m + 1 \le i \le m + s$,

$$\frac{1}{1+\rho_i} \le \frac{1}{1+\rho} \le \frac{1}{1+\rho\max\{\alpha_j, \beta_i\}} \le \sigma_{ij} \le \frac{1}{1+\rho\min\{\alpha_j, \beta_i\}} \le 1$$

and for $j = min$, $m + 1 \leq i \leq m + s$,

$$\frac{1}{1+\rho_i} \leq \frac{1}{1+\rho} \leq \frac{1}{1+\rho\max\{\alpha_j+\varepsilon,\beta_i\}} \leq \sigma_{ij} \leq \frac{1}{1+\rho\min\{\alpha_j+\varepsilon,\beta_i\}} \leq 1$$

To sum up, the conclusion can be verified. □

From the conclusions of Claim 1-5, we can get that this measure complies with all the properties described in section 2.

4 Case Study

To demonstrate the effectiveness of our measure, we evaluate the trustworthiness of a network software, its critical attributes include reliability y_1 and maintainability y_2. Its non-critical attributes are composed of portability y_3 and testability y_4. Reliability is described by error tolerance x_1, consistency x_2, simplicity x_3 and accuracy x_4. Maintainability is divided into consistency x_2, simplicity x_3, modularity x_5 and self-descriptiveness x_6. The decomposition of portability are modularity x_5, self-descriptiveness x_6, machine independence x_7 and software system independence x_8. Testability is associated with simplicity x_3, modularity x_5, self-descriptiveness $x6$ and instrumentation x_9. We suppose $(\alpha, \beta) = (0.7, 0.3)$, $(\alpha_1, \alpha_2, \beta_3, \beta_4) = (0.6, 0.4, 0.5, 0.5)$, and set

$$(w_{11}, w_{12}, w_{13}, w_{14}, w_{15}, w_{16}, w_{17}, w_{18}, w_{19}) = (0.3, 0.3, 0.2, 0.2, 0, 0, 0, 0, 0)$$
$$(w_{21}, w_{22}, w_{23}, w_{24}, w_{25}, w_{26}, w_{27}, w_{28}, w_{29}) = (0, 0.4, 0.2, 0, 0.2, 0.2, 0, 0, 0)$$
$$(w_{31}, w_{32}, w_{33}, w_{34}, w_{35}, w_{36}, w_{37}, w_{38}, w_{39}) = (0, 0, 0, 0, 0.25, 0.25, 0.3, 0.2, 0)$$
$$(w_{41}, w_{42}, w_{43}, w_{44}, w_{45}, w_{46}, w_{47}, w_{48}, w_{49}) = (0, 0, 0.1, 0, 0.3, 0.3, 0, 0, 0.3)$$
$$(\rho, \rho_1, \rho_2, \rho_3, \rho_4) = (0.4, 0.5, 0.6, 0.5, 0.7)$$

For convenience, we suppose the measures proposed in [16-17] are M_1 and M_2 separately. Three small simulations for M_1, M_2 and M_3 with parameters described above are given in Table 1. From the Sim. 2 in this table, we can observe that if the threshold y_0 decided by user is 8.48, then M_1 and M_2 do not satisfy expectability.

Table 1. Simulations for M_1, M_2 and M_3

	x_1	x_2	x_3	x_4	x_5	x_6	x_7	x_8	x_9	ϵ	y_1	y_2	y_3	y_4	M_3	M_1	M_2
1	8	8	9	8	9	8	8	8	8	0.1	8.2	8.4	8.2	8.4	8.1	5.6	5.2
	8	8	9	8	9	8	8	8	8	0.01	8.2	8.4	8.2	8.4	8.3	5.7	5.2
2	8	9	9	8	9	8	9	8	9	0.1	8.5	8.8	8.5	8.7	8.5	5.7	5.3
	8	9	9	8	9	8	9	8	9	0.01	8.5	8.8	8.5	8.7	8.6	5.8	5.3
3	7	8	9	8	7	8	8	7	8	0.1	7.9	8.0	7.5	8.0	7.7	5.5	5.0
	7	8	9	8	7	8	8	7	8	0.1	7.9	8.0	7.5	8.0	7.8	5.5	5.1

5 Comparative Study in Terms of the Properties Described in Section 2

In the following of this section we compare M_3 with M_1 [16], M_2 [17] and two other popular software trustworthiness measures: evidence theory based software trustworthiness measure (ERBSTM) [4] and fuzzy theory based software trustworthiness measure (FTBSTM) [7] through the properties given in section 2. The comparative results are summarized in Table 2, of which × represents the measure does not hold for the corresponding property and √ expresses the measure complies with the corresponding property. From the Table 2, we can obtain that M_3 is better than all the four measures through the properties introduced in section 2.

Table 2. Comparison of M_3 with 4 established measures in terms of properties presented in section 2

Measure \ Property	M_3	M_1	M_2	FTB STM	ERB STM
Monotonicity	√	√	√	√	√
Acceleration	√	√	√	×	×
Sensitivity	√	√	×	√	√
Substitutivity	√	√	√	×	×
Expectability	√	×	×	√	√

6 Conclusion and Future Work

In this paper, we use axiomatic approaches to measure software trustworthiness from the point of decomposition of trustworthy attributes and validate the measure M3 from theory. Compared with the model we once introduced in [16-17], ERBSTM [4] and FTBSTM [7], this model can measure software trustworthiness better in view of the properties described by us.

The properties given in section 2 should be considered as necessary sets but not sufficient ones, we will extend the set of properties presented here in the future. Those software trustworthiness measures based on the decomposition of trustworthy attributes that do not comply with the set of properties for software trustworthiness cannot be considered as reasonable ones. However, the models that do comply with the properties described here should only be considered as candidate measures, they still need to be better examined. Therefore, carrying out a thorough empirical validation of the model given in this paper to support its usefulness in practice is our future work too.

Acknowledgment. This article is supported from the National Natural Science Foundation of China (No. 91118007, 61472447), Shanghai Knowledge Service Platform Project (No. ZF1213).

References

1. Liu, K., Shan, Z., Wang, J., He, J., Zhang, Z., Qin, Y.: Overview On Major Research Plan of Trustworthy Software. Bull. of Nat. Nat. Sci. Foun. of Ch. 22(3), 145–151 (2008)
2. Hasselbring, W., Reussner, R.: Toward Trustworthy Software systems. IEEE Trans. on Comp. 39(4), 91–92 (2006)
3. Steffen, B., Wilhelm, H., Alexandra, P., et al.: Trustworthy Software Systems: a Discussion of Basic concepts and Terminology. ACM SIGSOFT Soft. Eng. Notes 31(6), 1–18 (2006)
4. Ding, S., Yang, S., Fu, C.: A Novel Evidential Reasoning Based Method for Software Trustworthiness Evaluation Under the Uncertain and Unreliable Environment. Exp. Sys. with App. 39, 2700–2709 (2012)
5. Li, X., Wang, X., Zhang, T., Yi, J.: Software Trustworthiness Evaluation Based on Weakness Analysis and Testing Assessment. J. Tsinghua Univ (Sci.& Tech.) 51(10), 1287–1293 (2011) (in Chinese)
6. Yuan, Y., Han, Q.: A Data Mining Based Measurement Method For Software Trustworthiness. Chin. J. of Elec. 21(1), 13–16 (2012)
7. Shi, H., Ma, J., Zou, F.: Software Dependability Evaluation Model Based on Fuzzy Theory. In: Zhou, J., Xie, Y., Chai, S., Wen, D. (eds.) Computer Science and Information Technology, pp. 102–106. IEEE Press, Piscataway (2008)
8. Fenton, N., Littlewood, B., Neil, M., Strigini, L., Sutcliffe, A., Wright, D.: Assessing Dependability of Safety Critical Systems Using Diverse Evidence. IEE Proc. Softw. 145(1), 35–39 (1998)
9. Wang, J., Chen, Y., Gu, B., Guo, X., Wang, B., Jin, S., Xu, J., Zhang, J.: An Approach to Measure and Grading Software Trust for Spacecraft Software. Sci. Sin. Tech. 45(2), 221–228 (2015)
10. Morasca, S.: Fundamental Aspects of Software Measurement. In: De Lucia, A., Ferrucci, F. (eds.) ISSSE 2009-2011. LNCS, vol. 7171, pp. 1–45. Springer, Heidelberg (2013)
11. Poels, G., Dedene, G.: Distance-based Software Measurement: Necessary and Sufficient Properties for Software Measures. Infor. and Soft. Tech. 42(1), 35–46 (2000)
12. Weyuker, E.: Evaluating Software Complexity Measures. IEEE Trans. on Soft. Eng. 14(9), 1357–1365 (1988)
13. Lakshmanan, K., Jayaprakash, S., Sinha, P.: Properties of Control-flow Complexity Measures. IEEE Trans. on Soft. Eng. 17(12), 1289–1295 (1991)
14. Briand, L., Morasca, S., Basili, V.: Property-based Software Engineering Measurement. IEEE Trans. on Soft. Eng. 22(1), 68–86 (1996)
15. Morasca, S.: Refining the Axiomatic Definition of Internal Software Attributes. In: Hans, D., Sebastian, G., Jürgen, M. (eds.) Empirical Software Engineering and Measurement, pp. 188–197. IEEE Press, Piscataway (2008)
16. Tao, H., Chen, Y.: A Metric Model for Trustworthiness of Softwares. In: Boldi, P., Vizzari, G. (eds.) Web Intelligence and Intelligent Agent Technology, vol. 3, pp. 69–72. IEEE Press, Piscataway (2009)
17. Tao, H., Chen, Y.: A New Metric Model for Trustworthiness of Softwares. Tele. Sys. 51(2-3), 95–105 (2012)

Review Relationship TPM as Mediator between TQM and Business Performance

M.F. Ahmad, A.H. Nor Aziati, Abdul Talib Bon, Y. Ngadiman, and Shiau Wei Chan

Department of Production & Operation Management
Faculty of Technology Management
Universiti Tun Hussein Onn Malaysia, Johor, Malaysia
{mohdfauzi,aziati,talib,yunos}@uthm.edu.my

Abstract. In today's highly competitive market, the demand for quality is the single most critical factor for companies to survive in the ever-expanding global marketplace. Total Quality Management (TQM) has been developed as a result of intense global competition. Most of the previous works show that TQM has significant relationship with business performance. However, the examining of moderators is less given in previous work, which mediators are known generally as general tools and techniques without specific focus on types of improvement. The purpose of this paper is to propose relationship between TQM practices and business performance with mediators effect of Total Productive Maintenance (TPM) based on extensive review of the literature. The main contribution of this paper is to identify the relationships among TQM, TPM practices and business performance as a conceptual model. This proposed conceptual model will help the academicians and industry players to have better understanding on the relationship between the practices.

1 Introduction

The concept of total quality management (TQM) has been developed as a result of intense global competition (MFB. Ahmad & Yusof, 2010; MFB Ahmad et al., 2008; Garvin, 1988). According to Garvin (1988), international competition requires higher levels of quality achievement to meet the customer satisfaction. TQM is a management philosophy that helps manage their organization to improve the effectiveness and performance to achieve world class status for the past two decades (M. F. Ahmad et al., 2012; M. F. Ahmad et al., 2014; Konecny & Thun, 2011). However, the study of mediators is neglected and is referred to less frequently in literature review. The fundamental systems-interactive paradigm of organisational analysis features the continual stages of input, processing, and output, which demonstrate the concept of openness and closeness. Processing is the process of changing from one "look" to another, or one culture to another (Rouse, 2005;M.F Ahmad et al., 2014). In this study, the author defines input as TQM; processing as application tools and techniques namely TPM; and output as business performance. Thus, one of the objectives of this study is to empirically analyse the impact of TPM between TQM and business performance.

M. Gen et al. (eds.), *Industrial Engineering, Management Science and Applications 2015*,
Lecture Notes in Electrical Engineering 349, DOI: 10.1007/978-3-662-47200-2_103

1.1 Literature Review

The first mediator that will be examined is total productive maintenance (TPM). The purpose of TPM is to minimise breakdown and to maximise equipment availability of production systems at minimal cost (Hans, 2000; Nakajima, 1988). TPM started in Japan in 1971 (Abdallah, 2013; Abdallah & Matsui, 2007). The word "total" means total effectiveness, total maintenance system, and total participation of all employees (Nakajima, 1988). On the other hand, preventive maintenance is a routine inspection to detect potential failures and to make minor adjustments or repairs that will prevent major breakdown maintenance at a premium cost (Konecny & Thun, 2011). Thus, TPM can be defined as an improvement programme that establishes a comprehensive productive-maintenance system throughout the entire life of the equipment with the participation of all employees through voluntary team-based activities (Dal et al., 2000). Furthermore, TPM can also be defined as an approach to achieve improvement of production processes by involving and empowering production employees, and by introducing an ongoing process improvement (Nakajima, 1988). The Japan Institute of Plant Maintenance (JIPM) defines TPM as maximising equipment effectiveness with a total system of preventive maintenance covering the entire life of the equipment involving everyone in all departments and at all levels (Imai, 1986). The Western definition of TPM is the philosophy in which all equipment in production are always kept in optimum condition and available for maximum output (Peter, 1994).

1.2 Methodology

An extensive literature search was designed to identify and retrieve primary empirical studies relevant to develop the hypotheses. The databases searched were SpringerLink, Emerald, Taylor & Francis, ScienceDirect, Elsevier, ProQuest and Google Scholar. The descriptor TQM and environment effect were used when possible; otherwise, it was searched as a keyword.

1.3 Hypotheses

H1: Relationship among TQM, TPM, and Business Performance

Most previous studies indicate a significant relationship between TQM practices and business performance (Jun et al., 2006; Bou & Beltrán, 2007; Gunday et al., 2011; Miyagawa & Yoshida, 2010;M. F. Ahmad et al., 2013). In contrast, other studies show that TQM does not improve business performance (Corredor & Goñi, 2011; Demirbag et al., 2006; M. F. Ahmad et al., 2013). Other findings show partial correlation between TQM practices and business performance (Demirbag et al., 2006; Feng et.al., 2006; Arumugam et al., 2008). Accordingly, the author proposes that:

H1: TQM practices are positively significant and have direct effects on business performance.

TPM is significantly supported by TQM for improving business performance (Konecny & Thun, 2011; Ahmad M.F , et al., 2014). Teeravaraprug et al., (2011) suggest that TQM and TPM should be implemented before lean production. Two sets of factors are critical for the effectiveness of TQM and TPM: (1) universally significant factors for all three approaches such as leadership, process management, and strategic planning; and (2) approach-specific factors such as equipment management and focus on customer satisfaction (Seth & Tripathi, 2005). TPM is comprehensive improvement originating from the concept of zero defects of TQM, which applies to control equipment performance (Seth & Tripathi, 2006). Teeravaraprug et al., (2011) reveals that TQM practices such as 5S, QC tools and Kaizen have significant impact on TPM. Abdallah (2013) indicates that TQM practices such as top management leadership, customer focus, training and continuous improvement have significantly impact on TPM implementation. Thus, TQM practices are positively correlated with TPM.
H2a: TQM practices are positively significance and direct effect on TPM.

Ahuja and Khamba (2008) note the critical success factors of TPM, such as top management leadership and involvement, maintenance practices, as well as holistic TPM initiatives, enhance business performance in Indian industry. Konecny and Thun (2011) indicate that TQM and TPM supported by human resource practices significantly improve business performance. Ahmad et al., (2012), Brah & Chong (2004) and Nakajima (1988) indicate that TPM practices have impact on business performance. Therefore, TPM practices are positively correlated with business performance. Accordingly, the author proposes that:
H2b: TPM practices are positively significance and direct effect on business performance.

There are lack of empirical evidences of TPM as mediator between TQM and business performance in previous work (Sadikoglu & Zehir, 2010). In this study, TPM acts as mediator between TQM and business performance. TQM provides soft and hard aspects such knowledge, skill, continuous improvement and work environment and culture to support adoption of TPM implementation (Konecny & Thun, 2011; Teeravaraprug et al., 2011). Meanwhile, TPM has positively impacts on business performance (Ahuja & Khamba, 2008; Brah & Chong, 2004; Nakajima, 1988). Thus, TQM will be helpful for effective and efficient in TPM adoption, in turn increasing business performance. Accordingly, the author proposes that:
H2: TPM is a mediator between TQM and business performance.

1.4 Conceptual Framework

Structural equation modeling (SEM) techniques are utilized to examine the relationships. Based on a comprehensive review of previous studies, a conceptual model has been proposed to understand the relationships as presented in Figure 1.

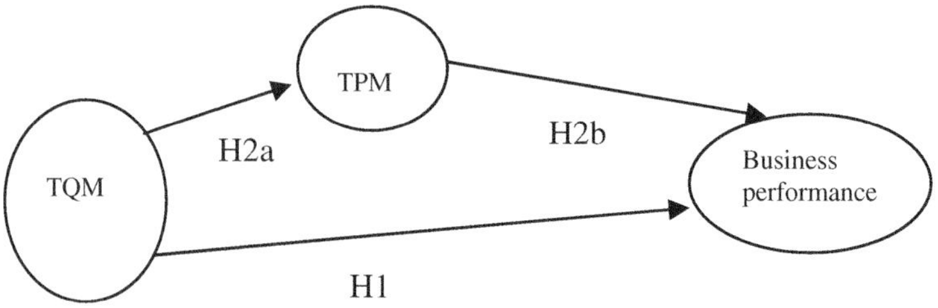

Fig. 1. Proposed conceptual model of TQM with mediators of TPM

2 Conclusion

The main objectives of this study are to investigate the relationships among TQM, TPM and business performances as a conceptual model. 3 hypotheses regarding the relations among TQM, TPM and business performance have been specified and conceptual framework have been proposed for future work.

Acknowledgement. The authors would like to thank Universiti Tun Hussein Onn (UTHM) for its support in this project.

References

Abdallah, A.B.: The Influence of "Soft" and "Hard" Total Quality Management (TQM) Practices on Total Productive Maintenance (TPM) in Jordanian Manufacturing Companies. International Journal of Business and Management 8(21), 1–13 (2013), doi:10.5539/ijbm.v8n21p1

Abdallah, A.B., Matsui, Y.: JIT and TPM: Their relationship and impact on JIT and competitive performances. TPM 1(9), 1–18 (2007)

Ahmad, M.F., Zakuan, N., Jusoh, A., Takala, J.: Relationship of TQM and Business Performance with Mediators of SPC, Lean Production and TPM. Procedia - Social and Behavioral Sciences 65(ICIBSoS), 186–191 (2012), doi:10.1016/j.sbspro.2012.11.109

Ahmad, M.F., Zakuan, N., Jusoh, A., Yusof, S.M., Takala, J.: Moderating Effect of Asean Free Trade Agreement between Total Quality Management and Business Performance. Procedia - Social and Behavioral Sciences 129, 244–249 (2014), doi:10.1016/j.sbspro.2014.03.673

Ahmad, M., Yusof, S.: Comparative study of TQM practices between Japanese and non-Japanese electrical and electronics companies in Malaysia: Survey results. Total Quality Management & Business Excellence 21(1), 11–20 (2010), doi:10.1080/14783360903492520

Ahmad, M.F.B., Yusof, S., Yusof, N.: Comparative study of quality practices between Japanese and non-Japanese base electrical and electronic companies in Malaysia: A Survey. Jurnal Teknologi 47(A), 75–89 (2008)

Ahmad, M.F., Arif, M., Zakuan, N., Rahman, S., Abdullah, T., Fadzil: The Effect of Demographics on Customer Satisfaction amongst Malaysia Hajj Pilgrims: Survey Result. Applied Mechanics and Materials 660, 1000–1004 (2014), http://www.scientific.net/AMM.660.1000
Ahmad, M.F., Zakuan, N., Jusoh, A., Takala, J.: Review of Relationship between TQM and Business Performance. Applied Mechanics and Materials 315, 166–170 (2013), doi:10.4028/www.scientific.net/AMM.315.166
Ahmad, M.F., Zakuan, N., Jusoh, A., Tasir, Z., Takala, J.: Meta-analysis of the relationship between TQM and Business Performance. In: IOP Conference Series: Materials Science and Engineering, vol. 46, p. 120. IOP Publishing, Bandung (2013), doi:10.1088/1757-899X/46/1/012020
Ahmad, M., Zakuan, N., Jusoh, A., Takala, J.: Relationship of TQM and Business Performance with Mediators of SPC, Lean Production and TPM. Procedia - Social and Behavioral Sciences 65(2012), 186–191 (2012), doi:10.1016/j.sbspro.2012.11.109
Ahmad, M.F., Zakuan, N., Jusoh, A., Yusof, S.M., Takala, J., Arif, M.S.: Comparative Study of TQM practices between Japanese and non-Japanese Companies: Proposed Conceptual Framework. Applied Mechanics and Materials 529(2014), 371–377 (2014), http://www.scientific.net/AMM.529.371
Ahuja, I.P.S., Khamba, J.S.: An evaluation of TPM initiatives in Indian industry for enhanced manufacturing performance. International Journal of Quality & Reliability Management 25(2), 147–172 (2008), doi:10.1108/02656710810846925
Ahuja, I.P.S., Khamba, J.S.: An evaluation of TPM initiatives in Indian industry for enhanced manufacturing performance. International Journal of Quality & Reliability Management 25(2), 147–172 (2008)
Arumugam, V., Ooi, K.-B., Fong, T.-C.: TQM practices and quality management performance: An investigation of their relationship using data from ISO 9001:2000 firms in Malaysia. The TQM Journal 20(6), 636–650 (2008), doi:10.1108/17542730810909383
Bou, J.C., Beltrán, I.: Total Quality Management, human resource strategy and firm performance: an empirical study. Total Quality Management 16(1), 37–41 (2005)
Brah, S.A., Chong, W.K.: Relationship between total productive maintenance and performance. Journal of Production Research 42(12), 2383–2401 (2004)
Corredor, P., Goñi, S.: TQM and performance: Is the relationship so obvious? Journal of Business Research 64(8), 830–838 (2011), doi:10.1016/j.jbusres.2010.10.002
Dal, B., Tugwell, P., Greatbanks, R.: Overall equipment effectiveness as a measure for operational improvement. Journal of Operations and Production Management 20(12), 1488–1502 (2000)
Demirbag, M., Tatoglu, E., Tekinkus, M., Zaim, S.: An analysis of the relationship between TQM implementation and organizational performance: Evidence from Turkish SMEs. Journal of Manufacturing Technology Management 17(6), 829–847 (2006), doi:10.1108/17410380610678828
Feng, J., Prajogo, D.I., Tan, K.C., Sohal, A.S.: The impact of TQM practices on performance: A comparative study between Australian and Singaporean organizations. European Journal of Innovation Management 9(3), 269–278 (2006), doi:10.1108/14601060610678149
Garvin, A.G.: Managing Quality. The Free Press, New York (1988)
Gunday, G., Ulusoy, G., Kilic, K., Alpkan, L.: Effects of innovation types on firm performance. International Journal of Production Economics 133(2), 662–676 (2011), doi:10.1016/j.ijpe.2011.05.014
Hans, L.: Measuring maintenance performance- in search for a maintenance productivity index. Int. J. Production Economics 63(8), 47 (2000)
Imai, M.: Kaizen: The Key to Japan's Competitive Success. Business. McGraw-Hill, New York (1986)

Jun, M., Cai, S., Shin, H.: TQM practice in maquiladora: Antecedents of employee satisfaction and loyalty. Journal of Operations Management 24(6), 791–812 (2006), doi:10.1016/j.jom.2005.09.006
Konecny, P.A., Thun, J.H.: Do it separately or simultaneously—An empirical analysis of a conjoint implementation of TQM and TPM on plant performance. International Journal of Production Economics 133(2), 496–507 (2011), doi:10.1016/j.ijpe.2010.12.009
Miyagawa, M., Yoshida, K.: TQM practices of Japanese-owned manufacturers in the USA and China. International Journal of Quality & Reliability Management 27(7), 736–755 (2010), doi:10.1108/02656711011062363
Nakajima, S.: Introduction to Total Productive Maintenance (TPM). Productivity Press, Cambridge (1988)
Peter, W.: Total Productive Maintenance. Butterworth-Heinemann Ltd., London (1994)
Rouse, W.B.: A theory of enterprise transformation. Systems Engineering 8(4), 279–295 (2005), doi:10.1002/sys.20035
Sadikoglu, E., Zehir, C.: Investigating the effects of innovation and employee performance on the relationship between total quality management practices and firm performance: An empirical study of Turkish firms. International Journal of Production Economics 127(1), 13–26 (2010), doi:10.1016/j.ijpe.2010.02.013
Seth, D., Tripathi, D.: Relationship between TQM and TPM implementation factors and business performance of manufacturing industry in Indian context. International Journal of Quality & Reliability Management 22(3), 256–277 (2005), doi:10.1108/02656710510582480
Seth, D., Tripathi, D.: Total Quality Management & Business A critical study of TQM and TPM approaches on business performance of Indian manufacturing industry A Critical Study of TQM and TPM Approaches on Business Performance of Indian Manufacturing Industry. Total Quality Management 17(7), 37–41 (2006)
Teeravaraprug, J., Kitiwanwong, K., SaeTong, N.: Relationship model and supporting activities of JIT, TQM and TPM. Songklanakarin Journal of Science and Technology 33(1), 101–106 (2011)

Theoretical Review of Critical Factors that Impact on Global Human Resource Practices: Case on Multinational Companies in Emerging Economies

Muhammad Mehmood Aslam[1], Syed Shaheer Hassan Rizvi[2], and Asif Hameed[3]

[1] Department of Management Sciences,
The Islamia University of Bahawalpur, Pakistan
[2] Islamia University of Bahawalpur, Pakistan
[3] Admin, Ministry of Education, Pakistan
urmehmood@yahoo.com, shaheer_rizvi@hotmail.com,
asifhameed59@gmail.com

Abstract.
Purpose

The aim of this review paper is to critically analyse those factors which influence upon the way human resource practices are worked-out in MNCs particularly those that operate in emerging markets like China & India. These factors comprises internal factors and also external factors that play their pivotal role in shaping the said HR practices like recruitment & selection, compensation, remuneration, training and development, etc. and therefore these determinants if truly taken into consideration by multinational corporations increase the likelihood of success that these companies know better ways in order to attract, develop and retain employees that keep potential, skills and capabilities with them. For firms whose strategic focus is on its employees do prosper and can earn competitive advantage against their counterpart even if they are operating in developed countries. Hence, in order to highlight those critical determinant/factors the conceptual framework completed by Mohan Thite, Adrian Wilkinson & Dhara Shah on the reserach subject of internationlization and Human resource strategies in MNCs of emerging markets. After having gone through the literature we come to know that the conceptual framework keeps some gaps in itself and doesn't include other important and critical factors which have been proposed by our research in form of extented concpetual framwork like for instance, strategic role of HR department, high performance work system (HPWS), corporate social responsibilty and its link with HR practices and many more.

Value

This research keeps significance for those MNCs which operate in emerging markets. Despite the fact that stiff competition, advance technological interventions, globalizaiton have made very much difficult environment for MNCs to earn success in form of competitive

M. Gen et al. (eds.), *Industrial Engineering, Management Science and Applications 2015*,
Lecture Notes in Electrical Engineering 349, DOI: 10.1007/978-3-662-47200-2_104

advantage in the industry, companies can lead from the front if they manage to utilize their workforce potential in the right direction. To achieve this end companies will have to deliberate upon internal and external factors that exert influnce over their HR practices as it is well known to everyone that if organizaitons want to gain competitive or even comparative advantage they need to take their human resource as a strategic asset not merely as an input for work process.

Conclusion

It has been concluded that if MNCs really want to outperform in emerging markets those should have to take all concerned critical factors into consideration which are significant for bringing success to them in form of outcomes that are to be generated by capable workforce MNCs have got with them. Thus, all depends on the way MNCs attract, develop and retain their human resource which is composed of diversity and get affected by multifarious challanges and from the organizational point of view it gives glimpse to the solution of problems that arise from forward or reverse diffusion to MNCs.

Keywords: Human Resource Practices, Competitive Advantage, Multinational Corporations, Emerging Markets, Capabilities Diffusion.

1 Introduction

By this time when every body is going through very fast pace to get ahead of ferocious competition prevailing in domestic and international markets. Every company is also putting forth best of it to come over all those concurrent and perplexing issues that could make them come by success and achievements irrespective of how big their competitors are. Amongst many factors that suitably enables an organization to have competitive advantage, that seems to be only source of survival today, those factors are also very much important that related directly or indirectly while managing human resources in international perspective.

There are many organizations which keep their human resources in better high value and also train and develop them accordingly. On the contrary, others do take care of profits only nothing except that. In this current research we are going to make an analysis as how organizations make their way through in building their human resource as a pivotal source of competitive advantage while taking into consideration various major and minor factors/variables that impact the policies and practices in human resource management when those organization trespass the geographical boundaries of their country in which they really exist. The research to be conducted is more likely to be qualitative in nature. However, it would be reviewed deeply from already researched work that what practices are involved in designing HR strategy and which factors do influence the way the said HR strategies are designed or implemented.

For gaining the research objective a profound look is required to have an understanding of the situations in which particularly Multinational Corporations performs their Human resource management policies and practices; thereupon it would be observed as which factors do curb the likelihood that MNCs will execute the said policies and practices successfully.

After having analyzed different researches made in response to know the arising importance of Human resource practices, it is observed that there has been little work done particularly to know about HR practices that are followed in MNCs which operate in emerging countries like India, China, etc. however by integrating different studies and analyzing the research made by Thite, Wilkinson & Shah on internationalization & HRM strategies across subsidiaries in MNC from emerging economies, we can figure out different categories of variables/factors that not only influence upon the employees working out there in those MNCs but also largely affect the way of their businesses. Having known that if MNC really want to have strategic competitive advantage in marketplace, it becomes possible if the organization is doing well in respect of giving importance to its human resource, its capabilities, skills and competencies and also number of factors that are related to corporate level of the organization. For instance, how much organization is giving importance to corporate social responsibility which is not merely the concept as how much firm does good with the society rather it also includes the way the organization utilizes and manages its human resource by having socially responsible behavior like for example, how to elevate the wellbeing standard of employees at work is a matter of paramount importance for the firm. Therefore, in this manner different factors have been observed throughout this research; which critically impact on HR practices of an MNC in emerging markets. These include internal and external factors; in which internal factors are said to be those which are someway or somehow in control of the organization and external factors would be those which are not under control of an MNC. For example, internal factors can be training and development of the human resource with strategic focus on it and its resultant relation with the performance of the organization, non-local competencies, alignment and realignment between HRM and organizational strategy, workforce diversity in the form of HCN, PCN, TCN and inpatriates, strategic role of human resource department, internal pressure within the organization and employees' related factors that is to say work, nature of work, remuneration and compensation, taxation, recruitment and selection, cross-cultural issues for expatriates, etc.

On the contrary there are also external factors like for instance economic situation in host countries, political pressures and the national culture prevailing in host countries, the level of control, coordination and communication that take places between subsidiary and its parent firm, pace of globalization, compliance with employee safety standards, technological advancements and its intervention into business processes, location of MNC from which it originates, etc.

It is to be notified that both internal and external factors that keep influence upon an MNC and its HR practices, forces the entire organization to do well in view of upgrading employees standards in terms of their working and also in terms of how much they are being given weight age as being employee of that MNC. Thus the ultimate goal of a successful MNC always remain to be an outperforming company by means of attracting, developing and retaining the pool of talented workforce at the workplace which can in turn become a likelihood of attaining completive advantage that is also to be sustainable for sure.

2 Literature Review

The phenomenon of globalization has genuinely led to creation of huge companies which aspire to expand their operational activities across multifarious countries

(Touron, 2008). Big companies do vary on wide spectrum in respect of their **strategically prevailing position**, **structure of the organization** and **processes of the management** itself. One of those existing type of organizations is multinational organization whose very primary and basic characteristic involves its sensitivity and the extent to which it tends to be responsive when it comes across different **environments particularly the national one**. It occasionally comprises multiple portfolios of national units which operate independently to each other (Bartlet & Ghoshal, 2000). MNCs in emerging economies seems to be small in **size along with least resources and experience** that is required internationally as compare to their counterparts prevailing in developed countries. (Thite, Wilkinson, & Shah). It is observed that 500 largest multinational organizations prove to be key leaders for being driving force in terms of globalization of multinational companies in future course of time, which also include firms situated at developing and emerging markets around the globe (Martelli & Abels, 2011). Study conducted before reveals a strong tendency from the side of students for preferring MNCs as compare to small and/or medium level firms when it come for them to make first choice in terms of their career (Teo & Poon, 1994; Moy & Lee, 2002 as cited in Gokuladas, 2010). Market perplexity as well as **trends related to internal organization** besides the development do have alter the manner of these firms when it comes to organize the activities globally or/and locally, in terms of their assets no matter tangible or intangible in nature. However apart from the whole discussions held before one thing turns out to be common is that as drastic changes occur in competition so as the effectiveness of the multinational companies' conventional approaches gets reduced (Michailova & Nielsen, 2006). Moreover, according to comparative institutional approach multinational corporations along with their subsidiaries are likely to be influenced by many factors like **way of thinking**, **behavior**, **structure** and **operating models** prevailing in those countries from which these corporations originate (Whitely, 2001 as sited in Chang, Smale, & Tsang, 2013).

Multinational companies are most of the time criticized for **stepping on cultures forcefully** and the **existent diversity** across the world and these seem to be embarking upon the journey to those countries where different standards are either weak or they even don't exist for instance **safety standards, governance, employees wellbeing standards, etc**. But at the same time these companies proved to be the transformers for having brought the learnt behavior, over a long span of time, to the new emerging markets encompassed with the ways to gain socially responsible status comparatively at **low cost** against what happens to be in developed economies. In economist the **social responsibility** is defined as how to generate wealth for their stakeholders, putting people towards work, uplifting the standards of living besides contributing their earnings to communities around them (Prout, 2006). In addition to that Baden et. Al., (2009) as cited in (Adebanjo, Ojadi, Laosirihongthong, & Tickle, 2013) highlighted **employees and staffing challenges/issues** an additional classified category so as far CSR is concerned. Human resource works as only the sole function in the whole organization that has considerable influence over the people at work and also impact over the whole organization for all times to come. HR management becomes quite handful in fostering the CSR culture into entire organization. Thus the **CSR-HR** has got connections between them. Looking at emerging evidence at present, if CSR is very well implemented, then CSR can make a significant and strong influence in maintaining the staff, developing the staff & retaining the staff as well.

Internationally HR managers are making an all-out effort in developing & also implementing appraisal system and incentive programs that bring forth sustainability in hiring people for work who also are to uphold those CSR values. (Strandberg, 2009).

Apart from economic downturn and unemployment, extremely advance countries like Japan and USA will be facing a shortage of talent over a long period of time due to the **retirement of the generation** called baby boomers and therefore, there seemed to be more people are retiring as compare to the people entering the job/labor market (Kapoor & Sherif, 2012).

Some economic factors also have an influence upon the propensity of Multinational Companies to invest abroad such as cost-related determinants, **factors relating to market** itself, how much **people of that organization keep local knowledge** besides **availability of technology** (Zvirgzde, Schiller, & Diez, 2013). While discussing the **strategic role of human resource department** of an MNC; the role of operations of its subsidiary Human Resource department in the host country be kept in consideration with enough focus on it (Sumelius, Smale, & Bjorkman, 2009). Similarly, many factors like **globalization into businesses**, emerging technologies, **volatile effects over nature of work and types of work**, **different sort of skill requirements**, **declining trend of unionization**, ever **rising professionalism of human resource practitioners** as well as researchers besides devolving employment severe conditions played a pivotal role in the development of HRM and its various practices and theories (Bach & Bordogna, 2011; Rau, 2012; Tissen et. al., 2010 as cited in Ananthram, Nankervis, & Chan, 2013). **Philosophy and the approach used by top management** of the organization assists designs and influence the overall IHRM system; exclusively the system that is destined to be utilized in abroad gets facilitated (Thite, Wilkinson, & Shah). It is further argued that **organizational pressure that are said to be internal** play a significant role by putting immense influence on Human resource recipes and the way those are delivered (Brewster, Sparrow, & Harris, 2005)**. Therefore, on one side the pressure on MNC is to respond locally and on the flipside of the coin there is a need of global integration along with the consistency within the framework of an MNC** (Myloni, Harzing, & Mirza, 2006)**.** However, merely confined to focus on strategic issues, sometimes human resource department leaves employees in isolation for not being in direct contact with them and therefore, this disconnection may bring them into a vulnerable situation due to the fact that Human Resource department becomes incapable of raising employee related issues in front of the top management of the company (Gilbert et. Al, 2011 as cited in Najeeb, 2013).

Amongst human resource management functions **training and development** is considered one the most integral function that influences upon the level of quality of human resource of the organization and resultantly also have **strong effect on the performance of the organization** (Losey, 1999; Mcdonnell, 2008 as cited in Sheehan, 2012). It is also argued that IHRM practices are mostly influenced by the **country of origin** in which MNC lies despite the fact it has got high or even low cultural context (Thite, Wilkinson, & Shah).

Sometimes few **non-local competencies**, that once developed in the parent company headquarter or even at subsidiary becomes **transferable**, enables multinational companies to maintain **competitive advantage** in the whole world (Borini & Fleury, 2011). Similarly, take resource based theory which transcribes that in order to gain and manage competitive advantage for the better interest of the MNC; **HRM needs to be in alignment** with the specifically defined organizational competencies (Barney, Conner, 1991 as cited in Myloni, Harzing, & Mirza, 2006).

Even in global marketing academic literature the burning issue that still continues related to the control of multinational companies over operations made in the foreign markets. Notwithstanding the fact that importance of having control on foreign operations that is kept in mind of global marketers, that is a **very costly** to do in form of initial investments & while sacrificing the autonomy of their foreign operations (Dong, Zou, & Taylor, 2008). It is further argued that HRM system makes use of **control mechanism** which is put in place to monitor that employees in overseas subsidiaries are working in the very best interest of the MNC and that the HRM system can be a **High performance work system (HPWS)** usually used for and considered important for control mechanism. Therefore, research evidence exhibits that both HPWS and strategic HRM orientation are said to be crucial for attaining the MNC goal while contributing its performance and very important aspect of both is that these create a leverage of Human resource capital of the firm and bring competitive advantage for the firm in definite terms (Foley, Ngo, & Loi, 2012). Many of the researches have taken into account corporate level context such as international strategy for competitiveness, experience of international business besides the influence of values that management places in line of HRM, and also spotlights the relationship that lie between parent and the subsidiary including many aspects like communication and the control (Beechler & Yang, 1994).

Hofstede, 2007 as cited in (Thite, Wilkinson, & Shah) argues that so as far emerging economies MNCs are concerned, *decision making style, culture of the organization & control over subsidiaries* is altogether very much different from that of those MNCs in developed countries because of their national culture & the difference that lies economically. Many academicians have noted that MNCs tend to adopt adaptive also called polycentric approach in the developed economies as compare to least developed economies due to the fact that MNC becomes successful in getting strong availability of skills required by managers (Bazeley & Richards, 2000 as cited in Thite, Wilkinson, & Shah).

Many academicans also went further to state that many factors at present are influential with respect to international human resource management for instance, **employee related issues** like visa-imigration, **cross cultural issues**, **firms strategic focus**, **dual couple career challenges**, **remunation, cost management, recruitment and selection issues and taxastion** (Cieri, Fencwich, & Hutchings, 2003). The complexity and perplexity while operating in many differet countries invovles the deployment of different categories of workers like expatriate, host coutnry nationals, third country nationals and inpatriates that fuels up **diversity in the organization** and these also make a line of damaracation between domestic & international HR management and that is quite different to the number of HR practices involved in HRM (Kiessling & Harvey, 2005).

The conceptual framework that has been propsed by *Mohan Thite, Adrian Wilkinson & Dhara Shah* discussed in literature review has been extracted (shown in Figure 1) and higlighted below for better comprehension of their work. However; the conceptual framework, afterwards, has been extented in our research to highlight critically significant factors both internal and external ones which impacts on HR policies and practices of multinational corporations that operate in emerging markets.

2.1 Figure No. 1

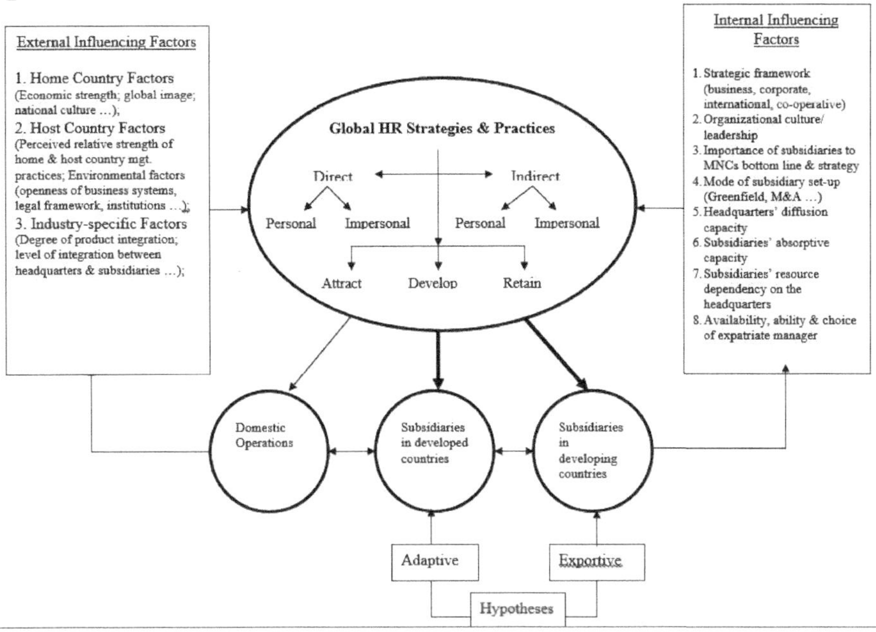

Fig. 1. Diffusion of Global HR Strategies & Practices across Subsidiaries in a Multinational Corporation from an Emerging Economy

3 Findings and Analysis

After having gone through the whole literature published by eminent academicians and research in different point of times we have analyzed various variables/factors that truly impact on the HR practices of Multinational companies particularly in emerging markets like China & India where power distance, for instance, is very much high and where due to cost constraint High performance work systems have not been put in place by MNCs. This review is a kind of critical analysis over the ***work of Mohan Thite, Adrian Wilkinson & Dhara Shah on the matter of internationlization and Human resource strategies in MNCs of emerging marktes*** and hence the conceptual framework they have been produced; particularly remained the focal point of interest during our research. The said conceptual framework seems to highlight different factors that influence upon HR practices of MNCs operating in emerging markets and the factors like internal ones and external ones play their role in designing the way through which MNCs attract, develope and retain their human resource capital which can lead the firm towards having competitive advantage.

However, after review the literaturre this research has brough forth some critical factors that has not been made as a part of that conceptual framework. Critical here refers to the sensitivity they hold with themselves. Like for instance, the role of corporate social responsibility seems to be very critical for the success of the MNC in the sense that if the firm considers its employees as an asset not merely work input this would be a strategic move for the firm to earn them competitive advantage over their counterparts. Likewise, the other factor is to implement high performance work system at subsidaries which comprises a full fledge control mechanisim for resouce monitoring for the parent company. American MNCs have been using that system at their subsidaries but unluckily due to its high cost mechanisim it has not been implemeted in Asian region MNCs on a wide spectrum. Next comes the role of HR department, which is to be considered more important so as far its strategic significance is concerned. If the subsidary HR department doesn't cordianate well with the parent company, this is supposedly called an isolated HR department with no significance at all. However, if the department is in direction of taking employee as an integral part of the whole process, it will raise the voice of employees in parent company meetings and sessions and thats how through this move the subsidary would have far reaching effects over its performance. Again the conceptual framwork lacked different varibles that particulary related to employees or group of employees working in MNCs of emerging markets domain. Employees concurrent issues like compensation, taxation, their safety and wellbeing standards do influence the way through which employees are developed & also retained at MNCs. On the contary, the experience and age of MNC also plays a pivotal role and even it exerts influence the way employees are taken care of. Suffice to say here that subsidaries of MNCs in emerging economies lack resource and capabilities that are not found to be witnessed in the parent country, though. The bigger the size of the company the more challanging it would become to deal with the diverse workforce and to address their arising challanging issues like for example as how to control that workforce. Likewise the role of employees' training and development is very much cruical due to the fact that organizations seek the capable, loyal employees at workplace who turn out to be very much helpful in bringing alignment in the firm's strategy and its various units strategies. Therefore, it is also well know and stated before as well that HR department adopt a pivotal role to play as the decisions it makes affect the whole organization entirely.

On the other side, the framework has less talked about external factors which though not under the control of the MNC while it is to operate overseas; these need to be put into consideration by the parent company. For instance, there is a rising trend of the phenomenon that particularly in USA and Japan the older generation called baby boomer is near to retirement stage and interestingly both the countries are hardly expecting their replacements with fresh blood as they are not in such number as well. Likewise, most of the multinational corporations do step up host country cultures forcefully which is not seen good and certainly not receive respect from the customers out there. Same is the case with ever changing technological advancements in the industry which put MNCs in a messy situation and all this come through the phenomenon of gloabization and the pace attached to it. Large western companies are making full use of being globalized and finding many ways to look for staff fueled up with skills and competencies which can bring the company a sustainable comparative advantage if not competitive advantage.

3.1 Figure No. 2

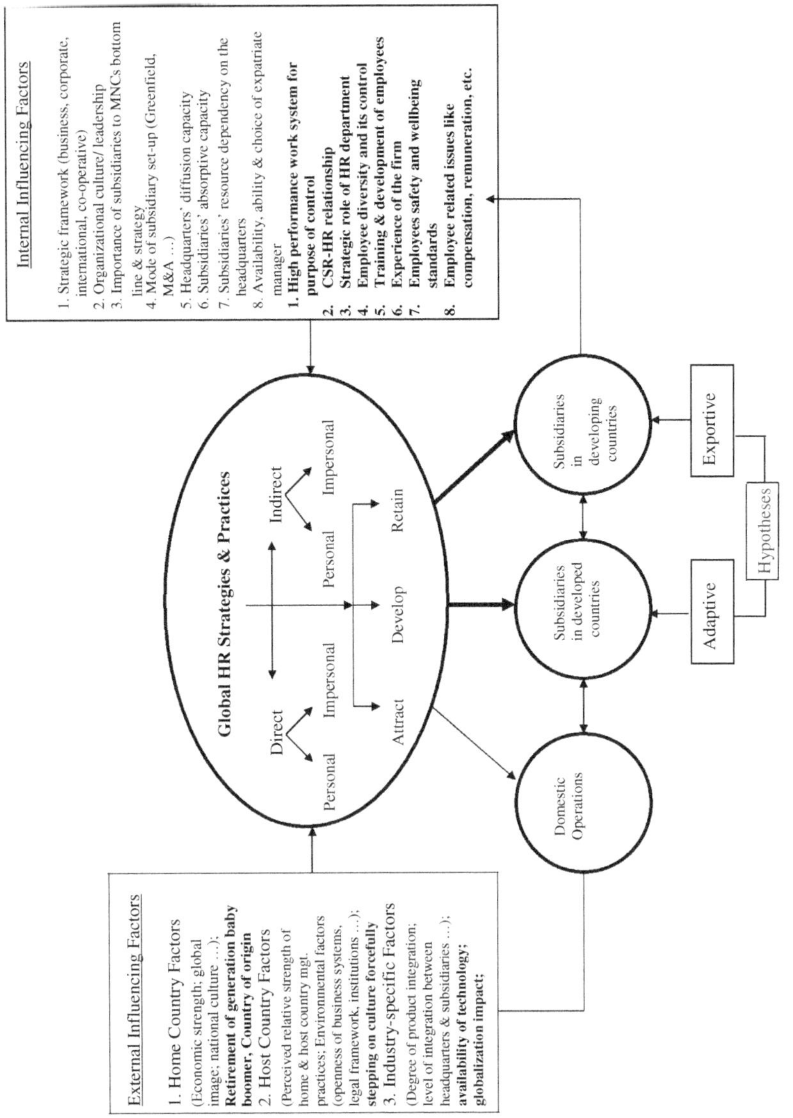

Fig. 2. Diffusion of Global HR Strategies & Practices across Subsidiaries in a Multinational Corporation from an Emerging Economy

Limitations
So as far the limitations are concerned to the extent of this review paper; we may go on to analyze these as below:

- Firstly, this review paper focuses solely on the conceptual framework made by Mohan Thite, Adrian Wilkinson & Dhara Shah and doesn't take into account other models having different concept involved into them.
- Secondly, there are certainly different perspectives to think about the topic in discussion but this review has been made focusing only the evolving natures of variables. Other different variables that are not obvious could not be identified.

4 Conclusion

If we put all said above together we may conclude that in order to be successful and to gain competitive advantge in the curent areana of globalization and technological advacements; multniationl corporations will have to strive for attaing capable workforce, though diverse in nature having diverse nature of works and tasks associated with them. For achieving this end the parent companies in emerging markets or the subsidaries that are operating in emerging markets and whose parent headquaters found to be in developed countries, MNCs have to take into consideration internal and external factors which do influence human resouce practices like recruitment and selection, compensation, managing expatriates, inpatriates, dual couple career challanages, minimizing costs related to managing human capital, etc. These factors are also of immense important in the sense that these *varibles exert a far reaching impact on human resouce practices of MNCs at times of attracting, developing and retaining capable, competitive workforce.*

Therefore, the conceptual framework has not included those critical factors that our reseach has proposed in the extended framework like high performacne work systems, strategic role of human resource department, the role of employee training and development, the linkage between corporate social responsibility and human resource practices, numerous employees related issues like cross cultural issues, firms strategic focus, dual couple career challenges, remuneration, cost management, recruitment and selection issues and taxastion. On the contrary our extented proposed model contains various external factors like globalization, technological advancement, country of origin of an MNC, cultural diversity and how to manage this diversity, retirement of baby boomer generation which would also affect the emering markets at times when "forward or reverse diffusion" of capability will be required.

In a nutshell, we may come up with the fact that MNCs operating in emerging markets like China & India lag behind in terms of resources and experience than those counterpart happened to be in developed economies; should have to take these cricital factors into their immediate consideration becuase this is the only way through from difficulty that MNCs face during their stay out there and these companies would therefore become able to better attract, develope and retain their workforce which could bring them competitive advantage if not more.

References

1. Adebanjo, D., Ojadi, F., Laosirihongthong, T., Tickle, M.: A case study of supplier selection in developing economies: a perspective on institutional theory and corporate social responsibility. Supply Chain Management: An International Journal 18(5), 553–566 (2013)
2. Ananthram, S., Nankervis, A., Chan, C.: Strategic human asset management: evidence from North America. Personnel Review 42(3), 281–299 (2013)
3. Bartlet, C., Ghoshal, S.: Managing the companies in global marketplace. The antidote from CSBC (25) (2000)
4. Beechler, S., Yang, J.Z.: The transfer of Japanese-style management to American Subsidiaries: Contingencies, constraints and competencies. Journal of International Business 25(3), 467–491 (1994)
5. Borini, F.M., Fleury, M.T.: Development of non-local competences in foreign subsidiaries of Brazilian multinationals. European Business Review 23(1), 106–119 (2011)
6. Chang, Y.Y., Smale, A., Tsang, S.S.: A diachronic analysis of HRM transfer: Taiwanese multinationals in the UK. Cross Cultural Management 20(3), 464–482 (2013)
7. Cieri, H.D., Fencwich, M., Hutchings, K.: International Human Resource Managment: Challanges for Teaching and Practices 59(3) (2003)
8. Dong, B., Zou, S., Taylor, C.R.: Factors That Influence Multinational Corporations' Control of Their Operations in Foreign Markets: An Empirical Investigation. Journal of International Marketing 16(1), 98–119 (2008)
9. Foley, S., Ngo, H.Y., Loi, R.: The adoption of high performance work systems in foreign subsidiaries. Journal of World Bsusiness, 106–113 (2012)
10. Gokuladas, V.K.: Factors that influence first-career choice of undergraduate engineers in software services companies. Career Development International 15(2), 144–165 (2010)
11. Kapoor, B., Sherif, J.: Global human resources (HR) information systems. Kybernetes 41(1/2), 229–238 (2012)
12. Kapoor, B., Sherif, J.: Global human resources (HR) information systems. Kybernetes 41(1/2), 229–238 (2012)
13. Kiessling, T., Harvey, M.: Strategic global human resource management research in the twenty-first century: an endorsement of the mixed-method research methodology. International Journal of Human Resource Management 16(1), 22–45 (2005)
14. Martelli, J.T., Abels, P.B.: Multinational corporations: the changing landscape. Management Research Review 34(8), 889–909 (2011)
15. Michailova, S., Nielsen, B.B.: MNCs and knowledge management:a typology and key features. Journal of Knowledge Management 10(1), 44–54 (2006)
16. Myloni, B., Harzing, A.-W., Mirza, H.: The effect of corporate-level organisational factors on the transfer of human resource management practices: European and US MNCs and their Greek subsidiaries. The International Journal of Human Resource Management (2006)
17. Najeeb, A.: The role of HR actors in designing and implementing HRM in tourist resorts in the Maldives. Employee Relations 35(6), 593–612 (2013)
18. Prout, J.: Corporate responsibility in the global economy: a business case. Society and Business Review 1(2), 184–191 (2006)
19. Sheehan, M.: Developing managerial talent; Exploring the link between management talent and perceived performance in multinational corporations (MNCs). European Journal of Training and Development 36(1), 66–85 (2012)

20. Sparrow, P., Brewster, C., Harris, H.: Globalizing Human Resource Management. Routledge, London (2004)
21. Strandberg, C.: The Role of Human Resource Management in Corporate Social Responsibility (2009)
22. Sumelius, J., Smale, A., Bjorkman, I.: The strategic role of HR in MNC subsidiaries in China between 1999 and 2006. Chinese Management Studies 3(4), 295–312 (2009)
23. Thite, M., Wilkinson, A., Shah, D. (n.d.): Internationalization & HRM Strategies across Subsideries in Multinational Corporation from Emerging Economies - A Conceptual Framework
24. Touron, M.M.: From an ethnocentric to ageocentric approach to IHRM, The case of a French multinational company. Cross Cultural Management 15(4), 335–352 (2008)
25. Zvirgzde, D., Schiller, D., Diez, J.R.: Location choices of multinational companies in transition economies: A literature review. Deliverable 2(1), 1–22 (2013)

Ranking Measures for Sustaining Quality Excellence Practices: An Empirical Investigation

Mehran Doulatabadi[1] and Sha'ri Mohd Yusof[2]

[1] Department of Business Administration,
Universiti Teknologi Malaysia, Johor Bahru Campus, Johor, Malaysia
[2] Department of Manufacturing and Industrial Engineering
Universiti Teknologi Malaysia, Johor Bahru Campus, Johor, Malaysia
dmehran2@live.utm.my, shari@utm.fkm.my

Abstract. Organizations regardless of their size and sector need to measure and review their quality performance strategically and tactically for gaining a competitive advantage. The primary purpose of this paper is to empirically identify and analyze the measures perceived as important for assessing business excellence practices within the United Arab Emirates context. Eight factors with a total of 52 practice sub-factors have been identified and derived from an extensive literature review followed by expert's panel assessment. A self-administered questionnaire survey is conducted to confirm the adaptability of the proposed measured with 138 companies from different sector and business. The SPSS software was employed to perform the required statistical analysis of the data collected from surveys. Based on analysis, five factors including 29 sub-factors have proposed as excellence practices measures for assessing business excellence practices. The proposed measures then become the main constructs and basis for developing a self-assessment tool in the future research.

1 Introduction

In this era of fast-changing business economy, there are now increasing number of commercial and public companies across the globe that are setting their vision and chief target on world class excellence. Over the years, quality subject has been regarded as a very important competitive priority for the long-term success among industries. In today business environment, it would be hard to find an organization that ignores the practice of quality management approaches. Within the increasing changing of the business economy, organizations have been forced to provide high quality consistently and pay more attention to quality of products and services. In light of the growing importance of quality management around the world, the UAE as the first developing country in the Middle East including North Africa (MENA) has enhanced competitiveness in the region by creating several significant quality and excellence award programs and schemes. Upon further analysis of the current literature and despite the numerous wealth of studies on quality management practices in various countries, very few studies have been observed with respect to the UAE context. As such, there is a need to conduct an empirical study to identify and analyze

M. Gen et al. (eds.), *Industrial Engineering, Management Science and Applications 2015*,
Lecture Notes in Electrical Engineering 349, DOI: 10.1007/978-3-662-47200-2_105

the factors that critical in sustaining quality management practices in the long run. As such, the main purpose of this survey is to match the quality excellence measures derived from the relevant literature with industry practices. This study is probably one of the first and the most extensive studies to date to examine quality management practices within the UAE context. This paper therefore seeks to investigate the critical factors in the practice of the quality management through surveying the past Dubai Quality Award recipients as an effective quality improvement programme in the region.

The paper is organized into six main sections. Following the introductory part, a description of the literature reporting the factors for quality management practices is presented. Then, it explains the methodology of study followed by results and discussion on the survey analysis. The paper ends with final conclusions, and suggestions for future research.

2 Literature Review

A comprehensive review of literature confirms that various pioneering researchers studies in developed countries empirically identified the measures with regards to quality management practices ([1],[2],[3],[4],[5],[6],[7],[8],[9],[10],[11],[12]), but very few studies have been observed with respect to the UAE context in general, particularly in the quality excellence context.

A plethora of literatures have been published to date devoted to quality management and its associated issues in the UAE context have been mostly descriptive, speculative, or theoretical in general [13]. Studies, however, use different methodology, instruments or constructs for their research such as SERVQUAL instrument modification [14], [15]; service quality measurement [16]; service quality and leadership [17], and the motives and benefits of ISO 9001: 2000 [18], but they focused mainly on the UAE banking industry. Surprisingly, only a limited set of empirical studies have thoroughly addressed to the concept of quality excellence practices [19], [20], and continuous quality improvement [21].

The first major research project on quality management was carried out by [22] to validate measurement of the critical factors of quality by focusing in UAE service and manufacturing industry. The research was replicated, but the study by [1] was conducted to prove the viability of the developed instrument in an international context. Their study revealed weaknesses in the instrument, in which there were elements that needed further clarification.

The research conducted by [13] attempted to identify soft elements critical to the successful TQM implementation process through the survey questionnaire in the UAE banking sector. They highlighted 16 factors that are critical for the successful implementation of TQM. However, this study has certain limitations such that its results may not be applicable to other industrial sectors.

[19] also undertook a survey to identify the most critical elements to achieve organizational excellence based on experience of nearly 250 participating engineering firms in the UAE and Saudi Arabia. They suggested fifteen traditional critical elements of excellence without focusing much on the soft elements of quality management practices. Finally, [21] conducted an exploratory survey with associated

to continuous quality improvement within the UAE industries. The author has indicated five key success factors by analysing only two case study organizations based in the UAE.

3 Methodology

In order to achieve the aim and objective of the study, an extensive literature review has been carried out to understand the barriers and difficulties to BX. In addition, a self-administered questionnaire survey was also used as primary instruments to elicit the required data and information. The questionnaire contained eight factors and 52 sub-factors that identified from an extensive literature review followed by expert's panel assessment. A five point Likert scale was used to measure the response with 1 (not important at all) to five (an extreme important).

To support the survey's objective all local private and public companies and institutions in various types of business were included as a target sample for the study. The selected companies are physically operating in the four largest land territories in the UAE including Abu Dhabi, Dubai, Sharjah, and Al Ain and mainly grouped in nine distinct industrial sectors and business clusters from small-to-medium enterprises to large companies with different degree of BX maturity. They were included manufacturing, service, construction, healthcare, higher education, trade, professional, finance, and tourism.

The questionnaire ensured to be completed by the person responsible for quality or business excellence in the company. They were mainly the Managing Directors and Quality Managers since they directly involved in the process and have first-hand knowledge of quality and business excellence practice. The final data collected from the survey instrument was mainly analyzed through some descriptive statistics using of Statistical Package for Social Sciences (SPSS) software Version 21.

With respect to the survey's objective, a research instrument was developed in the form of a questionnaire containing several items as an appropriate instrument to collect the required data and information. This questionnaire was designed based on a variety of inputs such as literature review, inputs from quality experts, a comparative study of quality award models and field visits. Then, the questionnaire was improved in terms of clarity based on a pilot survey.

The respondents were asked to determine the level of importance of the critical factors rating on a five-point interval scale ranging from 1 (being not important) to 5 (being an extremely important). The first page highlighted the objectives of the study. The survey questionnaire of the study is included two main sections as follows:

Section 1- Basic information about the organizational collection of demographic variables and Section 2- A number of 52 statements which resulted from an extensive review of literature on the critical factors to quality management practices as well as the feedback of academics and practitioners in the field.

The final data collected form survey questionnaires were mainly analyzed through Statistical Package for Social Sciences (SPSS) software Version 21. The descriptive analysis in the form of frequency analysis, percentage, cumulative frequency, and cumulative percentages were conducted to develop an overall understanding of the survey responses and a general picture of representative companies. Furthermore, the initial statistical techniques such as means and standard deviations were also used to analysis the research variables. In this process, all valid replies questionnaires were carefully transferred into the SPSS data file after ensuring that the respondents marked their questionnaires correctly.

3.1 Survey Sample and Response Rate

Of 138 sample companies that were potentially reachable, 92 companies returned the survey questionnaire completely. However, 9 companies were not reachable during the data collection process due to company move to another place or not operate. A total of 19 companies also did not give any respond due to reasons given such as lack of time, work pressures, company policy, and confidentiality issues. One (1) survey questionnaire was also unusable and did not use in data analysis process due to incomplete some questions and using the incorrect way in ranking the statements.

Therefore, with combining all the responses, a total of 91 completed questionnaires were used for analysis which is considered to be adequate for the survey method. In order to calculated response rate of the survey, the following formula was used as advocated by [23]:

$$\text{Active response rate} = \frac{\text{total number of responses}}{\text{total number in sample} - (\text{ineligible} + \text{unreachable})} \quad (1)$$

$$\text{Active response rate} = \frac{92}{138 - (1 + 26)} = 82.88\ \%$$

This response rate i.e. 82.88 percent is considered as very high response rate compare with previous studies around the topic since accessing the managers is usually difficult. Table 1 provides a summary of the survey responses' distribution and rate for this study.

Table 1. Survey responses' distribution and rate

Descriptions	Number/Percentage
Questionnaires distributed	138
Questionnaires completed and returned	92
Unreachable companies	27
Declined participation	19
Total response rate	82.88 %

3.2 Survey Responses' Background

The companies which took part in the survey were varied in terms of the job position of respondents, size, type, and ownership of company. With regard to the number of years organizations were operated, most of the respondents companies (58.2 percent) were established more than 20 years. About 18.7 percent of these companies have involved in business for more than 10 to 15 years followed by 11 percent for 5 to 10 years. Only a small number of organizations surveyed (4.4 percent) were newly established within UAE.

Regarding to type of business, the majority of respondents companies in this study were completely fully local companies (56 percent) followed by semi-governmental (20.9 percent) and joint venture companies (12.1 percent). The analysis indicated that there was no company that completely owned by foreigners, as this type of business was not included in the survey.

About type of industry, the analysis indicated that the distribution of business activity was quite different. As presented in Figure 3, service (22 percent), trade (16.5 percent) and manufacturing (14.3 percent) companies were as three main industries within the sample size, while healthcare (5.5 percent), professional and education (3.3 percent) were the lowest number of participants.

In terms of the size of company which was grouped based on the approximate number of full-time employees, this was not surprising as the majority of the respondents (71.4 percent) were part of large- sized companies with more than 250 employees. While the small and medium sized enterprises (13.2 percent) were the reminder with fewer than 250 employees.

With respect to the role of respondents, the highest percentages of participants were from middle management level (46.2 percent) followed by executive directors (22 percent) and senior management (18.7 percent). Of the respondents, only 7.7 percent hold positions in supervisory management (employees), since the researcher requested the questionnaire be completed by the persons in management level.

In terms of the respondents working experience, the analysis indicates that about 46.2 percent of the respondents had been working within the current company for 5 years or more, followed by 28.6 percent for 3 to 5 years.

Of the respondents, about 85.7 present had been working in the field of quality management for 5 to 7 years and more than 7 years. Only a few numbers of respondents (5.5 percent) were being worked for less than 3 year in the area of quality. This indicates that the participants were sufficiently experienced with quality management and BX practices. This further shows that underscoring the validity of inferences drawn. Table 2 presents the primary background of the companies and respondents who participated in the survey.

3.3 Survey Reliability and Validity Test

A specific reliability and validity tests were also conducted on the survey instrument for the purpose of this study. In order to validate item reliability, Cronbach's alpha test was performed as an indicator of internal consistency for each set of items in survey questionnaire. Table 3 shows the results of reliability analysis based on the eight measures for the sample collected in this study. As shown in Table, Cronbach's Alpha values of the eight main factors are ranged from 0.748 to 0.947, which was significantly greater than the general requirement (> 0.70). Therefore, form the results, it can be concluded that as the internal consistencies of each factors were deemed to be high the survey questionnaire are accepted as being reliable for the survey.

Table 2. Primary background of the companies and respondents

Demographics	Frequency (%)	Demographics	Frequency (%)
Years of operation		*Position*	
< 5 year	0 (0.0)	Senior quality coordinator	1 (6.7)
5.01 - 10 years	3 (20.0)	Quality manager/officer	6 (40.0)
10.01 - 15 years	5 (33.3)	Business Excellence manager	5 (33.3)
15.01 - 20 years	1 (6.7)	Operations manager/officer	3 (20.0)
> 20 years	6 (40.0)	General manager/director	0 (0.0)
Type of business		*Work Experience*	
Fully local-private	7 (46.7)	< 1 year	0 (0.0)
Joint Venture	1 (6.7)	1.01 - 3 years	0 (0.0)
Semi-Governmental	6 (40.0)	3.01 - 5 years	3 (20.0)
Governmental	1 (6.7)	5.01 - 7 years	4 (26.7)
Fully foreign-private	0 (0.0)	> 7 years	8 (53.3)
No. of employees		*Education*	
< 20	0 (0.0)	High School/Graduate	1 (6.7)
21 - 100	1 (6.7)	Technical degree	1 (6.7)
101 - 250	1 (6.7)	Bachelor degree	2 (13.3)
> 251	13 (86.7)	Master degree	8 (53.3)
		Doctoral degree	3 (20)

Table 3. Results of Construct reliability test

Item No.	Measures	No. of Items	Alpha (α) value
F01	Leadership and management commitment	7	0.870
F02	Strategy and planning	7	0.869
F03	Empowerment and involvement	7	0.784
F04	Education and training	7	0.838
F05	Teamwork and cooperation	5	0.748
F06	Recognition and reward	5	0.947
F07	Communication and relationship	7	0.839
F08	Work culture and climate	7	0.775
	Total	52	

With regards to the validity of items, various methods were employed for validity of constructs. After the choice of the variable from an extensive review of literature, the proposed survey was first reviewed by a group of practitioners and academic researchers. In addition, the questionnaire was tested separately through a pilot survey with the selected sample. It was believed that the final version of questionnaire was greatly improved and validated before conducting the actual survey.

4 Results and Analysis

Having analyzed of the survey result, it is appropriate to begin analyzing the main part of the study which is to match the quality excellence measures derived from the relevant literature with industry practices. Two aspects have been investigated in this study based on a total of 52 measurement items under eight critical factors including *Usefulness* and *Usage* levels of quality excellence measures. For this purpose, the respondents were asked to rate the usage and usefulness level of all the measures. A more detail description of the results are given in the next following sections.

4.1 Usefulness Level of Measures

The respondents were asked to rate the usefulness level of the quality excellence measures within their companies. A five-point Likert-type scale ranging from 1 (not useful at all) to 5 (very useful) was used to assess the perspective of respondents to the usefulness level of the measures. Table 4 shows the means of each measure.

Based on the findings, the overall mean values range from 4.07 to 4.40, which correspond between 'moderately important' and 'very important' (3 to 4 on the Likert scale). As ranked in Table 4.6, it is fairly clear that, of the eight critical factors, *Leadership and management commitment* was indicated by survey respondents to be the most important factor ones with an average mean of 4.40. *Recognition and Reward* (4.34), *Communication and relationship* (4.33) and *Work culture and climate* (4.33) were perceived to be the top three most important factors with a near equal importance. On the other hand, the results showed that *Teamwork and cooperation* (4.15), *Empowerment and involvement* (4.08), together with *Education and training* (4.07) are the least important critical factors perceived by respondent. These three factors are consistently at the bottom of the list which have separated themselves from the other factors.

Table 4. Mean usefulness level of the measures

Item no.	Measures	Overall mean	Rank
F01	Leadership and management commitment	4.4097	1
F06	Recognition and reward	4.3407	2
F07	Communication and relationship	4.3375	3
F08	Work culture and climate	4.3344	4
F02	Strategy and planning	4.2810	5
F05	Teamwork and cooperation	4.1538	6
F03	Empowerment and involvement	4.0879	7
F04	Education and training	4.0754	8

4.2 Usage Level of Measures

The respondents were also asked to rate the usage level of the quality excellence measures within their companies. A five-point Likert-type scale ranging from 1 (very low) to 5 (very high) was used to assess the extent to which a company uses each measure. Table 5 illustrates the result of ranking the extent of practice for the study. As can be seen in Table, the values range from 3.91 to 3.64, which correspond to a `moderate' to `high' level of practice (3 to 4 on the Likert scale).

Based on the results, *Communication and Relationship* (3.91) and *Strategy and Planning* (3.91) were the two highest practices which were indicated by the respondents in the survey. While *Empowerment and Involvement* (3.68) and *Education and Training* (3.64) were the bottom two the same as the perceived importance results. From this result, it can be observed that all the respondents rated at `moderate' to 'high' for degree of business excellence practices within the companies, indicating that companies are struggling to practice business excellence successfully.

Table 5. Mean usage level of the measures

Item no.	Measures	Mean	Rank
F07	Communication and relationship	3.9152	1
F02	Strategy and planning	3.9137	2
F06	Recognition and reward	3.9099	3
F01	Leadership and management commitment	3.9042	4
F08	Work culture and climate	3.7943	5
F05	Teamwork and cooperation	3.7341	6
F03	Empowerment and involvement	3.6860	7
F04	Education and training	3.6405	8

5 Discussions and Conclusions

From the survey, it is concluded that most of the measures match with industry practices. *Leadership and Management Commitment* and *Recognition and Reward* were indicated as the most important measures by respondents for evaluating quality excellence. They also indicated *Communication and Relationship* and *Work Culture and Climate* at a high level. For *Teamwork and Cooperation*, *Empowerment and Involvement*, and *Education and Training* the respondents placed them at a low usage and usefulness levels.

From the similarity of the usage and usefulness level, it can be concluded that the companies were considered the measures as key quality excellence practice that are the most important for their current and future success. It is indicated that the measures are suitable and appropriate for the companies and can help them to become more efficient, effective, and competitive.

Based on the results, the quality excellence measures have been modified. Due the low usage and the less importance, Teamwork and cooperation, Empowerment and Involvement, and Education and Training were removed from the measures. Finally, five factors with a total of 32 dimensions have proposed as the measures to assess quality excellence practice as presented in Table 6.

This paper has presented the results from a survey of the quality excellence measures within the United Arab Emirates context. Five constructs with a total of 32 dimensions have proposed as the measures to evaluate quality excellence practices. These measures will hopefully assist in developing an instrument to assess the level of quality excellence. As such, the further research will need to develop an instrument to evaluate the quality excellence practices.

Table 6. Proposed measures as perceived by respondents

Overall Rank	Item no.	Measure /sub-measure
1	F01	Leadership and mgmt. commitment (LMC)
		L1. Active participation and involvement L2. Continuous support and strong commitment L3. Effective communication among individuals L4. Team spirit and motivation approaches L5. Clear vision and mission L6. Personal commitment and involvement
2	F06	Recognition and reward (RER)
		R1. Recognition and appreciation system R2. Recognition as part of the motivation R3. Recognition as basic to increasing the involvement R4. Recognition via Senior executive R5. Continuous rewarded of employees
3	F07	Communication and relationship (COR)
		C1. Effective communication systems C2. Actual employee communication C3. Using ideas offered by junior staff C4. Employee relations and respects C5. Formal communication methods in all levels C6. 'Open door' policy
4	F08	Work culture and climate (WCC)
		W1. Management commitment to treating employees W2. Aware of the concept of 'excellence' W3. Continuously reinforce the culture of 'excellence' W4. Well-being of all employees W5. Continuous quality improvement W6. Excellence culture commitment
5	F02	Strategic quality planning (SQP)
		S1. Short and long term objectives S2. Clear strategic quality plan S3. Continuous improvement through strategies S4. Reviewing strategic plans and policies S5. Developing the plans and policies S6. Strategic planning activity at all levels

Acknowledgement. The authors would like to acknowledge all invited experts panel were completed and returned the survey questions as well as all the respondents companies that participated in this research project.

References

1. Saraph, J.V., Benson, G.P., Schroeder, R.G.: An instrument for measuring factors of quality management. Decision Sciences 20(4), 810–829 (1989)
2. Black, S.A., Porter, L.J.: Identification of the critical factors of TQM. Decision Sciences 27(1), 1–21 (1996)
3. Tamimi, N.: A second-order factor analysis of critical TQM factors. International Journal of Quality Science 3(1), 71–79 (1998)
4. Sohal, A.S., Terziovski, M.: TQM in Australian manufacturing: Factors critical to success. International Journal of Quality and Reliability Management 17(2), 158–167 (2000)
5. Dale, B.G., Zairi, M., Van der Wiele, A., Williams, A.R.T.: Quality is dead in Europe – long live excellence: True or false? Measuring Business Excellence 4(3), 4–10 (2000)
6. van der Wiele, A., Brown, A.: Quality Management Over a Decade –a longitudinal study. International Journal of Quality and Reliability Management 19(5), 508–523 (2002)
7. Zairi, M.: Total quality management sustainability: What it means and how to make it viable. International Journal of Quality and Reliability Management 19(5), 502–507 (2002)
8. Sebastianelli, R., Tamimi, N.: Understanding the obstacles to TQM success. The Quality Management Journal 10(3), 45–66 (2003)
9. Taylor, W.A., Wright, G.H.: A longitudinal study of TQM implementation: Factors influencing success and failure. Omega 31(2), 97–111 (2003)
10. Venkateswarlu, P., Nilakant, V.: Adoption and persistence of TQM programmes – case studies of five New Zealand organisations. Total Quality Management 16(7), 807–825 (2005)
11. Naor, M., Goldstein, S.M., Linderman, K.W., Schroeder, R.G.: The role of culture as driver of quality management and performance: Infrastructure versus core quality practices. Decision Sciences 39(4), 671–702 (2008)
12. Asif, M., de Bruijn, E.J., Douglas, A., Fisscher, O.A.M.: Why quality management programs fail: A strategic and operations management perspective. International Journal of Quality and Reliability Management 26(8), 778–794 (2009)
13. Al-Marri, K., Ahmed, A.M., Zairi, M.: Excellence in service: an empirical study of the UAE banking sector. International Journal of Quality & Reliability Management 24(2), 164–176 (2007)
14. Jabnoun, N., Al-Tamimi, H.: Measuring perceived service quality at UAE commercial banks. International Journal of Quality & Reliability Management 20(4), 458–472 (2003)
15. Badri, M., Abdulla, M., Al-Madani, A.: Information technology center service quality: Assessment and application of SERVQUAL. International Journal of Quality & Reliability Management 22(8), 819–848 (2005)
16. Jabnoun, N., Khalifa, A.: A customized measure of service quality in the UAE. Managing Service Quality 15(4), 374–388 (2005)
17. Jabnoun, N., Al Rasasi, A.: Transformational leadership and service quality in UAE hospitals. Managing Service Quality 15(1), 70–81 (2005)
18. Zaramdini, W.: An empirical study of the motives and benefits of ISO 9000 certification: the UAE experience. International Journal of Quality & Reliability Management 24(5), 472–491 (2007)

19. Magd, H., Hamza, S.: Critical Success Factors for Achieving Organizational Excellence in the Middle East: Engineering Firms Perspectives, e-TQM College Working Paper Series WP- 0102012007.wps.hbmsu.ac.ae/system/files/article03_0.pdf
20. Ahmed, A.M.M.B.: Staff suggestion scheme (3Ss) within the UAE context: Implementation and critical success factors. Education, Business and Society: Contemporary Middle Eastern Issues 2(2), 153–167 (2009)
21. Badri, M.A., Davis, D., Donna, D.: A study of measuring the critical factors of quality management. International Journal of Quality and Reliability Management 12(2), 36–53 (1995)

Software Reliability Analysis Considering the Fault Detection Trends for Big Data on Cloud Computing

Yoshinobu Tamura[1] and Shigeru Yamada[2]

[1] Yamaguchi University, Ube, Japan
tamura@yamaguchi-u.ac.jp
[2] Tottori University, Tottori, Japan
yamada@sse.tottori-u.ac.jp

Abstract. Recently, the cloud computing with big data is known as a next-generation software service paradigm. However, the effective method of software reliability assessment considering the big data and cloud computing has been only few presented. In particular, the big data on cloud computing is managed by using several software, i.e., Hadoop and NoSQL are used as the big-data-targeted processing software, OpenStack and Eucalyptus are well-known as the cloud computing software. In this paper, we propose the method of component-based reliability assessment for the software such as database and cloud. Moreover, we propose the method of system-wide reliability assessment considering the big data on cloud computing. In particular, we deeply analyze the software reliability based on two kinds of data set in terms of the background factors. Then, we analyze the software failure-occurrence time data and the cumulative number of detected faults data by applying the hazard rate model and stochastic differential equation one. Additionally, we show several numerical examples for the actual data.

1 Introduction

Many software reliability growth models (SRGM's)[1,2] and hazard rate models[3-6] have been applied to assess the reliability for quality management and testing-progress control of software development. At present, the cloud computing with big data is known as a next-generation software service paradigm. There are some interesting research papers in terms of the cloud computing, cloud service, mobile clouds, and cloud performance evaluation[7,8]. However, most of them have focused on the case studies of cloud service, cloud hardware, and cloud data storage technologies. The effective method of software reliability assessment considering the big data and cloud computing has been only few presented.

In particular, the big data on cloud computing is managed by using several software. Then, the database software and cloud software are generally-cited as the main software component. For example, Hadoop and NoSQL are used as the big-data-targeted processing software. Moreover, OpenStack and Eucalyptus are well-known as the cloud computing software.

M. Gen et al. (eds.), *Industrial Engineering, Management Science and Applications 2015*,
Lecture Notes in Electrical Engineering 349, DOI: 10.1007/978-3-662-47200-2_106

In this paper, we focus on these software in terms of the big data on cloud computing. First, we propose the method of component-based reliability assessment for the software such as database and cloud. Second, we propose the method of system-wide reliability assessment considering the big data on cloud computing. Also, we derive several assessment measures. In particular, we analyze actual software failure-occurrence time data and the cumulative numbers of detected faults data to show numerical examples of software reliability assessment for the big data on cloud computing. Then, we show that the proposed method may assist quality improvement for the software system management of the big data on cloud computing. Furthermore, we investigate a useful software reliability assessment method for the actual cloud computing system management.

2 Component-Based Reliability Analysis for Database and Cloud

The time-interval between successive software failures of $(k-1)$-th and k-th is represented as the random variable X_k $(k = 1, 2, \cdots)$. Therefore, we can define the hazard rate function $z_k(x)$ at time x during the operating-phase for X_k as follows[4]:

$$z_k(x) = \phi\{N - (k-1)\} \qquad (k = 1, 2, \cdots, N; N > 0, \phi > 0), \tag{1}$$

where we can define the quantities as follows:

$z_k(x)$: the hazard rate for each software of database and cloud,

N : the number of latent faults,

ϕ : the hazard rate per inherent fault.

Eq.(1) means the hazard rate for a software failure-occurrence phenomenon for each software of database and cloud.

In operating-phase of database and cloud software, the distribution function of X_k $(k = 1, 2, \cdots)$ representing the time-interval between successive software failures of $(k-1)$ th and k-th is defined as:

$$F_k(x) \equiv \Pr\{X_k \leq x\} \qquad (x \geq 0), \tag{2}$$

where $\Pr\{A\}$ represents the occurrence probability of event A. Therefore, the following function means the probability density function of X_k:

$$f_k(x) \equiv \frac{dF_k(x)}{dx}. \tag{3}$$

Also, the software reliability can be defined as the probability which a software failure does not occur during the time-interval $(0, x]$ in the operating-phase. The software reliability is given by

$$R_k(x) \equiv \Pr\{X_k > x\} = 1 - F_k(x). \tag{4}$$

From Eqs.(2) and (3), the hazard rate is given by the following equation:

$$z_k(x) \equiv \frac{f_k(x)}{1 - F_k(x)} = \frac{f_k(x)}{R_k(x)}, \tag{5}$$

where the hazard rate means the software failure rate in the operating-phase when the software failure does not occur during the time-interval $(0, x]$.

Therefore, we can obtain the software reliability assessment measures from our hazard rate model represented by Eq.(1). The probability density function can be derived as

$$f_k(x) = \phi\{N - k + 1\}\exp[-\{\phi(N - k + 1)\} \cdot x]. \tag{6}$$

Especially, we can give the following expressions as software reliability assessment measures derived from our hazard rate model:

- MTBF
  The mean time between software failures(MTBF) is useful to measure the property of the frequency of software failure-occurrence, and is given by

$$\mathrm{E}[X_k] = \frac{1}{\phi(N - k + 1)}. \tag{7}$$

- Software reliability
  Also, the software reliability can be defined as the probability which a software failure does not occur during the time-interval $(t, t + x]$ $(t \geq 0, x \geq 0)$ given that the debugging progress time of porting-phase is t. The software reliability is given as follows:

$$R_k(x) = \exp[-\phi(N - k + 1) \cdot x]. \tag{8}$$

3 System-Wide Reliability Analysis for Cloud Computing

Let $M(t)$ be the cumulative number of detected faults latent by operational time $t\,(t \geq 0)$. Suppose that $M(t)$ takes on continuous real values. Since latent faults in the software are detected and eliminated during the operation phase, $M(t)$ gradually increases as the operational procedures go on. Thus, under common assumptions for software reliability growth modeling[2], the following linear differential equation can be formulated:

$$\frac{dM(t)}{dt} = b(t)\{a - M(t)\}, \tag{9}$$

where $b(t)$ is the software fault-detection rate at operation time t and a non-negative function, a, means the number of latent faults in the environment on cloud computing. Then, we consider the big data in order to assess the reliability for cloud computing. Therefore, we extend Eq.(9) to the following stochastic differential equation modeling considering two Brownian motions[9,10]:

$$\frac{dM_1(t)}{dt} = \{b_1(t) + \sigma_1 \nu_1(t)\}\{a_1 - M_1(t)\}, \tag{10}$$

$$\frac{dM_2(t)}{dt} = \{b_2(t) + \sigma_2 \nu_2(t)\}\{a_2 - M_2(t)\}, \tag{11}$$

where σ_1 and σ_2 are a positive constant representing a magnitude of the irregular fluctuation, and $\nu_1(t)$ and $\nu_2(t)$ a standardized Gaussian white noise. We assume that $M_1(t)$ and a_1 are related with the software fault-detection rate $b_1(t)$ depending on the failure-occurrence phenomenon depending on the cloud computing. Also, $M_2(t)$ and a_2 are related with the software fault-detection rate $b_2(t)$ depending on the big data. Considering the independence of each noise, we can obtain the following integrated stochastic differential equation:

$$\frac{dM(t)}{dt} = \{b(t) + \sigma_1 \nu_1(t) + +\sigma_2 \nu_2(t)\}\{a - M(t)\}. \tag{12}$$

We extend Eqs.(10) and (11) to the following stochastic differential equation of an Ito type[11]:

$$dM_1(t) = \left\{b_1(t) - \frac{1}{2}\sigma_1{}^2\right\}\{a_1 - M_1(t)\}dt + \sigma_1\{a_1 - M_1(t)\}d\omega_1(t), \tag{13}$$

$$dM_2(t) = \left\{b_2(t) - \frac{1}{2}\sigma_2{}^2\right\}\{a_2 - M_2(t)\}dt + \sigma_2\{a_2 - M_2(t)\}d\omega_2(t), \tag{14}$$

where $\omega_i(t)$ is i-th one-dimensional Wiener process which is formally defined as an integration of the white noise $\nu_i(t)$ with respect to time t. Similarly, we can obtain the following integrated stochastic differential equation based on the independent noises $\nu_1(t)$ and $\nu_2(t)$.

$$dM(t) = \left\{ b(t) - \frac{1}{2}(\sigma_1{}^2 + \sigma_2{}^2) \right\} \{a - M(t)\} dt$$
$$+ \sigma_1 \{a - M(t)\} d\omega_1(t) + \sigma_2 \{a - M(t)\} d\omega_2(t). \tag{15}$$

We define the two dimensions processes $[\omega_1(t), \omega_2(t)]$ as follows[12]:

$$\tilde{\omega}(t) = (\sigma_1{}^2 + \sigma_2{}^2)^{-\frac{1}{2}} \{\sigma_1 \omega_1(t) + \sigma_2 \omega_2(t)\}. \tag{16}$$

Then, the Wiener processes, $\tilde{\omega}(t)$, are a Gaussian process. By using Ito's formula[9,10], we can obtain the solution of Eq.(15) under the initial condition $M(0) = 0$ as follows[11]:

$$M(t) = a\left[1 - \exp\left\{- \int_0^t b(s)ds - \sigma_1 \omega_1(t) - \sigma_2 \omega_2(t)\right\}\right]. \tag{17}$$

Using solution process $M(t)$ in Eq.(17), we can derive several software reliability measures. Moreover, we define the software fault-detection rate per fault in case of $b(t)$ defined as:

$$b(t) \cong \frac{\frac{dI(t)}{dt}}{a - I(t)} = \frac{b}{1 + c \cdot \exp(-bt)}, \tag{18}$$

where $I(t)$ means the mean value functions for the inflection S-shaped SRGM, based on a nonhomogeneous Poisson process (NHPP)[2], a the expected number of latent faults for SRGM, and b the fault-detection rate per fault. Generally, the parameter c is defined as $(1-l)/l$. We define the parameter l as the value of fault factor.

Therefore, the cumulative numbers of detected faults are obtained as follows, respectively:

$$M(t) = a\left[1 - \frac{1+c}{1 + c \cdot \exp(-bt)} \exp\{-bt - \sigma_1 \omega_1(t) - \sigma_2 \omega_2(t)\}\right]. \tag{19}$$

Since $M(t)$ is a random variable in the proposed model, its expected value can be a useful measure. It can be calculated from Eq.(17) as follows[11]:

$$\mathrm{E}[M(t)] = a\left[1 - \frac{1+c}{1 + c \cdot \exp(-bt)} \exp\left\{-bt + \frac{\sigma_1{}^2}{2} t + \frac{\sigma_2{}^2}{2} t\right\}\right], \tag{20}$$

where $\mathrm{E}[M(t)]$ is the expected cumulative number of faults detected up to time t.

The Wiener processes, $\tilde{\omega}(t)$, is composed of $\omega_1(t)$ and $\omega_2(t)$. Therefore, we can give each noise parameter σ_1 and σ_2 by considering the proportion of whole noise parameter $\tilde{\sigma}$ for $\tilde{\omega}(t)$. Then, we apply the proportion of the distribution function, $F_k(x) \equiv 1 - R_k(x)$, based on the hazard rate model to each noise parameter. Thereby, we can estimate σ_1 and σ_2 form $\tilde{\sigma}$, respectively.

4 Numerical Examples

We focus on Hadoop[13] and OpenStack[14] in order to evaluate the performance of our models. In this paper, we show numerical examples by using the data sets for Hadoop of database software and OpenStack of cloud software. The data sets used in this paper are collected in the bug tracking systems on the websites of Hadoop and OpenStack open source projects.

4.1 Results of Component-Based Reliability Assessment

Figs. 1 and 2 show the estimated MTBF for Hadoop database software and OpenStack cloud one. From Figs. 1 and 2, we found that the reliability of Hadoop grows more rapidly than one of OpenStack. Also, the estimated distribution function, $\hat{F}_k(x) \equiv 1 - \hat{R}_k(x)$, and the noise occurrence rate after the time-point of the end of fault detection are shown in Figs. 3 and 4. From Figs. 3 and 4, we can confirm that the software error and noise occurrence rate becomes large after the end of fault detection.

4.2 Results of System-Wide Reliability Assessment

Based on the results of component-based reliability assessment, we show the results of system-wide reliability assessment. The estimated sample paths of numbers of detected faults is shown in Fig. 5. From Fig. 5, we can confirm that the proposed method can estimate the individual cumulative numbers of detected faults for each database software and cloud software in terms of big data on cloud computing, respectively. In particular, we show the comparison results of the sample paths of the number of detected faults in Fig. 6. By comparing the Figs. 4 and 6, we found that the reliability of Hadoop becomes stable compared with OpenStack in the early operating-phase after the time-point of the end of fault detection.

The proposed method of reliability assessment are time-continuously-mixed by two different models. Moreover, two kinds of data sets are used in the proposed method of reliability assessment, i.e., the time-interval between software faults and the cumulative number of detected faults. Thereby, we can deeply analyze the software reliability based on the detailed data in terms of the background factors.

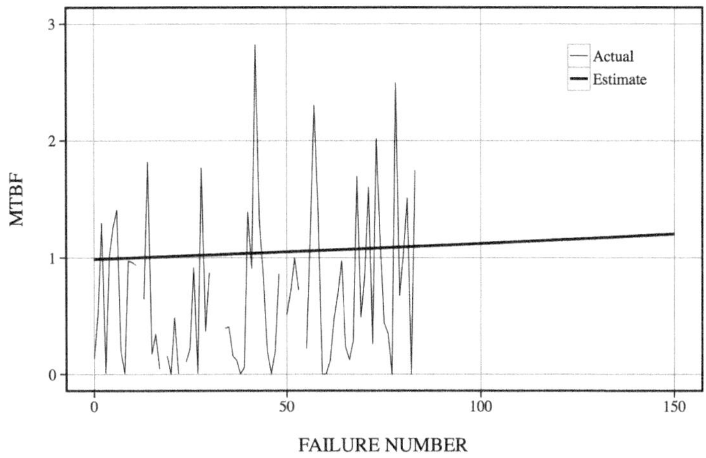

Fig. 1. The estimated MTBF for Hadoop database software

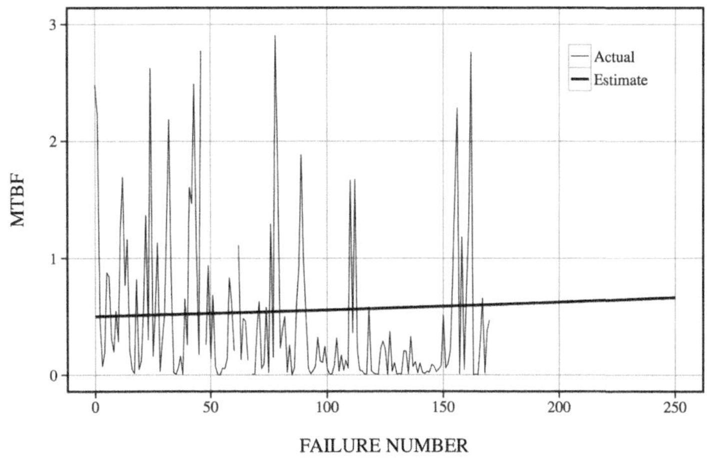

Fig. 2. The estimated MTBF for OpenStack cloud software

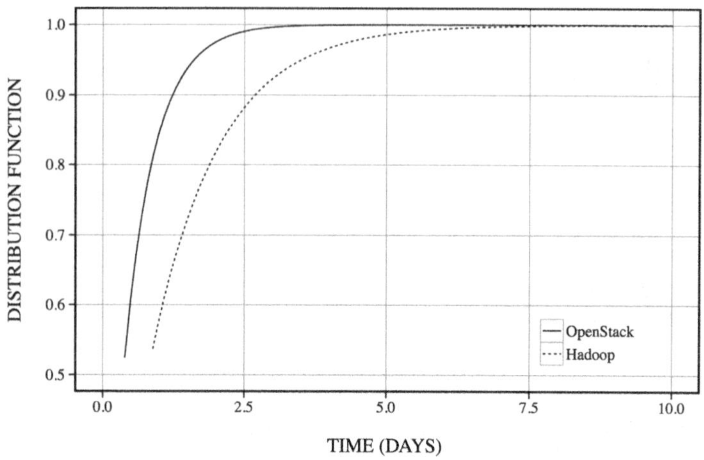

Fig. 3. The estimated distribution function, Hadoop $\hat{F}_{84}(x)$ and OpenStack $\hat{F}_{171}(x)$.

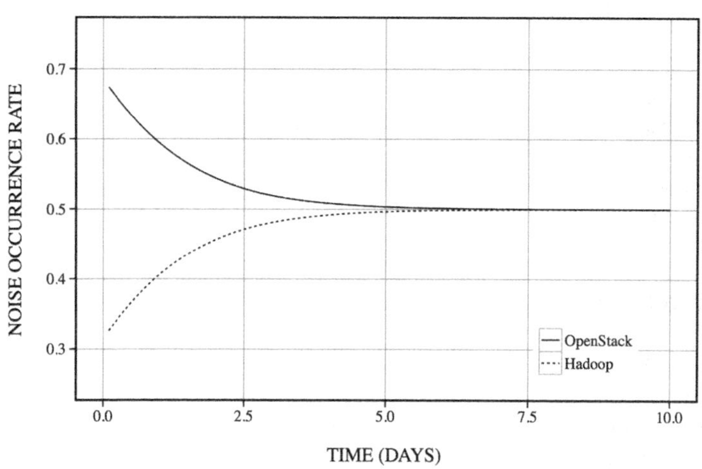

Fig. 4. The estimated noise occurrence rate

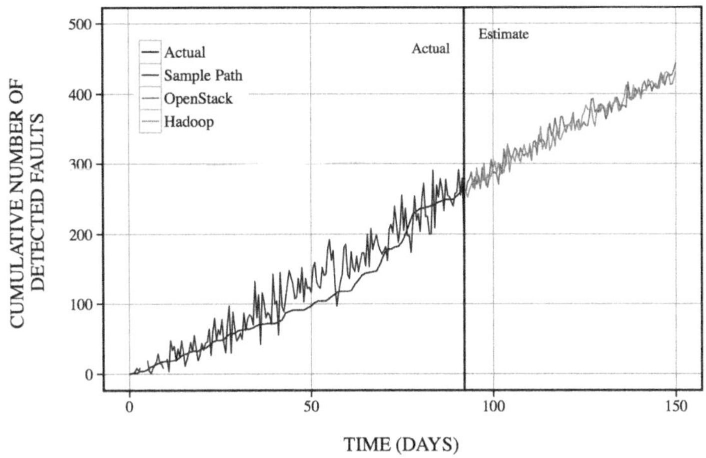

Fig. 5. The sample paths of the cumulative number of detected faults

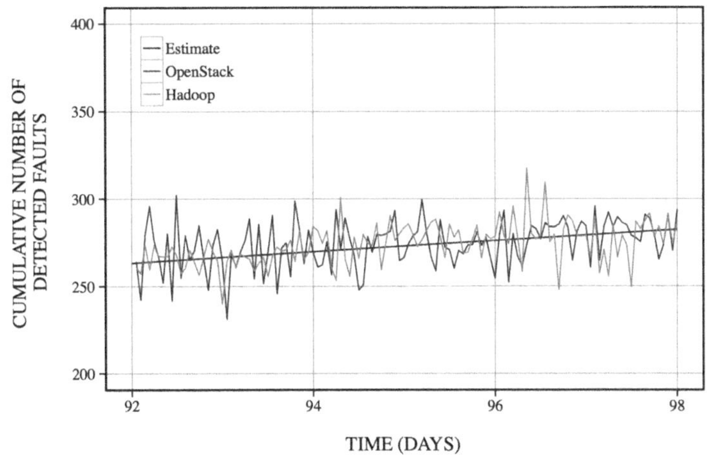

Fig. 6. Comparison of the sample paths of the cumulative number of detected faults

5 Concluding Remarks

It is important for software developers to control the operating-phase in cloud computing. At present, the cloud computing closely related to the big data. In this paper, we have proposed the method of software reliability assessment considering the big data and cloud computing. In particular, we have deeply analyzed the software reliability based on two kinds of data set in terms of the background factors. Then, we have used the software failure-occurrence time data and the cumulative number of detected faults data by applying the hazard rate model and stochastic differential equation one. Additionally, we have presented several numerical examples for the actual data.

The proposed method of reliability assessment will be useful for the software managers to deeply analyze the reliability of big data on cloud computing. In particular, the software managers will be able to understand the stability of the big data on cloud computing by using the sample path of the proposed model.

Acknowledgments. This work was supported in part by the Grant-in-Aid for Scientific Research (C), Grant No. 24500066 and No. 25350445 from the Ministry of Education, Culture, Sports, Science, and Technology of Japan.

References

1. Lyu, M.R. (ed.): Handbook of Software Reliability Engineering. IEEE Computer Society Press, Los Alamitos (1996)
2. Yamada, S.: Software Reliability Modeling: Fundamentals and Applications. Springer, Heidelberg (2013)
3. Schick, G.J., Wolverton, R.W.: An analysis of competing software reliability models. IEEE Trans. Software Engineering SE-4(2), 104–120 (1978)
4. Jelinski, Z., Moranda, P.B.: Software Reliability Research, in Statistical Computer Performance Evaluation, Freiberger, pp. 465–484. Academic Press, New York (1972)
5. Moranda, P.B.: Event–altered rate models for general reliability analysis. IEEE Trans. Reliability R-28(5), 376–381 (1979)
6. Xie, M.: On a generalization of the J-M model. In: Proc. Reliability 1989, 5. Ba/3/1-5 Ba/3/7 (1989)
7. Yang, B., Tan, F., Dai, Y.S.: Performance evaluation of cloud service considering fault recovery. Journal of Supercomputing (2011) (published online: February 23)
8. Khalifa, A., Eltoweissy, M.: Collaborative autonomic resource management system for mobile cloud computing. In: Proceedings of the Fourth International Conference on Cloud Computing, GRIDs, and Virtualization, Valencia, Spain, pp. 115–121 (2013)
9. Arnold, L.: Stochastic Differential Equations-Theory and Applications. John Wiley & Sons, New York (1974)
10. Wong, E.: Stochastic Processes in Information and Systems. McGraw-Hill, New York (1971)
11. Yamada, S., Kimura, M., Tanaka, H., Osaki, S.: Software reliability measurement and assessment with stochastic differential equations. IEICE Transactions on Fundamentals E77-A(1), 109–116 (1994)
12. Mikosch, T.: Elementary Stochastic Calculus, with Finance in View. Advanced Series on Statistical Science and Applied Probability, vol. 6. World Scientific, Singapore (1998)
13. The Apache Software Foundation, OpenStack, `http://www.openstack.org/`
14. The OpenStack project, Apache Hadoop, `http://hadoop.apache.org/`

Effect of Vibration Transmissibility on Fatigue Lifetime of Electronic Devices

Liu Yang[1,*], Ying Chen[1], Zenghui Yuan[1], and Liqun Chen[2]

[1] Science and Technology on Reliability and Environmental Engineering Laboratory, Beihang University, Beijing, China

[2] AVIC Aviation Motor Control System Institute Wuxi, Jiangsu, China

{yangliu92616,yuanzh6672,clq2272}@163.com, cheny@buaa.edu.cn

Abstract. Vibration fatigue is one of the main mechanisms which will cause the failures of electronic devices. If the natural frequency of a PCB and its case do not obey octave rule, the vibration of the PCB and the case will couple with each other, and stress applied on PCB will be amplified, resulting in early failure. With Steinberg vibration fatigue prediction models, this paper studies the effect of vibration transmissibility on fatigue lifetime of electronic devices with consideration of coupling. ADAMS software is used to simulate and analyze the vibration transmissibility of electronic devices. The correction of vibration transmissibility in Steinberg model is given. In case study, vibration fatigue lifetimes that compute with corrected Steinberg model and the model without consideration of vibration transmissibility are compared. Effect of vibration transmissibility on electronic devices' fatigue life is discussed.

Keywords: transmissibility, octave rule, vibration fatigue, coupling, lifetime.

1 Introduction

Vibration is one of main environmental conditions experienced by electronic devices, which will result in some failure mechanisms, such as random vibration fatigue, sinusoidal vibration fatigue, shock overstressing and so on. Analyzing vibration failure with Physics-of-Failure method and models has many advantages, including location of design weak location and formulation of improvement measures.

Vibration fatigue of electronic devices has been widely studied [1,2,3,4.5] by many researchers. The research on physics model of vibration fatigue failure can be traced back to 1970s. After many years of practical experience, Dave S. Steinberg proposed Steinberg model applied to lifetime estimation of electronic devices working under sinusoidal or random vibration conditions [6,7]. Although Manson model and other models [8] appeared in this field later, Steinberg model is still wildly used in engineering because of its obvious physical meaning. Dehbi.A et al. [9] studied the application of Steinberg model in tantalum capacitor. Through experiment, they

[*] Corresponding author.

M. Gen et al. (eds.), *Industrial Engineering, Management Science and Applications 2015*, Lecture Notes in Electrical Engineering 349, DOI: 10.1007/978-3-662-47200-2_107

provided the S-N curve in different sinusoidal sweeping-frequency vibration conditions. Marksteind et al. [10] proposed some principles of electronic systems to withstand high vibration and shock conditions.

Wu et al. [11] analyzed PCB's vibration with Steinberg model in CalcePWA. Steinberg model was spilt into two separate models, stress analysis model and fatigue damage model. They established a rapid test method which can directly test the effect of new structure and material on lifetime of PCB. Wu gave a suggestion that the Steinberg model in CalcePWA software required correction work for PCB with new structure and material. Chen et al. [12] estimated fatigue lifetime of electronic components in PBGA package by test of vibration damage and FEA. Liu et al. [13] studied the dynamic response and reliability of lead-free solder ball in BGA package under different G values and frequencies on the basis of Steinberg model. Urgueira et al. [14] used a variety of life prediction models including Steinberg model and evaluated lifetime of the position with maximum stress.

From the above discussion, Steinberg model has been widely used in engineering. In this paper, the effect of vibration transmissibility on fatigue lifetime of electronic devices is studied by Steinberg model. Transmissibility factor in Steinberg model represents the coupling state of PCB and its case. It has a significant impact on vibration fatigue lifetime of electronic devices.

2 Theoretical Basis

In Steinberg model, PCB can be approximated as a single degree of freedom system, when it vibrates under the fundamental resonance. In sinusoidal vibration environment, the actual dynamic single amplitude displacement of PCB's center is given by:

$$Z = \frac{9.8G_{out}}{f^2} = \frac{9.8G_{in}Q}{f_n^2} \quad . \tag{1}$$

Where Z is dynamic single amplitude displacement of PCB's center, f_n is resonant frequency of PCB, G_{out} is the root mean square acceleration of output , G_{in} is the root mean square acceleration of input, Q is transmissibility.

In random vibration environment, according to the stress level 3σ, the maximum dynamic single amplitude displacement of PCB's center is three times of the root mean square displacement which is as follows:

$$Z = 3\times\frac{9.8G_{RMS}}{f_n^2} . \tag{2}$$

Where G_{RMS} is the root mean square acceleration.

When the input PSD (Power Spectral Density) of random vibration is flat spectrum in resonance region, the root mean square acceleration response of a system is given by:

$$G_{out} = \sqrt{\frac{\pi}{2} P f_n Q}\,(RMS) . \tag{3}$$

Where P is the input PSD at resonant frequency.

Usually, electronic device can be simplified as two degrees of freedom spring-mass system which consists of spring, damping and mass block, as shown in Fig. 1 a). Changing stiffness ratio of spring can change the ratio of natural frequencies of PCB and case. Changing the mass of PCB and case will change the weight ratio. Changing damping ratio can make the ratio of uncoupling natural vibration transmissibility change. With the purpose of obtaining acceleration value G on PCB, we analyze the energy transmission from case to PCB in condition of different dynamic combinations.

In order to get the relation of vibration transmissibility in two degrees of freedom system, Adams software is used to model two degrees of freedom spring-mass system, as shown in Fig. 1 b) Mass block 1 in the figure represents PCB, and mass block 2 represents its case. The connection between mass block 1 and mass block 2 is a spring-damping system. Similarly the connection between mass 2 and ground is also a spring-damping system.

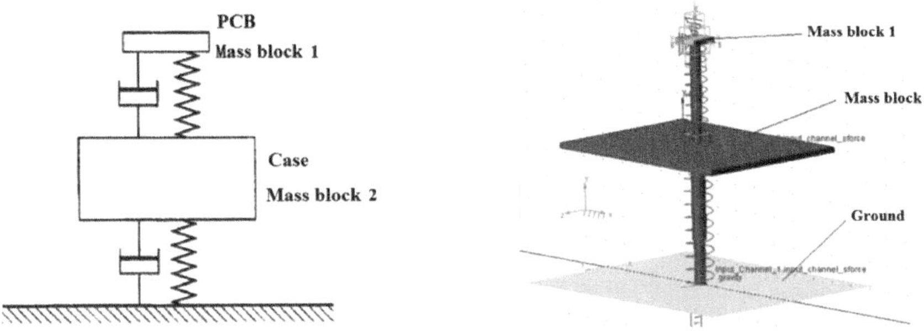

a) Two degrees of freedom dynamic spring-mass system b) ADAMS model

Fig. 1. Modeling PCB and the case

By regulating the mass of mass block 1 and mass block 2, the model is made to match the given condition of weight ratio between PCB and case. By regulating the elastic coefficient k and damping coefficient c of two spring-damping systems, the model is made to match the given condition of natural frequency ratio and natural transmissibility ratio between PCB and case. After that, the sinusoidal vibration load is applied on this system according to the given conditions. Stress response of mass block 1 under different conditions is shown in Fig. 2 a).

By studying two degrees of freedom spring-mass system under three conditions, the mass ratio of PCB and case is confirmed as 1:4. The ratio of PCB's uncoupling transmissibility q_2 and case's uncoupling transmissibility q_1 is determined as 0.25, 1 and 3. The ratio of PCB's coupling transmissibility Q_2 and case's uncoupling transmissibility q_1 varies with the ratio of two natural frequencies. (Fig.2 b))

The trend of Fig. 2 b) rises at first and then goes down. When the natural frequencies of PCB and case are very close, the vibration transmissibility reaches the peak. At this time, there is resonance phenomenon between PCB and case. A general equation of vibration transmissibility is shown in equation (4).

$$Q = A\left[\frac{f_n}{(G_{in})^{0.6}}\right]^{0.76} . \tag{4}$$

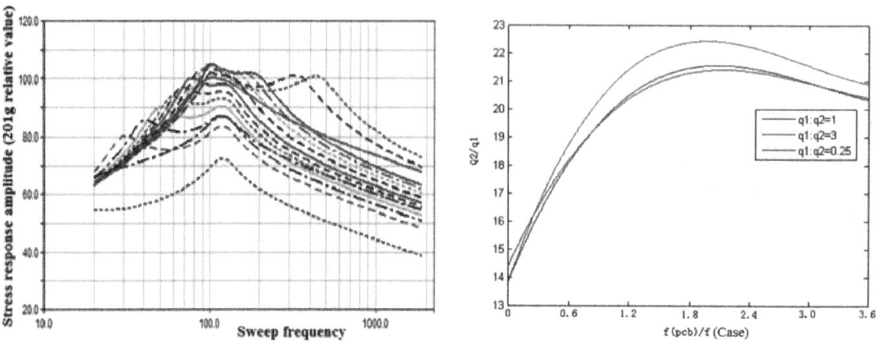

a) Stress response of mass block 1 under the condition of permanent mass ratio and transmissibility ratio (Different lines represent ratios of natural frequency)
b) Variation of vibration transmissibility

Fig. 2. ADAMS simulation results

The letter A is a constant related to structural support of electronic devices. When electronic device is girder structure, A=1.0. When it is periphery fixed PCB or plug-in mounting PCB, A=0.5. When it is a small sealed electronic case, A=0.2. f_n is resonant frequency. G_{in} is the root mean square acceleration of input. Q is transmissibility.

In the practical engineering calculation, Steinberg model can be simplified as shown in equation (5).

$$Q = c\sqrt{f_n}\ . \tag{5}$$

The value of c ranges from 0.5 to 2 which is a constant related to excitation amplitude and natural frequency. Generally for a PCB whose first-order natural frequency is in middle frequency band (200Hz-300Hz), the value of c is 1. And for an electronic case, the value is 0.5.

3 Case Study

3.1 PCB and Case Obey the Octave Rule in Design

An electronic device consists of two plug-in PCBs and a case. The device is conducted modal analysis by ANSYS Workbench. The first-order resonant frequency of case is 699.34Hz and the value of PCB-A is 86Hz, as shown in Fig. 3.

The frequency of random vibrational spectrum ranges from 10 Hz to 2000Hz and the power spectral density is 1.5g. In this case, prediction for failure time of components' interconnection due to vibration fatigue is given by CalcePWA, as shown in Fig. 4a). And the failure position is shown in Fig. 4 b).

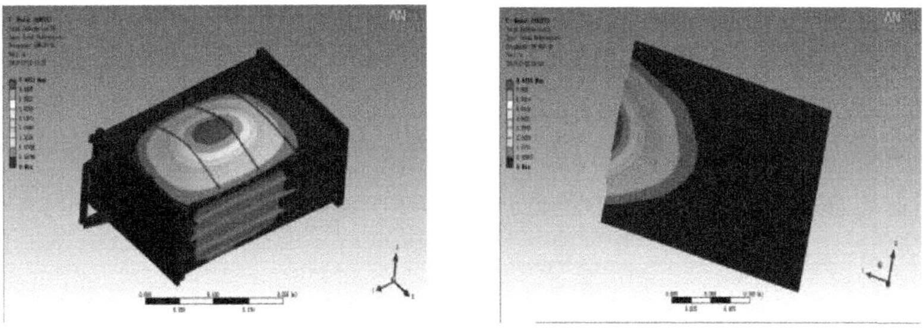

Fig. 3. The first-order modals of case and PCB-A

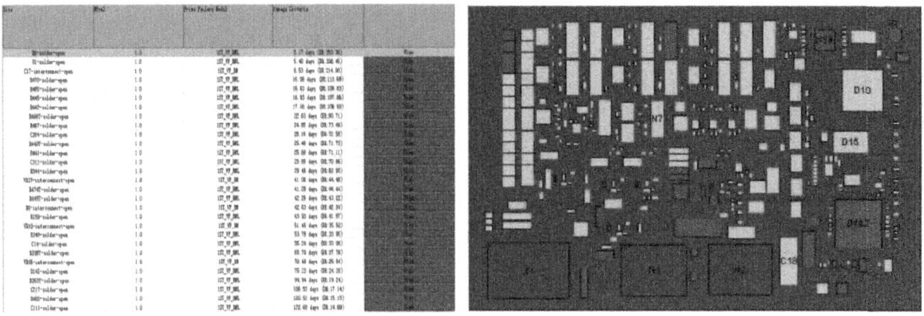

a) Failure prediction of PCB-A b) Potential failure location of PCB-A

Fig. 4. Failure prediction of PCB-A with CalcePWA

In Fig. 5, the ordinate axis represents the ratio between coupling transmissibility of PCB and uncoupling transmissibility of case. The abscissa axis represents the ratio between natural frequencies of PCB and case.

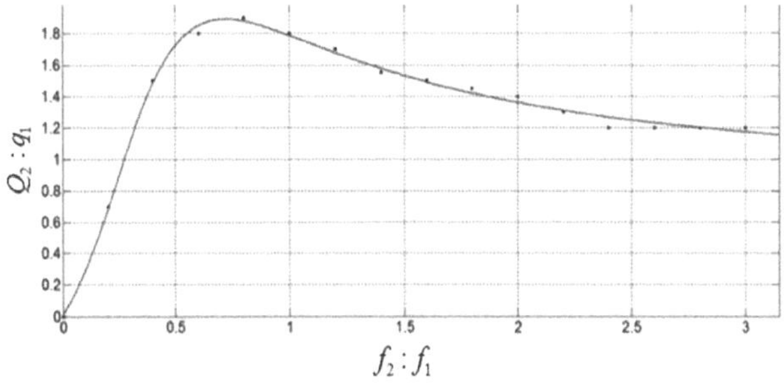

Fig. 5. The relation between two ratios

The fitting expression of above figure is equation (6).

$$y = \frac{6005x^2 + 7615x - 16.15}{x^3 + 8266x^2 - 3525x + 2805} . \tag{6}$$

When $f_2 : f_1 = 0.12$, the value of $Q_2 : q_1$ is about 0.39. By using sinusoidal vibration fatigue model in equation (1), component D10 in PCB is predicted its failure considering vibration coupling. If $Q_2 : q_1 = R$, substitute $Q_2 = R \cdot q_1$ into equation(1). Z can be expressed as shown in equation (7)

$$Z = \frac{9.8G_{out}}{f^2} = \frac{9.8Q_2G_{in}}{f^2} = \frac{9.8Rq_1G_{in}}{f^2} . \tag{7}$$

Relevant parameters of D10 and PCB are brought into sinusoidal vibration fatigue model. By calculation, the fatigue life is $N_2 = 4.47 \times 10^{13}$. However the result calculated by CalcePWA is 1.1645×10^{12}. Obviously the former is far greater than the latter. In other words, lifetime calculated without considering coupling is shorter than that considering coupling. The main reason is that the natural frequency of PCB is much smaller than case's, and they obey the octave rule. Therefore, this situation reflected on Fig. 5 is that the two frequencies have been away from dangerous area where serious coupling is much possible. In the circumstances, Q_2 calculated by $R \cdot q_1$ maybe be smaller than the approximate value which is square root of PCB's natural frequency.

When the input is random vibration, natural frequency of PCB is 86Hz, and natural frequency of case is 699.34Hz. According to the empirical formula, when PCB is excited at its natural frequency, the uncoupling transmissibility of PCB is $\sqrt{86} \approx 9.27$. The PCB should also be considered the additive energy which is gained from case coupling at 86Hz. And when forced frequency f_f is 86Hz, resonant frequency of case f_n is 699.34Hz, the ratio of forced frequency and resonant frequency of case R is 0.12, transmissibility of PCB can be calculated in equation (8)

$$Q = \frac{1}{1 - R^2} = 1.01 . \tag{8}$$

So the coupling transmissibility of PCB at 86Hz can be calculated by $Q_p = 1.01 \times 9.27 = 9.3627$, where Q_p represents coupling transmissibility.

The case's second resonance peak at 699.34Hz can be estimated. The uncoupling transmissibility of case is about 13.22. And the uncoupling transmissibility of PCB at 699.34Hz can be calculated by equation (8), the result is -0.0146. The minus means that the responses at 699.34Hz and 86Hz are in opposite direction. So coupling transmissibility of PCB at 699.34Hz is 0.193.

The PCB's resonance peaks at 86Hz and 699.34Hz are regarded as resonance peaks with single degree of freedom. Root mean square of the response in PCB's center is,

$$G_{RMS} = \sqrt{\frac{\pi}{2} \times 0.2 \times 86 \times 9.3627 + \frac{\pi}{2} \times 0.2 \times 699.34 \times 0.0146} = 16.005 . \tag{9}$$

Relevant parameters of D10 and PCB are put into random vibration fatigue model. Fatigue life of component D10 is obtained as 2.45×10^6 hours, which is smaller than 5.53×10^7 hours which is estimated by CalcePWA. The result can be explained by slight coupling of PCB and case in random vibration condition. Fortunately, the coupling degree is low enough. So displacement of PCB is not over enlarged, and the decrease of PCB's fatigue life is unconspicuous. Because the design of PCB-A and case is in strict conformance with octave rule.

3.2 PCB and Case Disobey the Octave Rule

Another electronic device has two power modules, two input output interface modules and two interface cards. Modal analysis of the electronic device is conducted by ANSYS workbench software. Through analyzing, the first resonant frequency of case is 579.6Hz. The first resonant frequency of module B is 219.8Hz. And other modules' first resonant frequencies are over 1000 Hz. Natural frequencies of module B and its case do not obey the octave rule, which may result in dynamic coupling. By CalcePWA, there is no weak link under sinusoidal vibration. When the input is random vibration, result of vibration fatigue life calculated by CalcePWA is shown in Fig. 6 a), and the failure position is shown in Fig. 6 b).

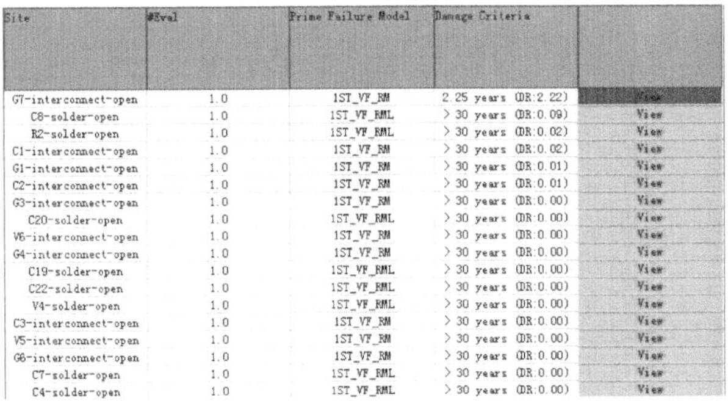

Site	#Eval	Prime Failure Model	Damage Criteria	
G7-interconnect-open	1.0	1ST_VF_RM	2.25 years (DR:2.22)	View
C8-solder-open	1.0	1ST_VF_RML	> 30 years (DR:0.09)	View
R2-solder-open	1.0	1ST_VF_RML	> 30 years (DR:0.02)	View
C1-interconnect-open	1.0	1ST_VF_RM	> 30 years (DR:0.02)	View
G1-interconnect-open	1.0	1ST_VF_RM	> 30 years (DR:0.01)	View
C2-interconnect-open	1.0	1ST_VF_RM	> 30 years (DR:0.01)	View
G3-interconnect-open	1.0	1ST_VF_RM	> 30 years (DR:0.00)	View
C20-solder-open	1.0	1ST_VF_RML	> 30 years (DR:0.00)	View
V6-interconnect-open	1.0	1ST_VF_RM	> 30 years (DR:0.00)	View
G4-interconnect-open	1.0	1ST_VF_RM	> 30 years (DR:0.00)	View
C19-solder-open	1.0	1ST_VF_RML	> 30 years (DR:0.00)	View
C22-solder-open	1.0	1ST_VF_RML	> 30 years (DR:0.00)	View
V4-solder-open	1.0	1ST_VF_RML	> 30 years (DR:0.00)	View
C3-interconnect-open	1.0	1ST_VF_RM	> 30 years (DR:0.00)	View
V5-interconnect-open	1.0	1ST_VF_RM	> 30 years (DR:0.00)	View
G6-interconnect-open	1.0	1ST_VF_RM	> 30 years (DR:0.00)	View
C7-solder-open	1.0	1ST_VF_RML	> 30 years (DR:0.00)	View
C4-solder-open	1.0	1ST_VF_RML	> 30 years (DR:0.00)	View

a) Failure prediction of module B

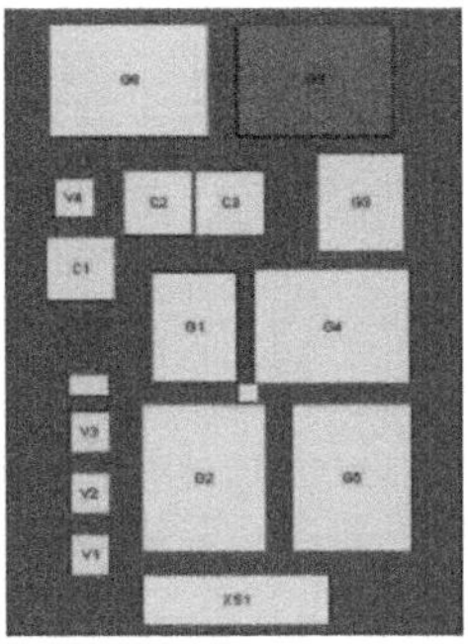

b) Potential failure location of module B

Fig. 6. Failure prediction of module B with CalcePWA

In Fig. 7, the ordinate axis represents the ratio between coupling transmissibility of module B and uncoupling transmissibility of case. The abscissa axis represents the ratio between natural frequencies of module B and case.

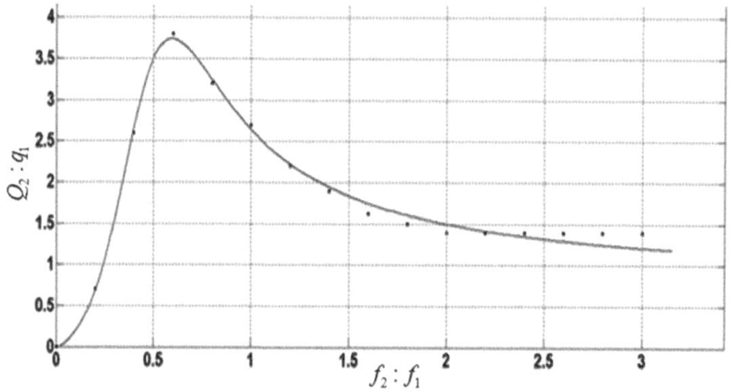

Fig. 7. The relation between two ratios

By using sinusoidal vibration fatigue model considering vibration coupling, fatigue life of G7 is calculated $N_2 = 1.15\times10^{12}$. It is slightly smaller than 1.5372×10^{12} which is the result from CalcePWA. Because the ratio of PCB's and case's natural frequency is about 1.5. They disobey the octave rule to some extent. The coupling effect can enlarge the displacement of PCB and shorten its fatigue lifetime. Seen from the Fig.7, frequencies' ratio 1.5 is at the edge of dangerous zone. Therefore, the reduction of fatigue lifetime is to a small extent.

Similarly, by using random vibration fatigue model, fatigue life of G7 considering vibration coupling is 411.37 hours. It is about a quarter of 1640 hours calculated by CalcePWA. PCB and case disobey the octave rule, so dynamic coupling effect quickly reduces components' fatigue life. If their natural frequencies are seriously contrary to octave rule, the effect of dynamic coupling will be more serious and the reducing of fatigue life will be more quickly.

4 Discussion and Conclusion

In this paper, fatigue lifetime of electronic devices is studied with Steinberg model considering transmissibility. The result is compared with that calculated by CalcePWA without considering vibration coupling. It is found that in sinusoidal vibration environment, lifetime considering vibration coupling is not always smaller than the lifetime without considering coupling. Only when the PCB and case seriously disobey octave rule, namely the ratio of their natural frequency is at dangerous zone, the calculation considering coupling is smaller. If the ratio of their natural frequency is away from dangerous zone, the calculation without considering coupling is smaller. In a general way, if the ratio of their natural frequency ranges from 0.75 to 1.25, it is defined as dangerous zone.

Compared with sinusoidal vibration, coupling in random vibration is more likely to enlarge the displacement of PCB. It is due to different characteristics of sinusoidal vibration and random vibration. In sinusoidal vibration, the natural frequencies of PCB and case are excited respectively. However in random vibration, they are excited simultaneously. If natural frequencies of PCB and case are more close to each other, they will disobey octave rule more seriously, and the effect of dynamic coupling will be more serious. Accordingly fatigue life is smaller than that without considering dynamic coupling, and the gap will be larger.

References

1. Dehbi, A., et al.: Vibration lifetime modelling of PCB assemblies using Steinberg model. Microelectronics Reliability 45(9-11), 1658–1661 (2005)
2. Markstein, W.H.: Designing electronics for high vibration and shock. Electron. Packag. Prod., 40(3) (1987)
3. Chin, I., et al.: A mechanical fatigue assessment methodology to study solder joint reliability. In: 2008 33rd IEEE/CPMT International. Electronic Manufacturing Technology Symposium (IEMT), Penang (2008)
4. Wu, M.L.: Vibration-induced fatigue life estimation of ball grid array packaging. Journal of Micromechanics and Microengineering 19(0650056) (2009)
5. Li, R.S.: A Methodology for Fatigue Prediction of Electronic Components Under Random Vibration Load. Journal of Electronic Packaging 123(4), 394–400 (1999)
6. Steinberg, D.S.: Vibration Analysis for Electronic Equipment, 3rd edn., p. 440. Wiley, New York (2000)
7. Chen, Y., Yang, L., Liu, B., Xue, D.: Applicability Study of Steinberg Vibration Fatigue Model in Electronic Products. In: IEEE 2014 Prognostics and System Health Management Conference, Zhangjiajie (2014)
8. Liu, F., Meng, G.: Random vibration reliability of BGA lead-free solder joint. Microelectronics Reliability 54(1), 226–232 (2014)
9. Chesné, S., Deraemaeker, A.: Damage localization using transmissibility functions: A critical review. Mechanical Systems and Signal Processing 38(2), 569–584 (2013)
10. Steenackers, G., Devriendt, C., Guillaume, P.: On the use of transmissibility measurements for finite element model updating. Journal of Sound and Vibration 303(3-5), 707–722 (2007)
11. Lim, J.H.: A correlation study of satellite finite element model for coupled load analysis using transmissibility with modified correlation measures. Aerospace Science and Technology
12. Manson, S.S.: Fatigue: a complex subject – some simple approximations. Exp. Mech. 5(3), 202 (1965)
13. Wang, D.: Study on simulation case of random vibration test at board level electronic products. Beihang University (2010)
14. Urgueira, A.P.V., Almeida, R.A.B., Maia, N.M.M.: On the use of the transmissibility concept for the evaluation of frequency response functions. Mechanical Systems and Signal Processing 25(3), 940–951 (2011)

Reliability Importance of the Channels in Safety Instrumented Systems

Yiliu Liu and Mary Ann Lundteigen

Department of Production and Quality Engineering,
Norwegian University of Science and Technology, Trondheim, 7491 Norway
yiliu.liu@ntnu.no

Abstract. Safety instrumented systems (SISs) are used in many application areas to reduce the likelihood of hazardous events and/or to mitigate the consequences of such events. A typical SIS has three subsystems, each with redundant channels that are vulnerable to common cause failures. This paper deals with the reliability importance of the channels of a SIS. Five different importance measures are determined with the average probability of failure on demand as reliability performance measure. A Markov method is used to link the failure rates of channels and the system unavailability. Common cause failures are also taken into account. An illustrative example is provided to describe the calculation process of the reliability importance of the channels.

Keywords: safety instrumented system, channel importance, Markov model, probability of failures on demand.

1 Introduction

The reliability importance of components in technical systems has been studied extensively by many authors, e.g. [1–3]. Such importance measures provide information on how the reliability of individual components influences the reliability or unreliability of the system of which they are a part. In general, the component's importance depends on the component's reliability and where it is located in the system structure.

This paper deals with the reliability importance of the various channels of a safety instrumented system (SIS) [4]. SISs are used in many application areas (e.g., process industry, transport, machinery) to reduce the likelihood of hazardous events and/or to mitigate the consequences of such events [4, 5]. A typical SIS has three subsystems: a sensor (S) subsystem, a logic solver (LS) subsystem, and a final element (FE) subsystem.

A simple SIS is illustrated in Fig. 1, where the sensor subsystem has three pressure transmitter channels that are voted 2-out-of-3 (2oo3), meaning that the subsystem is able to function when at least two of the three pressure transmitters (i.e., channels) are functioning. The pressure transmitters monitor the pressure in a pipeline and the measured value is sent to the logic solver. The logic solver compares the

M. Gen et al. (eds.), *Industrial Engineering, Management Science and Applications 2015*,
Lecture Notes in Electrical Engineering 349, DOI: 10.1007/978-3-662-47200-2_108

measurement with a specified threshold ("set point"), and sends a signal to close the final elements (i.e., valves) if two or more values have exceeded this set point value. The valves are, as illustrated in Fig. 1, voted as a 1-out-of-2 (1oo2) structure, such that the flow will be stopped when at least one of the valves closes. All the tree subsystems are required to perform the shutdown function when a demand occurs. The system to be protected, in this case the pipeline or an upstream or downstream system, is called the equipment under control (EUC) and may in Fig. 1, for example, be the pipeline or a vessel on the downstream side of the SIS. A failure of a channel to perform its safety function is called a dangerous (D) failure.

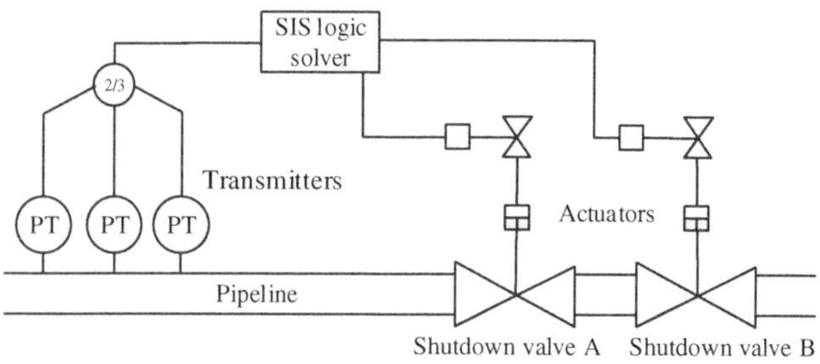

Fig. 1. A typical SIS

When the frequency of demands is less than once per year, the SIS is said to operate in low-demand mode, and when demands occur more often than once per year, the SIS is said to operate in high-demand mode. In this paper, we assume that the SIS is operated in low-demand mode. Further information about the different demand modes can be found in part 6 of IEC 61508 [4].

A low-demand SIS remains in a passive state until a demand occurs, and its safety function is only activated upon a demand. The (un)reliability of the SIS is therefore quantified by the *average probability of failure on demand* (PFD_{avg}). An automatic diagnostic function is implemented in most modern SISs and is able to perform self-tests of the channels almost continuously. The fraction of dangerous channel failures that can be detected by the diagnostic function is called the diagnostic coverage (DC). Two types of dangerous failures can therefore occur in a SIS: dangerous detected (DD) failures that are revealed almost immediately by the diagnostic function and dangerous undetected (DU) failures that can only be revealed by proof testing. It may be remarked that diagnostic functions are mainly associated with the electronic or programmable electronic components (i.e., pressure transmitters and logic solvers), while mechanical devices like valves would be, have limited or no diagnostic function.

The proof testing is often carried out periodically, e.g., once per year. If a channel has a DU failure, it is not able to function until the next proof testing. The safety function of the channel can therefore be unavailable for a rather long period before the fault is corrected.

The PFD_{avg} is a key parameter when making decision on whether a SIS is able to meet the specified safety integrity level (SIL) in accordance to standards like [4]. The PFD_{avg} calculated for a specific SIS function indicates to what degree the SIS is able to reduce the frequency of hazardous events arising in the EUC. The SIS in Fig. 1 may contribute to the necessary reduction in frequency of over pressuring the pipeline. In order to identify weaknesses of the system, it is useful to evaluate the contribution of the individual channels to the PFD_{avg} of a SIS. However, when analyzing the importance of these channels, some characteristics of SISs cannot be ignored:

- The PFD_{avg} is always used to measure the unavailability of a SIS in the long term. The importance of channels should therefore be examined based on this measure.
- SISs are responsive systems. Both DD and DU failures must be considered when determining the (safety) unavailability of the system, even if the contribution of DD failures is often small compared to the contribution from the DU failures.
- Common cause failures (CCFs) should be considered and modeled as part of safety analyses.

With these considerations, this paper will illustrate the analysis process for the reliability importance for channels of a SIS: Section II provides several general importance measures for SISs, and Section 3 discusses how to deal with the complex system structures. An illustrative example is included in Section 4, and finally, managerial implications and some research perspectives are outlined.

2 Measures of Channel Importance

2.1 Assumptions

Before the analysis, some assumptions for the SISs are necessary:

- All channels in the whole SIS are independent in terms of failures;
- Times to/between failures of these channels follow exponential distribution;
- The channels are repairable, and can be restored to an as-good-as-new state;
- Proof testings are carried out with a constant time interval, and in each proof testing, all channels in the SIS are tested;
- The interval between tests on different channels can be ignorable in one proof testing, and test durations are also very short;
- The repair times follow an exponential distribution with rate of 1/MTTR.

PFD_{avg} is a probability without unit. If, for example, $PFD_{avg} = 0.01$, this means that the SIS will, on the average, fail to respond adequately in 1 out of 100 demands. Such a measure also can be regarded as the percentage of time that the EUC is not protected by the SIS, e.g., when SIS is unavailable for 87.6 hour in one year, $PFD_{avg} = 0.01$ [5].

Several measures of reliability importance have been introduced (e.g., see [2, 3]). Here, we consider five of these measures. All of them are presented in detail and discussed in [5]. Here we introduce the five measures in the context of a SIS by using PFD_{avg} as the unavailability measure.

2.2 Birnbaum's Measure

Birnbaum's measure of reliability importance of a channel is a sensitivity measure, which indicates the effect a small change of the PFD_{avg} of the channel on the PFD_{avg} of the SIS.

$$I^{B}(i,j) = \frac{\partial PFD_{avg,SIS}}{\partial PFD_{avg(i,j)}} \tag{1}$$

where (i,j) denotes channel (j) that is part of subsystem (i).

By using the chain rule for derivatives, $I^{B}(i,j)$ may be written as

$$I^{B}(i,j) = \frac{\partial PFD_{avg,SIS}}{\partial PFD_{avg,(i)}} \cdot \frac{\partial PFD_{avg,(i)}}{\partial PFD_{avg(i,j)}} \tag{2}$$

The first factor in Eq. 1 is Birnbaum's measure of reliability importance of subsystem (i) for the SIS and the second factor is Birnbaum's measure of reliability importance of channel (i, j) for subsystem (i).

The three subsystems of the SIS in Fig. 1 (S, LS, and FE) operate as a series structure, and the unavailability of the SIS is therefore (e.g., see [5])

$$PFD_{avg,SIS} = PFD_{avg,S} + PFD_{avg,LS} + PFD_{avg,FE} \tag{3}$$

The average availability of the SIS is

$$A_{SIS} = (1 - PFD_{avg,S})(1 - PFD_{avg,LS})(1 - PFD_{avg,FE}) \tag{4}$$

Birnbaum's measure of reliability importance of subsystem i (i = S, LS or FE) can be written as

$$I^{B}(i) = \frac{\partial A_{SIS}}{\partial A_{i}} = A_{SIS}(1_{i} - p) - A_{SIS}(0_{i}, p) \tag{5}$$

where $A_{SIS}(1_i, p)$ denotes the average availability of the SIS when it is known that subsystem i is functioning ($PFD_{avg(i)} = 0$), and $A_{SIS}(0_i, p)$ is the availability of the SIS when subsystem i is known to be in a failed state ($PFD_{avg(i)} = 1$). Here p denotes the average availabilities of the other channels of the SIS. We can therefore obtain Birnbaum's measure for a subsystem, such as the sensor subsystem, as:

$$I^{B}(S) = (1 - PFD_{avg,LS})(1 - PFD_{avg,FE}) \approx 1 - PFD_{avg,LS} - PFD_{avg,LS} \tag{6}$$

The approximation is valid because the average PFDs of the subsystems of a SIS typically are much smaller than 0.1.

Furthermore, the probability of failure on demand of a channel comes from DD and DU failures, and we have ($\lambda_{i,j}$ is the dangerous failure rate)

$$\mathrm{PFD}_{\mathrm{avg},(i,j)} \approx \mathrm{DC}_{i,j} \cdot \lambda_{i,j} \cdot \mathrm{MTTR}_{i,j} + \frac{(1-\mathrm{DC}_{i,j})\lambda_{i,j}\tau}{2} \tag{7}$$

Such calculation is rather similar with the formula given in the standard of IEC 61508 [4], only ignoring the repair time, since this period is very short compared with the testing interval.

In this paper, we only consider the impact of channel reliability, and therefore the testing interval τ can be regarded as a constant for all channels. Thus, Eq. 1 can be expressed as

$$I^{B}(i,j) = K \cdot I^{B}(i) \cdot \frac{\partial \mathrm{PFD}_{\mathrm{avg},(i)}}{\partial \lambda_{ij}} \tag{8}$$

where

$$K = \frac{2}{2\mathrm{DC}_{i,j} \cdot \mathrm{MTTR}_{i,j} + (1-\mathrm{DC}_{i,j})\tau}$$

In most cases, DU failures are the main contributors of the $\mathrm{PFD}_{\mathrm{avg}}$ of a channel [5]. Eq. 8 can be simplified by ignoring the DD failures as

$$I^{B}(i,j) \approx \frac{2}{(1-\mathrm{DC}_{i,j})\tau} \cdot I^{B}(i) \cdot \frac{\partial \mathrm{PFD}_{\mathrm{avg}(i)}}{\partial \lambda_{i,j}} \tag{9}$$

Remark: The main condition for being able to ignore DD failures is that such failures are corrected within short time, often within a few hours.

2.3 Improvement Potential

The reliability importance measure improvement potential of a channel is used to measure the difference of the unavailability between the existing system and a system where the channel is replaced by a perfect channel. The improvement potential with respect to subsystem i is

$$I^{\mathrm{IP}}(i) = A_{\mathrm{SIS}}(1_i, p) - A_{\mathrm{SIS}}(p) = I^{\mathrm{B}}(i) \cdot \mathrm{PFD}_{\mathrm{avg},(i)} \tag{10}$$

Further, the improvement potential of channel (i,j) is

$$I^{\mathrm{IP}}(i,j) = I^{\mathrm{B}}(i,j) \cdot \mathrm{PFD}_{\mathrm{avg},(i)} \tag{11}$$

2.4 Risk achievement Worth

The risk achievement worth (RAW) of a channel is the ratio of the system unavailability if the channel is always faulty (PFD_{avg}≡1) with the actual system unavailability. Thus

$$I^{RAW}(i,j)=\frac{PFD_{avg,SIS}(0_{i,j},q)}{PFD_{avg,SIS}} \tag{12}$$

2.5 Risk Reduction Worth

The reliability reduction worth (RRW) is the ratio of the actual system unavailability with the unavailability of a system where channel (*i*, *j*) is replaced by a perfect channel (PFD_{avg}≡0). Thus

$$I^{RRW}(i,j)=\frac{PFD_{avg,SIS}}{PFD_{avg,SIS}(1_{i,j},q)} \tag{13}$$

2.6 Criticality Importance

The criticality importance of a channel expresses the probability that the channel has caused system failure, when we know that the system is failed. Such a measure is related to Birnbaum's measure, and according to the definition, it can be calculated as

$$I^{CI}(i,j)=\frac{I^{B}(i,j)PFD_{avg,(i,j)}}{PFD_{avg,SIS}}=\frac{I^{B}(i,j)\lambda_{i,j}}{PFD_{avg,SIS}}\cdot\frac{\partial PFD_{avg,(i)}}{\partial\lambda_{i,j}} \tag{14}$$

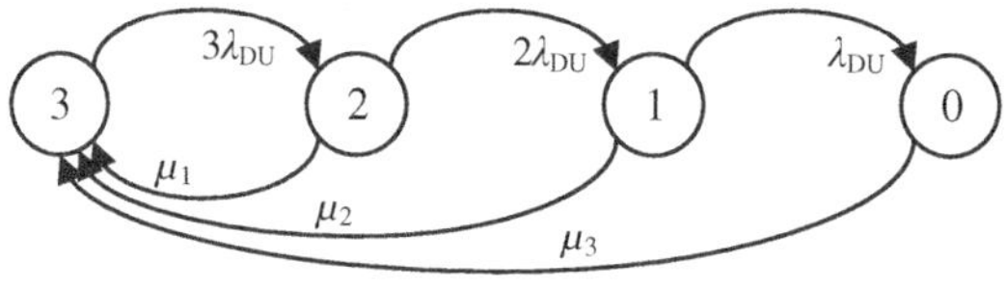

Fig. 2. A Markov model for the 2oo3 transmitter subsystem

3 Analysis for Complex Structures

3.1 Calculation of $PFD_{avg,SIS}$

The task is now to calculate the average PFD for the SIS based on the known failure rates of the channels. The $PFD_{avg,SIS}$ can be adequately approximated as the sum of average PFDs of the three subsystems as Eq. 2. We therefore need to calculate the average PFD for each of these subsystems.

The Markov approach is commonly used to model the behavior (e.g., failed states, repair actions) of a SIS subsystem and to calculate its PFD [4, 5, 6]. Fig. 2 shows the Markov model for the sensor subsystem voted 2oo3 in Fig. 1, and Table 1 describes the states of the model.

Table 1. Descriptions of states in Fig. 2

State	Description
0	3 DU
1	1 functional, 2 DU
2	2 functional, 1 DU
3	3 functional

The model in Fig. 2 considers three identical channels with the same DU failure rate, while ignoring possible DD failures and CCFs. It is straightforward to extend model to involve DD failures and CCFs, but this will increase the size of the model.

The sojourn probabilities of the subsystem in different states of a Markov model can be calculated (The details are skipped since the page limitation). A 2oo3 subsystem is unavailable both in states 2 and 3, and thus $PFD_{avg,S}$ is therefore equal to $P_2 + P_3$. The repair rate μ_k is the inverse of the mean downtime of the channel and is dependent on the mean occurrence time of the last DU failure when a dangerous subsystem fault is revealed in a proof test. For example, when we find two DU failures, we can show (see [5]) that the mean downtime until the fault is revealed is $\tau/3$. If several channels can be repaired together, the repair rate is

$$\mu_k = \frac{1}{\frac{\tau}{k+1} + \text{MTTR}} \tag{15}$$

where k is the number of channels that need to fail to get a dangerous subsystem failure.

The Markov approach can also be used to model a heterogeneous subsystem where the channels have different failure rates. Fig. 3 is such a model for the FE subsystem with shutdown valves A and B in Fig. 1, and the states are listed in Table 2.

Table 2. Descriptions of states in Fig. 3

State	Description
0	A and B DU
1	A functional, B DU
2	A DU, B functional
3	A and B functional

The repair rate μ_1 from state 1 or 2 to 3 in Fig. 3 is given by Eq. 14 as 1/(τ/2+MTTR), while μ_2 is 1/(τ/3+MTTR). Since the subsystem is voted 1oo2, $PFD_{avg,FE}$ can be calculated as P_0.

A more thorough introduction of Markov method can be found in [5, 7].

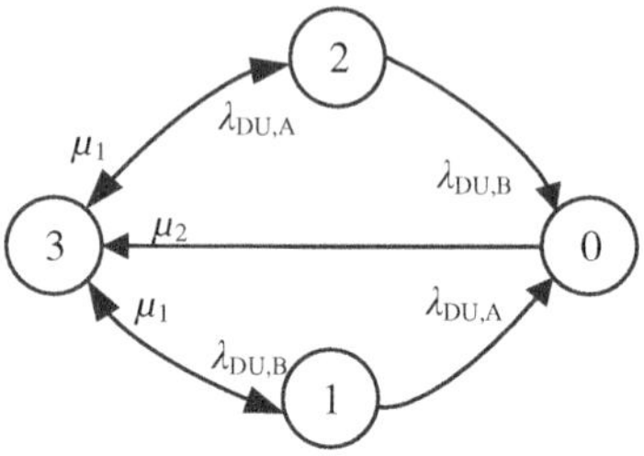

Fig. 3. A Markov model for the 1oo2 final element subsystem

3.2 Common Cause Failures

CCFs are defined in [4] as concurrent failures of two or more channels leading to system failure. The standard adopts the beta-factor model to quantify CCFs in SISs, where the beta-factor β is the fraction of CCFs among all failures of a channel. In this paper we adopt the multiple beta-factor model that is implemented in the PDS method [8]. For the 2oo3 subsystem in Fig. 2, if we assume channel 1 is failed, β can be regarded as the probability that another channel (2 or 3) is also failed.

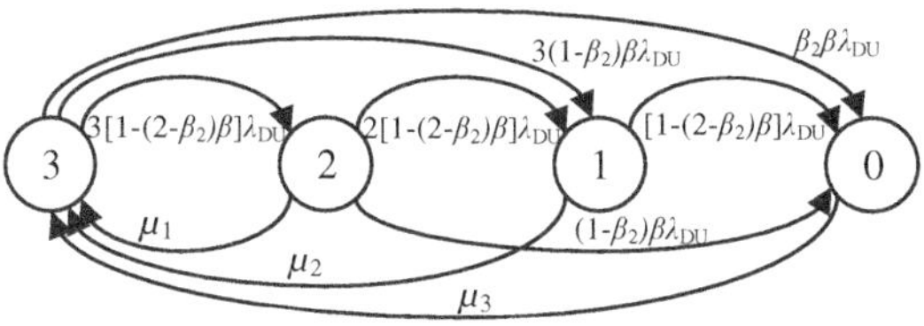

Fig. 4. A Markov model for the 2oo3 sensor subsystem with CCFs

And then, we can use β_2 to denote the conditional probability that the third channel fails. Thus, when a channel fault is revealed in a proof test, the probability that it is a triple fault is $\beta_2\beta$, the probability of a double fault is $3(1-\beta_2)\beta$, and the probability of an individual fault is $1-(2-\beta_2)\beta$. The details of the deduction can be found in Chapter 10 of [5]. Therefore, the model in Fig. 2 can be modified as in Fig. 4.

It is noted that the values of repair rates μ_2 and μ_3 are different from those in Fig. 2 considering the effects of CCFs. The failure rates are determined by the expected unavailable time of the SIS given that it has been found faulty in a proof test. The unavailability of the SIS can come from two contributors: individual failures in all channels and CCFs, which result in different unavailable time. We use v_1 to denote the probability of the SIS entering state 1, and $v_1=v_{1IND}+v_{1CCF}$, where v_{1IND} is the probability of the system entering state 1 due to individual failures, and v_{1CCF} is the probability of the system entering state 1 due to CCFs. According to the model in Fig. 4, we know v_{1IND}=6[1-(2-β_2)β]$^2\lambda_{DU}$, and v_{1CCF} is 3(1 − β_2)$\beta\lambda_{DU}$. Thus, μ_2 can be given by

$$\mu_2 = \frac{1}{v_1}\left(\frac{v_{1IND}}{\tau/3+\text{MTTR}}+\frac{v_{1CCF}}{\tau/2+\text{MTTR}}\right) \tag{16}$$

Similarly, we use v_0 to reflect the probability of the SIS entering state 0, which includes four probabilities: a CCF in 3 channels (v_{0CCF}), 3 individual failures (v_{0IND}), 1 CCF in 2 channel and then 1 individual failure (v_{0CI}), as well as 1 individual failure and then a CCF in the other two channels (v_{0IC}). Thus,

$$\mu_3 = \frac{1}{v_0}\left(\frac{v_{0IND}}{\tau/4+\text{MTTR}}+\frac{v_{0IC}+v_{0CI}}{\tau/3+\text{MTTR}}+\frac{v_{0CCF}}{\tau/2+\text{MTTR}}\right) \tag{17}$$

While for the 1oo2 subsystem in Fig. 3, the failure rates of the two channels are different. The geometric average of the two failure rates is

$$\lambda_{DU} = (\lambda_{DU,A}\cdot\lambda_{DU,B})^{\frac{1}{2}} \tag{18}$$

And then β can be determined as a fraction of this average value, thus the model in Fig. 3 can be modified as that in Fig. 5.

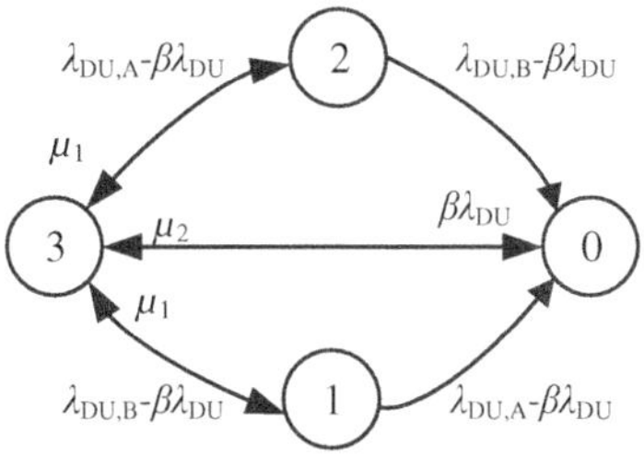

Fig. 5. A Markov model for the 1oo2 shutdown subsystem with CCFs

In Fig. 5, μ_2 is calculated in the similar way of Eq. 16. The probability of entering state 0 (v_0) is the sum of the probability due to individual failures (v_{0IND} = 2($\lambda_{DU,A}$− $\beta\lambda_{DU}$)($\lambda_{DU,B}$ − $\beta\lambda_{DU}$)) and that due to CCFs (v_{0CCF} = $\beta\lambda_{DU}$). So that, μ_2 in Fig. 5 is given by

$$\mu_2 = \frac{1}{v_0}\left(\frac{v_{0IND}}{\tau/3+\text{MTTR}}+\frac{v_{0CCF}}{\tau/2+\text{MTTR}}\right) \tag{19}$$

It should be noted that in many cases, since the individual failures contribute too little to the system unavailability, their effects on the value of the repair rate can be ignored, and so the repair rates can be approximated as $1/(\tau/2+\text{MTTR})$.

4 An Illustrative Example

4.1 Importance Measurements

The SIS in Fig. 1 is used to illustrate the calculation of the reliability importance of the channels. Total channel failure rates and diagnostic coverage factors from the PDS data handbook [9] are listed in Table 3.

Table 3. Failureratesofchannels

Channel	Failure rate (per hour)	DC
PT1	1.3x10−6	0.60
PT2	1.3x10−6	0.60
PT3	1.3x10−6	0.60
LS	1.1x10−7	0.00
SVA	2.1x10−6	0.10
SVB	0.7x10−6	0.15

Here we adopt a hardwired safety system to act as the logic solver. For the transmitters, the beta factor β is 0.06, and β_2 is 0.5. For the different shutdown valves, the beta factor is 0.05. In addition, let the proof testing interval be 8760 hour (one year) and assume that MTTR is 8 hours.

DD failures in transmitter and shutdown valve subsystems are ignored in the following calculation, since they have very little contribution on the system unavailability with being fixed as soon as possible while the redundant configurations are always functional during the repair. But for the logic solver, the DD failure also can result in the unavailability of the whole SIS, and so it is included in the calculation.

PFD_{avg} values of the subsystems can be calculated according to Eq. 2 and Fig. 4 and 5, and then PFD_{avg} of the whole SIS is the sum of these values as Eq. 11: $\text{PFD}_{\text{avg}}(\text{S}) = 0.2979 + 0.4818 + 0.2678 = 1.0475 \times 10^{-3}$ per hour.

The reliability importance of the three subsystems can be easily calculated based on the principle of Eq. 6 as: $I^{\text{B}}(\text{S})$ = 0.99925, $I^{\text{B}}(\text{LS})$ = 0.99943, and $I^{\text{B}}(\text{FE})$ = 0.99922. In order to check importance of channels with Birnbaum's measure, we can build functions of $\text{PFD}_{\text{avg}}(\text{SIS})$ with failure rates of channels based on the Markov models in Fig. 4 and 5 with regarding the failure rate of the studied channel as the variable, and the rates of others as constants. And then, based on Eq. 9, the Birnbaum's measures of all channels can be obtained as shown in Table 4. With the results of Birnbaum's measures, it is not hard to calculate the improvement potentials and critical importance of channels with Eq. 11 and Eq. 14.

Table 4. Reliability importance of channels

Channel	I^{B}	$I^{IP}(10^{-5})$	I^{CI}
PT1	0.1412	4.2076	0.3078
PT2	0.1412	4.2076	0.3078
PT3	0.1412	4.2076	0.3078
LS	0.9994	48.1525	0.4598
SVA	0.0180	0.4815	0.1423
SVB	0.0588	1.5751	0.1463

All of the three measures show that the logic solver is the most important channel in terms of performance of this SIS.

For the RAW measure, an always failed channel in some subsystem can result in a different structure and so the CCFs need to be dealt in different ways. For example, a 2oo3 subsystem will become a 2oo2 one which operates as shown in Fig. 6.

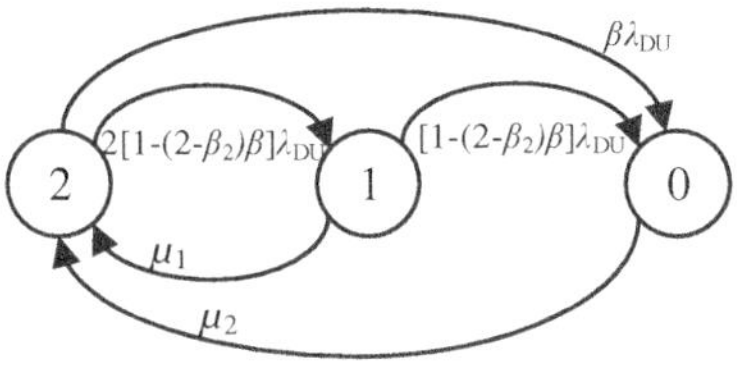

Fig. 6. A Markov model for the 2003 subsystem with one channel regarded as failed

We can understand the 2oo2 structure in Fig. 6 as a 2oo3 system while tolerating that the third channel stays in the faulty state due to an individual failure. If CCFs are regarded to be caused by some common events, and the probabilities of occurrences of such events are not changed when one of channels is in faulty, the rates of both individual and common failures in other channels should keep same with those for the model in Fig. 4. In Fig. 6, the unavailable states still are state 1 and 0.

For the 1oo2 subsystem, after the downgrade, there is only one channel SVA or SVB in consideration, and the failure rate is $\lambda_{DU,A}$ or $\lambda_{DU,B}$.

The RAW values of channels in the SIS are listed in Table 5.

Table 5. RAWs and RRWs of channels

Channel	I^{RAW}	I^{RRW}
PT1	4.8206	1.1712
PT2	4.8206	1.1712
PT3	4.8206	1.1712
LS	954.6539	1.8517
SVA	3.2331	1.3435
SVB	8.6488	1.3435

While for the RRW measure, a perfect channel can be introduced in the 2oo3 structure and the common events have impact on this channel. The model in Fig 6 also can be used to calculate PFD of the subsystem, but it should be noted that only the sojourn probability in state 0 is the contribution of PFD. For the other two subsystem, the introduction of a perfect channel will reduce PFD to 0. The results of RRW measures are listed in the right column of Table 5. According to all five measures, the logic solver obviously is the most important channel in this SIS. It is easy to understand since it has no redundant structure. While for channels in other two subsystems, rankings based on different measures are not same. With the Birnbaum's measure, improvement potential and critical importance method, the channels in the 2oo3 transmitter subsystem are more important, but with RAW and RRW method, the valves shows more important.

4.2 Discussions

We can consider to use a 1oo2 programmable logic controller (PLC) subsystem to replace the hardwired safety system. According to PDS data book [9], the failure rate of any of the two PLC channel is 1.24×10^{-5}, diagnostic coverage is 0.7 and the β-factor is 0.07. PFD_{avg} of the new SIS can be calculated as 0.2979 + 0.7853+ 0.2678 = 1.3510 $\times10^{-3}$ per hour, which is a little higher than that of the SIS mentioned in the last subsection.

And then, the reliability importance for channels in the new system is listed in Table 6.

Table 6. Reliability importance of channels in the new system

Channel	I^{B}	$I^{IP}(10^{-5})$	I^{RAW}	I^{RRW}	I^{CI}
PT1	0.1412	1.0736	3.9397	1.1278	0.2387
PT2	0.1412	1.0736	3.9397	1.1278	0.2387
PT3	0.1412	1.0736	3.9397	1.1278	0.2387
LS1	0.0870	6.8321	4.5050	2.3882	0.3555
LS2	0.0870	6.8321	4.5050	2.3882	0.3555
SVA	0.0180	1.3750	2.7174	1.2472	0.1103
SVB	0.0588	1.3750	6.9306	1.2472	0.1134

No channel has a very high criticality compared with others in such a system. In other words, the changes in reliability of any components has no crucial influence on the performance of the SIS, and therefore there is no obviously vulnerable part. In traditional reliability assessment of SISs, the system with lower PFD_{avg} is regarded more reliable. Criticality importance measures can help decision-makers with more information on how to prioritize reliability enhancement efforts.

Not all measures are same in effectiveness for the assessment of SISs. Consider the channels inside are usually with high reliability, the further increase in reliability maybe costly. In addition, since the main function of a SIS is to mitigate risks to EUC, the measures considering risks of channel failures can be recommended to evaluate the importance of channels in designing a SIS. These measures include Birnbaum's measure and RAW, which can help to locate the vital channel.

It is better for a SIS to be independent from individual failures, and we can examine such independence based on the importance measures, e.g., the sum up of $I_{i,j}^{\mathrm{RAW}}$, A smaller of the value of $\sum I_{i,j}^{\mathrm{RAW}}$, the SIS is inclined to be functional when one channel inside has failures.

It is also possible to check the "balance" in the system design, meaning that there is no super critical channel. We can, for example, measure the balance with the standard deviation of $I_{i,j}^{\mathrm{RAW}}$, When the deviation is small, it is more difficult to find one channel to attack so as to fail the whole system.

In the operational phase of the SIS, its performance can be maintained with periodical proof testings. Assume the safety function can be performed by a bypass channel during the test of a channel, the value of PFD_{avg} of the channel can be reduced with more frequent proof testings. IP and RRW measures can provide some clues to determine the priority of testings. For example, according to Table reftab6, we can increase the testing frequency on the logic servers to twice per year, so the PFD_{avg} of the SIS will be reduced to 0.2979+ 0.2052+0.2678 = 0.7709×10^{-3} per hour. While, if we increase the testing frequency of transmitters or final elements to twice per year, the PFD_{avg} becomes to 1.1957×10^{-3} and 1.2087×10^{-3} respectively.

5 Conclusion

This paper has demonstrated how channel importance measures may be used to locate weaknesses in SISs. Five different measures have been studied for this purpose. The main motivation for exploring the importance measures is to direct the focus to improve the basis for decision-making about design and operation of SIS, beyond the use of PFD_{avg} for this purpose. Based on the analysis, we find that the Birnbaum's measure and RAW (Risk achievement worth) is of particular use in the design phase, to identify the most important channel for which focus should be directed to improve the reliability. Measures like IP (Improvement potential) and RRW (Risk reduction worth) are more useful in the operational phase, in order to help prioritizing the effort of testing. Finally, the CI (criticality importance) approach has also its main use in the operational phase, and more particularly in the following up of SIS failures, to guide resources to components that are most likely to fail. It may be remarked that the application of the importance measures will be even more relevant, in the situation where the complexity of SIS functions is increasing, due to e.g., lack of proper independence of other SIS and non-SIS functions.

More complex structures of SIS modules, such as 2oo4, 3oo5, need to be studied in the future. New modeling methods should be adopted since the size of Markov model will increase explosively with more states of the system. Since this paper only considers two states of the channels, studies like [10] should be integrated to evaluate the importance of channels with multiple operational states. In addition, safe failures and safe failure fraction (SFF) also have been proved to be non-negligible for safety [11], thus they are needed to be considered in future studies. In this article, proof testings of all channels are simply supposed to carry out in short time, but more testing strategies, for example, staggered testings, different testing intervals, should be studied in order to examine their impacts on the importance of the channels.

References

1. Armstrong, M.J.: Joint reliability-importance of components. IEEE Trans. Relia. 44, 408–412 (1995)
2. Xing, L.: Maintenance-oriented fault tree analysis of component importance. In: 2004 Annual Symposium of Reliability and Maintainability, Los Angeles (2004)
3. Xing, L., Amari, S.V.: Effective component importance analysis for the maintenance of systems with common-cause failures. Inter. J. Relia. Qual. Safety Eng. 14(5), 459–478 (2007)
4. IEC 61508, Functional Safety of Electrical/Electronic/Programmable Electronic Safety-Related Systems: Part 1-7. International Electrotechnical Commission, Geneva: Switzerland (2009)
5. Rausand, M., Høyland, A.: System Reliability Theory: Models, Statistical Methods and Applications, 2nd ed. Wiley, Hoboken (2004)
6. Liu, Y., Rausand, M.: Reliability assessment of safety instrumented systems subject to different demand modes. J. Loss Prev. Proc. Ind. 24(1), 49–56 (2011)
7. Rausand, M.: Reliability of Safety-Critical Systems: Theory and Applications. Wiley, Hoboken (2014)
8. SINTEF, Reliability Prediction Method for Safety Instrumented Systems, SINTEF Safety Research, Trondheim, Norway (2013)
9. SINTEF, Reliability Data for Safety Instrumented Systems, SINTEF Safety Research, Trondheim, Norway (2013)
10. Si, S., Dui, H., Cai, Z., Sun, S.: The integrated importance measure of multi-state coherent systems for maintenance processes. IEEE Tran. Relia. 61, 266–273 (2012)
11. Yoshimura, I., Sato, Y.: Safety achieved by the safe failure fraction (SFF) in IEC 61508. IEEE Tran. Relia. 57, 662–669 (2008)

Fatigue Damage Ratios for Heavy Vehicles Operating in the Southern Part of Malaysian Peninsula

Osama Mahmoud Yassenn[1,*], Intan Rohani Endut[2], Mohamed Ahmed Hafez[3], Siti Zaharah Ishak[1], and Nurul Elma Kordi[1]

[1] Malaysia Institute of Transport, Universiti Teknologi MARA, 40450 Shah Alam, Selangor, Malaysia
[2] Faculty of civil engineering, Universiti Teknologi MARA, 40450 Shah Alam, Selangor, Malaysia
[3] Department of civil engineering, Universiti Tenaga Nasional, 43000 Kajang, Selangor, Malaysia
eng.osama.alhusyni@gmail.com

Abstract. In this study, fatigue damage ratios caused by different types of heavy vehicles operating in the southern part of Malaysian Peninsula were developed. More than 6 million heavy vehicles' traffic volumes and weights were collected from 3 Weight-In-Motion stations located at Tangkak, Yong Peng, and Sedenak. Through the analysis, an average fatigue damage ratios for each heavy vehicles types were developed to be used in flexible pavement design and rehabilitation purposes.

Keywords: Fatigue Damage Ratio, Flexible Pavement, Heavy Vehicles, Malaysia.

1 Introduction

It is well know that the economy of any country is highly influenced by the efficiency of its transportations system. It is not possible for any country to reach its potential economic growth without sufficient care and development given to its transportation system and its infrastructures.

The Malaysian economy is developing rapidly, thus it is important to keep the transportation system in the country at its highest possible efficiency. Although, transportation system consists of many modes, the most important transportation mode in Malaysia is the road transport.

Recently, it was observed that a large amount of premature damages were developed along the road network. This may be caused by overloading of heavy vehicles which was found to be forming a considerable percentage of heavy vehicles traffic volume in Malaysia [1-4]. Worth mentioning that, overloading of heavy vehicles is a common problem in many countries around the world [5-8].

* Corresponding author.

M. Gen et al. (eds.), *Industrial Engineering, Management Science and Applications 2015*,
Lecture Notes in Electrical Engineering 349, DOI: 10.1007/978-3-662-47200-2_109

Pavement deterioration is accelerated by several factors such as loading, environment, design consideration, material properties and quality of construction. However, the most important factor causing pavement deterioration is the effect of traffic loading especially the one of the heavy vehicle [9, 10]. It was reported that overloading could cause the damage of pavement to increase out of proportion and that an axle loaded twice the legal limits could cause the damage to increase up to sixty times as much damage as legally loaded axle [11].

There are several types of structural damages caused by loading. Fatigue damage is one the most common types of premature structural damages on flexible pavements along the road network in Malaysia. One way of describing fatigue damage is by using fatigue damage ratios. Fatigue damage ratios are used to represent the fatigue damage caused to flexible pavement by certain axle loading and configuration. It compares the fatigue damage caused by arbitrary axle loading and configuration to the fatigue damage caused by the standard axle load and configuration, which is single axle with four tires and 8.16 tons loading. In case of underestimation of these ratios due to overloading the pavements will encounter premature damages and high reduction of its service life. However, if these ratios were overestimated unnecessary expenditure will be spent due to overdesigning.

Thus, the main aim of this study is to develop fatigue damage ratios for each heavy vehicles type, operating in the southern portion of Malaysian peninsula, based on their actual loading.

2 Methodology

The heavy vehicles traffic volumes and weights were collected by PLUS expressway Berhad from several stations spread along the North-South Expressway, which is the biggest expressway in Malaysia. For the purpose of this study, the data collected from three stations located at the southern part of the expressway were used to represent the southern region of the Malaysian Peninsula.

In traffic volumes and weights data collection, Weight-In-Motion (WIM) systems are used. The Weight-In-Motion systems are used to measure the vehicles' gross weight or axle loading at normal traffic speed without the need of the vehicles to stop. When compared to static weighting system, this system is more effective.

The Weight-In-Motion systems have several advantages. One of the advantages is high processing rate. Since the systems measure vehicle weight at normal traffic speed, it is possible to process a large number of vehicles compared to static weighting system. Another advantage of this system is, due to the high processing rate, the systems make it possible to process data consciously, rather than sampling technique which is used in static weighting system. This will eliminate bias in the collected data. One more advantage of the systems is that, since the systems weight vehicles at their normal travelling speed and the vehicles do not need to stop like those in static weighting system, this will eliminate the accumulation of HVs heading to weighting station, which increases the safety of the weighting process and decreases the possibility of jams caused by the weighting station. A further advantage of this type of weighting systems is that the vehicles are weighted without alerting the driver, which results in a more trustworthy data, since the drivers will not try to avoid the weighting station.

On the other hand, there are a few disadvantages of using the WIM systems. Since the vehicles are weighted without stopping the vehicle, several data that would have been easily collected in the static weighting system will not be available with this method. This data includes fuel type, origin, and destination, however, for this study, the mentioned data are not required, which make this system an ideal choice.

The traffic volume and weight of each vehicle type were collected from three survey stations located along the North-South Expressway at Tangkak, Yong Peng, and Sedenak (Figure 1).

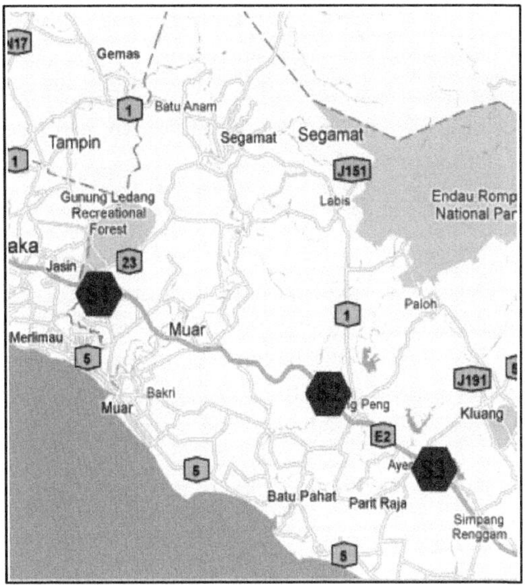

Fig. 1. Weight-In-Motion Approximate Locations. S1At Tangkak, S2 At Yong Peng, S3 At Sedenak.

Ten types of HVs, which form the majority of heavy vehicles operating at the southern part of the Malaysian Peninsula were investigated. The HVs types that were investigated in this study are as follows:

1- Single unit trucks with two axles
2- Single unit trucks with three axles
3- Single unit trucks with four axles
4- Single trailer trucks with four axles
5- Single trailer trucks with five axles
6- Single trailer trucks with six axles
7- Multi trailer trucks with five axles
8- Multi trailer trucks with six axles
9- Multi trailer trucks with seven axles
10- Buses

After collecting the traffic volume and weight data, flexible pavement of the North-South Expressway was analysed to estimate the fatigue life of the pavement under the current loading, and to compare it with its fatigue life under standard axle loading. To perform this analysis, it is necessary to identify pavement layer thicknesses and elasticity moduli. These two types of data were collected by PLUS Expressway Berhad every 250 meters. The data were collected using Falling Weight Deflectometer (FWD), Ground Penetration Radar (GPR), and Coring and Dynamic Cone Penetration (DCP).

The pavement cross-sections were analysed to determine the maximum tensile strain at each asphalt layer, which was used to determine the pavement fatigue life as shown by the following equation:

$$N_f = f_1(\varepsilon_t)^{-f_2}(E)^{-f_3}$$

Where:

N_f : Allowable number of load repetition before fatigue failure.

ε_t : Maximum tensile strain at the bottom of the Hot Mix Asphalt (HMA)

E : Elasticity modulus of HMA

f_1, f_2, f_3: Constants equal to 0.0796, 3.291, 0.854 based on American Asphalt Institute[12]

The same process was performed on pavement cross-sections using standard axle load. After that, fatigue damage ratio was calculated by dividing the pavement fatigue life under a specific HV load by the pavement fatigue life under standard axle load. This process was performed on more than 350km of the North-South Expressway, in which the WIM stations are located, then the average effect of each HV type was determined.

3 Results

3.1 Tangkak

Figure 2 shows a comparison between the percentages of weighted heavy vehicles at Tangkak Weight-In-Motion Station. A total of 2,074,116 heavy vehicles were weight at Tangkak Weight-In-Motion Station. It could be clearly seen that the majority of heavy vehicles traffic volumes (89.3%) is made of merely three types: single unit trucks with two axles (27.0%), single trailer truck with 4 axles (31.8%) and buses (30.5%).

All types of multi-trailer vehicles forms only 1.1%. The multi-trailers with five axles forms 1.1% and only 48 multi-trailer truck with six axles passed the weighting station (0.002%), also there is no records for any multi trailer vehicles with more than 6 axles passed the weighting station. All other types of heavy vehicles form 9.5%.

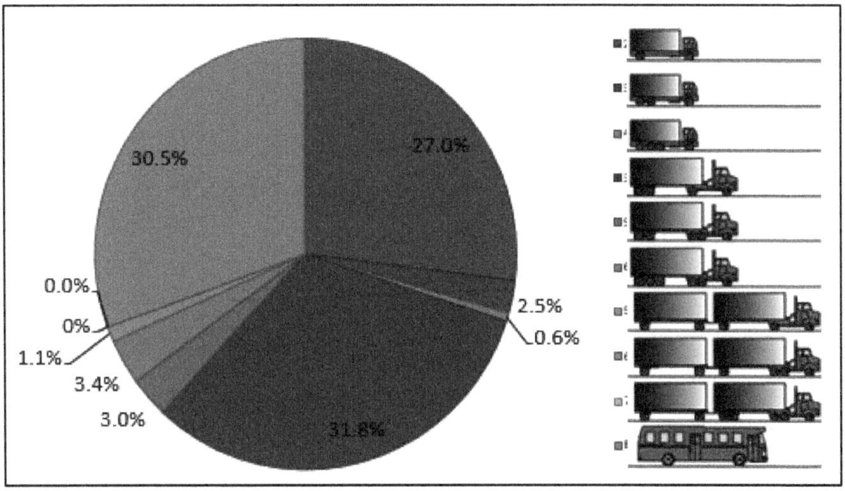

Fig. 2. Heavy Vehicles Traffic Volume at Tangkak WIM Station

Figure 3 compares between different fatigue damage ratios calculate for each vehicle class weighted by Tangkak WIM station. Based on the calculated fatigue damage ratio, it was found that multi-trailer trucks have the highest fatigue damage ratio. Multi-trailer trucks with five axles have a fatigue damage ratio of 7.93 followed by multi-trailer trucks with six axles (6.86). However, no multi-trailer trucks with seven axles were weighted in this WIM station.

Furthermore, the single trailer trucks have a medium fatigue damage ratio ranging between 5.91 and 5.19. Single trailer trucks with four axles have a fatigue damage ratio of 5.91 followed by the single trailer trucks with five axles (5.40) and the lowest is the single trailer trucks with six axles (5.19).

Moreover, the single unit trucks and busses have the lowest fatigue damage ratio comparing to all other heavy vehicles types. Arranged form the vehicle with highest to lowest fatigue damage ratio, busses (3.33), single unit trucks with two axles (3.08), single unit trucks with three axles (2.73), and single unit trucks with four axles (2.23).

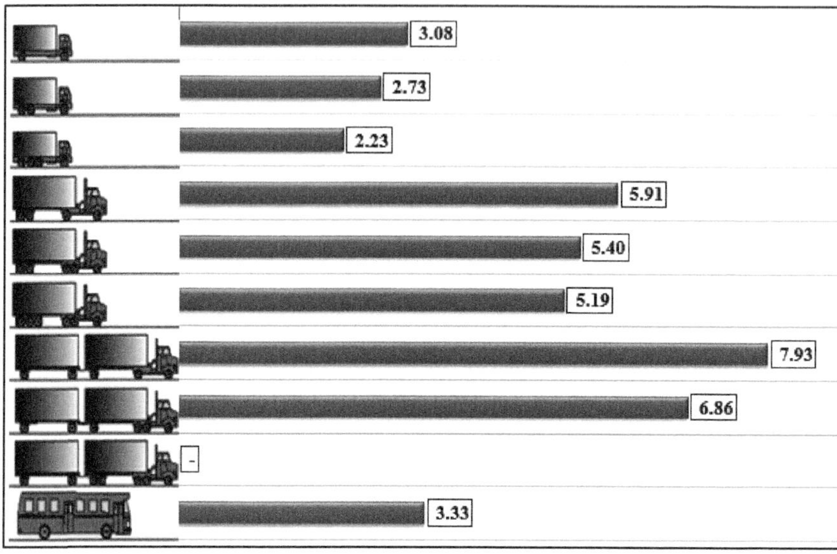

Fig. 3. Calculated Fatigue Damage Ratios for Heavy Vehicles at Tangkak WIM Station

3.2 Yong Peng

Figure 4 shows a comparison between the percentages of weighted heavy vehicles at Yong Peng Weight-In-Motion Station. A total of 1,850,915 heavy vehicles were weight at Yong Peng Weight-In-Motion Station. It could be clearly seen that the majority of heavy vehicles traffic volumes (87.5%) is made of merely three types: single unit trucks with two axles (17.0%), single trailer truck with 4 axles (35.2%) and buses (35.3%).

All types of multi-trailer vehicles forms only 1.6%. The multi-trailers with five axles forms 1.1% and the multi-trailer with seven axles forms only 0.5%. Also, only 33 multi-trailer truck with six axles passed the weighting station. All other types of heavy vehicles form 14.6%.

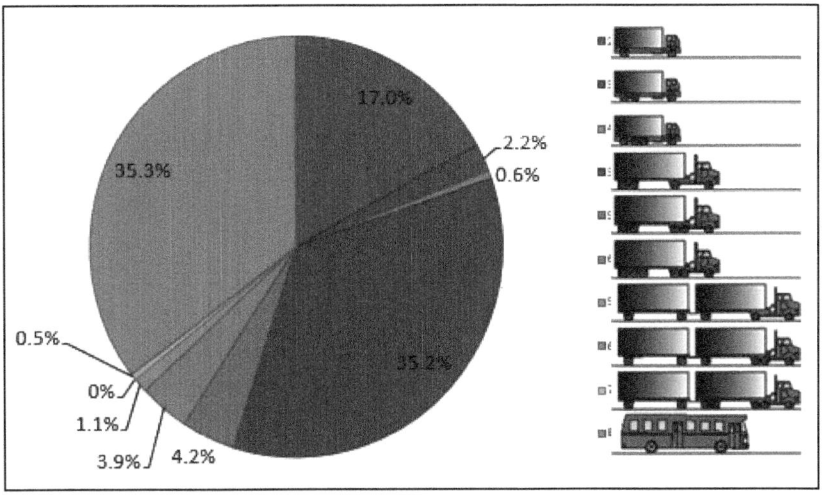

Fig. 4. Heavy Vehicles Traffic Volume at Yong Peng WIM station

Figure 5 compares between different fatigue damage ratios calculate for each vehicle class weighted by Yong Peng WIM station. Based on the calculated fatigue damage ratio, it was found that multi-trailer trucks have the highest fatigue damage ratio ranging between 6.74 for multi-trailer trucks with seven axles to 6.52 for multi-trailer trucks with five axles. Multi-trailer trucks with six axles have a fatigue damage ratio of 6.56.

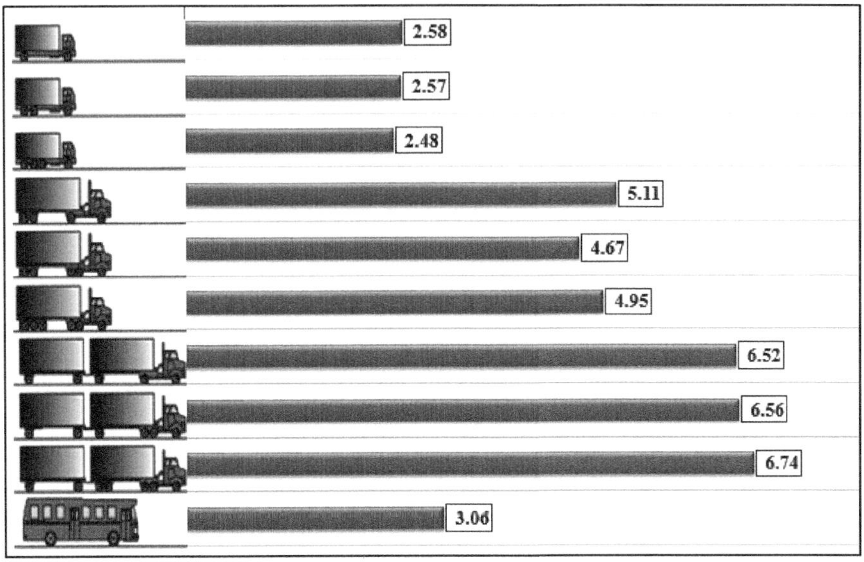

Fig. 5. Calculated Fatigue Damage Ratios for Heavy Vehicles at Yong Peng WIM Station

Furthermore, the single trailer trucks have a medium fatigue damage ratio ranging between 5.11 and 4.67. Single trailer trucks with four axles have a fatigue damage ratio of 5.11 followed by the single trailer trucks with six axles (4.95) and the lowest is the single trailer trucks with five axles (4.67).

Moreover, the single unit trucks and busses have the lowest fatigue damage ratio comparing to all other heavy vehicles types. Arranged form the vehicle with highest to lowest fatigue damage ratio, busses (3.06), single unit trucks with two axles (2.58), single unit trucks with three axles (2.57), and single unit trucks with four axles (2.48).

3.3 Sedenak

Figure 6 shows a comparison between the percentages of weighted heavy vehicles at Sedenak Weight-In-Motion Station. A total of 2,557,805 heavy vehicles were weight at Sedenak Weight-In-Motion Station. It could be clearly seen that the majority of heavy vehicles traffic volumes (86.6%) is made of merely three types: single unit trucks with two axles (20.0%), single trailer truck with 4 axles (35.5%) and buses (31.0%).

All types of multi-trailer vehicles forms only 1.5%. The multi-trailers with five axles forms 1.0% and the multi-trailer with seven axles forms only 0.5%. Also, only 36 multi-trailer truck with six axles passed the weighting station (0.001%). All other types of heavy vehicles form 11.9%.

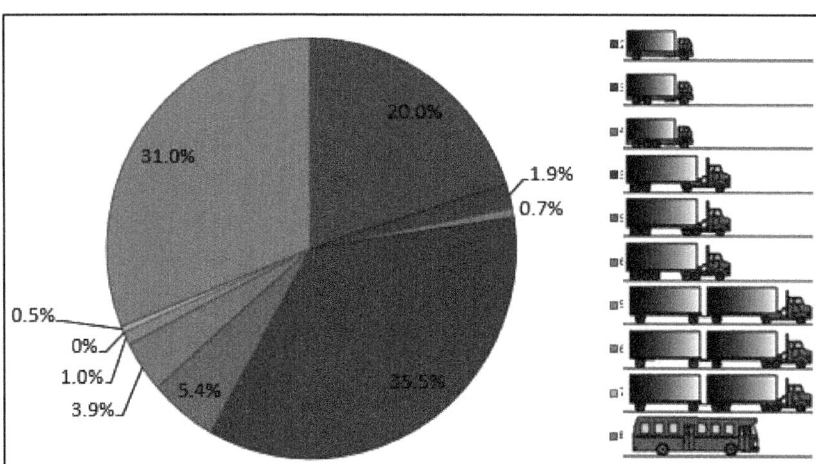

Fig. 6. Heavy Vehicles Traffic Volume at Sedenak WIM station

Figure 7 compares between different fatigue damage ratios calculate for each vehicle class weighted by Sedenak WIM station. Based on the calculated fatigue damage ratio, it was found that multi-trailer trucks have the highest fatigue damage ratio ranging between 6.47 for multi-trailer trucks with six axles to 6.04 for multi-trailer trucks with five axles. Multi-trailer trucks with seven axles have a fatigue damage ratio of 6.27.

Furthermore, the single trailer trucks have a medium fatigue damage ratio ranging between 4.90 and 3.67. Single trailer trucks with four axles have a fatigue damage ratio of 4.90 followed by the single trailer trucks with six axles (4.24) and the lowest is the single trailer trucks with five axles (3.67).

Moreover, the single unit trucks and busses have the lowest fatigue damage ratio comparing to all other heavy vehicles types. Arranged form the vehicle with highest to lowest fatigue damage ratio, single unit trucks with two axles (3.91), busses (2.93), single unit trucks with three axles (2.88), and single unit trucks with four axles (2.52).

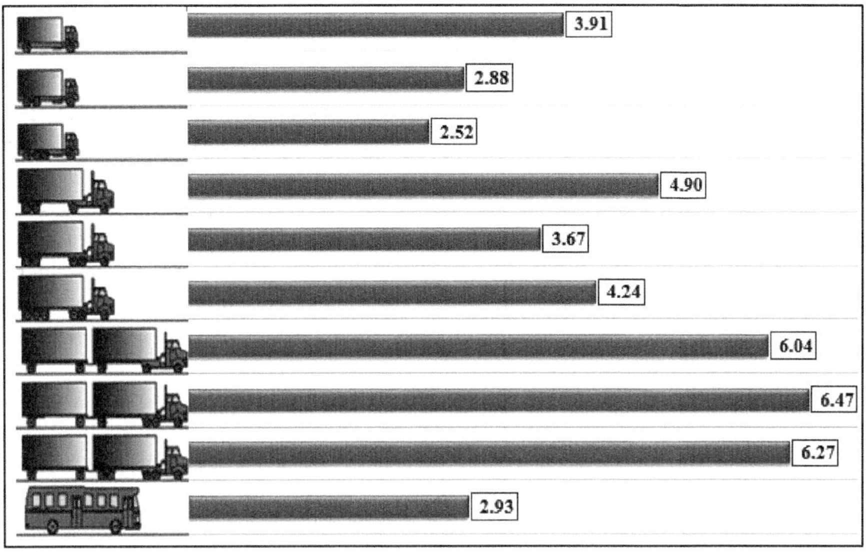

Fig. 7. Calculated Fatigue Damage Ratios for Heavy Vehicles at Sedenak WIM Station

4 Conclusion

For the pavements' new design and overlay design at the southern region of the Malaysian Peninsula, the average fatigue damage ratios based on actual weights for each vehicle type were calculated as follows:

1. Single unit trucks with two axles = 3.27
2. Single unit trucks with three axles = 2.74
3. Single unit trucks with four axles = 2.42
4. Single unit trailers with four axles = 5.26
5. Single unit trailers with five axles = 4.34
6. Single unit trailers with six axles = 4.73
7. Multi-trailer trucks with five axles = 6.82
8. Multi-trailer trucks with six axles = 6.66
9. Multi-trailer trucks with seven axles = 6.47
10. Buses = 3.09

In case of inaccurate traffic data, a weighted average of 4.03 can be used to represent fatigue damage ratio of heavy vehicles.

Acknowledgements. The authors would like to express their gratitude to the Fundamental Research Grant Scheme for initiating the grant awarded by the Ministry of Education (MOE), and also greatest support from Malaysia Institute of Transport (MITRANS), Research Management Institute (RMI), and Universiti Teknologi MARA (UiTM).

The writers would also like to express their gratitude to PLUS Expressway Bhd. and Soil Central Lab Sdn. Bhd. for providing WIM and FWD data.

References

1. Yassenn, O.M., et al.: Overloading At The Northern Part Of The Malaysian Expressway. In: 2011 IEEE Colloquium on Humanities, Science and Engineering Research, Penang, Malaysia (2011)
2. Yassenn, O.M., Hafez, M.A., Endut, I.R., Baharom, B., Wahab, M.Y.: Axle Load Distribution and Overloading at the Central Part of the North-South Expressway in Malaysia. European Journal of Scientific Research 79(2) (2012)
3. Yassenn, O.M., Hafez, M.A., Endut, I.R., Baharom, B., Wahab, M.Y.: Axle Overloading At The Southern Part of the North-South Expressway In Malaysia. In: 2012 IEEE Business Engineering and Industrial Applications Colloquium, Malaysia (2012)
4. Abdullah, A.S.: Development of Integrated Weigh-in-motion System and Analysis of Traffic Flow Characteristics considering Vehicle Weight. The University of Tokushima Japan (2011)
5. Wen, H., et al.: Site survey and analysis of highway trucks overloading status quo in Anhui (2005), http://eprints.qut.edu.au/27727/
6. Chan, Y.C.M.: Truck overloading study in developing countries and strategies to minimize its impact (2008)
7. Chaudry, R., Memon, A.B.: Effects of Variation in Truck Factor on Pavement Performance in Pakistan (2000)
.. Chou, C.-P.J.: Effect of overloaded heavy vehicles on pavement and bridge design. Transportation Research Record: Journal of the Transportation Research Board 1539(1), 58-. . . . .
.. Salama, H.K., Chatti, K., Lyles, R.W.: Effect of Heavy Multiple Axle Trucks on Flexible Pavement Damage Using In-Service Pavement Performance Data. Journal of Transportation Engineering 132(10), 763-. . . . .
10. Embong, H.M.B.H.: Kajian Kerosakan Permukaan Jalan Raya Dan Kerja-Kerja Membaiki Di Bawah Jabatan Kerja Raya Kuala Terengganu. In: Faculty of Civil Engineering. Universiti Teknologi Malaysia, Kuala Terengganu (2004)
11. CSIR and R.A.T. Technology. The Damaging Effects of Overloaded Heavy Vehicles on Roads (Pad27 - revised), Republic of South Africa (1997)
12. Huang, Y.H.: Pavement analysis and design. Pearson/Prentice Hall, United States of America (2004)

Process Reliability Modeling Based on Nonlinear Correlation Analysis

Chuanliang Zhang*, Wei Dai, and Yu Zhao

School of Reliability and Systems Engineering, Beihang University, Beijing 100191, China
zcls07@163.com, dw@buaa.edu.cn, lpds@126.com

Abstract. The analysis of the manufacturing process that formed the product inherent reliability and the determination of the key technological factors that have great influence on product inherent reliability can be the foundation of the process control and process improvement. And in order to improve the prediction accuracy and efficiency of the model, the input variables were selected firstly based on the analysis of input and output variables correlation. Firstly, the regularity of the product inherent reliability formed in manufacturing process was obtained through the analysis of production process. secondly, the estimation method of mutual information based on the copula entropy was given. thirdly, the steps choosing the key process factors were proposed based on the nonlinear correlation analysis. And finally the feasibility and practicability of the method were indicated by an application case.

1 Introduction

The product inherent reliability is decided by the design and machining process. Engineering practice shows that the inherent reliability of the same design standard of the products manufactured by different processing method will have great differences. Therefore, the process reliability can be reflected by evaluating the product inherent reliability of the same design standard. And the analysis of the manufacturing process that formed the product inherent reliability and the determination of the key technological factors that have great influence on product inherent reliability can be the foundation of the process control and process improvement.

If the input variables were not effectively selected in the establishment of inherent reliability prediction model, the role of the important variables may be obscured by the redundant or less affected variables contained in the model. The complexity of prediction model will be increased and the accuracy of prediction results will be decreased. Therefore, in order to improve the prediction accuracy and efficiency of the model, the input variables were selected firstly based on the analysis of input and output variables correlation. In the methods of the variable correlation analysis, to use fewer variables represent the main information of the original variables, the principal component analysis [1] and the independent component analysis [2] and canonical correlation analysis [3] mainly obtain a new set of variables through some transform methods, however the original physically meaning of the variables will be lost; the qualitative analysis but not the quantitative description can only be given by the

* Corresponding author.

M. Gen et al. (eds.), *Industrial Engineering, Management Science and Applications 2015*,
Lecture Notes in Electrical Engineering 349, DOI: 10.1007/978-3-662-47200-2_110

granger causality analysis [4,5]; the grey correlation degree[6] has great subjectivity because of its calculation has relation with the selection of the initial value. And many current processes are nonlinear and the correlation was often existed between variables, which makes the traditional linear correlation analysis methods are no longer applicable. While the mutual information [7] can effectively measure the nonlinear relationship of the variables, and it can reflect the amount of information shared between the two variables or more variables.

Firstly, the regularity of the product inherent reliability formed in manufacturing process was obtained through the analysis of production process; secondly, the estimation method of mutual information based on the copula entropy was given; thirdly, the steps choosing the key process factors were proposed based on the nonlinear correlation analysis; and finally the feasibility and practicability of the method were indicated by an application case.

2 Analysis of Production Process

The product reliability refers to the product properties of maintaining its own ability to work with time change in the use process. Thus the level of the product reliability can be reflected by the using properties of product, such as the wear resistance, the fatigue strength and the corrosion resistance performance. And the using properties of product were decided by the output parameters of process, such as the size, the surface roughness and other quality characteristics. And the quality characteristics are usually formed in the multi-channel processes, and each process has many process factors influenced on the quality characteristics. The using properties that influenced on the product reliability, the quality characteristics that influenced on the using properties and process factors that influenced on the quality characteristics can be determined by the analysis of FMEA and the process mechanism. The regularity of the product inherent reliability formed in manufacturing process can be expressed as figure 1.

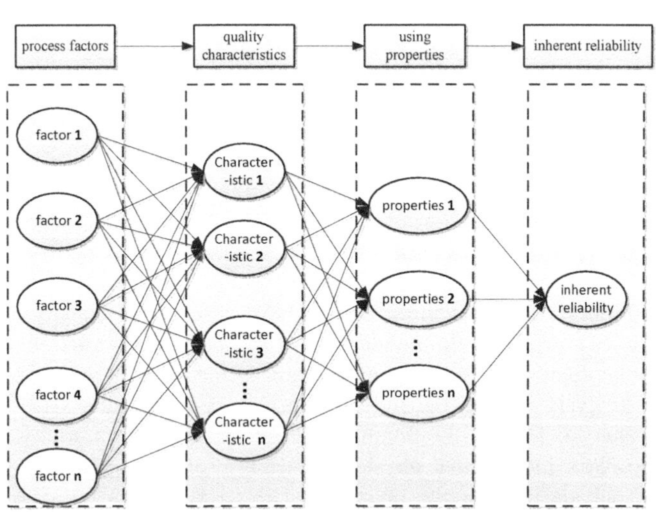

Fig. 1. Regularity of the product inherent reliability formed in manufacturing process

3 Estimation Method of Mutual Information Based on Copula entropy

The mutual information between the variables X and Y are defined as:

$$I(X,Y) = \iint_S f_{x,y}(x,y)\log\frac{f_{x,y}(x,y)}{f_x(x)f_y(y)}dxdy, \tag{1}$$

Where S was the definition domain of X and Y, $f_{x,y}(x,y)$ represented the joint probability density of X and Y, $f_x(x)$ and $f_y(y)$ were marginal probability density function of X and Y. The bigger mutual information reflect more information the variable X contains about the Y. The computation of mutual information was divided into three steps as the formula (1) show, firstly, using the data to calculate the joint distribution function; secondly, using the data to obtain the marginal distribution function; and then calculate the mutual information with using formula (1). However, the mutual information results will be affected by the joint probability density estimation accuracy with using this calculation method. Thus the accuracy of the mutual information estimation can be improved with avoiding the estimation of the joint probability density function. Copula entropy is an effective tool to measure the linear or nonlinear relationship, as it is no parametric and has no restrictions on the distribution of the variable function. The marginal distribution and joint distribution of variables were not need to be calculated with using the relationship between mutual information and Copula entropy.

The concrete steps of estimation method of mutual information based on the copula entropy were shown as follows [5] :

1) Normalize the data (X, Y) to [0,1], and the normalized formula was:

$$\overline{X} = \frac{X - X_{\min}}{X_{\max} - X_{\min}}, \quad \overline{Y} = \frac{Y - Y_{\min}}{Y_{\max} - Y_{\min}}, \tag{2}$$

where X_{min} is the minimum value of X, $X_{\max}$ is the maximum value of X, $Y_{\min}$ is the minimum value of Y, $Y_{\max}$ is the maximum value of Y.

2) Calculate the correlation coefficient of Kendall τ of $(\overline{X}, \overline{Y})$. Combined the same elements of $\overline{X}$ into s small collections, U_i express the number of elements contained in the small collection of I, Combined the same elements of $\overline{Y}$ into t small collections, V_i express the number of elements contained in the small collection of I, and

$$N_1 = \sum_{i=1}^{s}\frac{1}{2}U_i(U_i - 1), \quad N_2 = \sum_{i=1}^{t}\frac{1}{2}V_i(V_i - 1) \tag{3}$$

Thus the calculation formula of the correlation coefficient of Kendall τ of $(\overline{X}, \overline{Y})$ can be expressed as:

$$\tau = \frac{C - D}{\sqrt{(N_3 - N_1)(N_3 - N_2)}}, \tag{4}$$

Where C expressed the number of elements that have consistency of $(\overline{X}, \overline{Y})$ and D expressed the number of elements that have no consistency of $(\overline{X}, \overline{Y})$, and

$$N_3 = \frac{1}{2N(N-1)}, \tag{5}$$

3) Estimate the parameters of corresponding type of Copula function according to the correlation coefficient of Kendall τ.

4) Obtain the Copula density function $c(u, v)$ according to the parameters of Copula function.

5) Calculate the Copula entropy according to the following function:

$$H_c(u,v) = -\frac{1}{N}\sum_{i=1}^{N} \log c(u_i, v_i), \tag{6}$$

Where (u_i, v_i), $i = 1, \cdots, N$ is N sample points generated by Copula density function.

6) Thus the mutual information was:

$$I(X;Y) = -H_c(u,v) = \frac{1}{N}\sum_{i=1}^{N} \log c(u_i, v_i), \tag{7}$$

4 Selection of Key Process Factors Based on Nonlinear Analysis

Assumed R as the product inherent reliability, and through the analysis of FMEA and the process mechanism, t using properties that influenced on the product reliability were determined and denoted as z_1、z_2、…、z_t. The mutual information between the using properties and product reliability can be obtained with using the estimation method of mutual information based on the copula entropy and expressed as $I(z_i; R),\ i = 1,2,\cdots,t$. Assumed $l(l \le t)$ key using properties were selected finally, reorganized and marked as z_1、z_2、…、z_l. And through the analysis of FMEA and the process mechanism, p quality characteristics that influenced on the key using properties were determined and denoted as y_1、y_2、…、y_p. The mutual information between the quality characteristics and using properties can be obtained with using the estimation method of mutual information based on the copula entropy and expressed as:

$$\begin{bmatrix} & z_1 & z_2 & \cdots & z_l \\ y_1 & I(y_1; z_1) & I(y_1; z_2) & \cdots & I(y_1; z_l) \\ y_2 & I(y_2; z_1) & I(y_2; z_2) & \cdots & I(y_2; z_l) \\ \vdots & \vdots & \vdots & \ddots & \vdots \\ y_p & I(y_p; z_1) & I(y_p; z_2) & \cdots & I(y_p; z_l) \end{bmatrix}$$

The final mutual information of various quality characteristic was:

$$I(y_j; z) = \frac{1}{l}\sum_{s=1}^{l} I(y_j; z_s),\ j = 1,2,\cdots,p, \tag{8}$$

Assumed $m(m \le p)$ key quality characteristics were selected finally, reorganized and marked as y_1、y_2、…、y_m. And through the analysis of FMEA and the process mechanism, q process factors that influenced on the key quality characteristics were determined and denoted as x_1、x_2、…、x_q. The mutual information between the process factors and quality characteristics can be obtained with using the estimation method of mutual information based on the copula entropy and expressed as:

$$\begin{bmatrix} & y_1 & y_2 & \cdots & y_m \\ x_1 & I(x_1; y_1) & I(x_1; y_2) & \cdots & I(x_1; y_m) \\ x_2 & I(x_2; y_1) & I(x_2; y_2) & \cdots & I(x_2; y_m) \\ \vdots & \vdots & \vdots & \ddots & \vdots \\ x_q & I(x_q; y_1) & I(x_q; y_2) & \cdots & I(x_q; y_m) \end{bmatrix}$$

The final mutual information of various process factor was:

$$I(x_k; y) = \frac{1}{m}\sum_{f=1}^{m} I(x_k; y_m),\ k = 1,2,\cdots,q, \tag{9}$$

And the key process factors can be identified finally. Thus the prediction model of product reliability can be established on this basis, and product reliability and customer satisfaction can be improved through the effectively process monitoring.

5 Application Case

The coating process is a heat treatment increasing the reliability of the bearing. To establish the inherent reliability prediction model and obtain the foundation of the process control and process improvement, the key process factors were determined with using the proposed method. Firstly, the using properties that influenced on the inherent reliability, the quality characteristics that influenced on the using properties and process factors that influenced on the quality characteristics were determined by the analysis of FMEA and the process mechanism. The regularity of the product inherent reliability was expressed as figure 2.

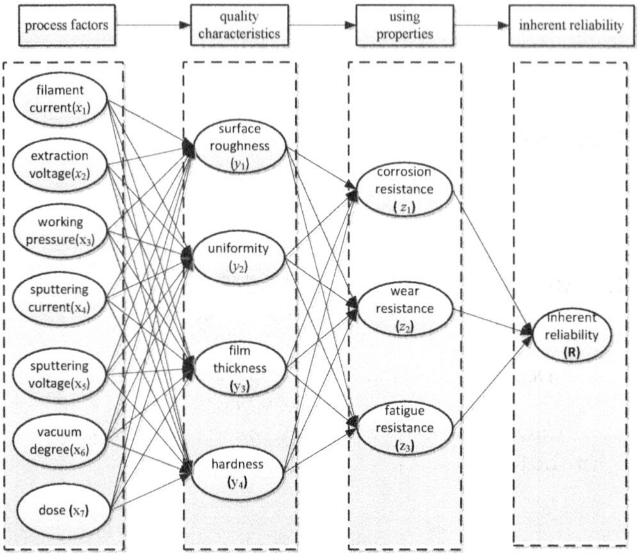

Fig. 2. Regularity of the bearing inherent reliability formed in coating process

The data of the process can be collected. And the mutual information between the using properties and product reliability was obtained with using the estimation method of mutual information based on the copula entropy as (0.8057, 0.8046, 0.8313). The result reflected that every using property has strong relationship with the inherent reliability. And then the mutual information between the quality characteristics and using properties can be obtained as Table 1.

Table 1. the mutual information between the quality characteristics and using properties

	z_1	z_2	z_3
y_1	0.441	0.525	0.412
y_2	0.173	0.102	0.112
y_3	0.423	0.631	0.512
y_4	0.621	0.544	0.613

The final mutual information of various quality characteristic was (0.459, 0.129, 0.522, 0.593), The result reflected that every quality characteristic has strong relationship with the using properties except film thickness. And then the mutual information between process factors and the key quality characteristics can be obtained as Table 2.

Table 2. The mutual information between process factors and the key quality characteristics

	y_1	y_2	y_4
x_1	0.731	0.643	0.765
x_2	0.832	0.841	0.752
x_3	0.121	0.312	0.214
x_4	0.132	0.214	0.141
x_5	0.113	0.315	0.238
x_6	0.621	0.758	0.832
x_7	0.732	0.653	0.792

The final mutual information of various process factor was (0.713, 0.808, 0.216, 0.162, 0.222, 0.737, 0.726), thus the key process factors can be determined as (x_1, x_2, x_6, x_7), and the prediction model of product reliability can be established on this basis.

6 Conclusion

The analysis of the regularity of the product inherent reliability formed in manufacturing process and the determination of the key technological factors that have great influence on product inherent reliability can be the foundation of the process control and process improvement. The intrinsic link between products inherent reliability and process factors was firstly reviewed; and then the correlation between variables was measured with using the estimation method of mutual information based on the copula entropy, and the key technological factors were identified; and finally the feasibility and practicability of the method were indicated by an application case.

References

1. Guo, W., Dai, L.R., Wang, R.H.: Feature mapping based on PCA. Acta Automatica Sinica 34(8), 876–879 (2008)
2. Yang, Z.Q., Li, Y., Hu, D.W.: Independent component analysis: a survey. Acta Automatica Sinica 28(5), 762–772 (2002)
3. Yamamoto, H., Yamaji, H., Fukusaki, E., Ohno, H., Fukuda, H.: Canonical correlation analysis for multivariate regression and its application to metabolic ngerprinting. Biochemical Engineering Journal 40(2), 199–204 (2008)
4. Barrett, A.B., Barnett, L., Seth, A.K.: Multivariate granger causality and generalized variance. Physical Review E 81(4), 041907 (2010)
5. Han, M., Liu, X.X.: Stepwise input variable selection based on mutual information for multivariate forecasting[J]. Acta Automatica Sinica 38(6), 999–1006 (2012)
6. Sallehuddin, R., Shamsuddin, S.M.H., Hashim, S.Z.M.: Application of grey relational analysis for multivariate time series. In: Proceedings of the 8th International Conference on Intelligent Systems Design and Applications, pp. 432–437. IEEE, Piscataway (2008)
7. Peng, H.C., Long, F., Ding, C.: Feature selection based on mutual information: criteria of max-dependency, max-relevance, and minredundancy. IEEE Transactions on Pattern Analysis and Machine Intelligence 27(8), 1226–1238 (2005)

Identification of Public Awareness in Preventive Maintenance for Personal Automobile

Y. Ngadiman, A.H. Nor Aziati, Abdul Talib Bon, M.F. Ahmad, Raja Zuraidah Raja Mohd Rasi, and Martin Yaw Swee Hock

Department of Production & Operation Management
Faculty of Technology Management and Business
Universiti Tun Hussein Onn Malaysia, Johor, Malaysia
{yunos,aziati,talib,mohdfauzi,rzuraida}@uthm.edu.my,
martinyaw06@gmail.com

Abstract. This study is to identify the awareness of public towards maintenance in their vehicle. The maintenance discuss in this study are narrowed and focused on two type of maintenance strategy particularly preventive maintenance. Maintaining vehicle have been an issue in Malaysia where we can see accidents happening in every seasonal in transportation sector. The lack of awareness towards maintenance may also lead to breakdown and malfunction. Therefore, the question comes whether the public has the awareness in maintaining their automobile, thus, this study will identify the public awareness in corrective and preventive maintenance towards personal Drivers. The study has been carried out at University of Tun Hussein Onn Malaysia (UTHM) and focuses on the Drivers age from early 20s to late 40s to identify their awareness or understanding towards maintenance. The study has been conducted with quantitative method and analyzed using statistical tools by cross tabulation. Studies on awareness in maintenance of vehicle have not been carrying out on individuals/users. Averagely, the respondents score a mean of 0.6475 which means that the respondents have more Yes answered in their sets of awareness questions. The study also concluded that experience and education on the respondents has the difference in affecting their awareness.

1 Introduction

The awareness of public users (excluding commercial users) is yet to be known its level towards preventive maintenance. As the knowledge on maintenance management is well practiced in the production sector and building maintenance sectors, we have yet to know the level of application of knowledge and preventive maintenance in each individual whom own vehicles in Malaysia. With such a large amount of vehicles moving in Malaysia, maintenance practicing has been introduced to the users to ensure that vehicles owned are well maintained [7]. The practice of maintenance comes in two very basic practices in the maintenance management practices such as corrective maintenance and preventive maintenance [6]. Vehicle is one of the main elements in transportation system in the industry [6], [7]. Vehicle can be use daily and

M. Gen et al. (eds.), *Industrial Engineering, Management Science and Applications 2015*,
Lecture Notes in Electrical Engineering 349, DOI: 10.1007/978-3-662-47200-2_111

the efficiency of the engines may drop gradually over time [5]. Owning a vehicle in Malaysia is considered as a very normal situation, whereby a report from JPJ Malaysia had registered a total of 628,239 vehicles from all sectors in Malaysia for the year 2012 [3].This study is to identify the level of awareness in public towards personal automobile. This study aims to be a guide or reference for students in future who are furthering their study in maintenance management. The study focus in UTHM Drivers which is divide into age group to further understand towards their awareness in corrective and preventive maintenance.

1.1 Research Background

UTHM is a public university located in Parit Raja, Batu Pahat, Johor of Malaysia which consists of various Drivers ranging from students, lecturers, technicians and administration workers. Besides, the study also aims to get different age group from UTHM to conclude their awareness in preventive maintenance in vehicle maintenance.

Maintenance management has been well practiced in the industry. The knowledge of maintenance management has contributed to the increase of efficiency/productivity and cost saving in the productivity line. However, this study is to identify the awareness of individual towards maintenance management in maintaining their automobile.

Through observation, most of the Drivers are practicing corrective maintenance. Corrective maintenance also known as reactive strategy [9] where task performed to identify, isolate, and rectify a fault so that the failed equipment, machine, or system can be restored to an operational condition within the tolerances or limits established for in-service operations.

Preventive maintenance also known as proactive strategy [9] is a servicing by personnel for the purpose of maintaining equipment and facilities in satisfactory operating condition by providing for systematic inspection, detection, and correction of incipient failures either before they occur or before they developed into major defects. This maintenance may cost amore but it is good for machineries in a long run.

Awareness is described as knowledge or perception of a situation or fact (Oxford Dictionary, 2013). Awareness in maintenance among public Drivers is a main factor that can relate to the issues of the accidents results and breakdowns. Therefore, the awareness of public Drivers towards preventive and corrective maintenance are yet to be concluded.

1.2 Problem Statement

The awareness towards maintenance might be narrow; most Drivers practice corrective maintenance rather than preventive maintenance [5]. There are a lot of issues regarding awareness on maintenance that leads to malfunction, accidents and higher cost of living. Machineries performance may deplete over time and it may lead to malfunction and fatal breakdown. Performing a maintenance might cost a lot especially comparing between preventive maintenance and corrective maintenance [5]. However, issues regarding typical Malaysian giving lack of understanding towards maintenance have brought up issues such as high rates of accidents during festive season [7]. Therefore, this study hope to give an insight and create more awareness by providing data previous studies that have been conducted by researchers such as [1], [5], and [10] have put the maintenance management field in the production line in manufacturing industry to evaluate their perspective and criteria. However, this study is different from their scope of studies. Instead of focusing on organization, the scope has been changed to identify the awareness on individual.

1.3 Research Objective

- To identify the public awareness Preventive maintenance for personal automobile in UTHM.
- To define the factors that associates with the public awareness in Preventive maintenance for personal automobile in UTHM.

2 Literature Review

In this section, we are going to discuss the relevant terms used in this study. There would be an introduction to the maintenance management field and the types of maintenance that are being discussed in the whole study. Besides, this section is going to discuss about the public awareness among the community in Malaysia. Lastly, previous studies and fact are going to be presented to support the framework of this study.

2.1 Maintenance Management

Maintenance management is a field that required certain aspects to keep a product, facility, and service to operate or function properly. The basic of the field is to reduce the operating cost and increase the efficiency in a long run. This practice has been executed in the manufacturing, building and service sector since the past few decades. The maintenance management can be divided into several categories. However in this study we are going to look into 2 categories which [13] are the Corrective maintenance and Preventive maintenance.

2.2 Public Awareness

Public has a variety of meaning and usage according to oxford dictionaries. Public are referred to ordinary people in general or a community. Awareness is referring to knowledge or perception of a situation or fact in Oxford dictionary. Therefore, the term public awareness in this context is referring to the knowledge and perception of the community. In the end of this study, the awareness is going to be label base on good and bad, depending on the result of the questionnaire.

2.3 Corrective maintenance

Corrective maintenance is known to be the oldest type and is ad hoc, under which maintenance is called upon only when the equipment fails. This could render poor yield when the equipment has been highly degraded and the repair time is also expected to be long due to a "complete non-functional" failure and "full system" overhaul [12]. Corrective maintenance is also known as reactive strategy [9] is a practice where task performed to identify, isolate, and rectify a fault so that the failed equipment, machine, or system can be restored to an operational condition within the tolerances or limits established for in-service operations.

2.4 Preventive Maintenance

Preventive maintenance is a comparatively more cost effective, requiring some simple statistical models to determine a constant optimal time interval for maintenance. However, a constant time-interval based maintenance schedule does not fit the actual maintenance requirements of the equipment as the required time to next maintenance will be too long when it is new or less used, and the time will be too short when it becomes "old" or ages. As a result, preventive maintenance can be cost-ineffective due to insufficient maintenance for aged equipment and/or a higher chance of maintenance-induced failure for new equipment as a consequence of "more than necessary" initial maintenance actions [12].

This routine, among others, include:

i. Changing engine lubricant after certain recommended kilometre-age (e.g.5,000 km);
ii. Replacing the timing belt (e.g. 100,000 km);
iii. Performing tyre rotation (e.g. every six months or 10,000 km); and
iv. Changing the transmission fluid (e.g. 40,000 km).

Preventive maintenance also known as proactive strategy [9] for the purpose of maintaining equipment and facilities in satisfactory operating condition by providing for systematic inspection, detection, and correction of incipient failures either before they occur or before they develop into major defects. This maintenance may cost more but it is good for machineries in a long run.

2.5 Personal Automobile

Personal is described by the meaning of Belonging to or affecting a particular person rather than anyone else in oxford dictionary. In this context Personal Automobile, carries the meaning by defining automobile that belongs to a person, it must not be a company vehicle nor belongings to an organization.

2.6 Drivers in Malaysia

Every year, Malaysia has a large amount of cars registered under the Road Transport Department of Malaysia [3]. According to a report by the JPJ, drivers in Malaysia have reached a total of 13,303,843 (cumulatively) in 2012 as the latest report. The mass amount of drivers in Malaysia has rose up a lot of issue such as traffic congestion. Particularly, during the festive season where traffics are struggling to move away from city. Despite that, new problem arose when there is an increase in the traffic accidents. The Figure 1 shows the statistic taken by the JPJ of Malaysia regarding the increase of drivers in Malaysia:

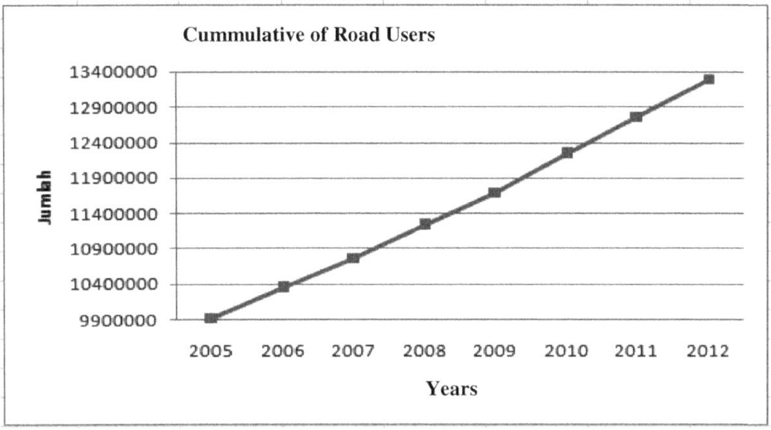

Fig. 1. Cumulative of road users

The increasing of drivers in Malaysia among the years has really gone to a point that maintenance should be put into consideration in each personal automobile owner to ensure their safety and daily life productivity. According to a research done by the Malaysian Institute of Road Safety (MIROS) in 2012, the total population of Malaysia is recorded at 29,300,000 while the total car registered was at 22,702,221 vehicles. This has concluded that each car in Malaysia has approximately 1.3 passengers. By rounding off, we can say that each Malaysian has own a car.

Malaysia has a different automotive ecosystem for the proportion of the types of vehicle on the road. Majority of the vehicle in Malaysia consist of locally made car such as Proton, Perodua and Naza. Local automotive producers cover up to 60% of the domestic vehicle sales.The remaining 40% of the vehicles are consisting of foreign manufactured cars. This situation has result both different drivers to experience different type of maintenance for their automobile. The drivers may experience the difference in maintenance cost, spare part availability and quality. However, this study is not focusing on the cost, spare part availability and quality of the maintenance but it is focused on how it affects their understanding/awareness of the need to maintain their own automobile.

2.7 Behaviour of Malaysia's Drivers Towards Automobile Maintenance

Previous studies conducted in the field of Maintenance Management focuses on industries such as manufacturing, building and IT system. Most of the studies related to manufacturing have concluded preventive maintenance is more beneficial compared to the corrective maintenance that is practiced conventionally. A study conducted in china for the manufacturing industries found out that most manufacturers in China that neglected the predictive maintenance and practices the corrective maintenance does not reduce the cost but rather have a lower machine availability that distort their production. While the manufacturers in china that are practicing predictive maintenance has more machine availability and lower bottleneck occurrence [10]. However, this scenario is applicable for organization that competes for continuous improvement and gain benefit from it. The concept of the paper that is being discuss are regarding on individual practices in understanding corrective maintenance and preventive maintenance. The previous idea from the study conducted by Pützr [10] regarding the maintenance strategies being use in the manufacturing industries in China can give a clear view of the maintenance strategies behaviour.

Another study conducted by [13] for the national review of maintenance issue for the national car in Malaysia also state that it is important to ensure a car is well-maintained to ensure the roadworthiness. Where roadworthiness is describe as fitting the vehicle to a accepted standard that it could be use on the open road. The study itself focuses between the relationship safety on the road with maintenance. The study has emphasized that maintenance management such corrective maintenance, preventive maintenance and predictive maintenance has play a vital part for the vehicle to be use on the road. However, the study didn't study on the perception or awareness of the road users in Malaysia towards those maintenance strategies. Somehow, the study has been identified [13] some factors that influence the users of personal automobile in performing maintenance for their vehicles:

1) Types of vehicles
   As mention above the ecosystem of vehicle in Malaysia comprises of 60% locally made cars while the remaining 40% are imported cars. Due to the availability of spare parts and cost. The decision made in maintaining the car will differ accordingly.

2) Cost of maintenance (financial)
   Normally, car owners' will attempt of perform preventive maintenance provided by the car dealer during warranty period. However, car owners' tend to switch and seek for a more flexible and reasonable price maintenance practices such as corrective maintenance. The study also stated the cost of maintenance can be influence by the type of spare part that is offer in the market.

3) Knowledge
   The study also concluded that the poor maintenance attitude can be influence by the lack of knowledge in the general knowledge on automobile maintenance. The study also highlighted the following:
   i. The inability to perform self-assessment on their cars;
   ii. Falling victim to irresponsible mechanics at service centres/workshops such as buying fake items, being charged unreasonably, etc;

iii. Failure to make the right decisions for maintenance work, especially in the case of a major breakdown; and
iv. Failure to appreciate the importance of maintenance equality/professional maintenance work.

Although, the study [13] have concluded that has been influencing the maintenance behaviour. The study did not come out with a recommendation on curbing the issue. The study also found out that the vehicle maintenance policy is implemented by the Malaysian government. It has indirectly affecting the behaviour of the automobile users nationally.

3 Methodology

3.1 Theoretical Framework

In this study, the perception or knowledge of the public towards corrective maintenance and preventive maintenance in automobile is taken as a dependent variable. It is expected to vary accordingly in each individual. Figure 2 is the framework of this study where 3 of the factors are being set to test whether it is affecting the public awareness. Knowledge background is the knowledge of the users in towards automobile maintenance. It can be consist of the users' education background.

Experience is the second factor that is put into the study to identify whether the users' experience will contribute to the level of public awareness. Experience can be label upon their exposure towards automobile maintenance and their encounter in car maintaining. Experience can be also mean by how long the users have been driving. As driving experience may be differ from every individual. It is expected that the problems encountered by the drivers may contribute to the drivers' awareness.

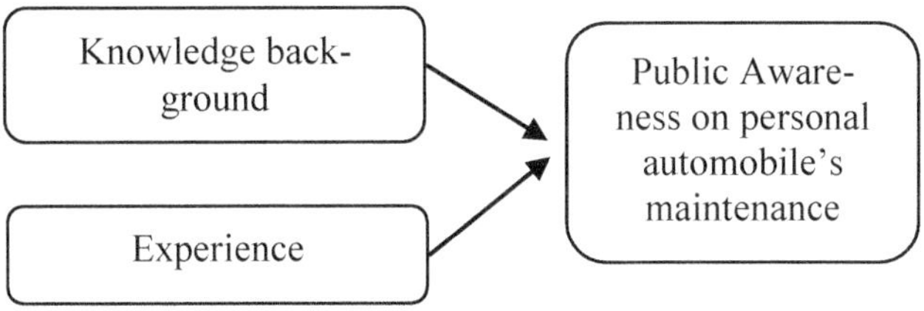

Fig. 2. Theoretical Framework

In a study conducted by Susan [10], one would expect, for instance that the less educated respondents would have higher don't know rates if they actually admitted when they didn't know the answer. On the other hand, the literature cautions us that persons with less education have more of a tendency to acquiesce. Even though, the report was conducted in 1995 but it is still relevant for the idea of linking knowledge background as a factor that influences the awareness.

In another study [2], where the study focused on the psychiatric knowledge. Jorm [2] mentioned a person affected by the symptom is being seen as a primary agent in symptom management. This perspective is important because it leads to bigger emphasis on increasing public knowledge and skills about mental health and empowering the experience of the people disabling the symptom. The study did mention that experience in a particular event may lead to better awareness in overcoming certain issues.

Therefore, in this studies the two main factors which is knowledge background and experience. However, the main purpose of the study is still to identify their awareness towards corrective maintenance and preventive maintenance in automobile maintenance.

3.2 Questionnaire

Questionnaire will be distributed to drivers in UTHM according to their age group. The questionnaire will ask about the details regarding the respondent's background and their understanding towards corrective maintenance and preventive maintenance. The questionnaire will be distributed to 80 respondents with 20 respondents in each age categories.

3.3 Data Collection

This research uses primary data collected through questionnaire. Questionnaire will be distributed to the sample. The content of the questionnaire will be comprises of sample's background, question regarding corrective maintenance and preventive maintenance. The data will then be collected and divide into categories such as age group.

3.4 Data Analysis

Data analysis is one of the important elements of the research. After the data being collected, analysis data process will be started in order to determine and find out the result and proves the hypothesis being accepted or rejected. The research data will be analyzed using Statistical Package for Science Social (SPSS). The descriptive analysis and correlation analysis were used to prove the hypotheses.

4 Data Analysis

This chapter will be discussing on the data analysis used in the study. The analysis will then be concluded into results. The data are being analyzed through IBM SPSS Data Editor.

This section will also discuss the reliability, normality, mean of the score and trend of scoring. It is important to note that each data here are taken primarily and computed through software. The result will then be discussed and detail later in this chapter.

4.1 Respondents' Demographic and Respond Rates

A total of 40 sets of questionnaire have been distributed through hard copies and Google drive. Fortunately, 40 of the respondents replied back to the questionnaire with full cooperation.

Table 1. Respondents' age and frequencies

Age	*Frequency*	*Cumulative Percent*
20-30	10	25.0
31-40	10	50.0
41-50	10	75.0
50 and above	10	100.0
Total	40	

Table 1 shows that the age of the respondents involves in this study. There are 4 classification of age namely 20-30, 31-40, 41-50, and 50 and above. Due to certain circumstances, it is unable for the study to be conducted to all the drivers available in UTHM.

During the distribution of the questionnaire in the targeted area, respondents of age 20-30 consists of 10 respondents. While respondents from 31-40, 41-50 and 50 and above have 10 respondents each. Respondent from group age 20-30 compare to others because the population in UTHM consist more drivers with age 20-30 years old.

Therefore, stratified sampling is used to identify the awareness from the respondents to ensure that there is similar size of sample from each age category. The respond rates are rather satisfied with 0% excluded rate.

Table 2. Respondents' gender and frequencies

Gender	**Frequency**	**Percent**	**Valid Percent**	**Cumulative Percent**
Male	19	47.5	47.5	47.5
Female	21	52.5	52.5	100.0
Total	40	100.0	100.0	

There are 19 males and 21 females involve in the questionnaire answering as in Table 2. Male has a 47.5% and female has a 52.5% as total of 100%. Both of the groups are almost equal in account.

In Table 3, it is almost balance between the imported and local cars. 18 respondents uses imported car while 22 respondents uses local cars. Imported cars cover up to 45% of the respondents while local cars cover up to 55% of the respondents' vehicle.

Table 3. Respondents' vehicle type and frequencies

Valid	Frequency	Percent	Valid Percent	Cumulative Percent
Imported	18	45.0	45.0	45.0
Local	22	55.0	55.0	100.0
Total	40	100.0	100.0	

Table 4. Respondents' education and frequencies

Education	Frequency	Percent
spm or equivalent	9	22.5
stpm or equivalent	9	22.5
bac degree	12	30.0
masters or PhD	10	25.0
Total	40	100.0

There are 9 respondents with SPM or equivalent and STPM or equivalent respectively as shown in Table 4. Bachelor degree has 12 respondents and there are 10 respondents with masters and PhD. The study defines bachelor degree and masters or PhD as a higher education level compare to SPM or equivalent and STPM or equivalent.

Table 5. Maintaining experience and frequencies.

Maintaining experience	Frequency	Percent
Yes	20	50.0
No	20	50.0
Total	40	100.0

Meanwhile Table 5 shows 20 respondents with maintaining experience and 20 respondents without maintaining experience. Which make an equal of both categories.

4.2 Reliability

Table 6 shows the reliability of all factors together which is an addition of 3 items namely driving experience, maintaining experience, and education background. The result of Cronbach alpha is at 0.714 which is a satisfactory level of reliability.

Table 6. Reliability Test's Cronbach Alpha

Cronbach's Alpha	Cronbach's Alpha Based on Standardized Items	N of Items
.714	.775	14

4.3 Skewness and Kurtosis

Table 7. Skewness and kurtosis from Question 1 to Question 5

Distributions	Type of Maintenance	Q2	Q3	Q4	Q5
Skewness	.777	1.200	.907	.315	1.200
Std. Error of Skewness	.374	.374	.374	.374	.374
Kurtosis	-1.473	-.592	-1.242	-2.003	-.592
Std. Error of Kurtosis	.733	.733	.733	.733	.733

Table 7 shows the descriptive data of the respondents' answers. 10 questions are put into the test to see whether the respondents have given a positive skewness or negative skewness. Bulmer, M.G., Principles of statistics stated that:-

- If a skewness is less than -1 or greater than +1 the distribution is highly skewed
- If the skewness is between -1 to -0.5 and +0.5 to +1 then the distribution is moderately skewed
- If the skewness is between -0.5 to +0.5 the distribution is approximately symmetrical

Table 8. Skewness and Kurtosis from Question 6 to Question 10

Distributions	Q6	Q7	Q8	Q9	Q10
Skewness	.907	.654	.424	.104	1.559
Std. Error of Skewness	.374	.374	.374	.374	.374
Kurtosis	-1.242	-1.658	-1.919	-2.097	.451
Std. Error of Kurtosis	.733	.733	.733	.733	.733

Therefore, there are 3 questions with skewness more than +1 in the sets of test as in Table 8. However, due to the questions has only 2 options, it has more chances for the answers to skew to one side. In this study the normality of the questions are not necessary important as the study focuses to look at their mean of the answers.

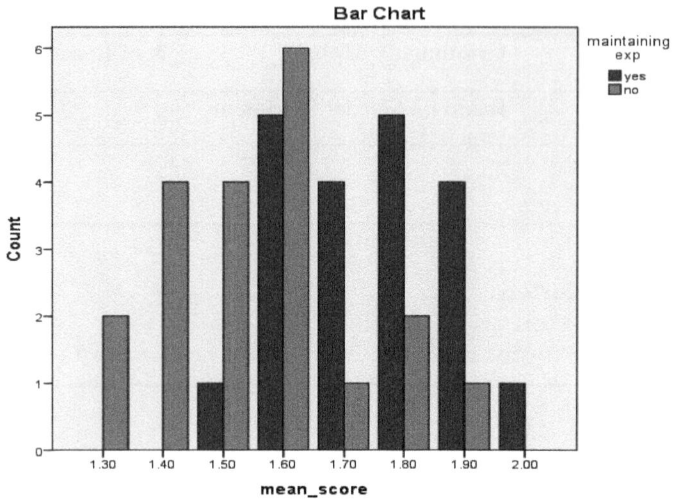

Fig. 3. Score by respondents according to education background

It is obviously seen in Figure 3, the bar chart that the mean score of the respondents with maintaining experience have their mean skewing to the right. However, respondents with no maintaining experience have their mean score distributed to the left of the bar chart. However, this does not imply that their awareness is low.

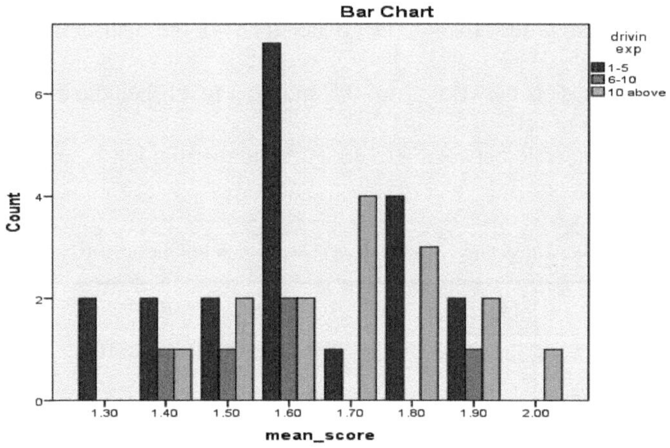

Fig. 4. Scored by respond according to education background

Figure 4 shows the bar chart between driving experience (driving exp) and the frequencies of their mean scores. Drivers with driving experience more than 10 years and above have their mean score distributed to the right. Drivers with driving experience 1-5 years have their mean score normally distributed throughout the chart.

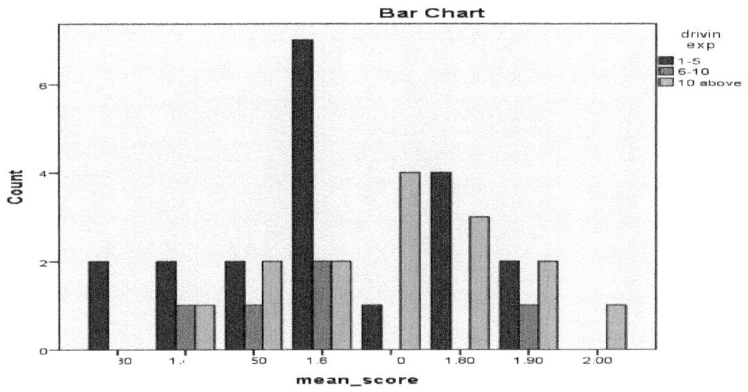

Fig. 5. Scored by respondents according to education background

Figure 5 shows the bar chart between driving experience (driving exp) and the frequencies of their mean scores. Drivers with driving experience more than 10 years and above have their mean score distributed to the right. Drivers with driving experience 1-5 years have their mean score normally distributed throughout the chart.

4.4 Score

Mean score is the score of mean according to each respondent. The questionnaire which is presented to obtain awareness has 2 options in each question to identify their awareness in Table 9. In Statistical Package for Science Social (SPSS), the value of 'Don't know' is 1 while the maximum value is 2. This indicates that each of the respondent with answered. Therefore, by interpreting the mean, the awareness will be identified through their mean at 1.6 and above will have a highest 'Yes'. The table shows that most of the respondents get a mean of 1.60. The lowest mean obtain by the respondents is 1.30. While the highest score obtain by the respondent is 2.00 with all answer 'Yes'.

Table 9. Respondents' average score

Valid	Frequency	Percent	Valid Percent	Cumulative Percent
0.30	2	5.0	5.0	5.0
0.40	4	10.0	10.0	15.0
0.50	5	12.5	12.5	27.5
0.60	11	27.5	27.5	55.0
0.70	5	12.5	12.5	67.5
0.80	7	17.5	17.5	85.0
0.90	5	12.5	12.5	97.5
1.00	1	2.5	2.5	100.0
Total	40	100.0	100.0	

Table 10. Descriptive statistic according to mean

Valid	40
Missing	0
Mean	0.6475
Mode	0.60
Skewness	-.057
Std. Error of Skewness	.374
Sum	65.90

Table 10 shows the mean of the mean score provided. The score of the score is at 0.6475. This means that the overall of their answer most of the respondents will have more 'Yes' answer rather than 'Don't know'. This set of data can conclude the level of awareness of the respondents in this study.

4.5 Conclusion

It is very conclusive that the data is reliable with a cronbach alpha of 0.714. It is at a satisfactory level for the cronbach alpha with 0.714 of internal consistency. The set has only 2 option as answer and the skewness of the set of questionnaire is being tested and it has most of the answer to be in normal. Besides, it is also found that respondent with higher level of education and with maintaining experience has a higher rate of Yes in their sets of answers. Lastly, it is conclusive that as a whole that the public awareness for the preventive maintenance is with a mean of 0.6475, which is moderately high.

5 Discussion

In this topic, we are going to discuss the result and relate it to our objectives. This study has 2 objectives:

a. To identify the public awareness in Preventive maintenance for personal automobile in UTHM.
b. To define the factors that is associates with the public awareness in Preventive maintenance for personal automobile in UTHM.

To make it easier for discussion, we will then discuss about the first objective which is to identify the public awareness in Preventive maintenance for personal automobile in UTHM. As the previous chapter accounts, the literature has support that the word "Public Awareness" carries the meaning of the knowledge of the public in one community.

Therefore, to identify the level of awareness the study has come out with a set of questions of questionnaire with 'Yes' and 'Don't know' as option to force the respondents to give out their true perceptions on the context of study. Their answer will then be summarized into mean to the rates of 'Yes' and 'Don't know'. In Statistical packages for Social Sciences (SPSS), the numerical value will be inserted 1- Don't know and 2- Yes.

This gives the study a value of mean with 0.5 as a median which indicates that the respondents with 0.5 score had actually answered the same amount of 'Yes' and 'Don't know' together. With the mean score above 0.5, then it is concluded that the respondent has answered more 'Yes' which means a higher level of awareness. This type of study is adapted from (Ciochetto& Haley, 2008) in their study 'How Do You Measure "Awareness"?' Experiences with The Lead-Based Paint Survey.

Table 11. Agregate mean result

N	Valid	40
	Missing	0
Mean		0.6475
Mode		0.60

The result in Table 11 has been collected from the respondents shows that the aggregate mean is at 0.6475. This means that by average of each respondents should have answer more 'Yes' than 'Don't know'. Conclusively, the awareness is averagely moderate. This has answer the first objective of the study.

The second objective of the study is to identify the factors affecting the public awareness in Preventive awareness. There are no previous studies regarding public awareness to Preventive maintenance. However, Preventive maintenance is described as a knowledge or practice which can be practice. Therefore, by adapting studies from psychological which implied that experience and education has an impact on awareness. Jorm mentioned experience in a particular event may lead to increase in awareness over a certain period of time in his study "Public knowledge and beliefs about mental disorders". In his study, he has concluded that people with experience in dealing with mental disorder patient will tend to have a higher significant of understanding towards the knowledge in that context.

In this study, the respondents with maintaining experience has the more 'Yes' answered.

Table 12.

		Maintaining experience		Total
		Yes	No	
Score	0.30	0	2	2
	0.40	0	4	4
	0.50	1	4	5
	0.60	5	6	11
	0.70	4	1	5
	0.80	5	2	7
	0.90	4	1	5
	1.00	1	0	1
Total		20	20	40

Table 12 shows the mean score of the respondents accordingly to those with maintaining experience. It is obviously seen that there are 19 respondents with maintaining experience with higher mean score of 0.6 and above. This analysis doesn't conclude the relationship but it is a study of trend to see how the respondents react towards a set of questions. The mean of different background has an impact towards their awareness. It is conclusive that the respondents with maintaining experience has higher mean compare to those with no maintaining experience

Table 13. Respondent educational level

	Educational spm or equivalent	stpm or equiva-lent	bachelor degree	masters or Phd	Total
0.3	0	0	1	1	2
0.4	1	1	2	0	4
0.5	2	2	1	0	5
0.6	3	2	4	2	11
0.7	2	1	0	2	5
0.8	1	1	1	4	7
0.9	0	1	3	1	5
1.0	0	1	0	0	1
Total	9	9	12	10	40

Table 13 shows the answer given by respondents according to their educational level. Respondents with higher education such as Masters or PhD has the mean score distributed over a higher mean score. However, respondents from STPM or equivalent has a high score of mean. Despite that, the result cant conclude that the higher the educational has a higher awareness in Preventive maintenance.

5.1 Limitation of the Study

With all the findings discussed in previous chapter, it will be a good measure to review the limitation of this study. There were several limitations occurred when conducting this study. The limitations met are shown as following.

5.2 Recommendation

There is no studies without minor flaws and misdesign in it. The study only concluded the awareness in Preventive maintenance for personal automobile in UTHM and how the factors can give different result in the awareness. There some several improvements that need to be done if there are future research carried on this study.

5.3 Increase in Sample Size

Due to some circumstances, the total amount of drivers registered in UTHM are still unable to obtain. Therefore, it is quite impossible to set a numerical value of respondents needed to obtain a qualified data. However, stratifying the respondents is a way to control variables of the respondents for maximal result.

5.4 Identify the Relationship between Factors and Awareness

This study only study the pattern and trend of the respondents background (factors) and their mean. However, chi square test can be apply by further modifying the questionnaire to understand the strength and correlation between awareness and the factors.

5.5 Provide Measure to Increase Awareness

Since the awareness hasn't been identify in the previous study, this study cant provide a set of solution or measure to improve the awareness. Therefore, it is better for the future study to provide measures and solution to improve the awareness at 0.6475.

5.6 Conclusion

In conclusion, the awareness for Preventive Maintenance for Personal Automobile in UTHM are moderate at 0.6475. Besides, the factors experience has a proportionate impact on the mean score of the awareness, it is conclusive that rspondents with maintaining experience have higher mean than those who have no experience. Other than that, we can also conclude that respondents with STPM or equivalent and above have a higher mean than SPM or equivalent.

References

[1] Jambekar, A.B.: A systems thinking perspective of maintenance, operations, and process quality. Journal of Quality in Maintenance Engineering 6(2), 123–132 (2000)
[2] Jorm, A.F.: Mental health literacy. Public knowledge and beliefs about mental disorders. The British Journal of Psychiatry: The Journal of Mental Science 177, 396–401 (2000)
[3] Jabatan Pengangkutan Jalan (JPJ). JumlahPendaftaran Kenderaan mengikut tahun,
[4] http://www.jpj.gov.my/pendaftaran-lain2-kenderaan
[5] (retrieved on March 11, 2014)
[6] Kamba, A.N., Rahmat, R.A.O.K., Ismail, A.: Why Do People Use Their Cars: A Case Study In Malaysia. Journal of Social Sciences 3(3), 117–122 (2007)
[7] Lo, H.: Management of industrial maintenance ± economic evaluation of maintenance policies (1994)
[8] Looi, S.C.: New Straight Times. "Accidents Death Involving Youth" (2013)
[9] http://www.nst.com.my/latest/accident-deaths-involving-youths-1.329433 (retrieved on March 11, 2014)
[10] Bulmer, M.G.: Principle of Statistic. Amazon Publication, UK (2003)
[11] New Straits Times (August 25, 2013), "Why Buses Turn into "Coffin", http://www.nst.com.my/nation/general/why-buses-turn-into-coffins-1.345135 (retrieved on March 11, 2014)
[12] Pintelon, L., Pinjala, S.K., Vereecke, A.: Evaluating the Effectiveness of Maintenance Strategies 12(1), 7–20 (2006)
[13] Pützr, F., Fischer, T., Wang, C., Lin, J.: Exploring maintenance strategies in Chinese product manufacturing companies. Management Research News 31(12), 941–950 (2008)
[14] Chiochetto, S.: How do you measure "awareness"? Experience with lead based paint survey. Bureau of Census, Center of survey method research Washington D.C (1995)
[15] Tan, C.M., Raghavan, N.: Root cause analysis based maintenance policy. International Journal of Quality & Reliability Management 24(2), 203–228 (2007)
[16] Voon, W.S.: Review of the National Automotive Policy on Car Maintenance Issues: Malaysia's Automotive Automative Ecosystem Explained (2012)

A Novel Analysis of Clinical Data and Image Processing Algorithms in Detection of Cervical Cancer

Abhishek Das[1], Avijit Kar[2], and Debasis Bhattacharyya[3]

[1] Dept. of Information Technology, Tripura University (A Central University), Agartala, India
[2] Dept. of Computer Sc. & Engineering, Jadavpur University, Kolkata, India
[3] Dept. of Gynecology, College of Medicine & Sagore Dutta Hospital, Kolkata, India
adas@tripurauniv.in

Abstract. Cervical Cancer is one of the ubiquitous forms of cancer afflicting the female population worldwide. A Digital Colposcope is a self-illuminated powerful microscope which acquires the image of the affected cervix. The raw cervix image acquired by the colposcope is known as a cervigram. The raw cervigram is preprocessed by removing the specular reflections and then the region of interest is sought. Different novel segmentation algorithms which were proposed in our earlier papers are applied to the cervigram to detect the acetowhite tissues. In the current paper, we try to draw a correlation between Pap smear results, colposcopic findings, histopathologic biopsy results and our proposed algorithmic analysis. The significance of association is assessed by chi-square testing and the strength of association by kappa statistic. The association between impression and histology was significant, the strength of correlation was also significant. The overall kappa statistic for the concordance of the algorithm with the human experts was 0.809, while that among raters, excluding the algorithm was 0.804. The proposed algorithm had better agreement with the expert decision.

Keywords: cervical cancer, segmentation algorithms, kappa statistic.

1 Introduction

Cervical Cancer is one of the ubiquitous forms of cancer afflicting the female population worldwide. According to World Health Organization (WHO) [1], every year in India alone there are 1 lakh 30 thousand women who are affected with cervical cancer. According to Chittaranjan National Cancer Research Institute's Cancer Registry Program [2], in the state of West Bengal alone, out of every 1000 women, aged between 30 years to 60 years old, 18 are affected by cervical cancer. To detect cervical cancer, the state does not possess the adequate infrastructure. Hence the real numbers of affected cervical cancer patients are still unknown. With lack of awareness and treatment, the deadly disease gets detected only in the Final Stage (i.e. CIN-3) of Cervical Cancer. Due to cervical cancer alone, 80,000 women dies every year in our Country. However, cervical cancer incidence and mortality remain high in resource-poor regions, where high quality Pap (Papanicolaou smear) screening programs

M. Gen et al. (eds.), *Industrial Engineering, Management Science and Applications 2015*,
Lecture Notes in Electrical Engineering 349, DOI: 10.1007/978-3-662-47200-2_112

[3],[4],[5],[6], often cannot be maintained because of inherent complexity and cost. A Digital Colposcope is a self-illuminated powerful microscope which acquires the image of the affected cervix. The raw cervix image acquired by the colposcope is known as a cervigram. The raw cervigram is preprocessed by removing the specular reflections and then the region of interest is sought. Before the image is made ready for implementing further image processing algorithms, illumination correction methods and intensity normalization methods are applied.

In our research, we have dealt with the image acquisition of the unhealthy cervix using a Digital Colposcope and then using various image processing techniques, we have tried to semi-automate the detection of cervical lesions and classify them into various Cervical Interepithelial Neoplasia (CIN) –I, II, III , Carcinoma-in-situ (CIS) stages or (LSIL and HSIL). The steps used in our techniques was removal of specular reflections, finding region of interest of the cervix, segmentation of AW tissues and further classification into mosaicism, punctuation and vasculatures. An automated characterization of the lesions borders along with the automated detection of the AW lesion is one the major goals of this work. Therefore, the understanding of the lesion margins and their relation to the staging of cervical cancer is a key point of such analysis and characterization. Lesion margins of the AW lesions on the cervix have a variety of forms and many visual interpretations. In order to categorize the AW lesions, the classification of Lesion margins can be grouped into two major categories, each one of those categories associated with low and high grades of cervical intraepithelial neoplasia (CIN). Low grade is known as mild dysplasia (CIN I), while high grade includes moderate dysplasia (CIN II), severe dysplasia (CIN III – cancer in situ), and cancer. Lower indexation by Reid's index (1-2 points) indicates low grade lesions (CIN I). Reid indices are cumulative for color, margins and vessel pattern. Reid's indexation yields a low score for 'indistinct borders,' known as diffuse borders, irregular lesions or 'satellite' lesions. Higher indexation (3-5 points) indicates high grade lesions. Reid's indexation for 'regular' ('smooth') lesions with a 'straight' outline is an indication for higher grade dysplasia and cancer (CIN II and CIN III). Strong demarcation lesion Margins is a well-defined characteristic of CIN III or sometimes as cancer in situ, while the diffusivity of the Lesion Margins indicates low grade lesions. The data from the colposcopic images has been graded accordingly.

2 Methods

In this paper, we performed a study of precancerous lesions and cervical cancer over a period of five years from 2008 to 2012 at SSKM Hospital, Kolkata who sought consultation for various gynecological disorders. Parity was from 0 to 6 with a medium of 3, and age was between 21 and 70 years. Colposcopy was performed with the help of B'orze® Colposcope on an outpatient basis. The cervigram data was later processed in a PC with Intel Core2 Duo processor, 2 GB Ram, 320 GB HDD using MatLab® software in Windows® XP operating system.

The patients having an unhealthy cervix were first administered Pap smear, and then Colposcopy images were taken. The colposcopy guided biopsy was also taken from all examined patients, by punch biopsy forceps from the most advanced part of lesion. The biopsy fragments were processed by the usual technique for inclusion in paraffin, Hemotoxylin-Eosin stained and interpreted in the Histo-pathology laboratory of SSKM Hospital, Kolkata which were further correlated with our algorithm findings.
Colposcopic examinations included:

- Direct examination of cervix with green filter and saline application;
- Examination of the cervix after test with 5% acetic acid, observing the junction of squamous cell, erosion, papillary lesions, aceto-white areas and vascular design;
- Examination of the cervix after Lugol test in which normal squamous epithelium, which contains glycogen, it turns brown.

The colposcopic findings were classified into nonmalignant and malignant categories. The non-malignant category included normal findings or viral wart changes, and the malignant category was divided into five groups: CIN I, CIN II, CIN III, micro-invasive carcinoma and CIS. Histological diagnosis was also classified into five groups: CIN I, CIN II, CIN III, micro-invasive carcinoma and CIS.

Using the model in our previous research [4],[5],[6],[7],[8],[9],[10] , the cervix tissues can be modelled as a mixture of Gaussians in 1-d feature space. The probability of occurrence of each Gaussian j is denoted α_j. For a set of n feature vectors: $\hat{y}_1, \ldots\ldots \hat{y}_n$, the maximum likelihood estimate of the different parameters can be found by maximizing $f(\hat{y}|\ \theta_j,c) = \Pi_i f(\hat{y}_i|\ \theta_j,c)$, using our proposed variant of EM algorithm [8].

Since we have the estimate of the parameter set, each iteration of the EM algorithm re-estimates the parameter set according to the following steps:

- Expectation step

$$p(T_i = j|\hat{y}_i, \theta, c) = \frac{\alpha_j f(\hat{y}_i|\theta_j, c)}{\sum_{t=1}^{J} \alpha_t f(\hat{y}_i|\theta_t, c)}$$

- Maximization step of the Mixture of Gaussian parameters:

$$\alpha_j = \frac{\sum_i p(T_i = j|\hat{y}_i, \theta, c)}{n},$$

$$\mu_j = \frac{\sum_i p(T_i = j|\hat{y}_i, \theta, c)(y_i - \sum_k c_k \phi_k(x_i))}{\sum_i p(T_i = j|\hat{y}_i, \theta, c)},$$

$$\sigma_j^2 = \frac{\sum_i p(T_i = j|\hat{y}_i, \theta, c)(y_i - \sum_k c_k \phi_k(x_i) - \mu_j)^2}{\sum_i p(T_i = j|\hat{y}_i, \theta, c)}$$

■ Maximization step of the illumination field parameters:

$$\begin{bmatrix} c_1 \\ c_2 \\ \vdots \\ c_K \end{bmatrix} = (A^T W A)^{-1} A^T W R; \quad A = \begin{bmatrix} \phi_1(x_1) & \phi_2(x_1) & \dots & \phi_K(x_1) \\ \phi_1(x_2) & \phi_2(x_2) & \dots & \phi_K(x_2) \\ \vdots & \vdots & \dots & \vdots \end{bmatrix}$$

$$W = diag(w_i); \quad w_i = \sum_j w_{ij}; \quad w_{ij} = \frac{p(T_i = j|\hat{y}_i, \theta, c)}{\sigma_j^2}$$

$$R = \begin{bmatrix} y_1 - \tilde{y}_1 \\ y_2 - \tilde{y}_2 \\ \vdots \end{bmatrix}; \quad \tilde{y}_i = \frac{\sum_j w_{ij}\mu_j}{\sum_j w_{ij}},$$

3 Results

From all 1050 patients examined, 555 patients (52.85%) presented non-malignant findings that do not require biopsy for histopathological examination, but required a full monitoring: bacteriological, cytological and colposcopic at different time intervals, and where not included in our study. 495 patients (47.14%) presented malignant findings and underwent colposcopic examination to further investigate a cytological abnormality on their pap smears – 345 cases (69.69%) or an abnormal appearance of the cervix – 150 patients (30.30%). These 345 patients represent our study group. We noticed that lesions were diagnosed in patients aged between 21 and 70 years, most cases encountered in the interval 31–50 years (Tables 1 and 2).

Age (years)	*Total*	*%*
21-30	**65**	**13.1**
31-40	**165**	**33.3**
41-50	**185**	**37.4**
51-60	**47**	**9.5**
61-70	**33**	**6.7**

Fig. 1. Table showing distribution of cases by age

Clinical and morphological parameters		No. of Cases	Percentage (%)
Colposcopic Findings	**CIN I**	28	5.6
	CIN II	142	28.7
	CIN III	279	56.4
	Micro-invasive carcinoma	26	5.3
	CIS	20	4.0
Histopathologic Findings	**Normal**	8	1.6
	CIN I	26	5.3
	CIN II	146	29.5
	CIN III	275	55.6
	Micro-invasive carcinoma	24	4.8
	CIS	16	3.2
	Total	495	100

Fig. 2. Table showing distribution of cases by clinical and morphological parameters

Colposcopic findings in our research were: 28 (5.6%) cases were CIN I, 142 (28.7%) cases were CIN II, 279 (56.4%) cases were CIN III, 26 (5.3%) cases were micro-invasive carcinoma and 20 (4%) cases were CIS (Figure 3).

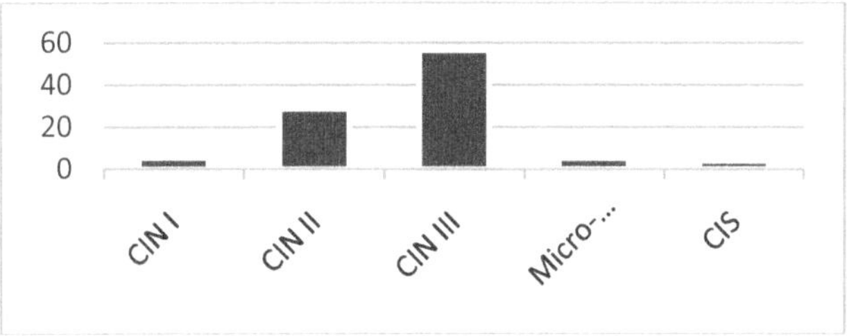

Fig. 3. Colposcopic findings in percent in our research

Colposcopic aspects of the selected cases showed a variety of issues that can be found isolated or associated:

- Mosaic: areas looking like pavement, composed of rectangular fields, square or polygonal, separated by dotted or red lines.
- Punctuation: thick acetowhite epithelium, with vascular spots on surface with variable distribution and size. Correspond to simple dysplasia, CIN I.
- Leukoplakia: simple shape, hypertrophic or warty. Colposcopic appearance is given by pearly white appearance, irregular surface, single or multiple plates. Correspond to CIN I, II, or III.

Histological results in the 495 examined cases were: 8 (1.6%) cases normal, 26 (5.3%) cases CIN I, 146 (29.5%) cases CIN II, 275 (55.6%) cases CIN III, 24 (4.8%) cases micro-invasive carcinoma and sixteen (3.2%) cases CIS.

CIN may be suspected after performing the colposcopic examination. The final diagnosis of CIN presence and the respective degree is made on cervical tissue specimen, dependent on the histological features concerned with differentiation, maturation and stratification of cells and nuclear abnormalities [3].

In CIN I, there is good maturation with minimal nuclear abnormalities and few mitotic figures. Undifferentiated cells are confined to the deeper layers (lower third) of the epithelium. Mitotic figures are present, but not very numerous [3].

CIN II is characterized by dysplastic cellular changes mostly restricted to the lower half or the lower two thirds of the epithelium, with more marked nuclear abnormalities than in CIN I. Mitotic figures may be seen throughout the lower half of the epithelium [3].

In CIN III, differentiation and stratification may be totally absent or present only in the superficial quarter of the epithelium with numerous mitotic figures. Nuclear abnormalities extend throughout the thickness of the epithelium. Many mitotic figures have abnormal forms [3].

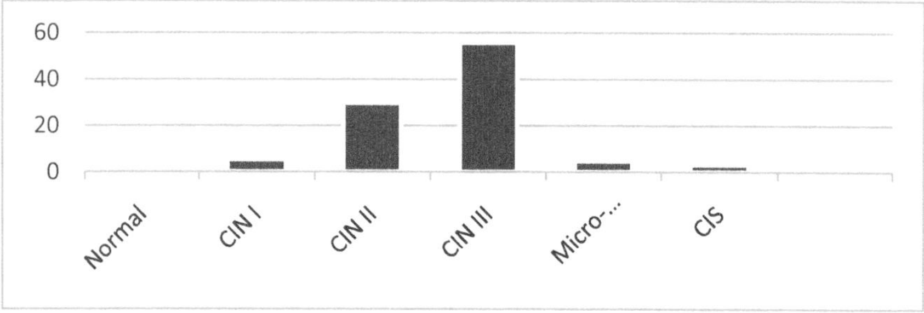

Fig. 4. Histopathologic findings in percent in our research

To determine the overall accuracy of the colposcopic impression and to compare our results with other published research [3], we used the cervical biopsy result as gold standard values because all probes were histologically analyzed. The colposcopic aspects of the lesions were compared with histologic results from colposcopic directed biopsy to asses a correlation between these two diagnostic methods. The correlation was calculated reporting the number of cases histologically confirmed to the number of cases colposcopic diagnosis for each lesion group separately.

Colposcopic Diagnosis

Histopathologic Diagnosis	Normal	CIN I	CIN II	CIN III	Micro-invasive carcinoma	CIS	Total
Normal	-	-	-	-	-	-	-
CIN I	6	22(78.5%)	-	-	-	-	28
CIN II	2	2	136(95.7%)	2	-	-	142
CIN III	-	2	7	265(94.9%)	5	-	279
Micro-invasive carcinoma	-	-	3	8	13(50%)	2	26
CIS	-	-	-	-	6	14(70%)	20
Total	8	26	146	275	24	16	495

Fig. 5. Table showing Correlation between Colposcopy and Histopathology

Table 5 shows the correlation between colposcopic findings and histological diagnosis. The correlation was 78.5% (22 out of 28 patients) in the CIN I category, 95.7% (136 out of 142 patients) in the CIN II category, 94.9% (265 out of 279 patients) in the CIN III category, 50% (thirteen out of 26 patients) for micro-invasive carcinoma and 70% (fourteen out of 20 patients) for CIS.

4 Conclusion and Discussion

As shown in Table 5, data obtained confirm that colposcopy performed better in differentiation of high-grade from low-grade lesions and that the positive predictive rate of the colposcopic impression is better as the cervical lesion is more severe: 265 cases confirmed as CIN III from 136 cases CIN II and 22 cases CIN I. The 28.4% (eight out of 28 cases) error rate in distinguishing between normal and LSIL, confirming it is difficult to distinguish between normal and abnormal. The colposcopy method incurred fewer false negatives (eight patients), giving a general accuracy rate of 98.38% (487 out of 495 biopsied patients). Sensitivity of colposcopic examination was 90.9% (450 out of 495 cases). As with cytology, some patients were under classified from CIN I to normal (six from 28 cases), from CIN II to CIN I (two out of 142 patients) and from CIN III to CIN II (seven out of 279 patients) using colposcopic examination. Evaluation of the colposcopic results yielded the following prevalence percentages based on the histologic diagnoses: 4.4% for CIN I, 27.5% for CIN II, 53.5% for CIN III, 2.6% for micro-invasive carcinoma and 2.8% for CIS.

	Pap smear Results	**Colposcopic findings**	**Histopathologic Results**	**Algorithmic analysis**
Pap Smear Results	-	K=0.710	K=0.780	K=0.802
Colposcopic findings	K=0.710	-	K=0.802	K=0.809
Histopathologic Results	K=0.780	K=0.802	-	K=0.882
Algorithmic analysis	K=0.802	K=0.809	K=0.882	-

Fig. 6. Table showing Correlation among the colposcopic findings, histopathologic results, pap results and our Algorithmic analysis using Kappa Values (K)

We further try to draw a correlation between Pap smear results, colposcopic findings, histopathologic biopsy results and our proposed algorithmic analysis [4],[5],[6],[7],[8],[9],[10], as shown in Table 6. The significance of association is assessed by chi-square testing and the strength of association by kappa statistic. Items such as physical exam findings or other diagnostic tests often rely on some degree of subjective interpretation by observers. Studies that measure the agreement between two or more observers should include a statistic that takes into account the fact that observers will sometimes agree or disagree simply by chance. The kappa statistic is the most prevalently used statistic for this purpose. Kappa K= $(p_o-p_e)/(1-p_e)$ where p_o is the observed agreement and p_e is the expected agreement . A kappa of 1 indicates perfect agreement, whereas a kappa of 0 indicates agreement equivalent to chance. The association between impression and histology was significant ($P < 0.001$), the strength of correlation was also significant. The overall kappa statistic for the concordance of the algorithm with the human experts was 0.809, while that among raters, excluding the algorithm was 0.804. The proposed algorithm had better agreement with the expert decision (kappa =0.882). Our algorithm could not be distinguished from the experts by this measure.

References

1. Massad, L.S., Jeronimo, J., Svhiffman, M.: Inter-observer agreement in the assessment of colposcopic grading. Obstet. Gynecol. 111, 1279–1280 (2008)
2. Craine, B.L., Craine, E.R., Engel, J.R., Wemple, N.T.: A clinical system for digital imaging colposcopy. In: Schneider, R.H., Dwyer III, S. (eds.) Medical Imaging II. Proc. SPIE, vol. 914, pp. 505–511 (1988)
3. Sankaranarayanan, R., Rajkumar, R., Theresa, R., Esmy, P.O., Mahe, C., Bagyalakshmi, K.R., Thara, S., Frappart, L., Lucas, E., Muwonge, R., Shanthakumari, S., Jeevan, D., Subbarao, T.M., Parkin, D.M., Cherian, J.: Initial results from a randomized trial of cervical visual screening in rural south India. Int. J. Cancer 109, 461–467 (2004)
4. Das, A., Kar, A., Bhattacharyya, D.: Elimination of Specular reflection and Identification of ROI: The First Step in Automated Detection of Uterine Cervical Cancer using Digital Colposcopy. In: Proc. of IEEE Imaging Systems & Techniques Penang, Malaysia, pp. 237–141 (2011), doi:1109/IST.2011.5962218
5. Das, A., Kar, A., Bhattacharyya, D.: Preprocessing for automatic detection of Cervical Cancer. In: Proc. of 15th International Conference on Information Visualisation, London, U.K., pp. 597–600 (2011), doi:10.1109/IV.2011.89, ISSN 1550-6037
6. Das, A., Kar, A., Bhattacharyya, D.: Implication of Technology on society in Asia: Automated detection of Cervical Cancer. In: Proc. of IEEE International Conference on Technology & Society, Singapore (2012)
7. Das, A., Kar, A., Bhattacharyya, D.: Detection of abnormal regions of precancerous lesions in Digitised Uterine Cervix images. In: Proc. of IEEE International Electrical Engineering Congress, at Pattaya city, Thailand (2014)
8. Das, A., Kar, A., Bhattacharyya, D.: Early Detection of Cervical Cancer using novel segmentation algorithms. Invertis Journal of Science & Technology (IJST) 7(2), 91–95 (2014) ISSN 0973-8940
9. Das, A., Kar, A., Bhattacharyya, D.: Q-Metrics for Early Detection of Cervical Cancer. Journal of Image Processing & Pattern Recognition Progress 14, 32–36 (2014)
10. Das, A., Kar, A., Bhattacharyya, D.: Probabilistic Segmentation Methods for Early Detection of Uterine Cervical Cancer. Journal of Applied Information Science (JAIS) 11, 28–31 (2013) ISSN 2321-6115

Author Index

Batch number: 10614928

Printed by Printforce, the Netherlands